CIRCUITS

Engineering Concepts and Analysis of Linear Electric Circuits

Preliminary Edition

CIRCUITS

Engineering Concepts and Analysis of Linear Electric Circuits

A. Bruce Carlson
Professor and Curriculum Chair
Electrical, Computer, and Systems Engineering
Rensselaer Polytechnic Institute

John Wiley & Sons, Inc.
New York Chichester Brisbane Toronto Singapore

ACQUISITIONS EDITOR Steven Elliot
MARKETING MANAGER Debra Riegert
SENIOR PRODUCTION EDITOR Cathy Ronda
DESIGNER Kevin Murphy
ILLUSTRATORS Wellington Studios
MANUFACTURING MANAGER Dorothy Sinclair
ILLUSTRATION COORDINATOR Rosa Bryant

This book was set in ITC Garamond Light by PRO-Image, and printed and bound by Hamilton Printing. The cover was printed by Hamilton Printing.

Recognizing the importance of preserving what has been written, it is a policy of John Wiley & Sons, Inc. to have books of enduring value published in the United States printed on acid-free paper, and we exert our best efforts to that end.

The paper in this book was manufactured by a mill whose forest management programs include sustained yield harvesting of its timberlands. Sustained yield harvesting principles ensure that the number of trees cut each year does not exceed the amount of new growth.

Library of Congress Cataloging in Publication Data:

Carlson, A. Bruce, 1937–
Circuits : engineering concepts and analysis of linear electric circuits / A. Bruce Carlson.
p. cm.
Includes index.
ISBN 0-471-15667-1 (pbk. : alk. paper)
1. Electric circuits, Linear. I. Title.
TK454.C357 1996
621.319'2—dc20 96-18043
CIP

Printed in the United States of America

10 9 8 7 6 5 4 3 2 1

Contents in Brief

Preface

This text is intended for the first course on linear electric circuits, usually taken in the second year by students majoring in electrical engineering and related fields. The length of the course might be one or two semesters or two or three quarters, depending upon the selected coverage.

FEATURES

Although several popular texts on linear circuits already exist, this one was written to provide an alternative with the following features.

1. **Focus on basic circuit concepts and analysis methods.** The text includes detailed coverage of node and mesh equations, phasors, the *s* domain, Fourier series, Laplace transforms, and state variables as useful circuit concepts and important analysis methods. However, topics such as Fourier transforms and flow graphs have been omitted because they more properly belong in subsequent courses on signal and system analysis. An introductory section on impulses and convolution provides a bridge to such courses. Similarly, the text coverage of resistive circuits, operational amplifiers, frequency response, and Bode plots has been designed to be self-contained as well as supporting subsequent courses on electronic circuits.
2. **Sequential development from elementary to more sophisticated topics.** The organization follows a "just-in-time" strategy that develops circuit concepts and methods at an introductory level before moving on to a higher level of abstraction. Specific examples of this strategy include:

- Treatment of duality in the context of resistive circuits, before capacitance and inductance;
- Superposition and Thévenin's theorem before systematic node and mesh analysis;
- Early introduction of operational amplifiers;
- Coverage of ac circuit analysis before second-order transients, so students become more comfortable with sinusoidal functions expressed as complex quantities;
- Ideal transformers before mutual inductance;
- Complex frequency and network functions before Laplace transforms but after natural response;
- Frequency response before Fourier series.

3. **Emphasis on modern engineering techniques.** While the text briefly discusses the classical solution of differential circuit equations, primary emphasis is given to more efficient approaches such as phasors, network functions, Laplace transforms, and state equations. Matrix notation is also introducted early on and used regularly for the representation and solution of simultaneous equations. When a particular technique involves several parts, it is presented as a procedural algorithm so students more clearly see the successive steps. Additionally, the text points out those tasks that are best done taking advantage of modern calculators or computer algebra programs.
4. **Stated instructional objectives.** Specific objectives are stated at the beginning of each chapter so the students know what they are expected to learn. These objectives also serve as a review guide.
5. **Examples of practical applications.** To stimulate student motivation, examples are often based on "real-world" applications. Several of these examples feature design considerations.
6. **Modular structure for flexibility.** The text has been structured with optional topics in self-contained sections or chapters to allow instructors to tailor the coverage for particular needs. Thus, for instance, instructors may elect to include ideal transformers even though the course length does not permit extensive coverage of mutual inductance. An appendix on PSpice is also provided so that instructors may choose to either integrate circuit simulation at appropriate points or to omit this topic. As a further aid for instructors, the table of contents identifies specific prerequisites for each section as well as parts that may be omitted without loss of continuity.

PREREQUISITE BACKGROUND

The text is written at a level suitable for students who have completed at least one term of college physics and one year of calculus. Familiarity with matrix notation is also assumed. Previous or concurrent courses on linear algebra and differential equations are helpful but not essential.

The prerequisite physics consists of elementary mechanics, dimensional analysis, and some knowledge of sinusoidal waves, electric charge and fields, and atomic structure. Conduction processes, electrical energy, and magnetic phenomena are discussed when introduced in the text.

The prerequisite mathematics consists of quadratic equations, simultaneous linear equations, trigonometric and exponential functions, simple series expansions, and elementary differential and integral calculus. A section of the text treats the solution of elementary differential circuit equations, and complex numbers are discussed in conjunction with ac circuits. An appendix at the back reviews matrix notation and Cramer's rule for sets of linear equations, plus more advanced concepts of linear algebra needed for the treatment of two-port networks and state variables. A table at the back summarizes the essential mathematical relations used in the text.

TEACHING AND LEARNING AIDS

As previously mentioned, each chapter starts with a list of instructional objectives to guide student work. More than 200 worked examples and 240 exercises are embedded within the chapters to help students master the objectives. Exercise solutions are given at the back of the book.

For further student practice, more than 1000 problems are provided at the ends of the chapters. These problems range from simple drills to more challenging analysis and design tasks. Special effort has been made to create problems that are educational, interesting, and "doable." However, instructors should note that the problems are organized by section, not necessarily in order of increasing difficulty.

Answers for about 20% of the problems are given at the back so students can check their work. These problems are marked with asterisks (*). Instructors may obtain a manual with complete problem solutions by contacting the publisher.

The notation has been chosen to avoid boldface symbols that would be difficult for instructors and students to reproduce by hand. In particular, phasors are denoted by an underbar, such as $\underline{V}$, while matrices are identified by brackets, such as $[A]$.

Most of the text material has been class-tested at Rensselaer Polytechnic Institute for several years. Feedback from instructors using this preliminary edition is also requested to further improve clarity and accuracy.

List of Reviewers

N.F. Audeh
University of Alabama–Huntsville

Calvin Finn
University of Idaho

Michael Georgiopoulos
University of Central Florida

Edwin L. Gerber
Drexel University

Edward W. Greeneich
Arizona State University

Ronald S. Gyurcsik
North Carolina State University

Ron Hollzeman
University of Pittsburgh

Deverl Hympherys
Brigham Young University

Derek Lile
Colorado State University

Paul J. Nahin
University of New Hampshire

Mahmood Nahvi
California Polytechnic State University

Clifford Pollack
Cornell University

Thaddeus Roppel
Auburn University

Mohammad R. Sayeh
Southern Illinois University at Carbondale

Rodney J. Soukup
University of Nebraska–Lincoln

Hal Tharp
University of Arizona

Mark Yoder
Rose-Hulman Institute of Technology

Reference Tables

Contents

The numbers in parentheses after each section name identify the prerequisite previous sections. The symbol † identifies sections within chapters that may be deferred or omitted without loss of continuity.

APPENDIX B CIRCUIT ANALYSIS WITH PSPICE 745

CHAPTER 1

Circuit Variables and Laws

We live in a "wired world" where the quality of daily life depends significantly on applications of electrical phenomena. Just look around and you'll probably see several electrical devices — the lamp that lights this page, a stereo system playing your favorite music, the alarm clock that woke you up, the radio or TV set that brings you news and entertainment, and of course your calculator or computer. All of these familiar products involve circuits designed by electrical engineers. And although electrical engineering includes many other topics, the underlying thread is the electric circuit.

The goal of this textbook is to help you learn the basic concepts and modern engineering methods of circuit analysis. Having learned those skills, you can put them to use in the study of specializations such as:

- Consumer and industrial electronics
- Power generation and distribution
- Semiconductor devices and integrated circuits
- Computer hardware
- Robotics and automation
- Control and communication systems

Further information about these and other facets of electrical engineering is contained in the publications of The Institute of Electrical and Electronics Engineers, commonly known as IEEE (I-triple-E). The magazines *IEEE Potentials* and *IEEE Spectrum* are particularly recommended for survey articles on recent advances and applications, as well as technological forecasts. You might also want to browse the IEEE homepage at http://www.ieee.org/i3e_hp.html.

But regardless of specialization, circuit analysis serves as an essential entry point for the wider scope of electrical engineering. Our investigation of electric circuits begins in this chapter with the variables and fundamental laws used throughout the book.

We'll define current and voltage, along with ideal electrical sources, and we'll develop their relationships to electric energy and power. Then we'll discuss Ohm's law and Kirchhoff's laws. Ohm's law describes individual resistance elements, while Kirchhoff's laws govern currents and voltages for interconnections of any types of elements. The joint use of these laws allows us to analyze simple circuits and even to tackle some design problems.

Like all the chapters that follow, this one contains two features intended to guide your study of the material:

- Stated at the beginning are the chapter *objectives,* a listing of specific concepts and skills to be learned;
- Within the chapter are several *exercises* that reinforce individual objectives.

You should read the objectives and work the exercises as you come to them. Then, for the additional practice needed to master the objectives, you should do some of the problems at the end of the chapter. Solutions for all of the exercises and answers to selected problems marked are given at the back of the book so you can check your work. Finally, you should reread the objectives for review purposes.

Objectives

After studying this chapter and working the exercises, you should be able to do each of the following:

1. Give the definitions and units for current, voltage, and electrical power (Section 1.1).
2. Express the value of an electrical quantity with an appropriate magnitude prefix (Section 1.1).

3. Draw the symbols and current–voltage curves for ideal sources and resistors (Sections 1.2 and 1.3).
4. Use Ohm's law to calculate current, voltage, resistance, and power (Section 1.3).
5. Identify nodes, supernodes, and loops in a given circuit, and apply Kirchhoff's laws to them (Section 1.4).
6. State the defining properties of series and parallel connections (Section 1.4).
7. Analyze a series or parallel circuit consisting of sources and resistors (Section 1.5).
8. Evaluate an unknown voltage, current, or resistance using branch variable analysis (Section 1.5).

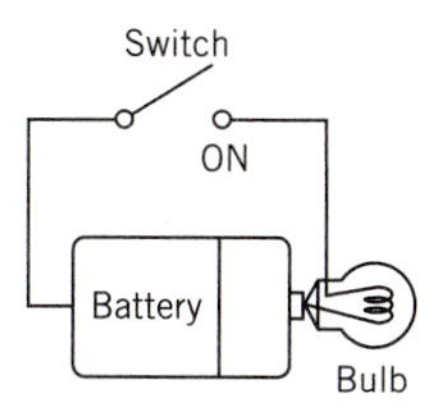

Figure 1.1
Flashlight circuit.

1.1 CURRENT, VOLTAGE, AND POWER

Figure 1.1 depicts a simple flashlight consisting of a battery, a bulb, a switch, and connecting wires. Each device here is a **two-terminal element**, having exactly two electrical contacts. Turning the switch to the ON position establishes a closed electrical path or **circuit**. The battery voltage then forces electric current in the form of moving charges to go from one battery terminal, through the bulb and wires, and back to the other battery terminal. The moving charges transfer energy from the battery to the bulb, where electrical power is converted to heat and light.

This section reviews the units and basic relationships of charge, current, energy, voltage, and electrical power. We'll also introduce magnitude prefixes used in conjunction with very large or very small electrical quantities.

Charge and Current

An electrical **charge** q has two attributes: amount and polarity. The *amount* of charge is expressed in **coulombs** (C) in the International System of Units (abbreviated SI). The *polarity* of the charge is either **positive** or **negative**. For example, the charge carried by an **electron** is

$$q_e = -1.60 \times 10^{-19} \text{ C} \tag{1.1}$$

A proton bears the same amount of charge but with positive polarity.

The polarity designations were assigned by early experimenters, and the negative polarity of an electron has no special significance by itself. What is significant is the fact that charges of opposite polarity attract and tend to neutralize each other, while like charges repel.

Electrical current exists whenever net charge flows past a given point. To illustrate, Fig. 1.2 shows a positive charge $q_1 = 9$ C and a negative charge $q_2 = -2$ C traveling to the right, while another positive charge $q_3 = 3$ C travels to the left. (Situations like this occur in semiconductor devices.) The charge transfer from left to right is given by

$$\Delta q = q_1 + q_2 - q_3 = 9 + (-2) - 3 = 4 \text{ C}$$

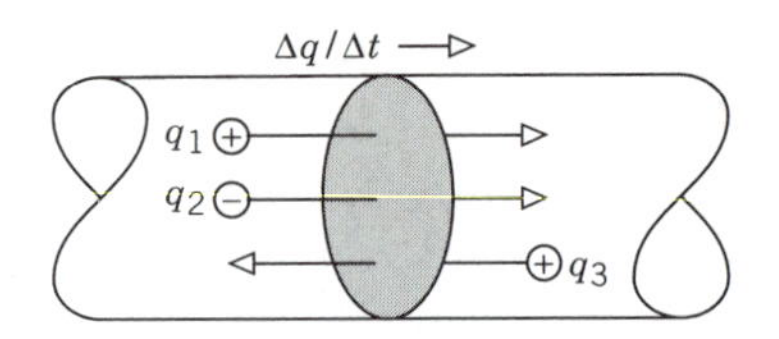

Figure 1.2 Positive and negative charges flowing past a point.

Note that since q_2 is a negative quantity, it has the same effect as a positive charge going the opposite way. If charge transfer Δq takes place in time Δt, then the flow rate or current is $\Delta q/\Delta t$. The arrow at the top of the drawing indicates the direction of flow when Δq is positive.

The study of conductors and electronic devices requires detailed knowledge of internal charge motion. But for purposes of circuit analysis, we only need to know the amount and direction of the resulting current through a wire or element. Specifically, the **instantaneous current** is defined as

$$i \stackrel{\Delta}{=} dq/dt \tag{1.2}$$

where dq/dt stands for the limiting value of the charge transfer rate $\Delta q/\Delta t$ as Δt becomes infinitesimally small.

The SI unit for current is the **ampere** (A), or **amp** for short, and Eq. (1.2) shows that

$$1 \text{ A} = 1 \text{ C/s}$$

Put in words, a one-amp current results from charge transfer at the rate of one coulomb per second.

We capitalize the abbreviations A for ampere and C for coulomb to honor the pioneering work of André Marie Ampère (1775–1836) and Charles Augustin de Coulomb (1763–1806). The letter i is used for current because Ampère originally gave it the French name *intensité.*

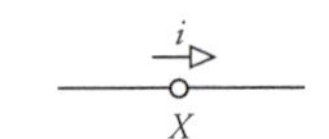

(a) Current reference arrow for equivalent positive charge transfer from left to right

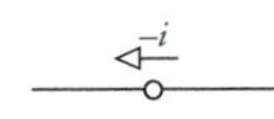

(b) Actual current direction when i is negative

Figure 1.3

The complete specification of the current at any point in a circuit involves *direction* as well as value. By standard convention

> The direction of a positive current i is the direction of equivalent positive charge transfer.

We'll stick with this convention even though current often consists of negatively charged electrons traveling the other way. The labeled arrow in Fig. 1.3*a* thus means that equivalent positive charge goes from left to right past point X at the rate i amps. Delta-tipped arrows such as the one shown here will be used exclusively to indicate current or charge flow.

Frequently, we must assume a **reference direction** for i in order to calculate its value. If the value turns out to be a negative quantity, then we know that the positive current $-i$ actually goes in the opposite of the reference direction. Figure 1.3*b* illustrates this point. In any case, the value label and reference direction must always appear together and be consistent — so the sign of the value must be changed if the direction of the reference arrow is reversed.

All of the circuit elements we'll consider have the property of **charge neutrality**, meaning that net charge cannot accumulate within them. Consequently, when current i enters the upper end of the two-terminal element in Fig. 1.4*a*, an equal current must exit from the bottom terminal. Figure 1.4*b* emphasizes this property with a single current arrow.

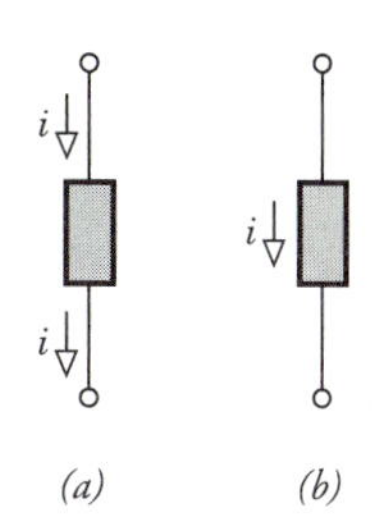

Figure 1.4
Current through a two-terminal element.

We have been speaking here of *instantaneous* current, implying that the value of i may vary with time t. Time dependence is made explicit by writing Eq. (1.2) in the functional notation

$$i(t) = dq(t)/dt$$

We'll use functional notation such as $i(t)$ when necessary to underscore time dependence. Otherwise,

> Lowercase letters represent instantaneous values of quantities that may vary with time.

Capital letters will be reserved for quantities that stay constant over time.

Now suppose we need to determine the **total charge** q_T transferred in some time interval of duration T seconds, say from $t = t_0$ to $t = t_0 + T$. If the current happens to be *constant* at $i = I$ amps throughout the interval, then $I = q_T/T$ so

$$q_T = IT \tag{1.3}$$

For a time-varying current, we draw upon Eq. (1.2) in the form $dq(t) = i(t)\,dt$ and integration yields

$$q_T = \int_{t_0}^{t_0+T} i(t)\,dt \tag{1.4}$$

Thus, q_T equals the *net area* under the plot of $i(t)$ versus t for the interval in question.

Another related quantity of interest is the **average current** over $t_0 \leq t \leq t_0 + T$, defined as

$$i_{av} \triangleq \frac{q_T}{T} = \frac{1}{T}\int_{t_0}^{t_0+T} i(t)\,dt \tag{1.5}$$

Since $q_T = i_{av}T$, comparison with Eq. (1.3) reveals that the charge transferred by a time-varying current with average value i_{av} equals the charge that would be transferred by a constant current of value $I = i_{av}$.

Example 1.1 *Charge Transfer and Average Current*

Suppose the current through some two-terminal element is known to be

$$i(t) = 10 \sin(\pi t/2) \text{ A} \qquad 0 \leq t \leq 6 \text{ s}$$

Figure 1.5 shows the plot of $i(t)$ versus t. We'll use this plot to calculate the charge transferred and average current over the time interval from $t = 0$ to $t = 6$ s.

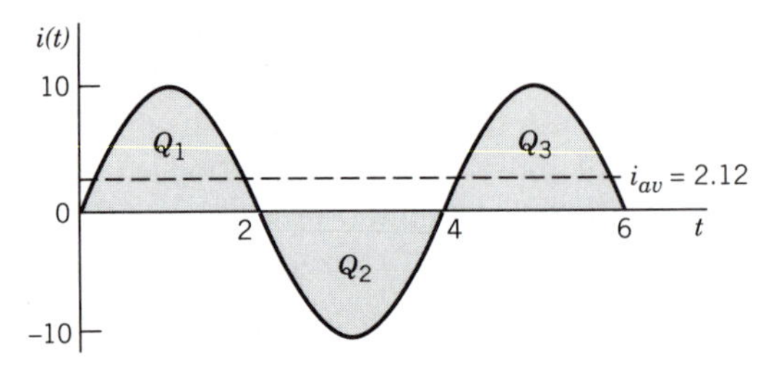

Figure 1.5 $i(t)$ and i_{av}.

During the subintervals $0 \leq t \leq 2$ and $4 \leq t \leq 6$, $i(t)$ is positive and equivalent positive charge goes in the direction of the reference arrow for i. The charge transferred in each of these intervals equals the shaded areas marked Q_1 and Q_3, which happen to have the same value. Hence,

$$Q_3 = Q_1 = \int_0^2 10 \sin \frac{\pi t}{2}\, dt = 10 \left. \frac{-\cos(\pi t/2)}{\pi/2} \right|_0^2 = \frac{40}{\pi} = 12.7 \text{ C}$$

But $i(t)$ takes on negative values throughout the middle interval $2 < t < 4$, so equivalent positive charge goes in the direction opposite the reference arrow. From symmetry, the charge transferred in this interval is $Q_2 = -Q_1$.

The net charge transferred over the entire interval $0 \leq t \leq 6$ s equals the sum

$$q_T = Q_1 + Q_2 + Q_3 = 12.7 \text{ C}$$

Accordingly, the average value of $i(t)$ is

$$i_{av} = 12.7 \text{ C}/6 \text{ s} = 2.12 \text{ A}$$

The dashed line in the figure compares i_{av} with the variations of $i(t)$. By definition of i_{av}, the area under the dashed line is $q_T = 12.7$ C.

Exercise 1.1

Suppose that 7×10^{15} electrons and 2×10^{15} protons travel to the right past a certain point in 0.005 seconds. If the reference direction for current is taken as left to right, then what is the value of i?

Exercise 1.2

Sketch $i(t)$ versus t and evaluate q_T and i_{av} over the interval $0 \leq t \leq 0.5$ s, given that

$$\begin{aligned} i(t) &= -8 \text{ A} \qquad 0 \leq t \leq 0.2 \\ &= +2 \text{ A} \qquad 0.2 < t \leq 0.5 \end{aligned}$$

Energy and Voltage

Current through a lightbulb produces heat and light because the moving charges give up energy to the bulb. Each charge therefore undergoes a change of potential energy, so a **potential difference** exists across the element.

The electrical variable associated with potential difference is **voltage**. Specifically, if charge dq gives up energy dw going from point X to point Y, then the voltage across those points is defined by

$$v \triangleq dw/dq \tag{1.6}$$

We measured voltage in **volts** (V), named after Alessandro Volta (1745–1827). With energy expressed in joules (J), the corresponding SI unit equation is

$$1 \text{ V} = 1 \text{ J/C}$$

Be careful to distinguish the unit abbreviation V from an instantaneous voltage denoted by v or a constant voltage denoted by V.

Polarity again enters the picture because the potential might decrease or increase in the direction of charge flow through an element. Figure 1.6 shows how we'll indicate voltage polarity with a plus sign (+) at the higher potential and a minus sign (−) at the lower potential. Like current reference arrows, the voltage reference marks must be labeled with a known or unknown value v. If v turns out to be negative quantity, then you can reverse the polarity marks *and* change the sign of the value.

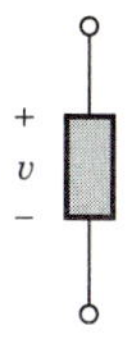

Figure 1.6
Polarity marks for voltage across a two-terminal element with the upper end assumed to be at the higher potential.

When the element under consideration absorbs energy, equivalent positive charge goes from higher to lower potential. Energy-absorbing elements are called **passive elements**, and Fig. 1.7*a* illustrates the **passive polarity convention** in which

> The current reference arrow for a passive element points in the direction from higher to lower potential.

We then say that we have a **voltage drop** across the element. But electrical sources have the ability to supply the energy that activates a circuit, so they are called **active elements**. Figure 1.7*b* illustrates the corresponding **active polarity convention**, in which

> The current reference arrow for an active element points in the direction from lower to higher potential.

Here we have a **voltage rise** across the element.

We use two polarity marks for voltage to emphasize that its definition always involves two points, whereas current is defined at a single point. Stated another way,

(a) Passive polarity convention *(b)* Active polarity convention

Figure 1.7

Voltage is an "across" variable (the potential difference across an element), while current is a "through" variable (the charge flow through a wire or element).

To bring out this difference more clearly, Fig. 1.8 diagrams a set-up for experimentally measuring voltage with a **voltmeter** (VM) and current with an **ammeter** (AM). The ammeter is inserted at one end of the element so that i goes *through* the meter and into the element. But the voltmeter's probes must touch the two ends of the element in order to sense the voltage v *across* it. Notice, by the way, that voltage is usually easier to measure because you can just touch the two terminals with the probes rather than making the disconnection and insertion needed to measure current.

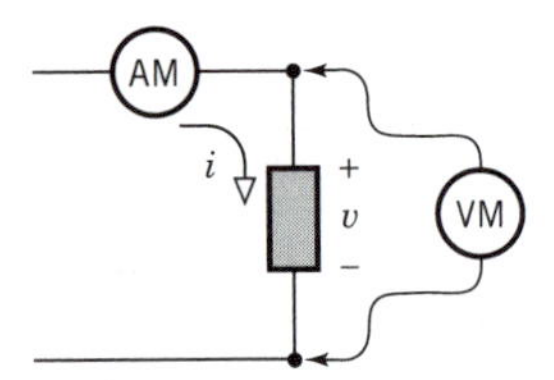

Figure 1.8 Measuring current with an ammeter (AM) and voltage with a voltmeter (VM).

Electric Power

Power is the rate of doing work or the rate of energy transfer, defined in general by

$$p \triangleq dw/dt \tag{1.7}$$

The SI unit for power is named the **watt** (W) in honor of James Watt (1736–1819).

The **electric power** consumed or supplied by a circuit element at any instant of time equals the voltage-current product, namely,

$$p = vi \tag{1.8}$$

so

$$1 \text{ W} = 1 \text{ V} \cdot \text{A}$$

Equation (1.8) follows directly from the definitions of i and v in Eqs. (1.2) and (1.6), since $v \times i = dw/dq \times dq/dt = dw/dt$.

When the reference polarities are drawn according to the passive convention, as in Fig. 1.7*a*, the product vi equals the power *consumed* by the element. When the reference polarities are drawn according to the active convention, as in Fig. 1.7*b*, vi equals the power *supplied* by the element. If vi turns out to have a negative value, then the active element is actually consuming power — as happens, for instance, when we charge a battery.

Electric power is the stock-in-trade of electric utility companies, and many electrical devices are characterized by their power ratings — a 100-W lightbulb, for instance. But your service bill is for energy usage, not power. When the power remains constant at P watts, the **total energy** w_T delivered in a time interval of duration T seconds is

$$w_T = PT \tag{1.9}$$

More generally, when the instantaneous power $p(t)$ varies over $t_0 \leq t \leq t_0 + T$, we must integrate Eq. (1.7) to get

$$w_T = \int_{t_0}^{t_0+T} p(t)\, dt \tag{1.10}$$

Equations (1.9) and (1.10) have the same mathematical form as the expressions for total charge in Eqs. (1.3) and (1.4) because power represents energy flow just as current represents charge flow.

Equation (1.9) brings out the fact that one joule equals one watt-second. However, utility companies generally measure energy in terms of the **kilowatt-hour** (kWh), which equals the energy supplied when $P = 1000$ W and $T = 1$ hour $= 3600$ seconds. Thus, from Eq. (1.9),

$$1 \text{ kWh} = 1000 \text{ W} \times 3600 \text{ s} = 3.6 \times 10^6 \text{ J} \tag{1.11}$$

The kilowatt-hour is not a standard SI unit.

Example 1.2 *Capacity of a Battery*

A typical 12-V automobile battery stores about 5×10^6 J of energy, or somewhat over 1 kWh. We'll calculate energy and charge transfer when a headlight is connected to the battery and draws a 4-A current.

With $v = 12$ V and $i = 4$ A, the instantaneous power consumed by the headlight will be

$$p = 12 \text{ V} \times 4 \text{ A} = 48 \text{ W}$$

If the voltage and current remain constant, then the energy supplied in one minute of operation is

$$w_T = 48 \text{ W} \times 60 \text{ s} = 2880 \text{ J}$$

Since $v = dw/dq$, we conclude that the total charge passing through the headlight during this interval is

$$q_T = w_T/v = 2880 \text{ J}/12 \text{ V} = 240 \text{ C}$$

This result agrees with $q_T = 4 \text{ A} \times 60 \text{ s} = 240$ C.

Incidentally, the maximum charge storage or "capacity" of an automobile battery is often rated in **ampere-hours** (Ah), where

$$1 \text{ Ah} = 1 \text{ C/s} \times 3600 \text{ s} = 3600 \text{ C}$$

For the battery at hand, the capacity is $q_{stored} = w_{stored}/v = 5 \times 10^6 \text{ J}/12 \text{ V} = 4.17 \times 10^5 \text{ C} = 116$ Ah. Hence, the battery would be "dead" after supplying 1 A for 116 hours or 4 A for $116/4 = 29$ hours.

Exercise 1.3

Let $i = 0.5$ A in Fig. 1.7*a*. Find the value of v when: (a) the element consumes $p = 4.5$ W; and (b) the element produces $p = 60$ W.

Exercise 1.4

A 60-W headlight is connected to the battery in Example 1.2. If the voltage stays constant, then how many hours will the headlight operate before the battery is completely discharged?

Magnitude Prefixes

Current, voltage, power, and other electrical quantities exist in very small to very large values. At one extreme, an electronic device might carry 10^{-6} amps and consume 10^{-5} watts. At the other extreme, a high-voltage transmission system might deliver 10^8 watts at 10^5 volts.

Instead of writing powers of 10 all the time, standard SI magnitude prefixes have been adopted to represent them. Table 1.1 lists the most common prefixes used in electrical engineering. The prefix abbreviations allow you to write 2×10^3 W more compactly as 2 kW (kilowatts), for example, or to write 80×10^{-6} A as 80 μA (microamps).

TABLE 1.1 SI Magnitude Prefixes

Prefix	Abbreviation	Magnitude
tera	T	10^{+12}
giga	G	10^{+9}
mega	M	10^{+6}
kilo	k	10^{+3}
milli	m	10^{-3}
micro	μ	10^{-6}
nano	n	10^{-9}
pico	p	10^{-12}
femto	f	10^{-15}

Example 1.3 *Magnitude Manipulations*

Consider a high-voltage device rated for a maximum of 20 W at 50 kV. The corresponding maximum current is

$$i_{max} = \frac{20 \text{ W}}{50 \text{ kV}} = \frac{20}{50 \times 10^3} = 0.4 \times 10^{-3} \text{ A} = 0.4 \text{ mA} = 400 \ \mu\text{A}$$

We could have obtained this result with fewer manipulations by observing that, as far as magnitudes are concerned, $(1 \text{ k})^{-1} = 10^{-3} = 1 \text{ m} = 1000 \ \mu$.

Exercise 1.5

Using appropriate magnitude prefixes, find the power drawn from a source with $v = 5$ kV when: **(a)** $i = 0.4$ μA; **(b)** $i = 10$ mA; and **(c)** $i = 300$ A.

1.2 SOURCES AND LOADS

A complete electric circuit generally includes at least one power-producing source element and one power-consuming load element. The simplest possible circuit thus consists of a source and a load, perhaps with connecting wires, as represented by Fig. 1.9*a*. Positive current i passes through the load from X to Y, and the potential at point X is v volts higher than the potential at point Y. A voltage drop therefore exists across the load, which consumes power $p = vi$. This power comes from the source, which has a voltage rise and supplies power $p = vi$.

Do you find this confusing? Perhaps a simple analogy will help. The hydraulic system in Fig. 1.9*b* acts in many ways like our source-load circuit. We'll describe its operation and indicate the analogous circuit concepts in parentheses. The pump (*source*) forces water flow (*current*) through pipes (*wires*) to drive the turbine (*load*). The water pressure (*potential*) is higher at the inlet port of the turbine than at the output. The turbine consumes power and a pressure drop (*voltage drop*) exists across it in the direction the water flows. The pump, on the other hand, produces a pressure rise (*voltage rise*) and delivers power.

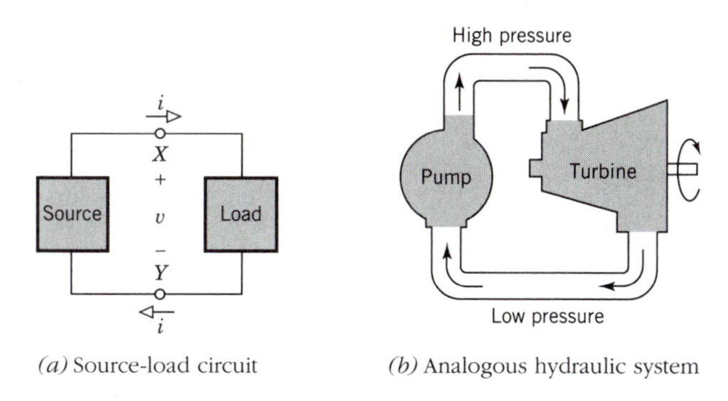

(*a*) Source-load circuit (*b*) Analogous hydraulic system

Figure 1.9

This section further develops the distinction between sources and loads, including the concept of ideal sources. We'll also discuss the role of idealized device models in circuit analysis. Our discussion begins with element descriptions in terms of i-v curves.

i-v Curves

Many two-terminal elements have the property of being **instantaneous** in the sense that they are completely described by the relationship between instantaneous voltage and current. That relationship might be expressed as an equation or plotted as a graph of current versus voltage called an ***i-v* curve**. Such curves can be obtained experimentally with the help of an adjustable source and the meter arrangement back in Fig. 1.8. An i-v curve provides valuable information about the nature and behavior of the element.

As an example, Fig. 1.10 shows an automobile headlight and its i-v curve when the reference polarities follow the passive convention. At every point along the curve we see that i and v have the same sign — both positive or both negative — so there is always a voltage drop across the headlight when $i \neq 0$. Hence, the headlight is a passive device that consumes power when $i \neq 0$ and cannot produce power. We intuitively knew this of course, but the i-v curve now confirms intuition.

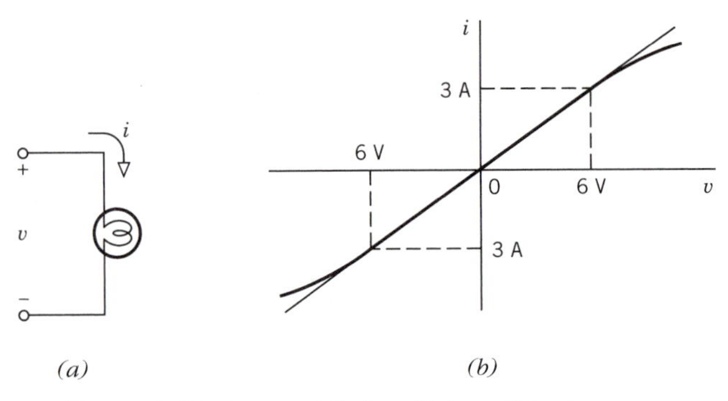

Figure 1.10 Automobile headlight and its i-v curve.

The curve also reveals a nearly straight-line or *linear* characteristic near the origin, with slope $\Delta i/\Delta v = 3\text{ A}/6\text{ V} = 0.5\text{ A/V}$. Accordingly, we have the simple but limited approximation

$$i \approx 0.5v \qquad -6\text{ V} < v < 6\text{ V}$$

At larger values of voltage, we would need to take account of the nonlinearity of the curve.

Plotting current versus voltage here suggests that i is a function of v in the sense that v is the *cause* and i is the *effect.* But this is not necessarily the case. Cause and effect generally depend upon the specific circuit configuration and the type of applied source. Furthermore, any information represented by an i-v curve could just as well be displayed in the form of a v-i curve.

Exercise 1.6

Consider a passive device having $i = 10^{-5}v^3$, with i in amps and v in volts. Sketch the i-v and v-i curves over the range $-30\text{ V} \leq v \leq 30\text{ V}$. (Save your results for use in Exercise 1.7.)

Ideal Sources

Batteries and wall outlets are familiar sources of electrical power. Both are usually classified as *voltage* sources. The voltage may be constant or time varying, but it does not change appreciably with current. Formalizing this concept, we say that:

> An **ideal voltage source** is a two-terminal element whose voltage is a specified constant or function of time, regardless of the current through it.

Source voltage is sometimes called **electromotive force (EMF)** to convey the notion that it provides the "force" that drives current.

Figure 1.11*a* shows the symbol we'll employ for an ideal voltage source with arbitrary time variation $v_s(t)$. We could use this symbol to represent the pickup on an electric guitar, for instance, or the output of a magnetic disk drive in a computer. The accompanying illustrative waveform depicts the voltage $v = v_s(t)$ that might be observed across the terminals.

The plus sign on the symbol for a voltage source identifies the terminal at the higher potential whenever v_s is positive. The reference current arrow has been drawn in accordance with the active convention, assuming that the source produces power and provides a voltage rise.

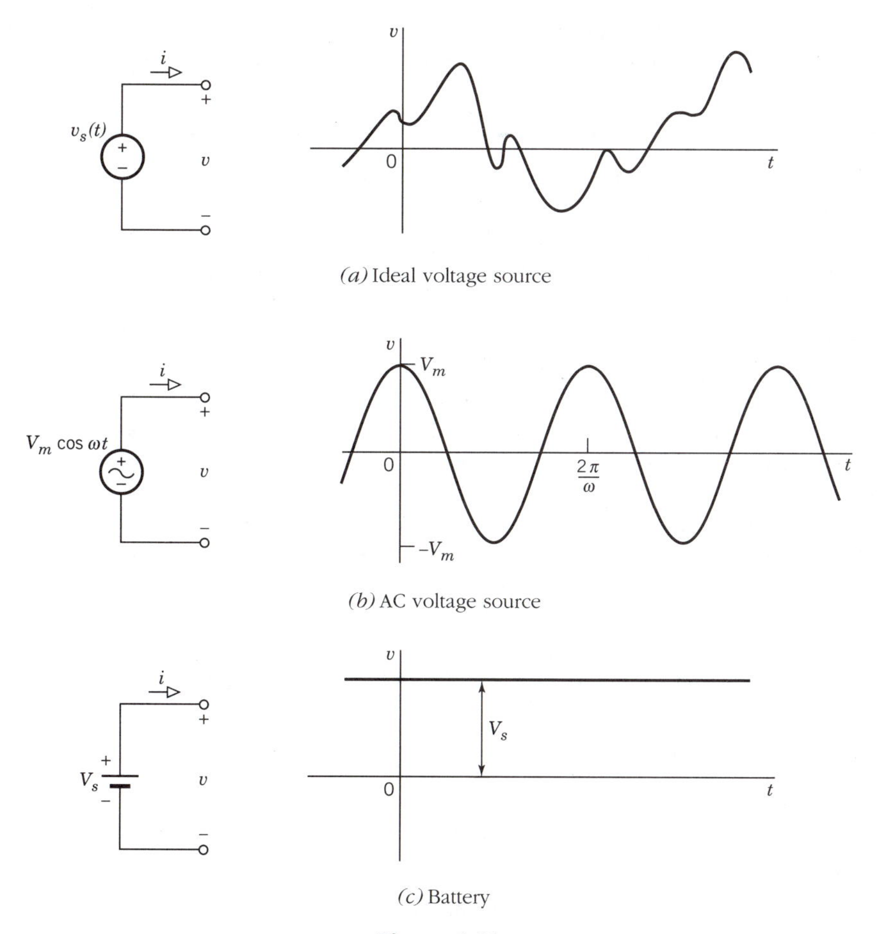

(a) Ideal voltage source

(b) AC voltage source

(c) Battery

Figure 1.11

Figure 1.11*b* shows the symbol and waveform for the special but important case of a **sinusoidal voltage source** with $v_s(t) = V_m \cos \omega t$. The sinusoidal voltage waveform continuously swings up and down between the peak values $+V_m$ and $-V_m$. During those intervals when $v_s(t)$ is negative, the terminal marked + is actually at a lower potential than the other terminal. This source usually supplies a sinusoidal or **alternating current**, abbreviated **ac**. Hence, a sinusoidal voltage source is often called an **ac voltage source**. We'll have much more to say about sinusoidal voltages and currents when we study ac circuits.

The **ideal battery** symbolized in Fig. 1.11*c* is a source of *constant* voltage with respect to time. The capitalized label V_s emphasizes this property, and we could omit the polarity marks here since the higher potential is at the longer of the two lines that make up the symbol. A battery normally supplies constant current $i = I$ whose value depends upon both V_s and the connected circuitry. Since I is a nonalternating or **direct current** (**dc**), we call the battery a **dc voltage source**.

Although most electrical generators act essentially as voltage sources, transistors and certain other devices act more like *current* sources. This means they tend to maintain a certain current rather than voltage. Formally stated,

> An **ideal current source** is a two-terminal element whose current is a specified constant or function of time, regardless of the voltage across it.

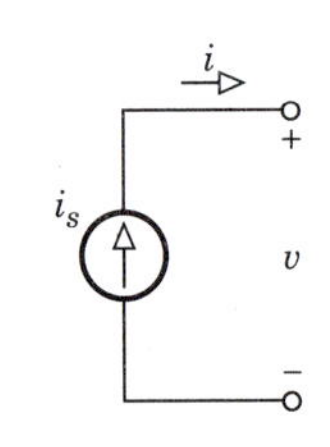

Figure 1.12
Ideal current source.

You should compare this definition with the previous one for an ideal voltage source.

Figure 1.12 represents an ideal current source that supplies current $i = i_s$ with arbitrary time variation. Again, the reference polarities comply with the active convention. We have no special symbol for a dc current source, which would be labeled with I_s instead of i_s. An ac current would be indicated by a sinusoidal "wiggle" drawn across the current arrow.

The difference between ideal voltage and current sources is underscored by their *i-v* curves in Fig. 1.13. The voltage across an ideal voltage source at some particular time is v_s, independent of i. Therefore, the *i-v* curve must be a vertical line at $v = v_s$. This vertical line runs from $i = -\infty$ to $i = +\infty$, and the particular value of i depends on what we connect to the voltage source. Conversely, the *i-v* curve for an ideal current source is a horizontal line at i

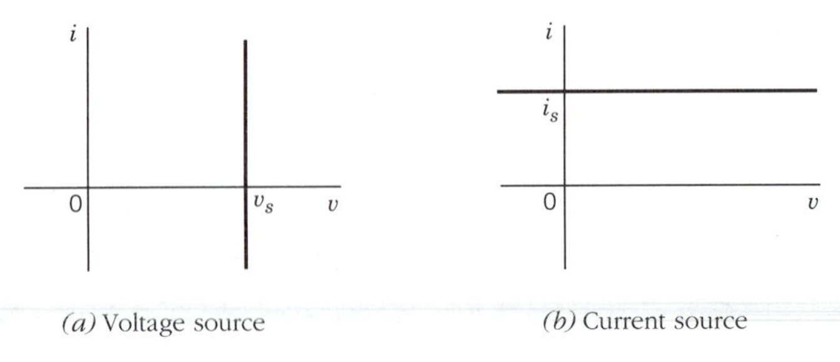

Figure 1.13

$= i_s$, and the value of v depends on what we connect to the current source. The reference polarities for both curves assume a voltage rise across the source in question, so either source produces power when both i and v have the same sign.

Example 1.4 ***i-v Curves for a Source and Load***

Suppose that an ideal 12-V battery is connected to the automobile headlight back in Fig. 1.10. We can find the resulting current and power graphically from Fig. 1.14, where the i-v curves for the battery and the headlight have been plotted on the same set of axes.

The intersection of the two curves is the only point that simultaneously satisfies the properties of both the source and load. Thus, although the ideal battery could supply any value of current, we see that this particular circuit operates only with $i = 5$ A. The battery therefore provides power $p = 12$ V $\times$ 5 A $= 60$ W, which is consumed by the headlight.

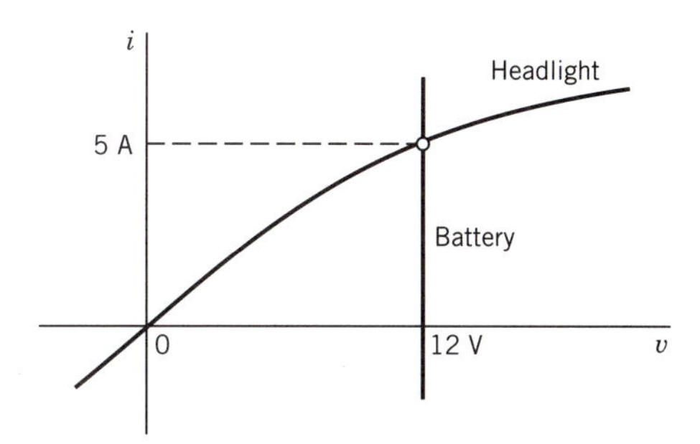

Figure 1.14 i-v curves for a battery and headlight.

Exercise 1.7

Let the current–voltage relationship for the load element in Fig. 1.9*a* be $i = 10^{-5}v^3$, with i in amps and v in volts. Calculate the power consumed by the load when connected to: **(a)** an ideal voltage source so that $v = -v_s = -20$ V; and **(b)** an ideal current source so that $i = i_s = 10$ mA.

Devices and Models

Theoretically, an ideal voltage or current source could produce an unlimited amount of power. Needless to say, such behavior is physically impossible, so ideal sources cannot exist. Why, then, do we bother defining them? The answer to this question is significant, for it relates to an important technique in electrical engineering, namely, the use of simplified representations or **models** of real devices.

Models of natural phenomena — expressed as mathematical relations, curves, etc. — are idealizations representing those aspects of the physical characteristics most relevant to a particular application. A good model allows you to predict with reasonable accuracy how the device will perform under the expected operating conditions. For instance, if you know that the headlight in Fig. 1.10 will have $|v| < 6$ V, then you can use the simplifying approximation $i \approx 0.5v$. Similarly, for a limited range of current, it might be quite reasonable to assume that a real battery acts like an ideal voltage source.

In general, a model helps you concentrate on significant factors and effects without getting bogged down in the details of a more accurate but more cumbersome description of the device. Of course, you must always bear in mind the assumptions and limitations of the model because predictions that go beyond a model's scope are likely to be erroneous. Consider, for example, the invalid conclusion that a battery could produce unlimited power.

Most branches of engineering and science involve mathematical models. (As a case in point, Newton's laws of motion are models that apply only when the velocity is small compared to the velocity of light.) Electrical engineering perhaps does more with models than some other fields. Indeed, every element symbol and circuit diagram is a model, often a very good one, but still a model.

From now on, it should go without saying that the various device representations are approximations of physical reality. Where appropriate, we'll discuss the significant limitations of these models. We'll also build better models of real devices by combining ideal elements.

1.3 OHM'S LAW AND RESISTORS

A kink in a garden hose restricts the flow of water, producing a pressure drop and the conversion of mechanical energy to heat. Analogously, the flow of charge in an electric current always encounters some *resistance,* resulting in a voltage drop and the conversion of electrical energy to heat.

Resistance may be desired in a circuit to produce a voltage drop or energy conversion, or it may be an unwanted but unavoidable part of a device or connecting wire. The properties of electrical resistance and resistors are examined here, along with a brief discussion of the resistivity of solid materials.

Resistors and Resistance

An **ideal linear resistor** is an energy-consuming element described by **Ohm's law**, named after Georg Simon Ohm (1797–1854). This law states that the voltage and current are directly proportional to each other, so we write

$$v = Ri \tag{1.12}$$

The proportionality constant R is the **resistance** measured in **ohms** (Ω).

Rewriting Ohm's law as $R = v/i$ yields the unit equation

$$1\ \Omega = 1\ \text{V/A}$$

Thus, resistance equals the ratio of voltage to current. Resistance values in typical circuits range from fractions of an ohm to **kilohms** (kΩ) or even as large as **megohms** (MΩ).

Figure 1.15 gives the symbol and i-v curve for an ideal linear resistor — henceforth simply called a resistor. The passive reference convention is used here because a resistor is a passive element. The slope of the i-v curve equals $1/R$ since, from Eq. (1.12),

$$i = \frac{v}{R} = \frac{1}{R}v$$

Clearly, i increases with v for a given value of R. However, i decreases as R increases because resistance *resists* current. This interpretation should help you keep track of the three different ways of writing Ohm's law — $v = Ri$, $R = v/i$, and $i = v/R$.

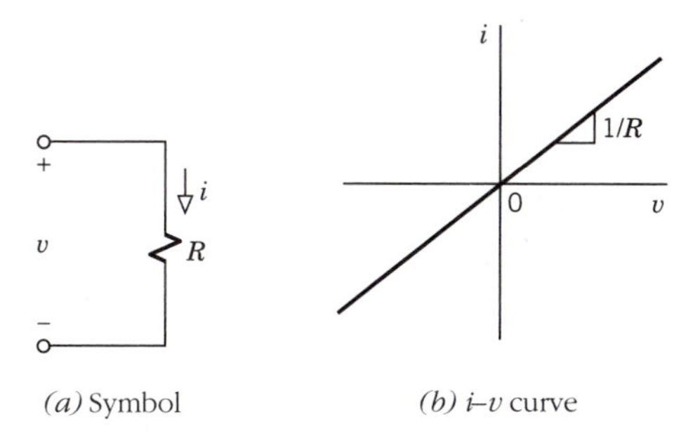

(*a*) Symbol (*b*) i–v curve

Figure 1.15 Ideal resistor.

Since the voltage and current are proportional to each other, a resistor in a circuit diagram may be labeled in terms of just one variable. Thus, Fig. 1.16*a* shows a resistor labeled with voltage v and current v/R, while Fig. 1.16*b* shows a resistor labeled with current i and voltage Ri. These labels expedite circuit analysis by directly incorporating Ohm's law. However, if one reference polarity happens to be reversed — counter to the passive convention — then Ohm's law must be written with a minus sign. In Fig. 1.16*c*, for example, the voltage is $v = R(-i) = -Ri$ because the current that goes in the assumed direction of the voltage drop is $-i$ rather than i. Consequently, either v or i will have a negative value.

For occasional use, it proves convenient to define **conductance** G as the reciprocal of resistance, so

$$G \triangleq 1/R \tag{1.13}$$

In recognition of the work of the brothers Werner and William Siemens, the SI unit for conductance is **siemens** (S). Nonetheless, you may still encounter the earlier and rather droll term **mho** (ohm spelled backwards), symbolized

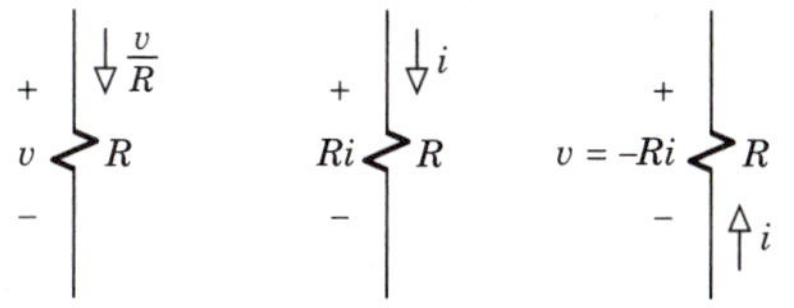

(*a*) (*b*) (*c*)

Figure 1.16 Resistor labeling for Ohm's law.

by an inverted omega (℧). Using conductance rather than resistance, Ohm's law becomes

$$i = Gv \tag{1.14}$$

With tongue in cheek, some people refer to Eq. (1.14) as "Mho's law."

Going back to the headlight in Fig. 1.10, we found that $i \approx 0.5v$ when $|v| < 6$ V. Thus, for small voltages, this *nonlinear* element can be modeled as a *linear* resistor with conductance $G \approx 0.5$ S — so $R = 1/G \approx 2\ \Omega$.

An ideal **ON–OFF switch** also may be modeled in terms of resistance or conductance, but with very extreme values. Specifically, in the closed or ON position of Fig. 1.17*a*, the switch creates a zero-resistance path ($R = 0$) such that $v = 0 \times i = 0$ for any i. This condition is known as a **short circuit**. In the open or OFF position of Fig. 1.17*b*, the path through the switch has zero conductance ($G = 0$) such that $i = 0 \times v = 0$ for any v. This condition is known as an **open circuit**.

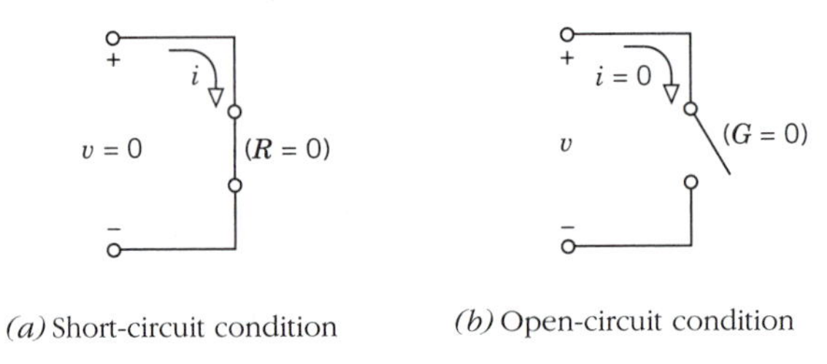

Figure 1.17 Ideal ON–OFF switch.

Exercise 1.8

If the current through a resistor is 50 mA when $v = 20$ V, then what are the values of R and G?

Power Dissipation

Combining Ohm's law with $p = vi$ gives two equivalent expressions for the power consumed by a resistor, namely:

$$p = (Ri)i = Ri^2 \tag{1.15a}$$
$$= v(v/R) = v^2/R \tag{1.15b}$$

Either expression can be used, depending upon whether you know the current through the resistor or the voltage across it.

In a process called **ohmic heating**, power consumed by a resistor is dissipated in the form of heat. Coils of resistive wire are used for precisely this purpose in toasters and other appliances. In the case of an incandescent lightbulb, the filament glows when heated by current and produces light. Ohmic heating also underlies the principle of the simple fuse: When the current gets too large, the heat melts the fuse element and the circuit is "broken."

Excessive ohmic heating may cause serious damage to connecting wires and electronic devices. For this reason, cooling fans are built into computers and other electronic systems.

Example 1.5 *Calculations with Consistent Units*

Figure 1.18 shows the circuit and i-v curves for a 4-mA current source connected to a 5-kΩ resistor. We'll find the resulting voltage across the resistor and the power dissipated by it.

From Ohm's law, the two i-v curves intersect at

$$v = Ri = 5\text{ k}\Omega \times 4\text{ mA} = 5 \cdot 10^3 \times 4 \cdot 10^{-3} = 20\text{ V}$$

We then have three ways of calculating power, since

$$p = vi = 20\text{ V} \times 4\text{ mA}$$
$$p = Ri^2 = 5\text{ k}\Omega \times (4\text{ mA})^2$$
$$p = v^2/R = (20\text{ V})^2/5\text{ k}\Omega$$

all of which yield

$$p = 80\text{ mW} = 0.08\text{ W}$$

Notice that voltage comes out directly in volts when we express resistance in kilohms and current in milliamps. Furthermore, with these same units, power comes out directly in milliwatts. We therefore say that volts, milliamps, kilohms, and milliwatts form a **consistent set of units**. Using consistent units eliminates the nuisance of converting magnitude prefixes back to powers of 10.

Now suppose the resistance increases to 5 MΩ, so $v = 5\text{ M}\Omega \times 4\text{ mA} = 20\text{ kV}$ and $p = 20\text{ kV} \times 4\text{ mA} = 80\text{ W}$. In the limit as $R \rightarrow \infty$, the voltage and power both become infinite, theoretically, because we have assumed an *ideal* current source.

Exercise 1.9

Calculate i and p when the current source in Fig. 1.18*a* is replaced by a 12-V battery with the upper terminal at the higher potential.

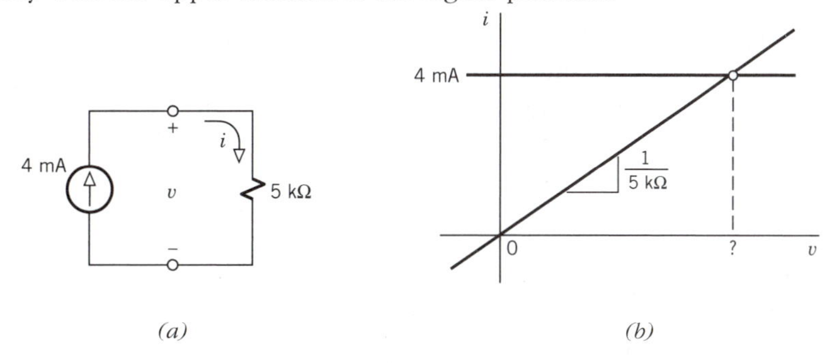

Figure 1.18 Circuit and i-v curves for a current source connected to a resistor.

Resistivity

As we have stated it, Ohm's law describes the *external* characteristics of a resistive element. But the resistance actually depends upon the material and shape of the element and its temperature.

Figure 1.19 depicts a voltage source connected across a uniform piece of solid material having length l and cross-sectional area A. The voltage v establishes an **electric field** $\mathcal{E}$ inside the material directed from higher to lower potential and of value

$$\mathcal{E} = v/l \tag{1.16}$$

This electric field exerts a force on any charged particles, and the resulting charge flow constitutes a current in the direction of the field given by

$$i = A\mathcal{E}/\rho \tag{1.17}$$

The constant ρ (rho) is an electrical property of the material called the **resistivity**, measured in **ohm-meters** ($\Omega \cdot$ m). The reciprocal of resistivity is the **conductivity**, denoted by $\sigma = 1/\rho$.

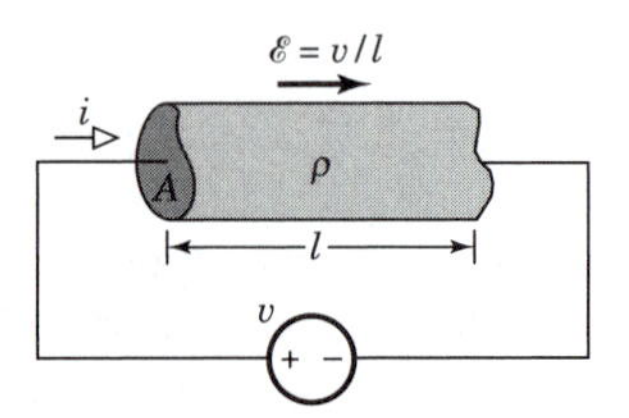

Figure 1.19 Current through a solid.

Although it may not appear to be so, Eq. (1.17) is just another version of Ohm's law. For when we insert Eq. (1.16), we get

$$i = \frac{A(v/l)}{\rho} = \frac{v}{(\rho l/A)} = \frac{v}{R}$$

where

$$R = \rho l/A \tag{1.18}$$

Equation (1.18) shows that the resistance of an element is proportional to resistivity and length, but inversely proportional to area. Hence, the wire used to make a large resistance should have high resistivity and small area. But a wire intended to carry large currents should have low resistivity and large area in order to minimize the ohmic heating $p = Ri^2$.

Table 1.2 lists values of ρ for some representative materials grouped into three general categories. At the one extreme are the electrical **conductors** — primarily metals — with very small resistivities. At the other extreme are the electrical **insulators** — rubber, plastics, etc. — with resistivities so large that ordinary voltages produce virtually no current. In between are the **semiconductors**, the basic materials for electronic devices and integrated circuits.

Thanks to the tremendous range of available resistivities, electricity is a convenient means for transporting energy. This is because good insulators

TABLE 1.2 Resistivities of Selected Materials at 20°C

Type	Material	ρ ($\Omega \cdot$ m)
Conductors	Copper	1.7×10^{-8}
	Aluminum	2.8×10^{-8}
	Nichrome	10^{-6}
	Carbon	3.5×10^{-5}
Semiconductors	Germanium	0.46
	Silicon	2300
Insulators	Rubber	10^{12}
	Polystyrene	10^{15}

keep the energy "contained" within the good conductors that waste little power in ohmic heating. Consider, for example, the 12-gauge copper wire that carries up to 20 A in residential wiring. This wire has a radius of about 1 mm so, from Eq. (1.18), the resistance per meter length is

$$R/l = \rho/\pi r^2 = (1.7 \times 10^{-8})/(\pi \times 10^{-6}) \approx 0.005\ \Omega/\text{m}$$

If $l = 10$ m and $i = 20$ A, then $R \approx 0.05\ \Omega$ and $p \approx 20$ W. When connecting a 120-V source to a load, the 10-m length of wire could handle as much as 120 V $\times$ 20 A = 2400 W while wasting less than 1% in ohmic heating.

An important factor not included in Table 1.2 is *temperature dependence*. The resistivity of a conductor generally increases with temperature, whereas the resistivity of an insulator decreases with temperature. Special care therefore must be taken when designing circuits for operation at extreme temperatures. But temperature dependence also lends itself to practical applications in devices such as the resistance-wire thermometer.

The temperature dependence of ρ explains in part why the *i-v* curve for many resistive elements becomes *nonlinear* at large values of current. An incandescent lamp filament, for instance, has the nonlinear characteristic previously seen in Fig. 1.10*b*, and its resistance at the operating temperature (around 2000°C) is more than 10 times as large as the "cold" resistance. Consequently, when first energized, the in-rush current will be substantially higher than the current drawn by the filament after it heats up to operating temperature.

Commercially manufactured resistors are designed to be essentially linear over their intended operating range. Such resistors are usually made of carbon, metal film, or wire, and they come in a variety of standard values, precision tolerances, and power ratings. Adjustable resistors called *potentiometers* will be described in Section 2.1.

Finally, we should at least mention **superconductors**. These special materials possess the unique property of *zero resistivity* and, hence, they completely eliminate the power loss associated with ohmic heating. Metals become superconducting only at temperatures near absolute zero, but some ceramic compounds show promise as superconductors at more reasonable temperatures for practical applications.

Example 1.6 *A Strain Gauge*

A **strain gauge** consists of many loops of thin, resistive wire glued to a flexible backing, as portrayed by Fig. 1.20. This device is used to measure the small fractional elongation or **strain** $\Delta l/l$ of a structural member to which it is attached. Straining the wire makes it somewhat longer and thinner, thereby increasing the resistance a small amount ΔR.

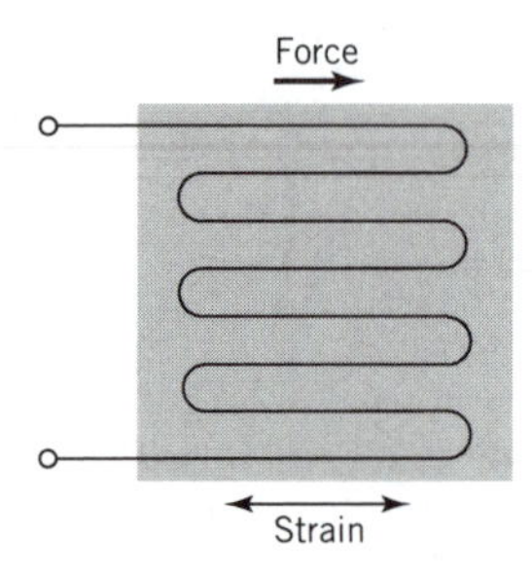

Figure 1.20 Resistive-wire strain gauge.

To derive an expression for ΔR, let the strained length be $l_{st} = l + \Delta l$. If the volume of the wire remains constant, then $A_{st}l_{st} = Al$ so its area under strained conditions becomes $A_{st} = Al/l_{st} = Al/(l + \Delta l)$. Thus, from Eq. (1.18), the strained resistance is

$$\begin{aligned} R_{st} &= \rho l_{st}/A_{st} = \rho(l + \Delta l)^2/Al \\ &= (\rho l/A)\,[1 + 2(\Delta l/l) + (\Delta l/l)^2] \end{aligned}$$

But with $(\Delta l/l)^2 \ll 1$, we can neglect the last term and write

$$R_{st} \approx (\rho l/A) + 2(\rho l/A)(\Delta l/l) = R + \Delta R \tag{1.19a}$$

where $R = \rho l/A$ is the unstrained resistance and

$$\Delta R = 2R(\Delta l/l) \tag{1.19b}$$

This result confirms that we can determine small mechanical strain by measuring the increased resistance of the strain gauge.

Exercise 1.10

Suppose a strain gauge is made from a 100-cm length of nichrome wire having 0.002-cm radius. Calculate ΔR when $\Delta l/l = 0.001$.

Lumped-Parameter Models

Since resistance is distributed along the entire length of conducting material, a more accurate pictorial representation of a resistor might take the form of Fig. 1.21*a*. In circuit analysis, however, we are concerned only with the voltage and current measured at the *terminals*. We therefore ignore internal effects and consider the resistance to be concentrated at a single point, as implied by Fig. 1.21*b*.

Figure 1.21*b* is called a **lumped-parameter model** because a spatially distributed property — resistance in this case — has been lumped entirely at one point. The lines connecting the element symbol to its terminals are then treated as ideal conductors having zero resistance. Almost all circuit diagrams consist of lumped-parameter symbols for the particular elements involved. This convention focuses attention on the terminal characteristics of the elements rather than on conditions within the elements.

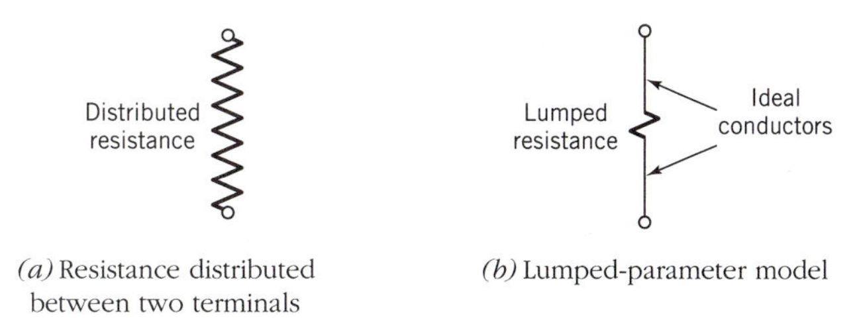

(*a*) Resistance distributed between two terminals

(*b*) Lumped-parameter model

Figure 1.21

1.4 KIRCHHOFF'S LAWS

The basic goal of circuit analysis is to determine particular voltage and current values, given a lumped-parameter diagram of the circuit in question. For this purpose we obviously need the *i-v* relationships of the individual elements. Additionally, we need the two laws first formulated in 1847 by Gustav Kirchhoff (pronounced Kear-koff).

Kirchhoff's laws are *circuit relationships* pertaining to the interconnection of elements, irrespective of the types of elements involved. This section presents and discusses those laws. We'll also introduce the distinctive properties of series and parallel connections.

Kirchhoff's Current Law

Electrical elements may be fastened together in various ways — using screw terminals, solder, or just twisted wires. Any such connection forms a **node**, defined by the property that

> A node is any connection point of two or more circuit elements.

Since charge must be conserved and does not accumulate at a node, the amount of charge flowing out of a node equals the amount flowing in at any instant. In other words, an electrical node acts like a junction of water pipes where the amount of water going out equals the amount coming in.

Kirchhoff's current law (abbreviated **KCL**) expresses conservation of charge in terms of currents entering and leaving a node. Specifically,

> The sum of the currents leaving any node equals the sum of the currents entering that node.

As an example, Fig. 1.22 shows a circuit fragment with one node marked by the solid dot. Currents i_1 and i_2 enter the node while i_3 leaves it, so

$$i_3 = i_1 + i_2$$

This relationship holds for each and every instant, regardless of what elements are connected at the node.

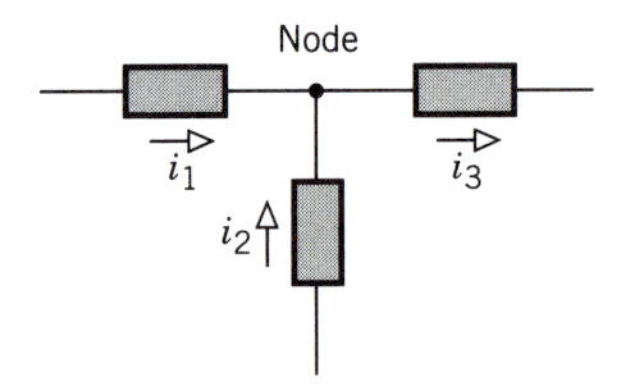

Figure 1.22 Node with three currents.

An alternative version of KCL is obtained by viewing currents leaving a node as *negative* currents entering the node. We can then say that

> The algebraic sum of all currents into any node equals zero.

This algebraic sum is written symbolically as

$$\sum_{\text{node}} i = 0 \tag{1.20}$$

Applying Eq. (1.20) to Fig. 1.22 yields

$$i_1 + i_2 + (-i_3) = 0$$

where i_3 bears a negative sign because its reference arrow points away from the node.

If you know the values of i_1 and i_3 in Fig. 1.22 and you want to evaluate i_2, then you would use KCL in the form $i_2 = i_3 - i_1$. We frequently need such expressions for circuit analysis, so you should practice writing them by inspection of the diagram.

A particularly simple but significant implication of KCL is obtained by considering Fig. 1.23, which is called a **series connection**. In general,

> Two or more elements are in series when each node connects just two elements.

The application of KCL in Fig. 1.23 yields $i_2 = i_1$ at node X and $i_3 = i_2$ at node Y, so

$$i_3 = i_2 = i_1$$

We therefore conclude that

Elements in series carry the same current.

This property holds at every instant of time.

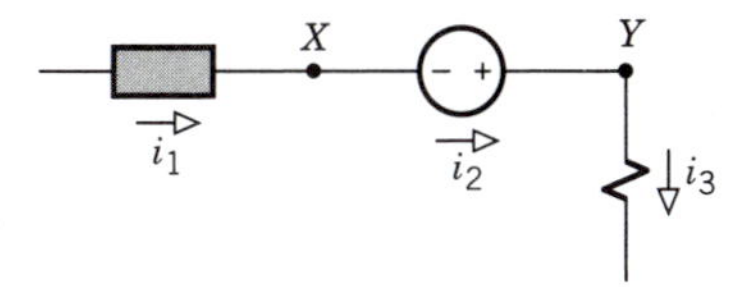

Figure 1.23 Series connection.

Given the foregoing conclusion, the series connection of ideal current sources in Fig. 1.24 appears to violate KCL. We'll eventually show that the apparent violation does not occur with *real* current sources. Meanwhile, we merely observe that series-connected current sources cannot reasonably be modeled as *ideal* sources.

Now consider the more complicated circuit fragment in Fig. 1.25. The dashed curve here surrounds a portion known as a **supernode**, where

A supernode is any closed region that contains two or more nodes and whose boundary intersects some connecting wires, each wire being intersected just once.

Since net charge cannot accumulate within a supernode, KCL requires that

The algebraic sum of all currents into any supernode equals zero.

For the supernode at hand we get

$$i_1 + i_2 - i_3 - i_4 = 0$$

This result is easily checked by noting at node A that $i_1 + i_2 - i_x = 0$ and that $i_x = i_3 + i_4$ at node B.

Figure 1.24 Series connection violating KCL.

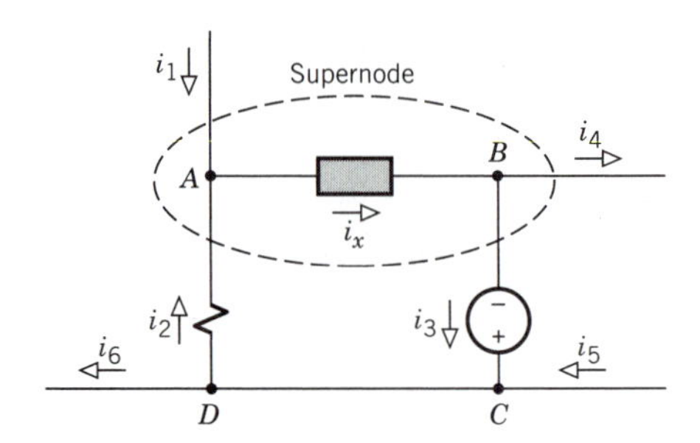

Figure 1.25 Circuit fragment marked with a supernode.

Supernodes save effort when you don't need to find a particular current such as i_x in Fig. 1.25. Furthermore, if you don't care about i_2, i_x, and i_3, then a supernode containing nodes *A*, *B*, *C*, and *D* leads to the conclusion that $i_1 - i_4 + i_5 - i_6 = 0$. You could also enclose nodes *C* and *D* to obtain $i_3 + i_5 = i_2 + i_6$. Actually, points *C* and *D* constitute a *single node* because they are joined by an ideal conductor.

Exercise 1.11

Find i_2 and i_6 in Fig. 1.25 given that $i_x = 4$ A, $i_1 = -2$ A, $i_4 = 6$ A, and $i_5 = 8$ A.

Kirchhoff's Voltage Law

Every electrical circuit contains at least one **loop**, defined by the property that

> A loop is any path that goes from node to node and returns to the starting node, passing just once through each node.

Since energy must be conserved when a charge goes around a loop, the energy given up by the charge equals the energy it gains. The same energy-conservation principle would apply if you carried a rock over a hill and back to where you started.

Kirchhoff's voltage law (KVL) expresses conservation of energy in terms of voltage rises and drops around a loop. Specifically,

> The sum of the voltage drops around any loop equals the sum of the voltage rises.

In this context, a voltage drop means that the potential decreases along the direction of travel around the loop while a voltage rise means that potential increases.

By way of example, consider the loop *ABEA* marked by the dashed curve in Fig. 1.26. If a positive charge travels in the clockwise direction as indicated, then v_1 and v_2 are voltage drops while v_x is a voltage rise, so

$$v_1 + v_2 = v_x$$

This relationship holds for each and every instant of time, regardless of what elements are in the loop.

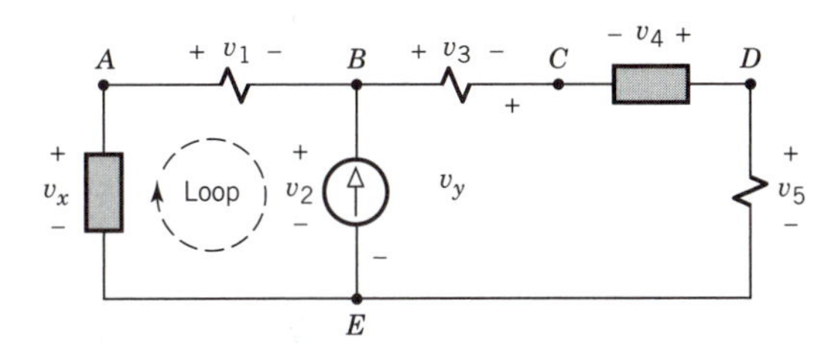

Figure 1.26 Circuit marked with a loop.

Like the current law, the voltage law can be stated more compactly as an algebraic sum by considering a voltage rise to be a *negative* voltage drop. Then

> The algebraic sum of all voltage drops around any loop equals zero.

Symbolically, this statement becomes

$$\sum_{\text{loop}} v = 0 \tag{1.21}$$

The algebraic sum for loop *ABEA* in Fig. 1.26 is

$$v_1 + v_2 + (-v_x) = 0$$

By the same method, going clockwise around the outer loop yields $v_1 + v_3 - v_4 + v_5 - v_x = 0$. Going counterclockwise would, of course, reverse all the signs. Or you might evaluate v_x by writing $v_x = v_1 + v_3 - v_4 + v_5$.

KVL also applies to loops containing an *open circuit* between two nodes. For instance, the voltage v_y in Fig. 1.26 certainly has some value that could be measured with a voltmeter. This value must satisfy KVL around any loop that includes the *CE* or *EC* jump. Thus, from the loop *BCEB*, we see that $v_y = v_2 - v_3$.

Figure 1.27 illustrates the simple but important configuration called a **parallel connection**. There are actually just two nodes here, but each one has been "split" to make a tidy rectilinear drawing. In general,

> Two or more elements are in parallel when their terminals are connected to the same pair of nodes.

The application of KVL in Fig. 1.27 yields $v_2 = v_1$ for the left-hand loop and $v_3 = v_2$ for the right-hand loop, so

$$v_3 = v_2 = v_1$$

We therefore conclude that

> Elements in parallel have the same voltage across each one of them.

This property holds at every instant of time.

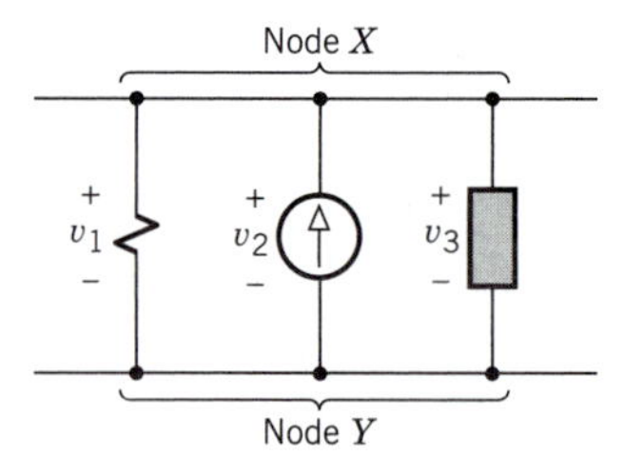

Figure 1.27 Parallel connection.

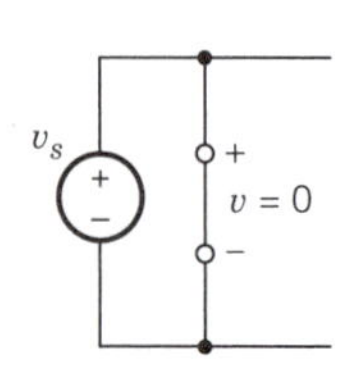

Figure 1.28 Parallel connection violating KVL.

Finally, Fig. 1.28 illustrates an apparent violation of KVL caused by a closed switch in parallel with an ideal voltage source. This arrangement is physically impossible because *ideal* voltage sources do not exist. (However, a short circuit across a *real* voltage source could draw enough current to damage or destroy the source!)

Example 1.7 *A Transistor Circuit*

Figure 1.29 diagrams a circuit containing a **bipolar junction transistor**. This transistor is a three-terminal electronic device connected at nodes *B*, *C*, and *E*, indicated by small circles for emphasis. Some of the voltages and currents have been measured, and their values are labeled on the diagram. We want to determine the remaining unknown values by applying Kirchhoff's laws.

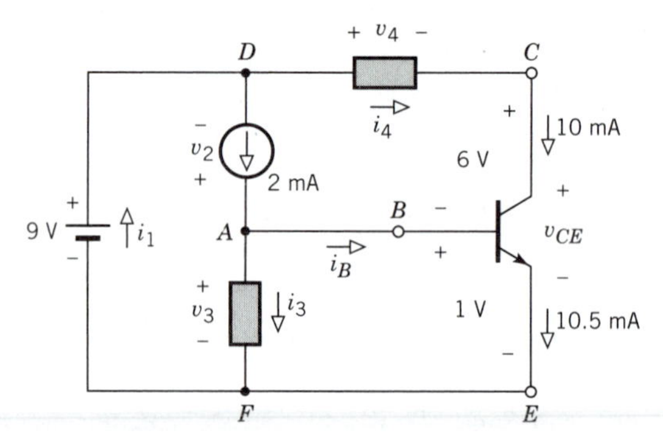

Figure 1.29 Circuit with a bipolar junction transistor.

First, we find i_B by mentally enclosing the transistor with a supernode. KCL then gives

$$i_B + 10 \text{ mA} - 10.5 \text{ mA} = 0 \quad \Rightarrow \quad i_B = 0.5 \text{ mA}$$

Next, from KVL for the loop *CEBC* around the transistor, we get

$$v_{CE} - 1 \text{ V} - 6 \text{ V} = 0 \quad \Rightarrow \quad v_{CE} = 7 \text{ V}$$

Further examination of the diagram reveals that

$$i_4 = 10 \text{ mA} \qquad v_3 = 1 \text{ V}$$

Additional node and loop sums yield the following results:

Node *D*	$i_1 = i_4 + 2 \text{ mA} = 12 \text{ mA}$
Node *A*	$i_3 = 2 \text{ mA} - i_B = 1.5 \text{ mA}$
Loop *DCEFD*	$v_4 = 9 \text{ V} - v_{CE} = 2 \text{ V}$
Loop *AFDA*	$v_2 = v_3 - 9 \text{ V} = -8 \text{ V}$

The negative value of v_2 means, of course, that node *A* is actually at a lower potential than node *D*. Consequently, the 2-mA source happens to absorb power in this circuit.

Exercise 1.12

What conditions must be satisfied to comply with KVL when two ideal voltage sources are connected in parallel?

Exercise 1.13

Find v_y and v_2 in Fig. 1.26, given that $v_3 = 20$ V, $v_4 = 3$ V, and $v_5 = -6$ V.

1.5 ELEMENTARY CIRCUIT ANALYSIS

With Ohm's and Kirchhoff's laws in hand, we're now prepared to analyze simple circuits consisting of resistors and ideal sources. We'll start with circuits in which all the elements are in series or in parallel. Then we'll tackle other configurations — including a design problem — using the branch variable method.

Series Circuits

A **series circuit** consists entirely of series-connected elements. Consequently, the same current i goes through each element. Figure 1.30 shows three such circuits, and the node dots have been omitted because we've already taken account of KCL. The analysis of these series circuits therefore involves the joint use of KVL and Ohm's law.

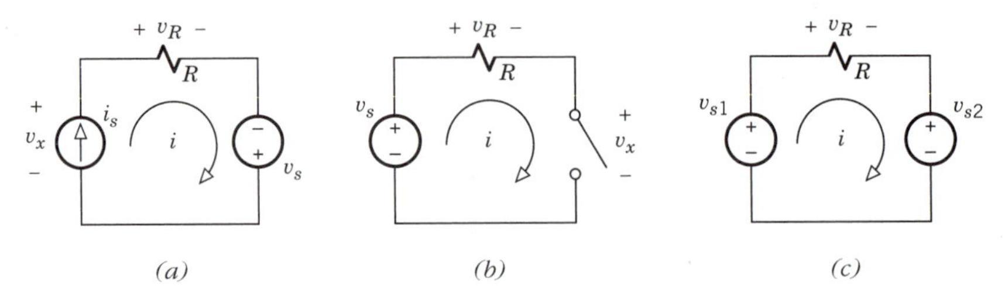

Figure 1.30 Series circuits.

The current source in Fig. 1.30*a* forces $i = i_s$, so $v_R = Ri_s$. Applying KVL around the loop in either direction yields

$$v_x = v_R - v_s = Ri_s - v_s \tag{1.22}$$

This result further illustrates the general property that the voltage across a current source is established by the elements connected to it.

The open circuit in Fig. 1.30*b* requires that $i = 0$, so $v_R = Ri = 0$. Thus, the resulting voltage across the open switch must be

$$v_x = v_s - v_R = v_s \tag{1.23}$$

Note carefully that v_x equals v_s because there is no voltage drop across the resistor when $i = 0$.

The current in Fig. 1.30*c* is not immediately known. However, KVL gives $v_R = v_{s1} - v_{s2}$ from which

$$i = v_R/R = (v_{s1} - v_{s2})/R \tag{1.24}$$

This expression leads to the conclusion that the effective source voltage across the resistor is the *algebraic sum* $v_{s1} - v_{s2}$. Hence, the two series voltage sources could be combined into one for purposes of analysis. By the way, Fig. 1.30*c* might represent a flashlight with two batteries, one of which has been turned the wrong way.

One final point can be noted here: Altering the *order* of the elements in these or any other series circuits would not affect the current or voltages since KVL around the loop still yields the same values.

Exercise 1.14

Let the circuit in Fig. 1.30*c* have $v_{s1} = 9$ V and $R = 60\ \Omega$. Find the values of v_R and i when: **(a)** $v_{s2} = 12$ V; **(b)** $v_{s2} = 9$ V; and **(c)** $v_{s2} = -9$ V.

Parallel Circuits

A **parallel circuit** consists entirely of parallel-connected elements. Consequently, from KVL, the same voltage appears across each element. Figure 1.31 shows three such circuits, each labeled with just one unknown voltage. The analysis of these parallel circuits therefore involves the joint use of KCL and

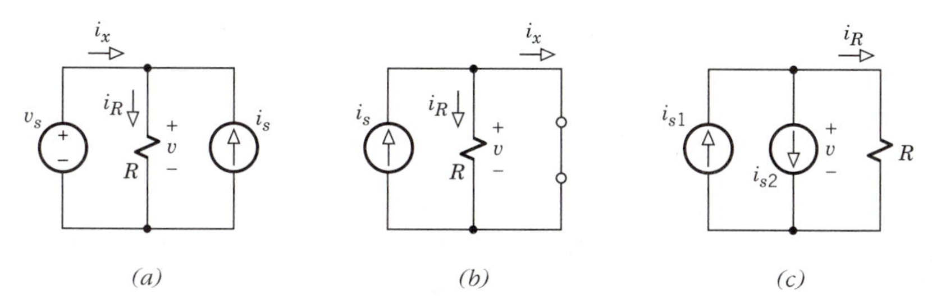

Figure 1.31 Parallel circuits.

Ohm's law. KCL may be applied at either the upper or lower node since both nodes involve the same currents.

The voltage source in Fig. 1.31*a* forces $v = v_s$, so $i_R = v_s/R = Gv_s$. Applying KCL at either node yields

$$i_x = i_R - i_s = Gv_x - i_s \tag{1.25}$$

This result further illustrates the general property that the current through a voltage source is established by the elements connected to it.

The short circuit in Fig. 1.31*b* requires that $v = 0$, so $i_R = v/R = 0$. Thus, the resulting current through the closed switch must be

$$i_x = i_s - i_R = i_s \tag{1.26}$$

Note carefully that i_x equals i_s because there is no current through the resistor when $v = 0$.

The voltage in Fig. 1.31*c* is not immediately known. However, KCL gives $i_R = i_{s1} - i_{s2}$ from which

$$v = Ri_R = R(i_{s1} - i_{s2}) \tag{1.27}$$

This expression leads to the conclusion that the effective source current through the resistor is the *algebraic sum* $i_{s1} - i_{s2}$. Hence, the two parallel current sources could be combined into one.

Finally, observe that altering the order of the elements in these or any other parallel circuits would not affect the voltage or currents since KCL at either node still yields the same values.

Example 1.8 ***Series and Parallel Source Connections***

To illustrate further the properties of series and parallel circuits, consider the following design task:

> A two-terminal passive electronic device is intended to operate with $v_x = 10$ V and $i_x = 2.5$ A. You want to supply these terminal values using either a 12-V voltage source or a 3-A current source.

Since the source values exceed the terminal values you must use at least one other element, which could be a resistor.

Figure 1.32*a* diagrams one possible design solution in the form of a series circuit with the voltage source. The role of the resistor R_{ser} is to introduce a voltage drop between the source and the device. Application of KVL around the loop shows that you want $v_R = 12 - v_x = 2$ V. But R_{ser} carries current i_x so $v_R = R_{ser} i_x = R_{ser} \times 2.5$ A. Hence,

$$R_{ser} \times 2.5 \text{ A} = 2 \text{ V} \quad \Rightarrow \quad R_{ser} = 2 \text{ V}/2.5 \text{ A} = 0.8\ \Omega$$

Alternatively, you could connect the current source in parallel with resistance R_{par} and device X, per Fig. 1.32*b*. The role of R_{par} is to bypass the excess current around X, and KCL at either node yields $i_R = 3 - i_x = 0.5$ A. But the voltage v_x appears across R_{par} so $v_x = R_{par} i_R = R_{par} \times 0.5$ A. Hence,

$$R_{par} \times 0.5 \text{ A} = 10 \text{ V} \quad \Rightarrow \quad R_{par} = 10 \text{ V}/0.5 \text{ A} = 20\ \Omega$$

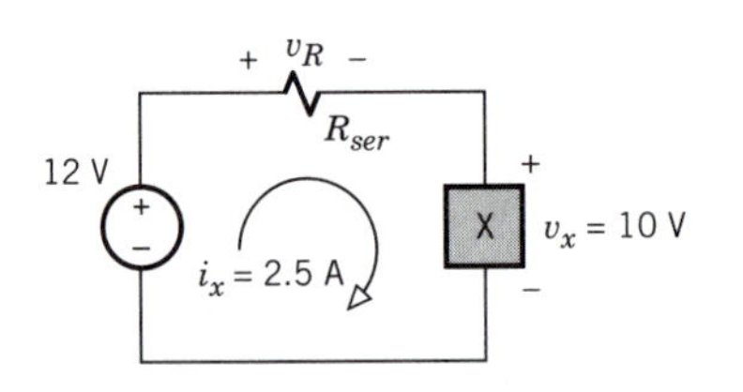

(*a*) Electronic device with series voltage source and resistor

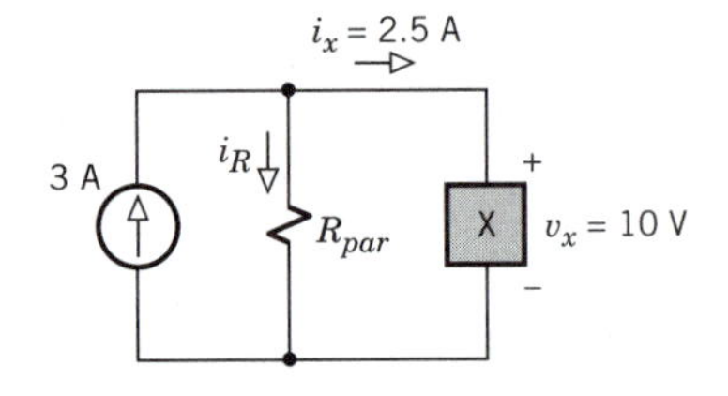

(*b*) Electronic device with parallel current source and resistor

Figure 1.32

Exercise 1.15

If the circuit in Fig. 1.31*a* has $i_s = 10$ mA, $R = 25$ kΩ, and $i_x = -4$ mA, then what is the value of v_s?

Exercise 1.16

Explain why the voltage source in Fig. 1.32*a* cannot be replaced by the current source from Fig. 1.32*b*, and vice versa.

Branch Variable Analysis

Circuit analysis becomes more difficult when the elements are not all series connected or parallel connected. One method for analyzing such circuits involves **branch variables**, which are the voltages or currents associated with individual elements. This method works best when you have *auxiliary information* about the circuit such as the current through a resistor or the voltage across it. More general techniques will be developed in subsequent chapters.

The key to branch variable analysis is careful labeling of the circuit diagram. The purpose of this labeling is to expedite the analysis by eliminating extraneous unknowns and directly incorporating Ohm's law. The procedure consists of four steps:

1. Label series-connected elements with the same current through them, and parallel-connected elements with the same voltage across them.

Choose any convenient reference polarities for unknown branch variables, provided that resistors are marked in accordance with the passive convention.

2. Incorporate the auxiliary information in the labels. Specifically, when the current i_R through a resistor R is known or is to be found, label the voltage $v_R = Ri_R$. Conversely, when v_R is known or is to be found, label the current $i_R = v_R/R$.
3. After all branch variables have been labeled, evaluate the unknown voltages and currents by successively applying Kirchhoff's laws to loops and nodes that involve just *one* unknown.
4. Check your results by putting them on the diagram to make sure that they satisfy Kirchhoff's laws.

Two examples will help illustrate the use of branch variables in circuit analysis and design.

But an additional comment is in order before we continue. The foregoing procedure is the first of several multiple-step techniques we will develop to help organize the work of circuit analysis. Accordingly, you should not attempt to memorize them blindly. Rather, you should concentrate on the underlying logic so that you can apply the procedure with understanding of the successive steps.

Example 1.9 *Calculating Branch Variables*

We are given the circuit diagrammed in Fig. 1.33*a*, with the additional information that $v_4 = 24$ V. The stated problem is to find the corresponding value of the source current i_s and the power supplied by the sources. We'll solve this problem by going through the steps of the branch variable analysis method.

Step 1: We start by labeling the circuit as shown in Fig. 1.33*b*, where i_1 and v_2 are the unknown branch variables needed to determine the supplied power.

Step 2: The branch current through the 8-Ω resistor is labeled $i_4 = v_4/8 = 3$ A (because we were given $v_4 = 24$ V) and it follows that $v_3 = 7i_4 = 21$ V. The labeled diagram now clearly brings out the things that we know and the unknowns to be found.

Step 3: Since v_2 is the only unknown voltage around the loop on the right, application of KVL yields

$$v_2 = v_3 + v_4 = 45 \text{ V} \qquad v_1 = 25 - v_2 = -20 \text{ V}$$

Hence,

$$i_2 = v_2/9 = 5 \text{ A} \qquad i_1 = v_1/10 = -2 \text{ A}$$

The negative value of i_1 means that the voltage source is actually absorbing power, so

$$p_v = 25 \text{ V} \times i_1 = -50 \text{ W}$$

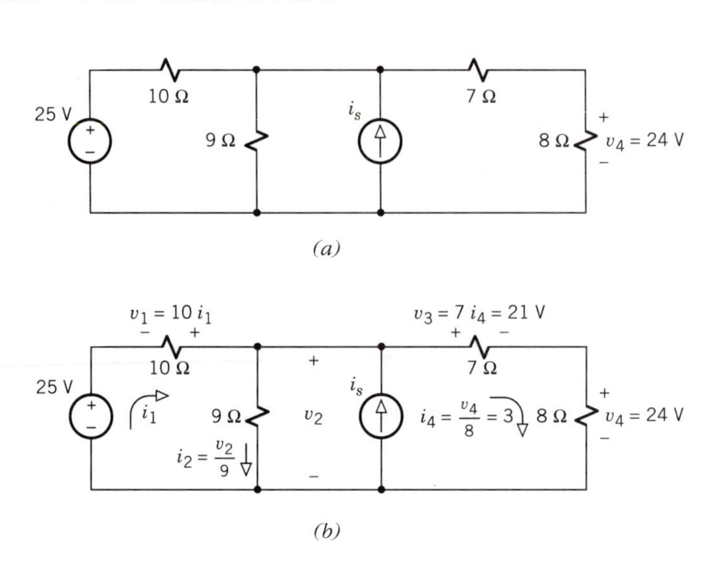

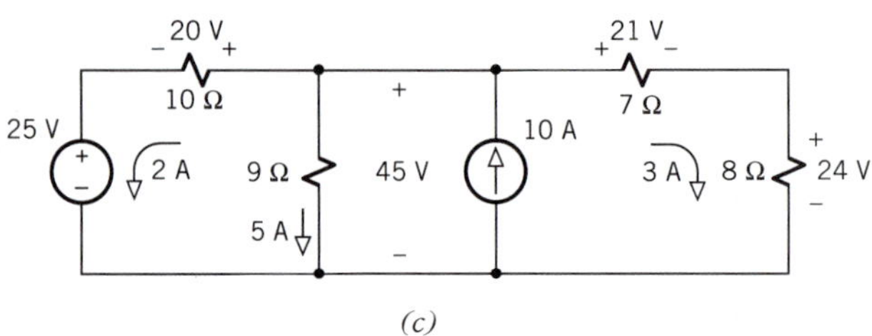

Figure 1.33 Illustration of branch variable analysis.

The current source must therefore supply all the power. The required current i_s is then determined by observing that the upper two nodes constitute a single node for KCL. Thus,

$$i_s = i_2 + i_4 - i_1 = 10 \text{ A}$$

and the current source supplies

$$p_i = v_2 \times i_s = 450 \text{ W}$$

Step 4: As a check on our results, Fig. 1.33*c* shows the same circuit labeled with all voltage and current values. Note that two reference polarities have been reversed here to account for the negative values $v_1 = -20$ V and $i_1 = -2$ A. This diagram confirms that all loop and node sums satisfy Kirchhoff's laws. Furthermore, using $p_R = Ri^2$, the total power dissipated by the resistors is found to be

$$p = 10 \times 2^2 + 9 \times 5^2 + (7 + 8) \times 3^2 = 400 \text{ W}$$

which agrees with $p_v + p_i = -50 + 450 = 400$ W.

Example 1.10 *Design of a Biasing Circuit*

Two passive electronic devices in an automobile are to be powered by the 12-V battery. However, the devices must operate under the following conditions:

	Voltage	Current
Device *A*	4 V	20 mA
Device *B*	5 V	16 mA

In view of the different voltage and current requirements, the devices cannot be connected directly in series or parallel with the battery. We therefore need to design a circuit configuration with additional elements.

Since the sum of the device voltages is less than 12 V, and since device *A* carries more current than device *B*, a suitable configuration might take the form of Fig. 1.34. The **biasing resistors** R_A and R_B are included here to satisfy

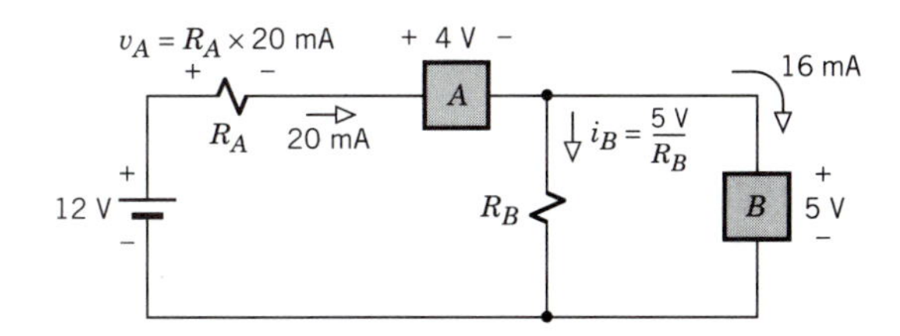

Figure 1.34 Circuit with two devices and biasing resistors.

Kirchhoff's laws. The design work now boils down to finding appropriate values for R_A and R_B.

The resistor in series with device A must introduce the additional voltage drop

$$v_A = 12 - 4 - 5 = 3 \text{ V}$$

Thus, we want

$$R_A = v_A/20 \text{ mA} = 0.15 \text{ k}\Omega = 150 \ \Omega$$

The resistor in parallel with device B must carry the surplus current

$$i_B = 20 - 16 = 4 \text{ mA}$$

Thus, we want

$$R_B = 5 \text{ V}/i_B = 1.25 \text{ k}\Omega$$

which completes the design.

Exercise 1.17

Given that $i_1 = 5$ A in Fig. 1.35, use branch variable analysis to find the corresponding values of v_2 and R_3.

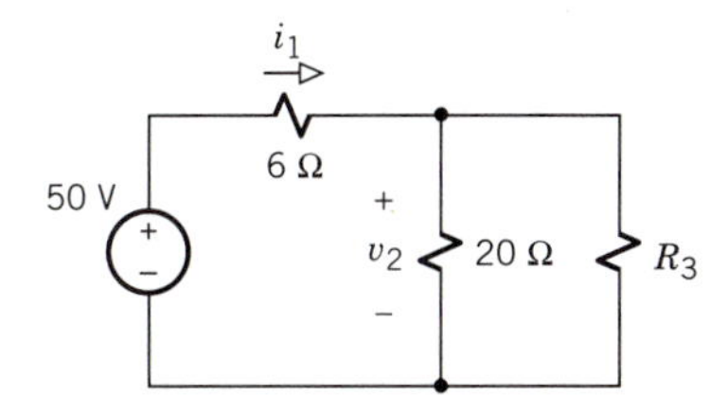

Figure 1.35

Exercise 1.18

Design a circuit with two biasing resistors so that the devices in Example 1.10 can be powered by a 6-V battery.

PROBLEMS

Section 1.1 Current, Voltage, and Power

1.1* Calculate the total charge transferred and the value of i_{av} over $0 \leq t \leq 20$ s when $i(t) = 0.3t$ A.

1.2 Calculate the total charge transferred and the value of i_{av} over $0 \leq t \leq 1$ min when $i(t) = 0.5 - 0.01t$ A with t in seconds.

1.3 Calculate the total charge transferred and the value of i_{av} over $0 \leq t \leq 1$ hour given that $i(t) = 0.1t - 200$ A with t in seconds.

1.4 Calculate the total charge transferred and the value of i_{av} over $0 \leq t \leq 1$ ms when $i(t) = e^{-500t}$ A with t in seconds.

1.5 A 12-V automobile battery is rated at 250 Ah. Find the stored charge and energy.

1.6* The charge stored by a new 9-V alkaline battery is 720 C. Find the stored energy and the ampere-hour rating.

1.7 The energy stored by a new 3-V lithium battery is 54 kJ. Find the stored charge and the ampere-hour rating.

1.8 If a utility company charges 5 cents/kWh, then how much does it cost to operate a 60-W light continuously for one week?

1.9 A load operating at 120 V draws constant power P and consumes 14.4 kWh in one day. Find P and the average current.

1.10 An automobile clock continuously draws 4 mA from the 12-V battery. How many kilowatt hours does it consume in one year?

Section 1.2 Sources and Loads

1.11 Suppose the source in Fig. P1.11 produces $v = 12 - 0.5i$ V. Find the supplied power when $i = 2$ A, 8 A, and -4 A. Compare each value with the power supplied by an ideal 12-V source.

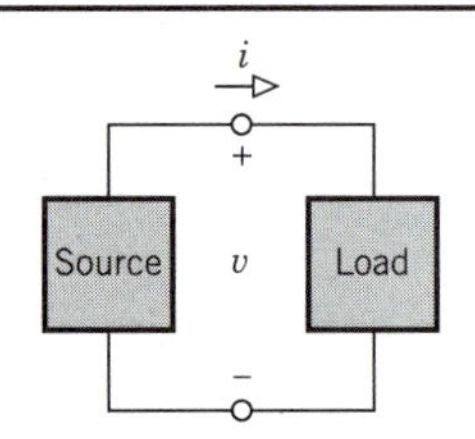

Figure P1.11

1.12 Do Problem 1.11 with $v = 12 - 1.5i$ V.

1.13 Suppose the source in Fig. P1.11 produces $i = 20 - 0.2v$ mA. Find the supplied power when $v = 10$ V, 50 V, and -25 V. Compare each value with the power supplied by an ideal 20-mA source.

1.14 Do Problem 1.13 with $i = 20 - 0.6v$ mA.

1.15 A certain source produces $v = 40 - 100i^2$ V. Justify the assertion that it acts like an ideal voltage source when $|i| \leq 0.2$ A.

1.16 A certain source produces the current $i = 0.5/\sqrt{400 + v^2}$ A. Justify the assertion that it acts like an ideal current source when $|v| \leq 10$ V.

Section 1.3 Ohm's Law and Resistors

1.17* Example 1.5 illustrated that volts, milliamps, kilohms, and milliwatts form a *consistent* set of units. Fill in the blanks in the table below so that each row forms another consistent set that does not require conversion of magnitude prefixes.

Voltage	Current	Resistance	Power
V	mA	kΩ	mW
kV	______	______	MW
V	______	MΩ	______

1.18 Add rows to the table in Problem 1.17 for:
(a) current in μA and power in nW;
(b) voltage in kV and resistance in MΩ.

1.19 Add rows to the table in Problem 1.17 for:
(a) current in kA and resistance in mΩ;
(b) resistance in Ω and power in pW.

1.20 A certain ideal source supplies 100 W when connected to a 4-Ω load resistor and 25 W when connected to a 16-Ω load resistor. Determine the type of source, and evaluate the voltage and current for each load.

1.21 A certain ideal source supplies 1.2 kW when connected to a 75-Ω load resistor and 4.8 kW when connected to a 300-Ω load resistor. Determine the type of source, and evaluate the voltage and current for each load.

1.22 A certain ideal source supplies 250 mW when connected to a 10-kΩ load resistor and 100 mW when connected to a 4-kΩ load resistor. Determine the type of source, and evaluate the voltage and current for each load.

1.23 A certain ideal source supplies 40 mW when connected to a 250-kΩ load resistor and 200 mW when connected to a 50-kΩ load resistor. Determine the type of source, and evaluate the voltage and current for each load.

1.24 A bar of silicon is 2 cm long with a square cross-section. If $R = 1$ MΩ at room temperature, then what are the cross-sectional dimensions in centimeters?

1.25 A bar of germanium is 5 cm long with a circular cross-section. If $R = 100$ Ω at room temperature, then what is the cross-sectional radius in centimeters?

1.26 A resistor is fabricated by depositing a thin film of carbon on a cylinder 10 mm long and 1 mm in diameter. Estimate the film's thickness t in micrometers such that $R = 1$ kΩ at room temperature.

1.27 Consider a 120-V lightbulb that dissipates 150 W at the operating temperature. Find the inrush current, given that the resistivity at the operating temperature is eight times the room-temperature resistivity.

1.28 The resistance of a **fuse** increases with temperature such that $R = R_c(1 + \alpha T)$, where R_c is the "cold" resistance, α is the temperature coefficient, and T the temperature rise above 20°C. The temperature rise is given by $T = kp$, where k is a constant and p is the power dissipated by the fuse. Obtain an expression for R in terms of the current i through the fuse, and show that it "blows out" ($R \rightarrow \infty$) at $i = 1/\sqrt{\alpha k R_c}$.

Section 1.5 Elementary Circuit Analysis

(See also PSpice problems B.1 and B.2.)

1.29 Let $R_2 = 10$ Ω in Fig. P1.29. Find v_1 and i_s, given that $i_4 = 2$ A.

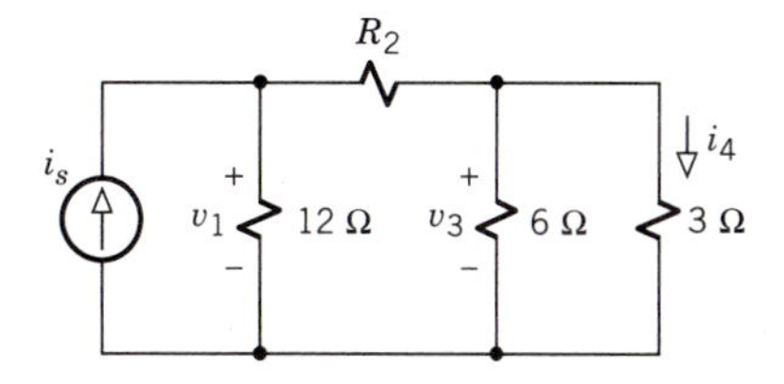

Figure P1.29

1.30 The circuit in Fig. P1.29 is found to have $v_1 = 18$ V and $v_3 = 9$ V. What are the values of i_s and R_2?

1.31 If the circuit in Fig. P1.31 has $v_2 = 15$ V and $R_3 = 24$ kΩ, then what are the values of i_3 and v_s?

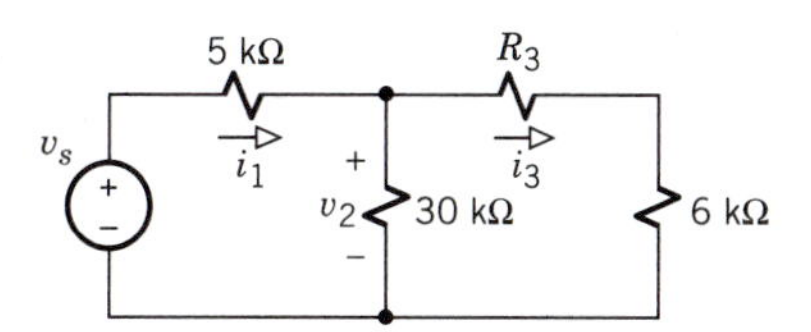

Figure P1.31

1.32 If the circuit in Fig. P1.31 has $v_s = 90$ V and $i_1 = 6$ mA, then what are the values of i_3 and R_3?

1.33 If the circuit in Fig. P1.31 has $v_s = 60$ V and $v_2 = 30$ V, then what are the values of i_1 and R_3?

1.34 Find v_b and R_3 in Fig. P1.34, given that $v_s = 12$ V and $v_a = 6$ V.

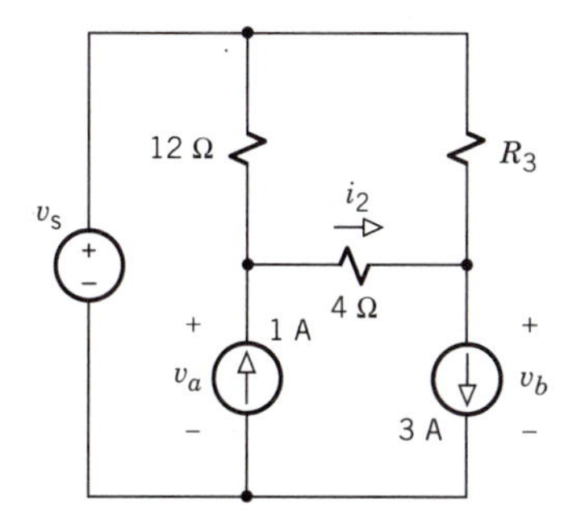

Figure P1.34

1.35 Find v_s and R_3 in Fig. P1.34, given that $v_a = 12$ V and $v_b = 8$ V.

1.36 Find v_b and R_3 in Fig. P1.34, given that $v_s = 18$ V and $i_2 = 2$ A.

1.37 The transistor circuit in Fig. P1.37 has $R_1 = 12.5$ kΩ, $R_2 = 25$ kΩ, $R_3 = 0$, $R_E = 700$ Ω, and $R_C = 1$ kΩ. If $v_1 = 5$ V and $i_E = 6$ mA, then what are the values of i_B, i_C, v_{CE}, and v_{BE}?

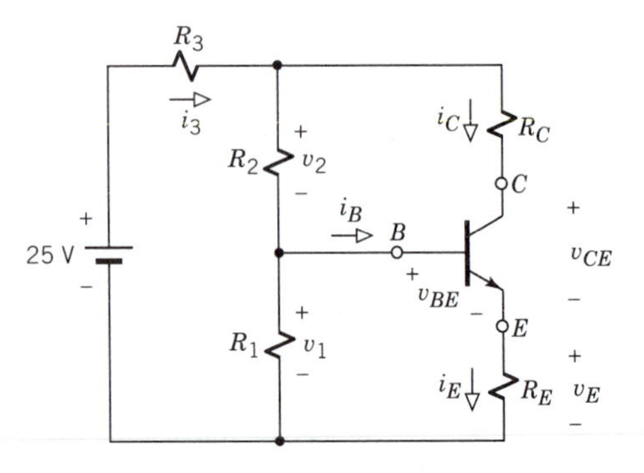

Figure P1.37

1.38 The transistor circuit in Fig. P1.37 has R_1 = 10 kΩ, R_2 = 13 kΩ, R_3 = 1 kΩ, R_E = 4 kΩ, and R_C = 5 kΩ. If i_B = 0.1 mA, i_C = 2 mA, and v_{BE} = 0.6 V, then what are the values of v_1, v_2, v_{CE}, and i_3?

1.39 The transistor circuit in Fig. P1.37 has R_1 = 2 kΩ, R_2 = 7.2 kΩ, R_3 = 1 kΩ, R_E = 500 Ω, and R_C = 1500 Ω. If v_1 = 3.6 V, v_{BE} = 1 V, and i_3 = 7 mA, then what are the values of i_B, i_C, v_{CE}, and v_2?

1.40 Let the circuit in Fig. P1.40 have i_s = 15 mA. Find v_s and the power supplied by each source when i_3 = 1 mA.

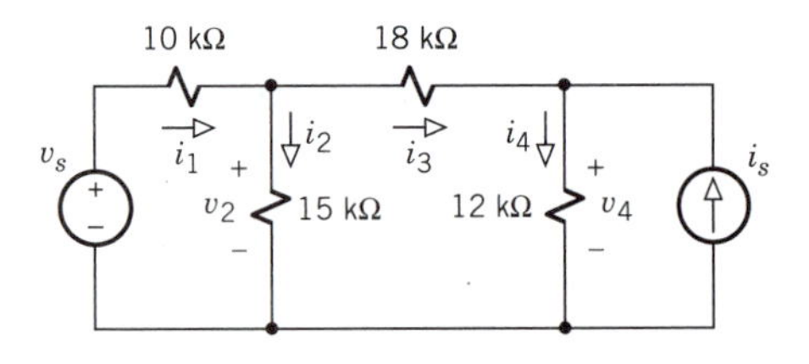

Figure P1.40

1.41 Let the circuit in Fig. P1.40 have i_s = 10 mA. Find v_s and the power supplied by each source when i_4 = 7 mA.

1.42 Let the circuit in Fig. P1.40 have v_s = 60 V. Find i_s and the power supplied by each source, given that i_2 = 4 mA.

1.43 Let the circuit in Fig. P1.40 have v_s = 40 V. Find i_s and the power supplied by each source, given that i_3 = 1 mA.

1.44 Suppose the transistor circuit in Fig. P1.37 has R_3 = 0. Determine values for the other resistors so that i_B = 0.1 mA, i_C = 2 mA, i_3 = 5.6 mA, v_{BE} = 0.7 V, v_{CE} = 6.7 V, and v_1 = 7 V.

1.45 Suppose the transistor circuit in Fig. P1.37 has R_E = 0. Determine values for the other resistors so that i_3 = 6 mA, i_B = 0.2 mA, i_C = 5 mA, v_{BE} = 0.6 V, v_{CE} = 6 V, and $v_1 + v_2$ = 16 V.

1.46 Suppose the transistor circuit in Fig. P1.37 has R_C = 0. Determine values for the other resistors so that i_B = 0.5 mA, i_E = 5 mA, i_3 = 6 mA, v_{BE} = 1 V, v_E = 3 V, and v_{CE} = 10 V.

1.47. Three devices have the following characteristics:

	Device A	*Device B*	*Device C*
Voltage	6 V	4 V	3 V
Current	15 mA	5 mA	10 mA

Design an appropriate circuit using two biasing resis tors and a 12-V source to supply these devices. Arrange your circuit to minimize the current drawn from the source.

1.48 Design a circuit using a 9-V source and two biasing resistors to supply the devices in Problem 1.47.

1.49 Design a circuit using a 7-V source and two biasing resistors to supply the devices in Problem 1.47.

1.50 Design a circuit using a 40-mA source and biasing resistors to supply the devices in Problem 1.47. Arrange your circuit to minimize the voltage across the source.

CHAPTER 2

Properties of Resistive Circuits

This chapter focuses on the properties of circuits consisting entirely of sources and resistors. Although resistive circuits might appear to lack glamour, there are two major reasons why they deserve further attention:

- The properties of resistive circuits will be put to use in the next chapter when we explore several important practical applications.
- The concepts and techniques developed here will be extended subsequently to include circuits containing other types of elements.

The study of resistive circuits thus becomes the foundation for all aspects of circuit analysis.

We'll begin with resistors connected in series and parallel, which leads us to the concepts of equivalent resistance and duality. Then we'll introduce controlled sources and generalize the definition of equivalent resistance. Finally, we'll develop handy analysis techniques based on proportionality, superposition, and Thévenin's and Norton's theorems.

All of our work here will be in terms of *instantaneous* voltages and currents, symbolized in general by v and i. Depending upon the situation, these symbols could stand for constant values, or for time-varying waveforms such as $v(t)$ and $i(t)$, or for the values of $v(t)$ and $i(t)$ at some particular time.

Objectives

After studying this chapter and working the exercises, you should be able to do each of the following:

1. Calculate the equivalent resistance of a two-terminal network of series and/or parallel resistors (Section 2.1).
2. Use voltage and current divider relations together with equivalent resistance to find all the voltages and currents in a resistive ladder network (Section 2.1).
3. Explain the meaning of duality, and list pairs of dual entities (Section 2.2).
4. Identify the symbols and properties of controlled sources (Section 2.3).
5. Determine the equivalent resistance of a resistive network with a controlled source (Section 2.3).
6. Analyze a ladder network by applying the proportionality principle (Section 2.4).
7. Invoke superposition to analyze a circuit with two or more independent sources (Section 2.4).
8. Find the Thévenin or Norton equivalent for a network containing at least one independent source (Section 2.5).
9. Simplify a circuit analysis problem using Thévenin's or Norton's theorem or source conversions (Section 2.5).

2.1 SERIES AND PARALLEL RESISTANCE

Electrical engineers frequently divide a complete circuit into smaller units to expedite analysis or design. These smaller units are called **networks**. A network may include any number of elements, but it must have at least two terminals available for connection to sources or other networks.

As a case in point, the load attached to a source is often a two-terminal network containing several resistors arranged in series and parallel. This section treats the analysis of series-parallel resistive networks using equivalent resistance and voltage and current dividers.

Series Resistance and Potentiometers

The simple load network in Fig. 2.1*a* consists of two series resistors and has two external terminals. The instantaneous terminal voltage v and current i are produced by some source not shown. We want to find the relationship between i and v and the values of the individual voltages v_1 and v_2.

Both resistors carry the same current i, so $v_1 = R_1 i$ and $v_2 = R_2 i$. Thus, from KVL,

$$v = v_1 + v_2 = R_1 i + R_2 i = (R_1 + R_2)i$$

This expression for the terminal v-i relationship has the form

$$v = R_{ser} i$$

with

$$R_{ser} = R_1 + R_2 \tag{2.1}$$

We call R_{ser} the **series equivalent resistance**.

The meaning of equivalent resistance is emphasized by Fig. 2.1*b* where a single resistor with resistance $R_{ser} = R_1 + R_2$ replaces the two resistors. The single equivalent resistor therefore draws current $i = v/R_{ser} = v/(R_1 + R_2)$, just like the series resistors. While we seldom make such a replacement physically, we often do it mentally for purposes of calculations. To illustrate, if $v =$ 20 V, $R_1 = 3\ \Omega$, and $R_2 = 2\Omega$, then $R_{ser} = 3 + 2 = 5\ \Omega$ and $i = 20\ \text{V}/5\ \Omega =$ 4A. Or if the voltage has a time variation so that $v(t) = 20 \cos \omega t$ V, then the current also varies with time and $i(t) = v(t)/R_{ser} = 4 \cos \omega t$ A.

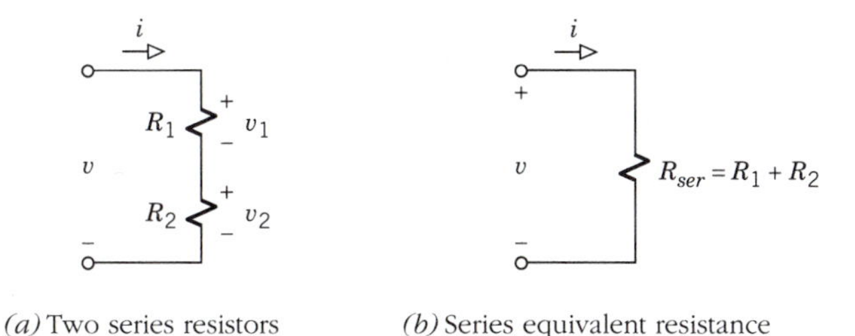

(*a*) Two series resistors (*b*) Series equivalent resistance

Figure 2.1

Series equivalent resistance is one aspect of the broader concept of **equivalent two-terminal networks**, and other aspects will emerge as we go along. In more general terms,

> Two-terminal networks are equivalent if they have exactly the same i-v characteristics at their terminals.

Consequently, the conditions at the terminals of a complicated network may be more easily calculated using a simpler equivalent network. However, the conditions *within* the networks might be quite different. For instance, the equivalent resistor in Fig. 2.1*b* clearly differs from the two series resistors in Fig. 2.1*a* because the individual voltages v_1 and v_2 exist in one network but not the other.

Now suppose you know the terminal voltage in Fig. 2.1*a* and you need the values of the voltages across R_1 and R_2. Since $v_1 = R_1 i$, $v_2 = R_2 i$, and $i = v/R_{ser} = v/(R_1 + R_2)$, you find that

$$v_1 = \frac{R_1}{R_1 + R_2} v \qquad v_2 = \frac{R_2}{R_1 + R_2} v \tag{2.2}$$

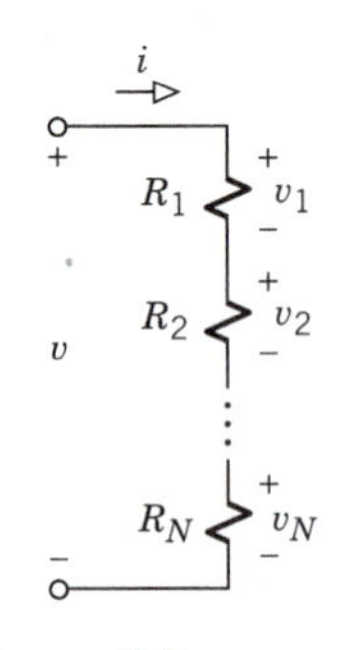

Figure 2.2
N series resistors.

These relations indicate that the total voltage v is "divided" between the two resistors, producing the smaller voltages v_1 and v_2. We therefore say that resistors carrying the same current form a **voltage divider**.

Voltage dividers keep popping up in circuits — especially electronic circuits. The voltage-divider relations in Eq. (2.2) save analysis time because they express v_1 and v_2 directly in terms of v, without the intermediate calculation of i. Equation (2.2) is also valuable for design work when you need to select R_1 and R_2 to get a specified value for v_1 or v_2.

Next, consider a network of N series resistors as represented by Fig. 2.2. Extrapolating from Eq. (2.1), the series equivalent resistance is

$$R_{ser} = R_1 + R_2 + \cdots + R_N \tag{2.3}$$

Thus,

> The value of R_{ser} will always be larger than the largest individual resistance.

The voltage v_n across any resistor R_n is found by generalizing Eq. (2.2) as

$$v_n = \frac{R_n}{R_{ser}} v = \frac{R_n}{R_1 + R_2 + \cdots + R_n} v \tag{2.4}$$

If all N resistors happen to have the same resistance R, then $R_{ser} = NR$ and $v_n = v/N$.

Another interesting resistive network is the **potentiometer**, or "pot" for short. Figure 2.3*a* gives the schematic symbol for this three-terminal device. The terminal labeled W is a movable contact point or **wiper** that intercepts a portion of the fixed total resistance R_{AB}. For a particular wiper position, the potentiometer acts like the two resistors R_{AW} and R_{WB} in Fig. 2.3*b*, where

$$R_{AW} + R_{WB} = R_{AB}$$

Potentiometers often come in a circular configuration with rotatable wiper

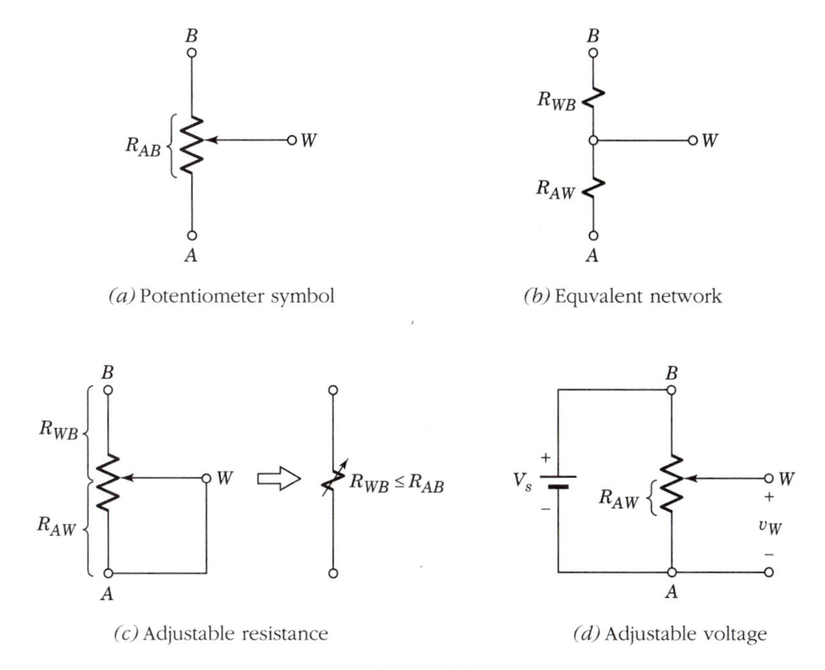

Figure 2.3

knobs or thumb wheels, and precision units have dials indicating the wiper position as accurately as $\pm 0.1\%$.

Potentiometers play vital roles for adjusting conditions in many electronic circuits. Connecting the wiper terminal to terminal A puts a short-circuit path around R_{AW} and creates the **adjustable resistance** $R_{WB} \leq R_{AB}$ depicted in Fig. 2.3*c*. Figure 2.3*d* shows how a potentiometer can be used to obtain an **adjustable voltage** from a battery. The potentiometer forms a voltage divider across V_s, resulting in

$$v_W = \frac{R_{AW}}{R_{AB}} V_s$$

However, this expression holds only when R_{AW} and R_{WB} remain effectively in series, so the wiper terminal must not carry any current.

Example 2.1 *Audio Volume Control*

Figure 2.4*a* depicts an audio amplifier with a 4-kΩ potentiometer at the input to provide volume control. The audio source generates a time-varying voltage $v_s(t)$ in series with a 1-kΩ resistance. The amplifier is an electronic circuit whose input voltage $v_{in}(t)$ produces the "bigger" output voltage $v_{out}(t) = 100v_{in}(t)$. We want to find the wiper setting of the potentiometer needed to get $v_{out}(t) = 60v_s(t)$, given that the amplifier draws no input current so $i_W = 0$.

Since the wiper terminal carries no current, the source resistance and potentiometer form three series resistors with total applied voltage $v_s(t)$, as dia-

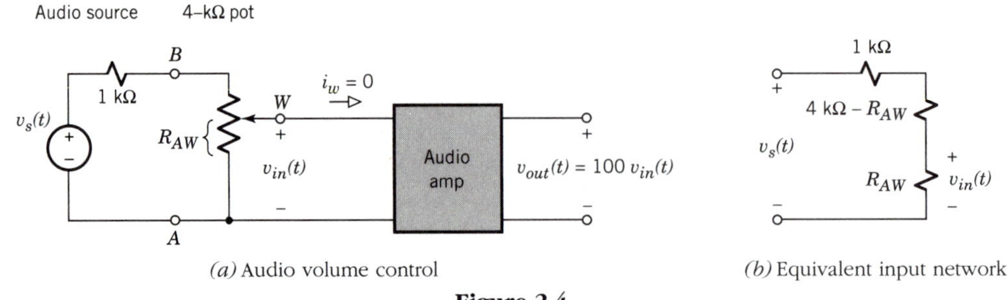

(a) Audio volume control (b) Equivalent input network

Figure 2.4

grammed in Fig. 2.4*b*. This diagram, representing just a part of original one, more clearly shows that $v_{in}(t)$ is related to $v_s(t)$ by the voltage-divider expression

$$v_{in}(t) = \frac{R_{AW}}{1\ \text{k}\Omega + (4\ \text{k}\Omega - R_{AW} + R_{AW})} v_s(t) = \frac{R_{AW}}{5\ \text{k}\Omega} v_s(t)$$

But if $v_{out}(t) = 100v_{in}(t) = 60v_s(t)$, then the potentiometer must be set such that $v_{in}(t) = 0.6v_s(t)$. Hence,

$$R_{AW}/5\ \text{k}\Omega = 0.6 \Rightarrow R_{AW} = 0.6 \times 5\ \text{k}\Omega = 3\ \text{k}\Omega$$

By varying the wiper position, the output voltage may be reduced to zero or increased to the maximum value

$$v_{out}(t) = 100 \frac{4\ \text{k}\Omega}{1\ \text{k}\Omega + 4\ \text{k}\Omega} v_s(t) = 80v_s(t)$$

A careful examination of Fig. 2.4*a* shows why the amplifier cannot deliver $v_{out}(t) = 100v_s(t)$ with the potentiometer in place.

Exercise 2.1

Consider three resistors in series, like Fig. 2.2. Obtain design values for R_1, R_2, and R_3 so that $v_1 = 0.2v$, $v_2 = 0.5v$, and $R_{ser} = 40\ \Omega$.

Parallel Resistance

Figure 2.5 represents N resistors in parallel, so voltage v appears across each one. Again we seek the i-v relationship at the network's terminals. To this end, we'll work with the conductances $G_1 = 1/R_1$, $G_2 = 1/R_2$, etc.

The current through R_1 is $i_1 = v/R_1 = G_1 v$, and likewise for all other currents. KCL then requires

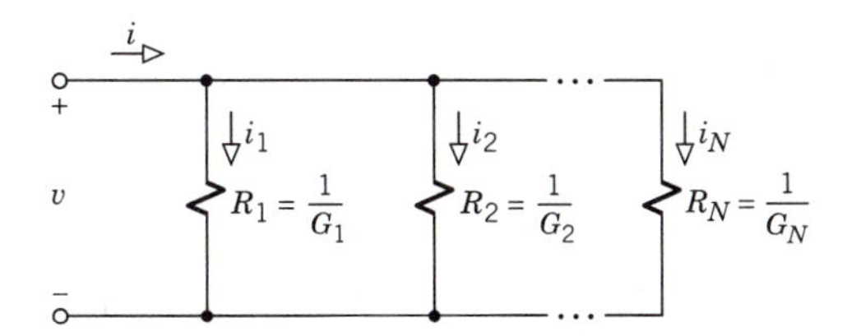

Figure 2.5 *N* parallel resistors.

$$\begin{aligned} i &= i_1 + i_2 + \cdots + i_N = G_1 v + G_2 v + \cdots + G_N v \\ &= (G_1 + G_2 + \cdots + G_N) v \end{aligned}$$

This expression for the terminal relationship has the form

$$i = G_{par} v$$

with

$$G_{par} = G_1 + G_2 + \cdots + G_N \tag{2.5}$$

We call G_{par} the **parallel equivalent conductance**.

The corresponding **parallel equivalent resistance** is $R_{par} = 1/G_{par}$, so Eq. (2.5) can be rewritten as

$$\frac{1}{R_{par}} = \frac{1}{R_1} + \frac{1}{R_2} + \cdots + \frac{1}{R_N} \tag{2.6}$$

A single resistor with resistance R_{par} has the same terminal *i-v* relationship as the *N* resistors in Fig. 2.5.

Parallel-connected resistors form a **current divider**, the total current i being "divided" into the smaller currents $i_1, i_2, \ldots, i_N$. The current through any resistor R_n is given by

$$i_n = \frac{G_n}{G_{par}} i = \frac{G_n}{G_1 + G_1 + G_2 + \cdots + G_N} i \tag{2.7}$$

Equation (2.7) follows from $i_n = G_n v$ with $v = R_{par} i = i/G_{par}$.

If all N resistors happen to have the same conductance $G = 1/R$, then $i_n = i/N$ and $G_{par} = NG$ — so $R_{par} = 1/NG = R/N$. In any case,

> The value of R_{par} will always be smaller than the smallest individual resistance.

This follows from the fact that putting resistors in parallel establishes additional current paths, thereby increasing the network's conductance and reducing its equivalent resistance.

Equations (2.5) and (2.7) have the minor disadvantage of being in terms of conductances, whereas we usually know resistance values. More convenient expressions are possible when there are just **two parallel resistors,** R_1 and R_2. The parallel equivalent resistance is then found from Eq. (2.6) to be

$$R_{par} = \left(\frac{1}{R_1} + \frac{1}{R_2}\right)^{-1} = \frac{R_1 R_2}{R_1 + R_2} \tag{2.8}$$

The "product-over-sum" quantity on the right-hand side occurs so often that we denote it by the symbol $R_1 \| R_2$, which stands for "R_1 in parallel with R_2." If $R_1 = 5\ \Omega$ and $R_2 = 20\ \Omega$, for instance, then the parallel equivalent value is $5\|20 = 100/25 = 4\ \Omega$. The product-over-sum formula may also be applied sequentially to calculate the equivalent resistance of three or more parallel resistors.

With $N = 2$ resistors, the current through R_1 is given by Eq. (2.7) as $i = (G_1/G_{par})i = (R_{par}/R_1)i$. Inserting R_{par} from Eq. (2.8) then yields

$$i_1 = \frac{R_2}{R_1 + R_2}\, i \tag{2.9}$$

This current-divider relation for two parallel resistors should be compared with the voltage-divider relation in Eq. (2.2). Note in particular that i_1 is proportional to the "opposite" resistance R_2 because a large value of R_2 forces more current through R_1.

Occasionally, you may encounter a design situation in which the resistance between two nodes must be reduced from its present value, say R_1, to a specified smaller value. The desired result is easily obtained by paralleling R_1 with another resistance R_2 to get $R_{par} = R_1 \| R_2 < R_1$. Solving Eq. (2.8) for R_2 yields

$$R_2 = \frac{R_1 R_{par}}{R_1 - R_{par}} \tag{2.10}$$

This "reverse parallel" formula also gives the value of R_2 when you know the values of R_1 and $R_1 \| R_2$.

Example 2.2 ***Parallel Resistance Calculations***

Given the parallel circuit in Fig. 2.6*a*, we want to find the value of R_x such that $i_x = 2$ A. We also want the resulting values of the terminal voltage v and the equivalent resistance.

Although Eq. (2.7) could be used to find $G_x = 1/R_x$, a more informative approach is to combine the three known resistances into the parallel equivalent

$$(1/12 + 1/24 + 1/8)^{-1} = 4\ \Omega$$

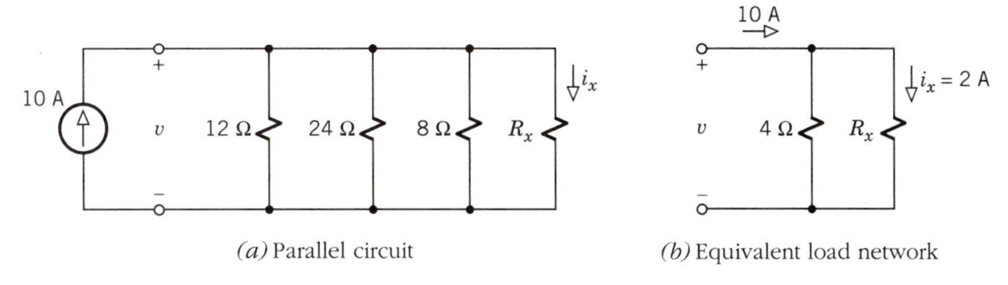

(a) Parallel circuit (b) Equivalent load network

Figure 2.6

Alternatively, noting that $24\|8 = 192/32 = 6\ \Omega$, we can use pairwise combinations to get

$$12\|(24\|8) = 12\|6 = 72/18 = 4\ \Omega$$

We then have the simplified network in Fig. 2.6*b*, where the branch-variable labels lead to

$$i_x = \frac{4}{4 + R_x} \times 10\text{ A} = 2\text{ A} \quad \Rightarrow \quad R_x = 16\ \Omega$$

Thus,

$$v = R_x i_x = 32\text{ V} \qquad R_{par} = 4\|16 = 3.2\ \Omega$$

These values agree with $v = R_{par} \times 10\text{ A} = 32$ V.

Example 2.3 *Electric Grill Unit*

Figure 2.7 depicts the type of cooking unit found in some electric grills. The unit consists of two resistive heating elements and a special switch that connects them to the source voltage individually, or in series, or in parallel. We'll determine the available ohmic heating powers when $R_1 = 12\ \Omega$, $R_2 = 24\ \Omega$, and $v_s = 120$ V.

The switch allows the selection of one of four increasing resistance values

$$R_1\|R_2 = 8\ \Omega \qquad R_1 = 12\ \Omega \qquad R_2 = 24\ \Omega \qquad R_1 + R_2 = 36\ \Omega$$

Since v_s is fixed and $p = v_s^2/R$, the power dissipation at the lowest and highest heat settings are

$$p_{min} = v_s^2/(R_1 + R_2) = 400\text{ W}$$
$$p_{max} = v_s^2/(R_1\|R_2) = 1800\text{ W}$$

Connecting the elements individually gives the intermediate heats $v_s^2/R_2 =$

600 W and $v_s^2/R_1 = 1200$ W. Four more heat settings are obtained when v_s has another possible value such as 240 V.

Exercise 2.2

Let three resistors in parallel be $R_1 = 4\ \Omega$, $R_2 = 36\ \Omega$, and $R_3 = 18\ \Omega$. Determine R_{par} using Eq. (2.6). Compare with the results obtained by calculating $R_1 \| (R_2 \| R_3)$ and $(R_1 \| R_2) \| R_3$.

Exercise 2.3

Let $v_s = 120$ V and $R_1 = 60\ \Omega$ in Fig. 2.7. Use Eq. (2.10) to find R_2 such that $p_{max} = 720$ W. Then calculate the other three heating powers.

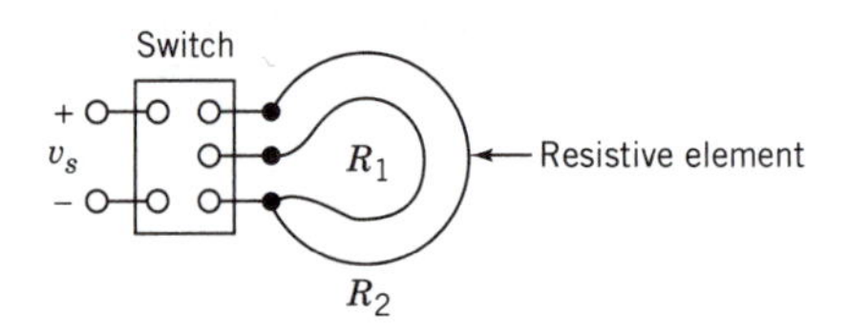

Figure 2.7 Electric grill unit.

Resistive Ladders

Now we're ready to tackle problems in which a source is connected to a network consisting entirely of series and parallel resistors. Such networks are often called **resistive ladders** because the network diagram looks rather like a ladder lying on its side.

Given a particular resistive-ladder circuit, we may need to find some or all of the voltages and currents. These branch variables can be determined by the **series-parallel reduction method**, as follows:

1. Start at a branch *farthest* from the source terminals and repeatedly combine series and parallel resistors until the entire network reduces to a *single equivalent resistance* R_{eq}.
2. Use the equivalent resistance to calculate the unknown voltage or current at the network's terminals.
3. Apply Kirchhoff's and Ohm's laws and/or divider relations at appropriate points in the partially reduced network diagram to find any other voltages or currents of interest.

An example will clarify this simple but valuable method.

Example 2.4 *Ladder Calculations*

Consider the circuit in Fig. 2.8*a*, where all resistances are in kilohms. We'll walk through the steps of the series-parallel reduction method to find the terminal current i, the total dissipated power p, and the voltages v_x and v_y.

Step 1: The resistive ladder contains two 20-kΩ resistors in parallel and a series connection of 4, 5, and 6 kΩ. We therefore replace these five elements by the two equivalent resistances

$$20\|20 = 10 \text{ k}\Omega \qquad 4 + 5 + 6 = 15 \text{ k}\Omega$$

The resulting partially reduced network in Fig. 2.8*b* has the 2-kΩ resistor in series with 10 kΩ‖15 kΩ. The entire ladder thus reduces to the single equivalent resistance

$$R_{eq} = 2 + 10\|15 = 2 + 6 = 8 \text{ k}\Omega$$

Step 2: From the terminal equivalent diagram in Fig. 2.8*c* we easily see that the load draws

$$i = 40 \text{ V}/8 \text{ k}\Omega = 5 \text{ mA} \qquad p = 40 \text{ V} \times i = 200 \text{ mW}.$$

Step 3: "Unfolding" some of the equivalent resistances takes us back to Fig. 2.8*b* where

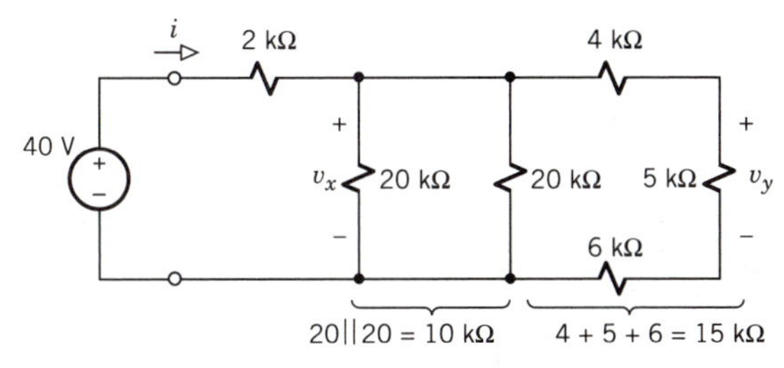

(a) Source connected to a ladder network

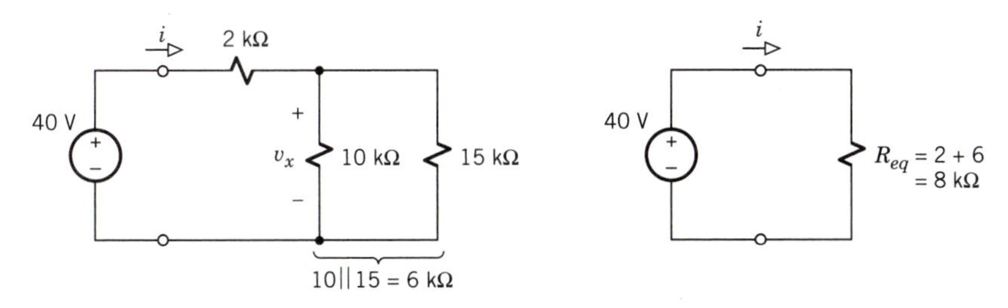

(b) Partitally reduced network

(c) Terminal equivalent diagram

Figure 2.8

$$v_x = 40\text{ V} - 2\text{ k}\Omega \times i = 30\text{ V}$$

But v_x also appears across the three-resistor voltage divider in Fig. 2.8*a*, so

$$v_y = \frac{5}{4 + 5 + 6} v_x = 10\text{ V}$$

Exercise 2.4

Find R_{eq} for the ladder network in Fig. 2.9 when $R_x = 4\ \Omega$ and $R_y = 9\ \Omega$. Then calculate v, p, and v_2.

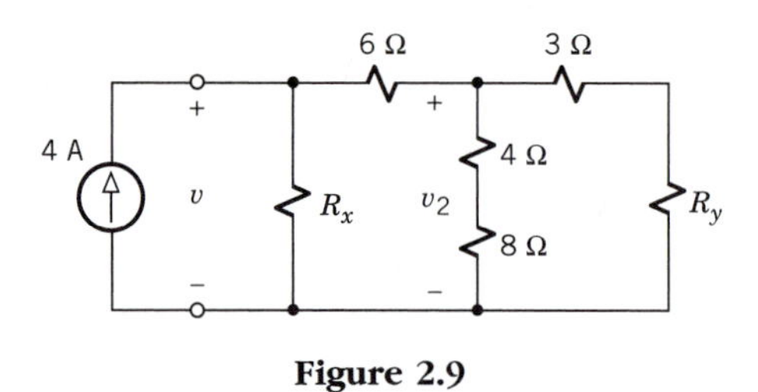

Figure 2.9

2.2 DUALITY

You may have spotted a resemblance between the equations for series-connected resistances and parallel-connected conductances. Those similarities were not accidental, but rather a consequence of the principle known as *duality.*

To bring out the duality principle more clearly, we'll do a side-by-side comparison of the series and parallel networks in Fig. 2.10. Table 2.1 lists the various relations using resistances for the series network and conductances for the parallel network.

Observe that each equation on one side of Table 2.1 is identical to the corresponding equation on the other side except for a change in symbols,

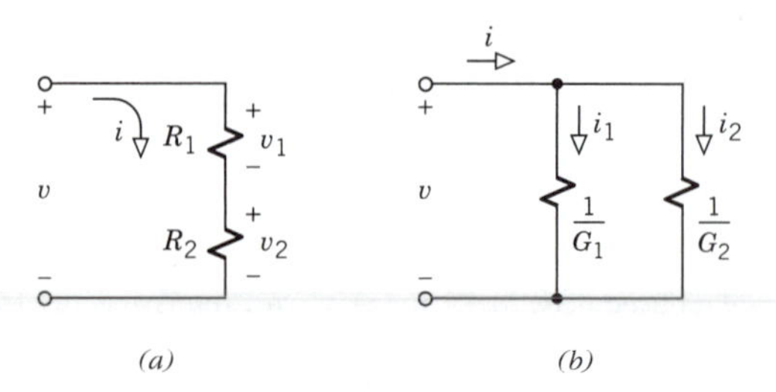

Figure 2.10 Dual networks.

TABLE 2.1

Series Network	Parallel Network
KVL loop equation:	KCL node equation:
$v = v_1 + v_2 = R_1 i + R_2 i$	$i = i_1 + i_2 = G_1 v + G_2 v$
Equivalent resistance:	Equivalent conductance:
$R_{ser} = R_1 + R_2$	$G_{par} = G_1 + G_2$
Voltage divider:	Current divider:
$v_1 = \dfrac{R_1}{R_1 + R_2} v$	$i_1 = \dfrac{G_1}{G_1 + G_2} i$
Open circuit:	Short circuit:
$v_1 = v$ when $R_1 \to \infty$	$i_1 = i$ when $G_1 \to \infty$

namely, v interchanged with i, and R interchanged with G. We therefore say that the networks are **duals**. Stated in general,

> Two different networks are duals when the i-v equations that describe one of them have the same mathematical form as the i-v equations for the other with voltage and current variables interchanged.

The two networks will be *exact* duals when their equations are numerically identical — a rare event. More commonly, we speak in terms of *structural* duals whose equations may have different numerical values. **Duality** refers to the comparable features of dual networks.

By definition, KCL and current are the respective duals of KVL and voltage. Furthermore, a review of Table 2.1 shows that conductance is the dual of resistance, parallel is the dual of series, and a short circuit is the dual of an open circuit. Table 2.2 lists these and other dual pairs, any entity on the right being the dual of the entity on the left, and vice versa. The last three pairs in the table will be discussed in later chapters but are included here for the sake of completeness.

Unlike equivalent resistance or divider relations, the duality principle is not a technique for analyzing particular circuits. Instead, duality is a conceptual aide that enhances intuition and underscores important relationships. For instance, starting from the general series resistance relation

$$v = (R_1 + R_2 + \cdots + R_N)i$$

you can invoke duality to write down the general parallel conductance relation

$$i = (G_1 + G_2 + \cdots + G_N)v$$

TABLE 2.2 Dual Pairs

KVL	KCL
Voltage	Current
Across	Through
Resistance	Conductance
Loop	Node
Series	Parallel
Open circuit	Short circuit
Node voltage	Mesh current
Capacitance	Inductance
Impedance	Admittance

Furthermore, duality expands your range of experience in that after you have analyzed the behavior of a particular circuit, you can draw immediate conclusions about the behavior of its dual.

Instructors also sometimes use duality to generate pairs of circuits for examples or problems. The following example illustrates the construction of a dual circuit.

Example 2.5 *Constructing a Dual Circuit*

Professor Bright has devised and analyzed the circuit in Fig. 2.11*a* as a lecture example. Now she wants to construct the exact dual circuit for an exam problem.

By interchanging voltage and current and resistance and conductance, Professor Bright knows that the dual circuit must contain a 12-A source, a 5-V source, and 2-S, 3-S, and 8-S conductances. Furthermore, since the left-hand section of Fig. 2.11*a* has a voltage source and two series resistances with current i_1 through each of them, Professor Bright concludes that the left-hand section of the dual will be a current source and two parallel conductances with voltage v_1 across each of them. Continuing to use dual pairs in this fashion, she arrives at the final diagram in Fig. 2.11*b*, where the resistors are labeled with their conductance values to emphasize the duality.

But the table of dual pairs does not automatically yield correct voltage polarities and current directions. These must be checked by writing the comparable equations. Putting KCL and KVL equations side-by-side, Professor Bright gets

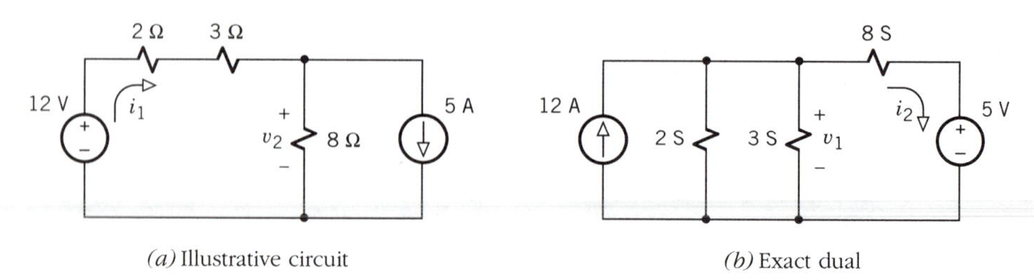

(*a*) Illustrative circuit

(*b*) Exact dual

Figure 2.11

Original circuit:

$$i_1 = \frac{v_2}{8\ \Omega} + 5\ \text{A}$$

$$v_2 = 12\ \text{V} - (2\ \Omega + 3\ \Omega)i_1$$

Dual circuit:

$$v_1 = \frac{i_2}{8\ \text{S}} + 5\ \text{V}$$

$$i_2 = 12\ \text{A} - (2\ \text{S} + 3\ \text{S})v_1$$

The pairs of equations are numerically identical, confirming that she has indeed constructed the exact dual.

Exercise 2.5

Construct the exact dual of the circuit in Fig. 2.12. Check your result by comparing the equations for v and i_2 with their duals.

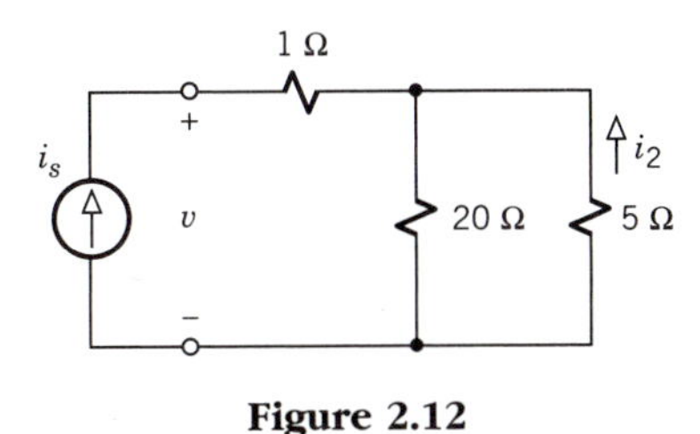

Figure 2.12

2.3 CIRCUITS WITH CONTROLLED SOURCES

The ideal sources defined back in Section 1.2 are classified as **independent sources** because the source voltage or current does not depend on any other voltage or current. But when modeling amplifiers and certain other electrical devices, we'll find it necessary to work with sources that do depend on some other voltage or current. They are therefore known as **dependent** or **controlled sources**.

This section defines and illustrates the properties of controlled sources. Then we'll examine the equivalent resistance of networks that contain controlled sources along with resistors.

Controlled Sources

To introduce the concept of controlled sources, Fig. 2.13*a* repeats the drawing of the amplifier with $v_{out} = 100v_{in}$ from Example 2.1. Although the amplifier no doubt consists of many elements, its effect at the output can be represented by the special source symbol in Fig. 2.13*b*. This diamond-shaped symbol stands for a source whose voltage v_{out} is controlled by another voltage v_{in}. You may find it helpful to think of the controlled source as an adjustable voltage generator operated by a friendly gnome, as depicted in Fig. 2.13*c*. The gnome has a pair of voltmeters to measure v_{in} and v_{out}, and he continuously readjusts the generator so that the controlled voltage always equals 100 times

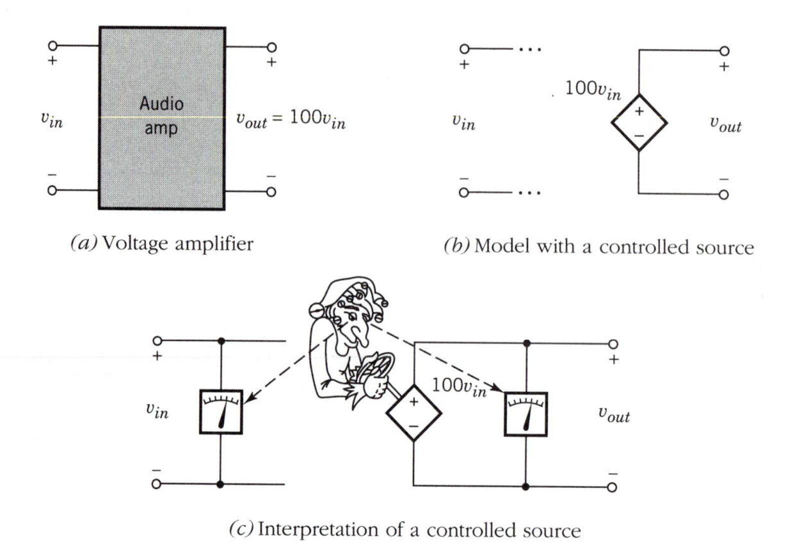

(a) Voltage amplifier

(b) Model with a controlled source

(c) Interpretation of a controlled source

Figure 2.13

the controlling voltage. If $v_{in} = 0.2 \cos \omega t$ V, for instance, then our gnome must make sure that $v_{out} = 20 \cos \omega t$ V at every instant of time.

Any voltage source that depends upon another voltage is termed a **voltage-controlled voltage source (VCVS)**. Figure 2.14*a* gives the general symbol for an ideal linear VCVS. The source voltage varies in direct proportion to some other voltage v_x, regardless of the current i through the source, so

$$v_c = \mu v_x \tag{2.11a}$$

We call v_x the **control variable**, and it must appear somewhere in conjunction

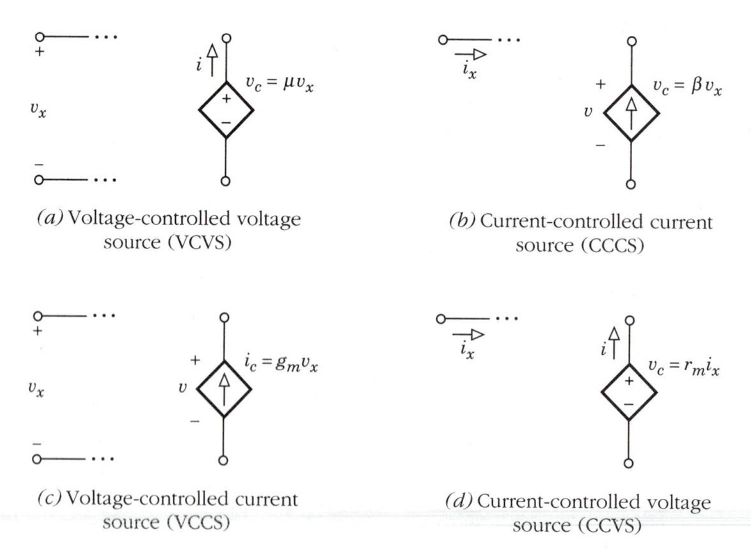

(a) Voltage-controlled voltage source (VCVS)

(b) Current-controlled current source (CCCS)

(c) Voltage-controlled current source (VCCS)

(d) Current-controlled voltage source (CCVS)

Figure 2.14

with the controlled source. The proportionality constant μ is a dimensionless quantity known as the **voltage gain**. With an appropriate value of μ, a VCVS may represent any linear voltage amplifier or voltage-amplifying device.

The dual of a VCVS is a **current-controlled current source (CCCS)**, symbolized by Fig. 2.14*b*. This current source produces

$$i_c = \beta i_x \tag{2.11b}$$

which is independent of v. Thus, i_x is the control variable and β is a dimensionless constant representing **current gain**. In electronic circuits, a CCCS often represents the current-amplifying property of a bipolar junction transistor.

But the current through a field-effect transistor depends upon a controlling voltage, symbolized by the **voltage-controlled current source (VCCS)** in Fig. 2.14*c*. This source produces

$$i_c = g_m v_x \tag{2.11c}$$

independent of the voltage v across the source. The constant g_m is called the **transconductance** because it represents a voltage-to-current "transfer" effect and has the same units as conductance.

Finally, the dual of a VCCS is the **current-controlled voltage source (CCVS)** in Fig. 2.14*d*, where

$$v_c = r_m i_x \tag{2.11d}$$

which is independent of the current i through the source. The constant r_m is called the **transresistance** because we now have a current-to-voltage transfer effect with the same units as resistance.

The four controlled sources in Fig. 2.14 account for all possible combinations of controlled and controlling variables. Each source is a *linear* element in that the controlled voltage or current is directly proportional to the control variable. As a consequence of the controlling effect, they all have the property that

> A controlled source produces voltage or current only when an independent source activates the control variable.

Thus, if a network includes controlled sources but no independent sources, then all voltages and currents will be zero. Nonetheless, a controlled source often represents a device capable of supplying electrical energy when "turned on" by the control variable.

Simple circuits containing linear controlled sources can be analyzed using the branch variable method, bearing in mind that the value of the controlled source depends upon the value of the controlling variable. Hence,

An essential step in the analysis of circuits with controlled sources is relating the control voltage or current to other quantities of interest.

Some of our results for series and parallel resistance may be used for this purpose, provided that the control variable does not become lost in the process.

Example 2.6 *Amplifier with a Field-Effect Transistor*

Figure 2.15 is the model of a simple voltage amplifier built with a field-effect transistor. We'll develop an expression for the output voltage in terms of the input voltage, given the transconductance value $g_m = 5 \times 10^{-3}$ S = 5 mS.

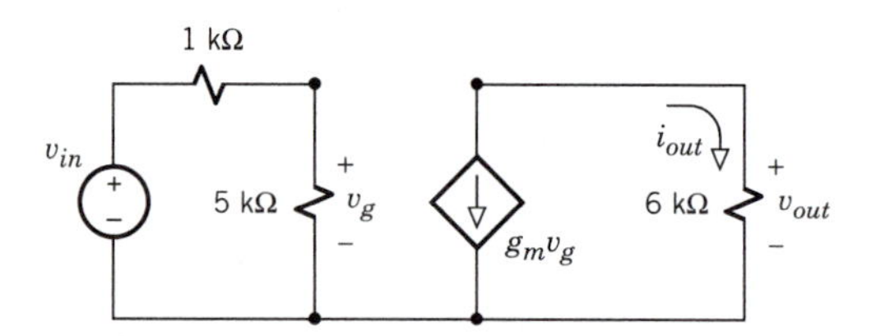

Figure 2.15 Model of a voltage amplifier with a field-effect transistor.

The input voltage produces the control voltage v_g via a voltage divider, so $v_g = (5/6)v_{in}$. The VCVS then generates the output current $i_{out} = -g_m v_g$, which, in turn, develops the output voltage $v_{out} = 6\text{ k}\Omega \times i_{out}$. Thus,

$$v_{out} = 6\text{ k}\Omega \times [-(5\text{ mS}) \times (5/6)v_{in}] = -25v_{in}$$

The negative value here means that v_{out} will be negative when v_{in} is positive, and vice versa, so we call this circuit an *inverting* amplifier.

Example 2.7 *Analysis with a VCVS*

The circuit in Fig. 2.16*a* poses more of a challenge because the control voltage v_2 interacts with the controlled voltage v_c. To illustrate the analysis technique, we'll find the current i_1 produced by an arbitrary driving voltage v_s.

First, we observe that the control voltage appears across the 6-Ω resistor, whereas i_1 goes through the 2-Ω and 10-Ω resistors. We may therefore combine the two series resistors and redraw the circuit, as shown in Fig. 2.16*b*. Next, focusing on v_2 and i_1, we apply KVL to the right-hand loop to get

$$v_2 + 3v_2 = 12i_1 \quad \Rightarrow \quad v_2 = 3i_1$$

This is the needed relationship between the control variable and the variable being sought.

But relating i_1 to v_s involves the unknown current i since, from KVL around the left-hand loop,

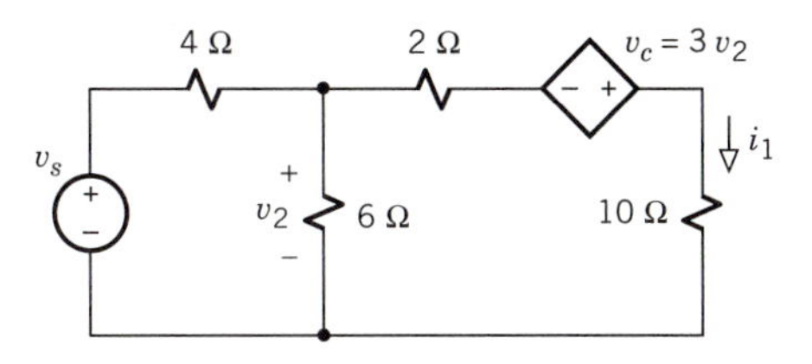

(a) Circuit with a VCVS

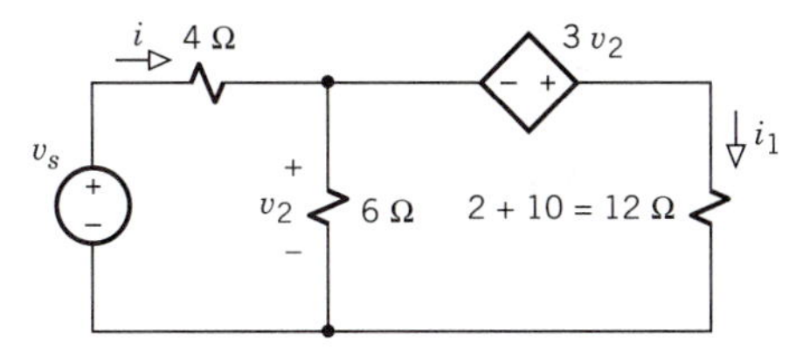

(b) Simplified diagram

Figure 2.16

$$v_s - 4i = v_2 = 3i_1$$

We therefore eliminate i with the help of KCL at the upper node, which yields

$$i = i_1 + v_2/6 = 1.5i_1$$

Thus, $v_s = 4i + v_2 = 4(1.5i_1) + 3i_1 = 9i_1$ so

$$i_1 = v_s/9\ \Omega$$

which is the desired result.

If the driving voltage is turned off, then $v_s = 0$, $i_1 = 0$, $v_2 = 0$, and $i = 0$. Hence, *all* voltages and currents will be zero — a property we've asserted for any controlled-source circuit when no independent source activates the control variable.

Exercise 2.6

The CCCS in Fig. 2.17 represents a bipolar junction transistor with $\beta = 9$. Find i_1 in terms of i by writing two expressions for v. (Save your results for use in Exercise 2.8).

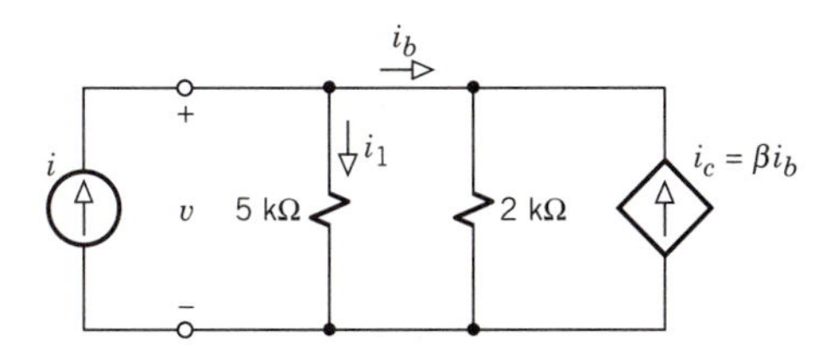

Figure 2.17

Generalized Equivalent Resistance

We previously showed how resistive ladder networks could be reduced to single equivalent resistances. Now we'll go further along this line by considering networks containing controlled sources.

Figure 2.18*a* represents an arbitrary **load network**, which we define as follows:

> A load network is any two-terminal network that contains no independent sources. If controlled sources are included, then the control variables must be within the same network.

Independent sources are excluded from load networks to be consistent with our viewpoint that we complete a circuit by connecting an independent source to a load. Controlled sources are allowed since they remain "off" until an independent source activates the control variables. The control variables must remain in the load network because they cannot be separated physically from the devices they control.

Relative to Fig. 2.18*a*, the **equivalent resistance theorem** states that

> When a load network consists entirely of resistances or resistances and controlled sources, the terminal voltage and current are related by $v = R_{eq}i$, where R_{eq} is a constant.

Consequently, at its terminals, the load network acts like the single equivalent resistance R_{eq} in Fig. 2.18*b*.

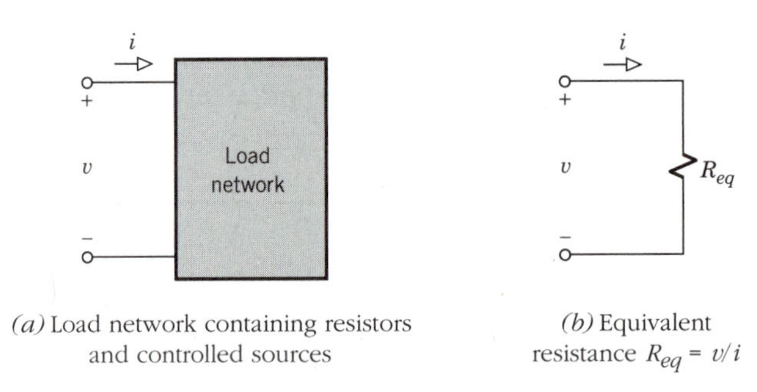

(*a*) Load network containing resistors and controlled sources

(*b*) Equivalent resistance $R_{eq} = v/i$

Figure 2.18

But how do you find the equivalent resistance when controlled sources are involved? The answer to this question comes indirectly from the theorem by noting that

$$R_{eq} = v/i \tag{2.12}$$

Thus, if you have obtained numerical values for v and i, or an expression for

v in terms of i or for i in terms of v, then you can determine R_{eq} by forming the ratio v/i. Of course, the reference polarities for v and i must be drawn in accordance with the passive convention, as in Fig. 2.18*a*.

Our next example illustrates the use of Eq. (2.12).

Example 2.8 *Equivalent Resistance with a VCCS*

The two-terminal network in Fig. 2.19 contains a VCCS, and the control variable happens to be the terminal voltage v. The reference direction for the terminal current i goes from higher to lower potential, assuming $v > 0$. Lacking numerical values, we'll find R_{eq} by writing i in terms of v.

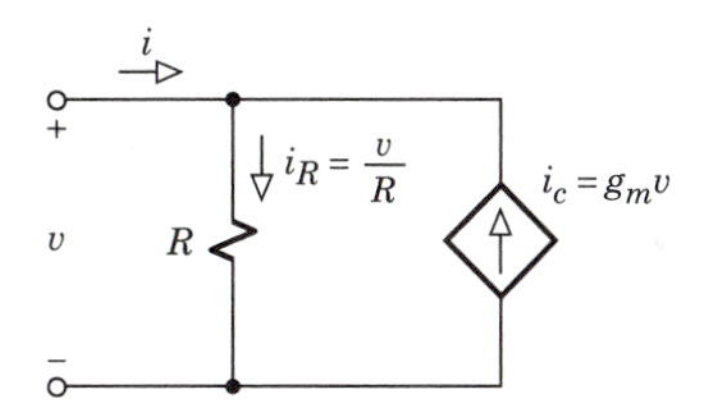

Figure 2.19 Load network with a VCCS.

KCL at the upper node gives $i = i_R - i_c$, where $i_R = v/R$ and $i_c = g_m v$ so

$$i = i_R - i_c = \frac{v}{R} - g_m v = \frac{1 - g_m R}{R} v$$

Forming the ratio v/i now yields

$$R_{eq} = \frac{v}{i} = \frac{R}{1 - g_m R}$$

As predicted by the equivalent resistance theorem, R_{eq} is indeed a constant with the units of resistance.

However, the value of R_{eq} here depends critically upon the value of the product $g_m R$. If $g_m R = 1/2$, for instance, then $1 - g_m R = 1/2$ and $R_{eq} = 2R$. Or if $g_m R = 1$, then $1 - g_m R = 0$ and $R_{eq} = \infty$ — like an open circuit. The equivalent resistance is greater than R in both of these cases because the controlled source provides part or all of i_R, thereby reducing the input current for a given value of v. Thus, with $i < v/R$, R_{eq} must be greater than R.

Now suppose that $g_m R = 2$, so $1 - g_m R = -1$ and $R_{eq} = -R$ — a *negative* equivalent resistance! This is an unusual but not impossible result because controlled sources may supply energy. The negative value of R_{eq} means that the controlled source in Fig. 2.19 actually pumps current *out* of the terminal at the higher potential. The terminal i-v characteristic is still a straight line through the origin with slope $1/R_{eq}$, even though R_{eq} happens to be negative.

Exercise 2.7

Let R be removed from Fig. 2.19, leaving only the controlled source. Obtain the expression for R_{eq} in this case.

Exercise 2.8

Using your results from Exercise 2.6, calculate the equivalent resistance $R_{eq} = v/i$ for the load network in Fig. 2.17 with $\beta = 9$.

2.4 LINEARITY AND SUPERPOSITION

This section develops two new analysis techniques based on fundamental properties of linear circuits. First, we'll exploit the proportionality property to analyze circuits with one independent source. Then we'll apply the superposition property to circuits with two or more independent sources.

Linear Elements and Circuits

Previously we said that resistors obeying Ohm's laws are *linear elements* because the voltage and current vary in direct proportion to each other. The controlled sources discussed in Section 2.2 are also *linear elements* because their source values vary in direct proportion to the control variables. We now introduce the important concept of **linear circuits** by stating that

> A circuit is linear when it consists entirely of linear elements and independent sources.

This definition allows for the inclusion of other types of linear elements considered in later chapters. Here, we'll explore the consequences of linearity.

To begin with a simple case, consider any linear load network with just one pair of external terminals. The external source applied to the network could be either a voltage or a current, so we'll let x stand for the instantaneous value of the source. Likewise, we'll let y stand for the instantaneous value of any branch variable of interest — be it a voltage or a current. This generalized notation allows us to focus on the *excitation* x and the resulting *response* y.

The cause-and-effect relationship between x and y can be represented mathematically by some function

$$y = f(x)$$

Since we're dealing with a linear circuit, $f(x)$ must be a *linear function* having the properties of proportionality and superposition. The **proportionality property** means that if the excitation is increased or decreased by some constant multiplying factor, say K, then the response becomes

$$f(Kx) = Kf(x) \tag{2.13a}$$

The **superposition property** means that if the excitation consists of the sum of two components, say $x_a + x_b$, then the response becomes

$$f(x_a + x_b) = f(x_a) + f(x_b) \tag{2.13b}$$

We combine these two properties by writing

$$f(K_a x_a + K_b x_b) = K_a f(x_a) + K_b f(x_b) \tag{2.14}$$

where K_a and K_b are arbitrary constants.

Equation (2.14) describes precisely those properties we would want from an amplifier. For example, suppose a voltage amplifier has input $x = v_{in}$ and output $y = v_{out} = 100v_{in}$, so $f(x) = 100x$. The input might come from an audio mixer combining the voltage v_1 from a keyboard with the voltage v_2 from a microphone to produce $v_{in} = 0.5v_1 + 3v_2$. The resulting amplifier output will then be

$$v_{out} = f(0.5v_1 + 3v_2) = 0.5f(v_1) + 3f(v_2)$$

But $f(v_1) = 100v_1$ and $f(v_2) = 100v_2$ so

$$v_{out} = 50v_1 + 300v_2$$

which is the amplified version of the composite input.

Equation (2.14) also serves as a test for linearity of individual elements. Take, for instance, a resistance governed by the voltage–current relationship $v = Ri$, so $v = f(i)$ with $f(i) = R \times i$. Letting $i = K_a i_a + K_b i_b$ yields

$$\begin{aligned} f(K_a i_a + K_b i_b) &= R \times (K_a i_a + K_b i_b) \\ &= K_a R i_a + K_b R i_b = K_a f(i_a) + K_b f(i_b) \end{aligned}$$

This expression agrees with Eq. (2.14) and affirms that the resistance is a linear element. However, power dissipation by a resistance is *not* a linear function of the current since $p = Ri^2$ and $R \times (K_a i_a + K_b i_b)^2 \neq K_a R i_a^2 + K_b R i_b^2$.

Exercise 2.9

Show that the current–voltage relationship $i = Gv$ is a linear function of v.

Proportionality Principle

Now consider a linear circuit driven by a single independent source whose instantaneous value x produces the response $y = f(x)$. Suppose the excitation changes to a new value denoted by

$$\hat{x} = Kx \tag{2.15}$$

The proportionality property in Eq. (2.13a) tells us that the instantaneous value of the reponse changes to

$$\hat{y} = f(\hat{x}) = f(Kx) = Kf(x) = Ky \tag{2.16}$$

Thus, *any* branch variable in a linear circuit varies in direct proportion to the source. Equation (2.16) is called the **proportionality principle**.

This principle leads to a clever technique for the analysis of ladder networks. The ladder structure makes it easy to find all branch variables and the source value when you happen to know the value of a branch variable *farthest* from the source. Drawing upon Eq. (2.16), the same "backwards" process helps you determine the branch variables when the source value is known.

The procedure for analyzing a ladder network using the proportionality principle goes as follows:

1. Temporarily ignore the actual source value and *assume* any convenient value y_1 for a branch variable farthest from the source.
2. Work toward the source by calculating y_2 from y_1, calculating y_3 from y_1 and y_2, and so on, until you obtain the source value x that would produce y_1.
3. Evaluate $K = \hat{x}/x$, where $\hat{x}$ is the actual source value.
4. Compute the *actual* branch variables of interest via $\hat{y}_1 = Ky_1$, $\hat{y}_2 = Ky_2$, etc.

This procedure is simple and direct for resistive ladders. It may also be effective when a controlled source is included, provided that the control variable appears in a branch adjacent to the controlled source.

Example 2.9 *Analysis of a Ladder Network*

The circuit in Fig. 2.20 was the subject of Example 2.7, where we used branch-variable analysis to obtain $i_1 = v_s/9\ \Omega$. Here, we're given that $\hat{v}_s = 72$ V, and we want to evaluate $\hat{\imath}$, $\hat{v}_2$, and $\hat{\imath}_1$ by applying the proportionality principle — a valid approach since the load network has a ladder configuration and the control variable v_2 is in a branch adjacent to the VCVS.

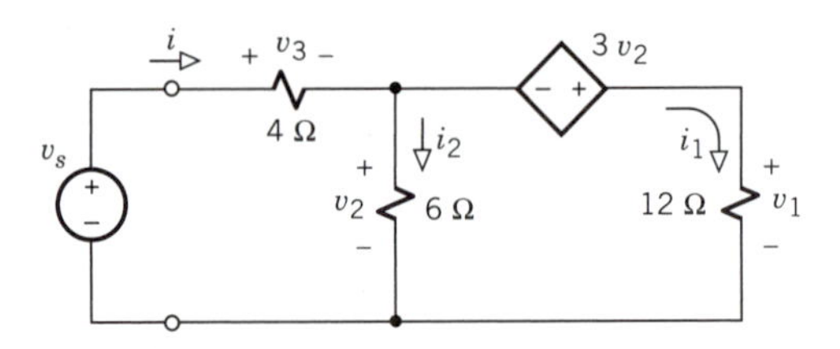

Figure 2.20 Ladder network for analysis by the proportionality principle.

Step 1: The variables i_1 and v_1 are farthest from the source, so we start with the convenient assumption

$$i_1 = 1\text{ A} \quad \Rightarrow \quad v_1 = 12i_1 = 12\text{ V}$$

Step 2: Working backwards toward the source using Ohm's and Kirchhoff's laws, we get

$$\begin{array}{ll} v_2 + 3v_2 = v_1 \quad \Rightarrow & v_2 = v_1/4 = 3\text{ V} \\ i_2 = v_2/6 = 0.5\text{ A} & i = i_2 + i_1 = 1.5\text{ A} \\ v_3 = 4i = 6\text{ V} & v_s = v_3 + v_2 = 9\text{ V} \end{array}$$

Step 3: The actual source voltage is $\hat{v}_s = 72$ V, so

$$K = \hat{v}_s / v_s = 72/9 = 8$$

Step 4: Finally, the actual values of the variables we seek are

$$\hat{\imath} = Ki = 12\text{ A} \quad \hat{v}_2 = Kv_2 = 24\text{ V} \quad \hat{\imath}_1 = Ki_1 = 8\text{ A}$$

The value of $\hat{\imath}_1$ agrees with our previous result that $i_1 = v_s/9$. We also note that the equivalent resistance of the ladder network is

$$R_{eq} = \hat{v}_s / \hat{\imath} = v_s / i = 6\ \Omega$$

Exercise 2.10

Let $\beta = 9$ and $\hat{\imath} = 3$ mA in Fig. 2.17 (p. 57). Find $\hat{v}$ by assuming that $i_c = 9$ mA.

Superposition Theorem

The proportionality method works only for networks driven by a *single* independent source. For circuits with *multiple sources,* superposition often expedites analysis. The underlying **superposition theorem** states that

> When a linear circuit contains two or more independent sources, the value of any branch variable equals the algebraic sum of the individual contributions from each and every independent source with all other independent sources set to zero.

A complicated analysis problem may thereby reduce to several easier problems — one for each independent source. Summing the results from these easier problems yields the final answer.

To cast superposition in mathematical form, consider initially the case of two independent sources. Let y_n stand for the value of any branch variable, and let the values of the independent sources be x_1 and x_2. Then we can write

$$y_n = f(x_1, x_2)$$

where $f(x_1, x_2)$ is a linear function of both x_1 and x_2. The individual contributions from each source alone are given by

$$y_{n-1} \triangleq f(x_1, 0) \qquad y_{n-2} \triangleq f(0, x_2) \tag{2.17}$$

With both sources acting together, the theorem tells us that

$$y_n = f(x_1, 0) + f(0, x_2) = y_{n-1} + y_{n-2} \tag{2.18}$$

This algebraic sum readily generalizes for the case of three or more independent sources.

When evaluating the contribution from one independent source, you do not bodily remove the other sources from the circuit. Instead, you suppress all other independent sources by setting their values to zero. The **suppressed sources** then behave like the following replacements:

Ideal independent voltage source → Short circuit
Ideal independent current source → Open circuit

These replacements preserve the property that an ideal voltage source allows any amount of current through it, even if $v_s = 0$, while an ideal current source allows any amount of voltage across it, even if $i_s = 0$. We say that the suppressed sources are deactivated or "dead," while the unsuppressed source remains active or "alive."

Special care must be taken if the circuit also includes controlled sources. In general, a controlled source affects the individual contribution of each independent source. Consequently,

> Controlled sources are not suppressed during analysis by superposition.

For this reason, superposition may require a little more effort when controlled sources are present.

Example 2.10 *Superposition Calculations*

Given the circuit in Fig. 2.21*a*, we want to calculate the value of i_1 using superposition. Since there are three independent sources and no controlled sources, superposition involves the analysis of three single-source circuits.

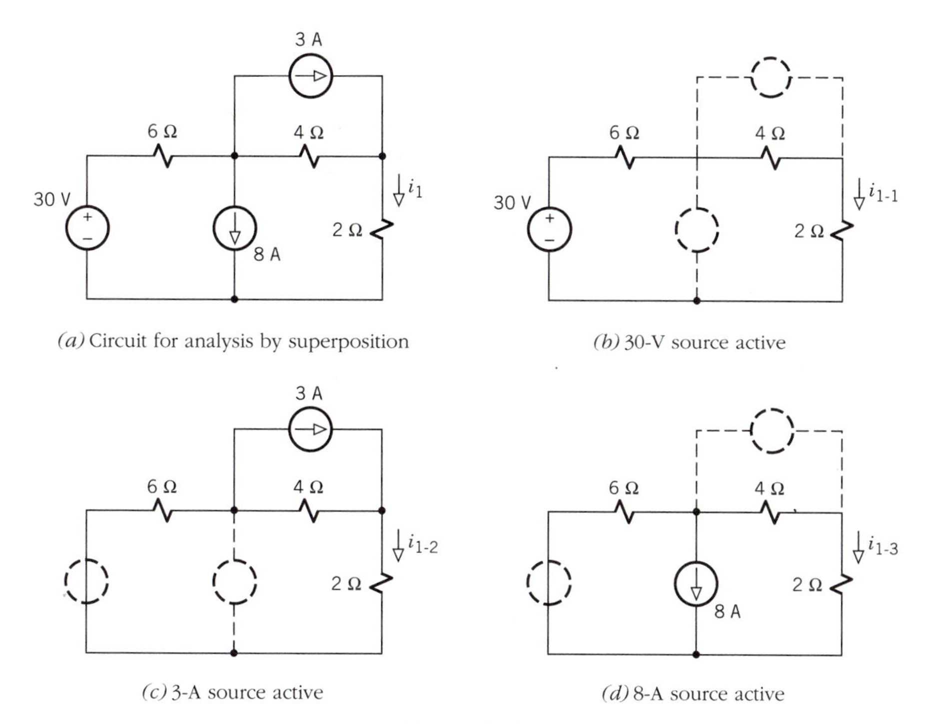

(a) Circuit for analysis by superposition

(b) 30-V source active

(c) 3-A source active

(d) 8-A source active

Figure 2.21

First, we find the contribution to i_1 from the 30-V source by replacing the current sources with open circuits. The resulting diagram in Fig. 2.21*b* includes dashed lines to show where suppressed sources were located. All remaining elements are in series, so we immediately get

$$i_{1-1} = 30/(6 + 4 + 2) = 2.5 \text{ A}$$

Second, we find the contribution from the 3-A source by replacing the voltage source with a short circuit and the other current source with an open circuit. Careful examination of the resulting diagram in Fig. 2.21*c* reveals that the 4-Ω resistor is now in parallel with the series combination of the 6-Ω and 2-Ω resistors. Thus, from the current–divider relation,

$$i_{1-2} = \frac{4}{(6 + 2) + 4} \times 3 = 1 \text{ A}$$

Third, we find the contribution from the 8-A source by suppressing the other two sources. The resulting diagram in Fig. 2.21*c* is another current divider, but the 8-A source drives current in the opposite direction of the reference arrow for i_{1-3}. Thus,

$$i_{1-3} = \frac{6}{6 + (2 + 4)}(-8) = -4 \text{ A}$$

Finally, with all sources active, the value of i_1 is given by the algebraic sum

$$i_1 = i_{1-1} + i_{1-2} + i_{1-3} = 2.5 + 1 - 4 = -0.5 \text{ A}$$

Example 2.11 *Superposition with a Controlled Source*

To illustrate superposition with a controlled source, let the 8-A independent source in the foregoing example be changed to a CCCS having $i_c = 8i_1$. The new diagram in Fig. 2.22*a* contains two independent sources, and again we seek the value of i_1. However, the controlled source is not suppressed for superposition analysis, so we now investigate just two reduced circuits — each circuit including the CCCS along with one active independent source.

The contribution of the 30-V source is found from Fig. 2.22*b*, where $i_c = 8i_{1-1}$. The total current through the 6-Ω resistor is $i_c + i_{1-1} = 9i_{1-1}$, so KVL requires

$$30 - 6(9i_{1-1}) - (4 + 2)i_{1-1} = 0$$

from which $i_{1-1} = 0.5$ A.

The contribution of the 3-A source is found from Fig. 2.22*c*, where $i_c = 8i_{1-2}$. The current through the 6-Ω resistor is $9i_{1-2}$, while the current through the 4-Ω resistor is $i_{1-2} - 3$ A. Thus,

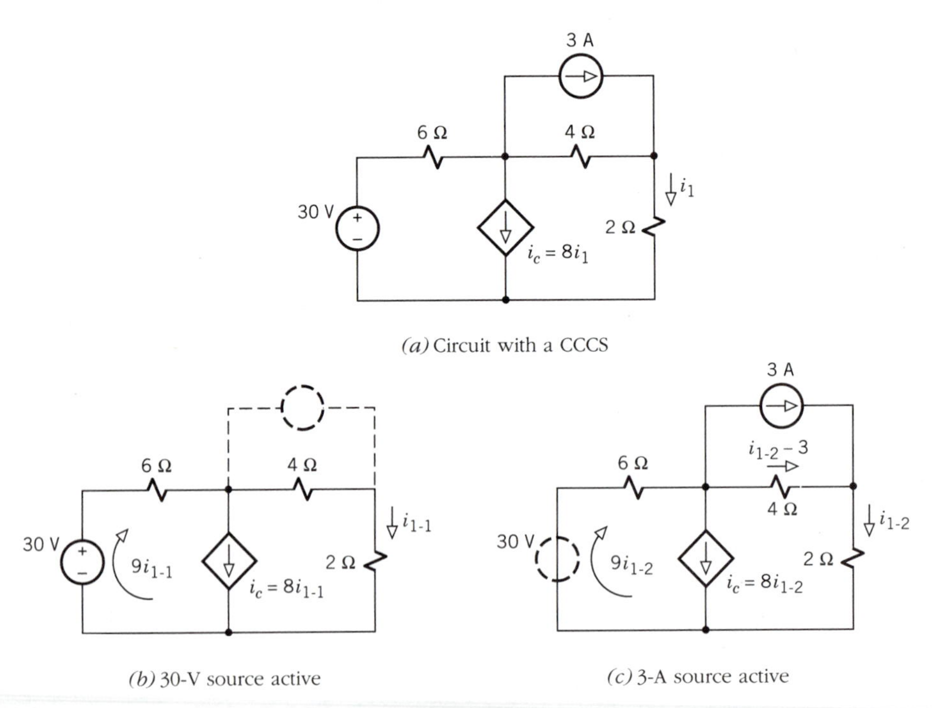

Figure 2.22

$$6(9i_{1-2}) + 4(i_{1-2} - 3) + 2i_{1-2} = 0$$

from which $i_{1-2} = 0.2$ A.

With both independent sources active, the value of i_1 is given by the sum

$$i_1 = i_{1-1} + i_{1-2} = 0.5 + 0.2 = 0.7 \text{ A}$$

Exercise 2.11

Use superposition to find v_1 in Fig. 2.23 when $R = 24\ \Omega$.

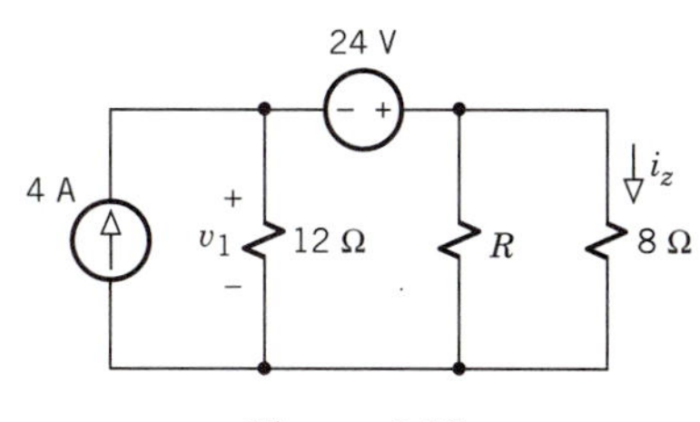

Figure 2.23

Exercise 2.12

What would you say to someone who suggests using superposition to find i when $v_s = 18$ V in Fig. 2.20 (p. 62)?

2.5 THÉVENIN AND NORTON NETWORKS

As a companion of equivalent resistance, this closing section expands the concept of terminal equivalence for networks that include *independent sources.* In fact, we'll discuss two equivalent source networks based on Thévenin's and Norton's theorems. These equivalent networks can be put to use for a variety of tasks in circuit analysis. Additionally, Thévenin-Norton equivalence leads to handy source conversions that simplify many analysis and design problems.

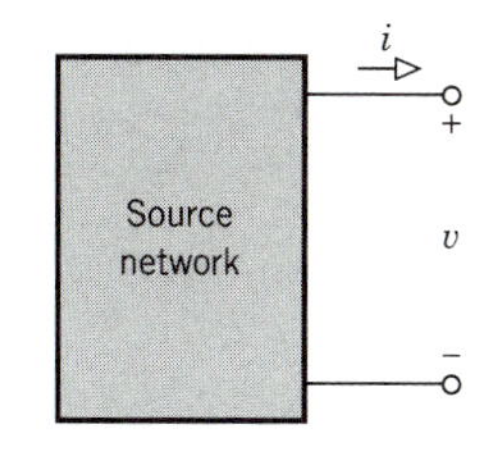

Figure 2.24
Source network containing a least one independent source.

Thévenin's and Norton's Theorems

Thévenin's and Norton's theorems pertain to the terminal behavior of networks containing independent sources. Before stating the theorems, we need to introduce some additional terminology.

Figure 2.24 represents an arbitrary **source network**, which we define as follows:

> A source network is any two-terminal network that consists entirely of linear resistances and sources, including at least one independent source. If controlled sources are present, then the control variables must be within the same network.

The reference polarities for the terminal variables v and i in Fig. 2.24 have been drawn assuming that the source network supplies power when connected to a load. We'll develop a general expression for the terminal v-i relationship by considering two special cases.

At one extreme, let the terminals be left open as in Fig. 2.25*a*. Then $i = 0$, and we define the resulting **open-circuit voltage** by

$$v_{oc} \triangleq v|_{i=0} \tag{2.19}$$

At the other extreme, let the terminals be shorted together as in Fig. 2.25*b*. Then $v = 0$, and we define the resulting **short-circuit current** by

$$i_{sc} \triangleq i|_{v=0} \tag{2.20}$$

Since the network is linear, its v-i curve will be the straight line in Fig. 2.25*c*, which goes through the point v_{oc} at $i = 0$ and the point i_{sc} at $v = 0$.

The slope of the v-i line has been labeled $-R_t$, where

$$R_t \triangleq v_{oc}/i_{sc} \tag{2.21}$$

When $v_{oc} = 10$ V and $i_{sc} = 2$ mA, for instance, Eq. (2.21) yields the value $R_t = 10\text{ V}/2\text{ mA} = 5\text{ k}\Omega$. We call R_t the network's **Thévenin resistance**, named in honor of Charles Leon Thévenin (1857–1926). A physical interpretation of R_t will be given shortly.

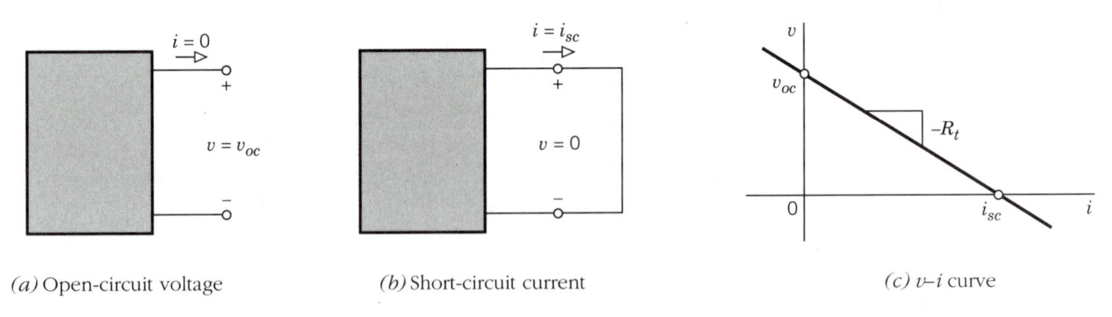

(a) Open-circuit voltage   *(b)* Short-circuit current   *(c)* v–i curve

Figure 2.25

Meanwhile, let the source network be connected to any other two-terminal network. For purposes of analyzing the conditions at the terminals, **Thévenin's theorem** states that

> Any linear source network acts at its terminals like an ideal voltage source of value v_{oc} in series with a resistor having resistance R_t.

Figure 2.26*a* diagrams the corresponding **Thévenin equivalent network**. With any value of *i*, we see that the terminal voltage of the Thévenin network is given by

$$v = v_{oc} - R_t i \tag{2.22}$$

Setting $i = 0$ then yields $v = v_{oc}$. Furthermore, setting $v = 0$ and using Eq. (2.21) yields $i = v_{oc}/R_t = i_{sc}$. Hence, Eq. (2.22) exactly matches the v-i curve in Fig. 2.25*c*.

A half-century after Thévenin, another terminal equivalent network was pointed out by the American engineer Edward L. Norton (1989–1983). **Norton's theorem** states that

> Any linear source network acts at its terminals like an ideal current source of value i_{sc} in parallel with a resistor having resistance R_t.

Figure 2.26*b* diagrams the corresponding **Norton equivalent network**, whose structure is the dual of the Thévenin network. With any value of *v*, the terminal current of the Norton network is given by

$$i = i_{sc} - v/R_t \tag{2.23}$$

This equation also matches our original v-i curve since $i = i_{sc}$ when $v = 0$ while setting $i = 0$ yields $v = R_t i_{sc} = v_{oc}$.

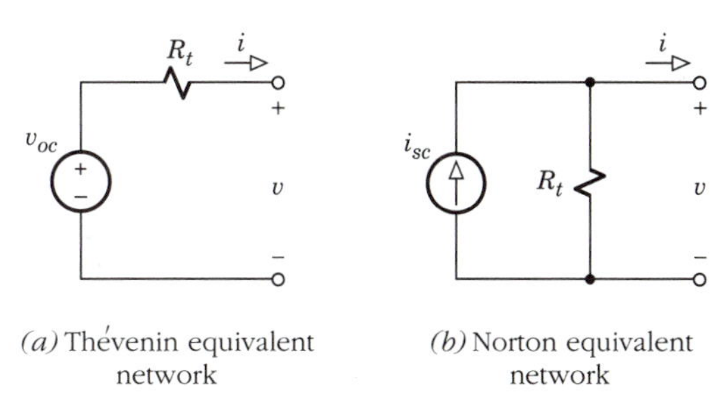

(a) Thévenin equivalent network

(b) Norton equivalent network

Figure 2.26

We'll refer to v_{oc}, i_{sc}, and R_t collectively as the **Thévenin parameters**. Having found the Thévenin parameters of a source network, you can use either the Thévenin or Norton equivalent to predict the terminal behavior of the source network. This strategy has particular value when you need to investigate what happens at the terminals as the external circuitry changes, or when you are designing another network for connection to the source network.

Keep in mind, however, that the equivalence holds only at the terminals of the source network — not *within* the network. As a case in point, we see from Fig. 2.26 that the Thévenin network dissipates no power when $i = 0$, whereas the Norton network dissipates $p = R_t i_{sc}^2$ when $i = 0$.

Example 2.12 *Thévenin Parameters from a v-i Curve*

Figure 2.27*a* diagrams an inverting voltage amplifier whose Thévenin parameters are sought. For this purpose a numerical stimulation using PSpice was performed with a variable load resistance R_L. (Appendix B gives information about PSpice simulations, but the details are not needed here.)

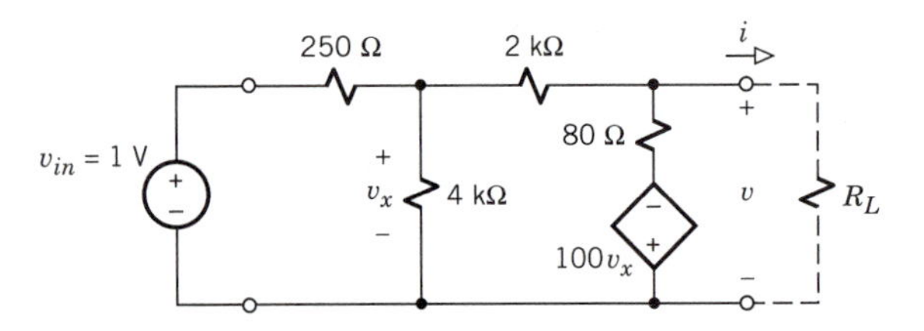

(a) Circuit diagram of an inverting voltage amplifier

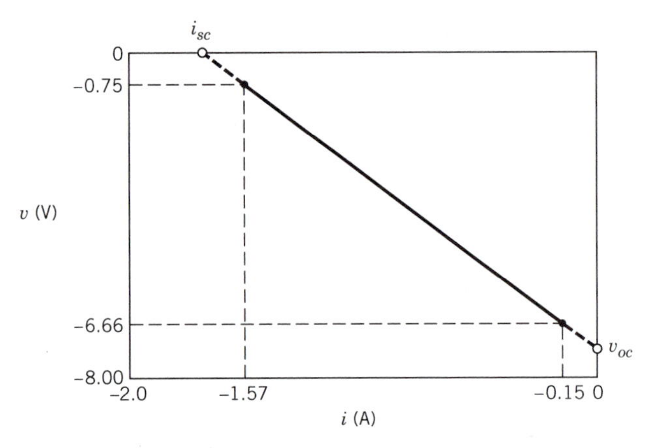

(b) v-i curve obtained by PSpice simulation

Figure 2.27

The open-circuit voltage and short-circuit current could be found by setting $R_L = \infty$ and $R_L = 0$, respectively. However, since PSpice does not allow these extreme resistance values, R_L was varied from 0.5 Ω to 50 Ω to trace the *v-i* curve in Fig. 2.27*b*. This linear curve shows that $v = -0.75$ V when $i = -1.57$ A, and $v = -6.66$ V when $i = -0.15$ A. We'll use these values to determine v_{oc}, i_{sc}, and R_t.

Equation (2.22) says that v and i are always related by $v_{oc} - R_t i = v$.

Hence, the data from the v-i curve yields two equations with two unknowns, namely

$$v_{oc} - R_t(-1.57\ \text{A}) = -0.75\ \text{V}$$
$$v_{oc} - R_t(-0.15\ \text{A}) = -6.66\ \text{V}$$

from which

$$v_{oc} = -7.28\ \text{V} \qquad R_t = 4.16\ \Omega$$

Equation (2.20) then yields

$$i_{sc} = v_{oc}/R_t = -1.75\ \text{A}$$

Note that the inverting effect of the amplifier causes both v_{oc} and i_{sc} to have negative values.

Example 2.13 *Equivalent Source Networks*

Figure 2.28*a* shows another source network connected to a load resistor R_L. The application in question calls for two different design values of R_L. One value must be such that $v = 24$ V, while the other must be such that $i = 8$ A. We also want to evaluate the current i_1 supplied by the voltage source in each case.

Problems like this involving variable load conditions are best tackled using an equivalent network for the source. Accordingly, we'll first determine the network's Thévenin parameters.

We calculate v_{oc} from the open-circuit conditions in Fig. 2.28*b*. Since the two resistors carry the same current here, they form a voltage divider with

$$v_{oc} = \frac{20}{(5 + 20)} \times 50 = 40\ \text{V}$$

We then calculate i_{sc} from the short-circuit conditions in Fig. 2.28*c*. The short circuit bypasses current around the 20-Ω resistor, so

$$i_{sc} = 50/5 = 10\ \text{A}$$

Equation (2.21) then yields the Thévenin resistance

$$R_t = v_{oc}/i_{sc} = 40/10 = 4\ \Omega$$

Now we can find the design value of R_L that produces $v = 24$ V by substituting the Thévenin equivalent for the source network, as shown in Fig. 2.28*d*. We thus have a new voltage divider, from which

$$v = \frac{R_L}{4 + R_L} \times 40 = 24 \text{ V} \quad \Rightarrow \quad R_L = 6 \; \Omega$$

Then we find the design value of R_L that produces $i = 8$ A by substituting the Norton equivalent network, as shown in Fig. 2.28*e*. We thus have a current divider, from which

$$i = \frac{4}{4 + R_L} \times 10 = 8 \text{ A} \quad \Rightarrow \quad R_L = 1 \; \Omega$$

Although both of these results could have been obtained using either equivalent network, the series structure of the Thévenin network generally expedites voltage calculations while the parallel structure of the Norton network generally expedites current calculations.

But neither equivalent network mimics the *internal* conditions of the actual source network. To evaluate the internal current i_1, we must therefore go back to the original diagram in Fig. 2.28*a* where $i_1 = (50 - v)/5$. With $R_L = 6\;\Omega$ and $v = 24$ V, $i_1 = 26/5 = 5.2$ A — in contrast to the loop current $i = v/R_L = 4$ A in Fig. 2.28*d*. With $R_L = 1\;\Omega$ and $i = 8$ A, $v = R_L i = 8$ V so $i_1 = 42/5 = 8.4$ A — in contrast to the source current $i_{sc} = 10$ A in Fig. 2.28*e*.

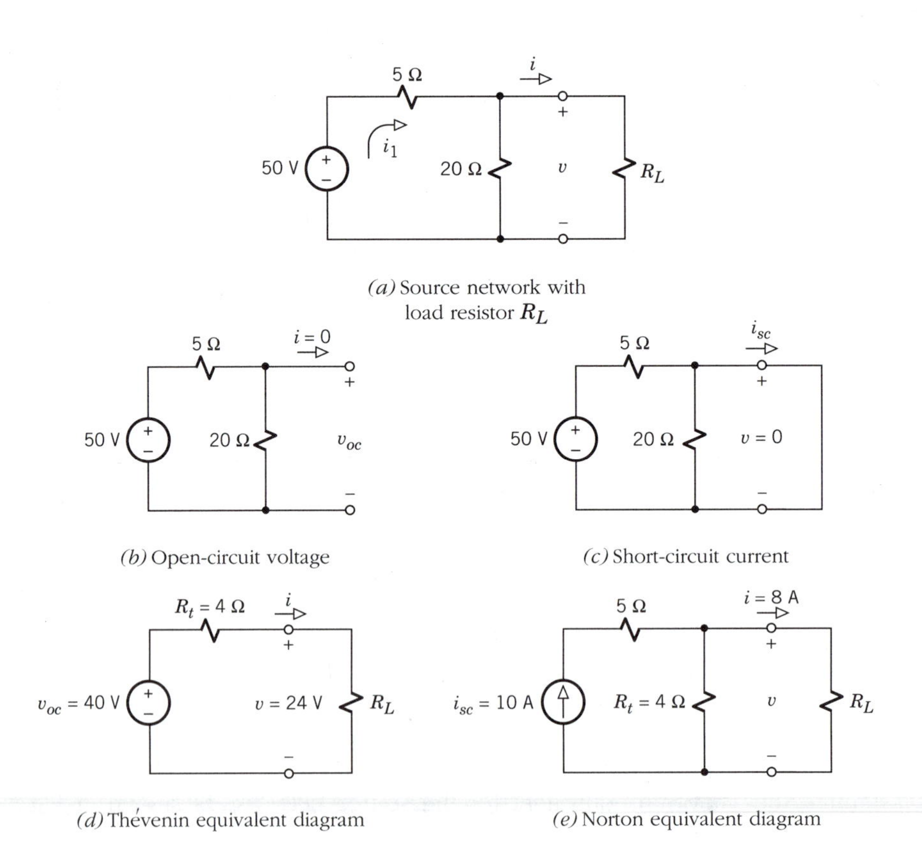

(a) Source network with load resistor R_L

(b) Open-circuit voltage

(c) Short-circuit current

(d) Thévenin equivalent diagram

(e) Norton equivalent diagram

Figure 2.28

Exercise 2.13

Measurements at the output terminals of a linear amplifier show that v = 30 V when i = 5 mA, and v = −30 V when i = −15 mA. Find the amplifier's Thévenin parameters.

Exercise 2.14

Obtain the Thévenin and Norton equivalents for the network in Fig. 2.29. Then find the value of R_L that would draw i = 4 A when connected to the network's terminals.

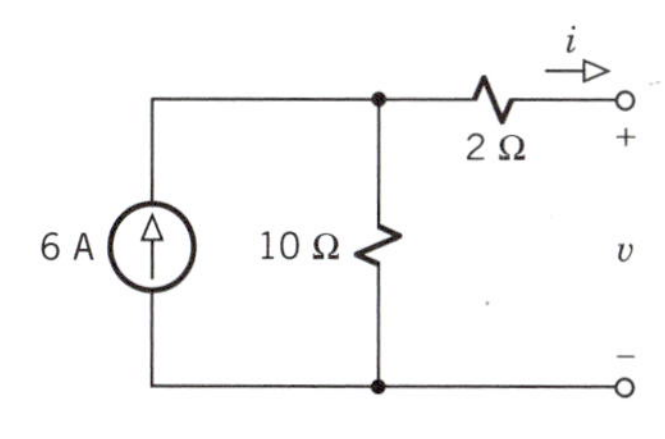

Figure 2.29

Thévenin Resistance

The parameters v_{oc} and i_{sc} have clear physical interpretations that help us calculate their values by considering open-circuit and short-circuit conditions. The Thévenin resistance R_t also has a physical interpretation that we'll develop here. Besides promoting insight, the physical interpretation helps us calculate Thévenin resistance without resorting to the formula $R_t = v_{oc}/i_{sc}$.

Suppose we set $v_{oc} = 0$ in the Thévenin network (Fig. 2.26*a*) or $i_{sc} = 0$ in the Norton network (Fig. 2.26*b*). Since a zero voltage source becomes a short circuit and a zero current source becomes an open circuit, both equivalent networks thereby reduce to a source-free network with *equivalent resistance* R_t. The very same property must hold for the original source network in question. Specifically,

> The Thévenin resistance of a source network equals the equivalent resistance of that network when all independent sources have been suppressed.

Independent sources are suppressed by making the replacements previously used for superposition calculations. We say that the source network is **dead** when all independent sources have been suppressed. As before, any controlled sources are not suppressed because they may affect the value of R_t.

When the source network has a ladder structure and contains no controlled sources, R_t is easily found by series-parallel reduction of the dead network. For instance, the network back in Fig. 2.28*b* reduces to two parallel resistors when the voltage source is replaced by a short circuit. We thus get $R_t = 5\|20 = 4\ \Omega$, in agreement with $v_{oc}/i_{sc} = 4\ \Omega$ from Example 2.13.

When the source network contains controlled sources, the Thévenin resistance can be found using the method represented by Fig. 2.30. Here, the dead source network has been connected to an external **test source**. This test source may be any independent voltage or current source that establishes v_t and i_t at the terminals, with i_t referenced *into* the higher-potential terminal. Since the dead network contains only resistors and controlled sources, and since R_t equals the equivalent resistance of the dead network, the equivalent resistance theorem tells us that

$$R_t = v_t/i_t \tag{2.24}$$

Accordingly, R_t can be found by forming the ratio v_t/i_t.

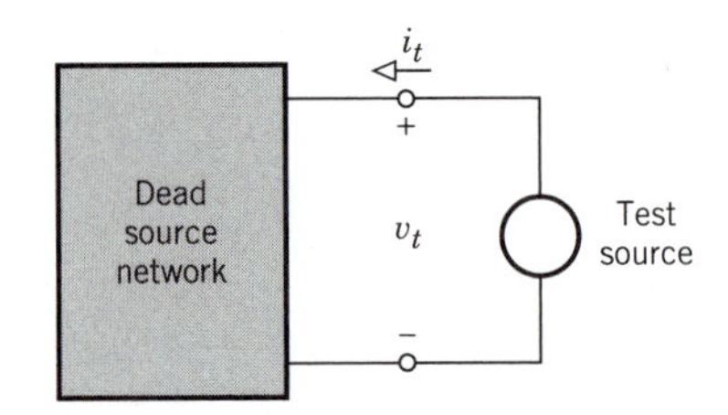

Figure 2.30 Dead source network with test source to find $R_t = v_t/i_t$.

We now know how to determine R_t without calculating the open-circuit voltage or short-circuit current of the unsuppressed source network. But we also know that $R_t = v_{oc}/i_{sc}$. Therefore, if you have found R_t and i_{sc}, then you can evaluate v_{oc} via

$$v_{oc} = R_t i_{sc} \tag{2.25a}$$

Or if you have found R_t and v_{oc}, then you can evaluate i_{sc} via

$$i_{sc} = v_{oc}/R_t \tag{2.25b}$$

These relations provide welcome flexibility when analyzing a source network because the network will be fully characterized by any two of the three Thévenin parameters.

Example 2.14 *Calculating Thévenin Parameters*

Suppose you need the Thévenin parameters for the source network in Fig. 2.31*a*, which includes a VCVS. An examination of the diagram suggests that finding v_{oc} might be rather tedious since there are no simplifications under open-circuit conditions. You therefore shift your attention to i_{sc} and R_t.

First, for i_{sc}, you draw the short-circuited diagram in Fig. 2.31*b* where the 40-kΩ resistor has become bypassed. The control voltage must satisfy the KVL equation

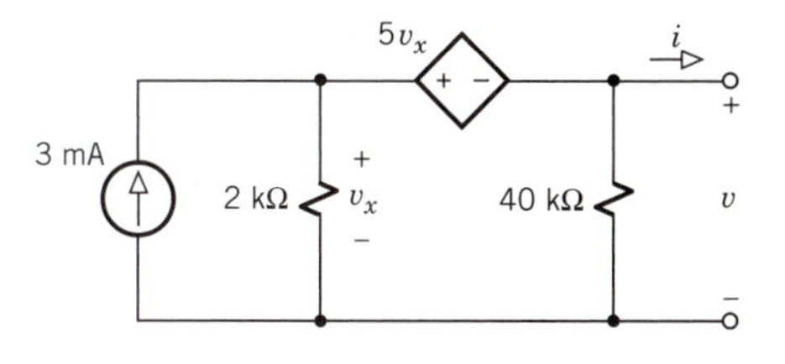

(a) Source network with a VCVS

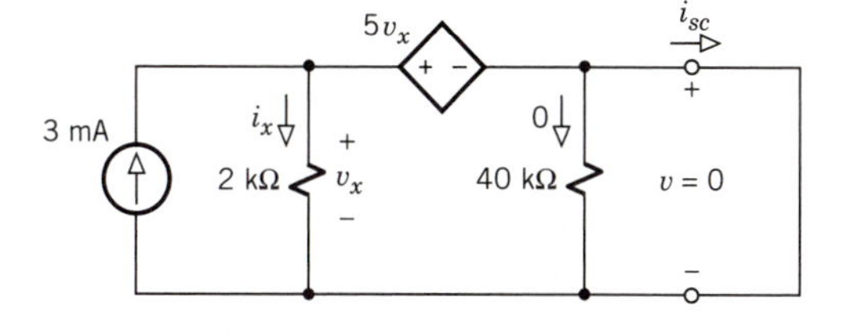

(b) Short circuit at the output

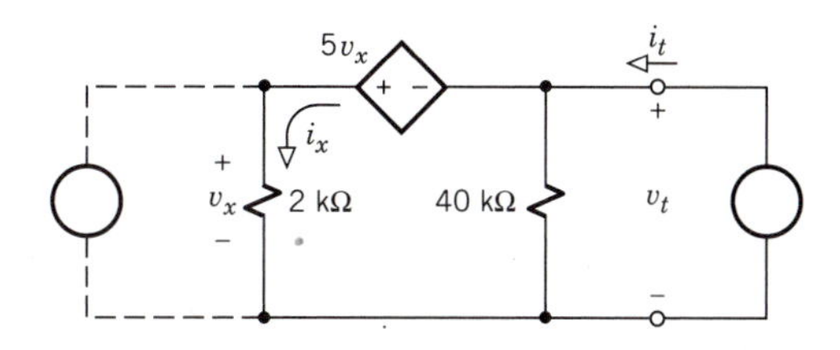

(c) Dead network with test source

Figure 2.31

$$v_x - 5v_x = v = 0 \quad \Rightarrow \quad v_x = 0$$

Hence, $i_x = 0$ and KCL yields

$$i_{sc} = 3 \text{ mA}$$

Next, for R_t, you seek the equivalent resistance of the dead network in Fig. 2.31c by attaching an arbitrary test source. The control voltage must now satisfy

$$v_x - 5v_x = v_t \Rightarrow v_x = -0.25v_t$$

Thus, working with kilohms and milliamps,

$$i_x = v_x/2 = -0.125v_t$$

You then find i_t in terms of v_t by writing

$$i_t = i_x + v_t/40 = -0.125v_t + 0.025v_t = -0.1v_t$$

from which

$$R_t = v_t/i_t = v_t/(-0.1v_t) = -1/0.1 = -10 \text{ k}\Omega$$

The controlled source in this network causes the Thévenin resistance to have a negative value.

Finally, you use Eq. (2.25a) to get the remaining Thévenin parameter

$$v_{oc} = R_t i_{sc} = -10\ \text{k}\Omega \times 3\ \text{mA} = -30\ \text{V}$$

Exercise 2.15

Obtain the Thévenin parameters for the network in Fig. 2.32 by determining v_{oc} and R_t. Then check your calculated value of i_{sc} by direct analysis.

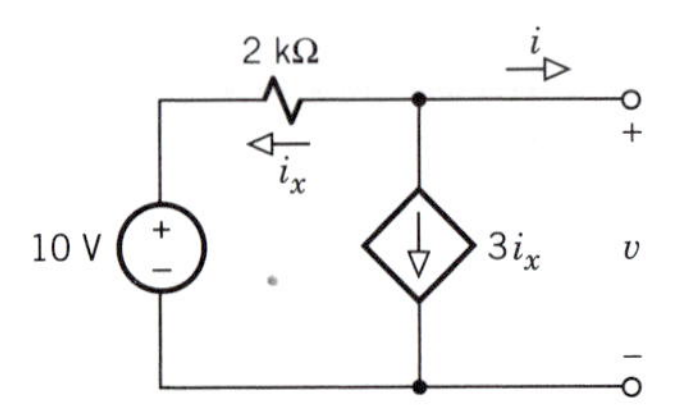

Figure 2.32

Source Conversions

A highly valuable byproduct of Thévenin's and Norton's theorems is the technique of source conversions. Source conversions are based upon the observation that if a Thévenin network and a Norton network are both equivalent to a particular source network, then they must also be *equivalent to each other.* This observation allows you to simplify an analysis or design problem by converting a voltage source with series resistance to an equivalent current source with parallel resistance, or vice versa.

To generalize notation for source conversions, suppose that part of a circuit contains a voltage source v_s in series with finite nonzero resistance R_s. Since this part has the same structure as a Thévenin network, we could conve rt it to a Norton network having R_s in parallel with a current source i_s. The value of i_s must equal the short-circuit current from v_s in series with R_s, so

$$i_s = v_s / R_s \tag{2.26a}$$

Conversely, suppose that part of a circuit has a Norton structure consisting of a current source i_s in parallel with finite nonzero resistance R_s. Then we could convert this part to a Thévenin structure having R_s in series with v_s, where

$$v_s = R_s i_s \tag{2.26b}$$

Figure 2.33 summarizes the source conversions in both directions. Note carefully that

> The arrow for the current source points in the direction of the higher-potential end of the voltage source.

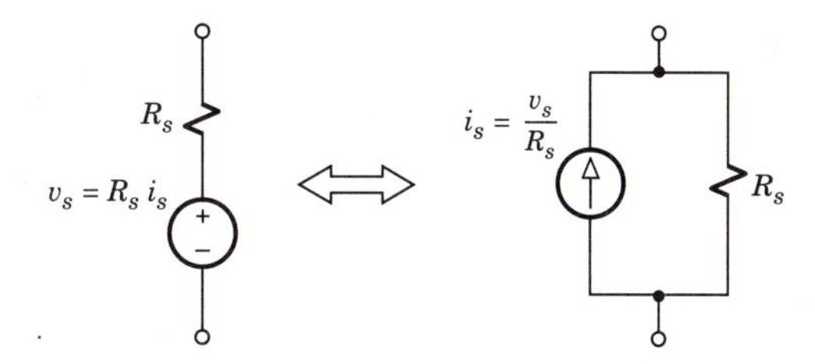

Figure 2.33 Equivalent source conversions.

This ensures that both networks have the same sign for the open-circuit voltage and short-circuit current.

Source conversions may be applied to portions of a circuit or network to simplify intermediate calculations. Going even further, repeated source conversions may reduce a circuit to all-series or all-parallel form. Conversions may also be applied to controlled sources as well as to independent sources. However,

> When controlled sources are present, source conversions must not obliterate the identity of any control variables.

Usually, you also want to preserve the identity of the particular variables being sought.

Example 2.15 *Circuit Reduction by Source Conversions*

Figure 2.34*a* repeats the circuit from Example 2.9, where we got $v_2 = 24$ V using the proportionality principle. However, if v_2 is the only quantity of interest, then source conversions yield an all-parallel circuit for easier analysis.

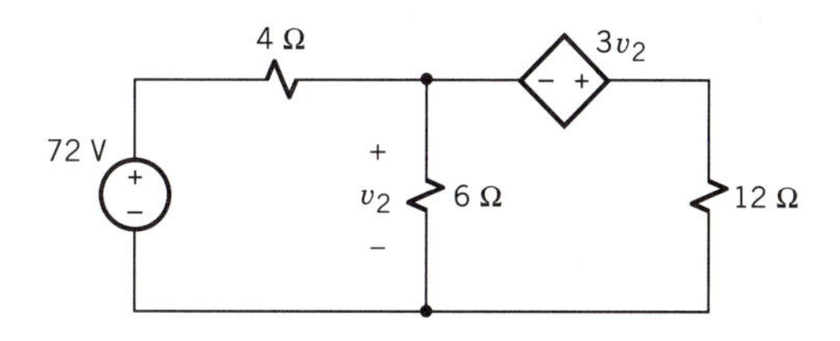

(a) Circuit for analysis by source convertion

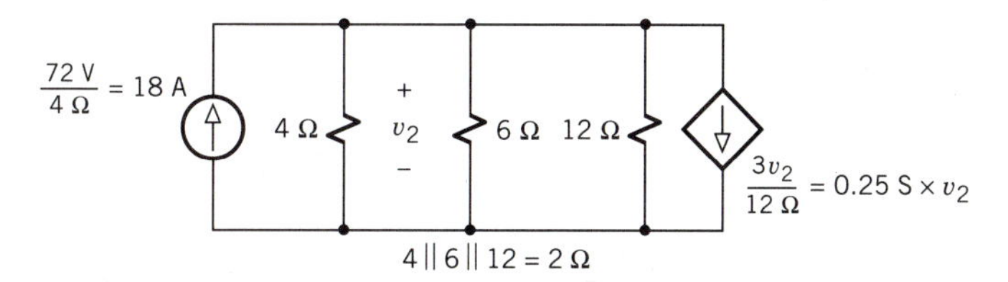

(b) Equivalent parallel circuit

Figure 2.34

The 72-V source and 4-Ω series resistance convert to a parallel structure with source current 72 V/4 Ω = 18 A. The VCVS and 12-Ω series resistance likewise convert to a parallel structure with source current $3v_2/12\ \Omega = 0.25\ \text{S} \times v_2$ — so the VCVS becomes a VCCS whose transconductance is $g_m = 0.25$ S. The resulting diagram in Fig. 2.34*b* still has the control voltage v_2 in place, as required.

But now all three resistances can be combined as $4\|6\|12 = 2\ \Omega$. This parallel equivalent resistance carries the net current from the sources, so

$$v_2 = 2(18 - 0.25v_2)$$

Solving for v_2 yields

$$v_2 = 36/1.5 = 24\text{ V}$$

which agrees with our previous result.

Example 2.16 ***Thévenin Network via Source Conversions***

Consider the source network in Fig. 2.35*a*. In Example 2.10, we invoked superposition and found that a 2-Ω resistor connected to the terminals carried $i = -0.5$ A. Here, we'll gain insight on that result by applying source conversions to obtain the Thévenin equivalent network.

Our ultimate objective is a series configuration carrying current *i*. We therefore begin by converting the 3-A source and 4-Ω parallel resistance to get Fig. 2.35*b*. Next, we convert the 30-V source and 6-Ω series resistance as shown in Fig. 2.35*c*. This step yields two parallel current sources, which can be summed algebraically and converted with the 6-Ω resistance to obtain the reduced diagram in Fig. 2.35*d*. Since the order of series elements does not matter, we now add the two resistances and algebraically sum the source voltages to get the final result in Fig. 2.35*e*.

Figure 2.35*e* has the form of a Thévenin network with $v_{oc} = -6$ V and $R_t = 10\ \Omega$. Consequently, a 2-Ω load resistor connected to the terminals would draw $i = -6/(10 + 2) = -0.5$ A.

Exercise 2.16

Let $R = 8\ \Omega$ in Fig. 2.23 (p. 67). Find v_1 using source conversions to get an all-parallel circuit.

Exercise 2.17

Find i_1 in Fig. 2.22*a* (p. 66) by using source conversions to get an all-series circuit.

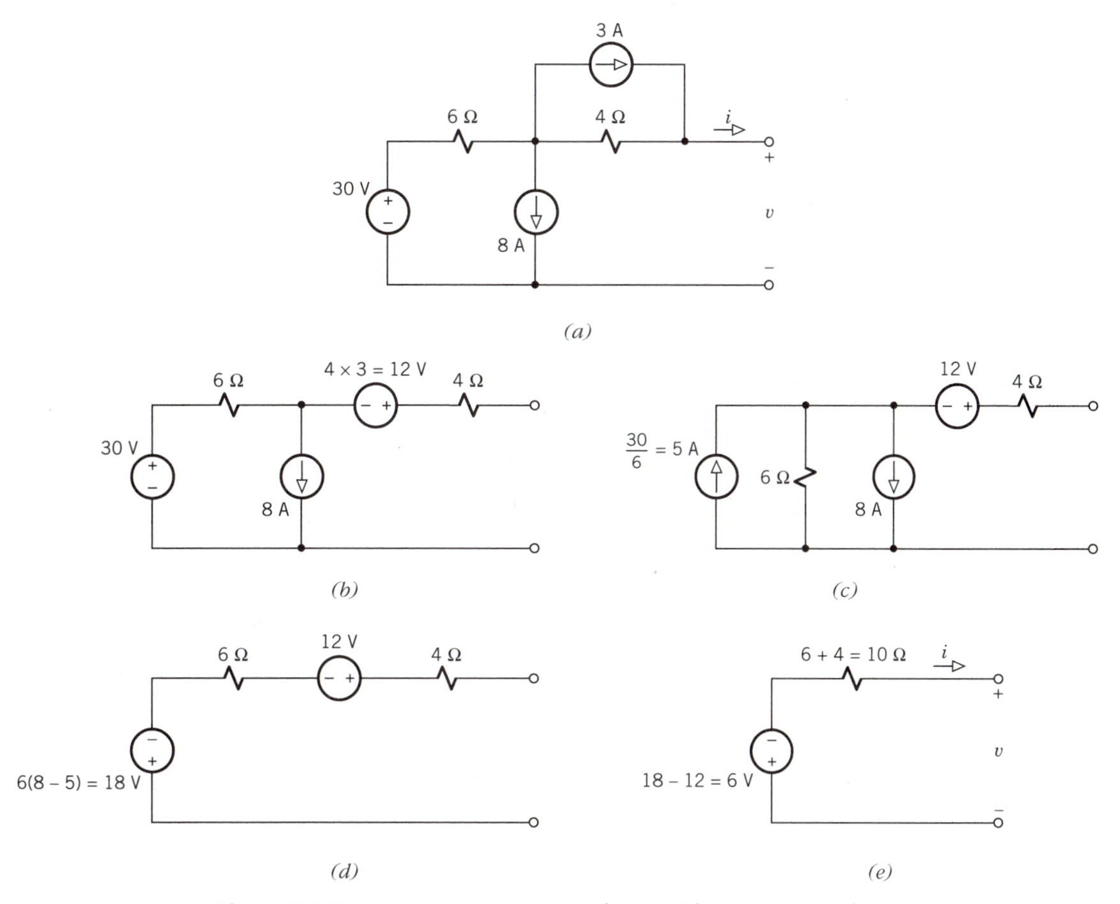

Figure 2.35 Source conversions to obtain a Thévenin network.

PROBLEMS

Section 2.1 Series and Parallel Resistance

2.1 Obtain an expression similar to Eq. (2.8) for three parallel resistors.

2.2 Obtain an expression similar to Eq. (2.8) for four parallel resistors.

2.3* Suppose an electric grill unit like Fig. 2.7 (p. 48) has v_s = 120 V and R_1 = 8 Ω. Using Eq. (2.10), design R_2 such that p_{max} = 2400 W. Then calculate the power for the other three heat settings.

2.4 Do Problem 2.3 with R_1 = 10 Ω.

2.5 Obtain the condition on R_1 and R_2 in Fig. 2.7 (p. 48) such that $p_{max} = 4p_{mim}$. Why would this be undesirable?

2.6 Show how you would connect some available 20-Ω resistors to reduce the resistance between two nodes from 60 Ω to R_{eq} = 24 Ω.

2.7 Do Problem 2.6 for R_{eq} = 20 Ω.

2.8 Do Problem 2.6 for R_{eq} = 12 Ω.

2.9* If v = 48 V and R_y = 33 Ω in Fig. P2.9, then what is the value of R_x?

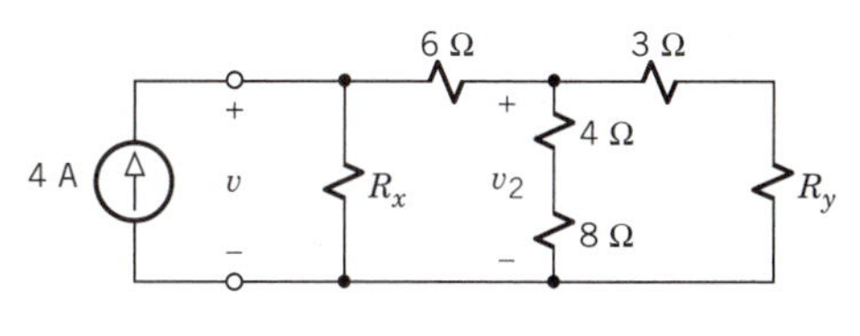

Figure P2.9

2.10 If v = 24 V and R_y = 3 Ω in Fig. P2.9, then what is the value of R_x?

2.11 If v = 12 V and v_2 = 4 V in Fig. P2.9, then what is the value of R_y?

2.12 If $v = 28$ V and $v_2 = 16$ V in Fig. P2.9, then what is the value of R_y?

2.13 Let $v_s = 75$ V and $R_3 = 54$ kΩ in Fig. P2.13. Find i_1, i_3, and the power supplied by the source.

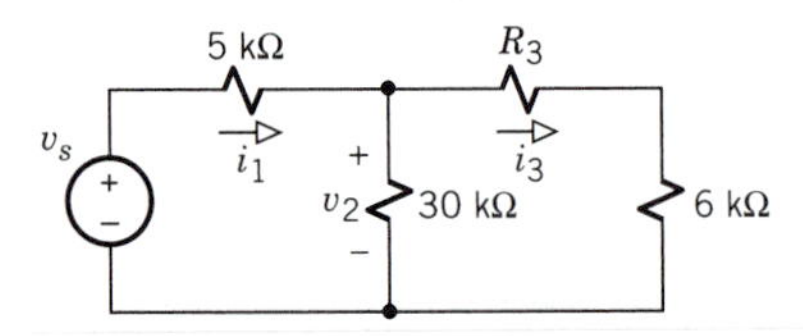

Figure P2.13

2.14 Let $v_s = 25$ V and $R_3 = 4$ kΩ in Fig. P2.13. Find i_1, i_3, and the power supplied by the source.

2.15* Find v_1, i_2, i_3, and the power supplied by the source in Fig. P2.15 when $R_2 = 19$ kΩ and $R_3 = 30$ kΩ.

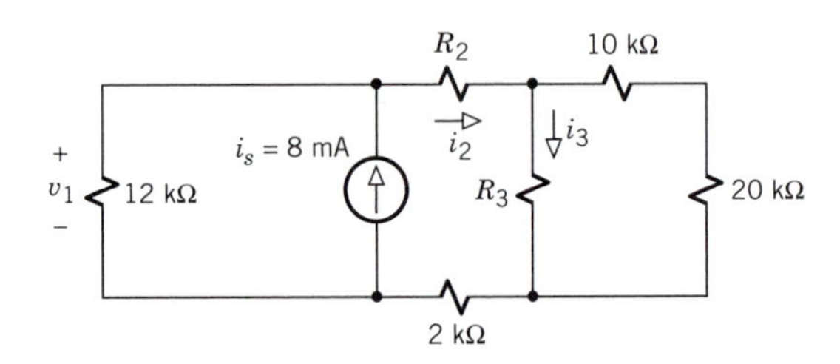

Figure P2.15

2.16 Find v_1, i_2, i_3, and the power supplied by the source in Fig. P2.15 when $R_2 = 8$ kΩ and $R_3 = 15$ kΩ.

2.17 Find v_1, i_2, i_3, and the power supplied by the source in Fig. P2.15 when $R_2 = 2.5$ kΩ and $R_3 = 10$ kΩ.

2.18 Calculate i, v_1, and v_2 in Fig. P2.18 when both wipers are set at the middle of the potentiometers.

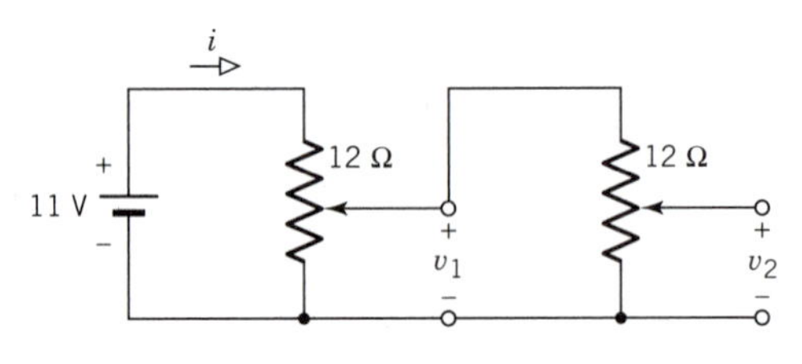

Figure P2.18

2.19 Calculate i, v_1, and v_2 in Fig. P2.18 when both wipers are set one-third from the bottom of the potentiometers.

2.20 Calculate i, v_1, and v_2 in Fig. P2.18 when both wipers are set one-third from the top of the potentiometers.

2.21 The circuit in Fig. P2.21 is intended to measure temperature. It features a **temperature-dependent resistor** having $R_T = R_0(1 + \alpha T)$, where T is the temperature in degrees Celsius and α is the temperature coefficient. (a) Find R_0 and α given that $v = 6$ V at $T = 0$°C and $v = 15$ V at $T = 100$°C. Hint: Write an equation for i in terms of v. (b) What value of v will be measured when $T = 350$°C?

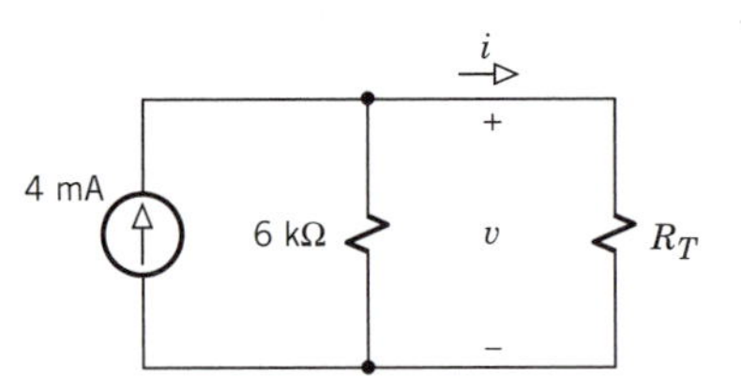

Figure P2.21

2.22 Let the temperature measuring circuit described in Problem 2.21 have $R_0 = 3$ kΩ and $\alpha = 0.02/$°C. By considering G_{par}, derive an expression for the temperature T in terms of v. Then find the values of T corresponding to $v = 12$ V and $v = 18$ V.

Section 2.3 Circuits with Controlled Sources

2.23* Use KVL to determine i_x in Fig. P2.23 when $R_y = 3$ Ω and $i_c = 4 i_x$.

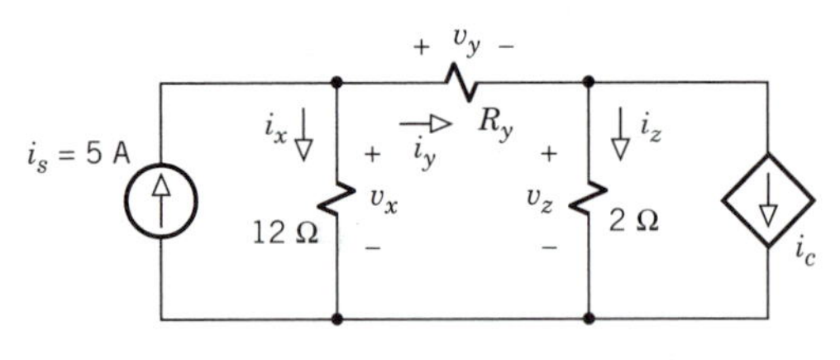

Figure P2.23

2.24 Use KVL to determine i_y in Fig. P2.23 when $R_y = 10$ Ω and $i_c = v_x/(4\ \Omega)$.

2.25 Use KVL to determine v_x in Fig. P2.23 when $R_y = 20$ Ω and $i_c = 2 i_y$.

2.26 Use KCL to determine i_y in Fig. P2.26 when $R_x = 10$ kΩ and $v_c = 25$ kΩ × i_y.

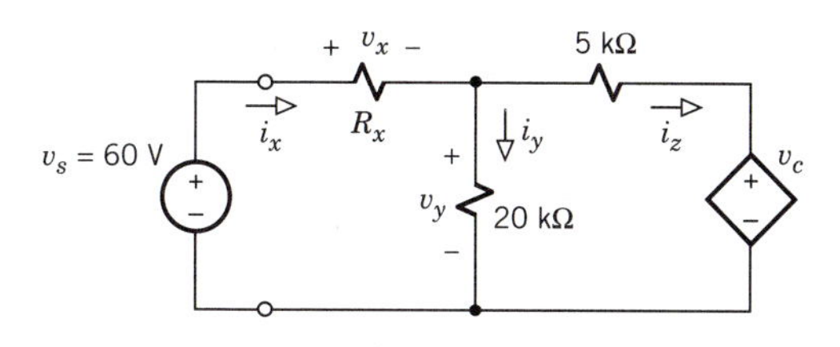

Figure P2.26

2.27 Use KCL to determine v_y in Fig. P2.26 when $R_x = 4\ \text{k}\Omega$ and $v_c = 5v_x$.

2.28 Use KCL to determine v_x in Fig. P2.26 when $R_x = 12\ \text{k}\Omega$ and $v_c = 10\ \text{k}\Omega \times i_x$.

2.29* Find $R_{eq} = v/i$ for Fig. P2.29 when $v_c = 4v$.

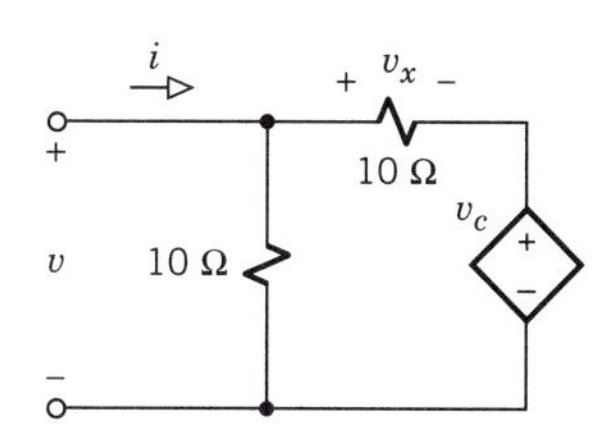

Figure P2.29

2.30 Find $R_{eq} = v/i$ for Fig. P2.29 when $v_c = 3v_x$.

2.31 Find $R_{eq} = v/i$ for Fig. P2.29 when $v_c = 20\ \Omega \times i$.

2.32 Find $R_{eq} = v/i$ for Fig. P2.32 when $i_c = 3i$.

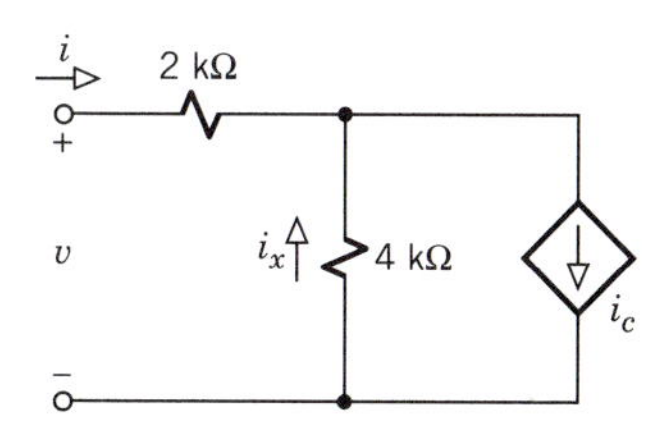

Figure P2.32

2.33 Find $R_{eq} = v/i$ for Fig. P2.32 when $i_c = 3i_x$.

2.34 Find $R_{eq} = v/i$ for Fig. P2.32 when $i_c = v/(4\ \text{k}\Omega)$.

2.35 The value of R_1 in Fig. P2.35 is unknown, but measurements reveal that $v_3 = 36$ V. (a) Find i_3, v_4, i_2, and i_1. (b) Determine the value of R_1 and the equivalent resistance connected to the 48-V source. (c) Calculate the total power supplied by the two sources. Then make a separate calculation of the total power dissipated by the four resistors.

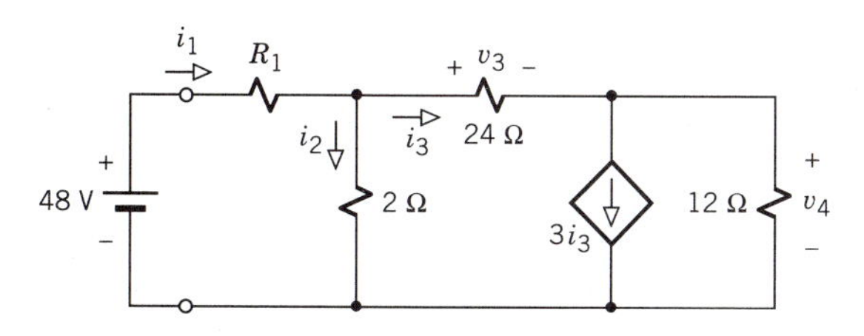

Figure P2.35

2.36 Figure P2.36 is the model of an amplifier with two transistors, each having current gain β. (a) Show that $i_x = -i_{in}$, regardless of R_L. (b) Use the result from (a) to obtain an expression for $R_{in} = v_{in}/i_{in}$ in terms of β. Do your work with resistance in kilohms and current in milliamps. (c) Let $v_s = 2$ V and $R_L = 5\ \text{k}\Omega$. Make a table listing the values of R_{in}, i_{in}, and v_{out} for β = 8, 10, 12, and 14. For what value(s) of β would the circuit model be invalid? Explain your answer.

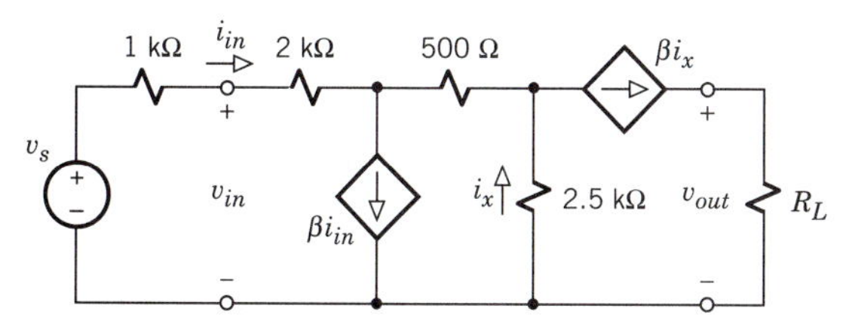

Figure P2.36

Section 2.4 Linearity and Superposition

2.37* Let $R_3 = 4\ \text{k}\Omega$ and $v_s = 20$ V in Fig. P2.13. Assume $\hat{i}_3 = 3$ mA to calculate i_1 and i_3.

2.38 Let $R_2 = 6\ \Omega$ and $i_s = 4$ A in Fig. P2.38. Assume $\hat{v}_3 = 6$ V to calculate v_1 and i_4.

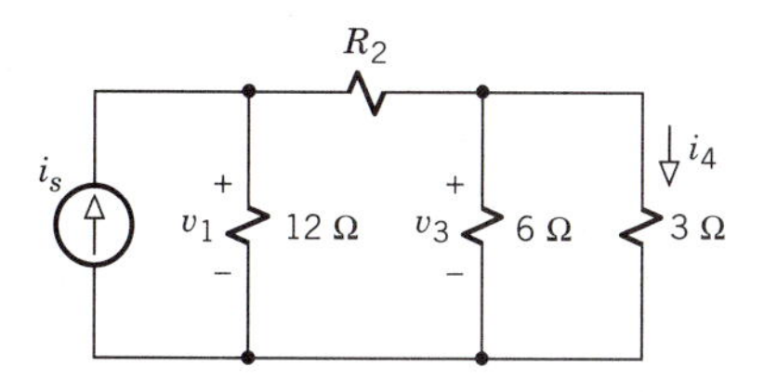

Figure P2.38

2.39 Let $R_2 = 0.2\ \text{k}\Omega$, $R_3 = 6\ \text{k}\Omega$, and $i_s = 24$ mA in Fig. P2.15. Assume $\hat{\imath}_3 = 5$ mA to calculate v_1 and i_3.

2.40 Let $R_x = 5\ \text{k}\Omega$ and $v_c = 15\ \text{k}\Omega \times i_y$ in Fig. P2.26. Find v_x, v_y, and v_c by assuming $\hat{\imath}_y = 1$ mA.

2.41 Let $R_x = 5\ \text{k}\Omega$ and $v_c = 1.5 v_y$ in Fig. P2.26. Find v_x, v_y, and v_c by assuming $\hat{v}_y = 20$ V.

2.42 Let $R_x = 6\ \text{k}\Omega$ and $v_c = -25\ \text{k}\Omega \times i_z$ in Fig. P2.26. Find v_x, v_y, and v_c by assuming $\hat{\imath}_z = 1$ mA.

2.43 Let $R_y = 6\ \Omega$ and $i_c = 4 i_y$ in Fig. P2.23. Find i_x, i_z, and i_c by assuming $\hat{\imath}_y = 1$ A.

2.44 Let $R_y = 14\ \Omega$ and $i_c = v_y/(7\ \Omega)$ in Fig. P2.23. Find i_x, i_z, and i_c by assuming $\hat{v}_y = 14$ V.

2.45 Let $R_y = 2\ \Omega$ and $i_c = -4 i_z$ in Fig. P2.23. Find i_x, i_z, and i_c by assuming $\hat{\imath}_z = 1$ A.

2.46* Find v_a in Fig. P2.46 using superposition.

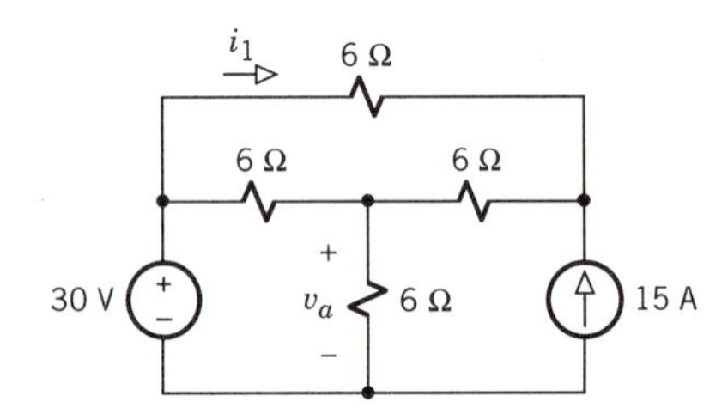

Figure P2.46

2.47 Find i_1 in Fig. P2.46 using superposition.

2.48 Let $v_s = 10$ V and $R_3 = 8\ \Omega$ in Fig. P2.48. Apply superposition to calculate i_2.

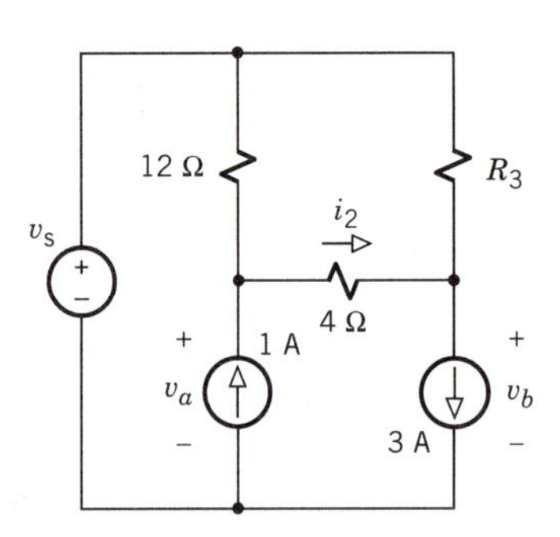

Figure P2.48

2.49 Let $v_s = 12$ V and $R_3 = 20\ \Omega$ in Fig. P2.48. Apply superposition to calculate v_a.

2.50 Let $v_s = 16$ V and $R_3 = 56\ \Omega$ in Fig. P2.48. Apply superposition to calculate v_b.

2.51 Given $i_s = 5$ mA in Fig. P2.51, use superposition to find v_s such that $v_2 = 30$ V.

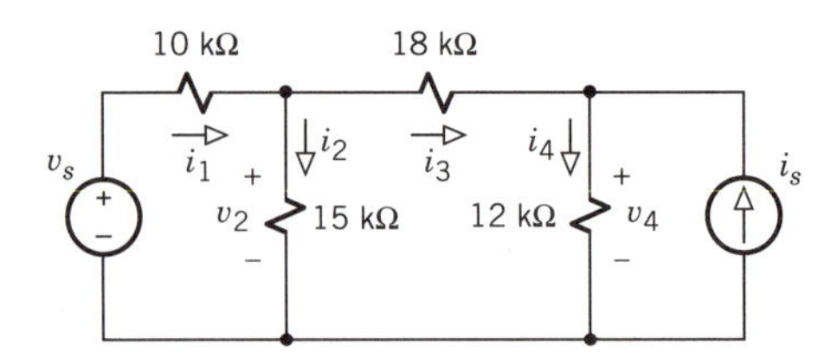

Figure P2.51

2.52 Given $i_s = 3$ mA in Fig. P2.51, use superposition to find v_s such that $v_4 = 36$ V.

2.53 Given $v_s = 20$ V in Fig. P2.51, use superposition to find i_s such that $i_3 = 2$ mA.

2.54 Given $v_s = 30$ V in Fig. P2.51, use superposition to find i_s such that $i_2 = 3$ mA.

2.55 The VCVS in Fig. P2.55 has variable gain μ. (a) Relate v_3 to v_1 to show that $v_2 = (4 - \mu)v_1/4$. (b) Assume $v_1 = 20$ V. Use proportionality and the result from (a) to obtain an expression for $R_{eq} = v/i$ in terms of μ. (c) Evaluate R_{eq} when $\mu = 0$. Check your result directly from the circuit diagram for this case. (d) Let $v_s = 60$ V and let μ vary over the following values: 1, 5, 6, 7, 9, 10. Make a table listing the resulting values of R_{eq}, i, v, and the power $p = vi$. For what value(s) of μ would the circuit model be invalid? Explain your answer.

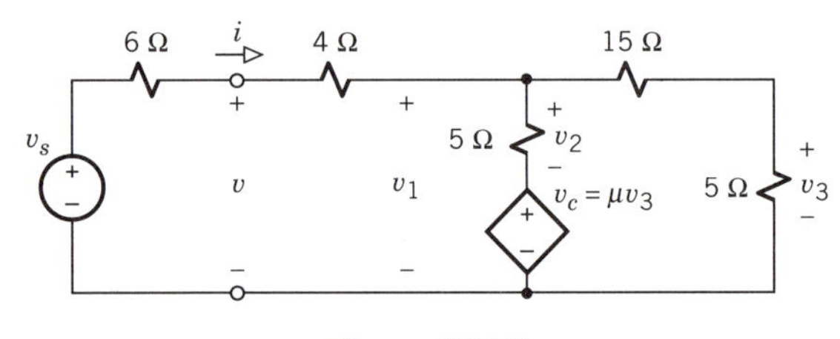

Figure P2.55

2.56 The VCVS in Fig. P2.56 has variable gain μ. (a) Relate v_2 to v_1 to show that $v = (5 - \mu)v_2$. (b) Assume $v_2 = 4$ V. Use proportionality and the result from (a) to obtain an expression for $R_{eq} = v/i$ in terms of μ. Do your work with resistance in kilohms and current in milliamps. (c) Evaluate R_{eq} when $\mu = 0$. Check your result directly from the circuit diagram for this case. (d) Let $v_s = 30$ V and let μ vary over the following values: 0, 5, 10, 15, 20, 30. Make a table listing the resulting values of R_{eq}, i, v, and the power $p = vi$. For what value(s) of μ would the circuit model be invalid? Explain your answer.

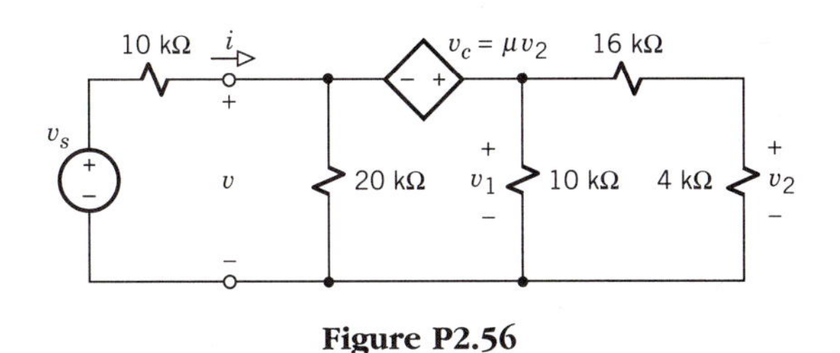

Figure P2.56

Figure P2.61

Section 2.5 Thévenin and Norton Networks

(See also PSpice problems B.5 and B.6.)

2.57* Let $R_x = 0$ in Fig. P2.57. Find v_{oc} and i_{sc} using superposition. Then confirm that v_{oc}/i_{sc} equals the equivalent resistance of the dead network.

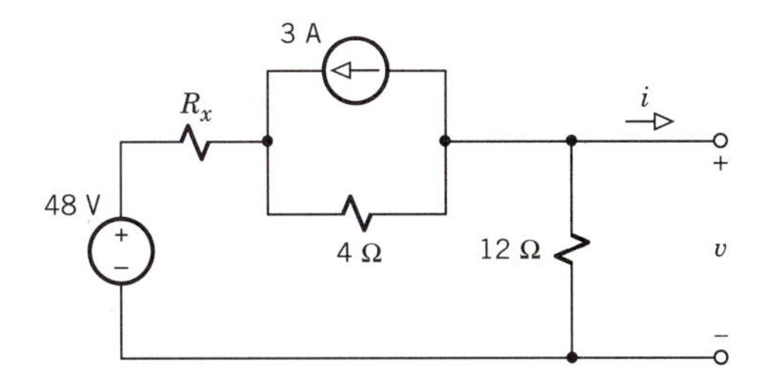

Figure P2.57

2.58 Do Problem 2.57 with $R_x = 2\ \Omega$.

2.59 Let $R_x = 0$ in Fig. P2.59. Find v_{oc} and i_{sc} using superposition. Then confirm that v_{oc}/i_{sc} equals the equivalent resistance of the dead network.

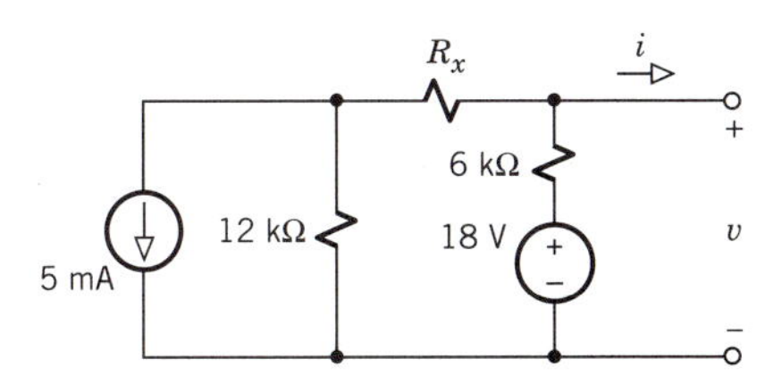

Figure P2.59

2.60 Do Problem 2.59 with $R_x = 18\ \text{k}\Omega$.

2.61 Let $i_c = 4i$ in Fig. P2.61. Determine R_t and either v_{oc} or i_{sc} expressed in terms of v_s. Then calculate the remaining Thévenin parameter.

2.62 Do Problem 2.61 with $i_c = v/(10\ \text{k}\Omega)$.

2.63 Do Problem 2.61 with $i_c = v_x/(1\ \text{k}\Omega)$.

2.64* Let $v_c = 2v$ in Fig. P2.64. Determine R_t and either v_{oc} or i_{sc} expressed in terms of i_s. Then calculate the third Thévenin parameter.

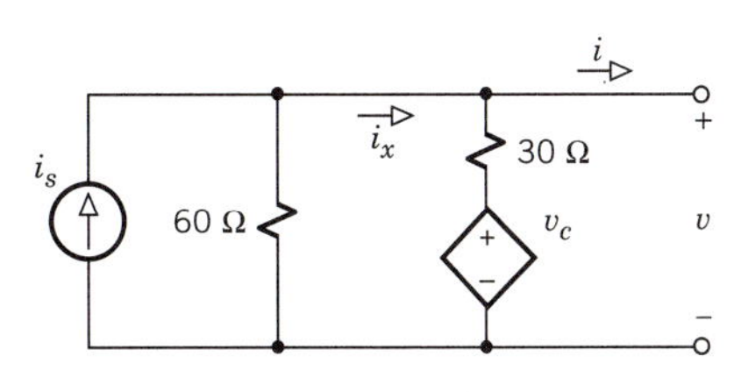

Figure P2.64

2.65 Do Problem 2.64 with $v_c = 90\ \Omega \times i_x$.

2.66 Do Problem 2.64 with $v_c = 45\ \Omega \times i$.

2.67 Perform source conversions to obtain the Thévenin parameters for Fig. P2.57 with $R_x = 0$.

2.68 Perform source conversions to obtain the Thévenin parameters for Fig. P2.57 with $R_x = 8\ \Omega$.

2.69 Perform source conversions to obtain the Thévenin parameters for Fig. P2.59 with $R_x = 0$.

2.70 Perform source conversions to obtain the Thévenin parameters for Fig. P2.59 with $R_x = 12\ \text{k}\Omega$.

2.71* Let $v_s = 24$ V and $i_s = 1$ mA in Fig. P2.51. Find i_1 with the help of source conversions.

2.72 Let $v_s = 20$ V and $i_s = 4$ mA in Fig. P2.51. Find v_4 with the help of source conversions.

2.73 Let $v_s = 30$ V and $i_s = 3$ mA in Fig. P2.51. Find i_3 with the help of source conversions.

2.74 Let $v_s = 30$ V and $i_s = 2$ mA in Fig. P2.51. Find v_2 with the help of source conversions.

2.75 Use source conversions to find v_y in Fig. P2.23 when $R_y = 6\ \Omega$ and $i_c = v_y/(0.2\ \Omega)$.

2.76 Use source conversions to find i_y in Fig. P2.26 when $R_x = 8\ \text{k}\Omega$ and $v_c = 30\ \text{k}\Omega \times i_y$.

2.77 Figure P2.77 is the model of a transistor power amplifier with variable load resistance R_L. Let $\beta = 49$ and $R = 36\ \Omega$. (a) Find i_{out}/v_{in} when $R_L = 0$ and v_{out}/v_{in} when $R_L = \infty$. (b) Use the results from (a) to calculate the Thévenin resistance seen looking back from R_L. Check your value of R_t by supressing v_{in} and replacing R_L with a test source. (c) Draw the equivalent output circuit when $v_{in} = 10$ V and $R_L = 8\ \Omega$. Find v_{out} and the power p_{out} delivered to R_L. Then find i_{in} and calculate the power gain p_{out}/p_{in}, where $p_{in} = v_{in}i_{in}$.

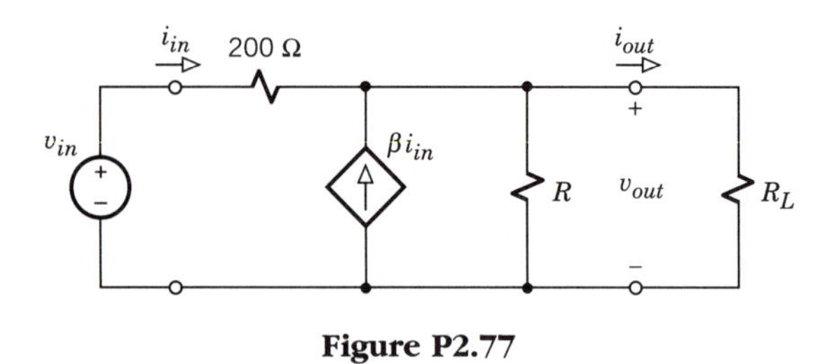

Figure P2.77

2.78 Do Problem 2.77 with $\beta = 99$ and $R = 2000\ \Omega$.

CHAPTER 3

Applications of Resistive Circuits

This chapter discusses some important practical applications involving resistive circuits. We'll begin with the limitations of real sources, as distinguished from ideal sources, and we'll investigate the related topic of power transfer. Then we'll introduce circuit models of electronic amplifiers and develop corresponding analysis methods for amplifier circuits. Special consideration will be given to the versatile operational amplifier, or "op-amp," whose properties expedite the design of linear amplification systems. The last section deals with dc meters and measurement techniques.

Objectives

After studying this chapter and working the exercises, you should be able to do each of the following:

1. Distinguish between real and ideal sources, and explain the practical difference between a real voltage source and a real current source (Section 3.1).
2. State the condition for maximum power transfer from a source to a load (Section 3.1).
3. Model a voltage or current amplifier using controlled sources and resistors (Section 3.2).
4. Given the circuit model, calculate the amplification of an amplifier or cascaded amplifiers (Section 3.2).
5. Describe the major features of an operational amplifier (Section 3.2).
6. Analyze op-amp circuits by invoking the properties of the ideal op-amp (Section 3.2).
7. Design an op-amp system with noninverting, inverting, and/or summing amplifiers (Section 3.2).
8. Explain why the internal resistances of an op-amp usually have negligible effect (Section 3.3).†
9. Draw and analyze the basic circuits used for dc ammeters and voltmeters (Section 3.4).†
10. Explain the operating principles of the Wheatstone bridge (Section 3.4).†

3.1 REAL SOURCES AND POWER TRANSFER

Real sources are physical devices such as batteries and generators capable of supplying electrical energy. These devices differ from *ideal* sources in two respects: (1) real sources cannot deliver unlimited amounts of power, and (2) they dissipate power internally. To take account of those differences, this section develops models for real sources and explores the limitations on power transfer from a real source to a load.

Source Models and Loading

A real voltage source is characterized, in part, by the instantaneous terminal voltage $v = v_s$ measured under open-circuit conditions. But as increasing current i is drawn from the source, v might typically decrease along a curve like the one in Fig. 3.1*a*. Such behavior is, of course, nonlinear.

However, we usually operate a real voltage source with sufficiently small current that its v-i curve can be approximated by the straight dashed line shown in the figure. The slope of this line has been denoted $-R_s$, so the linearized v-i relationship becomes

$$v = v_s - R_s i \qquad (3.1)$$

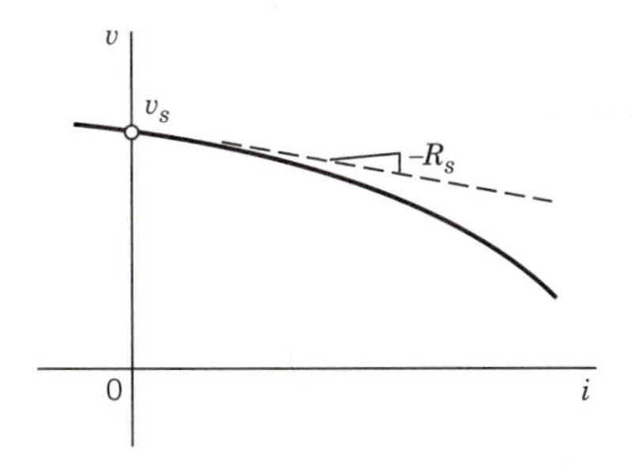

(a) v-i curve of a real voltage source

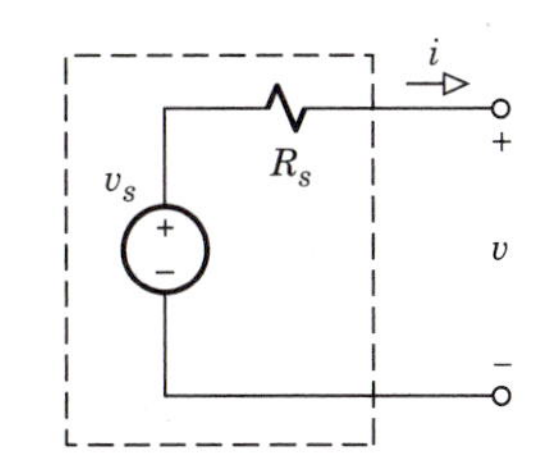

(b) Lumped-element model for $|i| \ll v_s/R_s$

Figure 3.1

which holds for $|i| \ll v_s/R_s$. We call v_s the **source voltage**, meaning the *open-circuit* voltage, and we call R_s the **source resistance**.

Based upon Eq. (3.1), an appropriate lumped-element model for a real voltage source takes the form of Fig. 3.1*b*. This model consists of an ideal voltage source v_s with series resistance R_s, similar to a Thévenin network. And like a Thévenin network, this model represents only the terminal behavior of a real voltage source. Thus, v_s and R_s are modeling elements that do not necessarily exist as separate entities within the source. Nonetheless, all real voltage sources have internal resistance that causes an internal voltage drop and power dissipation when $i \neq 0$.

A real current source also exhibits nonlinear behavior, like the typical *i-v* curve in Fig. 3.2*a*. But when the source operates at sufficiently small voltage, we can use a straight-line approximation to write the linearized *i-v* relationship as

$$i = i_s - \frac{v}{R_s} \tag{3.2}$$

which holds for $|v| \ll R_s i_s$. The corresponding lumped-element model in Fig. 3.2*b* consists of an ideal current source i_s with parallel resistance R_s, similar to a Norton network. We call i_s the **source current**, meaning the *short-circuit*

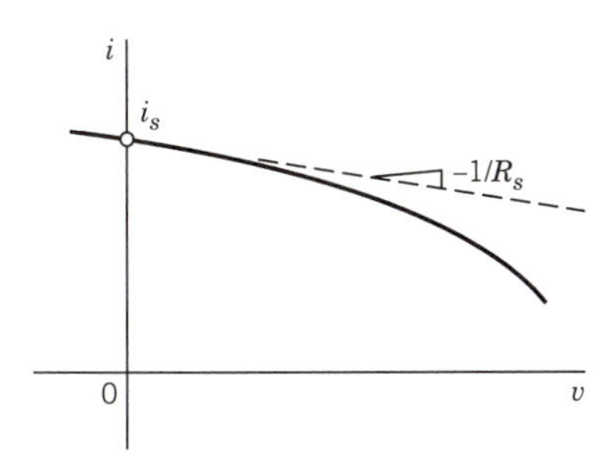

(a) i-v curve of a real current source

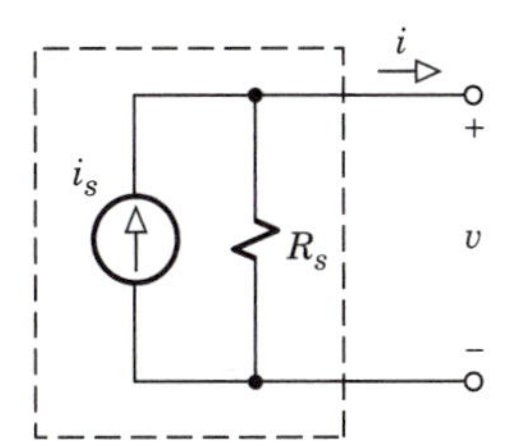

(b) Lumped-element model for $|v| \ll R_s i_s$

Figure 3.2

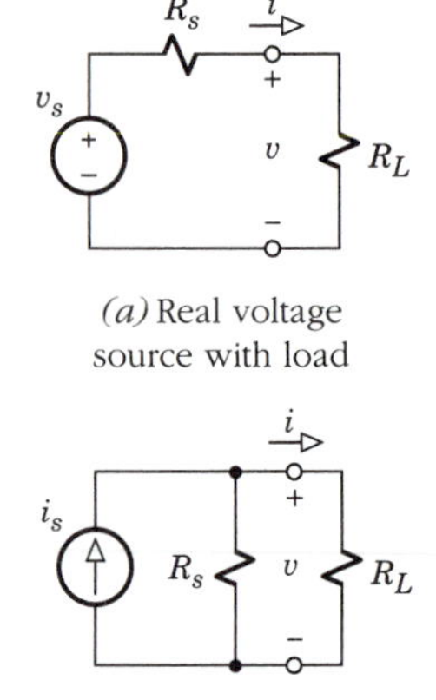

(*a*) Real voltage source with load

(*b*) Real current source with load

Figure 3.3

current, and the source resistance R_s again represents internal power dissipation.

Now recall from Section 2.5 that a current-source network like Fig. 3.2*b* could be converted into a voltage-source network like Fig. 3.1*b*, or vice versa. Furthermore, since neither v nor i remains constant at the terminals of a real source, there is no *theoretical* distinction between the two types of sources operating in their linear regions. The *practical* distinction emerges when we examine the **loading effects** that occur with a variable load connected to a real voltage or current source.

Consider first what happens when we attach a load resistance R_L to a real voltage source, as represented by Fig. 3.3*a*. We then have a voltage divider with

$$v = v_s - R_s i = \frac{R_L}{R_s + R_L} v_s \tag{3.3}$$

The internal voltage drop $R_s i$ thus causes the load voltage v to be less than the open-circuit source voltage v_s. A familiar demonstration of this effect occurs when you start an automobile engine with the headlights on; the headlights dim because the internal resistance of the battery and the large current drawn by the starter combine to decrease the voltage at the battery terminals.

Now let R_L be connected to a real current source, as represented by Fig. 3.3*b*. We then have a current divider with

$$i = i_s - \frac{v}{R_s} = \frac{R_s}{R_s + R_L} i_s \tag{3.4}$$

The internal current v/R_s thus causes the load current i to be less than the short-circuit source current i_s.

Comparing the loading effects in Eqs. (3.3) and (3.4), we see that a practical voltage source has $v \approx v_s$ when $R_s \ll R_L$, whereas a practical current source has $i \approx i_s$ when $R_s \gg R_L$. Hence, the key distinction between practical sources is the value of R_s relative to R_L. A "good" voltage source has *small internal resistance* compared to R_L, so that $v \approx v_s$ over the expected operating range. Conversely, a "good" current source has *large internal resistance* compared to R_L, so that $i \approx i_s$ over the expected operating range. In such cases, we are justified in using ideal sources as models to simplify analysis. The same observations hold for controlled sources, which also have internal resistance that may or may not play a significant role.

Example 3.1 *Model of a Battery*

Terminal measurements on a new battery yield $v = 6.0$ V when $i = 0$ and $v = 5.8$ V when $i = 0.05$ A. We'll use these results to model the battery and to determine the conditions under which it acts essentially like an ideal voltage source.

Clearly, the open-circuit source voltage is $v_s = 6.0$ V. The source resistance is then calculated from the slope of the v-i curve via

$$R_s = -\frac{\Delta v}{\Delta i} = -\frac{(6.0 - 5.8)\text{V}}{(0 - 0.05)\text{A}} = 4\ \Omega$$

Thus, the model in Fig. 3.3*a* holds for $|i| \ll v_s/R_s = 1.5$ A.

The battery acts essentially like an ideal voltage source with $v \approx v_s = 6$ V if $R_L \gg R_s$. But for a more precise condition on R_L, we might specify that the terminal voltage should be at least 90% of the open-circuit value. Setting $v \geq 0.9v_s = 5.4$ V and $R_s = 4\ \Omega$ in Eq. (3.3) gives

$$v = \frac{R_L}{4 + R_L} \times 6 \geq 5.4 \quad \Rightarrow \quad R_L \geq 36\ \Omega$$

This is the condition on R_L for treating the battery as an ideal voltage source.

With continued operation, however, the battery "runs down" and its internal resistance increases. The internal voltage drop thus increases and the terminal voltage decreases.

Example 3.2 *A Paradox Resolved*

If the current sources in Fig. 3.4*a* were ideal, then the configuration would violate KCL because the terminal current i cannot be both 8 mA and 2 mA at the same time. This paradox is resolved by noting that the sources are real and have internal resistance.

The circuit model in Fig. 3.4*b* includes the source resistances, and source conversions lead to the series circuit in Fig. 3.4*c*. Here we see that

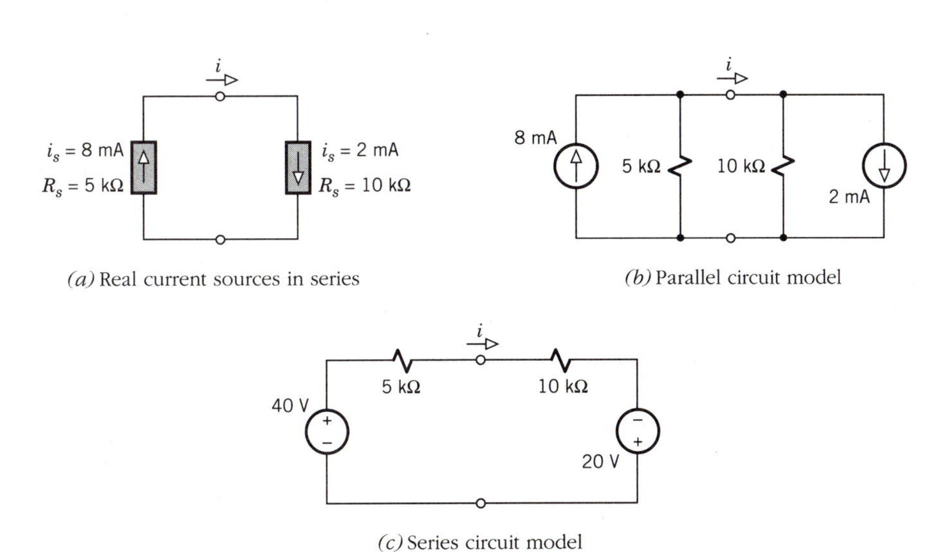

(a) Real current sources in series

(b) Parallel circuit model

(c) Series circuit model

Figure 3.4

$$i = (40 + 20)\text{V}/(5 + 10)\text{k}\Omega = 4 \text{ mA}$$

which falls between the two short-circuit current values.

Exercise 3.1

Find i_s and R_s for the current-source model in Fig. 3.3*b* given that $i = 100$ mA when $R_L = 0$ and $i = 80$ mA when $R_L = 1$ kΩ.

Exercise 3.2

What are the terminal values of i and v for the parallel-connected batteries in Fig. 3.5?

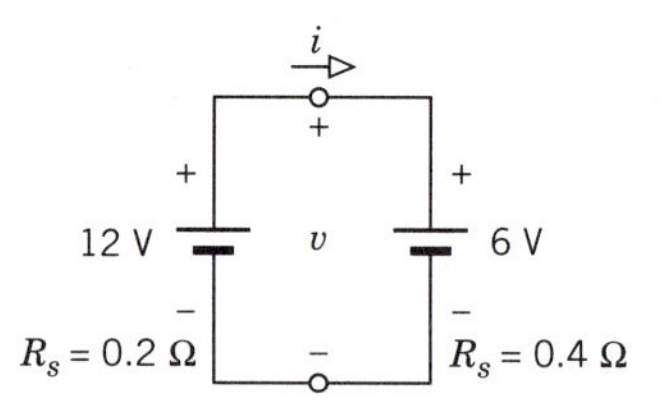

Figure 3.5

Power Transfer and Efficiency

Besides causing loading effect, internal resistance limits the amount of power that can be transferred from a source to a load resistance. We'll investigate this situation to determine the maximum power transfer and the power-transfer efficiency.

Consider a load resistance R_L connected to a voltage source v_s with internal resistance R_s. Power p_L is delivered to the load via $i = v_s/(R_s + R_L)$, so

$$p_L = R_L i^2 = \frac{R_L}{(R_s + R_L)^2} v_s^2 \tag{3.5}$$

The solid curve in Fig. 3.6 shows p_L plotted versus the resistance ratio R_L/R_s. We use this ratio to display how p_L varies with R_L in the usual case when R_s has a fixed value. As seen from the plot, the curve goes through a distinct *maximum* at $R_L/R_s = 1$ or

$$R_L = R_s \tag{3.6}$$

The corresponding maximum value of p_L is

$$p_{max} = v_s^2/4R_s \tag{3.7}$$

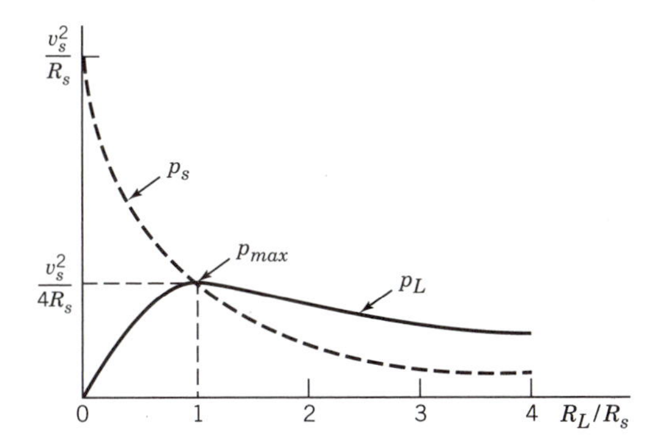

Figure 3.6 Load power p_L and internal power dissipation p_s versus R_L/R_s.

We call p_{max} the **maximum available power** since a source with fixed values of v_s and R_s cannot deliver more power than p_{max}.

The results in Eqs. (3.6) and (3.7) are derived by differentiating Eq. (3.5) with respect to R_L, which yields

$$\frac{dp_L}{dR_L} = \frac{(R_s + R_L)^2 - 2(R_s + R_L)R_L}{(R_s + R_L)^4} v_s^2 = \frac{R_s - R_L}{(R_s + R_L)^3} v_s^2$$

But p_L is maximum when $dp_L/dR_L = 0$, so $R_s - R_L = 0$ or $R_L = R_s$. Substituting $R_L = R_s$ in Eq. (3.5) gives the corresponding expression $p_{max} = R_s v_s^2/(2R_s)^2 = v_s^2/4R_s$.

Had R_L been connected to a current source i_s with R_s in parallel, we would have obtained the dual of Eq. (3.7), namely,

$$p_{max} = R_s i_s^2/4 \tag{3.8}$$

As in the case of a voltage source, the maximum value occurs when $R_L = R_s$.

Taking both cases into account, the **maximum power transfer theorem** states that

> If a source has fixed nonzero resistance R_s, then maximum power transfer to a load resistance requires $R_L = R_s$.

When $R_L = R_s$, the load is said to be **matched** to the source for maximum power transfer. This theorem also applies to any *linear source network* with Thévenin resistance $R_t = R_s$, since the source models in Fig. 3.3 could represent any such network.

Finally, observe that power transfer from a real source always produces ohmic heating in the source resistance. Calculations of such *internal* effects require information about the actual structure and cannot, in general, be based upon Thévenin or Norton equivalent networks. However, the entire load current i usually passes through the internal resistance of a real source, so we'll represent the internal conditions by the lumped-parameter voltage-source model in Fig. 3.3*a*. The resulting internal power dissipated by R_s is then

$$p_s = R_s i^2 = \frac{R_s}{(R_s + R_L)^2} v_s^2 \tag{3.9}$$

The dashed curve in Fig. 3.6 shows that p_s steadily decreases as R_L increases and that $p_s = p_L$ when $R_L/R_s = 1$.

Since the total power generated by the source is $p_L + p_s$, the wasted internal power p_s should be small compared to p_L for efficient operation. Formally, we define the **power-transfer efficiency**

$$\text{Eff} \triangleq \frac{p_L}{p_L + p_s} \tag{3.10}$$

which is often expressed as a percentage. If the load has been matched for maximum power transfer, then $p_s = p_L$ so Eff $= 0.5 = 50\%$. Moreover, with $R_L = R_s$, the terminal voltage drops to $v = v_s/2$. Clearly, electric utility companies would not, and should not, strive for maximum power transfer. They seek instead higher power-transfer efficiency by making p_s as small as possible.

When do we want maximum power transfer? Primarily in applications where voltage or current signals are used to convey *information* rather than to deliver large amounts of power. For instance, the first stage of a radio or television receiver should get as much power as possible out of the information-bearing signals that arrive via antenna or cable. Those tiny signals account for only a small fraction of the total power consumed by the receiver, so power-transfer efficiency is not a significant concern.

Example 3.3 *Calculating Power Transfer and Efficiency*

Figure 3.7 represents a power tool connected by a long extension cord to a 120-V source. Electrically, the power tool acts like a variable load whose resistance depends upon the amount of work being done. We'll calculate power and efficiency values for various values of R_L.

The equivalent source applied at the terminals of the power tool has $v_s =$ 120 V and $R_s = 0.2 + 4.8 = 5\ \Omega$, which includes the resistance of the extension

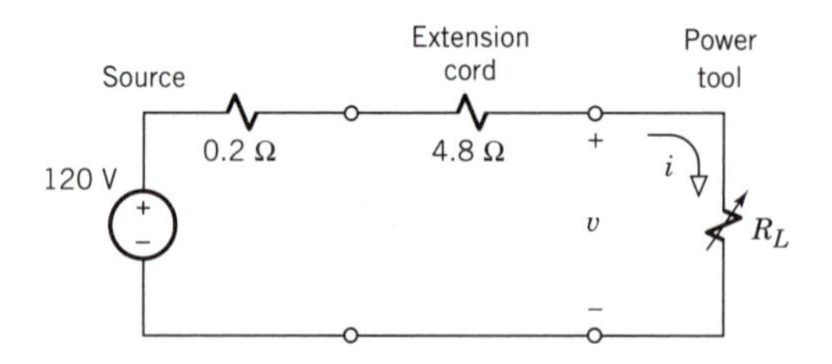

Figure 3.7 Model of a power tool connected to a source by an extension cord.

cord because power dissipated here is wasted power. Maximum power transfer therefore occurs when $R_L = 5\ \Omega$, and the resulting power delivered to the tool will be

$$p_{max} = (120\ \text{V})^2/(4 \times 5\ \Omega) = 720\ \text{W}$$

This value corresponds to about one **horsepower** (hp), since

$$1\ \text{hp} = 746\ \text{W}$$

But with $R_L = R_s$, the terminal voltage at the tool drops to $v_s/2 = 60$ V, the 0.2-Ω and 4.8-Ω resistances together dissipate $p_s = p_{max} = 720$ W, and the power-transfer efficiency is 50%. Furthermore, the extension cord may be destroyed by excessive heating since it must carry $i = 120\ \text{V}/(5 + 5)\ \Omega = 12$ A.

A more realistic load might be $R_L = 35\ \Omega$, in which case $i = 120/40 = 3$ A, $v = 120 - 5i = 105$ V, $p_L = 35i^2 = 315\ \text{W} \approx 0.4$ hp, $p_s = 5i^2 = 45$ W, and Eff $= 315/(315 + 45) \approx 88\%$. Table 3.1 lists these and a few other calculated results for comparison purposes. The numbers here illustrate the significance of the curves in Fig. 3.6.

TABLE 3.1 Power Transfer with $v_s = 120$ V and $R_s = 5\ \Omega$

$R_L(\Omega)$	i(A)	v(V)	p_L(W)	p_s(W)	Eff(%)
1	20	20	400	2000	17
5	12	60	720	720	50
35	3	105	315	45	88
115	1	115	115	5	96

Exercise 3.3

Consider the source network in Fig. 3.8. What value of R_L should be used to obtain the maximum available power? Calculate the corresponding values of v, i, p_s, and Eff. Hint: Note that p_s must be determined from the network diagram, not the Thévenin model.

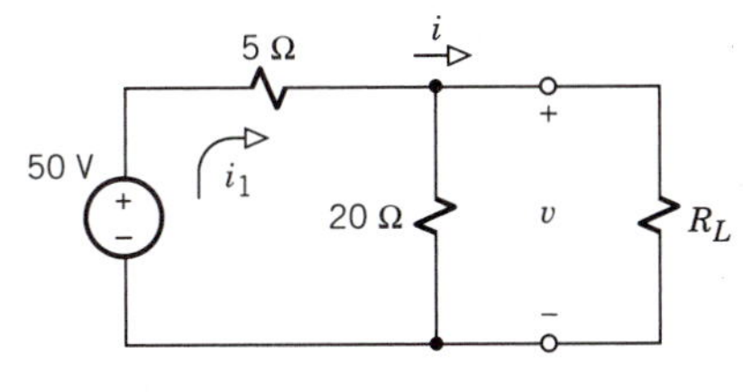

Figure 3.8

3.2 AMPLIFIERS AND OP-AMPS

Amplifiers generally "enlarge" the variations of an electrical signal represented by a voltage or current waveform. Amplification is achieved with the help of electronic devices, primarily transistors, having the ability to control a large voltage or current in proportion to a smaller input variable. Of course, the large voltage or current being controlled must come from something other than the input source, so every amplifier also needs an appropriate power supply — usually a battery or an electronic circuit for ac to dc conversion.

But we are not concerned here with the details of electronic amplification. Rather, this section focuses on equivalent models of amplifiers and how they are used to determine the input–output relationships. We'll give particular attention to circuits built with operational amplifiers, and we'll discuss some important applications and design techniques as well as analysis methods.

Amplifier Models

Figure 3.9*a* represents a **voltage amplifier** as a "black box" with a pair of input terminals and a pair of output terminals. The source connected at the input generates a time-varying voltage $v_s(t)$, and the amplifer produces a larger output voltage waveform $v_{out}(t)$ across the attached load. The source might be the pickup on an electric guitar, for instance, and the load might actually be a loudspeaker. Ideally, the action of the amplifier is expressed by the simple linear relation

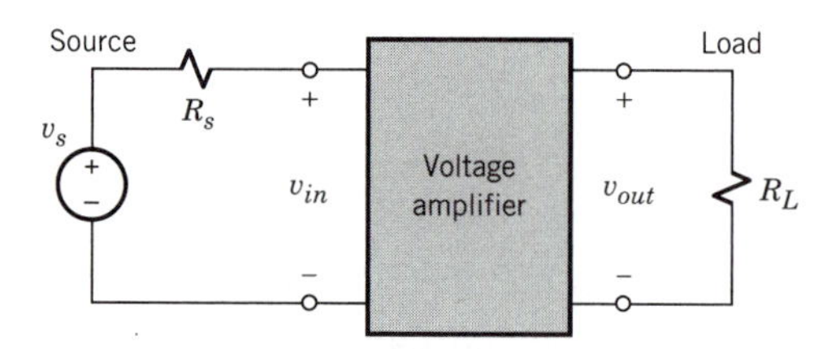

(a) Voltage amplifier with input source and output load

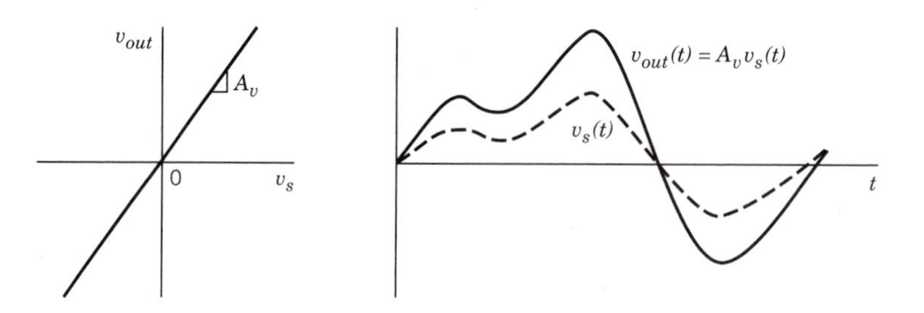

(b) Transfer curve and typical waveforms

Figure 3.9

$$v_{out}(t) = A_v v_s(t)$$

where A_v denotes the overall voltage amplification.

Since $v_{out}(t)/v_s(t) = A_v$ at any instant of time, the plot of v_{out} versus v_s will be a straight line through the origin with slope A_v. Such plots are called **transfer curves**. Figure 3.9*b* shows the transfer curve and typical waveforms for a linear voltage amplifier with $A_v > 1$. The output signal has the same relative variations as the input, but with bigger excursions. Some amplifiers also *invert* the output waveform, which corresponds to $A_v < -1$.

Although we cannot lift the lid on the "black box" and discuss the internal electronics, we can describe the terminal relations using the equivalent circuit model in Fig. 3.10. This generic model contains an **input resistance** R_i to reflect the property that the amplifier may draw current from the applied source. Consequently, the terminal voltage v_{in} often differs from the source voltage v_s. Amplification of v_{in} is modeled by a VCVS with gain μ, which would be a positive quantity for a noninverting amplifier or a negative quantity for an inverting amplifier. In either case, the VCVS represents the **open-circuit output voltage** $v_c = \mu v_{in}$.

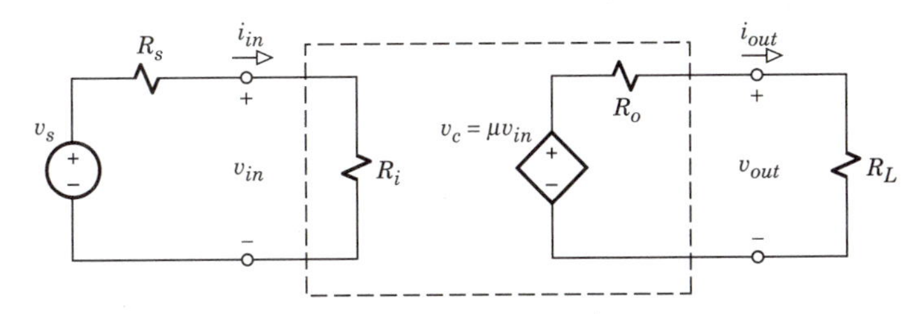

Figure 3.10 Model of a voltage amplifier.

But the model also includes an **output resistance** R_o, accounting for power dissipation within the amplifier. The output side thus takes the form of a Thévenin network, and R_o equals the *Thévenin equivalent resistance* seen looking back into the output terminals with the input source suppressed. The amplifier's power supply does not appear in the model because it is represented indirectly by the controlled source whose power comes from the power supply.

The impact of internal resistances on the overall amplification is brought out by calculating $A_v = v_{out}/v_s$, noting that loading at input and output result in $v_{in} \neq v_s$ and $v_{out} \neq v_c$. To include those loading effects, we first write the ratio v_{out}/v_s as the "chain" expansion

$$\frac{v_{out}}{v_s} = \frac{v_{in}}{v_s} \times \frac{v_c}{v_{in}} \times \frac{v_{out}}{v_c} \tag{3.11}$$

This expansion has the advantage that we can easily determine each of the three voltage ratios. Indeed, we already know that $v_c/v_{in} = \mu$, while voltage–divider relations at the input and output yield

$$\frac{v_{in}}{v_s} = \frac{R_i}{R_s + R_i} \qquad \frac{v_{out}}{v_c} = \frac{R_L}{R_o + R_L}$$

Substituting these ratios into Eq. (3.11) yields

$$A_v = \frac{v_{out}}{v_s} = \frac{R_i}{R_s + R_i} \times \mu \times \frac{R_L}{R_o + R_L} \tag{3.12}$$

Equation (3.12) indicates that $|A_v| < |\mu|$, and we clearly see that loading effects reduce the overall voltage gain.

Input loading will be negligible when the amplifier's input resistance is large compared to the source resistance, while output loading will be negligible when the output resistance is small compared to the load resistance. Thus, a "good" voltage amplifier has $R_i \gg R_s$ and $R_o \ll R_L$, so that $A_v \approx \mu$.

Certain electronic circuits inherently amplify *current* rather than voltage. The generic model of a **current amplifier** in Fig. 3.11 also has two pairs of terminals, but the control variable is now the input current through R_i. Accordingly, we show a current source applied at the input, and we represent amplification by a CCCS with gain β producing the **short-circuit output current** $i_c = \beta i_{in}$. The output side then takes the form of a Norton network consisting of the CCCS in parallel with the Thévenin equivalent output resistance R_o.

Using another chain expansion, the overall current amplification in Fig. 3.11 is found to be

$$A_i = \frac{i_{out}}{i_s} = \frac{R_s}{R_s + R_i} \times \beta \times \frac{R_o}{R_o + R_L} \tag{3.13}$$

Thus, a "good" current amplifier with $A_i \approx \beta$ has small input resistance ($R_i \ll R_s$) and large output resistance ($R_o \gg R_L$). These observations agree with the fact that Fig. 3.11 is the dual of Fig. 3.10.

Some applications require more voltage or current gain than can be achieved with a single amplifier. A common solution to this problem is connecting two or more amplifiers in **cascade** — the output of the first amplifier driving the input of the second, and so forth. The overall amplification of the cascade is then proportional to the product of the individual controlled-source gains. The following example illustrates the analysis procedure for a cascade of voltage amplifiers.

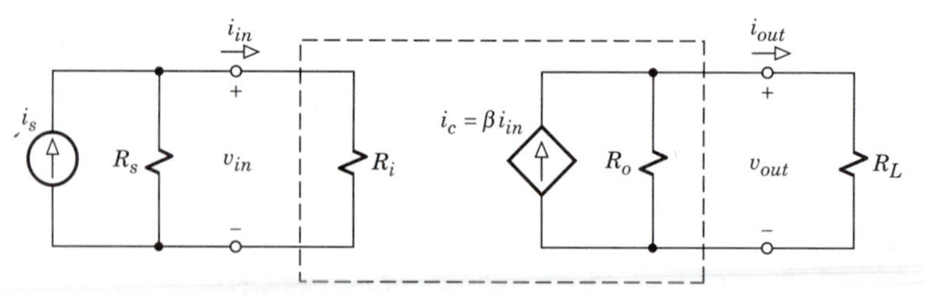

Figure 3.11 Model of a current amplifier.

Example 3.4 *Cascade Voltage Amplifier*

A certain medical transducer attached to a patient produces the oscillating voltage signal $v_s = 0.1 \sin \omega t$ V, which provides information about the patient's condition. The transducer has $R_s = 8$ kΩ and a maximum current limitation of ±10 μA. The signal v_s is to be amplified for plotting on a strip-chart recorder. The recorder has a ±10-V scale, and it acts like a 5-kΩ load resistance at its input terminals. The amplifiers available for this application are inverting circuits with $R_i = 24$ kΩ, $R_o = 1$ kΩ, and $\mu = -10$. Since one amplifier would only produce $|A_v| < 10$ and $|v_{out}| < 1$ V, we'll examine the suitability of a cascade built with two amplifiers.

Figure 3.12 shows the complete equivalent circuit for the two-stage cascade. The output of the first stage is denoted here by v_x, which becomes the input to the second stage. Accordingly, the second VCVS has been labeled $v_{c2} = -10v_x$.

We calculate the overall amplification of the cascade using the chain expansion

$$A_v = \frac{v_{out}}{v_s} = \frac{v_{in}}{v_s} \times \frac{v_{c1}}{v_{in}} \times \frac{v_x}{v_{c1}} \times \frac{v_{c2}}{v_x} \times \frac{v_{out}}{v_{c2}}$$

$$= \frac{24}{8+24}(-10)\frac{24}{1+24}(-10)\frac{5}{1+5} = 0.6 \times (-10)^2 = 60$$

This calculation reveals that loading effects reduce the overall amplification by a factor of 0.6, and that the second voltage inversion counteracts the first.

The resulting output will be $v_{out} = A_v v_s = 6 \sin \omega t$ V, which oscillates over ±6 V and nicely suits the recorder's range. The current drawn from the transducer is $i_{in} = v_s/(8 + 24 \text{ k}\Omega) = 3.125 \sin \omega t$ μA, which falls within the transducer's limitation. We therefore conclude that the cascade amplifier meets the stated specifications.

Exercise 3.4

Use the dual of Eq. (3.11) to derive Eq. (3.13) from Fig. 3.11.

Exercise 3.5

Using the model in Fig. 3.11, draw a diagram similar to Fig. 3.12 for a two-stage cascade of identical current amplifiers with a real current source at the

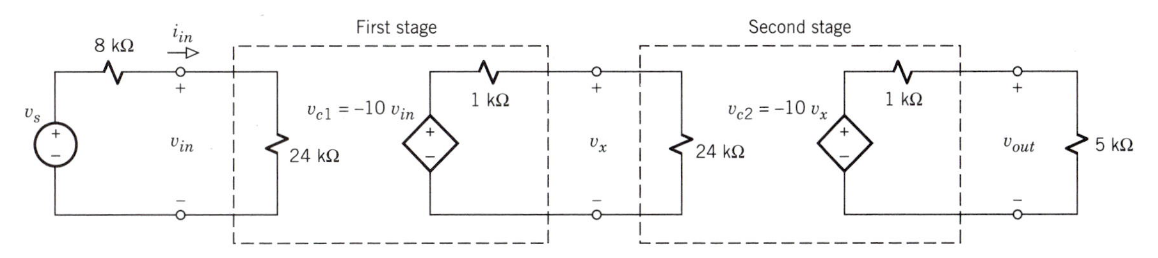

Figure 3.12 Cascade amplifier with two stages.

input. Then obtain an expression for $A_i = i_{out}/i_s$. What are the conditions on the resistances such that $A_i \approx \beta^2$?

Operational Amplifiers

Before the advent of microelectronics, an electrical engineer who needed an amplifier had to design the entire circuit using a dozen or more components. But that chore has been largely eliminated thanks to the modern integrated-circuit **operational amplifier** — commonly known as the "**op-amp**." Now you can buy an op-amp off the shelf and connect a few elements to it to construct a more reliable, less expensive amplifier than one built from scratch. We'll demonstrate these points after an introduction to op-amp characteristics.

The name "operational amplifier" actually refers to a large family of general-purpose and special-purpose units having the distinctive feature that

> Op-amps provide high-gain amplification of the difference between two input voltages.

The two input voltages v_p and v_n and the output voltage v_{out} are shown on the op-amp schematic diagram in Fig. 3.13*a*. These three voltages are measured with respect to a single reference point identified by the **ground** emblem (⏚). The ground point is established external to the op-amp by the power-supply connections, shown as dashed lines.

Most op-amps operate with two equal dc supply voltages, here labeled $+V_{PS}$ and $-V_{PS}$. The ground point is then taken at the node that connects the lower potential of the positive supply to the higher potential of the negative supply. That arrangement allows the output voltage to have any value in the range from $-V_{PS}$ to $+V_{PS}$.

Although the dc supply voltages are essential for operation, we'll be primarily concerned with the variable *signal voltages*. Accordingly, we'll use the less cluttered op-amp symbol in Fig. 3.13*b*, where the terminal labels v_p, v_n, and v_{out} denote signal voltages measured with respect to ground. The input terminal marked + is called the **noninverting input**, and the terminal

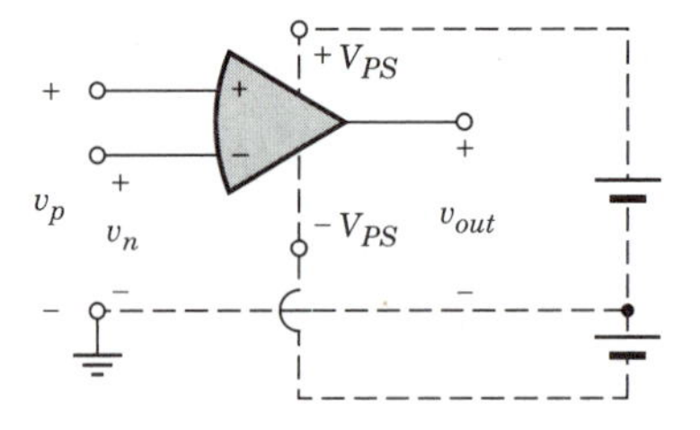

(a) Schematic diagram of an op-amp with two power supplies

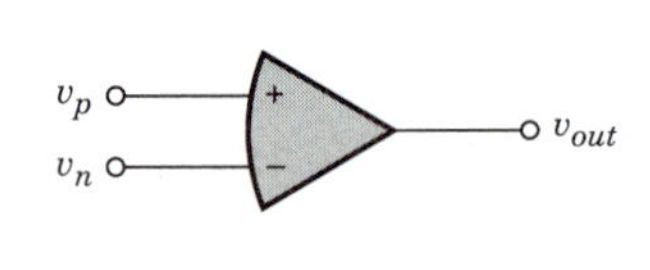

(b) Op-amp symbol

Figure 3.13

marked − is the **inverting input**. Of course, the voltages v_p and v_n applied to these terminals may be of either polarity. The resulting voltage across the input terminals is the **difference voltage**

$$v_d = v_p - v_n$$

This difference voltage entirely controls the output voltage.

The transfer curve relating v_{out} to v_d typically has the shape plotted in Fig. 3.14*a*. The curve consists of a middle **linear region** bounded by positive and negative **saturation regions**. Saturation occurs if $|v_d|$ exceeds a critical value v_{max}. The illustrative waveforms in Fig. 3.14*b* emphasize the fact that the output becomes "pinned" at $v_{out} \approx +V_{PS}$ whenever $v_d > v_{max}$, or pinned at $v_{out} \approx -V_{PS}$ whenever $v_d < -v_{max}$. At either extreme, the op-amp no longer acts as a linear amplifier and the output waveform is obviously distorted.

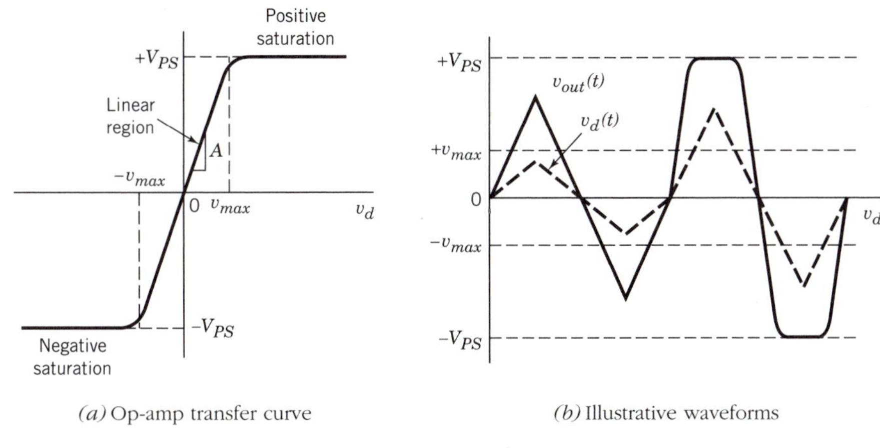

(a) Op-amp transfer curve

(b) Illustrative waveforms

Figure 3.14

Undistorted linear amplification takes place in the middle region where the transfer curve has slope *A*, so

$$v_{out} = Av_d = A(v_p - v_n) \tag{3.14}$$

Increasing the voltage v_p at the *noninverting* terminal causes v_{out} to move in the *positive* direction, whereas increasing the voltage v_n at the *inverting* terminal causes v_{out} to move in the *negative* direction.

When operating within its linear region, an op-amp behaves essentially like the simplified model in Fig. 3.15. This model has an open circuit at the input terminals, so $i_p = i_n = 0$, and a grounded VCVS produces Av_d at the output terminal. (The effects of internal input and output resistance will be considered later.) Aside from the difference–voltage input, the distinctive characteristic of an op-amp is its gigantic gain A — usually 10,000 or more!

But our linear op-amp model holds only over the output range

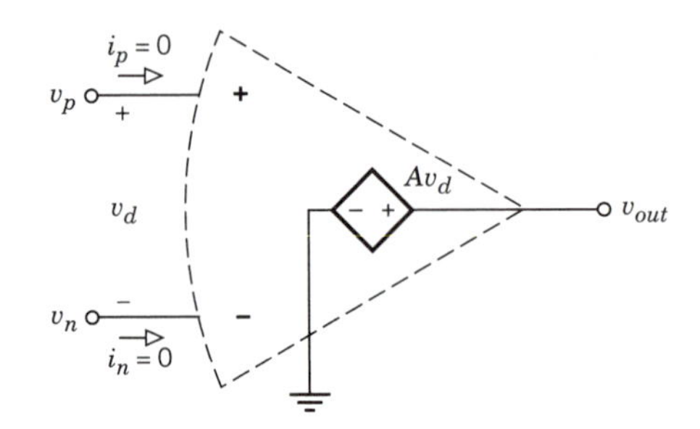

Figure 3.15 Simplified op-amp model.

$$-V_{PS} < v_{out} < +V_{PS} \tag{3.15}$$

The corresponding input range is

$$-v_{max} < v_d < v_{max} \tag{3.16a}$$

where, from Fig. 3.14*a*,

$$v_{max} \approx V_{PS}/A \tag{3.16b}$$

Since the typical op-amp gain is $A \geq 10^4$, and since the supply voltage seldom exceeds 30 V, the variations of the difference voltage must be limited to

$$|v_d| < v_{max} \leq 3 \text{ mV}$$

Otherwise, with $|v_d| \geq v_{max}$, the op-amp saturates and the output waveform becomes distorted.

In view of the limitation on v_d, an op-amp alone is not a practical amplifier because it would be driven into saturation by any signal voltage whose variations exceed a few millivolts. Consequently,

> Linear op-amp circuits always include **negative feedback** connecting the output terminal to the inverting terminal.

Properly designed feedback results in a greatly reduced difference voltage, thereby preventing saturation. Additionally, the amplification with feedback becomes nearly independent of the op-amp gain A — an important consideration since op-amp gains tend to fluctuate unpredictably, sometimes increasing or decreasing by a factor of 2 or greater. We'll subsequently show how negative feedback turns an op-amp into a useful and reliable amplifier.

Exercise 3.6

Calculate v_{max} for an op-amp with $A = 10^5$ and $V_{PS} = 25$ V. Then let $v_p = 0.5$ mV and make a table listing the values of v_d and v_{out} when $v_n = 0$, 0.2, 0.4, 0.6, 0.8, and 1.0 mV.

Noninverting Op-Amp Circuits

The op-amp circuit in Fig. 3.16*a* functions as a **noninverting voltage amplifier**, meaning that v_{out} has the same polarity as v_{in}. The input is applied directly to the noninverting terminal, so $v_p = v_{in}$ and $i_{in} = i_p = 0$. The resistors R_F and R_1 constitute a feedback connection from the output to the inverting terminal and then to the ground point. A detailed analysis of this configuration will bring out the feedback effects.

We begin by replacing the op-amp symbol with the model from Fig. 3.15, so the rearranged diagram takes the form of Fig. 3.16*b*. Since $i_n = 0$, the external resistors act as a voltage divider producing the **feedback voltage**

$$v_n = Bv_{out} \qquad B = \frac{R_1}{R_F + R_1} \tag{3.17}$$

where B stands for the **feedback factor**.

Next, we note that $v_n + v_d = v_{in}$ and $v_{out} = Av_d$, so the resulting difference in voltage is

$$v_d = v_{in} - v_n = v_{in} - Bv_{out} = v_{in} - ABv_d$$

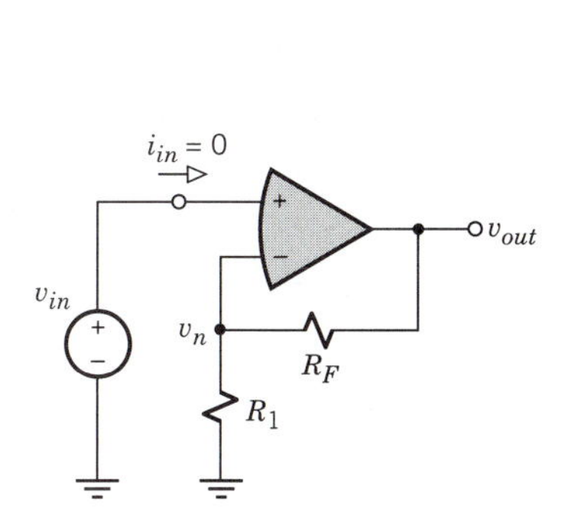

(*a*) Diagram of a noninverting voltage amplifier

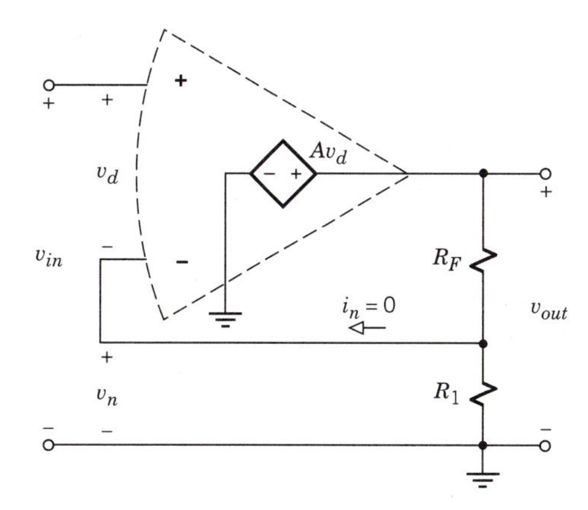

(*b*) Simplified circuit model

Figure 3.16

We thus have *negative* feedback in that Bv_{out} is subtracted from v_{in} to form v_d. Solving for v_d and using $v_{out} = Av_d$ now yields

$$v_d = \frac{1}{1 + AB} v_{in} \qquad v_{out} = \frac{A}{1 + AB} v_{in} \tag{3.18}$$

The product AB appearing here is called the **loop gain** from v_d back to v_n because $v_n/v_d = Bv_{out}/v_d = AB$. The overall amplification v_{out}/v_{in} is known as the **closed-loop gain**.

Most op-amps have sufficiently large gain A that the loop gain satisfies the condition

$$AB \gg 1 \tag{3.19}$$

Then $1 + AB \approx AB$ in Eq. (3.18) and the input–output relation simplifies to

$$v_{out} \approx \frac{1}{B} v_{in} = \frac{R_F + R_1}{R_1} v_{in} \tag{3.20}$$

This expression shows that the closed-loop gain depends almost entirely upon the feedback factor B associated with the resistive voltage divider, as distinguished from the unreliable gain A of the op-amp itself. Furthermore, the magnitude of the difference voltage becomes

$$|v_d| \approx \frac{1}{AB} |v_{in}| \ll |v_{in}|$$

which confirms that feedback does, indeed, reduce v_d as required for linear amplification of v_{in}.

Figure 3.17 summarizes our results in the form of an equivalent circuit for the noninverting amplifier when $AB \gg 1$. There are no loading effects at input or output, so $v_{in} = v_s$ and

$$A_v = \frac{v_{out}}{v_s} = \frac{R_F + R_1}{R_1} \tag{3.21}$$

Hence, you can design for any "moderate" amplification A_v by picking appro-

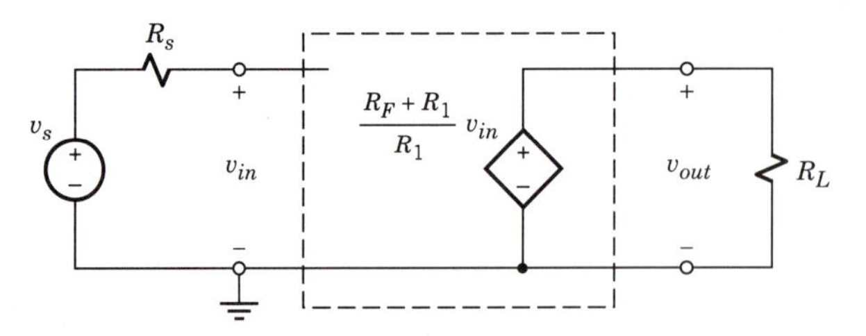

Figure 3.17 Equivalent circuit for a noninverting amplifier with $AB \gg 1$.

priate values of R_F and R_1. The op-amp's gain is unimportant, provided that $A \gg A_v$ to satisfy the loop-gain condition.

In theory, R_1 and R_F may have any desired values. In practice, however, the internal electronics of an op-amp generally require that

> The external resistances should be in the range from 1 kΩ to 100 kΩ.

The corresponding amplification in Eq. (3.21) is limited to $A_v \leq 101$, but larger values can be obtained by connecting two op-amp circuits in cascade. When adjustable amplification is needed, a potentiometer may be used for the R_1-R_F voltage divider.

A special type of noninverting amplifier known as a **voltage follower** is shown in Fig. 3.18*a*. This stripped-down circuit emerges from Fig. 3.16*a* when $R_F \rightarrow 0$ and $R_1 \rightarrow \infty$. Thus,

$$B = v_n / v_{out} = 1$$

and Eq. (3.18) yields

$$\frac{v_{out}}{v_{in}} = \frac{A}{1 + A} \approx 1 \tag{3.22}$$

Since $v_{out} \approx v_{in}$, we say that the output voltage "follows" the input.

You may well wonder why we bother with a circuit that only provides unity voltage amplification. The explanation comes from Fig. 3.18*b*, where $v_{out} \approx v_{in} = v_s$ regardless of R_s and R_L. A voltage follower therefore functions as a **buffer** that eliminates loading effects between a high-resistance voltage source and a low-resistance load.

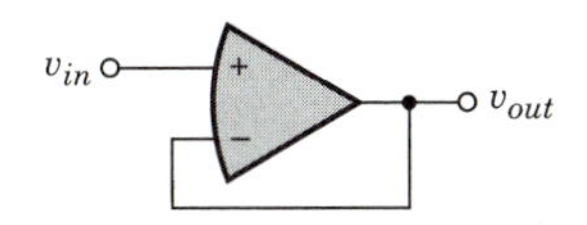

(a) Voltage follower

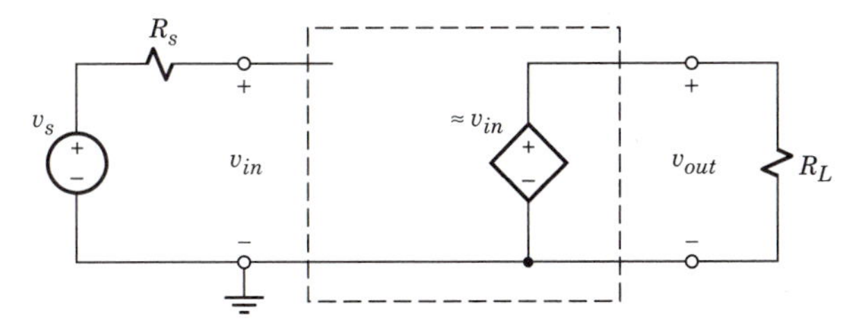

(b) Buffer between source and load

Figure 3.18

Example 3.5 *Designing a Noninverting Amplifier*

Suppose you want to build a noninverting amplifier to provide as much amplification as possible for a time-varying signal $v_{in}(t)$ with $|v_{in}|_{max} = 0.2$ V. Further suppose that the components available to you include an op-amp with $A \approx 10^5$ and a dual-voltage supply with $V_{PS} = 12$ V. You carry out the design calculations and analysis as follows.

First, you determine the maximum allowable amplification for linear operation with $|v_{out}| < V_{PS} = 12$ V. Since $|v_{in}|_{max} = 0.2$ V,

$$|v_{out}|_{max} = A_v \times 0.2 \text{ V} < 12 \text{ V} \Rightarrow A_v < 60$$

You therefore decide that $A_v \approx 50$ would provide ample safety margin to avoid saturation and output distortion.

Next, setting $A_v = 50$ in Eq. (3.21), you get the needed design equation $(R_1 + R_F)/R_1 = 50$. To keep $R_F \leq 100$ kΩ, you take $R_1 = 2$ kΩ and calculate R_F via

$$R_F + R_1 = 50R_1 \Rightarrow R_F = 49R_1 = 98 \text{ k}\Omega$$

The corresponding feedback factor and loop gain are

$$B = \frac{2}{98 + 2} = \frac{1}{50} \qquad AB \approx \frac{10^5}{50} = 2000 \gg 1$$

Inserting these values into Eq. (3.18) gives

$$\frac{v_d}{v_{in}} \approx \frac{1}{2001} \approx 0.0005 \qquad \frac{v_{out}}{v_{in}} \approx \frac{10^5}{2001} = 49.975 \approx 50$$

Thus, the difference voltage v_d will have a maximum value of about 0.0005×0.2 V $= 0.1$ mV, whereas $|v_{out}|_{max} \approx 50 \times 0.2$ V $= 10$ V.

Finally, you check out the impact of unreliable op-amp gain A, noting that the feedback factor remains fixed at $B = 1/50$. If the gain decreases to $A = 0.5 \times 10^5$, then

$$AB = \frac{50{,}000}{50} = 1000 \qquad \frac{v_{out}}{v_{in}} = \frac{50{,}000}{1001} = 49.950 \approx 50$$

If the gain increases to $A = 2 \times 10^5$, then

$$AB = \frac{200{,}000}{50} = 4000 \qquad \frac{v_{out}}{v_{in}} = \frac{200{,}000}{4001} = 49.988 \approx 50$$

You therefore conclude that your design will work successfully despite possible variations of op-amp gain.

Exercise 3.7

Consider a noninverting amplifier with $R_1 = 4\ \text{k}\Omega$ and $R_F = 76\ \text{k}\Omega$.

(a) What's the condition on V_{PS} to avoid saturation if $|v_{in}| \leq 0.8$ V?

(b) Find the minimum and maximum values of v_d/v_{in} and v_{out}/v_{in} when the gain of the op-amp varies from 10^4 to 10^6. Give your results with six significant figures.

Ideal Op-Amps

As we have seen, an inverting amplifier circuit with loop gain $AB \gg 1$ exhibits four significant properties:

- The magnitude of the difference voltage is very small;
- The value of the output voltage is essentially independent of the op-amp gain;
- The currents into the inverting and noninverting terminals are neglible; and
- The equivalent output resistance is negligible.

These effects also appear in any linear op-amp circuit, and they lead to the handy concept of the *ideal* op-amp.

An **ideal op-amp** is a fictitious device having *infinite gain*, so

$$A = \infty$$

Since $v_{out} = Av_d$, you might jump to the conclusion that $|v_{out}|$ becomes unbounded. But negative feedback forces $v_n \rightarrow v_p$ as $A \rightarrow \infty$, so $v_d = v_p - v_n \rightarrow 0$ and v_{out} remains finite. As a case in point, letting $A \rightarrow \infty$ in Eq. (3.18) shows that $v_d \rightarrow 0$ while $v_{out} \rightarrow v_{in}/B$. Thus, negative feedback keeps an ideal op-amp in the linear region despite the infinite gain.

When an ideal op-amp has a negative-feedback connection, its input terminals act as a **virtual short** — meaning that $v_d = 0$ (like a short circuit) while $i_p = i_n = 0$ (like an open circuit). Accordingly, we'll represent an ideal op-amp symbolically by Fig. 3.19, where the double-headed arrow at the input stands for the properties

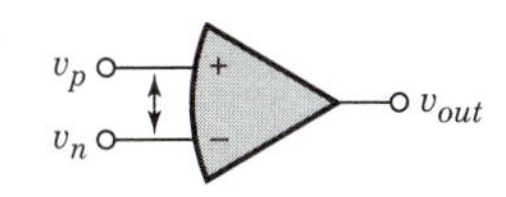

Figure 3.19 Ideal op-amp symbol with virtual short.

$$v_p = v_n \qquad i_p = i_n = 0 \tag{3.23}$$

But these properties do not provide an explicit relationship for v_{out}. Instead,

> The value of v_{out} will be whatever is needed to satisfy the conditions of the virtual short.

This simple rule eliminates calculations involving the loop gain, and thereby expedites the analysis and design of op-amp circuits. However, two precautions should be observed.

First, the virtual-short conditions depend upon negative feedback, so

> When using the ideal op-amp model, the circuit under consideration must have a feedback connection from the output terminal to the inverting input terminal.

Second, the output current comes from the power supplies via paths not shown in Fig. 3.19, so

> You cannot apply KCL at the output terminal of an ideal op-amp.

This restriction raises no problems because the virtual-short conditions pertain only to the *input* terminals of the ideal op-amp.

Real op-amps have finite gain, of course, but negative feedback ensures that v_d will be very small when the loop gain satisfies $AB \gg 1$. Thus, results obtained using the ideal op-amp are sufficiently accurate for nearly all practical purposes. Given that most op-amps have $A > 10^4$, we'll hereafter assume that $A = \infty$ to take advantage of the ease of virtual-short analysis. The following example illustrates the technique.

Example 3.6 ***A Noninverting Current Amplifier***

The op-amp circuit in Fig. 3.20 has a grounded current source at the input and a grounded load resistance at the output. We'll apply the virtual-short conditions of an ideal op-amp to determine v_{out} and the current gain i_{out}/i_s.

Since zero current enters the op-amp's input terminals, we quickly find that

$$v_p = R_s i_s \qquad v_n = \frac{R_1}{R_1 + R_F} v_{out} = \frac{v_{out}}{K}$$

But the virtual short also requires $v_n = v_p$, so

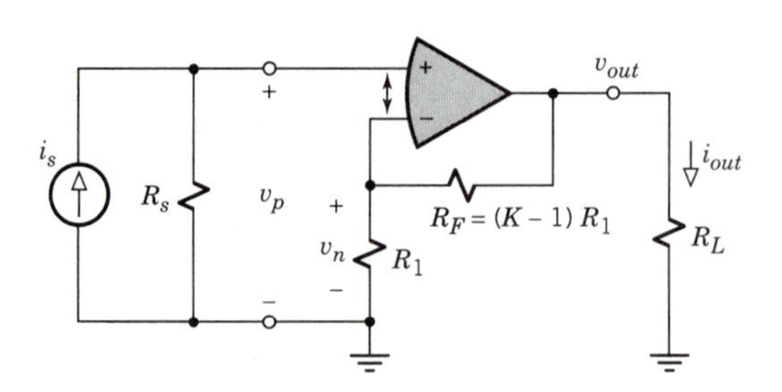

Figure 3.20 Noninverting current amplifier.

$$v_{out} = Kv_n = Kv_p = KR_s i_s$$

Writing $i_{out} = v_{out}/R_L$ then gives the current gain

$$A_i = i_{out}/i_s = KR_s/R_L$$

Hence, this circuit performs as a **noninverting current amplifier**. However, it has the disadvantage that A_i depends upon the values of R_s and R_L.

Exercise 3.8

Use virtual-short analysis of Fig. 3.16*a* to show that $v_{out}/v_{in} = (R_F + R_1)/R_1$.

Inverting and Summing Op-Amp Circuits

Many op-amp circuits function in an *inverting* mode, the input being connected to the inverting terminal. For such cases we usually draw the op-amp symbol with its inverting terminal at the top. The feedback path likewise appears across the top.

Figure 3.21*a* diagrams an **inverting voltage amplifier**. The input voltage is applied through R_1 to the inverting terminal, while the noninverting terminal is grounded so $v_p = 0$. The virtual short now establishes a "virtual ground" such that $v_n = v_p = 0$, as indicated in Fig. 3.21*b*. Feedback takes the form of the current i_f through R_F. Since $i_{in} + i_f = 0$, we must have

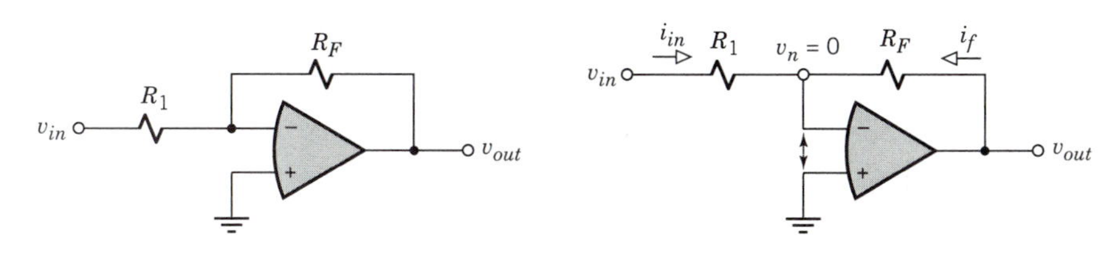

(*a*) Diagram of an inverting voltage amplifier

(*b*) Analysis using the virtual short

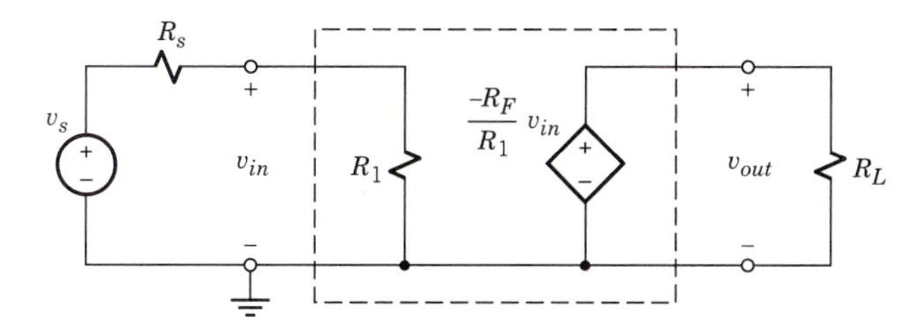

(*c*) Equivalent circuit

Figure 3.21

$$i_f = -i_{in}$$

KVL around the left-hand and right-hand loops through the virtual short yields

$$i_{in} = \frac{v_{in} - v_n}{R_1} = \frac{v_{in}}{R_1} \qquad i_f = \frac{v_{out} - v_n}{R_F} = \frac{v_{out}}{R_F}$$

Therefore, $v_{out}/R_F = -v_{in}/R_1$ and

$$v_{out} = -\frac{R_F}{R_1} v_{in} \tag{3.24}$$

This circuit thus amplifies the input voltage by the factor R_F/R_1 and inverts the output. When $R_1 = R_F$, the circuit becomes a **unity-gain inverter** with $v_{out} = -v_{in}$.

Unlike a noninverting amplifier, an inverting amplifier circuit has finite input resistance because it draws input current through R_1 (but not into the op-amp). Since the virtual ground puts the right-hand end of R_1 at ground potential, the input resistance is just

$$R_i = v_{in}/i_{in} = R_1$$

Accordingly, we must now give attention to source loading by using the circuit model in Fig. 3.21*c*. If the source has open-circuit voltage v_s and internal resistance R_s, then

$$A_v = \frac{v_{out}}{v_s} = \frac{R_1}{R_s + R_1}\left(-\frac{R_F}{R_1}\right) = -\frac{R_F}{R_s + R_1} \tag{3.25}$$

But we usually design inverting amplifiers with $R_1 \gg R_s$ to minimize loading and get $A_v \approx -R_F/R_1$. Alternatively, a voltage follower may be inserted to provide buffering between the source and the inverting amplifier.

Connecting an additional input resistor to the inverting terminal creates the **inverting summing amplifier** in Fig. 3.22. We can invoke superposition here to find v_{out} by suppressing v_2 and then v_1. With $v_2 = 0$, no current goes through R_2 and $v_{out-1} = -(R_F/R_1)v_1$. Similarly, with $v_1 = 0$, we get $v_{out-2} = -(R_F/R_2)v_2$. The total output with both sources active is $v_{out} = v_{out-1} + v_{out-2}$, so

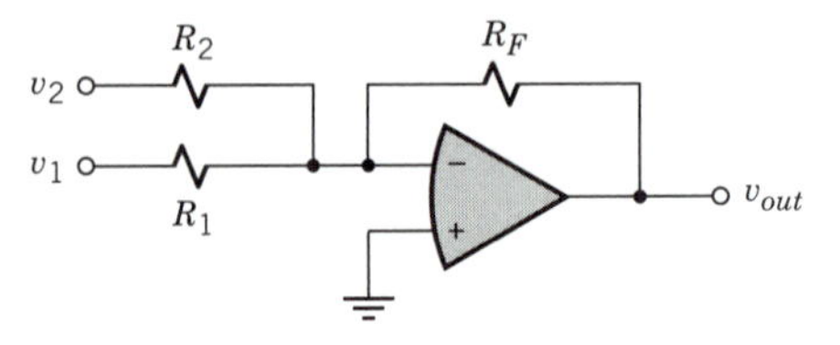

Figure 3.22 Inverting summing amplifier with two inputs.

$$v_{out} = -\left(\frac{R_F}{R_1} v_1 + \frac{R_F}{R_2} v_2\right) \tag{3.26a}$$

$$= -(R_F/R_1)(v_1 + v_2) \qquad R_2 = R_1 \tag{3.26b}$$

Hence, this circuit sums, amplifies, and inverts the two input signals. Equations (3.26a) and (3.26b) readily generalize for the case of three or more inputs.

If you were puzzled about the name *operational* amplifier, its meaning should now be clear, because we have shown how different op-amp circuits perform the operations of *amplification, inversion,* and *summation* of voltage signals. The operation of *subtraction* can also be performed by a system with two or more op-amps. As illustrated in our next example, the basic strategy for subtraction involves an inverting amplifier to pre-invert one of the signals applied to an inverting summing amplifier so the two inversions cancel each other.

Example 3.7 ***Design of an Op-Amp System***

We are given two signal sources, each having $R_s = 20\ \Omega$. One source generates $v_a(t)$ with $|v_a(t)| \leq 25$ mV, and the other generates $v_b(t)$ with $|v_b(t)| \leq 150$ mV. We want to design an op-amp system that produces a combined output signal

$$v_{out}(t) \approx 200v_a(t) - 40v_b(t)$$

There are a number of solutions to this problem, but at least two op-amps are required to carry out the subtraction and to get the gain of 200.

One design solution is based on rewriting the desired output in the form

$$v_{out} \approx -40[(-5v_a) + v_b] = -40(v_1 + v_2)$$

where

$$v_1 = -5v_a \qquad v_2 = v_b$$

A comparison of the expression for v_{out} with Eq. (3.26b) suggests an inverting summing amplifier with $R_2 = R_1$ and $R_F/R_1 = 40$. The input $v_1 = -5v_a$ can then be obtained by applying v_a to an inverting amplifier with $R_F/R_1 = 5$. Figure 3.23 gives the complete system diagram taking $R_1 = 1$ kΩ for both amplifiers so that $R_i = R_1 \gg R_s$.

Finally, we must determine the supply voltage required for the maximum output voltage. We compute $|v_{out}|_{max}$ by considering *worst-case conditions,* with v_a at its positive peak when v_b is at its negative peak, or vice versa. Then

$$|v_{out}|_{max} = 200|v_a|_{max} + 40|v_b|_{max}$$
$$= 200 \times 25 \text{ mV} + 40 \times 150 \text{ mV} = 11 \text{ V}$$

Therefore, the system must have $V_{PS} > 11$ V to keep operation within the linear region.

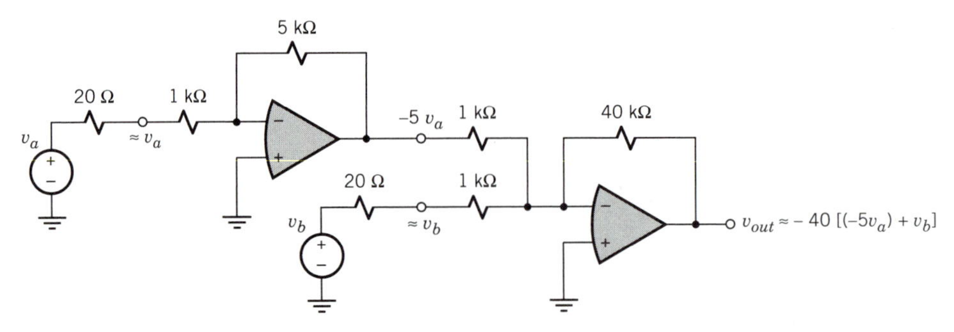

Figure 3.23 Op-amp system with two signal sources.

Exercise 3.9

An inverting amplifier like Fig. 3.21 is to be designed such that $v_{out}/v_{in} = -20$. Given our usual limitation on the values of R_1 and R_F, find the largest possible value of the input resistance R_i and the corresponding values of R_1 and R_F.

Exercise 3.10

Let the circuit in Fig. 3.22 have $R_1 = 1$ kΩ and $V_{PS} = 12$ V, and let $|v_1| \leq 0.2$ V and $|v_2| \leq 0.6$ V. Determine values for R_F and R_2 to get $v_{out} = -K(3v_1 + v_2)$, where K is as large as possible.

Example 3.8 *Difference Amplifier*

Another op-amp circuit for voltage subtraction is the **difference amplifier** in Fig. 3.24. The subtraction effect can be seen from the following qualitative observations. If $v_1 = 0$, then R_1 becomes grounded and v_p is proportional to v_2, so we have a *noninverting* amplifier with v_{out} proportional to v_2. But if $v_2 = 0$, then $v_p = 0$, so we have an *inverting* amplifier with v_{out} proportional to $-v_1$. Thus, in general, we expect the output to be a difference voltage.

For a quantitative analysis, we assume an ideal op-amp and note that R_2 and R_3 form a voltage divider such that

$$v_p = R_3 v_2/(R_2 + R_3)$$

Then, since an ideal op-amp has $v_n = v_p$ and $i_n = 0$,

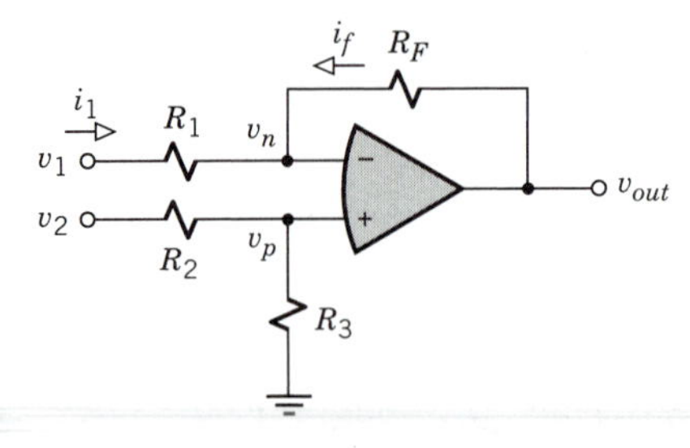

Figure 3.24 Difference amplifier.

$$i_f = -i_1 = -\frac{v_1 - v_n}{R_1} = \frac{R_3}{R_2 + R_3}\frac{v_2}{R_1} - \frac{v_1}{R_1}$$

Therefore,

$$v_{out} = v_n + R_F i_f = \frac{R_3}{R_2 + R_3} v_2 + \frac{R_3}{R_2 + R_3}\frac{R_F}{R_1} v_2 - \frac{R_F}{R_1} v_1$$

$$= \frac{R_F}{R_1}\frac{1 + R_1/R_F}{1 + R_2/R_3} v_2 - \frac{R_F}{R_1} v_1 \qquad (3.27a)$$

Finally, if $R_2/R_3 = R_1/R_F$, then Eq. (3.27a) reduces to

$$v_{out} = (R_F/R_1)(v_2 - v_1) \qquad R_3/R_2 = R_F/R_1 \qquad (3.27b)$$

which emphasizes the difference operation.

Exercise 3.11

Derive Eq. (3.27b) by applying superposition to Fig. 3.24 with $R_3/R_2 = R_F/R_1$.

3.3 INTERNAL OP-AMP RESISTANCES†

We previously omitted the input and output resistances of an operational amplifier because they usually can be ignored. This section takes account of those internal resistances to develop quantitative measures of their effects and to justify their omission in most practical applications.

Figure 3.25 shows a more complete model for a nonideal op-amp, including a large but finite input resistance r_i and a small but nonzero output resistance r_o. (The lowercase letters distinguish these quantities from the input and output resistances of complete amplifier circuits built with op-amps.) Most op-amps have $r_i > 100$ kΩ and $r_o < 200$ Ω. However, negative feedback in an op-amp circuit makes the equivalent input resistance much greater than r_i

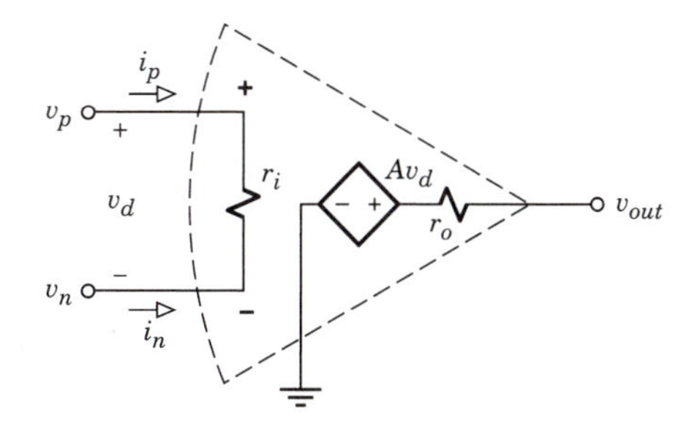

Figure 3.25 Model of a nonideal op-amp with input and output resistance.

and the equivalent output resistance much less than r_o. We'll demonstrate this point by re-examining the characteristics of a noninverting amplifier.

Figure 3.26*a* diagrams the noninverting amplifier circuit with the op-amp model from Fig. 3.25. We also include a source having resistance R_s at the input and a load resistance R_L at the output. The equivalent input resistance is $R_i = v_{in}/i_{in}$, as distinguished from r_i for the op-amp alone. We calculate R_i by noting that

$$v_{in} = v_d + v_n \qquad v_d = r_i i_{in} \qquad v_n = R_1(i_{in} + i_f)$$

$$v_{out} = R_L i_{out} = R_F i_f + v_n = Av_d - r_o(i_f + i_{out})$$

Some algebraic manipulation of these relations yields the rather cumbersome expression

$$R_i = \frac{v_{in}}{v_{in}} = r_i + \frac{R_1(Ar_i + R_F + r_o + R_F r_o/R_L)}{R_1 + R_F + r_o + (R_1 + R_F)r_o/R_L}$$

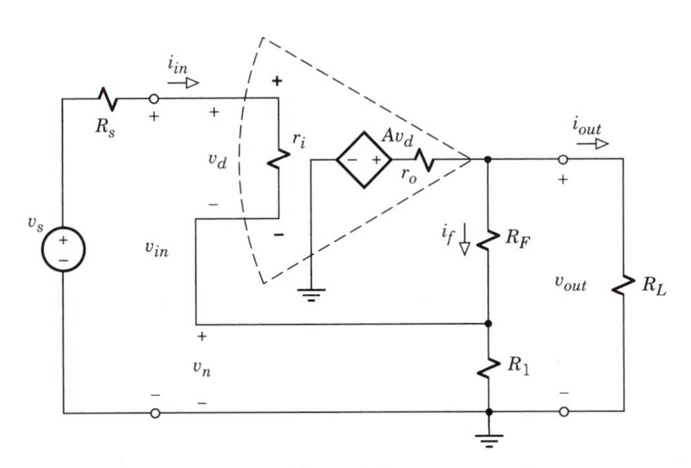

(*a*) Noninverting amplifier including op-amp resistances

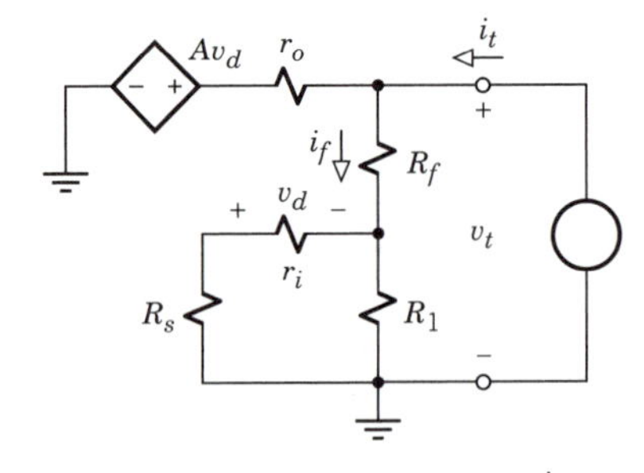

(*b*) Diagram for determining the Thévenin equivalent output resistance

Figure 3.26

Thus, in general, the input resistance R_i depends on the load resistance R_L at the output.

But practical op-amp circuits usually have

$$R_L \gg r_o \qquad R_1 + R_F \gg r_o \qquad Ar_i \gg R_F + r_o + R_F r_o / R_L$$

Accordingly, our result for R_i reduces to the useful approximation

$$R_i \approx (1 + AB)r_i \tag{3.28}$$

where, as before,

$$B = R_1/(R_1 + R_F) \tag{3.29}$$

Equation (3.28) more clearly indicates that the feedback creates a very large equivalent input resistance when $AB \gg 1$.

To calculate the Thévenin equivalent output resistance, we suppress v_s and replace R_L with a test source. The resulting diagram in Fig. 3.26*b* has

$$i_t = i_f + \frac{v_t - Av_d}{r_o} \qquad i_f = \frac{v_t}{R_F + R_1'} \qquad v_d = -\frac{r_i}{R_s + r_i} R_1' i_f$$

where we have introduced

$$R_1' = R_1 \| (r_i + R_s)$$

Solving for i_t in terms of v_t yields

$$\frac{1}{R_o} = \frac{i_t}{v_t} = \frac{1}{r_o} + \frac{AR_1 r_i + r_o(r_i + R_1 + R_s)}{r_o[R_1(r_i + R_s) + R_F(r_i + R_1 + R_s)]}$$

Thus, in general, the output resistance R_o depends on the source resistance R_s at the input.

But we can clean up our expression via the reasonable assumptions that

$$r_i \gg R_1 \qquad AR_1 \gg r_o$$

We then get the more informative approximation

$$R_o \approx \frac{r_o}{1 + AB} \tag{3.30}$$

Thus, feedback greatly reduces R_o compared to the output resistance of the op-amp alone.

Since a noninverting amplifier should have $AB \gg 1$, the input resistance R_i usually will be large enough and the output resistance R_o usually will be small enough that we can ignore loading effects at the input and output. These considerations justify the simplified equivalent circuit for a noninverting amplifier in Fig. 3.17. Similar arguments justify ignoring op-amp resistances in inverting and summing amplifiers.

The following example illustrates calculations that include R_i and R_o.

Example 3.9 *Source Buffering with a Voltage Follower*

The voltage variations from a temperature sensor are to be displayed on a chart recorder. The sensor has a source resistance of 2400 Ω and generates $v_s = 1\text{–}4$ V, depending on the temperature. The recorder has a full-scale range of 5 V and a 600-Ω input resistance.

If the sensor is connected directly to the recorder, then loading effect reduces the terminal voltage to

$$v = 600v_s/(2400 + 600) = 0.2\text{–}0.8 \text{ V}$$

which would be just a small fraction of the 5-V chart range. To get a better chart record, a voltage follower is inserted as a buffer between the sensor and the recorder. This is the configuration we want to analyze.

The voltage follower has $R_F = 0$ and $R_1 = \infty$, so $B = 1$. We'll conservatively assume an op-amp with $A = 10^4$, $r_i = 100$ kΩ, and $r_o = 200$ Ω. Thus, since $1 + AB \approx 10^4$, the follower's input and output resistances are

$$R_i \approx 10^4 \times 10^5 = 10^9 \ \Omega \qquad R_o \approx 200/10^4 = 0.02 \ \Omega$$

We also need the follower's open-circuit voltage amplification given by Eq. (3.22) as

$$v_{out-oc}/v_{in} = A/(1 + A) = 0.99990$$

Figure 3.27 shows the resulting complete equivalent circuit. By chain expansion,

$$\frac{v_{out}}{v_s} = \frac{10^9}{2400 + 10^9} \times 0.99990 \times \frac{600}{0.02 + 600} = 0.99986$$

so

$$v_{out} = 0.99986v_s \approx 1\text{–}4 \text{ V}$$

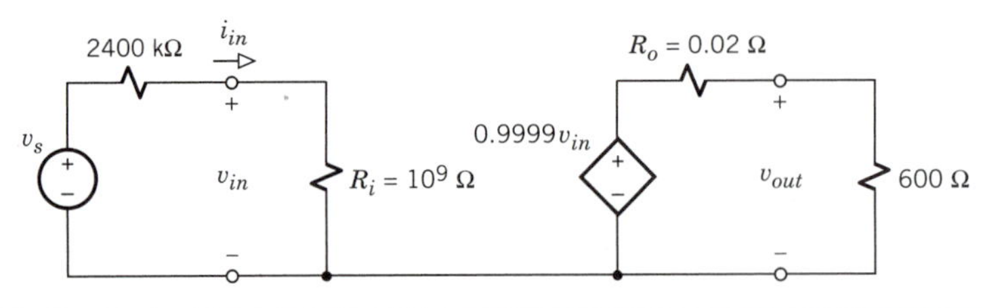

Figure 3.27 Equivalent circuit of a voltage follower with input and output resistances.

which makes much fuller use of the chart range. Also note that the maximum input current is only $4\ \text{V}/(2400 + 10^9)\Omega \approx 4 \times 10^{-9}$ A. Thus, we might just as well have assumed that $R_i = \infty$ and $R_o = 0$.

Exercise 3.12

Calculate R_i and R_o for a noninverting amplifier with $R_1 = 4\ \text{k}\Omega$, $R_F = 76\ \text{k}\Omega$, $A = 10^4$, $r_i = 3\ \text{M}\Omega$, and $r_o = 100\ \Omega$.

3.4 DC METERS AND MEASUREMENTS†

This chapter concludes with an introduction to dc measuring instruments. Measurement techniques and error considerations will also be discussed.

Routine electrical measurements are usually made with a *multimeter*. These versatile instruments contain a sensing device, auxiliary circuitry, and switches that allow you to measure voltage, current, or resistance in several ranges. Most modern multimeters have an electronic sensing device with a *digital* display. But prior to the development of integrated-circuit electronics, the sensing device was electromechanical with an *analog* display. Analog meters have the advantage of lower cost, and they are still widely used for single-function panel meters and inexpensive multimeters.

We'll begin with the analysis and design of simple analog dc meters, and we'll describe the similarities and differences of digital meters as we go along. Then we'll discuss specialized measuring techniques that exploit the principle of null measurements. As a reminder that we are dealing here with constant or slowly varying voltages and currents, the capitalized symbols V and I will be used throughout.

Voltmeters and Ammeters

The heart of an analog meter is usually a **d'Arsonval movement**, whose structural features are depicted in Fig. 3.28. This electromechanical sensing device consists of a flat coil suspended in the magnetic field between a

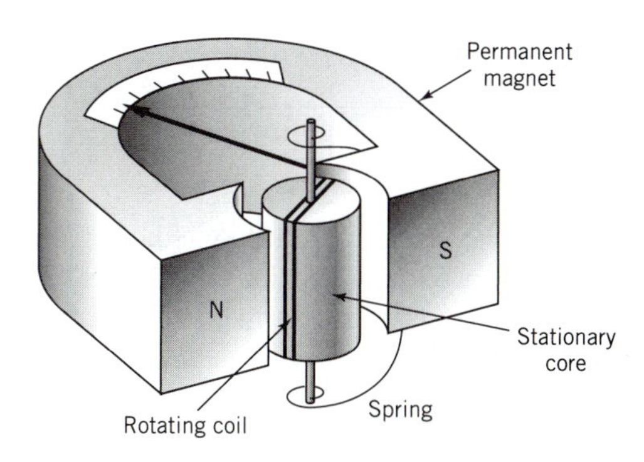

Figure 3.28 d'Arsonval movement for analog dc meters.

horseshoe-shaped permanent magnet and a stationary core. Current through the coil interacts with the magnetic field to create torque that rotates the coil about its vertical axis. Opposing torque comes from a pair of spiral springs, and the resulting angular deflection is displayed by a pointer attached to the coil assembly.

When the coil carries constant current, the steady-state deflection of the pointer is proportional to the amount of current. The pointer's maximum or **full-scale deflection** corresponds to the current I_{fs}. Smaller deflections are thus proportional to currents over the range $0 \leq I \leq I_{fs}$. Some movements have a *zero-center* configuration to sense currents of either direction within the range $-I_{fs} \leq I \leq I_{fs}$.

Viewed from its terminals, a d'Arsonval movement in the steady state is equivalent to the resistance R_m in Fig. 3.29. This resistance represents the fact that the movement draws power from the source that supplies I. Therefore, the terminal voltage at full-scale deflection is

$$V_{fs} = R_m I_{fs} \tag{3.31}$$

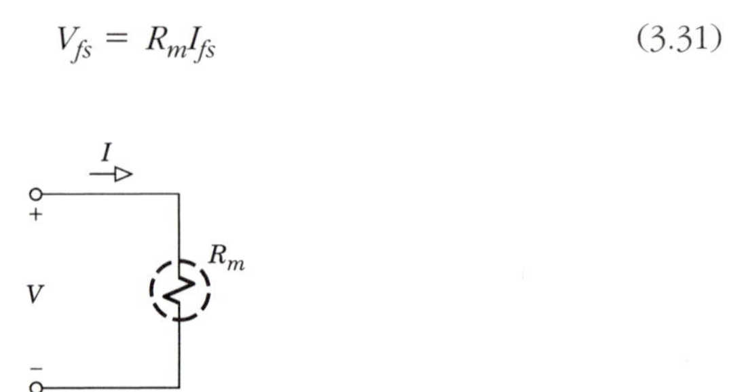

Figure 3.29 Equivalent resistance of a d'Arsonval movement.

The characteristics of a given movement are specified by any two of the three parameters I_{fs}, R_m, and V_{fs}. Table 3.2 lists typical values for a low-current movement and a high-current movement. Low-current movements are said to have greater *sensitivity* because full-scale deflection requires less current.

TABLE 3.2 Typical d'Arsonval Movement Parameters

Type	I_{fs}	R_m	V_{fs}
Low current	50 μA	5 kΩ	250 mV
High current	10 mA	1 Ω	10 mV

With the help of additional resistors and switches, the d'Arsonval movement can measure dc current or voltage in various ranges. In particular, a **multirange dc voltmeter** consists of a movement and switchable *series* resistors. Figure 3.30 diagrams the circuitry needed to provide two voltage ranges for measuring the unknown voltage V_u. The voltage V across the movement is produced by a voltage divider consisting of R_m and either R_1 or R_2.

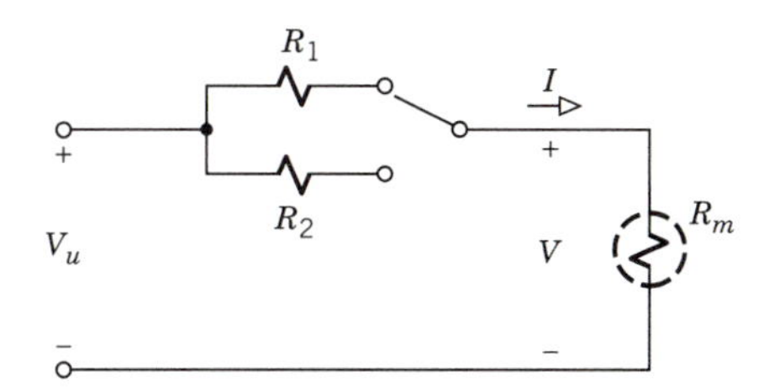

Figure 3.30 Multirange DC voltmeter with two multipliers.

Consequently, the range-setting resistors R_1 and R_2 are known as **multipliers**, since they allow V_u to be greater than V.

We'll analyze Fig. 3.30 taking the switch to be in the upper position, which puts R_1 in series with R_m. Then $V = R_m V_u/(R_m + R_1)$ so $V_u = (R_m + R_1)V/R_m$ and full-scale deflection corresponds to

$$V_{ufs} = (1 + R_1/R_m)V_{fs} \tag{3.32}$$

Replacing R_1 in Eq. (3.32) with R_2 gives the full-scale relationship for the other switch position. By way of example, if $V_{fs} = 250$ mV, $R_m = 5$ kΩ, $R_1 = 15$ kΩ, and $R_2 = 95$ kΩ, then the two full-scale voltages are $(1 + 15/5) \times 250$ mV = 1 V and $(1 + 95/5) \times 250$ mV = 5 V.

A **multirange dc ammeter** employs switchable *parallel* resistors, as illustrated by Fig. 3.31. The range-setting resistors here are known as **shunts** because they bypass part of the unknown current I_u around the movement. When R_1 is connected, full-scale deflection corresponds to

$$I_{ufs} = (1 + R_m/R_1)I_{fs} \tag{3.33}$$

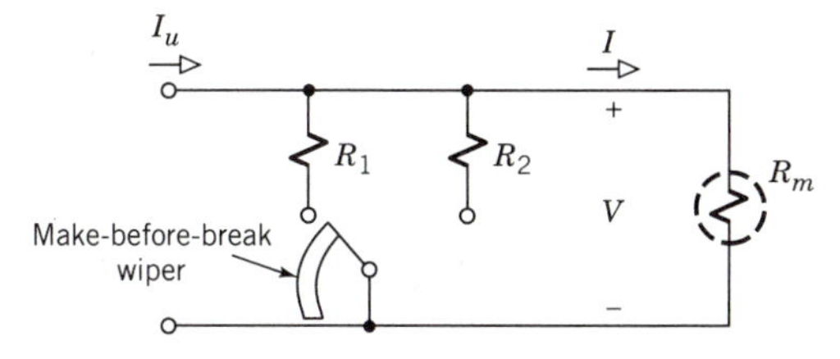

Figure 3.31 Multirange DC ammeter with two shunts.

The derivation of this expression is left for an exercise.

As illustrated in Fig. 3.31, the switch in a multirange ammeter must have a special *make-before-break wiper*. This wiper has an extended tab that "makes" contact with R_2 before it "breaks" contact with R_1, so there is always some resistance in parallel with the movement. Were this not the case, the entire current I_u could go through the movement and the resulting ohmic heating might destroy the coil. Voltmeters do not require a special switch, but all meter movements should have a series fuse for protection from excessive current.

The foregoing voltmeter and ammeter circuitry can be combined in an **analog multimeter**. Multimeters have a rotary selector switch with mechanically coupled wipers to select the proper connection of multipliers and shunts. For instance, Fig. 3.32 gives the arrangement of a simple multimeter with one voltage range and one current range. The two wipers are coupled to move together, and the contact marked "NC" has no connection to it. Redraw the circuit, if necessary, to convince yourself that R_v serves as a multiplier when the wipers are in the upper position while R_a serves as a shunt when the wipers are in the lower position.

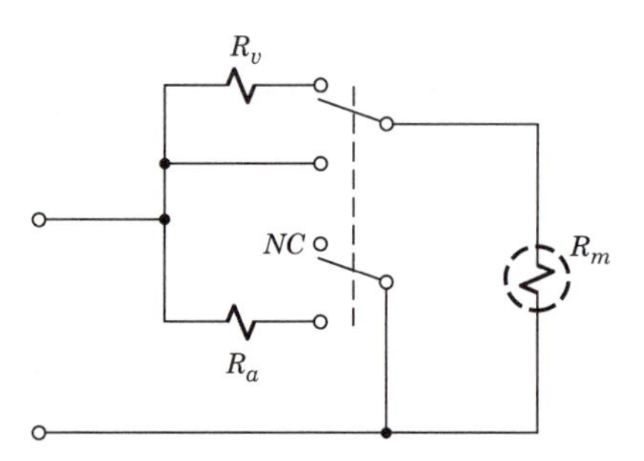

Figure 3.32 Analog multimeter with one voltage range and one current range.

A **digital multimeter (DMM)** differs from a d'Arsonval instrument in that the sensing unit is an electronic **digital voltmeter (DVM)**. Besides providing an easy-to-read digital display, a DVM draws essentially no current — like the input of an op-amp. Figure 3.33a illustrates how multipliers are connected as a voltage-divider "string" at the DVM input to obtain N voltage ranges. Figure 3.33b shows how shunts are connected in parallel with the DVM to provide N current ranges. The smallest full-scale voltage and current readings of a typical DMM are 200 mV and 200 μA, respectively.

Whether analog or digital, most multimeters also include provision for measuring ac voltage and current. AC measurements are made possible by switching in a rectifying network that performs ac to dc conversion.

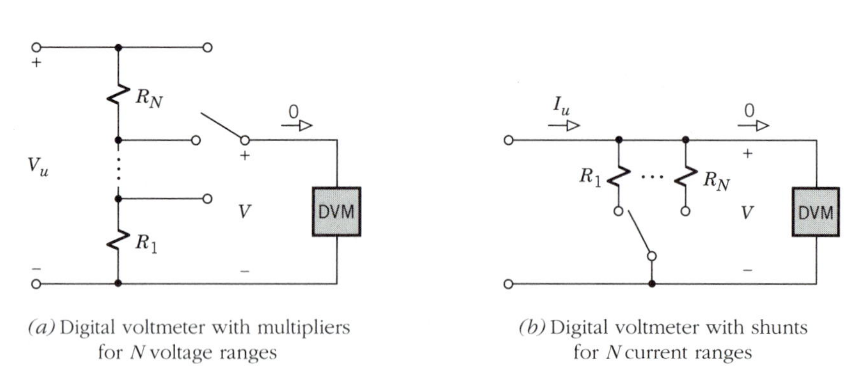

(a) Digital voltmeter with multipliers for N voltage ranges

(b) Digital voltmeter with shunts for N current ranges

Figure 3.33

Example 3.10 ***Analog Multimeter Design***

Suppose you want to build a multimeter like Fig. 3.32 with $V_{ufs} = 5$ V and $I_{ufs} = 30$ mA using a d'Arsonval movement having $V_{fs} = 120$ mV and $I_{fs} = 200$ μA. You carry out the design calculations for R_v and R_a as follows.

First, you find the movement's resistance to be

$$R_m = V_{fs}/I_{fs} = 120 \text{ mV}/200 \text{ μA} = 600 \ \Omega$$

Next, you determine the value of the multiplier resistor R_v by rewriting Eq. (3.32) in the form

$$1 + R_v/600 = 5 \text{ V}/120 \text{ mV} \Rightarrow R_v = 24{,}400 \ \Omega$$

Finally, you determine the value of the shunt resistor R_a by rewriting Eq. (3.33) in the form

$$1 + 600/R_a = 30 \text{ mA}/200 \text{ μA} \Rightarrow R_a = 4.027 \ \Omega$$

Exercise 3.13

Derive Eq. (3.33) from Fig. 3.31.

Exercise 3.14

Rework the design calculations in Example 3.10 for a meter movement with $I_{fs} = 2$ mA and $R_m = 15 \ \Omega$.

Measurement Errors

Measurements made with a voltmeter or ammeter involve two general types of errors. One type of error stems from the limited **accuracy** of the meter itself, which depends upon the design and the quality of the components. Manufacturers usually state meter accuracy as a percentage of the actual reading or the full-scale reading. But the voltage or current measured by the meter may also differ from the actual quantity you want to measure. This second type of error stems from interaction effects between the meter and the circuit under test.

Interaction error in voltage measurements takes the form of **circuit loading**. We'll analyze this effect using the diagram in Fig. 3.34, where R_{vm} represents the total input resistance of the voltmeter. The voltage to be measured appears across resistance R, which could be a Thévenin equivalent resistance. The voltmeter loads the circuit by drawing current I, and the measured voltage is

$$V_u = (R \| R_{vm})I_s = RI_s/(1 + R/R_{vm})$$

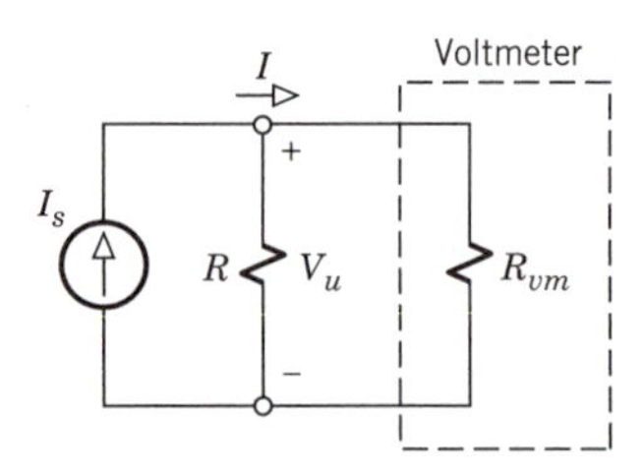

Figure 3.34 Voltage measurement with circuit loading.

Since the actual voltage without the meter connected is $V_{act} = RI_s$, we can write

$$V_{act} = (1 + R/R_{vm})V_u \tag{3.34}$$

This expression brings out the fact that circuit loading causes the measured value V_u to be *less* than V_{act}.

In principle, Eq. (3.34) could also be used to calculate V_{act} if we know the values of R and R_{vm}. In practice, however, determining R usually would require complicated additional measurements. We therefore rely instead upon the assumption that $R_{vm} \gg R$ so $V_{act} \approx V_u$. For this reason, a "good" voltmeter should have *large input resistance.*

Going back to Fig. 3.33*a*, we see that the input resistance of a multirange digital voltmeter is independent of the range setting and that $R_{vm} = R_1 + R_2 + \cdots + R_N$. Thus, the multiplier string can be designed to minimize circuit loading, and R_{vm} usually equals or exceeds 10 MΩ.

An analog voltmeter has smaller input resistance because the d'Arsonval movement always draws current. Furthermore, R_{vm} varies with range setting and is usually expressed as *ohms per volt* of the full-scale range. From the diagram back in Fig. 3.30, we see that $V_u = V_{ufs}$ when $I = I_{fs}$. Hence, $R_{vm} = V_{ufs}/I_{fs}$ or

$$R_{vm}/V_{ufs} = 1/I_{fs} \tag{3.35}$$

Most analog voltmeters are built with low-current d'Arsonval movements having $I_{fs} = 50\ \mu A$, and the corresponding input resistance is $R_{vm}/V_{ufs} = 1/(50\ \mu A) = 20{,}000$ ohms per volt. Thus, for instance, the 5-V range would have $R_{vm} = 20\ k\Omega/V \times 5\ V = 100\ k\Omega$.

Current measurements produce another interaction effect when an ammeter is inserted in a circuit branch. We'll analyze this effect using the diagram in Fig. 3.35, where R_{am} represents the ammeter's input resistance. The current to be measured comes through resistance R, which again could be a Thévenin equivalent resistance. The nonzero ammeter resistance introduces a **burden voltage** V, and the measured current is

$$I_u = V_s/(R + R_{am}) = V_s/[R(1 + R_{am}/R)]$$

Since the actual current is $I_{act} = V_s/R$, we can write

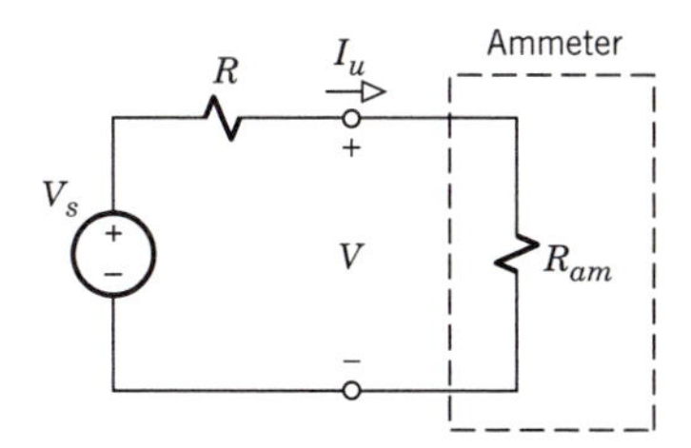

Figure 3.35 Current measurement with burden voltage.

$$I_{act} = (1 + R_{am}/R)I_u \tag{3.36}$$

The measured value I_u is therefore *less* than I_{act}.

Equation (3.36) reveals that a "good" ammeter should have *small input resistance*, so $R_{am} \ll R$ and $I_u \approx I_{act}$. But the input resistance of a multirange ammeter depends upon the selected range, so R_{am} is typically expressed as *ohms times milliamps* of the full-scale range. From both the analog and digital ammeter diagrams back in Figs. 3.31 and 3.33*b* we see that $V = V_{fs}$ when $I_u = I_{ufs}$. Hence, $R_{am} = V_{fs}/I_{ufs}$ or

$$R_{am} \times I_{ufs} = V_{fs} \tag{3.37}$$

If $V_{fs} = 250$ mV, then $R_{am} \times I_{ufs} = 250$ ohms times milliamps, and the 10-mA range would have $R_{am} = 250\ \Omega\text{-mA}/10\ \text{mA} = 25\ \Omega$.

For comparison purposes, Table 3.3 lists the properties relevant to measurement errors with typical analog and digital multimeters.

TABLE 3.3 Typical Multimeter Properties

Type	Accuracy	R_m	V_{fs}	R_{vm}	R_{am}
Analog	±3% of full scale	5 kΩ	250 mV	20 kΩ/V	250 Ω-mA
Digital	±0.5% of reading	∞	200 mV	10 MΩ	200 Ω-mA

Example 3.11 *Estimating Voltage Measurement Error*

Consider the situation depicted in Fig. 3.36, where the voltage in question appears across a 5-kΩ resistor. A reading of 7.40 V has been obtained using the 10-V scale of an analog multimeter. We'll estimate the measurement error, given that the meter has the properties listed in Table 3.3.

First, since the meter's accuracy is ±3% of full scale, the measured voltage is anywhere in the range

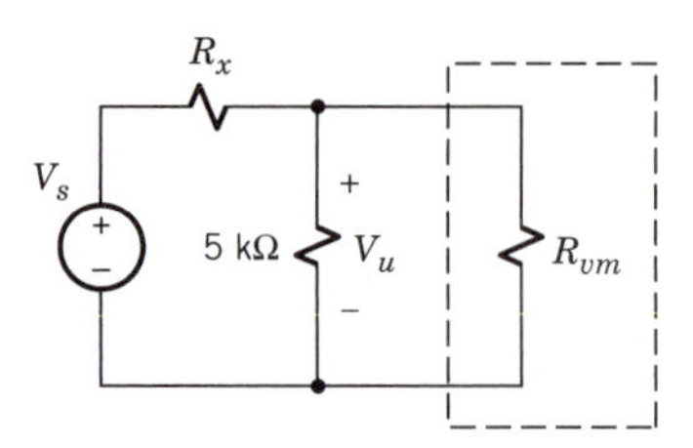

Figure 3.36 Measuring voltage across a resistor.

$$V_u = 7.40 \text{ V} \pm 0.03 \times 10 \text{ V} = 7.10\text{–}7.70 \text{ V}$$

Next, to account for circuit loading, Eq. (3.34) shows that V_u should be multiplied by $(1 + R/R_{vm})$. We know that the meter has $R_{vm} = 20 \text{ k}\Omega/\text{V} \times 10 \text{ V} = 200 \text{ k}\Omega$, but the resistance R_x is unknown. However, the Thévenin equivalent resistance in parallel with the meter is $R = R_x \| (5 \text{ k}\Omega) \leq 5 \text{ k}\Omega$ so

$$R/R_{vm} \leq 5 \text{ k}\Omega/200 \text{ k}\Omega = 0.025$$

Thus, the upper bound on the actual voltage is

$$V_{act} \leq (1 + 0.025) \times 7.70 = 7.89 \text{ V}$$

The lower bound corresponds to $R/R_{vm} = 0$ so

$$V_{act} \geq (1 + 0) \times 7.10 = 7.10 \text{ V}$$

Exercise 3.15

A reading of 1.60 mA is obtained on the 2-mA scale of a digital multimeter inserted in series with a 1-kΩ resistor. Estimate the upper and lower bounds on I_{act}.

Ohmmeters

In contrast to voltmeters and ammeters, an **ohmmeter** must supply current through the element whose resistance is being measured. Consequently, both analog and digital ohmmeters contain a battery or some other constant source.

A simplified analog ohmmeter is shown in Fig. 3.37*a*, where R_u is the unknown resistance to be measured and V_s provides the test current sensed by the d'Arsonval movement. The values of the range resistor R_r and the parallel combination $R_{am} = R_a \| R_m$ are chosen to get full-scale deflection when $R_u = 0$, so

$$\frac{R_{am}}{R_r + R_{am}} V_s = V_{fs}$$

With $R_u \neq 0$, the voltage across the movement becomes

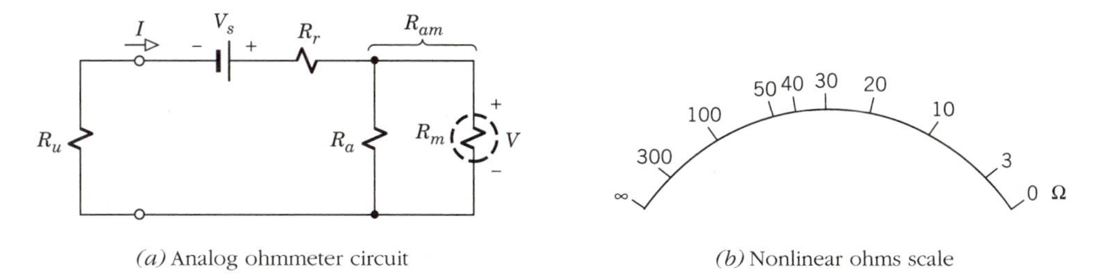

(a) Analog ohmmeter circuit

(b) Nonlinear ohms scale

Figure 3.37

$$V = \frac{R_{am}}{R_u + R_r + R_{am}} V_s = \frac{R_r + R_{am}}{R_u + R_r + R_{am}} V_{fs} \tag{3.38}$$

Equation (3.38) reveals that $V = 0.5V_{fs}$ when $R_u = R_r + R_{am}$, whereas $V = 0$ when $R_u = \infty$ (open-circuited terminals). Hence, the ohms scale is *backwards* and *nonlinear.*

Figure 3.37*b* illustrates the scale of an analog ohmmeter with $R_r + R_{am} = 30\ \Omega$. We see that the most easily read resistance values fall around *half-scale* deflection, and readablilty deteriorates when R_u is much smaller or larger than $R_r + R_{am}$. In view of the nonlinearity, accuracy is typically stated as $\pm 2.0°$ of arc.

Most analog multimeters include ohmmeter circuitry similar to Fig. 3.37*a*, but with switched range resistors to provide multiple ranges with different half-scale values. Additionally, a variable resistance labeled "Ohms adjust" compensates for the changing value of V_s as the battery ages. A multimeter with scales for volts, ohms, and milliamps is known by the acronym VOM.

Digital multimeters employ different techniques for resistance measurements using a DVM. One technique involves a precise current source to produce a voltage across R_u. Another technique based on the *voltage ratio method* is diagrammed in Fig. 3.38. The unknown R_u is connected here in series with a selectable reference resistance R_{ref} and a voltage source, thereby establishing the two voltages $V_u = R_u V_s/(R_u + R_{ref})$ and $V_{ref} = R_{ref} V_s/(R_u + R_{ref})$. The DVM measures both voltages and a built-in calculator computes R_u from the ratio $V_u/V_{ref} = R_u/R_{ref}$, so

$$R_u = (V_u/V_{ref})R_{ref}$$

Typical accuracy is $\pm 0.5\%$ of the ohms reading.

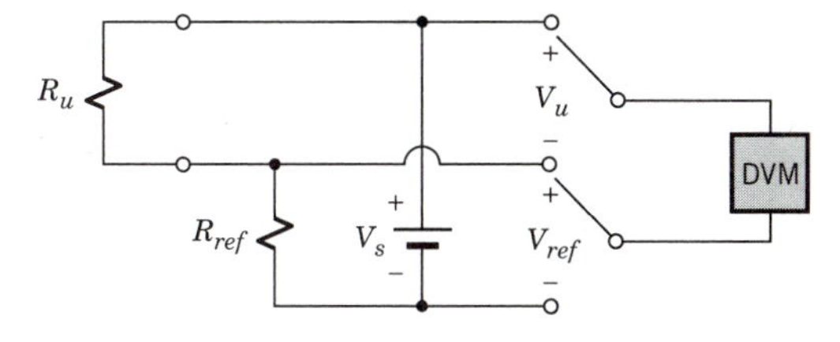

Figure 3.38 Digital ohmmeter using voltage ratio.

Whether using an analog or digital meter, two precautions should be observed. First, an ohmmeter can only measure isolated resistors or the equivalent resistance of a resistive network, but not the value of a resistor embedded within a network. Second, ohmmeters should not be connected to energized circuits nor to any electronic components that might be damaged by the current from the internal source.

Exercise 3.16

An ohmmeter circuit like Fig. 3.37*a* is to be built using a d'Arsonval movement with $R_m = 3\ \text{k}\Omega$. Explain why R_a is needed by considering the half-scale deflection when R_a is removed.

Null Measurements and Bridges

Null measurements differ from direct measurements in that the quantity being measured is compared with a known reference quantity, analogous to weight measurement using a balancing scale. The balancing strategy avoids unwanted interaction effects and generally results in greater precision than a direct measurement.

To introduce the principle of null measurements, consider the **potentiometric voltage measurement system** in Fig. 3.39. This system employs a calibrated potentiometer and a zero-center meter to compare an unknown source voltage V_u with the adjustable voltage V_W derived from a standard reference voltage V_{ref}. You adjust the potentiometer's wiper until the meter indicates that $V_x = V_u - V_W = 0$, so

$$V_u = V_W = (R_{AW}/R_{AB})V_{ref}$$

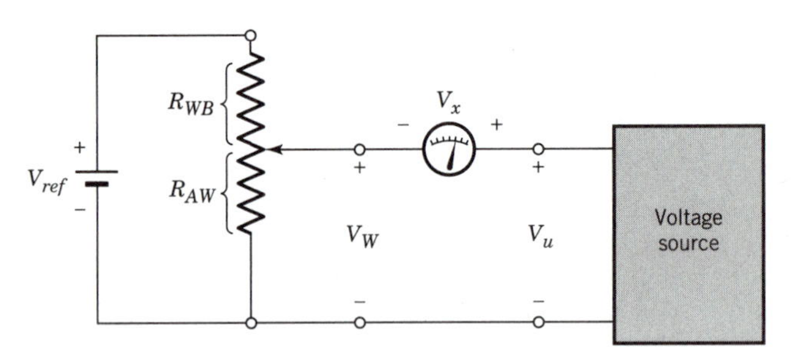

Figure 3.39 Potentiometric voltage measurement system.

The value of V_u can be determined by reading the calibrated dial of the potentiometer. No current is drawn from the unknown source in the balanced condition, making this method ideal for measuring sensitive electrochemical processes and the like.

An automated **self-balancing potentiometer** incorporates a small motor driven by the difference voltage $V_u - V_W$. The motor moves the wiper until balance is achieved and $V_u - V_W = 0$. The self-balancing system will continuously follow a time-varying voltage, if not too rapid, and is often used for strip-chart and X–Y recorders.

Another null-measurement circuit called the **Wheatstone bridge** is diagrammed in Fig. 3.40*a*. This circuit is designed for precise measurement of an unknown resistance R_u with the help of a calibrated potentiometer R plus two known resistances R_1 and R_2. We adjust the potentiometer until $V_x = 0$, so the bridge is said to be *balanced.* Since no current flows through the meter under balanced conditions, we can redraw the circuit as shown in Fig. 3.40*b* where

$$V_1 = \frac{R_1}{R_1 + R} V_s = \frac{V_s}{1 + (R/R_1)} \qquad V_2 = \frac{R_2}{R_2 + R_u} V_s = \frac{V_s}{1 + (R_u/R_2)}$$

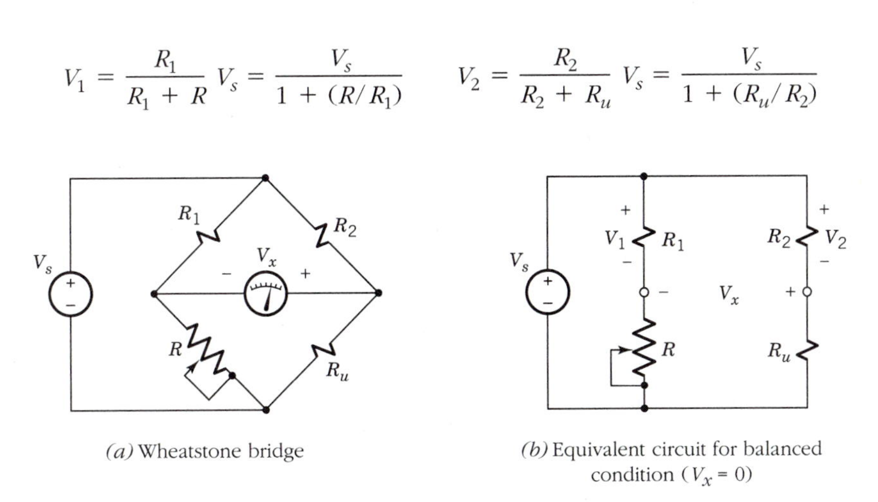

(*a*) Wheatstone bridge

(*b*) Equivalent circuit for balanced condition ($V_x = 0$)

Figure 3.40

But $V_x = V_1 - V_2 = 0$, so $V_2 = V_1$ and $R_u/R_2 = R/R_1$. Thus,

$$R_u = (R_2/R_1)R \tag{3.39}$$

which gives R_u in terms of known quantities, independent of V_s.

Commercial Wheatstone bridges have switchable resistors for R_1 and R_2, and the ratio R_2/R_1 can be set for various powers of 10 — 0.1, 1, 10, 100, etc. The value of R is then read from the dial of the potentiometer and multiplied by the appropriate power of 10 to get R_u. This arrangement allows the use of a single potentiometer to measure a wide range of unknown resistances.

Some bridges include two potentiometers to measure small *increases* of resistance, which occur in instrumentation transducers such as resistance thermometers and strain gauges. The circuit is shown in Fig. 3.41, where the second potentiometer is labeled ΔR and the resistance to be measured increases from R_u to $R_u + \Delta R_u$. Initially, ΔR_u equals zero and the bridge is balanced using R, with $\Delta R = 0$, so $R_u = (R_2/R_1)R$. Then, after the subject resistance

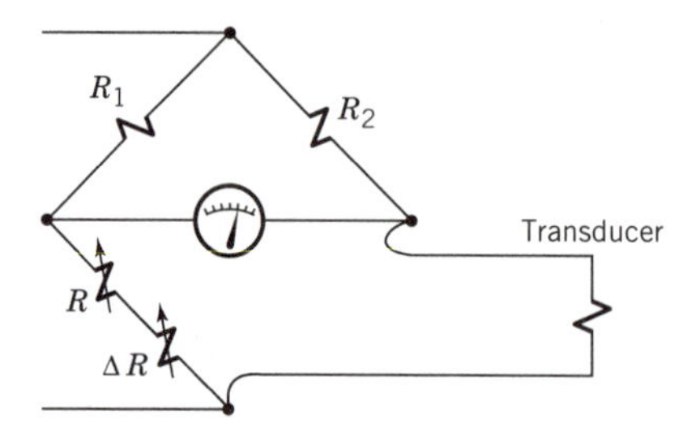

Figure 3.41 Wheatstone bridge network for strain measurement.

increases by ΔR_u, the bridge is rebalanced using ΔR alone. Thus, $(R_u + \Delta R_u) = (R_2/R_1)(R + \Delta R)$ and

$$\Delta R_u = (R_2/R_1)\Delta R \tag{3.40}$$

If ΔR has much smaller resistance increments than R, then very small resistance changes can be measured with great accuracy.

Example 3.12 ***Strain Measurement with a Wheatstone Bridge***

Suppose the unknown resistance in Fig. 3.41 represents a strain gauge with unstrained resistance R_u. We showed in Example 1.6 that the mechanical strain $\Delta l/l$ increases the gauge resistance to $R_u + \Delta R_u$, where $\Delta R_u = 2R_u(\Delta l/l)$. Now we'll calculate the strain given the following bridge measurements.

Before strain is applied, the bridge is balanced with $R_1 = 1000\ \Omega$, $R_2 = 100\ \Omega$, $R = 625\ \Omega$, and $\Delta R = 0$. Thus, from Eq. (3.39),

$$R_u = (100/1000) \times 625 = 62.5\ \Omega$$

After strain is applied, the bridge is rebalanced by adjusting the second potentiometer to $\Delta R = 2.4\ \Omega$, so Eq. (3.40) yields

$$\Delta R_u = (100/1000) \times 2.4 = 0.24\ \Omega$$

The corresponding strain is

$$\Delta l/l = \Delta R_u/2R_u = 0.24/125 = 0.00192$$

Exercise 3.17

Suppose the potentiometer in Fig. 3.40 can be varied from 0 to 1000 Ω with a reading accuracy of 1 Ω, and the resistance ratio R_2/R_1 can be set for any of the following values: 0.001, 0.01, 0.1, 1, 10, 100, 1000. Also suppose that $R_u \approx 50\ \Omega$.

(a) With what values of R_2/R_1 can the bridge be balanced?

(b) What value of R_2/R_1 should be used for maximum accuracy?

PROBLEMS

Section 3.1 Real Sources and Power Transfer

3.1* Let $v_s = 120$ V and $R_s = 3\ \Omega$ in Fig. P3.1. Calculate p_{max} and determine the condition on R_L such that $v \geq 110$ V.

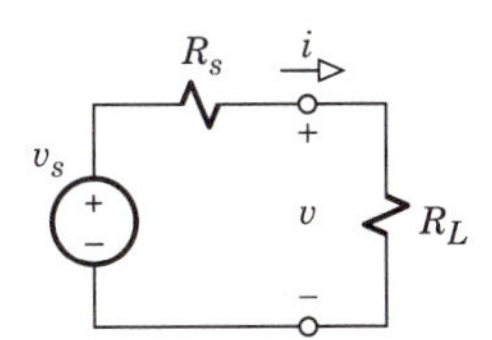

Figure P3.1

3.2 Let $v_s = 20$ V, $R_s > 0$, and $R_L = 2\ \text{k}\Omega$ in Fig. P3.1. Calculate p_{max} and determine the condition on R_s such that $v \geq 19$ V.

3.3 Let $i_s = 5$ A and $R_s = 80\ \Omega$ in Fig. P3.3. Calculate p_{max} and determine the condition on R_L such that $i \geq 4.5$ A.

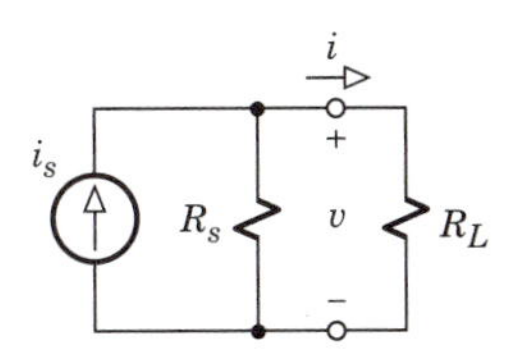

Figure P3.3

3.4 Let $i_s = 40$ mA, $R_s < \infty$, and $R_L = 5\ \text{k}\Omega$ in Fig. P3.3. Find p_{max} and the condition on R_s such that $i \geq 38$ mA.

3.5 Evaluate p_L/p_{max}, p_s/p_{max} and Eff for Fig. P3.1 when $R_L = R_s/2$ and when $R_L = 4R_s$.

3.6 Evaluate p_L/p_{max} and Eff for the circuits in Figs. P3.1 and P3.3 when $R_L = 2R_s$.

3.7 Do Problem 3.6 with $R_L = R_s/3$.

3.8 Obtain an expression for p_L in Fig. P3.3. Then derive Eq. (3.8) by solving $dp_L/dR_L = 0$.

3.9* Let a matched load resistor be connected to the terminals of Fig. 2.29 (p. 73). Use the results of Exercise 2.14 to find p_L, p_s, and Eff.

3.10 Let a matched load resistor be connected to the terminals of Fig. 2.32 (p. 75). Use the results of Exercise 2.15 to find p_L. Then calculate Eff by finding the power supplied by the sources.

3.11 Let a matched load resistor be connected to the terminals of Fig. 2.35*a* (p. 78). Use the results of Example 2.16 to find p_L. Then calculate Eff by finding the power supplied by the sources.

Section 3.2 Amplifiers and Op-Amps

(See also PSpice problems B.7 and B.8.)

3.12 Find μ needed to get $A_v = 20$ in Fig. P3.12 when $R_s = 1\ \text{k}\Omega$, $R_i = 4\ \text{k}\Omega$, $R_o = 100\ \Omega$, and $R_L = 500\ \Omega$. Then calculate the corresponding value of the current gain $A_i = i_{out}/(v_s/R_s)$.

3.13 Find μ needed to get $A_v = 1$ in Fig. P3.12 when $R_s = 1\ \text{k}\Omega$, $R_i = 3\ \text{k}\Omega$, $R_o = 4\ \Omega$, and $R_L = 8\ \Omega$. Then calculate the corresponding value of the current gain $A_i = i_{out}/(v_s/R_s)$.

3.14 Find β needed to get $A_i = 60$ in Fig. P3.14 when $R_s = 300\ \Omega$, $R_i = 200\ \Omega$, $R_o = 5\ \text{k}\Omega$, and $R_L = 1\ \text{k}\Omega$. Then calculate the corresponding value of the voltage gain $A_v = v_{out}/(R_s i_s)$.

3.15 Find β needed to get $A_i = 1$ in Fig. P3.14 when $R_s = 50\ \Omega$, $R_i = 10\ \Omega$, $R_o = 2.5\ \text{k}\Omega$, and $R_L = 0.5\ \text{k}\Omega$. Then calculate the corresponding value of the voltage gain $A_v = v_{out}/(R_s i_s)$.

3.16* Suppose the amplifier in Fig. P3.12 has $R_i = 1\ \text{k}\Omega$ and $R_o = 0$, and a **feedback resistor** R_F is connected from the upper input terminal to the upper output terminal. Find $R_i' = v_{in}/i_{in}$ and $A_v = v_{out}/v_s$ with $R_s = 1\ \text{k}\Omega$, $R_F = 12\ \text{k}\Omega$, and $\mu = 10$.

3.17 Do Problem 3.16 with $R_s = 4\ \text{k}\Omega$, $R_F = 10\ \text{k}\Omega$, and $\mu = 16$.

3.18 Do Problem 3.16 with $R_s = 0.5\ \text{k}\Omega$, $R_F = 15\ \text{k}\Omega$, and $\mu = -14$.

3.19 Let both stages in Fig. P3.19 have $R_i = 5\ \text{k}\Omega$, $\mu = -40$, and $R_o = 200\ \Omega$. What value of R_L yields $A_v = 500$ when $R_s = 3\ \text{k}\Omega$?

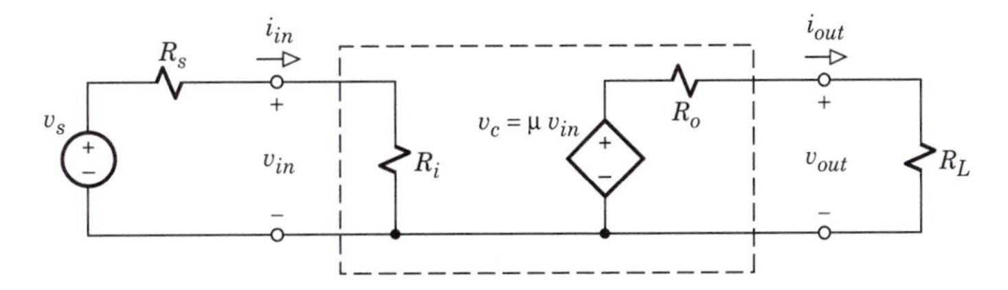

Figure P3.12

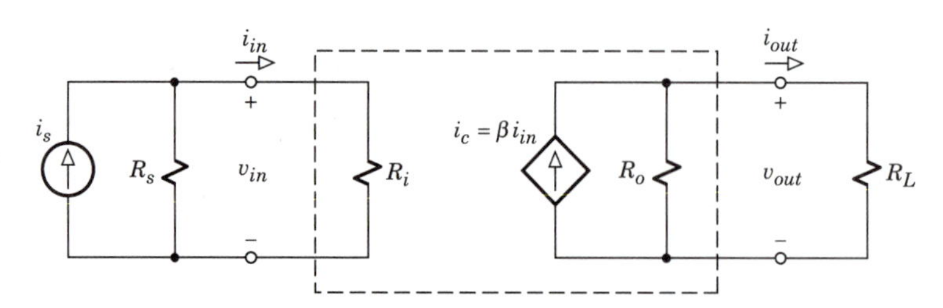

Figure P3.14

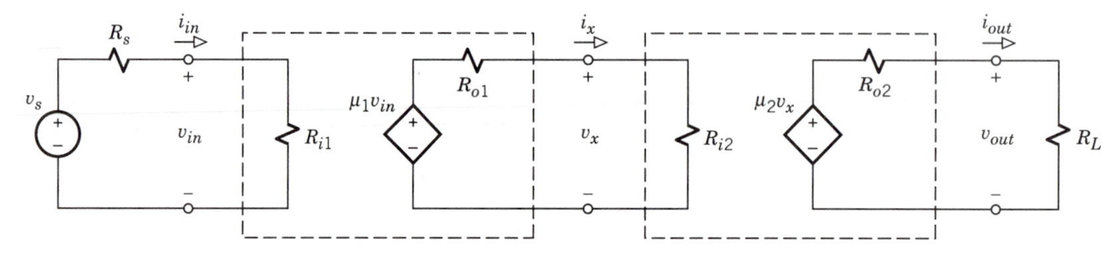

Figure P3.19

3.20 Let the first stage in Fig. P3.19 be replaced by a current amplifier with $R_{i1} = 400\ \Omega$, $\beta = 25$, and $R_{o1} = 1\ k\Omega$. Write a chain expression to find the value of μ_2 that yields $A_v = -300$ when $R_{i2} = 1\ k\Omega$, $R_{o2} = 2\ \Omega$, $R_s = 100\ \Omega$, and $R_L = 8\ \Omega$.

3.21 Let the second stage in Fig. P3.19 be replaced by a current amplifier with $R_{i2} = 500\ \Omega$, $\beta = -50$, and $R_{o2} = 300\ \Omega$. Write a chain expression to find the value of μ_1 that yields $A_v = -80$ when $R_{i1} = 20\ k\Omega$, $R_{o1} = 1\ k\Omega$, $R_s = 5\ k\Omega$, and $R_L = 100\ \Omega$.

3.22 Sketch v_{out} versus v_{in} for a noninverting op-amp circuit with $R_1 = 4\ k\Omega$, $R_F = 76\ k\Omega$, $V_{PS} = 10$ V, and $A = \infty$.

3.23 Consider a noninverting op-amp circuit with $R_1 = 5\ k\Omega$, $R_F = 45\ k\Omega$, $V_{PS} = 30$ V, and $A = \infty$. Sketch v_{out} versus t when v_{in} has the waveform in Fig. P3.23.

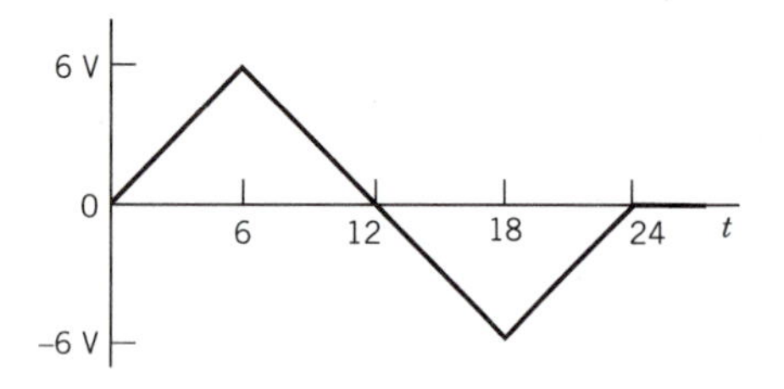

Figure P3.23

3.24* A noninverting op-amp circuit has $A = 10^4$, $v_{out}/v_{in} = 25$, and $R_F = 100\ k\Omega$. Find the corresponding values of v_d/v_{in} and R_1.

3.25 Consider an inverting op-amp circuit with $R_1 = 5\ k\Omega$, $R_F = 15\ k\Omega$, and $V_{PS} = 12$ V. Sketch v_{out} versus t when v_{in} has the waveform in Fig. P3.23.

3.26 Consider an inverting amplifier with $R_1 = 5\ k\Omega$, $R_F = 25\ k\Omega$, and $V_{PS} = 10$ V. Sketch v_{out} versus t when v_{in} has the waveform in Fig. P3.23.

3.27 Plot v_{out} versus v_1 for an inverting summing amplifier with $R_1 = R_2 = 4\ k\Omega$, $R_F = 16\ k\Omega$, $V_{PS} = 12$ V, and $v_2 = -3$ V.

3.28 Plot v_{out} versus v_1 for an inverting summing amplifier with $R_1 = 4\ k\Omega$, $R_2 = 6\ k\Omega$, $R_F = 12\ k\Omega$, $V_{PS} = 30$ V and $v_2 = 3$ V.

3.29 Figure P3.29 shows the external terminals available on a certain integrated circuit. The input voltage is applied to pin 1 and the output is taken at pin 5. List the other pin connections needed to get $v_{out}/v_{in} = -1$, 3, and -3.

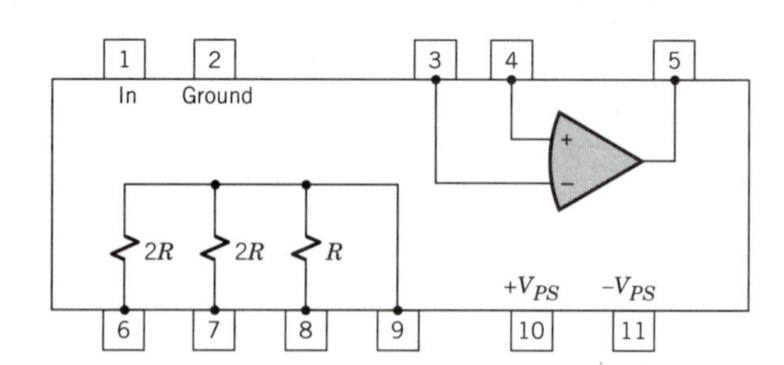

Figure P3.29

3.30 Do Problem 3.29 for $v_{out}/v_{in} = -2$, 2, and 4.

3.31 Design a system with two op-amps and input v_a that produces $v_{out} = -100v_a$. The input must see $R_i = \infty$, and all resistors must be within the range 5 kΩ–50 kΩ. Specify the minimum supply voltage given that $|v_a| \le 180$ mV.

3.32 Design a system with three op-amps and inputs v_a and v_b that produces $v_{out} = -2v_a - 100v_b$. The two inputs must see $R_i = \infty$, and all resistors must be in the range 5 kΩ–50 kΩ. Specify the minimum supply voltage given that $|v_a| \le 1.5$ V and $|v_b| \le 50$ mV.

3.33 Design a system with two op-amps and inputs v_a, v_b, and v_c that produces $v_{out} = 50v_a + 50v_b - 5v_c$. The three inputs must see $R_i = 20$ kΩ, and all resistors must be in the range 5 kΩ–100 kΩ. Specify the minimum supply voltage given that $|v_a| \le 30$ mV, $|v_b| \le 80$ mV, and $|v_c| \le 300$ mV.

3.34 Design a system with two op-amps and inputs v_a, v_b, and v_c that produces $v_{out} = 80v_a - 8v_b - 8v_c$. The three inputs must see $R_i = 10$ kΩ, and all resistors must be in the range 10 kΩ–100 kΩ. Specify the minimum supply voltage given that $|v_a| \le 50$ mV, $|v_b| \le 500$ mV, and $|v_c| \le 250$ mV.

3.35 Design a system with three op-amps and inputs v_a, v_b, and v_c that produces $v_{out} = -40v_a + 20v_b - 4v_c$. The three inputs must see $R_i = 5$ kΩ, and all resistors must be in the range 5 kΩ–50 kΩ. Specify the minimum supply voltage given that $|v_a| \le 0.1$ V, $|v_b| \le 0.2$ V, and $|v_c| \le 1$ V.

3.36 Design a system with five op-amps and inputs v_a, v_b, and v_c that produces $v_{out} = 100v_a + 50v_b - 30v_c$. The three inputs must see $R_i = \infty$, and all resistors must be in the range 5 kΩ–25 kΩ. Specify the minimum supply voltage given that $|v_a| \le 60$ mV, $|v_b| \le 100$ mV, and $|v_c| \le 300$ mV.

3.37 An inverting summing amplifier with three inputs is to act as a **digital-to-analog converter**. Obtain the design equations for R_1, R_2, and R_3 in terms of R_F to satisfy the following input–output table. Then write the general expression for v_{out}.

v_1	0	1	0	1	0	1	0	1
v_2	0	0	1	1	0	0	1	1
v_3	0	0	0	0	1	1	1	1
v_{out}	0	−1	−2	−3	−4	−5	−6	−7

3.38 Figure P3.38 is called a **leap-frog circuit**. (a) Show that $v_1 = -v_{in}$, $v_2 = -K_1v_{in}$, and $v_3 = K_2K_1v_{in}$. Hint: Note that each of the first two stages are inverting summing amplifiers with two inputs. (b) Given that all resistors must fall in the range 1 kΩ–100 kΩ, determine the maximum allowable value of K_1. Then find the corresponding value of R and the maximum value of K_2.

3.39 Figure P3.39 is a **negative-resistance converter**. Assume an ideal op-amp to show that $v_{in}/i_{in} = -R_L$.

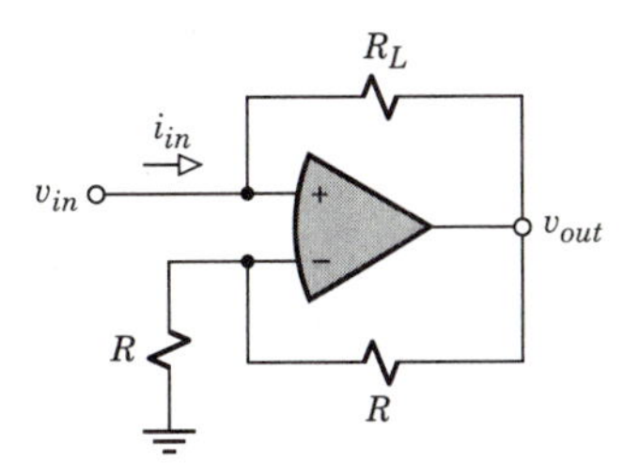

Figure P3.39

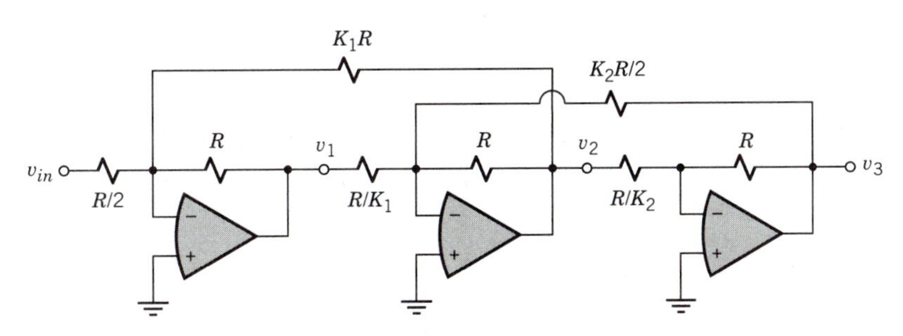

Figure P3.38

3.40 Figure P3.40 is an **inverting/noninverting amplifier**. Assume an ideal op-amp to find v_{out}/v_{in} in terms of α, and show that adjusting the potentiometer changes the output from $v_{out} = -v_{in}$ to $v_{out} = +v_{in}$.

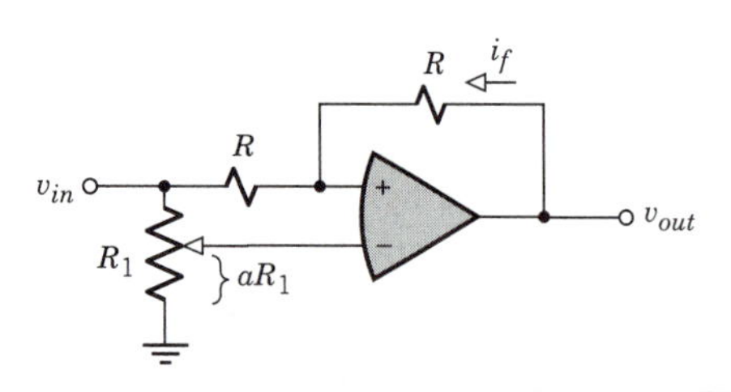

Figure P3.40

3.41 Assume an ideal op-amp to show that Fig. P3.41 is an **inverting current amplifier** with $A_i = i_{out}/i_s = -(R_F + R_1)/R_1$, independent of R_L.

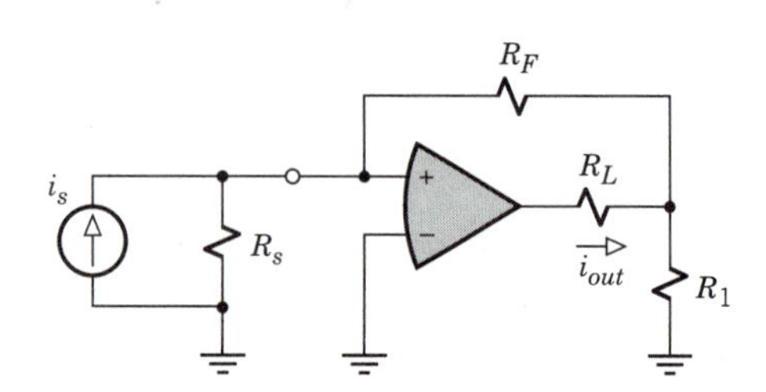

Figure P3.41

3.42 Figure P3.42 is a **current amplifier with a grounded load**. Assume an ideal op-amp to show that $i_{out} = -Ki_{in}$ and find $R_i = v_{in}/i_{in}$. Then obtain an expression for $A_i = i_{out}/i_s$ and simplify for the extreme cases $R_s >> KR_L$ and $R_s >> KR_L$.

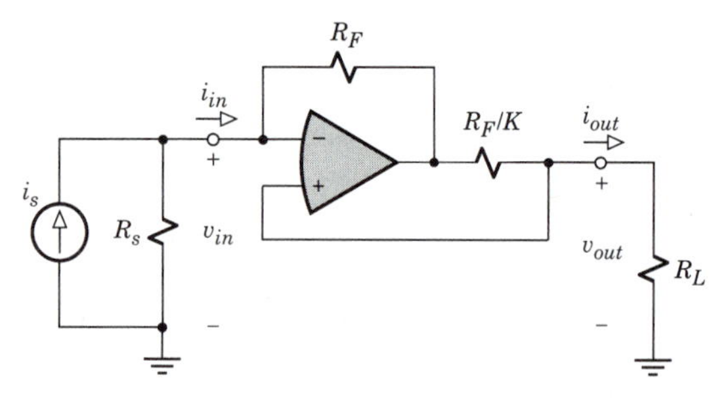

Figure P3.42

3.43 Figure P3.43 is a **noninverting summing amplifier** with two inputs. Assume an ideal op-amp and use superposition to find v_p and v_{out}. Then show that your result reduces to $v_{out} = (R_2v_1 + R_1v_2)/R$ when $R_1 + R_2 = R_F + R$.

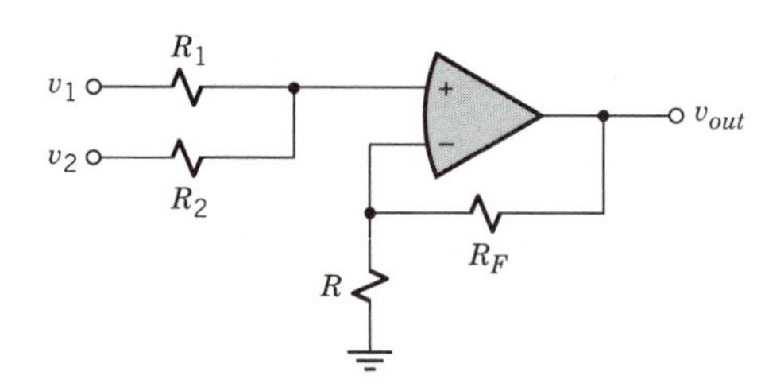

Figure P3.43

3.44 Let a third input v_3 and resistor R_3 be connected at the noninverting op-amp terminal in Fig. P3.43. Assume an ideal op-amp and use superposition to find v_p and v_{out}. Then show that your result reduces to $v_{out} = (R_F + R)(v_1 + v_2 + v_3)/3R$ when $R_1 = R_2 = R_3$.

3.45 Figure P3.45 is a **voltage-to-current converter**, with $i_{out} = gv_{in}$ where g is independent of R_L. (a) Show that $v_a = v_b - v_{in}$ by expressing v_p in terms of v_b. Then confirm that $i_{out} = gv_{in}$ and obtain the expression for g. (b) For a specified value of $|i_{out}|_{max}$, find the limitation on R_L needed to keep the op-amps in their linear region when they both have the supply voltages $\pm V_{PS}$.

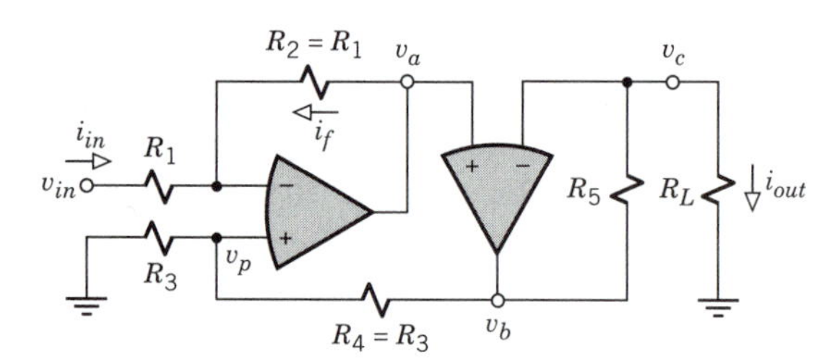

Figure P3.45

3.46* Consider the effect of finite op-amp gain A on the inverting amplifier in Fig. 3.21*a* (p. 107). Find v_{out}/v_{in} and i_{in}/v_{in} using the model from Fig. 3.15 (p. 100).

3.47 Let the circuit in Fig. P3.39 be built with an op-amp having finite gain A. Find v_{out}/v_{in} and v_{in}/i_{in} using the model from Fig. 3.15 (p. 100).

3.48 Consider the effect of finite op-amp gain A on the inverting summing amplifier in Fig. 3.22 (p. 108). Obtain an expression for v_{out} using the model from Fig. 3.15 (p. 100).

3.49 Consider the effect of finite op-amp gain A on the difference amplifier in Fig. 3.24 (p. 110). Taking $R_F = KR_1$ and $R_3 = KR_2$, apply superposition to obtain an expression for v_{out} using the model from Fig. 3.15 (p. 100).

Section 3.3 Internal Op-Amp Resistances

(See also PSpice problems B.9 and B.10.)

3.50* A voltage follower is built with an op-amp having $A = 10^3$, $r_i = 500\ \text{k}\Omega$, and $r_o = 100\ \Omega$. Taking account of the input and output resistances, find i_{in}, v_d, and v_{out} when $v_{in} = 10$ V and $R_L = 1\ \text{k}\Omega$.

3.51 A noninverting amplifier is built with $R_1 = 1\ \text{k}\Omega$, $R_F = 49\ \text{k}\Omega$, and an op-amp having $A = 10^4$, $r_i = 1\ \text{M}\Omega$, and $r_o = 200\ \Omega$. Taking account of the input and output resistances, find i_{in}, v_d, and v_{out} when $v_{in} = 0.2$ V and $R_L = 2\ \text{k}\Omega$.

3.52 The exact input resistance of a voltage follower can be derived by taking $R_1 = \infty$ and $R_F = 0$ in Fig. 3.26*a* (p. 112). Analyze the resulting circuit to obtain

$$R_i = r_i + R_L(Ar_i + r_o)/(R_L + r_o)$$

3.53 The exact output resistance of a voltage follower can be derived by taking $R_1 = \infty$ and $R_F = 0$ in Fig. 3.26*b* (p. 112). Analyze the resulting circuit to obtain

$$R_o = r_o(r_i + R_s)/[(1 + A)r_i + r_o + R_s]$$

Section 3.4 DC Meters and Measurements

3.54* Suppose a voltmeter is connected to the output of the potentiometer circuit in Fig. P3.54. If the meter reads $v_W = 4.0$ V when $R_{AW} = R_{WB} = 300\ \Omega$ and $V_s = 10$ V, then what is the value of the voltmeter's resistance R_i?

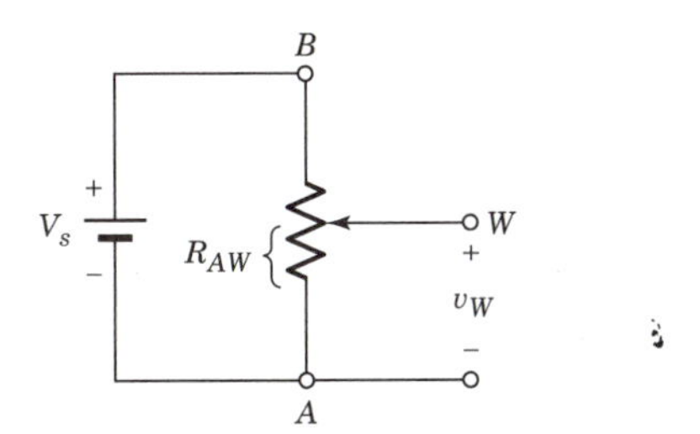

Figure P3.54

3.55 Suppose a voltmeter is connected to the output of the potentiometer circuit in Fig. P3.54. If the meter reads $v_W = 4.0$ V when $R_{AW} = R_{WB} = 500\ \Omega$ and $V_s = 9$ V, then what is the value of the voltmeter's resistance R_i?

3.56 An analog multimeter has a circuit similar to Fig. 3.32 (p. 118). Specify the required shunt and multiplier resistors to get full-scale ranges of 1 mA, 1 A, 0.2 V, and 20 V when the meter movement has $I_{fs} = 1$ mA and $R_m = 20\ \Omega$.

3.57 Do Problem 3.56 with $I_{fs} = 100\ \mu\text{A}$ and $R_m = 2\ \text{k}\Omega$.

3.58 A analog multimeter has a circuit similar to Fig. 3.32 (p. 118). Specify the required shunt and multiplier resistors to get full-scale ranges of 50 mA, 1 A, 0.5 V, and 10 V when the meter movement has $V_{fs} = 50$ mV and $R_m = 5\ \Omega$.

3.59 Do Problem 3.58 with $V_{fs} = 120$ mV and $R_m = 600\ \Omega$.

3.60 The movement in a certain analog multimeter has $I_{fs} = 100\ \mu\text{A}$ and $R_m = 200\ \Omega$. Find the input resistance R_i for the 5-V scale and the 10-mA scale.

3.61 Do Problem 3.60 for a movement with $V_{fs} = 0.1$ V and $R_m = 50\ \Omega$.

3.62* Suppose the voltmeter in Fig. P3.62 has $1/I_{fs} = 5\ \text{k}\Omega/\text{V}$. What is the value of R if the ammeter reads 0.62 mA and the voltmeter reads 6.0 V on the 10-V scale?

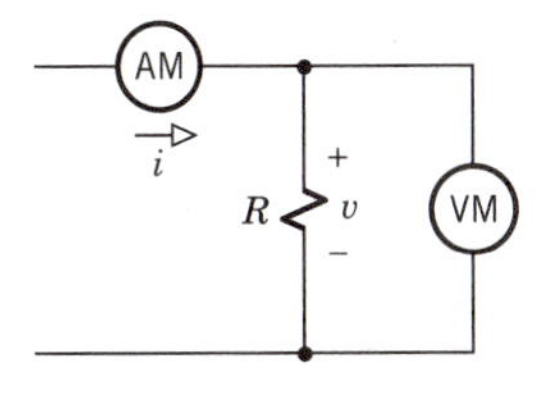

Figure P3.62

3.63 Suppose the voltmeter in Fig. P3.62 has $1/I_{fs} = 20\ \text{k}\Omega/\text{V}$. What is the value of R if the ammeter reads 0.43 mA and the voltmeter reads 1.2 V on the 2-V scale?

3.64 Consider the ohmmeter in Fig. 3.37*a* (p. 123). If $V_s = 3$ V, $V_{fs} = 0.2$ V, and $R_{am} = 4\ \text{k}\Omega$, then what values of R_u correspond to readings of 20%, 50%, and 80% of full scale?

3.65 Do Problem 3.64 with $V_s = 6$ V, $V_{fs} = 30$ mV, and $R_{am} = 10\ \Omega$.

3.66 Consider the Wheatstone bridge in Fig. 3.40 (p. 125). Suppose the values of R_1 and R_2 are known to an accuracy of $\pm 0.5\%$ and the potentiometer dial can be read with an accuracy of $\pm 1\%$. Find the maximum and minimum possible values of R_u to three significant figures when the bridge is balanced with the following nominal settings: $R_1 \approx 400\ \Omega$, $R_2 \approx 10\ \Omega$, $R \approx 200\ \Omega$.

3.67 Do Problem 3.66 with the following nominal settings: $R_1 \approx 100\ \Omega$, $R_2 \approx 6000\ \Omega$, $R \approx 40\ \Omega$.

3.68 Consider the transducer bridge in Fig. 3.41 (p. 126). Suppose the transducer is a strain gauge, and $R_1 = 500\ \Omega$ and $R_2 = 100\ \Omega$. (a) Calculate R_u, ΔR_u, and $\Delta l/l$, given that the bridge is initially balanced with $R = 275\ \Omega$ and rebalanced with $\Delta R = 4.4\ \Omega$ under strain. (b) If the stated values of R and ΔR have an accuracy of $\pm 0.5\%$, then what are minimum and maximum values of $\Delta l/l$?

3.69 Do Problem 3.68 with $R = 1500\ \Omega$ and $\Delta R = 13.5\ \Omega$.

3.70 One method of measuring small resistance changes involves a standard Wheatstone bridge with $R_2 = R_1$ and a DVM that draws negligible current. The bridge is initially balanced with $R = R_u$ and R is not readjusted when R_u changes to some new value $R_u = R + \Delta R$, where $|\Delta R| \ll R$. (a) Use the approximation $(A + a)^{-1} \approx (1 - a/A)/A$ for $|a/A| \ll 1$ to show that the meter reading after R_u changes will be $V_x \approx [R_1 \Delta R/(R_1 + R)^2]V_s$. (b) Suppose R_u represents a resistance thermometer having $R_u = 50(1 + 0.004T)\ \Omega$, so the bridge is balanced at temperature $T = 250°C$ when $R = R_u = 100\ \Omega$. Let the temperature change to $T = 250°C + \Delta T$. Taking $R_1 = 400\ \Omega$ and $V_s = 25$ V, find the maximum and minimum values of $|\Delta T|$ that can be measured by a meter with $V_{fs} = 200$ mV and a three-digit display.

CHAPTER 4

Systematic Analysis Methods

Any resistive circuit can, in principle, be analyzed using the techniques previously developed. However, the systematic methods presented in this chapter are better suited to circuit problems that involve several unknown branch variables. These systematic methods have three important advantages for the analysis of resistive circuits:

- They provide efficient ways of translating circuit diagrams into matrix equations with a minimum number of unknowns;
- The matrix equations are in standard form for solution by hand or with a calculator or computer program;

- Any or all other variables of interest can be obtained from the solution of the matrix equation.

Furthermore, we'll subsequently generalize these methods to incorporate other elements in addition to resistors and sources.

Much of the efficiency of our systematic approach stems from matrix notation. A concise summary of this topic is given in the first section of Appendix A at the back of the book. You may find it helpful to read that section before going on here.

The appendix also covers the use of determinants and Cramer's rule for solving simultaneous equations by hand. If a matrix equation involves symbolic quantities, then it must be solved either by hand or with the help of a computer algebra system such as *Maple* or *Mathematica*. But numerical equations may be solved using a programmed calculator, which is recommended for the exercises and the problems at the end of this chapter when they involve more than two unknowns.

Objectives

After studying this chapter and working the exercises, you should be able to do each of the following:

1. Given a circuit consisting of resistors and independent sources, identify an appropriate set of node voltages and write the matrix node equation by inspection (Section 4.1).
2. Write the node equations for a circuit containing a floating voltage source, with or without series resistance (Section 4.1).
3. Calculate all other branch variables in a circuit from either the node voltages or mesh currents (Sections 4.1 and 4.2).
4. Given a circuit consisting of resistors and independent sources, identify an appropriate set of mesh currents and write the matrix mesh equation by inspection (Section 4.2).
5. Write the mesh equations for a circuit containing an interior current source, with or without parallel resistance (Section 4.2).
6. Carry out systematic analysis of a circuit that includes controlled sources (Sections 4.3).
7. Choose a suitable systematic method to determine the equivalent resistance or Thévenin parameters of a network (Section 4.4).
8. Use node equations to analyze a circuit with ideal op-amps (Section 4.5).†
9. Perform delta-wye transformations to simplify resistive networks for analysis (Section 4.6).†

4.1 NODE ANALYSIS

The complete analysis of a resistive circuit entails finding *all* branch variables. When the circuit has several branches, direct application of Ohm's and

Kirchhoff's laws generally produces an unorganized set of simultaneous equations with many unknowns. Such brute-force analysis has serious drawbacks because formulating the equations is an error-prone process, and solving the equations may be a tedious chore.

This section presents a systematic method involving the concept of *node voltages*. For many circuits, an appropriate set of node voltages constitutes the *minimum* number of unknowns from which all other branch variables can be determined. Notational and computational efficiency will be enhanced by putting the node equations in matrix form, which also facilitates error checking.

We'll temporarily confine our attention to circuits driven by independent sources. Systematic analysis with controlled sources will be postponed to Section 4.3.

Node Voltages

To introduce the concept of node voltages, Fig. 4.1*a* depicts a typical circuit configuration with four nodes. The node at the bottom is marked with the ground symbol (⏚) to indicate that it has been chosen as the **reference node**. The location of the reference node is arbitrary, and the circuit need not actually be grounded.

The voltages v_1, v_2, and v_3 in Fig. 4.1*a* are called **node voltages** because they equal the potential differences between the other nodes and the reference. In general,

> The node voltages of a circuit are the potentials at nonreference nodes with respect to a specified reference node.

Thus, if one lead of a voltmeter is clipped to the reference node, then the node voltages can be measured by probing the other nodes. Of course some of the node voltages may turn out to be negative quantities.

Having chosen and marked a reference, we'll hereafter omit node-voltage polarities and label the nonreference nodes directly with node voltages, as

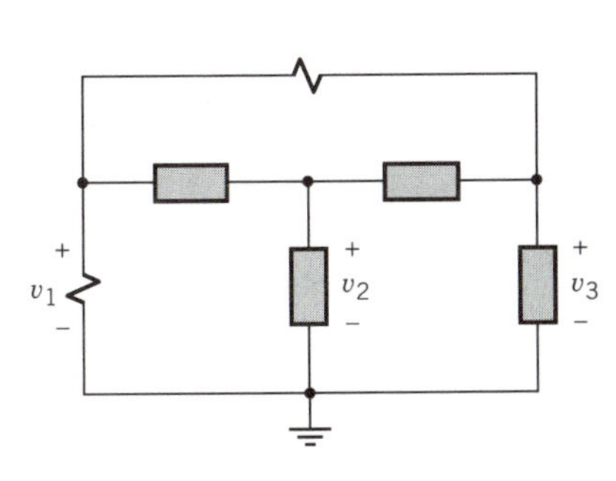

(*a*) Circuit with four nodes and three node voltages

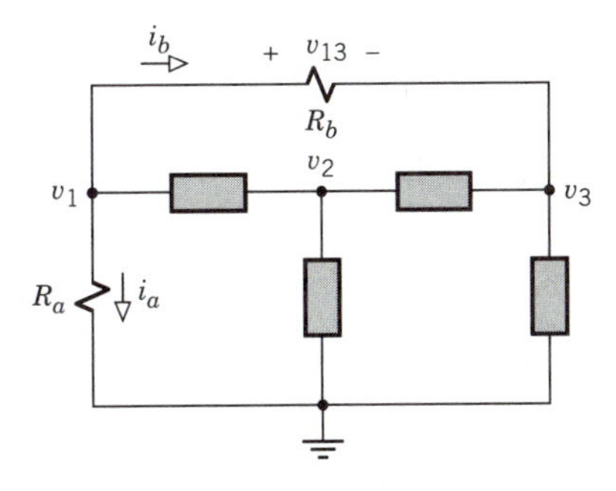

(*b*) Relating other branch variables to node voltages

Figure 4.1

shown in Fig. 4.1*b*. We'll also identify the nonreference nodes by the subscripts of the node voltages, so node 1 is associated with v_1, node 2 with v_2, etc.

Polarity marks may be needed for other voltages such as v_{13} in Fig. 4.1*b*. The double-subscript notation reflects the fact that v_{13} is expressible in terms of v_1 and v_3 since KVL requires $v_1 - v_{13} - v_3 = 0$ so $v_{13} = v_1 - v_3$. In fact, any voltage appearing between two nonreference nodes may be written as

$$v_{nm} = v_n - v_m \tag{4.1}$$

where v_n and v_m are the voltages at nodes n and m, respectively.

We have now seen that every branch voltage in a circuit is either a node voltage or the difference between two node voltages. Furthermore, all unknown branch currents in a resistive circuit can be found from the branch voltages using Ohm's law. For example, resistance R_a connects node 1 directly to the reference node in Fig. 4.1*b*, so the current through that resistance is just

$$i_a = v_1/R_a = G_a v_1$$

where $G_a = 1/R_a$. Similarly, R_b connects node 1 directly to node 3, so the current from node 1 to node 3 through that resistance is

$$i_b = v_{13}/R_b = G_b(v_1 - v_3)$$

We'll have frequent use for expressions like those for i_a and i_b. We'll usually write such expressions with conductances, even though our diagrams are always labeled with resistances.

Extrapolating from the foregoing observations leads us to the important conclusion that

> All branch variables in a circuit can be determined from the values of the node voltages.

This is the crucial property of node voltages for circuit analysis.

As a simple illustration of that property, consider the circuit in Fig. 4.2*a*. Taking the reference at the bottom leaves v_1 as the only *unknown* node voltage because the voltage source connected from the reference to node 2 forces

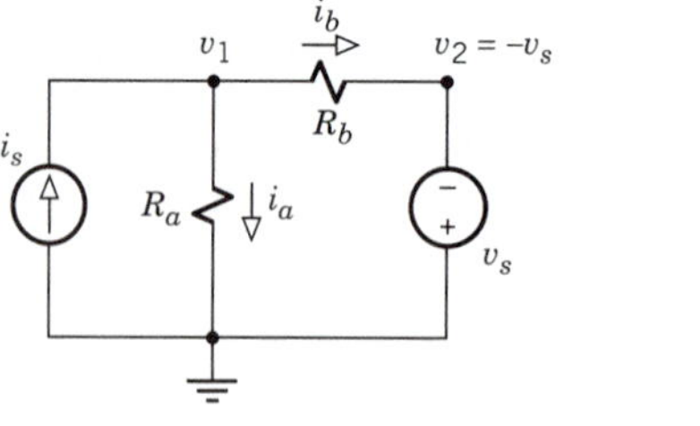

(*a*) Circuit with one unknown node voltage

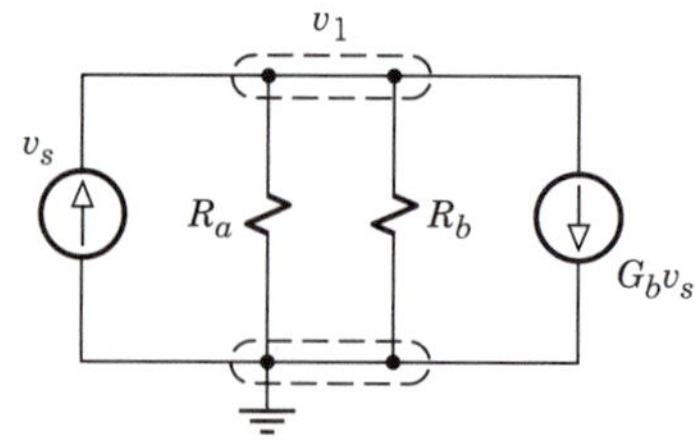

(*b*) Equivalent parallel circuit

Figure 4.2

$v_2 = -v_s$. If we know the value of v_1, then we can calculate the branch currents via

$$i_a = G_a v_1 \qquad i_b = G_b(v_1 - v_2) = G_b(v_1 + v_s)$$

The most convenient way of finding v_1 is to algebraically sum the currents that leave node 1. Accordingly, we write

$$i_a + i_b - i_s = 0$$

or, in terms of v_1,

$$G_a v_1 + G_b(v_1 + v_s) - i_s = 0 \tag{4.2a}$$

Rearrangement then yields the **node equation**

$$(G_a + G_b)v_1 = i_s - G_b v_s \tag{4.2b}$$

which can be solved for v_1 given the element values.

Comparing Eq. (4.2b) with the circuit diagram, we see that the quantity $G_a + G_b$ on the left side equals the sum of conductances connected to node 1. The quantity $i_s - G_b v_s$ on the right side of Eq. (4.2b) also has a physical interpretation, namely, the net equivalent source current *into* node 1. We justify that interpretation by converting the voltage source with series resistance into a current source with parallel resistance, as shown in Fig. 4.2*b*. This diagram clearly reveals that the equivalent source current $G_b v_s$ leaves node 1, so the net source current into node 1 is $i_s - G_b v_s$.

Now consider any circuit with just one unknown node voltage v_1. The node equation for v_1 takes the general form

$$G_{11} v_1 = i_{s1} \tag{4.3}$$

where

G_{11} = Sum of conductances connected to node 1
i_{s1} = Net equivalent source current into node 1

These interpretations allow you to write the node equation easily by *inspection* of the circuit diagram. The only restriction on the inspection technique is that any voltage sources must have one terminal at the reference node so you can determine the corresponding equivalent source current.

Example 4.1 ***Node Analysis with One Unknown***

The circuit in Fig. 4.3 is to be analyzed using a node equation. In particular, we want to find the three indicated branch currents.

Taking the reference node at the bottom, v_1 is the only unknown node voltage since the potentials at the other two nodes are +18 V and −60 V with respect to the reference. We omit the node dot between the 5-Ω and 7-Ω

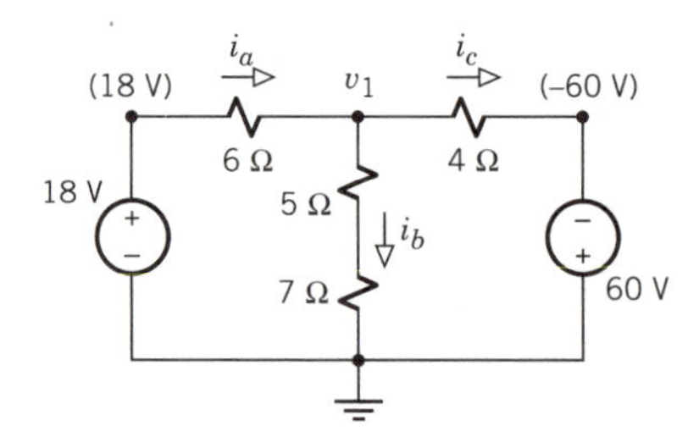

Figure 4.3 Circuit for node analysis with one unknown.

resistors in series because they act like a single resistance of 5 + 7 = 12 Ω between node 1 and the reference node. Hence, the sum of the conductances connected at node 1 is

$$G_{11} = \frac{1}{6} + \frac{1}{5+7} + \frac{1}{4} = \frac{6}{12} = \frac{1}{2}\text{ S}$$

Next, we note that the 18-V source in series with 6 Ω and the 60-V source in series with 4 Ω could be converted into current sources. Without actually doing those conversions, we conclude that the net equivalent source current into node 1 is

$$i_{s1} = \frac{18}{6} + \frac{-60}{4} = -12\text{ A}$$

We thus obtain the node equation

$$\frac{1}{2}v_1 = -12 \quad\Rightarrow\quad v_1 = -24\text{ V}$$

The negative value of v_1 means that node 1 is at a lower potential than the reference node.

Finally, we use v_1 to calculate the branch currents via

$$i_a = \frac{18 - v_1}{6} = 7\text{ A} \qquad i_b = \frac{v_1}{5+7} = -2\text{ A} \qquad i_c = \frac{v_1 - (-60)}{4} = 9\text{ A}$$

These values agree with the KCL relation $i_a = i_b + i_c$.

Exercise 4.1

Use a node equation to find v_1 and i_a in Fig. 4.4.

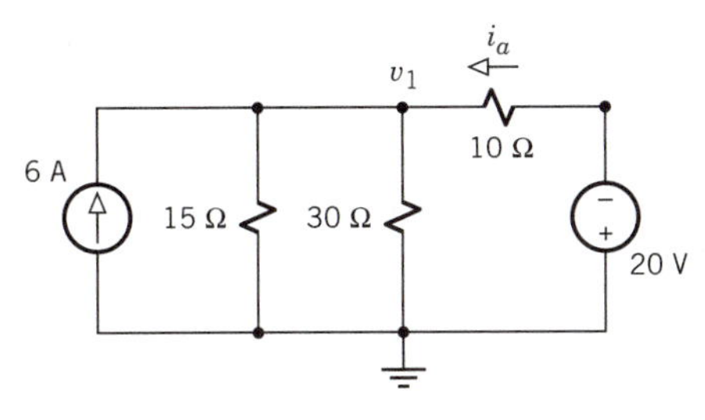

Figure 4.4

Matrix Node Equations

If a circuit contains more than one unknown node voltage, then you must take account of branch currents going from one nonreference node to another. You must also formulate a set of node equations that can be solved for the unknowns. The matrix method developed here expedites these tasks and saves considerable labor.

Before presenting the matrix method, we'll establish the necessary background by analyzing the circuit in Fig. 4.5 from scratch. This circuit has three unknown node voltages, so we'll need three independent equations to evaluate v_1, v_2, and v_3. Since our node-voltage labels automatically satisfy KVL, and since branch currents can be expressed in terms of node voltages via Ohm's law, we turn immediately to KCL. The required independence of the equations is ensured by applying KCL at each nonreference node having an unknown voltage.

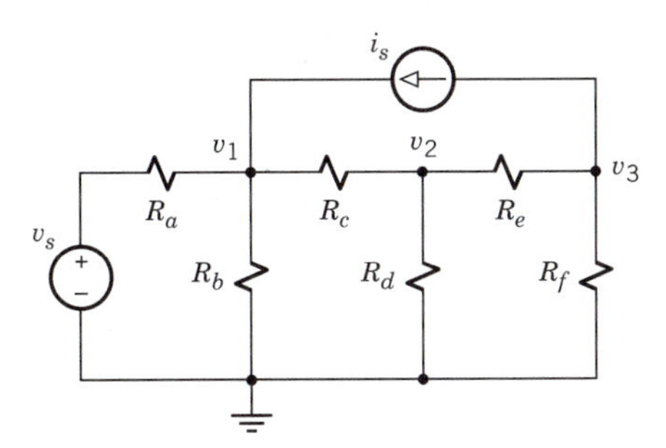

Figure 4.5 Circuit with three unknown node voltages.

Proceeding with this task, we again use conductances to write the sum of the branch currents leaving node 1 as

$$G_a(v_1 - v_s) + G_b v_1 + G_c(v_1 - v_2) - i_s = 0 \tag{4.4a}$$

All terms involving v_1 appear in this sum as positive quantities because the currents leaving node 1 through R_a, R_b, and R_c will be positive when this node is at a higher potential than the other nodes. Furthermore, the number of terms on the left of Eq. (4.4a) equals the number of branches connected at node 1. For the second equation, we sum the three branch currents leaving node 2 to get

$$G_c(v_2 - v_1) + G_d v_2 + G_e(v_2 - v_3) = 0 \tag{4.4b}$$

so all terms involving v_2 are now positive quantities. Similarly, at node 3,

$$G_e(v_3 - v_2) + G_f v_3 + i_s = 0 \tag{4.4c}$$

We have thereby produced a set of three simultaneous equations that can be solved for the unknown node voltages. And after solving for the node voltages, we can evaluate any other branch variables of interest.

But simultaneous solution of our three node equations would be awkward as they stand because we have an unorganized mix of known and unknown quantities. We therefore rearrange them to obtain the standard form:

$$(G_a + G_b + G_c)v_1 - G_c v_2 + 0 \times v_3 = i_s + G_a v_s \tag{4.5a}$$

$$-G_c v_1 + (G_c + G_d + G_e)v_2 - G_e v_3 = 0 \tag{4.5b}$$

$$0 \times v_1 - G_e v_2 + (G_e + G_f)v_3 = -i_s \tag{4.5c}$$

Note carefully that G_c appears twice on the left side of Eq. (4.5a) because the branch current $G_c(v_1 - v_2)$ in Eq. (4.4a) has now been separated into two parts. For the same reason, G_c and G_e appear twice in Eq. (4.5b), and G_e appears twice in Eq. (4.5c).

A comparison of Eqs. (4.5a)–(4.5c) with the circuit diagram in Fig. 4.5 shows that each term has a distinct physical interpretation. In particular, compare Eq. (4.5a) with the connections at node 1. The coefficient $(G_a + G_b + G_c)$ multiplying v_1 equals the sum of the conductances connected to node 1, while the coefficient of v_2 is the conductance connected directly between nodes 1 and 2. But no conductance connects node 1 to node 3, so the coefficient of v_3 equals zero in Eq. (4.5a). Furthermore, the quantity $i_s + G_a v_s$ on the right of Eq. (4.5a) equals the net equivalent source current *into* node 1. All the terms in Eqs. (4.5b) and (4.5c) have corresponding interpretations from the connections at nodes 2 and 3.

For a more convenient display of Eqs. (4.5a)–(4.5c), we put them together in the **matrix node equation**

$$\begin{bmatrix} G_a + G_b + G_c & -G_c & 0 \\ -G_c & G_c + G_d + G_e & -G_e \\ 0 & -G_c & G_e + G_f \end{bmatrix} \begin{bmatrix} v_1 \\ v_2 \\ v_3 \end{bmatrix} = \begin{bmatrix} G_a v_s + i_s \\ 0 \\ -i_s \end{bmatrix} \tag{4.6}$$

The first row of the matrix equation represents the KCL equation at node 1, and likewise for the second and third rows. Hence, the matrix quantities have the same interpretation as before.

Now consider any circuit consisting of resistors and independent sources and having N unknown node voltages. By generalization of Eq. (4.6), the unknown node voltages can be found from a matrix equation having the form

$$[G]\,[v] = [i_s] \tag{4.7}$$

The vectors $[v]$ and $[i_s]$ stand for the $N \times 1$ column arrays

$$[v] = \begin{bmatrix} v_1 \\ v_2 \\ \vdots \\ v_N \end{bmatrix} \qquad [i_s] = \begin{bmatrix} i_{s1} \\ i_{s2} \\ \vdots \\ i_{sN} \end{bmatrix}$$

where

v_n = Unknown node voltage at node n
i_{sn} = Net equivalent source current into node n

The elements of the equivalent source-current vector $[i_s]$ may be positive, negative, or zero, depending upon the circuit configuration.

The **conductance matrix** $[G]$ in Eq. (4.7) stands for the $N \times N$ square array

$$[G] = \begin{bmatrix} G_{11} & -G_{12} & \cdots & -G_{1N} \\ -G_{21} & G_{22} & \cdots & -G_{2N} \\ \vdots & \vdots & & \vdots \\ -G_{N1} & -G_{N2} & \cdots & G_{NN} \end{bmatrix} \tag{4.8}$$

where

G_{nn} = Sum of conductances connected to node n
$G_{nm} = G_{mn}$ = Equivalent conductance directly connecting nodes n and m

The quantity G_{nm} equals G_{mn} because the subscript sequence does not matter when we calculate the equivalent conductance directly connecting any two nonreference nodes. Further study of Eq. (4.8) reveals that:

- The conductance matrix has symmetry about the main diagonal (upper left to lower right) since $G_{nm} = G_{mn}$;
- Each element G_{nn} on the main diagonal is positive, whereas each off-diagonal element is negative unless it happens to be zero;
- Any off-diagonal quantity G_{nm} also appears as part of the diagonal terms G_{nn} and G_{mm} because G_{nm} connects to nodes n and m so it must be part of the connections to node n and to node m.

These properties serve as simple tests for error checking.

Drawing on the physical interpretations of the conductance matrix and the source-current vector, we can determine their element values by *inspection* of a circuit diagram. Thus, the systematic matrix method of node analysis goes as follows:

1. Mark the reference node, indentify any known node voltages fixed by voltage sources, and label the remaining unknown node voltages. Also label all other variables of interest.
2. Fill in the conductance matrix $[G]$ using the properties that G_{nn} equals the sum of conductances connected to node n while G_{nm} equals the equivalent conductance directly connecting nodes n and m.
3. Check $[G]$ to make sure that $G_{mn} = G_{nm}$ and that these off-diagonal elements carry a negative sign. Also check that each off-diagonal quantity G_{nm} appears as part of the diagonal terms G_{nn} in the same row and G_{mm} in the same column.
4. Fill in the source-current vector $[i_s]$ using the property that i_{sn} equals the net equivalent source current into node n. If resistance R_s connects node n to a node fixed at voltage $\pm v_s$ by a voltage source, then i_{sn} includes the equivalent source current $\pm G_s v_s$.
5. Solve the resulting matrix equation $[G]\,[v] = [i_s]$ for the unknown node voltages.

All other quantities of interest may then be calculated from the node voltages.

But before you start a node analysis, you should be sure of the *minimum number* of unknown node voltages needed. Since voltage sources establish known node voltages, whereas current sources have no effect on the number of unknowns, you can

> Determine the minimum number of unknown node voltages by suppressing all sources, counting the remaining nodes, and subtracting one for the reference node.

The location of the reference node may also be an important consideration because our inspection method for node analysis requires that all voltage sources have a common terminal, which must be taken as the reference to find the equivalent source currents. We'll revisit this limitation subsequently.

Example 4.2 *Matrix Node Analysis with Two Unknowns*

The circuit in Fig. 4.6*a* is to be analyzed to find the power supplied by each of the three sources. We should be able to do this problem with two unknown node voltages because source suppression leaves the three nodes in Fig. 4.6*b*. However, the two voltage sources in Fig. 4.6*a* do not have a common terminal to take as the reference node. Recalling that the order of elements in series is not important, we redraw the circuit per Fig. 4.6*c* where the positions of the 30-V source and 6-Ω resistor have been interchanged.

Step 1: The reference node and the known and unknown node voltages have been labeled on Fig. 4.6*c*. The branch variables v_{12}, i_a, and i_b also have been indicated for the eventual purpose of power calculations.

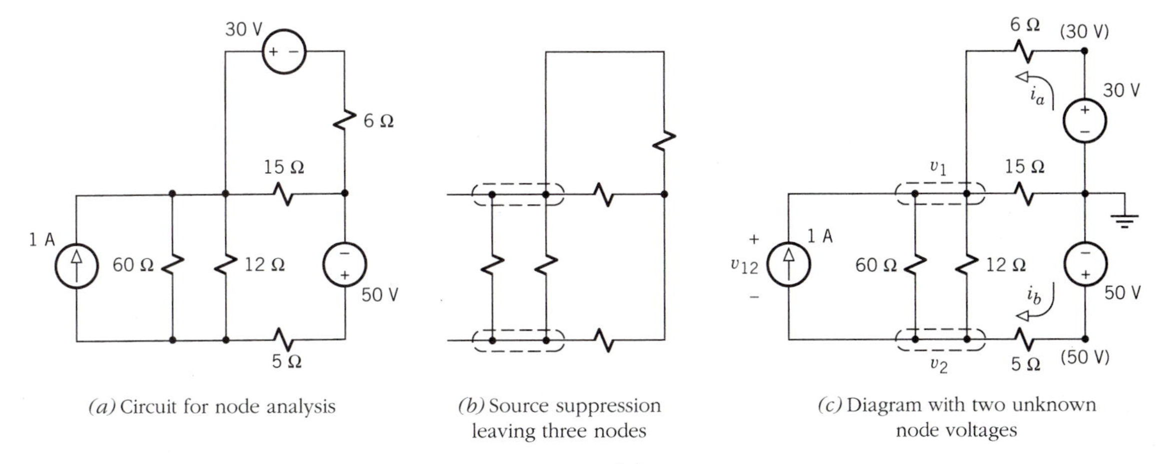

(*a*) Circuit for node analysis

(*b*) Source suppression leaving three nodes

(*c*) Diagram with two unknown node voltages

Figure 4.6

Step 2: By inspection of the diagram, the sum of the conductances connected to nodes 1 and 2 are

$$G_{11} = \frac{1}{6} + \frac{1}{15} + \frac{1}{12} + \frac{1}{60} \qquad G_{22} = \frac{1}{5} + \frac{1}{12} + \frac{1}{60}$$

The equivalent conductance connecting nodes 1 and 2 is

$$G_{12} = G_{21} = \frac{1}{12} + \frac{1}{60}$$

Thus,

$$[G] = \begin{bmatrix} \frac{1}{6} + \frac{1}{15} + \frac{1}{12} + \frac{1}{60} & -\frac{1}{12} - \frac{1}{60} \\ -\frac{1}{12} - \frac{1}{60} & \frac{1}{5} + \frac{1}{12} + \frac{1}{60} \end{bmatrix} = \begin{bmatrix} \frac{20}{60} & -\frac{6}{60} \\ -\frac{6}{60} & \frac{18}{60} \end{bmatrix}$$

Step 3: Our conductance matrix has the stated properties, and the off-diagonal term appears as part of G_{11} and G_{22} on the main diagonal.

Step 4: For the elements of $[i_s]$, we note that the 1-A source current goes out of node 2 and into node 1. Additionally, the upper voltage source provides an equivalent source current 30 V/6 Ω into node 1, while the lower voltage source provides 50 V/5 Ω into node 2. Thus,

$$[i_s] = \begin{bmatrix} \frac{30}{6} + 1 \\ \frac{50}{5} - 1 \end{bmatrix} = \begin{bmatrix} 6 \\ 9 \end{bmatrix}$$

Step 5: Using $[G]$ and $[i_s]$ as we have found them, the matrix node equation could be solved directly with a programmed calculator. For hand solution, however, we prefer to clear the fractions in $[G]$ by multiplying every element by 30 — so each element of $[i_s]$ must also be multiplied by 30. The matrix equation then becomes

$$\begin{bmatrix} 10 & -3 \\ -3 & 9 \end{bmatrix} \begin{bmatrix} v_1 \\ v_2 \end{bmatrix} = \begin{bmatrix} 180 \\ 270 \end{bmatrix}$$

Cramer's rule for the evaluation of v_1 and v_2 involves the determinants

$$\Delta = \begin{vmatrix} 10 & -3 \\ -3 & 9 \end{vmatrix} = 81 \qquad \Delta_1 = \begin{vmatrix} 180 & -3 \\ 270 & 9 \end{vmatrix} = 2430 \qquad \Delta_2 = \begin{vmatrix} 10 & 180 \\ -3 & 270 \end{vmatrix} = 3240$$

from which

$$v_1 = \Delta_1/\Delta = 30 \text{ V} \qquad v_2 = \Delta_2/\Delta = 40 \text{ V}$$

Now we can calculate the branch variables

$$v_{12} = v_1 - v_2 = -10 \text{ V} \quad i_a = \frac{30 - v_1}{6} = 0 \quad i_b = \frac{50 - v_2}{5} = 2 \text{ A}$$

Hence, the 50-V source supplies 50 V $\times\ i_b$ = 100 W, while the 30-V source supplies no power since 30 V $\times\ i_a = 0$, and the 1-A source actually absorbs power since $v_{12} \times 1$ A $= -10$ W.

Example 4.3 ***Matrix Node Analysis with Three Unknowns***

Problem: Find v_1, v_2, and v_3 by node analysis of Fig. 4.7.

Solution: Suppressing all sources confirms that we need three unknown node voltages, and the reference node indicated on the diagram is at one

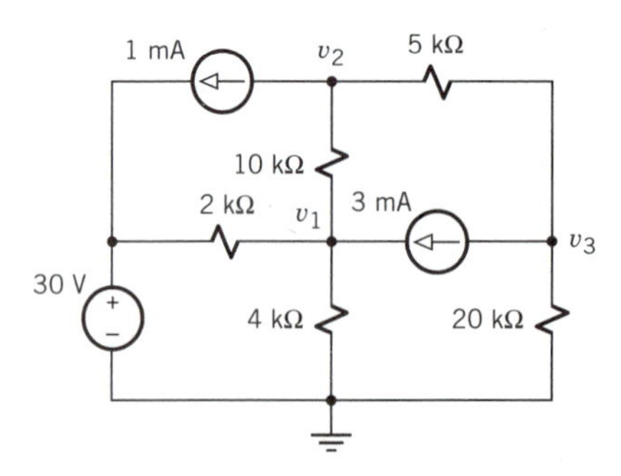

Figure 4.7 Circuit for node analysis with three unknowns.

terminal of the only voltage source. Since all resistances are in kilohms and both source currents are in milliamps, we'll work with a consistent set of units by taking conductances in millisiemens (the reciprocal of kilohms).

The conductance matrix follows by inspection, noting that there is no resistive path between nodes 1 and 3. Thus,

$$[G] = \begin{bmatrix} \frac{1}{4}+\frac{1}{2}+\frac{1}{10} & -\frac{1}{10} & 0 \\ -\frac{1}{10} & \frac{1}{10}+\frac{1}{5} & -\frac{1}{5} \\ 0 & -\frac{1}{5} & \frac{1}{5}+\frac{1}{20} \end{bmatrix} = \begin{bmatrix} 0.85 & -0.1 & 0 \\ -0.1 & 0.3 & -0.2 \\ 0 & 0 & 0.25 \end{bmatrix}$$

which passes our error checks. For the equivalent source-current vector, we observe that the 30-V source connects to node 1 through 2 kΩ. Thus,

$$[i_s] = \begin{bmatrix} \frac{30}{2}+3 \\ -1 \\ -3 \end{bmatrix} = \begin{bmatrix} 18 \\ -1 \\ -3 \end{bmatrix}$$

We thereby obtain the matrix node equation

$$\begin{bmatrix} 0.85 & -0.1 & 0 \\ -0.1 & 0.3 & -0.2 \\ 0 & -0.2 & 0.25 \end{bmatrix} \begin{bmatrix} v_1 \\ v_2 \\ v_3 \end{bmatrix} = \begin{bmatrix} 18 \\ -1 \\ -3 \end{bmatrix}$$

from which

$$v_1 = 20\ \text{V} \qquad v_2 = -10\ \text{V} \qquad v_3 = -20\ \text{V}$$

The fact that the 30-V source supplies equivalent source current into node 1 (but not node 2) can be justified two ways. One way is to write out the KCL equations

$$\textit{Node 1:}\ \frac{v_1 - 30}{2} + \frac{v_1}{4} + \frac{v_1 - v_2}{10} - 3 = 0$$

$$\textit{Node 2:}\ \frac{v_2 - v_1}{10} + \frac{v_2 - v_3}{5} + 1 = 0$$

Rearrangement then yields the first two rows of the matrix equation with $i_{s1} = (30/2) + 3$ and $i_{s2} = -1$.

A second, more pictorial justification comes from the "source node splitting" shown in Fig. 4.8*a*. Here we have put a second 30-V source in parallel with the first to split the 30-V node into two nodes, leaving the rest of the

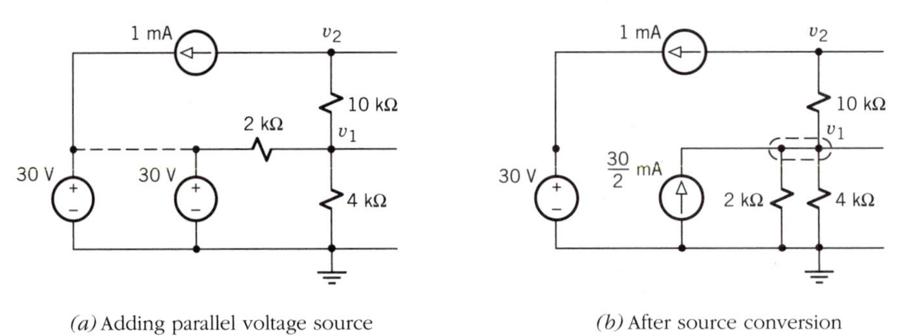

(*a*) Adding parallel voltage source (*b*) After source conversion

Figure 4.8 Source node splitting.

circuit unaffected. Source conversion then yields Fig. 4.8*b*, where an equivalent source current of 30 V/2 kΩ goes into node 1, whereas the 1-mA source clearly takes current out of node 2.

Exercise 4.2

Determine by inspection of Fig. 4.9 the conductance and current values for the matrix node equation. Then solve for v_1 and v_2. (Hint: The 20-V source supplies equivalent source current for both nodes 1 and 2.)

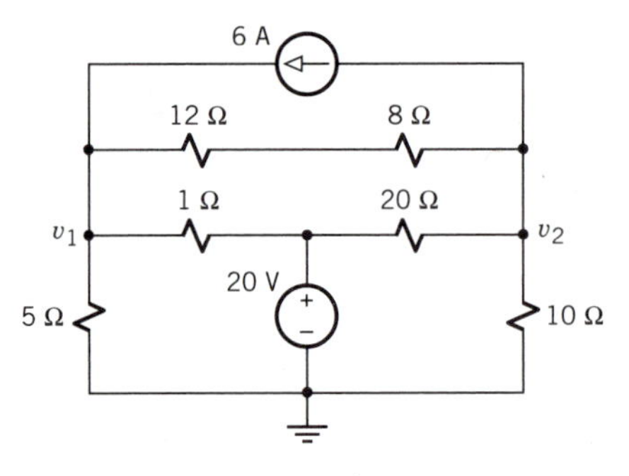

Figure 4.9

Exercise 4.3

The conductance matrix for a certain circuit includes $G_x = 1/R_x$ in G_{11}. What do you conclude about the location of R_x when:
(**a**) G_x does not appear anywhere else in $[G]$ or $[i_s]$?
(**b**) G_x also appears in i_{s1} but nowhere else in $[G]$?

Exercise 4.4

Suppose Fig. 4.7 is modified by inserting resistance R in series with the 3-mA source. Explain why R should *not* be included in the conductance matrix, even though it is connected to either node 1 or node 3.

Floating Voltage Sources

Consider any circuit containing two or more voltage sources without a common terminal. If the reference node is taken at a terminal of one source, then the other sources lack direct connections to the reference and are said to be **floating**. Node analysis with floating voltage sources requires extra care, and there are two different cases to examine.

Case I: Floating Sources with Series Resistance Figure 4.10*a* shows a voltage source with series resistance connected between two nonreference nodes, n and m. The intervening node, marked X, may be ignored because the voltage at X is $v_x = v_n + v_s$ — so v_x will be known immediately after finding v_n. We are therefore free to perform a source conversion giving the parallel structure in Fig. 4.10*b*. This figure reveals that the equivalent source current v_s/R_s goes out of node n into node m. Additionally, we now see that the conductance $1/R_s$ affects both nodes, a fact that was not evident before the conversion.

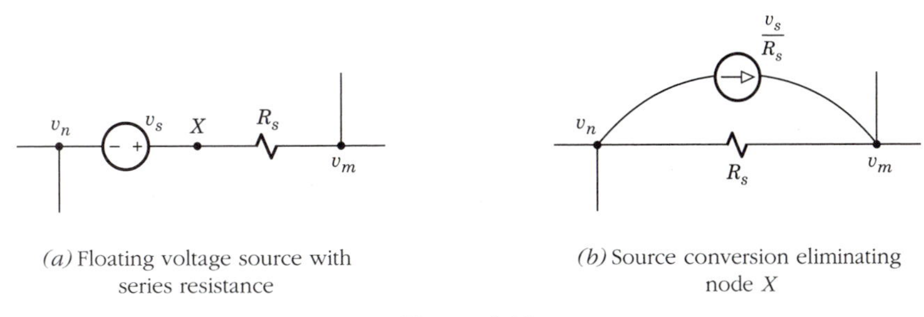

(*a*) Floating voltage source with series resistance

(*b*) Source conversion eliminating node X

Figure 4.10

When a circuit includes a floating source like Fig. 4.10*a*, you can carry out node analysis by inspection after redrawing the diagram per Fig. 4.10*b*. But keep in mind that the source conversion is only equivalent at the terminal nodes n and m. Thus, to find the actual branch current through R_s, for instance, you must calculate it from the original configuration in Fig. 4.10*a* using the values of v_m and $v_x = v_n + v_s$.

Also keep in mind that careful selection of the reference node eliminates unnecessary source conversions. Accordingly, we conclude that

> The reference node should be chosen such that it ties to one terminal of as many voltage sources as possible.

Any remaining floating voltage sources with series resistance can then be converted to equivalent current sources.

Example 4.4 *Matrix Node Analysis with Source Conversion*

Node analysis is to be used to find the currents i_a and i_b in Fig. 4.11*a*. Source suppression reduces the circuit to three nodes, so we can do the analysis with two unknown node voltages.

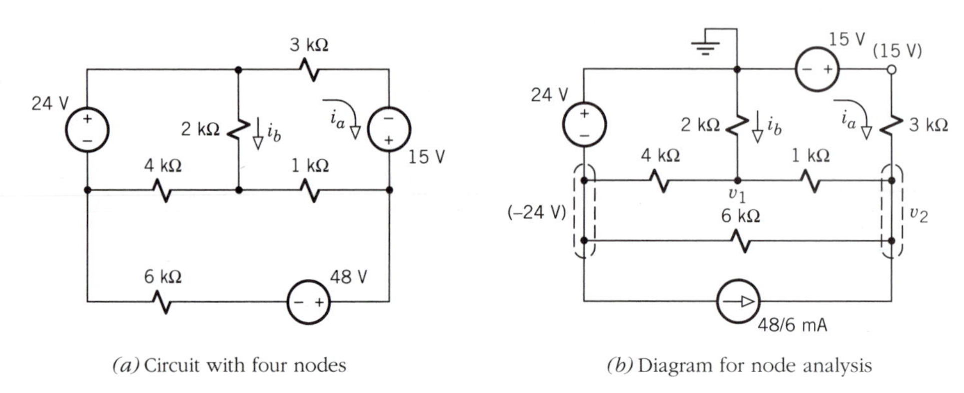

(a) Circuit with four nodes

(b) Diagram for node analysis

Figure 4.11

At first glance, the obvious choice for the reference node is at the far right, which is a common terminal for the 15-V and 48-V sources. However, that selection would leave the 24-V source floating *without* series resistance for source conversion. We therefore take the reference at the upper node, since both currents in question leave to this node.

Figure 4.11*b* shows the diagram redrawn for node analysis with the following modifications:

- The positions of the 15-V source and the 3-kΩ resistor have been interchanged to tie the source to the reference node.
- The floating 48-V source with series resistance has been converted into a current source with parallel resistance.

Note that the 15-V source has not been converted because it is not floating, and we want to preserve the identity of i_a. Also note that the 24-V source now connects to node 2 via the 6-kΩ resistor as well as to node 1 via the 4-kΩ resistor.

Working with consistent units, inspection of Fig. 4.11*b* yields

$$[G] = \begin{bmatrix} \frac{1}{4}+\frac{1}{2}+\frac{1}{1} & -\frac{1}{1} \\ -\frac{1}{1} & \frac{1}{3}+\frac{1}{1}+\frac{1}{6} \end{bmatrix} = \frac{1}{12}\begin{bmatrix} 21 & -12 \\ -12 & 18 \end{bmatrix}$$

$$[i_s] = \begin{bmatrix} \frac{-24}{4} \\ \frac{15}{3}+\frac{-24}{6}+\frac{48}{6} \end{bmatrix} = \frac{1}{12}\begin{bmatrix} -72 \\ 108 \end{bmatrix}$$

Thus, after canceling the common factor 1/12, we have

$$\begin{bmatrix} 21 & -12 \\ -12 & 18 \end{bmatrix} \begin{bmatrix} v_1 \\ v_2 \end{bmatrix} = \begin{bmatrix} -72 \\ 108 \end{bmatrix}$$

from which

$$v_1 = 0 \text{ V} \qquad v_2 = 6 \text{ V}$$

Finally, the currents in question are

$$i_a = \frac{15 - v_2}{3 \text{ k}\Omega} = 3 \text{ mA} \qquad i_b = \frac{-v_1}{2 \text{ k}\Omega} = 0 \text{ mA}$$

Exercise 4.5

The sources in Fig. 4.12 have arbitrary values. Write the $[G]$ and $[i_s]$ matrices needed to find v_1, v_2, and v_3.

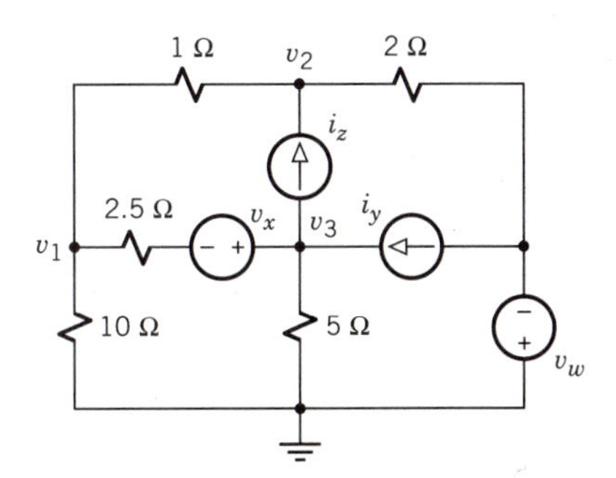

Figure 4.12

Case II: Floating Ideal Sources When a circuit has been modeled with *ideal* voltage sources, the absence of series resistance occasionally precludes the source conversions needed to do node analysis by inspection. We'll investigate two ways of handling such cases.

Consider the circuit in Fig. 4.13, where v_y is a floating ideal source and v_1 and v_2 are unknown node voltages. But v_3 is not an *independent* unknown because the floating source forces $v_3 = v_2 + v_y$. The fact that there are just two unknown node voltages agrees with our source-suppression test. However, we cannot write node equations directly because we cannot convert the floating source and we do not know the current i_y through the floating source.

One approach to this problem is to treat i_y like a *fictitious source current* and write three individual node equations with $v_3 = v_2 + v_y$. Summing conductances in the usual manner gives

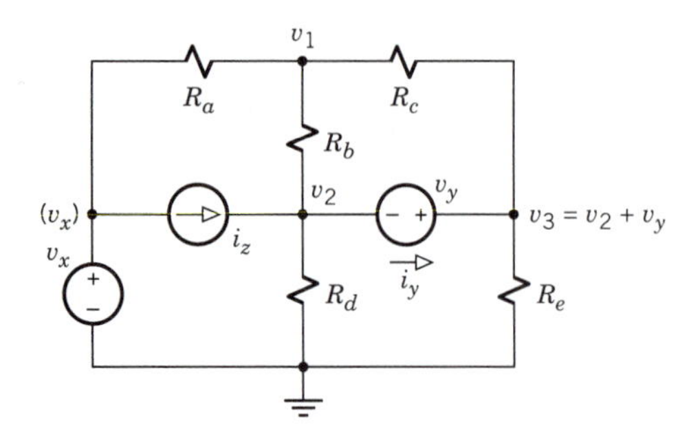

Figure 4.13 Circuit with floating ideal voltage source.

$$(G_a + G_b + G_c)v_1 - G_b v_2 - G_c(v_2 + v_y) = G_a v_x \tag{4.9a}$$

$$-G_b v_1 + (G_b + G_d)v_2 - 0(v_2 + v_y) = i_z - i_y \tag{4.9b}$$

$$-G_c v_1 - 0 \times v_2 + (G_c + G_e)(v_2 + v_y) = i_y \tag{4.9c}$$

We then have three equations with three unknowns, namely, v_1, v_2, and i_y. But i_y can be eliminated immediately by adding Eqs. (4.9b) and (4.9c) to get

$$-(G_b + G_c)v_1 + (G_b + G_d)v_2 + (G_c + G_e)(v_2 + v_y) = i_z \tag{4.9d}$$

Equations (4.9a) and (4.9d) now constitute two independent equations with unknowns v_1 and v_2.

Alternatively, instead of working with i_y as an additional unknown, suppose we enclose both terminals of the floating source within the *supernode* shown in Fig. 4.14. The unknown current through the voltage source stays entirely inside the supernode, and summing the currents leaving the supernode yields

$$G_b(v_2 - v_1) + G_d v_2 - i_z + G_c(v_2 + v_y - v_1) + G_e(v_2 + v_y) = 0 \tag{4.10}$$

Closer examination reveals that Eq. (4.10) is identical to Eq. (4.9d), so the supernode equation is equivalent to combining the two equations at the terminal nodes of the floating source. Thus, we can get the pair of equations for

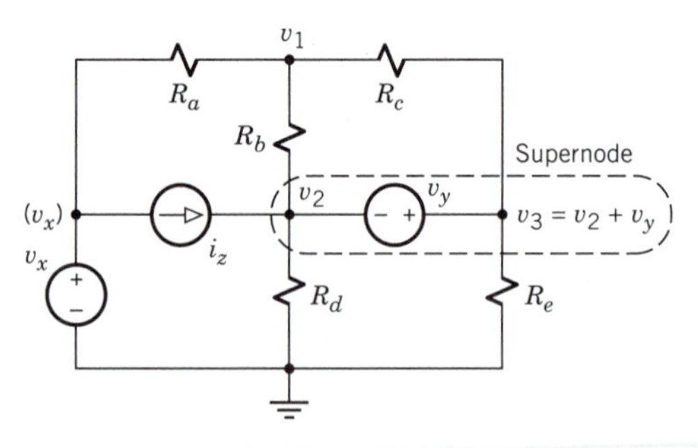

Figure 4.14 Supernode around floating ideal source.

v_1 and v_2 from the supernode equation together with the equation at node 1, as previously given in Eq. (4.9a).

Example 4.5 *Node Analysis with a Supernode*

Node analysis is to be used to determine the values of v_1 and v_2 in Fig. 4.15. We'll do this problem by enclosing both terminals of the floating source in a supernode, as shown in the figure.

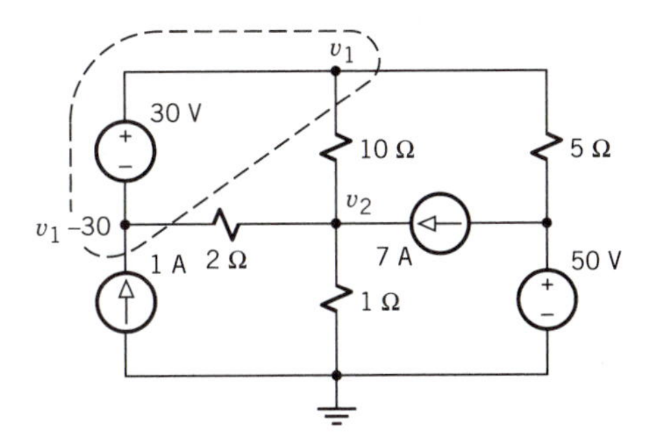

Figure 4.15 Circuit for analysis with a supernode.

The voltage at the lower terminal of the floating source is $v_1 - 30$, so summing the currents leaving the supernode gives the supernode equation

$$\frac{(v_1 - 30) - v_2}{2} - 1 + \frac{v_1 - v_2}{10} + \frac{v_1 - 50}{5} = 0$$

There are four terms on the left of this equation because the supernode cuts across four branches. Next, we write a conventional equation for the currents leaving node 2 to get

$$\frac{v_2 - v_1}{10} + \frac{v_2 - (v_1 - 30)}{2} + \frac{v_2}{1} - 7 = 0$$

Rearrangement then yields a pair of equations in standard form as

$$\begin{aligned} 0.8v_1 - 0.6v_2 &= 26 \\ -0.6v_1 + 1.6v_2 &= -8 \end{aligned}$$

from which

$$v_1 = 40 \text{ V} \qquad v_2 = 10 \text{ V}$$

Exercise 4.6

Find v_1 and v_2 in Fig. 4.16 by writing a single supernode equation.

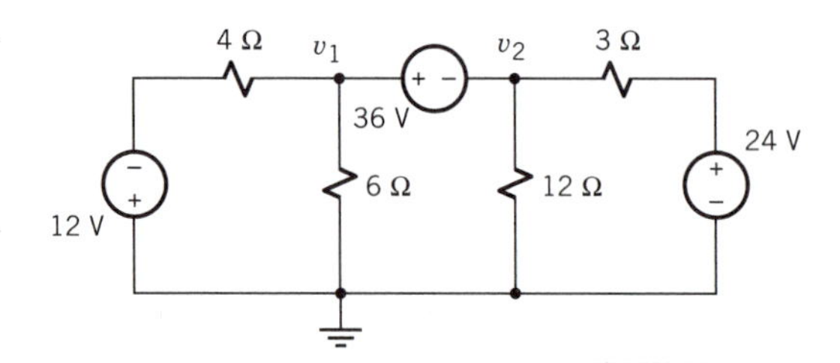

Figure 4.16

4.2 MESH ANALYSIS

This section presents our other systematic analysis method, which is the *dual* of node analysis and involves the concept of *mesh currents*. For some circuits, the use of mesh currents rather than node voltages yields the minimum number of unknown variables. And even when the number of unknown mesh currents equals the number of unknown node voltages, mesh currents may be more appropriate variables in certain problems.

Mesh Currents

Unlike node analysis, mesh analysis is restricted to the category called **planar circuits**. The defining feature of this category is that

> The diagram of a planar circuit can be drawn on a plane without "hop-overs" for crossing branches.

Figure 4.17*a* illustrates a *nonplanar* circuit, whereas Fig. 4.17*b* is actually a planar circuit since the diagram could be redrawn without the hop-over. Most practical circuits satisfy the planar condition, so the restriction on mesh analysis is not severe.

The absense of crossing branches ensures that any planar circuit consists of a number of unique **meshes**, defined by the property that

> A mesh is a closed current path that contains no closed paths within it.

In other words, each mesh constitutes the border of a "hole" or "window pane" in the circuit diagram.

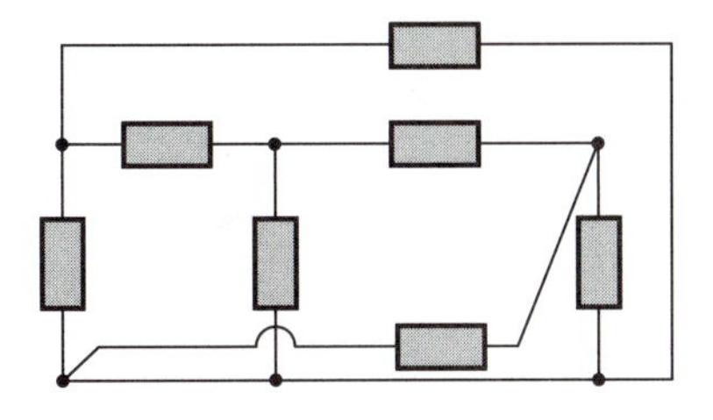

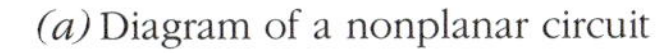

(a) Diagram of a nonplanar circuit

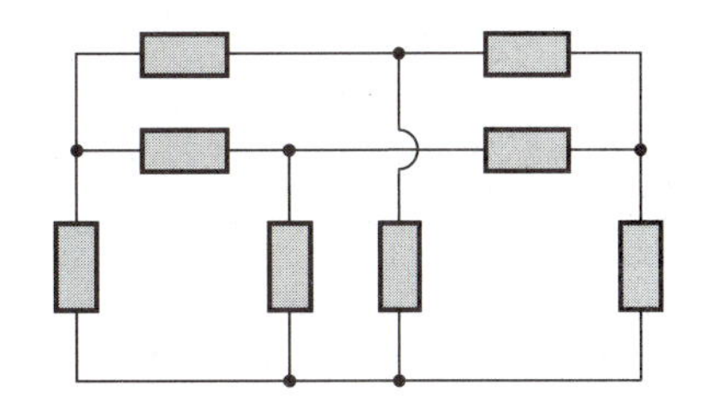

(b) Diagram of a planar circuit that could be redrawn without a hopover

Figure 4.17

Figure 4.18*a* depicts a circuit configuration comprising three meshes and labeled with three currents, i_1, i_2, and i_3. Each of these currents goes through a branch on the perimeter of one of the meshes. But the three interior branches are labeled differently because they form the boundaries between meshes. From KCL, the current through these interior branches are $i_{12} = i_1 - i_2$, $i_{23} = i_2 - i_3$, and $i_{13} = i_{12} + i_{13} = i_1 - i_3$.

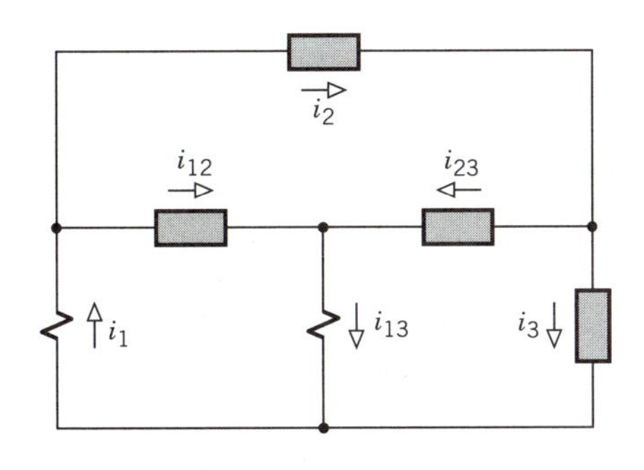

(a) Circuit with three meshes

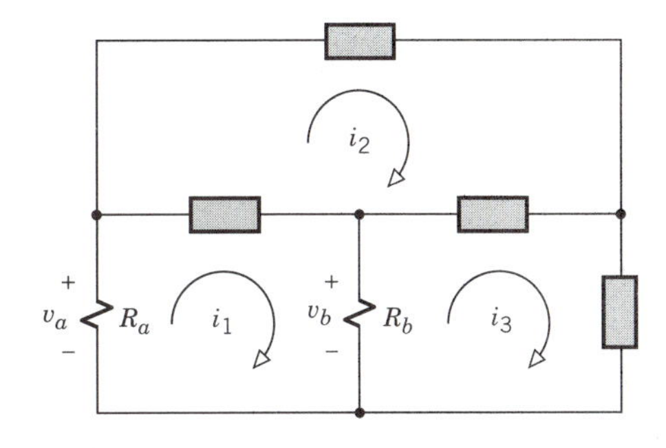

(b) Mesh currents

Figure 4.18

Alternatively, we could draw just three current arrows, as shown in Fig. 4.18*b*. Here we visualize i_1, i_2, and i_3 as currents circulating completely around their respective meshes. We therefore say that i_1, i_2, and i_3 are **mesh currents**. In general,

> The mesh currents of a circuit are the currents that circulate completely around each of the meshes.

Any branch on the perimeter of a circuit carries a mesh current, whereas any interior branch carries two mesh currents.

Hereafter, we'll number the meshes of a circuit according to the subscripts of the mesh currents. We'll also adopt the **reference convention** that

> All unknown mesh currents circulate in the same rotational sense, either clockwise or counterclockwise.

Although the choice of the reference rotation is arbitrary, consistent use of this convention allows us to write interior branch currents in the form

$$i_{nm} = i_n - i_m \tag{4.11}$$

where i_n and i_m are the unknown currents around meshes n and m, respectively.

Since every branch current is either a mesh current or consists of two mesh currents, any unknown branch voltages can be found from the mesh currents. In Fig. 4.18*b*, for example, Ohm's law yields

$$v_a = -R_a i_1 \qquad v_b = R_b i_{13} = R_b(i_1 - i_3)$$

We thus conclude that

> All branch variables in a circuit can be determined from the values of the mesh currents.

Hence, mesh currents provide another way of doing the complete analysis of any planar circuit.

Turning from generalities to specifics, consider the circuit in Fig. 4.19*a*. There are two meshes here, but i_s is a source current so i_1 is the only unknown mesh current. The most convenient way of finding i_1 is to follow the direction of i_1 and algebraically sum the voltage *drops* around mesh 1. We thereby obtain

$$R_a(i_1 - i_s) + R_b i_1 - v_s = 0 \tag{4.12a}$$

Rearrangement then yields the **mesh equation**

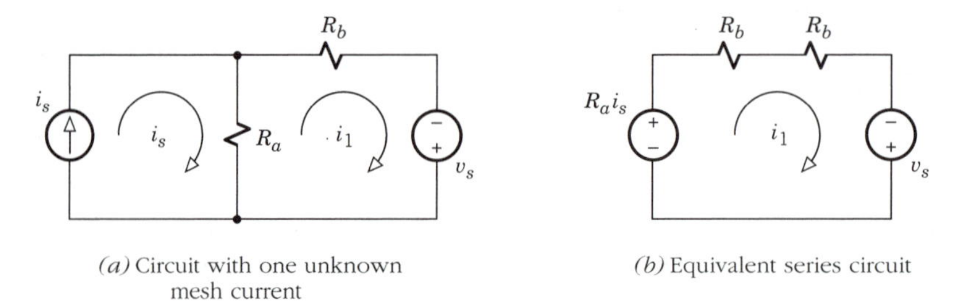

(*a*) Circuit with one unknown mesh current

(*b*) Equivalent series circuit

Figure 4.19

$$(R_a + R_b)i_1 = v_s + R_a i_s \tag{4.12b}$$

which can be solved for i_1 given the element values.

Comparing Eq. (4.12b) with the circuit diagram, we see that the quantity $R_a + R_b$ on the left side equals the *sum of resistances around mesh* 1. The quantity $v_s + R_a i_s$ on the right side of Eq. (4.12b) also has a physical interpretation, namely, the *net equivalent source voltage driving* i_1. We justify that interpretation by converting the current source with parallel resistance into a voltage source with series resistance, as shown in Fig. 4.19*b*. This diagram clearly reveals that the net equivalent source voltage $v_s + R_a i_s$ drives i_1 in its clockwise sense.

Exercise 4.7

Write and solve a mesh equation to find i_1 and v_a in Fig. 4.20.

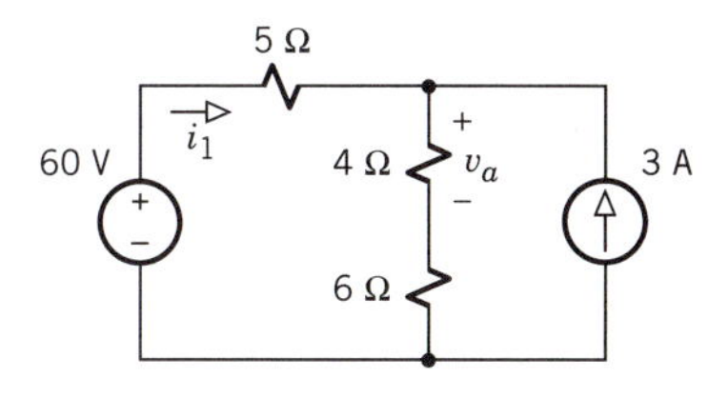

Figure 4.20

Matrix Mesh Equations

When a circuit contains more than one unknown mesh current, the matrix method developed here can be used to save labor and provide error checking. We'll establish the background for that method by carrying out a detailed mesh analysis with three unknowns.

The circuit in Fig. 4.21 has four meshes and one current source. The unknown mesh currents are i_1, i_2, and i_3, and their reference arrows have been

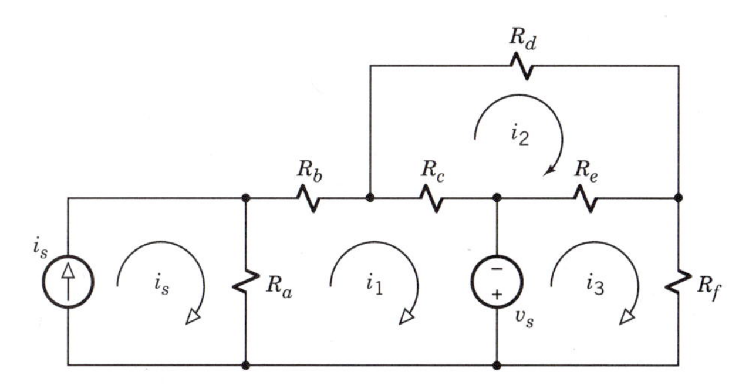

Figure 4.21 Circuit with three unknown mesh currents.

drawn with the same clockwise sense. The mesh-current arrows automatically satisfy KCL, and any branch voltages can be expressed in terms of mesh currents via Ohm's law, so we'll use KVL to obtain three independent equations. The required independence is ensured by applying KVL around each mesh having an unknown current.

Following the direction of i_1 to sum the voltage drops around mesh 1 we get

$$R_a(i_1 - i_s) + R_b i_1 + R_c(i_1 - i_2) - v_s = 0 \tag{4.13a}$$

All terms involving i_1 appear in this sum as positive quantities, and the number of terms on the left equals the number of branches around mesh 1. For the second equation, we sum the three voltage drops around mesh 2 to get

$$R_c(i_2 - i_1) + R_d i_2 + R_e(i_2 - i_3) = 0 \tag{4.13b}$$

so all terms involving i_2 are now positive quantities. Similarly, for mesh 3,

$$R_e(i_3 - i_2) + R_f i_3 + v_s = 0 \tag{4.13c}$$

Finally, we put Eqs. (4.13a)–(4.13c) together in the **matrix mesh equation**

$$\begin{bmatrix} R_a + R_b + R_c & -R_c & 0 \\ -R_c & R_c + R_d + R_e & -R_e \\ 0 & -R_e & R_e + R_f \end{bmatrix} \begin{bmatrix} i_1 \\ i_2 \\ i_3 \end{bmatrix} = \begin{bmatrix} R_a i_s + v_s \\ 0 \\ -v_s \end{bmatrix} \tag{4.14}$$

which can be solved for i_1, i_2, and i_3.

Each row of the matrix in Eq. (4.14) corresponds to a KVL equation around a mesh with an unknown current, and each quantity has a physical interpretation. In particular, a comparison of the first row with the circuit diagram shows that the coefficient $R_a + R_b + R_c$ multiplying i_1 equals the sum of the resistances around mesh 1, while the coefficient $-R_c$ multiplying i_2 is the negative of the resistance in the branch shared by meshes 1 and 2. Resistance R_c appears twice here because the branch voltage $R_c(i_1 - i_2)$ in Eq. (4.13a) has now been separated into two parts. But no resistance is in the branch shared by meshes 1 and 3, so the coefficient of i_3 equals zero in the first row of the matrix. Furthermore, the quantity $R_a i_s + v_s$ on the right equals the net equivalent source voltage driving i_1. All the terms in the second and third rows of the matrix have like interpretations from the branches in meshes 2 and 3.

Clearly, the matrix mesh equation has many similarities to a matrix node equation. In fact, if you compare Eq. (4.14) with Eq. (4.6), then you'll see that they have exactly the same form with voltage and current interchanged and resistance and conductance interchanged. Hence, the circuit in Fig. 4.21 is the *structural dual* of the circuit back in Fig. 4.5.

Now consider any planar circuit that has N unknown mesh currents and contains only resistors and independent sources. The unknown mesh currents can be found from a matrix equation in the general form

$$[R]\,[i] = [v_s] \tag{4.15}$$

The vectors $[i]$ and $[v_s]$ stand for the $N \times 1$ column arrays

$$[i] = \begin{bmatrix} i_1 \\ i_2 \\ \vdots \\ i_N \end{bmatrix} \qquad [v_s] = \begin{bmatrix} v_{s1} \\ v_{s2} \\ \vdots \\ v_{sN} \end{bmatrix}$$

where

i_n = Unknown current around mesh n
v_{sn} = Net equivalent source voltage driving i_n

The elements of the equivalent source-voltage vector $[v_s]$ may be positive, negative, or zero, depending upon the circuit configuration.

The **resistance matrix** $[R]$ in Eq. (4.15) stands for the $N \times N$ square array

$$[R] = \begin{bmatrix} R_{11} & -R_{12} & \cdots & -R_{1N} \\ -R_{21} & R_{22} & \cdots & -R_{2N} \\ \vdots & \vdots & & \vdots \\ -R_{N1} & -R_{N2} & \cdots & R_{NN} \end{bmatrix} \tag{4.16}$$

where

R_{nn} = Sum of resistances around mesh n
$R_{nm} = R_{mn}$ = Equivalent resistance shared by meshes n and m

Thus, $[R]$ has the same symmetry and error-checking properties as the conductance matrix $[G]$ in a node equation.

With the help of the interpretations of the resistance matrix and the source-voltage vector, we can determine their element values directly by *inspection* of a circuit diagram. Thus, the systematic matrix method of mesh analysis goes as follows:

1. Identify any known mesh currents fixed by current sources, select the reference rotational direction, and label the remaining unknown mesh currents. Also label all other variables of interest.
2. Fill in the resistance matrix $[R]$ using the properties that R_{nn} equals the sum of resistances around mesh n while R_{nm} equals the equivalent resistance shared by meshes n and m.
3. Check $[R]$ to make sure that $R_{mn} = R_{nm}$ and that these off-diagonal elements carry a negative sign. Also check that each off-diagonal quantity R_{nm} appears as part of the diagonal terms R_{nn} in the same row and R_{mm} in the same column.
4. Fill in the source-voltage vector $[v_s]$ using the property that v_{sn} equals the net equivalent source voltage driving current i_n. If resistance R_s in

mesh n also carries current $\pm i_s$ fixed by a current source, then v_{sn} includes the equivalent source voltage $\pm R_s i_s$.

5. Solve the resulting matrix equation $[R]\,[i] = [v_s]$ for the unknown mesh currents.

All other variables of interest may then be calculated from the mesh currents.

Since current sources establish known mesh currents, whereas voltage sources have no effect on the number of unknowns, you can

> Determine the minimum number of unknown mesh currents by suppressing all sources and counting the remaining meshes.

But keep in mind that the circuit in question must be planar and that our inspection method for mesh analysis requires all current sources to be on the perimeter of the circuit diagram to find the equivalent source voltages. We'll relax this requirement subsequently.

Example 4.6 *Matrix Mesh Analysis with Two Unknowns*

In Example 4.2 we used matrix node analysis with two unknowns to find the power supplied by the sources in Fig. 4.22*a*. Now we'll use matrix mesh analysis with i_1 and i_2 as the unknowns. We need just two unknown mesh currents because, if we combine the two parallel resistors and suppress all sources, then we get the two-mesh pattern in Fig. 4.22*b*.

Step 1: The diagram in Fig. 4.22*a* has been labeled with the known 1-A mesh current, and the counterclockwise rotational sense has been selected for i_1 and i_2.

Step 2: Noting that $60\|12 = 10\ \Omega$, inspection of the diagram gives the resistance matrix

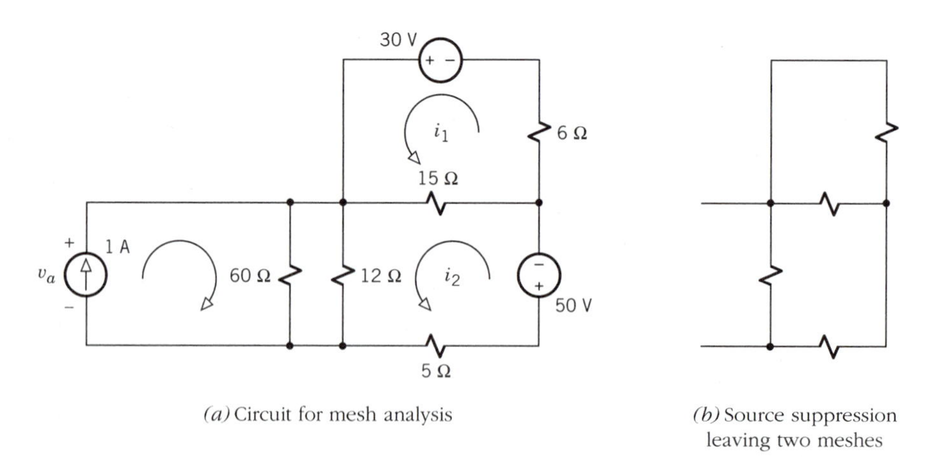

(a) Circuit for mesh analysis

(b) Source suppression leaving two meshes

Figure 4.22

$$[R] = \begin{bmatrix} 6 + 15 & -15 \\ -15 & 15 + 60\|12 + 5 \end{bmatrix} = \begin{bmatrix} 21 & -15 \\ -15 & 30 \end{bmatrix}$$

Step 3: Our resistance matrix has the stated properties, and the 15-Ω off-diagonal term appears as part of R_{11} and R_{22} on the main diagonal.

Step 4: For the elements of $[v_s]$, we note that the 30-V source drives i_1 while i_2 is opposed by the 50-V source and by the equivalent source voltage (60‖12) Ω × 1 A. Thus,

$$[i_s] = \begin{bmatrix} 30 \\ -50 - (60\|12) \times 1 \end{bmatrix} = \begin{bmatrix} 30 \\ -60 \end{bmatrix}$$

Step 5: The resulting matrix mesh equation is

$$\begin{bmatrix} 21 & -15 \\ -15 & 30 \end{bmatrix} \begin{bmatrix} i_1 \\ i_2 \end{bmatrix} = \begin{bmatrix} 30 \\ -60 \end{bmatrix}$$

which is easily solved to get

$$i_1 = 0 \qquad i_2 = -2 \text{ A}$$

The power supplied by the 50-V source is $50 \times (-i_2) = 100$ W, while the 30-V source supplies no power since $30 \times i_1 = 0$. We finally calculate the branch voltage

$$v_a = (60\|12)(1 + i_2) = -10 \text{ V}$$

so the power supplied by the 1-A source is $v_a \times 1 = -10$ W. All of these results agree with our previous node-voltage calculations.

Example 4.7 ***Matrix Mesh Analysis with Three Unknowns***

Problem: Find i_1, i_2, and i_3 by mesh analysis of Fig. 4.23.

Solution: Suppressing all sources confirms that we need three unknown mesh currents, and the clockwise rotational sense has been chosen. We'll work

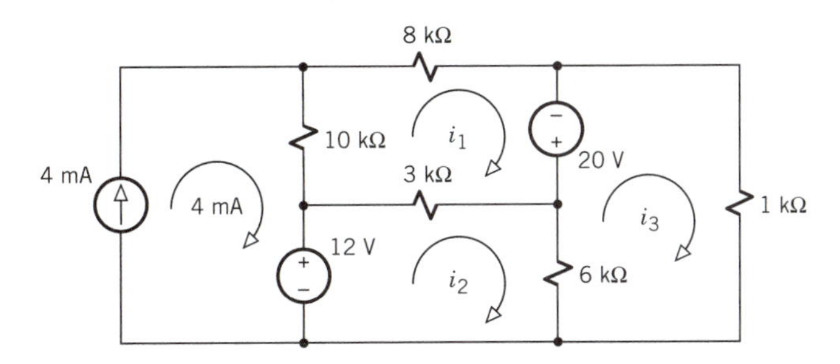

Figure 4.23 Circuit for mesh analysis with three unknowns.

with a consistent set of units, taking resistances in kilohms and currents in milliamps.

The resistance matrix is found by inspection, noting that the branch shared by meshes 1 and 3 contains no resistance. Thus,

$$[R] = \begin{bmatrix} 10 + 8 + 3 & -3 & 0 \\ -3 & 3 + 6 & -6 \\ 0 & -6 & 6 + 1 \end{bmatrix}$$

which passes our error checks. For the equivalent source-voltage vector, we note that the 20-V source drives i_1 but opposes i_3, the 12-V source drives i_2, and the equivalent voltage 10 kΩ × 4 mA drives i_1 and has no affect on i_2. Thus,

$$[v_s] = \begin{bmatrix} 20 + 10 \times 4 \\ 12 \\ -20 \end{bmatrix}$$

We now have the matrix mesh equation

$$\begin{bmatrix} 21 & -3 & 0 \\ -3 & 9 & -6 \\ 0 & -6 & 7 \end{bmatrix} \begin{bmatrix} i_1 \\ i_2 \\ i_3 \end{bmatrix} = \begin{bmatrix} 60 \\ 12 \\ -20 \end{bmatrix}$$

from which

$$i_1 = 3 \text{ mA} \qquad i_2 = 1 \text{ mA} \qquad i_3 = -2 \text{ mA}$$

The fact that the 4-mA source supplies an equivalent source voltage in mesh 1 (but not mesh 2) can be justified by writing out the KVL equations. But a diagrammatic justification comes from the "source mesh splitting" shown in Fig. 4.24*a*. Here we have put a second 4-mA source in series with the first to split the 4-mA mesh into two meshes, leaving the rest of the circuit unaffected. Source conversion then yields Fig. 4.24*b*, where an equivalent source voltage of 10 kΩ × 4 mA drives i_1, whereas only the 12-V source drives i_2.

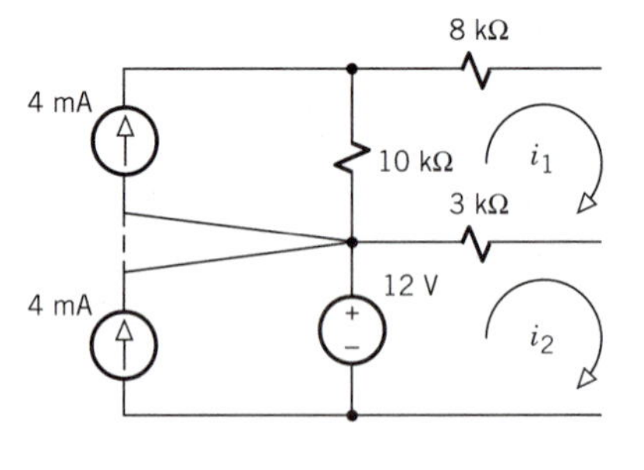

(*a*) Adding series current source

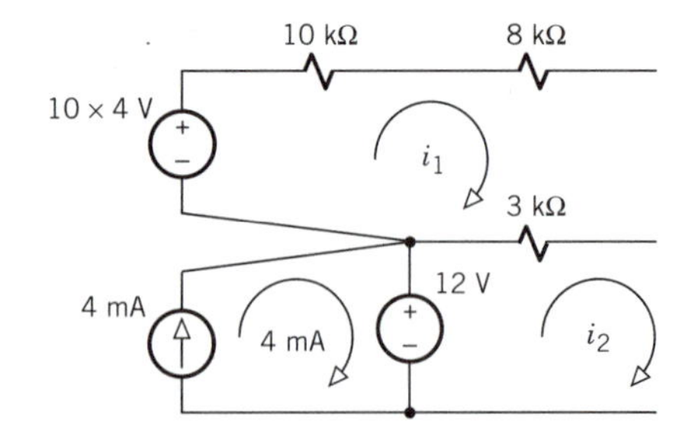

(*b*) After source conversion

Figure 4.24 Source mesh splitting.

Exercise 4.8

Determine by inspection of Fig. 4.25 the resistance and voltage values for the matrix mesh equation. Then solve for i_1 and i_2. (Hint: The 3-A source supplies equivalent source voltages for both meshes 1 and 2.)

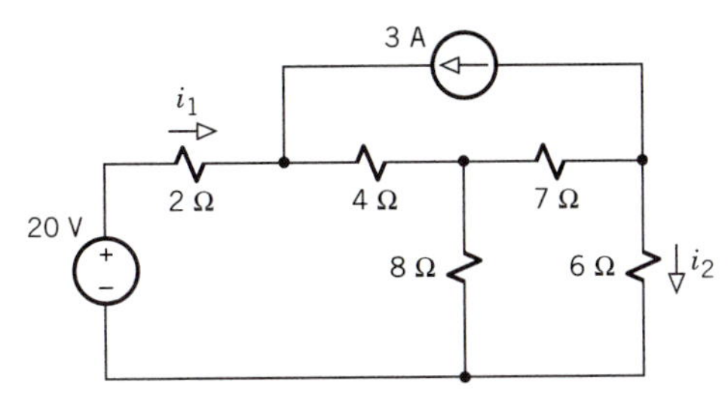

Figure 4.25

Exercise 4.9

The resistance matrix for a certain circuit includes R_x in R_{22}. What do you conclude about the location of R_x when:

(a) R_x does not appear anywhere else in $[R]$ or $[v_s]$?

(b) R_x also appears in v_{s2} but nowhere else in $[R]$?

Exercise 4.10

Suppose Fig. 4.23 is modified by putting resistance R in parallel with the 20-V source. Explain why R should *not* be included in the resistance matrix.

Interior Current Sources

The inspection method for writing mesh equations always works when all current sources are on the perimeter of the circuit. Interior current sources require additional consideration, and there are two different cases to examine.

Case I: Interior Sources with Parallel Resistance Figure 4.26*a* shows a current source with parallel resistance between two meshes, n and m. The intervening mesh current i_x may be ignored because the source requires $i_x - i_n = i_s$ or $i_x = i_n + i_s$ — so i_x will be known immediately after finding i_n. We are therefore free to perform a source conversion giving the series structure in Fig. 4.26*b*. This figure reveals that the equivalent source voltage $R_s i_s$ drives current i_m and opposes i_n. Additionally, we now see that the resistance R_s affects both meshes, a fact that was not evident before the conversion.

When a circuit includes an interior source like Fig. 4.26*a*, you can carry out mesh analysis by inspection *after* redrawing the diagram per Fig. 4.26*b*. But keep in mind that the source conversion is only equivalent at the terminal

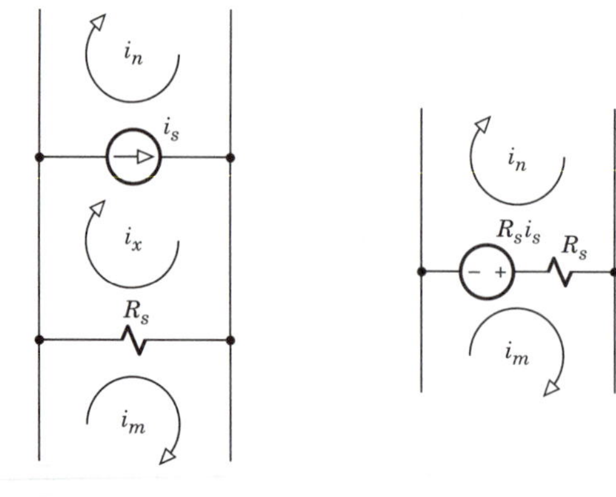

(a) Interior current source with parallel resistance

(b) Source conversion eliminating i_x

Figure 4.26

nodes. Thus, to find the actual branch current through R_s, for instance, you must calculate it from the original configuration in Fig. 4.26*a* using the values of i_m and $i_x = i_n + i_s$.

Example 4.8 *Matrix Mesh Analysis with Source Conversion*

Suppose you want to find v_a and i_b in Fig. 4.27*a*. Suppressing all sources gives the pattern in Fig. 4.27*b*, which has four nodes and two meshes. You therefore decide that mesh analysis will be easier than node analysis.

Converting the 3-A interior source gives the diagram in Fig. 4.27*c*, where the 7-A perimeter source current goes through resistors in mesh 1 and mesh 2. Although the identity of i_b has been lost here, the partial diagram in Fig. 4.27*d* reveals that

$$i_b = 7 + i_x \qquad i_x + i_2 = 3$$

Thus, both v_a and i_b can be found from the mesh currents.

Having disposed of the preliminaries, the rest of the work goes quickly. By inspection of Fig. 4.27*c* you write the matrix mesh equation

$$\begin{bmatrix} 4+2+8 & -2 \\ -2 & 6+2+10 \end{bmatrix} \begin{bmatrix} i_1 \\ i_2 \end{bmatrix} = \begin{bmatrix} 20 + 8 \times 7 \\ 24 + 6 \times 7 \end{bmatrix}$$

from which

$$i_1 = 6 \text{ A} \qquad i_2 = 4 \text{ A}$$

You then calculate

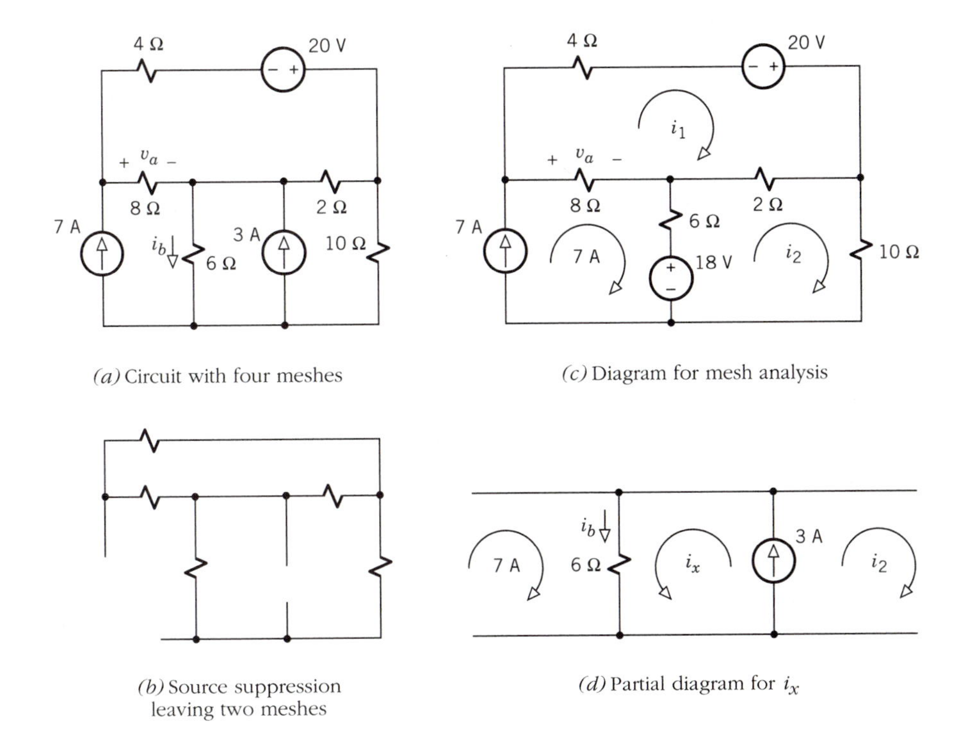

(a) Circuit with four meshes

(b) Source suppression leaving two meshes

(c) Diagram for mesh analysis

(d) Partial diagram for i_x

Figure 4.27

$$v_a = 8(7 - i_1) = 8 \text{ V}$$

$$i_x = 3 - i_2 = -1 \text{ A} \qquad i_b = 7 + i_x = 6 \text{ A}$$

which completes the work.

Had you opted for node analysis, a careful choice of the reference node would eliminate the need for any source conversions. However, you would have to write and solve a matrix equation with three unknowns.

Exercise 4.11

Redraw the diagram in Fig. 4.28*a* to put both sources on the perimeter. Then write and solve one mesh equation for i_1.

Exercise 4.12

Redraw the diagram in Fig. 4.28*b* and write a matrix mesh equation for the unknown currents i_1, i_2, and i_3.

Case II: Interior Ideal Sources When a circuit has been modeled with *ideal* current sources, the absence of parallel resistance may preclude the source conversions needed to do mesh analysis by inspection. We'll investigate two

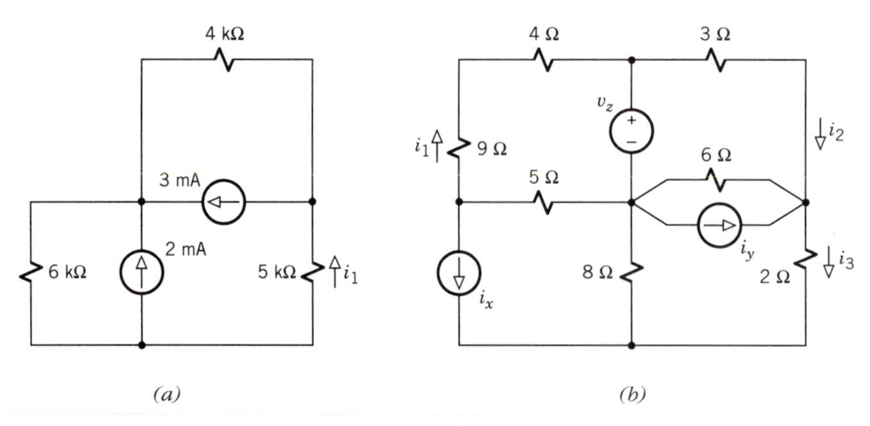

Figure 4.28

ways of handling such cases. Not surprisingly, they are the duals of our two methods for node analysis with floating ideal voltage sources.

Consider the circuit in Fig. 4.29. The interior ideal source i_y establishes a relationship between the mesh currents labeled i_2 and i_3, namely, $i_3 - i_2 = i_y$. If we take i_2 as an unknown, then i_3 is not an *independent* unknown since $i_3 = i_2 + i_y$. This conclusion agrees with the source-suppression test, which leaves just two meshes when i_y is replaced with an open circuit. But direct analysis by inspection is impossible because we do not know the voltage v_y across the interior source.

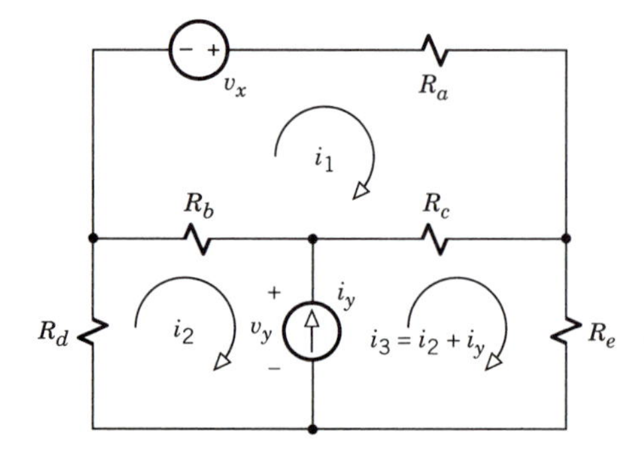

Figure 4.29 Circuit with interior ideal current source.

One approach to this problem is to treat v_y like a *fictitious source voltage* and write three individual mesh equations with $i_3 = i_2 + i_y$. We thereby obtain

$$(R_a + R_b + R_c)i_1 - R_b i_2 - R_c(i_2 + i_y) = v_x \tag{4.17a}$$

$$-R_b i_1 + (R_b + R_d)i_2 - 0(i_2 + i_y) = -v_y \tag{4.17b}$$

$$-R_c i_1 - 0 \times i_2 + (R_c + R_e)(i_2 + i_y) = v_y \tag{4.17c}$$

We then have three equations with three unknowns, i_1, i_2, and v_y. But v_y can be eliminated immediately by adding Eqs. (4.17b) and (4.17c) to get

$$-(R_b + R_c)i_1 + (R_b + R_d)i_2 + (R_c + R_e)(i_2 + i_y) = 0 \quad (4.17d)$$

Equations (4.17a) and (4.17d) now constitute two independent equations with unknowns i_1 and i_2.

Alternatively, instead of working with v_y as an additional unknown, suppose we apply KVL to the loop indicated by the dashed line in Fig. 4.30. This loop is known as a **supermesh** because it encloses more than one mesh. The unknown voltage v_y does not appear as we go around the supermesh, and summing the voltage drops yields

$$R_d i_2 + R_b(i_2 - i_1) + R_c(i_2 + i_y - i_1) + R_e(i_2 + i_y) = 0 \quad (4.18)$$

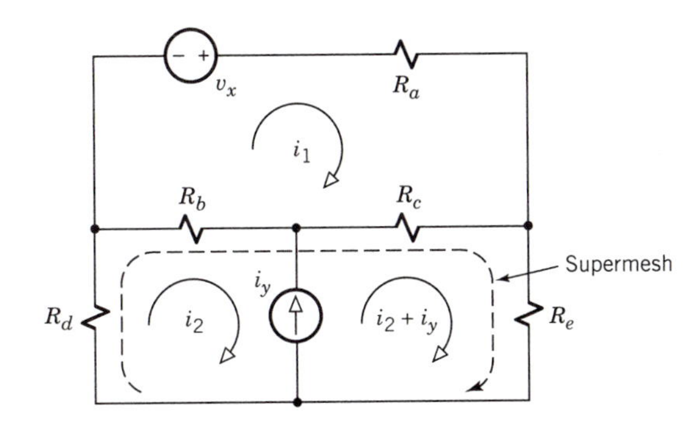

Figure 4.30 Supermesh around interior ideal source.

Closer examination confirms that Eq. (4.18) is identical to Eq. (4.17d), so the supermesh equation is equivalent to writing and combining the two mesh equations that involve the interior current source. Thus, we can get the pair of equations for i_1 and i_2 from the supermesh equation together with the equation for mesh 1, as previously given in Eq. (4.17a).

Example 4.9 *Mesh Analysis with a Supermesh*

The value of i_1 in Fig. 4.31 is to be found. Source suppression reveals that there are two unknown node voltages but just one unknown mesh current.

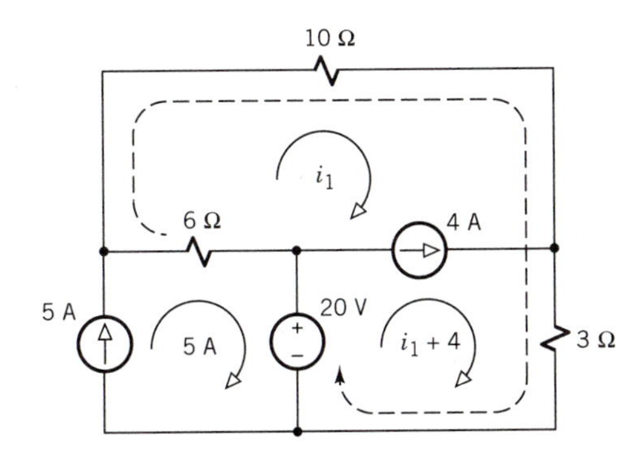

Figure 4.31 Circuit for analysis with a supermesh.

We therefore choose mesh analysis with i_1 as the unknown, which we'll find via a single supermesh equation.

The current around the lower portion of the supermesh is $i_1 + 4$, so

$$6(i_1 - 5) + 10i_1 + 3(i_1 + 4) - 20 = 0$$

Rearrangement gives

$$(6 + 10 + 3)i_1 = 6 \times 5 - (3 \times 4) + 20$$

so

$$19i_1 = 38 \quad \Rightarrow \quad i_1 = 2\text{ A}$$

Exercise 4.13

Use a supermesh equation to obtain the matrix equation for i_1 and i_2 in Fig. 4.32.

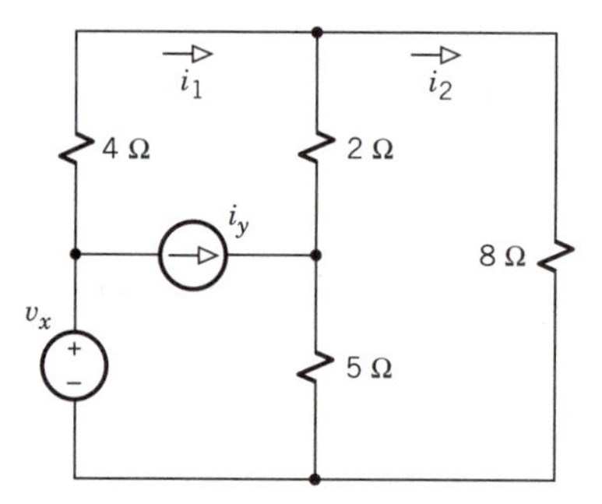

Figure 4.32

4.3 SYSTEMATIC ANALYSIS WITH CONTROLLED SOURCES

This section extends mesh and node analysis to deal with linear electronic circuits or any other circuits that include controlled sources. We'll show that the essential new step involves writing additional equations for the control variables, and we'll present systematic procedures for incorporating those equations into our matrix format.

Mesh Analysis

The circuit in Fig. 4.33 provides a simple introduction to mesh analysis with controlled sources. There is only one unknown current here, and the mesh equation for i_1 is

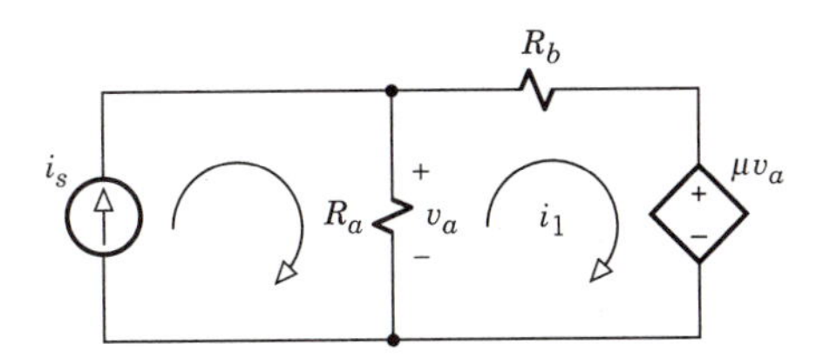

Figure 4.33 Circuit with a controlled source.

$$(R_a + R_b)\,i_1 = R_a i_s - \mu v_a \tag{4.19}$$

But the control variable v_a is also unknown, so we need another equation relating v_a to i_1. Further examination of the diagram shows that

$$v_a = R_a(i_s - i_1) \tag{4.20}$$

Equations (4.19) and (4.20) thus constitute a set of simultaneous equations with two unknowns, i_1 and v_a. We call Eq. (4.20) a **constraint equation** because it expresses the control variable in terms of known constants and the unknown mesh current.

If we only want to find i_1, then we substitute the constraint equation into the mesh equation to get

$$\begin{aligned}(R_a + R_b)\,i_1 &= R_a i_s - \mu R_a(i_s - i_1)\\ &= (1 - \mu)\,R_a i_s + \mu R_a i_1\end{aligned}$$

We'll recast this expression in the form

$$R_{11}\,i_1 = \tilde{v}_s + \tilde{R} i_1 \tag{4.21}$$

where $R_{11} = R_a + R_b$ and

$$\tilde{v}_s = (1 - \mu)\,R_a i_s \qquad \tilde{R} = \mu R_a$$

The tilde (˜) thus identifies quantities affected by the controlled source. Since Eq. (4.21) has i_1 on both sides, we regroup terms as

$$(R_{11} - \tilde{R})\,i_1 = \tilde{v}_s \tag{4.22}$$

which can be solved for i_1.

Now consider any planar circuit that has two or more unknown mesh currents and includes at least one controlled source. We begin by writing the usual matrix mesh equation

$$[R]\,[i] = [v_s]$$

However, the control variables will appear as extraneous unknowns in the source-voltage vector, so we must formulate constraint equations for all control variables and insert them into $[v_s]$. The resulting voltage vector then takes on the general form

$$[v_s] = \begin{bmatrix} \tilde{v}_{s1} + \tilde{R}_{11}i_1 + \tilde{R}_{12}i_2 + \cdots \\ \tilde{v}_{s2} + \tilde{R}_{21}i_1 + \tilde{R}_{22}i_2 + \cdots \\ \vdots \end{bmatrix} \tag{4.23}$$

where the $\tilde{v}$'s and $\tilde{R}$'s are constants.

Solving the matrix mesh equation requires separating out of Eq. (4.23) those terms that involve the unknown mesh currents $i_1, i_2, \ldots$ To that end, we expand $[v_s]$ as a vector of constants plus a matrix-vector product so that

$$[v_s] = \begin{bmatrix} \tilde{v}_{s1} \\ \tilde{v}_{s2} \\ \vdots \end{bmatrix} + \begin{bmatrix} \tilde{R}_{11} & \tilde{R}_{12} & \cdots \\ \tilde{R}_{21} & \tilde{R}_{22} & \cdots \\ \vdots & \vdots & \vdots \end{bmatrix} \begin{bmatrix} i_1 \\ i_2 \\ \vdots \end{bmatrix}$$

Symbolically, we now have

$$[v_s] = [\tilde{v}_s] + [\tilde{R}][i]$$

where $[\tilde{v}_s]$ includes all the independent sources and $[\tilde{R}][i]$ accounts for the mesh currents in the constraint equations. Although the elements of $[\tilde{R}]$ have the units of resistance, this matrix generally differs from the resistance matrix $[R]$ and does not necessarily possess any symmetry.

After carrying out the expansion of $[v_s]$, the matrix mesh equation becomes

$$[R][i] = [\tilde{v}_s] + [\tilde{R}][i]$$

Regrouping terms finally yields the modified matrix equation

$$[R - \tilde{R}][i] = [\tilde{v}_s]$$

where

$$[R - \tilde{R}] = [R] - [\tilde{R}]$$

Note that the matrix $[R - \tilde{R}]$ is obtained by subtracting the elements of $[\tilde{R}]$ from the corresponding elements of $[R]$. This means that we should not expect $[R - \tilde{R}]$ to have symmetry about the main diagonal.

To summarize, a systematic procedure for mesh analysis with controlled sources consists of the following steps:

1. Identify the unknown mesh currents and write the resistance matrix $[R]$ by inspection.
2. For each controlled source, formulate a constraint equation expressing the control variable in terms of known constants and/or unknown mesh currents.

3. Write the source-voltage vector by inspection, insert the constraint equations, and expand $[v_s]$ in the form

$$[v_s] = [\tilde{v}_s] + [\tilde{R}]\,[i] \tag{4.24}$$

4. Calculate $[R - \tilde{R}]$, and solve for the unknown mesh currents from

$$[R - \tilde{R}]\,[i] = [\tilde{v}_s] \tag{4.25}$$

The following examples illustrate the techniques used in this procedure.

Example 4.10 *Mesh Analysis with a CCCS*

The circuit in Fig. 4.34 has two unknown mesh currents and a current-controlled current source. We'll carry out the steps of our systematic procedure to evaluate i_1 and i_2.

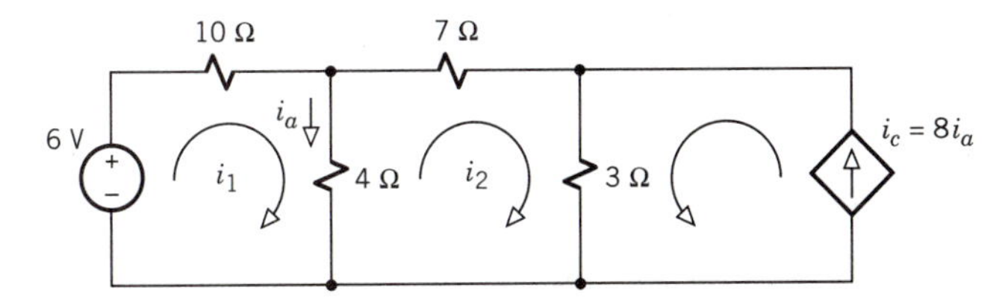

Figure 4.34 Circuit for mesh analysis with a CCS.

Step 1: We find the resistance matrix by inspection of the diagram to be

$$[R] = \begin{bmatrix} 10+4 & -4 \\ -4 & 4+7+3 \end{bmatrix} = \begin{bmatrix} 14 & -4 \\ -4 & 14 \end{bmatrix}$$

Step 2: The constraint equation relating i_a to the mesh currents is

$$i_a = i_1 - i_2$$

Step 3: We write the source-voltage vector and insert the constraint equation to get

$$[v_s] = \begin{bmatrix} 6 \\ -3 \times 8i_a \end{bmatrix} = \begin{bmatrix} 6 \\ -24i_1 + 24i_2 \end{bmatrix}$$

Expansion then yields

$$[v_s] = \begin{bmatrix} 6 \\ 0 \end{bmatrix} + \begin{bmatrix} 0 \\ -24i_1 + 24i_2 \end{bmatrix} = \begin{bmatrix} 6 \\ 0 \end{bmatrix} + \begin{bmatrix} 0 & 0 \\ -24 & 24 \end{bmatrix} \begin{bmatrix} i_1 \\ i_2 \end{bmatrix}$$

Thus,

$$[\tilde{v}_s] = \begin{bmatrix} 6 \\ 0 \end{bmatrix} \qquad [\tilde{R}] = \begin{bmatrix} 0 & 0 \\ -24 & 24 \end{bmatrix}$$

Step 4: We calculate

$$[R - \tilde{R}] = \begin{bmatrix} 14 - 0 & -4 - 0 \\ -4 - (-24) & 14 - 24 \end{bmatrix} = \begin{bmatrix} 14 & -4 \\ 20 & -10 \end{bmatrix}$$

which does not exhibit any symmetry. The modified matrix mesh equation finally becomes

$$\begin{bmatrix} 14 & -4 \\ 20 & -10 \end{bmatrix} \begin{bmatrix} i_1 \\ i_2 \end{bmatrix} = \begin{bmatrix} 6 \\ 0 \end{bmatrix}$$

from which $i_1 = 1$ A and $i_2 = 2$ A.

Example 4.11 ***Mesh Analysis of a Current Amplifier***

Problem: Find the current gain $A_i = i_{out}/i_s$ of the amplifier circuit in Fig. 4.35, which contains two voltage-controlled current sources.

Solution: Mesh analysis is the best choice since there are four unknown node voltages but three unknown mesh currents. Working with resistances in kilohms and currents in milliamps, the analysis proceeds as follows:

$$[R] = \begin{bmatrix} 37 & -1 & 0 \\ -1 & 41 & -4 \\ 0 & -4 & 19 \end{bmatrix}$$

$$v_a = 6i_1 \qquad v_b = 4(i_3 - i_2)$$

$$[v_s] = \begin{bmatrix} 30i_s \\ -36 \times 4v_a \\ -12 \times 2v_b \end{bmatrix} = \begin{bmatrix} 30i_s \\ -864i_1 \\ 96i_2 - 96i_3 \end{bmatrix}$$

$$= \begin{bmatrix} 30i_s \\ 0 \\ 0 \end{bmatrix} + \begin{bmatrix} 0 & 0 & 0 \\ -864 & 0 & 0 \\ 0 & 96 & -96 \end{bmatrix} \begin{bmatrix} i_1 \\ i_2 \\ i_3 \end{bmatrix}$$

Thus,

$$\begin{bmatrix} 37 - 0 & -1 - 0 & 0 - 0 \\ -1 - (-864) & 41 - 0 & -4 - 0 \\ 0 - 0 & -4 - 96 & 19 - (-96) \end{bmatrix} \begin{bmatrix} i_1 \\ i_2 \\ i_3 \end{bmatrix} = \begin{bmatrix} 30i_s \\ 0 \\ 0 \end{bmatrix}$$

Since the only independent source i_s has an arbitrary value, A_i can be calculated by drawing upon the proportionality principle. Setting $i_s = 1$ mA, for convenience, the matrix equation yields

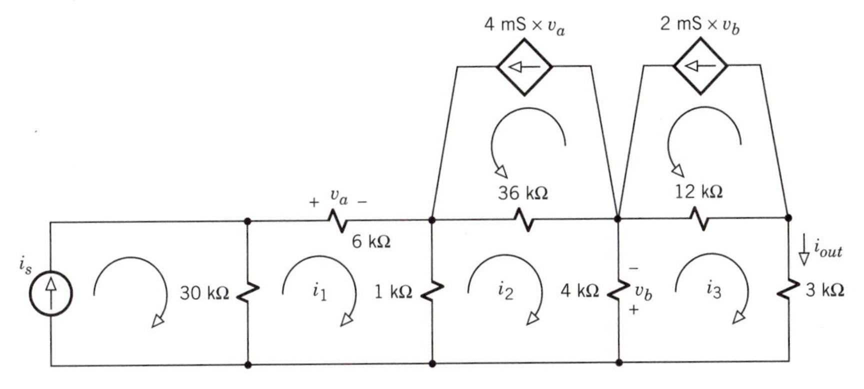

Figure 4.35 Current amplifier with two controlled sources.

$$i_1 = 0.5 \text{ mA} \qquad i_2 = -11.5 \text{ mA} \qquad i_3 = -10 \text{ mA}$$

Hence,

$$A_i = i_{out}/i_s = i_3/i_s = -10$$

Exercise 4.14

Given that $v_c = 20\ \Omega \times i_a$ in Fig. 4.36, show that

$$[v_s] = \begin{bmatrix} 42 \\ -60 \end{bmatrix} + \begin{bmatrix} 20 & 0 \\ -20 & 0 \end{bmatrix} \begin{bmatrix} i_1 \\ i_2 \end{bmatrix}$$

Then obtain the modified matrix mesh equation. (Save $[R]$ for use in Exercise 4.15.)

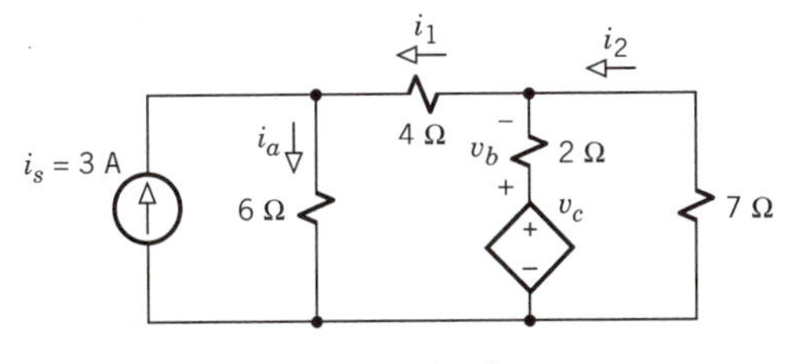

Figure 4.36

Exercise 4.15

Use mesh analysis to find i_2 in Fig. 4.36 when $v_c = 6v_b$.

Node Analysis

Having seen how to modify mesh analysis to include controlled sources, we'll exploit duality to derive the modifications for node analysis with controlled sources.

Consider any circuit that includes at least one controlled source. The systematic procedure for node analysis starts with the usual conductance matrix $[G]$ and source-current vector $[i_s]$. Since all control variables appear in $[i_s]$, we must write a constraint equation for each controlled source that relates the control variable to the node voltages. After inserting the constraint equations, $[i_s]$ will be the dual of Eq. (4.23), namely,

$$[i_s] = \begin{bmatrix} \tilde{i}_{s1} + \tilde{G}_{11}v_1 + \tilde{G}_{12}v_2 + \cdots \\ \tilde{i}_{s2} + \tilde{G}_{21}v_1 + \tilde{G}_{22}v_2 + \cdots \\ \vdots \end{bmatrix}$$

where the $\tilde{i}$'s and $\tilde{G}$'s are constants. We therefore expand $[i_s]$ as

$$[i_s] = \begin{bmatrix} \tilde{i}_{s1} \\ \tilde{i}_{s2} \\ \vdots \end{bmatrix} + \begin{bmatrix} \tilde{G}_{11} & \tilde{G}_{12} & \cdots \\ \tilde{G}_{21} & \tilde{G}_{22} & \cdots \\ \vdots & \vdots & \vdots \end{bmatrix} \begin{bmatrix} v_1 \\ v_2 \\ \vdots \end{bmatrix}$$

so that

$$[i_s] = [\tilde{i}_s] + [\tilde{G}]\,[v]$$

The matrix node equation now becomes $[G]\,[v] = [\tilde{i}_s] + [\tilde{G}]\,[v]$. Hence, the node voltages can be found from $[G - \tilde{G}]\,[v] = [\tilde{i}_s]$, where $[G - \tilde{G}] = [G] - [\tilde{G}]$.

A systematic procedure for node analysis with controlled sources consists of the following steps:

1. Identify the unknown node voltages and write the conductance matrix $[G]$ by inspection.
2. For each controlled source, formulate a constraint equation expressing the control variable in terms of known constants and/or unknown node voltages.
3. Write the source-current vector by inspection, insert the constraint equations, and expand $[i_s]$ in the form

$$[i_s] = [\tilde{i}_s] + [\tilde{G}]\,[v] \tag{4.26}$$

4. Calculate $[G - \tilde{G}]$, and solve for the unknown node voltages from

$$[G - \tilde{G}]\,[v] = [\tilde{i}_s] \tag{4.27}$$

As in the case of mesh analysis, we can't expect $[G - \tilde{G}]$ to have symmetry about the main diagonal.

Example 4.12 *Node Analysis of an Inverting Amplifier*

Figure 4.37 is the equivalent circuit of an inverting amplifier constructed with an op-amp. This circuit would have $v_{out}/v_{in} = -R_F/R_1 = -50$ if the op-amp were ideal (i.e., $r_i = \infty$, $r_o = 0$, and $A = \infty$). But here we'll use node analysis to find v_{out}/v_{in} taking account of the op-amp's input and output resistances and finite gain A.

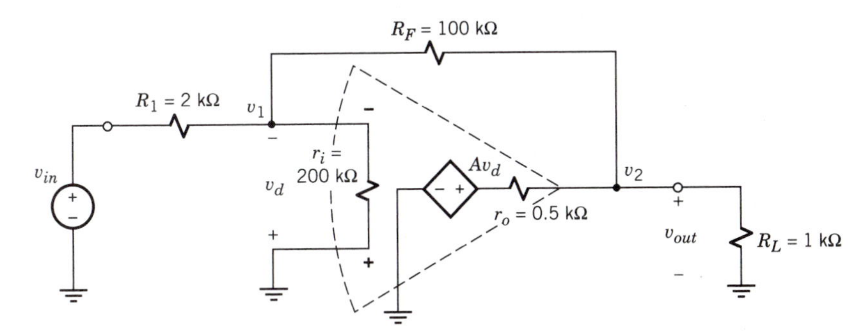

Figure 4.37 Model of an op-amp inverting amplifier.

The unknown node voltages are $v_1 = -v_d$ and $v_2 = v_{out}$, as marked on the diagram. The corresponding conductance matrix in millisiemens is

$$[G] = \begin{bmatrix} \frac{1}{2} + \frac{1}{100} + \frac{1}{200} & -\frac{1}{100} \\ -\frac{1}{100} & \frac{1}{100} + \frac{1}{0.5} + \frac{1}{1} \end{bmatrix} = \begin{bmatrix} 0.515 & -0.01 \\ -0.01 & 3.01 \end{bmatrix}$$

The constraint equation for v_d is just

$$v_d = -v_1$$

so the source-current vector in milliamps is

$$[i_s] = \begin{bmatrix} v_{in}/2 \\ A(-v_1)/0.5 \end{bmatrix} = \begin{bmatrix} 0.5v_{in} \\ 0 \end{bmatrix} + \begin{bmatrix} 0 & 0 \\ -2A & 0 \end{bmatrix}\begin{bmatrix} v_1 \\ v_2 \end{bmatrix}$$

Subtracting $[\tilde{G}]$ from $[G]$ yields the modified matrix equation

$$\begin{bmatrix} 0.515 & -0.01 \\ 2A - 0.01 & 3.01 \end{bmatrix}\begin{bmatrix} v_1 \\ v_2 \end{bmatrix} = \begin{bmatrix} 0.5v_{in} \\ 0 \end{bmatrix}$$

Since both A and v_{in} have arbitrary values, we solve for v_2 by applying Cramer's rule to obtain the determinants

$$\Delta = 1.55 + 0.02A \qquad \Delta_2 = (0.05 - A)v_{in}$$

Thus, with $v_{out} = v_2 = \Delta_2/\Delta$,

$$\frac{v_{out}}{v_{in}} = \frac{\Delta_2/\Delta}{v_{in}} = \frac{0.05 - A}{1.55 + 0.02A}$$

Table 4.1 lists values of v_{out}/v_{in} for selected values of the op-amp gain A. We see from these results that $v_{out}/v_{in} \approx -50$ when $A \geq 10^4$, so the circuit performs essentially as though the op-amp were ideal.

TABLE 4.1

A	10^2	10^3	10^4	10^5
v_{out}/v_{in}	−28.15	−46.40	−49.2	−49.96

Exercise 4.16

Obtain the modified matrix node equation for v_1 and v_2 in Fig. 4.38.

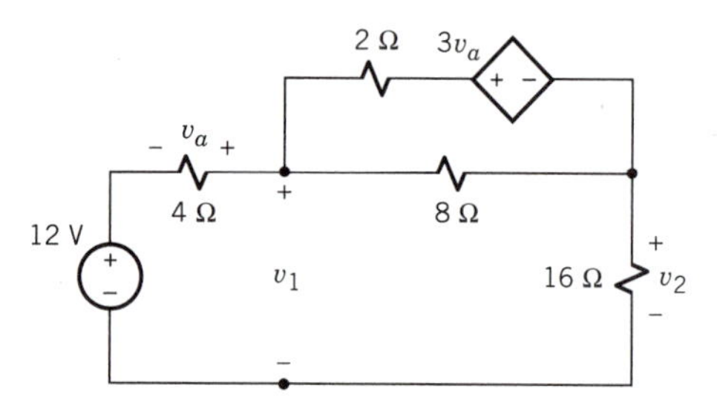

Figure 4.38

4.4 APPLICATIONS OF SYSTEMATIC ANALYSIS

This section applies systematic analysis methods to important problems of the type that often occur in the study of electronic circuits. Specifically, we'll show how node and mesh equations can be used to calculate the equivalent resistance of load networks and the Thévenin parameters of source networks. Our work here will also shed further light on the choice between node or mesh analysis for particular problems.

Equivalent Resistance

In Section 2.3, we introduced the concept of generalized equivalent resistance for load networks that consist entirely of resistors or resistors and controlled sources. Now we'll apply systematic analysis to the task of determining equivalent resistance when a nonsystematic method would require excessive labor.

Our general strategy is represented by Fig. 4.39, where an arbitrary load network has been connected to a test source that establishes v or i at the terminals. We know that the terminal variables are related by $v = R_{eq} i$, so

$$R_{eq} = v / i \tag{4.28}$$

Hence, the equivalent resistance R_{eq} can be calculated by finding v in terms of i or i in terms of v.

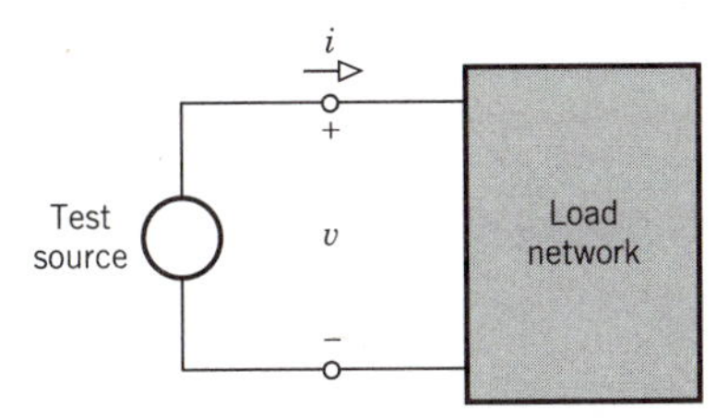

Figure 4.39 Test source for determining $R_{eq} = v/i$.

We have considerable freedom here since the test source may be either a voltage source or a current source, and we have the option of using either node or mesh analysis. Accordingly, we should choose the combination of test source and analysis method that leads most easily to the desired relationship between v and i for a given network. This usually means that we seek to minimize the number of simultaneous unknowns. The following example illustrates these considerations.

Example 4.13 *Equivalent Resistance of a Bridge Network*

Figure 4.40*a* shows a **resistive bridge**. Although this network consists entirely of resistors, it does not have a ladder structure that allows series-parallel reduction. We therefore turn to systematic analysis to find R_{eq}.

Suppose we assume a voltage test source and draw the circuit in Fig. 4.40*b*. Since we want to find the resulting current i, mesh analysis might appear to be the best choice. But mesh analysis would involve three unknown currents, whereas there are only two unknown node voltages and we can easily find i from v_1 and v_2. Pursuing the node analysis, we have

$$\begin{bmatrix} \frac{1}{12} + \frac{1}{4} + \frac{1}{3} & -\frac{1}{4} \\ -\frac{1}{4} & \frac{1}{8} + \frac{1}{4} + \frac{1}{2} \end{bmatrix} \begin{bmatrix} v_1 \\ v_2 \end{bmatrix} = \begin{bmatrix} \frac{v}{12} \\ \frac{v}{8} \end{bmatrix}$$

from which $v_1 = v_2 = v/5$. KCL at the reference node then yields

$$i = \frac{v_1}{3} + \frac{v_2}{2} = \frac{v}{15} + \frac{v}{10} = \frac{v}{6}$$

Hence, $R_{eq} = v/i = 6\ \Omega$.

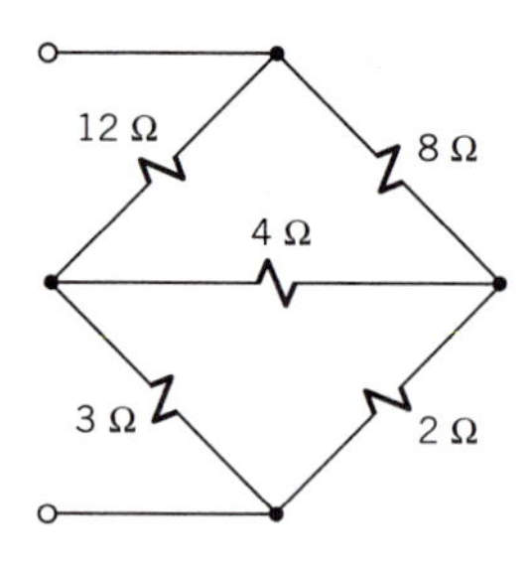

(a) Resistive bridge network

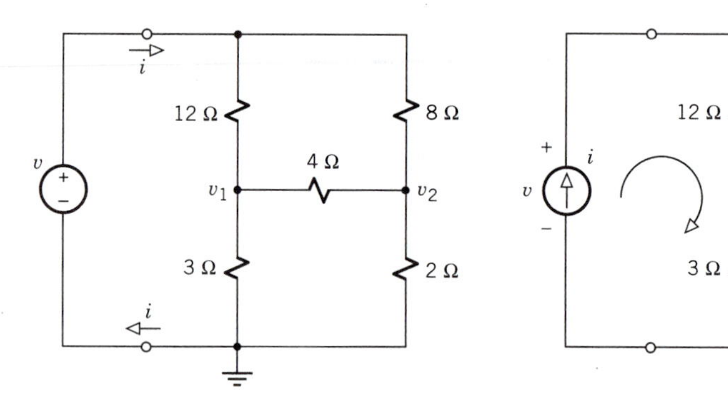

(b) Test voltage source for node analysis

(c) Test current source for mesh analysis

Figure 4.40

Alternatively, we could assume a test current source as shown in Fig. 4.40*c*. Now node analysis for v would involve three unknown voltages, whereas there are only two unknown mesh currents and we can easily find v from i_1 and i_2. The matrix mesh equation is

$$\begin{bmatrix} 12 + 8 + 4 & -4 \\ -4 & 3 + 4 + 2 \end{bmatrix} \begin{bmatrix} i_1 \\ i_2 \end{bmatrix} = \begin{bmatrix} 12i \\ 3i \end{bmatrix}$$

from which $i_1 = i_2 = 3i/5$. KCL around the outer loop then yields

$$v = 8i_1 + 2i_2 = \frac{24i}{5} + \frac{6i}{5} = 6i$$

Hence, $R_{eq} = v/i = 6\ \Omega$ as before.

For this particular network, either node or mesh analysis leads to a minimum of two unknowns, provided that we choose the appropriate test source.

Exercise 4.17

Consider the load network in Fig. 4.41. Find R_{eq} by choosing a test source and method that involve just one unknown.

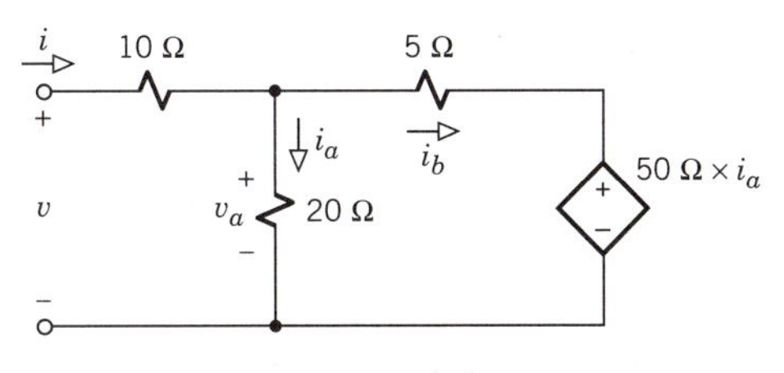

Figure 4.41

Thévenin Parameters

Recall from Section 2.5 that a two-terminal network consisting of resistors, controlled sources, and at least one independent source can be reduced to a Thévenin or Norton equivalent network. Here we discuss systematic methods for determining the Thévenin parameters of source networks.

If the network under consideration has a complicated structure, then node or mesh analysis may be used to find the open-circuit voltage v_{oc} or the short-circuit current i_{sc}. Systematic analysis may also be used to find the Thévenin resistance R_t, which is the equivalent resistance of the dead network. We need any two of these three parameters, since they are related by

$$v_{oc} = R_t i_{sc} \tag{4.29}$$

The direct approach therefore requires analyzing two different circuit configurations.

An indirect approach called the **one-step method** yields the Thévenin parameters from the analysis of just *one* circuit configuration. This method involves an external current or voltage source connected to the terminals of the source network.

Figure 4.42*a* shows the Thévenin model for an arbitrary source network, and an external current source has been connected to establish the terminal current *i*. The resulting terminal voltage will be

$$v = v_{oc} - R_t i \tag{4.30}$$

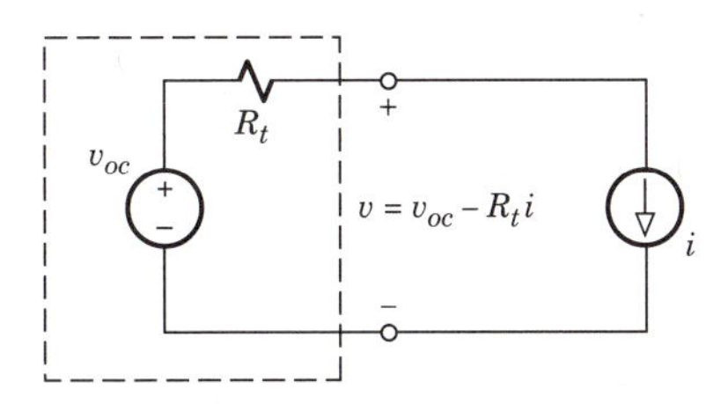

(a) Thévenin model with external current source

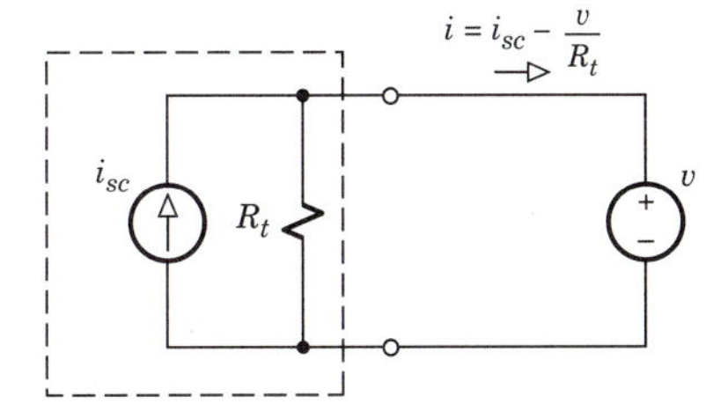

(b) Norton model with external voltage source

Figure 4.42 One-step method for Thévenin parameters.

Thus, v_{oc} and R_t can be found by using node analysis to obtain an expression for v in terms of i.

The dual situation in Fig. 4.42*b* employs the Norton model with an external voltage source establishing the terminal voltage v. The resulting terminal current will be

$$i = i_{sc} - \frac{v}{R_t} \tag{4.31}$$

Thus, i_{sc} and R_t can be found by using mesh analysis to obtain an expression for i in terms of v.

Example 4.14 *Thévenin Parameters of a Current Amplifier*

Figure 4.43*a* is the circuit model of a current amplifier. We want to find the Thévenin parameters looking back into the output terminals. Node analysis with an external current source would involve three unknown voltages, while mesh analysis with a voltage source would involve two unknown currents. We therefore choose the latter.

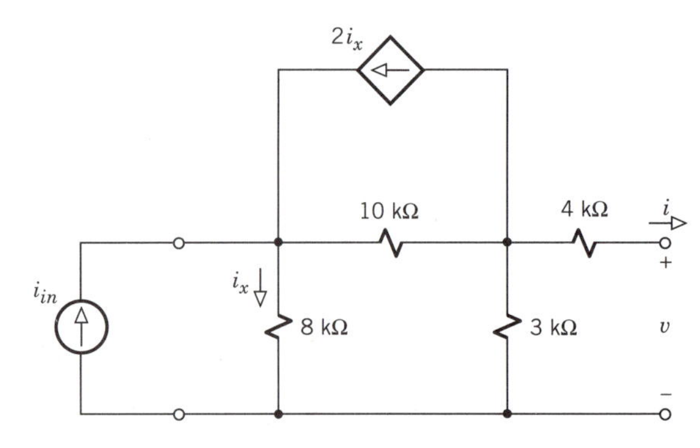

(*a*) Circuit model of a current amplifier

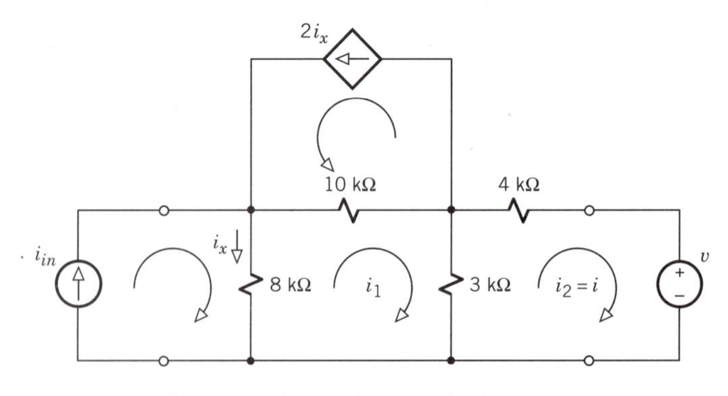

(*b*) Mesh analysis with external voltage source

Figure 4.43

Redrawing the circuit as shown in Fig. 4.43*b* and applying systematic inspection we obtain

$$[R] = \begin{bmatrix} 21 & -3 \\ -3 & 7 \end{bmatrix} \qquad i_x = i_{in} - i_1$$

$$[v_s] = \begin{bmatrix} 8i_{in} - 10 \times 2i_x \\ -v \end{bmatrix} = \begin{bmatrix} -12i_{in} \\ -v \end{bmatrix} + \begin{bmatrix} 20 & 0 \\ 0 & 0 \end{bmatrix} \begin{bmatrix} i_1 \\ i \end{bmatrix}$$

Thus,

$$\begin{bmatrix} 1 & -3 \\ -3 & 7 \end{bmatrix} \begin{bmatrix} i_1 \\ i \end{bmatrix} = \begin{bmatrix} -12i_{in} \\ -v \end{bmatrix}$$

Solving for i in terms of i_{in} and v yields

$$i = \frac{-v - 36i_{in}}{-2} = 18i_{in} - \frac{v}{-2}$$

Hence, by comparison with Eq. (4.31),

$$i_{sc} = 18i_{in} \qquad R_t = -2 \text{ k}\Omega$$

The open-circuit ouput voltage is then $v_{oc} = R_t i_{sc} = -36 \text{ k}\Omega \times i_{in}$.

Exercise 4.18

Use the one-step method with an external current source to find v_{oc} and R_t for the source network in Fig. 4.44. Note that there will be just one unknown node voltage.

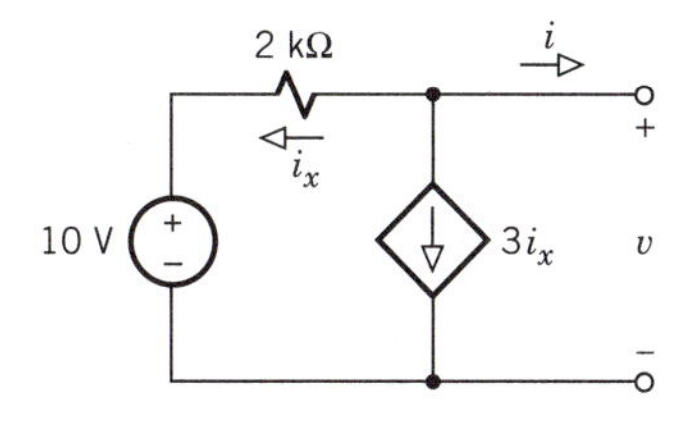

Figure 4.44

4.5 NODE ANALYSIS WITH IDEAL OP-AMPS†

We showed in Section 3.2 that real op-amps act essentially like ideal op-amps with infinite gain when the circuit provides negative feedback. Then we used the corresponding virtual short to analyze some basic op-amp circuits. Here, we combine virtual-short conditions with node analysis for the study of more complicated op-amp circuits.

Consider a circuit that includes one or more op-amps with negative feedback, so we can make the assumption of ideal op-amps. The virtual shorts at the op-amp inputs will determine the values of some voltages. However, further analysis may be needed to find other voltages and the resulting overall performance of the circuit.

Since the voltage at the output of an op-amp comes from a controlled source, it must be regarded as a *dependent* variable whose value will be whatever is needed to satisfy the virtual-short conditions. Consequently, a node equation cannot be written at the output terminal of an ideal op-amp. Instead, we proceed as follows:

1. Mark the virtual shorts at op-amp inputs, label all node voltages except those at any virtual grounds, and identify all unknown node voltages other than op-amp outputs.
2. Write a node equation for each unknown voltage, treating op-amp outputs as controlled source voltages.
3. In lieu of constraint equations, write the voltage relationships required by the virtual-short conditions.
4. Use the node equations and virtual-short relations to solve for the voltages of interest.

This procedure usually goes along quite rapidly because many op-amp circuits have just one unknown voltage.

Example 4.15 *Noninverting/Inverting Op-Amp Circuit*

Professor Bright tells you that the circuit in Fig. 4.45 may act either as a noninverting or inverting amplifier. You want to test the professor's claim by deriving an expression for v_{out}/v_{in}. After checking for negative feedback, you assume that both op-amps are ideal and you carry out the following analysis steps.

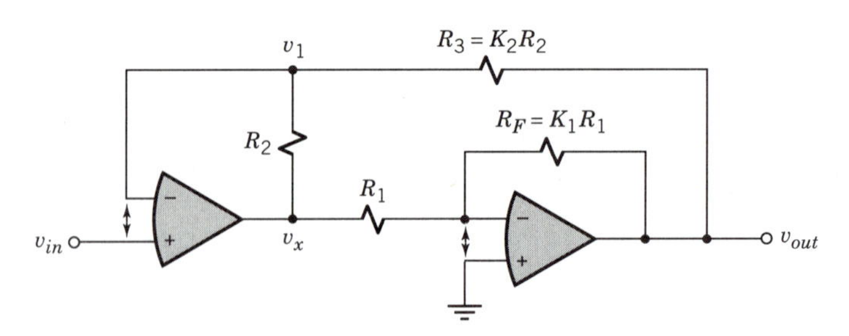

Figure 4.45 Noninverting/inverting op-amp circuit.

Step 1: You mark the virtual shorts and observe that the input to the second op-amp is a virtual ground. You then label all other voltages as shown, noting that v_1 is the only unknown node voltage that is not an op-amp output.

Step 2: The wire connecting v_1 to the inverting terminal of the first op-amp carries no current, so you write the single node equation for v_1 as

$$\left(\frac{1}{R_2} + \frac{1}{K_2 R_2}\right) v_1 = \frac{v_x}{R_2} + \frac{v_{out}}{K_2 R_2}$$

Step 3: The virtual-short condition at the input of the first op-amp requires

$$v_1 = v_{in}$$

The virtual ground at the input of the second op-amp requires $v_x/R_1 + v_{out}/R_F = 0$, so

$$v_{out} = -(R_F/R_1)\, v_x = -K_1 v_x$$

This expression agrees, of course, with the fact that the second op-amp is part of a standard inverting amplifier configuration.

Step 4: Now you eliminate v_1 and v_x from the node equation by substituting $v_1 = v_{in}$ and $v_x = -v_{out}/K_1$. Hence,

$$\left(1 + \frac{1}{K_2}\right) v_{in} = \frac{-v_{out}}{K_1} + \frac{v_{out}}{K_2} = \frac{K_1 - K_2}{K_1 K_2} v_{out}$$

from which

$$\frac{v_{out}}{v_{in}} = \frac{K_1(K_2 + 1)}{K_1 - K_2}$$

Your result thus confirms the interesting property that the circuit serves as a noninverting amplifier with $v_{out}/v_{in} > 0$ when $K_1 > K_2$, whereas it becomes an inverting amplifier with $v_{out}/v_{in} < 0$ when $K_1 < K_2$. You also see that K_1 and K_2 should be chosen with care because $|v_{out}/v_{in}| \rightarrow \infty$ when $K_2 = K_1$ and $|v_{out}|$ can be very large when $K_2 \approx K_1$.

Exercise 4.19

Let the op-amp circuit in Fig. 4.46 have

$$R_1 + R_2 = R_4 \qquad R_1 = R_4/K_1 \qquad R_3 = R_4/K_2$$

Assume an ideal op-amp to show that

$$v_{out}/v_{in} = K_1(K_2 + 2)$$

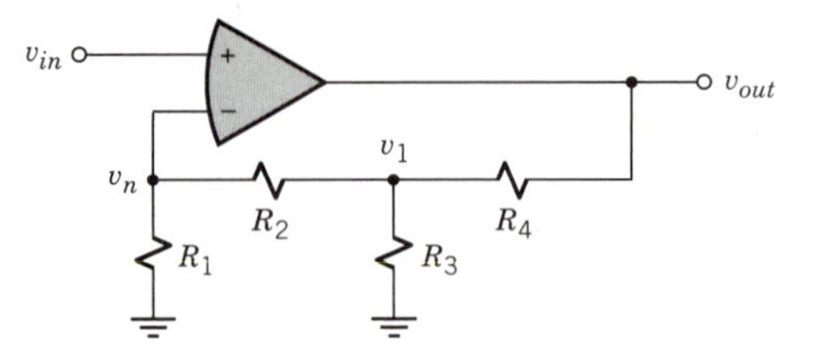

Figure 4.46

Hint: Do not take v_n as an unknown node voltage since it is easily related to v_1 by a voltage divider.

4.6 DELTA-WYE TRANSFORMATIONS†

We close this chapter with an investigation of resistive networks having *three* external terminals. Such networks play a major role in many AC power systems, and they also appear as part of other networks such as resistive bridges. We'll use systematic analysis to derive expressions for the equivalent resistance between any pair of terminals in a three-terminal resistive network. These expressions then lead to handy network transformations.

There are two basic configurations of three-terminal resistive networks, the **wye (Y) network** in Fig. 4.47*a* and the **delta (Δ) network** in Fig. 4.47*b*. The names of these networks reflect their distinctive structures. We want to determine the relationships between the resistances that allow us to transform a given delta network into an equivalent wye network, or vice versa.

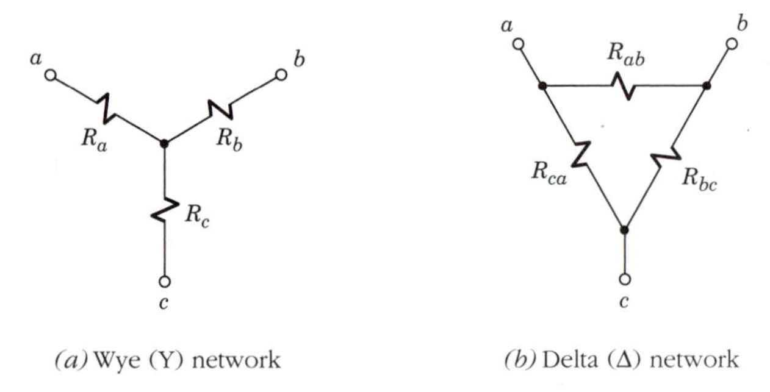

Figure 4.47 Three-terminal resistive networks.

To begin the investigation, we take the bottom node as the reference and redraw each network in a four-terminal configuration with input and output terminal pairs. The wye structure then becomes the **tee (T) network** in Fig. 4.48*a*, while the delta structure becomes the **pi (Π) network** in Fig. 4.48*b*.

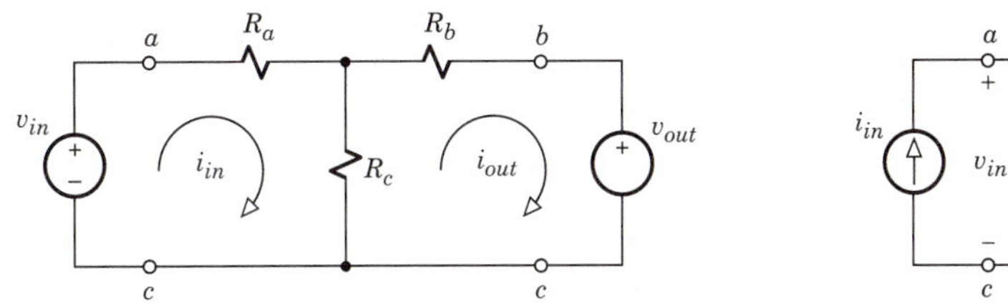

(a) Wye network connected as a tee (T) network

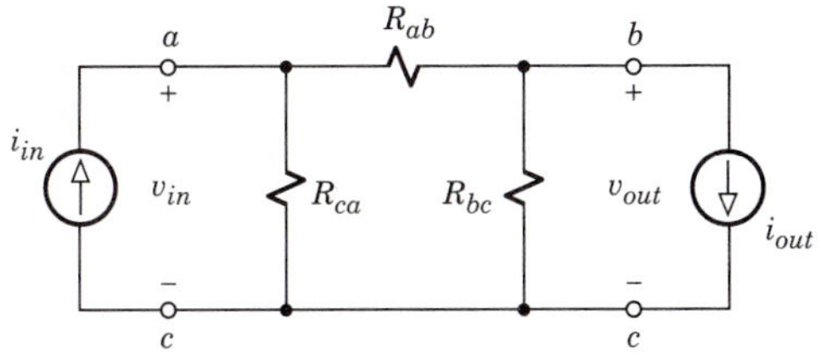

(b) Delta network connected as a pi (Π) network

Figure 4.48

External sources have also been connected at the terminal pairs for purposes of analysis. If the networks are equivalent at the input and output terminal pairs, then the input and output voltage-current relationships of the wye/tee network must be the same as the input and output voltage-current relationships of the delta/pi network.

The mesh equations for the wye/tee network are easily obtained by inspection of Fig. 4.48*a* to be

$$\begin{aligned}(R_a + R_c)\, i_{in} - R_c i_{out} &= v_{in} \\ -R_c i_{in} + (R_b + R_c)\, i_{out} &= -v_{out}\end{aligned} \tag{4.32}$$

Solving for the currents in terms of the voltages then gives

$$\begin{aligned}i_{in} &= \frac{(R_b + R_c) v_{in} - R_c v_{out}}{R_a R_b + R_b R_c + R_c R_a} \\ i_{out} &= \frac{R_c v_{in} - (R_a + R_c) v_{out}}{R_a R_b + R_b R_c + R_c R_a}\end{aligned} \tag{4.33}$$

The node equations for the delta/pi network in Fig. 4.48*b* are

$$\begin{aligned}\left(\frac{1}{R_{ca}} + \frac{1}{R_{ab}}\right) v_{in} - \frac{1}{R_{ab}} v_{out} &= i_{in} \\ -\frac{1}{R_{ab}} v_{in} + \left(\frac{1}{R_{bc}} + \frac{1}{R_{ab}}\right) v_{out} &= -i_{out}\end{aligned} \tag{4.34}$$

and solving for the voltages yields

$$\begin{aligned}v_{in} &= \frac{(R_{ab} R_{ca} + R_{ca} R_{bc})\, i_{in} - R_{ca} R_{bc} i_{out}}{R_{ab} + R_{bc} + R_{ca}} \\ v_{out} &= \frac{R_{ca} R_{bc} i_{in} - (R_{bc} R_{ab} + R_{ca} R_{bc})\, i_{out}}{R_{ab} + R_{bc} + R_{ca}}\end{aligned} \tag{4.35}$$

Equations (4.32) and (4.35) express v_{in} and v_{out} in terms of i_{in} and i_{out} for each network. Upon equating both expressions for v_{in} and both expressions for v_{out}, we find that the wye/tee network described by Eq. (4.32) will be equivalent to the delta/pi network described by Eq. (4.35) when

$$R_a = \frac{R_{ab}R_{ca}}{\Sigma R} \qquad R_b = \frac{R_{bc}R_{ab}}{\Sigma R} \qquad R_c = \frac{R_{ca}R_{bc}}{\Sigma R} \tag{4.36a}$$

where

$$\Sigma R \triangleq R_{ab} + R_{bc} + R_{ca} \tag{4.36b}$$

These relationships constitute the **delta-to-wye transformation**, meaning that a given delta network acts at its terminals like a wye network whose resistances are related to the delta resistances by Eqs. (4.36a) and (4.36b). Observe that the equivalent wye resistance R_a equals the product of the two delta resistances connected at node a divided by the sum of the delta resistances. Like interpretations hold for R_b and R_c.

Conversely, from Eqs. (4.33) and (4.34) we find that the delta/pi network will be equivalent at its terminals to a wye/tee network with

$$R_{ab} = \frac{R^2}{R_c} \qquad R_{bc} = \frac{R^2}{R_a} \qquad R_{ca} = \frac{R^2}{R_b} \tag{4.37a}$$

where

$$R^2 \triangleq R_aR_b + R_bR_c + R_cR_a \tag{4.37b}$$

These relationships thus constitute the **wye-to-delta transformation**. Observe that the equivalent delta resistance R_{ab} connecting nodes a and b equals the sum of the pairwise products of the wye resistances divided by the wye resistance at the opposite node. Like interpretations hold for R_{bc} and R_{ca}.

The following example illustrates how delta-wye transformations may be used to analyze resistive networks that do not have a ladder structure.

Example 4.16 *Transformation of a Resistive Bridge*

Figure 4.49*a* shows the resistive bridge whose equivalent resistance was determined by systematic analysis in Example 4.13. But now we see that portions of the bridge may be viewed as wye or delta networks. This viewpoint is advantageous because a wye-to-delta or delta-to-wye transformation will reduce the bridge to a series-parallel structure. Then we can find the equivalent resistance without having to solve node or mesh equations.

As indicated by the labeling in Fig. 4.49*b*, we'll transform the wye network on the left into an equivalent delta network. The common numerator term for Eq. (4.37a) is

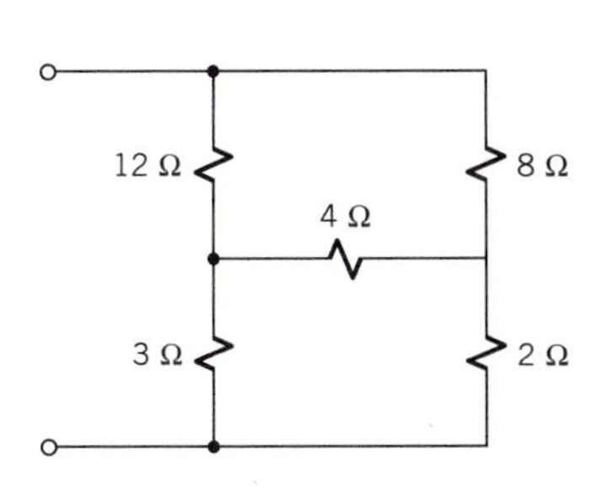

(a) Resistive bridge network

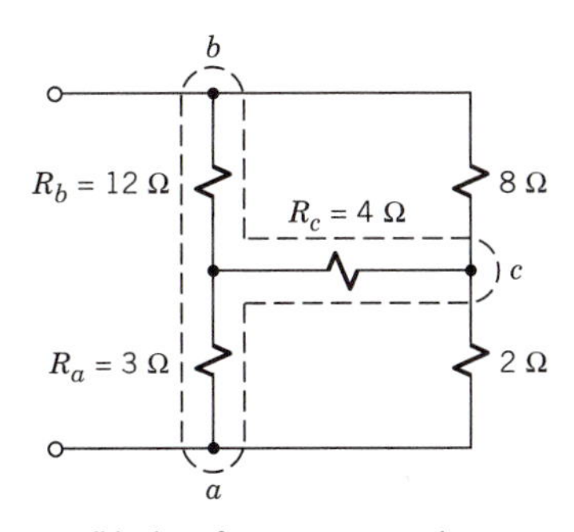

(b) Identifying an internal wye network

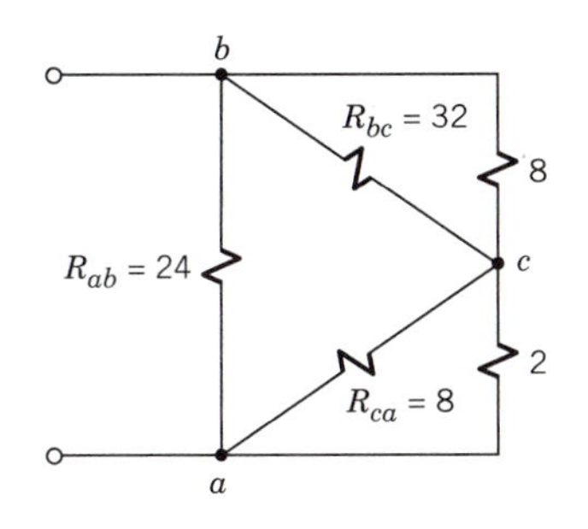

(c) After wye-to-delta transformation

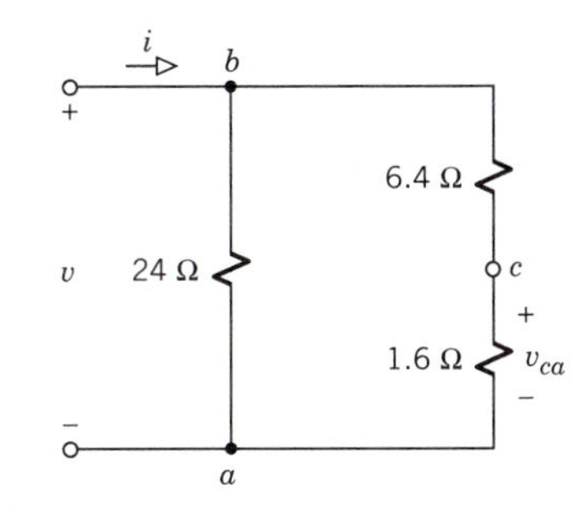

(d) Reduced equivalent network

Figure 4.49

$$R^2 = R_a R_b + R_b R_c + R_c R_a = 96\ \Omega^2$$

so the equivalent delta resistances are

$$R_{ab} = \frac{96}{4} = 24\ \Omega \qquad R_{bc} = \frac{96}{3} = 32\ \Omega \qquad R_{ca} = \frac{96}{12} = 8\ \Omega$$

The resulting transformed network is given in Fig. 4.49*c*.

Now we can combine the parallel resistances $R_{bc} \| 8 = 6.4\ \Omega$ and $R_{ca} \| 2 = 1.6\ \Omega$. The diagram thus reduces to Fig. 4.49*d*, where

$$R_{eq} = v/i = 24 \| (6.4 + 1.6) = 6\ \Omega$$

which agrees with our result from Example 4.13. We also get a bonus from this diagram, namely, the voltage–divider relationship given by $v_{ca}/v = 1.6/(6.4 + 1.6) = 0.2$.

Exercise 4.20

Calculate R_{eq} for Fig. 4.49*a* by performing a delta-to-wye transformation, treating the upper three resistors as an upside down delta/pi network.

PROBLEMS

Note: *A programmed calculator is recommended for those problems that require numerical solution of three simultaneous equations. Problems that involve two simultaneous equations may be solved by hand or by calculator.*

Section 4.1 Node Analysis

4.1* Use node analysis to find v_1, v_2, v_3, and i_1 in Fig. P4.1 with i_s = 2 A and R = 20 Ω.

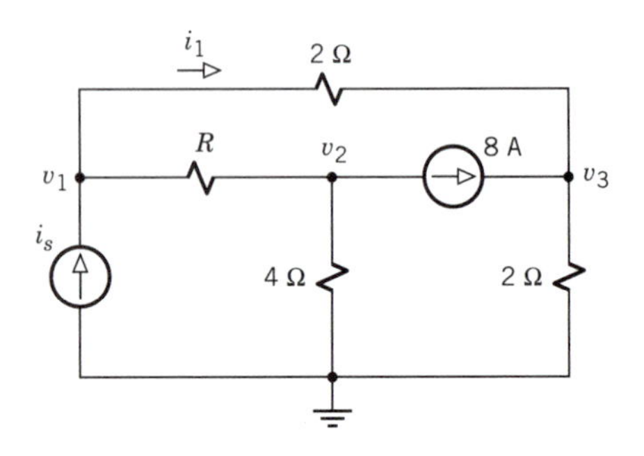

Figure P4.1

4.2 Use node analysis to find v_1, v_2, v_3, and i_1 in Fig. P4.1 with i_s = 6 A and R = 10 Ω.

4.3 Calculate v_1, v_2, and v_3 in Fig. P4.3, given that an external current source supplies i_1 = 3 A.

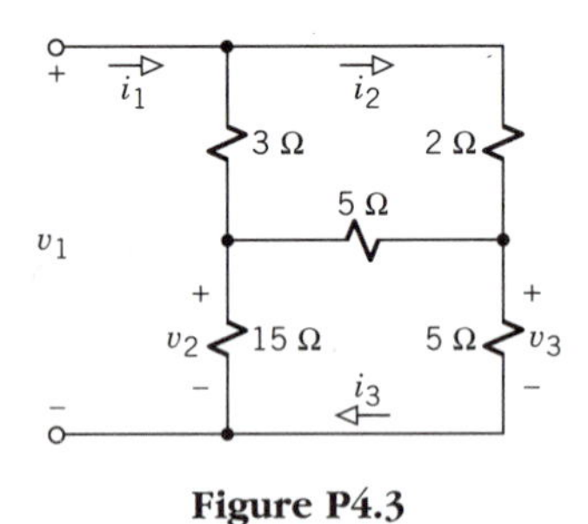

Figure P4.3

4.4 Calculate v_1, v_2, and v_3 in Fig. P4.4, given that an external current source supplies i_1 = 5 A.

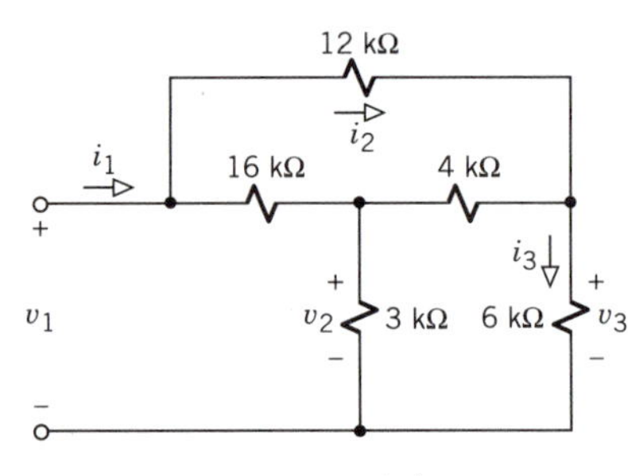

Figure P4.4

4.5 Let R_x = 18 kΩ and i = 0 in Fig. P4.5. Find v and the power supplied by the current source.

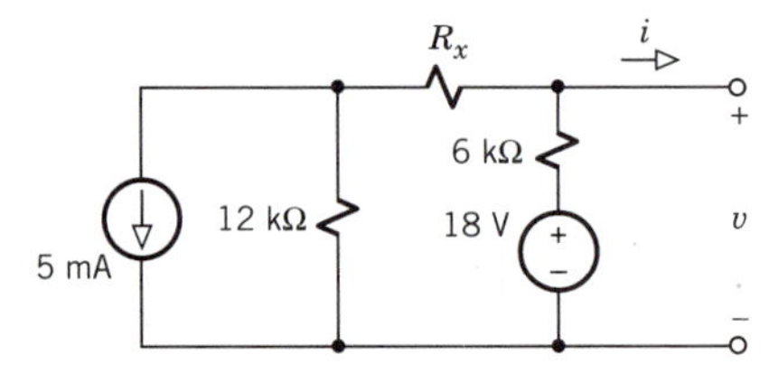

Figure P4.5

4.6* Let v_s = 60 V and i_s = 3 mA in Fig. P4.6. Find v_2, v_4, and the power supplied by each source.

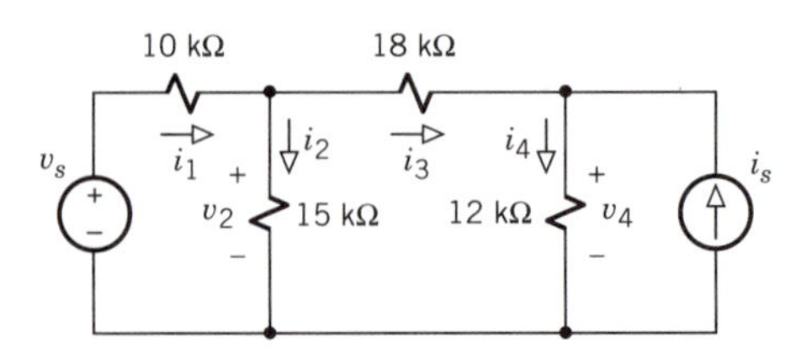

Figure P4.6

4.7 Let $v_s = 10$ V and $i_s = 5$ mA in Fig. P4.6. Use node analysis to find v_2, v_4, and the power supplied by each source.

4.8 If $v_s = 12$ V and $R_3 = 8\ \Omega$ in Fig. P4.8, then what are the values of v_a, v_b, and i_2?

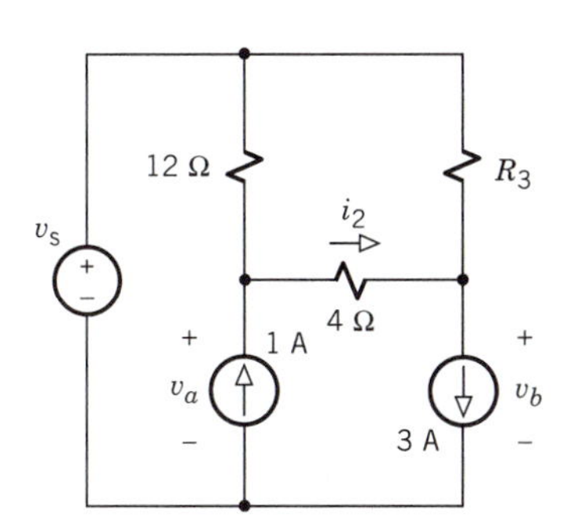

Figure P4.8

4.9 If $v_s = 18$ V and $R_3 = 20\ \Omega$ in Fig. P4.8, then what are the values of v_a, v_b, and i_2?

4.10 Using node analysis, determine the power supplied by each source in Fig. P4.10.

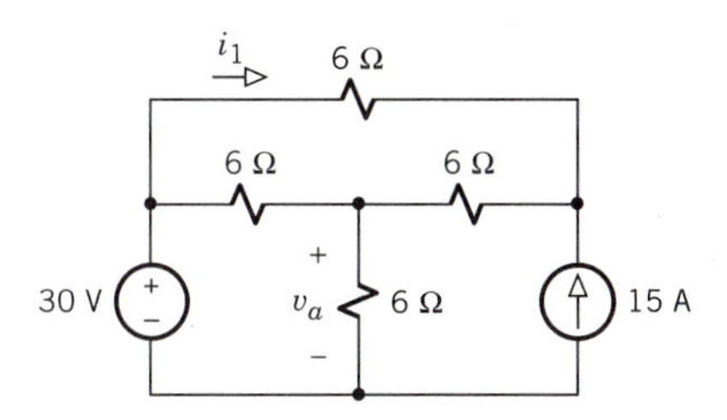

Figure P4.10

4.11* Apply node analysis to find v_1, v_2, v_3, i_1, and i_2 in Fig. P4.11 when $v_s = 42$ V and $R = 5\ \Omega$.

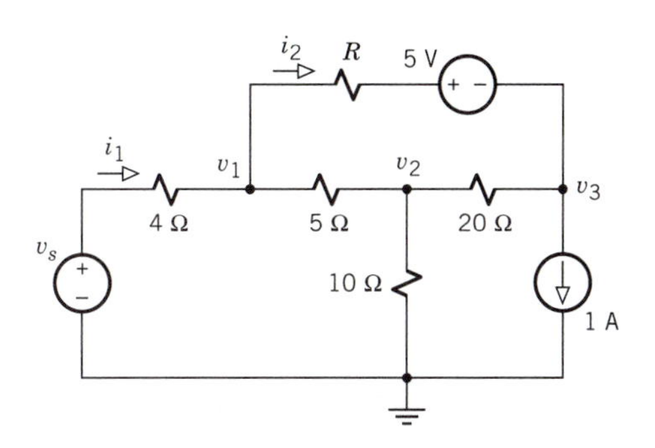

Figure P4.11

4.12 Apply node analysis to find v_1, v_2, v_3, i_1, and i_2 in Fig. P4.11 when $v_s = 16$ V and $R = 20\ \Omega$.

4.13 Let $v_s = 10$ V and $R = 4$ kΩ in Fig. P4.13. Find v_1, v_2, i_1, i_2, and i_3.

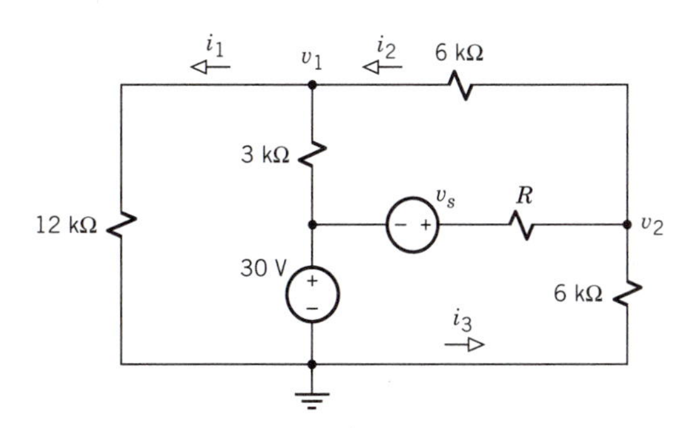

Figure P4.13

4.14 Let $v_s = 25$ V and $R = 1$ kΩ in Fig. P4.13. Find v_1, v_2, i_1, i_2, and i_3.

4.15* Let $v_s = 22$ V and $R = 3\ \Omega$ in Fig. P4.15. Use supernode analysis to calculate v_1, v_2, i_1, i_2, and i_3

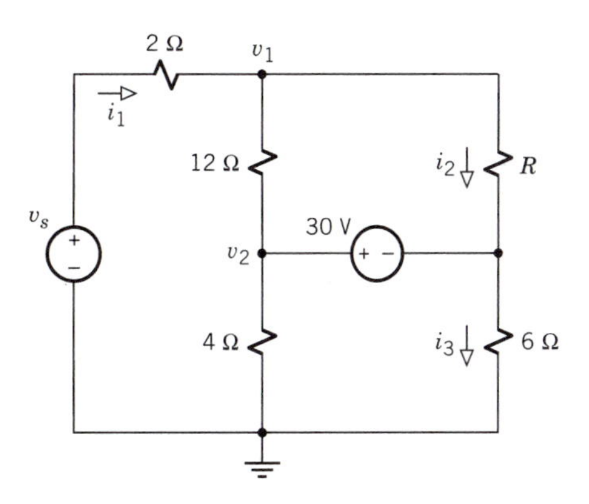

Figure P4.15

4.16 Let $v_s = 23$ V and $R = 12\ \Omega$ in Fig. P4.15. Use supernode analysis to calculate v_1, v_2, i_1, i_2, and i_3.

4.17 Evaluate v_1, v_2, i_1, i_2, and i_3 by node analysis of Fig. P4.17 with $i_s = 3.6$ A and $R_a = R_b = 10\ \Omega$.

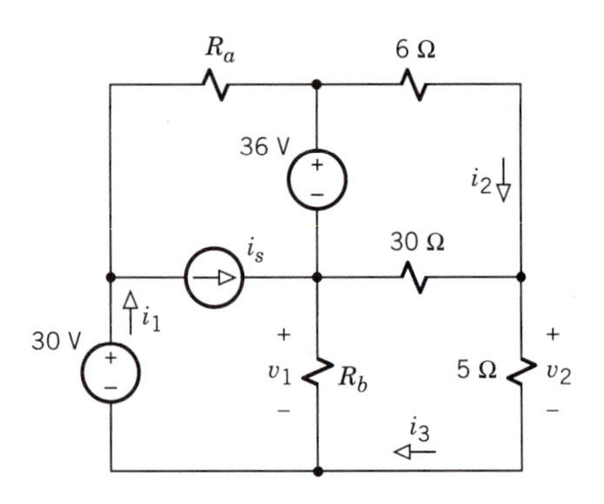

Figure P4.17

4.18 Evaluate v_1, v_2, i_1, i_2, and i_3 by node analysis of Fig. P4.17 with $i_s = 2$ A, $R_a = 3\ \Omega$, and $R_b = 15\ \Omega$.

4.19 Given that $i_s = 5$ mA and $R = 4$ kΩ in Fig. P4.19, find v_1, v_2, i_1, and i_2 by node analysis.

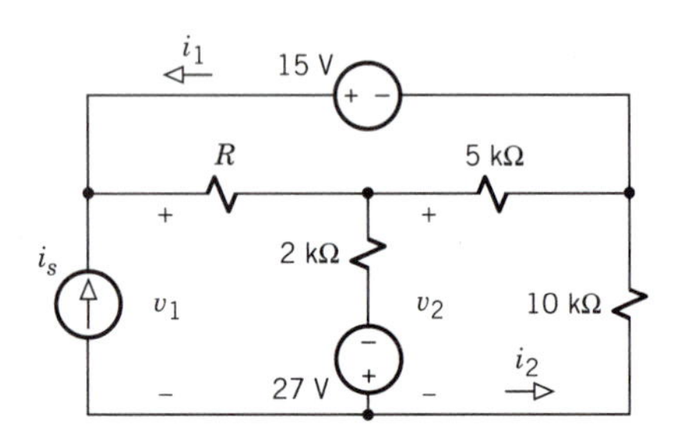

Figure P4.19

4.20 Given that $i_s = 3$ mA and $R = 60$ kΩ in Fig. P4.19, find v_1, v_2, i_1, and i_2 by node analysis.

4.21 Let $R = 3\ \Omega$ and $i_s = 4$ A in Fig. P4.21. Use node analysis to calculate v_1, v_2, i_1, and i_2.

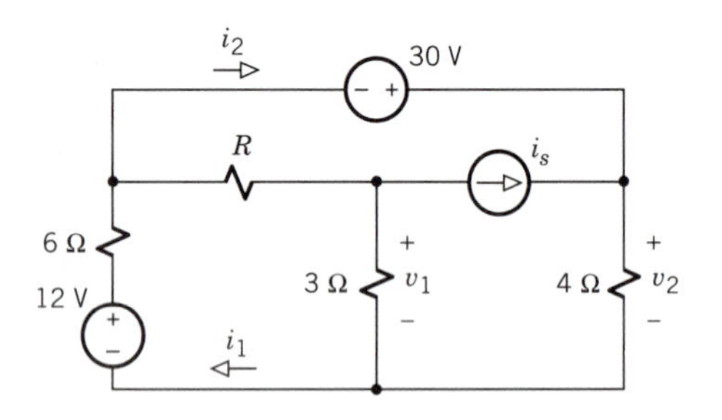

Figure P4.21

4.22 Let $R = 11\ \Omega$ and $i_s = 7$ A in Fig. P4.21. Use node analysis to calculate v_1, v_2, i_1, and i_2.

Section 4.2 Mesh Analysis

4.23* Let $v_s = 22$ V and $R = 3\ \Omega$ in Fig. P4.15. Use mesh analysis to calculate i_1, i_2, i_3, v_1, and v_2.

4.24 Let $v_s = 23$ V and $R = 12\ \Omega$ in Fig. P4.15. Use mesh analysis to calculate i_1, i_2, i_3, v_1, and v_2.

4.25 Let $v_s = 10$ V and $R = 4$ kΩ in Fig. P4.13. Find i_1, i_2, i_3, v_1, and v_2 by mesh analysis.

4.26 Let $v_s = 25$ V and $R = 1$ kΩ in Fig. P4.13. Find i_1, i_2, i_3, v_1, and v_2 by mesh analysis.

4.27 Calculate i_1, i_2, and i_3 in Fig. P4.3, given that an external voltage source supplies $v_1 = 30$ V.

4.28 Calculate i_1, i_2, and i_3 in Fig. P4.4, given that an external voltage source supplies $v_1 = 9$ V.

4.29* Let $v_s = 40$ V and $i_s = 5$ mA in Fig. P4.6. Find i_1, i_3, and the power supplied by each source.

4.30 Let $v_s = 20$ V and $i_s = 10$ mA in Fig. P4.6. Find i_1, i_3, and the power supplied by each source.

4.31 Let $R_x = 0$ in Fig. P4.5, and let a 12-kΩ load resistor be connected across the output terminals. Calculate i and the power supplied by the current source.

4.32 Let $R_x = 18$ kΩ in Fig. P4.5, and let a short circuit be connected across the output terminals. Calculate i and the power supplied by the current source.

4.33 Using mesh analysis, determine the power supplied by each source in Fig. P4.10.

4.34 Apply mesh analysis to find i_1, i_2, v_1, v_2, and v_3 in Fig. P4.11 when $v_s = 42$ V and $R = 5\ \Omega$.

4.35 Apply mesh analysis to find i_1, i_2, v_1, v_2, and v_3 in Fig. P4.11 when $v_s = 16$ V and $R = 20\ \Omega$.

4.36 Given that $i_s = 5$ mA and $R = 4$ kΩ in Fig. P4.19, find i_1, i_2, v_1, and v_2 by mesh analysis.

4.37 Given that $i_s = 3$ mA and $R = 60$ kΩ in Fig. P4.19, find i_1, i_2, v_1, and v_2 by mesh analysis.

4.38* Use supermesh analysis to find i_1 in Fig. P4.1 with $i_s = 6$ A and $R = 10\ \Omega$. Then calculate v_1, v_2, and v_3.

4.39 Use supermesh analysis to find i_1 in Fig. P4.1 with $i_s = 2$ A and $R = 20\ \Omega$. Then calculate v_1, v_2, and v_3.

4.40 Let $R = 3\ \Omega$ and $i_s = 4$ A in Fig. P4.21. Use mesh analysis to calculate i_1, i_2, v_1, and v_2.

4.41 Let $R = 11\ \Omega$ and $i_s = 7$ A in Fig. P4.21. Use mesh analysis to calculate i_1, i_2, v_1, and v_2.

4.42 Evaluate i_1, i_2, i_3, v_1, and v_2 by mesh analysis of Fig. P4.17 with $i_s = 3.6$ A and $R_a = R_b = 10\ \Omega$.

4.43 Evaluate i_1, i_2, i_3, v_1, and v_2 by mesh analysis of Fig. P4.17 with $i_s = 2$ A, $R_a = 3\ \Omega$, and $R_b = 15\ \Omega$.

Section 4.3 Systematic Analysis with Controlled Sources

4.44* Let $R_x = 4$ kΩ and $v_c = 5$ kΩ × i_x in Fig. P4.44. Find i_x and i_z.

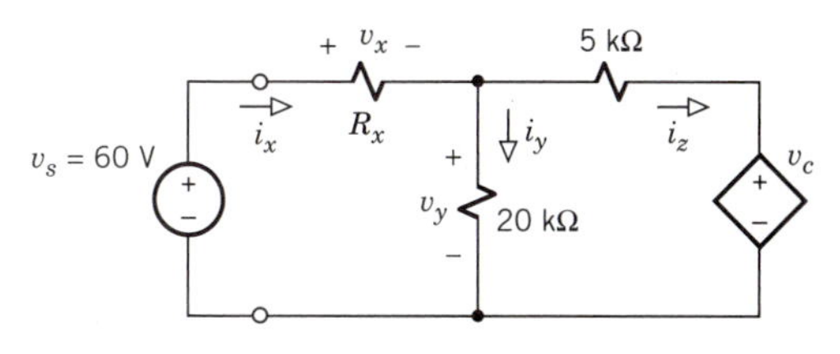

Figure P4.44

4.45 Let $R_x = 8\ \text{k}\Omega$ and $v_c = 30\ \text{k}\Omega \times i_y$ in Fig. P4.44. Find i_x and i_z.

4.46 Find i_x and i in Fig. P4.46, given that $i_s = 3$ A, $v_c = 90\ \Omega \times i_x$, and a 10-Ω load resistor is connected across the output terminals.

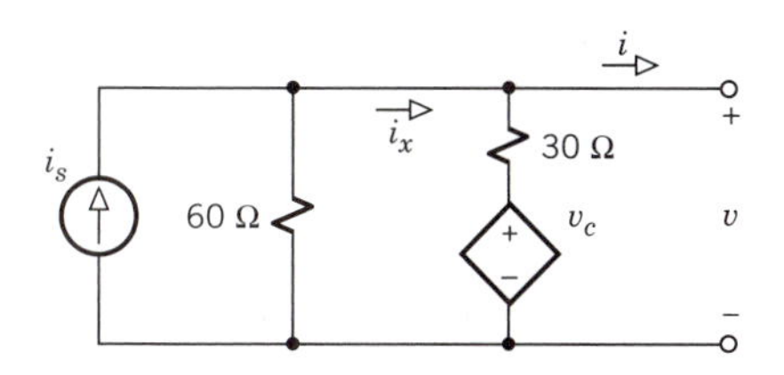

Figure P4.46

4.47 Find i_x and i in Fig. P4.46, given that $i_s = 1$ A, $v_c = 2v$, and an 80-Ω load resistor is connected across the output terminals.

4.48 If $i_c = v_2/(50\ \Omega)$ and $v_c = 4\ \Omega \times i_x$ in Fig. P4.48, then what are the values of i_1 and i_2?

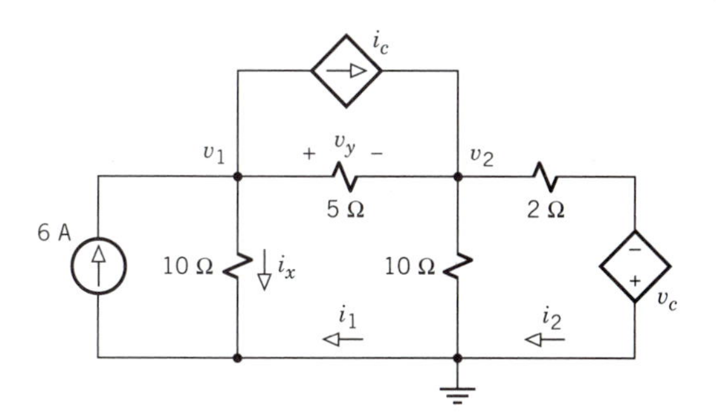

Figure P4.48

4.49 If $v_c = 14\ \text{k}\Omega \times i_1$ and $i_c = v_x/(25\ \text{k}\Omega)$ in Fig. P4.49, then what are the values of i_1 and i_2?

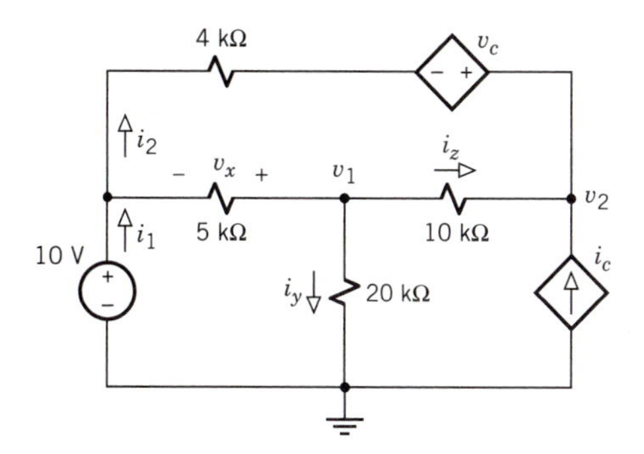

Figure P4.49

4.50 If $v_c = 4v_x$ and $i_c = 3i_y$ in Fig. P4.49, then what are the values of i_1 and i_2?

4.51 Let the circuit in Fig. P4.51 have $v_s = 24$ V, $i_c = 3i_y$, $v_c = 1\ \Omega \times i_x$, and $R = \infty$ (an open circuit). (a) Redraw the circuit for mesh analysis with i_1 and i_2 as the unknowns. Then form and solve the matrix mesh equation for i_1 and i_2. (b) Use the results of (a) to calculate v_1, v_2, and the equivalent input resistance $R_{eq} = v_s/i_1$.

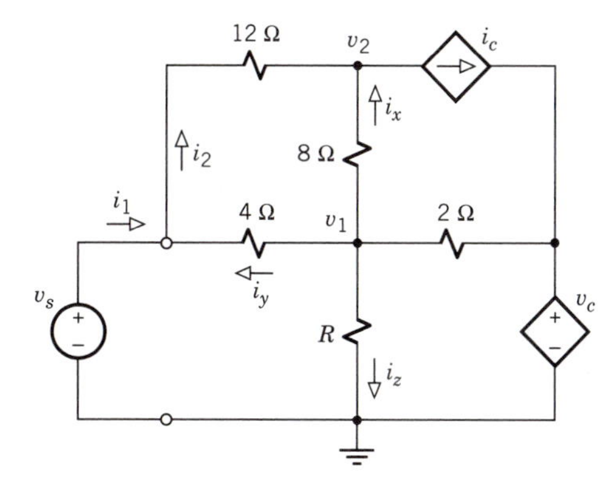

Figure P4.51

4.52* Find v_x and v_z in Fig. P4.52 when $R_y = 3\ \Omega$ and $i_c = v_x/(3\ \Omega)$.

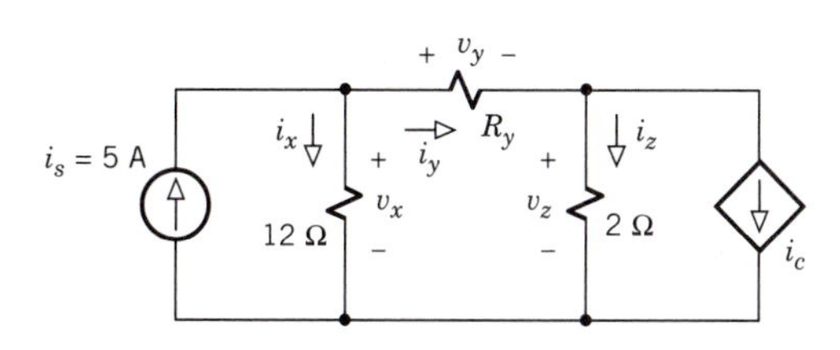

Figure P4.52

4.53 Find v_x and v_z in Fig. P4.52 when $R_y = 6\ \Omega$ and $i_c = 4i_y$.

4.54 Find v_x and v in Fig. P4.54, given that $v_s = 8$ V, $i_c = v_x/(1\ \text{k}\Omega)$, and a 12-k$\Omega$ load resistor is connected across the output terminals.

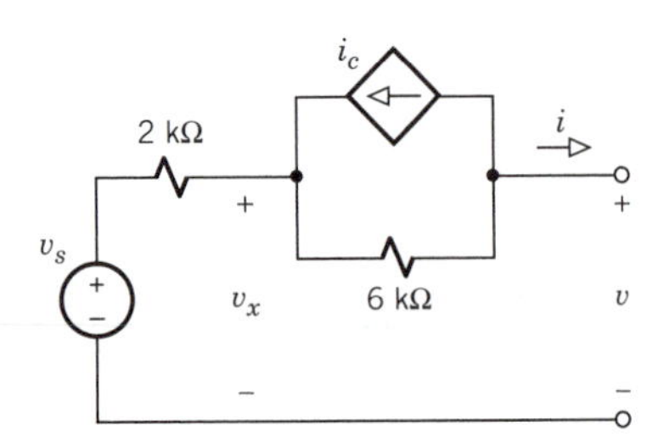

Figure P4.54

4.55 Find v_x and v in Fig. P4.54, given that $v_s = 25$ V, $i_c = 4i$, and an 18-kΩ load resistor is connected across the output terminals.

4.56 If $v_c = 3v_x$ and $i_c = v_2/(8\ \text{k}\Omega)$ in Fig. P4.49, then what are the values of v_1 and v_2?

4.57 If $v_c = 18\ \text{k}\Omega \times i_z$ and $i_c = 8i_y$ in Fig. P4.49, then what are the values of v_1 and v_2?

4.58 If $i_c = 3i_x$ and $v_c = v_y$ in Fig. P4.48, then what are the values of v_1 and v_2?

4.59 Let the circuit in Fig. P4.51 have $v_s = 12$ V, $i_c = 2i_x$, $v_c = 6\ \Omega \times i_z$, and $R = 3\ \Omega$. (a) Redraw the circuit for node analysis with v_1 and v_2 as the unknowns. Then form and solve the matrix node equation for v_1 and v_2. (b) Use the results of (a) to calculate i_1, i_2, and the equivalent input resistance $R_{eq} = v_s/i_1$.

4.60 Figure P4.60 is the model of a current amplifier with $g_m = 2$ mS and load resistance R_L. The goal of this problem is to find the input resistance and the Thévenin parameters of the output. (a) Obtain the matrix mesh equation for i_1 and i_{out} with i_{in} and R_L being arbitrary. Then solve for i_1 and i_{out} in terms of i_{in} and R_L. (b) Use the results of (a) to evaluate i_{out}/i_{in}, i_1/i_{in}, and $R_i = v_{in}/i_{in}$ when $R_L = 0$, 1 kΩ, and ∞. (c) Use the results of (a) to find the output Thévenin parameters v_{out-oc}, i_{out-sc}, and $R_o = v_{out-oc}/i_{out-sc}$. (d) Check your value of R_o by suppressing i_{in} and replacing R_L with a test current source so $i_{out} = -i_t$. Then write a single node equation for $v_{out} = v_t$ and calculate $R_o = v_t/i_t$.

4.61 Figure P4.61 is the model of a voltage amplifier with load resistance R_L. The goal of this problem is to find the input resistance and the Thévenin parameters of the output. (a) Obtain the matrix node equation for v_1 and v_{out} with v_{in} and $G_L = 1/R_L$ being arbitrary. Then solve for v_1 and v_{out} in terms of v_{in} and G_L. (b) Use the results of (a) to evaluate v_{out}/v_{in}, v_1/v_{in}, and $R_i = v_{in}/i_{in}$ when $R_L = 0$, 1 kΩ, and ∞. (c) Use the results of (a) to find the output Thévenin parameters v_{out-oc}, i_{out-sc}, and $R_o = v_{out-oc}/i_{out-sc}$. (d) Check your value of R_o by suppressing v_{in} and replacing R_L with a test current source so $i_{out} = -i_t$. Then write a single node equation for $v_{out} = v_t$ and calculate $R_o = v_t/i_t$.

Section 4.4 Applications of Systematic Analysis

(See also PSpice problems B.11 and B.12.)

4.62* Evaluate R_{eq} for the network in Fig. P4.3 by solving a pair of node equations.

4.63 Evaluate R_{eq} for the network in Fig. P4.4 by solving a pair of node equations.

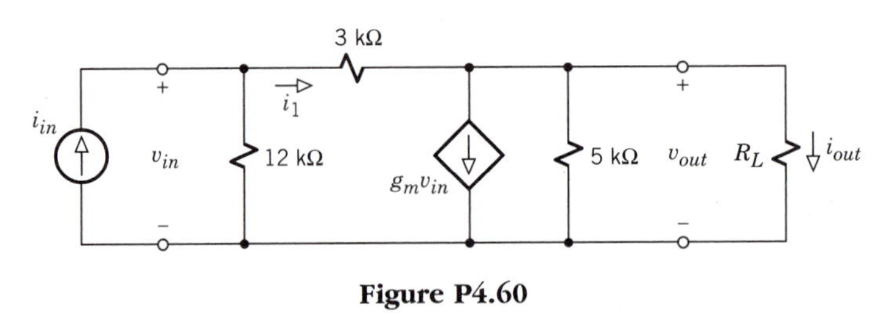

Figure P4.60

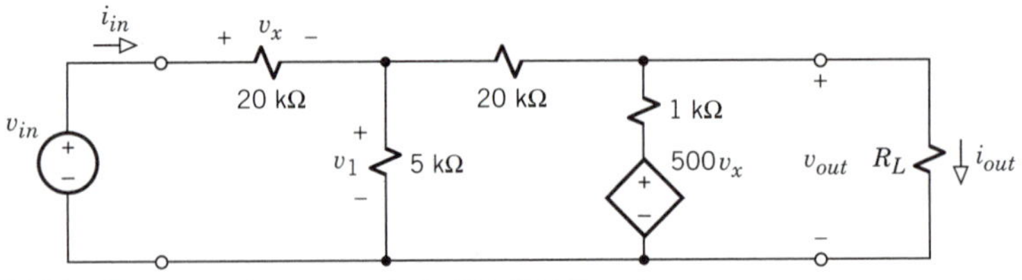

Figure P4.61

4.64 Evaluate R_{eq} for the network in Fig. P4.3 by solving a pair of mesh equations.

4.65 Evaluate R_{eq} for the network in Fig. P4.4 by solving a pair of mesh equations.

4.66 Using mesh analysis, find R_{eq} for Fig. P4.66, given that $R = 6\ \Omega$ and $i_c = 4i$.

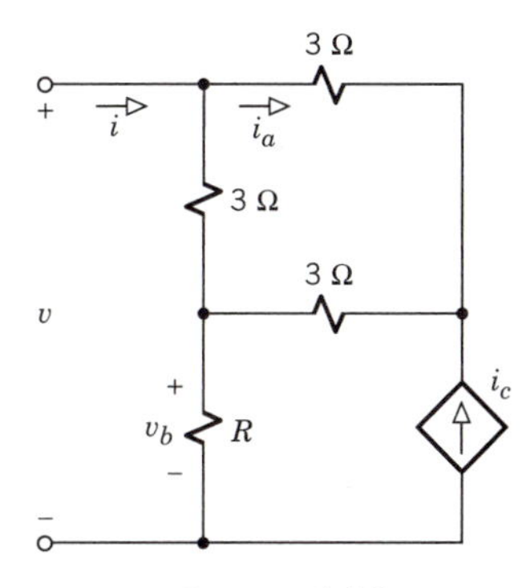

Figure P4.66

4.67 Using mesh analysis, find R_{eq} for Fig. P4.66, given that $R = 9\ \Omega$ and $i_c = 2i_a$.

4.68 Using node analysis, find R_{eq} for Fig. P4.68, given that $R = 4$ kΩ and $v_c = 5v$.

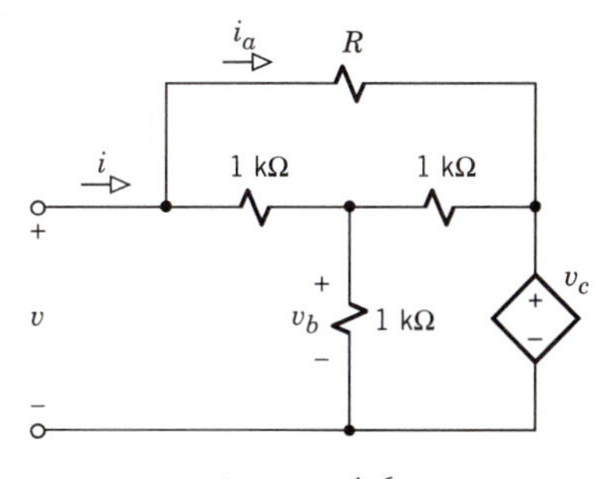

Figure P4.68

4.69 Using node analysis, find R_{eq} for Fig. P4.68, given that $R = 10$ kΩ and $v_c = 4v_b$.

4.70 Solve one node equation to determine v_{oc} and R_t for the network in Fig. P4.46 with $i_s = 0.5$ A and $v_c = 2v$.

4.71 Solve one mesh equation to determine i_{sc} and R_t for the network in Fig. P4.54 with $v_s = 8$ V and $i_c = v_x/(1\ \text{k}\Omega)$.

4.72* Use the one-step method and node analysis to find v_{oc} and R_t for the network in Fig. P4.54 when $v_s = 6$ V and $i_c = 4i$.

4.73 Use the one-step method and mesh analysis to find i_{sc} and R_t for the network in Fig. P4.46 when $i_s = 1$ A and $v_c = 45\ \Omega \times i$.

4.74 Figure P4.74 is the model of a voltage amplifier. Find v_{oc} and R_t with $R_x = 1$ kΩ, $R_1 = 50$ kΩ, and $v_c = 1000v_1$.

4.75 Do Problem 4.74 with $R_x = 50$ kΩ, $R_1 = 2$ kΩ, and $v_c = 1000v_x$.

Section 4.5 Node Analysis with Ideal Op-amps

(See also PSpice problems B.13 and B.14.)

4.76* Consider the op-amp circuit in Fig. P3.43 (p. 130). Use node analysis to obtain an expression for v_{out} in terms of v_1 and v_2 when $R_F = (K - 1)R$.

4.77 Use node analysis to find v_{out}/v_{in} for the op-amp circuit in Fig. P3.40 (p. 130).

4.78 Find v_{out}/v_{in} for the op-amp circuit in Fig. P4.78.

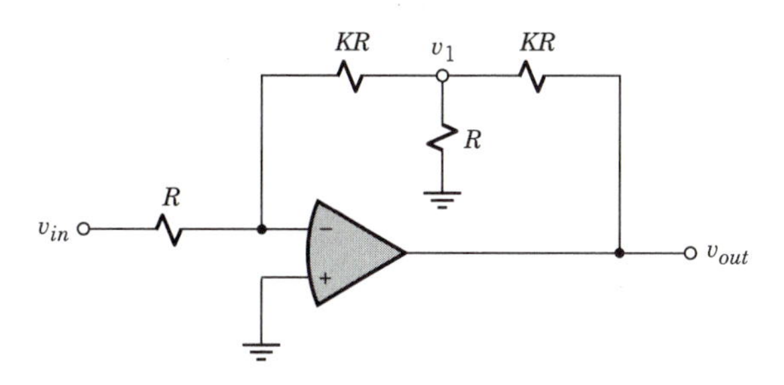

Figure P4.78

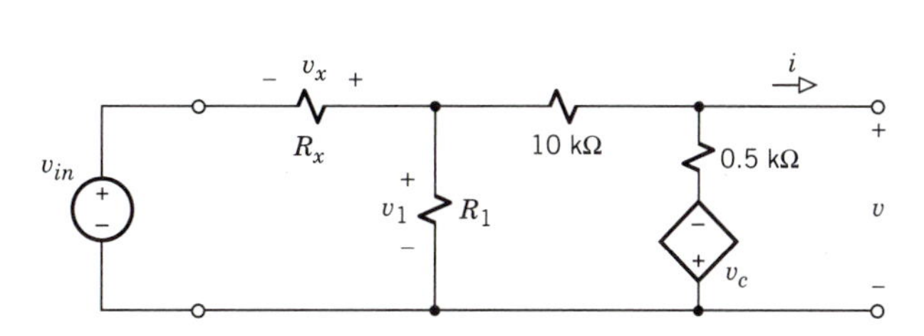

Figure P4.74

4.79 Consider the circuit in Fig. P4.79. Obtain an expression for v_{out} in terms of v_1 and v_2 when $R_1 = 2$ kΩ, $R_2 = 4$ kΩ, $R_3 = 6$ kΩ, $R_F = 8$ kΩ, and $R_4 = \infty$ (an open circuit).

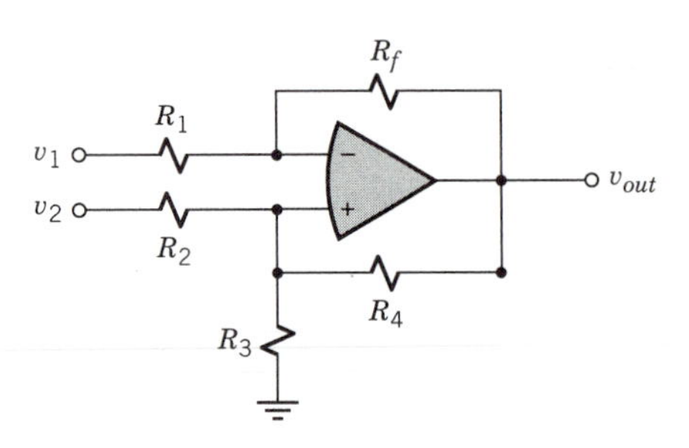

Figure P4.79

4.80 Consider the circuit in Fig. P4.79. Write two node equations to obtain an expression for v_{out} in terms of v_1 and v_2 when $R_1 = 2$ kΩ, $R_2 = 1$ kΩ, $R_4 = 6$ kΩ, $R_F = 5$ kΩ, and $R_3 = \infty$ (an open circuit).

4.81 Find v_{out}/v_{in} for the op-amp circuit in Fig. P4.81.

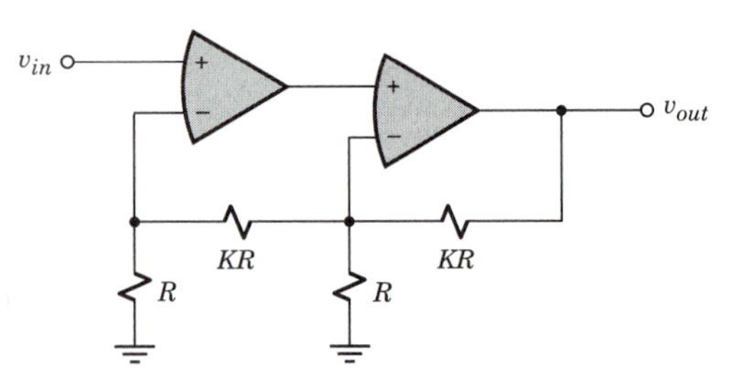

Figure P4.81

4.82 Let the op-amp circuit in Fig. P4.82 have $R_1 = R_2 = R$ and $R_3 = R/K$. Show that

$$v_a = (2K + 2)(v_1 - v_2)$$

4.83 Obtain expressions for both outputs of the op-amp circuit in Fig. P4.82. Put your results in the form $v_b = K_{b2}v_2 - K_{b1}v_1$ and $v_a = K_{a1}v_1 - K_{a2}v_2$.

4.84 Circuits like Fig. P4.84 are sometimes used in automobile sound systems to produce $v_{out} = 2Kv_{in}$. Derive the corresponding design equation for R_a by finding expressions for v_a and v_b.

Section 4.6 Delta-Wye Transformations

4.85* Apply a delta-to-wye transformation to Fig. P4.3 to calculate i_2/i_1 and R_{eq}.

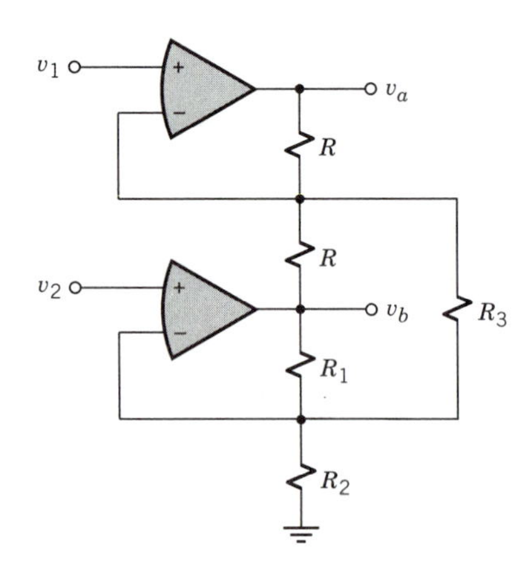

Figure P4.82

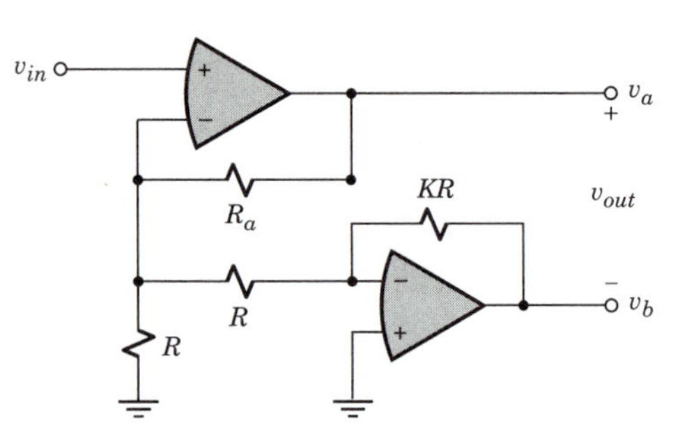

Figure P4.84

4.86 Apply a delta-to-wye transformation to Fig. P4.4 to calculate i_3/i_1 and R_{eq}.

4.87 Apply a wye-to-delta transformation to Fig. P4.3 to calculate v_2/v_1 and R_{eq}.

4.88 Apply a wye-to-delta transformation to Fig. P4.4 to calculate v_2/v_1 and R_{eq}.

4.89 Use a wye-to-delta transformation and a node equation to find the power supplied by the current source in Fig. P4.10.

4.90 Use a delta-to-wye transformation and a mesh equation to find the power supplied by the voltage source in Fig. P4.10.

4.91 Use an appropriate delta-wye transformation in Fig. P4.66 to find v_b/i and R_{eq} when $R = 10$ Ω and $i_c = 8i$.

4.92 Use an appropriate delta-wye transformation in Fig. P4.68 to find v/i_a and R_{eq} when $R = 5$ kΩ and $v_c = 2v$.

4.93 Let $R_y = 6$ Ω and $i_c = v_x/1.2$ Ω in Fig. P4.52. Use an appropriate delta-wye transformation to find v_x and v_z.

CHAPTER 5

Energy Storage and Dynamic Circuits

If circuits consisted only of sources and resistors, then life might be simpler but certainly much duller for electrical engineers. However, many circuits also contain capacitors or inductors, whose properties are introduced in this chapter. Unlike sources or resistors, these elements neither produce nor dissipate power. Instead, they possess the distinctive ability to absorb electrical energy from a source, store the energy temporarily, and then return the energy to a circuit.

Capacitors and inductors exhibit electrical "memory" effects in the sense that energy stored at an earlier time may contribute to the present value of a voltage or current. Consequently, circuits with energy storage behave in a markedly different way compared to "memory-less" resistive circuits. When time-varying currents or voltages are involved, the behavior of an energy-storage circuit is said to be *dynamic*. Dynamic circuit behavior leads to important practical applications in electronics, filtering, waveform generation, and electric power systems.

Our study of electrical energy storage begins here with the structure and properties of capacitors and inductors. Then we'll formulate and investigate differential equations that describe the dynamic response of circuits with energy-storage elements. These initial investigations pave the way for the in-depth treatment of dynamic circuits in later chapters.

Objectives

After studying this chapter and working the exercises, you should be able to do each of the following:

1. Describe the physical structure and energy storage of capacitors and inductors (Sections 5.1 and 5.2).
2. Draw the symbols for ideal capacitors and inductors, and write and apply the i–v and v–i relationships (Sections 5.1 and 5.2).
3. State the continuity conditions for capacitance and inductance (Sections 5.1 and 5.2).
4. Analyze a circuit with stored energy under dc steady-state conditions (Sections 5.1 and 5.2).
5. Calculate equivalent capacitance and inductance for ladder networks of capacitors or inductors (Sections 5.1 and 5.2).
6. Formulate the differential equation for simple circuits with one or two energy-storage elements (Section 5.3).
7. Distinguish between the natural response, forced response, and complete response of a dynamic circuit (Section 5.3).
8. Calculate the complete response of a first-order circuit with a simple excitation function (Section 5.3).

5.1 CAPACITORS

Capacitors store energy in an *electric field*. The electric field is created by displacing positive and negative charges such that they tend to attract each other. The familiar effect of "static cling" illustrates the force from the energy stored by an electric field. In electric circuits, capacitors are used to store energy for flash lamps, to provide a tuning mechanism for radios, and to perform various other useful tasks.

This section describes the physical structure of capacitors and discusses the related concepts of displaced charge, displacement current, and stored energy. We'll also develop the voltage–current relationships for capacitors individually or in series or parallel connections.

Capacitance and Displacement Current

Figure 5.1a depicts the essential parts of a **capacitor**. This device consists of two metallic surfaces or **plates** separated by a **dielectric**. Ideally, the dielectric

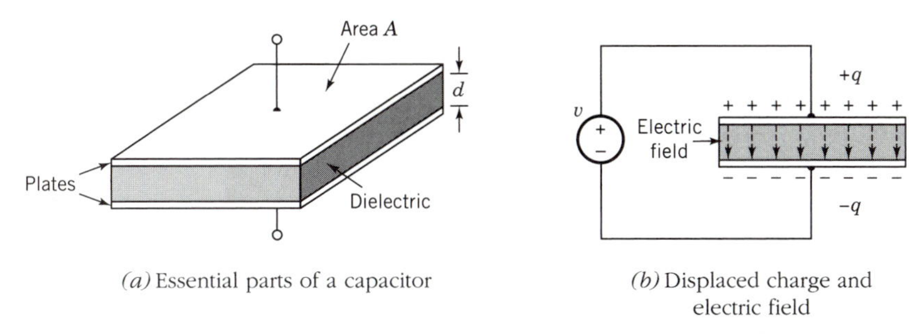

(a) Essential parts of a capacitor

(b) Displaced charge and electric field

Figure 5.1

is a perfect insulator, so it prevents charge flow inside the capacitor. Conduction must therefore take place via external circuitry attached to the plates.

When a voltage source is connected to a capacitor, the source draws free electrons away from the plate at the higher potential and deposits an equal number of additional electrons on the plate at the lower potential. Figure 5.1*b* illustrates the resulting condition, with equivalent positive charge $+q$ on one plate and excess negative charge $-q$ on the other. Some people refer to q as "*stored charge*," even though the *net* charge on the capacitor equals zero. To be more accurate, we'll call q the **displaced charge** because the source actually displaces q from the lower-potential plate to the higher-potential plate.

Charge displacement creates an **electric field** between the plates, as indicated in Fig. 5.1*b*. This field holds the energy supplied by the source when it moved the charges. Hence,

> A capacitor stores energy in an electric field produced by displaced charge on the plates.

We'll calculate the stored energy after further consideration of charge displacement.

Theory and experiments indicate that q is proportional to the instantaneous voltage v across a capacitor, so we write $q = Cv$. The proportionality constant C is called the **capacitance**, defined by

$$C \triangleq q/v \tag{5.1}$$

Capacitance is measured in **farads** (F), and Eq. (5.1) indicates that 1 farad equals 1 coulomb per volt. Alternatively, 1 coulomb equals 1 ampere-second, so we can express the unit equation for capacitance as

$$1 \text{ F} = 1 \text{ A} \cdot \text{s/V}$$

However, a capacitor with $C = 1$ F would be an exceedingly large device, as we'll soon show.

The capacitance of a parallel-plate structure like Fig. 5.1*a* is given by the formula

$$C = \epsilon_r \epsilon_0 A/d \qquad (5.2)$$

Here, A is the area of the plates, d is their spacing, and ϵ_r is the **dielectric constant** or **relative permittivity** compared to the free-space permittivity ϵ_0. The permittivity of free space is expressed in terms of the velocity of light c as

$$\epsilon_0 = 10^7/4\pi c^2 = 8.854 \text{ pF/m} \qquad (5.3)$$

Free space has $\epsilon_r = 1$, by definition, whereas typical solid dielectrics have $2 < \epsilon_r < 100$.

Now suppose you wanted to build a 1-F capacitor using a dielectric with $\epsilon_r \approx 10$ and $d = 0.1$ mm. From Eq. (5.2), you would need plates having area $A = Cd/\epsilon_r\epsilon_0 \approx 10^6 \text{ m}^2$ — larger than 100 football fields! Consequently, most practical capacitance values range from a few thousand **microfarads** ($1\ \mu\text{F} = 10^{-6}$ F) down to a few **picofarads** ($1 \text{ pF} = 10^{-12}$ F). Small capacitances also occur naturally whenever two conducting surfaces come into proximity, like the wires in a cable or the leads of an electronic device. This effect is called **stray** or **parasitic capacitance**.

Next, consider what happens when the voltage across a capacitor varies with time, so q must likewise vary with time. Varying the charge on each plate requires a *current* $i = dq/dt$ entering the higher-potential terminal and an equal current leaving the lower-potential terminal. The resulting current–voltage relationship follows by differentiating $q = Cv$, noting that C is a constant. Thus, $dq/dt = d(Cv)/dt = C\, dv/dt$ so

$$i = C\frac{dv}{dt} \qquad (5.4)$$

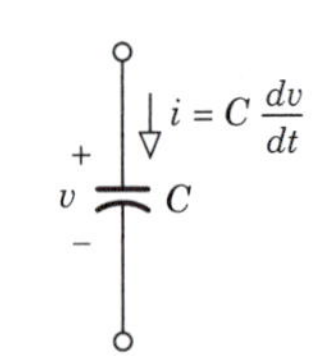

Figure 5.2
Capacitor symbol.

We call i a **displacement current** because it represents the change of displaced charge on the plates of the capacitor rather than internal charge flow through the dielectric. Nonetheless, charges do flow through the connecting wires at the same rate i. Since i enters the terminal at the higher potential and exits the terminal at the lower potential, the symbol for a capacitor given in Fig. 5.2 includes polarity marks following the usual passive convention.

Equation (5.4) states that instantaneous capacitor current is proportional to the rate of change of the voltage. But if the voltage remains *constant*, then $dv/dt = 0$ and the current equals zero. Constant voltage corresponds to a dc voltage source, so we say that a capacitor acts as a **dc open circuit**. The "gap" in the capacitor symbol reflects this open-circuit property.

Any source that changes the voltage across a capacitor must supply a nonzero current $i = C\, dv/dt$ and instantaneous power

$$p = vi = Cv\frac{dv}{dt} \qquad (5.5)$$

The power does not get dissipated as heat or some other form of energy, as it would in a resistor, because an ideal capacitor contains no mechanism for power dissipation. Instead, p represents energy transfer from the source to the capacitor's electric field.

The energy transfer takes place at the rate $dw/dt = p$, and we can rewrite Eq. (5.5) as $dw = p\,dt = Cv\,dv$. Integrating this expression now gives

$$w = \int Cv\,dv = \tfrac{1}{2}Cv^2 + K_{int}$$

where K_{int} is the constant of integration. Since there is no electric field to store energy when $v = 0$, it follows that $K_{int} = 0$. Therefore, the **instantaneous stored energy** is just

$$w = \tfrac{1}{2}Cv^2 \tag{5.6}$$

For example, a 1-μF capacitor with $v = 20$ V stores $w = \frac{1}{2} \times 10^{-6} \times 20^2 = 200\ \mu$J.

If we disconnect the source after charging an ideal capacitor to voltage v, then we still find $v = q/C$ across the capacitor because the displaced charge cannot leave the plates when $i = 0$. Furthermore, the displaced charge and stored energy remain indefinitely within an ideal capacitor as long as its terminals are left open. The energy may be recovered at a later time by connecting a load to the charged capacitor, allowing the charges to leave the plates and transfer energy from the field to the load. This store-and-discharge ability is exploited in electronic flash lamps, heart defibrillators, and similar pulsed applications that require large bursts of energy. The energy is built up and stored during the time intervals between bursts.

Example 5.1 *Capacitor Waveforms*

Suppose the voltage across a 3-μF capacitor has the waveform plotted in Fig. 5.3*a*. The time axis here is in milliseconds (ms), and

$$\begin{aligned} v &= 20 \sin 50\pi t \text{ V} && 0 < t \le 10 \text{ ms} \\ &= 20 \text{ V} && 10 < t \le 30 \text{ ms} \\ &= -20 \sin 50\pi t \text{ V} && 30 < t \le 60 \text{ ms} \end{aligned}$$

The displaced charge q will vary in the same manner, with $q_{max} = Cv_{max} = 60\ \mu$C.

The resulting current is found using Eq. (5.4) and $d(\sin 50\pi t)/dt = 50\pi \cos 50\pi t$. Thus, $i_{max} = C|dv/dt|_{max} = (3 \times 10^{-6}) \times 20 \times 50\pi = 3\pi$ mA and

$$\begin{aligned} i &= 3\pi \cos 50\pi t \text{ mA} && 0 < t \le 10 \text{ ms} \\ &= 0 && 10 < t \le 30 \text{ ms} \\ &= -3\pi \cos 50\pi t \text{ mA} && 30 < t \le 60 \text{ ms} \end{aligned}$$

This waveform is plotted in Fig. 5.3*b*.

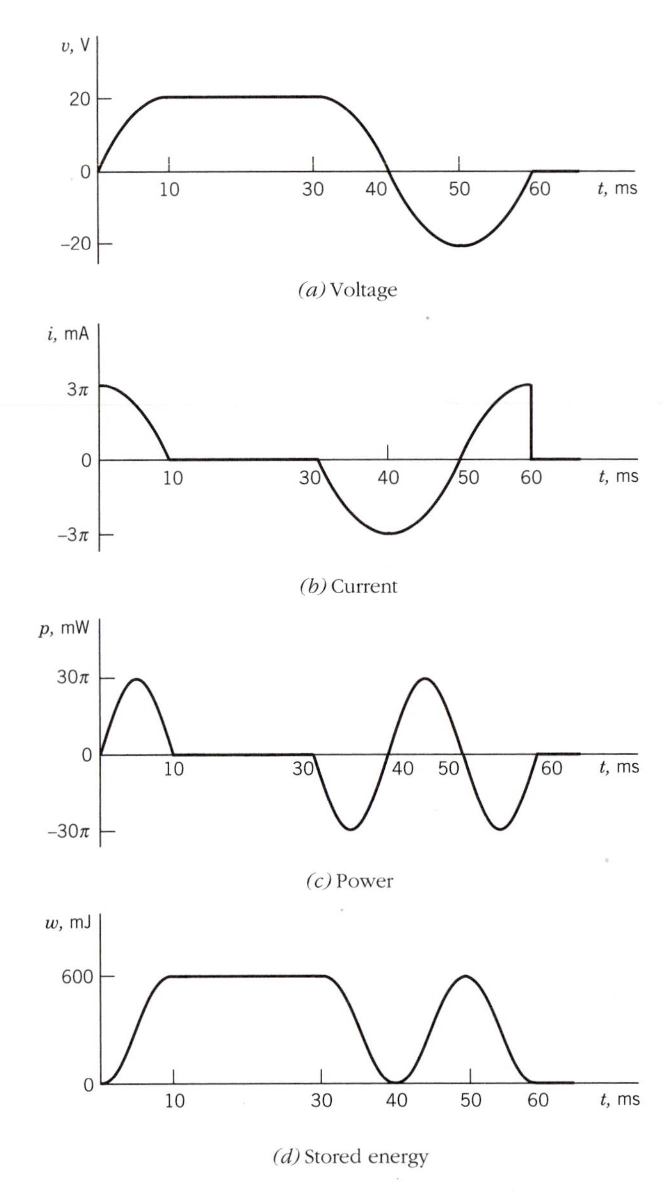

Figure 5.3 Illustrative capacitor waveforms.

The plots of v and i emphasize the fact that a capacitor's current and voltage generally have *different waveforms* — in contrast to a resistor whose current and voltage always vary proportionally. Also note that i must be positive when v increases and negative v when decreases, regardless of the sign of v. Of course i equals zero when v remains constant at any value.

Multiplying the voltage and current waveforms gives the instantaneous power $p = vi$ plotted in Fig. 5.3*c*. This waveform should be compared with the stored energy $w = \frac{1}{2}Cv^2$ in Fig. 5.3*d*. We see that w increases when $p > 0$, so the source is transferring energy to the capacitor. Conversely, when $p < 0$, the capacitor returns energy to the source and w decreases.

Exercise 5.1

The horizontal sweep circuit in a TV set is driven by a voltage waveform like Fig. 5.4, where the time axis is in microseconds (μs). Evaluate dv/dt for the two time intervals and plot the current waveform when v appears across a 200-pF capacitor. Then sketch p and w versus time, labeling the extreme values.

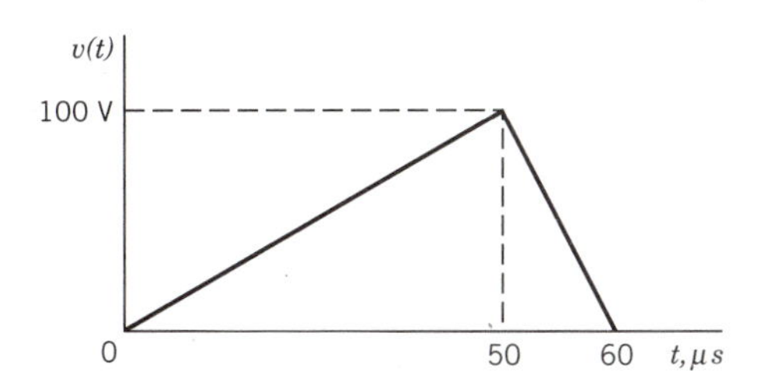

Figure 5.4

Voltage–Current Relationships

The relationship $i = C\,dv/dt$ tells us how to calculate capacitor current i when we know the voltage derivative dv/dt. Conversely, the v–i relationship will involve an integral, and we must take care when evaluating the constants of integration. For this reason, we'll work with *definite integrals* over specified intervals of time. We'll also indicate the time dependence explicitly by writing $i = i(t)$, $v = v(t)$, etc.

Suppose we start at some initial time t_0 when the displaced charge is $q(t_0)$. If the capacitor draws current $i(t) = dq/dt$, then the charge at a later time $t_0 + T$ will be

$$q(t_0 + T) = q(t_0) + \int_{t_0}^{t_0+T} i(t)\,dt$$

To recast this expression for any instant $t > t_0$, we let $t = t_0 + T$ and we introduce the dummy integration variable λ to avoid confusion with the new upper limit t. Hence,

$$q(t) = q(t_0) + \int_{t_0}^{t} i(\lambda)\,d\lambda \qquad t > t_0$$

where the function $i(\lambda)$ is identical to $i(t)$ with t replaced by the dummy variable λ. Finally, since $v(t) = q(t)/C$, we obtain

$$v(t) = v(t_0) + \frac{1}{C}\int_{t_0}^{t} i(\lambda)\,d\lambda \qquad t > t_0 \tag{5.7}$$

This result gives the instantaneous voltage $v(t)$ for $t > t_0$ in terms of the **initial voltage** $v(t_0)$ and the subsequent current.

Alternatively, we let $t_0 \rightarrow -\infty$ and assume that $v(-\infty) = 0$ — a reasonable assumption since any capacitor would have no initial voltage at the time it was built. Equation (5.7) then becomes

$$v(t) = \frac{1}{C}\int_{-\infty}^{t} i(\lambda)\, d\lambda \tag{5.8}$$

This expression brings out the fact that a capacitor possesses electrical **memory**, the voltage at any time t being dependent upon the entire past behavior of the current. Physically, capacitor memory takes the form of the resultant displaced charge q delivered by i over all previous time.

The compact form of Eq. (5.8) lends itself to general analysis problems. But Eq. (5.7) is better suited to practical calculations because the initial voltage $v(t_0)$ incorporates the effect of i for all $t < t_0$. In any case, we must know either the initial voltage or the entire past behavior of i in order to find $v(t)$.

As a simple illustration using Eq. (5.7), take the case of a capacitor with $v = V_0$ at $t = 0$ and subsequent current

$$i(t) = At^2 \qquad t > 0$$

Upon setting $t_0 = 0$, $v(t_0) = V_0$, and $i(\lambda) = A\lambda^2$, we get

$$v(t) = V_0 + \frac{1}{C}\int_{0}^{t} A\lambda^2\, d\lambda = V_0 + \frac{At^3}{3C} \qquad t > 0$$

Thus, again we find that the voltage waveform differs from the current waveform.

Now assume any initial voltage and let the current be constant over some time interval, say $i = I$ from $t = t_1$ to $t = t_2 > t_1$. Taking $t_0 = t_1$ and $i = I$ in Eq. (5.7) yields the resulting voltage

$$v(t) = v(t_1) + \frac{I}{C}(t - t_1) \qquad t_1 < t \leq t_2 \tag{5.9}$$

Figure 5.5 illustrates this waveform known as a **voltage ramp**. The ramp begins at $v(t_1)$ and increases linearly with slope $dv/dt = I/C$ up to $v(t_2) = v(t_1) + (I/C)(t_2 - t_1)$. If the current drops to zero at time t_2 — as it does

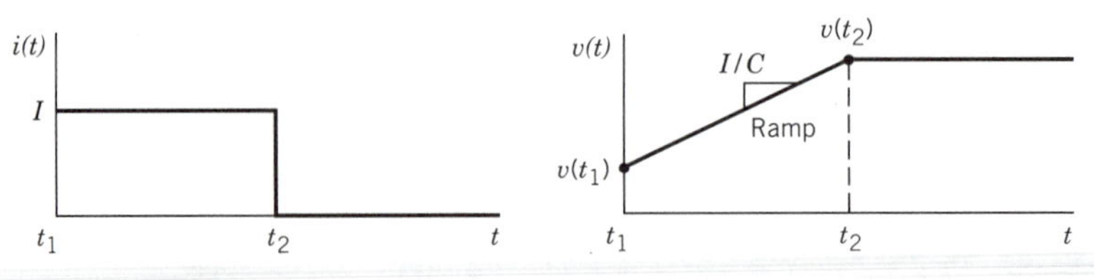

Figure 5.5 Capacitor waveforms with constant current.

in the figure — then the voltage remains constant at $v(t) = v(t_2)$ for $t > t_2$. This behavior follows from Eq. (5.7) with t_2 as the new initial time and $i = 0$ for $t > t_2$. Notice that the current discontinuity or *jump* at $t = t_2$ does not produce a voltage jump.

To explore the general possibility of voltage jumps, suppose the current jumps at an arbitrary time t_j, as shown in Fig. 5.6. Since $i(t)$ is discontinuous at t_j, we must introduce the notation t_j^- and t_j^+ to stand for the instants just before and just after the jump. These instants are defined mathematically by letting Δt be a positive quantity and taking the limits

$$t_j^- = \lim_{\Delta t \to 0} t_j - \Delta t \qquad t_j^+ = \lim_{\Delta t \to 0} t_j + \Delta t \tag{5.10}$$

Hence, the numerical values of t_j^- and t_j^+ equal t_j, even though they have different physical meanings.

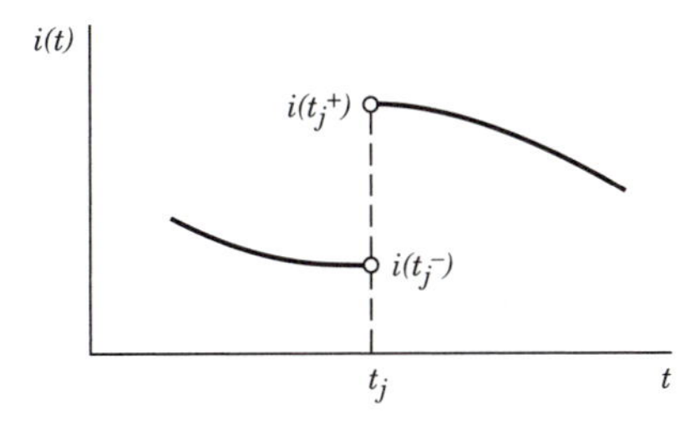

Figure 5.6 Current waveform with discontinuity.

The voltage just before the current jumps can be denoted in general by $v(t_j^-)$, and Eq. (5.7) gives the voltage just after the jump as

$$v(t_j^+) = v(t_j^-) + \frac{1}{C}\int_{t_j^-}^{t_j^+} i(\lambda)\, d\lambda$$

But when i is *finite,* the area under the current waveform from t_j^- to t_j^+ must be vanishingly small. Therefore,

$$v(t_j^+) = v(t_j^-) \qquad \text{if } |i| < \infty \tag{5.11}$$

which means that v does not jump at time t_j. Stated another way, there is **continuity of capacitor voltage** in that

> The voltage across a capacitor cannot change discontinuously when the current remains finite.

Voltage continuity jibes with energy considerations, for a voltage jump would imply a sudden change of the stored energy $w = \frac{1}{2}Cv^2$, which, in turn, would require infinite instantaneous power $p = dw/dt$.

Example 5.2 *Waveform Generation in a Hazard Blinker*

Some waveform generators take advantage of the voltage ramp produced across a capacitor by a constant current. An interesting example is the circuit in Fig. 5.7*a*, a simplified version of the portable blinkers placed around traffic hazards. The flash lamp here is an idealized two-state device having the unusual i–v characteristics in Fig. 5.7*b*, so the lamp is always OFF when $v <$ 40 V, and always ON when $v >$ 80 V. But it may be either OFF or ON in the range 40 V $< v <$ 80 V, depending on whether the voltage is increasing or decreasing.

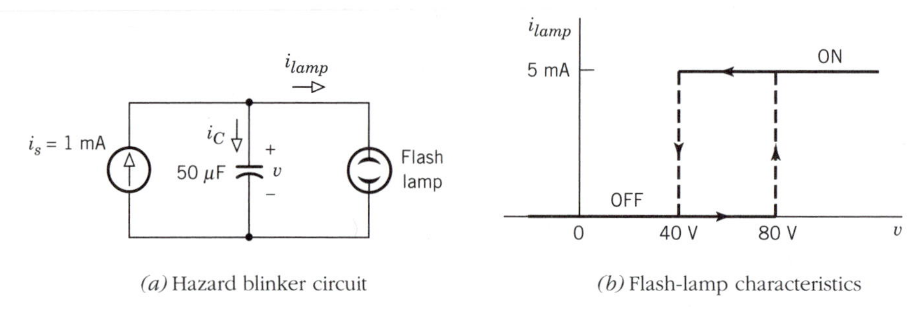

(*a*) Hazard blinker circuit (*b*) Flash-lamp characteristics

Figure 5.7

We'll use the flash-lamp characteristics to analyze the operation of the blinker starting with $v = 0$ at $t = 0$. Our analysis leads to the waveforms plotted in Fig. 5.8.

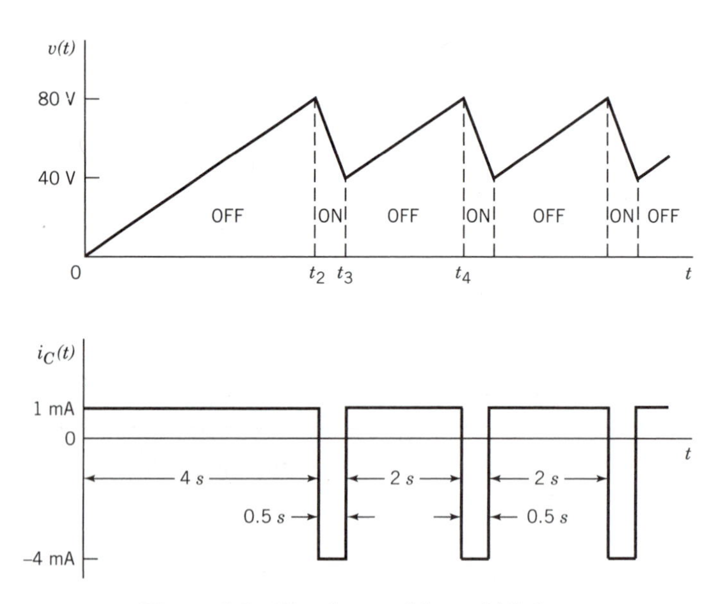

Figure 5.8 Waveforms of hazard blinker.

Since the lamp is OFF initially and draws no current as long as $v < 80$ V, the initial capacitor current is constant at $i_C = i_s = 1$ mA. This current charges the capacitor, and $v(t)$ is given by Eq. (5.9) with $v(0) = 0$, so

$$v(t) = \frac{1 \text{ mA}}{50 \ \mu\text{F}} t = 20t \qquad 0 < t \leq t_2$$

The lamp then goes ON at $t = t_2$ when $v(t_2) = 80$ V. Hence, $20t_2 = 80$ and $t_2 = 4$ s.

The lamp now draws $i_{lamp} = 5$ mA and the capacitor current becomes $i_C = i_s - i_{lamp} = -4$ mA, so the voltage begins to decrease as the capacitor discharges through the lamp. Inserting the new initial condition $v(t_2) = 80$ V into Eq. (5.9) yields

$$v(t) = 80 + \frac{-4 \text{ mA}}{50 \ \mu\text{F}}(t - t_2) = 80 - 80(t - t_2) \qquad t_2 < t \leq t_3$$

The discharge continues until $t = t_3$ when $v(t_3) = 40$ V and the lamp goes OFF. Hence, $80 - 80(t_3 - t_2) = 40$ and $t_3 = t_2 + 0.5$ s.

With the lamp OFF, the capacitor is again charged by $i_C = i_s$ starting from $v(t_3) = 40$ V. Hence,

$$v(t) = 40 + \frac{1 \text{ mA}}{50 \ \mu\text{F}}(t - t_3) = 40 + 20(t - t_3) \qquad t_3 < t \leq t_4$$

The lamp goes ON again when $v(t_4) = 80$ V, so $40 + 20(t_4 - t_3) = 80$ and $t_4 = t_3 + 2$ s.

The situation at t_4 is now identical to that at t_2. Operation thus continues periodically thereafter with the lamp blinking ON for 0.5 s and OFF for 2 s, as shown in Fig. 5.8. Although the current jumps at $t = t_2, t_3, \ldots$, the voltage waveform exhibits continuity throughout.

Example 5.3 *Op-Amp Integrator*

The form of Eq. (5.8) suggests that a capacitor might be used for electronic integration of time-varying electrical signals.

Figure 5.9 diagrams one popular circuit of this type, employing an op-amp with capacitance in the feedback path. The output voltage is

$$v_{out}(t) = -v_C(t) = -\frac{1}{C}\int_{-\infty}^{t} i_C(\lambda)\, d\lambda$$

But the virtual short at the op-amp's input requires $i_C = i_{in} = v_{in}/R$, so

$$v_{out}(t) = -\frac{1}{RC}\int_{-\infty}^{t} v_{in}(\lambda)\, d\lambda \qquad (5.12)$$

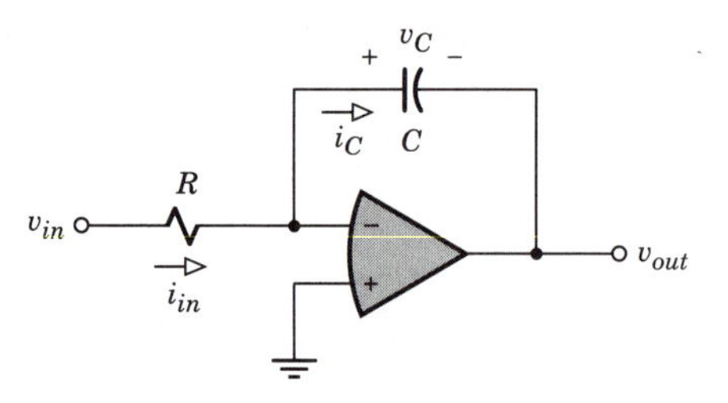

Figure 5.9 Op-amp integrator.

Hence, this **op-amp integrator** produces an output voltage proportional to the integral of the input voltage.

Exercise 5.2

How long will it take to charge a 10-μF capacitor to 200 V starting at $v(0) = 0$ and using a 5-mA constant current source?

Exercise 5.3

Obtain an expression for the voltage $v(t)$ across capacitor C when $v(0^-) = 0$ and $i(t) = Ie^{-at}$ for $t \geq 0$. Confirm that $v(0^+) = 0$, so $v(t)$ has continuity at $t = 0$. Hint: Note from Eq. (5.10) that 0^+ numerically equals 0 so $e^{-at} = 1$ at $t = 0^+$.

Parallel and Series Capacitance

While the capacitance element equations look quite different from Ohm's law, they are nonetheless *linear* relationships. We can show that the v–i equation satisfies both the proportionality and superposition properties by letting $i(t) = K_a i_a(t) + K_b i_b(t)$ in Eq. (5.8). The voltage is then

$$
\begin{aligned}
v(t) &= \frac{1}{C}\int_{-\infty}^{t} [K_a i_a(\lambda) + K_b i_b(\lambda)]\, d\lambda \\
&= \frac{K_a}{C}\int_{-\infty}^{t} i_a(\lambda)\, d\lambda + \frac{K_b}{C}\int_{-\infty}^{t} i_b(\lambda)\, d\lambda
\end{aligned}
$$

We thus have the linear expression

$$v(t) = K_a v_a(t) + K_b v_b(t)$$

with

$$v_a(t) = \frac{1}{C}\int_{-\infty}^{t} i_a(\lambda)\, d\lambda \qquad v_b(t) = \frac{1}{C}\int_{-\infty}^{t} i_b(\lambda)\, d\lambda$$

The linearity of the i–v equation is similarly confirmed by letting $v = K_a v_a + K_b v_b$ in $i = C\,dv/dt$.

As a direct consequence of linearity, any two-terminal network consisting entirely of capacitors acts at its terminals like a single *equivalent* capacticance. In particular, the network of N parallel capacitors in Fig. 5.10 has $i = i_1 + i_2 + \cdots + i_N$, with $i_1 = C_1\,dv/dt$, etc., so the terminal i–v relationship is

$$i = C_1 \frac{dv}{dt} + C_2 \frac{dv}{dt} + \cdots + C_N \frac{dv}{dt} = (C_1 + C_2 + \cdots + C_N)\frac{dv}{dt}$$

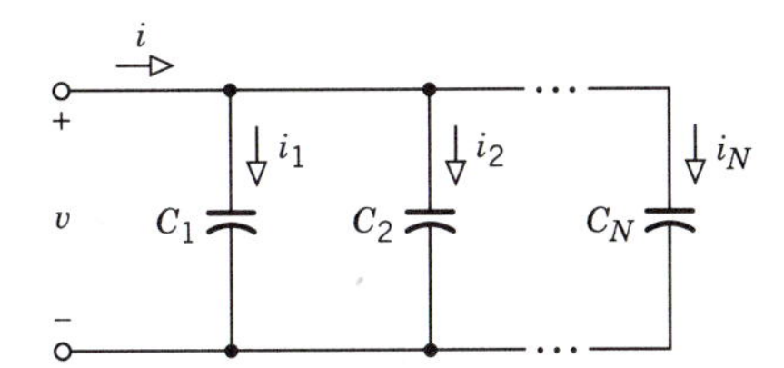

Figure 5.10 Capacitors in parallel.

This expression clearly has the form $i = C_{par}\,dv/dt$, where the **parallel equivalent capacitance** is

$$C_{par} = C_1 + C_2 + \cdots + C_N \tag{5.13}$$

Note carefully that *parallel capacitance adds,* like parallel *conductance* rather than resistance. Thus, large capacitance "banks" for energy storage often consist of several capacitors connected in parallel.

Working from Fig. 5.10, you can also show that the current i_n through capacitor C_n is related to the terminal current i via

$$i_n = (C_n / C_{par})i \tag{5.14}$$

This is the **current–divider relation** for parallel capacitance.

If N capacitors are connected in series, as in Fig. 5.11, then $v = v_1 + v_2 + \cdots + v_N$ and $i = C_1\,dv_1/dt = C_2\,dv_2/dt = \ldots$, so

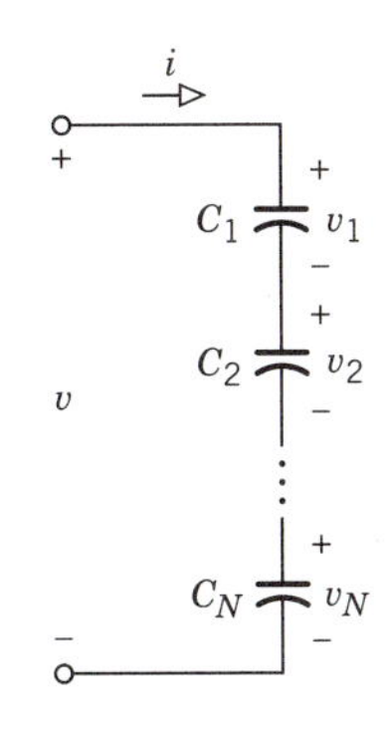

Figure 5.11 Capacitors in series.

$$\frac{dv}{dt} = \frac{dv_1}{dt} + \frac{dv_2}{dt} + \cdots + \frac{dv_N}{dt} = \left(\frac{1}{C_1} + \frac{1}{C_2} + \cdots + \frac{1}{C_N}\right) i$$

Rewriting this expression as $i = C_{ser}\,dv/dt$ yields the **series equivalent capacitance** in the form

$$\frac{1}{C_{ser}} = \frac{1}{C_1} + \frac{1}{C_2} + \cdots + \frac{1}{C_N} \tag{5.15}$$

For the special case of just *two* capacitors in series, Eq. (5.15) simplifies to

$$C_{ser} = \left(\frac{1}{C_1} + \frac{1}{C_2}\right)^{-1} = \frac{C_1 C_2}{C_1 + C_2}$$

The equivalent capacitance for a ladder network of capacitors can be found by series-parallel reduction using Eqs. (5.13) and (5.15).

Finally, suppose you know the terminal voltage v in Fig. 5.11 and you want to find the voltage v_n across capacitor C_n. Since the current through C_n is $i = C_{ser}\, dv/dt$, Eq. (5.7) yields

$$v_n(t) = v_n(t_0) + \frac{1}{C_n}\int_{t_0}^{t} C_{ser}\frac{dv}{d\lambda}\, d\lambda$$

from which

$$v_n(t) = v_n(t_0) + \frac{C_{ser}}{C_n}[v(t) - v(t_0)] \qquad t \geq t_0 \tag{5.16}$$

This result expresses the **voltage-divider relation** for series capacitors. Note that you must know the initial voltage $v_n(t_0)$ as well as the terminal voltage waveform $v(t)$.

Example 5.4 *Calculations for Series Capacitors*

Figure 5.12*a* depicts a 3-μF capacitor and a 6-μF capacitor in series with the following conditions at $t = 0$:

$$v_1(0) = 10 \text{ V} \qquad q_1(0) = C_1 v_1(0) = 30\ \mu\text{C}$$
$$v_2(0) = -10 \text{ V} \qquad q_2(0) = C_2 v_2(0) = -60\ \mu\text{C}$$
$$v(0) = v_1(0) + v_2(0) = 0 \text{ V}$$

We'll calculate the new conditions at $t_1 > 0$ when a source connected to the terminals establishes $v(t_1) = 30$ V.

The equivalent capacitance is $C_{ser} = (3 \times 6)/(3 + 6) = 2\ \mu$F. Thus, from the voltage divider in Eq. (5.16), we have

$$v_1(t_1) = 10 + \frac{2}{3}[30 - 0] = 30 \text{ V} \qquad q_1(t_1) = 90\ \mu\text{C}$$

Likewise,

$$v_2(t_1) = -10 + \frac{2}{6}[30 - 0] = 0 \text{ V} \qquad q_2(t_1) = 0\ \mu\text{C}$$

These results are summarized in Fig. 5.12*b*.

Note that the values of $v_1(t_1)$ and $v_2(t_1)$ agree with the fact that $v_1(t_1) + v_2(t_1) = v(t_1) = 30$ V. Also note that the source has removed 60 μC of charge from each of the lower plates and added 60 μC to each of the upper plates.

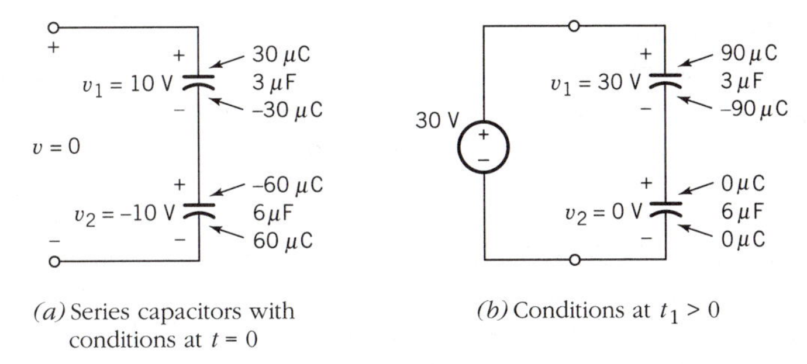

(a) Series capacitors with conditions at $t = 0$

(b) Conditions at $t_1 > 0$

Figure 5.12

Exercise 5.4

Derive Eq. (5.14) from Fig. 5.10.

Exercise 5.5

Confirm that the equivalent capacitance of the network in Fig. 5.13 is 4 μF. Then find $v_1(t_1)$ and $v_2(t_1)$ given that $v_1(0) = 0$, $v_2(0) = 10$ V, and $v(t_1) = 0$.

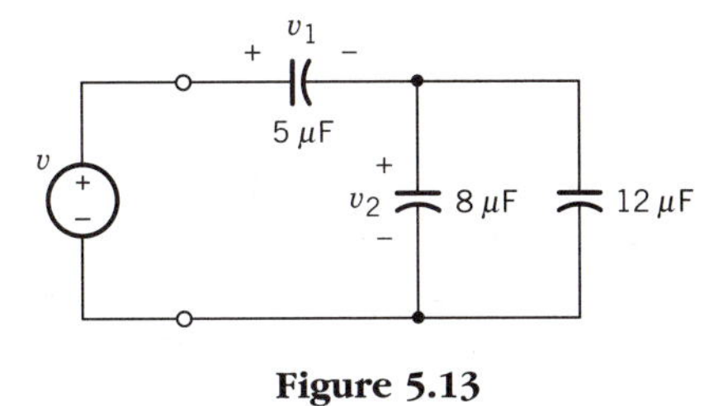

Figure 5.13

Types of Capacitors

Capacitors come in a wide variety of shapes and sizes. Particular families are categorized according to the type of dielectric material.

Small capacitors often consist of thin **ceramic discs** with metal coatings on the flat surfaces to form the plates. Larger capacitances are achieved by rolling sheets of metal foil and flexible dielectric into a tubular shape. The dielectric is usually a **plastic film** such as Mylar.

Electrolytic capacitors provide even more capacitance per unit volume because the dielectric is an extremely thin oxide layer formed on aluminum or tantalum foil. Using this technology, a 1-F capacitor can be crammed into a cylinder 8 cm in diameter and 22 cm high. However, the oxide can be destroyed by excessive voltage or the wrong voltage polarity, so electrolytic capacitors are restricted to relatively low voltages of one polarity. (If you violate these restrictions, then the electrolytic capacitor may explode and shower you with bits of foil!)

Variable capacitors usually have movable plates with air as the dielectric. The relatively wide spacing between plates and the low dielectric permittivity ($\epsilon_r \approx 1$) limit these devices to small values of capacitance.

Ideally, a capacitor is lossless in the sense that there is no internal conduction path or power dissipation. But some current may leak through the dielectric of a real capacitor. When dielectric leakage is a significant factor, it can be modeled by a large **leakage resistance** R_l in parallel with the capacitor. Leakage increases with capacitance, and the product R_lC tends to be a constant whose value depends primarily on the dielectric material. Table 5.1 lists typical capacitance ranges, maximum voltages, and R_lC products for some common types of capacitors.

TABLE 5.1 Typical Capacitor Characteristics

Type	Available Capacitances	Maximum Voltage	R_lC (Ω-F)
Air-variable	10–500 pF	500	∞
Ceramic disc	5 pf–50 nF	600–1000	1000
Plastic film	1 nF–5 μF	100–600	100,000
Electrolytic	1 μF–1 F	6–250	50–500

5.2 INDUCTORS

Inductors store energy in a *magnetic field*. The magnetic field is created by passing current through a coil of wire, as is done in an electromagnet such as those used to lift scrap iron. In electric circuits, inductors often play the role of "smoothing" devices that reduce current variations. But when the current is forced to change abruptly, the inductor generates a large induced voltage — large enough to cause a spark in the extreme case of an automobile ignition circuit.

This section describes the physical structure of inductors and discusses the related concepts of flux linkage and induced voltage. We'll then show that inductance is the dual of capacitance, and we'll exploit this duality to develop the current–voltage relationships for inductors individually or in series or parallel connections.

Inductance and Induced Voltage

Figure 5.14*a* depicts the essential parts of an **inductor**. This two-terminal device consists of a wire **coil** wound around a **core**. The core may be cylindrical, as shown, or it might be doughnut shaped. The coil has N **turns**, insulated from each other and from the core. Ideally, there is no resistance in the wire that makes up the coil.

When the wire carries current i, a **magnetic field** is created in the space around the coil and becomes concentrated within the core, as illustrated in Fig. 5.14*b*. Like an electric field, the magnetic field holds energy that comes from the source that established the field. Hence,

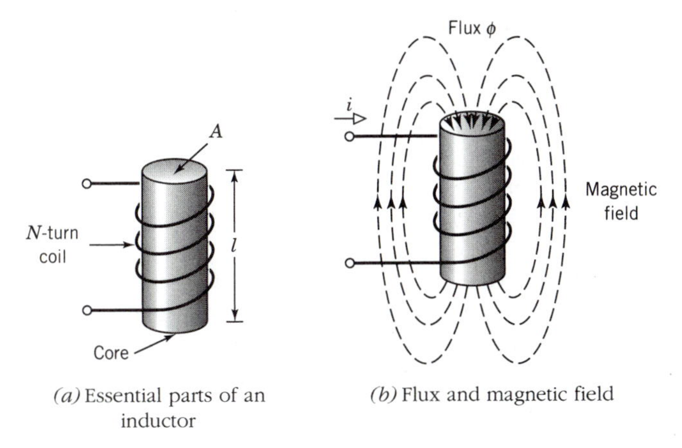

(a) Essential parts of an inductor

(b) Flux and magnetic field

Figure 5.14

> An inductor stores energy in a magnetic field produced by current through a wire coil.

We'll calculate the stored energy after further consideration of the magnetic field.

The strength of the field is expressed in terms of the **magnetic flux** ϕ through the core, and the product $N\phi$ is known as the **flux linkage**. Experiments and theory indicate that flux linkage is directly proportional to the current, so we write $N\phi = Li$. The proportionality constant L is the **inductance**, defined by

$$L \triangleq N\phi / i \tag{5.17}$$

which is measured in **henrys** (H). Since the SI unit for flux is the **weber**, Eq. (5.17) shows that 1 henry equals 1 weber per ampere. But 1 weber is equivalent to 1 volt-second, and we can express the unit equation for inductance as

$$1 \text{ H} = 1 \text{ V} \cdot \text{s/A}$$

Practical values of inductance range from several henrys down to a few microhenrys, depending upon the coil size and core material. Every circuit also has very small amounts of **stray inductance** because current through any conductor — even a straight wire — creates a magnetic field around it.

The inductance of the structure in Fig. 5.14*a* is given approximately by the formula

$$L = \mu_r \mu_0 N^2 A / l \tag{5.18}$$

Here, A is the area of core, l is its length, and μ_r is the **relative permeability** of the core material compared to the free-space permeability μ_0. The permeability of free space is

$$\mu_0 = 4\pi \times 10^{-7} = 1.26\ \mu\text{H/m} \tag{5.19}$$

Free space has $\mu_r = 1$, by definition, and nonmagnetic materials have $\mu_r \approx 1$, whereas ferromagnetic materials have $\mu_r \gg 1$.

Large inductances require coils with hundreds of turns wrapped on large steel or iron cores. However, if the coil current gets too large, then the magnetic properties of ferromagnetic cores become nonlinear and the current is no longer proportional to the flux. We'll therefore limit our consideration to inductors operating with small enough currents to be in the linear magnetic region.

Now suppose the current through an inductor varies with time, so ϕ likewise varies with time. **Faraday's law** tells us that the time-varying flux *induces* a voltage in the coil given by $v = N\,d\phi/dt$. But $\phi = (L/N)i$, and if L and N are constant, then $d\phi/dt = (L/N)\,di/dt$ so the **induced voltage** is

$$v = L\frac{di}{dt} \tag{5.20}$$

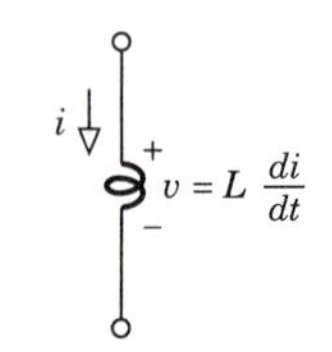

Figure 5.15
Inductor symbol.

This voltage appears across the terminals of the coil, and it tends to oppose current entering the coil. Accordingly, the symbol for an inductor shown in Fig. 5.15 includes polarity marks with the usual passive convention.

Equation (5.20) states that the instantaneous voltage across an inductor is proportional to the rate of change of the current. If the current stays constant, then $di/dt = 0$ and the voltage equals zero — provided that the coil has no resistance. Hence, we say that an ideal inductor acts as a **dc short circuit**.

Since these properties are just the "opposite" of the properties of an ideal capacitor, we conclude that

Inductance and capacitance are duals.

Indeed, the inductor equation $v = L\,di/dt$ is precisely the dual of the capacitor equation $i = C\,dv/dt$. Thus, any set of capacitor waveforms can be relabeled via duality to apply for an inductor, or vice versa. Furthermore, any capacitance relationship also holds for inductance when we replace C with L and interchange i and v.

Recall, in particular, that the instantaneous stored energy in a capacitor was found to be $w = \frac{1}{2}Cv^2$. Applying duality, we immediately obtain the inductor's **instantaneous stored energy**

$$w = \tfrac{1}{2}Li^2 \tag{5.21}$$

When the current and stored energy changes with time, the instantaneous power delivered to the inductor is

$$p = iv = Li\frac{di}{dt} \qquad (5.22)$$

Equation (5.22) follows from Eq. (5.20), and it is also the dual of the capacitor equation $p = vi = Cv\,dv/dt$.

Despite the dualism between inductance and capacitance, inductors are seldom used to hold stored energy for long time intervals. Since w is proportional to i^2, an inductor cannot just be disconnected from the source. Rather, the current through the inductor must be sustained by short-circuiting the terminals until the energy is needed. But every real inductor has some **winding resistance** in the coil, and the resistance soon dissipates the energy. For this reason, inductors are usually switched directly from the source to the energy-consuming load — as is done in ignition systems and similar applications that require high-voltage pulses. (For the same reason, engineers have special interest in the new superconducting materials that exhibit zero resistance at temperatures above absolute zero. A superconducting coil could thus store energy for a much longer time than a coil with resistance.)

Example 5.5 *Inductor Waveforms*

Suppose that a 50-mH inductor is switched from a 2-A source to a load that causes the current to decrease linearly to zero in 100 μs. The inductor current therefore varies as shown in Fig. 5.16*a*.

The decreasing current produces the negative voltage pulse in Fig. 5.16*b*, where $v = 50\text{ mH} \times (-2\text{ A}/100\ \mu\text{s}) = -1\text{ kV}$ over the interval in question. The instantaneous power delivered to the load is the negative of $p = iv$ plotted in Fig. 5.16*c*. If the inductor has negligible resistance, then the total energy transferred to the load equals the initial stored energy $w = \frac{1}{2} \times 50\text{ mH} \times (2\text{ A})^2 = 100\text{ mJ}$.

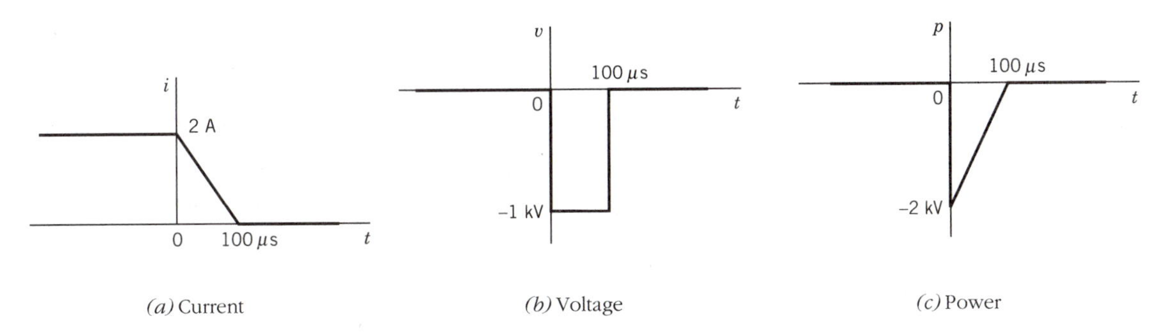

(a) Current *(b)* Voltage *(c)* Power

Figure 5.16 Illustrative inductor waveforms.

Example 5.6 *DC Steady-state Analysis*

The circuit in Fig. 5.17*a* is known to be in the **dc steady state** — meaning that all voltages and currents are constant. The inductor therefore acts as a short circuit while the capacitor acts as an open circuit. We want to calculate the resulting stored energy.

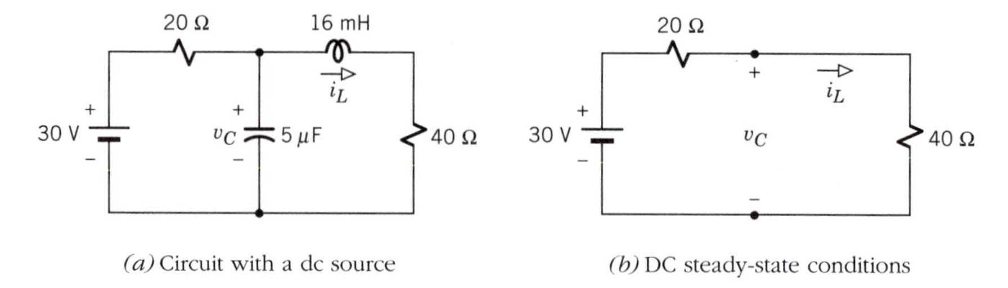

(a) Circuit with a dc source (b) DC steady-state conditions

Figure 5.17

Using the dc properties, we can redraw the circuit as shown in Fig. 5.17*b* where we clearly see that

$$i_L = \frac{30}{20 + 40} = 0.5 \text{ A} \qquad v_C = \frac{40}{20 + 40} 30 = 20 \text{ V}$$

Hence, the individual stored energies are

$$w_L = \tfrac{1}{2}(16 \times 10^{-3})0.5^2 = 2 \text{ mJ} \qquad w_C = \tfrac{1}{2}(5 \times 10^{-6})20^2 = 1 \text{ mJ}$$

The total energy stored in the circuit is $w = w_L + w_C = 3$ mJ.

Exercise 5.6

Find the total stored energy in Fig. 5.17*a* when the positions of the capacitor and inductor are interchanged.

Exercise 5.7

Derive Eq. (5.21) using Eq. (5.22) and $dw = p\,dt$.

Current–Voltage Relationships

Having established the duality between inductance and capacitance, we now draw upon our previous voltage–current relationships for capacitors to determine the corresponding current–voltage relationships for inductors.

When you know the **initial current** $i(t_0)$ at time $t = t_0$, the instantaneous current at any later time is related to the voltage $v(t)$ by

$$i(t) = i(t_0) + \frac{1}{L}\int_{t_0}^{t} v(\lambda)\, d\lambda \qquad t > t_0 \tag{5.23}$$

Alternatively, letting $t_0 \to -\infty$ and assuming that $i(-\infty) = 0$, Eq. (5.23) simplifies to

$$i(t) = \frac{1}{L}\int_{-\infty}^{t} v(\lambda)\, d\lambda \tag{5.24}$$

This expression shows that an inductor also possesses electrical **memory**, the current at any time t being dependent on the entire past behavior of the voltage. Physically, inductor memory takes the form of the resultant flux linkage $N\phi$ created by the voltage $v = N\,d\phi/dt$ over all previous time. Thus, to calculate $i(t)$, you must know either the initial current or the past behavior of v.

If the voltage stays constant at $v = V$ from $t = t_1$ to $t = t_2$, then the resulting current is

$$i(t) = i(t_1) + \frac{V}{L}(t - t_1) \qquad t_1 < t \leq t_2 \tag{5.25}$$

Constant voltage therefore produces a **current ramp** that begins at $i(t_1)$ and increases linearly with slope $di/dt = V/L$.

If the voltage jumps at some time t_j but has a finite value, then the current cannot jump. Stated mathematically,

$$i(t_j^+) = i(t_j^-) \qquad \text{if } |v| < \infty \tag{5.26}$$

In other words, there is **continuity of inductor current** in that

> The current through an inductor cannot change discontinuously when the voltage remains finite.

Thus, inductors tend to smooth out current variations, while capacitors tend to smooth out voltage variations.

To summarize, Table 5.2 gives a side-by-side comparison of the capacitor and inductor properties covered so far. The table also includes series and parallel equivalent inductance, which we discuss next.

TABLE 5.2 Summary of Capacitor and Inductor Properties

Capacitors	Inductors
$i = C\dfrac{dv}{dt}$	$v = L\dfrac{di}{dt}$
DC open circuit	DC short circuit
$w = \frac{1}{2}Cv^2$	$w = \frac{1}{2}Li^2$
$v(t) = v(t_0) + \dfrac{1}{C}\displaystyle\int_{t_0}^{t} i\,d\lambda$	$i(t) = i(t_0) + \dfrac{1}{L}\displaystyle\int_{t_0}^{t} v\,d\lambda$
$\quad = \dfrac{1}{C}\displaystyle\int_{-\infty}^{t} i\,d\lambda$	$\quad = \dfrac{1}{L}\displaystyle\int_{-\infty}^{t} v\,d\lambda$
Continuity of voltage when $\lvert i \rvert$ is finite.	Continuity of current when $\lvert v \rvert$ is finite.
$C_{par} = C_1 + C_2 + \cdots$	$L_{ser} = L_1 + L_2 + \cdots$
$\dfrac{1}{C_{ser}} = \dfrac{1}{C_1} + \dfrac{1}{C_2} + \cdots$	$\dfrac{1}{L_{par}} = \dfrac{1}{L_1} + \dfrac{1}{L_2} + \cdots$

Exercise 5.8

Consider the capacitor waveforms in Fig. 5.5 (p. 200). Using duality, relabel the drawing to correspond to inductor waveforms.

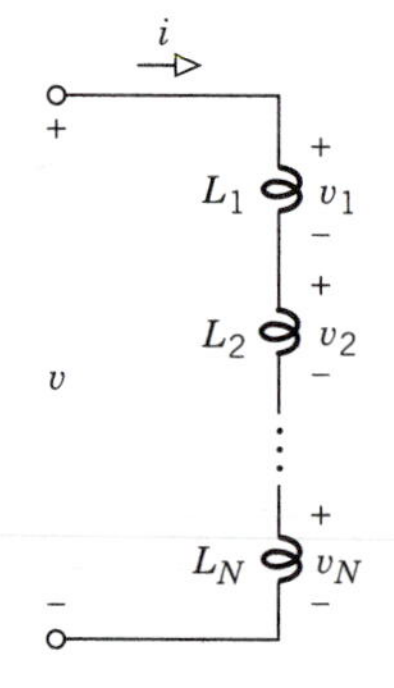

Figure 5.18 Inductors in series.

Series and Parallel Inductance

In the absence of magnetic nonlinearity, inductors are *linear* elements. Consequently, any two-terminal network consisting entirely of ideal inductors acts at its terminals like a single *equivalent* inductance.

When N inductors are connected in series, as in Fig. 5.18, the terminal voltage is $v = v_1 + v_2 + \ldots + v_N$ with $v_1 = L_1\, di/dt$, etc. Thus,

$$v = L_1 \frac{di}{dt} + L_2 \frac{di}{dt} + \cdots + L_N \frac{di}{dt} = (L_1 + L_2 + \cdots + L_N) \frac{di}{dt}$$

so the **series equivalent inductance** is

$$L_{ser} = L_1 + L_2 + \cdots + L_N \tag{5.27}$$

The resulting voltage v_n across L_n is given by the **voltage-divider relation**

$$v_n = (L_n/L_{ser})v \tag{5.28}$$

Note that these two equations have exactly the same form as those for resistors in series.

If N inductors are connected in parallel, as in Fig. 5.19, then $i = i_1 + i_2 + \cdots + i_N$ and $v = L_1\, di_1/dt = L_2\, di_2/dt, = \ldots$, so

$$\frac{di}{dt} = \frac{di_1}{dt} + \frac{di_2}{dt} + \cdots + \frac{di_N}{dt} = \left(\frac{1}{L_1} + \frac{1}{L_2} + \cdots + \frac{1}{L_N}\right) v$$

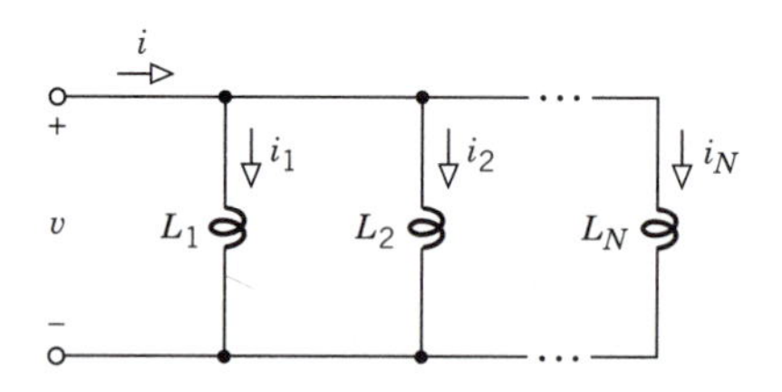

Figure 5.19 Inductors in parallel.

Hence, the **parallel equivalent inductance** can be calculated via

$$\frac{1}{L_{par}} = \frac{1}{L_1} + \frac{1}{L_2} + \cdots + \frac{1}{L_N} \tag{5.29}$$

If $N = 2$, then Eq. (5.29) simplifies to

$$L_{par} = \left(\frac{1}{L_1} + \frac{1}{L_2}\right)^{-1} = \frac{L_1 L_2}{L_1 + L_2}$$

The equivalent inductance for a ladder network of inductors can be found by series-parallel reduction using Eqs. (5.27) and (5.29).

The voltage across each inductor in Fig. 5.19 is $v = L_{par}\, di/dt$. Hence, the current i_n through L_n is given by Eq. (5.23) as

$$i_n(t) = i_n(t_0) + \frac{1}{L_n}\int_{t_0}^{t} L_{par}\frac{di}{d\lambda}\, d\lambda$$

from which

$$i_n(t) = i_n(t_0) + \frac{L_{par}}{L_n}[i(t) - i(t_0)] \qquad t \geq t_0 \tag{5.30}$$

This result expresses the **current-divider relation** for parallel inductors, including initial currents.

Equations (5.27)-(5.30) are, of course, the duals of our previous results for parallel and series capacitors. However, the inductor equations become invalid when the coils are close together and their magnetic fields interact. Such interaction creates a coupling effect known as **mutual inductance**. Mutual inductance is the physical phenomenon underlying the action of **transformers**, and it will be discussed more fully in Chapter 8. Capacitors seldom interact because their electric fields are almost completely confined within their respective dielectrics.

Example 5.7 ***LR Network Calculations***

Suppose you know that the network in Fig. 5.20*a* has no mutual-inductance effects and that

$$i(0) = 0 \qquad i_a(t) = 5t \qquad t \geq 0$$

You want to find the resulting terminal variables $v(t)$ and $i(t)$ for $t \geq 0$.

To expedite matters, you observe that the two 0.4-H inductors can be combined into the parallel equivalent $L_{par} = (0.4 \times 0.4)/(0.4 + 0.4) = 0.2$ H. You also observe that the other two inductors carry the same current, so they can be combined into the series equivalent $L_{ser} = 0.1 + 0.5 = 0.6$ H. Figure 5.20*b* shows the resulting simplified diagram. No further series-parallel reductions are possible because you cannot combine different types of elements.

From the simplified diagram you now see that

$$v(t) = 2i_a(t) + 0.6\frac{di_a}{dt} = 10t + 3$$

$$i(t) = i_a(t) + i_b(t) = 5t + i_b(t)$$

Next, you obtain $i_b(t)$ from $v(t)$ by setting $v(\lambda) = 10\lambda + 3$ in Eq. (5.23), so

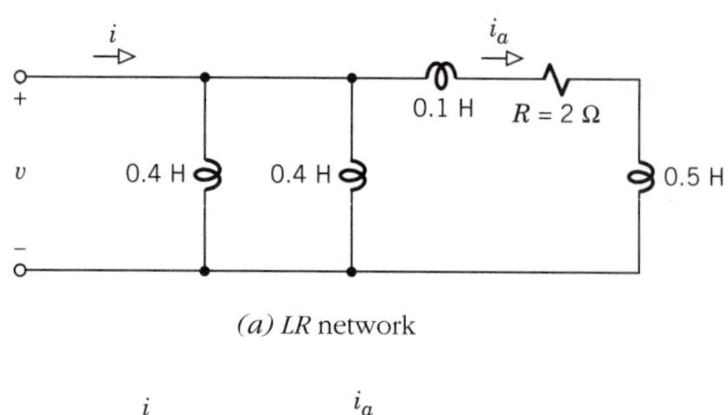

(a) LR network

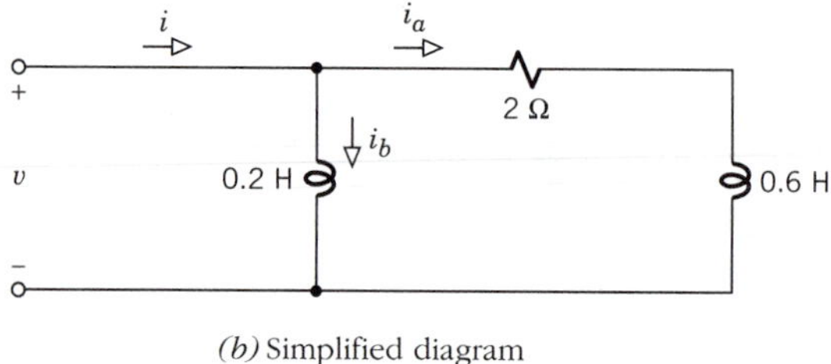

(b) Simplified diagram

Figure 5.20

$$i_b(t) = i_b(0) + \frac{1}{0.2}\int_0^t (10\lambda + 3)\, d\lambda = i_b(0) + 25t^2 + 15t$$

Finally, you evaluate $i_b(0)$ from

$$i(0) = i_a(0) + i_b(0) = i_b(0) = 0$$

Hence,

$$i(t) = 5t + 0 + 25t^2 + 15t = 25t^2 + 20t$$

Exercise 5.9

Let the network in Fig. 5.20*a* have $R = 0$. Find $v(t)$ and $i_a(t)$ for $t \geqslant 0$ when $i(t) = 4 \sin 50t$ A and $i_a(0) = 2$ A.

5.3 DYNAMIC CIRCUITS

Dynamic circuits include at least two different types of elements — capacitance and resistance, or inductance and capacitance, for instance. Since the element equations for capacitors and inductors involve time derivatives, the voltage–current relationships for such circuits usually take the form of *differential equations*. Obtaining the solutions of those equations is an essential part of circuit analysis because dynamic circuit behavior appears in numerous applications.

But our purpose here is not a comprehensive study of differential equations. Rather, we'll identify important properties of dynamic circuits to establish a framework for more efficient analysis methods developed in subsequent chapters. This section thereby provides the underlying concepts of the engineering approach to dynamic circuits.

We'll begin by formulating differential equations for some typical circuits. Then we'll show how the equations can be solved by focusing on three particular types of behavior: the *natural response,* the *forced response,* and the *complete response.* To avoid undue complications, we'll devote most of our attention to circuits containing just one energy-storage element and one source. Circuits with multiple sources can be treated by invoking superposition and analyzing two or more single-source circuits.

Differential Circuit Equations

The analysis of any dynamic circuit starts with two fundamental types of information:

- Kirchhoff's laws for the circuit configuration
- Device equations for the individual elements

The combination of device equations with Kirchhoff's laws leads to relationships between the branch variables and the applied sources. Sometimes the result is a direct expression relating the branch variable to the applied source. More often, however, the relationship is an indirect one in the form of a differential equation.

As a simple example, consider the series *RL* network driven by a current source in Fig. 5.21*a*. The resulting terminal voltage v might be the variable of interest here, and Kirchhoff's voltage law requires of the element voltages that

$$v_L + v_R = v$$

The element voltages, in turn, are given by the device equations

$$v_L = L\,di/dt \qquad v_R = Ri$$

Inserting these devices equations into the KVL equation yields

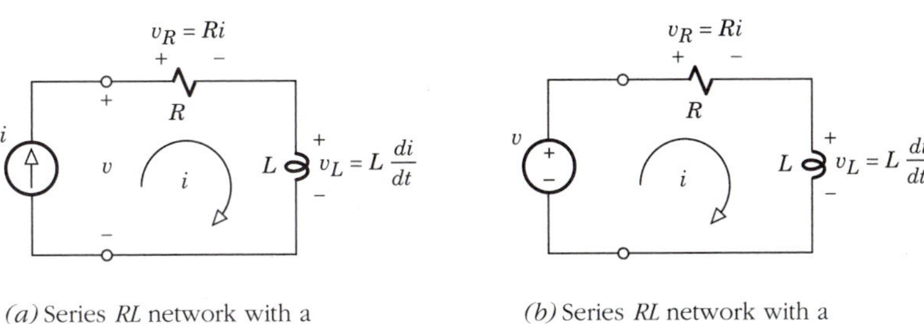

(a) Series *RL* network with a current source

(b) Series *RL* network with a voltage source

Figure 5.21

$$v = L\frac{di}{dt} + Ri$$

This relationship is *direct* in that you can find $v(t)$ by inserting the expression for the source current $i(t)$.

But suppose the same network is driven by a voltage source, as in Fig. 5.21*b*. Since KVL and the device equations are the same as before, our previous relationship between v and i still holds. However, v is now the applied source and i is the unknown branch variable, so di/dt is also unknown. We therefore put the unknown terms on the left and write

$$L\frac{di}{dt} + Ri = v \qquad (5.31)$$

Although this expression is the desired result, it is an *indirect* relationship that cannot be solved for $i(t)$ just by inserting the expression for $v(t)$.

Equation (5.31) is called an **inhomogeneous linear first-order differential equation** — quite a mouthful! It is a *differential* equation because both the unknown and its derivatives are present. It is a *first-order* equation because no higher derivatives of the unknown are included. It is a *linear* equation because only linear functions of the unknown are involved. It is an *inhomogeneous* equation because it contains a nonzero excitation term on the right-hand side. Such equations occur time and again in conjunction with dynamic circuits.

By way of further illustration, consider the voltage across the parallel *RC* circuit in Fig. 5.22*a*. This circuit is driven by a current source, and KCL requires

$$i_C + i_R = i$$

where

$$i_C = C\,dv/dt \qquad i_R = v/R$$

Hence, v is related to i via

$$C\frac{dv}{dt} + \frac{1}{R}v = i \qquad (5.32)$$

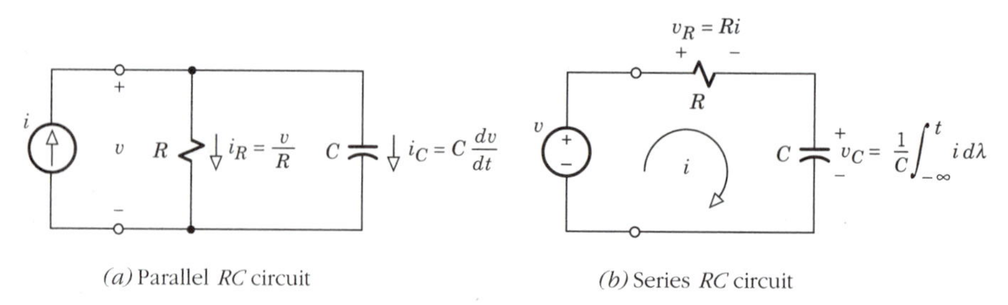

(*a*) Parallel *RC* circuit

(*b*) Series *RC* circuit

Figure 5.22

This equation has the same mathematical form as Eq. (5.31), with v being the unknown and i the excitation. Indeed, Eq. (5.32) is the *dual* of Eq. (5.31) because Fig. 5.22*a* is the dual of Fig. 5.21*b*.

The series *RC* circuit in Fig. 5.22*b* offers more of a challenge. KVL and the device equations immediately yield the *integral equation*

$$Ri + \frac{1}{C}\int_{-\infty}^{t} i(\lambda)\, d\lambda = v$$

We then convert this into a differential equation by differentiating both sides with respect to time, using the general property that

$$\frac{d}{dt}\int_{-\infty}^{t} x(\lambda)\, d\lambda = x(t)$$

Thus,

$$R\frac{di}{dt} + \frac{1}{C}i = \frac{dv}{dt} \tag{5.33a}$$

Again we have the same form as Eq. (5.31), except that the term on the right-hand side is now the derivative of the excitation. Alternatively, we could take v_C as the unknown and note that $i = C\, dv_C/dt$ so

$$RC\frac{dv_C}{dt} + v_C = v \tag{5.33b}$$

Differentiating both sides of Eq. (5.33b) and substituting $dv_C/dt = i/C$ leads to Eq. (5.33a).

Let's summarize the foregoing results by using $y(t)$ to stand for any unknown branch variable of interest. Then Eqs. (5.31)–(5.33) may be put in the generic form

$$a_1\frac{dy}{dt} + a_0 y = f(t) \tag{5.34}$$

The constants a_1 and a_0 incorporate element values, while $f(t)$ represents the effect of the excitation and is known as the **forcing function**. Most linear circuits consisting of resistance and one energy-storage element are described by first-order differential equations like Eq. (5.34). They are therefore called **first-order circuits**.

Turning to circuits with more than one energy-storage element, consider the series *LRC* circuit in Fig. 5.23. To find the relationship between the current i and the applied voltage v, we combine the device equations with KVL and get the *integral-differential equation*

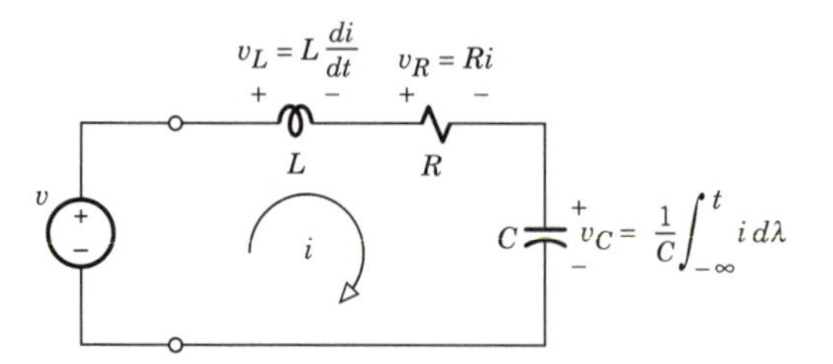

Figure 5.23 Series *LRC* circuit.

$$L\frac{di}{dt} + Ri + \frac{1}{C}\int_{-\infty}^{t} i\,d\lambda = v$$

Differentiating both sides then yields

$$L\frac{d^2i}{dt^2} + R\frac{di}{dt} + \frac{1}{C}i = \frac{dv}{dt} \tag{5.35}$$

Here d^2i/dt^2 is the derivatives of di/dt or the *second derivative* of i. Thus, Eq. (5.35) is a **second-order differential equation**.

Second-order equations arise whenever a circuit has two independent energy-storage elements. Accordingly, such circuits are called **second-order circuits**. Any second-order circuit is described by a differential equation in the form

$$a_2\frac{d^2y}{dt^2} + a_1\frac{dy}{dt} + a_0y = f(t) \tag{5.36}$$

Equation (5.36) is the logical extension of Eq. (5.34). Furthermore, as you might suspect, the differential equation of a circuit with n energy-storage elements usually includes all derivatives of the unknown up to the nth-derivative.

Example 5.8 *Second-Order Amplifier Circuit*

Figure 5.24 diagrams an amplifier with a CCCS. We'll derive the relationship between v_{in} and v_{out}, which we expect to be a second-order differential equation because the circuit contains two energy-storage elements.

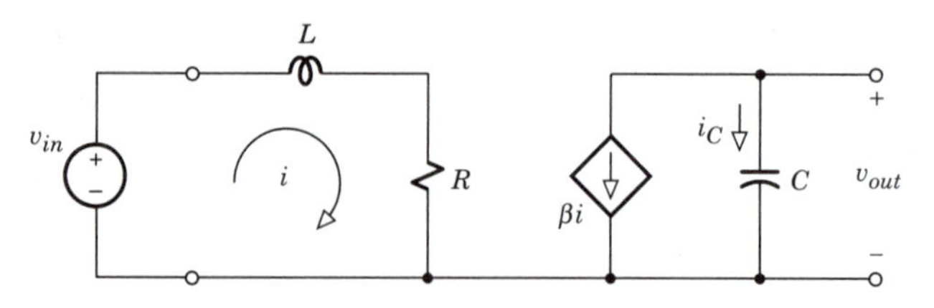

Figure 5.24 Second-order amplifier circuit.

The CCCS drives the current through the capacitor to produce the output voltage, so $C\, dv_{out}/dt = i_C = -\beta i$ and

$$i = -\frac{C}{\beta}\frac{dv_{out}}{dt}$$

The control variable i on the input side is related to the input voltage by

$$v_{in} = L\frac{di}{dt} + Ri$$

Substituting for i yields

$$v_{in} = -\frac{LC}{\beta}\frac{d}{dt}\left(\frac{dv_{out}}{dt}\right) - \frac{RC}{\beta}\frac{dv_{out}}{dt}$$

or, after rearrangement,

$$LC\frac{d^2 v_{out}}{dt^2} + RC\frac{dv_{out}}{dt} = -\beta v_{in}$$

This is indeed a second-order equation, but the term directly proportional to v_{out} is missing on the left-hand side — like Eq. (5.36) with $a_0 = 0$.

Exercise 5.10

Derive a differential equation for the circuit in Fig. 5.23 with v_C as the unknown. Then differentiate your result to obtain Eq. (5.35).

Natural Response

The task of solving a differential circuit equation can be approached by dealing first with the **natural response**. This special case is more easily handled because

> The natural response is the solution of the circuit equation with the forcing function set to zero.

Mathematicians call the natural response the **complementary solution**. Physically, natural response occurs by itself when the only excitation comes from internal stored energy.

Suppose, for example, that the parallel *RC* circuit in Fig. 5.25 has no applied source and hence no input current. Even so, if the capacitor has been charged to some initial voltage, then current i_R flows through R and the voltage

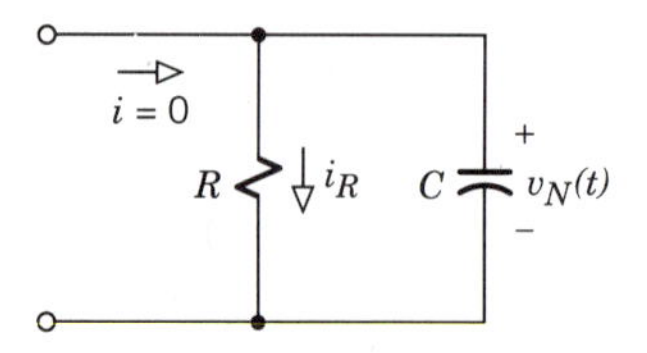

Figure 5.25 *RC* network with natural-response voltage.

changes with time as the capacitor discharges. The label $v_N(t)$ emphasizes that this voltage is a natural response.

For more generality, consider any branch variable $y(t)$ in any first-order circuit described by our generic relationship $a_1\, dy/dt + a_0 y = f(t)$. Since we are concerned only with the natural response, denoted $y_N(t)$, we set $y = y_N$ and $f(t) = 0$ to obtain

$$a_1 \frac{dy_N}{dt} + a_0 y_N = 0 \tag{5.37}$$

Now we have a **homogeneous differential equation**, as distinguished from equations with nonzero forcing functions.

But even a homogeneous equation cannot be solved by algebraic manipulation because you would need to know the solution y_N to evaluate dy_N/dt and vice versa. Instead, we'll make an informed *guess* at the solution and then test our guess. While this guessing method may not seem to be very "scientific," it is certainly valid if it leads to the correct result. The key is making an *informed* guess based on Eq. (5.37) rewritten as

$$y_N = -(a_1/a_0)\, dy_N/dt$$

This expression shows that $y_N(t)$ must be *proportional to its own derivative,* and the sole class of time functions possessing that property are *exponential* waveforms. We therefore assume that

$$y_N(t) = Ae^{st} \tag{5.38}$$

where A and s are constants to be determined. If $y_N(t)$ is as given by Eq. (5.38), then $dy_N/dt = sAe^{st}$ — so $y_N(t)$ is indeed proportional to dy_N/dt.

To test our assumed solution, we substitute Eq. (5.38) and its derivative back into Eq. (5.37) to obtain $a_1 sAe^{st} + a_0 Ae^{st} = 0$. Factoring the common term Ae^{st} then yields

$$(a_1 s + a_0)Ae^{st} = 0$$

which holds if either $a_1 s + a_0 = 0$ or $Ae^{st} = 0$. The latter option corresponds to the trivial case $y_N = Ae^{st} = 0$, which holds little interest. We're therefore left with

$$a_1 s + a_0 = 0$$

This relationship is known as the **characteristic equation** because it dictates the value of s, the result being $s = -a_0/a_1$. Hence, the nontrivial solution of Eq. (5.37) is

$$y_N(t) = Ae^{st} \qquad s = -a_0/a_1 \tag{5.39}$$

An alternative derivation of Eq. (5.39) that avoids guessing starts with Eq. (5.37) rewritten as

$$dy_N/dt = -(a_0/a_1)y_N$$

from which

$$\frac{dy_N}{y_N} = -\frac{a_0}{a_1}dt$$

Indefinite integration of both sides then yields

$$\ln y_N = -(a_0/a_1)t + K_{int}$$

where K_{int} is the constant of integration. Now recall that if $\ln a = b$ then $a = e^b$, so

$$y_N = e^{[-(a_0/a_1)t + K_{int}]} = e^{K_{int}}e^{-(a_0/a_1)t}$$

Letting $e^{K_{int}} = A$ finally yields

$$y_N(t) = Ae^{-(a_0/a_1)t} \tag{5.40}$$

in agreement with Eq. (5.39).

We have now shown by two methods that Eq. (5.40) satisfies the homogeneous equation with any value of the constant A. Evaluating A therefore requires additional information. The most useful information for that purpose is the **initial condition** on the natural response. In particular, if we know that $y_N(0^+) = Y_0$, then we can set $t = 0^+$ in Eq. (5.40) to obtain

$$y_N(0^+) = Ae^{0+} = A = Y_0$$

Thus,

$$y_N(t) = Y_0 e^{-(a_0/a_1)t} \qquad t > 0$$

This result indicates that the natural response of a first-order circuit with $a_0/a_1 > 0$ starts at the initial value Y_0 and *decays exponentially* toward zero as $t \rightarrow \infty$.

Decaying exponentials may also appear in the natural response of circuits containing more than one energy-storage element. For example, take the second-order circuit described by

$$4\frac{d^2y}{dt^2} + 32\frac{dy}{dt} + 60y = f(t)$$

so the related homogeneous equation is

$$4\frac{d^2y_N}{dt^2} + 32\frac{dy_N}{dt} + 60y_N = 0$$

This second-order equation cannot be solved by integration, as we did in the first-order case. However, the assumption that $y_N(t) = Ae^{st} \neq 0$ yields

$$4s^2Ae^{st} + 32sAe^{st} + 60Ae^{st} = 0$$

which leads to the characteristic equation

$$4s^2 + 32s + 60 = 0$$

This quadratic equation factors as

$$4(s + 3)(s + 5) = 0$$

Hence, there are two different roots, namely

$$s = -3 \qquad s = -5$$

Since both values of s satisfy the homogeneous equation, and since superposition holds in a linear circuit, the natural response consists of *two exponentials functions* in the form

$$y_N(t) = A_1e^{-3t} + A_2e^{-5t}$$

Evaluating the constants A_1 and A_2 now requires *two initial conditions*, consistent with the fact that a second-order circuit stores energy in two separate elements. But the main point to notice here is that $y_N(t)$ decays exponentially to zero — just as it does in a first-order circuit with $a_0/a_1 > 0$.

Besides exponentials, other types of waveforms may be found in the natural response of circuits with two or more energy-storage elements. Nonetheless, regardless of the number of capacitors and inductors, it is always true that

> If a circuit consists of resistors and energy-storage elements, but no controlled sources, then the natural response dies away and
>
> $$y_N(t) \rightarrow 0 \text{ as } t \rightarrow \infty$$

Circuits having this property are said to be **stable**. Stable behavior occurs because resistance eventually dissipates all the initial stored energy and there is no other energy source to sustain $y_N(t)$. Circuits containing controlled sources may or may not be stable, depending upon the particular configuration.

Example 5.9 *Capacitor Discharge*

Suppose the network back in Fig. 5.25 actually represents a lossy capacitor with capacitance $C = 300\ \mu\text{F}$ and leakage resistance $R = 2\ \text{M}\Omega$. Further suppose that the capacitor was initially charged to 1000 V at $t = 0^-$ and it discharges through its own leakage resistance for $t > 0$.

The natural-response voltage $v_N(t)$ obeys the homogeneous equation obtained from Eq. (5.32), namely

$$C\,dv_N/dt + v_N/R = 0$$

Multiplying through by R gives the more convenient expression

$$RC\,dv_N/dt + v_N = 0$$

The assumption that $v_N(t) = Ae^{st}$ then yields

$$RCsAe^{st} + Ae^{st} = (RCs + 1)Ae^{st} = 0$$

so the characteristics equation is

$$RCs + 1 = 0$$

Hence,

$$s = -1/RC = -1/600$$

and

$$v_N(t) = Ae^{-t/600}$$

The constant A equals the initial value $v_N(0^+)$, and continuity of capacitor voltage requires that $v_N(0^+) = v(0^-) = 1000$ V. Thus, $A = 1000$ V and

$$v_N(t) = 1000e^{-t/600}\ \text{V} \qquad t > 0$$

which is plotted in Fig. 5.26.

Although $v_N(t)$ is a decaying exponential function, the plot reveals that the voltage at $t = 600$ s has only decayed to $v_N(600) = 1000e^{-1} = 368$ V. However, it does become quite small for $t \geq 5 \times 600$ s when $v_N(t) \leq 1000e^{-5} \approx 7$ V. The slow rate of decay is consistent with the tiny leakage current $i_R = v_N/R \leq 0.5$ mA.

The numerical values here happen to be typical for a capacitor in the high-voltage section of a TV set. Consequently, you might get a nasty "zap"

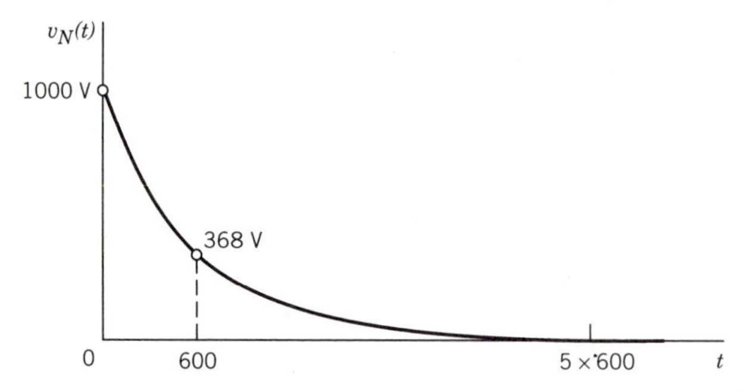

Figure 5.26 Decaying exponential waveform $v_N(t) = 1000e^{-t/600}$ V.

from the capacitor even though the power has been off for several minutes. People who repair TV sets often speed up the decay by putting a small external resistance R_x across the capacitor's terminals so C discharges through the parallel combination $R_x \| R \approx R_x$. If $R_x = 20\ \Omega$, for instance, then $s \approx -1/R_xC = -1/0.006$ and $v_N(t) \leqslant 7$ V for $t \geqslant 5 \times 0.006 = 0.03$ s.

Exercise 5.11

Let Fig. 5.21*b* represent a coil with inductance $L = 50$ mH and winding resistance $R = 0.5\ \Omega$. Taking $v = 0$ to correspond to a short across the terminals, find $i_N(t)$ for $t > 0$ when $i(0^-) = 2$ A. Then evaluate $i_N(t)$ at $t = 0.1$, 0.5, and 1.0 s.

Forced Response

In contrast to natural response, **forced response** occurs when the circuit has a nonzero forcing function established by an applied source. More specifically,

> The forced response is the solution of the circuit's inhomogeneous differential equation, independent of any initial conditions.

Mathematicians call this response the **particular solution** because there is only one such solution for a given differential equation with a given forcing function.

We'll let $y_F(t)$ denote the forced response of an arbitrary branch variable. For second-order circuits described by Eq. (5.36), the behavior of $y_F(t)$ is governed by the inhomogeneous equation

$$a_2 \frac{d^2 y_F}{dt^2} + a_1 \frac{dy_F}{dt} + a_0 y_F = f(t) \tag{5.41}$$

This equation also holds for first-order circuits by the simple expedient of letting $a_2 = 0$.

The forced response must satisfy Eq. (5.41) for all values of t with the forcing function in question. As we did for the natural response, we'll attempt to find the forced response by making an informed guess of the particular solution.

In many circuit problems, $f(t)$ is a simple time function having a finite number of different derivatives. Equation (5.41) then implies that $y_F(t)$ involves a sum of functions like $f(t)$ and its derivatives, each term being multiplied by some constant coefficient. This observation suggest the following solution technique.

Method of undetermined coefficients — The method consists of two steps:

1. Select a trial form for $y_F(t)$. For that purpose, Table 5.3 lists selected forcing functions and the corresponding trial solutions with coefficients K_0, K_1, . . . If $f(t)$ equals a sum or product of the listed functions, then $y_F(t)$ should be taken as a sum or product of the trial solutions.
2. Evaluate the coefficients by substituting $y_F(t)$ into the differential equation along with the forcing function. The values of the coefficients are then determined from the property that the resulting expression must hold for all t. Hence, there will be no unknown constants in the final result.

TABLE 5.3 Selected Trial Solutions for Forced Response

$f(t)$	$y_F(t)$
k_0 (a constant)	K_0 (a constant)
$k_1 t$	$K_1 t + K_0$
$k_2 e^{at}$	$K_2 e^{at}$
$k_3 \cos \omega t + k_4 \sin \omega t$	$K_3 \cos \omega t + K_4 \sin \omega t$

Example 5.10 *Sinusoidal Forced Response*

Let the voltage-driven series RL circuit back in Fig. 5.21b have $R = 4\ \Omega$ and $L = 0.1$ H. From Eq. (5.31), the behavior of the current is governed by

$$0.1\, di/dt + 4i = v$$

We'll find the forced response i_F produced by the sinusoidal source voltage $v = 25 \sin 30t$ V. The inhomogeneous equation thus becomes

$$0.1\, di_F/dt + 4i_F = 25 \sin 30t$$

Step 1: The forcing function on the right has the form $k_3 \cos \omega t + k_4 \sin \omega t$ with $k_3 = 0$ and $\omega = 30$, so the appropriate particular solution from Table 5.3 is

$$i_F(t) = K_3 \cos 30t + K_4 \sin 30t$$

Step 2: We evaluate the coefficients K_3 and K_4 by substitution in the inhomogeneous equation. The derivative of $i_F(t)$ is obtained by recalling that

$$\frac{d}{dt}(\cos \omega t) = -\omega \sin \omega t \qquad \frac{d}{dt}(\sin \omega t) = \omega \cos \omega t$$

Hence,

$$di_F/dt = -30K_3 \sin 30t + 30\, K_4 \cos 30t$$

Substituting i_F and di_F/dt into the inhomogeneous equation and regrouping terms gives

$$(4K_3 + 3K_4) \cos 30t + (-3K_3 + 4K_4) \sin 30t = 25 \sin 30t$$

This relationship holds for all values of t only if

$$4K_3 + 3K_4 = 0 \qquad -3K_3 + 4K_4 = 25$$

Simultaneous solution then yields

$$K_3 = -3 \qquad K_4 = 4$$

Hence, our result for the forced response is

$$i_F(t) = -3 \cos 30t + 4 \sin 30t \text{ A}$$

The waveform $i_F(t)$ and its two components are plotted in Fig. 5.27. Notice that the sum turns out to be another sinusoidal waveform with the same repetition period as the forcing function. We'll take advantage of that property for the study of ac circuits in the next chapter.

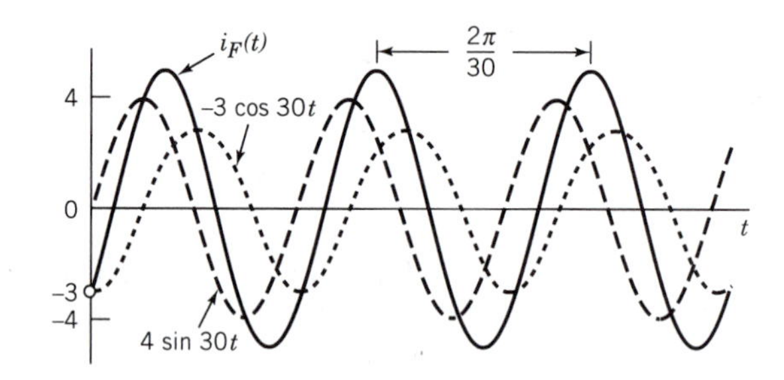

Figure 5.27 Forced-response waveform $i_F(t) = -3 \cos 30t + 4 \sin 30t$ A.

Exercise 5.12

Find the forced response for the voltage in Fig. 5.22*a* when $R = 5$ kΩ, $C = 100$ μF, and $i = 2t$ mA. (Save your work for use in Exercises 5.13 and 5.14.)

Excitation at a natural frequency is a special case that sometimes occurs when $f(t)$ includes an exponential function. The exponential trial solution may then lead to an impossible condition on the coefficients. This case therefore requires special treatment as follows:

> If the trial form for $y_F(t)$ contains any term proportional to a component of the natural response, then that term must be multiplied by t before evaluating the coefficients.

We need this modification because any component of the natural response would be a solution of the *homogeneous* equation rather than the inhomogeneous equation.

Example 5.11 *Exponential Forced Response*

In Example 5.10, we found the sinusoidal forced response of the current governed by

$$0.1\ di/dt + 4i = v$$

This time we'll find the forced response when the source voltage has the exponential waveform $v = 10e^{-bt}$ V. We'll consider two different values of the constant b, namely, $b = 20$ and $b = 40$.

Since the exponential source function might correspond to excitation at a natural frequency, we must first find the form of the natural response. Setting $v = 0$ and $i = i_N$ gives the related homogeneous equation

$$0.1\ di_N/di + 4i_N = 0$$

The assumption that $i_N = Ae^{st}$ then yields the characteristic equation

$$0.1s + 4 = 0 \quad \Rightarrow \quad s = -40$$

The natural response is therefore $i_N(t) = Ae^{-40t}$.

Now, for the forced response, we replace i with i_F to obtain the inhomogeneous equation

$$0.1\ di_F/dt + 4i_F = 10e^{-bt} \qquad b = 20 \text{ or } 40$$

The forcing function on the right has the form k_2e^{at} with $a = -b$, and the corresponding trial solution in Table 5.3 is $K_2e^{at} = K_2e^{-bt}$.

With $b = 20$, the trial solution differs from the form of $i_N(t)$ so we just take $i_F(t) = K_2e^{-20t}$, whose derivative is $di_F/dt = -20K_2e^{-20t}$. We then evaluate K_2 by substituting for i_F and di_F/dt in the inhomogeneous equation to get

$$0.1(-20K_2e^{-20t}) + 4K_2e^{-20t} = 10e^{-20t}$$

or, after simplification,

$$2K_2e^{-20t} = 10e^{-20t}$$

This expression holds for all t only when $K_2 = 5$. Thus,

$$i_F(t) = 5e^{-20t} \text{ A}$$

which is our final result when $b = 20$.

But when $b = 40$, the trial solution K_2e^{-40t} is proportional to i_N, so we must multiply it by t and take

$$i_F(t) = K_2te^{-40t}$$

Then, using the chain rule for differentiating the product te^{-40t}, the inhomogeneous equation becomes

$$0.1(K_2e^{-40t} - 40K_2te^{-40t}) + 4K_2te^{-40t} = 10e^{-40t}$$

so

$$0.1K_2e^{-40t} = 10e^{-40t} \quad \Rightarrow \quad K_2 = 100$$

Thus,

$$i_F(t) = 100te^{-40t} \text{ A}$$

which is our result when $b = 40$.

Had we neglected the modification of multiplying by t when $b = 40$, we would have assumed that $i_F(t) = K_2e^{-40t}$. Substitution into the inhomogeneous equation then gives

$$0.1(-40K_2e^{-40t}) + 4K_2e^{-40t} = 10e^{-40t}$$

which reduces to the impossible condition on K_2 that

$$0 \times K_2e^{-40t} = 10e^{-40t}$$

This clearly incorrect result illustrates why we must modify the form of the particular solution when it contains a term proportional to the natural response.

Exercise 5.13

Find $v_N(t)$ and $v_F(t)$ for the circuit in Exercise 5.12 with $i = 5e^{-2t}$ mA. (Save v_N for use in Exercise 5.14).

Complete Response

Although a forced response $y_F(t)$ satisfies the inhomogeneous equation, it does not necessarily constitute the *complete* response $y(t)$ of a dynamic circuit. The difference between $y(t)$ and $y_F(t)$ hinges upon *initial conditions,* which, by definition, do not affect the forced response.

Accounting for initial conditions generally requires another component in the complete response. Since this component will be independent of the forcing function, it should satisfy the *homogeneous* differential equation rather than the inhomogeneous equation. Accordingly, we conclude that the additional component must take the form of the *natural response.* Hence, the **complete response** is given by

$$y(t) = y_F(t) + y_N(t) \tag{5.42}$$

where $y_N(t)$ includes arbitrary constants to be evaluated from the initial conditions on $y(t)$.

Equation (5.42) obviously holds when the circuit contains initial stored energy that would produce the natural response in absence of other excitation. But Eq. (5.42) also holds for a circuit *without* initial stored energy because the energy-storage elements cannot adjust immediately to the forced response. The readjustment behavior of the complete response thus gives rise to a natural-response component, with or without initial stored energy.

To summarize, the "classical" method for the complete solution of dynamic circuit equations involves the following steps:

1. Find the natural response $y_N(t)$ that satisfies the homogeneous equation and includes one arbitrary constant for each energy-storage element.
2. Select a trial solution for the forcing function in question, and use the inhomogeneous equation to determine the coefficients in $y_F(t)$.
3. Write the complete response as $y(t) = y_F(t) + y_N(t)$ and evaluate the constants in $y_N(t)$ from the initial conditions on $y(t)$.

Note carefully that the third step must be done at the end to incorporate the initial conditions in the entire expression for the complete response. For a first-order circuit, there is just one constant in $y_N(t)$ and we need just one initial condition.

Regardless of the number of energy-storage elements, $y_N(t)$ will die away with time if the circuit under consideration is *stable.* Consequently, the complete response behaves such that

$$y(t) = y_F(t) \qquad \text{as} \qquad t \to \infty \tag{5.43}$$

We say that the circuit has reached the **steady state** when $y_N(t)$ becomes negligible compared to $y_F(t)$. Before arriving at the steady state, the circuit undergoes a readjustment known as the **transient response**, which involves both the natural and forced response.

Since most circuits of practical interest are stable, special analysis techniques have been developed for the study of steady-state and transient response. Several of these specialized techniques will be presented in later chapters. We close here with an example of the complete solution of a first-order circuit using the classical method.

Example 5.12 *Complete Response Calculation*

Again consider a voltage-driven series RL circuit whose current is governed by

$$0.1\ di/dt + 4i = v$$

We'll find the complete response given that $i(t) = 0$ for $t < 0$ and $v(t) = 400 \sin 280t$ V for $t > 0$.

Step 1: The form of the natural response was previously found in Example 5.11 to be

$$i_N(t) = Ae^{-40t}$$

where A is an arbitrary constant.

Step 2: The forced response must satisfy

$$0.1\ di_F/dt + 4i_F = 400 \sin 280t$$

We therefore assume that $i_F(t) = K_3 \cos 280t + K_4 \sin 280t$, and we evaluate K_3 and K_4 by the same method used in Example 5.10 to obtain

$$i_F(t) = -14 \cos 280t + 2 \sin 280t$$

This expression has $i_F(0^+) = -14$, whereas continuity of inductor current requires that $i(0^+) = i(0^-) = 0$. We must therefore add $i_N(t)$ to the forced response to get the correct initial value for the complete response.

Step 3: The complete response is the sum

$$i(t) = i_F(t) + i_N(t) = -14 \cos 280t + 2 \sin 280t + Ae^{-40t}$$

Now we can evaluate A from the initial condition by setting $t = 0^+$ to get

$$i(0^+) = -14 + A = 0 \quad \Rightarrow \quad A = 14$$

Hence, our final result is

$$i(t) = -14 \cos 280t + 2 \sin 280t + 14e^{-40t} \qquad t > 0$$

Figure 5.28 shows the complete response along with its components $i_F(t)$ and $i_N(t)$.

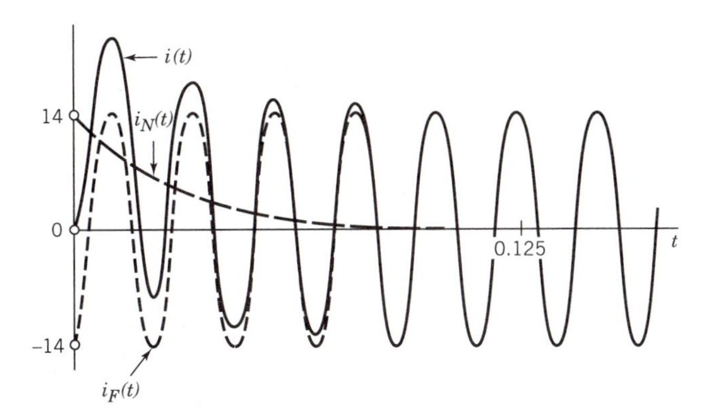

Figure 5.28 Complete response $i(t)$ and its components $i_F(t)$ and $i_N(t)$.

We see that the natural-response component dies away rather quickly, leaving the circuit in the steady state with $i(t) \approx i_F(t)$ for $t \geqslant 5/40 = 0.125$ s. The transient response corresponds to the behavior for $t < 0.125$ s, where $i_N(t)$ is a significant part of $i(t)$.

Exercise 5.14

Find the complete response $v(t)$ for the circuit in Exercise 5.12 when the source current starts at $t = 0^+$ and $v(0) = -30$ V. Also estimate the duration of the transient interval.

PROBLEMS

Section 5.1 Capacitors

5.1* The voltage across a 10-μF capacitor is given by $v = 4t^2$ V. Evaluate w and i at $t = 5$ s.

5.2 Do Problem 5.1 with $v = 40e^{-0.2t}$ V.

5.3 A 2-μF capacitor has $v(0) = -50$ V and $i = 3$ mA for $t \geqslant 0$. Find v and w at $t = 0.1$ s.

5.4 A 50-μF capacitor has $v(3) = 60$ V and $i = -10$ mA for $t \geqslant 3$. Find v and w at $t = 3.5$ s.

5.5* A 40-μF capacitor has $v(0) = 0$. What constant current $i > 0$ results in $w = 5$ J at $t = 4$ s?

5.6 A 200-μF capacitor has $v(0) = 50$ V. What constant current $i > 0$ results in $w = 4$ J at $t = 3$ s?

5.7 The voltage waveform across a 2-μF capacitor is plotted in Fig. P5.7. Sketch the waveforms of $i(t)$, $w(t)$, and $p(t)$, labeling extreme values.

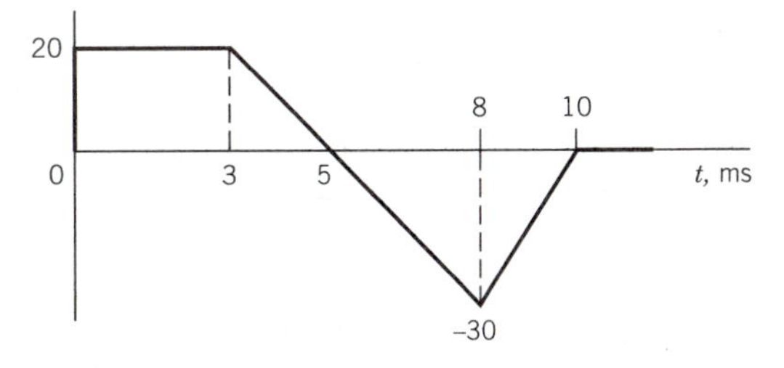

Figure P5.7

5.8 The voltage across a 30-μF capacitor is given by $v(t) = 4 \cos 50t$ V. Sketch the resulting waveforms of $i(t)$, $p(t)$, and $w(t)$ for $0 \leqslant t \leqslant \pi/25$ s, labeling all extreme values.

5.9 An 8-μF capacitor has $v(0) = 0$ and the current waveform in milliamps is plotted in Fig. P5.9. Sketch the waveforms of $v(t)$, $w(t)$, and $p(t)$, labeling all extreme values.

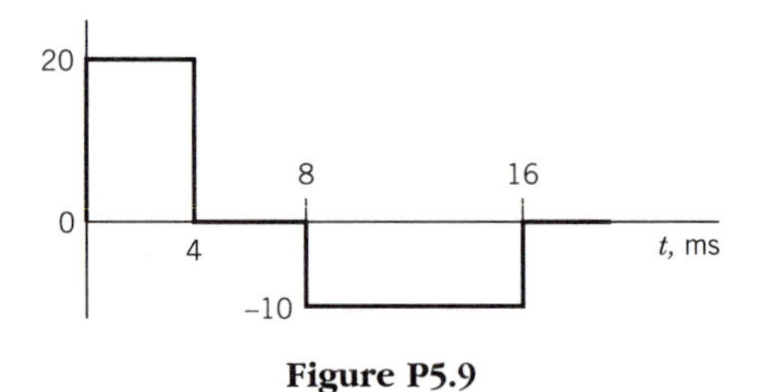

Figure P5.9

5.10 A 5-μF capacitor has $v(0) = 80$ V and the current waveform in milliamps is plotted in Fig. P5.10. Sketch the waveforms of $v(t)$, $w(t)$, and $p(t)$, labeling extreme values.

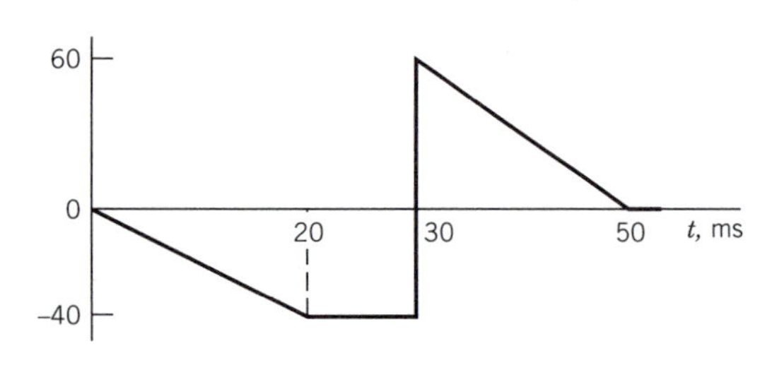

Figure P5.10

5.11 Show that the op-amp circuit in Fig. 5.9 (p. 204) becomes a **differentiator** when the resistor and the capacitor are interchanged.

5.12* Find the equivalent capacitance when a 10-μF capacitor is put in series with parallel-connected 40-μF and 50-μF capacitors.

5.13 Find the equivalent capacitance when a 9-μF capacitor is put in parallel with series-connected 70-μF and 30-μF capacitors.

5.14 Find the equivalent capacitance when series-connected 5-μF and 20-μF capacitors are put in parallel with series-connected 6-μF and 30-μF capacitors.

5.15 Find the equivalent capacitance when series-connected 8-μF and 10-μF capacitors are put in series with parallel-connected 15-μF and 25-μF capacitors.

5.16* The network in Fig. P5.16 has $v_1 = -6$ V, $v_2 = 8$ V, and $v_3 = 6$ V at $t = 0$. If the terminals are connected by a short circuit at $t_1 > 0$, then what are the values of v_1, v_2, and v_3 at t_1?

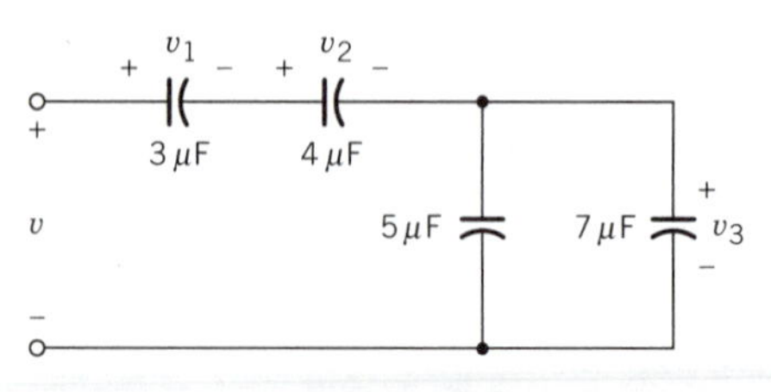

Figure P5.16

5.17 The network in Fig. P5.16 has $v_1 = -10$ V, $v_2 = 10$ V, and $v_3 = 8$ V at $t = 0$. A source is then connected to the terminals such that $v = 40$ V at $t_1 > 0$. What are the values of v_1, v_2, and v_3 at t_1?

5.18 The network in Fig. P5.18 has $v_1 = 8$ V and $v_2 = v_3 = 12$ V at $t = 0$. If the terminals are connected by a short circuit at $t_1 > 0$, then what are the values of v_1, v_2, and v_3 at t_1?

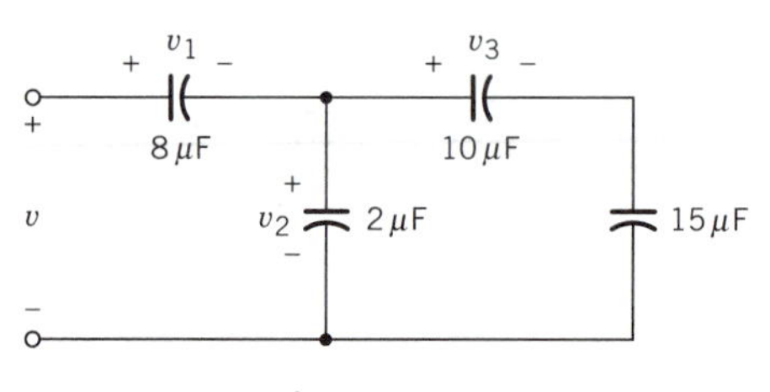

Figure P5.18

5.19 The network in Fig. P5.18 has $v_1 = -20$ V and $v_2 = v_3 = 0$ at $t = 0$. A source is then connected to the terminals such that $v = 30$ V at $t_1 > 0$. What are the values of v_1, v_2, and v_3 at t_1?

5.20 Design a capacitance bank for maximum possible energy storage when connected to a 1400-V source, given a supply of ten 20-μF capacitors rated for up to 800 V. [Hint: Consider Eq. (5.16) for the case of N equal capacitors with zero initial voltages.] How many joules will it store? If the leakage resistance of each capacitor is $R_l = 1$ MΩ, then how much current does the bank draw from the source when the capacitors are fully charged?

5.21 Do Problem 5.20 with an 1800-V source.

5.22 Do Problem 5.20 with a 2000-V source.

5.23 Consider the capacitor in Fig. 5.1a (p. 195). The electric field strength between the plates is $\mathcal{E} = v/d$, and the dielectric breaks down if $\mathcal{E}$ exceeds a certain value. Calculate the maximum possible stored energy for a capacitor having volume $Ad = 1$ cm^3 when: (a) the dielectric is air, for which $\mathcal{E}_{max} \approx 3 \times 10^6$ V/m and $\epsilon_r \approx 1$; (b) the dielectric is a solid with $\mathcal{E}_{max} \approx 2 \times 10^8$ V/m and $\epsilon_r \approx 6$.

Section 5.2 Inductors

5.24* The current through a 0.4-H inductor is given by $i = 10te^{-0.5t}$. Evaluate w and v at $t = 2$ s.

5.25 A 0.8-H inductor has $i(0) = -5$ A and $v = 16$ V for $t \geq 0$. Find i and w at $t = 0.5$ s.

5.26 A 0.5-H inductor has $i(0) = 0$. How long will it take to get $w = 1$ J if $v = 10$ V for $t \geq 0$?

5.27 The current in milliamps through a 0.4 H inductor is plotted in Fig. P5.7. Sketch the waveforms of $v(t)$, $w(t)$, and $p(t)$, labeling extreme values.

5.28 A 40-mH inductor has $i(0) = -1$ A, and the voltage waveform is plotted in Fig. P5.9. Sketch the waveforms of $i(t)$, $w(t)$, and $p((t)$, labeling all extreme values.

5.29 A 1-mH inductor has $i(0) = 0.2$ A, and the voltage waveform is plotted in Fig. P5.10. Sketch the waveforms of $i(t)$, $w(t)$, and $p((t)$, labeling all extreme values.

5.30* Find v_{in}, i_1, and i_2 in Fig. P5.30 under dc steady-state conditions with $i_{in} = 2$ A. Then calculate the total stored energy when $L = 5$ mH and $C = 25$ μF.

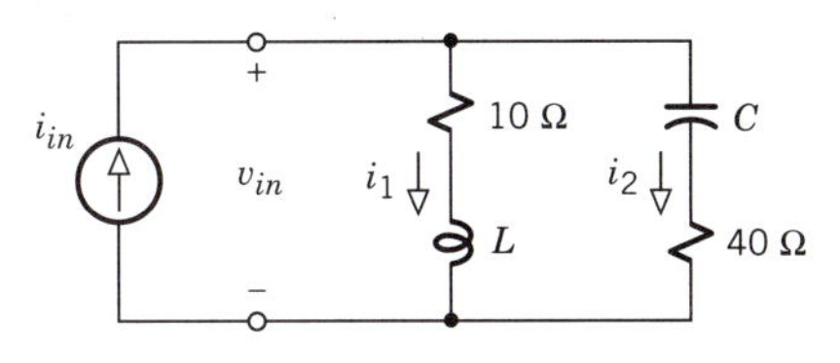

Figure P5.30

5.31 Find i_{in}, v_1, and v_2 in Fig. P5.31 under dc steady-state conditions with $v_{in} = 15$ V. If $L = 16$ mH and the total stored energy is 5 mJ, then what is the value of C?

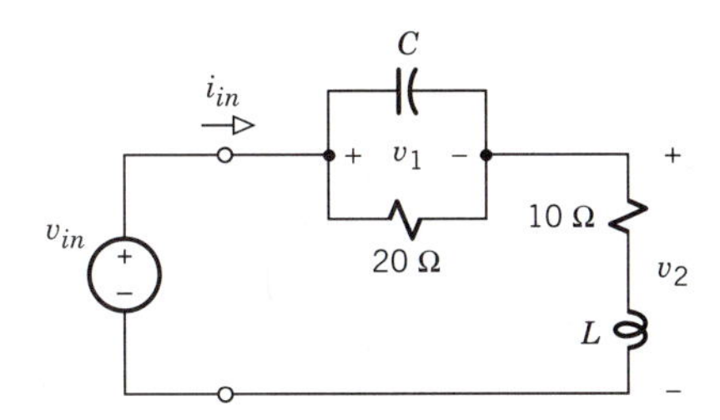

Figure P5.31

5.32 Find v_{in}, i_1, and i_2 in Fig. P5.32 under dc steady-state conditions with $i_{in} = 1$ A. If $C = 80$ μF and the total stored energy is 80 mJ, then what is the value of L?

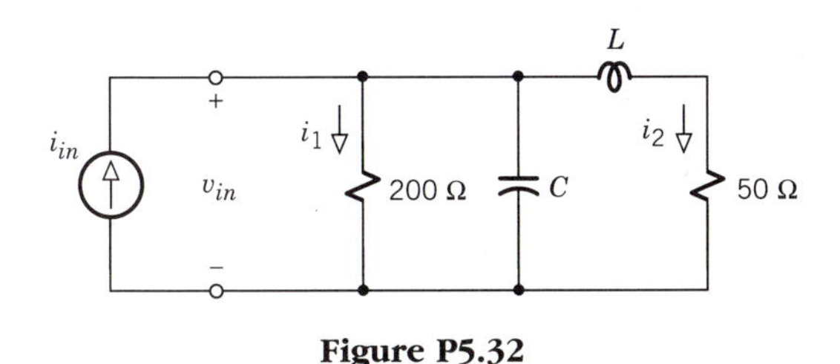

Figure P5.32

5.33 Let the input voltage source in Fig. P5.31 be replaced by a dc current source $i_s = 3$ A and parallel resistance $R_s = 60\ \Omega$. Find i_{in} and v_1 under steady-state conditions. If $C = 25$ μF and the total stored energy is 30 mJ, then what is the value of L?

5.34 Let the input current source in Fig. P5.32 be replaced by a dc voltage source $v_s = 25$ V and series resistance $R_s = 10\ \Omega$. Find v_{in} and i_2 under steady-state conditions. If $L = 25$ mH and the total stored energy is 8 mJ, then what is the value of C?

5.35 A network consists of a 1-μF capacitor in series with parallel-connected 3-H and 0.6-H inductors. Taking the passive convention for polarities, find the total voltage v across the network when the voltage across the capacitor is $v_C = 4 \cos 2000t$.

5.36 A network consists of a 250-μF capacitor in parallel with series-connected 4-H and 6-H inductors. Taking the passive convention for polarities, find the total current i entering the network when the current through the inductors is $i_L = te^{-20t}$.

5.37 A 3-H inductor and a 20-μF capacitor are in series with an unknown inductor. If the voltage across the capacitor is $v_C = 5 \sin 100t$ and the total voltage across the network equals zero, then what is the value of the unknown inductance L_x?

5.38 A 6-H inductor and a 100-μF capacitor are in parallel with an unknown inductor. If the total current entering the network is $i = 1.5e^{-50t}$ and the current entering the capacitor $i_C = 0.5e^{-50t}$, then what is the value of the unknown inductance L_x?

5.39* For the network in Fig. P5.39, find $i_1(t)$ for $t \geq 0$ given that $L_1 = 0.1$ H, $L_2 = 0.15$ H, $L_3 = 0.12$ H, $i_1(0) = -4$, and $v(t) = 3t$ for $t \geq 0$.

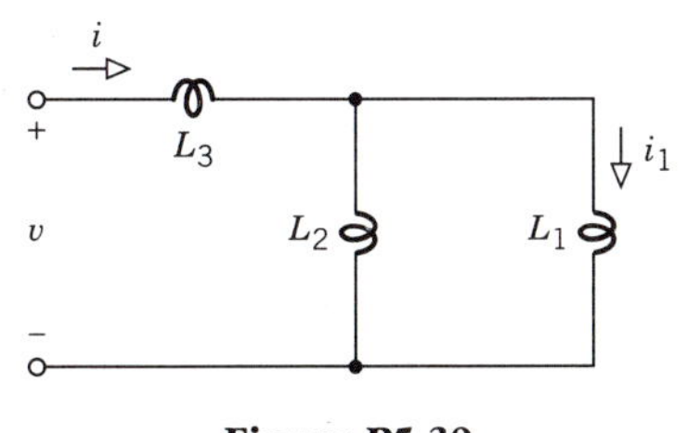

Figure P5.39

5.40 For the network in Fig. P5.39, find $i_1(t)$ for $t \geq 0$ given that $L_1 = 40$ mH, $L_2 = 10$ mH, $L_3 = 16$ mH, $i_1(0) = 0.5$, and $v(t) = 60e^{-1000t}$ for $t \geq 0$.

5.41 For the network in Fig. P5.39, find $i_1(t)$ for $t \geq 0$ given that $L_1 = 4$ mH, $L_2 = 12$ mH, $L_3 = 3$ mH, $i_1(0) = -2$, and $v(t) = 60 \sin 500t$ for $t \geq 0$.

Section 5.3 Dynamic Circuits

5.42* Let element X in Fig. P5.42 be a resistor R_1. Obtain an expression relating i_L to v in the form $a_1\, dy/dt + a_0 y = f(t)$.

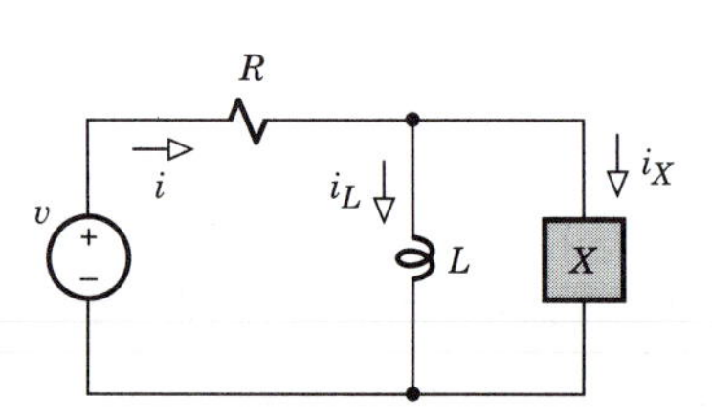

Figure P5.42

5.43 Obtain an expression in the form $a_1\, dy/dt + a_0\, y = f(t)$ relating v_C to v for the circuit in Fig. P5.43.

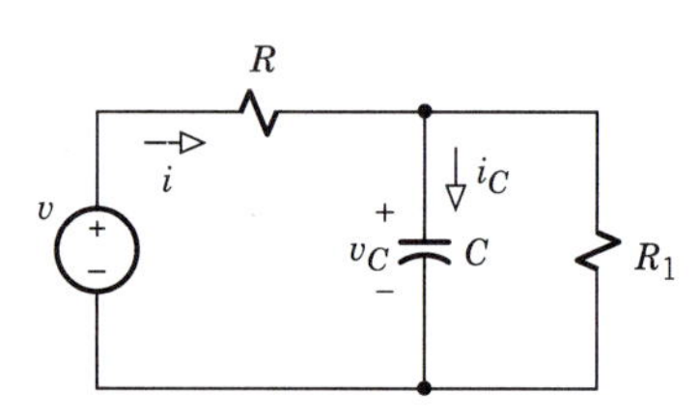

Figure P5.43

5.44 Let the op-amp circuit in Fig. P5.44 have a capacitor C added in parallel with R_F. Derive the differential equation relating v_{out} to v_{in} assuming an ideal op-amp.

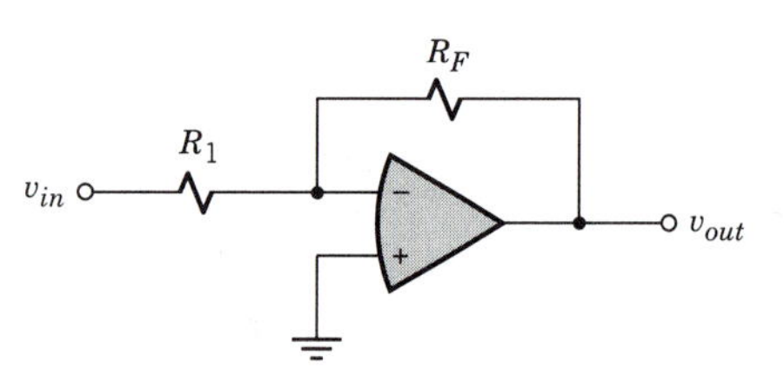

Figure P5.44

5.45 Derive the differential equation relating v_{out} to v_{in} for the ideal op-amp circuit in Fig. P5.44 when a capacitor C is put in series with R_1.

5.46 Obtain a differential equation relating v_C to i in Fig. P5.46 when element X is a short circuit and element Y is an inductor L.

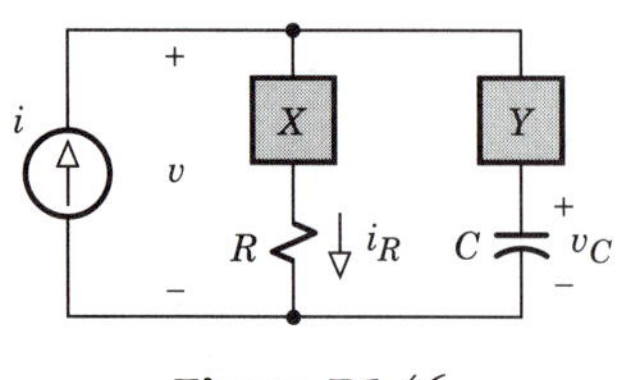

Figure P5.46

5.47 Obtain a differential equation relating i_L to v in Fig. P5.42 when element X is a capacitor C.

5.48* Let element Y in Fig. P5.46 be a short circuit and let element X be an inductor L. Derive the differential equation relating i_R to i. Then differentiate your result to relate v_L to i.

5.49 Let inductance L be added in series with R in Fig. P5.43. Derive the differential equation relating v_C to v. Then differentiate your result to relate i_C to v.

5.50 Consider the series LRC circuit described by $L\, d^2i/dt^2 + R\, di/dt + i/C = dv/dt$. Find the characteristic equation and $i_N(t)$ when $L = 1$ H, $R = 120$ Ω, and $C = 500$ μF. Is the circuit stable?

5.51 Do Problem 5.50 with $L = 1$ H, $C = 2000$ μF, and $R = -60$ Ω (a negative resistance).

5.52 Consider the amplifier circuit in Example 5.8 (p. 220). Find the characteristic equation and the natural response of $v_{out}(t)$ when $L = 5$ mH, $R = 100$ Ω, $C = 2$ μF, and $\beta = 100$. Is the circuit stable?

5.53 A second-order circuit with a controlled source is described by $2\, d^2y/dt^2 + 6\, dy/dt - 20y = f(t)$. Find the characteristic equation and $y_N(t)$. Is the circuit stable?

5.54* Consider the series RC circuit described by $R\, di/dt + i/C = dv/dt$. Find the forced response i_F when $R = 40$ kΩ, $C = 100$ μF, and $v = 50t^2 + 200t$.

5.55 Consider the parallel CR circuit described by $C\, dv/dt + v/R = i$. Find the forced response v_F when $C = 10$ μF, $R = 2$ kΩ, and $i = 10^{-3}(1 - e^{-25t})$.

5.56 Consider the series LR circuit described by $L\, di/dt + Ri = v$. Find the forced response i_F when $L = 0.3$ H, $R = 5$ Ω, and $v = 10te^{-20t}$.

5.57 A series RLC circuit is described by $L\, d^2i/dt^2 + R\, di/dt + i/C = dv/dt$. Find the forced response i_F when $L = 0.2$ H, $R = 0$, $C = 10$ μF, and $v = 40 \sin 1000t$.

5.58 Do Problem 5.57 with $L = 2$ H, $R = 3$ Ω, $C = 50$ μF, and $v = 60 \cos 100t$.

5.59 A certain second-order circuit is described by $d^2y/dt^2 + 5\,dy/dt = x$. Find the forced response y_F when $x = 20 + 18e^{-2t}$. (Hint: First find the form of y_N.)

5.60 Do Problem 5.55 with $C = 20\ \mu\text{F}$.

5.61 Do Problem 5.56 with $L = 0.25$ H.

5.62* Given that $L\,di/dt + Ri = v$, find the complete response for $i(t)$ with $L = 0.3$ H and $R = 6\ \Omega$ when $v = 12t$ for $t \geq 0$ and $i(0^-) = 0.4$ A.

5.63 Given that $C\,dv/dt + v/R = i$, find the complete response for $v(t)$ with $C = 5\ \mu\text{F}$ and $R = 250\ \Omega$ when $i = 0.5e^{200t}$ for $t \geq 0$ and $v(0^-) = 60$ V.

5.64 Given that $R\,di/dt + i/C = dv/dt$, find the complete response $i(t)$ with $R = 400\ \Omega$ and $C = 100\ \mu\text{F}$ when $v = 200t + 40e^{25t}$ for $t \geq 0$ and $i(0^-) = 0$.

5.65 Find the complete response for $i(t)$ from Eq. (5.33a) with $R = 4\ \Omega$ and $C = 25\ \mu\text{F}$ when $v = 300 \sin 20t$ for $t \geq 0$ and $i(0^-) = 0.05$ A.

CHAPTER 6

AC Circuits

The abbreviation *ac* stands for *alternating current,* which could have any type of alteration. But the specific type of alternating waveform most often encountered in circuit analysis is the *sinusoid.* And one of the most important tasks of circuit analysis is finding the steady-state response forced by a sinusoidal excitation after the natural response has died away. Consequently, in agreement with common practice, we'll say that

> An *ac circuit* is a stable linear circuit operating in the steady state with sinusoidal excitation.

This chapter investigates the properties of such circuits.

AC circuits have long been the bread and butter of electrical engineering — in power distribution, lightning, consumer products, and industrial systems. Moreover, an understanding of ac circuit concepts is an essential prerequisite for myriad topics ranging from electronics and rotating machinery to automatic control, communications, and signal processing. Thus, our study of ac circuits has both immediate and subsequent applications.

Fundamentally, ac circuit analysis involves the sometimes daunting tasks of formulating differential circuit equations and then calculating particular solutions with sinusoidal excitations. However, a much more convenient technique was developed in the 1890s by Charles Proteus Steinmetz (1865–1923), a German-Austrian immigrant working at the General Electric Company. Steinmetz revolutionized electrical engineering with his startling discovery that the use of complex numbers reduces ac circuit problems to relatively simple algebraic manipulations!

Following Steinmetz's lead, we begin our investigation here with phasor notation, complex algebra, and impedance as the keys to efficient ac circuit analysis. These tools are then applied to calculate the steady-state response of circuits containing energy-storage elements as well as resistance. The combination of phasors and impedance has come to be known as the *phasor transform method*, and it allows us to treat ac circuits in a manner similar to resistive circuits.

Even so, the behavior of ac circuits with energy storage differs from that of resistive circuits in several respects. We'll explore some of those differences, including the important and interesting phenomenon known as resonance. The closing section discusses ac bridge circuits for the measurement of unknown capacitance and inductance.

Objectives

After studying this chapter and working the exercises, you should be able to do each of the following:

1. Represent a sinusoidal waveform by a phasor, and use phasor addition to sum sinusoids at the same frequency (Section 6.1).
2. Convert a complex number from polar to rectangular form, or vice versa, and carry out the operations of addition, subtraction, multiplication, and division with complex numbers (Section 6.1).
3. State Ohm's law for ac circuits, and find the impedance or admittance of a resistor, inductor, and capacitor (Section 6.2).
4. Define and calculate ac resistance, reactance, ac conductance, and susceptance (Section 6.2).

5. Draw the frequency-domain diagram of an ac ladder network, find the equivalent impedance or admittance, and evaluate any branch variable (Section 6.2).
6. Apply the phasor transform method to analyze a circuit in the ac steady state using proportionality. Thévenin's or Norton's theorem, or systematic node or mesh equations (Section 6.3).
7. Construct phasor diagrams for the voltages and currents in *RC, RL*, and *RLC* circuits (Section 6.4).
8. Evaluate and explain the significance of the resonant frequency and quality factor of a series or parallel *RLC* circuit (Section 6.4).
9. Use superposition to analyze a circuit with ac sources at different frequencies (Section 6.5).
10. Determine unknown element values given the balanced conditions of an impedance bridge (Section 6.6).†

6.1 PHASORS AND THE AC STEADY STATE

Suppose a stable linear circuit is driven by a sinusoidal source applied sufficiently long ago that the natural response has died away to zero. Then, as we saw in Section 5.3, the complete response equals the forced response and we say that the circuit has reached the **ac steady state**.

This section further examines ac steady-state conditions. As preparation, we'll first review the properties of sinusoidal waveforms — or **sinusoids** for short. Then we'll introduce two valuable tools: the phasor concept for representing sinusoids, and the algebra of complex numbers for manipulating phasors.

Sinusoids and Phasors

Figure 6.1 illustrates two familiar time functions: the cosine wave $X_m \cos \omega t$, and the sine wave $X_m \sin \omega t$. Both waveforms exhibit the same oscillatory behavior that continues indefinitely. The **amplitude** X_m equals the peak ver-

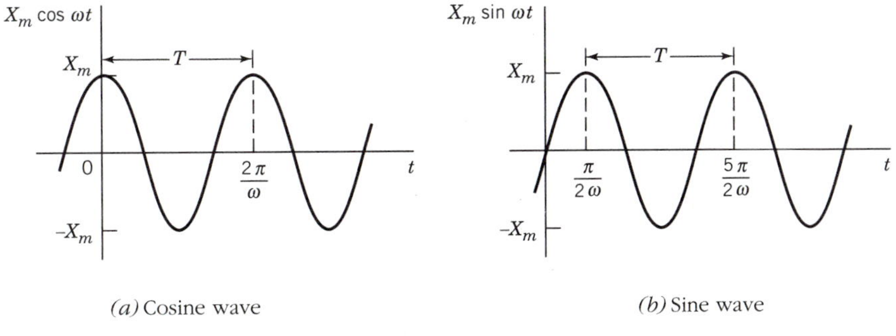

(a) Cosine wave (b) Sine wave

Figure 6.1

tical excursions, and the waveforms swing between $+X_m$ and $-X_m$. Hence, in the circuit context, the unit of X_m would be volts or amps. The **angular** or **radian frequency** ω (omega) equals the oscillation rate in **radians per second** (rad/s), and the waveforms repeat themselves when t increases by $2\pi/\omega$ so the sinusoidal argument ωt increases by 2π rad.

The rate of oscillation may also be specified two other ways. The time duration of one full cycle is the **period**, defined by

$$T \triangleq 2\pi/\omega \tag{6.1a}$$

The reciprocal of the period is the **cyclical frequency**, defined by

$$f \triangleq 1/T = \omega/2\pi \tag{6.1b}$$

This quantity corresponds to the number of oscillation cycles per second, and its unit has been named the **hertz** (Hz) to honor Heinrich Hertz (1857–1894) who first confirmed the existence of oscillatory electromagnetic waves. Suppose, for instance, that the frequency of an oscillation is $f = 60$ Hz. Then the period is $T = 1/f \approx 16.7$ ms and the radian frequency is $\omega = 2\pi f \approx 377$ rad/s. Although engineers most often speak of frequency in hertz, we use radian frequency to write sinusoidal arguments compactly as ωt rather than the equivalent expressions $2\pi ft$ or $2\pi t/T$.

The only difference between the two waveforms in Fig. 6.1 is the point where $t = 0$. To allow for *any* position of the time origin, we'll work with sinusoids having the general form

$$x(t) = X_m \cos(\omega t + \phi) \tag{6.2}$$

The additional term ϕ (phi) is called the **phase angle**, or just the **phase**, with respect to the reference cosine function. The phase should be expressed in radians when evaluating the instantaneous total angle $\omega t + \phi$, but ϕ by itself is usually measured in degrees.

Adding ϕ in the argument of a cosine function shifts the central positive peak from $t = 0$ to $t = t_0$, as shown in Fig. 6.2. If ϕ is positive, then t_0 will be negative because $\omega t_0 + \phi = 0$ at the peak. Thus, for any ϕ,

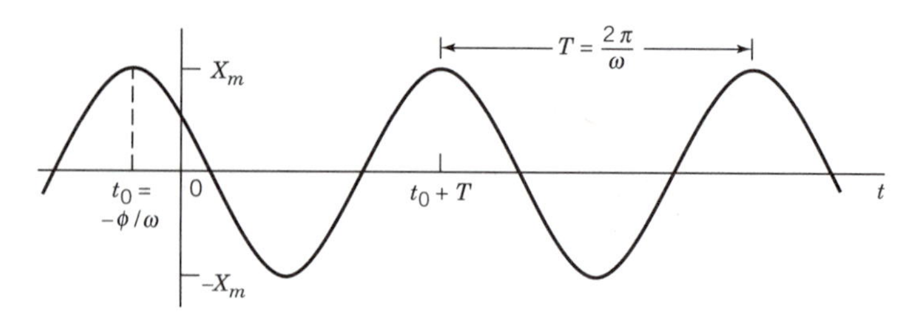

Figure 6.2 The sinusoidal wave $X_m \cos(\omega t + \phi)$.

$$t_0 = -\frac{\phi(\text{rad})}{\omega} = -\frac{\phi(\text{deg})}{360^\circ}T \tag{6.3}$$

where we have noted that 2π rad $= 360^\circ$ so $1/\omega = T/(2\pi \text{ rad}) = T/360^\circ$.

Don't be distressed by our choice of the *cosine* function in Eq. (6.2) to represent an arbitrary *sinusoid.* This is just one of our standard conventions for ac circuit analysis. To put a sine wave in cosine form, we invoke the trigonometric identity

$$\sin \alpha = \cos(\alpha - 90^\circ) \tag{6.4a}$$

Another standard convention is that we regard amplitude as being a *positive* quantity. When negative signs occur, they may be absorbed in the phase angle via

$$-\cos \alpha = \cos(\alpha \pm 180^\circ) \tag{6.4b}$$

To illustrate these points, the sinusoidal expression $-6 \sin(\omega t + 50^\circ)$ can be rewritten as

$$-6 \sin(\omega t + 50^\circ) = -6 \cos(\omega t + 50^\circ - 90^\circ) = 6 \cos(\omega t - 40^\circ \pm 180^\circ)$$

Thus, the amplitude is $X_m = 6$ and the phase can be taken to be either $\phi = -40^\circ + 180^\circ = 140^\circ$ or $\phi = -40^\circ - 180^\circ = -220^\circ$. In such cases, we usually take the value of ϕ that falls in the range $-180^\circ \leq \phi \leq 180^\circ$.

Our generic sinusoid $x(t) = X_m \cos \omega t + \phi)$ involves three parameters: amplitude X_m, frequency ω, and phase ϕ. These same three parameters also characterize the two-dimensional picture in Fig. 6.3*a*, called a **rotating phasor**. Here, a directed line of length X_m forms the instantaneous total angle $\omega t + \phi$ relative to the horizontal axis. Thus, as time goes by, the line rotates counterclockwise about the origin at the cyclical rate $f = \omega/2\pi$, and its horizontal projection traces out the values of $x(t) = X_m \cos(\omega t + \phi)$. We may therefore view any sinusoid as the *horizontal projection of a rotating phasor.*

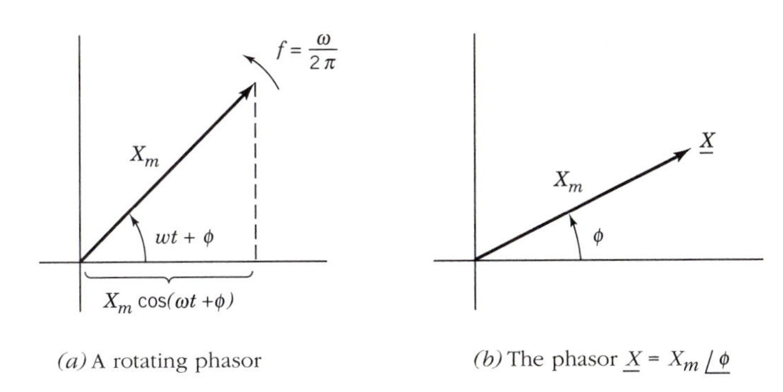

(a) A rotating phasor

(b) The phasor $\underline{X} = X_m \angle \phi$

Figure 6.3

A snapshot of the rotating phasor taken at $t = 0$ yields the *nonrotating* **phasor** in Fig. 6.3*b*. This phasor has just two parameters, length and angle, and we'll represent it by the notation

$$\underline{X} = X_m \,\underline{/\phi} \tag{6.5}$$

The symbol $\underline{X}$ denotes the phasor associated with the sinusoidal waveform $x(t) = X_m \cos(\omega t + \phi)$, and the symbols $X_m \,\underline{/\phi}$ mean that $\underline{X}$ has length X_m at angle ϕ measured counterclockwise from the horizontal axis. (Many texts indicate phasors by boldface letters such as **X**, but the underlined letter $\underline{X}$ is easier to reproduce in handwriting.)

Phasor notation does not include information about frequency, which was eliminated by taking $t = 0$ in the rotating-phasor picture. However, if you know the phasor associated with some voltage or current at a stated frequency, then you can immediately write down the corresponding sinusoidal function.

The advantage of phasor notation for the study of ac circuits stems from the following property:

> When a linear circuit has a sinusoidal excitation at some frequency ω, the forced response of any branch variable will be another sinusoid in the general form $y(t) = Y_m \cos(\omega t + \phi_y)$.

Proof of this property will be given later. Meanwhile, we observe that the response oscillates at exactly the same frequency as the excitation but with different amplitude Y_m and phase ϕ_y. Hence, ac circuit analysis involves *two unknowns* for each branch variable of interest. Representing $y(t)$ by the non-rotating phasor $\underline{Y} = Y_m \,\underline{/\phi_y}$ emphasizes that Y_m and ϕ_y are the unknowns to be found, whereas the frequency ω is not an unknown.

One final point deserves mention. Although a phasor looks like a two-dimensional *vector*, we avoid the latter terminology because vectors usually refer to spatial orientation of forces or movement. Furthermore, we'll soon consider phasor operations that would be undefined for vectors.

Example 6.1 *From Phasor to Sinusoid*

Suppose that analysis of an ac circuit with frequency $f = 200$ Hz yields the phasor current $\underline{I} = 7\ mA \,\underline{/-45°}$. You now want to write the corresponding sinusoidal expression for $i(t)$.

The radian frequency of $i(t)$ is $\omega = 2\pi f = 400\pi$ rad/s, and the phasor has amplitude $I_m = 7$ mA and angle $\phi_i = -45°$. Hence,

$$i(t) = I_m \cos(\omega t + \phi_i) = 7 \cos(400\pi t - 45°) \text{ mA}$$

Exercise 6.1

Write the phasors associated with the following sinusoids:
(a) $v(t) = 100 \sin(\omega t - 60°)$ V;
(b) $i(t) = -0.2 \cos(\omega t + 80°)$ A.

Exercise 6.2

Calculate T and t_0 and sketch the sinusoids represented by:
(a) $\underline{I} = 30$ mA $\angle 90°$ with $f = 250$ kHz;
(b) $\underline{V} = 1.5$ kV $\angle -144°$ with $\omega = 100\pi$ rad/s.

Complex Numbers

Steinmetz's method for ac circuit analysis involves manipulating phasors, and we'll later encounter other two-dimensional quantities that also play important roles in electrical engineering. Such quantities may be viewed either as directed lines or as complex numbers. Both viewpoints will be combined here by treating complex numbers as points in a complex plane.

The **complex plane** laid out in Fig. 6.4*a* is a two-dimensional space wherein any point A may be specified using two coordinates. By convention, the horizontal axis of the complex plane is called the **real axis** (abbreviated Re) while the vertical axis is called the **imaginary axis** (abbreviated Im). Correspondingly, the horizontal coordinate of any point is known as the **real part** of A, and the vertical coordinate is known as the **imaginary part** of A. These two **rectangular coordinates** are denoted by

$$\text{Re}[A] = a_r \qquad \text{Im}[A] = a_i$$

But there's nothing "imaginary" about Im[A], nor is there anything particularly "real" about Re[A]. The terms "real" and "imaginary" merely label the rectangular coordinates of the point A.

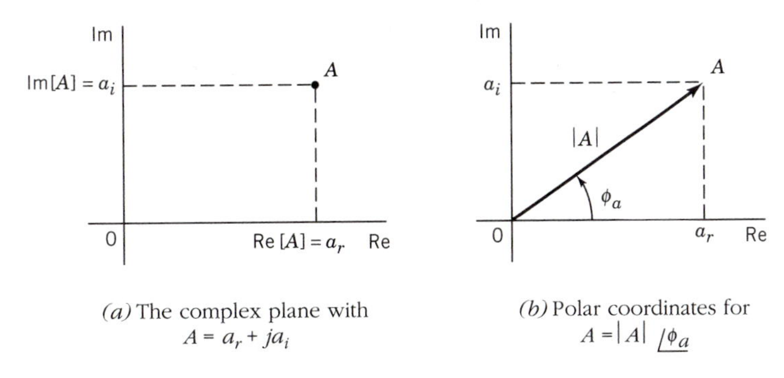

(a) The complex plane with $A = a_r + ja_i$

(b) Polar coordinates for $A = |A| \angle \phi_a$

Figure 6.4

In the same vein, we identify vertical coordinates by introducing the **imaginary unit**

$$j \triangleq \sqrt{-1} \tag{6.6}$$

(We use j here instead of the mathematician's i to avoid confusion with our general symbol for current.) The imaginary unit allows us to express any point A as a **complex number** in the form

$$A = \text{Re}[A] + j\text{Im}[A] = a_r + ja_i \tag{6.7}$$

This expression actually stands for *two* separate relations, namely, $\text{Re}[A] = a_r$ and $\text{Im}[A] = a_i$. In general, any complex equation requires equality of the real and imaginary parts.

Either part of a complex number may be positive, negative, or zero. If one part is zero, then we often omit it entirely, as illustrated by $0 + j8 = j8$ or $-3 + j0 = -3$. If the imaginary part is negative, then we usually put the minus sign before j, as illustrated by $2 + j(-5) = 2 - j5$. Although both parts are always *real* numbers, the quantity A in Eq. (6.7) is *complex* because its imaginary part has been multiplied by j. Thus, the term "complex" does not necessarily imply something "complicated." Rather, it only means that the quantity in question has two distinct parts.

Any point in the complex plane may also be specified by the **polar coordinates** shown in Fig. 6.4*b*, where we have added a directed line from the origin to the point A. The length of the line is called the **magnitude** of A, symbolized by $|A|$. The angle of the line measured counterclockwise relative to the positive real axis is called the **angle** of A, symbolized by $\angle A = \phi_a$. Then, borrowing from phasor notation, we can emphasize the polar interpretation of a complex number by writing

$$A = |A| \underline{/\phi_a}$$

We don't underline A here because symbols such as $\underline{X}$ will be reserved exclusively for phasors, which are always associated with sinusoidal time functions. However, we can now say that *a phasor is a complex number*, and it may be expressed in either polar or rectangular notation.

To convert any complex number from polar to rectangular form, we observe in Fig. 6.4*b* that

$$a_r = |A| \cos \phi_a \qquad a_i = |A| \sin \phi_a \tag{6.8}$$

These relations also hold when ϕ_a is negative, by virtue of the general properties that $\cos(-\alpha) = \cos \alpha$ and $\sin(-\alpha) = -\sin \alpha$. For conversion from rectangular to polar form, we see that

$$|A| = \sqrt{a_r^2 + a_i^2} \qquad \phi_a = \tan^{-1}(a_i/a_r) \tag{6.9}$$

The formula for ϕ_a in Eq. (6.9) means the angle formed by the rectangular coordinates, so it must be used with care when a_r is negative. An alternative version for that case is

$$\phi_a = \pm 180° - \tan^{-1}[a_i/(-a_r)] \qquad a_r < 0 \tag{6.10}$$

Most scientific calculators have built-in routines for polar-to-rectangular conversion (P ⟶ R) and for rectangular-to-polar conversion (R ⟶ P), including the case of $a_r < 0$. Nonetheless, as a guard against keystroke errors, you should always estimate your expected results by making a quick sketch of the quantity being converted.

Now we have some of the tools needed to carry out various operations on complex numbers or phasors. **Addition** and **subtraction** must be performed in rectangular form, because you add or subtract the respective horizontal and vertical components. Specifically, if $A = a_r + ja_i$ and $B = b_r + j\,b_i$, then

$$A \pm B = (a_r + ja_i) \pm (b_r + jb_i) = (a_r \pm b_r) + j(a_i \pm b_i)$$

so

$$\text{Re}[A \pm B] = \text{Re}[A] \pm \text{Re}[B] \qquad \text{Im}[A \pm B] = \text{Im}[A] \pm \text{Im}[B] \tag{6.11}$$

Figure 6.5 illustrates the geometric picture of $A + B$ using the familiar parallelogram construction for vector addition in two dimensions. This construction also works for subtraction by writing $A - B = A + (-B)$.

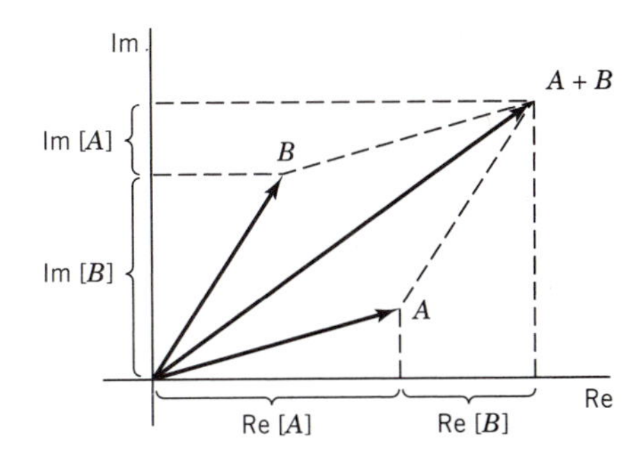

Figure 6.5 Parallelogram construction for $A + B$.

Multiplication can be carried out in rectangular form by noting from Eq. (6.6) that $j^2 = -1$. Thus,

$$\begin{aligned} AB &= (a_r + ja_i)(b_r + jb_i) \\ &= a_r b_r + ja_r b_i + ja_i b_r + j^2 a_i b_i \\ &= (a_r b_r - a_i b_i) + j(a_r b_i + a_i b_r) \end{aligned}$$

so

$$\text{Re}[AB] = a_r b_r - a_i b_i \qquad \text{Im}[AB] = a_r b_i + a_i b_r \tag{6.12}$$

As a special case, let $A = k$ with k being any *real* quantity. We then have

$$\text{Re}[kB] = k\,\text{Re}[B] \qquad \text{Im}[kB] = k\,\text{Im}[B]$$

which follows by setting $a_r = k$ and $a_i = 0$ in Eq. (6.12).

Before tackling complex division, we introduce the **conjugate** operation defined as changing the sign of the imaginary part of a complex number. In particular, if

$$A = a_r + ja_i = |A| \angle\phi_a$$

then its **complex conjugate** is

$$A^* \triangleq a_r - ja_i = |A| \angle -\phi_a \qquad (6.13)$$

Figure 6.6 depicts the relationship between A and A* along with the sum

$$A + A^* = 2\,\text{Re}[A] = 2a_r \qquad (6.14a)$$

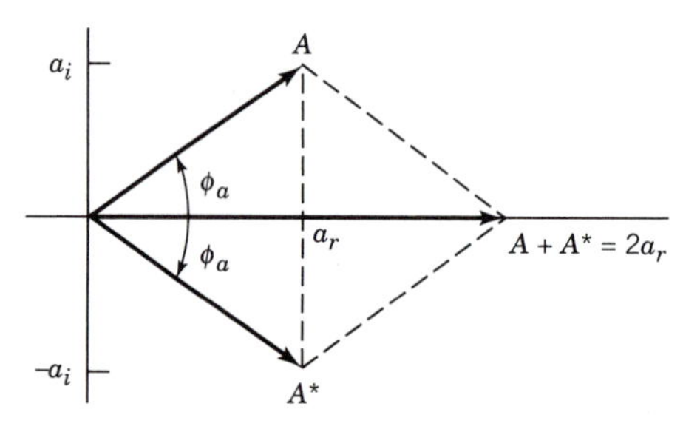

Figure 6.6 The number A and its complex conjugate A^*.

More importantly for our current purposes, multiplying A by A^* yields

$$AA^* = a_r^2 + a_i^2 = |A|^2 \qquad (6.14b)$$

which is always a positive real quantity.

Division of complex numbers in rectangular form is less direct than multiplication because we must eliminate the complex denominator. For that purpose, we take advantage of Eq. (6.14b) and multiply both numerator and denominator by the complex conjugate of the denominator. Thus,

$$\frac{B}{A} = \frac{BA^*}{AA^*} = \frac{(b_r + jb_i)(a_r - ja_i)}{|A|^2} = \frac{b_r a_r + b_i a_i}{a_r^2 + a_i^2} + j\frac{b_r a_i - b_i a_r}{a_r^2 + a_i^2} \qquad (6.15)$$

This process is known as **rationalization**, and it yields the real and imaginary parts of B/A.

Some calculators have the capability of adding, subtracting, multiplying, or dividing complex numbers expressed in any form, and the result may be obtained in either polar or rectangular form. If you have such a calculator, then you can save considerable time in numerical problems by learning how to use those features. However, you must still know the basic methods to do symbolic manipulations.

Example 6.2 *Calculations with Complex Numbers*

Suppose you want the polar form of the complex number

$$D = A + B/C$$

given that

$$A = 8 + j3 \qquad B = 100\,\underline{/90^\circ} \qquad C = 3 - j4$$

For problems with multiple operations like this one, you should carefully plan your attack before proceeding with the calculations.

In the case at hand, you need B/C in rectangular form to add to A. Since the rectangular form of C is given, you first convert B to get

$$B = 100 \cos 90^\circ + j100 \sin 90^\circ = 0 + j100 = j100$$

Division by rationalization then yields

$$\frac{B}{C} = \frac{j100}{3 - j4} = \frac{j100(3 + j4)}{(3 - j4)(3 + j4)} = \frac{j300 + j^2 400}{3^2 + 4^2} = -16 + j12$$

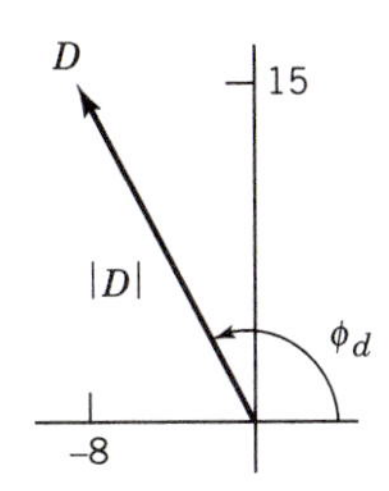

Figure 6.7

Now you can do the addition

$$D = A + B/C = (8 + j3) + (-16 + j12) = -8 + j15$$

which is sketched in the complex plane in Fig. 6.7.

You see from the sketch that $|D|$ must fall in the range $15 < |D| < (8 + 15)$, while ϕ_d must fall in the range $90^\circ < \phi_d < 135^\circ$. Performing the rectangular to polar conversion, you obtain

$$|D| = \sqrt{(-8)^2 + 15^2} = 17 \qquad \phi_d = \pm 180^\circ - \tan^{-1}(15/8) = 118.1^\circ$$

so your final answer is $D = 17\,\underline{/118.1^\circ}$.

Exercise 6.3

Given $A = -4 + j3$, find $A - A^*$, jA, and $A + 25/A$ in rectangular form and convert to polar form.

Euler's Formula

The mathematical link between the rectangular and polar form of complex numbers comes from a famous relationship by Euler (pronounced Oiler). **Euler's formula** states that, for any angle α,

$$e^{\pm j\alpha} = \cos\alpha \pm j\sin\alpha \tag{6.16}$$

Thus, the quantity $e^{j\alpha}$ is a complex number with $\mathrm{Re}[e^{j\alpha}] = \cos\alpha$ and $\mathrm{Im}[e^{j\alpha}] = \sin\alpha$. The corresponding polar coordinates are readily found to be

$$|e^{j\alpha}| = 1 \qquad \angle e^{j\alpha} = \alpha$$

Accordingly, we can write

$$e^{j\alpha} = 1\,\underline{/\alpha}$$

This expression brings out the fact that $e^{j\alpha}$ is a unit-length line at angle α, as diagrammed in Fig. 6.8. Similarly, $e^{-j\alpha} = 1\,\underline{/-\alpha}$.

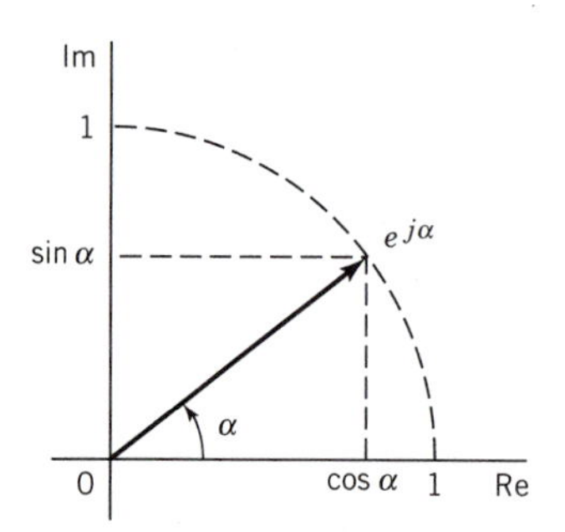

Figure 6.8 Euler's formula in the complex plane.

Drawing upon the relation $1\,\underline{/\alpha} = e^{j\alpha}$, an arbitrary complex number having magnitude $|A|$ and angle ϕ_a can be put into the **exponential form**

$$A = |A|e^{j\phi_a} \tag{6.17}$$

Equation (6.17) provides the mathematical meaning behind our polar notation $A = |A|\,\underline{/\phi_a}$. Likewise, the exponential version of the phasor notation $\underline{X} = X_m\,\underline{/\phi}$ is $\underline{X} = X_m e^{j\phi}$.

Multiplication and division of complex numbers are easily done using the exponential form, recalling that $e^a e^b = e^{(a+b)}$ while $e^a/e^b = e^{(a-b)}$. Thus,

$$AB = |A|e^{j\phi_a}|B|e^{j\phi_b} = |A||B|e^{j(\phi_a+\phi_b)} = |A||B|\,\underline{/\phi_a + \phi_b} \tag{6.18}$$

and

$$\frac{A}{B} = \frac{|A|e^{j\phi_a}}{|B|e^{j\phi_b}} = \frac{|A|}{|B|}e^{j(\phi_a - \phi_b)} = \frac{|A|}{|B|}\,\underline{/\phi_a - \phi_b} \tag{6.19}$$

Hereafter, you can skip over the intermediate steps by noting that you multiply or divide the magnitudes but you add or subtract the angles.

With the help of Eqs. (6.18) and (6.19), we next develop geometric interpretations of j and explain why the vertical component of a complex number is called the *imaginary* part. First observe that j is a complex number with zero real part and unit imaginary part, so

$$j = 0 + j1 = 1\,\underline{/90^\circ}$$

Multiplying A by j then yields

$$jA = 1\,\underline{/90^\circ} \times |A\,\underline{/\phi_a} = |A|\,\underline{/\phi_a + 90^\circ}$$

which is equivalent to a 90° counterclockwise rotation. Taking $A = j$ confirms that

$$j^2 = j \times j = 1\,\underline{/180^\circ} = -1$$

Consistency therefore requires the "imaginary" property $j = \sqrt{-1}$. A related property is the reciprocal

$$1/j = 1\,\underline{/-90^\circ} = -j$$

which agrees with Eq. (6.19) when $A = 1$ and $B = j$. Figure 6.9 summarizes these characteristics of j.

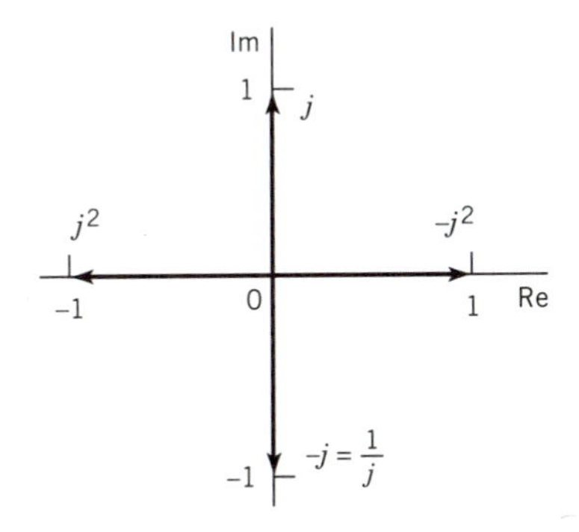

Figure 6.9

Finally, we use the property $-1 = j^2$ to verify Euler's formula by letting

$$E = \cos\alpha + j\sin\alpha$$

Differentiating E with respect to α yields

$$dE/d\alpha = -\sin\alpha + j\cos\alpha = j(j\sin\alpha + \cos\alpha) = jE$$

But differentiating $e^{j\alpha}$ yields

$$d(e^{j\alpha})/d\alpha = j(e^{j\alpha})$$

Hence, by comparison, $e^{j\alpha} = E = \cos\alpha + j\sin\alpha$. Replacing α with $-\alpha$ then shows that $e^{-j\alpha} = \cos\alpha + j\sin(-\alpha) = \cos\alpha - j\sin\alpha$.

Exercise 6.4

(a) Given $A = 3 + j4$, find A^2 and $1/A$ in polar form.
(b) Using the polar form, show that $A^*/B^* = (A/B)^*$.

AC Steady-State Conditions

Having introduced phasors and developed complex algebra for phasor calculations, we're almost ready to derive some important conclusions about circuits in the ac steady state. What we're missing is the mathematical relationship between phasors and waveforms for sinusoids and sums of sinusoids.

Again consider the phasor $\underline{X} = X_m \underline{/\phi} = X_m e^{j\phi}$ associated with our generic sinusoid $x(t) = X_m \cos(\omega t + \phi)$. Multiplying $\underline{X}$ by $e^{j\omega t}$ gives

$$\underline{X}e^{j\omega t} = X_m e^{j\phi} e^{j\omega t} = X_m e^{j(\omega t + \phi)}$$

The complex quantity $X_m e^{j(\omega t+\phi)}$ has magnitude X_m and instantaneous angle $\omega t + \phi$, so it represents a *rotating phasor* just like the one back in Fig. 6.3*a*. Furthermore, from Euler's theorem.

$$\mathrm{Re}[\underline{X}e^{j\omega t}] = X_m \cos(\omega t + \phi) \tag{6.20}$$

where

$$X_m = |\underline{X}| \qquad \phi = \angle\underline{X}$$

Equation (6.20) provides an explicit formula for obtaining a sinusoid from its associated phasor.

Now suppose a waveform consists of two sinusoids at the *same frequency*, say $X_1 \cos(\omega t + \phi_1) + X_2 \cos(\omega t + \phi_2)$. We apply Eq. (6.20) to this case by introducing the two associated phasors

$$\underline{X}_1 = X_1 \underline{/\phi_1} \qquad \underline{X}_2 = X_2 \underline{/\phi_2}$$

so that

$$X_1 \cos(\omega t + \phi_1) + X_2 \cos(\omega t + \phi_2) = \mathrm{Re}[\underline{X}_1 e^{j\omega t}] + \mathrm{Re}[\underline{X}_2 e^{j\omega t}]$$

But

$$\mathrm{Re}[\underline{X}_1 e^{j\omega t}] + \mathrm{Re}[\underline{X}_2 e^{j\omega t}] = \mathrm{Re}[\underline{X}_1 e^{j\omega t} + \underline{X}_2 e^{j\omega t}] = \mathrm{Re}[(\underline{X}_1 + \underline{X}_2) e^{j\omega t}]$$

Thus,

$$X_1 \cos(\omega t + \phi_1) + X_2 \cos(\omega t + \phi_2) = X_m \cos(\omega t + \phi) \qquad (6.21a)$$

with

$$X_m = |\underline{X}_1 + \underline{X}_2| \qquad \phi = \angle(\underline{X}_1 + \underline{X}_2) \qquad (6.21b)$$

Equation (6.21a) tells us that a sum of sinusoids having the same frequency equals another sinusoid at that frequency. Equation (6.21b) tells us that the amplitude and phase of the sum can be found by *adding the phasors* associated with the individual sinusoids. And since phasors are two-dimensional quantities, the sum $\underline{X}_1 + \underline{X}_2$ is calculated just like the sum of two complex numbers. Identical results hold for a sum of three or more sinusoids with the same frequency.

Getting back to ac circuit analysis, recall our investigation of natural and forced response in Section 5.3. There we found that the natural response of a stable circuit dies away with time, so the steady-state response equals the forced response. We also found that if the forcing function is

$$f(t) = A_1 \cos \omega t + A_2 \sin \omega t$$

then the forced response of any branch variable has the form

$$y(t) = B_1 \cos \omega t + B_2 \sin \omega t$$

Since $\sin \omega t = \cos(\omega t - 90°)$, we now see from Eq. (6.21a) that both of these expressions represent single sinusoids. We thereby conclude that

> The steady-state response of any branch variable in a stable circuit with a sinusoidal excitation will be another sinusoid at the same frequency as the source.

The response may differ from the excitation only in amplitude and phase.

Another important conclusion for ac circuit analysis relates to Kirchhoff's laws. Since all steady-state voltages and currents are sinusoids at the same frequency, and since the amplitude and phase of a sum of sinusoids can be found by adding the associated phasors, it follows that

> Kirchhoff's laws hold in phasor form, with each sinusoid replaced by its associated phasor.

Phasor notation thereby allows us to concentrate on the unknown quantities — amplitude and phase — without having to write down the full-blown sinusoidal expressions for each step.

To summarize, for any stable linear circuit operating in the ac steady state, we have shown that:

- All voltages and currents are sinusoids at the source frequency.
- Phasor notation provides a convenient way of representing the sinusoidal amplitudes and phase angles.
- Calculations involving Kirchhoff's laws can be performed using phasors in place of sinusoids.

The following examples further illustrate the value of phasors for ac steady-state circuit analysis.

Example 6.3 *Parallel Network with an AC Voltage Source*

Figure 6.10*a* depicts a parallel RC network driven by the ac source voltage

$$v(t) = 30 \cos (4000t + 20°) \text{ V}$$

The source was applied some time ago, and we want to find the resulting steady-state input current $i(t) = i_R(t) + i_C(t)$. Since all the steady-state currents are sinusoids at frequency $\omega = 4000$, we'll determine $i(t)$ from the phasor sum $\underline{I} = \underline{I}_R + \underline{I}_L$.

With $R = 5\ \Omega$, the current through the resistor is easily found via Ohm's law to be

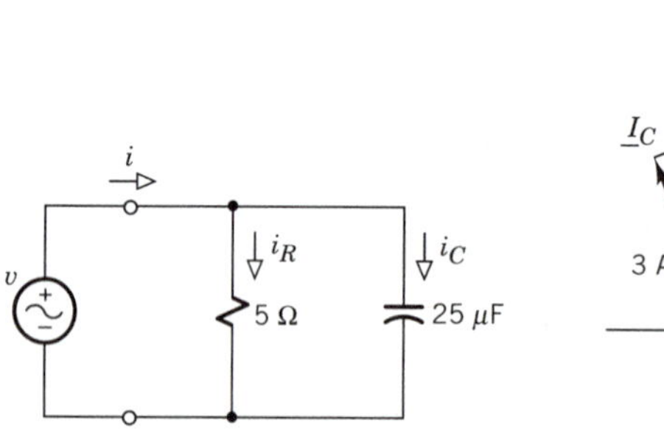

(a) RC circuit in the ac steady state

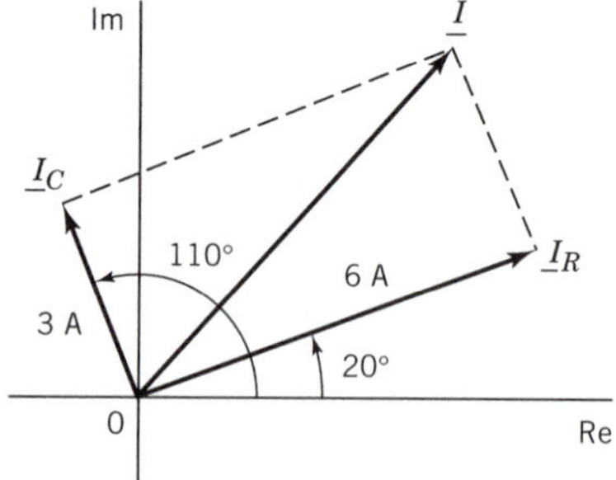

(b) Phasor diagram for the currents

Figure 6.10

$$i_R(t) = v(t)/R = 6 \cos(4000t + 20°) \text{ A}$$

so its associated phasor is

$$\underline{I}_R = 6 \text{ A } \underline{/20°} = 5.64 + j2.05$$

For the capacitor's current, we insert $C = 25\ \mu\text{F} = 25 \times 10^{-6}$ F and differentiate $v(t)$ to get

$$\begin{aligned} i_C(t) &= C\,dv(t)/dt = 25 \times 10^{-6}[-4000 \times 30 \sin(4000t + 20°)] \\ &= -3 \sin(4000t + 20°) = -3 \cos(4000t + 20° - 90°) \\ &= 3 \cos(4000t - 70° \pm 180°) \text{ V} \end{aligned}$$

Thus, taking the phase angle to be $-70° + 180° = 110°$.

$$\underline{I}_C = 3 \text{ A } \underline{/110°} = -1.03 + j2.82$$

These calculations confirm that both $i_R(t)$ and $i_C(t)$ are sinusoids with $\omega = 4000$ and may therefore be represented by the associated phasors.

Now we obtain $\underline{I}$ from the phasor sum diagrammed in Fig. 6.10*b*. Adding the rectangular components and converting to polar form yields

$$\underline{I} = \underline{I}_R + \underline{I}_C = 4.61 + j4.87 = 6.71 \text{ A } \underline{/46.6°}$$

Hence, $I_m = |\underline{I}| = 6.71$ A, $\phi_i = \angle\underline{I} = 46.6°$, and

$$i(t) = I_m \cos(\omega t + \phi_i) = 6.71 \cos(4000t + 46.6°) \text{ A}$$

which is our final result.

Example 6.4 *Parallel Network with an AC Current Source*

The previous example was straightforward because we knew the voltage across each element. A more challenging problem is posed in Fig. 6.11, where the same *RC* network operates in the ac steady state with input current

$$i(t) = 3 \cos 4000t \text{ A}$$

Now we seek the resulting voltage $v(t)$ at the source terminals.

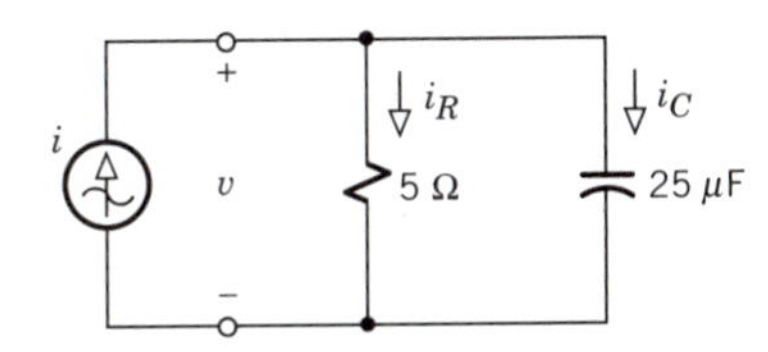

Figure 6.11 Parallel network with an AC current source.

The brute-force approach to this problem would require finding the particular solution of the differential equation

$$i_R(t) + i_C(t) = \frac{1}{R}v(t) + C\frac{dv(t)}{dt} = i(t)$$

But we'll take advantage of phasor concepts by writing

$$i(t) = 3\cos(4000t + 0°) = \text{Re}[\underline{I}e^{j4000t}] \qquad \underline{I} = 3\text{ A }\angle 0°$$

We also know that the steady-state voltage must be another sinusoid having the form

$$v(t) = V_m\cos(4000t + \phi_v) = \text{Re}[\underline{V}e^{j4000t}] \qquad \underline{V} = V_m \angle \phi_v$$

To evaluate V_m and ϕ_v, we'll solve $\underline{I}_R + \underline{I}_C = \underline{I}$ with $\underline{I}_R$ and $\underline{I}_C$ expressed in terms of $\underline{V}$. The KCL phasor equation thereby takes the place of the differential equation.

The current phasors $\underline{I}_R$ and $\underline{I}_C$ are found the same way as in Example 6.3, except that V_m and ϕ_v are unknowns. Thus,

$$i_R(t) = v(t)/R = 0.2V_m\cos(4000t + \phi_v)$$

and

$$\underline{I}_R = 0.2V_m \angle \phi_v = 0.2\underline{V}$$

We then differentiate the expression for $v(t)$ to get

$$\begin{aligned} i_C(t) &= C\,dv/dt = -0.1V_m\sin(4000t + \phi_v) \\ &= 0.1V_m\cos(4000t + \phi_v + 90°) \end{aligned}$$

so

$$\underline{I}_C = 0.1V_m \angle \phi_v + 90° = 0.1 \angle 90° \times V_m \angle \phi_v = j0.1\underline{V}$$

where we used the fact that $j = 1 \angle 90°$.

After substituting for $\underline{I}_R$ and $\underline{I}_C$, the phasor equation becomes

$$\underline{I}_R + \underline{I}_C = 0.2\underline{V} + j0.1\underline{V} = (0.2 + j0.1)\underline{V} = \underline{I}$$

from which

$$\underline{V} = \frac{\underline{I}}{0.2 + j0.1} = \frac{3 \angle 0°}{0.224 \angle 26.6°} = 13.4 \angle -26.6°$$

Hence, $V_m = |\underline{V}| = 13.4$ V, $\phi_v = \angle\underline{V} = -26.6°$, and

$$v(t) = 13.4 \cos(4000t - 26.6°) \text{ V}$$

Exercise 6.5

Find $\underline{I}_R$ and $\underline{I}_C$ when the circuit in Fig. 6.10*a* has $v(t) = 60 \sin 8000t$ A. Then evaluate $\underline{I} = \underline{I}_R + \underline{I}_C$ to find $i(t)$.

6.2 IMPEDANCE AND ADMITTANCE

Besides phasor notation and complex algebra, we need one more part of Steinmetz's method for ac circuit analysis, namely, the concept of *impedance* for elements and networks. This section introduces impedance and its reciprocal, admittance.

Starting with individual elements, we'll obtain simple equations for the voltage and current phasors in terms of impedance. Then we'll expand our scope to the equivalent impedance of networks. We'll show that the joint use of phasors and impedance allows us to handle networks with energy-storage elements in a fashion similar to our previous treatment of resistive networks.

Impedance and Admittance of Elements

Under ac steady-state conditions, both the voltage across a circuit element and the current through that element are sinusoids at the same frequency. Accordingly, we'll write these sinusoids in general as

$$\begin{aligned} v(t) &= V_m \cos(\omega t + \phi_v) = \text{Re}[\underline{V}e^{j\omega t}] \\ i(t) &= I_m \cos(\omega t + \phi_i) = \text{Re}[\underline{I}e^{j\omega t}] \end{aligned} \qquad (22a)$$

where the associated phasors are

$$\underline{V} = V_m \underline{/\phi_v} \qquad \underline{I} = I_m \underline{/\phi_i} \qquad (22b)$$

Our objective here is to find the relationships between $\underline{V}$ and $\underline{I}$ for resistors, inductors, and capacitors.

Resistors: A resistor is described by $v = Ri$, suggesting that $\underline{V} = R\underline{I}$. For confirmation, we substitute $v(t) = \text{Re}[\underline{V}e^{j\omega t}]$ and $i(t) = \text{Re}[\underline{I}e^{j\omega t}]$ into Ohm's law to get

$$\text{Re}[\underline{V}e^{j\omega t}] = R \times \text{Re}[\underline{I}e^{j\omega t}]$$

We then draw upon the general property that $k\,\text{Re}[A] = \text{Re}[kA]$ when k is a real quantity, so

$$\text{Re}[\underline{V}e^{j\omega t}] = \text{Re}[R\underline{I}e^{j\omega t}]$$

Since this relation must hold for any value of t, the phasor equation for a resistor is just

$$\underline{V} = R\underline{I} \tag{23}$$

as expected.

Like all phasor relationships, Eq. (23) actually stands for two separate equations relating the amplitudes and phase angles. Specifically, putting $\underline{V}$ and $\underline{I}$ in polar form yields

$$V_m \,\underline{/\phi_v} = RI_m \,\underline{/\phi_i}$$

and we clearly see that

$$V_m = RI_m \qquad \phi_v = \phi_i$$

Thus, the amplitudes differ by the factor R, but the phase angles are the same. Figure 6.12*a* shows the phasor diagram and the corresponding waveforms, the latter plotted versus ωt to bring out the phase relationship. The phasors $\underline{V}$ and $\underline{I}$ are **colinear** or **In phase** with each other — meaning that the waveforms $v(t)$ and $i(t)$ differ only in amplitude.

Inductors: We derive the phasor equation for an inductor from the element equation $v = L\,di/dt$. With i expressed in terms of $\underline{I}$, the time derivative becomes

$$\frac{di}{dt} = \frac{d}{dt}\text{Re}[\underline{I}e^{j\omega t}] = \text{Re}\left[\underline{I}\frac{de^{j\omega t}}{dt}\right] = \text{Re}[\underline{I}j\omega e^{j\omega t}]$$

Thus,

$$\text{Re}[\underline{V}e^{j\omega t}] = L\,\text{Re}[\underline{I}j\omega e^{j\omega t}] = \text{Re}[j\omega L\underline{I}e^{j\omega t}]$$

from which

$$\underline{V} = j\omega L\underline{I} \tag{6.24}$$

Writing $j\omega L\underline{I} = \omega L \,\underline{/90^\circ} \times I_m \,\underline{/\phi_i} = \omega LI_m \,\underline{/\phi_i + 90^\circ}$ then yields the amplitude and phase equations

$$V_m = \omega LI_m \qquad \phi_v = \phi_i + 90^\circ$$

The phasor diagram in Fig. 6.12*b* shows that $\underline{I}$ **lags** $\underline{V}$ by 90°. Correspondingly, each peak of the waveform $i(t)$ occurs *after* a peak of $v(t)$.

Capacitors: The phasor equation for a capacitor is the dual of the inductor relationship in Eq. (6.24). We thus have $\underline{I} = j\omega C\underline{V}$, or

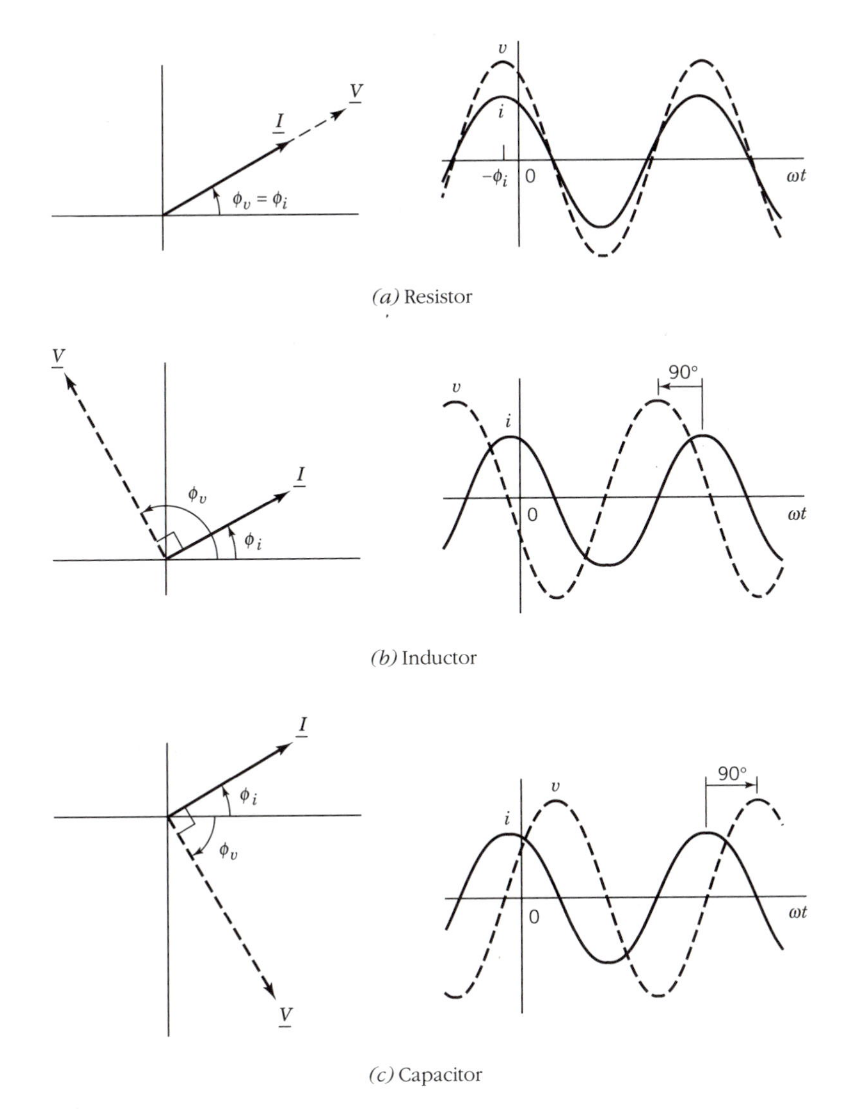

Figure 6.12 Phasor diagram and waveforms for elements.

$$\underline{V} = \frac{1}{j\omega C}\underline{I} = -\frac{j}{\omega C}\underline{I} \tag{6.25}$$

It then follows that

$$V_m = \frac{I_m}{\omega C} \qquad \phi_v = \phi_i - 90°$$

The phasor diagram in Fig. 6.12*c* shows that $\underline{I}$ **leads** $\underline{V}$ by 90°, so each peak of $i(t)$ occurs before a peak of $v(t)$.

Looking back at the phasor relations in Eqs. (6.23)–(6.25), we see that each one falls under the general form

$$\underline{V} = Z\underline{I} \tag{6.26}$$

The quantity Z is called the **impedance**. Impedance is measured in *ohms* (Ω) because $Z = \underline{V}/\underline{I}$ and the ratio $\underline{V}/\underline{I}$ carries the units of volts per amp. Since impedance has the same unit as resistance, and since Eq. (6.26) has the same form as $v = Ri$, the phasor equation $\underline{V} = Z\underline{I}$ may be called **Ohm's law for ac circuits**. The name "impedance" conveys the notion that Z "impedes" phasor current $\underline{I}$, just as resistance R "resists" instantaneous current i. Keep in mind, however, that impedance is defined only in conjunction with phasors.

The impedance of a specific element will be identified by attaching the subscript R, L, or C. From Eqs. (6.23)–(6.25) we find that

$$Z_R = R \tag{6.27a}$$

$$Z_L = j\omega L = \omega L \underline{/90^\circ} \tag{6.27b}$$

$$Z_C = 1/j\omega C = -j/\omega C = 1/\omega C \underline{/-90^\circ} \tag{6.27c}$$

Thus, the impedance of a resistor just equals R, whereas the impedance of an energy-storage element involves the frequency ω as well as L or C. The angles of Z_L and Z_C embody the phase differences previously seen in Figs. 6.12*b* and 6.12*c*.

Figure 6.13 provides a pictorial summary of our results. The upper part of the figure consists of **time-domain diagrams** showing each element with its general v–i or i–v relationship. The lower part of the figure assumes ac steady-state conditions and shows the phasor relationships with the element impedances. Such pictures are called **frequency-domain diagrams** because Z_L and Z_C depend upon ω. Note carefully that the frequency-domain diagrams

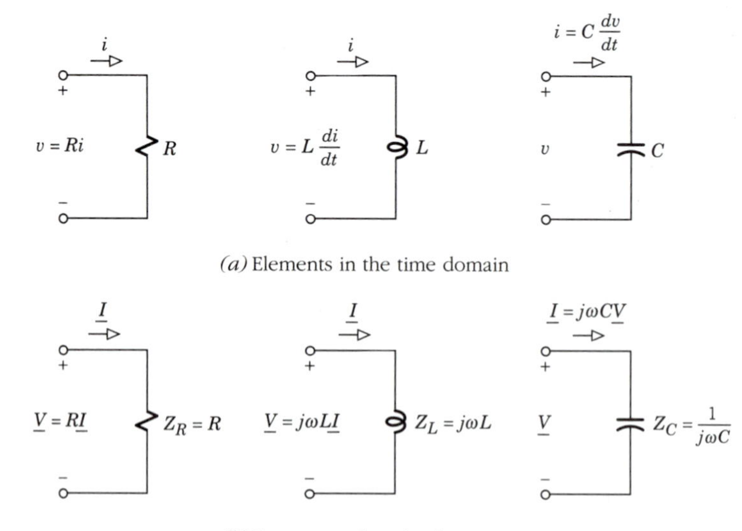

Figure 6.13

are labeled with *phasors* and *impedances* because we always use them together.

The frequency dependances of Z_L and Z_C reflect the rate-of-change effects in the time-domain relationships. At the low-frequency extreme, $Z_L \rightarrow 0$ and $Z_C \rightarrow \infty$ when $\omega \rightarrow 0$ — which corresponds to *dc* steady-state conditions. Thus, we have agreement with our earlier observation that an inductor acts as a dc short circuit while a capacitor acts as a dc open circuit. At the high-frequency extreme, $Z_C \rightarrow 0$ and $Z_L \rightarrow \infty$ when $\omega \rightarrow \infty$, so a capacitor acts as a high-frequency short circuit and an inductor acts as a high-frequency open circuit (sometimes called a "choke").

Besides impedance, another useful quantity is the **admittance** of an element defined by

$$Y \triangleq 1/Z \tag{6.28}$$

Admittance is therefore measured in *siemens* (S), the reciprocal of ohms. In terms of admittance, our ac version of Ohm's law becomes $\underline{V} = \underline{I}/Y$ or

$$\underline{I} = Y\underline{V} \tag{6.29}$$

Hence, you can use admittance to find $\underline{I}$ when you know $\underline{V}$.

Comparing Eqs. (6.29) and (6.26) reveals that admittance is the *dual* of impedance, just as conductance is the dual of resistance. Indeed, the admittance of a resistor equals its conductance since $Y_R = 1/Z_R = 1/R = G$. The admittance of an inductor is $Y_L = 1/j\omega L = -j/\omega L$, and the admittance of a capacitor is $Y_C = j\omega C$. These relations are listed in Table 6.1, together with the impedance relations.

TABLE 6.1 Impedance and Admittance of Elements

Element	Impedance	Admittance
Resistor	R	$1/R = G$
Inductor	$j\omega L$	$1/j\omega L = -j/\omega L$
Capacitor	$1/j\omega C = -j/\omega C$	$j\omega C$

Example 6.5 *Capacitor Calculations*

In Example 6.3, we had to use differentiation to find the current phasor $\underline{I}_C$ when the voltage across a 25-μF capacitor was $v = 30 \cos (4000t + 20°)$ V. Now we can determine $\underline{I}_C$ via Ohm's law for ac circuits.

With $C = 25 \times 10^{-6}$ and $\omega = 4000$, the impedance and admittance of the capacitor are

$$Z_C = -j/\omega C = -j10\ \Omega = 10\ \Omega\ \underline{/-90°}$$
$$Y_C = 1/Z_C = +j0.1\ \text{S} = 0.1\ \text{S}\ \underline{/+90°}$$

Then, since $\underline{V} = 30\ \text{V}\ \underline{/20°}$,

$$\underline{I}_C = Y_C\underline{V} = 0.1 \text{ S } \angle 90° \times 30 \text{ V } \angle 20° = 3 \text{ A } \angle 110°$$

which agrees with the previous result.

But if the source frequency increases to $\omega = 20{,}000$, then the impedance decreases to $Z_C = 2\ \Omega\ \angle{-90°}$ while the admittance increases to $Y_C = 0.5$ S $\angle 90°$. The current amplitude likewise increases to $|\underline{I}_C| = 15$ A if $\underline{V}$ remains fixed.

Exercise 6.6

Derive Eq. (6.25) from $i = C\,dv/dt$ *with* $v = Re\,[\underline{V}e^{j\omega t}]$.

Exercise 6.7

Evaluate the impedance and admittance of a 25-mH inductor when $\omega = 800$ rad/s. Then find $\underline{V}$ and $v(t)$ when the current through the inductor is $i(t) = 4\cos(800t - 50°)$ A.

Equivalent Impedance and Admittance

In general, different types of elements cannot be combined into a single equivalent element. However, under ac steady-state conditions, a network consisting of resistors, inductors, and capacitors acts at its terminals like an *equivalent impedance*. We'll prove this important statement here, and we'll examine the properties of equivalent impedance and admittance.

Figure 6.14*a* is the frequency-domain diagram of a series network operating in the ac steady state at some frequency ω. Regardless of the particular types of elements, the phasor voltage and current associated with each element are related by the element's impedance — $\underline{V}_1 = Z_1\underline{I}$, $\underline{V}_2 = Z_2\underline{I}$, etc. Since KVL holds for the voltage phasors, the terminal voltage phasor is

$$\begin{aligned}\underline{V} &= \underline{V}_1 + \underline{V}_2 + \cdots + \underline{V}_N = Z_1\underline{I} + Z_2\underline{I} + \cdots + Z_N\underline{I} \\ &= (Z_1 + Z_2 + \cdots + Z_N)\underline{I} = Z_{ser}\underline{I}\end{aligned}$$

We thereby obtain the **series equivalent impedance**

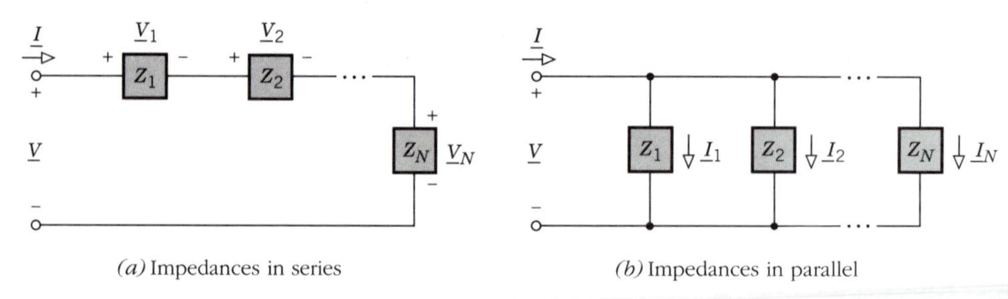

(*a*) Impedances in series

(*b*) Impedances in parallel

Figure 6.14

$$Z_{ser} = Z_1 + Z_2 + \cdots + Z_N \tag{6.30}$$

Thus, series impedances add like series resistances.

The parallel network in Fig. 6.14*b* is the dual of the series network. Consequently, $\underline{I} = Y_{par}\underline{V}$ with the **parallel equivalent admittance** being

$$Y_{par} = Y_1 + Y_2 + \cdots + Y_N \tag{6.31}$$

Thus, parallel admittances add like parallel conductances. For the special case of $N = 2$ elements, the parallel equivalent impedance is

$$Z_{par} = \frac{1}{Y_{par}} = \frac{1}{Y_1 + Y_2} = \frac{1}{(1/Z_1) + (1/Z_2)} = \frac{Z_1 Z_2}{Z_1 + Z_2} \tag{6.32}$$

We can therefore adapt our parallel-resistance notation and write $Z_{par} = Z_1 \| Z_2$ for the product-over-sum expression.

As an extension of the series and parallel cases, let the **load network** in Fig. 6.15 consist of arbitrarily connected resistors, inductors, capacitors, and controlled sources — but no independent sources. The terminal voltage and current in the ac steady state are sinusoids having the associated phasors $\underline{V}$ and $\underline{I}$, and the ration $\underline{V}/\underline{I}$ will be a constant. Therefore, we define the equivalent impedance in general by

$$Z_{eq} \triangleq \underline{V}/\underline{I} \tag{6.33}$$

Equation (6.33) is the impedance version of our equivalent resistance theorem.

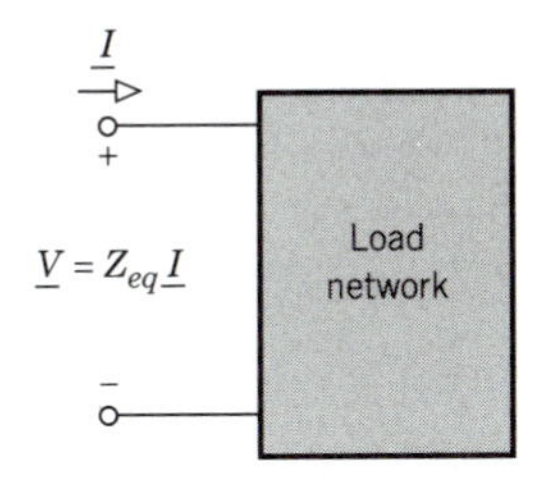

Figure 6.15 Load network with $Z_{eq} = \underline{V}/\underline{I}$.

But equivalent impedance differs from equivalent resistance in two critical respects: It may be a complex quantity, and its value may vary with frequency. Moreover, in contrast to the impedances of individual elements, both the real and imaginary parts of an equivalent impedance may be nonzero. For example, the equivalent impedance of a resistor in series with an inductor is $Z_{ser} = R + j\omega L$, which follows from Eq. (6.30) with $Z_1 = Z_R = R$ and $Z_2 = Z_L = j\omega L$.

To emphasize the complex and frequency-dependent properties, we often write *any* impedance in the form

$$Z = Z(j\omega) = R(\omega) + jX(\omega) \tag{6.34a}$$

where

$$R(\omega) \triangleq \text{Re}[Z] \qquad X(\omega) \triangleq \text{Im}[Z] \tag{6.34b}$$

The real part $R(\omega)$ is known as the **ac resistance**, and it may or may not vary with ω. The imaginary part $X(\omega)$ is known as the **reactance**, and it always varies with ω because it reflects the presence of energy storage. Reactance $X(\omega)$ is measured in ohms, like impedance $Z(j\omega)$ and ac resistance $R(\omega)$.

Equation (6.34a) also holds for each of the individual elements, since $Z_R = R = R + j0$, $Z_L = j\omega L = 0 + j\omega L$, and $Z_C = 1/j\omega C = 0 + j(-1/\omega C)$. Thus, a resistor has no reactance, while the energy-storage elements have no ac resistance. For this reason, inductors and capacitors are called **reactive elements**. Furthermore, for an inductor or capacitor alone,

$$X_L = \text{Im}[Z_L] = \omega L \qquad X_C = \text{Im}[Z_C] = -1/\omega C \tag{6.35}$$

We therefore say that

Inductive reactance is positive, and capacitive reactance is negative.

AC resistance can be negative only when a network includes a controlled source.

Figure 6.16 diagrams an arbitrary impedance in the complex plane with horizontal component $R(\omega)$ and vertical component $X(\omega)$. We call this figure the **impedance triangle**. The complex number Z is *not* a phasor because impedance does not represent a sinusoidal time function. Nonetheless, we can express Z in the polar notation

$$Z = |Z| \underline{/\theta} \tag{6.36a}$$

with

$$|Z| = \sqrt{R^2(\omega) + X^2(\omega)} \qquad \theta = \angle Z = \tan^{-1}\frac{X(\omega)}{R(\omega)} \tag{6.36b}$$

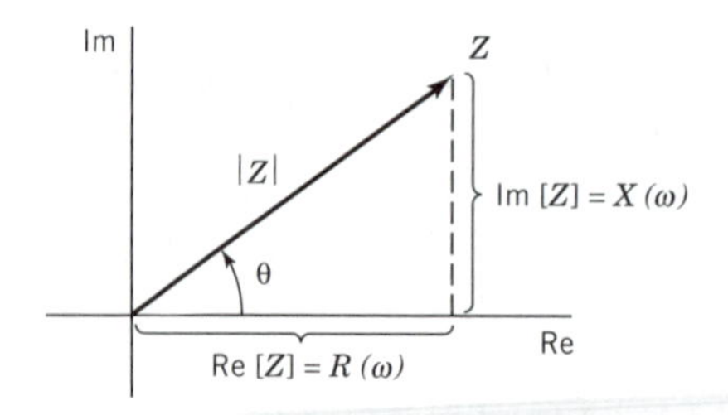

Figure 6.16 Impedance triangle for $Z(j\omega) = R(\omega) + jX(\omega)$.

Hereafter, all impedance angles will be symbolized by θ, with or without subscripts.

Finally, we note that any admittance is also a complex, frequency-dependent quantity. Hence, we can write

$$Y = Y(j\omega) = G(\omega) + jB(\omega) \tag{6.37a}$$

where

$$G(\omega) \triangleq \text{Re}[Y] \qquad B(\omega) \triangleq \text{Im}[Y] \tag{6.37b}$$

The real part $G(\omega)$ is known as the **ac conductance**, and the imaginary part is known as the **susceptance**. Although $G(\omega)$ and $B(\omega)$ are measured in siemens and are the duals of $R(\omega)$ and $X(\omega)$, respectively, they do not necessarily equal the reciprocals of $R(\omega)$ and $X(\omega)$. To determine the ac conductance and susceptance given the ac resistance and reactance, you must first find $Y = 1/Z$ and then use Eq. (6.37b). Table 6.2 summarizes the nomenclature and definitions for ac impedance and admittance.

TABLE 6.2 Nomenclature for Impedance and Admittance

Impedance	$Z(j\omega) = V/I$	Admittance	$Y(j\omega) = I/V$
Resistance	$R(\omega) = \text{Re}[Z]$	Conductance	$G(\omega) = \text{Re}[Y]$
Reactance	$X(\omega) = \text{Im}[Z]$	Susceptance	$B(\omega) = \text{Im}[Y]$

Example 6.6 *Impedance Analysis of a Parallel RC Circuit*

Again consider the parallel *RC* circuit with

$$v = 30 \cos(4000t + 20°) \text{ V} \quad R = 5\ \Omega \quad C = 25\ \mu\text{F}$$

As in Example 6.3, we want to determine the phasor current $\underline{I}$ supplied by the source. But now we can apply all of Steinmetz's phasor transform method to eliminate many intermediate steps.

The capacitor's impedance at $\omega = 4000$ was found in Example 6.5 to be $Z_C = -j10\ \Omega$, so Fig. 6.17*a* is our frequency-domain diagram. We put the

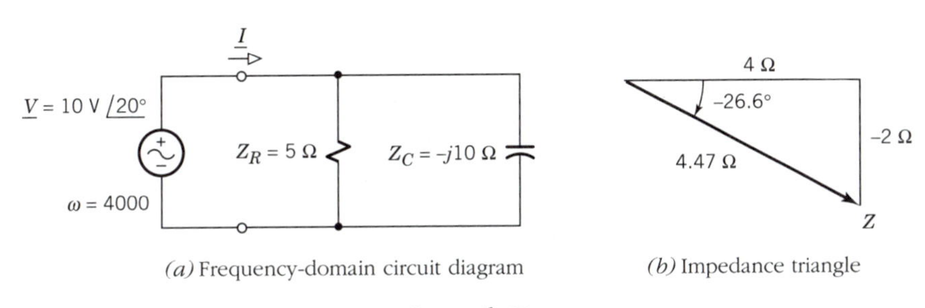

(*a*) Frequency-domain circuit diagram (*b*) Impedance triangle

Figure 6.17

value of ω beside the source as a reminder that this diagram holds only for the specified frequency. The equivalent impedance connected to the source is

$$Z = 5\|(-j10) = \frac{-j50}{5 - j10} = 4 - j2\ \Omega = 4.47\ \Omega\ \underline{/-26.6^\circ}$$

Figure 6.17b shows the corresponding impedance triangle with $R(\omega) = \text{Re}[Z] = 4\ \Omega$ and $X(\omega) = \text{Im}[Z] = -2\ \Omega$. Alternatively, we could determine the equivalent admittance.

$$Y = \frac{1}{5} + \frac{1}{-j10} = 0.2 + j0.1\ \text{S} = 0.224\ \text{S}\ \underline{/26.6^\circ}$$

from which $G(\omega) = \text{Re}[Y] = 0.2\ \text{S} \neq 1/R(\omega)$ and $B(\omega) = \text{Im}[Y] = 0.1\ \text{S} \neq 1/X(\omega)$.

Finally, we use $\underline{V} = 30\ \text{V}\ \underline{/20^\circ}$ and either Z or Y to calculate the phasor current

$$\underline{I} = \underline{V}/Z = Y\underline{V} = 6.71\ \text{A}\ \underline{/46.6^\circ}$$

Exercise 6.8

Evaluate the ac resistance, reactance, ac conductance, and susceptance of a parallel RL network when $R = 10\ \Omega$, $L = 25$ mH, and $\omega = 800$.

Frequency Dependence. There are two distinct types of ac analysis problems, depending upon whether the frequency is a known constant or is to be treated as a variable.

If the value of ω is specified, then the impedance or admittance of each element is a complex number and you can proceed by combining those complex numbers in the appropriate manner. But to investigate questions of *frequency dependence*, you must manipulate complex quantities in which ω appears as a variable. This type of problem is illustrated in the following example.

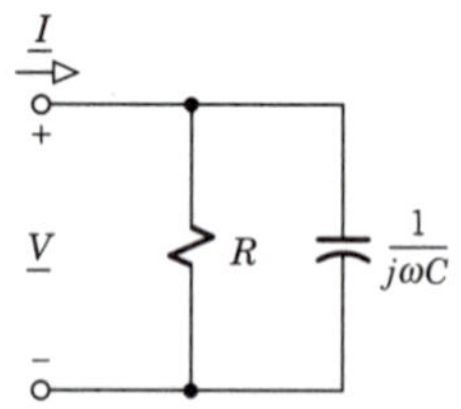

Figure 6.18 Parallel RC network.

Example 6.7 *Frequency Dependence of a Parallel RC Network*

The diagram in Fig. 6.18 represents a parallel RC network operating at an arbitrary frequency. We'll obtain expressions for the impedance, the ac resistance, and the reactance in terms of R, C, and ω.

We start with the parallel equivalent impedance

$$Z(j\omega) = \frac{Z_R Z_C}{Z_R + Z_C} = \frac{R/j\omega C}{R + 1/j\omega C} = \frac{R}{j\omega CR + 1}$$

Rationalization then yields the rectangular form

$$Z(j\omega) = \frac{R}{1 + j\omega CR}\frac{1 - j\omega CR}{1 - j\omega CR} = \frac{R - j\omega CR^2}{1 + (\omega CR)^2}$$

Hence, the ac resistance and reactance are

$$R(\omega) = \text{Re}[Z] = \frac{R}{1 + (\omega CR)^2} \qquad X(\omega) = \text{Im}[Z] = -\frac{\omega CR^2}{1 + (\omega CR)^2}$$

We see here that $R(\omega)$ is positive, but it does not equal the dc resistance R unless $\omega C = 0$. We also see that $X(\omega)$ is negative, but it does not equal X_C unless $R = \infty$.

If $\omega \rightarrow 0$, then the capacitor becomes an open circuit, so $R(0) = R$, $X(0) = 0$, and $Z(j0) = R + j0$. If $\omega \rightarrow \infty$, then the capacitor becomes a short circuit, so $R(\infty) = 0$, $X(\infty) = 0$, and $Z(j\infty) = 0 + j0$.

Exercise 6.9

Obtain expressions for $R(\omega)$ and $X(\omega)$ for a parallel RL network with arbitrary frequency ω.

AC Ladder Networks

An ac ladder network consists of resistors, inductors, and capacitors connected in series and/or parallel, with two terminals available for an external source. When the source frequency is specified, a ladder network is easily analyzed by **series-parallel reduction**. The method goes along the lines of the technique developed for resistive ladders back in Section 2.1, except that now *impedance replaces resistance.*

To find the branch variables in a ladder network, you proceed as follows:

1. Evaluate the element impedance at the source frequency and draw the frequency-domain diagram labeled with phasors and impedances.

2a. Combine series and parallel impedance to reduce the entire network to an equivalent impedance.

2b. Calculate the phasors of interest using the impedance versions of Ohm's law and voltage or current dividers.

3. Convert the resulting phasors into sinusoidal time functions.

Example 6.8 *AC Ladder Calculations*

The current driving the ladder network in Fig. 6.19*a* is

$$i(t) = 10 \cos 50{,}000t \text{ mA}$$

We want to find the steady-state voltages $v(t)$, $v_L(t)$, and $v_C(t)$.

Step 1: The impedances of the reactive elements at $\omega = 50$ krad/s are $Z_L = j10{,}000\ \Omega$ and $Z_C = -j10{,}000\ \Omega$. Figure 6.19*b* shows our frequency-domain diagram with all impedances in kilohms.

Step 2a: The parallel *RC* section has impedance

$$5 \| (-j10) = 4 - j2 \text{ k}\Omega$$

The total equivalent impedance is

$$Z = 40 \| [j10 + (4 - j2)] = 4.8 + j6.4 \text{ k}\Omega = 8 \text{ k}\Omega \underline{/53.1^\circ}$$

Thus, $R(\omega) = \text{Re}[Z] = 4.8$ kΩ and $X(\omega) = \text{Im}[Z] = +6.4$ kΩ. The reduced diagram in Fig. 6.19*c* emphasizes the fact that this network has an ac resistance of 4.8 kΩ and an inductive (positive) reactance of 6.4 kΩ at $\omega = 50$ krad/s.

Step 2b: With $\underline{I} = 10 \text{ mA } \underline{/0^\circ}$, the resultant terminal voltage is

$$\underline{V} = Z\underline{I} = 48 + j64 = 80 \text{ V } \underline{/53.1^\circ}$$

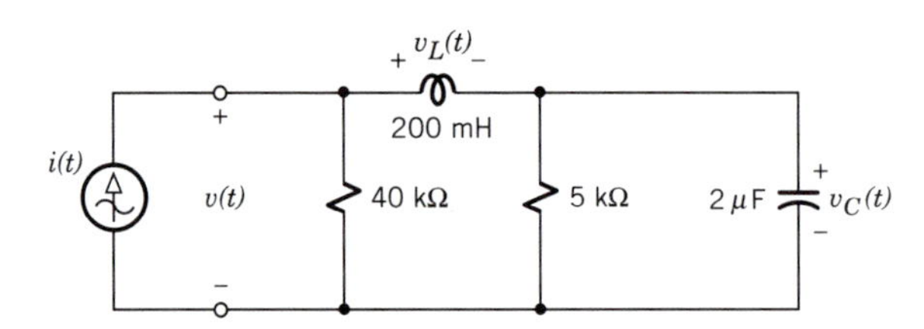

(a) AC ladder network with a current source

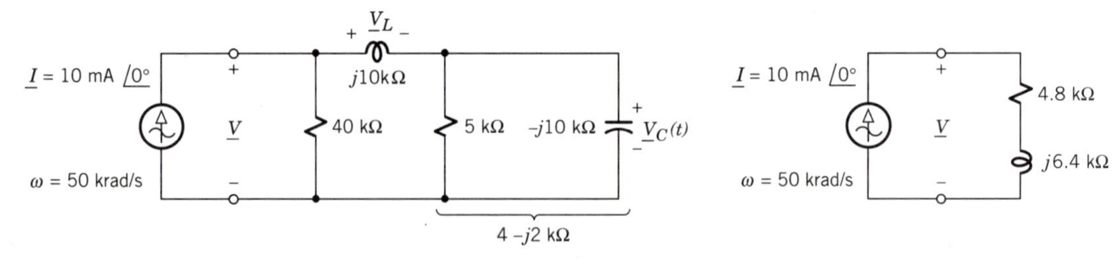

(b) Frequency-domain diagram

(c) Equivalent impedances

Figure 6.19

Then, going back to Fig. 6.19*b*, we calculate $\underline{V}_L$ and $\underline{V}_C$ from the voltage dividers

$$\underline{V}_L = \frac{j10}{(4 - j2) + j10}\underline{V} = 3.2 + j88 = 89.4 \text{ V} \angle 79.7^\circ$$

$$\underline{V}_C = \frac{4 - j2}{(4 - j2) + j10}\underline{V} = 32 - j24 = 40 \text{ V} \angle -36.9^\circ$$

Step 3: Converting phasors to time function, we get the voltage waveforms

$$v(t) = 80 \cos (50{,}000t + 53.1^\circ) \text{ V}$$
$$v_L(t) = 89.4 \cos (50{,}000t + 79.7^\circ) \text{ V}$$
$$v_C(t) = 40 \cos (50{,}000t - 36.9^\circ) \text{ V}$$

Notice that $v_L(t)$ has a larger amplitude (89.4 V) than the terminal voltage (80 V). This surprising result comes from the partial cancellation of inductive and capacitive reactances in the denominator of the voltage divider for $\underline{V}_L$. The related phenomenon of *resonance* will be explored in Section 6.4. Meanwhile, Fig. 6.20 shows the phasor construction $\underline{V}_L + \underline{V}_C = \underline{V}$. The accompanying plot of the waveforms confirms that $v_L(t) + v_C(t)$ equals $v(t)$ at every instant of time — even though the amplitude of $v_L(t)$ is greater than the amplitude of $v(t)$.

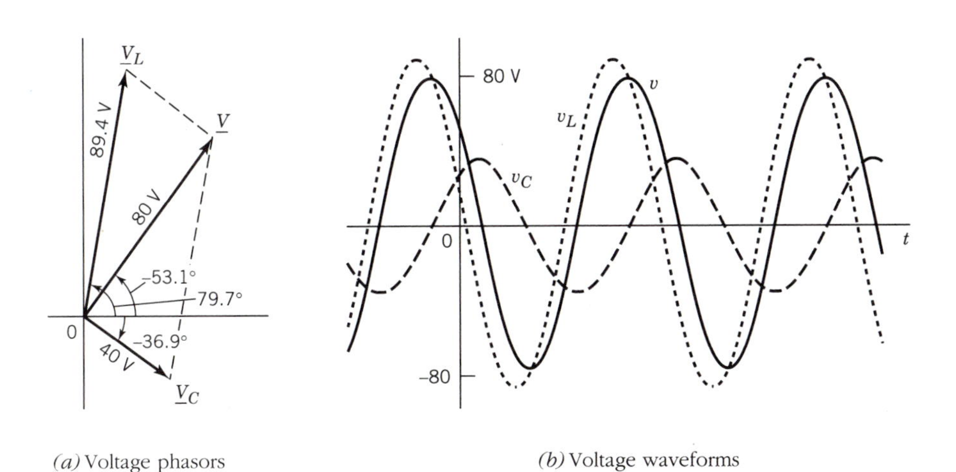

(*a*) Voltage phasors (*b*) Voltage waveforms

Figure 6.20

Exercise 6.10

Let the circuit in Fig. 6.21 have $v(t) = 60 \cos 1000t$ V, $R_1 = 8\ \Omega$, $L = 14$ mH, $C = 25\ \mu$F, and $R_2 = 80\ \Omega$. Calculate the equivalent impedance in rectangular

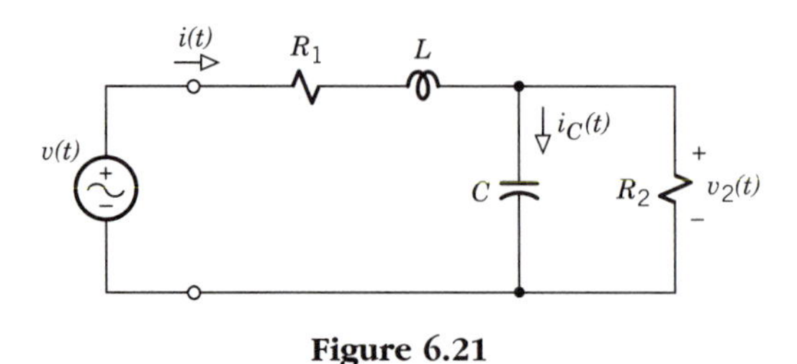

Figure 6.21

and polar form, and find $i(t)$ and $i_L(t)$. (Save your diagram for use in Exercise 6.11.)

6.3 AC CIRCUIT ANALYSIS

This section further exploits the concepts of phasors and impedance to find any steady-state response of interest in any stable linear circuit with sinusoidal excitation. The excitation may come from one or more independent sources, provided that all sources operate at the *same frequency* so the impedances have unique values. We'll also encounter design-oriented problems in which the frequency or an element value is to be determined from specified circuit characteristics.

The starting point in every case will be the frequency-domain diagram. Then we draw upon the property that the phasor equation $\underline{V} = Z\underline{I}$ has the same form as Ohm's law, so

> All resistive-circuit analysis techniques also hold for frequency-domain diagrams, with phasors in place of time functions, impedance in place of resistance, and admittance in place of conductance.

Accordingly, the **phasor transform method** consists of three general steps:

1. Go from the time domain to the frequency domain by representing sinusoids as phasors and evaluating element impedances at the source frequency.
2. Apply any appropriate techniques to analyze the frequency-domain diagram for the response phasors of interest.
3. Return to the time domain by converting the resulting response phasors to sinusoidals.

We call this procedure a *transform* method because it transforms difficult time-domain problems into more easily solved frequency-domain problems.

Since we already know how to go to and from the frequency domain, we'll concentrate here on analysis techniques in the frequency domain. In particular, we'll develop and apply the frequency-domain versions of:

- Proportionality
- Thévenin's and Norton's theorems and source conversions
- Systematic node and mesh equations

Superposition in the frequency domain is deferred to a later section, where we'll also use it to tackle circuits with more than one source frequency.

To simplify numerical work, some of the circuits considered here will have impractical element values or frequencies. (However, such circuits could be scaled models of practical circuits.) Additionally, when there is just one independent source, its phase angle will usually be taken to be zero. (Any other angle would simply produce a time shift of all waveforms.) Often, we'll stop after the second step of the phasor transform method because the resulting phasors adequately define the corresponding sinusoidal time functions.

Proportionality

Since the phasor equation $\underline{V} = Z\underline{I}$ is a *linear* relationship, we can analyze ac circuits using the **proportionality method** developed back in Section 2.4. This method is simple and direct for ladder networks driven by a single source. It may also be effective when the network includes a controlled source.

To analyze such networks, you first assume a convenient value for a voltage or current phasor farthest from the source. Next, you work toward the source by calculating other voltage and current phasors until you obtain the terminal conditions that would produce those branch variables. Finally, if the value of the source phasor is given, you invoke the proportionality principle to determine the actual values of the branch variables.

Example 6.9 ***AC Network with a Controlled Source***

Figure 6.22 shows the time-domain and frequency-domain diagrams of a network that contains a VCVS and operates at $\omega = 1000$. You want to find the input impedance $Z = \underline{V}/\underline{I}$ and the resulting capacitor current i_1 when $v = 20 \cos 1000t$ V.

The proportionality method is made to order for this problem because ratios such as $\underline{V}/\underline{I}$ are independent of the actual source phasor. Accordingly, you assume for convenience that

$$\underline{I}_2 = 1\text{ A}\,\underline{/0^\circ} = 1 + j0$$

in which case

$$\underline{V}_x = j8\underline{I}_2 = j8$$

Going back toward the source yields

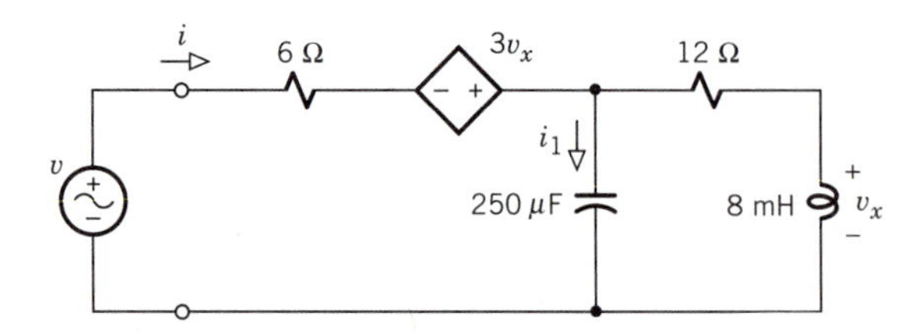

(a) AC network with a controlled source

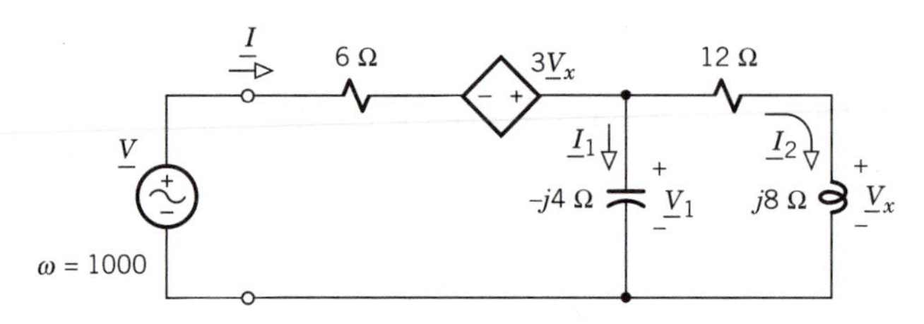

(b) Frequency-domain diagram

Figure 6.22

$$\underline{V}_1 = (12 + j8)\underline{I}_2 = 12 + j8$$
$$\underline{I}_1 = \underline{V}_1/(-j4) = -2 + j3$$
$$\underline{I} = \underline{I}_1 + \underline{I}_2 = -1 + j3$$
$$\underline{V} = 6\underline{I} - 3\underline{V}_x + \underline{V}_1 = 6 + j2$$

Hence, the input impedance is

$$Z = \underline{V}/\underline{I} = 0 - j2 = 2\ \Omega\ \angle -90^\circ$$

Since Re$[Z] = 0$, the controlled source causes the network to have zero ac resistance when $\omega = 1000$.

Finally, to calculate i_1, you note from your previous results that $\underline{I}_1/\underline{I} = (-2 + j3)/(-1 + j3) = 1.14\ \angle 15.3^\circ$. Then, with $\underline{V} = 20\text{ V}\ \angle 0^\circ$.

$$\underline{I} = \underline{V}/Z = 10\text{ A}\ \angle 90^\circ \qquad \underline{I}_1 = (\underline{I}_1/\underline{I})\underline{I} = 11.4\text{ A}\ \angle 105.3^\circ$$

so

$$i_1 = 11.4 \cos(8000t + 105.3^\circ)\text{ A}$$

Example 6.10 *Phase-Shift Oscillator*

An **oscillator** is an electronic circuit that generates a sinusoidal output without an independent input source. Some oscillator circuits have a structure like Fig. 6.23*a*, which consists of a phase-shift network and a VCVS with *positive feedback* from the output to the input, shown here as a dashed line.

The VCVS represents a voltage amplifier that produces $v_{out} = Kv_x$ (which could be implemented using a noninverting op-amp circuit). The phase-shift

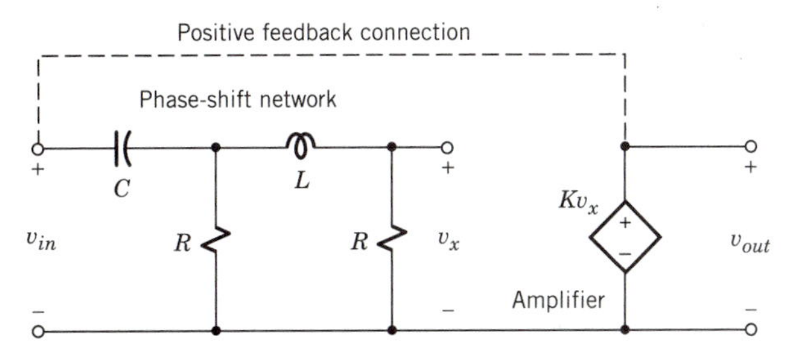

(a) Phase-shift oscillator circuit

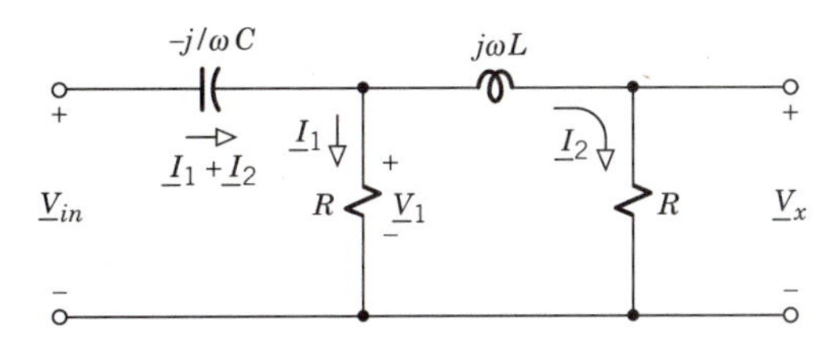

(b) Phase-shift network in the frequency domain

Figure 6.23

network is designed such that $v_x = v_{in}/K$ at one and only one frequency. Consequently, at the design frequency,

$$v_{out} = (v_{out}/v_x)(v_x/v_{in})v_{in} = K(1/K)v_{in} = v_{in}$$

This means that the input can come directly from the output, via the feedback connection, so an independent source is not required. Instead, we only need some initial stored energy in the capacitor or inductor, and the circuit will thereafter generate a sinusoidal output. The oscillation amplitude is dictated by operating limits of the VCVS.

For purposes of oscillator design, our task here is to analyze the phase-shift network redrawn in the frequency domain in Fig. 6.23*b*. Specifically, we want to find the value of ω at which $\underline{V}_x/\underline{V}_{in} = 1/K$ or, equivalently, $\underline{V}_{in}/\underline{V}_x = K\underline{/0°}$. We also want to find the corresponding value of K.

Exploiting proportionality, let's assume that $\underline{I}_2 = 1\underline{/0°} = 1$. Then

$$\underline{V}_x = R\underline{I}_2 = R \qquad \underline{V}_1 = (R + j\omega L)\underline{I}_2 = R + j\omega L$$

$$\underline{I}_1 = \underline{V}_1/R = 1 + j\omega L/R \qquad \underline{I}_1 + \underline{I}_2 = 2 + j\omega L/R$$

$$\underline{V}_{in} = (-j/\omega C)(\underline{I}_1 + \underline{I}_2) + \underline{V}_1 = R + (L/CR) + j(\omega L - 2/\omega C)$$

so

$$\underline{V}_{in}/\underline{V}_x = 1 + (L/CR^2) + j(\omega L - 2/\omega C)/R$$

Since $\angle(\underline{V}_{in}/\underline{V}_x) = 0°$ only when $\mathrm{Im}[\underline{V}_{in}/\underline{V}_x] = 0$, we see that oscillation requires $\omega L - 2/\omega C = 0$ or $\omega L = 2/\omega C$. Solving for the oscillation frequency yields

$$\omega_{osc} = \sqrt{2/LC}$$

And when $\omega = \omega_{osc}$, $\underline{V}_{in}/\underline{V}_x = 1 + (L/CR^2)$, so the VCVS must have

$$K = 1 + (L/CR^2)$$

Exercise 6.11

Assume $\underline{V}_2 = 80\text{ V}\ \underline{/0^\circ}$ to find the ratio $\underline{V}_2/\underline{V}$ for the circuit in Exercise 6.10 (p. 269).

Exercise 6.12

Let the oscillator in Example 6.10 have $R = 1\text{ k}\Omega$ and $L = 20\text{ mH}$. Find the design values of C and K so that $f_{osc} = \omega_{osc}/2\pi = 10\text{ kHz}$.

Thévenin and Norton Networks

Thévenin's and Norton's theorems allow us to replace ac source networks with simpler equivalent networks. The frequency-domain versions of these theorems differ only slightly from our original statements.

An ac source network is any two-terminal network that consists entirely of linear elements and sources, including at least one independent ac source. But if there are two or more independent sources, then each of them must be at the *same frequency* so we can represent the network by an equivalent one containing a single source.

The frequency-domain Thévenin parameters of such a network are the **open-circuit voltage phasor** $\underline{V}_{oc}$, the **short-circuit current phasor** $\underline{I}_{sc}$, and the **Thévenin impedance** $Z_t = \underline{V}_{oc}/\underline{I}_{sc}$. For purposes of analyzing the conditions at the terminals, Thévenin's and Norton's theorems together state that

> An ac source network is equivalent in the frequency domain to a phasor voltage source $\underline{V}_{oc}$ in series with impedance Z_t, or a phasor current source $\underline{I}_{sc}$ in parallel with impedance Z_t.

Figure 6.24 diagrams the corresponding Thévenin and Norton equivalent networks.

Since the Thévenin and Norton networks are equivalent to each other, the three parameters must be related by

$$\underline{V}_{oc} = Z_t \underline{I}_{sc} \tag{6.38}$$

It also follows from Fig. 6.24 that Z_t equals the equivalent impedance of the source network after the independent sources have been suppressed.

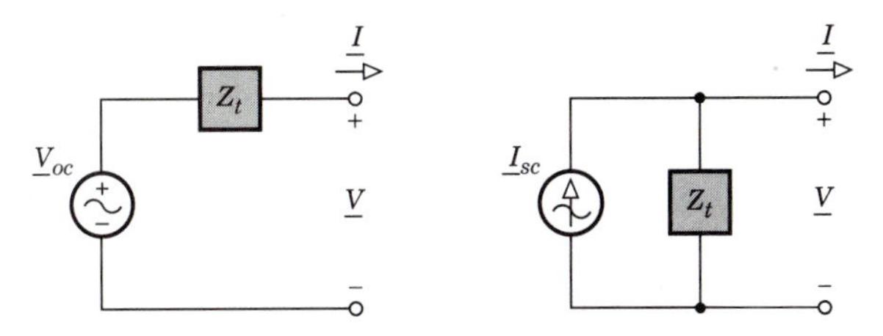

Figure 6.24 Thévenin and Norton networks in the frequency domain.

Finally, note that any source network having a Thévenin structure like Fig. 6.24*a* may be converted to a Norton structure like Fig. 6.24*b*, and vice versa. Equation (6.38) provides the general rule for ac source conversions.

Example 6.11 *Application of an AC Norton Network*

Given the frequency-domain source network in Fig. 6.25*a*, we want to maximize the amplitude of the terminal voltage $\underline{V}$ by selecting appropriate resistors and/or reactive elements for the load impedance Z. To solve this problem, we'll first find the Norton equivalent source network.

The Thévenin impedance is easily obtained since the network contains just one source. After suppressing this independent voltage source by substituting a short circuit, the inductor and resistor will be in parallel so

$$Z_t = j40\|280 - j20 = 5.6 + j19.2 = 20\ \Omega\ \underline{/73.7^\circ}$$

Next, the open-circuit voltage is calculated using the voltage divider

$$\underline{V}_{oc} = \frac{280}{280 + j40}\,10 = 9.90\ \text{V}\ \underline{/-8.13^\circ}$$

Therefore, the short-circuit current is

$$\underline{I}_{sc} = \underline{V}_{oc}/Z_t = 0.495\ \text{A}\ \underline{/-81.8^\circ}$$

Figure 6.25*b* diagrams the resulting Norton network with attached load.

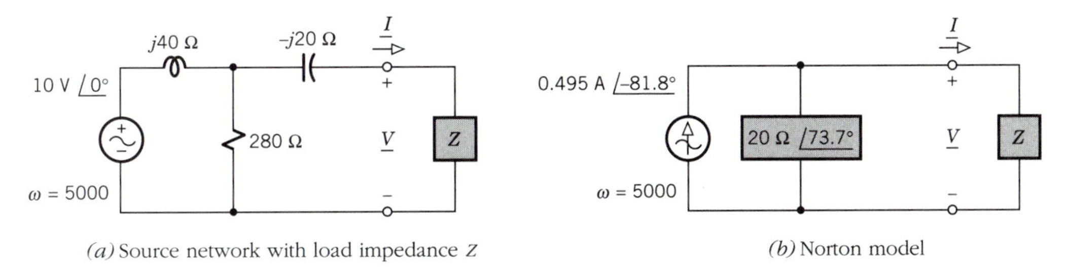

(a) Source network with load impedance Z *(b)* Norton model

Figure 6.25

The equivalent admittance connected to the current source is $Y_{eq} = Y_t + 1/Z$, where

$$Y_t = 1/Z_t = 0.05 \angle{-73.7°} = 0.014 - j0.048 \text{ S}$$

Since $\underline{V} = \underline{I}_{sc}/Y_{eq}$, the amplitude $|\underline{V}|$ will be maximum when $|Y_{eq}|$ is as small as possible. We find the required load impedance by letting $1/Z = Y = G + jB$ so

$$Y_{eq} = Y_t + Y = (0.014 + G) + j(B - 0.048)$$

and

$$|Y_{eq}| = \sqrt{(0.014 + G)^2 + (B - 0.048)^2}$$

The load's ac conductance G cannot be negative, but its susceptance B may be positive or negative. Consequently, $|Y_{eq}|$ is minimum when $G = 0$ and $B - 0.048 = 0$. We therefore want the load to have

$$Y = 0 + j0.048 \text{ S} \qquad Z = 1/Y = 0 - j20.8\ \Omega$$

Observing that Z consists entirely of *negative* reactance, the load should be a capacitor whose value satisfies the condition $-j/\omega C = Z = -j20.8\ \Omega$. Hence, we want $C = 1/20.8\omega = 9.62\ \mu\text{F}$.

Finally, let's calculate the resulting terminal voltage. The minimized admittance is $Y_{eq} = 0.014 + j0$, and

$$\underline{V} = \underline{I}_{sc}/Y_{eq} = 35.4 \text{ V} \angle{-81.8°}$$

whose magnitude is considerably larger than $|\underline{V}_{oc}| = 9.90$ V.

Exercise 6.13

Let the network in Fig. 6.26 have $i_s = 2 \cos 5t$ A. Find Z_t. $\underline{V}_{oc}$, and $\underline{I}_{sc}$ without using Eq. (6.38). Then confirm that your results satisfy Eq. (6.38). (Save your results for use in Exercise 6.19.)

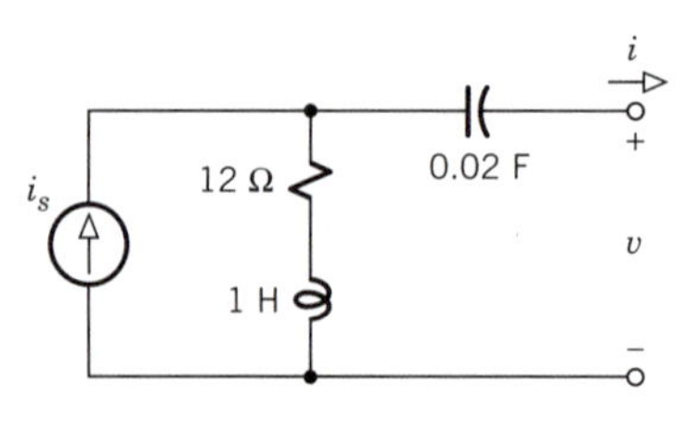

Figure 6.26

Node and Mesh Equations

When an ac circuit contains more than one source or has a complicated configuration, the use of systematic node or mesh equations often proves to be the most efficient analysis method. The frequency-domain techniques mimic those we developed for resistive circuits, except that we now work with phasors and impedance or admittance. Also remember that if there is more than one source, then *all independent sources must have the same frequency.*

AC Mesh Analysis. The matrix mesh equation for a single-frequency ac circuit takes the general form

$$[Z][\underline{I}] = [\underline{V}_s] \tag{6.39}$$

Here, $[\underline{I}]$ is the vector of unknown current phasors, while $[\underline{V}_s]$ is the vector of equivalent source-voltage phasors. The **impedance matrix** $[Z]$ has the same symmetry properties as the resistance matrix $[R]$ for a resistive circuit, except that its elements may be complex quantities.

Equation (6.39) can be solved directly for the unknown current phasors when there are no controlled sources. But if controlled sources are present, we expand $[\underline{V}_s]$ as

$$[\underline{V}_s] = [\underline{\tilde{V}}_s] + [\tilde{Z}][\underline{I}]$$

and the matrix mesh equation becomes

$$[Z - \tilde{Z}][\underline{I}] = [\underline{\tilde{V}}_s] \tag{6.40}$$

where $[Z - \tilde{Z}] = [Z] - [\tilde{Z}]$.

Writing and solving mesh equations for ac circuits follows the same principles we developed for resistive circuits. The only significant difference is that the ac case involves complex numbers. To expedite calculations, all complex numbers should be expressed in rectangular form using rationalization where necessary. The matrix equations can then be solved via Cramer's rule or, better yet, using a calculator that handles matrices with complex coefficients.

Example 6.12 *Systematic AC Mesh Analysis*

The circuit in Fig. 6.27*a* is driven by two independent sources

$$v = 30 \cos (10t + 60°) \text{ V} \qquad i = 1 \cos 10t \text{ A}$$

so both sources have $\omega = 10$. Figure 6.27*b* shows the frequency-domain diagram labeled for mesh analysis. Matrix notation will not be needed because $\underline{I}_1$ is the only unknown mesh current. However, we'll still follow our systematic process to determine Z and $\underline{V}_s$ for the single mesh equation $Z\underline{I}_1 = \underline{V}_s$.

We begin by combining the two parallel impedances to get $10 \| j20 = 8 + j4$. Thus, the sum of the impedances around the mesh is

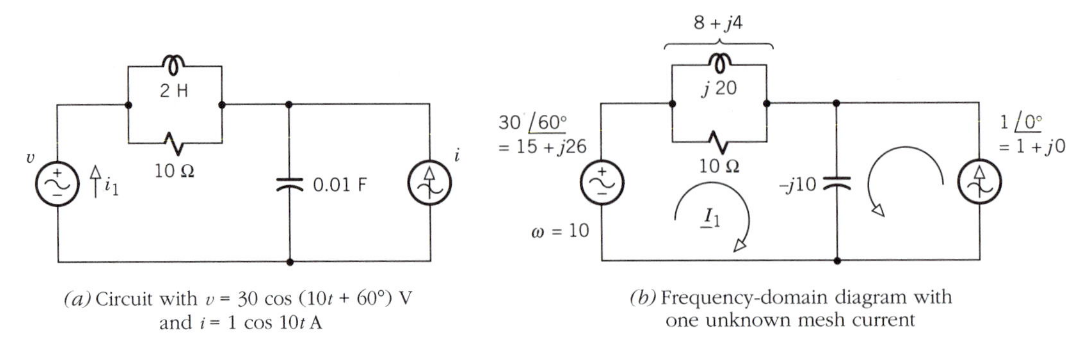

(a) Circuit with $v = 30 \cos(10t + 60°)$ V and $i = 1 \cos 10t$ A

(b) Frequency-domain diagram with one unknown mesh current

Figure 6.27

$$Z = (8 + j4) - j10 = 8 - j6\ \Omega$$

Next, we use the rectangular form of the source phasors to find the net voltage driving $\underline{I}_1$. The voltage source has $\underline{V} = 30\ \underline{/60°} = 15 + j26$, and the Norton structure on the right-hand side has open-circuit voltage $-j10 \times 1$ that opposes $\underline{I}_1$, so

$$\underline{V}_s = (15 + j26) - (-j10) = 15 + j36 \text{ V}$$

Hence, our mesh equation is

$$(8 - j6)\underline{I}_1 = 15 + j36$$

and converting to polar form yields

$$\underline{I}_1 = (39 \text{ V } \underline{/67.4°})/(10\ \Omega\ \underline{/-36.9°}) = 3.9 \text{ A } \underline{/104.3°}$$

Any other branch variable can now be found from the diagram and the value of $\underline{I}_1$.

AC Node Analysis. The matrix node equation for an ac circuit takes the general form

$$[Y][\underline{V}] = [\underline{I}_s] \tag{6.41}$$

where $[Y]$ is the **admittance matrix**, $[\underline{V}]$ is the vector of unknown voltage phasors, and $[\underline{I}_s]$ is the vector of equivalent source-current phasors.

If controlled sources are present, then $[\underline{I}_s]$ can be expanded as

$$[\underline{I}_s] = [\underline{\tilde{I}}_s] + [\tilde{Y}][\underline{V}]$$

and the matrix node equation becomes

$$[Y - \tilde{Y}][\underline{V}] = [\tilde{\underline{I}}_s] \tag{6.42}$$

where $[Y - \tilde{Y}] = [Y] - [\tilde{Y}]$.

Example 6.13 *Systematic AC Node Analysis*

The frequency-domain circuit diagram in Fig. 6.28 contains a CCCS. We'll use node analysis to find the voltage $\underline{V}_1$ and the equivalent impedance $Z_1 = \underline{V}_1/\underline{I}$ connected to the source and inductor.

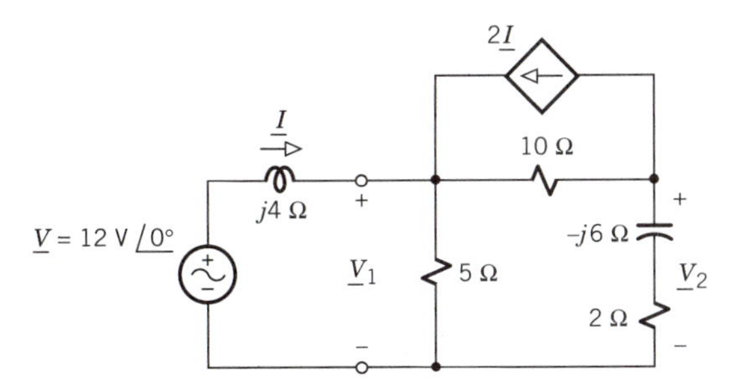

Figure 6.28 Frequency-domain diagram with two unknown node voltages.

First, we write the admittance matrix in rationalized form. The inductor's admittance is $1/j4 = -j/4$ and the admittance of the series RC section is $1/(2 - j6) = (1 + j3)/20$, so

$$[Y] = \begin{bmatrix} \frac{1}{10} + \frac{1}{5} - \frac{j}{4} & -\frac{1}{10} \\ -\frac{1}{10} & \frac{1}{10} + \frac{1 + j3}{20} \end{bmatrix} = \frac{1}{20}\begin{bmatrix} 6 - j5 & -2 \\ -2 & 3 + j3 \end{bmatrix}$$

The fractions have been cleared by multiplying each term by 20 and dividing the entire matrix by 20.

Second, we formulate the constraint equation relating the control variable $\underline{I}$ to the node voltages. Inspection of the diagram shows that

$$\underline{I} = (\underline{V} - \underline{V}_1)/j4 = -j3 + j0.25\underline{V}_1$$

Third, we write the source-current vector and expand it using the constraint equation. Thus,

$$[\underline{I}_s] = \begin{bmatrix} 2\underline{I} + \underline{V}/j4 \\ -2\underline{I} \end{bmatrix} = \begin{bmatrix} -j9 + j0.5\underline{V}_1 \\ j6 - j0.5\underline{V}_1 \end{bmatrix} = \begin{bmatrix} -j9 \\ j6 \end{bmatrix} + \begin{bmatrix} j0.5 & 0 \\ -j0.5 & 0 \end{bmatrix}\begin{bmatrix} \underline{V}_1 \\ \underline{V}_2 \end{bmatrix}$$

Again multiplying and dividing by 20, we have

$$[\tilde{\underline{I}}_s] = \frac{1}{20}\begin{bmatrix} -j180 \\ j120 \end{bmatrix} \qquad [\tilde{Y}] = \frac{1}{20}\begin{bmatrix} j10 & 0 \\ -j10 & 0 \end{bmatrix}$$

Now we subtract $[\tilde{Y}]$ from $[Y]$ and cancel the common factor of 1/20 to get the matrix node equation

$$\begin{bmatrix} 6 - j15 & -2 \\ -2 + j10 & 3 + j3 \end{bmatrix} \begin{bmatrix} \underline{V}_1 \\ \underline{V}_2 \end{bmatrix} = \begin{bmatrix} -j180 \\ j120 \end{bmatrix}$$

Solving for $\underline{V}_1$ via determinants yields

$$\Delta = 59 - j7 \qquad \Delta_1 = 540 - j300$$
$$\underline{V}_1 = \Delta_1/\Delta = 9.62 - j3.94 = 10.4 \text{ V } \underline{/-22.3^\circ}$$

Finally, inserting the value of $\underline{V}_1$ into our constraint equation gives

$$\underline{I} = 0.986 - j0.595 = 1.15 \text{ A } \underline{/-31.1^\circ}$$

so

$$Z_1 = \underline{V}_1/\underline{I} = 9.03\ \Omega\ \underline{/8.8^\circ} = 8.92 + j1.39\ \Omega$$

Observe that Z_1 has positive reactance even though the network to the right of $\underline{V}_1$ contains no inductance. In this case, then, the controlled source makes capacitive reactance look like inductive reactance.

Exercise 6.14

Write a single node equation to find $\underline{V}_1$ in Fig. 6.29.

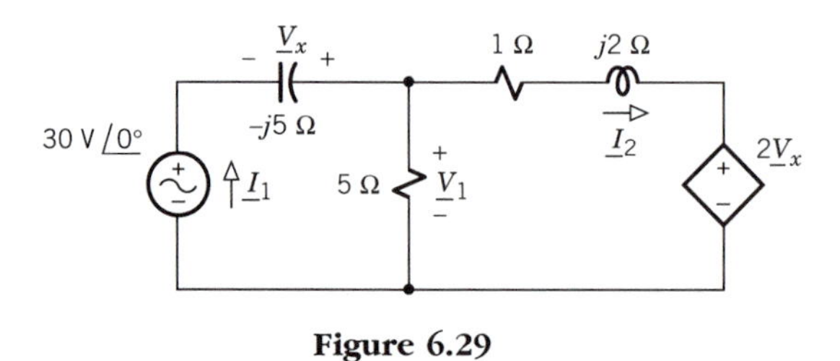

Figure 6.29

Exercise 6.15

Obtain the matrix mesh equation for $\underline{I}_1$ and $\underline{I}_2$ in Fig. 6.29. You need not solve for the unknowns.

6.4 PHASOR DIAGRAMS AND RESONANCE

This section delves further into the behavior of circuits under ac steady-state conditions. We'll start with the construction of phasor diagrams for simple circuits. Then we'll examine resonance in series and parallel circuits, drawing upon phasor diagrams as an aid to understanding resonance effects. Practical applications of resonance will be discussed later in conjunction with power systems and filters.

Phasor Diagrams

Phasor diagrams provide an informative picture of ac voltage and current relationships. Since these diagrams are used primarily for qualitative rather than quantitative purposes, we'll emphasize construction techniques that require only a few calculations.

Consider the series *RL* network in Fig. 6.30*a*. The network's impedance is $Z = R + j\omega L$ and

$$\theta = \angle Z = \tan^{-1}(\omega L/R)$$

If both L and R are positive and nonzero, then the impedance angle falls in the range

$$0 < \theta < 90^\circ$$

This property is the essential information needed to construct all three voltage phasors along with the current phasor.

For convenience, we'll take $\angle \underline{I} = 0$ to establish the phase reference because there is just one current in this network. The corresponding angles of the voltage phasors are then determined as follows:

$$\begin{aligned} \underline{V} &= Z\underline{I} && \Rightarrow \quad \angle \underline{V} = \angle Z + \angle \underline{I} = \theta \\ \underline{V}_R &= R\underline{I} && \Rightarrow \quad \angle \underline{V}_R = \angle \underline{I} = 0 \\ \underline{V}_L &= j\omega L\underline{I} && \Rightarrow \quad \angle \underline{V}_L = 90^\circ + \angle \underline{I} = 90^\circ \end{aligned}$$

Furthermore, we know that

$$\underline{V}_L + \underline{V}_R = \underline{V}$$

Figure 6.30*b* shows the parallelogram based on the foregoing considerations.

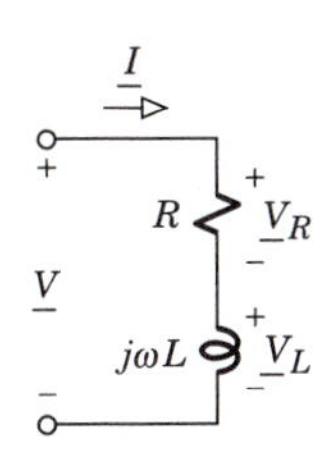

(*a*) Series *RL* network

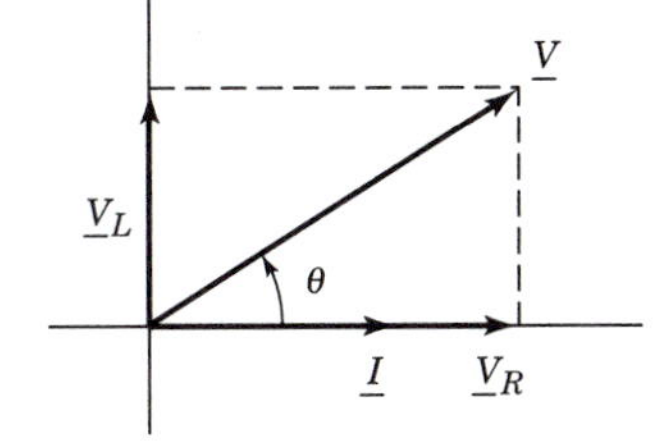

(*b*) Phasor diagram

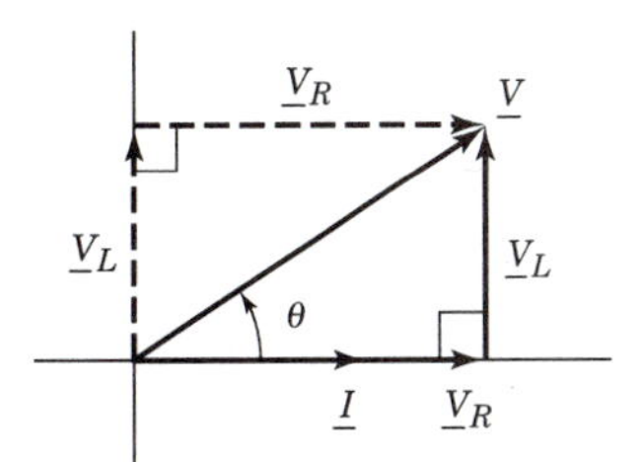

(*c*) Head-to-tail constructions for $\underline{V}$

Figure 6.30

Alternatively, we could use either of the two "head-to-tail" constructions in Fig. 6.30*c*. On the one hand, the solid triangle is formed by putting the tail of $\underline{V}_L$ at the head of $\underline{V}_R$ to obtain $\underline{V}$. On the other hand, the dashed triangle has the tail of $\underline{V}_R$ at the head of $\underline{V}_L$.

Although our diagrams do not include calibration for the phasor lengths, the voltage phasors are drawn to scale and in correct relationship to the current phasor for a particular value of θ. Of course the scale for the current phasor may differ from the scale for the voltage phasors because they have different units. To account for $\angle \underline{I} \neq 0$, the entire diagram may be rotated so that $\underline{V}$ and $\underline{I}$ appear at the proper angles.

Next, consider the series *RC* network in Fig. 6.31*a*. The impedance is $Z = R - j/\omega C$ so

$$\theta = \tan^{-1}(-1/\omega CR) = -\tan^{-1}(1/\omega CR)$$

If both *C* and *R* are positive and nonzero, then

$$-90° < \theta < 0$$

Again taking $\underline{I}$ as our phase reference, we find that $\angle \underline{V} = \theta$ and $\angle \underline{V}_R = 0$, whereas

$$\underline{V}_C = -j\underline{I}/\omega C \quad \Rightarrow \quad \angle \underline{V}_C = -90° + \angle \underline{I} = -90°$$

The resulting phasor diagram showing $\underline{I}$ and the two triangular constructions for $\underline{V}_C = \underline{V}_R = \underline{V}$ is given in Fig. 6.31*b*. Note that the direction arrow for θ points clockwise to denote a *negative* angle.

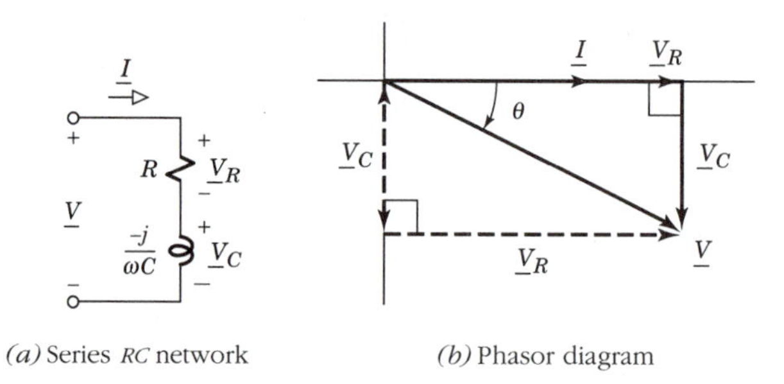

(*a*) Series *RC* network (*b*) Phasor diagram

Figure 6.31

Turning to parallel networks, let jX in Fig. 6.32*a* stand for the impedance of an inductor or capacitor. In the parallel *RL* case, the admittance is $Y = G - j/\omega L$ and $\angle Z = -\angle Y$. Hence, the impedance angle falls in the positive range $0 < \theta < 90°$. Now we'll take $\angle \underline{V} = 0$ for the phase reference, so

$$\underline{I} = \underline{V}/Z \quad \Rightarrow \quad \angle \underline{I} = \angle \underline{V} - \angle Z = -\theta$$
$$\underline{I}_R = \underline{V}/R \quad \Rightarrow \quad \angle \underline{I}_R = \angle \underline{V} = 0$$
$$\underline{I}_L = \underline{V}/j\omega L \quad \Rightarrow \quad \angle \underline{I}_L = \angle \underline{V} - 90° = -90°$$

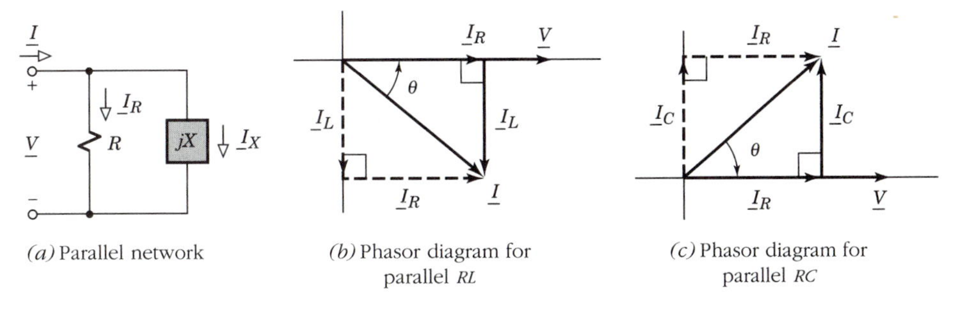

(*a*) Parallel network

(*b*) Phasor diagram for parallel *RL*

(*c*) Phasor diagram for parallel *RC*

Figure 6.32

Figure 6.32*b* shows the diagram with $\underline{V}$ and $\underline{I}_R + \underline{I}_L = \underline{I}$. Finally, in the parallel *RC* case, the impedance angle is negative and

$$\underline{I}_C = j\omega C\underline{V} \quad \Rightarrow \quad \angle\underline{I}_C = 90° + \angle\underline{V} = 90°$$

The phasor construction is given in Fig. 6.32*c*.

A review and comparison of our phasor diagrams leads to four general conclusions about the properties of networks consisting of one resistor and one reactive element:

- For an inductive network, either series or parallel, $\underline{I}$ lags $\underline{V}$ by an angle that falls between 0° and 90°.
- For a capacitive network, either series or parallel, $\underline{I}$ leads $\underline{V}$ by an angle that falls between 0° and 90°.
- For a series network, either inductive or capacitive, the voltage phasors form a right triangle with $\underline{V}$ as the hypotenuse and $\underline{V}_R$ in phase with $\underline{I}$.
- For a parallel network, either inductive or capacitive, the current phasors form a right triangle with $\underline{I}$ as the hypotenuse and $\underline{I}_R$ in phase with $\underline{V}$.

Phasor diagrams for more complicated circuits can be constructed using these properties and a few additional calculations.

Example 6.14 ***Phasor Diagram Construction***

The terminal impedance of the network in Fig. 6.33*a* is $Z = (-j5)\|(4 + j2) = 4 - j2\ \Omega$, so $\theta = -26.6°$. Figure 6.33*b* shows all voltage and current phasors drawn to scale taking $\angle\underline{I} = 0$. This diagram was constructed by the following steps.

First, we draw $\underline{I}$ and $\underline{V}$ using the fact that the entire network is capacitive and $\underline{I}$ leads $\underline{V}$ by 26.6°. However, the lengths of $\underline{I}$ and $\underline{V}$ are arbitrary since they have different scales.

Second, we construct $\underline{V}_R$ and $\underline{V}_L$ to scale by noting that

$$\underline{V}_R = 4\underline{I}_2 \qquad \underline{V}_L = j2\underline{I}_2 \qquad \underline{V}_R + \underline{V}_L = \underline{V}$$

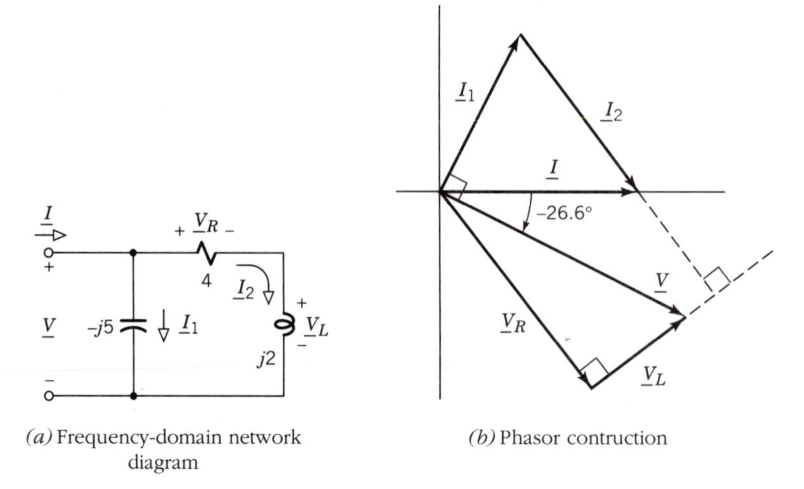

(*a*) Frequency-domain network diagram

(*b*) Phasor contruction

Figure 6.33

Thus, $\angle \underline{V}_L = \angle \underline{V}_R + 90°$, $|\underline{V}_L| = |\underline{V}_R|/2$, and the voltage phasors thus form a right triangle with $\underline{V}$ as the hypotenuse.

Third, we construct $\underline{I}_1$ and $\underline{I}_2$ to scale by noting that

$$\underline{I}_1 = \underline{V}/(-j5) \qquad \underline{I}_2 = \underline{V}_R/4 = \underline{V}_L/j2 \qquad \underline{I}_1 + \underline{I}_2 = \underline{I}$$

The current phasors thus form a triangle, but not a right triangle. Instead, $\angle \underline{I}_1 = \angle \underline{V} + 90°$ while $\angle \underline{I}_2 = \angle \underline{V}_R = \angle \underline{V}_L - 90°$.

Exercise 6.16

The terminal impedance of the network in Fig. 6.34 is $Z = 2 + j0$. Construct a diagram showing all voltage and current phasors drawn to scale with $\angle \underline{V} = 0$. Hint: Start with the current phasors.

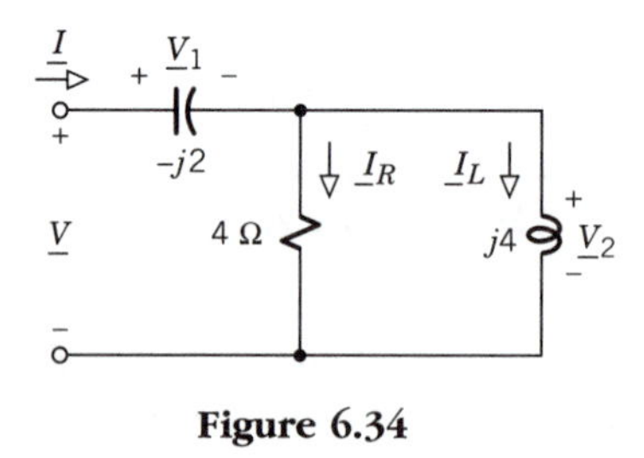

Figure 6.34

Series Resonance

Inductors and capacitors have "opposite" properties in two respects: Inductive reactance ($X_L = \omega L$) is positive and increases with frequency, whereas capacitive reactance ($X_C = -1/\omega C$) is negative and decreases with frequency. These properties lead to important and sometimes unexpected effects in circuits that contain both types of reactive elements. Depending upon the excitation frequency, either the inductance or the capacitance may dominate, or the two reactances may cancel out and produce the phenomenon known as *resonance.*

Consider, in particular, the series *RLC* network in Fig. 6.35*a*. Its terminal impedance is the sum

$$Z(j\omega) = R + j\omega L - j/\omega C = R + jX(\omega) \tag{6.43a}$$

where

$$X(\omega) = \omega L - 1/\omega C \tag{6.43b}$$

On the one hand, the capacitance dominates at low frequencies and the net reactance is negative. On the other hand, the inductance dominates at high frequencies and the net reactance is positive. The borderline between these two cases occurs at $\omega = \omega_0$ when

$$X(\omega_0) = \text{Im}[Z(j\omega_0)] = 0 \tag{6.44}$$

Equation (6.44) defines the **series resonance condition**. The corresponding **resonant frequency** for an *RLC* network must satisfy $\omega_0 L - 1/\omega_0 C = 0$, so $\omega_0{}^2 = 1/LC$ and

$$\omega_0 = 1/\sqrt{LC} \tag{6.45}$$

The network appears to be purely resistive at resonance, since $Z(j\omega_0) = R + jX(\omega_0) = R$.

Figure 6.35*b* displays the frequency variation of the magnitude and angle of $Z(j\omega)$ of this network, calculated from

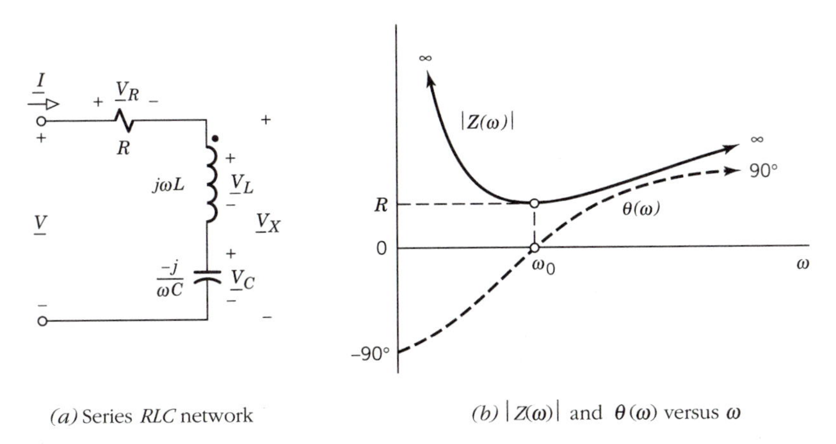

(*a*) Series *RLC* network

(*b*) $|Z(\omega)|$ and $\theta(\omega)$ versus ω

Figure 6.35

$$|Z(\omega)| = \sqrt{R^2 + (\omega L - 1/\omega C)^2} \qquad \theta(\omega) = \tan^{-1}\frac{\omega L - 1/\omega C}{R}$$

We see that $|Z(\omega)|$ has a unique *minimum* at ω_0, and that $\theta(\omega)$ goes from $-90°$ to $+90°$ as ω increases.

The phasor diagrams in Fig. 6.36 provide another view of the frequency variation for a series resonant network. Here we take $\angle \underline{V} = 0$ and let $\underline{V}_X = \underline{V}_L + \underline{V}_C$ to construct $\underline{V}_R + \underline{V}_X = \underline{V}$. Regardless of ω, $\underline{I}$ always lags $\underline{V}_L$ by 90°, leads $\underline{V}_C$ by 90°, and is in phase with $\underline{V}_R$. But the network is capacitive below resonance ($\omega < \omega_0$), so the diagram on the left has $|\underline{V}_L| < |\underline{V}_C|$ and $\underline{I}$ leads $\underline{V}$. Conversely, the network is inductive above resonance ($\omega > \omega_0$), so the diagram on the right has $|\underline{V}_L| > |\underline{V}_C|$ and $\underline{I}$ lags $\underline{V}$. The middle diagram depicts the resonance situation ($\omega = \omega_0$) when $|\underline{V}_L| = |\underline{V}_C|$, so $\underline{V}_X = 0$, $\underline{V}_R = \underline{V}$, and $\underline{I}$ is in phase with $\underline{V}$.

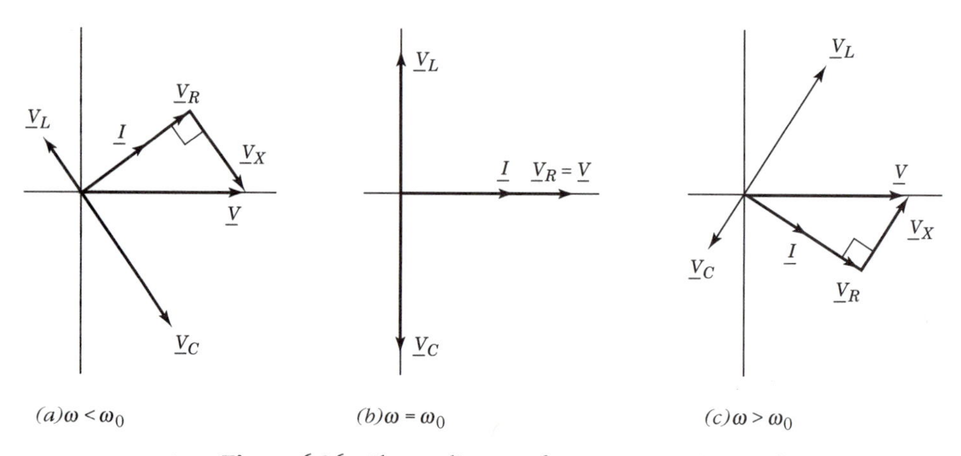

Figure 6.36 Phasor diagrams for a series *RLC* network.

Although $\underline{V}_X = 0$ at resonance, the individual reactive voltages $v_L(t)$ and $v_C(t)$ may actually have large amplitudes — perhaps even larger than the amplitude of the terminal voltage! To explore this possibility, we note that the terminal current and voltage at ω_0 are related by $\underline{I} = \underline{V}/Z(j\omega_0) = \underline{V}/R$. Then, with $\omega = \omega_0 = 1/\sqrt{LC}$,

$$\underline{V}_L = j\omega_0 L\underline{I} = \frac{j}{R}\sqrt{\frac{L}{C}}\underline{V} \qquad \underline{V}_C = \frac{\underline{I}}{j\omega_0 C} = -\frac{j}{R}\sqrt{\frac{L}{C}}\underline{V}$$

Now we define the **quality factor** of a series *RLC* network to be

$$Q_{ser} \triangleq \frac{\omega_0 L}{R} = \frac{1}{\omega_0 CR} = \frac{1}{R}\sqrt{\frac{L}{C}} \tag{6.46}$$

Thus, at resonance.

$$\underline{V}_L = jQ_{ser}\underline{V} \qquad \underline{V}_C = -jQ_{ser}\underline{V} \tag{6.47}$$

If $Q_{ser} > 1$, then the amplitudes $|\underline{V}_L|$ and $|\underline{V}_C|$ will exceed $|\underline{V}|$ — an effect known as **resonant voltage rise**.

When first encountered, this effect seems paradoxical. But Fig. 6.36*b* shows that $|\underline{V}_L|$ and $|\underline{V}_C|$ may be arbitrarily large compared to $|\underline{V}|$ since $\underline{V}_X = \underline{V}_L + \underline{V}_C = 0$. The physical explanation of resonant voltage rise relates to oscillating energy storage, in that $|Z(\omega_0)|$ is minimum and the large current $|\underline{I}|$ transfers a large amount of energy back and forth between the inductor and capacitor. In fact, an alternative, more general expression for quality factor is

$$Q = 2\pi \frac{\text{Maximum energy stored}}{\text{Energy lost per cycle}}$$

Since the lost energy is dissipated as heat, a high-Q series circuit must have small resistance, as seen explicitly in Eq. (6.46).

Analogous resonance effects occur in certain low-loss mechanical systems. For instance, when you "pump" a playground swing at its resonant frequency, the swing goes much higher than when you pump it at some other frequency. And as the swing goes back and forth, the stored energy changes from maximum potential energy when the swing changes direction at its highest point to maximum kinetic energy when the swing has greatest velocity at its lowest point.

Resonant voltage rise in electrical circuits is sometimes exploited to obtain voltage amplitudes much greater than the available source voltage. (It may also deliver a shocking surprise when you're working with a series circuit that you didn't know was resonant!) Equation (6.46) shows that we generally want large L and small R to get $Q_{ser} \gg 1$. However, real inductors always include some winding resistance, and the resulting value of $\omega_0 L/R$ may be less than required to achieve the desired Q_{ser} at low resonant frequencies.

Example 6.15 ***Series Resonance Design***

Suppose you want to design a circuit that produces a 1−kV sinusoid, given a source with $|\underline{V}| = 100$ V and $\omega = 5000$. You decide to take advantage of series resonance with $\omega_0 = 5000$ and $Q_{ser} = 10$, so $|\underline{V}_C| = Q_{ser}|\underline{V}| = 1000$ V.

Available for this purpose is a 0.4–henry coil, and you determine the series capacitance and resistance from Eqs. (6.45) and (6.46) as follows:

$$C = 1/\omega_0^2 L = 10^{-7} = 0.1\ \mu\text{F} \qquad R = \omega_0 L/Q_{ser} = 200\ \Omega$$

But when you build and test the circuit, you find that $|\underline{V}_C| = 800$ V, implying that $Q_{ser} = 8$ instead of 10.

After pondering this result, you conclude that the coil must have significant winding resistance R_w that decreases the quality factor. Since the total series resistance is $R_w + 200\ \Omega$, and since $Q_{ser} = 8$, you infer that

$$R_w + 200 = \omega_0 L/Q_{ser} = 250\ \Omega \quad \Rightarrow \quad R_w = 50\ \Omega$$

Accordingly, you replace the 200-Ω resistor with 150 Ω, so $R_w + 150 = 200\ \Omega$, $Q_{ser} = 10$, and $|\underline{V}_C| = 1000$ V as required.

Exercise 6.17

Evaluate ω_0 and Q_{ser} for the network in Fig. 6.35*a* with $R = 5\ \Omega$, $L = 0.1$ mH, and $C = 0.25\ \mu$F. Then construct a phasor diagram showing all the voltages when $\omega = 0.8\omega_0$ and $\underline{I} = 2\text{ A }\angle 0°$. Use your diagram to calculate $\underline{V}$ in polar form.

Parallel Resonance

The dual of a series *RLC* network is the parallel *RCL* network in Fig. 6.37*a*. Its admittance is

$$Y(j\omega) = G + j\omega C - j/\omega L = G + jB(\omega) \tag{6.48a}$$

where $G = 1/R$ and

$$B(\omega) = \omega C - 1/\omega L \tag{6.48b}$$

The susceptance $B(\omega)$ changes sign as frequency increases, and we define the **parallel resonance condition** by

$$B(\omega_0) = \text{Im}[Y(j\omega_0)] = 0 \tag{6.49}$$

But our previous expression for the resonant frequency still holds, since $\omega_0 C - 1/\omega_0 L = 0$ requires

$$\omega_0 = 1/\sqrt{LC}$$

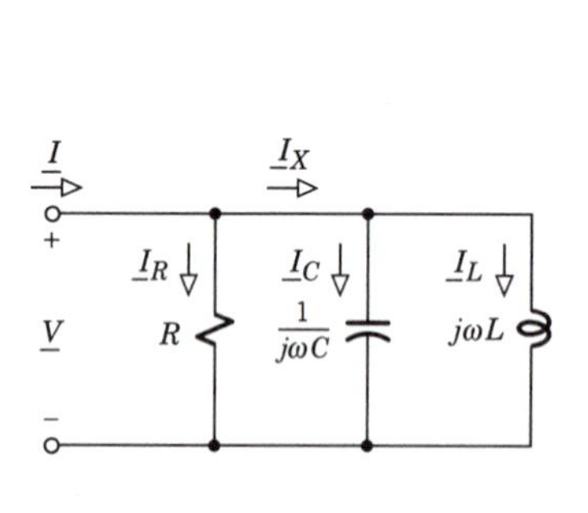

(*a*) Parallel *RCL* network

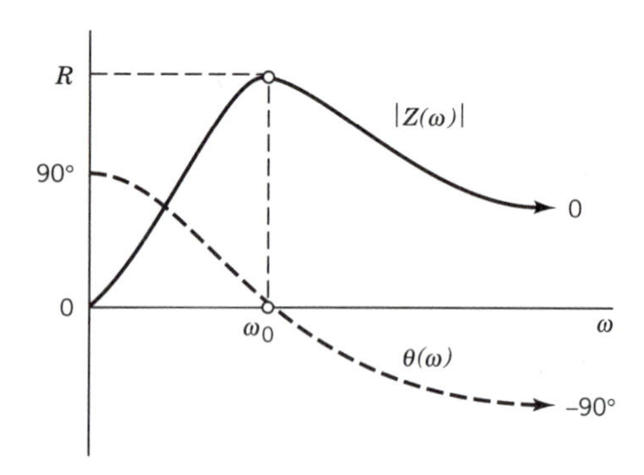

(*b*) $|Z(\omega)|$ and $\theta(\omega)$ versus ω

Figure 6.37

Moreover, the parallel network also appears to be purely resistive at resonance since $Y(j\omega_0) = G = 1/R$.

To bring out the difference between series and parallel resonance, we'll calculate the magnitude and angle of the impedance of the parallel network. Substituting Eq. (6.48a) into $Z(j\omega) = 1/Y(j\omega)$ yields

$$|Z(\omega)| = \frac{1}{\sqrt{G^2 + (\omega C - 1/\omega L)^2}} \qquad \theta(\omega) = -\tan^{-1}\frac{\omega L - 1/\omega C}{G}$$

Figure 6.37*b* plots the frequency variation of $|Z(\omega)|$ and $\theta(\omega)$. We see that $|Z(\omega)|$ has a unique *maximum* at ω_0, and that $\theta(\omega)$ goes from $+90°$ to $-90°$ as ω increases — just the opposite of the series case back in Fig. 6.35*b*.

A related difference between parallel and series resonance is that parallel networks may exhibit **resonant current rise**. Although $\underline{I}_X = \underline{I}_C + \underline{I}_L = 0$ when $\omega = \omega_0$, the reactive currents are

$$\underline{I}_C = jQ_{par}\underline{I} \qquad \underline{I}_L = -jQ_{par}\underline{I} \tag{6.50}$$

Equation (6.50) is the dual of Eq. (6.47), and the quality factor Q_{par} is given by the dual of Eq. (6.46), namely,

$$Q_{par} \triangleq \omega_0 CR = \frac{R}{\omega_0 L} = R\sqrt{\frac{C}{L}} \tag{6.51}$$

Thus, a high-Q parallel resonant network must have *large* resistance to reduce energy loss, whereas a high-Q series resonant network has small resistance.

Lastly, to take account of the winding resistance R_w associated with a real inductor, consider the modified network diagrammed in Fig. 6.38*a*. Here, R_w is the only resistance, and the admittance of the network is

$$Y(j\omega) = j\omega C + \frac{1}{R_w + j\omega L} = \frac{j\omega CR_w - \omega^2 LC + 1}{R_w + j\omega L}$$

Since R_w should be small, we'll simplify the analysis by assuming that $R_w + j\omega L \approx j\omega L$. Then

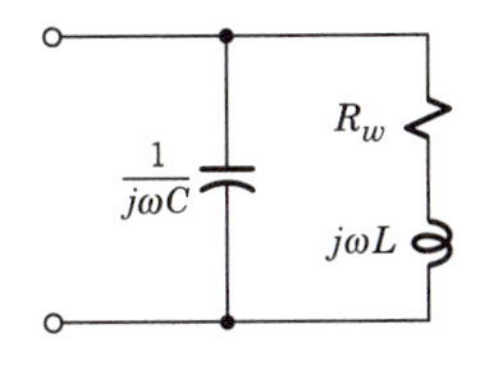

(*a*) Parallel network with winding resistance

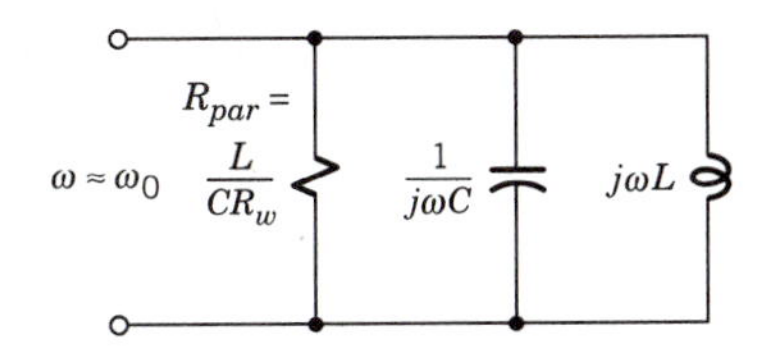

(*b*) Equivalent network for $\omega \approx \omega_0$

Figure 6.38

$$Y(j\omega) \approx (CR_w/L) + j(\omega C - 1/\omega L)$$

so $\text{Im}[Y(j\omega_0)] = 0$ at $\omega_0 \approx 1/\sqrt{LC}$. Furthermore, comparison with Eq. (6.48a) reveals that R_w has essentially the same effect as a parallel resistor

$$R_{par} = L/CR_w \tag{6.52}$$

which holds for

$$R_w \ll \omega_0 L$$

Figure 6.38*b* shows the approximate equivalent network for frequencies near ω_0.

Example 6.16 ***Parallel Resonance Calculations***

The circuit in Fig. 6.39*a* is driven at $\omega = 5000$ by a voltage source with internal resistance, and the inductor has winding resistance $R_w = 2.5\ \Omega$. The capacitance is to be designed for parallel resonance at the source frequency, and the resulting amplitudes and angles of v, i_1, i_C, and i_2 are to be evaluated.

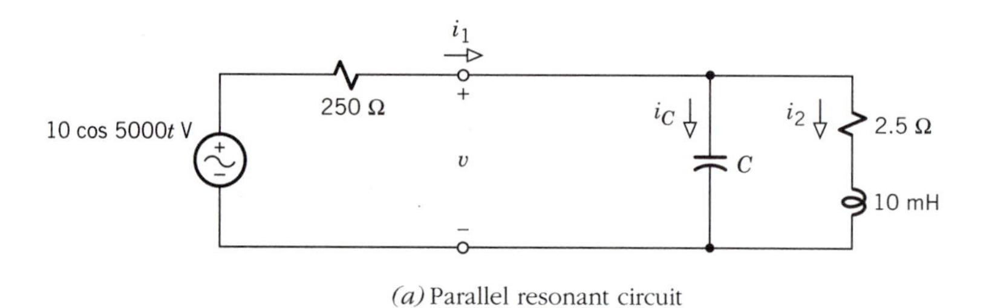

(a) Parallel resonant circuit

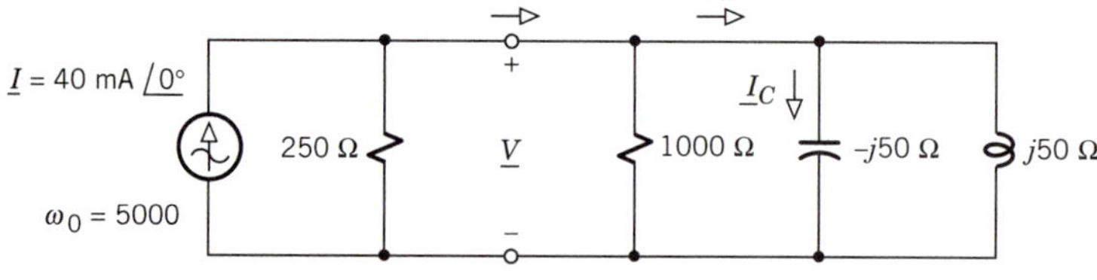

(b) Equivalent frequency-domain diagram

Figure 6.39

We first observe that if $\omega_0 = 5000$, then $\omega_0 L = 50\ \Omega$ and $R_w = 2.5\ \Omega \ll \omega_0 L$. Therefore, the network connected to the source will indeed act like a parallel *RCL* network. Accordingly, the required capacitance value is $C = 1/\omega_0^2 L = = 4\ \mu\text{F}$.

For the remaining analysis, we convert the voltage source into its Norton equivalent, and we use Eq. (6.52) to convert R_w into

$$R_{par} = L/CR_w = 1000\ \Omega$$

We thereby obtain the frequency-domain diagram in Fig. 6.39*b*. The phasor $\underline{I}_2$ does not exist in this diagram because R_w has been converted and removed from the inductance branch. However, the phasors $\underline{V}$, $\underline{I}_1$, and $\underline{I}_C$ still represent the branch variables in the time-domain diagram.

We now see that the equivalent parallel resistance is $R = 250\|1000 = 200\ \Omega$, and we know that $\underline{I}_X = 0$ at resonance. Thus, with $\underline{I} = 40\text{ mA }\angle 0°$,

$$\underline{V} = R\underline{I} = 8\text{ V }\angle 0° \qquad \underline{I}_1 = \underline{V}/1000 = 8\text{ mA }\angle 0°$$

Next, we calculate $\underline{I}_C$ using

$$Q_{par} = R/\omega_0 L = 200/50 = 4 \qquad \underline{I}_C = jQ_{par}\underline{I} = 160\text{ mA }\angle 90°$$

The time-domain diagram then indicates that

$$\underline{I}_2 = \underline{I}_1 - \underline{I}_C = 8 - j160 = 160.2\text{ mA }\angle -87.1°$$

Hence, both $\underline{I}_C$ and $\underline{I}_2$ exhibit resonant current rise.

Exercise 6.18

Derive Eq. (6.50) from Fig. 6.37*a* with $\omega = \omega_0$.

6.5 AC SUPERPOSITION

Consider a linear circuit operating in the steady state with two or more independent sinusoidal sources. If all sources have the *same* frequency, then superposition provides an alternative to other analysis techniques such as node or mesh equations. You simply find the phasor response of interest produced by each source acting alone, and then you add the phasors to obtain the total response.

More importantly, superposition is the only practical analysis method when a circuit contains sources at *different* frequencies. The superposition principle thus plays a key role in audio and communications engineering because many signal waveforms may be treated as a sum of two or more sinusoids at different frequencies.

But you cannot directly add phasors representing sinusoids with different frequencies. Instead, you must return to the time domain and add the sinusoidal components. Consequently, when two or more frequencies are involved, the superposition procedure goes as follows:

1. Draw the frequency-domain diagram for one of the source frequencies, suppressing all independent sources at different frequencies. Analyze

this diagram to find the corresponding phasor response and convert it to a sinusoidal function.

2. Repeat the first step for each of the other source frequencies.
3. Add the sinusoidal components at different frequencies to obtain the total response produced by all sources acting together.

This procedure remains valid when dc sources are included, since a dc voltage or current may be viewed as a sinusoid with $\omega = 0$.

Example 6.17 ***AC Superposition Calculations***

The two sources in Fig. 6.40*a* operate at different frequencies. We'll apply superposition to find the steady-state voltage $v_C(t)$. For this purpose we'll let $v_C(t) = v_{C-1}(t) + v_{C-2}(t)$, where $v_{C-1}(t)$ is the component at one frequency and $v_{C-2}(t)$ is the component at the other frequency.

Step 1. With the voltage source active, we have $\omega = 5$ and Fig. 6.40*b* is the corresponding frequency-domain diagram. Although a voltage divider might be used here, a node equation is simpler, so we write

$$\left(\frac{1}{50} + \frac{j}{10} - \frac{j}{20}\right) \underline{V}_{C-1} = \frac{60}{j20}$$

Clearing the fractions then yields $\underline{V}_{C-1} = -j300/(2 + j5) = 55.7 \text{ V} \angle -158.2°$.

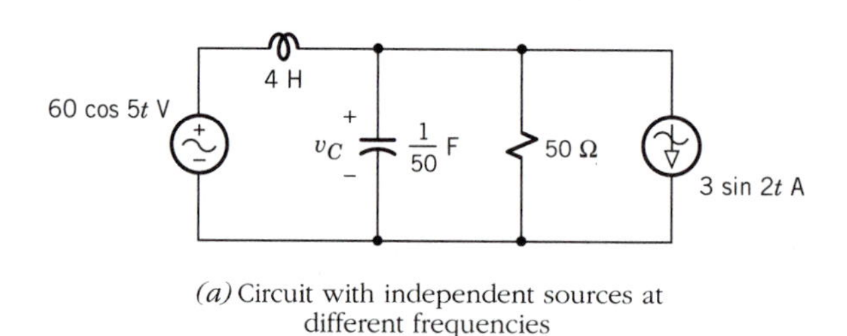

(a) Circuit with independent sources at different frequencies

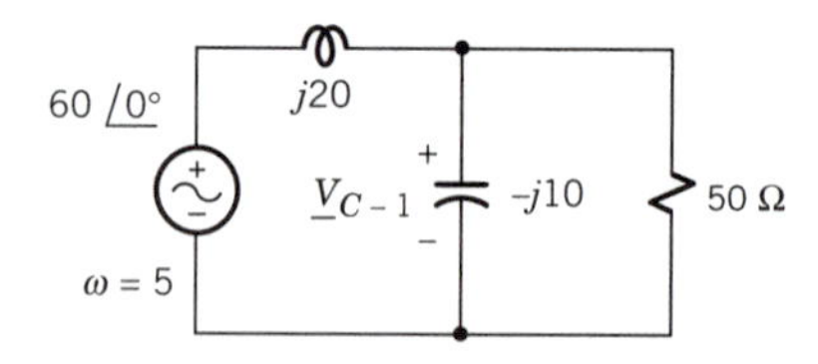

(b) Frequency-domain diagram with the voltage source active

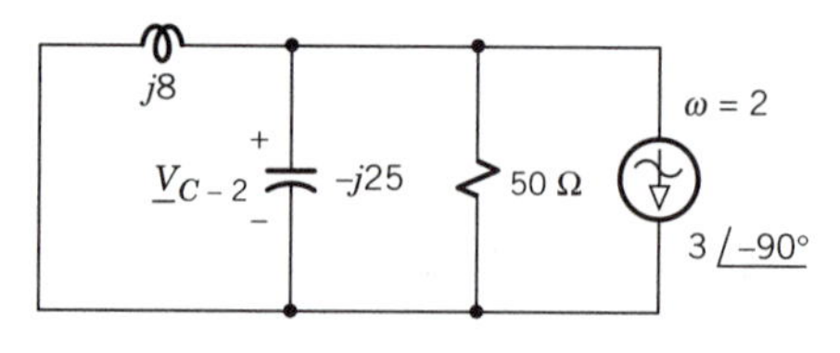

(c) Frequency-domain diagram with the current source active

Figure 6.40

Step 2. With the current source active, we have $\omega = 2$ and Fig. 6.40*c* is the corresponding frequency-domain diagram. Note that the source current phasor is $3\,\underline{/-90^\circ}$ since the time-domain current is $3 \sin 2t = 3 \cos (2t - 90^\circ)$. The node equation for this case is

$$\left(\frac{1}{50} + \frac{j}{25} - \frac{j}{8}\right) \underline{V}_{C-2} = -3\,\underline{/-90^\circ} = j3$$

from which $\underline{V}_{C-2} = j600/(4 - j17) = 34.4 \text{ V } \underline{/166.8^\circ}$.

Step 3. After converting the two phasors to sinusoids, we sum them to obtain the final result

$$v_C(t) = 55.7 \cos (5t - 158.2^\circ) + 34.4 \cos (2t + 166.8^\circ) \text{ V}$$

Exercise 6.19

Consider the network in Fig. 6.26 (p. 276). Find the current i when $i_s = 2 \cos 5t$ A (as in Exercise 6.13) and a voltage source is connected to the terminals so that $v = 52 \cos 10t$ V.

6.6 IMPEDANCE BRIDGES†

Impedance bridges exploit the same balancing strategy as a Wheatstone bridge. However, an impedance bridge must contain both resistors and reactive elements because an unknown impedance may have a resistive and a reactive component. Specific bridge configurations are used to measure inductance with winding resistance and capacitance with or without leakage resistance.

Figure 6.41 shows the general structure of an impedance bridge in the frequency domain, with Z_u being the unknown. The other three impedances

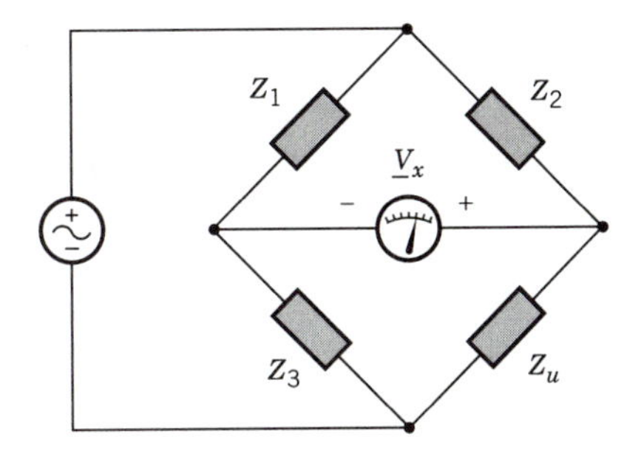

Figure 6.41 Impedance bridge.

are known, and two of them must be adjustable. The bridge is balanced by varying the adjustable impedances until $\underline{V}_x = 0$. This voltage null may be sensed by an ac voltmeter, as shown, or by displaying $v_x(t)$ on an oscilloscope.

Under balanced conditions, the unknown impedance is related to the other three by the impedance version of the Wheatstone bridge equation, namely,

$$Z_u = Z_2 Z_3 / Z_1 \tag{6.53}$$

But since Eq. (6.53) involves complex quantities, it actually represents *two* equalities. An impedance bridge thereby allows for the measurement of resistance and reactance simultaneously. We'll demonstrate this capability by examining two particular types of bridges.

The **parallel-comparison bridge** in Fig. 6.42*a* is intended for parallel *RC* unknowns. The parallel structure suggests analysis via the admittance Y_u and Y_3 rather than impedances, so we first rewrite Eq. (6.53) as

$$Y_u = 1/Z_u = Z_1/Z_2Z_3 = (Z_1/Z_2)Y_3$$

Inserting the corresponding element expressions yields

$$\frac{1}{R_u} + j\omega C_u = \frac{R_1}{R_2}\left(\frac{1}{R_3} + j\omega C_3\right)$$

We then equate the real and imaginary parts on both sides to obtain

$$R_u = (R_2/R_1)R_3 \qquad C_u = (R_1/R_2)C_3 \tag{6.54}$$

Note the ω does not appear in Eq. (6.54), which means that we do not need

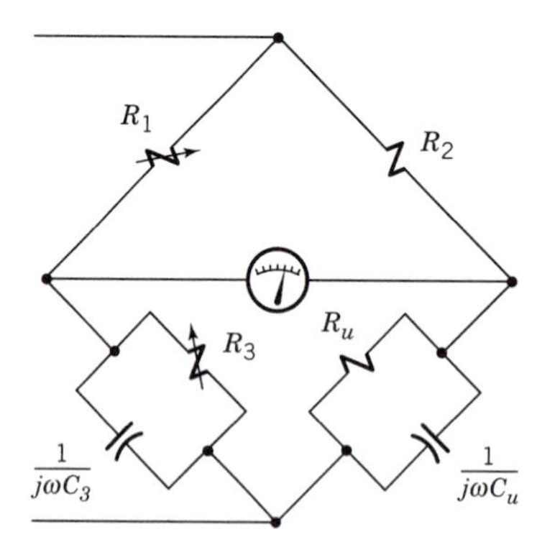

(*a*) Parallel-comparison bridge with parallel *RC* unknown

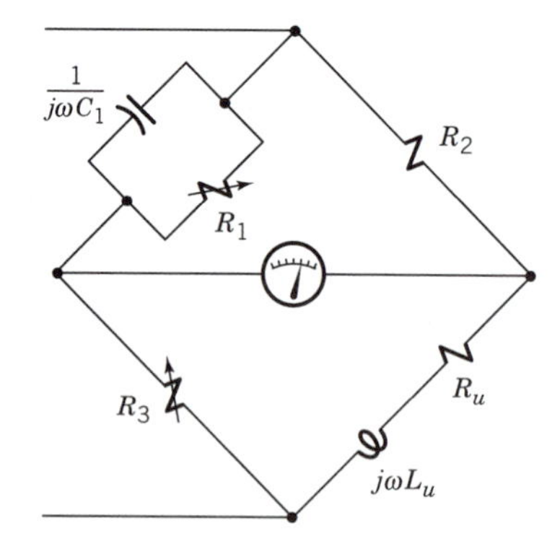

(*b*) Maxwell bridge with series *RL* unknown

Figure 6.42

the value of ω. Also note that capacitance without parallel resistance can be measured by taking $R_3 = \infty$ (an open circuit) to agree with $R_u = \infty$

Some parallel-comparison bridges include a dial for reading the **dissipation factor** of capacitors with leakage. When R_u represents parallel leakage resistance, the dissipation factor is defined by

$$D \triangleq |Z_C|/|Z_R| = 1/\omega C_u R_u \tag{6.55}$$

A good capacitor should have $D \ll 1$ so that $R_u \gg 1/\omega C$. If an inductor is put in parallel with R_u and C_u to form a resonant circuit with $\omega_0 = \omega$, then the quality factor will be $Q_{par} = \omega C_u R_u = 1/D$.

The **Maxwell bridge** is Fig. 6.42*b* is intended for series *RL* unknowns, but the circuit employs a parallel *RC* balancing impedance to avoid the problems of obtaining calibrated inductance. The bridge equation for this case becomes

$$R_u + j\omega L_u = R_2 R_3 (1 + j\omega C_1 R_1)/R_1$$

from which

$$R_u = (R_2/R_1)R_3 \qquad L_u = C_1 R_2 R_3 \tag{6.56}$$

Again, we see the desirable feature that ω does not appear in these relations.

Since R_u may represent the winding resistance of a coil with inductance L_u, some Maxwell bridges include a dial for reading the coil's **quality factor** defined by

$$Q_{coil} \triangleq |Z_L|/|Z_R| = \omega L_u/R_u \tag{6.57}$$

This definition corresponds to Q_{ser} for a resonant network formed by connecting the coil in series with a capacitor such that $\omega_0 = \omega$.

Table 6.3 lists information about six different types of impedance bridges, including the two we have discussed. Several of these configurations together with a Wheatstone bridge are built into a **universal impedance bridge**. This flexible measuring instrument usually contains a dc source, a 1−kHz ac source, and terminals for external sources and meters or oscilloscopes.

TABLE 6.3 Impedance bridges with $Z_2 = R_2$

Type	Z_1	Z_3	Z_u
Parallel-comparison	R_1	$R_3 \| (1/j\omega C_3)$	Capacitive
Series-comparison	R_1	$R_3 + 1/j\omega C_3$	Capacitive
Schering	$R_1 \| (1/j\omega C_1)$	$1/j\omega C_3$	Capacitive
Maxwell	$R_1 \| (1/j\omega C_1)$	R_3	Inductive
Hay	$R_1 + 1/j\omega C_1$	R_3	Inductive
Owen	$1/j\omega C_1$	$R_3 + 1/j\omega C_3$	Inductive

Exercise 6.20

Suppose Z_u contains inductance L_u and capacitance C_u. Explain why both reactive elements cannot be determined from a single-frequency measurement.

Exercise 6.21

Show that Eq. (6.54) also holds for a series-comparison bridge when $Z_u = R_u + 1/j\omega C_u$.

PROBLEMS

Section 6.1 Phasors and the AC Steady State

6.1* Let x and α be positive quantities, and let

$$\sqrt{1+x^2} = y \quad \tan^{-1} x = \phi \quad \cos\alpha = a \quad \sin\alpha = b$$

Obtain rectangular expressions for each of the following quantities. Then rework the calculations in polar form to obtain polar expressions.
(a) $-(3\,\angle\alpha)^*$ (b) $(1 - jx)(2\,\angle 30°)$ (c) $j/(1 - jx)^*$

6.2 Do Problem 6.1 for
(a) $j(6\,\angle{-\alpha})^*$ (b) $91 + jx)(2\,\angle{-60°})$
(c) $-1/(-1 - jx)^*$

6.3 Do Problem 6.1 for
(a) $-(1 - jx)^*$ (b) $(-5 + j12)(2\,\angle\phi)$ (c) $j/(-1 + jx)$

6.4 Do Problem 6.1 for
(a) $j(-1 + jx)^*$ (b) $(15 - j8)(2\,\angle{-\phi})$
(c) $-1/(1 - jx)$

6.5 Do Problem 6.1 for
(a) $(-5\,\angle\alpha)^*$ (b) $(1 - jx)(\sqrt{2}\,\angle{-45°})$ (c) $(1 + jx)^{-2}$

6.6 Do Problem 6.1 for
(a) $j(4\,\angle\alpha)^*$ (b) $(-1 + jx)(\sqrt{2}\,\angle 45°)$ (c) $(1 - jx)^{-2}$

6.7 Carry out the division $(1 + jx)/(1 + jy)$ in both rectangular and polar form to show that

$$\tan^{-1}[(x - y)/(1 + xy)] = \tan^{-1} x - \tan^{-1} y$$

6.8 Show from Euler's theorem that

$$\cos\alpha = (e^{j\alpha} + e^{-j\alpha})/2 \qquad \sin\alpha = (e^{j\alpha} - e^{-j\alpha})/2j$$

6.9 Use the relations in Problem 6.8 to derive the trigonometric identity

$$\cos\alpha\cos\beta = \tfrac{1}{2}\cos(\alpha - \beta) + \tfrac{1}{2}\cos(\alpha + \beta)$$

6.10 Use the relations in Problem 6.8 to derive the trigonometric identity

$$\sin\alpha\sin\beta = \tfrac{1}{2}\cos(\alpha - \beta) - \tfrac{1}{2}\cos(\alpha + \beta)$$

6.11 Use the relations in Problem 6.8 to derive the trigonometric identity

$$\sin\alpha\cos\beta = \tfrac{1}{2}\sin(\alpha - \beta) + \tfrac{1}{2}\sin(\alpha + \beta)$$

6.12* Apply phasor analysis to evaluate the sum

$$v = 150\cos(\omega t + 60°) + 55\cos(\omega t - 90°)\text{ V}$$

6.13 Apply phasor analysis to evaluate the sum

$$i = 16\cos(\omega t + 45°) + 30\cos(\omega t - 45°)\text{ mA}$$

6.14 Apply phasor analysis to evaluate the sum

$$v = 17\cos(\omega t - 45°) - 17\cos\omega t\text{ V}$$

6.15 Apply phasor analysis to evaluate the sum

$$i = 2\cos(\omega t + 20°) - 2\cos(\omega t - 40°)\text{ A}$$

6.16 Apply phasor analysis to evaluate the sum

$$v = 50\cos(\omega t + 36.9°) - 15\cos\omega t + 90\cos(\omega t - 90°)\text{ V}$$

6.17 Apply phasor analysis to evaluate the sum

$$i = 2.6\cos(\omega t - 22.6°) - 9\cos(\omega t - 90°) - 12\cos(\omega t - 66.4°)\text{ A}$$

Section 6.2 Impedance and Admittance

6.18* At a certain frequency, a 100–Ω resistor in parallel with inductance L has the same admittance as a 20–Ω resistor in series with a 2–H inductor. What are the values of ω and L?

6.19 At a certain frequency, a 9–Ω resistor in series with inductance L has the same impedance as a 10–Ω resistor in parallel with a 5–H inductor. What are the values of ω and L?

6.20 At a certain frequency, a 25–Ω resistor in parallel with capacitance C has the same impedance as a 20–Ω resistor in series with a 0.01–F capacitor. What are the values of ω and C?

6.21 At a certain frequency, a 2–Ω resistor in series with capacitance C has the same impedance as a 4–Ω resistor in parallel with a 0.05–F capacitor. What are the values of ω and C?

6.22* Let the network in Figure P6.22 have $R_1 = 5$ Ω, $L = 10$ mH, $R_2 = 0$, $C = 200$ μF, and $i = 4 \cos 500t$ A. Calculate Z and v. Then find i_1 and i_2, and confirm that $\underline{I}_1 + \underline{I}_2 = \underline{I}$.

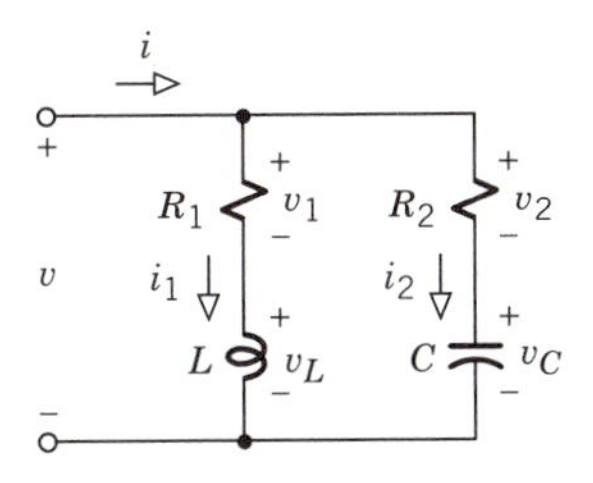

Figure P6.22

6.23 Do Problem 6.22 with $R_1 = 0$, $L = 20$ mH, $R_2 = 150$ Ω, $C = 0.5$ μF, and $i = 30 \cos 10{,}000t$ mA.

6.24 Let the network in Fig. P6.24 have $C = 50$ μF, $R_1 = 20$ Ω, $R_2 = 4$ Ω, $L = 4$ mH, and $v = 10 \cos 2000t$ V. Calculate Z and i. Then find v_C and v_1, and confirm that $\underline{V}_C + \underline{V}_1 = \underline{V}$.

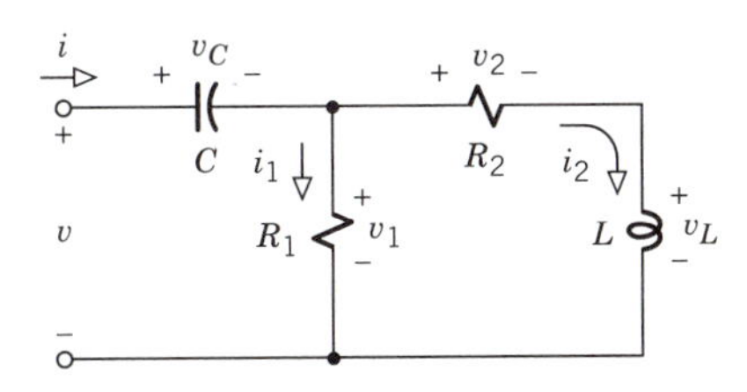

Figure P6.24

6.25 Do Problem 6.24 with $C = 0.05$ μF, $R_1 = 3$ kΩ, $R_2 = 1$ kΩ, L = 40 mH, and $v = 26 \cos 50{,}000t$ V.

6.26 Let the network in Fig. P6.26 have $L = 9$ mH, $R_1 = 6$ Ω, $R_2 = 30$ Ω, $C = 50$ μF, and $i = 2 \cos 2000t$ A. Calculate Z and v. Then find i_L and i_1, and confirm that $\underline{I}_L + \underline{I}_1 = \underline{I}$.

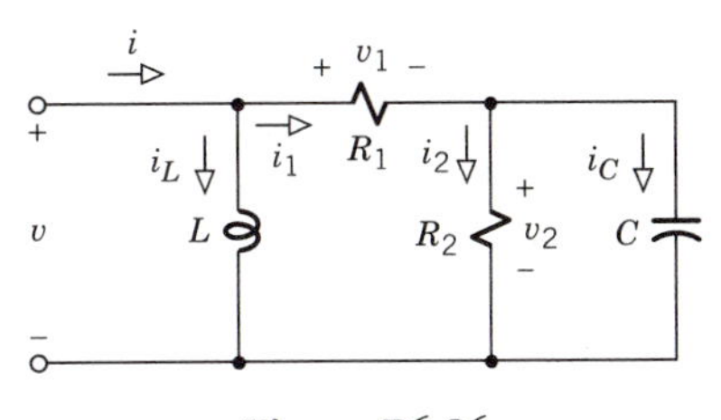

Figure P6.26

6.27 Do Problems 6.26 with $L = 0.8$ H, $R_1 = 1$ kΩ, $R_2 = 10$ kΩ, $C = 0.02$ μF, and $i = 5 \cos 10{,}000t$ mA.

6.28* Obtain expressions for $R(\omega)$ and $X(\omega)$ when the network in Fig. P6.22 has $R_1 = 5$ Ω, $L = 5$ H, $R_2 = 0$, and $C = 1/10$ F.

6.29 Obtain expressions for $G(\omega)$ and $B(\omega)$ when the network in Fig. P6.22 has $R_1 = 0$, $L = 2$ H, $R_2 = 10$ Ω, and $C = 1/100$ F.

6.30 Obtain expressions for $R(\omega)$ and $X(\omega)$ when the network in Fig. P6.24 has $C = 1/10$ F, $R_1 = 6$ Ω, $R_2 = 4$ Ω, and $L = 2$ H.

6.31 Obtain expressions for $G(\omega)$ and $B(\omega)$ when the network in Fig. P6.26 has $L = 4$ H, $R_1 = 30$ Ω, $R_2 = 10$ Ω, and $C = 1/10$ F.

6.32* The network in Fig. P6.32 has $R = 50$ Ω but the reactive element is unknown. Measurements with $|\underline{V}|$ held constant reveal that $|\underline{V}_R| = ½\,|\underline{V}|$ when $\omega = 200$ and that $|\underline{V}_X|$ decreases when ω increases. Determine the type and value of the reactive element.

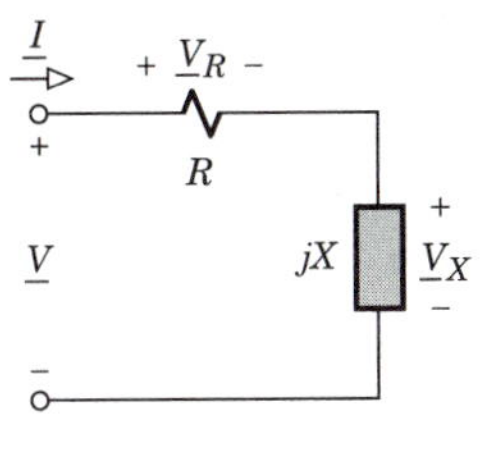

Figure P6.32

6.33 The network in Fig. P6.32 has $R = 20$ Ω, but the reactive element is unknown. Measurements with $|\underline{V}|$ held constant reveal that $|\underline{V}_X| = ½\,|\underline{V}|$ when $\omega = 100$ and that $|\underline{V}_X|$ increases when ω increases. Determine the type and value of the reactive element.

6.34 The resistance and the reactance of the network in Fig. P6.32 are unknown. Measurements with $|\underline{V}| = 20$ V reveal that $|\underline{I}| = 4$ A when $\omega = 100$ and $|\underline{I}| = 2$ A when $\omega = 400$. Determine the type and value of the reactive element and the value of the resistance.

6.35 The resistance and the reactance of the network in Fig. P6.32 are unknown. Measurements with $|\underline{V}| = 60$ V reveal that $|\underline{I}| = 1$ A when $\omega = 1000$ and $|\underline{I}| =$

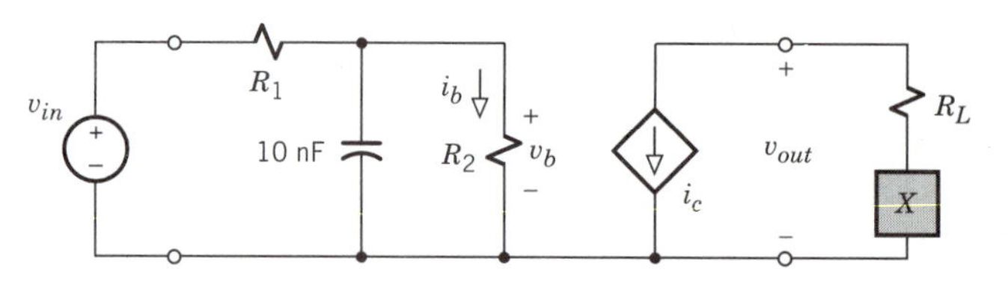

Figure P6.36

1.5 A when $\omega = 2000$. Determine the type and value of the reactive element and the value of the resistance.

6.36 Figure P6.36 is the model of a transistor amplifier with $R_1 = 1.5\ \text{k}\Omega$, $R_2 = 1\ \text{k}\Omega$, $R_L = 2\ \text{k}\Omega$, and $i_c = 100 i_b$. The input voltage is sinusoidal at an arbitrary frequency ω. The output load includes an unspecified energy-storage element X whose impedance is $Z_X = jX(\omega)$. (a) Find $\underline{I}_c / \underline{V}_{in}$ and show that the phasor voltage amplification has the form

$$\underline{V}_{out}/\underline{V}_{in} = -K[1 + jaX(\omega)]/(1 + jb\omega)$$

where K, a, and b are known constants. (b) What element (type and value) should be used for X so that $\underline{V}_{out}/\underline{V}_{in} = K$, independent of frequency.

6.37 Do Problem 6.36 with $R_1 = 1\ \text{k}\Omega$, $R_2 = 4\text{k}\Omega$, $R_L = 500\ \Omega$, and $i_c = 0.1\ \text{S} \times v_b$.

Section 6.3 AC Circuit Analysis

(See also PSpice problems B.15—B.18.)

6.38* Assume a value for $\underline{I}_L$ in Fig. P6.38 to calculate $\underline{I}_L/\underline{I}$ and $\underline{V}/\underline{I}$ in polar form when $R_1 = 19\ \Omega$, $R_2 = 20\ \Omega$, and $1/\omega C = \omega L = 10\ \Omega$.

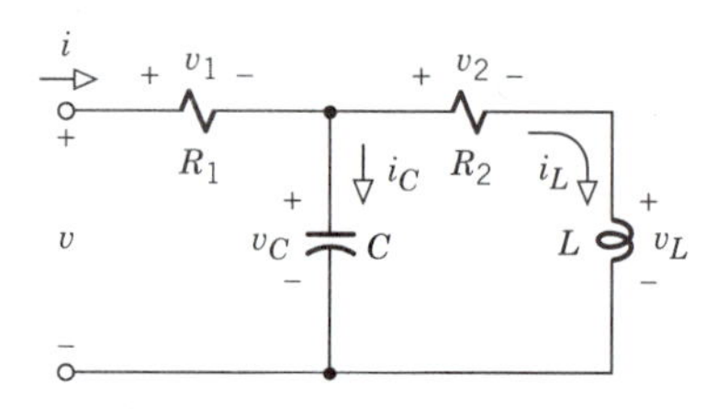

Figure P6.38

6.39 Assume a value for $\underline{V}_L$ in Fig. P6.24 to calculate $\underline{V}_1/\underline{V}$ and $\underline{V}_L/\underline{I}$ in polar form when $R_1 = R_2 = 4\ \Omega$, $1/\omega C = 2\ \Omega$, and $\omega L = 8\ \Omega$.

6.40 Assume a value for $\underline{I}_2$ in Fig. P6.22 to calculate $\underline{V}_C/\underline{V}_L$ and $\underline{V}/\underline{I}$ in polar form when $R_1 = R_2 = 2\ \Omega$ and $\omega L = 1/\omega C = 6\ \Omega$.

6.41 Assume a value for $\underline{V}_R$ in Fig.P6.41 to calculate $\underline{V}_R/\underline{V}$ and $\underline{V}/\underline{I}$ in polar form when $R = 4\ \Omega$ and $\underline{V}_x = \underline{V}_R$.

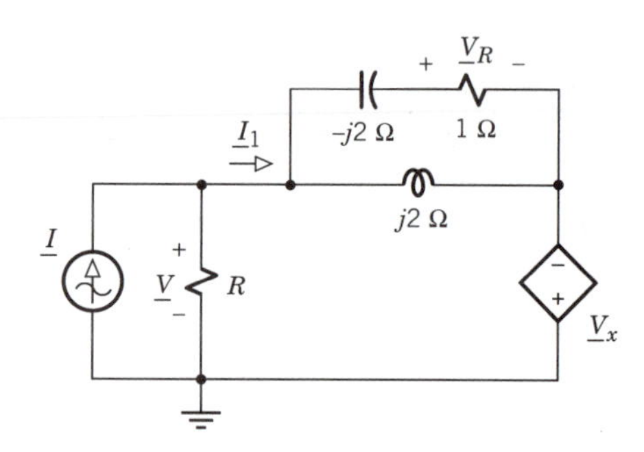

Figure P6.41

6.42 Assume a value for $\underline{I}_1$ in Fig. P6.42 to calculate $\underline{V}_1/\underline{V}$ and $\underline{V}/\underline{I}$ in polar form when $R = 6\ \Omega$ and $\underline{I}_x = 2\underline{I}_1$.

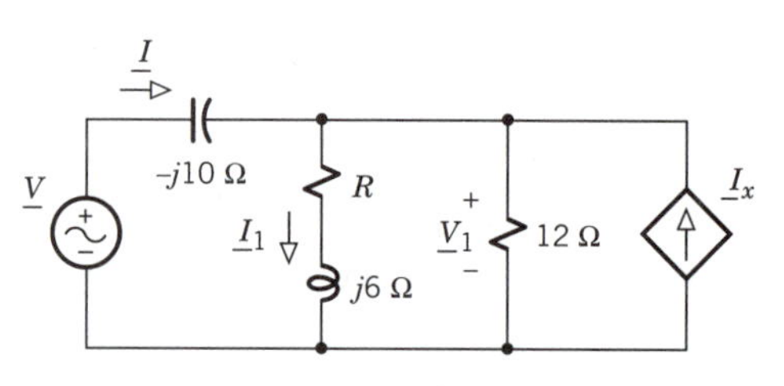

Figure P6.42

6.43 Assume a value for $\underline{V}_2$ in Fig. P6.43 to calculate $\underline{I}_2/\underline{I}$ and $\underline{V}_1/\underline{I}$ in polar form when $R = 5\ \Omega$ and $\underline{V}_x = 2\underline{V}_2$.

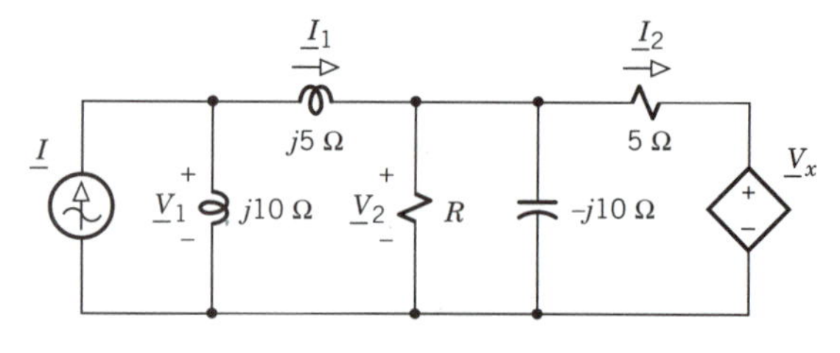

Figure P6.43

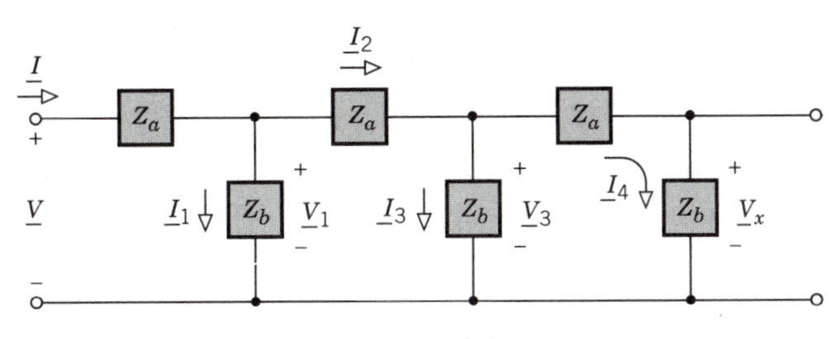

Figure P6.44

6.44 Some oscillator designs avoid the need for inductors by employing the phase-shift network in Fig. P6.44, where $Z_a = 1/j\omega C$ and $Z_b = R$. find $\underline{V}_x$ and $\underline{V}$ when $\underline{I}_4 = 1\ \text{A}\ \angle 0°$, and use your results to determine ω_{osc} at which $\angle(\underline{V}_x/\underline{V}) = -180°$. Then show that $\underline{V}_x/\underline{V} = -1/29$ when $\omega = \omega_{osc}$. Hints: Work with $X = -1/\omega C$ for convenience, and note that $\angle(\underline{V}_x/\underline{V}) = \angle \underline{V}_x - \angle \underline{V}$.

6.45 Do Problem 6.44 with $Z_a = R$ and $Z_b = 1/j\omega C$.

6.46 Apply Norton's theorem to find v_2 in Fig. P6.46 when $C = 1/500$ F, $R = 50\ \Omega$, $L = 4$ H, $v = 80 \cos 5t$ V, and $i = 0.4 \cos 5t$ A.

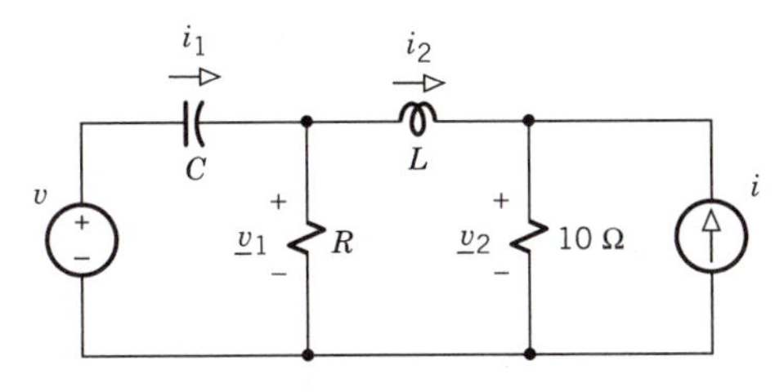

Figure P6.46

6.47 Using source conversions, find i_1 in Fig. P6.47 when $C = 1/200$ F, $L = 0.5$ H, $R = 20\ \Omega$, $i = 2 \cos (20t + 180°)$ A, and $v = 80 \cos 20t$ V.

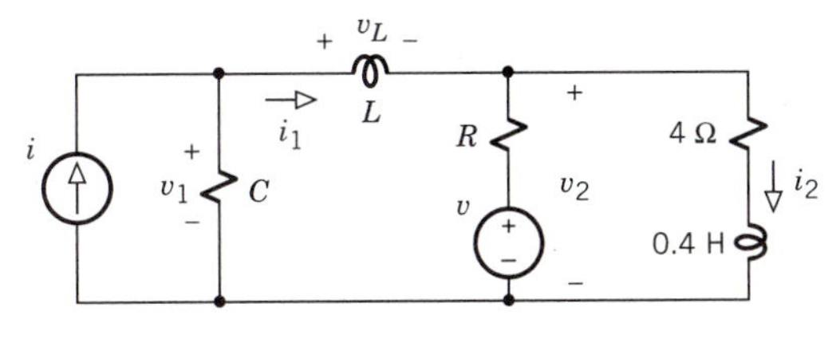

Figure P6.47

6.48 The circuit in Fig. P6.48 has $R_1 = 75\ \Omega$, $L = 9$ H, $v = 75 \cos 4t$ V, and $i = 3 \cos 4t$ A. Apply Thévenin's theorem to determine the values of R_2 and $|\underline{I}_a|$ given that $\angle \underline{I}_a = 45°$.

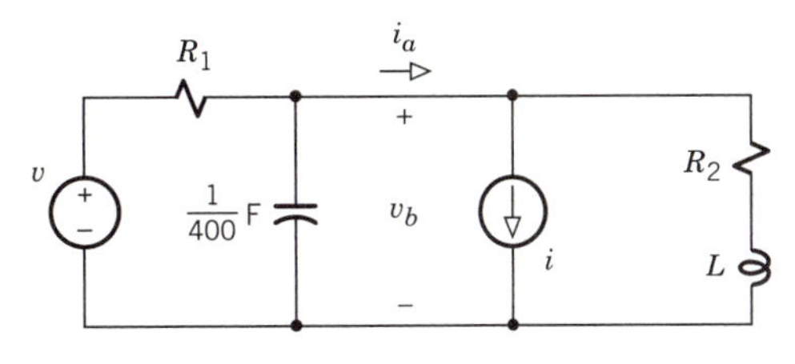

Figure P6.48

6.49 The circuit in Fig. P6.48 has $R_1 = 10\ \Omega$, $R_2 = 4\ \Omega$, $v = 5 \cos 40t$ V, and $i = 0.2 \cos 40t$ A. Apply Norton's theorem to determine the values of L and $|\underline{V}_b|$ given that $\angle \underline{V}_b = 0$ and $L > 0.1$ H.

6.50* Write a single node equation to evaluate $\underline{V}_1$ and $\underline{I}$ in Fig. P6.42 when $R = 2\ \Omega$, $\underline{I}_x = \underline{I}_1/3$, and $\underline{V} = 7\ \text{V}\ \angle 0°$.

6.51 Write a single node equation to evaluate $\underline{V}$ and $\underline{I}_1$ in Fig. P6.41 when $R = 4\ \Omega$, $\underline{V}_x = -2\underline{V}$, and $\underline{I} = 1\ \text{A}\ \angle 0°$.

6.52 Use node analysis to find $\underline{V}_1$ and $\underline{V}_2$ in Fig. P6.46 when $C = 1/50$ F, $R = 10\ \Omega$, $L = 4$ H, $v = 15 \cos 5t$ V, and $i = 3 \cos 5t$ A.

6.53 Use node analysis to find $\underline{V}_1$ and $\underline{V}_2$ in Fig. P6.47 when $C = 1/200$ F, $L = 0.25$ H, $R = 20\ \Omega$, $i = 4 \cos 20t$ A, and $v = 40 \cos 20t$ V.

6.54 Use node analysis to find $\underline{V}_1$ and $\underline{V}_2$ in Fig. P6.43 when $R = 5\ \Omega$, $\underline{V}_x = 2v_2$, and $\underline{I} = 3\ \text{A}\ \angle 0°$.

6.55 Use node analysis to find $\underline{V}_1$ and $\underline{V}_2$ in Fig. P6.47 when $C = 1/80$ F, $L = 0.8$ H, $R = 2\ \Omega$, $i = 5 \cos 10t$ A, and the voltage source is replaced by a VCVS with $v = 0.75 v_L$.

6.56* Write a single mesh equation to evaluate $\underline{I}$ and $\underline{V}_1$ in Fig. P6.42 when $R = 0$, $\underline{I}_x = 1.5\underline{I}$, and $\underline{V} = 20\ \text{V}\ \angle 0°$.

6.57 Write a single mesh equation to evaluate $\underline{I}_1$ and $\underline{V}$ in Fig. P6.41 when $R = 2\ \Omega$, $\underline{V}_x = 4\ \Omega \times \underline{I}_1$, and $\underline{I} = 6\ \text{A}\ \angle 0°$.

6.58 Use mesh analysis to find $\underline{I}_1$ and $\underline{I}_2$ in Fig. P6.46 when $C = 1/50$ F, $R = 10\ \Omega$, $L = 4$ H, $v = 15 \cos 5t$ V, and $i = 3 \cos 5t$ A.

6.59 Use mesh analysis to find $\underline{I}_1$ and $\underline{I}_2$ in Fig. P6.47 when $C = 1/80$ F, $L = 0.8$ H, $R = 2\ \Omega$, $i = 2 \cos 10t$ A, and $v = -10 \cos 10t$ V.

6.60 Use mesh analysis to find $\underline{I}_1$ and $\underline{I}_2$ in Fig. P6.43 when $R = 10\ \Omega$, $\underline{V}_x = \underline{V}_1$, and $\underline{I} = 2\text{ A}\ \underline{/0°}$.

6.61 Use mesh analysis to find $\underline{I}_1$ and $\underline{I}_2$ in Fig. P6.46 when $C = 1/40$ F, $R = 2\ \Omega$, $L = 0.7$ H, $v = 20 \cos 10t$ V, and the current source is replaced by a VCCS with $i = 0.6\text{ S} \times v_1$.

Section 6.4 Phasor Diagrams and Resonance

(See also PSpice problems B.19–B.22.)

6.62* Let the network in Fig. P6.24 have $Z = 5 + j0\ \Omega$, $1/\omega C = 5\ \Omega$, and $\omega L = 2R_2$. Construct a diagram showing all voltage phasors to scale when $\underline{I} = 1\text{ A}\ \underline{/0°}$.

6.63 Let the network in Fig. P6.38 have $Z = 8 - j6\ \Omega$, $R_1 = 5\ \Omega$ and $R_2 = 2\omega L$. Construct a diagram showing all voltage phasors to scale when $\underline{I} = 1\text{ A}\ \underline{/0°}$.

6.64 Let the network in Fig. P6.22 have $Y = 0.25 - j0.25$ S, $\omega L = R_1$, and $R_2 = 0$. Construct a diagram showing all voltage and current phasors to scale when $\underline{V} = 4\text{ V}\ \underline{/0°}$. Take the scale such that 1 A = 1 V.

6.65 Let the network in Fig. P6.24 have $Z = 10 + j6\ \Omega$, $1/\omega C = 4\ \Omega$, and $R_2 = 0$. Construct a diagram showing all voltage and current phasors to scale when $\underline{I} = 1\text{ A}\ \underline{/0°}$. Take the scale such that 1 A = 10 V.

6.66 Let the network in Fig. P6.26 have $Z = 15 + j0\ \Omega$, $\omega L = 20\ \Omega$, and $1/\omega C = R_2/3$. Construct a diagram showing all voltage and current phasors to scale when $\underline{V} = 60\text{ V}\ \underline{/0°}$. Take the scale such that 1 A = 10 V.

6.67 Let the network in Fig. P6.26 have $Z = 1 + j2\ \Omega$, $\omega L = 2\ \Omega$, and $1/\omega C = R_2$. Construct a diagram showing all voltage and current phasors to scale when $\underline{I} = 4\text{ A}\ \underline{/0°}$. Take the scale such that 1 V = 0.5 A.

6.68* Suppose the series network in Fig. P6.68 has $\omega_0 = 10^4$, $R = 200\ \Omega$, and $L = 50$ mH. Find the corresponding values of C and Q_{ser}. Then calculate $|\underline{V}_L|$ and $|\underline{V}_R + \underline{V}_L|$ when $\omega = \omega_0$ and $|\underline{V}| = 8$ V.

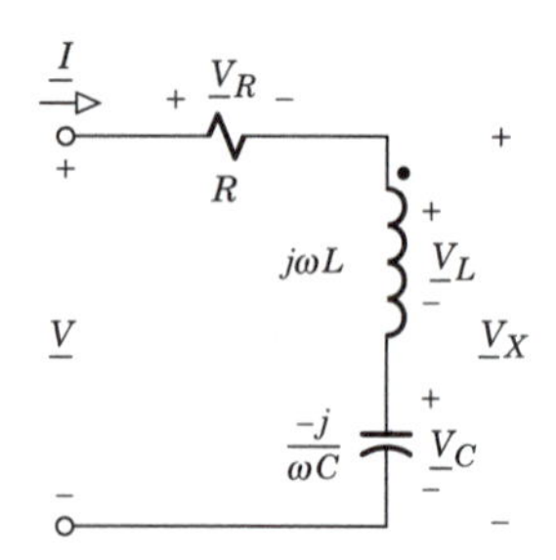

Figure P6.68

6.69 Suppose the series network in Fig. P6.68 has $\omega_0 L = 50\ \Omega$ and $Q_{ser} = 25$. If the network is driven at resonance by a 10–V source with internal resistance $R_s = 3\ \Omega$, then what are the values of $|\underline{V}_C|$ and $|\underline{V}|$?

6.70 A coil with inductance L and resistance R_w is connected in series to a 1–μF capacitor and a 5–V ac source. Measurements show that $|\underline{V}_C|_{max} = 100$ V at $\omega = 5 \times 10^4$. What are the values of L and R_w?

6.71 The resistance of a certain coil is given by $R_w = 100\sqrt{L}\ \Omega$, where L is the coil's inductance in henrys. The coil is connected to a capacitor to form a series network like Fig. P6.58 with $R = R_w$. Determine the values of C, L, and R_w if $\omega_0 = 10^5$ and $Q_{ser} = 50$.

6.72 Suppose the circuit in Fig. P6.72 satisfies the series resonance condition $X(\omega_0) = 0$ with $\omega_0 = 2000$. If $R_1 = 14\ \Omega$, $C = 5\ \mu$F, and $R_2 = 75\ \Omega$, then what are the values of $Z(j\omega_0)$ and L? Does ω_0^2 equal $1/LC$?

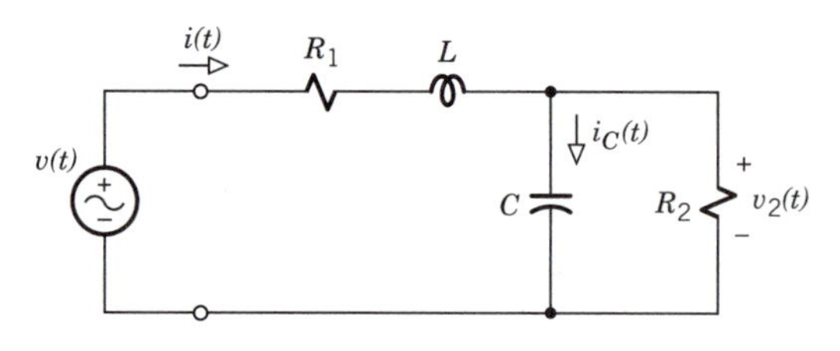

Figure P6.72

6.73 Suppose the circuit in Fig. P6.24 satisfies the series resonance condition $X(\omega_0) = 0$ with $\omega_0 = 5000$. If $R_1 = 30\ \Omega$, $R_2 = 10\ \Omega$, and $L = 2$ mH, then what are the values of $Z(j\omega_0)$ and C? Does ω_0^2 equal $1/LC$?

6.74 The network in Fig. P6.72 may exhibit series resonance with $R_1 = R_2 = R$. Set $X(\omega_0) = 0$ to obtain an expression for ω_0^2/ω_{LC}^2 where $\omega_{LC}^2 = 1/LC$. Then determine the condition on L/C such that $\omega_0^2 > 0$, so resonance can occur.

6.75 Do Problem 6.74 for the network in Fig. P6.24 with $R_1 = R_2 = R$.

6.76 Suppose the parallel network in Fig. P6.76 has $\omega_0 = 5 \times 10^4$, $R = 15$ kΩ, and $C = 0.4\mu$F. Find the corresponding values of L and Q_{par}. Then calculate $|\underline{V}|$ and $|\underline{I}_L|$ when $\omega = \omega_0$ and $|\underline{I}| = 2$ mA.

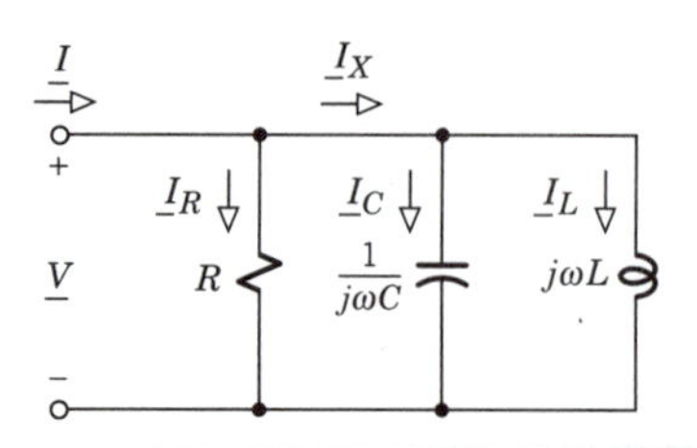

Figure P6.76

6.77 Suppose the parallel network in Fig. P6.76 has $\omega_0 L = 200\ \Omega$ and $Q_{par} = 50$. If the network is driven at resonance by a 1–mA source with parallel internal resistance $R_s = 15\ \text{k}\Omega$, then what are the values of $|\underline{I}_L|$ and $|\underline{I}_R|$?

6.78 Given a parallel *RCL* network with $L = 20\ \mu\text{H}$, $\omega_0 = 5 \times 10^5$, and $Q_{par} = 100$, what external capacitor and resistor should be connected in parallel to get the lower values $\omega_0{}' = 2.5 \times 10^5$ and $Q_{par}{}' = 40$?

6.79 A certain coil has $L = 0.5$ mH and $R_w = 10\ \Omega$. What capacitor should be connected in parallel with the coil to get $\omega_0 = 10^6$, and what is the resulting value of Q_{par}?

6.80 A certain coil has $L = 0.1$ mH and $R_w = 0.2\ \Omega$. What capacitor and resistor should be connected in parallel with the coil to get $\omega_0 = 2 \times 10^5$ and $Q_{par} = 20$?

6.81 The network in Fig. P6.22 may exhibit parallel resonance when $R_1 = 0$ and $R_2 = R$. Set $B(\omega_0) = 0$ to obtain an expression for ω_0^2/ω_{LC}^2 where $\omega_{LC}^2 = 1/LC$. Then determine the condition on L/C such that $\omega_0^2 > 0$, so resonance can occur.

6.82 Do Problem 6.81 with $R_1 = 2R$ and $R_2 = R$.

Section 6.5 AC Superposition

6.83* Apply superposition to find $i_1(t)$ in Fig. P6.22 when $R_1 = R_2 = 5\ \text{k}\Omega$, $L = 0.25$ H, $C = 10$ nF, and $i = 4 + 2 \cos 20{,}000t$ mA.

6.84 Apply superposition to find $i(t)$ in Fig. P6.38 when $R_1 = 2\ \text{k}\Omega$, $R_2 = 8\ \text{k}\Omega$, $C = 2$ nF, $L = 0.12$ H, and $v = 20 + 13 \cos 50{,}000t$ V.

6.85 Find $v_2(t)$ in Fig. P6.46 when $C = 1/400$ F, $L = 0.1$ H, $i = 3$ A, and $v = 20 \cos 100t$ V.

6.86 Find $i_d(t)$ in Fig. P6.48 when $R_1 = 2\ \Omega$, $C = 1/400$ F, $R_2 = 4\ \Omega$, $L = 5$ mH, $v = 30$ V, and $i = 4 \cos 200t$ A.

6.87 Find $i_1(t)$ in Fig. P6.47 when $C = 1/400$ F, $L = 0.25$ H, $R = 20\ \Omega$, $i = 3 \cos 20t$ A, and $v = 60 \cos 40t$ V.

6.88 Find $v_b(t)$ in Fig. P6.48 when $R_1 = 5\ \Omega$, $R_2 = 2\ \Omega$, $L = 0.02$ H, $v = 18 \cos (100t + 45°)$ V, and $i = 3 \cos 200t$ A.

6.89 Find $v_1(t)$ in Fig. P6.46 when $C = 1/100$ F, $R = 30\ \Omega$, $L = 0.5$ H, $v = 10 \cos 40t$ V, and $i = 2 \cos 10t$ A.

Section 6.6 Impedance Bridges

6.90* Derive the relationships for R_u and L_u when an Owen bridge is balanced with $Z_u = R_u + j\omega L_u$. What elements should be adjustable so that the real and reactive parts can be balanced independently?

6.91 Derive the relationships for R_u and C_u when a Scherling bridge is balanced with $Z_u = R_u + 1/j\omega C_u$. What elements should be adjustable so that the real and reactive parts can be balanced independently?

6.92 Derive the relationships for R_u and L_u when a Hay bridge is balanced with $Z_u = R_u + j\omega L_u$. Simplify your result for the case of $\omega C_1 R_1 \ll 1$.

6.93 A Schering bridge is balanced with $Z_u = R \| (1/j\omega C)$. Determine the values of R and C given what $R_1 = 1\ \text{k}\Omega$, $C_1 = 2\ \mu\text{F}$, $R_2 = 2\ \text{k}\Omega$, $C_3 = 1\ \mu\text{F}$, and $\omega = 1000$. Hint: Use $Y_u = (Z_2 Z_3 / Z_1)^{-1}$.

6.94 An Owen bridge is balanced with $Z_u = R \| j\omega L$. Determine the values of R and L given that $C_1 = 0.05\ \mu\text{F}$, $R_2 = 20\ \Omega$, $R_3 = 25\ \text{k}\Omega$, $C_3 = 0.2\ \mu\text{F}$, and $\omega = 1000$. Hint: Use $Y_u = (Z_2 Z_3 / Z_1)^{-1}$.

6.95 A series-comparison bridge is balanced with $Z_u = R \| (1/j\omega C)$, and it is assumed that $R \gg 1/\omega C$ so $\omega CR \gg 1$. Find the approximate values of R and C given that $R_1 = 2\ \text{k}\Omega$, $R_2 = 10\ \Omega$, $R_3 = 40\ \Omega$, $C_3 = 0.05\ \mu\text{F}$, and $\omega = 1000$. Hint: See Exercise 6.21.

CHAPTER 7

AC Power and Three-Phase Circuits

The basic purpose of many circuits operating in the ac steady state is *energy transfer*. Power enters the picture as the rate of energy transfer, which ranges from the picowatt level in radio receivers to the megawatt level in high-voltage transmission systems.

This chapter begins with important concepts of ac power, including average power, rms values, and maximum power transfer. Then we introduce the viewpoint and techniques of electric power engineering as applied to single-phase and three-phase power systems. A major concern throughout our investigation is the crucial task of efficient power delivery from source to load. Other practical considerations such as ac power measurement and residential circuits and wiring will be discussed as we go along.

Objectives

After studying this chapter and working the exercises, you should be able to do each of the following:

1. Calculate the average power dissipated in an ac circuit, using either the peak or rms values of voltage and current (Section 7.1).
2. Design an ac circuit for maximum power transfer (Section 7.1).
3. Define and calculate real, reactive, complex, and apparent power (Section 7.2).
4. Use the power triangle to analyze an ac power system and determine the power factor (Section 7.2).
5. State the purpose of power-factor correction, and design the correction needed for a given combination of loads (Section 7.2).
6. Identify the properties of three-phase sources and the advantages of three-phase systems for power transfer (Section 7.3).
7. Analyze a three-phase circuit with a balanced wye or delta load (Section 7.3).
8. Explain how power can be measured in a three-phase circuit (Section 7.3).
9. Analyze a three-phase circuit with an unbalanced load (Section 7.4).†

7.1 POWER IN AC CIRCUITS

Under ac steady-state conditions, all voltages and currents in a circuit are sinusoids. The instantaneous power therefore has periodic oscillatory fluctuations. Our objectives here are to find the resulting average power and determine the conditions for maximum power transfer. We'll also introduce the use of rms values in expressions for average power.

Average Power

Fundamentally, the concept of power relates to energy transfer. When a load draws instantaneous power $p(t)$ over some time interval t_1 to t_2, the total energy transferred to the load during that interval is

$$w = \int_{t_1}^{t_2} p(t)\, dt$$

We therefore define the **average power** P as

$$P \triangleq \frac{w}{t_2 - t_1} = \frac{1}{t_2 - t_1}\int_{t_1}^{t_2} p(t)\, dt \tag{7.1}$$

which represents the average rate of energy transfer over t_1 to t_2.

Usually, we want the *long-term* average obtained by letting $t_2 \rightarrow \infty$. But when $p(t)$ varies periodically, as it does in the ac steady state, the long-term average equals the average over one or more periods. Accordingly, letting T stand for any integer multiple of the period of $p(t)$, we take $t_2 = t_1 + T$ so Eq. (7.1) becomes

$$P = \frac{w}{T} = \frac{1}{T}\int_{t_1}^{t_1+T} p(t)\, dt \tag{7.2}$$

This long-term average of any periodic function $p(t)$ has four significant properties:

- The average is independent of the starting time t_1.
- When $p(t)$ consists of two or more components, the average of the sum equals the sum of the averages.
- If $p(t)$ contains a constant component, then the average of that component just equals the constant.
- Any sinusoidal component in $p(t)$ averages to zero.

We'll illustrate these properties by considering a particular but important case.

Suppose the instantaneous power has the form

$$p(t) = P_0 + P_a \cos(n\omega t + \phi) \tag{7.3a}$$

This function consists of a constant plus a sinusoid with period $2\pi/n\omega = T/n$, where $T = 2\pi/\omega$. Then, from the previously stated properties, we expect that

$$P = P_0 \tag{7.3b}$$

Hence, the average power just equals the constant component of $p(t)$. To check our result, we substitute Eq. (7.3a) into Eq. (7.2) to get the sum of the averages

$$\begin{aligned} P &= \frac{1}{T}\int_{t_1}^{t_1+T} [P_0 + P_a \cos(n\omega t + \phi)]\, dt \\ &= \frac{P_0}{T}\int_{t_1}^{t_1+T} dt + \frac{P_a}{T}\int_{t_1}^{t_1+T} \cos(n\omega t + \phi)\, dt \end{aligned}$$

The first term immediately reduces to P_0 since

$$\int_{t_1}^{t_1+T} dt = (t_1 + T) - t_1 = T$$

We evaluate the second integral by making the change of variable $\lambda = n\omega t + \phi$, so $d\lambda = n\omega\, dt = (2n\pi/T)\, dt$ and the integration limits become $n\omega t_1 + \phi = \lambda_1$ and $n\omega(t_1 + T) + \phi = \lambda_1 + 2n\pi$. Then

$$\int_{t_1}^{t_1+T} \cos(n\omega t + \phi)\, dt = \frac{T}{2n\pi}\int_{\lambda_1}^{\lambda_1+2n\pi} \cos\lambda\, d\lambda$$
$$= \frac{T}{2n\pi}[\sin(\lambda_1 + 2n\pi) - \sin(\lambda_1)]$$

But $\sin(\lambda_1 + 2n\pi) = \sin(\lambda_1)$ for any integer n, so

$$\int_{t_1}^{t_1+T} \cos(n\omega t + \phi)\, dt = 0$$

which confirms that any sinusoidal component with period T/n averages to zero.

Now consider a resistor in a circuit operating under ac steady-state conditions, as represented by the frequency-domain diagram in Fig. 7.1*a*. If $\underline{I} = I_m \angle\phi_i$, then $i(t) = I_m \cos(\omega t + \phi_i)$ and the instantaneous power dissipated by R is

$$p_R(t) = Ri^2(t) = RI_m^2 \cos^2(\omega t + \phi_i) \tag{7.4a}$$

Figure 7.1*b* plots the waveforms of the instantaneous current and power, showing that $p_R(t)$ repeats every $T/2$ seconds when $i(t)$ has period T. We also see from the vertical excursions of $p_R(t)$ that the instantaneous power oscillates between $p_{max} = RI_m^2$ and $p_{min} = 0$. Since the variations of $p_R(t)$ are symmetric relative to the dashed line at the middle, it appears that the average value is

$$P_R = \tfrac{1}{2}(p_{max} - p_{min}) = \tfrac{1}{2}RI_m^2$$

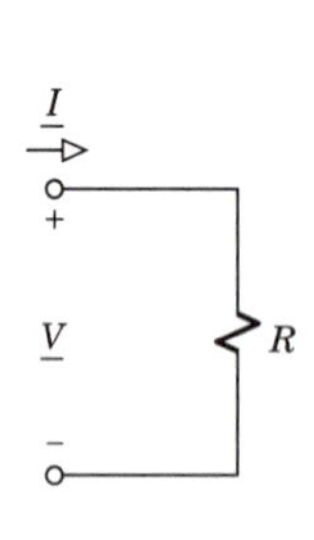

(*a*) Resistor in the ac steady state

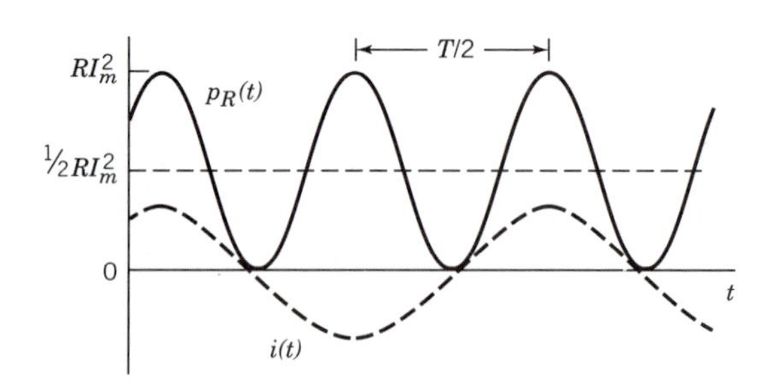

(*b*) Waveforms of $i(t)$ and $p_R(t) = Ri^2(t)$

Figure 7.1

To verify our inference, we expand Eq. (7.4a) using the trigonometric identity

$$\cos^2 \alpha = \tfrac{1}{2} + \tfrac{1}{2} \cos 2\alpha$$

so

$$p_R(t) = \tfrac{1}{2} R I_m^2 [1 + \cos(2\omega t + 2\phi_i)] \tag{7.4b}$$

This expansion shows that $p_R(t)$ consists of a constant plus a sinusoid, so the average power P_R equals the constant component $\tfrac{1}{2}RI_m^2$. We can also express P_R in terms of the voltage amplitude by noting from $\underline{I} = \underline{V}/R$ that $I_m = V_m/R$. Thus,

$$P_R = \tfrac{1}{2} R I_m^2 = \frac{V_m^2}{2R} \tag{7.5}$$

which holds regardless of phase angles. Equation (7.5) thus gives the average power dissipated by any resistor in any ac circuit when I_m is the peak current through R or when V_m is the peak voltage across R.

Next, to broaden our scope, let Fig. 7.2*a* represent an arbitrary load network containing any number of resistors and reactive elements. If $\underline{I} = I_m \angle \phi_i$ and if the load has impedance $Z = |Z| \angle \theta$, then

$$\underline{V} = Z\underline{I} = |Z| I_m \angle \theta + \phi_i$$

Thus,

$$v(t) = |Z| I_m \cos(\omega t + \theta + \phi_i) \qquad i(t) = I_m \cos(\omega t + \phi_i)$$

and

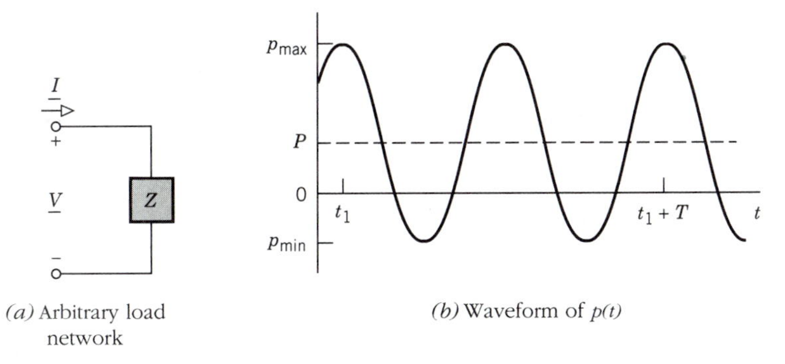

(a) Arbitrary load network

(b) Waveform of *p(t)*

Figure 7.2

$$p(t) = v(t)i(t) = |Z|I_m^2 \cos(\omega t + \theta + \phi_i) \cos(\omega t + \phi_i)$$

We then apply the identity

$$\cos \alpha \cos \beta = \tfrac{1}{2} \cos(\alpha - \beta) + \tfrac{1}{2} \cos(\alpha + \beta)$$

which yields the more informative relation

$$p(t) = \tfrac{1}{2}|Z|I_m^2 [\cos \theta + \cos(2\omega t + \theta + 2\phi_i)] \tag{7.6}$$

Since $\cos \theta$ stays constant while $\cos(2\omega t + \theta + 2\phi_i)$ varies from -1 to $+1$, the extreme values of $p(t)$ are

$$p_{max} = \tfrac{1}{2}|Z|I_m^2(\cos \theta + 1) \qquad p_{min} = \tfrac{1}{2}|Z|I_m^2(\cos \theta - 1) \tag{7.7}$$

Figure 7.2*b* shows the waveform of $p(t)$ for the normal case of $0 < |\theta| < 90°$, so $p_{min} < 0$ and the instantaneous power has both positive and negative values. Physically, the source supplies energy to the load whenever $p(t) > 0$, whereas the load returns energy back to the source whenever $p(t) < 0$ — the returned energy having been stored temporarily in the reactive elements.

The average power delivered to the load equals the constant component in Eq. (7.6). Thus,

$$P = \tfrac{1}{2}|Z|I_m^2 \cos \theta$$

which falls precisely halfway between p_{max} and p_{min}. But $|Z|\cos \theta = \text{Re}[Z] = R(\omega)$ and $I_m = V_m/|Z|$, so we can write

$$P = \tfrac{1}{2}R(\omega)I_m^2 = \frac{R(\omega)}{2|Z|^2} V_m^2 \tag{7.8}$$

Equation (7.8) allows direct calculation of the power dissipated by *all* the resistors in a network.

Comparing Eq. (7.8) with Eq. (7.5) reveals that the network's *ac resistance* $R(\omega)$ accounts for the combined effect of the resistors and reactive elements. However, note carefully in Eq. (7.8) that $P \neq V_m^2/2R(\omega)$ because V_m does not necessarily equal the peak voltage across the ac resistance. If the network consists entirely of resistors, then $Z = R(\omega) = R_{eq}$ and Eq. (7.8) reduces to Eq. (7.5) with $R = R_{eq}$. On the other hand, if the network contains only reactive elements, then $R(\omega) = 0$ and $P = 0$, consistent with the property that ideal inductors and capacitors never dissipate average power.

Example 7.1 *AC Power Calculations*

Figure 7.3 repeats the circuit studied in Example 6.8. We previously found that

$$Z = 4.8 + j6.4 \text{ k}\Omega \qquad |\underline{V}| = 80 \text{ V} \qquad |\underline{V}_C| = 40 \text{ V}$$

Now we'll calculate average powers.

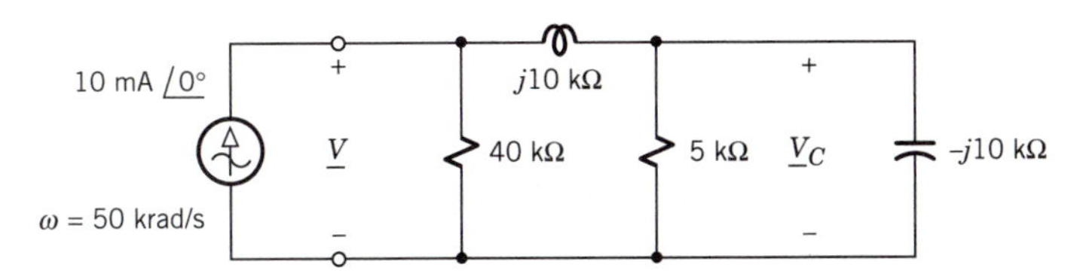

Figure 7.3 Frequency-domain circuit diagram.

The total average power supplied by the source is found from Eq. (7.8) with $R(\omega) = 4.8$ kΩ and $I_m = 10$ mA, so

$$P = \tfrac{1}{2}R(\omega)I_m^2 = \tfrac{1}{2} \times 4.8 \times 10^2 = 240 \text{ mW}$$

This power is actually dissipated by the 40-kΩ and 5-kΩ resistors, and we can put $|\underline{V}|$ and $|\underline{V}_C|$ into Eq. (7.5) to get the individual values

$$P_{40} = \frac{80^2}{2 \times 40} = 80 \text{ mW} \qquad P_5 = \frac{40^2}{2 \times 5} = 160 \text{ mW}$$

Thus, $P_{40} + P_5 = 240 \text{ mW} = P$, as expected.

Exercise 7.1

A parallel *RL* network with $R = 10\ \Omega$ and $L = 25$ mH is driven by $i(t) = 5 \sin 200t$ A. Calculate P, p_{max}, and p_{min}.

RMS Values

Equations for average power often have the voltage or current couched in terms of their *effective* or *rms* values. We'll introduce this concept here and then apply it to ac power calculations.

Consider a resistance R carrying a time-varying current $i(t)$ with period T. Since $p(t) = Ri^2(t)$, the average power dissipation is

$$P = \frac{1}{T}\int_{t_1}^{t_1+T} p(t)\, dt = R\left[\frac{1}{T}\int_{t_1}^{t_1+T} i^2(t)\, dt\right] \tag{7.9a}$$

But if the same resistance carries a *constant* current I, then $p(t)$ is constant and

$$P = RI^2 \tag{7.9b}$$

Comparing Eqs. (7.9a) and (7.9b), we see that $i(t)$ produces the same power dissipation as the constant current I when

$$I^2 = \frac{1}{T}\int_{t_1}^{t_1+T} i^2(t)\,dt \tag{7.9c}$$

Hence, I is the *effective value* of $i(t)$ with respect to power dissipation.

The effective value of any periodic current $i(t)$ is also called its **root-mean-square (rms) value**. This name follows from Eq. (7.9c) rewritten as the definition

$$I_{rms} \triangleq \sqrt{\frac{1}{T}\int_{t_1}^{t_1+T} i^2(t)\,dt} \tag{7.10a}$$

Thus, the root-mean-square value equals the square-*root* of the *mean* (average) of the *square* of $i(t)$. Likewise, the rms value of a periodic voltage $v(t)$ is

$$V_{rms} \triangleq \sqrt{\frac{1}{T}\int_{t_1}^{t_1+T} v^2(t)\,dt} \tag{7.10b}$$

For the ac case with sinusoids $i(t) = I_m \cos(\omega t + \phi_i)$ and $v(t) = V_m \cos(\omega t + \phi_v)$, Eqs. (7.10a) and (7.10b) yield

$$I_{rms} = \frac{I_m}{\sqrt{2}} \approx 0.707 I_m \qquad V_{rms} = \frac{V_m}{\sqrt{2}} \approx 0.707 V_m \tag{7.11}$$

These values are independent of the phase angles ϕ_i and ϕ_v.

Most ac voltmeters and ammeters are calibrated for rms values, and we normally speak of the rms voltage and current supplied by ac power sources. Standard residential ac circuits in the United States have $V_{rms} \approx 120$ V, so the peak voltage is actually $V_m = \sqrt{2}\, V_{rms} \approx 170$ V.

The rationale behind the widespread use of rms values emerges when we substitute I_{rms} and V_{rms} back into Eq. (7.8) for the average ac power. The factors of ½ then disappear and we get the tidier expression

$$P = R(\omega) I_{rms}^2 = R(\omega) V_{rms}^2/|Z|^2 \tag{7.12}$$

Hereafter, we'll always do average power calculations in terms of rms values. For a given impedance Z, V_{rms} and I_{rms} are related by

$$V_{rms} = |Z| I_{rms} \tag{7.13}$$

This relation follows from $\underline{V} = Z\underline{I}$, so $V_m = |Z|I_m$ and dividing both sides by $\sqrt{2}$ yields Eq. (7.13).

Example 7.2 *RMS Value of a Half-Rectified Wave*

Figure 7.4 plots a half-rectified current $i(t)$ found in some electronic power supplies. This waveform consists of a sine wave with the negative excursions removed, so $i(t) = I_m \sin \omega t$ for $0 \leq t \leq T/2$ but $i(t) = 0$ for $T/2 < t < T$, and so forth. We'll determine the corresponding rms value.

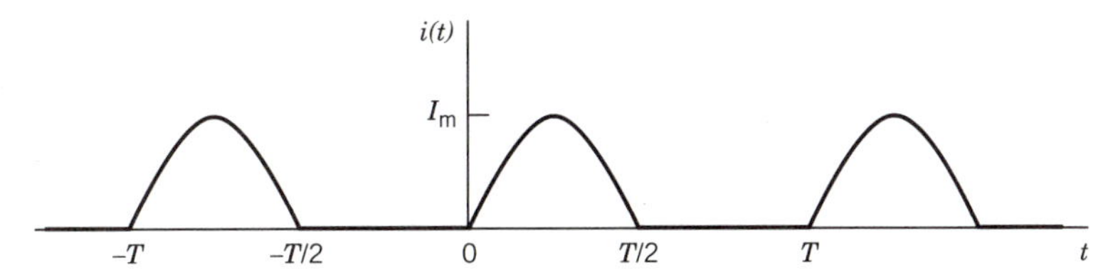

Figure 7.4 Half-rectfied current waveform.

Taking $t_1 = 0$ in Eq. (7.10a) and using the expansion $\sin^2 \alpha = ½(1 - \sin 2\alpha)$, we get

$$I_{rms}^2 = \frac{1}{T}\int_0^T i^2(t)\, dt = \frac{1}{T}\int_0^{T/2} I_m^2 \sin^2 \frac{2\pi t}{T}\, dt$$
$$= \frac{I_m^2}{2T}\left(\int_0^{T/2} dt - \int_0^{T/2} \sin \frac{4\pi t}{T}\, dt\right) = \frac{I_m^2}{4}$$

Hence,

$$I_{rms} = I_m/2$$

If resistance R carries a half-rectified current, then $P_R = RI_{rms}^2 = RI_m^2/4$, whereas an ac current with the same peak value would produce $P_R = R(I_m/\sqrt{2})^2 = RI_m^2/2$. The difference, of course, stems from the fact that the half-rectified wave is "off" half of the time.

Exercise 7.2

Let $v(t) = V_m \cos(\omega t + \phi_v)$. Show from Eq. (7.10b) that $V_{rms}^2 = V_m^2/2$. Hint: Use a trigonometric expansion to take advantage of the property that a sinusoid averages to zero.

Maximum Power Transfer

In Section 3.1 we showed that if a voltage source had fixed internal resistance R_s, then maximum power transfer to a load resistance R required $R = R_s$. The frequency-domain diagram in Fig. 7.5 depicts a similar situation. Here we have an ac voltage source with impedance Z_s, and we want to investigate the average power P transferred to a load impedance Z.

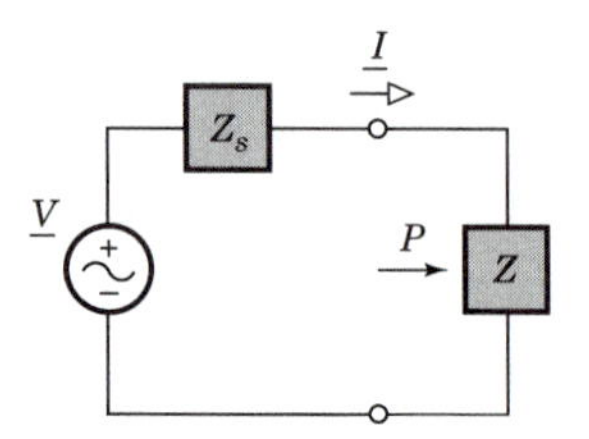

Figure 7.5 AC power transfer from source to load.

Both impedances may consist of ac resistance plus reactance, so we'll write

$$Z_s = R_s + jX_s \qquad Z = R + jX$$

The dependence on ω is understood but omitted for convenience in manipulations. If I_{rms} denotes the rms current, then the load power is

$$P = RI_{rms}^2$$

Likewise, the power dissipated in the source is

$$P_s = R_s I_{rms}^2$$

The corresponding power transfer efficiency is

$$\text{Eff} = \frac{P}{P + P_s} = \frac{R}{R + R_s} \tag{7.14}$$

Hence, efficiency can be calculated directly from the ac resistances.

But to determine the conditions for maximum power transfer, we must express P in terms of the fixed rms source voltage. Since $\underline{I} = \underline{V}/(Z_s + Z)$, $I_{rms} = V_{rms}/|Z_s + Z|$ and

$$P = \frac{RV_{rms}^2}{|Z_s + Z|^2} = \frac{RV_{rms}^2}{(R_s + R)^2 + (X_s + X)^2} \tag{7.15}$$

Presumably, R_s and either X_s or X are fixed, while the other two components are variable. Of course, the load must have $R > 0$; otherwise, it could not absorb power from the source. Under these conditions, we conclude from Eq. (7.15) that the maximum value of P occurs when $X_s + X = 0$ or

$$X = -X_s$$

Thus, X should be inductive if X_s is capacitive, and vice versa. Since the reactive components cancel out — similar to series resonance — the equivalent circuit reduces to a voltage source in series with R_s and R. Hence, just like a purely resistive circuit, maximum power transfer additionally requires that

$$R = R_s$$

This condition follows from Eq. (7.15) by setting $X_s + X = 0$ and solving $dP/dR = 0$ for R.

The two foregoing requirements are combined with the help of complex-conjugate notation in the form

$$Z = R + jX = R_s - jX_s = Z_s^* \tag{7.16}$$

We then say that the impedances are **matched** for maximum power transfer at the source frequency ω. The corresponding value of P is called the **maximum available power**, given by

$$P_{max} = V_{rms}{}^2/4R_s \tag{7.17}$$

However, when $R = R_s$, Eq. (7.14) indicates that maximum power transfer corresponds to 50% efficiency, meaning that equal amounts of power are dissipated in the source and load. We tolerate this low efficiency only when the objective is to get as much power as possible from a feeble source such as an electronic oscillator or a radio antenna.

Sometimes the ratio X/R is fixed but the *magnitude* of the load impedance can be adjusted. (We'll see in Chapter 8 how this is done with the help of a transformer.) To determine the optimum value of $|Z|$ for this case, we substitute $R = |Z| \cos \theta$ and $X = |Z| \sin \theta$ into Eq. (7.15), which becomes

$$P = \frac{|Z| \cos \theta \; V_{rms}{}^2}{|Z_s|^2 + 2|Z|(R_s \cos \theta + X_s \sin \theta) + |Z|^2}$$

Differentiating P with respect to $|Z|$ yields

$$\frac{dP}{d|Z|} = \frac{(|Z_s|^2 - |Z|^2) \cos \theta \; V_{rms}{}^2}{[|Z_s|^2 + 2|Z|(R_s \cos \theta + X_s \sin \theta) + |Z|^2]^2}$$

so $dP/d|Z| = 0$ when $|Z_s|^2 - |Z|^2 = 0$. Hence, the optimum value of $|Z|$ is

$$|Z| = |Z_s| \tag{7.18}$$

The resulting value of P will be less than P_{max} but greater than P when $|Z| \neq |Z_s|$.

Example 7.3 *Power Transfer from an Oscillator*

A certain high-frequency electronic oscillator has

$$V_{rms} = 1.2\text{ V} \qquad Z_s = 6 + j8\text{ k}\Omega = 10\text{ k}\Omega\ \underline{/53.1^\circ}$$

If we are free to connect a matched load impedance, then we should take

$$Z = Z_s^* = 6 - j8\text{ k}\Omega$$

so that

$$P = P_{max} = 1.2^2/(4 \times 6) = 0.06\text{ mW} = 60\ \mu\text{W}$$

But if the load has the fixed ratio $X/R = -7/24$, then we must use

$$Z = (24 - j7)c = 25c\ \underline{/16.3^\circ}$$

where c is an adjustable constant. Since Eq. (7.18) calls for $|Z| = |Z_s|$ in this case, we should take $25c = 10\text{ k}\Omega$ so that

$$Z = 10\text{ k}\Omega\ \underline{/16.3^\circ} = 9.6 - j2.8\text{ k}\Omega$$

The resulting load power is found from Eq. (7.15) to be

$$P = \frac{9.6 \times 1.2^2}{(6 + 9.6)^2 + (8 - 2.8)^2} = 0.0511\text{ mW} = 51.1\ \mu\text{W}$$

The transfer efficiency in this case is Eff $= 9.6/(9.6 + 6) \approx 62\%$.

Exercise 7.3

Suppose that the only variable component in Eq. (7.15) is the load resistance R. Solve $dP/dR = 0$ to show that the optimum value of R is

$$R = \sqrt{R_s^2 + (X_s + X)^2}$$

7.2 POWER SYSTEMS

We've seen that instantaneous power in an ac circuit has an oscillatory behavior. The implications of that behavior are explored more fully here, with particular reference to ac systems that handle large amounts of power — say a kilowatt or more.

We'll first examine the difference between real and reactive power, and we'll introduce the useful concept of complex power. Then we'll define the power factor of a load and show how power-factor correction improves the efficiency of power transfer. We'll also describe wattmeters for ac power measurements.

Real and Reactive Power

Two conventions commonly used in electric power engineering need to be stated at the outset. These conventions simplify notation, and they will be employed throughout the remainder of this chapter.

Our first convention reflects the fact that ac power systems operate at a *fixed frequency*, which, in the US, is usually

$$\omega = 2\pi \times 60 \text{ Hz} \approx 377 \text{ rad/s}$$

We may therefore ignore frequency-dependent effects and write any load impedance as

$$Z = |Z| \angle\theta = R + jX \tag{7.19a}$$

where

$$R = |Z| \cos\theta \qquad X = |Z| \sin\theta \tag{7.19b}$$

Figure 7.6 shows our load model and the corresponding impedance triangle. This series model holds equally well for a parallel load or a combination of series and parallel elements, with R and X being interpreted as the series equivalent values calculated from Eq. (7.19b).

Our second convention combines phasor notation with root-mean-square values by expressing all *phasor magnitudes in rms units*. Specifically, for the terminal voltage and current phasors in Fig. 7.6a we write

$$|\underline{V}| = V_{rms} \qquad |\underline{I}| = I_{rms}$$

Ohm's law still relates $\underline{V}$ and $\underline{I}$ by $\underline{V} = Z\underline{I}$ because we use rms values for both

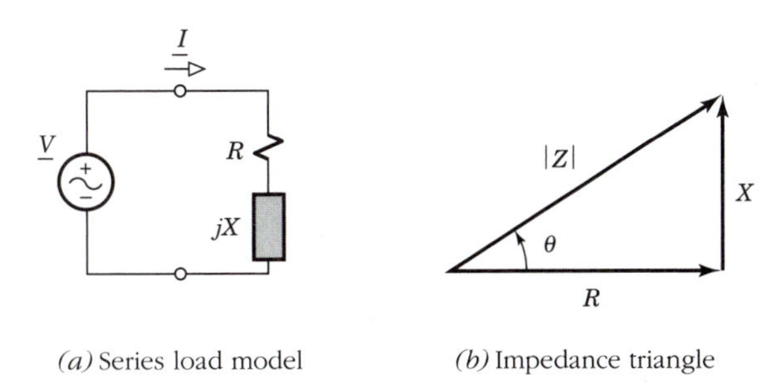

(a) Series load model　　*(b)* Impedance triangle

Figure 7.6

voltage and current. If we take the angle ϕ_v of the terminal voltage as the phase reference, then $\phi_i = \angle \underline{I} = \angle(\underline{V}/Z) = \phi_v - \theta$ so

$$\underline{V} = V_{rms} \angle \phi_v \qquad \underline{I} = I_{rms} \angle \phi_v - \theta \tag{7.20}$$

However, the *peak* values of the associated sinusoids must now be calculated via

$$V_m = \sqrt{2} V_{rms} = \sqrt{2} |\underline{V}| \qquad I_m = \sqrt{2} I_{rms} = \sqrt{2} |\underline{I}| \tag{7.21}$$

Having stated our conventions, we're ready to study the instantaneous power delivered to the load. Recall from Eq. (7.6) that if $\phi_i = \phi_v - \theta$ then

$$p(t) = \tfrac{1}{2} |Z| I_m^2 [\cos \theta + \cos (2\omega t + 2\phi_v - \theta)]$$

But now we'll substitute $\frac{1}{2}|Z|I_m^2 = |Z|I_{rms}^2 = V_{rms}I_{rms}$, and we'll expand $\cos (2\omega t + 2\phi_v - \theta)$ via

$$\cos (\alpha - \beta) = \cos \alpha \cos \beta + \sin \alpha \sin \beta$$

Thus,

$$p(t) = V_{rms}I_{rms} \cos \theta \, [1 + \cos 2(\omega t + \phi_v)] + V_{rms}I_{rms} \sin \theta \sin 2(\omega t + \phi_v)$$

Finally, we let

$$P = V_{rms}I_{rms} \cos \theta \qquad Q = V_{rms}I_{rms} \sin \theta$$

and write

$$p(t) = p_R(t) + p_X(t) \tag{7.22a}$$

where

$$p_R(t) = P[1 + \cos 2(\omega t + \phi_v)] \qquad p_X(t) = Q \sin 2(\omega t + \phi_v) \tag{7.22b}$$

We have thereby decomposed $p(t)$ into the two components $p_R(t)$ and $p_X(t)$ sketched in Fig. 7.7, where $\phi_v = 0$ for clarity.

The component $p_R(t)$ consists of a constant plus a sinusoid, so its average value is thus $P = V_{rms}I_{rms} \cos \theta$. To justify the symbol P for this quantity, note that $V_{rms} = |Z|I_{rms}$ and the impedance triangle in Fig. 7.6 shows that $\cos\theta = R/|Z|$. Hence,

$$P = V_{rms}I_{rms} \cos \theta = RI_{rms}^2 \tag{7.23}$$

so the average value of $p_R(t)$ equals the *average power* absorbed by the load's resistance. We'll call P the **real power**, for reasons that will become clear subsequently. Of course, P is measured in watts (W).

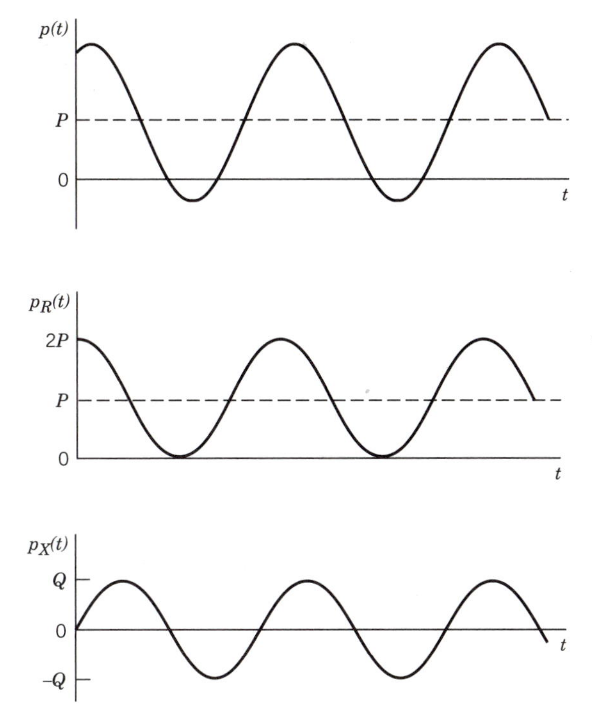

Figure 7.7 Instantaneous ac power and its real and reactive components (with $\phi_v = 0$).

The other component of $p(t)$ is $p_X(t)$, which oscillates sinusoidally between $+Q$ and $-Q$ and has zero average value. Consequently, $p_X(t)$ does not contribute to average power absorbed by the load. Instead, we note from Fig. 7.6 that $\sin\theta = X/|Z|$, so

$$Q = V_{rms} I_{rms} \sin\theta = X I_{rms}^2 \tag{7.24}$$

The presence of X indicates that $p_X(t)$ represents the *rate of energy exchange* between the source and the load's reactance. Accordingly, we call Q the **reactive power**. (Be careful not to confuse the symbol Q used here with the quality factor of resonant circuits defined in Section 6.4.) To emphasize that reactive power represents alternating energy storage rather than power transfer, we express Q in **volt-amperes reactive (VAr)**.

If the load happens to be entirely resistive, then $X = 0$, $Q = 0$, and $p(t) = p_R(t)$. But if the load happens to be entirely reactive, then $R = 0$, $P = 0$, and $p(t) = p_X(t)$. In the case of a single inductor with rms current $|\underline{I}_L|$, rms voltage $|\underline{V}_L|$, and $X = \omega L$, Eq. (7.24) becomes

$$Q_L = \omega L |\underline{I}_L|^2 = |\underline{V}_L|^2/\omega L \tag{7.25a}$$

Similarly, in the case of a single capacitor with $X = -1/\omega C$,

$$Q_C = -|\underline{I}_C|^2/\omega C = -\omega C|\underline{V}_C|^2 \tag{7.25b}$$

The negative value of Q_C simply means that the waveform $p_X(t)$ would be inverted in Fig. 7.7. More generally, a load with inductive reactance ($X > 0$) has $Q > 0$, while a load with capacitive reactance ($X < 0$) has $Q < 0$. We therefore say that an inductive load "consumes" reactive power while a capacitive load "produces" reactive power.

Reactive power represents an oscillatory energy exchange and it does not contribute to power transfer. On the contrary, reactive power often reduces power-transfer efficiency by increasing the rms current required to deliver a specified average power from the source to the load. The increased current then results in power wasted as ohmic heating in the generator's internal resistance and the resistance of the wires connecting source to load. An example will help illustrate this effect.

Example 7.4 *Power-Transfer Efficiency*

Figure 7.8 diagrams a parallel *RC* load connected to an ac voltage generator with $V_{rms} = 300$ V. The combined source and wire resistance are represented by $R_s = 2\ \Omega$. We want to find the resulting rms current, the real and reactive power, and the power-transfer efficiency.

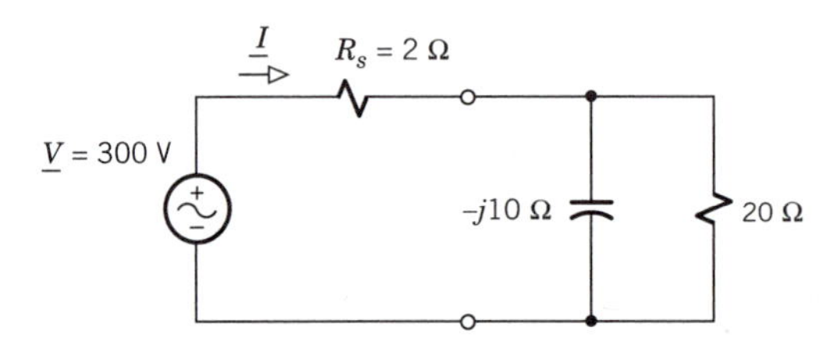

Figure 7.8 Circuit for power transfer.

The load impedance is $Z_{RC} = (-j10)\|20 = 4 - j8\ \Omega$, so the source sees the total impedance

$$Z = R_s + Z_{RC} = 6 - j8\ \Omega = 10\ \Omega\ \underline{/-53.1^\circ}$$

Hence, $I_{rms} = V_{rms}/|Z| = 300\ \text{V}/10\ \Omega = 30$ A, and Eqs. (7.23) and (7.24) give

$$P = 6 \times 30^2 = 5400\ \text{W} = 5.4\ \text{kW}$$
$$Q = (-8) \times 30^2 = -7200\ \text{VAr} = -7.2\ \text{kVAr}$$

The power dissipated by R_s is $P_s = R_s I_{rms}{}^2 = 1.8$ kW and the load actually receives $P_L = \text{Re}[Z_{RC}] \times I_{rms}{}^2 = 3.6$ kW. The power-transfer efficiency therefore equals the resistance ratio

$$P_L/P = 4/(2 + 4) \approx 67\%$$

a rather poor efficiency.

But suppose the capacitive reactance could be disconnected from the load so that $I_{rms} = 300\ \text{V}/(2 + 20)\Omega = 13.6$ A. Then the ohmic heat loss would fall to $R_s I_{rms}^2 = 372$ W while the load power would increase slightly to $RI_{rms}^2 =$ 3.72 kW. The efficiency would thus becomes

$$P_L/P = 20/(2 + 20) \approx 91\%$$

and we have a considerable improvement.

Exercise 7.4

Find I_{rms}, P, Q, and P_L/P for the circuit in Fig. 7.8 when an inductor with impedance $j12.5\ \Omega$ is inserted in series with R_s.

Complex Power and Power Factor

The product of the rms voltage and current at the terminals of a load is called the **apparent power**, expressed in **volt-amperes (VA)**. Although the quantity $V_{rms}I_{rms}$ is easily measured with the help of simple ac meters, it does not necessarily equal the actual power absorbed by the load. Instead, by squaring and adding Eqs. (7.23) and (7.24), we find that $P^2 + Q^2 = (V_{rms}I_{rms})^2 \cos^2\theta + (V_{rms}I_{rms})^2 \sin^2\theta$ or

$$V_{rms}I_{rms} = \sqrt{P^2 + Q^2}$$

The form of this equation suggests a triangular relationship between real, reactive, and apparent power.

To develop the geometric picture, we'll use the complex conjugate of the current phasor in Eq. (7.20), namely,

$$\underline{I}^* = (I_{rms}\ \underline{/\phi_v - \theta})^* = I_{rms}\ \underline{/\theta - \phi_v}$$

Since $\underline{V} = V_{rms}\ \underline{/\phi_v}$, multiplying $\underline{V}$ by $\underline{I}^*$ eliminates ϕ_v and yields the **complex power**

$$S \triangleq \underline{V}\,\underline{I}^* = V_{rms}I_{rms}\ \underline{/\theta} \tag{7.26a}$$

Thus, in rectangular form,

$$S = V_{rms}I_{rms}\cos\theta + jV_{rms}I_{rms}\sin\theta = P + jQ \tag{7.26b}$$

The real power P therefore equals Re[S] and the reactive power Q equals Im[S].

Figure 7.9 shows the resulting complex-plane picture known as the **power triangle**, a right triangle with hypotenuse of length

$$|S| = \sqrt{P^2 + Q^2} = V_{rms}I_{rms} \tag{7.27}$$

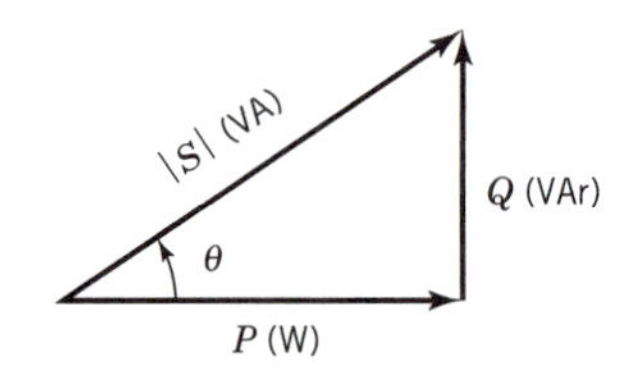

Figure 7.9 The power triangle.

This triangle has exactly the same shape as the impedance triangle back in Fig. 7.6*b* because

$$S = P + jQ = RI_{rms}^2 + jXI_{rms}^2 = ZI_{rms}^2$$

Consequently, the power triangle equals the impedance triangle multiplied by I_{rms}^2.

A valuable property of complex power for the study of power systems is the **conservation law**, which states that:

> When several loads are connected to the same source, the total complex power from the source equals the sum of the complex powers of the individual loads.

For instance, if two loads draw $S_1 = P_1 + jQ_1$ and $S_2 = P_2 + jQ_2$, respectively, then the source must supply

$$S = S_1 + S_2 = (P_1 + P_2) + j(Q_1 + Q_2) \tag{7.28}$$

The real part of Eq. (7.28) agrees with the familiar observation that if a 60-W lamp and a 100-W lamp are connected to the same source, then the source must supply $P = 60 + 100 = 160$ W. And the same summation effect holds for reactive power. This conservation property, together with the power triangle, eliminates the need for finding combined impedances and thereby simplifies many power calculations.

Loads that consume large amounts of power are often characterized in terms of the **power factor**, defined by

$$\text{pf} \triangleq P/|S| = \cos\theta \tag{7.29}$$

which is the ratio of real power to apparent power. Clearly, the power factor of a passive load always falls in the range $0 \leq \text{pf} \leq 1$ since $P \geq 0$ and $|S| \geq P$. Unity power factor simply means that $|S| = P$, which occurs when $Q = 0$ and $\theta = 0$.

A load with pf = 1 has zero equivalent series reactance and draws the *minimum source current* $I_{rms} = P/V_{rms}$ needed for a specified value of P. Otherwise, any load with $Q \neq 0$ draws

$$I_{rms} = \frac{|S|}{V_{rms}} = \frac{P/\text{pf}}{V_{rms}} > \frac{P}{V_{rms}}$$

In view of the increased current that must be supplied when pf < 1, electric utilities usually charge higher rates to large industrial consumers whose plants operate at low power factors.

Given the power rating and power factor of a load, you can easily compute the apparent power via $|S| = P/\text{pf}$. Then you can compute the *magnitude* of the reactive power from Eq. (7.27) rewritten as

$$Q = \pm\sqrt{|S|^2 - P^2} = \pm|S|\sqrt{1 - \text{pf}^2} \tag{7.30}$$

However, the *sign* of Q depends upon the nature of the load's reactance. If the load is *inductive,* so $Q > 0$ and $\theta > 0$, then we say that it has a **lagging** power factor — meaning that the current phasor lags the voltage phasor. Conversely, if the load is *capacitive,* so $Q < 0$ and $\theta < 0$, then we say that it has a **leading** power factor — meaning that the current phasor leads the voltage phasor.

For reference purposes, Table 7.1 lists our various ac power quantities, while Table 7.2 summarizes power-factor terminology.

Most industrial loads include inductive components, such as motor windings, and therefore have lagging power factors. Capacitors are often used in conjunction with those loads for the purpose of **power-factor correction**. The capacitor is connected in *parallel* with the load to avoid an unwanted voltage drop. But an appropriate capacitor in parallel with an inductive load cancels out the reactive power and the combined load has pf = 1, thereby minimizing current drawn from the source. Sometimes a special rotating machine known as a **synchronous capacitor** must be employed for power-factor correction because a conventional capacitor would be impractically large. Furthermore, the capacitance of a synchronous capacitor may be adjusted electrically to achieve the desired correction.

TABLE 7.1 AC Power Quantities

For load $Z = R + jX = \|Z\| \angle\theta$			
Quantity	**Relations**	**Unit**	**Meaning**
Real power	$P = V_{rms}I_{rms}\cos\theta$ $= RI_{rms}^2$	W	Average power delivered to the load
Reactive power	$Q = V_{rms}I_{rms}\sin\theta$ $= XI_{rms}^2$	VAr	Rate of reactive energy exchange
Complex power	$S = \underline{V}\,\underline{I}^* = P + jQ$ $= ZI_{rms}^2$		Two-dimensional combination of P and Q
Apparent power	$\|S\| = V_{rms}I_{rms}$ $= \sqrt{P^2 + Q^2}$	VA	Magnitude of complex power
Power factor	$\text{pf} = P/\|S\|$ $= \cos\theta$		Ratio of real power to apparent power

TABLE 7.2 Power-Factor Terminology

Power Factor	Type of Load	Conditions
Unity	Resistive	$X = 0,\ \theta = 0,\ Q = 0$
Lagging	Inductive	$X > 0,\ \theta > 0,\ Q > 0$
Leading	Capacitive	$X < 0,\ \theta < 0,\ Q < 0$

Example 7.5 *Designing Power-Factor Correction*

Figure 7.10*a* represents an industrial plant with a parallel capacitor to be designed for power-factor correction. The plant operates from a 500-V(rms) source at 60 Hz and the plant consists of two loads having

$$P_1 = 48 \text{ kW} \qquad \text{pf}_1 = 0.60 \text{ lagging}$$
$$P_2 = 24 \text{ kW} \qquad \text{pf}_2 = 0.96 \text{ leading}$$

We infer from this data that load 1 is highly inductive, while load 2 is slightly capacitive. The additional capacitor has been included to correct the plant's overall power factor. We'll first use conservation of complex power and the power triangle to combine the two loads. Then we'll determine the value of C needed to get unity power factor.

The apparent and reactive power for load 1 are found from Eqs. (7.29) and (7.30) to be

$$|S_1| = 48 \text{ kW}/0.6 = 80 \text{ kVA} \qquad Q_1 = +\sqrt{80^2 - 48^2} = +64 \text{ kVAr}$$

where we know that $Q_1 > 0$ because pf_1 is lagging. The rms current drawn by this load is

$$|\underline{I}_1| = 80 \text{ kVA}/500 \text{ V} = 160 \text{ A}$$

Similarly, for load 2 we find that

$$|S_2| = 25 \text{ kVA} \qquad Q_2 = -7 \text{ kVAr} \qquad |\underline{I}_2| = 50 \text{ A}$$

Now we can construct the combined power triangle shown in Fig. 7.10*b* with

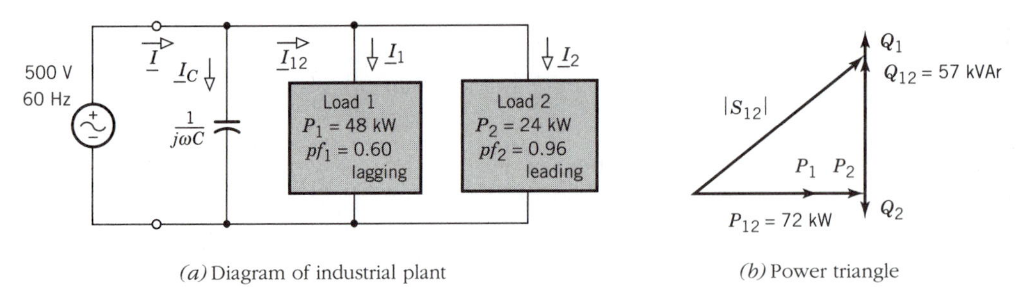

(*a*) Diagram of industrial plant

(*b*) Power triangle

Figure 7.10

$$P_{12} = P_1 + P_2 = 72 \text{ kW} \qquad Q_{12} = Q_1 + Q_2 = +57 \text{ kVAr}$$

Thus,

$$|S_{12}| = \sqrt{72^2 + 57^2} = 91.8 \text{ kVA}$$
$$|\underline{I}_{12}| = 91.8 \text{ kVA}/500 \text{ V} = 184 \text{ A}$$

so $\text{pf}_{12} = P_{12}/|S_{12}| = 0.784$, which is still lagging since $Q_{12} > 0$.

Table 7.3 summarizes our results so far. You should note carefully those quantities that add directly (P and Q) and those that do not ($|S|$ and I_{rms}). The last two lines of the table pertain to power-factor correction using an ideal lossless capacitor with $P_C = 0$ and $Q_C = -57$ kVAr. The corresponding value of C is calculated from Eq. (7.25b) as

$$C = -Q_C/\omega|\underline{V}_C|^2 = 605 \ \mu\text{F}$$

since $|\underline{V}_C| = |\underline{V}| = 500$ V and $\omega = 377$ rad/sec.

TABLE 7.3

Load	P (kW)	Q (kVAr)	$\|S\|$ (kVA)	I_{rms} (A)
1	48	+64	80	160
2	24	− 7	25	50
1 & 2	72	+57	91.8	184
C	0	−57	57	114
Plant	72	0	72	144

The entire plant with its corrected power factor has

$$P = P_{12} + P_C = 72 \text{ kW} \qquad Q = Q_{12} + Q_C = 0$$

so pf $= 1$ and $|S| = P$. Accordingly, the rms current from the source will be

$$I_{rms} = |\underline{I}| = 72 \text{ kVA}/500 \text{ V} = 144 \text{ A}$$

which is less than the current for load 1 alone! In this situation, the source sees a plant that appears to be entirely resistive and draws the minimum current $I_{rms} = P/V_{rms}$. But the plant actually contains inductive and capacitive reactances exchanging stored energy via "reactive currents." These reactive currents circulate entirely between the capacitor and the loads, similar to a parallel resonant circuit, which accounts for the larger values of $|\underline{I}_C|$, $|\underline{I}_{12}|$, and $|\underline{I}_1|$.

The diagram of current phasors in Fig. 7.11 provides further insight on power-factor correction. This diagram is constructed taking $\angle\underline{V} = 0$, so $\angle\underline{I} = 0$ (since the corrected plant is purely resistive), whereas $\angle\underline{I}_C = +90°$. We then

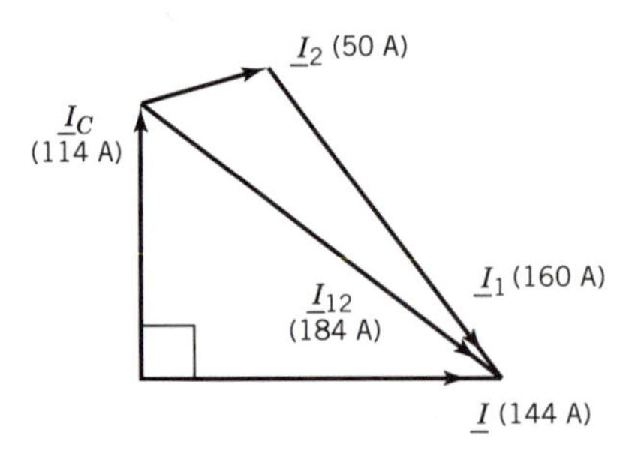

Figure 7.11 Phasor diagram for power-factor correction.

note that $\underline{I} = \underline{I}_C + \underline{I}_{12}$ and $\underline{I}_{12} = \underline{I}_1 + \underline{I}_2$, where $\angle\underline{I}_1 < 0$ (since load 1 has a lagging power factor) while $\angle\underline{I}_2 > 0$ (since load 2 has a leading power factor).

Example 7.6 *Improving Power-Transfer Efficiency*

Power-factor correction becomes particularly important when resistance in the source and/or transmission line wastes a significant amount of power. We'll illustrate this point using the circuit in Fig. 7.12, where the load impedance is $20 + j40\ \Omega$ and the high-voltage source and transmission line have a combined impedance of $4 + j15\ \Omega$.

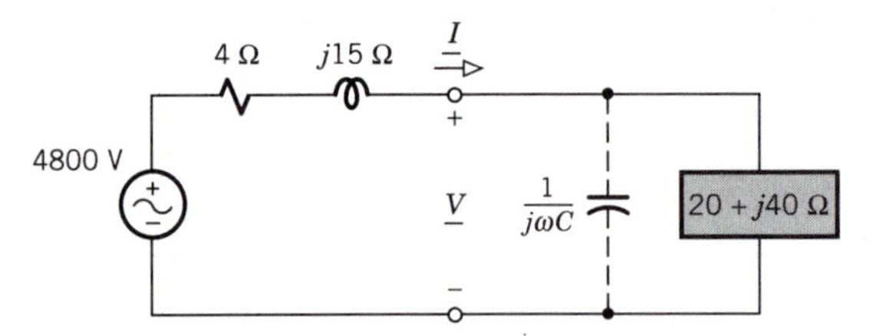

Figure 7.12 Circuit for power transfer via a transmission line.

Without power-factor correction, the total series impedance is the sum $24 + j55\ \Omega$. The resulting rms load current and voltage and the complex power supplied by the source are

$$I_{rms} = |\underline{I}| = 4800/|24 + j55| = 80.0 \text{ A}$$
$$V_{rms} = |\underline{V}| = |20 + j40|\,|\underline{I}| = 3580 \text{ V}$$
$$S = (24 + j55)|\underline{I}|^2 = 154 \text{ kW} + j352 \text{ kVAr}$$

From the resistance ratio, the power-transfer efficiency is

$$P_L/P = 20/(4 + 20) \approx 83\%$$

Appropriate power-factor correction should decrease the current and increase the load voltage and efficiency.

As indicated on the diagram, we'll reduce $|\underline{I}|$ and the power wasted in the 4-Ω resistance by putting a capacitor across the load terminals. Our strategy here is to choose C such that corrected load becomes purely resistive and therefore has unity power factor. Although we don't know the resulting load voltage, the value of ωC is easily found from the equivalent admittance

$$Y_{eq} = j\omega C + \frac{1}{20 + j40} = \frac{20}{2000} + j\left(\omega C - \frac{40}{2000}\right)$$

We should thus take $\omega C = 40/2000$ so the corrected load impedance becomes $Z_{eq} = (20/2000)^{-1} = 100 + j0\ \Omega$. Then the impedance seen by the source is $104 + j15\ \Omega$, so

$$\begin{aligned} I_{rms} &= |\underline{I}| = 4800/|104 + j15| = 45.7\text{ A} \\ V_{rms} &= |\underline{V}| = 100|\underline{I}| = 4570\text{ V} \\ S &= (104 + j15)|\underline{I}|^2 = 217\text{ kW} + j31\text{ kVAr} \\ P_L/P &= 100/(4 + 100) \approx 96\% \end{aligned}$$

A comparison with the previous values shows substantial improvement on all counts.

Exercise 7.5

An inductive ac motor operates from a 200-V(rms) source at 60 Hz. The motor draws $P = 8$ kW and $I_{rms} = 85$ A(rms). Calculate pf and Q. Then find the motor's equivalent impedance Z in rectangular form. (Save your work for use in the following exercise.)

Exercise 7.6

An 8-Ω resistive heating element and a capacitor C are connected in parallel with the motor in Exercise 7.5. The entire plant has unity power factor. Make a table like Table 7.3 for this plant, and find the design value of C.

Wattmeters

Electrical power can be measured using the meter shown in Fig. 7.13. The heart of this instrument is an **electrodynamometer movement** with a current-sensing coil carrying $i_i(t)$ and a voltage-sensing coil carrying $i_v(t)$. The movement is designed such that the steady-state deflection angle γ_{ss} of the rotating pointer is proportional to the average of the product of the currents $i_v(t)$ and $i_i(t)$. Thus,

$$\gamma_{ss} = \frac{K_M}{T}\int_{t_1}^{t_1+T} i_v(t)i_i(t)\,dt \tag{7.31}$$

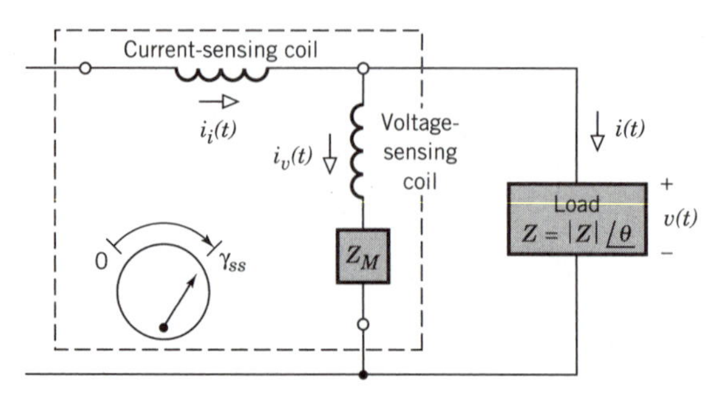

Figure 7.13 Wattmeter with electrodynamometer movement.

where K_M is a proportionality constant and T is the period of the current waveforms.

To measure the ac power delivered to the load, the impedance Z_M in series with the voltage-sensing coil must be a large resistance $R_M \gg |Z|$. If the load has

$$v(t) = \sqrt{2}|\underline{V}| \cos(\omega t + \phi_v) \qquad i(t) = \sqrt{2}|\underline{I}| \cos(\omega t + \phi_i)$$

then

$$i_v(t) = \frac{\sqrt{2}|\underline{V}|}{R_M} \cos(\omega t + \phi_v) \qquad i_i(t) \approx \sqrt{2}|\underline{I}| \cos(\omega t + \phi_i)$$

The corresponding steady-state pointer angle becomes

$$\begin{aligned} \gamma_{ss} &= \frac{2K_M|\underline{V}||\underline{I}|}{R_M T} \int_{t_1}^{t_1+T} \cos(\omega t + \phi_v) \cos(\omega t + \phi_i)\, dt \\ &= \frac{K_M}{R_M} |\underline{V}||\underline{I}| \cos(\phi_v - \phi_i) = \frac{K_M}{R_M} P \end{aligned} \qquad (7.32)$$

where we have used the fact that $\phi_v - \phi_i = \angle Z = \theta$. Therefore, with appropriate dial calibration, we have an ac **wattmeter**.

Actually, the meter reading will differ slightly from the predicted value because $i_i(t) = i(t) + i_v(t) \neq i(t)$. To obtain a more accurate result, you disconnect the load and subtract the no-load reading from the original value. More sophisticated instruments have a **compensating winding** that automatically cancels the no-load term.

Hereafter, we'll assume an **ideal ac wattmeter** represented in Fig. 7.14 where, by convention, the current input terminal is labeled ±. In terms of the voltage and current phasors, the meter reads

$$P = |\underline{V}|\, |\underline{I}| \cos(\angle \underline{V} - \angle \underline{I}) = \text{Re}\,[\underline{V}\underline{I}^*] \qquad (7.33)$$

This reading equals the power delivered through the meter since an ideal

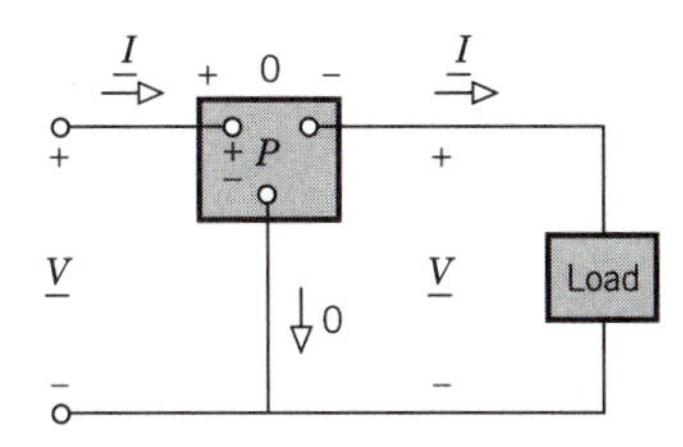

Figure 7.14 Ideal AC wattmeter measuring $P = \text{Re}[\underline{V}\underline{I}^*]$.

meter has zero voltage drop across the current-sensing coil and draws zero current through the voltage-sensing coil.

Exercise 7.7

By considering the magnitude and phase of $\underline{I}_v$, modify Eq. (7.32) to show that Fig. 7.13 becomes a **varmeter** measuring Q when $Z_M = j\omega L_M$ with $\omega L_M \gg Z$.

7.3 BALANCED THREE-PHASE CIRCUITS

A power system supplied by just one ac source is called a **single-phase** (1-ϕ) **circuit**. At large power levels, the oscillatory behavior of the instantaneous power in a single-phase circuit puts severe pulsating strain on the generating and load equipment. **Polyphase circuits** with multiple sources were developed in response to this problem.

The most common polyphase configuration is the **balanced three-phase** (3-ϕ) **circuit**, which has three ac sources arranged to achieve constant instantaneous power. As a further benefit, this type of circuit delivers more watts per kilogram of conductor than an equivalent single-phase circuit. For these reasons, almost all bulk electric power generation, distribution, and consumption take place via three-phase systems. The operating principles of balanced three-phase circuits thus deserve our attention here.

Three-Phase Sources and Symmetrical Sets

Figure 7.15*a* depicts the essential components of a three-phase generator. The stationary frame or **stator** holds three identical windings equally spaced around the cylindrical inner surface by $360°/3 = 120°$. The **rotor** is an electromagnet that rotates inside the stator at the rate $f = \omega/2\pi$ revolutions per second. A steam turbine or some other prime mover provides the mechanical energy needed to turn the rotor.

As the rotor sweeps past each winding, its magnetic field induces a sinusoidal voltage with frequency ω and rms value V_ϕ. If we take $v_a(t)$ as the phase reference, then $v_b(t)$ and $v_c(t)$ differ only by 120° phase shifts and

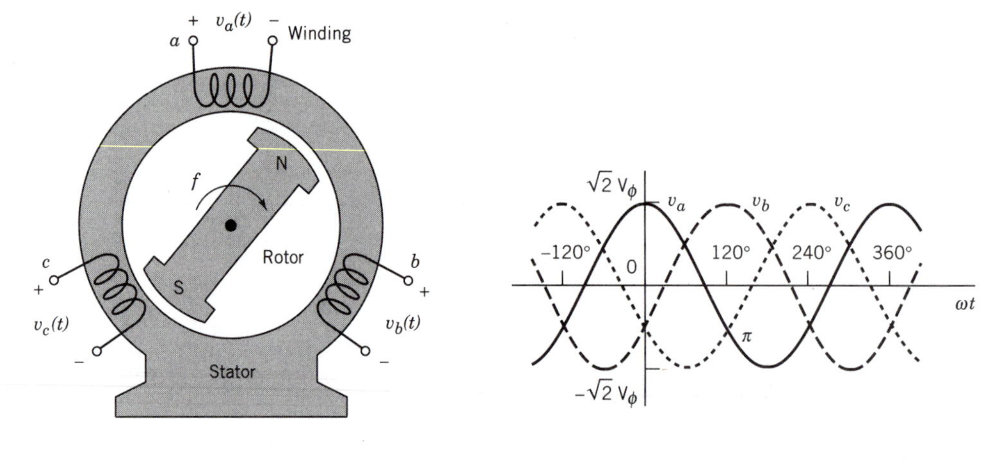

(a) Three-Phase generator *(b)* Voltage waveforms

Figure 7.15

$$\begin{aligned} v_a(t) &= \sqrt{2}V_\phi \cos \omega t \\ v_b(t) &= \sqrt{2}V_\phi \cos(\omega t - 120^\circ) \\ v_c(t) &= \sqrt{2}V_\phi \cos(\omega t + 120^\circ) \end{aligned} \tag{7.34}$$

These waveforms are plotted versus ωt in Fig. 7.15*b*.

Equation (7.34) defines a **symmetrical three-phase set** with **phase sequence** *a-b-c*. The phase sequence indicates the time sequence of the waveform peaks, and we'll assume an *a-b-c* sequence throughout. But regardless of phase sequence, the sum of the voltages is

$$v_a(t) + v_b(t) + v_c(t) = 0 \tag{7.35}$$

which holds at each and every instant of time. This property can be inferred from the waveforms in Fig. 7.15*b*. Its proof follows from the trigonometric relationship

$$\cos \alpha + \cos(\alpha - 120^\circ) + \cos(\alpha + 120^\circ) = 0 \tag{7.36}$$

for any angle α.

To bring out the significance of Eq. (7.35), consider the simple three-phase circuit in Fig. 7.16. Here, each winding of the generator has been represented by our standard source symbol, and the reference terminals have been joined at node n, called the **neutral**. The remaining three source terminals and the neutral are connected by four wires to a load consisting of three equal resistances tied together at the load's neutral N. Such loads having three *identical* branches are said to be **balanced**.

At a casual glance, Fig. 7.16 merely looks like three *single-phase* circuits with the neutral wire common to all three. But note that the balanced load

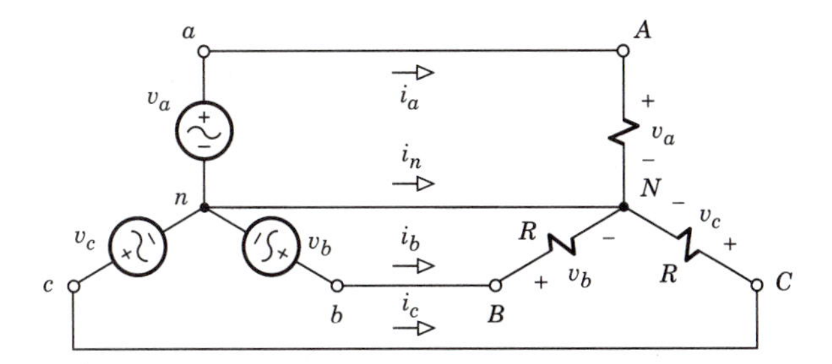

Figure 7.16 Simple three-phase circuit.

draws $i_a(t) = v_a(t)/R$, $i_b(t) = v_b(t)/R$, and $i_c(t) = v_c(t)/R$, so these currents also form a symmetrical three-phase set. Hence, the total instantaneous current through the neutral wire is

$$i_n(t) = -[i_a(t) + i_b(t) + i_c(t)] = -[v_a(t) + v_b(t) + v_c(t)]/R = 0$$

Since $i_n(t) = 0$ when the load is balanced, the neutral wire might as well be omitted! The three-phase circuit therefore needs just *three* wires from the generator to the load.

The instantaneous power supplied to the load via current $i_a(t)$ is $p_a(t) = Ri_a^2(t) = v_a^2(t)/R$. Figure 7.17 shows this waveform along with $p_b(t) = v_b^2(t)/R$ and $p_c(t) = v_c^2(t)/R$. Using point-by-point addition, we find that the total instantaneous power transferred from generator to load is

$$p(t) = p_a(t) + p_b(t) + p_c(t) = 3V_\phi^2/R$$

This result will be proved mathematically later on. Meanwhile, observe that $p(t)$ has a *constant* value!

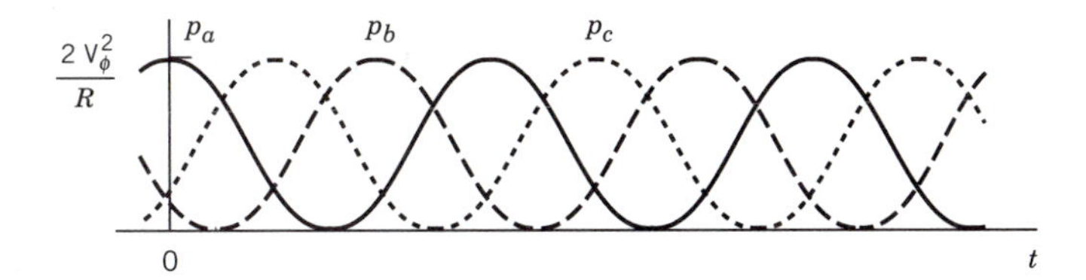

Figure 7.17 Instantaneous power waveforms in a three-phase circuit.

We are now in a position to make two statements of profound significance about this balanced three-phase circuit:

- The three-phase circuit requires fewer conductors than three single-phase circuits handling the same total power.
- The total instantaneous power is constant, rather than pulsating as in a single-phase circuit.

The latter property implies less vibration and mechanical strain at the generator and smoother power delivery to the load equipment — somewhat analogous to the advantages of a six-cylinder engine over a one-cylinder engine.

For a more detailed study of three-phase sources we turn to the frequency-domain diagram in Fig. 7.18 — descriptively named a **wye (Y) generator**. The neutral point n may or may not be externally available, or safety considerations may call for it to be grounded. The generator is labeled with two sets of voltage phasors: the **phase voltages** $\underline{V}_a$, $\underline{V}_b$, and $\underline{V}_c$, defined with respect to the neutral; and the **line voltages** $\underline{V}_{ab}$, $\underline{V}_{bc}$, and $\underline{V}_{ca}$, defined across pairs of terminals. Consistent with double-subscript notation, the line voltages are related to the phase voltages via

$$\underline{V}_{ab} = \underline{V}_a - \underline{V}_b \qquad \underline{V}_{bc} = \underline{V}_b - \underline{V}_c \qquad \underline{V}_{ca} = \underline{V}_c - \underline{V}_a$$

The name *line voltage* reflects the fact that $\underline{V}_{ab}$, $\underline{V}_{bc}$, and $\underline{V}_{ca}$ appear between the wires or "lines" connecting the generator to the load. These voltages are easily measured, and three-phase generators are commonly rated in terms of line voltage because phase voltages can only be measured when the neutral point is accessible.

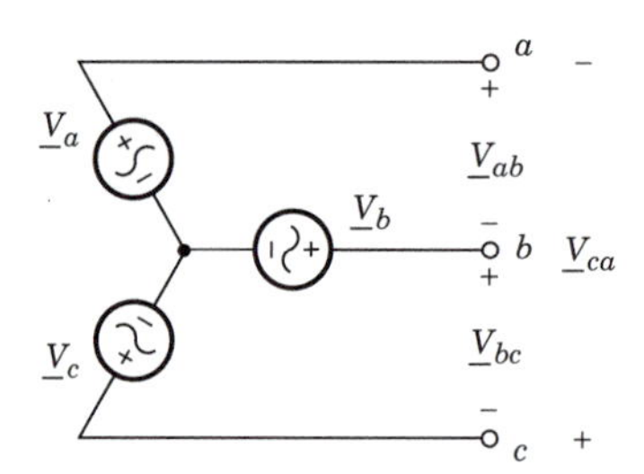

Figure 7.18 Frequency-domain diagram of a wye generator.

But the phase voltages of a wye generator correspond directly to the source voltages. Accordingly, we call V_ϕ the **rms phase voltage**, and we draw upon Eq. (7.34) to write

$$\underline{V}_a = V_\phi \angle 0^\circ \qquad \underline{V}_b = V_\phi \angle -120^\circ \qquad \underline{V}_c = V_\phi \angle 120^\circ \tag{7.37}$$

which is the phasor version of a symmetrical three-phase set. Figure 7.19*a* diagrams these phasors, and we'll use this diagram to find the resulting line voltages.

The phasor construction for $\underline{V}_{ab} = \underline{V}_a - \underline{V}_b$ in Fig. 7.19*b* immediately reveals that $\angle \underline{V}_{ab} = 30^\circ$. Then, letting $|\underline{V}_{ab}| = V_l$, the right triangle shows that

$$V_\phi^2 = (V_l/2)^2 + (V_\phi/2)^2$$

Solving for V_l yields the **rms line voltage**

$$V_l = \sqrt{3}V_\phi \approx 1.73 V_\phi \tag{7.38}$$

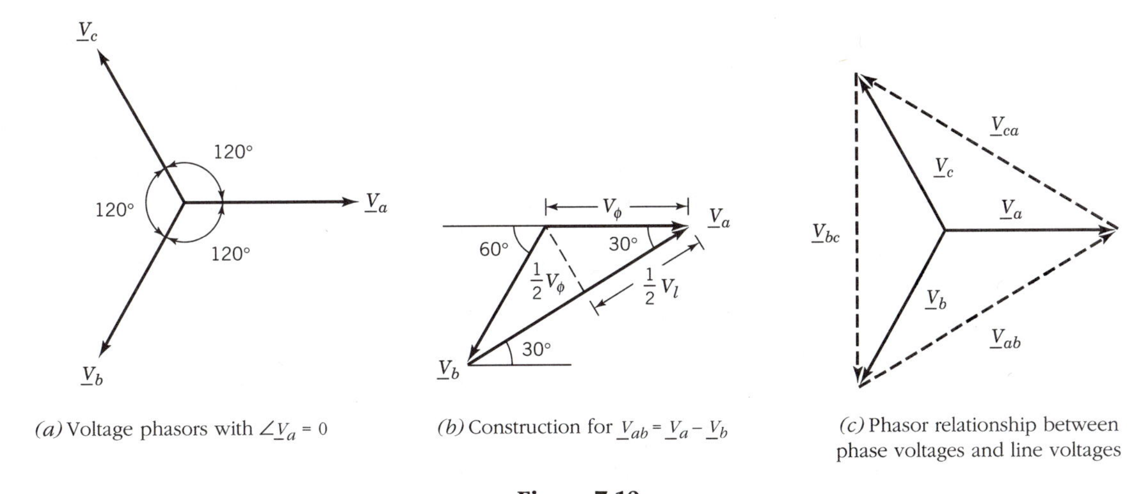

(a) Voltage phasors with $\angle \underline{V}_a = 0$

(b) Construction for $\underline{V}_{ab} = \underline{V}_a - \underline{V}_b$

(c) Phasor relationship between phase voltages and line voltages

Figure 7.19

By symmetry, the other line voltages have the same rms value but are shifted by $\pm 120°$ relative to $\underline{V}_{ab}$. Thus,

$$\underline{V}_{ab} = V_l \underline{/30°} \qquad \underline{V}_{bc} = V_l \underline{/-90°} \qquad \underline{V}_{ca} = V_l \underline{/150°} \tag{7.39}$$

which is another symmetrical three-phase set.

Figure 7.19*c* shows the relationships between the phase and line voltages. Clearly, $\underline{V}_{ab} + \underline{V}_{bc} + \underline{V}_{ca}$ equals zero, as it should to satisfy both KVL and the requirement for a symmetrical three-phase set. If we rearranged $\underline{V}_a$, $\underline{V}_b$, and $\underline{V}_c$ head to tail, then they too would form an equilateral triangle, confirming that $\underline{V}_a + \underline{V}_b + \underline{V}_c = 0$.

Three-phase windings may also be connected per Fig. 7.20, called a **delta (Δ) generator**. The line voltages equal the source voltages here, but we'll continue to use V_l for the rms line voltage. The neutral point and the phase

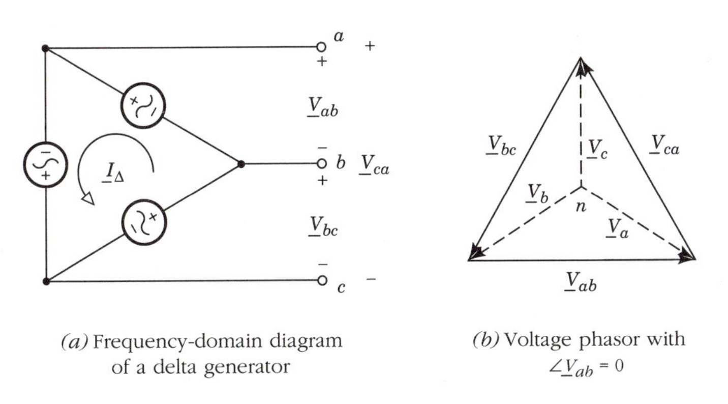

(a) Frequency-domain diagram of a delta generator

(b) Voltage phasor with $\angle \underline{V}_{ab} = 0$

Figure 7.20

voltages do not physically exist in a delta generator. Nonetheless, the equivalent neutral point and phase voltages are defined by the accompanying phasor diagram, where $\underline{V}_{ab}$ has been taken as the reference. Thus, $\underline{V}_{ab} = V_l \angle 0°$, $\underline{V}_a = (V_l/\sqrt{3}) \angle -30°$, and so forth.

Except for a possible phase shift of all voltages, a delta generator acts externally just like a three-terminal wye generator with the same rms line voltage. Delta generators differ internally from wye generators by the absence of the neutral point and the presence of the delta mesh. Theoretically, the interior mesh current $\underline{I}_\Delta$ equals zero because $\underline{V}_{ab} + \underline{V}_{bc} + \underline{V}_{ca} = 0$. However, any deviation from that voltage condition would produce a large and unwanted circulating current. Consequently, delta generators are usually found only in special applications.

Example 7.7 *Calculating Line Voltages*

Problem: Find $v_{ab}(t)$, $v_{bc}(t)$, and $v_{ca}(t)$ for a three-phase generator with $\underline{V}_a = 15$ kV(rms) $\angle 90°$ and *a-b-c* sequence.

Solution: Since $V_\phi = 15$ kV, the rms line voltage is

$$V_l = \sqrt{3} \times 15 \text{ kV} \approx 26 \text{ kV(rms)}$$

Then, from Fig. 7.19*c*, $\angle \underline{V}_{ab} = \angle \underline{V}_a + 30°$, $\angle \underline{V}_{bc} = \angle \underline{V}_{ab} - 120°$, and $\angle \underline{V}_{ca} = \angle \underline{V}_{ab} + 120° = \angle \underline{V}_{ab} - 240°$, so

$$\underline{V}_{ab} = 26 \text{ kV} \angle 120° \qquad \underline{V}_{bc} = 26 \text{ kV} \angle 0° \qquad \underline{V}_{ca} = 26 \text{ kV} \angle -120°$$

Finally, converting to the peak voltages $\sqrt{2} \times 26$ kV $= 36.8$ kV, we obtain

$$v_{ab}(t) = 36.8 \cos(\omega t + 120°) \text{ kV}$$
$$v_{bc}(t) = 36.8 \cos \omega t \text{ kV}$$
$$v_{ca}(t) = 36.8 \cos(\omega t - 120°) \text{ kV}$$

Exercise 7.8

Use the expansion for $\cos(\alpha \pm \beta)$ to prove Eq. (7.36).

Exercise 7.9

Find $\underline{V}_a$, $\underline{V}_b$, and $\underline{V}_c$ in Fig. 7.18 when $\underline{V}_{ab} = 52$ kV $\angle 0°$. (Save your results for use in Exercise 7.10.)

Balanced Wye Loads

A three-phase load consists of three impedance branches. These branches might correspond to three separate devices or to a three-phase device such as a motor with three windings. When the impedances are connected in a wye configuration like Fig. 7.21*a*, they form a **wye load**. The load is *balanced* if each branch has the *same* impedance $Z_Y = |Z_Y| \angle\theta$.

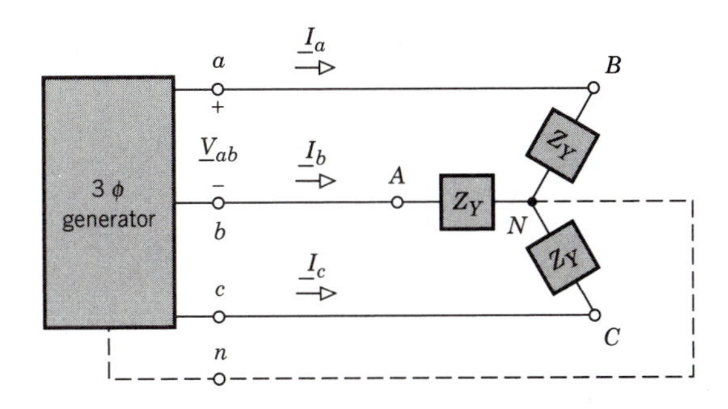

(a) Three-phase circuit with balanced wye load

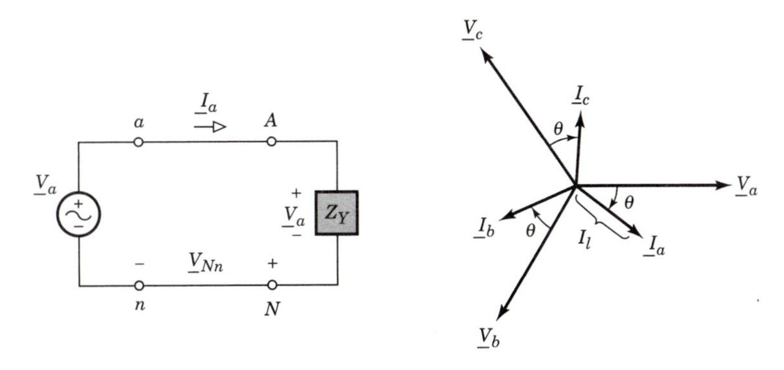

(b) Equivalent circuit for phase *a*

(c) Voltage and current phasors

Figure 7.21

The balanced condition has practical importance because the total instantaneous power to the load will be constant, even when the load includes energy-storage elements as well as resistance. Consequently, three-phase loads are usually designed to be balanced. As an added bonus, the symmetry of the balanced condition simplifies our work because

> A three-phase circuit with a balanced wye load can be analyzed using an equivalent circuit for just one phase.

We'll develop this technique to find the currents and average power drawn by a balanced wye load. Then we'll confirm that the total instantaneous power is constant.

For purposes of analysis, we assume that the lines from the generator are ideal conductors. We also temporarily assume a wye generator whose neutral point is connected to the load neutral *N* through an ideal conductor indicated by the dashed line. These assumptions allow us to draw phase *a* by itself, as shown in Fig. 7.21*b*. Here, $\underline{V}_{Nn} = 0$ and the source voltage $\underline{V}_a$ appears across the branch impedance, so $\underline{I}_a = \underline{V}_a/Z_Y$. Diagrams for the other two phases would be identical except for the subscripts on the voltages and currents. We thus deduce from the phase-*a* diagram alone that

$$\underline{I}_a = \underline{V}_a/Z_Y \qquad \underline{I}_b = \underline{V}_b/Z_Y \qquad \underline{I}_c = \underline{V}_c/Z_Y$$

Since $|\underline{I}_a| = |\underline{I}_b| = |\underline{I}_c| = V_\phi/|Z_Y|$ and $V_\phi = V_l/\sqrt{3}$, all three lines carry the same **rms line current** given by

$$I_l = \frac{V_\phi}{|Z_Y|} = \frac{V_l}{\sqrt{3}|Z_Y|} \tag{7.40}$$

Then, taking $\angle\underline{V}_a = 0°$, we have

$$\underline{I}_a = I_l \underline{/-\theta} \qquad \underline{I}_b = I_l \underline{/-120° - \theta} \qquad \underline{I}_c = I_l \underline{/120° - \theta} \tag{7.41}$$

Figure 7.21*c* gives the phasor diagram for the line currents and phase voltages.

Clearly, the currents form a symmetrical three-phase set, regardless of the value of Z_Y. Hence, the neutral wire again carries no current and it may be removed from Fig. 7.21*a*. Nonetheless, the symmetry still ensures that

$$\underline{V}_{Nn} = 0 \tag{7.42}$$

We can also drop our assumption about the generator's configuration because a delta generator acts at its terminals just like a three-terminal wye generator. Equations (7.40) and (7.41) therefore hold for either type of generator.

Having determined the voltages and currents, we turn to the power drawn by the load. Each phase has rms voltage V_ϕ, rms current I_l, and impedance Z_Y, so the real and reactive power *per phase* are

$$\begin{aligned} P_\phi &= \text{Re}[Z_Y]I_l^2 = V_\phi I_l \cos\theta \\ Q_\phi &= \text{Im}[Z_Y]I_l^2 = V_\phi I_l \sin\theta \end{aligned} \tag{7.43}$$

Since the generator supplies three identical phases, the *total* real and reactive power from the generator are

$$\begin{aligned} P &= 3P_\phi = 3V_\phi I_l \cos\theta = \sqrt{3}V_l I_l \cos\theta \\ Q &= 3Q_\phi = 3V_\phi I_l \sin\theta = \sqrt{3}V_l I_l \sin\theta \end{aligned} \tag{7.44}$$

where we have noted that $3V_\phi = \sqrt{3}V_l$. The total apparent power is

$$|S| = \sqrt{P^2 + Q^2} = \sqrt{3}V_l I_l \tag{7.45}$$

from which

$$\text{pf} = P/|S| = \cos\theta$$

Thus, the three-phase power factor exactly equals the power factor of a single branch impedance.

Although the instantaneous power in each branch exhibits the same oscillating behavior found in single-phase circuits, the total instantaneous power will be constant. This key property follows from

$$p(t) = v_a(t)i_a(t) + v_b(t)i_b(t) + v_c(t)i_c(t)$$

where $v_a(t) = \sqrt{2}V_\phi \cos \omega t$, $i_a(t) = \sqrt{2}I_l \cos (\omega t - \theta)$, and the other voltages and currents have $\pm 120°$ phase shifts. Thus,

$$\begin{aligned} p(t) &= 2V_\phi I_l [\cos \omega t \cos (\omega t - \theta) \\ &\quad + \cos (\omega t - 120°) \cos (\omega t - 120° - \theta) \\ &\quad + \cos (\omega t + 120°) \cos (\omega t + 120° - \theta)] \\ &= V_\phi I_l [3 \cos \theta + \cos (2\omega t - \theta) \\ &\quad + \cos (2\,\omega t - \theta - 120°) + \cos (2\,\omega t - \theta + 120°)] \end{aligned}$$

The last three terms of this expression sum to zero per Eq. (7.36), leaving the constant value

$$p(t) = 3V_\phi I_l \cos \theta = 3P_\phi = P$$

If $Z_Y = R$, then $\theta = 0$, $I_l = V_\phi / R$, and $p(t) = 3V_\phi^2/R$, as was asserted for the balanced wye load consisting of three identical resistors back in Fig. 7.16.

Example 7.8 ***Three-Phase Circuit with Line Impedances***

Figure 7.22*a* depicts a three-phase transmission line connecting a high-voltage generator to a load. Each phase of the line has impedance Z_l, and the load is a balanced wye with branch impedance Z. We are given that

$$|\underline{V}_{ab}| = 45 \text{ kV} \qquad Z_l = 0.5 + j3\ \Omega \qquad Z = 4.5 + j9\ \Omega$$

Our task is to find the rms line current and the various powers. Since the load plus transmission line presents a balanced wye condition to the generator, we can expedite the analysis using an equivalent circuit for one phase.

First, we calculate the phase voltage at the generator from Eq. (7.38) with $V_l = |\underline{V}_{ab}|$, so

$$V_\phi = 45 \text{ kV}/\sqrt{3} \approx 26 \text{ kV}$$

Second, we take $\underline{V}_a$ as the reference and draw the equivalent phase-*a* loop in Fig. 7.22*b*. The total phase impedance is

$$Z_Y = Z_l + Z = 5 + j12\ \Omega = 13\ \Omega \underline{/67.4°}$$

from which

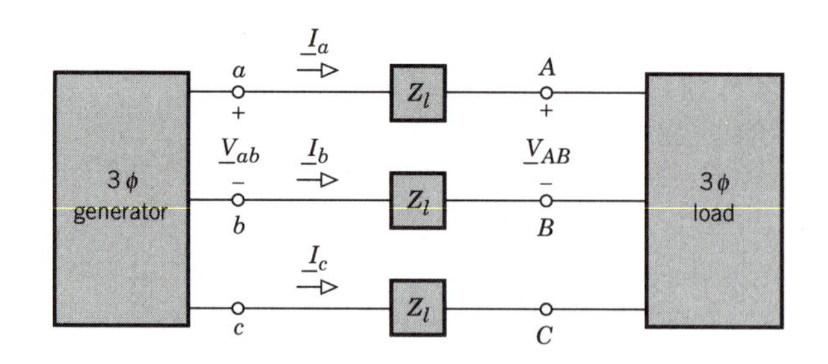

(a) Three-phase circuit with line impedances

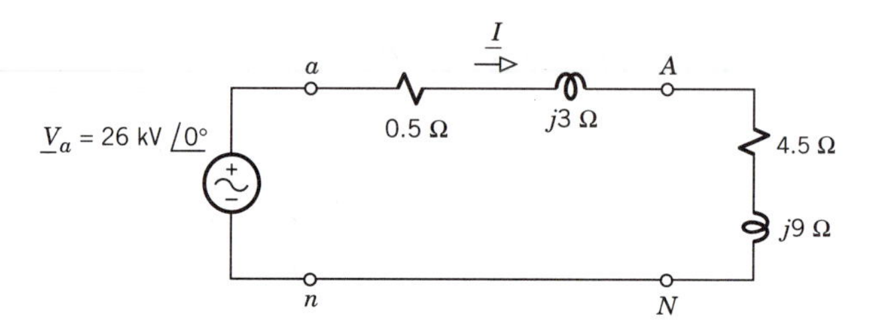

(b) Equivalent circuit for phase *a*

Figure 7.22

$$I_l = |\underline{I}_a| = 26\ \text{kV}/13\ \Omega = 2\ \text{kA(rms)}$$

Thus,

$$P = 3 \times 5\ \Omega \times (2\ \text{kA})^2 = 60\ \text{MW}$$
$$Q = 3 \times 12\ \Omega \times (2\ \text{kA})^2 = 144\ \text{MVAr}$$

Notice that powers come out in *mega* values when we work with kilovolts, kiloamps, and ohms.

Of the 60 MW supplied by the generator, the power that reaches the load is $P_L = 3 \times 4.5\ \Omega \times (2\ \text{kA})^2 = 54$ MW, so

$$P_L/P = 54/60 = 90\%$$

Although this would appear to be a reasonable efficiency, the transmission line still dissipates $P_L - P = 6$ MW! Power-factor correction should therefore be employed to reduce the amount of wasted power.

Exercise 7.10

Suppose the generator in Fig. 7.21*a* is wye connected and has $\underline{V}_{ab} = 52\ \text{kV}\ \underline{/0^\circ}$. If $\underline{I}_a = 1\ \text{kA}\ \underline{/-90^\circ}$, then what are the values of Z_Y, P, and Q? (Save your results for use in Exercise 7.12).

Balanced Delta Loads

Three-phase loads may also be arranged in a delta configuration, and Fig. 7.23 diagrams a **balanced delta load** having equal branch impedances $Z_\Delta = |Z_\Delta| \underline{/\theta}$. This load may be driven by either a delta or wye generator, even though there is no point for a neutral connection. If the lines from the generator are ideal conductors, then a line voltage appears across each impedance. Hence, we could analyze this configuration starting with the *branch* currents $\underline{I}_{ab} = \underline{V}_{ab}/Z_\Delta$, $\underline{I}_{bc} = \underline{V}_{bc}/Z_\Delta$, and $\underline{I}_{ca} = \underline{V}_{ca}/Z_\Delta$.

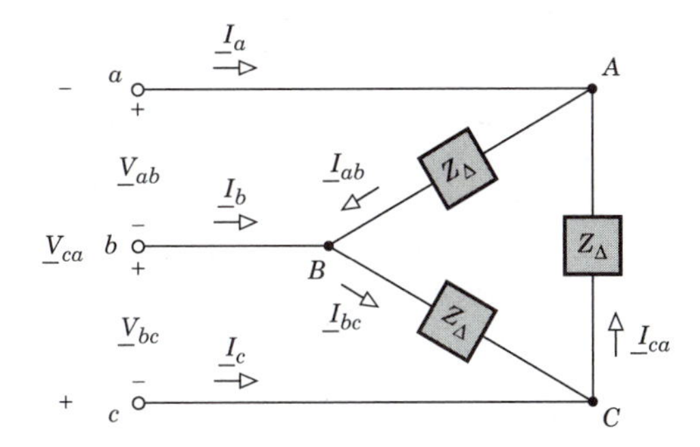

Figure 7.23 Balanced delta load.

But rather than beginning again from scratch, we'll take advantage of our previous results by finding a balanced *wye* load that draws the same line currents as a balanced delta load. For this purpose we note that

$$\underline{I}_a = \underline{I}_{ab} - \underline{I}_{ca} = (\underline{V}_{ab} - \underline{V}_{ca})/Z_\Delta$$

where

$$\begin{aligned}\underline{V}_{ab} - \underline{V}_{ca} &= (\underline{V}_a - \underline{V}_b) - (\underline{V}_c - \underline{V}_a)\\ &= 2\underline{V}_a - \underline{V}_b - \underline{V}_c = 3\underline{V}_a - (\underline{V}_a + \underline{V}_b + \underline{V}_c)\end{aligned}$$

Then, since $\underline{V}_a + \underline{V}_b + \underline{V}_c = 0$,

$$\underline{I}_a = 3\underline{V}_a/Z_\Delta = \underline{V}_a/(Z_\Delta/3)$$

It likewise follows that $\underline{I}_b = \underline{V}_b/(Z_\Delta/3)$ and $\underline{I}_c = \underline{V}_c/(Z_\Delta/3)$. A balanced delta load therefore acts like a balanced wye load with

$$Z_Y = Z_\Delta/3 \tag{7.46}$$

which we call the **equivalent wye impedance**.

Combining Eqs. (7.40) and (7.46) yields the rms line current

$$I_l = \frac{3V_\phi}{|Z_\Delta|} = \frac{\sqrt{3}V_l}{|Z_\Delta|} \tag{7.47}$$

as contrasted with the rms branch currents $|\underline{I}_{ab}| = |\underline{I}_{bc}| = |\underline{I}_{ca}| = V_l/|Z_\Delta|$. The power relations hold as they stand back in Eqs. (7.43) and (7.44) with $Z_Y = Z_\Delta/3$ and $\theta = \angle Z_\Delta$. Consequently, a delta load draws three times the current and power of a wye load built with the same impedances.

To summarize, we have seen that the symmetry of a balanced three-phase circuit leads to simple expressions for voltage, current, and power with either a wye or delta load. More complicated circuits involving line impedances or combined loads are also easy to analyze as long as symmetry is preserved so we can work with an equivalent circuit for one phase alone.

Example 7.9 *Power-Factor Correction with Parallel Loads*

Figure 7.24*a* represents a three-phase system with **parallel loads**. The generator has $V_l = 780$ V and $f = 60$ Hz. Load 1 is a balanced delta with branch impedance $Z_{\Delta 1} = 30 + j60\ \Omega$, while load 2 is a balanced wye with branch impedance $Z_{Y2} = 30 + j0\ \Omega$. You want to add capacitors to correct the system's power factor, and predict the resulting values of P and I_l.

The equivalent wye impedance of the delta load is $Z_{Y1} = Z_{\Delta 1}/3 = 10 + j20\ \Omega$, and you know that the potential at its equivalent neutral $N1$ equals the potential at $N2$ in the wye load. Accordingly, you draw the equivalent phase-a impedance in Fig. 7.24*b*, which brings out the fact that

$$Z_Y = Z_{Y1} \| Z_{Y2} = (10 + j20) \| 30 = 12 + j9\ \Omega = 15\ \Omega\ \underline{/36.9^\circ}$$

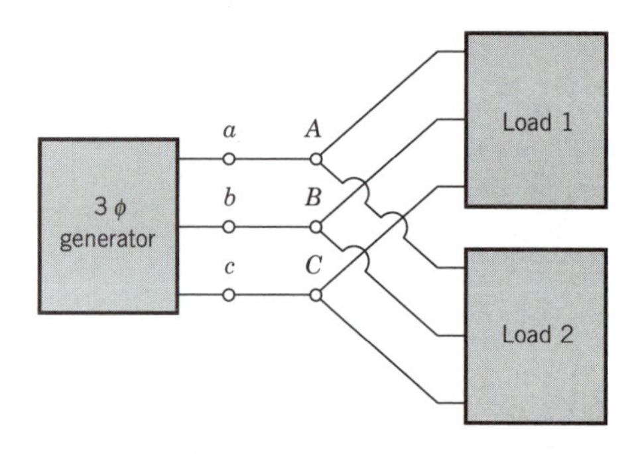

(a) Three-phase system with parallel loads

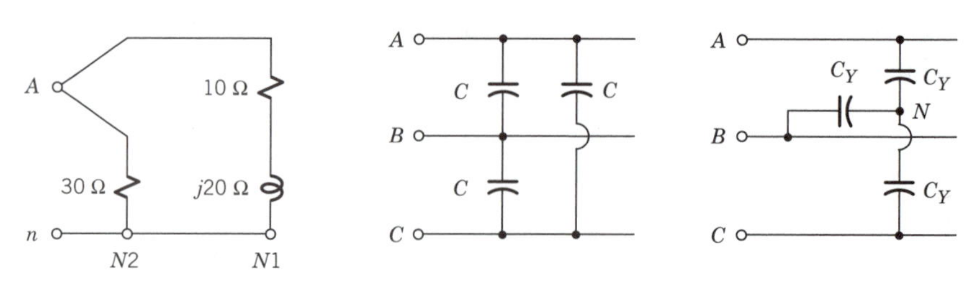

(b) Equivalent phase *a* impedances

(c) Capacitors for power-factor correction

Figure 7.24

Since $V_\phi = 780/\sqrt{3} \approx 450$ V, the rms current for the combined loads is 450 V/15 Ω = 30 A.

Next, to determine the required power-factor correction, you calculate the reactive power per phase via

$$Q_\phi = 9\ \Omega \times (30\ \text{A})^2 = +8.1\ \text{kVar}$$

You therefore want to connect a capacitor to each phase such that $Q_C = -\omega C|\underline{V}_C|^2 = -Q_\phi$. To preserve balanced conditions, the three capacitors must be equal and arranged in a delta or wye configuration, as shown in Fig. 7.24*c*. For the delta configuration, $|\underline{V}_C| = V_l$ and

$$C_\Delta = Q_\phi/2\pi f V_l^2 = 63.9\ \mu\text{F}$$

For the wye configuration, $|\underline{V}_C| = V_\phi = V_l/\sqrt{3}$ so

$$C_Y = 3C_\Delta = 192\ \mu\text{F}$$

Thus, the delta configuration involves less capacitance, but the capacitors must be rated for the full line voltage (780 V) rather than the smaller phase voltage (450 V).

Finally, since ideal capacitors draw no power, the total average power supplied by the generator is still given by

$$P = 3 \times 12\ \Omega \times (30\ \text{A})^2 = 32.4\ \text{kW}$$

However, with $Q_\phi + Q_C = 0$, $P = |S| = \sqrt{3}V_l I_l$ and the rms line current will be

$$I_l = P/\sqrt{3}V_l = 24\ \text{A}$$

as compared to the 30-A current drawn by the combined load.

Exercise 7.11

A balanced delta load with $Z_\Delta = 4.5 + j6\ \Omega$ is supplied by a generator having $\underline{V}_a = 500\ \text{V}\ \angle 0°$ and $f = 50$ Hz. Three equal capacitors are added in parallel with the load branches to get unity power factor. Evaluate the rms line current before and after power-factor correction, and determine the required value of C_Δ.

Power Measurements

Be it balanced or unbalanced, a three-phase load consists of three branches capable of consuming power. It would therefore seem that *three* separate measurements are needed to find the total power delivered to the load. This is indeed true for *unbalanced wye-wye systems* because such systems have four wires that carry nonzero current. But any other three-phase system has just three current-carrying wires. We are therefore free to choose one of the wires for the voltage reference and measure the power flow through the other two wires, getting by with only only *two* measurements.

As a case in point, take a balanced or unbalanced delta load with two wattmeters arranged as shown in Fig. 7.25. The complex power delivered to Z_{ab} is $S_{ab} = \underline{V}_{ab}\underline{I}_{ab}^*$, so we can express the real power consumed by this branch as $P_{ab} = \text{Re}[S_{ab}]$. Adding like expressions for P_{bc} and P_{ca} gives the total load power

$$P = \text{Re}[S_{ab}] + \text{Re}[S_{bc}] + \text{Re}[S_{ca}] = \text{Re}[S] \tag{7.48a}$$

where, from the conservation property,

$$S = \underline{V}_{ab}\underline{I}_{ab}^* + \underline{V}_{bc}\underline{I}_{bc}^* + \underline{V}_{ca}\underline{I}_{ca}^* \tag{7.48b}$$

We then draw upon Eq. (7.33) to write the two wattmeter readings as

$$P_1 = \text{Re}[\underline{V}_{ab}\underline{I}_a^*] \qquad P_2 = \text{Re}[\underline{V}_{cb}\underline{I}_c^*] \tag{7.49}$$

in which $\underline{V}_{cb} = -\underline{V}_{bc}$. We'll prove that $P_1 + P_2$ equals the total power P.

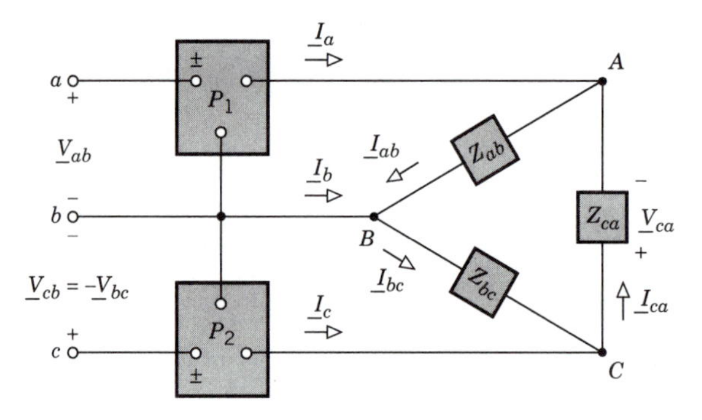

Figure 7.25 Two wattmeters connected to a delta load.

Our proof starts by observing from Fig. 7.25 that $\underline{I}_{ab} = \underline{I}_a + \underline{I}_{ca}$ and $\underline{I}_{bc} = \underline{I}_{ca} - \underline{I}_c$. Substituting these relations into Eq. (7.48b) yields

$$\begin{aligned} S &= \underline{V}_{ab}(\underline{I}_a + \underline{I}_{ca})^* + \underline{V}_{bc}(\underline{I}_{ca} - \underline{I}_c)^* + \underline{V}_{ca}\underline{I}_{ca}^* \\ &= \underline{V}_{ab}\underline{I}_a^* - \underline{V}_{bc}\underline{I}_c^* + (\underline{V}_{ab} + \underline{V}_{bc} + \underline{V}_{ca})\underline{I}_a^* \\ &= \underline{V}_{ab}\underline{I}_a^* + \underline{V}_{cb}\underline{I}_c^* \end{aligned}$$

where we have noted that $-\underline{V}_{bc} = \underline{V}_{cb}$ and $\underline{V}_{ab} + \underline{V}_{bc} + \underline{V}_{ca} = 0$ under any conditions. Now, using Eqs. (7.48a) and (7.49),

$$\begin{aligned} P &= \text{Re}[S] = \text{Re}[\underline{V}_{ab}\underline{I}_a^* + \underline{V}_{cb}\underline{I}_c^*] \\ &= \text{Re}[\underline{V}_{ab}\underline{I}_a^*] + \text{Re}[\underline{V}_{cb}\underline{I}_c^*] = P_1 + P_2 \end{aligned} \tag{7.50}$$

Hence, the sum of the wattmeter readings equals the total power to a balanced or unbalanced delta load.

Additionally, if the load happens to be *balanced,* then the two wattmeter readings provide sufficient data to determine the magnitude of the total *reactive* power. We'll develop the relationship by taking an *a-b-c* phase sequence with $\angle \underline{V}_a = 0°$, so

$$\underline{V}_{ab} = V_l \underline{/30°} \qquad \underline{V}_{cb} = -\underline{V}_{bc} = V_l \underline{/90°}$$
$$\underline{I}_a^* = I_l \underline{/\theta} \qquad \underline{I}_c^* = I_l \underline{/\theta - 120°}$$

The *difference* between the two readings in Eq. (7.49) is then

$$\begin{aligned} P_2 - P_1 &= \mathrm{Re}[V_l I_l \underline{/\theta - 30°}] - \mathrm{Re}[V_l I_l \underline{/\theta + 30°}] \\ &= V_l I_l [\cos(\theta - 30°) - \cos(\theta + 30°)] \end{aligned}$$

But

$$\cos(\theta - 30°) - \cos(\theta + 30°) = 2 \sin\theta \sin 30° = \sin\theta$$

so

$$P_2 - P_1 = V_l I_l \sin\theta = Q/\sqrt{3}$$

Repeating this analysis with an *a-c-b* phase sequence yields $P_2 - P_1 = -Q/\sqrt{3}$, and we account for both possibilities by writing

$$|Q| = \sqrt{3}|P_2 - P_1| \tag{7.51}$$

Thus, you can calculate $|Q|$ from P_1 and P_2 even when you don't know the phase sequence.

Equations (7.50) and (7.51) also hold for a balanced or unbalanced wye load, provided that the unbalanced load does not have a neutral connection. Furthermore, the voltage reference for the two wattmeters may be any one of the three current-carrying lines. In all cases, however, the individual readings P_1 and P_2 have no particular physical significance. Indeed, one of the readings may even turn out to be zero or negative.

Example 7.10 *Calculation of Wattmeter Readings*

Again consider the three-phase system in Fig. 7.24*a*, which has $V_l = 780$ V and parallel balanced loads. Prior to power-factor correction, we found in Example 7.9 that $\theta = \angle Z_Y = 36.9°$ and $I_l = 30$ A. Now we'll calculate the readings of two wattmeters inserted at terminals *a* and *c*, with terminal *b* being voltage reference. Thus, from Eq. (7.49), the meters read $P_1 = \mathrm{Re}[\underline{V}_{ab}\underline{I}_a^*]$ and $P_2 = \mathrm{Re}[\underline{V}_{cb}\underline{I}_c^*]$.

Taking $\angle \underline{V}_a = 0°$, the phasors needed for the calculation are

$$\underline{V}_{ab} = 780 \text{ V } \underline{/30^\circ} \qquad \underline{I}_a = 30 \text{ A } \underline{/-36.9^\circ}$$
$$\underline{V}_{cb} = 780 \text{ V } \underline{/90^\circ} \qquad \underline{I}_c = 30 \text{ A } \underline{/83.1^\circ}$$

Hence,

$$P_1 = \text{Re}[\underline{V}_{ab}\underline{I}_a^*] = \text{Re}[780 \underline{/30^\circ} \times 30 \underline{/36.9^\circ}] = 23{,}400 \cos 66.9^\circ = 9.2 \text{ kW}$$
$$P_2 = \text{Re}[\underline{V}_{cb}\underline{I}_c^*] = \text{Re}[780 \underline{/90^\circ} \times 30 \underline{/-83.1^\circ}] = 23{,}400 \cos 6.9^\circ = 23.2 \text{ kW}$$

so

$$P_1 + P_2 = 32.4 \text{ kW} \qquad \sqrt{3}|P_2 - P_1| = 24.3 \text{ kVAr}$$

These readings agree with our previous results that $P = 32.4$ kW and $Q = 3Q_\phi = 24.3$ kVAr.

Exercise 7.12

Let two wattmeters be inserted into the system in Exercise 7.10 such that $P_1 = \text{Re}[\underline{V}_{ac}\underline{I}_a^*]$ and $P_2 = \text{Re}[\underline{V}_{bc}\underline{I}_b^*]$. Evaluate P_1 and P_2 to verify that Eqs. (7.50) and (7.51) yield correct results.

7.4 UNBALANCED THREE-PHASE CIRCUITS†

Much of the simplifying symmetry goes out the window when a three-phase circuit is *unbalanced*. A circuit sometimes becomes unbalanced because the generator voltages do not form a symmetrical three-phase set. But the unbalanced state usually results from loads with unequal impedances. This section briefly examines analysis methods for common types of unbalanced loads.

Composite Loads

Three-phase sources often supply the power for single-phase loads along with balanced three-phase loads. We'll refer to this situation as a **composite load**.

Figure 7.26 depicts a simple composite load with one single-phase load impedance Z_{cn} connected line-to-neutral. If the wires have negligible impedance, then the voltages at the three-phase load remain unaffected by the single-phase load, and vice versa. We may therefore use our previous methods to find $\underline{I}_A$, $\underline{I}_B$, and $\underline{I}_C$, and we know that $\underline{I}_N = 0$. Thus, the currents from the generator are

$$\underline{I}_a = \underline{I}_A \qquad \underline{I}_b = \underline{I}_B \qquad \underline{I}_c = \underline{I}_C + \underline{I}_{cn} \qquad \underline{I}_n = -\underline{I}_{cn}$$

where $\underline{I}_{cn} = \underline{V}_c / Z_{cn}$.

Similar arguments hold when there are two or more single-phase loads, which may be connected line-to-neutral or line-to-line. The total real power

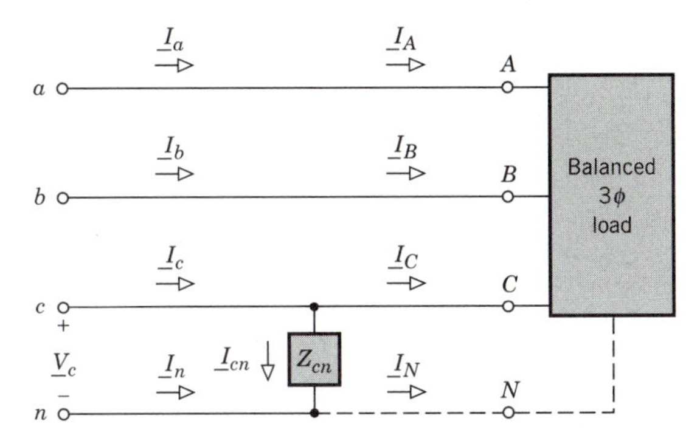

Figure 7.26 Composite load with three-phase and single-phase impedances.

P and reactive power Q are calculated by adding the single-phase powers to the three-phase load powers. The equivalent power factor seen by the generator can then be found from the basic definition

$$\text{pf} \triangleq P/|S| = P/\sqrt{P^2 + Q^2} \tag{7.52}$$

In general, this power factor differs from the power factor of the three-phase load by itself.

Exercise 7.13

Find $\underline{I}_a$, $\underline{I}_b$, $\underline{I}_c$, and $\underline{I}_n$ in polar form when the three-phase load in Fig. 7.26 has $Z_Y = 12 + j0\ \Omega$, $Z_{cn} = -j5\ \Omega$, and $\underline{V}_a = 600\text{ V}\ \underline{/0°}$. Then calculate P, Q, and pf.

Unbalanced Wye and Delta Loads

Occasionally, a balanced three-phase load becomes unbalanced because one of the branches has developed an **open-circuit fault**. Since the fault eliminates a current path, such cases are easily analyzed by direct application of Kirchhoff's laws to the resulting frequency-domain diagram.

Loads with finite but unequal impedances call for further consideration. In particular, the analysis of **unbalanced wye loads** requires special attention regarding the presence or absence of a neutral path.

On the one hand, consider the unbalanced *four-wire* circuit in Fig. 7.27*a*. The neutral connection is only possible with a wye generator, but the path may be indirect via grounds at the generator and load. If all current paths have zero impedance, then the neutral connection ensures that $\underline{V}_{Nn} = 0$ and the corresponding phase voltage appears across each branch impedance. Hence,

$$\underline{I}_a = \underline{V}_a/Z_a \qquad \underline{I}_b = \underline{V}_b/Z_b \qquad \underline{I}_c = \underline{V}_c/Z_c \tag{7.53}$$

and

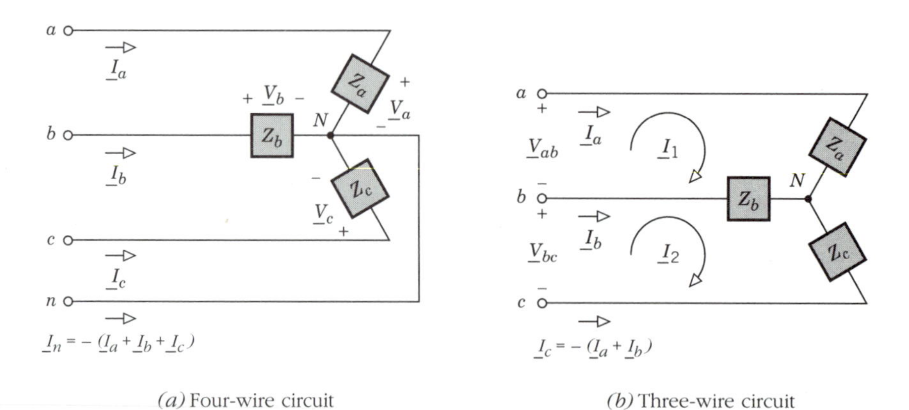

(a) Four-wire circuit (b) Three-wire circuit

Figure 7.27 Unbalanced wye loads.

$$\underline{I}_n = -(\underline{I}_a + \underline{I}_b + \underline{I}_c)$$

whereas a balanced system would have $\underline{I}_n = 0$.

On the other hand, consider the *three-wire* circuit in Fig. 7.27*b*. Here, the ungrounded or "floating" load neutral N forces $\underline{I}_a + \underline{I}_b + \underline{I}_c = 0$. But the phase voltages no longer appear across the branch impedances, so we now work with mesh currents. Specifically, we let $\underline{I}_1 = \underline{I}_a$ and $\underline{I}_2 = -\underline{I}_c$ to write the matrix mesh equation

$$\begin{bmatrix} Z_a + Z_b & -Z_b \\ -Z_b & Z_b + Z_c \end{bmatrix} \begin{bmatrix} \underline{I}_1 \\ \underline{I}_2 \end{bmatrix} = \begin{bmatrix} \underline{V}_{ab} \\ \underline{V}_{bc} \end{bmatrix} \tag{7.54a}$$

After solving for $\underline{I}_1$ and $\underline{I}_2$, we can calculate the line currents

$$\underline{I}_a = \underline{I}_1 \qquad \underline{I}_b = \underline{I}_2 - \underline{I}_1 \qquad \underline{I}_c = -\underline{I}_2 \tag{7.54b}$$

If desired, the voltage at the load's neutral is determined from any of the three phases via

$$\underline{V}_{Nn} = \underline{V}_a - Z_a\underline{I}_a = \underline{V}_b - Z_b\underline{I}_b = \underline{V}_c - Z_c\underline{I}_c$$

A balanced system would, of course, have $\underline{V}_{Nn} = 0$.

The effect of generator impedances and/or line impedances may be incorporated into Eqs. (7.53) and (7.54). You simply let Z_a be the total impedance in the path from $\underline{V}_a$ to N, and likewise for Z_b and Z_c. Keep in mind, however, that Eq. (7.53) assumes a zero-impedance path for the neutral current.

Circuits with **unbalanced delta loads** cannot have a neutral connection, and the branch currents are easily calculated when known line voltages appear directly across the branch impedances. Thus, in Fig. 7.28 we have

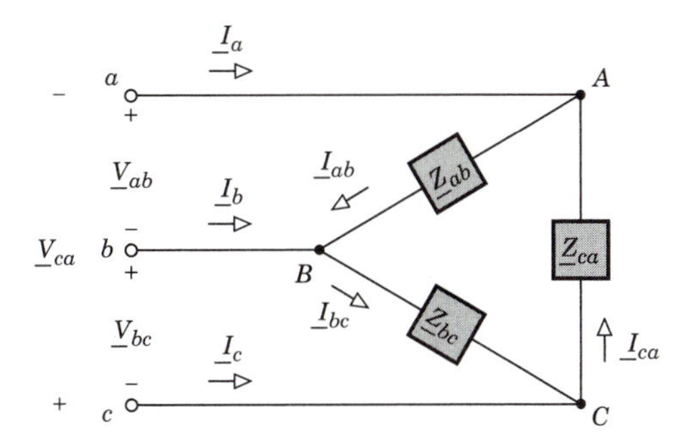

Figure 7.28 Unbalanced delta load.

$$\underline{I}_{ab} = \underline{V}_{ab}/Z_{ab} \qquad \underline{I}_{bc} = \underline{V}_{bc}/Z_{bc} \qquad \underline{I}_{ca} = \underline{V}_{ca}/Z_{ca} \tag{7.55a}$$

and

$$\underline{I}_a = \underline{I}_{ab} - \underline{I}_{ca} \qquad \underline{I}_b = \underline{I}_{bc} - \underline{I}_{ab} \qquad \underline{I}_c = \underline{I}_{ca} - \underline{I}_{bc} \tag{7.55b}$$

all obtained by inspection of the diagram.

But the problem becomes more complicated when generator and/or line impedances cause an intervening voltage drop, as represented in Fig. 7.29*a*. We handle this configuration by calling upon the **delta-to-wye transformation** derived back in Section 4.6 for resistive networks. In our present context, the unbalanced delta load can be converted to an **equivalent wye load** whose branch impedances at terminals *A*, *B*, and *C*, are given by

$$Z_A = \frac{Z_{AB}Z_{CA}}{\Sigma Z} \qquad Z_B = \frac{Z_{BC}Z_{AB}}{\Sigma Z} \qquad Z_C = \frac{Z_{CA}Z_{BC}}{\Sigma Z} \tag{7.56a}$$

where

$$\Sigma Z \triangleq Z_{AB} + Z_{BC} + Z_{CA} \tag{7.56}$$

For the special case of a balanced delta, $Z_{AB} = Z_{BC} = Z_{CA} = Z_\Delta$ and $\Sigma Z = 3Z_\Delta$, so Eq. (7.56a) reduces to our previous result that $Z_Y = Z_\Delta/3$.

For the unbalanced delta in Fig. 7.29*a*, we use Eq. (7.56a) to replace it by the equivalent unbalanced wye, as shown in Fig. 7.29*b*. The resulting unbalanced three-wire wye system is then analyzed by applying Eq. (7.54) with $Z_a = Z_{aA} + Z_A$, etc. An example will help clarify this procedure.

Example 7.11 ***Calculations for an Unbalanced Delta Load***

Suppose the circuit in Fig. 7.29*a* has the following parameter values:

$$V_l = 900 \text{ V} \qquad Z_{aA} = Z_{bB} = Z_{cC} = 5\ \Omega$$
$$Z_{AB} = 30\ \Omega \qquad Z_{BC} = 60 + j60\ \Omega \qquad Z_{CA} = j30\ \Omega$$

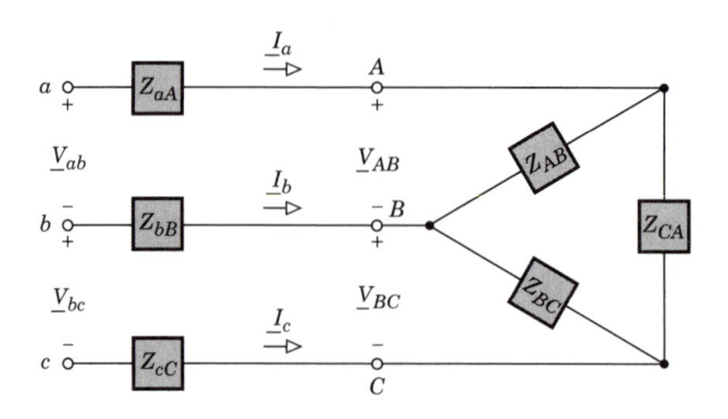

(a) Unbalanced delta load with line impedances

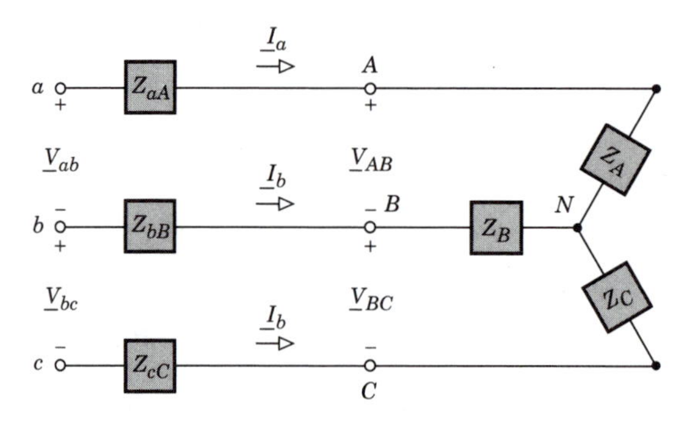

(b) Equivalent wye load

Figure 7.29

We want to find the line currents and the total real and reactive powers supplied by the generator.

First, we use the delta-to-wye transformation in Eq. (7.56a) with $\Sigma Z = 90 + j90$ to obtain

$$Z_A = 5 + j5 \qquad Z_B = 20 + j0 \qquad Z_C = 0 + j20$$

Adding the 5-Ω series resistance in each line gives the total equivalent wye impedances

$$Z_a = 10 + j5 \qquad Z_b = 25 + j0 \qquad Z_c = 5 + j20$$

Next, we take $\underline{V}_{ab}$ as the phase reference so

$$\underline{V}_{ab} = 900 \text{ V} \underline{/0^\circ} = 900 \qquad \underline{V}_{bc} = 900 \underline{/-120^\circ} = -450 - j780$$

Equation (7.54a) thus becomes

$$\begin{bmatrix} 35 + j5 & -25 \\ -25 & 30 + j20 \end{bmatrix} \begin{bmatrix} \underline{I}_1 \\ \underline{I}_2 \end{bmatrix} = \begin{bmatrix} 900 \\ -450 - j780 \end{bmatrix}$$

from which $\underline{I}_1 = 4.6 - j16.8$ and $\underline{I}_2 = -26.2 - j22.5$, so Eq. (7.54b) yields

$$\underline{I}_a = 4.6 - j16.8 = 17.3 \text{ A } \angle -74.5°$$
$$\underline{I}_b = -30.8 - j5.7 = 31.3 \text{ A } \angle -169.4°$$
$$\underline{I}_c = 26.2 + j22.5 = 34.5 \text{ A } \angle 40.7°$$

Finally, we calculate the total real and reactive powers directly as

$$P = R_a|\underline{I}_a|^2 + R_b|\underline{I}_b|^2 + R_c|\underline{I}_c|^2 = 33.5 \text{ kW}$$
$$Q = X_a|\underline{I}_a|^2 + X_b|\underline{I}_b|^2 + X_c|\underline{I}_c|^2 = 25.4 \text{ kVAr}$$

The impedance angle of the total load is not defined for the unbalanced case because each branch has a different impedance angle. Nonetheless, Eq. (7.52) gives the equivalent power factor

$$\text{pf} = P/\sqrt{P^2 + Q^2} = 0.79$$

which is a lagging power factor since $Q > 0$.

Exercise 7.14

Find all four line currents in Fig. 7.27*a* when $\underline{V}_{ab} = 208 \text{ V } \angle 0°$, $Z_a = j10\ \Omega$, $Z_b = 6 + j8\ \Omega$, and $Z_c = 10\ \Omega$.

7.5 RESIDENTIAL CIRCUITS AND WIRING†

No discussion of ac power would be complete without at least some coverage of the wiring arrangements that distribute electric power to the outlets of a typical home. This section, therefore, describes several of the major features of residential circuits. Since we are concerned only with voltage and current magnitudes, capital letters are used throughout to denote rms values.

A word of warning before we begin: This material is intended only for illustrative purposes and does not qualify you to be an electrician!

Dual-Voltage Service Entrance

Most homes in the United States have a **dual voltage** ac supply provided by a three-line entrance cable like that of Fig. 7.30*a*. One line, called the **neutral**, is connected to an earth ground. The other two lines, labeled *B* and *R*, are

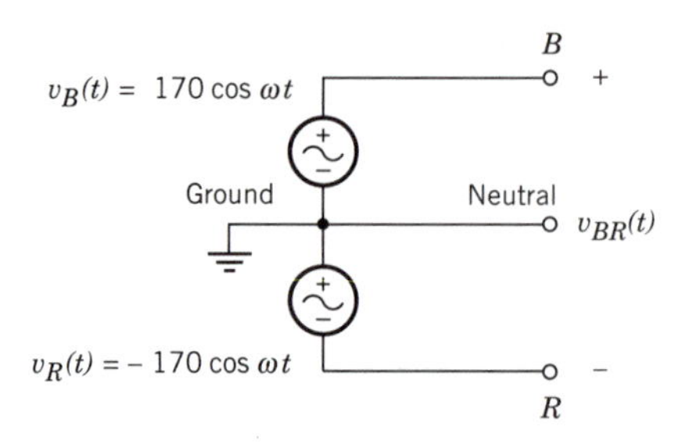

(a) Dual voltage ac supply

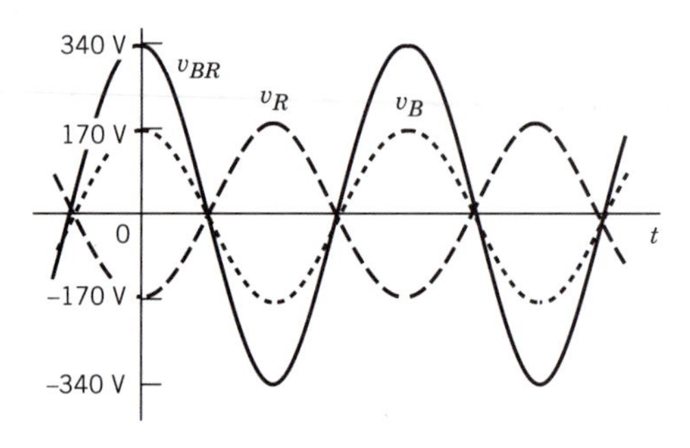

(b) Voltage waveforms

Figure 7.30

"hot" in the sense that they have 170-V peak sinusoidal potential variations relative to ground potential, with one sinusoid inverted in polarity compared to the other. The two sources shown actually correspond to voltages at the output of a transformer connected to a three-phase power line.

Formally, we write $v_B = 170 \cos \omega t$ and $v_R = -v_B = 170 \cos(\omega t \pm 180°)$, with rms values $V_B = V_R = 170/\sqrt{2} = 120$ V. The line-to-line voltage is then $v_{BR} = v_B - v_R = 340 \cos \omega t$, as shown in Fig. 7.30*b*, and it has the rms value $v_{BR} = 340/\sqrt{2} = 240$ V.

After passing through the electric meter that measures energy consumption, the entrance cable terminates at the **main panel**. Here, the hot lines connect to individual circuits for lighting, appliances, and so forth, while the neutral connects to a **busbar** and then to the local earth ground.

Figure 7.31 diagrams a main panel with **breakers** serving the joint role of disconnecting switches and overcurrent protection. The modern thermal-magnetic breaker is a spring-loaded switch having a bimetallic element that opens under small but continuous current overload, and a magnetic coil that trips the switch instantly under heavy overloads. (Older equipment would have separate switches and fuses instead of breakers.) The breaker labeled GFCI has additional features described later.

The figure also shows four different types of circuits going out of the panel. Each circuit has a minimum of *three* wires, and they have been labeled in accordance with standard insulation color codes as follows:

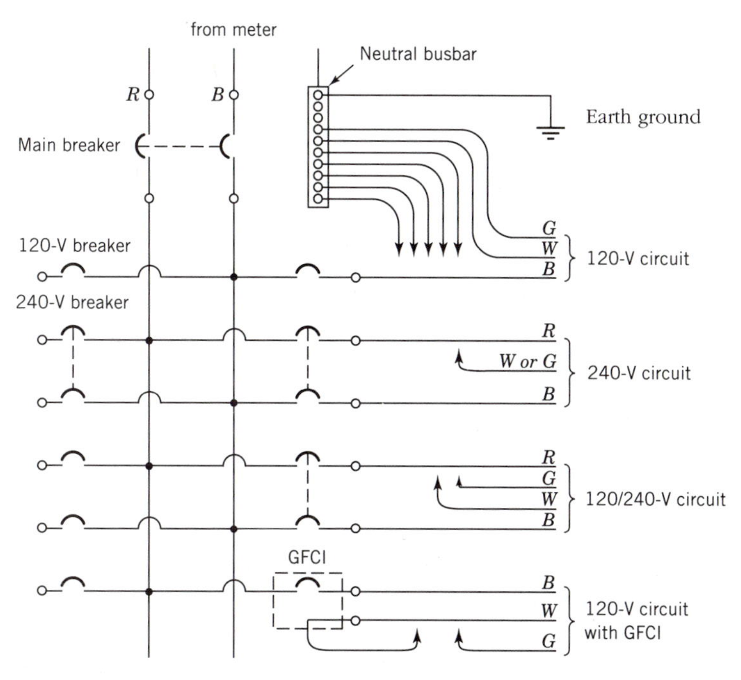

Figure 7.31 Main panel with breakers and typical circuit connections.

Hot = B (black) or R (red)

Neutral = W (white)

Ground = G (green) or uninsulated

Every outgoing hot wire must connect to a breaker, whereas every neutral wire and ground wire must be tied directly to earth ground at the neutral busbar. The functional difference between neutral and ground wires will be brought out by considering circuit wiring.

Wiring and Grounding

Figure 7.32 depicts the wiring and wire resistances from the panel to a 120-V grounded outlet. Several other outlets and lights would typically be connected in parallel on the same circuit. Under normal conditions, current flows to the load (not shown) through only the hot and neutral wires. Hence, the ground terminal at the outlet is at zero volts with respect to earth ground, despite the resistance R_G. However, current I through R_W and R_B causes the neutral terminal to be at $R_W I$ volts with respect to earth ground, and the available load voltage becomes $120 - (R_W + R_B)I$. Resistance in the entrance cable increases this loading effect even further, so home appliances are usually designed to operate over a range of 110–120 V.

From the viewpoint of an electronics engineer, the ground wire provides a valuable reference potential for voltage measurements, independent of the

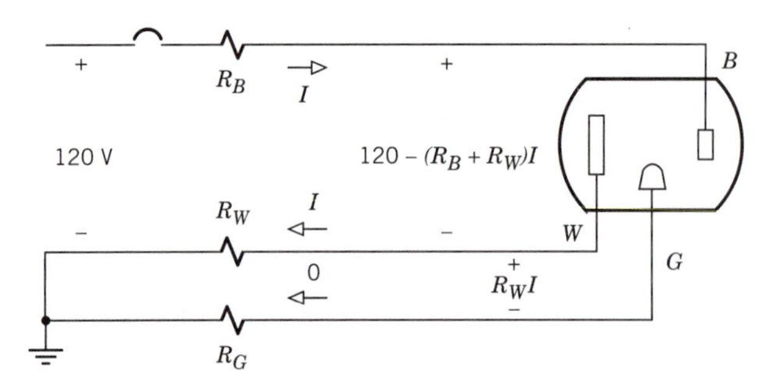

Figure 7.32 Wiring resistance between panel and outlet.

neutral and its voltage offset. In addition, electrical interference and ground-loop problems are minimized by connecting all instruments to one ground point. But vastly more important is the ground wire's role in protecting human life against electrical shock.

Table 7.4 lists effects of various levels of 60-Hz ac current on the human body. The 100–300-mA range turns out to pose the greatest electrocution hazard. Larger currents may cause serious or life-threatening burns, but they induce a temporary heart contraction that actually protects it from fatal damage. As Table 7.4 implies, the amount of current rather than voltage is the key factor in electrical shock. Voltage enters the picture when we take account of body resistance, which ranges from around 500 kΩ for dry skin down to 1 kΩ for wet skin. Thus, a person with wet skin risks electrocution from ac voltages as low as 100 V.

TABLE 7.4 Effects of AC Electrical Shock

RMS Current	Effects
1–5 mA	Threshold of sensation
10–20 mA	Involuntary muscle contractions ("can't-let-go")
20–100 mA	Pain, breathing difficulties
100–300 mA	Ventricular fibrilation, **possible death**
>300 mA	Respiratory paralysis, burns, unconsciousness

Now suppose you touch the metal frame of an ungrounded appliance that has an internal wiring fault between the frame and the hot line, as in Fig. 7.33*a*. Your body then provides a possible conducting path to ground, and you may experience a serious shock — especially if you're standing on a damp concrete floor or other grounded surface so that the current passes through your chest. The circuit breaker offers no help in this situation, because it is designed to protect the circuit — not you — against excess currents of 15 A or more.

An internal connection from the frame to neutral significantly improves the situation, but even better is the grounded frame shown in Fig. 7.33*b*. This arrangement keeps the frame at ground potential in absence of a wiring fault or, at worst, a few volts from ground if a fault results in current through the

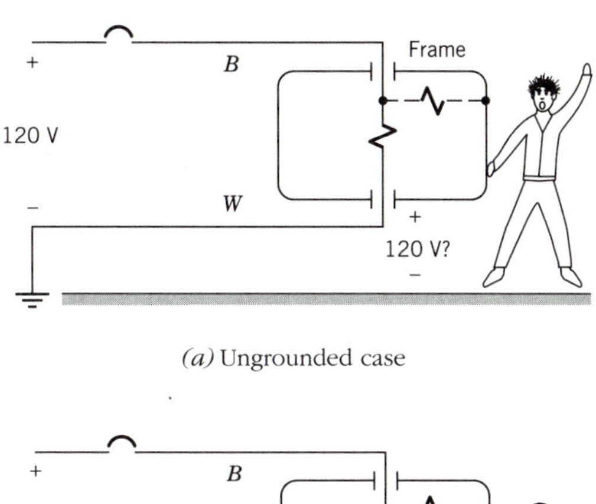

(a) Ungrounded case

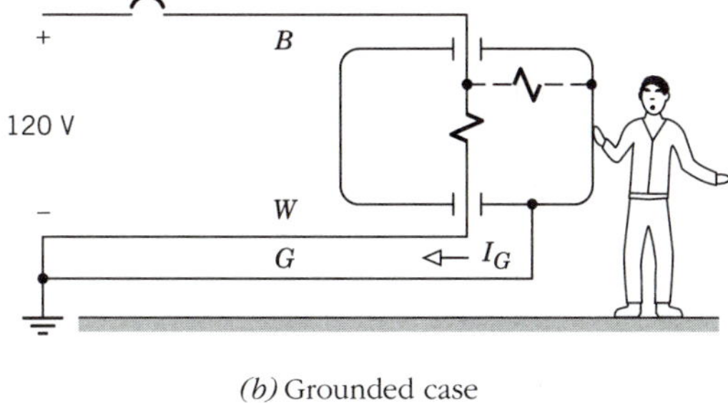

(b) Grounded case

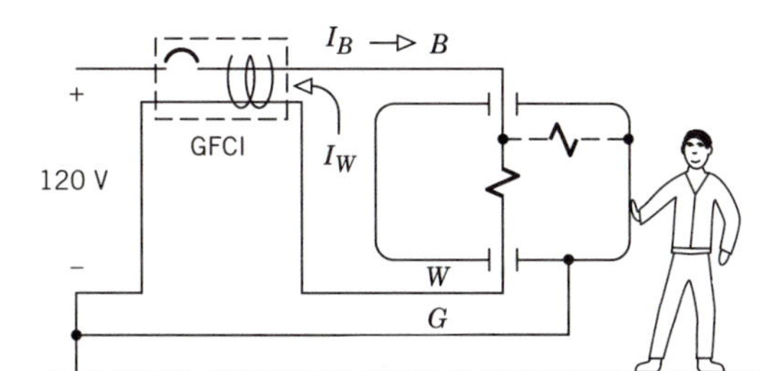

(c) With ground-fault circuit interrupter

Figure 7.33 Appliance with wiring fault.

ground wire. But the best possible shock protection is afforded by the **ground-fault circuit interrupter** (GFCI) depicted by Fig. 7.33*c*. The GFCI has a sensing coil around the hot and neutral wires, and any ground-fault current — through you or the ground wire — that results in an imbalance of $|I_B - I_W| > 5$ mA induces a current in the sensing coil and opens the circuit. The GFCI may be located at an outlet, or it may be part of a circuit breaker at the main panel.

Figure 7.34 diagrams three residential wiring patterns involving 240 V: (a) a 240-V load, such as an electric heating unit; (b) a dual-voltage load, such an electric range with 120-V and 240-V elements; (c) a "three-wire circuit" with two 120-V loads sharing common neutral and ground wires. In this last case (intended solely to reduce wire costs), $I_W = I_B - I_R$ and the neutral current

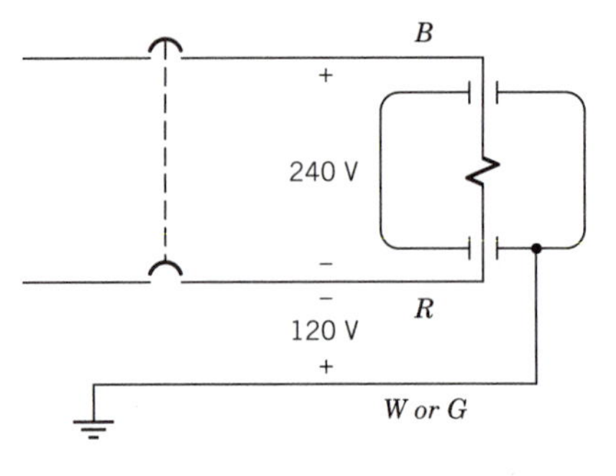

(a) 240-V load

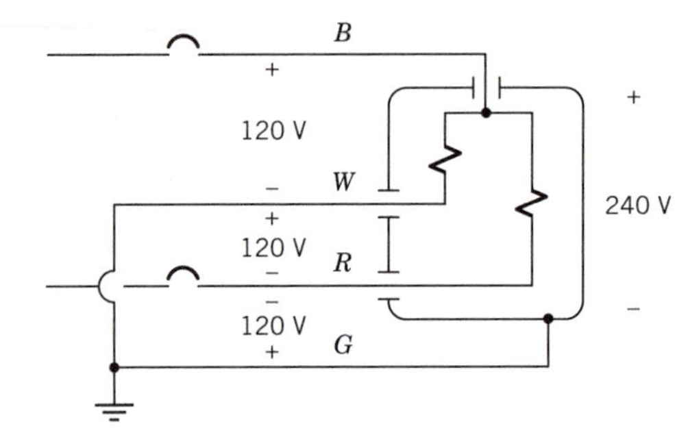

(b) Dual-voltage load

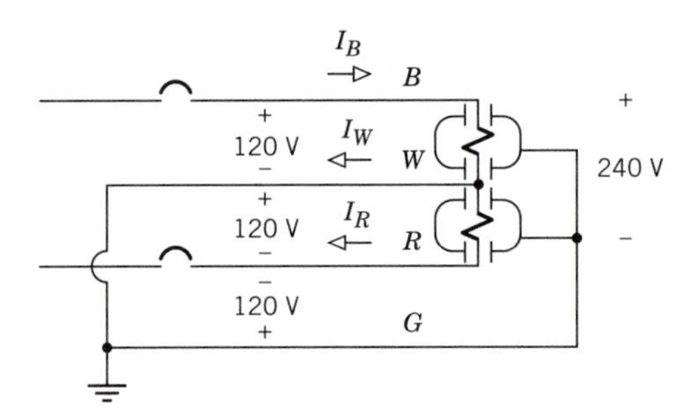

(c) Three-wire circuit for two 120-V loads

Figure 7.34

will be small if the loads are nearly equal. Hence, the neutral wire need not be bigger than the hot wires.

Finally, Fig. 7.35 illustrates how a light or some other device can be controlled independently from two different locations using single-pole double-throw (SPDT) switches, commonly known as "three-way" switches. The hot wire is switched between two "travelers" at the first switch and from the travelers to the light at the second. Therefore, we have a complete circuit only when both switches are either up or down, and flipping either switch opens the circuit. Control at three or more locations is also possible using "four-way" switches that interchange the travelers. In any case, neutral and ground wires are never switched.

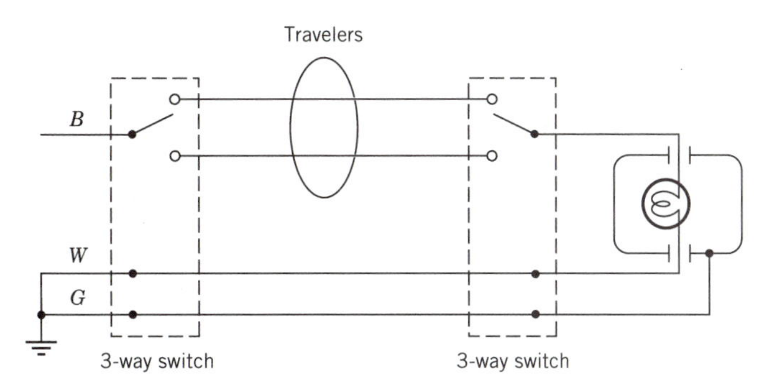

Figure 7.35 Wiring for three-way switches.

PROBLEMS

Section 7.1 Power in AC Circuits

7.1* Let the circuit in Fig. P7.1 have $R_1 = 20\ \Omega$, $L = 1$ H, $R_1 = 0$, $C = 1/100$ F, and $i = 6 \cos 10t$ A. Find the resulting rms value of v, and use Eq. (7.8) to calculate the average power dissipated. Then confirm that $P = \frac{1}{2}R_1|I_1|^2$.

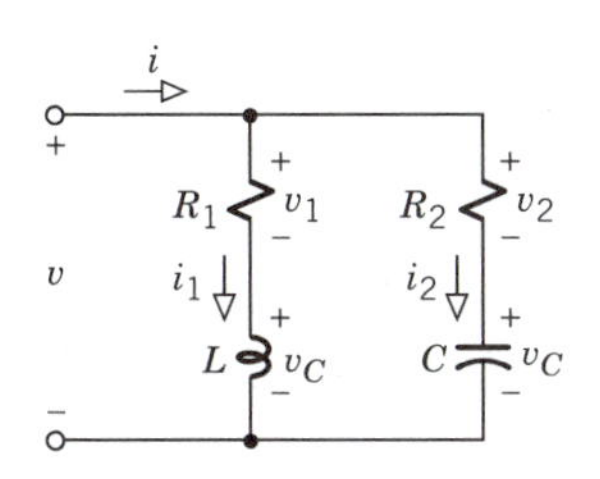

Figure P7.1

7.2 Do Problem 7.1 with $R_1 = 0$, $L = 1$ H, $R_2 = 4$ kΩ, $C = 0.1$ μF, and $i = 5 \cos 5000t$ mA.

7.3 Let the circuit in Fig. P7.3 have $C = 1/50$ F, $R_1 = 5\ \Omega$, $R_2 = 0$, $L = 1$ H, and $v = 15 \cos 10t$ V. Find the resulting rms value of i, and use Eq. (7.8) to calculate the average power dissipated. Then confirm that $P = \frac{1}{2}R_1|I_1|^2$.

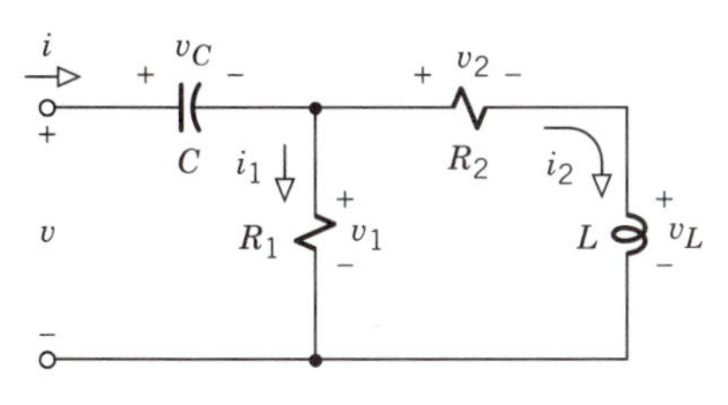

Figure P7.3

7.4 Let the circuit in Fig. P7.3 have $C = 0.2\ \mu$F, $R_1 = 20$ kΩ, $R_2 = 4$ kΩ, $L = 8$ H, and $v = 40 \cos 1000t$ V. Find the resulting rms value of i, and use Eq. (7.8) to calculate the average power dissipated. Then calculate the power dissipated by each resistor to confirm that $P = P_1 + P_2$.

7.5 Let the circuit in Fig. P7.5 have $L = 2$ H, $R_1 = 2$ kΩ, $R_2 = 4$ kΩ, $C = 0.25\ \mu$F, and $i = 20 \cos 1000t$ mA. Find the resulting rms value of v, and use Eq. (7.8) to calculate the average power dissipated. Then calculate the power dissipated by each resistor to confirm that $P = P_1 + P_2$.

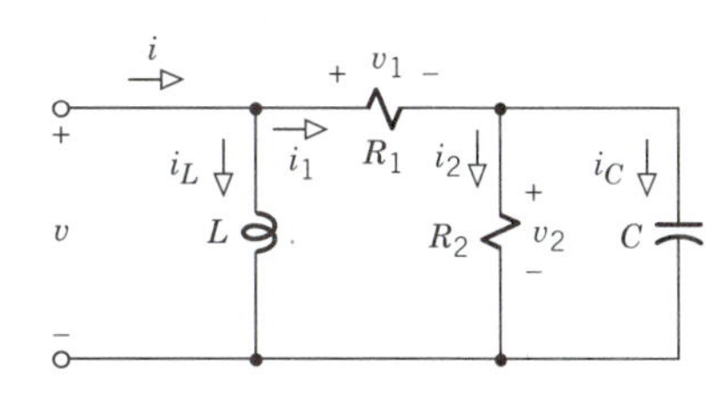

Figure P7.5

7.6 Do Problem 7.5 with $L = 5$ mH, $R_1 = 5\ \Omega$, $R_2 = 10\ \Omega$, C = 50 μF, and $i = 5 \cos 2000t$ A.

7.7 A series *RLC* circuit driven by $v = 20 \cos \omega t$ V has $R = 50\ \Omega$, $L = 1$ H, and $C = 100\ \mu$F. Obtain an expression for the average power as a function of frequency, and plot P versus ω by evaluating P at $\omega = 0^+$, 50, 100, 200, and ∞.

7.8 A parallel *RLC* circuit driven by $i = 0.2 \cos \omega t$ *A* has $R = 500\ \Omega$, $L = 0.25$ H, and $C = 1\ \mu$F. Obtain an expression for the average power as a function of frequency, and plot P versus ω by evaluating P at $\omega = 0^+$, 1000, 2000, 4000, and ∞.

7.9 The current $i = 1 \cos \omega t$ A drives a network consisting of a 100-μF capacitor in parallel with a coil having $R = 50\ \Omega$ and $L = 0.25$ H. Obtain an expression for the average power as a function of frequency, and plot P versus ω by evaluating P at $\omega = 0^+$, 100, 200, 400, and ∞.

7.10* Find V_{rms} when $v(t)$ has the periodic waveform in Fig. P7.10 with $\tau = 3T/4$, $A = 40$ V, and $B = -20$ V.

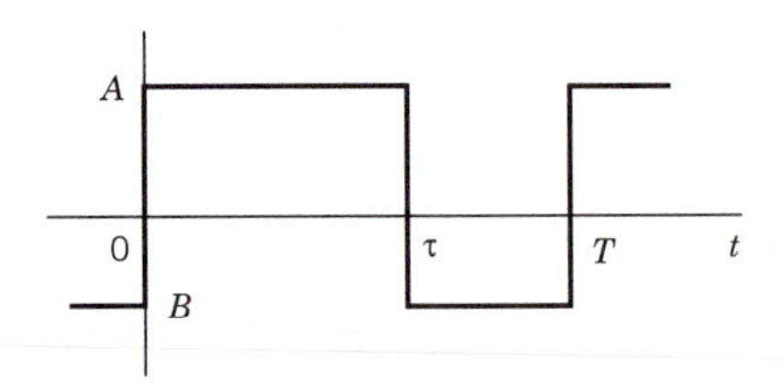

Figure P7.10

7.11 Find I_{rms} when $i(t)$ has the periodic waveform in Fig. P7.10 with $\tau = T/3$, $A = 4$ A, and $B = -7$ A.

7.12 Find I_{rms} when $i(t)$ has the periodic waveform in Fig. P7.12 with $A = 26$ A and $B = 0$.

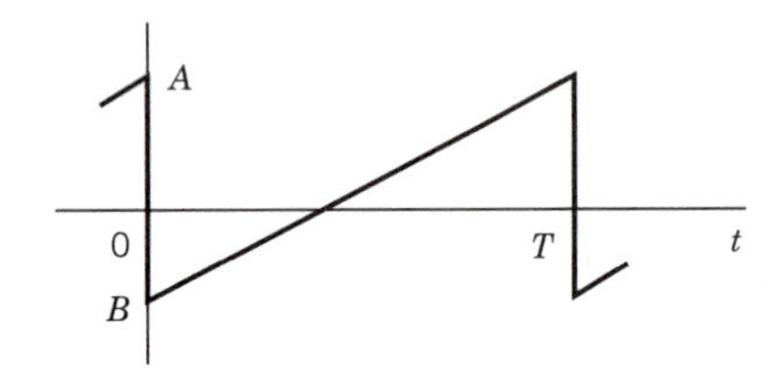

Figure P7.12

7.13 Find V_{rms} when $v(t)$ has the periodic waveform in Fig. P7.12 with $A = 45$ V and $B = -45$ V.

7.14 Suppose $Z_s = 100\ \Omega\ \angle 45°$ and $\omega = 10^4$ in Fig. 7.5 (p. 312). By considering admittances, determine the two elements that should be connected in parallel to form Z for maximum power transfer.

7.15 Do Problem 7.14 with $Z_s = 2\ \text{k}\Omega\ \angle -30°$ and $\omega = 10^6$.

7.16* Let the source network in Fig. P7.16 have $R = 500\ \Omega$, $L = 0$, $C = 0.4\ \mu$F, and $v_s = 10 \cos 10^4 t$ V. Find the load impedance Z that draws maximum power, and calculate the resulting value of P_{max}.

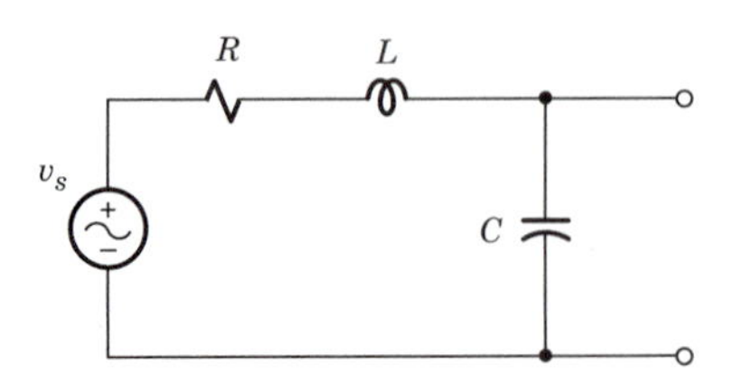

Figure P7.16

7.17 Do Problem 7.16 with $R = 1\ \text{k}\Omega$, $L = 2$ mH, $C = 500$ pF, and $v_s = 4 \cos 10^6 t$ V.

7.18 Do Problem 7.16 with $R = 2.5\ \text{k}\Omega$, $L = 50$ mH, $C = 1$ nF, and $v_s = 1 \cos 10^5 t$.

7.19 A certain source has $V_{rms} = 2.4$ V and $Z_s = 24 - j7\ \text{k}\Omega$. Find the load impedance Z for maximum power transfer, and the resulting values of P and Eff for each of the following cases: (a) Z is fully adjustable; (b) $|Z|$ is adjustable but $X/R = +3/4$; (c) R is adjustable but $X = -10\ \text{k}\Omega$. Hint: For case (c), see Exercise 7.3.

7.20 Do Problem 7.19 with $Z_s = 12 + j5\ \text{k}\Omega$.

Section 7.2 Power Systems

(See also PSpice problems B.23 and B.24.)

7.21 Obtain expressions for P and Q in terms of $|\underline{I}|$ when impedances $Z_R = R$ and $Z_X = jX$ are connected in parallel.

7.22 A certain load draws $P = 720$ W and $|\underline{I}| = 10$ A at 120 V and 60 Hz. The load current decreases when the frequency increases. Find the element values to represent the load as: (a) two elements in parallel; (b) two elements in series.

7.23 Do Problem 7.22 with $P = 960$ W.

7.24 A certain load draws $P = 1100$ W and $|\underline{I}| = 5$ A at 240 V and 60 Hz. The load current increases when the frequency increases. Find the element values to represent the load as: (a) two elements in parallel; (b) two elements in series.

7.25 Do Problem 7.24 with $P = 600$ W.

7.26* An ac motor having impedance $Z_M = 3.6 + j4.8\ \Omega$ is supplied from a 240-V, 60-Hz source. (a) Calculate pf, $|\underline{I}|$, P, and Q. (b) The power factor is corrected to unity by connecting a capacitor in parallel with the motor. Find C and the power P and current $|\underline{I}|$ supplied by the source. (c) Suppose the power factor is corrected to unity by connecting a capacitor in *series* with the motor. Find C and the current $|\underline{I}|$ and power P supplied by the source. Also calculate the voltage $|\underline{V}_M|$ across the motor, and comment on your result.

7.27 Do Problem 7.26 with $Z_M = 3.0 + j5.2\ \Omega$.

7.28 Do Problem 7.26 with $Z_M = 1.68 + j5.76\ \Omega$.

7.29* Two inductive loads are connected in parallel with a 600-V source. The power and current drawn by each load are:

$$P_1 = 23\ \text{kW} \qquad |\underline{I}_1| = 150\ \text{A}$$
$$P_2 = 60\ \text{kW} \qquad |\underline{I}_2| = 200\ \text{A}$$

(a) Find the complex power and rms current supplied by the source, and calculate the power factor of the combined loads. (b) A capacitor is now added in parallel, increasing the power factor to 0.9 lagging. Find the current drawn from the source and the capacitor's reactive power.

7.30 Do problem 7.29 for the case of:

$$P_1 = 42 \text{ kW} \qquad |\underline{I}_1| = 250 \text{ A}$$
$$P_2 = 75 \text{ kW} \qquad |\underline{I}_2| = 400 \text{ A}$$

7.31 Two loads are connected in parallel with a 2000-V source. The individual power factors and currents are:

$$\text{pf}_1 = 0.5 \text{ lagging} \qquad |\underline{I}_1| = 40 \text{ A}$$
$$\text{pf}_2 = 0.8 \text{ leading} \qquad |\underline{I}_2| = 15 \text{ A}$$

(a) Find the total power and current from the source, and calculate the power factor of the combined loads. (b) A capacitor is now added in parallel, increasing the power factor to 0.95 lagging. Find the current drawn from the source and the capacitor's reactive power.

7.32 Do problem 7.31 for the case of:

$$\text{pf}_1 = 0.75 \text{ leading} \qquad |\underline{I}_1| = 20 \text{ A}$$
$$\text{pf}_2 = 0.28 \text{ lagging} \qquad |\underline{I}_2| = 50 \text{ A}$$

7.33 An industrial dryer operates at 600 V and requires 50 A. The unit consists of a fan in parallel with a heater. The fan draws 20 kW and has a lagging power factor of 0.8. Use a power triangle to find the resistance of the heater, assuming that it has unity power factor.

7.34 An industrial furnace operates at 1300 V and requires 100 A. The unit consists of a blower in parallel with a heater. The blower draws 50 kW and has a power factor of 0.707 lagging. Use a power triangle to find the resistance of the heater, assuming that it has unity power factor.

7.35 An inductive motor draws 24 kW and 40 A from a 1000-V, 60-Hz source. The total current drops to 25 A when a capacitor is connected in parallel with the motor. Show from a power triangle that there are two possible values of C, and find those values.

7.36 An inductive motor draws 16 kW and 68 A from a 500-V, 60-Hz source. The total current drops to 40 A when a capacitor is connected in parallel with the motor. Show from a power triangle that there are two possible values of C, and find those values.

7.37 A motor is connected to a 240-V, 60-Hz source by a transmission line with resistance $R_l = 1\ \Omega$. The motor has pf = 0.5 lagging, and it would draw 4.8 kW if operated at 240 V. However, the resistance of the transmission line reduces the voltage at the motor and wastes power.
(a) Find the equivalent impedance of the motor to determine the capacitance that should be connected in parallel with it to minimize the line loss.
(b) Calculate the resulting values of the current and power from the source and the power-transfer efficiency. Compare these with the values without the capacitor.

7.38 Do Problem 7.37 for a motor with pf = 0.4 lagging that would draw 6 kW if operated at 240 V.

7.39 A capacitor is to be connected in parallel with a motor that draws $P = 10$ kW and $Q = 6$ kVAr from a 240-V, 60-Hz source. Derive an expression for C in terms of the resulting power factor. Then find the cost of obtaining pf = 0.9, 0.95, and 1.0, given that 240-V capacitors are priced at $0.25 per microfarad.

7.40 When the source voltage and frequency are known, the real and reactive powers drawn by a load can be determined using an ac ammeter and a known capacitor. First the rms current is measured without the capacitor; call this value $|\underline{I}_1|$. Then the current is measured with the capacitor connected in parallel; call this value $|\underline{I}_2|$. By considering $|\underline{V}\underline{I}_1|^2 - |\underline{V}\underline{I}_2|^2$, derive an expression for Q in terms of known and measured quantities. Then write an expression for P in terms of Q and other known quantities.

Section 7.3 Balanced Three-Phase Circuits

(See also PSpice problems B.25 and B.26.)

7.41* A balanced wye load is connected to a 60-Hz three-phase source with $\underline{V}_{ab} = 208 \text{ V} \angle 0°$. The load has pf = 0.5 lagging, and each phase draws P_ϕ = 6 kW. (a) Calculate I_l and find Z_Y, $\underline{I}_a$, $\underline{I}_b$, and $\underline{I}_c$ in polar form. (b) What value of C should be put in parallel with each load element to minimize the current from the source, and what is the resulting line current?

7.42 Do Problem 7.41 with pf = 0.8 lagging and $P_\phi = 7.2$ kW.

7.43 A balanced delta load is connected to a 60-Hz three-phase source with $\underline{V}_a = 120 \text{ V} \angle 0°$. The load has pf = 0.6 lagging, and each phase draws P_ϕ = 9 kW. (a) Calculate I_l and find Z_Δ, $\underline{I}_{ab}$, $\underline{I}_{bc}$, and $\underline{I}_{ca}$ in polar form. (b) What value of C should be put in parallel with each load element to minimize the current from the source, and what is the resulting line current?

7.44 Do Problem 7.43 with pf = 0.4 lagging and $P_\phi = 9.6$ kW.

7.45 Consider the system in Fig. P7.45 with $\underline{V}_{ab}$ = 2600 V $\angle 0°$ and a delta-connected load. Find $\underline{I}_a$, $\underline{V}_{AN}$, $\underline{V}_{AB}$, and the total real and reactive powers supplied by the generator when

$$Z_l = 0 + j6\ \Omega \qquad Z_\Delta = 45 + j60\ \Omega$$

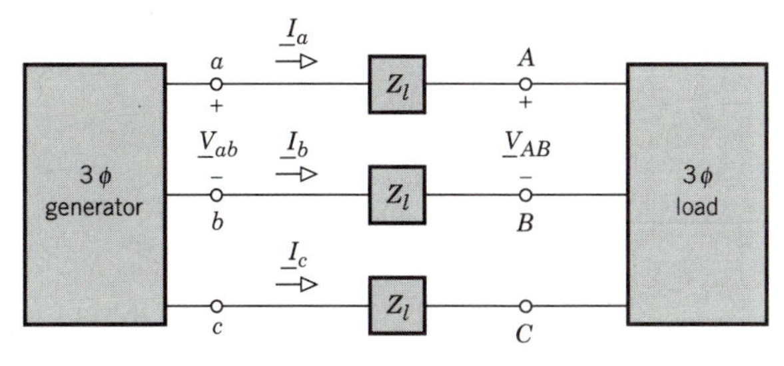

Figure P7.45

7.46 Do Problem 7.45 with

$$Z_l = 4 + j0\ \Omega \qquad Z_\Delta = 15 + j36\ \Omega$$

7.47 Do Problem 7.45 with

$$Z_l = 8 + j25\ \Omega \qquad Z_\Delta = 48 + j90\ \Omega$$

7.48* Load 1 in Fig. P7.48 has a balanced wye configuration, while load 2 has a balanced delta configuration. Find $|\underline{I}_a|$, $|\underline{V}_{AB}|$, the rms current and average power drawn by each load, and the power factor seen by the generator when

$$|\underline{V}_{ab}| = 658\ \text{V} \qquad Z_l = 6 + j0\ \Omega$$
$$Z_{Y1} = 20 + j0\ \Omega \qquad Z_{\Delta 2} = 36 + j72\ \Omega$$

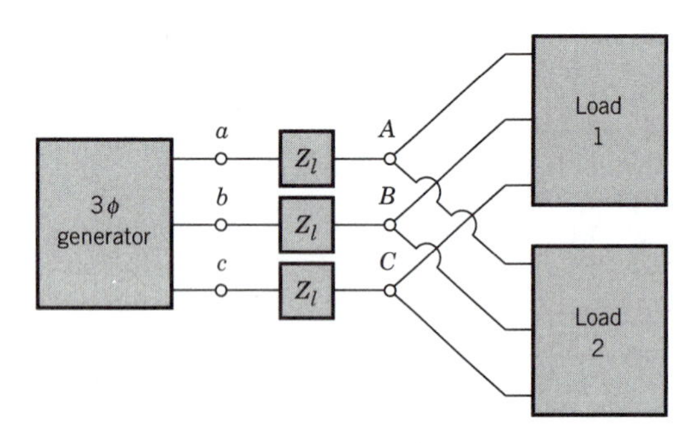

Figure P7.48

7.49 Do Problem 7.48 with

$$|\underline{V}_{ab}| = 1040\ \text{V} \qquad Z_l = 0 + j23\ \Omega$$
$$Z_{Y1} = 120 + j0\ \Omega \qquad Z_{\Delta 2} = 45 + j135\ \Omega$$

7.50 Do Problem 7.48 with

$$|\underline{V}_{ab}| = 520\ \text{V} \qquad Z_l = 3 + j1\ \Omega$$
$$Z_{Y1} = 20 + j0\ \Omega \qquad Z_{\Delta 2} = 12 + j24\ \Omega$$

7.51 Let the system in Fig. P7.45 have $|\underline{V}_{ab}| = 45\ \text{k}\Omega$ and $f = 60$ Hz. Three capacitors are to be wye connected at the load terminals to minimize the current from the generator. Find C_Y and the resulting values of $|\underline{I}_a|$, P, Q, and the power-transfer efficiency, given that $Z_l = 14 + j30\ \Omega$ and the load has $Z_\Delta = 75 + j228\ \Omega$.

7.52 Do Problem 7.51 with $Z_l = 20 + j50\ \Omega$ and $Z_\Delta = 90 + j349\ \Omega$.

7.53* Two wattmeters are used to measure the properties of a balanced three-phase motor operating at $V_l = 1300$ V. The motor is known to be inductive, and the meters read $P_1 = 149$ kW and $P_2 = 59$ kW. Find P, Q, and I_l. Then calculate Z_Y in rectangular form.

7.54 Do Problem 7.53 for the case of $P_1 = 114$ kW and $P_2 = 186$ kW.

7.55 Find P_1 and P_2 in Fig. P7.55 when $\underline{V}_{ab} = 208\ \text{V}\ \underline{/0^\circ}$ and the load is balanced with $Z_\Delta = 9 + j12\ \Omega$. Then use direct power calculations to confirm that Eqs. (7.50) and (7.51) give correct results to within round-off error.

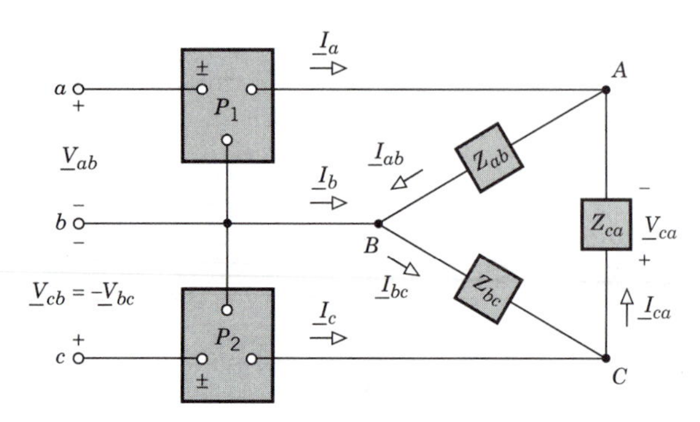

Figure P7.55

7.56 Do Problem 7.55 with $Z_\Delta = 2.1 + j7.2\ \Omega$.

7.57 Wattmeters are inserted in a balanced three-phase circuit such that $P_1 = \text{Re}[\underline{V}_{ba}\underline{I}_b^*]$ and $P_2 = \text{Re}[\underline{V}_{ca}\underline{I}_c^*]$. Find P_1 and P_2 when $\underline{V}_a = 300\ \text{V}\ \underline{/0^\circ}$ and the load has $Z_Y = 1.5 + j2.6\ \Omega$. Then use direct power calculations to confirm that Eqs. (7.50) and (7.51) give correct results to within round-off error.

7.58 Do Problem 7.57 with $Z_Y = 2.4 + j5.5\ \Omega$.

Section 7.4 Unbalanced Three-Phase Circuits

(See also PSpice problems B.27 and B.28.)

7.59* Consider the composite system in Fig. 7.26 (p. 343) with $\underline{V}_a = 120$ V and a balanced wye load having $Z_Y = 3 + j4\ \Omega$. Let Z_{cn} represent a capacitor chosen such that $Q = 0$ for the entire system. Find $\underline{I}_a$, $\underline{I}_b$, $\underline{I}_c$, and $\underline{I}_n$ in rectangular and polar forms.

7.60 Do Problem 7.59 with the addition of a single-phase motor connected from b to n so that $Z_{bn} = 8 + j6\ \Omega$.

7.61 Figure P7.61 depicts a three-wire composite system with two single-phase loads and two wattmeters. The given values are

$$\underline{V}_{ab} = 1040\ \text{V}\ \underline{/0^\circ} \qquad Z_{ab} = Z_{bc} = 104\ \Omega$$
$$Z_Y = 30 + j52 \text{ for the three-phase load}$$

(a) Find $\underline{I}_a$, $\underline{I}_b$, and $\underline{I}_c$ in rectangular and polar form. (b) Find the wattmeter readings and use direct power calculations to confirm that $P_1 + P_2 = P$ to within round-off error.

7.62 Do Problem 7.61 with

$$\underline{V}_a = 3\ \text{kV}\ \underline{/0^\circ} \qquad Z_{ab} = 520\ \Omega$$
$$Z_{bc} = -j520\ \Omega$$
$$Z_Y = 300\ \Omega\ \underline{/30^\circ} \text{ for the three-phase load}$$

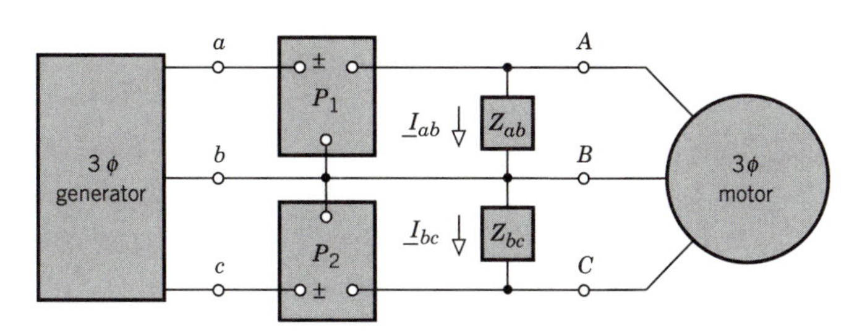

Figure P7.61

7.63 Do Problem 7.61 with

$$\underline{V}_{ab} = 580 \text{ V} \angle 0^\circ \qquad Z_{ab} = 8 - j6\ \Omega$$

$$Z_{bc} = 10\ \Omega$$

$$Z_\Delta = 10\ \Omega \angle 45^\circ \text{ for the three-phase load}$$

7.64 Suppose a delta load has an open-circuit fault so that $Z_{ca} = \infty$ while $Z_{ab} = Z_{bc} = 5 + j12\ \Omega$. Calculate $\underline{I}_a$, $\underline{I}_b$, P, and Q when $\underline{V}_{ab} = 208 \text{ V} \angle 0^\circ$. Then construct a phasor diagram to find $\underline{I}_b$.

7.65 Suppose the wye load in a four-wire system has an open-circuit fault so that $Z_c = \infty$ while $Z_a = Z_b = 4 + j3\ \Omega$. Calculate $\underline{I}_a$, $\underline{I}_b$, P, and Q when $\underline{V}_{ab} = 208 \text{ V} \angle 0^\circ$ and an ideal conductor connects the load and generator neutrals. Then construct a phasor diagram to find $\underline{I}_n$.

7.66 Suppose the wye load in a three-wire system has an open-circuit fault so that $Z_c = \infty$ while $Z_a = Z_b = 12 + j5\ \Omega$. Calculate $\underline{I}_a$, $\underline{I}_b$, P, and Q when $\underline{V}_a = 120 \text{ V} \angle 0^\circ$. Then construct a phasor diagram to find the voltage $\underline{V}_{Nn}$ across the two neutral points.

7.67* Consider the four-wire system in Fig. 7.27*a* (p. 344). Find $\underline{I}_a$, $\underline{I}_b$, $\underline{I}_c$, $\underline{I}_n$, P, and Q when $\underline{V}_{ab} = 520 \text{ V} \angle 0^\circ$, $Z_a = 5\ \Omega \angle 0^\circ$, and $Z_b = Z_c = 10\ \Omega \angle 60^\circ$.

7.68 Consider the three-wire wye load in Fig. 7.27*b* (p. 344). Find $\underline{I}_a$, $\underline{I}_b$, $\underline{I}_c$, and P when $\underline{V}_{ab} = 240 \text{ V} \angle 0^\circ$, $Z_a = Z_b = 10\ \Omega$, and $Z_c = 5\ \Omega$.

7.69 Consider the unbalanced delta load in Fig. 7.28 (p. 345). Find $\underline{I}_a$, $\underline{I}_b$, $\underline{I}_c$, and P when $\underline{V}_a = 580 \text{ V} \angle 0^\circ$, $Z_{ab} = Z_{bc} = 50\ \Omega$, and $Z_{ca} = 25\ \Omega$.

7.70 Consider the unbalanced system in Fig. 7.29*a* (p. 346). Find $\underline{I}_a$, $\underline{I}_b$, $\underline{I}_c$, P, and Q when $\underline{V}_{ab} = 600 \text{ V} \angle 0^\circ$, $Z_{aA} = Z_{bB} = Z_{cC} = 1 + j0\ \Omega$, $Z_{AB} = 9 + j9\ \Omega$, $Z_{BC} = 9 + j0\ \Omega$, and $Z_{CA} = 9 - j9\ \Omega$.

7.71 Consider the unbalanced system in Fig. 7.29*a* (p. 346). Find $\underline{I}_a$, $\underline{I}_b$, $\underline{I}_c$, P, and Q when $\underline{V}_{ab} = 300 \text{ V} \angle 0^\circ$, $Z_{aA} = Z_{bB} = Z_{cC} = 0 + j1\ \Omega$, $Z_{AB} = Z_{CA} = 4 + j0\ \Omega$, and $Z_{BA} = 8 + j0\ \Omega$.

CHAPTER 8

Transformers and Mutual Inductance

During the 1880s, the two founders of the electric industry waged the "battle of the currents" in the United States. On one side stood Thomas Edison, promoting his system based on direct current. On the other side, George Westinghouse argued for the superiority of alternating current. Westinghouse's arguments eventually prevailed, and ac became the dominant vehicle for transmission and distribution of electric power. Two major factors account for the advantage of ac over dc: the simplicity of ac generators and motors, and the availability of transformers for adjusting ac voltage or current amplitudes at various points to improve power-transfer efficiency and to meet the needs of different loads.

Transformers consist of magnetically coupled coils that respond only when driven by time-varying excitations. Such coils can be arranged to "step up" or "step down" a voltage or current waveform into another waveform with larger or smaller excursions. This chapter introduces transformer properties and develops analysis methods for ac circuits containing transformers.

We'll begin with ideal transformers, which approximate real transformers in many practical situations but allow for simplified analysis. Then we'll turn to real transformers by examining the mutual-inductance effect that occurs when coils are coupled magnetically. Our study of circuits with mutual inductance will also explore the conditions under which a real transformer acts essentially like an ideal transformer. Finally, the closing section considers losses and efficiency of power transformers.

Objectives

After studying this chapter and working the exercises, you should be able to do each of the following:

1. List the properties of an ideal transformer, and draw the controlled-source model (Section 8.1).
2. Use a referred source or load network to analyze an ac circuit coupled by an ideal transformer (Section 8.1).
3. Design a circuit with impedances matched by an ideal transformer (Section 8.1).
4. State the relationship between the mutual inductance and the self-inductances of magnetically coupled coils, and explain the significance of unity coupling (Section 8.2).
5. Write the voltage–current equations and phasor relationships for magnetically coupled coils (Sections 8.2 and 8.3).
6. Identify the conditions under which a real transformer behaves approximately like an ideal transformer (Sections 8.2 and 8.3).
7. Use an appropriate equivalent network to analyze an ac circuit with mutual inductance (Section 8.3).
8. Given the parameters of a power transformer, evaluate its performance when connected to a specified load (Section 8.4).†

8.1 IDEAL TRANSFORMERS

Many ac circuits involve a source network magnetically coupled to a load network by a transformer. The physical principles underlying transformer action will be covered in the next section. Here, we'll state the properties of *ideal* transformers, and we'll derive from those properties a simple method of analysis using referred source or load networks. We'll also see how transformers can be employed to improve power-transfer efficiency or maximize power transfer from an ac source to a given load.

Properties of Ideal Transformers

Figure 8.1 symbolizes an ideal transformer, a four-terminal device consisting of two wire coils. One coil has N_1 turns and is called the **primary winding**, while the other has N_2 turns and is called the **secondary winding**. Either winding may serve as the input of the transformer, so we'll take the reference direction for both currents into the upper terminals.

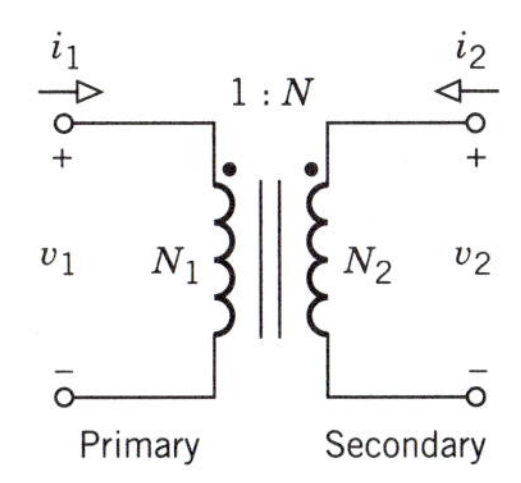

Figure 8.1 Symbol for an ideal transformer.

Magnetic coupling provides interaction between the coils, but no direct electrical connection exists from primary to secondary. Consequently, we indicate the relative voltage polarities by the following **dot convention**:

> The potential difference from the dotted to undotted end of the secondary winding has the same polarity as the potential difference from the dotted to undotted end of the primary winding.

The reference polarity signs in Fig. 8.1 agree with this convention, taking the dotted ends to be at the higher potential.

Both coils ideally have zero resistance, and they act like short circuits under dc steady-state conditions. A constant current through one winding therefore produces no effect in the other winding. But the transformer springs to life when either winding has a *time-varying excitation,* which activates the magnetic coupling and produces an *induced voltage* in the opposite winding. The behavior of an ideal transformer with time-varying excitation depends entirely upon its **turns ratio**

$$N \triangleq N_2/N_1 \tag{8.1}$$

Specifically, the terminal voltages and currents are related by

$$v_2(t) = Nv_1(t) \qquad i_2(t) = -i_1(t)/N \tag{8.2}$$

The negative sign in the current equation simply means that i_2 goes in the opposite direction of its reference arrow when i_1 is positive.

If v_1 varies with time and if $N > 1$, then $|v_2| = N|v_1| > |v_1|$ so we have a **step-up transformer** whose instantaneous output voltage magnitude will be greater than the input. But Eq. (8.2) indicates that $|i_2|$ will be less than $|i_1|$ when $N > 1$. Conversely, a **step-down transformer** with $N < 1$ produces $|v_2| < |v_1|$ and $|i_2| > |i_1|$. A transformer with $N_1 \neq N_2$ may operate in either the step-up or step-down mode, depending upon which winding is used for the input.

A transformer with $N = 1$ also has practical value as an **isolation transformer** that decouples the dc potential levels on either side without affecting the time-varying quantities. Thus, the sole purpose of an isolation transformer is to eliminate direct electrical connection between the primary and secondary sides.

Regardless of the value of N, an ideal transformer never dissipates power. We confirm this property by noting that the total instantaneous power into the transformer is $p = v_1 i_1 + v_2 i_2$. Substituting for v_2 and i_2 from Eq. (8.2) then yields

$$p = v_1 i_1 + v_2 i_2 = v_1 i_1 + (Nv_1)(-i_1/N) = 0 \tag{8.3}$$

Hence, an ideal transformer transfers power without any internal loss. Furthermore, although the winding symbols in Fig. 8.1 look like inductors, an ideal transformer never stores any energy.

The foregoing properties of an ideal transformer can be represented by **controlled-source models.** The model in Fig. 8.2*a* has a VCVS on the secondary side to establish $v_2 = Nv_1$, while the CCCS on the primary side incorporates the fact that $i_1 = -Ni_2$ when $i_2 = -i_1/N$. Similar reasoning leads to the model in Fig. 8.2*b*, which has a CCCS on the secondary side and a VCVS on the primary side.

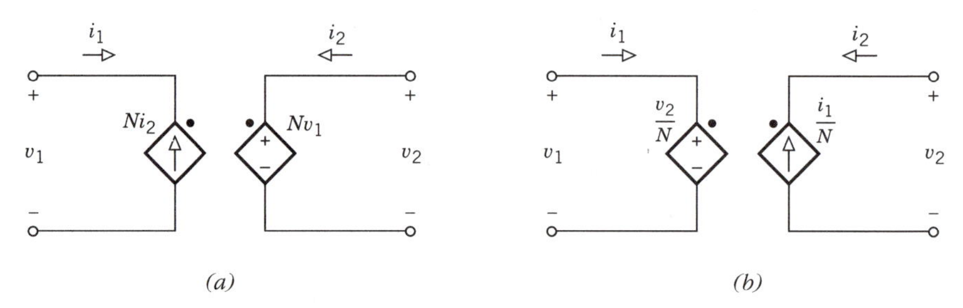

Figure 8.2 Controlled-source models of an ideal transformer.

Sometimes a transformer consists of three or more windings, or one winding has additional terminals known as **taps.** Such units provide multiple voltages from one source, as needed in applications such as an electronic power supply or a dual-voltage ac system.

For instance, Fig. 8.3*a* shows an ideal transformer with a tap dividing the secondary into two segments having N_2 and N_3 turns. Each secondary segment is magnetically coupled to the primary so that

$$v_2 = \frac{N_2}{N_1} v_1 \qquad v_3 = \frac{N_3}{N_1} v_1 \tag{8.4a}$$

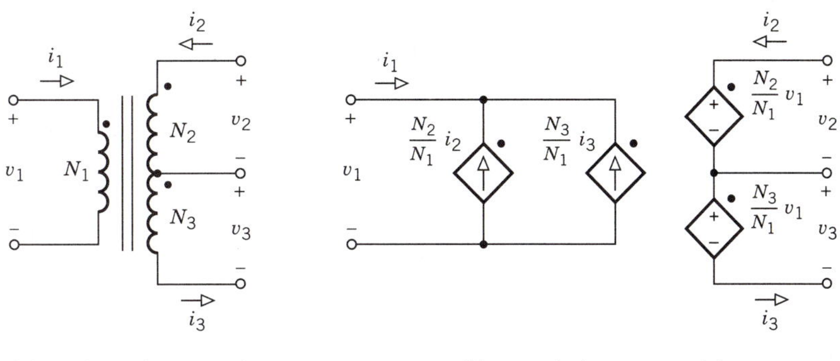

(a) Ideal transformer with tapped secondary

(b) Controled–source model

Figure 8.3

The resulting primary current is found from the lossless property $p = v_1 i_1 + v_2 i_2 + v_3 i_3 = 0$, so

$$i_1 = -\frac{v_2}{v_1} i_2 - \frac{v_3}{v_1} i_3 = -\frac{N_2}{N_1} i_2 - \frac{N_3}{N_1} i_3 \tag{8.4b}$$

Figure 8.3*b* shows the corresponding controlled-source model with two CCCSs and two VCVSs. Other tapped configurations are handled in a similar manner.

Example 8.1 *Analysis of a Transformer Circuit*

To illustrate analysis with a controlled-source model, consider the transformer circuit in Fig. 8.4*a*. Here we have an ac source coupled to a resistive load by an ideal step-up transformer with $N = 3$. But there is also an electrical connection between source and load via a capacitor. We seek the current and average power supplied by the source given that $v = 60 \cos 5000t$ V.

We begin by drawing the frequency-domain diagram in Fig. 8.4*b*. Since $\underline{V}_1 = \underline{V}$, the transformer is represented using a VCVS in the secondary to produce

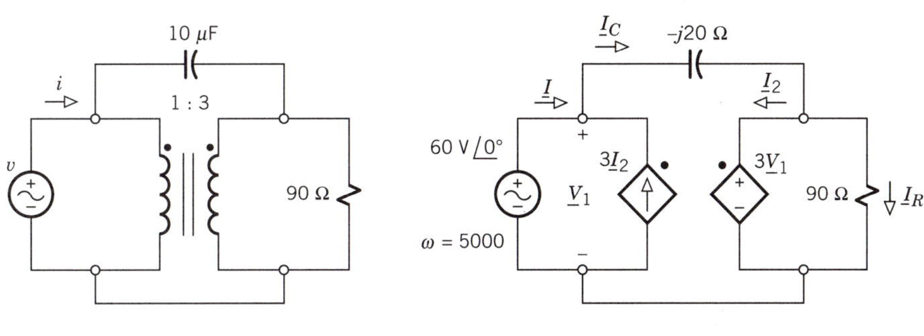

(a) Circuit with an ideal transformer

(b) Frequency–domain diagram

Figure 8.4

$$\underline{V}_2 = 3\underline{V}_1 = 3\underline{V} = 180 \text{ V } \angle 0°$$

Inspection of the diagram shows that $\underline{I}_2 = \underline{I}_C - \underline{I}_R$ where

$$\underline{I}_R = \frac{3\underline{V}_1}{90} = 2 \text{ A} \qquad \underline{I}_C = \frac{\underline{V}_1 - 3\underline{V}_1}{-j20} = -j6 \text{ A}$$

Thus,

$$\underline{I} = \underline{I}_C - 3\underline{I}_2 = \underline{I}_C - 3(\underline{I}_C - \underline{I}_R) = 6 + j12 = \sqrt{180} \text{ A } \angle 63.4°$$

$$Z = \underline{V}/\underline{I} = 2 - j4 \ \Omega \qquad P = \tfrac{1}{2}\text{Re}[Z]|\underline{I}|^2 = 180 \text{ W}$$

As a check on these results, we note that the average power dissipated by the 90-Ω resistor is $P_R = \frac{1}{2} \times 90|\underline{I}_R|^2 = 180 \text{ W} = P$ — confirming that all the power from the source is transferred to the resistor.

Exercise 8.1

Find the polar form of $\underline{I}_1$ and $\underline{I}_2$ for the transformer circuit in Fig. 8.5, given that $N_1 = 50$, $N_2 = 20$, and $v_1 = 25 \cos 100t$ V.

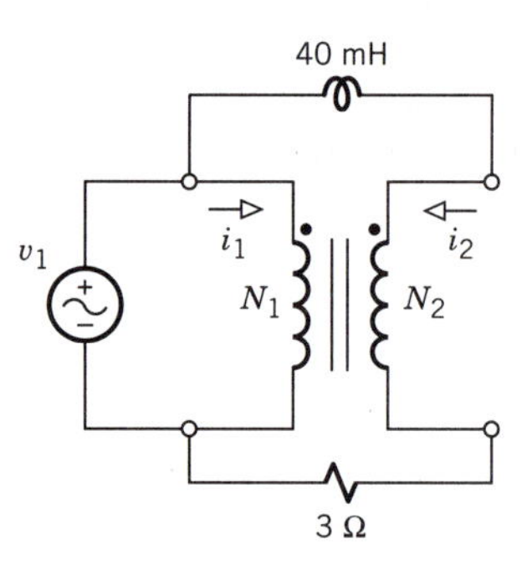

Figure 8.5

Referred Networks

Far and away the most common type of transformer circuit has the structure of Fig. 8.6*a*, where an ideal transformer serves as an *interface* between an ac source and a load. We emphasize the interface role by reversing the arrow for the secondary current, so $i_{out} = -i_2$.

Since any ac source network can be represented by its Thévenin model, and since any passive load network can be represented by its equivalent impedance, the corresponding frequency-domain diagram becomes as shown in Fig. 8.6*b*. Instead of using a controlled-source model here, the properties of the ideal transformer are displayed as the phasor relations

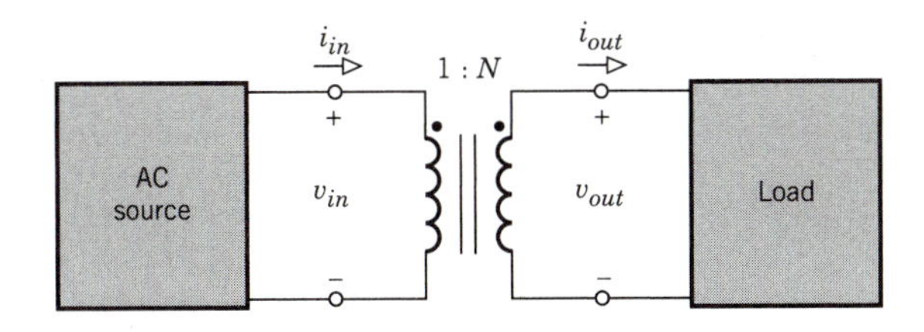

(a) Ideal transformer interfacing a source and a load

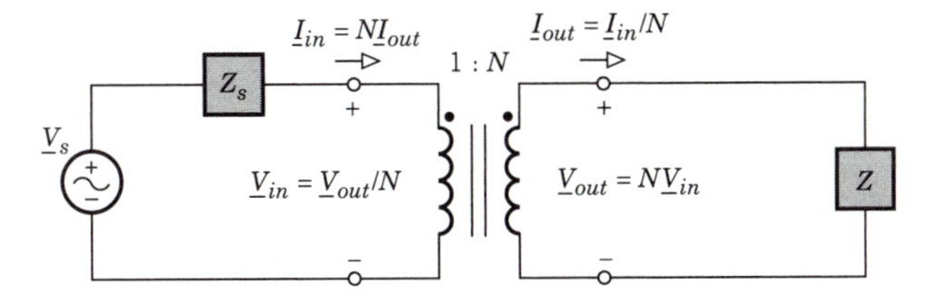

(b) Frequency–domain diagram

Figure 8.6

$$\begin{aligned} \underline{I}_{in} &= N\underline{I}_{out} & \underline{I}_{out} &= \underline{I}_{in}/N \\ \underline{V}_{in} &= \underline{V}_{out}/N & \underline{V}_{out} &= N\underline{V}_{in} \end{aligned} \tag{8.5}$$

We'll draw upon these relations to derive two other circuits better suited to further analysis.

First, we calculate the equivalent impedance seen looking into the primary. Since $\underline{I}_{out} = \underline{V}_{out}/Z = N\underline{V}_{in}/Z$, the primary current is

$$\underline{I}_{in} = N\underline{I}_{out} = N(N\underline{V}_{in}/Z) = N^2\underline{V}_{in}/Z$$

and therefore

$$Z_{in} = \underline{V}_{in}/\underline{I}_{in} = Z/N^2 \tag{8.6}$$

We call Z/N^2 the **referred load impedance** seen in the primary. Figure 8.7*a*

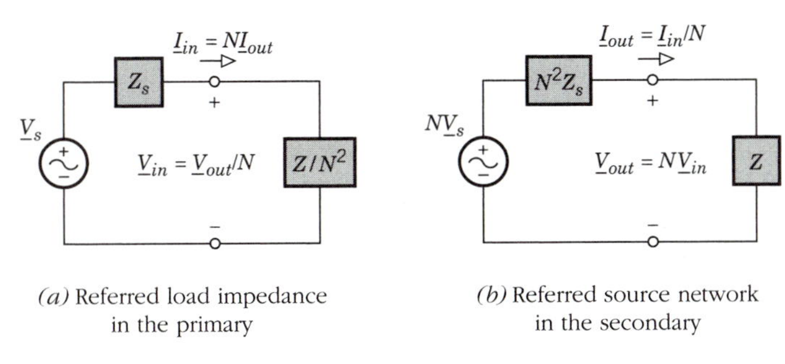

(a) Referred load impedance in the primary

(b) Referred source network in the secondary

Figure 8.7 Equivalent transformer circuits.

diagrams the equivalent circuit with the referred load impedance replacing the transformer and load.

To bring out an important implication of Eq. (8.6), let the load be a series *RLC* network with $Z = R + j\omega L - j/\omega C$. The referred load impedance can then be written as

$$\frac{Z}{N^2} = \frac{R}{N^2} + j\omega\left(\frac{L}{N^2}\right) - \frac{j}{\omega(N^2C)}$$

which corresponds to a series network composed of resistance R/N^2, inductance L/N^2, and capacitance N^2C. We thus conclude in general that

> A referred load network has the same structure as the load network with each R and L divided by N^2 and each C multiplied by N^2.

This property allows you to refer the *entire* load network to the primary, element by element, without having to calculate the equivalent load impedance Z. But keep in mind that referred impedances, not element values, are divided by N^2. Thus, referred capacitance must be *multiplied* by N^2 because its impedance is inversely proportional to capacitance.

Next, we focus on the output terminals in Fig. 8.6*b* and seek the Thévenin equivalent network seen looking back into the secondary. This is easily done in an indirect fashion by observing that

$$\underline{V}_{in} = \underline{V}_s - Z_s\underline{I}_{in} = \underline{V}_s - Z_s(N\underline{I}_{out})$$
$$\underline{V}_{out} = N\underline{V}_{in} = N[\underline{V}_s - Z_s(N\underline{I}_{out})]$$

so the secondary voltage and current are related by

$$\underline{V}_{out} = N\underline{V}_s - (N^2Z_s)\underline{I}_{out} \tag{8.7}$$

The corresponding equivalent circuit in Fig. 8.7*b* has the **referred source network** based on Eq. (8.7) instead of the original source network and the transformer.

Since $N\underline{V}_s$ is the open-circuit voltage referred to the secondary, and since N^2Z_s is the referred source impedance, the referred short-circuit current would be $N\underline{V}_s/N^2Z_s = \underline{V}_s/NZ_s$. Thus, in general,

> A referred source network has the same structure as the source network with each voltage multiplied by N, each impedance multiplied by N^2, and each current divided by N.

This property allows you to refer the *entire* source network into the secondary, element by element.

Either of the referred networks in Fig. 8.7 may be used to analyze the original circuit in Fig. 8.6*b*. The advantage of doing so is that a referred network does not contain a transformer and, consequently, can be analyzed by any of our standard methods. Which referred network you choose generally depends upon the unknowns of interest since the primary voltage and current are more easily found by referring the load to the primary (Fig. 8.7*a*), while the secondary voltage and current are more easily found by referring the source to the secondary (Fig. 8.7*b*). In either case, if the load happens to have a parallel configuration, then it may be advantageous to put the source network in Norton form.

However, note carefully that our referral methods work only when the transformer separates the circuit into two parts that interact exclusively via the transformer's magnetic coupling. As a counterexample, the circuit back in Fig. 8.4*a* does not satisfy this condition because it has current paths around the transformer. Such cases must be analyzed by direct application of the transformer relations, as previously illustrated in Example 8.1.

Example 8.2 *Power Transmission with Transformers*

Examples in Chapter 7 brought out the point that resistance in transmission lines may dissipate significant amounts of power, resulting in low power-transfer efficiency. One of the most important applications of transformers is increasing power-transfer efficiency by increasing the voltage on a transmission line, thereby reducing the current and the line loss. This example employs referral methods to demonstrate that effect.

To establish a baseline for comparison, consider first the situation in Fig. 8.8*a*. This transformerless circuit consists of an ac voltage source with negligible impedance connected to a remote 3-Ω load resistance via a transmission line with 2-Ω resistance. The objective here is to deliver $P = 15$ kW to the load, so

$$|\underline{I}_{out}| = \sqrt{2P/3\ \Omega} = 100\ \text{A} \qquad |\underline{V}_{out}| = 3\ \Omega \times |\underline{I}_{out}| = 300\ \text{V}$$

The corresponding line current and source voltage are

$$|\underline{I}| = |\underline{I}_{out}| = 100\ \text{A} \qquad |\underline{V}_s| = (2 + 3)\ \Omega \times |\underline{I}| = 500\ \text{V}$$

From the resistance ratios, the power-transfer efficiency has the discouraging value

$$\text{Eff} = 3/(3 + 2) = 60\%$$

which reflects the fact that the transmission line wastes $\frac{1}{2} \times 2|\underline{I}|^2 = 10$ kW.

Now consider the modified circuit in Fig. 8.8*b*, where a 1:*N* step-up transformer has been inserted at the source and a 4:1 step-down transformer at the load. We still want $P = 15$ kW delivered to the load, so, taking $\angle\underline{I}_{out} = 0°$ for simplicity, we want

$$\underline{I}_{out} = 100\ \text{A}\ \underline{/0°} \qquad \underline{V}_{out} = 300\ \text{V}\ \underline{/0°}$$

We'll find the required value of *N*, and we'll calculate the power-transfer efficiency of this new arrangement. These tasks are easily accomplished by re-

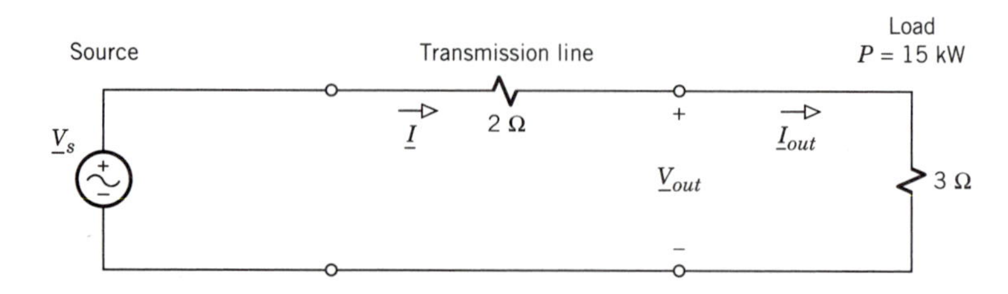

(a) Power transfer via transmission line

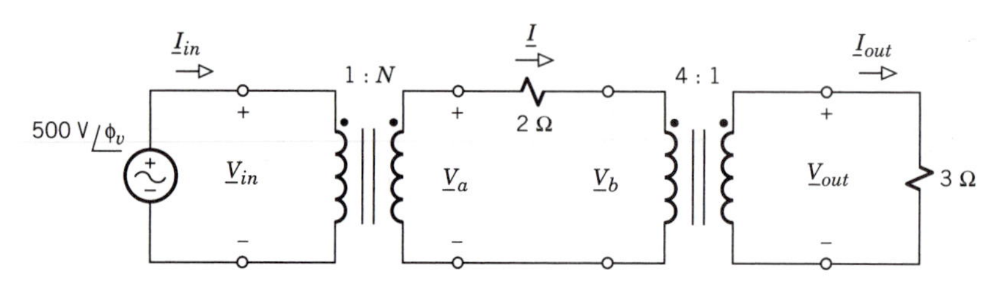

(b) Circuit with transformers at each end

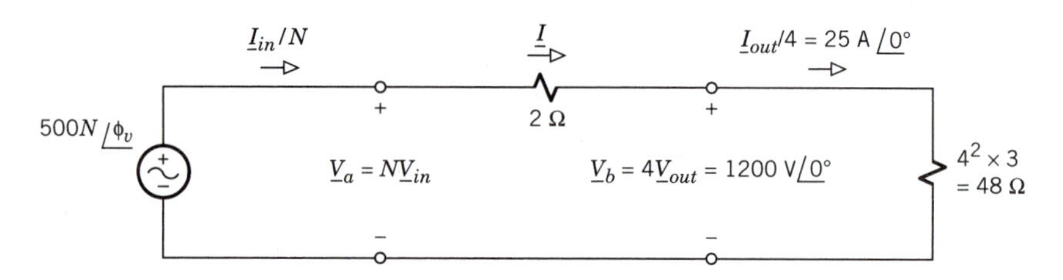

(c) Frequency–domain diagram with both ends referred into the middle

Figure 8.8

ferring both the source and the load into the middle section with the transmission line.

Since the source has negligible impedance, the source network referred to the secondary of the step-up transformer is just $N\underline{V}_s = 500N\,\angle\phi_v$. Then, at the load, the 4:1 turns ratio of the step-down transformer corresponds to 1:N with $N = 1/4$, so the load resistance referred to the primary is $3\ \Omega/(1/4)^2 = 4^2 \times 3 = 48\ \Omega$. Figure 8.8$c$ diagrams our resulting equivalent circuit, including labels for the referred quantities $\underline{I}_{in}/N$, $\underline{I}_{out}/4 = 25\text{ A }\angle 0°$, $\underline{V}_a = N\underline{V}_{in}$, and $\underline{V}_b = 4\underline{V}_{out} = 1200\text{ V }\angle 0°$.

Recalling that ideal transformers never dissipate power, we immediately see that the power-transfer efficiency has been increased to

$$\text{Eff} = 48/(48 + 2) = 96\%$$

a much more satisfactory result! To find the required value of N, we note that the line current will be

$$\underline{I} = 500N\,\angle\phi_v/(2 + 48) = 10N\,\angle\phi_v = \underline{I}_{out}/4 = 25\text{ A }\angle 0°$$

so $\phi_v = 0°$ and

$$N = 25/10 = 2.5$$

The power waste by the transmission line has thus been reduced from 10 kW to ½ × $2|\underline{I}|^2$ = 625 W. The remaining quantities of interest are

$$\underline{I}_{in} = N\underline{I} = 62.5 \text{ A } \angle 0^\circ \qquad \underline{V}_a = N\underline{V}_{in} = \underline{V}_b + 2\underline{I} = 1250 \text{ V } \angle 0^\circ$$

Example 8.3 *Transformer-Coupled Oscillator*

Another type of transformer application is illustrated by Fig. 8.9*a*, which represents an electronic oscillator circuit. The oscillator generates a 12-V sinusoid at ω = 50,000 and has 1-kΩ source resistance. A 20-V battery supplies dc current and power to the oscillator through the primary of an ideal 2:1 step-down transformer, which also couples the oscillator to an *RC* load. The transformer's isolation property has been exploited here to ground the undotted end of the secondary, even though the grounded battery puts the undotted end of the primary at 20 V.

By superposition, the primary current i_{in} consists of an ac component i_{in-1} plus a dc component i_{in-2} = −20 V/1 kΩ = −20 mA. The dc component produces no effect in the secondary, so i_{out} will be a sinusoid at ω = 50,000. We'll find this current by referring the ac source to the secondary using N = ½. Then we'll find the ac primary component via $i_{in-1} = Ni_{out}$.

The frequency-domain circuit with the referred source network is shown in Fig. 8.9*b*, where we have converted the Thévenin structure to Norton form with parallel resistance $N^2 \times R_s$ = 250 Ω and source current (N × 12 V)/

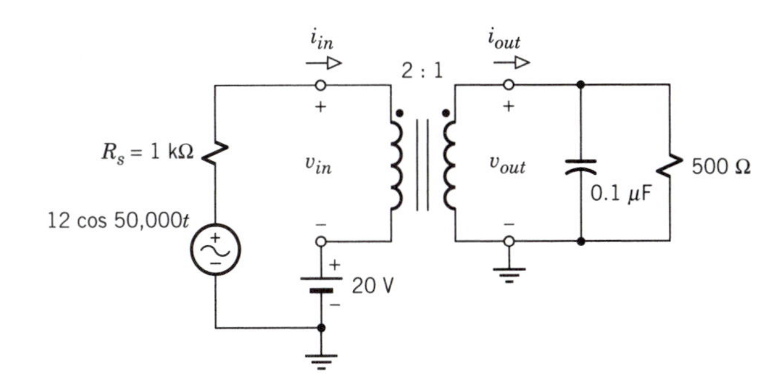

(*a*) Transformer circuit with ac and dc sources

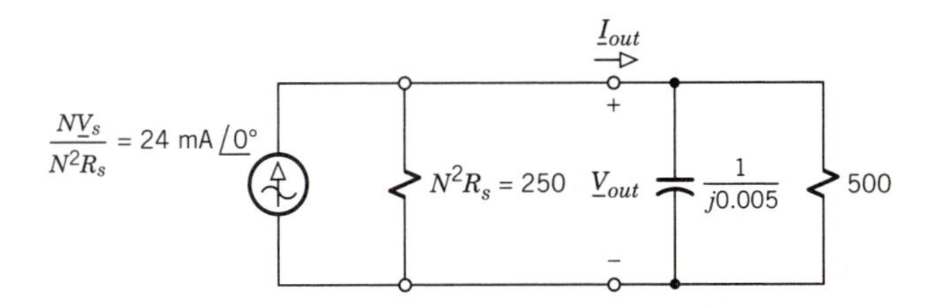

(*b*) Frequency–domain diagram with referred source network in the secondary

Figure 8.9

$(N^2 \times R_s) = 24$ mA $\angle 0°$. The resulting voltage across the secondary is calculated from the node equation

$$\left(\frac{1}{250} + j0.005 + \frac{1}{500}\right) \underline{V}_{out} = 24 \times 10^{-3}$$

which yields $\underline{V}_{out} \approx 3$ V $\angle -40°$. Now we can find $\underline{I}_{out}$ via

$$\underline{I}_{out} = \left(j0.005 + \frac{1}{500}\right) \underline{V}_{out} \approx 16 \text{ mA } \angle 20°$$

Hence, $\underline{I}_{in-1} = N\underline{I}_{out} \approx 8$ mA $\angle 20°$, and the total primary current is $i_{in} = i_{in-1} + i_{in-2} \approx 8 \cos(50{,}000t + 20°) - 20$ mA.

Exercise 8.2

Let the circuit in Fig. 8.6*b* have $\underline{V}_s = 25$ V $\angle 0°$, $Z_s = 3 + j4\ \Omega$, $N = 2$, and $Z = 12 - j9\ \Omega$. Calculate $\underline{I}_{in}$, $\underline{V}_{in}$, $\underline{I}_{out}$, and $\underline{V}_{out}$ by referring the secondary to the primary.

Exercise 8.3

Rework Exercise 8.2 by referring the primary to the secondary.

Impedance Matching

Recall our investigation of ac power transfer in Section 7.1. There we found that maximum power transfer from a source with fixed resistance R_s and reactance X_s occurs when the load has the *matched impedance*

$$Z = Z_s^* = R_s - jX_s$$

Thus, if $Z = R + jX$, then maximum power transfer requires $R = R_s$ and $X = -X_s$. The resulting maximum available power delivered to the load is

$$P_{max} = V_{rms}{}^2/4R_s$$

where V_{rms} is the open-circuit rms source voltage.

But if R_s and R have unequal fixed values, then impedance matching requires the help of a transformer between the source and load, as shown back in Fig. 8.6*b*. The load impedance referred to the primary becomes Z/N^2, so power transfer is maximized by taking

$$Z/N^2 = Z_s^* \tag{8.8}$$

The transformer's turns ratio N should thus be chosen to satisfy $R/N^2 = R_s$, and the load reactance should be adjusted such that $X/N^2 = -X_s$.

If neither R nor X can be adjusted, then we cannot get the maximum available power. In this case, the largest possible power is delivered to the load when

$$|Z|/N^2 = |Z_s| \tag{8.9}$$

which follows from the derivation of Eq. (7.18).

Example 8.4 *Impedance Matching with a Transformer*

The left side of Fig. 8.10*a* is the frequency-domain model for the output of an amplifier operating at $\omega = 10^5$. The load has a fixed resistance of 500 Ω, and we want to find the load reactance X and the transformer's turns ratio N to achieve maximum power transfer.

Performing a Norton-to-Thévenin conversion on the source yields

$$Z_s = 400\|(-j200) = 80 - j160\ \Omega$$
$$|\underline{V}_s| = |Z_s\underline{I}_s| \approx 18\ \text{V}$$

Figure 8.10*b* gives the equivalent circuit with the load impedance referred to the primary. Impedance matching to get $Z/N^2 = Z_s$ requires an inductive load reactance and a turns ratio such that

$$500/N^2 = R_s = 80\ \Omega \qquad \omega L/N^2 = -X_s = +160\ \Omega$$

We therefore need

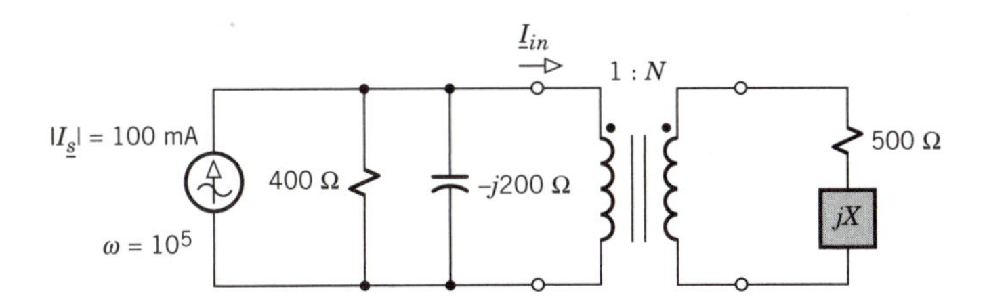

(a) Model of amplifier with impedance–matching transformer

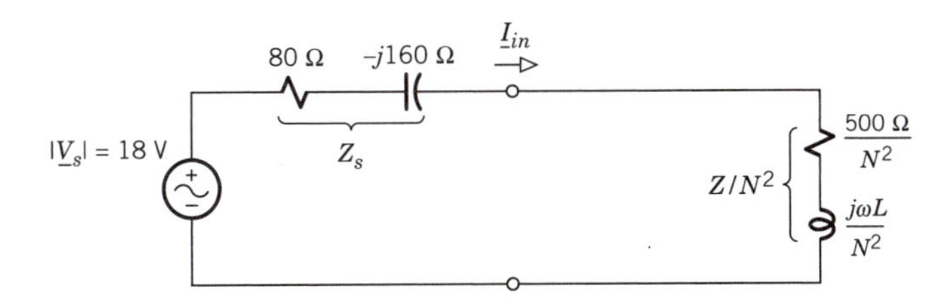

(b) Referred load in the primary

Figure 8.10

$$N = \sqrt{500/80} = 2.5 \qquad L = 160N^2/\omega = 10 \text{ mH}$$

Since the ideal transformer is lossless, the load receives

$$P_{max} = (18/\sqrt{2} \text{ V})^2/(4 \times 80\ \Omega) \approx 510 \text{ mW}$$

Exercise 8.4

Suppose the reactance X cannot be included in the load in Fig. 8.10a. What should be the value of N, and what's the resulting average power delivered to the 500-Ω resistor?

8.2 MAGNETIC COUPLING AND MUTUAL INDUCTANCE

Magnetic coupling exists whenever current-carrying coils are sufficiently close that their magnetic fields interact. But transformers usually require a high degree of coupling that can be achieved only by winding the coils around a single core, thereby creating a *magnetic circuit*. This section begins with a brief description of magnetic circuits, from which we develop the concept of *mutual inductance*. We'll give close scrutiny to mutual inductance because it is the underlying phenomenon for transformer action.

Magnetic Circuits

Our discussion of inductance in Section 5.2 began by considering a magnetic device like Fig. 8.11a. Here, a coil has N turns wrapped around a cylindrical core that provides a path for magnetic flux ϕ produced by current i. The direction of ϕ is given by the **right-hand rule:**

> When you grasp the coil in your right hand with your fingers curled in the direction of the current, your extended thumb points in the direction of the flux inside the coil.

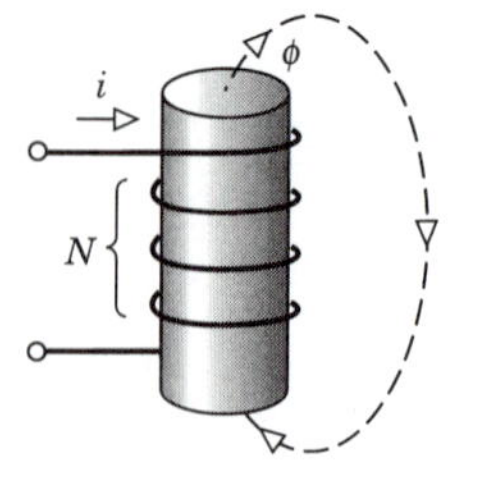

(*a*) Coil with cylindrical core

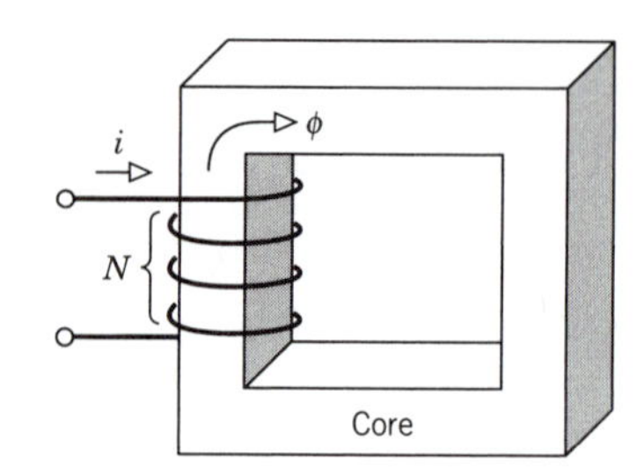

(*b*) Magnetic circuit

Figure 8.11

Thus, when i enters the upper terminal of the coil, as shown, the direction of ϕ is from bottom to top through the core. The flux then exits the core at the top, passes through space, and enters the core at the bottom.

The amount of flux is greatly increased if we use a closed core shape like Fig. 8.11b and if the core material has a large relative permeability μ_r. These conditions ensure that almost no flux leaves the core. The resulting value of ϕ is given approximately by

$$\phi = \mu_r \mu_0 NiA/l \tag{8.10}$$

where A is the average cross-sectional area of the core and l is the average length of the flux path through the core.

We call Fig. 8.11b a **magnetic circuit** because the flux circulates around a closed magnetic path, just as current circulates around a closed electrical path. The analogy between electric and magnetic circuits is further enhanced by introducing two additional terms.

First, we express the flux-producing effect of the coil in terms of **magnetomotive force (mmf),** defined by

$$\mathcal{F} \triangleq Ni \tag{8.11}$$

Second, we express the magnetic properties of the flux path in terms of **reluctance,** defined by

$$\mathcal{R} \triangleq l/\mu_r \mu_0 A \tag{8.12}$$

Combining Eqs. (8.10)–(8.12) then yields the simple expression

$$\mathcal{F} = \mathcal{R}\phi \tag{8.13}$$

which is the magnetic version of $v = Ri$. The corresponding magnetic circuit is diagrammed in Fig. 8.12, and we clearly see that the mmf source $\mathcal{F}$ applied to reluctance $\mathcal{R}$ results in flux $\phi = \mathcal{F}/\mathcal{R}$.

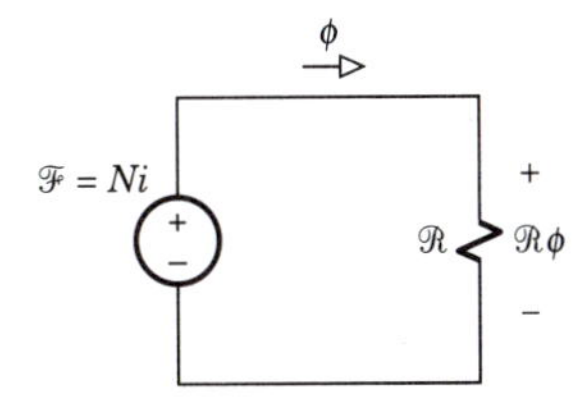

Figure 8.12 Diagram of a magnetic circuit.

Figure 8.13a shows a more complicated core configuration with two windings and two closed flux paths. The coil on the left links both ϕ_1 and ϕ_2, but the coil on the right only links ϕ_2. The magnetic circuit diagram in Fig. 8.13b therefore includes two mmf sources. From the right-hand rule, the source $\mathcal{F}_1 = N_1 i_1$ drives both ϕ_1 and ϕ_2 while the source $\mathcal{F}_2 = N_2 i_2$ opposes ϕ_2. The

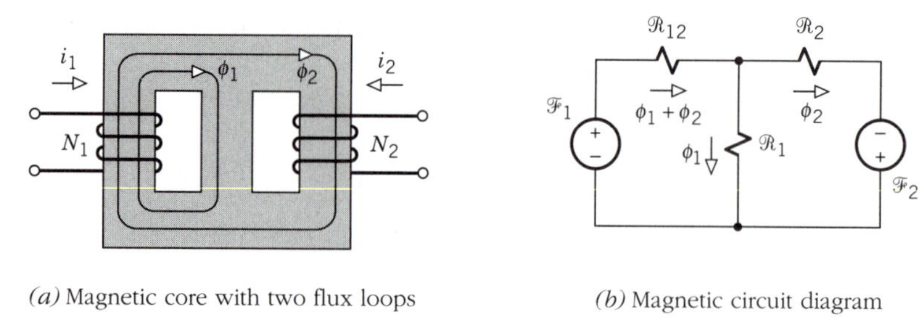

(a) Magnetic core with two flux loops

(b) Magnetic circuit diagram

Figure 8.13

diagram has reluctance $\mathcal{R}_{12}$ representing the path of ϕ_1 and ϕ_2 together, $\mathcal{R}_1$ representing the path of ϕ_1 alone, and $\mathcal{R}_2$ representing the path of ϕ_2 alone. Given the mmf and reluctance values, we could apply standard methods of circuit analysis to evaluate ϕ_1 and ϕ_2.

Of course, our standard methods work only for a *linear* magnetic circuit. But high-permeability ferromagnetic materials may exhibit *nonlinear* behavior when the mmf gets too large. As a further complication, these materials have a *hysteresis* phenomenon in that the variations of ϕ lag behind the variations of $\mathcal{F}$. Since nonlinearity and hysteresis produce unwanted effects in transformers (as well as complicating the analysis), we'll continue to assume that the cores of magnetic circuits operate in the linear region and that hysteresis effects are negligible. These assumptions justify the use of linear magnetic circuit diagrams.

Self-Inductance and Mutual Inductance

Figure 8.14*a* depicts a magnetic circuit of the type commonly used for real transformers. Our objective here is to derive the relationships between the terminal voltages and currents of the two coils. Those relationships will lead us to the concepts of self-inductance and mutual inductance.

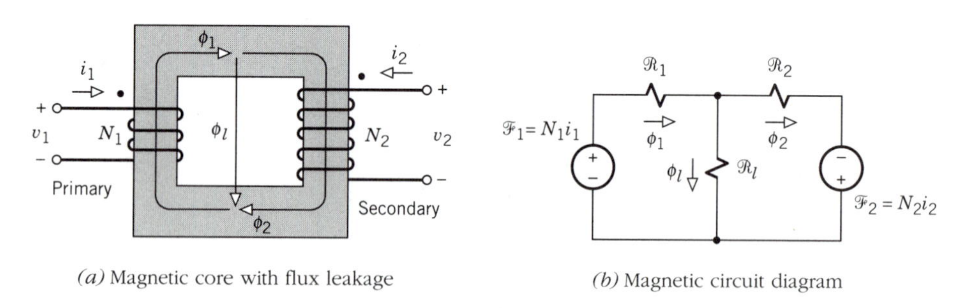

(a) Magnetic core with flux leakage

(b) Magnetic circuit diagram

Figure 8.14

As before, the coil with N_1 turns has been designated as the primary winding and the coil with N_2 turns has been designated as the secondary winding. Current i_1 entering the dotted terminal of the primary establishes $\mathcal{F}_1 = N_1 i_1$, which drives flux in the clockwise direction around the core. Current i_2 likewise establishes $\mathcal{F}_2 = N_2 i_2$, which also drives flux in the clockwise direction. Thus, the dots here mean that:

> Current entering the dotted end of one winding produces flux in the same direction as the flux produced by current entering the dotted end of the other winding.

We'll subsequently show that these polarity dots agree with our previous convention for ideal transformers.

Although most of the flux stays within the core, there is always some **leakage flux** ϕ_l that escapes. Consequently, the total flux ϕ_1 linking the primary differs from the total flux ϕ_2 linking the secondary. The magnetic circuit diagram therefore takes the form of Fig. 8.14*b*, where $\mathcal{R}_l$ represents the reluctance of the leakage path. Treating ϕ_1 and ϕ_2 like mesh currents in this diagram, we easily get the matrix equation

$$\begin{bmatrix} \mathcal{R}_1 + \mathcal{R}_l & -\mathcal{R}_l \\ -\mathcal{R}_l & \mathcal{R}_2 + \mathcal{R}_l \end{bmatrix} \begin{bmatrix} \phi_1 \\ \phi_2 \end{bmatrix} = \begin{bmatrix} N_1 i_1 \\ N_2 i_2 \end{bmatrix}$$

Simultaneous solution then yields

$$\phi_1 = \frac{N_1 i_1}{\mathcal{R}_{11}} + \frac{N_2 i_2}{\mathcal{R}_M} \qquad \phi_2 = \frac{N_1 i_1}{\mathcal{R}_M} + \frac{N_2 i_2}{\mathcal{R}_{22}} \tag{8.14}$$

where we have introduced

$$\mathcal{R}_{11} = \mathcal{R}_1 + \mathcal{R}_2 \| \mathcal{R}_l \qquad \mathcal{R}_{22} = \mathcal{R}_2 + \mathcal{R}_1 \| \mathcal{R}_l$$
$$\mathcal{R}_M = \mathcal{R}_1 + \mathcal{R}_2 + (\mathcal{R}_1 \mathcal{R}_2 / \mathcal{R}_l) \tag{8.15}$$

The quantities $\mathcal{R}_{11}$ and $\mathcal{R}_{22}$ are just the equivalent reluctances seen by the mmfs $\mathcal{F}_1$ and $\mathcal{F}_2$, respectively. The mutual reluctance $\mathcal{R}_M$ accounts for the fact that $\mathcal{F}_1$ contributes to ϕ_2 and $\mathcal{F}_2$ contributes to ϕ_1.

Next, we assume that the coils have negligible resistance, so the terminal voltages are given directly by Faraday's law. Specifically, from Eq. (8.14), we get

$$v_1 = N_1 \frac{d\phi_1}{dt} = \frac{N_1^2}{\mathcal{R}_{11}} \frac{di_1}{dt} + \frac{N_1 N_2}{\mathcal{R}_M} \frac{di_2}{dt}$$
$$v_2 = N_2 \frac{d\phi_2}{dt} = \frac{N_1 N_2}{\mathcal{R}_M} \frac{di_1}{dt} + \frac{N_2^2}{\mathcal{R}_{22}} \frac{di_2}{dt}$$

These two equations can be cleaned up and made more understandable by defining

$$L_1 \triangleq \frac{N_1^2}{\mathcal{R}_{11}} \qquad L_2 \triangleq \frac{N_2^2}{\mathcal{R}_{22}} \qquad M \triangleq \frac{N_1 N_2}{\mathcal{R}_M} \tag{8.16}$$

so we have

$$v_1 = L_1 \frac{di_1}{dt} + M \frac{di_2}{dt} \qquad v_2 = M \frac{di_1}{dt} + L_2 \frac{di_2}{dt} \tag{8.17}$$

Notice that v_1 and v_2 depend upon both i_1 and i_2 because ϕ_1 and ϕ_2 in Eq. (8.14) depend upon both currents.

To interpret the quantities L_1 and M, let the secondary be open-circuited so $di_2/dt = 0$ and Eq. (8.17) reduces to

$$v_1 = L_1\, di_1/dt \qquad v_2 = M\, di_1/dt$$

Clearly, L_1 corresponds to the inductance of the primary by itself, known as the **self-inductance** of the primary. But the primary current also affects the secondary, and M represents the magnetic coupling from primary to secondary. This coupling arises because part of flux ϕ_1 produced by $N_1 i_1$ links the secondary and induces the open-circuit voltage $v_2 = M\, di_1/dt$. Conversely, if $di_1/dt = 0$, then

$$v_1 = M\, di_2/dt \qquad v_2 = L_2\, di_2/dt$$

Hence, L_2 is the self-inductance of the secondary, and M now represents magnetic coupling from secondary to primary. Since the coupling is the same in either direction, M is named the **mutual inductance** between the coils.

Equation (8.16) indicates that the self-inductances L_1 and L_2 are proportional to N_1^2 and N_2^2, respectively, whereas the mutual inductance is proportional to $N_1 N_2$. From Eqs. (8.15) and (8.16), M is related to L_1 and L_2 by

$$M = k\sqrt{L_1 L_2} \tag{8.18a}$$

where

$$k = \frac{\mathcal{R}_l}{\sqrt{(\mathcal{R}_1 + \mathcal{R}_l)(\mathcal{R}_2 + \mathcal{R}_l)}} \leq 1 \tag{8.18b}$$

We call k the **coupling coefficient.** Maximum coupling ($k = 1$) occurs when $\mathcal{R}_l = \infty$, so there is no leakage and all of the flux links both coils. The mutual inductance then has the largest possible value $M_{max} = \sqrt{L_1 L_2}$. Otherwise, leakage reduces the amount of flux linking both coils, resulting in $k < 1$ and $M < \sqrt{L_1 L_2}$.

The foregoing equations were based on a linear magnetic circuit with both coils wound in the same sense. If one coil in Fig. 8.14*a* had been wound in the opposite sense, then its mmf would oppose the flux produce by the other coil. Consequently, the terminal relationships would be

$$v_1 = L_1 \frac{di_1}{dt} - M \frac{di_2}{dt} \qquad v_2 = -M \frac{di_1}{dt} + L_2 \frac{di_2}{dt} \tag{8.19}$$

Equation (8.19) differs from Eq. (8.17) in that M has been replaced by $-M$, so reversing one winding simply changes the sign of the magnetic coupling in both directions.

Although our derivations have been somewhat lengthy, our results can be summarized in a simple and concise fashion. Hereafter, we'll represent magnetically coupled coils by the schematic symbols in Fig. 8.15, which indicate that mutual inductance has a bidirectional effect and always occurs in conjunction with two self-inductances. The coils in part (*a*) have dots at the same end to denote the same winding sense, while the coils in part (*b*) have dots at the opposite ends to denote opposite winding sense. These schematic symbols stand for the voltage–current relationships

$$v_1 = L_1 \frac{di_1}{dt} \pm M \frac{di_2}{dt} \qquad v_2 = \pm M \frac{di_1}{dt} + L_2 \frac{di_2}{dt} \tag{8.20}$$

where the upper sign for M goes with Fig. 8.15*a* and the lower sign goes with Fig. 8.15*b*. In either case, the mutual inductance is related to the self-inductances by

$$M = k\sqrt{L_1 L_2} \qquad 0 \le k \le 1 \tag{8.21}$$

Equations (8.20) and (8.21) constitute the essential information needed to analyze circuits with magnetic coupling, provided that the cores have negligible nonlinear effects. Our prior assumption of negligible coil resistance is not essential, since we can always model winding resistance by adding appropriate resistors in series with the primary and secondary.

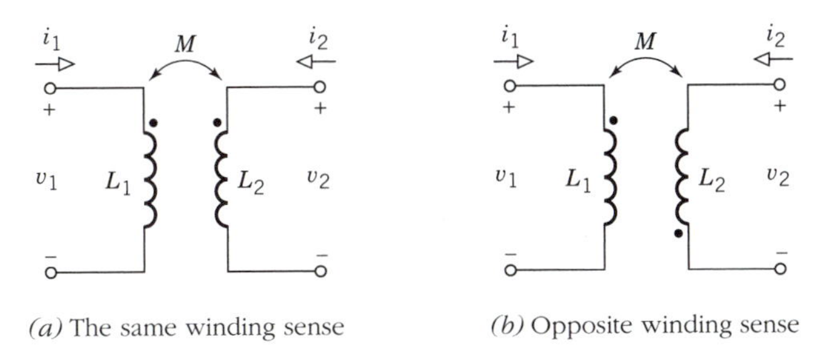

(*a*) The same winding sense (*b*) Opposite winding sense

Figure 8.15 Symbols for magnetically coupled coils.

Example 8.5 *Series Equivalent Inductance*

Figure 8.16*a* shows two coils connected in series to form a two-terminal network. In absence of mutual inductance, we would write $v = L_{eq} di/dt$ with $L_{eq} = L_1 + L_2$. But what is the effect of mutual inductance on the equivalent inductance?

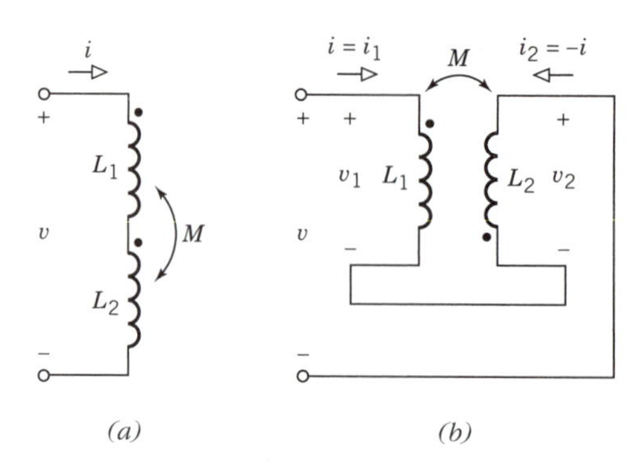

Figure 8.16 Coils connected in series.

To answer this question, we redraw and label the network per Fig. 8.16*b* so we clearly see that $i_1 = i$, $i_2 = -i$, and $v = v_1 - v_2$. Since the coils have the opposite winding sense, we take the lower signs in Eq. (8.20) to get

$$v_1 = L_1 \frac{di_1}{dt} - M\frac{di_2}{dt} = L_1 \frac{di}{dt} - M\frac{d(-i)}{dt} = (L_1 + M)\frac{di}{dt}$$

$$v_2 = -M\frac{di_1}{dt} + L_2\frac{di_2}{dt} = -M\frac{di}{dt} + L_2\frac{d(-i)}{dt} = -(M + L_2)\frac{di}{dt}$$

Thus, $v = v_1 - v_2 = (L_1 + 2M + L_2)\, di/dt = L_{eq}\, di/dt$ where

$$L_{eq} = L_1 + L_2 + 2M \leq L_1 + L_2 + 2\sqrt{L_1 L_2}$$

The mutual inductance therefore increases the series equivalent inductance. If the coils had the same winding sense, or if we reversed the connections to one coil, then we would have obtained $L_{eq} = L_1 + L_2 - 2M$.

Exercise 8.5

By taking $M = M_{max}$, show that $L_1 + L_2 - 2M$ cannot be a negative quantity.

Exercise 8.6

Suppose the secondary in Fig. 8.17 is short-circuited, so $v_2 = 0$. Show that $v_1 = L_{eq}\, di_1/dt$, and express L_{eq} in terms of L_1 and k.

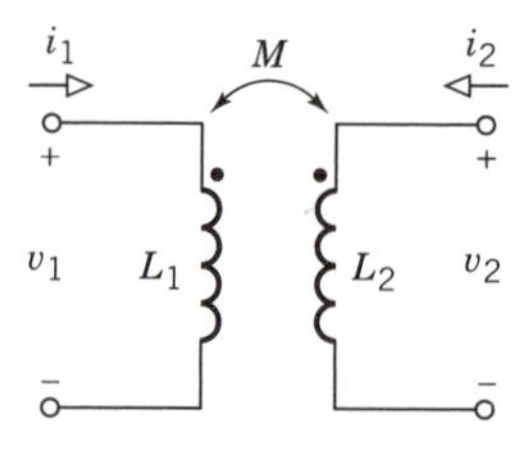

Figure 8.17

Energy Storage and Unity Coupling

We close this section with an examination of energy storage in magnetically coupled coils, giving special attention to unity coupling. For simplicity, we'll work with coils having the same winding sense since we can always replace M by $-M$ to account for the opposite sense.

The rate of energy transfer into the coils in Fig. 8.17 is $dw/dt = p = v_1 i_1 + v_2 i_2$. Expanding v_1 and v_2 per Eq. (8.17) and noting that $Mi_1\, di_2/dt + Mi_2\, di_1/dt = M\, d(i_1 i_2)/dt$, we get

$$\frac{dw}{dt} = L_1 i_1 \frac{di_1}{dt} + M \frac{d(i_1 i_2)}{dt} + L_2 i_2 \frac{di_2}{dt}$$

To find the instantaneous stored energy, we integrate dw/dt over time from $-\infty$ to t. Thus, using the dummy variable λ,

$$\begin{aligned} w(t) &= L_1 \int_{-\infty}^{t} i_1(\lambda)\, di_1(\lambda) + M \int_{-\infty}^{t} d[i_1(\lambda) i_2(\lambda)] \\ &\quad + L_2 \int_{-\infty}^{t} i_2(\lambda)\, di_2(\lambda) \\ &= \tfrac{1}{2} L_1 i_1^2(t) + M i_1(t) i_2(t) + \tfrac{1}{2} L_2 i_2^2(t) \end{aligned} \tag{8.22}$$

where we have made the usual assumption that the stored energy and the currents equal zero at $t = -\infty$.

If there is no magnetic coupling, then $M = 0$ and Eq. (8.22) reduces to $w = \frac{1}{2} L_1 i_1^2 + \frac{1}{2} L_2 i_2^2$ — precisely what you would expect for the total energy stored by two isolated inductors. The magnetic coupling increases or decreases the stored energy, depending upon the winding sense and the signs of the currents. When $Mi_1 i_2$ is a positive quantity, w is increased. But w is decreased when $Mi_1 i_2$ is a negative quantity.

Additional information about the stored energy can be gained from Eq. (8.22) by keeping one of the currents constant. In particular, Fig. 8.18 plots w versus i_2 with i_1 held at a fixed positive value. This curve exhibits a unique minimum at $i_2 = -(M/L_2) i_1$ where

$$w_{min} = \frac{L_1 L_2 - M^2}{2 L_2} i_1^2 \tag{8.23}$$

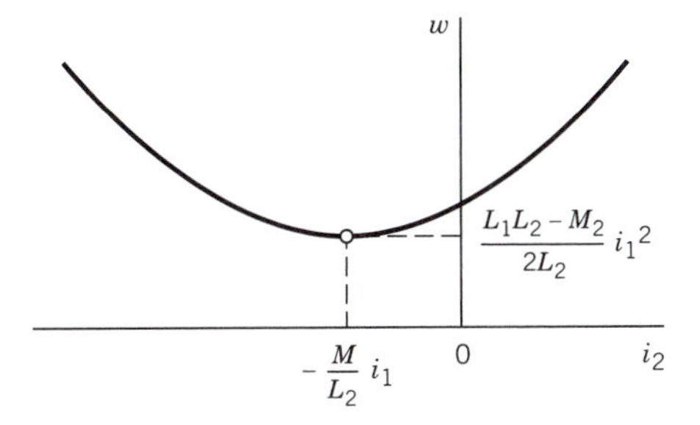

Figure 8.18 Stored energy versus i_2 with i_1 held constant.

Since w cannot be negative, it follows that $L_1L_2 - M^2 \geqslant 0$. This conclusion agrees with the property $M^2 = k^2L_1L_2 \leqslant L_1L_2$. We also see that if $k = 1$, then $M^2 = L_1L_2$ and $w_{min} = 0$.

Let's give closer attention to the case of *unity coupling* ($k = 1$), which occurs when there is no flux leakage. Correspondingly, setting $\mathcal{R}_l = \infty$ in Eq. (8.15) yields $\mathcal{R}_{11} = \mathcal{R}_{22} = \mathcal{R}_M = \mathcal{R}_1 + \mathcal{R}_2$, and Eq. (8.16) simplifies to

$$L_1 = \frac{N_1^2}{\mathcal{R}_1 + \mathcal{R}_2} \qquad L_2 = \frac{N_2^2}{\mathcal{R}_1 + \mathcal{R}_2} \qquad M = \frac{N_1N_2}{\mathcal{R}_1 + \mathcal{R}_2} \tag{8.24a}$$

which confirms that $M = \sqrt{L_1L_2}$ and also shows that

$$L_2/L_1 = N_2^2/N_1^2 \tag{8.24b}$$

The voltage relationship is then obtained from Eq. (8.17) with $M = \sqrt{L_1L_2}$, so

$$v_1 = L_1\frac{di_1}{dt} + \sqrt{L_1L_2}\frac{di_2}{dt}$$

$$\begin{aligned} v_2 &= \sqrt{L_1L_2}\frac{di_1}{dt} + L_2\frac{di_2}{dt} = \sqrt{\frac{L_2}{L_1}}\left(L_1\frac{di_1}{dt} + \sqrt{L_1L_2}\frac{di_2}{dt}\right) \\ &= \sqrt{L_2/L_1}\,v_1 = (N_2/N_1)v_1 = Nv_1 \end{aligned}$$

where we have identified the *turns ratio* $N = N_2/N_1$. Hence, insofar as the voltages are concerned, a pair of coils with unity coupling acts just like an *ideal transformer.* Furthermore, from Fig. 8.18, unity-coupled coils store zero energy when

$$i_2 = -\frac{M}{L_2}i_1 = -\frac{N_1}{N_2}i_1 = -\frac{i_1}{N}$$

which is the same current relationship as an ideal transformer. We'll pursue these observations in the next section.

Exercise 8.7

Taking i_1 to be constant in Eq. (8.22), show that $w_{min} = (L_1L_2 - M^2)i_1^2/2L_2$ at $i_2 = -(M/L_2)i_1$.

8.3 CIRCUITS WITH MUTUAL INDUCTANCE

Having completed the necessary background, we're ready to analyze circuits with mutual inductance. Specifically, we'll extend the method of impedance analysis to include mutual inductance, and we'll use impedance analysis to quantify the conditions under which real transformers act nearly like ideal transformers. Then we'll derive useful equivalent networks for the self-inductances and mutual inductance of real transformers.

Impedance Analysis

In principle, any circuit with mutual inductance can be analyzed starting from the basic voltage–current equations

$$v_1 = L_1 \frac{di_1}{dt} \pm M \frac{di_2}{dt} \qquad v_2 = \pm M \frac{di_1}{dt} + L_2 \frac{di_2}{dt} \tag{8.25}$$

Alternatively, we may use the circuit model in Fig. 8.19, where the effects of the mutual inductance appear as voltage sources controlled by the *derivatives* di_1/dt and di_2/dt. The upper signs on the controlled sources correspond to coils with the same winding sense, while the lower signs correspond to coils with the opposite sense. The advantage of the controlled-source model is that it explicitly represents the relationships between the terminal voltages and currents.

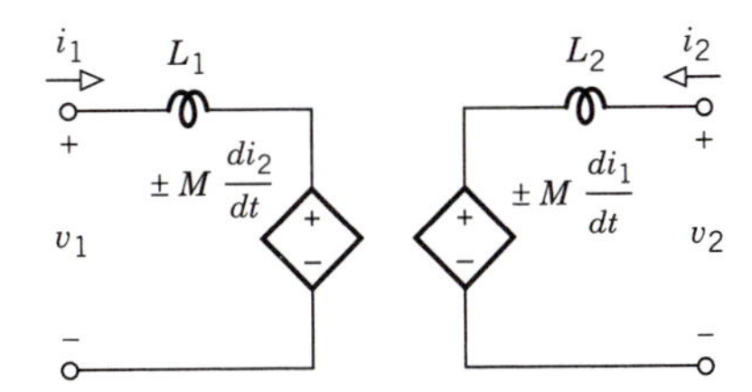

Figure 8.19 Mutual inductance represented by controlled sources.

Under ac steady-state conditions, analysis is further simplified by working with phasors and impedances. Clearly, any self-inductance term such as $L_1\, di_1/dt$ should be replaced by $j\omega L_1 \underline{I}_1$ in the frequency domain. Likewise, a mutual-inductance term such as $M\, di_2/dt$ should be replaced by $j\omega M \underline{I}_2$, even though M does not exist as a distinct inductance element. The ac version of Eq. (8.25) therefore becomes

$$\underline{V}_1 = j\omega L_1 \underline{I}_1 \pm j\omega M \underline{I}_2 \qquad \underline{V}_2 = \pm j\omega M \underline{I}_1 + j\omega L_2 \underline{I}_2 \tag{8.26}$$

Figure 8.20 shows the resulting frequency-domain model. The value of ωM is related to the reactances ωL_1 and ωL_2 by

$$\omega M = k\sqrt{\omega L_1 \omega L_2} \leq \sqrt{\omega L_1 \omega L_2} \tag{8.27}$$

since $M = k\sqrt{L_1 L_2}$ and $k \leq 1$.

Let's put our frequency-domain model immediately to use by analyzing the circuit in Fig. 8.21*a*. Here we again have a transformer serving as the interface between an ac source and a load. But now we're dealing with a *real*

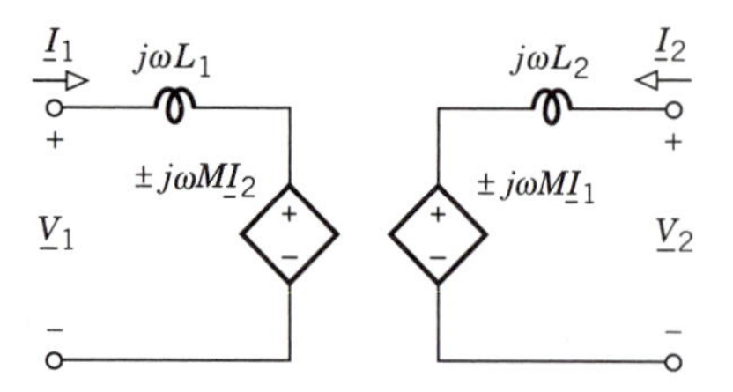

Figure 8.20 Frequency-domain model of self- and mutual inductances.

transformer consisting of magnetically coupled coils. The coils have the same winding sense, and $i_{out} = -i_2$, so the frequency-domain diagram becomes as shown in Fig. 8.21*b*. (A more complete diagram might also include series resistors representing the winding resistance of the primary and secondary.)

The quantities of interest in Fig. 8.21*b* are the ratios $\underline{V}_{out}/\underline{V}_{in}$ and $\underline{I}_{out}/\underline{I}_{in}$ and the input impedance $Z_{in} = \underline{V}_{in}/\underline{I}_{in}$.

We begin the analysis by writing the loop equations

$$j\omega L_1 \underline{I}_{in} - j\omega M \underline{I}_{out} = \underline{V}_{in} \qquad (j\omega L_2 + Z)\underline{I}_{out} = -j\omega M \underline{I}_{in}$$

from which

$$\frac{\underline{I}_{out}}{\underline{I}_{in}} = \frac{j\omega M}{j\omega L_2 + Z} \tag{8.28a}$$

$$Z_{in} = j\omega L_1 + \frac{(\omega M)^2}{j\omega L_2 + Z} = \frac{j\omega L_1 Z + (\omega M)^2 - \omega L_1 \omega L_2}{j\omega L_2 + Z} \tag{8.28b}$$

Then, since $\underline{V}_{out} = Z\underline{I}_{out}$,

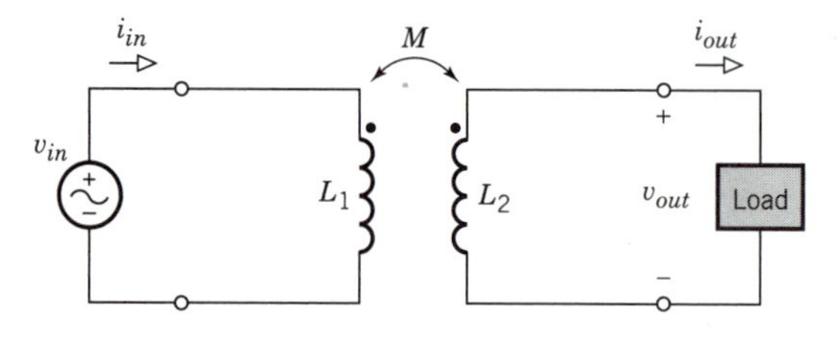

(a) Transformer interfacing a source and a load

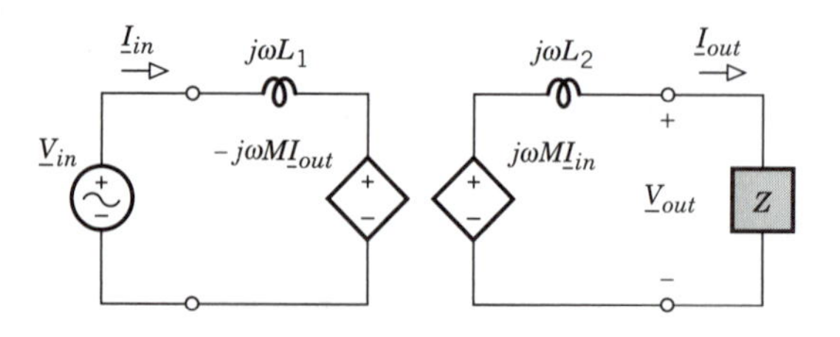

(b) Frequency–domain diagram with controlled sources

Figure 8.21

$$\frac{\underline{V}_{out}}{\underline{V}_{in}} = \frac{j\omega MZ}{j\omega L_1 Z + (\omega M)^2 - \omega L_1 \omega L_2} \tag{8.28c}$$

These expressions hold for any load impedance Z with any coupling coefficient k.

But if $k = 1$, then $(\omega M)^2 - \omega L_1 L_2 = 0$ and we can also draw upon the corresponding expressions for L_1, L_2, and M in Eq. (8.24a). Equations (8.28a)–(8.28c) thereby simplify to

$$\frac{\underline{I}_{out}}{\underline{I}_{in}} = \frac{M}{L_2}\frac{1}{1 + Z/j\omega L_2} = \frac{1}{N}\frac{1}{1 + Z/j\omega L_2} \tag{8.29a}$$

$$Z_{in} = \frac{L_1}{L_2}\frac{Z}{1 + Z/j\omega L_2} = \frac{1}{N^2}\frac{Z}{1 + Z/j\omega L_2} \tag{8.29b}$$

$$\frac{\underline{V}_{out}}{\underline{V}_{in}} = \frac{M}{L_1} = \frac{N_2}{N_1} = N \tag{8.29c}$$

Equation (8.29c) restates our previous conclusion that a unity-coupled transformer produces the same voltage ratio as an *ideal* transformer. And although Eqs. (8.29a) and (8.29b) differ from the current ratio and input impedance of an ideal transformer, there are conditions such that they too become more familiar. In particular, suppose the self-inductance of the secondary is large enough that $|Z/j\omega L_2| \ll 1$. Then $1 + Z/j\omega L_2 \approx 1$ and

$$\underline{I}_{out}/\underline{I}_{in} \approx 1/N \qquad Z_{in} \approx Z/N^2$$

Hence, a real transformer mimics the behavior of an ideal transformer when

$$k \approx 1 \qquad \omega L_2 \ll |Z| \tag{8.30}$$

These conditions often hold in practice.

Example 8.6 *Comparison of a Real and Ideal Transformer*

Let the circuit in Fig. 8.21 have a resistive load, and let

$$\omega L_1 = 100\ \Omega \qquad \omega M = 490\ \Omega \qquad \omega L_2 = 2500\ \Omega \qquad Z = 200\ \Omega$$

Since $\omega L_2 \gg |Z|$ and since $k = \omega M/\sqrt{\omega L_1 \omega L_2} = 0.98$, we suspect that the transformer will act almost like an ideal transformer with $N = N_2/N_1 = \sqrt{L_2/L_1} = \sqrt{\omega L_2/\omega L_1} = 5$.

Inserting numerical values into Eqs. (8.28a)–(8.28c) yields

$$\frac{\underline{I}_{out}}{\underline{I}_{in}} = \frac{j4900}{j2500 + 200} = 0.195\ \underline{/4.5^\circ}$$

$$Z_{in} = \frac{j20{,}000 - 9900}{j2500 + 200} = 8.90\ \Omega\ \underline{/30.9^\circ}$$

$$\frac{\underline{V}_{out}}{\underline{V}_{in}} = \frac{j98{,}000}{j20{,}000 - 9900} = 4.39\ \underline{/-26.3^\circ}$$

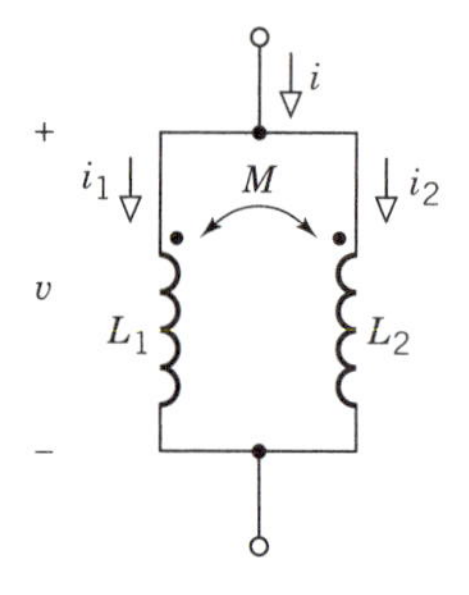

Figure 8.22

For comparison, an ideal transformer would result in $\underline{I}_{out}/\underline{I}_{in} = 0.2\ \underline{/0°}$, $Z_{in} = 200/5^2 = 8\ \Omega\ \underline{/0°}$, and $\underline{V}_{out}/\underline{V}_{in} = 5\ \underline{/0°}$.

Exercise 8.8

Suppose the transformer in Fig. 8.22 has $k = 0.75$, and the network operates in the ac steady state with $\omega L_1 = 8\ \Omega$ and $\omega L_2 = 2\ \Omega$. Show that $\underline{I}_2 = -5\underline{I}_1$, and use this relationship to evaluate $Z_{eq} = \underline{V}/\underline{I}$.

Exercise 8.9

Use Eq. (8.28b) to obtain an expression for Z_{in} in rectangular form when $Z = R + jX$. Under what conditions is Z_{in} entirely capacitive?

Equivalent Tee and Pi Networks

Many applications involve a transformer with the primary and secondary connected as shown in Fig. 8.23*a*, where the added wire from node n_1 to node n_2 ensures that $v_n = 0$. In such circumstances, we may replace the transformer by the **equivalent tee (T) network** in Fig. 8.23*b*.. The upper sign associated with M still corresponds to coils having the same winding sense, so the three inductances are $L_1 - M$, M, and $L_2 - M$. The lower sign corresponds to coils having the opposite winding sense, in which case the three inductances are $L_1 + M$, $-M$, and $L_2 + M$.

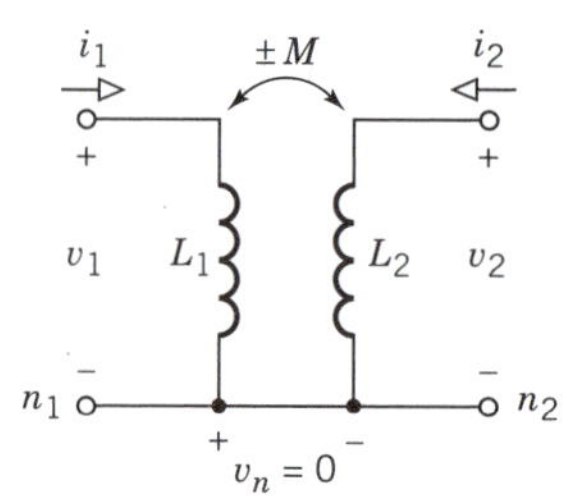

(*a*) Magnetically coupled coils

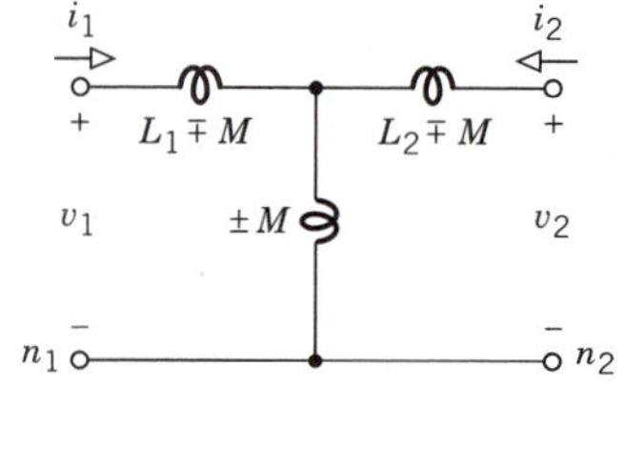

(*b*) Equivalent tee (T) network

Figure 8.23

The tee network consists of *fictitious* inductances, some having *negative* values. Nonetheless, the equivalence is easily established by writing the terminal voltage–current relationships. The current going down through the middle branch in Fig. 8.23*b* is $i_1 + i_2$, so

$$v_1 = (L_1 \mp M)\frac{di_1}{dt} \pm M\frac{d}{dt}(i_1 + i_2) = L_1\frac{di_1}{dt} \pm M\frac{di_2}{dt}$$

$$v_2 = (L_2 \mp M)\frac{di_2}{dt} \pm M\frac{d}{dt}(i_1 + i_2) = \pm M\frac{di_1}{dt} \pm L_2\frac{di_2}{dt}$$

which are identical to the relationships back in Eq. (8.25). The tee network therefore correctly represents the terminal conditions of the transformer.

Analyzing transformer circuits becomes easier when we can substitute an equivalent tee network because it directly incorporates the voltage–current relationships without controlled sources. Furthermore, having identified the proper sign for M, we are free to take any convenient reference polarities and labels for the voltages and currents.

The tee-network structure clearly suits systematic mesh analysis to find the currents i_1 and i_2. But sometimes we want to find v_1 and v_2 via systematic node analysis. The middle node of the tee network then introduces a third unknown voltage having no physical significance. For such cases, we can use the **equivalent pi (Π) network** in Fig. 8.24, where

$$L_{11} = \frac{L_1 L_2 - M^2}{L_2 \mp M} \qquad L_{12} = \frac{L_1 L_2 - M^2}{\pm M} \qquad L_{22} = \frac{L_1 L_2 - M^2}{L_1 \mp M} \tag{8.31}$$

provided that

$$k \neq 1$$

Like the tee network, the pi network consists of fictitious inductances arranged to yield the correct terminal relationships. The pi network is derived from the tee network by applying the impedance version of the wye-to-delta transformation developed in Section 4.6. However, the transformation involves division by $L_1 L_2 - M^2$, so the pi network does not exist when $k = 1$.

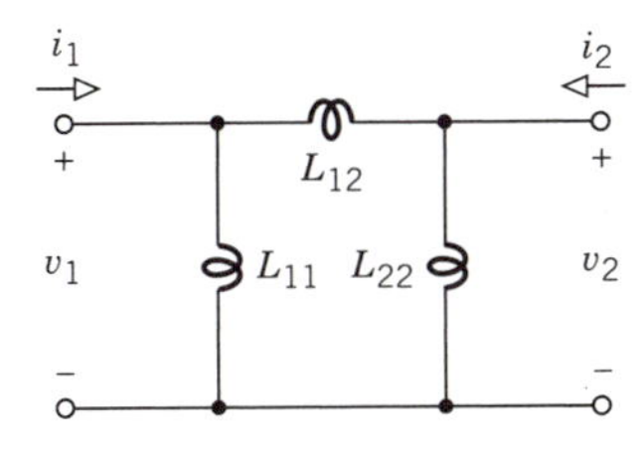

Figure 8.24 Equivalent pi (Π) network for magnetically coupled coils.

Finally we note that an equivalent tee or pi network may be invoked for analysis even when the transformer lacks a direct connection between n_1 and n_2 in Fig. 8.23*a*. The essential condition is that the voltage v_n must have no time-varying component that would be shorted out by the bottom wire of the tee or pi network.

Example 8.7 *Transformer Circuit Analysis with a Tee Network*

We want to find $\underline{I}_{in}$, $\underline{I}_{out}$, and $\underline{V}_{out}$ when the transformer circuit in Fig. 8.25*a* operates under steady-state conditions. We can use a tee network for the transformer since the bottom wire will not short-out any time-varying voltage. The

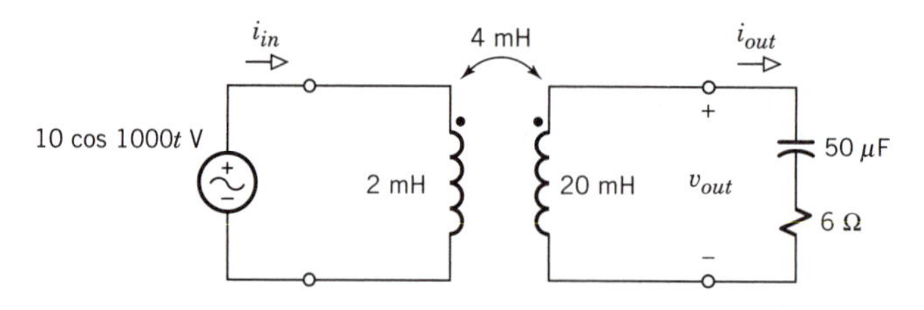

(a) Circuit with magnetic coupling

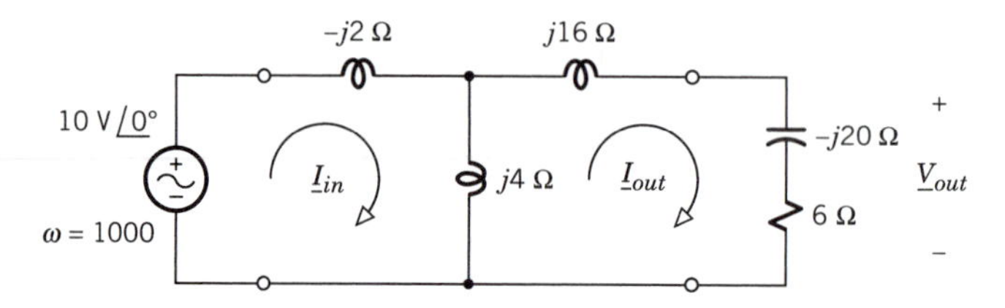

(b) Frequency–domain diagram with the equivalent tee network

Figure 8.25

dots show that the coils have the same winding sense, so the parameters for the tee network are

$$L_1 - M = 2\text{ mH} - 4\text{ mH} = -2\text{ mH} \qquad L_2 - M = 20\text{ mH} - 4\text{ mH} = 16\text{ mH}$$

in addition to $M = 4$ mH.

Multiplying the tee inductances by $\omega = 1000$ leads to the frequency-domain diagram in Fig. 8.25*b*, where the negative inductance has impedance $j\omega(L_1 - M) = -j2\ \Omega$. Routine mesh analysis now yields

$$\begin{bmatrix} -j2 + j4 & -j4 \\ -j4 & j4 + j16 - j20 + 6 \end{bmatrix} \begin{bmatrix} \underline{I}_{in} \\ \underline{I}_{out} \end{bmatrix} = \begin{bmatrix} 10 \\ 0 \end{bmatrix}$$

from which

$$\underline{I}_{in} = 3\text{ A }\underline{/-36.9^\circ} \qquad \underline{I}_{out} = 2\text{ A }\underline{/53.1^\circ}$$

The resulting output voltage is

$$\underline{V}_{out} = (6 - j20)\underline{I}_{out} = 41.8\text{ V }\underline{/-20.2^\circ}$$

whereas $\underline{V}_{in} = 10\text{ V }\underline{/0^\circ}$. The transformer thus has a step-up effect.

Example 8.8 *Step-Up Autotransformer*

Figure 8.26*a* shows an ac source connected to a load via the coil configuration known as a **step-up autotransformer.** We'll determine the ratios $\underline{I}_{out}/\underline{I}_{in}$ and $\underline{V}_{out}/\underline{V}_{in}$ with the help of the equivalent tee network.

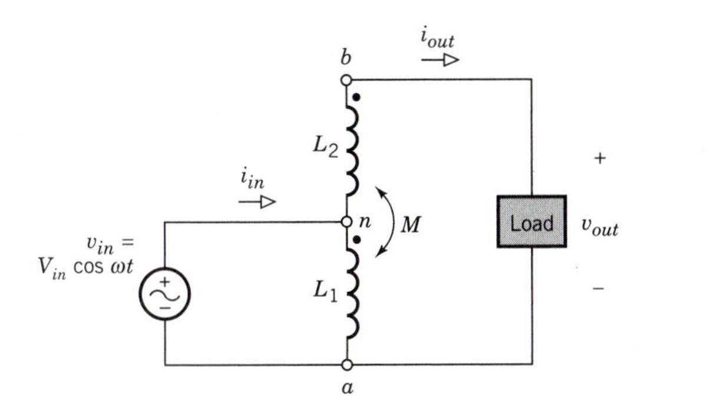

(a) Circuit with step–up autotransformer

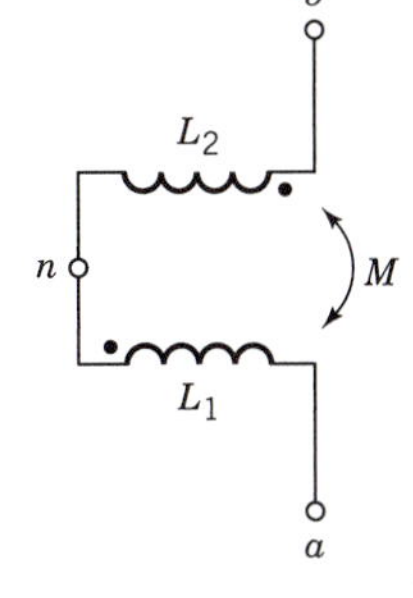

(b) Coils redrawn for clarity

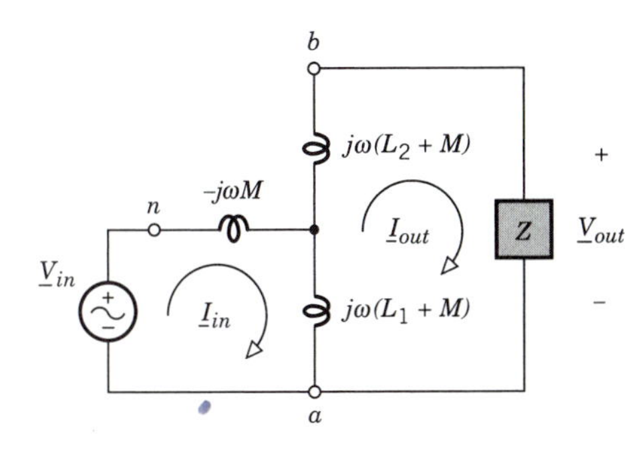

(d) Frequency–domain diagram

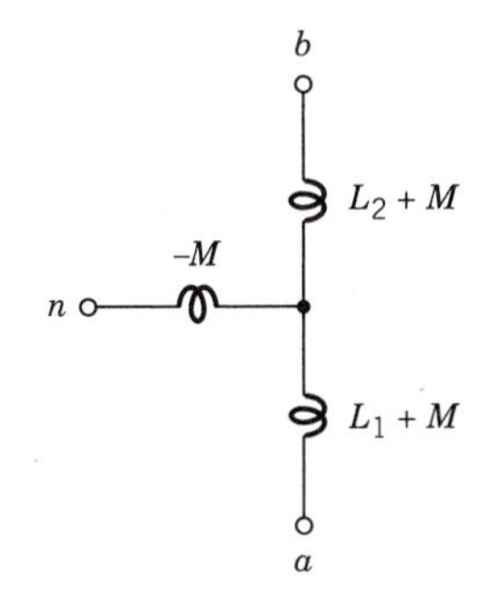

(c) Equivalent tee network

Figure 8.26

Since the coils appear in an unusual orientation, we first redraw them alone, as shown in Fig. 8.26*b*. This drawing reveals that the coils have the opposite winding sense and that node n combines nodes n_1 and n_2. Thus, Fig. 8.26*c* is the corresponding tee network.

Figure 8.26*d* gives our resulting frequency-domain diagram with $j\omega(-M) = -j\omega M$. By inspection of the diagram, we obtain the impedance matrix

$$[Z] = \begin{bmatrix} -j\omega M + j\omega(L_1 + M) & -j\omega(L_1 + M) \\ -j\omega(L_1 + M) & j\omega(L_1 + M) + j\omega(L_1 + M) + Z \end{bmatrix}$$

Thus, the matrix mesh equation is

$$\begin{bmatrix} j\omega L_1 & -j\omega(L_1 + M) \\ -j\omega(L_1 + M) & j\omega(L_1 + 2M + L_2) + Z \end{bmatrix} \begin{bmatrix} \underline{I}_{in} \\ \underline{I}_{out} \end{bmatrix} = \begin{bmatrix} \underline{V}_{in} \\ 0 \end{bmatrix}$$

Solving for $\underline{I}_{in}$ and $\underline{I}_{out}$ yields

$$\frac{\underline{I}_{out}}{\underline{I}_{in}} = \frac{j\omega(L_1 + M)}{j\omega(L_1 + 2M + L_2) + Z} \tag{8.32a}$$

$$\frac{\underline{V}_{out}}{\underline{V}_{in}} = \frac{Z\underline{I}_{out}}{\underline{V}_{in}} = \frac{j\omega(L_1 + M)Z}{j\omega L_1 Z - \omega^2(L_1 L_2 - M^2)} \tag{8.32b}$$

If $|j\omega(L_2 + 2M + L_2)| \gg |Z|$, and if the coils have unity coupling, then Eqs. (8.32a) and (8.32b) reduce to

$$\frac{\underline{I}_{out}}{\underline{I}_{in}} \approx \frac{L_1 + M}{L_1 + 2M + L_2} = \frac{N_1}{N_1 + N_2} \qquad \frac{\underline{V}_{out}}{\underline{V}_{in}} \approx \frac{L_1 + M}{L_1} = \frac{N_1 + N_2}{N_1} \tag{8.33}$$

Thus, the autotransformer behaves like an ideal transformer with turns ratio $N = (N_1 + N_2)/N_1$. The practical advantage of the autotransformer configuration is that it requires fewer secondary turns to achieve a specified value for the effective turns ratio N. However, the load has a direct electrical connection to the source, rather than being isolated as in conventional transformer circuits.

Exercise 8.10

Use the equivalent tee network to show that the network in Fig. 8.22 has $L_{eq} = (L_1L_2 - M^2)/(L_1 + L_2 - 2M)$.

Exercise 8.11

Use the equivalent pi network and one node equation to evaluate $\underline{V}_{out}/\underline{V}_{in}$ in Fig. 8.26*a* when $\omega = 100$, $L_1 = 0.1$ H, $L_2 = 1$ H, $M = 0.3$ H, and the load impedance is $Z = 50\ \Omega$.

Other Equivalent Networks

Several other equivalent networks for mutual inductance involve an ideal transformer. The inclusion of an ideal transformer represents the inherent electrical isolation of the primary and secondary windings, and it leads to simpler equivalent networks for applications ranging from power transformers to high-frequency amplifiers.

Consider the modified tee network in Fig. 8.27*a*, where an ideal transformer with turns ratio N_0 has been put on the secondary side. Routine analysis confirms that the terminal variables of this network are related by our usual transformer equations, regardless of the value of N_0. But judicious choice of N_0 yields valuable simplifications.

If we take $N_0 = \pm L_2/M$, then $\mp M/N_0 = -M^2/L_2 = -k^2L_1$ and thus $(L_2/N_0^2) \mp (M/N_0) = 0$. We thereby obtain the *two*-inductor network in Fig. 8.27*b*. Or if we let $N_0 = \pm M/L_1$, then $L_1 \mp M/N_0 = 0$ and $L_2/N_0^2 = L_1/k^2$, so we get another two-inductor network shown in Fig. 8.27*c*. The negative turns ratios $-L_2/M$ and $-M/L_1$ in these two networks represent coils with opposite winding sense, which reverses the polarities of v_2 and i_2.

Two more two-inductor networks are easily derived from Figs. 8.27*b* and 8.27*c* by referring the elements to the secondary side of the ideal transformer. We might also construct equivalent pi networks with ideal transformers, but they have little practical value.

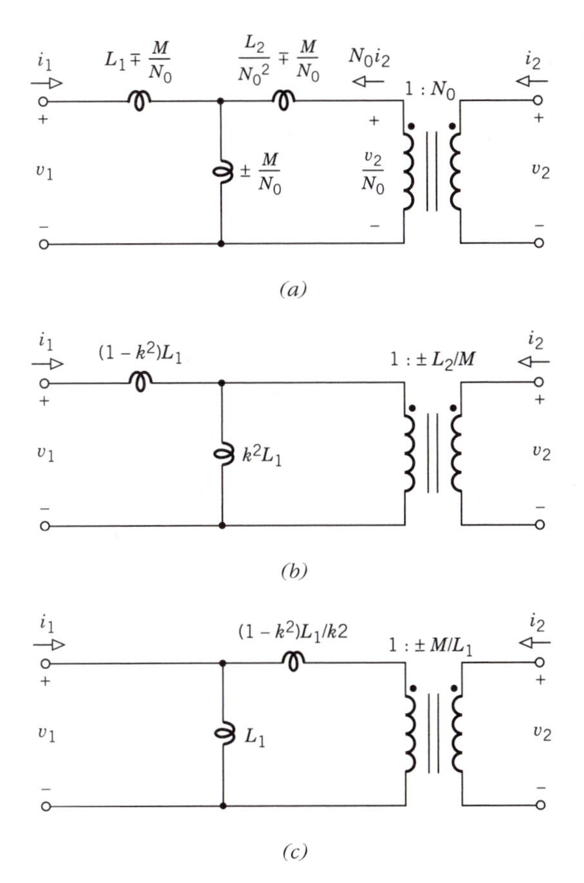

Figure 8.27 Equivalent networks with an ideal transformer to represent mutual inductance.

Example 8.9 *Design of a Tuned Amplifier*

The current source in Fig. 8.28*a* represents the output of a transistor amplifier. Transformer coupling to the *RC* load creates a parallel resonance effect, and we say that the amplifier is "tuned." The design task in this particular case is to determine values for C and R so that the circuit resonates at $\omega_0 = 10^6$ with quality factor $Q_{par} = 25$.

To bring out the parallel structure, we'll work with the equivalent network in Fig. 8.27*b* for the transformer. The network's parameters are

$$k^2 = \frac{M^2}{L_1 L_2} = 0.8 \qquad (1 - k^2)L_1 = 1\ \mu H \qquad k^2 L_1 = 4\ \mu H \qquad \frac{M}{L_1} = 2$$

Figure 8.28*b* shows the resulting equivalent circuit. We then refer the primary of the 1:2 ideal transformer into the secondary to get Fig. 8.28*c*. This diagram confirms that v_{out} appears across a parallel *RLC* network, since the other inductance in series with the current source just introduces a small voltage drop at ω_0.

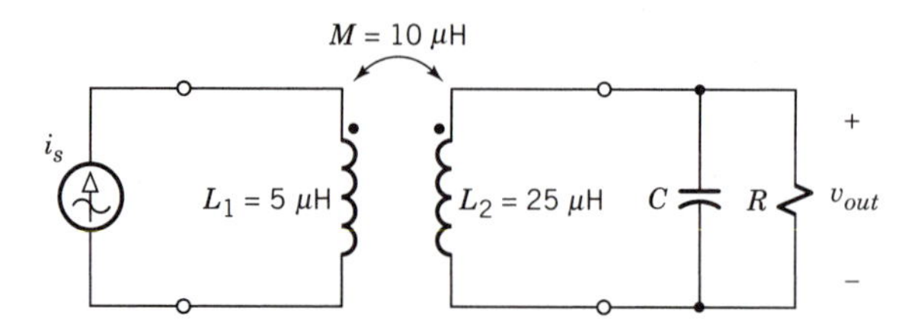

(a) Model of "tuned" transistor amplifier

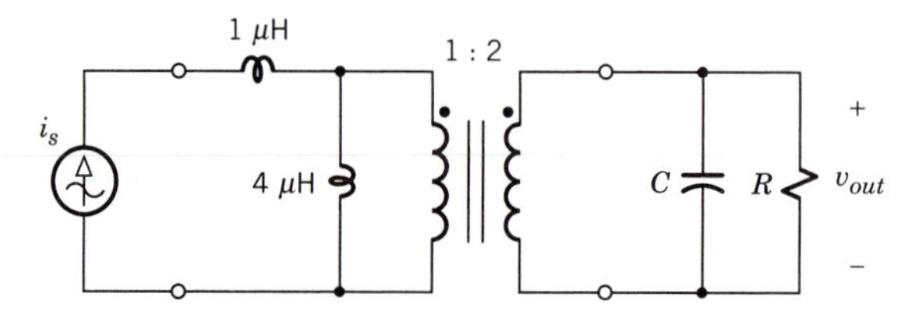

(b) Equivalent circuit with an ideal transformer

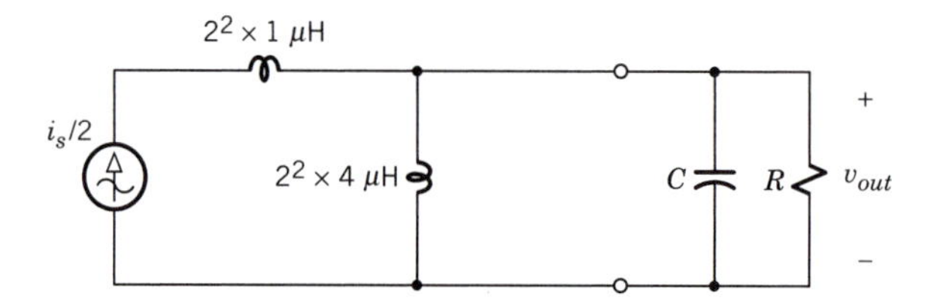

(c) Referred source network in the secondary

Figure 8.28

Finally, we calculate C and R from $\omega_0^2 = 1/LC$ and $Q_{par} = R/\omega_0 L$, where $L = 2^2 \times 4 = 16\ \mu\text{H}$. Thus,

$$C = 1/(\omega_0^2 L) = 62.5\text{ nF} \qquad R = \omega_0 L Q_{par} = 400\ \Omega$$

which completes the design.

Exercise 8.12

Verify that the terminal voltages and currents in Fig. 8.27*a* are related by Eq. (8.25).

8.4 POWER TRANSFORMERS†

This closing section focuses on power considerations in ac circuits with transformers. We'll discuss the types of losses that occur in power transformers, and we'll develop a model to represent those power losses along with the magnetic coupling. We'll also investigate techniques for measuring the parameters of a power transformer.

Losses and Efficiency

Transformers designed for ac power applications invariably have high-permeability ferromagnetic cores with coils wrapped one over the other, which all but eliminates flux leakage and ensures that $k \approx 1$. Furthermore, in normal operation, the load impedance is usually small compared to the impedance of the secondary's self-inductance. A power transformer therefore behaves approximately like an ideal transformer.

However, since large amounts of power are involved, we must also consider losses within the transformer. The most obvious cause of power loss is ohmic heating in the winding resistance. The average power dissipated in the windings is called the **copper loss** to distinguish it from power dissipated within the core.

But **core losses** also occur whenever a magnetic circuit has a time-varying mmf, and there are two major sources of heating within ferromagnetic cores. One source is the **hysteresis effect** that causes the flux variations to lag behind the mmf variations. For ac operation at frequency f, the average hysteresis power loss is given by Steinmetz's empirical formula

$$P_h = K_h f \phi_{max}{}^n \tag{8.34}$$

where ϕ_{max} is the maximum flux while K_h and n are characteristics of the core. The value of the Steinmetz exponent n ranges from 1.5 to 2.0.

The other major source of core heating is **eddy currents.** To explain this effect, Fig. 8.29a depicts a time-varying flux $\phi(t)$ passing through the cross

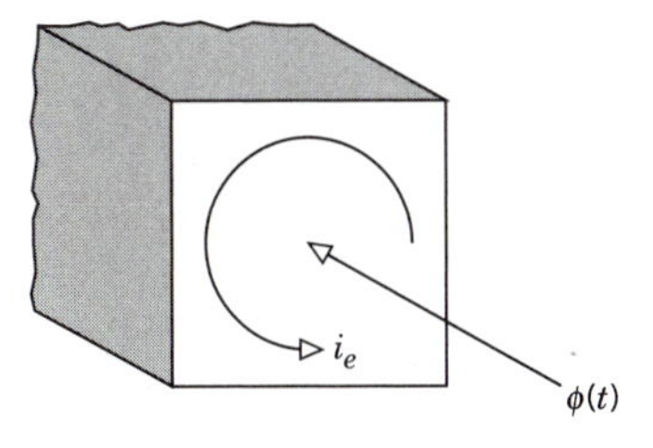

(a) A solid core

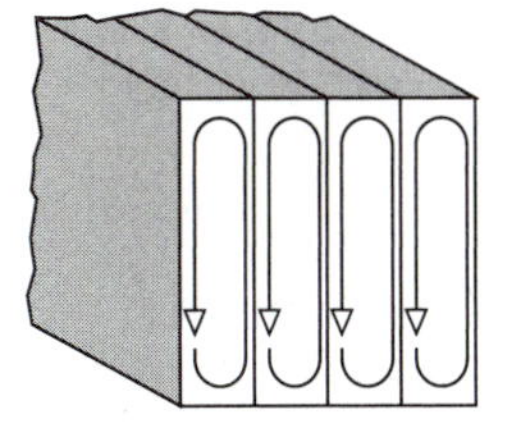

(b) A laminated core

Figure 8.29 Eddy currents.

section of a ferromagnetic core. According to Faraday's law, $\phi(t)$ induces a voltage around the perimeter of the cross section which, in turn, produces the eddy current i_e. As i_e circulates around the path perpendicular to the flux, it encounters resistance R_e in the core material and thus dissipates power $R_e i_e^2$. The resulting average eddy-current power loss is

$$P_e = K_e f^2 \phi_{max}^2 / R_e \tag{8.35}$$

where K_e is another core characteristic. Unless special steps are taken, P_e may account for more than half of the total power supplied to a transformer.

Fortunately, we can minimize eddy-current losses simply by building the core with thin sheets called **laminations**, as illustrated in Fig. 8.29*b*. The laminations must be parallel to the flux path and insulated from each other by varnish coatings. If there are m laminations, then each one carries ϕ_{max}/m and has a resistance of about mR_e since the width of the conducting path has been reduced by about $1/m$. The eddy-current power loss in a laminated core than becomes

$$m \times \frac{K_e f^2 (\phi_{max}/m)^2}{mR_e} = \frac{P_e}{m^2}$$

Thus, the eddy-current loss in a core with 10 laminations will be only 1% of the loss in a comparable solid core.

To analyze losses along with magnetic coupling in a power transformer, we'll use the frequency-domain model in Fig. 8.30. As indicated by the in/out subscripts, we assume that a source is connected to the primary side and a load to the secondary side. We also assume that the coils have the same winding sense. Consistent with our objective of power calculations, all phasor magnitudes will be expressed in *rms* values.

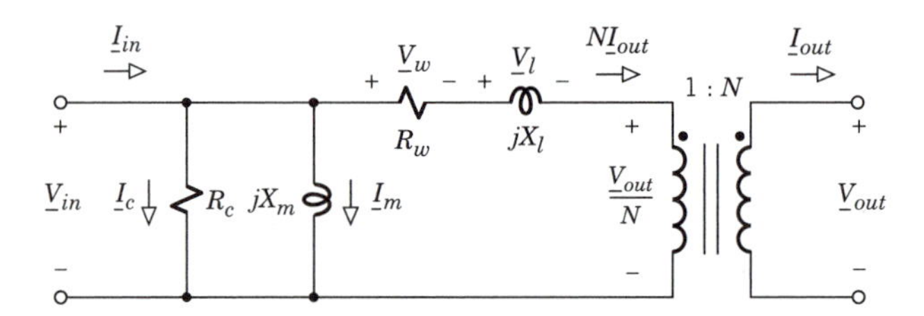

Figure 8.30 Frequency-domain model of a power transformer.

The transformer's magnetic coupling is represented here by the two inductors and ideal transformer from Fig. 8.27*c*, with the additional provision that $k \approx 1$. Accordingly, the turns ratio of the ideal transformer is $M/L_1 = k\sqrt{L_2 L_1} \approx N_2/N_1 = N$. The two inductors appear as the **magnetizing reactance** X_m and the **leakage reactance** X_l, where

$$X_m = \omega L_1 \qquad X_l = \omega(1 - k^2)L_1/k^2$$

Note that the leakage reactance is much smaller than the magnetizing reactance since $X_l/X_m = (1 - k^2)/k^2 \ll 1$ when $k \approx 1$.

Two resistors have been added to the model to account for the transformer's internal power dissipation. Core losses are represented by a large resistance R_c in parallel with jX_m, so the value of R_c must satisfy

$$P_h + P_e = |\underline{V}_{in}|^2/R_c$$

The location of R_c reflects the fact that core losses are essentially independent of the load current $\underline{I}_{out}$. Copper losses are represented by a small resistance R_w in series with jX_l. Of course, the two windings actually have separate resistances. But we can get by with the single resistance, as shown, if it satisfies

$$P_w = R_w|N\underline{I}_{out}|^2$$

where P_w is the power dissipated by both windings. Thus, R_w models the total effect rather than the individual windings.

Figure 8.30 has also been labeled with several variables such as $\underline{I}_c$ and $\underline{V}_w$ that do not exist as distinct physical quantities. Nonetheless, they can be used to calculate measurable quantities such as $\underline{I}_{in}$ and $\underline{V}_{out}$. By way of illustration, Fig. 8.31 gives the phasor construction for $\underline{V}_{in}$ and $\underline{I}_{in}$ taking $\underline{V}_{out}$ as the reference. The construction assumes an inductive load with impedance angle $\theta > 0$, so $\underline{I}_{out}$ lags $\underline{V}_{out}$. The voltage phasors depict the KVL relation $\underline{V}_{in} = \underline{V}_{out}/N + \underline{V}_w + \underline{V}_l$, while the current phasors depict the KCL relation $\underline{I}_{in} = N\underline{I}_{out} + \underline{I}_c + \underline{I}_m$.

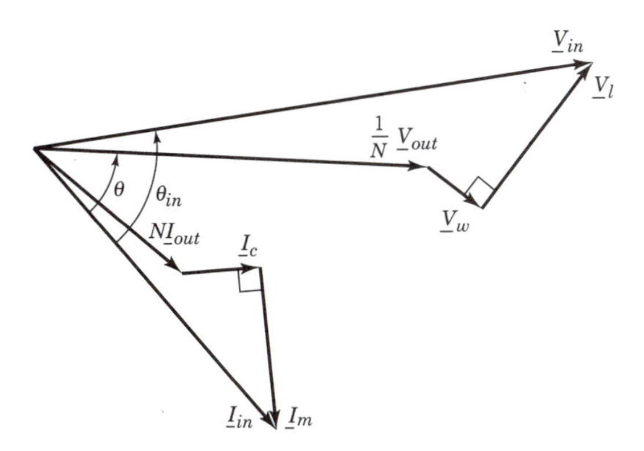

Figure 8.31 Phasor diagram for a power transformer when $\underline{I}_{out}$ lags $\underline{V}_{out}$.

An important conclusion from the phasor diagram concerns the angle of the equivalent impedance at the terminals of the primary, namely, $\theta_{in} = \angle\underline{V}_{in} - \angle\underline{I}_{in}$. Since $\theta_{in} > \theta$ in this case, the source sees a *lower* power factor than the power factor of the load impedance alone. This observation supports the fact that the source must supply reactive power to the transformer as well as to the inductive load.

If we know the values of the parameters R_c, X_m, R_w, and X_l, then we can predict how the transformer performs when connected to a specified load. In particular, from Fig. 8.30 we see that the power dissipated in the transformer will be

$$P_{dis} = R_c|\underline{I}_c|^2 + R_w|N\underline{I}_{out}|^2 \tag{8.36}$$

Hence, the power-transfer efficiency is

$$\text{Eff} = \frac{P_{out}}{P_{out} + P_{dis}} \tag{8.37}$$

where P_{out} is the power absorbed by the load. The following example illustrates these calculations.

Example 8.10 *Efficiency of a Power Transformer*

A certain step-up transformer with $N = 5$ has the following parameter values at $f = 60$ Hz:

$$R_c = 40\ \Omega \qquad X_m = 24\ \Omega \qquad R_w = 0.08\ \Omega \qquad X_l = 0.5\ \Omega$$

We'll investigate the performance of this transformer when it couples a 120-V(rms) ac source to a 36-Ω resistive load. For comparison purposes, we note in advance that an ideal transformer would produce $|\underline{V}_{out}| = 5 \times 120 = 600$ V and deliver $P_{out} = 600^2/36 = 10{,}000$ W.

Figure 8.32 shows the equivalent circuit diagram incorporating our model with the load resistance referred to the primary as $36/5^2 = 1.44\ \Omega$. Routine analysis yields the rms values

$$\underline{I}_c = 120/40 = 3\text{ A} \qquad \underline{I}_m = 120/j24 = -j5\text{ A}$$

$$5\underline{I}_{out} = \frac{120}{0.08 + j0.5 + 1.44} = 71.2 - j23.4 = 75.0\text{ A }\underline{/-18.2^\circ}$$

$$\underline{I}_{in} = \underline{I}_c + \underline{I}_m + 5\underline{I}_{out} = 74.2 - j25.4 = 79.5\text{ A }\underline{/-21.2^\circ}$$

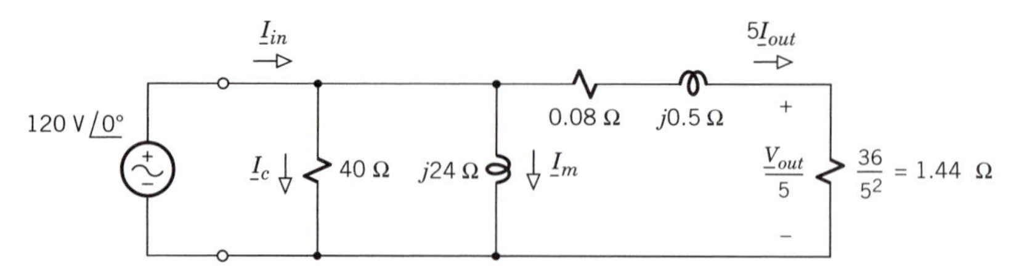

Figure 8.32 Equivalent circuit of a power transformer with load resistance referred to the primary.

Thus,

$$\underline{I}_{out} = 15.0 \text{ A} \,\underline{/-18.2^\circ} \qquad \underline{V}_{out} = 36\underline{I}_{out} = 540 \text{ V} \,\underline{/-18.2^\circ}$$

$$P_{out} = 36 \times 15.0^2 = 8100 \text{ W}$$

$$P_{dis} = 40 \times 3^2 + 0.08 \times 75.0^2 = 810 \text{ W}$$

$$\text{Eff} = 8100/8910 \approx 91\%$$

We see that the transformer provides respectable power-transfer efficiency, even though its losses and reactances have significantly reduced $|\underline{V}_{out}|$ and P_{out} compared to an ideal transformer.

Exercise 8.13

Suppose the transformer described in Example 8.10 is connected to a load having impedance $Z = 58 + j125\ \Omega$. Find $|\underline{V}_{out}|$, P_{out}, and the efficiency when $|\underline{V}_{in}| = 240$ V(rms).

Parameter Measurements

Our transformer model in Fig. 8.30 readily lends itself to *parameter measurements* of real transformers — a necessary and important task when the manufacturer does not provide all the information needed to analyze the performance of a power transformer.

The data listed on the nameplate of a transformer usually consists of the rms voltage ratings and the rated (maximum) apparent power at the specified frequency. From these you can calculate the turns ratio and the current ratings. For instance, suppose that a particular nameplate bears the legend

60 Hz, 720/240 V, 36 kVA

This means that the transformer has been designed for operation at $f = 60$ Hz with

$$|\underline{V}_{in}|_{max} = 720 \text{ V} \qquad |\underline{V}_{out}|_{max} = 240 \text{ V}$$

$$N = |\underline{V}_{out}|_{max}/|\underline{V}_{in}|_{max} = 240/720 = 1/3$$

$$|\underline{I}_{in}|_{max} = 36 \text{ kVA}/720 \text{ V} = 50 \text{ A}$$

$$|\underline{I}_{out}|_{max} = 36 \text{ kVA}/240 \text{ V} = 150 \text{ A}$$

We could also turn this transformer around and operate it in the step-up mode with $N = 3$, $|\underline{V}_{in}|_{max} = 240$ V, etc.

Given the nameplate information, the values of the model parameters R_c, X_m, R_w, and X_l may be determined experimentally using a wattmeter and rms ammeters and voltmeters. The procedure involves the following two steps.

1. Open-circuit measurements. Use the set-up in Fig. 8.33*a* with the secondary left open. Adjust the primary voltage at or below its rated value,

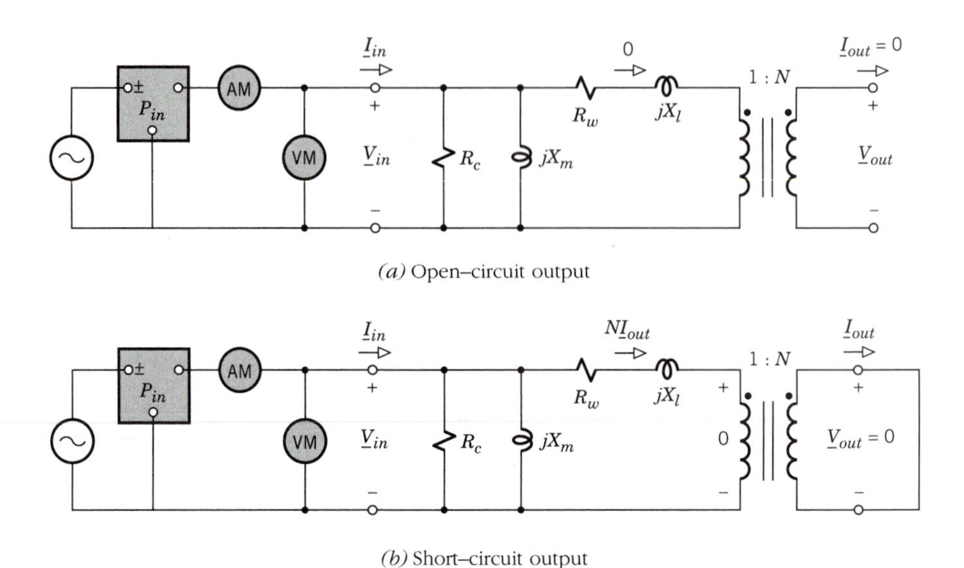

(a) Open–circuit output

(b) Short–circuit output

Figure 8.33 Set-ups for power-transformer measurements.

and record the values of $|\underline{V}_{in}|$, $|\underline{I}_{in}|$, and $P_{in} = P_{oc}$. The real power P_{oc} is dissipated entirely by R_c, while X_m absorbs all the reactive power Q_{oc}. Hence,

$$R_c = |\underline{V}_{in}|^2/P_{oc} \qquad X_m = |\underline{V}_{in}|^2/Q_{oc} \tag{8.38}$$

where

$$Q_{oc} = \sqrt{|\underline{V}_{in}|^2|\underline{I}_{in}|^2 - P_{oc}^{\ 2}}$$

A measurement of the open-circuit output voltage would also give the turns ratio via $N = |\underline{V}_{out}|/|\underline{V}_{in}|$.

2. Short-circuit measurements. Use the set-up in Fig. 8.33*b* with the secondary shorted. Adjust the primary current at or below its rated value, and record the values of $|\underline{V}_{in}|$, $|\underline{I}_{in}|$ and $P_{in} = P_{sc}$. The real power P_{sc} is now dissipated primarily by R_w, while X_l absorbs most of the reactive power Q_{sc}. Thus,

$$R_w \approx P_{sc}|\underline{I}_{in}|^2 \qquad X_l \approx Q_{sc}/|\underline{I}_{in}|^2 \tag{8.39}$$

where

$$Q_{sc} = \sqrt{|\underline{V}_{in}|^2|\underline{I}_{in}|^2 - P_{sc}^{\ 2}}$$

Equation (8.39) has sufficient accuracy for most purposes, provided that

$$|(R_c \| jX_m)| >> |R_w + jX_l| \tag{8.40}$$

If the values obtained from Eqs. (8.38) and (8.39) do not satisfy Eq. (8.40), then the short-circuit output current also must be measured and R_w and X_l are calculated from

$$R_w = \frac{P_{sc} - |\underline{V}_{in}|^2/R_c}{N^2|\underline{I}_{out}|^2} \qquad X_l = \frac{Q_{sc} - |\underline{V}_{in}|^2/X_m}{N^2|\underline{I}_{out}|^2} \tag{8.41}$$

Exercise 8.14

Derive Eq. (8.41) from Fig. 8.33b.

PROBLEMS

Section 8.1 Ideal Transformers

(See also PSpice problems B.29 and B.30.)

8.1 Let capacitance C be connected across the secondary in Fig. 8.1 (p. 361). Show that

$$i_1 = N^2 C\, dv_1/dt$$

8.2 Let inductance L be connected across the secondary in Fig. 8.1 (p. 361). Show that

$$v_1 = (L/N^2)\, di_1/dt.$$

8.3* A time-varying source is connected to the network in Fig. P8.3. What are the values of v_{in}/i_{in} and v_x/v_{in} when $R = 0$ and $N = 1/5$?

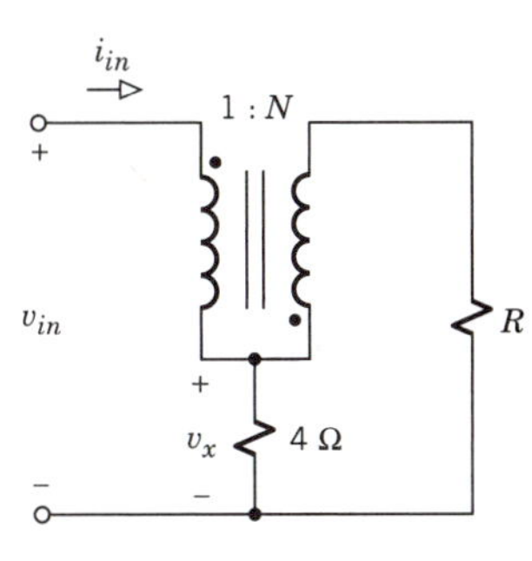

Figure P8.3

8.4 Do Problem 8.3 with $R = 3\ \Omega$ and $N = 1/2$.

8.5 Do Problem 8.3 with $R = 24\ \Omega$ and $N = 2$.

8.6 Find i_1, i_2, and i_3 at $t = 0$ in Fig. P8.6 when $R_1 = 2\ \Omega$, $R_2 = 4\ \Omega$, $R_3 = 0$, and $N = 4$.

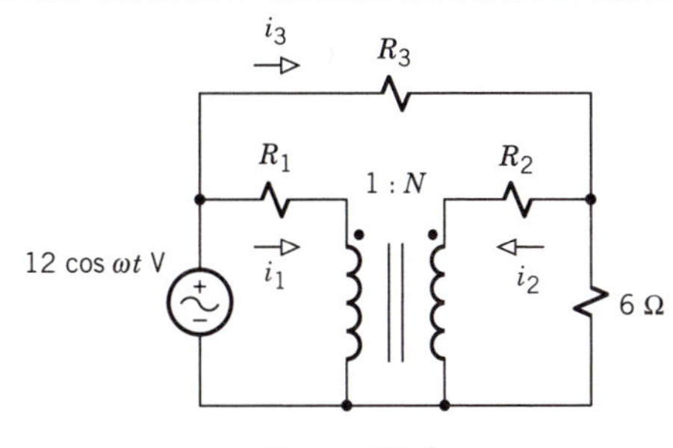

Figure P8.6

8.7 Find i_1, i_2, and i_3 at $t = 0$ in Fig. P8.6 when $R_1 = 0$, $R_2 = 1\ \Omega$, $R_3 = 18\ \Omega$, and $N = 3$.

8.8 Find i_1, i_2, and i_3 at $t = 0$ in Fig. P8.6 when $R_1 = 30\ \Omega$, $R_2 = 0$, $R_3 = 1\ \Omega$, and $N = 1/3$.

8.9 Let a parallel RLC network be the load in Fig. 8.6 (p. 365). Verify that the referred load admittance corresponds to a parallel network consisting of R/N^2, L/N^2, and N^2C.

8.10* By referring the source and load into the middle section of Fig. P8.10, find $\underline{I}$, $\underline{I}_{in}$, and all the voltage phasors when $N_a = 2$, $Z = 30 + j30\ \Omega$, $N_b = 4$, and $Z_{out} = 800 - j32\ \Omega$.

8.11 Do Problem 8.10 with $N_a = 5$, $Z = 60 + j20\ \Omega$, $N_b = 1/4$, and $Z_{out} = 15 - j20\ \Omega$.

8.12 Do Problem 8.10 with $N_a = 0.5$, $Z = 5 - j5\ \Omega$, $N_b = 3$, and $Z_{out} = 180 + j180\ \Omega$.

8.13 Do Problem 8.10 with $N_a = 1/4$, $Z = 0.75 + j0.5\ \Omega$, $N_b = 1/2$, and $Z_{out} = 3 - j2\ \Omega$.

8.14 Figure P8.14 is a parallel resonant circuit. Calculate ω_0 and Q_{par}, given that $N = 2$, $R_x = 60\ \text{k}\Omega$, and $C_x = 0.25\ \mu\text{F}$.

8.15 Let the parallel resonant circuit in Fig. P8.14 have $R_x = 10\ \text{k}\Omega$ and $C_x = 1\ \mu\text{F}$. Find N needed to

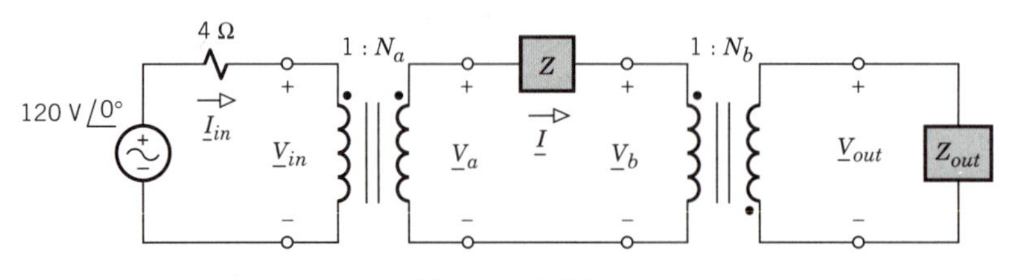

Figure P8.10

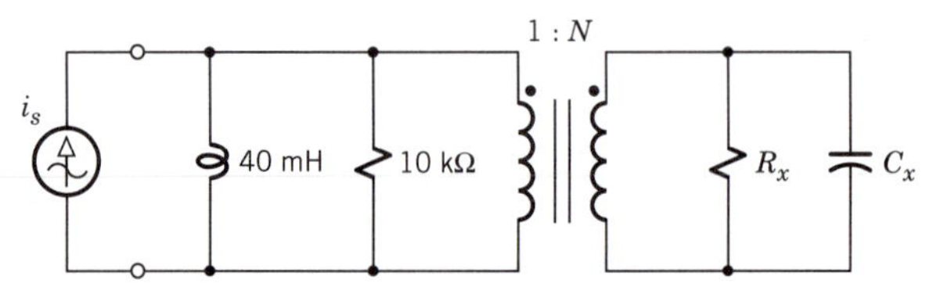

Figure P8.14

get $\omega_0 = 10{,}000$, and determine the resulting value of Q_{par}.

8.16 Let the parallel resonant circuit in Fig. P8.14 have $R_x = 160$ kΩ. Find N and C_x such that $Q_{par} = 50$ and $\omega_0 = 2500$.

8.17 Let the parallel resonant circuit in Fig. P8.14 have $C_x = 1\ \mu$F. Find N and R_x such that $\omega_0 = 500$ and $Q_{par} = 100$.

8.18* Evaluate $\underline{I}$, $\underline{V}_a$, and $\underline{V}_b$ in Fig. P8.18 given that $N_a = 4$, $N_b = 2$, and $\underline{I}_x = 2\underline{I}$.

8.19 Evaluate $\underline{I}$, $\underline{V}_a$, and $\underline{V}_b$ in Fig. P8.18 given that $N_a = 8$, $N_b = 1/4$, and $\underline{I}_x = 6\underline{I}_a$.

8.20 Evaluate $\underline{I}$, $\underline{V}_a$, and $\underline{V}_b$ in Fig. P8.18 given that $N_a = 1/5$, $N_b = 1/2$, and $\underline{I}_x = \underline{V}_a/32\ \Omega$.

8.21* Let $Z_s = 60\ \Omega\ \angle{-30°}$ and $Z = R + j270\ \Omega$ in Fig. P8.21. Find the values of N and R needed to get maximum power transfer.

8.22 Let $Z_s = 600\ \Omega + jX_s$ and $Z = 48\ \Omega\ \angle 60°$ in Fig. P8.21. Find the values of N and X_s needed to get maximum power transfer.

8.23 The source in Fig. P8.21 operates at $\omega = 40{,}000$ and has $Z_s = 5 + j10\ \Omega$. If the load is a parallel RC network with $C = 0.02\ \mu$F, then what values of N and R are needed for maximum power transfer?

8.24 The source in Fig. P8.21 operates at $\omega = 20{,}000$ and has $Z_s = 2 + j2$ kΩ. The load is a parallel RC network with $R = 250\ \Omega$. What values of N and C are needed for maximum power transfer?

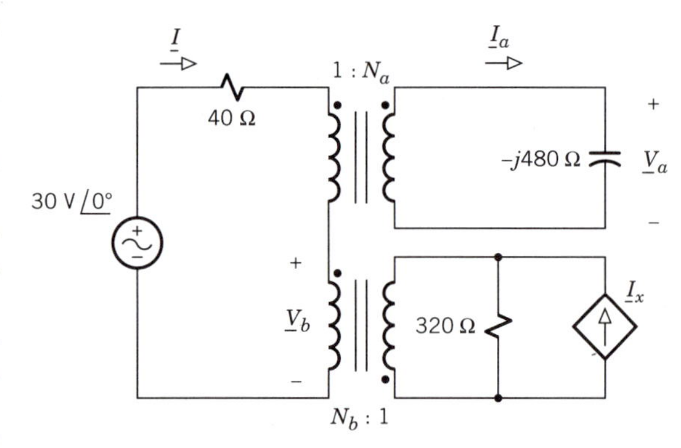

Figure P8.18

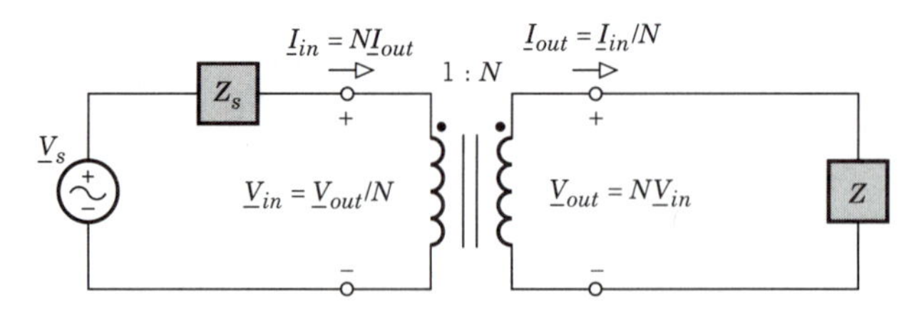

Figure P8.21

Section 8.2 Magnetic Coupling and Mutual Inductance

8.25* A certain core has $\Re_1 = \Re_2 = 10^5$ H^{-1} and $\Re_l = 19\,\Re_1$. Find N_1 and N_2 such that $L_1 = 8$ mH and $L_2 = 50$ mH. Then calculate k and M.

8.26 A certain core has $\Re_1 = \Re_2 = 10^4$ H^{-1} and $\Re_l = 9\,\Re_1$. Find N_1 and N_2 such that $L_1 = 1$ H and $M = 0.1$ H. Then calculate L_2.

8.27 A certain core has $\Re_1 = \Re_2 = 10^5$ H^{-1} and $\Re_l = 49\,\Re_1$. Find N_1 such that $L_1 = 10$ mH. Then calculate L_2, k, and M when $N_2 = 5N_1$.

8.28 A certain core has $\Re_1 = \Re_2 = 10^6$ H^{-1} and $\Re_l = 4\,\Re_1$. Find N_1 and N_2 such that $M = 240\ \mu$H and $N_1 = 3N_2$. Then calculate L_1 and L_2.

8.29 Suppose the coils in Fig. P.8.29 have $L_1 = 1$ mH, $L_2 = 25$ mH, and $k = 0.8$. Find $i_1(t)$ and $v_2(t)$ under ac steady-state conditions with $v_1(t) = 20 \sin 1000t$ V and $i_2(t) = 5 \cos 1000t$ A.

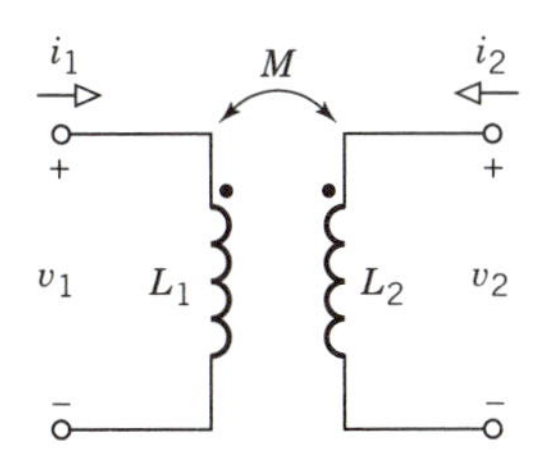

Figure P8.29

8.30 Suppose the coils in Fig. P8.30 have $L_1 = 200$ mH, $L_2 = 8$ mH, and $k = 0.5$. Find $i_2(t)$ and $v_2(t)$ given that $v_1(t) = 5$ V, $i_1(t) = t^2$ A, and $i_2(0) = 0$.

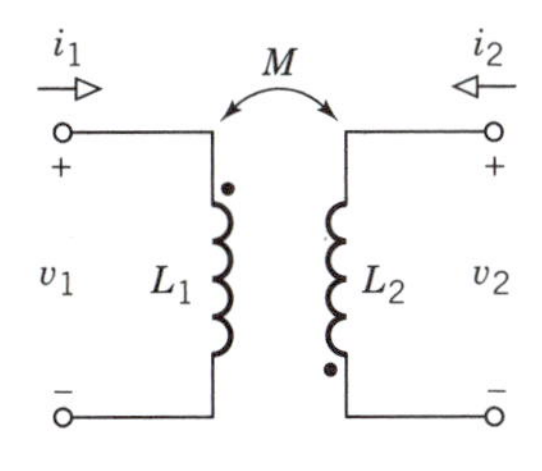

Figure P8.30

8.31 Measurements are made on the coils in Fig. P8.29 under steady-state conditions with $v_1 = 10 \cos 200t$ V. With the secondary open, it is found that $i_1 = 0.1 \sin 200t$ A and $v_2 = 6 \cos 200t$ V. With the secondary shorted, it is found that $i_1 = 1 \sin 200t$ A. What are the values of L_1, L_2, M, and k?

8.32 Measurements are made on the coils in Fig. P8.30 under steady-state conditions with $v_1 = 5 \cos 100t$ V. With the secondary open, it is found that $i_1 = 0.5 \sin 100t$ A and $v_2 = -10 \cos 100t$ V. With the secondary shorted, it is found that $i_1 = 1.5 \sin 100t$ A. What are the values of L_1, L_2, M, and k?

Section 8.3 Circuits with Mutual Inductance
(See also PSpice problems B.31 and B.32.)

8.33* Substitute the ac controlled-source model for the transformer in Fig. P8.33 to find $\underline{V}_1$ and $\underline{I}_2$ given that $\omega L_1 = 8\ \Omega$, $\omega L_2 = 2\ \Omega$, $\omega M = 4\ \Omega$, the load is a short circuit, and the coils have the opposite winding sense.

8.34 Substitute the ac controlled-source model for the transformer in Fig. P8.33 to find $\underline{V}_1$ and $\underline{I}_2$ given that $\omega L_1 = 2\ \Omega$, $\omega L_2 = 5\ \Omega$, $\omega M = 3\ \Omega$, the load impedance is $Z = -j5\ \Omega$, and the coils have the same winding sense.

8.35 Substitute the ac controlled-source model for the transformer in Fig. P8.33 to find $\underline{V}_1$ and $\underline{I}_2$ given that $\omega L_1 = 2\ \Omega$, $\omega L_2 = 6\ \Omega$, $\omega M = 3\ \Omega$, the load impedance is $Z = 3\ \Omega$, and the coils have the opposite winding sense.

8.36 Using the ac controlled-source model for the transformer, calculate $\underline{I}_1$ and $\underline{V}_2$ in Fig. P8.36 given that $R = 10\ \Omega$, $M = 0.01$ H, $L_2 = 0.05$ H, the load is a capacitor with $C = 1/500$ F, and the coils have the opposite winding sense.

8.37 Using the ac controlled-source model for the transformer, calculate $\underline{I}_1$ and $\underline{V}_2$ in Fig. P8.36 given that $R = 20\ \Omega$, $M = 0.1$ H, $L_2 = 0.2$ H, the load is a capacitor with $C = 1/1000$ F, and the coils have the same winding sense.

8.38 Using the ac controlled-source model for the transformer, calculate $\underline{I}_1$ and $\underline{V}_2$ in Fig. P8.36 given that $R = 10\ \Omega$, $M = 0.1$ H, $L_2 = 0.3$ H, the load is a 20-Ω resistor, and the coils have the opposite winding sense.

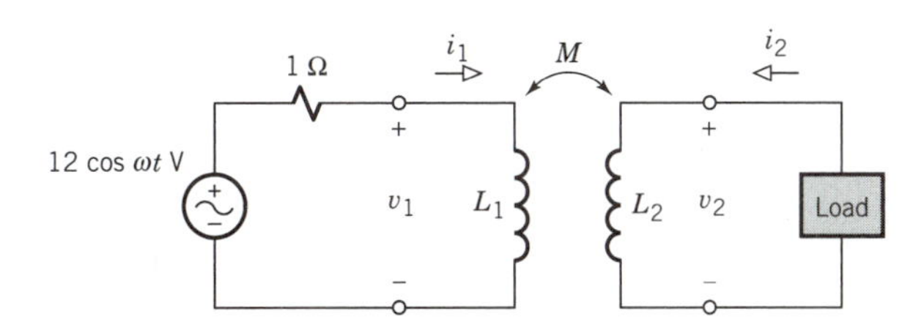

Figure P8.33

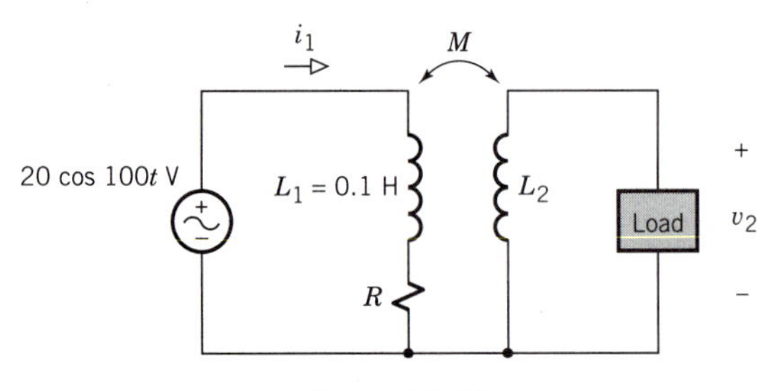

Figure P8.36

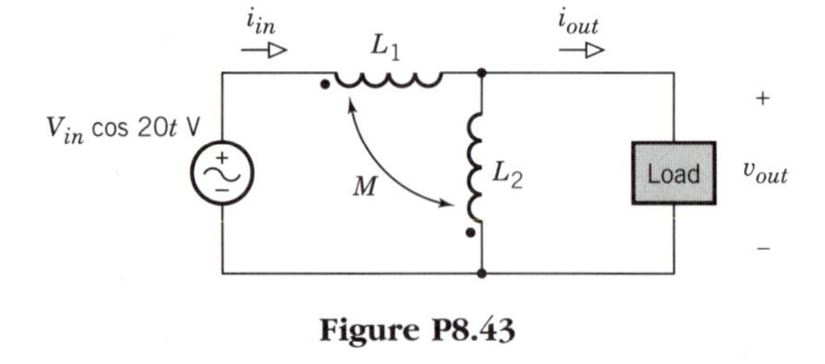

Figure P8.43

8.39 Using the ac controlled-source model for the transformer, calculate $\underline{I}_1$ and $\underline{V}_2$ in Fig. P8.36 given that $R = 30\ \Omega$, $M = 0.2$ H, $L_2 = 0.4$ H, the load is a 40-Ω resistor, and the coils have the same winding sense.

8.40* Replace the transformer in Fig. P8.40 by its equivalent tee network to find $\underline{I}_1$ and $\underline{I}_2$ when $L_1 = 0.6$ H, $L_2 = M = 0.4$ H, and $R_2 = 4\ \Omega$.

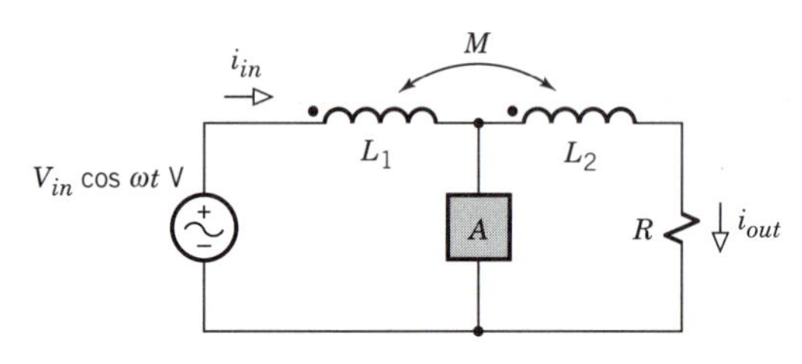

Figure P8.46

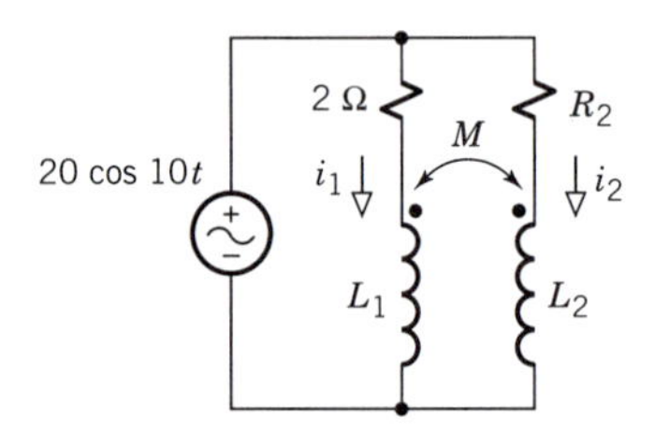

Figure P8.40

8.41 Replace the transformer in Fig. P8.40 by its equivalent tee network to find $\underline{I}_1$ and $\underline{I}_2$ when $L_1 = 0.8$ H, $L_2 = 4.8$ H, $M = 1.6$ H, and $R_2 = 0$.

8.42 Replace the transformer in Fig. P8.40 by its equivalent tee network to find $\underline{I}_1$ and $\underline{I}_2$ when $L_1 = 0.9$ H, $L_2 = 0.6$ H, $M = 0.7$ H, and $R_2 = 1\ \Omega$.

8.43 Using mesh analysis with an equivalent tee network, evaluate $\underline{I}_{out}/\underline{I}_{in}$ and $\underline{V}_{out}/\underline{V}_{in}$ in Fig. P8.43 when $L_1 = 0.2$ H, $L_2 = 0.15$ H, $M = 0.1$ H, and the load impedance is $Z = -j2\ \Omega$.

8.44 Do Problem 8.43 with $L_1 = 0.05$ H, $L_2 = 0.2$ H, $M = 0.1$ H, and $Z = 3\ \Omega$.

8.45 Do Problem 8.43 with $L_1 = 2.2$ H, $L_2 = 0.2$ H, $M = 0.6$ H, and $Z = 2 - j4\ \Omega$.

8.46* Find $\underline{I}_{out}/\underline{I}_{in}$ in polar form and $Z_{in} = \underline{V}_{in}/\underline{I}_{in}$ in rectangular form when the circuit in Fig. P8.46 has $\omega L_1 = 25\ \Omega$, $\omega L_2 = 4\ \Omega$, $\omega M = 10\ \Omega$, $R = 3\ \Omega$, and let the impedance of element A be $Z_A = 0\ \Omega$.

8.47 Do Problem 8.46 with $\omega L_1 = 1\ \Omega$, $\omega L_2 = 6\ \Omega$, $\omega M = 2\ \Omega$, $R = 4\ \Omega$, and $Z_A = -j6\ \Omega$.

8.48 Do Problem 8.46 with $\omega L_1 = 2.6\ \Omega$, $\omega L_2 = 20\ \Omega$, $\omega M = 4\ \Omega$, $R = 12\ \Omega$, and $Z_A = 3\ \Omega$.

8.49 A **step-down autotransformer** has the source and load interchanged in Fig. 8.26 (p. 387). (a) Derive expressions for $\underline{I}_{out}/\underline{I}_{in}$ and $\underline{V}_{out}/\underline{V}_{in}$. (b) Simplify your results for the case of $k = 1$ and $\omega L_1 >> |Z|$.

8.50 Using node analysis with an equivalent pi network, evaluate $\underline{V}_1$ and $\underline{V}_2$ in Fig. P8.50 given that $R = 3\ \Omega$, $L_1 = 0.4$ H, $L_2 = 1.2$ H, $M = 0.6$ H, the load is an open circuit, and the coils have the opposite winding sense.

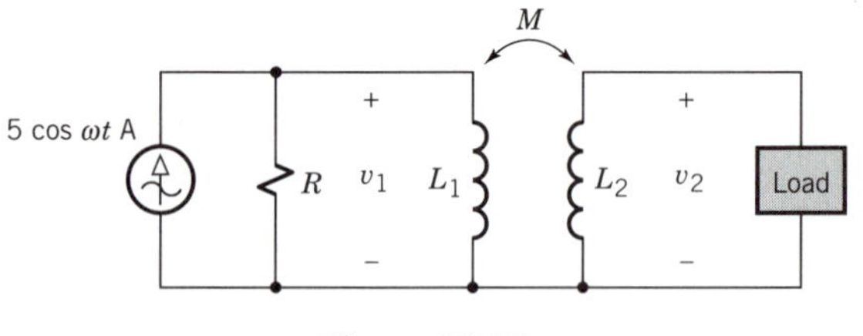

Figure P8.50

8.51 Do Problem 8.50 given that $R = 10\ \Omega$, $L_1 = 0.4$ H, $L_2 = 3$ H, $M = 1$ H, the load impedance is

$Z = -j5\ \Omega$, and the coils have the same winding sense.

8.52 Do Problem 8.50 given that $R = 8\ \Omega$, $L_1 = 0.5$ H, $L_2 = 0.5$ H, $M = 0.3$ H, the load impedance is $Z = 2\ \Omega$, and the coils have the same opposite winding sense.

8.53 Suppose the transformer in Fig. P8.33 has $\omega L_1 = 0.75\ \Omega$, $\omega L_2 = 3\ \Omega$, $k = 0.5$, and the coils have the opposite winding sense. Use the equivalent network in Fig. 8.27c (p. 389) to find $\underline{I}_1$ and $\underline{V}_2$ when the load is an open circuit.

8.54 Suppose the transformer in Fig. P8.33 has $\omega L_1 = 5\ \Omega$, $\omega L_2 = 1\ \Omega$, $k = 1$, and the coils have the opposite winding sense. Use the equivalent network in Fig. 8.27b (p. 389) to find $\underline{I}_1$ and $\underline{V}_2$ when the load is a 2-Ω resistor.

8.55 Suppose the transformer in Fig. P8.33 has $\omega L_1 = 5\ \Omega$, $\omega L_2 = 25\ \Omega$, $k^2 = 0.8$, and the coils have the same winding sense. Use the equivalent network in Fig. 8.27*b* (p. 389) to find $\underline{I}_1$ and $\underline{V}_2$ when the load is a 16-Ω resistor.

8.56 Suppose the transformer in Fig. P8.33 has $\omega L_1 = 2\ \Omega$, $\omega L_2 = 5\ \Omega$, $k^2 = 0.5$, and the coils have the same winding sense. Use the equivalent network in Fig. 8.27*c* (p. 389) to find $\underline{I}_1$ and $\underline{V}_2$ when the load impedance is $Z = 5 - j15\ \Omega$.

Section 8.4 Power Transformers

8.57* A power transformer is connected to a load with impedance Z. Find $\underline{V}_{out}$, $\underline{I}_{in}$, P_{in}, Q_{in}, and Eff when

$$R_c = 40\ \Omega \quad X_m = 30\ \Omega \quad R_w = 0.8\ \Omega \quad X_l = 6.4\ \Omega$$
$$N = 2.5 \quad \underline{V}_{in} = 240\ \text{V(rms)}\ \underline{/0^\circ} \quad Z = 25\ \Omega$$

8.58 Do Problem 8.57 with

$$R_c = 900\ \Omega \quad X_m = 300\ \Omega \quad R_w = 4\ \Omega \quad X_l = 10\ \Omega$$
$$N = 0.5 \quad \underline{V}_{in} = 1800\ \text{V(rms)}\ \underline{/0^\circ} \quad Z = 12 + j5\ \Omega$$

8.59 A load impedance Z is connected to the secondary of a power transformer, and a capacitor is put across the primary terminals along with a 60-Hz ac source that establishes $\underline{V}_{in} = 240$ V(rms) $\underline{/0^\circ}$. Determine the value of C for unity power factor, find the resulting current $\underline{I}$ from the source, and calculate the transformer currents $\underline{I}_{in}$ and $\underline{I}_{out}$ when

$$R_c = 10\ \Omega \quad X_m = 20\ \Omega \quad R_w = 0.2\ \Omega \quad X_l = 1\ \Omega$$
$$N = 5 \quad Z = 45\ \Omega$$

8.60 Do Problem 8.59 with

$$R_c = 100\ \Omega \quad X_m = 300\ \Omega \quad R_w = 1\ \Omega \quad X_l = 3\ \Omega$$
$$N = 0.4 \quad Z = 0.8 + j2.4\ \Omega$$

8.61 Suppose the transformer described in Problem 8.57 is operated in the step-down mode by connecting the primary to a 24-Ω load resistance and driving the secondary with $\underline{V}_{in} = 2000$ V(rms) $\underline{/0^\circ}$. Find $\underline{V}_{out}$, $\underline{I}_{in}$, P_{in}, Q_{in}, and Eff.

8.62 Suppose the transformer described in Problem 8.58 is operated in the step-up mode by connecting the primary to a 300-Ω load resistance and driving the secondary with $\underline{V}_{in} = 600$ V(rms) $\underline{/0^\circ}$. Find $\underline{V}_{out}$, $\underline{I}_{in}$, P_{in}, Q_{in}, and Eff.

8.63 A source voltage $\underline{V}_s$ with series resistance R_s is connected to the primary in Fig. 8.30 (p. 392). Derive approximate expressions for the open-circuit voltage and Thévenin equivalent impedance seen looking back into the secondary when $R_s \ll R_c$, $R_w \ll R_c$, and $X_l \ll X_m$.

8.64 Calculate the parameter values for the transformer model in Fig. 8.30 (p. 392), given the following measurement data:

Secondary Open	Secondary Shorted
$\lvert\underline{V}_{in}\rvert = 120$ V	$\lvert\underline{V}_{in}\rvert = 60$ V
$\lvert\underline{I}_{in}\rvert = 0.5$ A	$\lvert\underline{I}_{in}\rvert = 10$ A
$P_{in} = 36$ W	$P_{in} = 240$ W
$\lvert\underline{V}_{out}\rvert = 600$ V	$\lvert\underline{I}_{out}\rvert = 2$ A

Use the value of $\lvert\underline{I}_{out}\rvert$ only if Eq. (8.40) is not satisfied.

8.65 Do Problem 8.64 with:

Secondary Open	Secondary Shorted
$\lvert\underline{V}_{in}\rvert = 240$ V	$\lvert\underline{V}_{in}\rvert = 5$ V
$\lvert\underline{I}_{in}\rvert = 5.2$ A	$\lvert\underline{I}_{in}\rvert = 25$ A
$P_{in} = 480$ W	$P_{in} = 750$ W
$\lvert\underline{V}_{out}\rvert = 60$ V	$\lvert\underline{I}_{out}\rvert = 10$ A

8.66 Do Problem 8.64 with:

Secondary Open	Secondary Shorted
$\lvert\underline{V}_{in}\rvert = 240$ V	$\lvert\underline{V}_{in}\rvert = 240$ V
$\lvert\underline{I}_{in}\rvert = 2.5$ A	$\lvert\underline{I}_{in}\rvert = 7.3$ A
$P_{in} = 480$ W	$P_{in} = 1400$ W
$\lvert\underline{V}_{out}\rvert = 240$ V	$\lvert\underline{I}_{out}\rvert = 4.8$ A

CHAPTER 9

Transient Response

When a circuit contains resistance and one or more energy-storage elements, the complete response of any branch variable consists of the natural response as well as the forced response. We distinguished between those two components in Chapter 5 by noting that the natural response depends upon the initial stored energy whereas the forced response depends upon the excitation. We also observed that the natural response usually dies away with time, leaving only the forced response. That property allowed us to develop ac steady-state analysis in terms of the forced response of stable circuits with sinusoidal excitation.

Now we return to the natural response and the resulting transients that appear in the complete response of dynamic circuits. We'll begin with first-order transients in circuits containing just one inductor or capacitor. Then we'll examine second-order transients in circuits containing two energy-storage elements.

The study of these transients is important because, on the one hand, they are often put to practical use for applications such as wave-shaping circuits and timing devices. But, on the other hand, transients often impose limits on the performance of digital systems and other high-speed electrical circuits.

Objectives

After studying this chapter and working the exercises, you should be able to do each of the following:

1. Evaluate the time constant of a first-order circuit, and sketch the zero-input response of any voltage or current in the circuit (Section 9.1).
2. Find and sketch the step response and pulse response of a first-order circuit (Section 9.1).
3. Find and sketch the transient response of any variable in a first-order circuit with a switched dc excitation (Section 9.1).
4. Find and sketch the transient response of any variable in a first-order circuit with a switched ac excitation (Section 9.2).†
5. Calculate the characteristic values for a second-order circuit, and determine whether the circuit is overdamped, underdamped, or critically damped (Section 9.3).
6. Identify the general form of the natural response of an overdamped, underdamped, or critically damped circuit (Section 9.3).
7. Find the characteristic polynomial of a second-order circuit (Section 9.3).
8. Determine the initial value and slope of any variable in a second-order circuit (Section 9.4).
9. Obtain the transient response of any variable in a second-order circuit with a switched dc excitation (Section 9.4).

9.1 FIRST-ORDER TRANSIENTS

This section examines transients in circuits containing just one inductor or capacitor, so their behavior is governed by first-order differential equations. We'll start with a review and elaboration of the natural response of *RC* and *RL* circuits in the context of the *zero-input response.* Then we'll study the complete response of first-order circuits with switched dc sources, including the important special cases of the step and pulse response.

Zero-Input Response

Figure 9.1*a* represents a situation often encountered in circuits with mechanical switches or electronic switching devices. The switch has been in the upper position for a long time prior to some instant $t = t_0$, so the capacitor charges to $v_C = V_0$ and stores energy $w_C = \frac{1}{2}CV_0^2$. The charged capacitor acts as a dc block under steady-state conditions, and $i_C = 0$ for $t < t_0$. The switch then

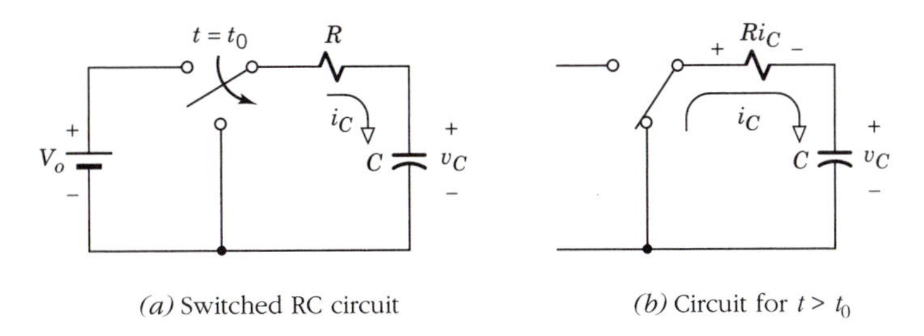

(a) Switched RC circuit

(b) Circuit for $t > t_0$

Figure 9.1

goes to the lower position at $t = t_0$, putting a short across the series *RC* network. We want to study the circuit's behavior after the switching instant.

The equivalent circuit for $t > t_0$ is diagrammed in Fig. 9.1*b*, where $v_C \neq 0$ and

$$Ri_C + v_C = 0$$

Thus, i_C must be nonzero for $t > t_0$, and v_C must decrease as the capacitor discharges through the resistor. The resulting behavior of i and v_C is called the **zero-input response** because the circuit has no applied source for $t > t_0$. This zero-input response will be of the same form as the natural response since they both satisfy the homogeneous differential equation.

To express the differential equation compactly, we'll introduce the primed notation for time derivatives defined in general by

$$y' \triangleq dy/dt \tag{9.1}$$

Then we can write the capacitor current as

$$i_C = C\,dv_C/dt = Cv_C'$$

Substituting for i_C in the KVL equation yields

$$RCv_C' + v_C = 0 \tag{9.2}$$

which is a first-order homogeneous differential equation.

We know from Section 5.3 that the solution of Eq. (9.2) is an exponential function of the form $v_C = Ae^{st}$, whose derivative is $v_C' = sAe^{st}$. We determine the value of s by inserting the assumed solution into Eq. (9.2) to get

$$RCsAe^{st} + Ae^{st} = 0$$

Since $v_C = Ae^{st} \neq 0$, we can divide both sides by Ae^{st} and thus obtain the **characteristic equation**

$$RCs + 1 = 0$$

Hence,

$$s = -1/RC = -1/\tau$$

where we've introduced

$$\tau = RC \tag{9.3}$$

We call τ the **time constant** because it carries the unit of time. The natural response of v_C is then

$$v_C(t) = Ae^{-t/\tau} \tag{9.4}$$

with A being an arbitrary constant.

But the zero-input response differs from the natural response in that A must be a specific value, namely, the value that satisfies the *initial condition.* To evaluate A, recall that the voltage across a capacitor cannot jump as long as the current remains finite. Therefore, at the switching instant, v_C obeys the continuity condition

$$v_C(t_0^+) = v_C(t_0^-) = V_0$$

Setting $t = t_0^+$ in Eq. (9.4) then yields

$$v_C(t_0^+) = Ae^{-t_0/\tau} = V_0$$

from which

$$A = V_0 e^{+t_0/\tau}$$

so the zero-input response of the voltage is

$$v_C(t) = (V_0 e^{+t_0/\tau})e^{-t/\tau} \qquad t > t_0$$

The corresponding zero-input response of the current is obtained by noting that $i_C = -v_C/R$ in Fig. 9.1*b*. Thus, for $t > t_0$, we have

$$v_C(t) = V_0 e^{-(t-t_0)/\tau} \qquad i_C(t) = -\frac{V_0}{R}e^{-(t-t_0)/\tau} \tag{9.5}$$

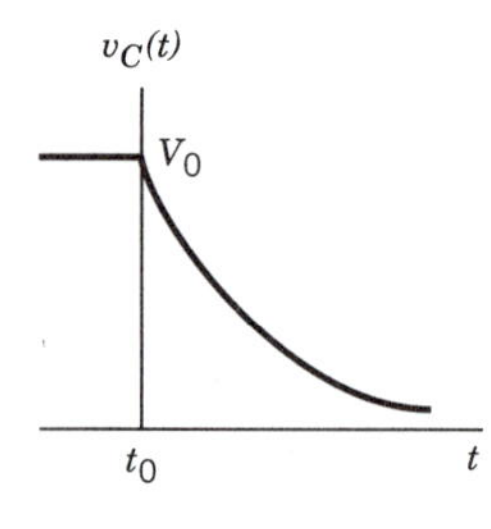

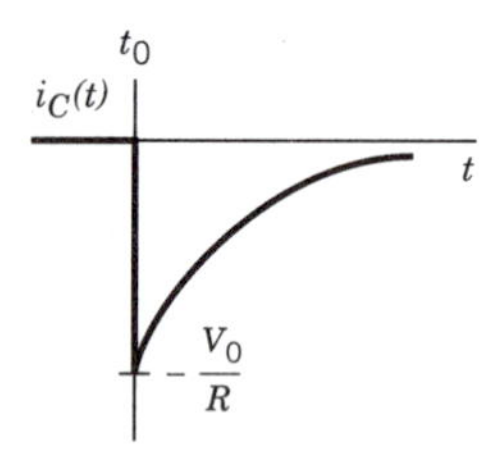

Figure 9.2

The term $(t - t_0)$ appearing here reflects the fact that these waveforms start at the switching instant $t = t_0$.

The decaying exponential waveforms $v_C(t)$ and $i_C(t)$ are sketched in Fig. 9.2, with the dc steady state for $t < t_0$. Although the voltage has continuity at the switching instant, the current jumps from $i_C(t_0^-) = 0$ to $i_C(t_0^+) = -V_0/R$. The negative value of i_C for $t > t_0$ simply means that the current transfers energy from C to R. And, in fact, the resistance eventually absorbs all of the energy stored by the capacitor.

To explore this energy transfer, we note that the resistance dissipates instantaneous power

$$p_R = Ri_C^2 = (V_0^2/R)e^{-2(t-t_0)/RC} \qquad t > t_0$$

As $t \to \infty$, the total energy absorbed by R is

$$w_R = \int_{t_0}^{\infty} p_R\, dt = \frac{V_0^2}{R} e^{2t_0/RC} \int_{t_0}^{\infty} e^{-2t/RC}\, dt$$
$$= \frac{V_0^2}{R} e^{2t_0/RC} \left(\frac{-RC}{2}\right)\left(e^{-\infty} - e^{-2t_0/RC}\right) = \frac{V_0^2 C}{2}$$

Thus, as expected, w_R equals the capacitor's initial stored energy $w_C = \frac{1}{2}CV_0^2$.

Decaying exponential waveforms also occur in switched *RL* circuits like Fig. 9.3. Here we have $v_L = 0$ in the dc steady state for $t < t_0$, so the constant current source establishes $i_L = I_0$ and $w_L = \frac{1}{2}LI_0^2$. When the switch goes to the lower position at $t = t_0$, the source is disconnected and KVL around the *RL* loop requires $Ri_L + v_L = 0$ with $v_L = L\, di_L/dt = Li_L'$. The differential equation is therefore $Li_L' + Ri_L = 0$ or

$$(L/R)i_L' + i_L = 0$$

The assumption that $i_L = Ae^{st}$ then yields the characteristic equation

$$(L/R)s + 1 = 0$$

from which

$$s = -R/L = -1/\tau$$

where

$$\tau = L/R \tag{9.6}$$

Hence, the natural response of i_L is

$$i_L(t) = Ae^{-t/\tau}$$

with A being an arbitrary constant.

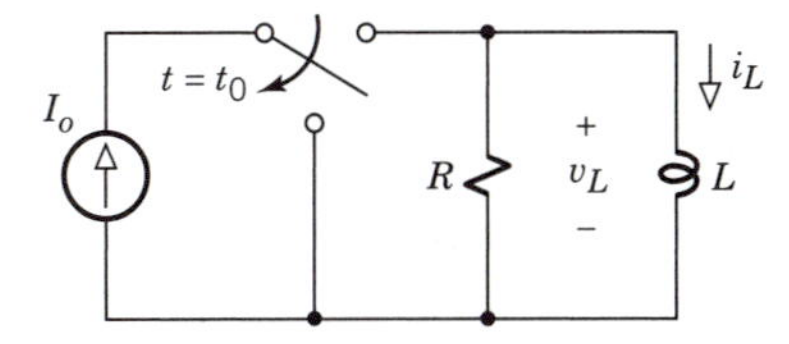

Figure 9.3 Switched *RL* circuit.

But the current through the inductor must have continuity at the switching instant, so

$$i_L(t_0^+) = i_L(t_0^-) = I_0$$

which requires that $i_L(t_0^+) = Ae^{-t_0/\tau} = I_0$ or $A = I_0 e^{+t_0/\tau}$. Therefore, noting that $v_L = -Ri_L$ for $t > t_0$, the zero-input response of $i_L(t)$ and $v_L(t)$ are

$$i_L(t) = I_0 e^{-(t-t_0)/\tau} \qquad v_L(t) = -RI_0 e^{-(t-t_0)/\tau} \tag{9.7}$$

These expressions show that $i_L(t)$ and $v_L(t)$ decay to zero for $t > t_0$, as the resistor dissipates the energy initially stored by the inductor. Closer examination of Eqs. (9.6) and (9.7) also reveals that they are the *duals* of Eqs. (9.3) and (9.5). The reason, of course, is that the *RL* circuit is the structural dual of the previous *RC* circuit.

Now we'll generalize our results by letting $y(t)$ be *any* zero-input response in *any* network consisting of resistance and one energy-storage element — either capacitance or inductance — but no independent sources. Viewed from the energy-storage element, the rest of the network acts as a single *equivalent resistance* R_{eq}, so the time constant can be calculated from Eq. (9.3) or (9.6) written as

$$\tau = \begin{cases} R_{eq}C \\ L/R_{eq} \end{cases} \tag{9.8}$$

The zero-input response is given in general by

$$y(t) = Y_0 e^{-(t-t_0)/\tau} \qquad t > t_0 \tag{9.9}$$

where

$$Y_0 = y(t_0^+)$$

If $y(t)$ is a capacitor voltage or inductor current, then continuity requires that $y(t_0^+) = y(t_0^-)$. For other variables, the value of Y_0 must be determined from the continuity of v_C or i_L at the initial instant t_0.

Since Eq. (9.9) describes the zero-input response of any *RC* or *RL* network, we should give further attention to the corresponding waveform plotted in Fig. 9.4. The dashed line indicates the **initial slope**

$$y'(t_0^+) \triangleq \left.\frac{dy}{dt}\right|_{t=t_0^+} = -\frac{Y_0}{\tau} \tag{9.10}$$

This line shows that $y(t)$ initially heads toward zero at $t = t_0 + \tau$. But the slope progressively decreases for $t > t_0$, and $y(t)$ actually equals about 37% of its initial value after an interval equal to one time constant has elapsed. Although the zero-input response never completely reaches zero in finite time, for most purposes $y(t)$ becomes negligibly small after five time constants when

$$y(t_0 + 5\tau) = Y_0 e^{-5} \approx 0.007\,Y_0$$

which is less than 1% of the initial value.

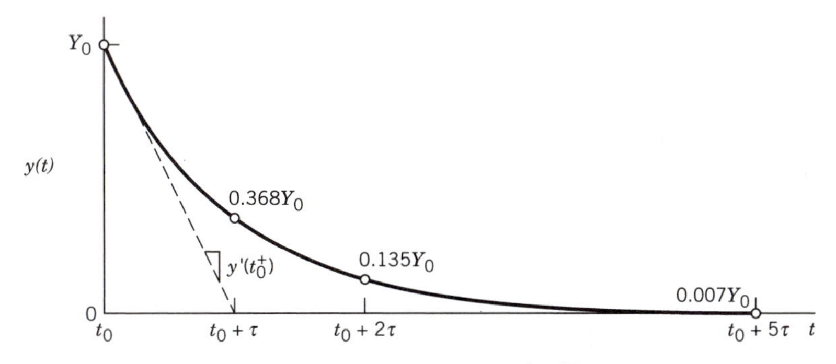

Figure 9.4 $y(t) = Y_0 e^{-(t-t_0)/\tau}$

As a final point, we observe that Eqs. (9.8)–(9.10) hold even when the network contains *controlled sources.* However, controlled sources may result in *negative equivalent resistance.* In such cases, the time constant τ will be negative so the zero-input response grows with time and the circuit is *unstable.* Further discussion of this effect is deferred to Chapter 10.

Example 9.1 *Zero-Input Response of an RL Circuit*

Suppose you know that the circuit in Fig. 9.5 has reached the dc steady state when the switch is opened at $t = 0$. You want to find $i(t)$ and $v(t)$ for $t > 0$.

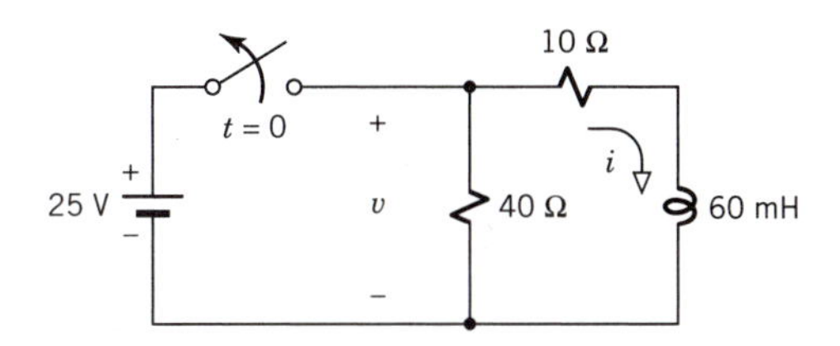

Figure 9.5 Circuit for Example 9.1.

First, you calculate the time constant by observing that the equivalent resistance connected to the inductor with the switch open in $R_{eq} = 40 + 10 = 50\ \Omega$, so

$$\tau = L/R_{eq} = 60\text{ mH}/50\ \Omega = 1.2\text{ ms}$$

Next, since the current i has continuity here, you invoke the steady-state condition at $t = 0^-$ to get

$$i(0^+) = i(0^-) = 25\text{ V}/10\ \Omega = 2.5\text{ A}$$

Hence, using Eq. (9.9) with $t_0 = 0$,

$$i(t) = 2.5e^{-t/\tau}\text{ A} \qquad t > 0$$

from which

$$v(t) = -40i(t) = -100e^{-t/\tau}\text{ V} \qquad t > 0$$

Continuity of current in this circuit thus generates a negative voltage "spike" whose 100-V peak value substantially exceeds the 25-V source voltage. The magnitude of the spike decays to about 37 V at $t = \tau = 1.2$ ms and to $|v(t)| < 0.7$ V for $t > 5\tau = 6$ ms.

Exercise 9.1

Let the inductor in Fig. 9.5 be replaced by a 2-μF capacitor. Show that $v(0^+) = 20$ V. Then find $v(t)$ and $i(t)$ for $t > 0$.

Step Response

We now investigate transients that occur in stable first-order circuits having no initial stored energy at the time when a dc source is applied. This situation is just the opposite of the zero-input response in that the capacitor voltage or inductor current starts at zero and builds up to a nonzero value in the dc steady state. The transitional behavior is known as the **step response** since the circuit's excitation switches abruptly from OFF to ON.

For notational convenience, we'll express switched dc sources using the **unit step function**

$$u(t) \triangleq \begin{cases} 0 & t < 0 \\ 1 & t > 0 \end{cases} \tag{9.11a}$$

As shown in Fig. 9.6*a*, $u(t)$ jumps from 0 to 1 at $t = 0$. More generally, multiplying $u(t)$ by any constant K and replacing t by $t - t_0$ yields

$$Ku(t - t_0) = \begin{cases} 0 & t - t_0 < 0 \\ K & t - t_0 > 0 \end{cases} \tag{9.11b}$$

This function, plotted in Fig. 9.6*b*, makes a step of "height" K at time t_0.

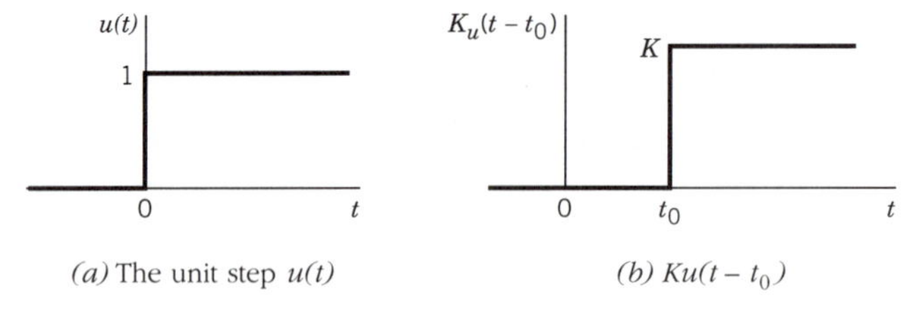

(*a*) The unit step $u(t)$ (*b*) $Ku(t - t_0)$

Figure 9.6

Our study of step response begins with an arbitrary *RC* circuit driven by a dc source that goes on at $t = t_0$. We'll focus on the capacitor voltage because any other variable of interest can be expressed in terms of $v_C(t)$. Relative to the capacitor, the rest of the circuit acts like a Thévenin source network having resistance R_{eq} and a switched open-circuit voltage

$$v_{oc}(t) = V_{ss}u(t - t_0) = \begin{cases} 0 & t < t_0 \\ V_{ss} & t > t_0 \end{cases}$$

Figure 9.7 diagrams the equivalent circuit.

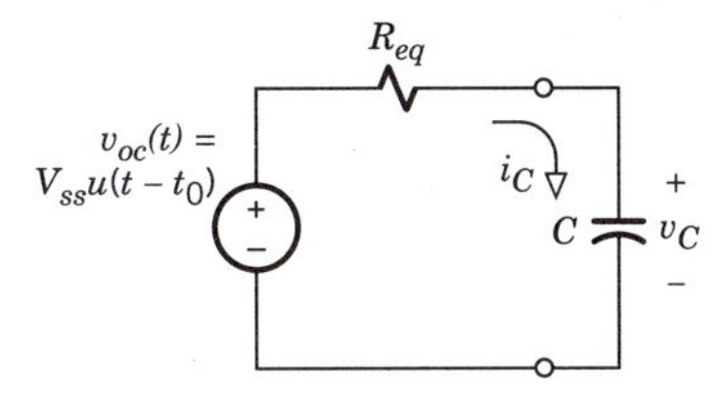

Figure 9.7 Circuit model for step response with a capacitor.

The capacitor has no initial stored energy, so $v_C(t_0^+) = v_C(t_0^-) = 0$, but the source drives the circuit toward dc steady-state conditions with $v_C = V_{ss}$ as $t \rightarrow \infty$. Since $i_C = Cv_C'$ and $R_{eq}i_C + v_C = v_{oc}(t) = V_{ss}$ for $t > t_0$, the transient behavior of $v_C(t)$ is governed by the nonhomogeneous differential equation

$$\tau v_C' + v_C = V_{ss} \qquad t > t_0$$

with

$$\tau = R_{eq}C$$

The general solution of the differential equation has the form

$$v_C(t) = V_{ss} + Ae^{-t/\tau}$$

which consists of the forced response V_{ss} plus the natural response $Ae^{-t/\tau}$. The natural response must be included to satisfy the initial condition $v_C(t_0^+) = 0$, and we evaluate the constant A via

$$v_C(t_0^+) = V_{ss} + Ae^{-t_0/\tau} = 0 \quad \Rightarrow \quad A = -V_{ss}e^{+t_0/\tau}$$

Hence, the step response of the capacitor voltage is

$$v_C(t) = V_{ss} + (-V_{ss}e^{+t_0/\tau})e^{-t/\tau} = V_{ss}[1 - e^{-(t-t_0)/\tau}] \qquad t > t_0$$

We'll discuss this type of waveform after considering *RL* circuits.

To find the step response of an *RL* circuit, we focus on the current through the inductor and we replace the rest of the circuit by a Norton network. The resulting diagram in Fig. 9.8 then has the stepped short-circuit current

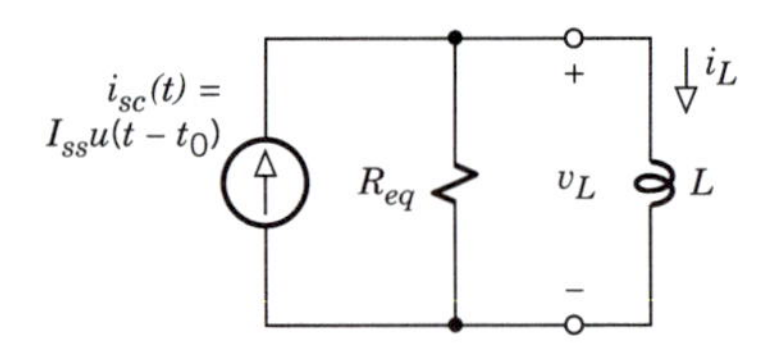

Figure 9.8 Circuit model for step response with an inductor.

$$i_{sc}(t) = I_{ss}u(t - t_0)$$

But Fig. 9.8 is the dual of Fig. 9.7, so we immediately conclude from our previous results that the step response of the inductor current is given by the dual expression

$$i_L(t) = I_{ss}[1 - e^{-(t-t_0)/\tau}] \qquad t > t_0$$

with $\tau = L/R_{eq}$.

Now let $y(t)$ stand for either $v_C(t)$ or $i_L(t)$ and let Y_{ss} denote the value of the dc forced response. The step response may then be written in the generic form

$$y(t) = Y_{ss}[1 - e^{-(t-t_0)/\tau}] \qquad t > t_0 \tag{9.12}$$

The plot of $y(t)$ in Fig. 9.9 shows how the response goes through a transient interval on its way to the steady state with $y(t) = Y_{ss}$ as $t \rightarrow \infty$. The initial slope is

$$y'(t_0^{+}) = Y_{ss}/\tau$$

and the response starts out as the ramp waveforms

$$y(t) \approx y'(t_0^{+})(t - t_0) = \frac{Y_{ss}}{\tau}(t - t_0)$$

Using this property, the transient portion is easily sketched by drawing a straight line from the starting point to the point $y = Y_{ss}$ at $t = t_0 + \tau$. The transient is essentially over after five time constants, since $y(t) \approx Y_{ss}$ for $t > t_0 + 5\tau$.

Based on Fig. 9.9, which represents either $v_C(t)$ or $i_L(t)$, we can henceforth assume that any stable first-order circuit is in the dc steady state if at least five time constants have elapsed after the source was applied. Accordingly, when we say that a switch has been in a certain position for a "long time," we mean at least five time constants.

Example 9.2 ***Step Response of an RC Circuit***

The switch in Fig. 9.10*a* has been open for a long time before $t = 0$, so the capacitor has no initial stored energy when the switch closes at $t = 0$. We

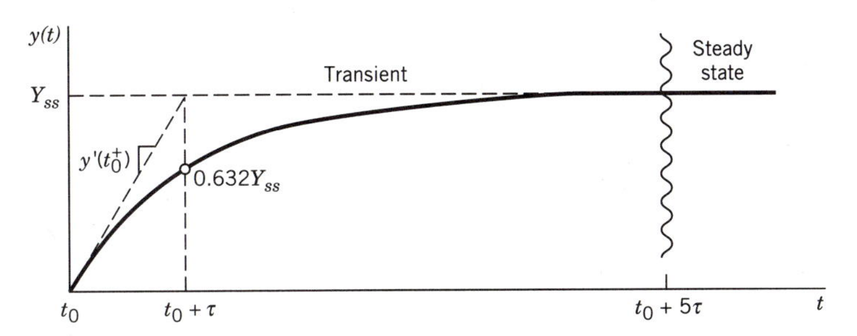

Figure 9.9 $y(t) = Y_{ss}[1 - e^{-(t-t_0)/\tau}]$

want to find the resulting step-response voltage $v(t)$ and current $i(t)$, given that $C = 50\ \mu F$.

With the switch closed, the Thévenin parameters of the source network connected to the capacitor are

$$R_{eq} = 3\|6 = 2\ k\Omega \qquad v_{oc} = V_{ss} = \frac{6}{3+6} \times 12 = 8\ V$$

Hence,

$$\tau = R_{eq}C = 2\ k\Omega \times 50\ \mu F = 0.1\ s$$

and Eq. (9.12) yields

$$v(t) = 8(1 - e^{-10t})\ V \qquad t > 0$$

We then find the current from the equivalent circuit in Fig. 9.10b, where

$$i(t) = \frac{8\ V - v(t)}{2\ k\Omega} = 4e^{-10t}\ mA \qquad t > 0$$

The transient interval is essentially over at $t = 5\tau = 0.5$ s, when $v(t) = 8(1 - e^{-5}) \approx 8$ V and $i(t) = 4e^{-5}$ mA $\approx 27\ \mu A$.

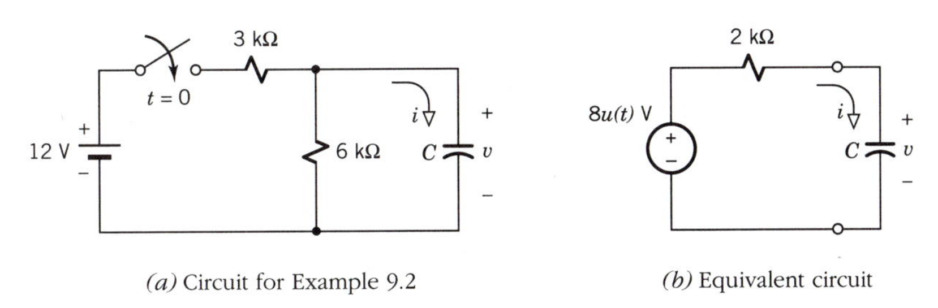

(a) Circuit for Example 9.2 *(b)* Equivalent circuit

Figure 9.10

Exercise 9.2

Find $i(t)$ and $v(t)$ in Fig. 9.10*a* when a 20–mH inductor replaces the capacitor.

Pulse Response

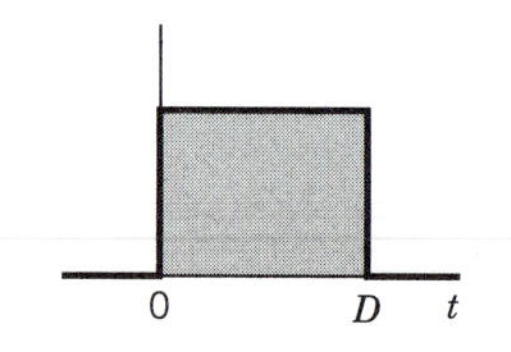

Figure 9.11
Rectangular pulse.

The **pulse response** occurs when a circuit has no initial stored energy and the excitation is a **rectangular pulse** like Fig. 9.11. The pulse starts at $t = 0$ and ends at $t = D$, where D is the pulse **duration**. We are interested in the corresponding response because pulsed waveforms are used to convey information in computers and other types of digital systems.

The pulse begins just like a step, so the pulse response at first will be identical to the *step response*. We therefore write

$$y(t) = Y_{ss}(1 - e^{-t/\tau}) \qquad 0 < t \leq D \tag{9.13a}$$

Here, $y(t)$ still denotes the capacitor voltage or inductor current in a first-order circuit, and Y_{ss} is the steady-state value that $y(t)$ would approach if the pulse lasted forever. But the pulse ends at $t = D$ when $y(t)$ has the value

$$y(D) = Y_{ss}(1 - e^{-D/\tau}) \tag{9.13b}$$

Since the circuit's excitation equals zero after the pulse ends, the remainder of the pulse response will be identical to the *zero-input response*. Hence,

$$y(t) = y(D)e^{-(t-D)/\tau} \qquad t > D \tag{9.13c}$$

obtained from Eq. (9.9) with $t_0 = D$ and $Y_0 = y(D^+) = y(D)$.

Figure 9.12 shows the pulse-response waveforms for three different values of the time constant τ relative to the pulse duration D. With a short time constant $\tau \ll D$, the response rises quickly and flattens off so that $y(D) \approx Y_{ss}$. The response then decays quickly at the end of the pulse. With a longer time constant $\tau \approx D$, the rise and decay are less rapid, and the value of $y(D)$ must be computed from Eq. (9.13b). With an even longer time constant $\tau \gg D$, the rise approximates the linear ramp $y(t) \approx Y_{ss}t/\tau$, so $y(D) \approx Y_{ss}D/\tau \ll Y_{ss}$. We therefore conclude that the pulse response of a first-order circuit de-

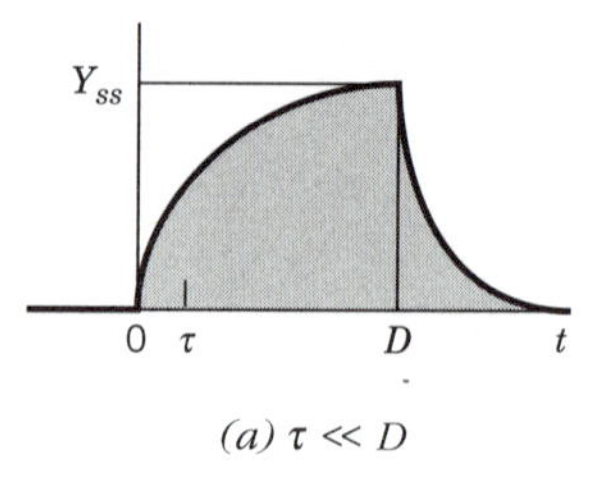

(a) $\tau \ll D$

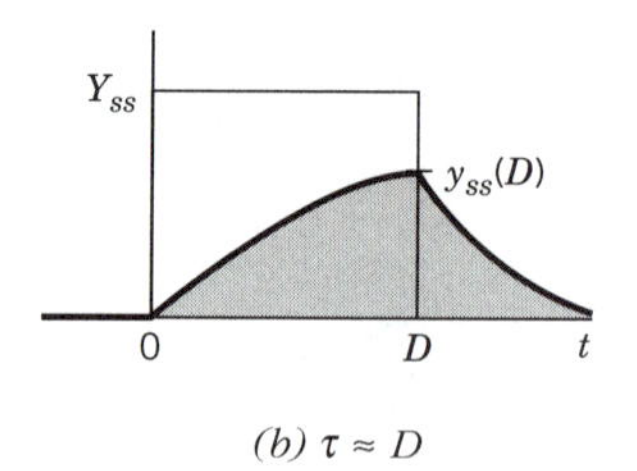

(b) $\tau \approx D$

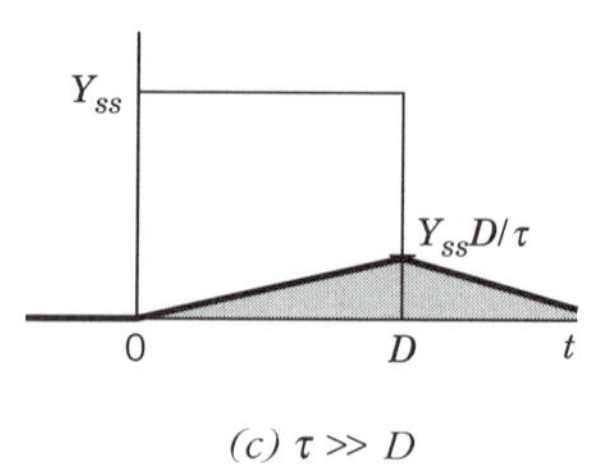

(c) $\tau \gg D$

Figure 9.12 Pulse response.

pends critically upon the relative values of τ and D. Preserving the pulse shape requires a *short* time constant, say

$$\tau \leq D/5$$

Otherwise, $y(t)$ will not have a "flat" top and the pulse shape becomes significantly distorted.

Example 9.3 *Analysis of a Relay Driver*

Figure 9.13*a* represents a **relay driver** consisting of a pulsed voltage source and a relay coil. Circuits of this type are used to control high-current switches and various other electromechanical devices. You have found that the relay is actuated when i_L reaches the 150-mA "pull-in" current, and it becomes deactuated when i_L falls to the 40-mA "drop-out" current. Your task now is to determine how long the relay stays actuated when the source produces a 5-V pulse with 30-ms duration.

Inspection of the diagram shows that $R_{eq} = 10 + 15 = 25\ \Omega$, so the time constant is $\tau = L/R_{eq} = 400\ \text{mH}/25\ \Omega = 16$ ms. With the pulse applied, i_L approaches the steady-state value $I_{ss} = 5\ \text{V}/25\ \Omega = 200$ mA. Hence, expressing time in milliseconds, you write

$$i_L(t) = 200(1 - e^{-t/16})\ \text{mA} \qquad 0 < t \leq 30\ \text{ms}$$

which is sketched in Fig. 9.13*b*. The relay closes at time t_1 when $i_L(t_1) = 150$ mA, so

$$200(1 - e^{-t_1/16}) = 150$$

and

$$t_1 = -16 \ln(1 - 150/200) = 22.2\ \text{ms}$$

You also note that the value of current at the end of the pulse is $i_L(D) = 200(1 - e^{-30/16}) = 169$ mA.

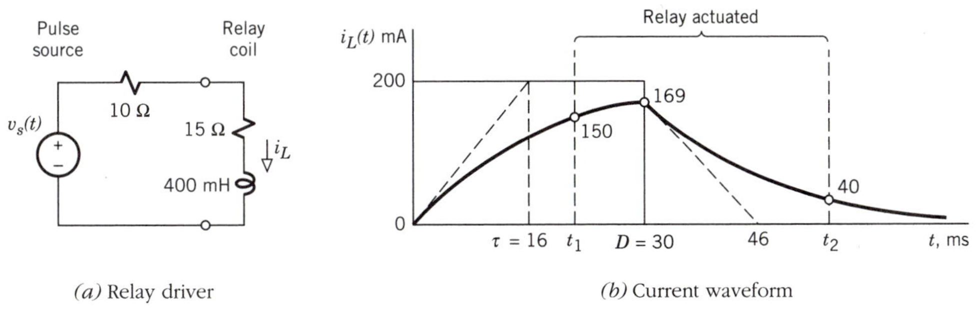

(*a*) Relay driver (*b*) Current waveform

Figure 9.13

After the pulse is over, the exponential decay of the current becomes

$$i_L(t) = 169e^{-(t-30)/16} \qquad t > 30 \text{ ms}$$

The relay then opens at time t_2 when $i_L(t_2) = 40$ mA, so

$$169e^{-(t_2-30)/16} = 40$$

from which you obtain

$$t_2 = 30 - 16 \ln (40/169) = 53.1 \text{ ms}$$

Hence, the relay remains closed over the interval

$$t_2 - t_1 = 30.9 \text{ ms}$$

The waveform in Fig. 9.13*b* shows that this inverval is considerably delayed compared to the beginning and end of the driving pulse.

Exercise 9.3

Suppose the switch in Fig. 9.10*a* closes at $t = 0$ and opens again at $t = 0.4$ s. (**a**) What's the condition on C so that $v(t)$ has a recognizable pulse shape? (**b**) Find and sketch $v(t)$ when $C = 200\ \mu$F.

Switched DC Transients

Switched dc transients occur when the source applied to a network jumps from one constant value to another. Accordingly, the voltages and currents undergo transitional readjustments to new steady-state values. We'll perform a general analysis of such transients in first-order circuits by letting $y(t)$ be any voltage or current of interest. We'll also let $\tau = R_{eq}C$ or $\tau = L/R_{eq}$, where R_{eq} is the equivalent resistance connected to the energy-storage element.

Suppose the switching takes place at time t_0, and $y(t)$ then approaches a new steady-state value Y_{ss}. The resulting transient behavior is governed by the differential equation

$$\tau y' + y = Y_{ss} \qquad t > t_0$$

which is the same as for the step response. The solution will thus have the form

$$y(t) = Y_{ss} + Ae^{-t/\tau} \qquad t > t_0$$

The new feature here is that A must satisfy a nonzero initial condition.

Given the initial value $y(t_0^{+}) = Y_0$, we can evaluate A by setting $t = t_0^{+}$ to get

$$y(t_0^+) = Y_{ss} + Ae^{-t_0/\tau} = Y_0$$

Hence, $A = (Y_0 - Y_{ss})e^{+t_0/\tau}$ and

$$y(t) = Y_{ss} + (Y_0 - Y_{ss})e^{-(t-t_0)/\tau} \qquad t > t_0 \tag{9.14}$$

With appropriate values of Y_{ss}, Y_0, and τ, Eq. (9.14) represents the switched dc transient response of *any* variable in *any* first-order circuit. This expression also includes both the zero-input response and the step response as special cases in which $Y_{ss} = 0$ or $Y_0 = 0$.

An organized procedure for the application of Eq. (9.14) involves four steps, as follows:

1. Draw the circuit diagram for $t = t_0^-$ to find $y(t_0^-)$ and $v_C(t_0^-)$ or $i_L(t_0^-)$.
2. Draw the circuit diagram for $t > t_0$, and find the initial value $Y_0 = y(t_0^+)$ from the continuity of v_C or i_L.
3. Find the steady-state value Y_{ss} via dc analysis of the circuit diagram for $t > t_0$.
4. Find the time constant τ by calculating R_{eq} connected to the energy-storage element when all independent sources have been suppressed.

The following example illustrates how this procedure can be used when there are two or more switching instants.

Example 9.4 ***Sequential Switched Transients***

Both switches in Fig. 9.14*a* have been open for a long time. The switch on the right closes at $t = 0$, and the switch on the left closes at $t = 1$ s. We want to find the resulting transient waveforms $v(t)$ and $i(t)$. We'll start by following our procedure for the first switching instant.

Step 1: We note from Fig. 9.14*a* that $v(0^-) = 0$ and $i(0^-) = 0$.

Step 2: For the interval $0 < t \leq 1$ s, we use the diagram in Fig. 9.14*b*. Since $v(t)$ has continuity, the initial values are

$$V_0 = v(0^+) = v(0^-) = 0 \qquad I_0 = i(0^+) = \frac{-16 - v(0^+)}{8} = -2 \text{ mA}$$

Step 3: By dc steady-state analysis, $v(t)$ and $i(t)$ head for the values

$$V_{ss} = \frac{24}{8 + 24}(-16) = -12 \text{ V} \qquad I_{ss} = \frac{-16}{8 + 24} = -0.5 \text{ mA}$$

Step 4: Suppressing the 16-V source shows that $R_{eq} = 8\|24 = 6 \text{ k}\Omega$, so

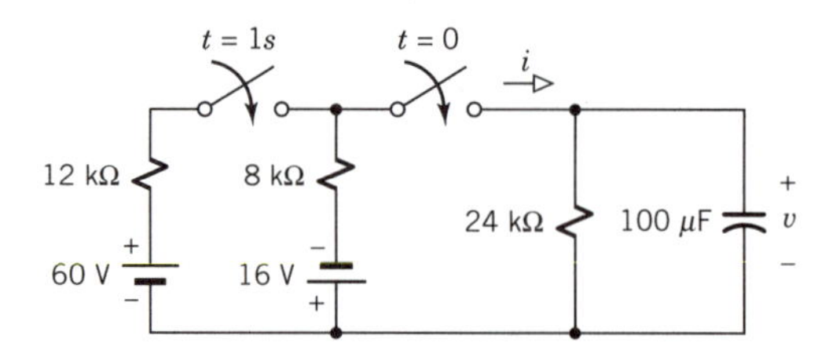

(a) Circuit with sequential switching

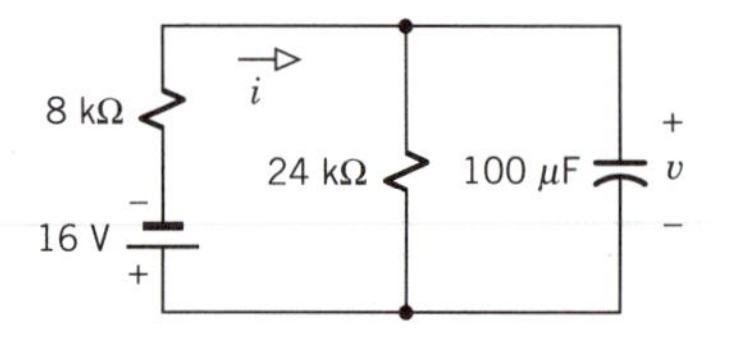

(b) Equivalent circuit for $0 < t \leq 1s$

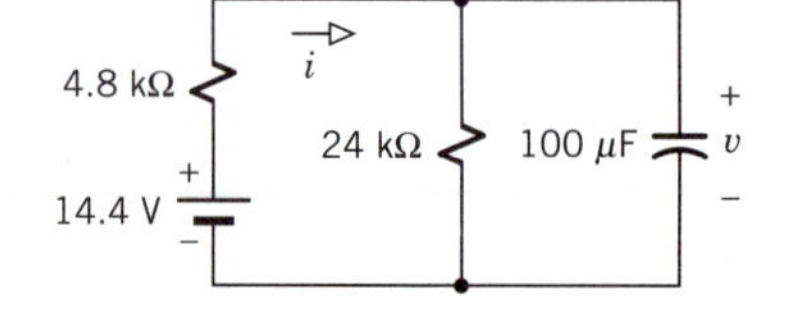

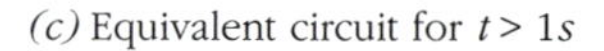

(c) Equivalent circuit for $t > 1s$

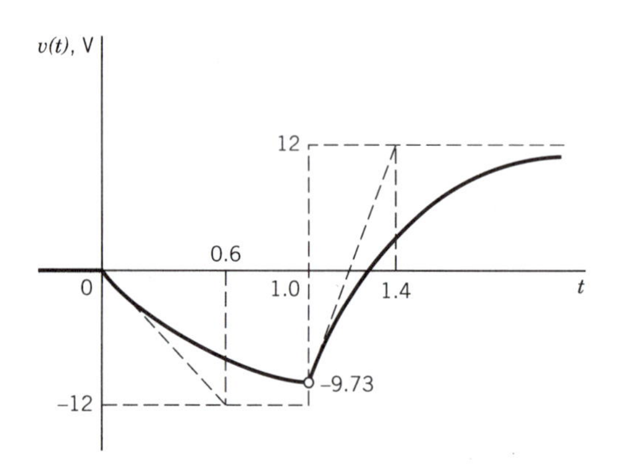

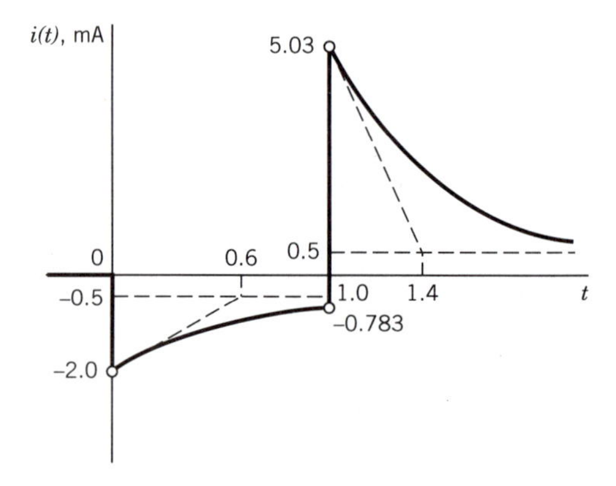

(d) Switched waveforms

Figure 9.14

$$\tau = 6 \text{ k}\Omega \times 100 \text{ }\mu\text{F} = 0.6 \text{ s}$$

Inserting our results into Eq. (9.14) with $t_0 = 0$ we obtain

$$v(t) = V_{ss} + (V_0 - V_{ss})e^{-t/\tau} = -12 + 12e^{-t/0.6} \text{ V}$$
$$i(t) = I_{ss} + (I_0 - I_{ss})e^{-t/\tau} = -0.5 - 1.5e^{-t/0.6} \text{ mA}$$

which hold for $0 < t \leq 1$ s. Setting $t = 1$ s gives

$$v(1) = -9.73 \text{ V} \qquad i(1) = -0.783 \text{ mA}$$

which are the values at the end of this interval.

The second interval begins when both switches are closed, so the circuit is now driven by two dc sources. We combine these by applying source conversions to the left of the 24-kΩ resistor in Fig. 9.14*a*, which yields the new equivalent circuit in Fig. 9.14*c*. The capacitor voltage again has continuity, so the initial values at $t = 1^+$ s are

$$V_0 = v(1^+) = v(1) = -9.73 \text{ V} \qquad I_0 = i(1^+) = \frac{14.4 - v(1^+)}{4.8} = 5.03 \text{ mA}$$

But $v(t)$ and $i(t)$ now head for

$$V_{ss} = \frac{24}{4.8 + 24} 14.4 = 12 \text{ V} \qquad I_{ss} = \frac{14.4}{4.8 + 24} = 0.5 \text{ mA}$$

We also have a new time constant since

$$R_{eq} = 4.8 \| 24 = 4 \text{ k}\Omega \qquad \tau = 4 \text{ k}\Omega \times 100 \ \mu\text{F} = 0.4 \text{ s}$$

Thus, with $t_0 = 1$ s in Eq. (9.14), we have

$$v(t) = 12 - 21.73e^{-(t-1)/0.4} \text{ V} \qquad i(t) = 0.5 + 4.53e^{-(t-1)/0.4} \text{ mA}$$

which hold for $t > 1$ s.

These waveforms are plotted in Fig. 9.14*d*, including the steady-state values for $t < 0$. The waveforms are easily sketched by drawing straight lines for the initial slopes at $t = 0$ and $t = 1$ s. The current jumps at both switching instants because $v(t)$ cannot jump.

Exercise 9.4

Find $i(t)$ for $t > 0$ in Fig. 9.15 given that

$$i_s(t) = -4 \text{ A} \qquad t < 0$$
$$= 2 \text{ A} \qquad t > 0$$

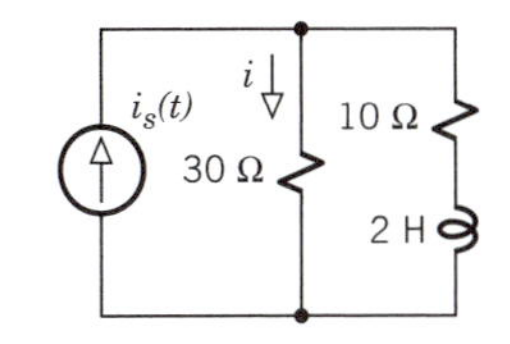

Figure 9.15

9.2 SWITCHED AC TRANSIENTS†

Switched ac transients occur when a sinusoidal excitation goes on at $t = t_0$, or when some parameter of an ac source undergoes an abrupt change. We investigate the resulting transients here for the case of stable first-order circuits. The analysis method is similar to that of switched dc transients, except that the forced response will be a sinusoid rather than a constant.

The forced response of any variable $y(t)$ in a stable circuit can be found by assuming ac steady-state conditions and using phasor analysis to obtain the phasor $\underline{Y} = Y_m \angle \phi$. Accordingly, we write

$$y_F(t) = Y_m \cos(\omega t + \phi) \qquad (9.15)$$

The complete response of a first-order circuit with time constant τ is then

$$y(t) = y_F(t) + Ae^{-t/\tau} \qquad t > t_0$$

Given the initial value $y(t_0^+) = Y_0$, we can evaluate the constant A by setting $t = t_0^+$ to get

$$y(t_0^+) = y_F(t_0^+) + Ae^{-t_0/\tau} = Y_0$$

Hence, $A = [Y_0 - y_F(t_0^+)]e^{+t_0/\tau}$ and

$$y(t) = y_F(t) + [Y_0 - y_F(t_0^+)]e^{-(t-t_0)/\tau} \qquad t > t_0 \tag{9.16}$$

This expression shows that $y(t)$ approaches the ac steady state as $t \to \infty$, since $y(t) \approx y_F(t)$ for $t > t_0 + 5\tau$, provided that $\tau > 0$.

Example 9.5 *Transients in an AM Radio Signal*

The applied voltage in Fig. 9.16*a* is a sinusoid with $\omega = 15$, but its amplitude jumps from 6 V to 12 V at $t = 0$, so

$$\begin{aligned} v_s(t) &= 6 \cos 15t \text{ V} \qquad && t < 0 \\ &= 12 \cos 15t \text{ V} && t > 0 \end{aligned}$$

Waveforms like this occur in radio signals with **amplitude modulation (AM)**. We want to find $i(t)$ and $v(t)$.

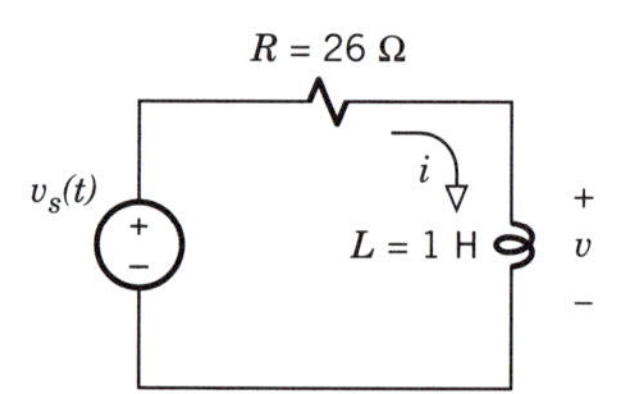

(*a*) Circuit with switched ac source

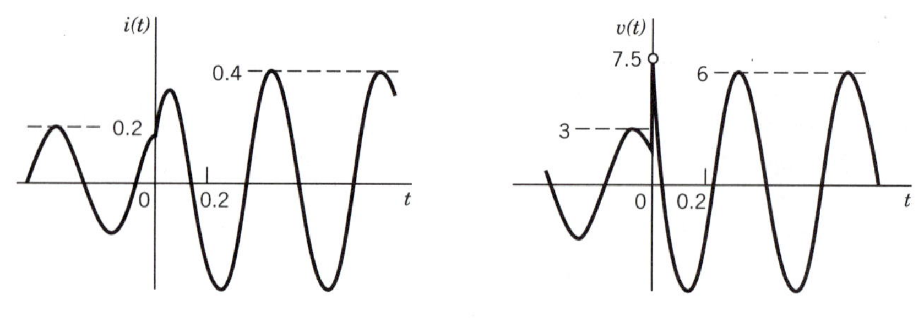

(*b*) Waveforms

Figure 9.16

First, we determine the initial values at $t = 0^+$ by assuming ac steady-state conditions for $t < 0$. Routine phasor analysis gives

$$\underline{I} = 6/(26 + j15) = 0.2 \text{ A } \underline{/-30°} \qquad \underline{V} = j15\underline{I} = 3 \text{ V } \underline{/60°}$$

so the waveforms for $t < 0$ are

$$i(t) = 0.2 \cos(15t - 30°) \qquad v(t) = 3 \cos(15t + 60°)$$

Since $i(t)$ has continuity,

$$I_0 = i(0^+) = i(0^-) = 0.2 \cos(-30°) = 0.173 \text{ A}$$

The initial value of $v(t)$ is then calculated via

$$V_0 = v(0^+) = v_s(0^+) - 26\,i(0^+) = 7.50 \text{ V}$$

Next, phasor analysis for $t > 0$ yields

$$i_F(t) = 0.4 \cos(15t - 30°) \qquad v_F(t) = 6 \cos(15t + 60°)$$

from which

$$i_F(0^+) = 0.346 \text{ A} \qquad v_F(0^+) = 3 \text{ V}$$

Finally, we observe from the circuit diagram that

$$\tau = 1 \text{ H}/26\ \Omega = 1/26 \text{ s}$$

After substituting our results into Eq. (9.16), we obtain the complete response for $t > 0$ as

$$i(t) = 0.4 \cos(15t - 30°) - 0.173e^{-26t} \text{ A}$$
$$v(t) = 6 \cos(15t + 60°) + 4.50e^{-26t} \text{ V}$$

The waveforms plotted in Fig. 9.16*b* show that the continuity of $i(t)$ produces a smooth transition to the new ac steady state, whereas $v(t)$ has an initial jump discontinuity. The transient interval is over rather quickly because the exponential terms nearly vanish after $5\tau \approx 0.2$ s, which is about one-half of the sinusoidal period $2\pi/\omega \approx 0.4$ s.

Exercise 9.5

Consider the circuit in Fig. 9.15 (p. 419). Find $i(t)$ for $t > 0$ given that

$$i_s(t) = -4 \text{ A} \qquad t < 0$$
$$= 4 \cos 10t \text{ A} \qquad t > 0$$

9.3 SECOND-ORDER NATURAL RESPONSE

When a circuit contains two or more energy-storage elements, the transient response may include time functions other than decaying exponentials. In particular, circuits with at least one capacitor and one inductor have three possible types of behavior known as *overdamped*, *underdamped*, and *critically damped*. This section examines the basic properties and natural response of second-order circuits, in preparation for our study of second-order transients in the next section.

Second-Order Circuit Equations

We begin with the definition that

> A second-order circuit contains two independent energy-storage elements.

The stipulation of *independent* energy-storage elements means that the energy stored by one of them must be independent of the energy stored by the other. As a simple counterexample, two inductors in series carry the same current i_L and would not be independent because the energies $\frac{1}{2}L_1 i_L^2$ and $\frac{1}{2}L_2 i_L^2$ are related.

When energy-storage elements are independent, transient analysis is expedited by focusing on the *capacitor voltages* and *inductor currents*. These variables play a crucial role for two reasons:

- The total stored energy depends entirely upon the values of v_C and i_L.
- Any other variable of interest can be expressed in terms of v_C, i_L, and the excitation.

In view of these important properties, we say that

> The capacitor voltages and inductor currents constitute the **state variables** of any circuit having independent energy-storage elements.

The name *state* variable emphasizes the fact that the values of v_C and i_L, together with the excitation, fully determine the state of the circuit at any instant of time.

If the circuit contains two independent energy-storage elements, then the response of any variable y is governed by a *second-order* differential equation. Such equations involve the second derivative d^2y/dt^2, written more compactly in the double-prime notation

$$y'' \triangleq d^2y/dt^2 = dy'/dt \tag{9.17}$$

We'll use state variables to find differential equations for any variables of interest in representative second-order circuits.

Series *LRC* circuit. The inductor and capacitor in Fig. 9.17 are independent, so i_L and v_C are the state variables. If we know their values along with the excitation v_s, then we can calculate the remaining variables using Kirchhoff's and Ohm's laws. In particular, we see that

$$i_C = i_L \qquad v_R = Ri_L \qquad v_L = v_s - v_R - v_C = v_s - Ri_L - v_C$$

But formulating the differential circuit equation also requires the energy-storage element relations

$$i_C = C\,dv_C/dt = Cv_C' \qquad v_L = L\,di_L/dt = Li_L'$$

Inserting Cv_C' in the KCL equation and Li_L' in the KVL equation yields the pair of *first*-order differential equations

$$Cv_C' = i_L \tag{9.18a}$$

$$Li_L' = v_s - Ri_L - v_C \tag{9.18b}$$

These equations are *coupled* in that v_C' depends upon i_L and i_L' depends upon v_C.

Although a pair of coupled first-order equations provides one description of a second-order circuit, we seek here a single equation relating either state variable to the excitation. For that purpose, we differentiate Eq. (9.18a) to obtain

$$i_L' = di_L/dt = C\,dv_C'/dt = Cv_C''$$

Then, after substituting for i_L and i_L', Eq. (9.18b) becomes

$$LCv_C'' = v_s - RCv_C' - v_C$$

or, after rearrangement,

$$v_C'' + \frac{R}{L}v_C' + \frac{1}{LC}v_C = \frac{1}{LC}v_s \tag{9.19}$$

This is the desired relationship between v_C and v_s. The fact that Eq. (9.19) is a second-order differential equation confirms that the two energy-storage elements are independent.

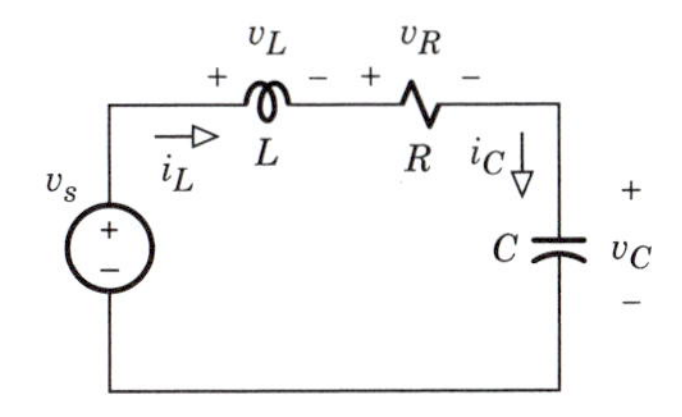

Figure 9.17 Series *LRC* circuit.

We can also relate i_L to v_s by the same strategy. We use the derivative of Eq. (9.18b) to eliminate v_C' from Eq. (9.18a), thereby obtaining

$$i_L'' + \frac{R}{L} i_L' + \frac{1}{LC} i_L = \frac{1}{L} v_s' \tag{9.20}$$

This is another second-order differential equation, and the coefficients on the left-hand side are exactly the same as those in Eq. (9.19). However, the right-hand side now includes the source derivative v_s'.

Note that Eqs. (9.19) and (9.20) have been arranged so the coefficient of the highest derivative of v_C or i_L equals unity. That arrangement expedites the task of obtaining differential equations for other variables. As a case in point, suppose we want to relate the inductor voltage v_L to the excitation v_s. Having previously found that $v_L = v_s - (Ri_L + v_C)$, we multiply Eq. (9.20) by R and add it to Eq. (9.19) to get

$$Ri_L'' + v_C'' + \frac{R}{L}(Ri_L' + v_C') + \frac{1}{LC}(Ri_L + v_C) = \frac{R}{L} v_s' + \frac{1}{LC} v_s$$

Substituting $Ri_L + v_C = v_s - v_L$ then yields the desired result

$$v_L'' + \frac{R}{L} v_L' + \frac{1}{LC} v_L = v_s''$$

The coefficients on the left-hand side are exactly the same as those in Eqs. (9.19) and (9.20). Indeed, this property holds for *all* variables in any given circuit.

Parallel *CRL* circuit. The circuit in Fig. 9.18 is the dual of the series *LRC* circuit. Consequently, i_L and v_C are related to the source current i_s by the duals of Eqs. (9.19) and (9.20), namely,

$$i_L'' + \frac{1}{RC} i_L' + \frac{1}{LC} i_L = \frac{1}{LC} i_s \tag{9.21}$$

$$v_C'' + \frac{1}{RC} v_C' + \frac{1}{LC} v_C = \frac{1}{C} i_s' \tag{9.22}$$

Again, the left-hand sides have the same coefficients.

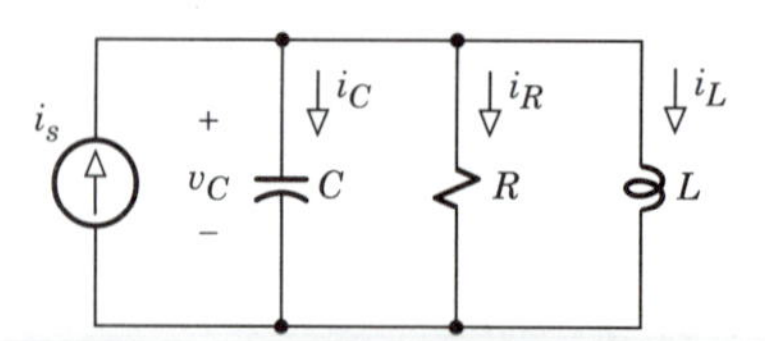

Figure 9.18 Parallel *CRL* circuit.

A review of the results so far reveals that all of our second-order differential equations take the general form

$$y'' + 2\alpha y' + \omega_0^2 y = f(t) \tag{9.23}$$

Here, α and ω_0^2 are constants incorporating element values, while $f(t)$ represents the effect of the excitation. In Eq. (9.22), for instance, we have $y = v_C$, $2\alpha = 1/RC$, $\omega_0^2 = 1/LC$, and $f(t) = i_s'/C$. We use the symbol ω_0 in Eq. (9.23) because it equals the *resonant frequency* when the circuit has resonant properties. Otherwise, ω_0^2 just stands for some combination of element values. The reason why we include the factor 2 that multiplies α in Eq. (9.23) will emerge from our subsequent discussion of second-order natural response.

But before going on, let's summarize the method we've used to obtain second-order differential circuit equations. Our method employs capacitor voltages and inductor currents as state variables for the following steps:

1. Apply Ohm's and Kirchhoff's laws to express i_C and v_L in terms of v_C, i_L, and the excitation.
2. Set $i_C = Cv_C'$ and $v_L = Li_L'$ to obtain a pair of coupled first-order differential equations.
3. Use one first-order equation to eliminate one state variable from the other equation, thereby obtaining a second-order equation for v_C or i_L.
4. When needed, repeat step 3 for the other state variable.
5. When needed, obtain the equation for any other variable using the relations from steps 3 and 4.

Two examples will further illustrate this method.

Example 9.6 *Second-Order Circuit with Two Inductors*

Suppose we need the equation relating i_2 to i_s in Fig. 9.19. There are no capacitors here, so we take the inductor currents i_1 and i_2 as the state variables and proceed as follows.

Step 1: Clearly, $i = i_s - i_1$ and $i_x = i_s - i_1 - i_2$, so KVL and Ohm's law gives

$$v_2 = R_x i_x = R_x(i_s - i_1 - i_2)$$
$$v_1 = Ri + v_2 = R(i_s - i_1) + R_x(i_s - i_1 - i_2)$$

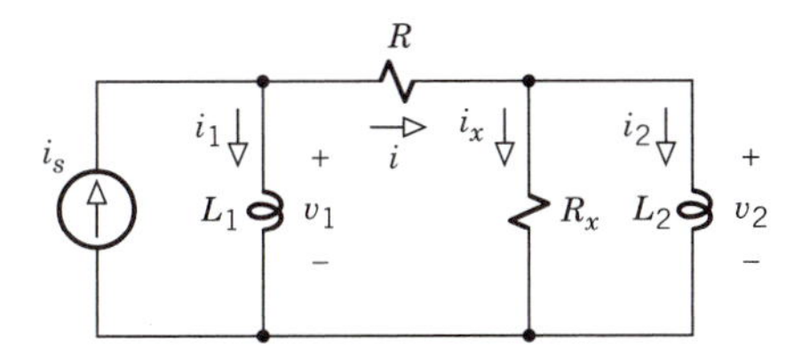

Figure 9.19 Circuit with two inductors.

Step 2: Setting $v_2 = L_2 i_2{}'$ and $v_1 = L_1 i_1{}'$, we obtain the pair of equations

$$L_2 i_2{}' = R_x i_s - R_x i_1 - R_x i_2$$
$$L_1 i_1{}' = (R + R_x) i_s - (R + R_x) i_1 - R_x i_2$$

Step 3: To eliminate i_1 and $i_1{}'$ from the second equation, we rewrite the first equation as

$$i_1 = i_s - i_2 - (L_2/R_x) i_2{}'$$

Substitution and rearrangement then yields

$$\frac{L_1 L_2}{R_x} i_2'' + \left(L_1 + \frac{R + R_x}{R_x} L_2\right) i_2' + R i_2 = L_1 i_s'$$

This result has two interesting implications.

First, if $R_x \to \infty$, then the term involving i_2'' disappears and we have

$$(L_1 + L_2) i_2' + R i_2 = L_1 i_s'$$

which is a *first*-order equation. This means that the two inductors are not independent when R_x becomes an open circuit. We check this conclusion by setting $i_x = 0$ in the circuit diagram, so $i_2 = i = i_1 - i_s$ and the energy stored by L_2 is related to the energy stored by L_1.

Second, with finite R_x, multiplying through by $R_x/L_1 L_2$ puts the second-order equation in our standard form

$$i_2'' + \left(\frac{R_x}{L_2} + \frac{R + R_x}{L_1}\right) i_2' + \frac{R R_x}{L_1 L_2} i_2 = \frac{R_x}{L_2} i_s' \qquad (9.24a)$$

Comparison with Eq. (9.23) then shows that

$$\alpha = \frac{R_x}{2L_2} + \frac{R + R_x}{2L_1} \qquad \omega_0{}^2 = \frac{R R_x}{L_1 L_2} \qquad (9.24b)$$

Example 9.7 *Second-Order Circuit with a Controlled Source*

Figure 9.20 diagrams the **phase-shift oscillator** previously studied in Example 6.10. We'll use the state variables v_C and i_L to obtain the differential equation governing the behavior of $v_{out} = K v_x$ with an arbitrary value of K.

Combining the first two steps of our procedure, we have

$$v_x = R i_L \qquad v_1 = K v_x - v_C \qquad i_1 = v_1/R$$
$$C v_C' = i_C = i_1 + i_L = (K + 1) i_L - v_C/R$$
$$L i_L' = v_L = v_1 - v_x = (K - 1) R i_L - v_C$$

Eliminating v_C then gives

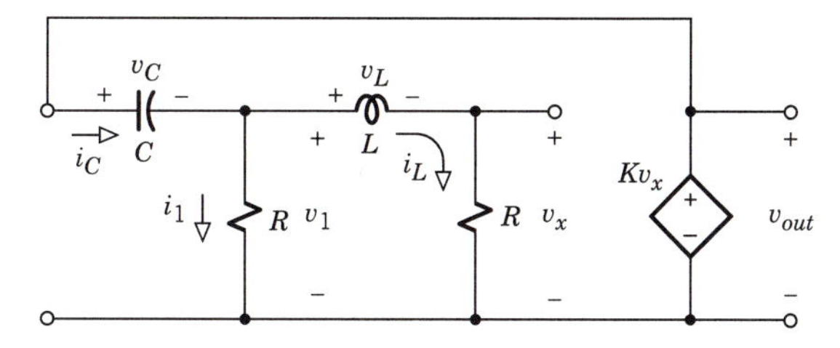

Figure 9.20 Phase-shift oscillator.

$$LCi_L'' + (RC - KRC + L/R)i_L' + 2i_L = 0$$

But $i_L = v_x/R = v_{out}/KR$, so

$$v_{out}'' + \frac{R}{L}\left(\frac{L}{R^2C} + 1 - K\right) v_{out}' + \frac{2}{LC} v_{out} = 0 \qquad (9.25a)$$

Hence,

$$\alpha = \frac{R}{2L}\left(\frac{L}{R^2C} + 1 - K\right) \qquad \omega_0^2 = \frac{2}{LC} \qquad (9.25b)$$

The right-hand side of Eq. (9.25a) has $f(t) = 0$ because the circuit has no applied excitation.

Exercise 9.6

Show that i_L in Fig. 9.21 is related to v_s by

$$i_L'' + \frac{1}{RC} i_L' + \frac{1}{LC} i_L = \frac{1}{L} v_s' + \frac{1}{LRC} v_s$$

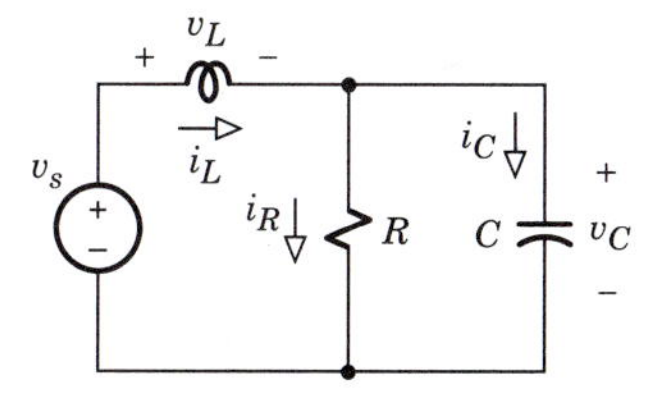

Figure 9.21

Overdamped Response

The natural response of any variable in a second-order circuit is governed by the homogeneous equation obtained from Eq. (9.23). Specifically, with $y = y_N$ and $f(t) = 0$ we have

$$y_N'' + 2\alpha y_N' + \omega_0^2 y_N = 0 \tag{9.26}$$

The coefficients α and ω_0^2 depend only upon the particular circuit elements and configuration, regardless of the variable under consideration. Hence, *all* variables in a given circuit will have the same type of natural behavior.

If we assume the usual exponential solution $y_N = Ae^{st}$, then $y_N' = sAe^{st}$ and $y_N'' = s^2Ae^{st}$. Inserting these into Eq. (9.26) we have $s^2Ae^{st} + 2\alpha sAe^{st} + \omega_0^2Ae^{st} = 0$ with $Ae^{st} \neq 0$, so factoring out Ae^{st} yields the **second-order characteristic equation**

$$s^2 + 2\alpha s + \omega_0^2 = 0 \tag{9.27}$$

Note that Eq. (9.27) is identical to Eq. (9.26) with the derivatives of y_N replaced by powers of s, i.e.,

$$y_N'' \to s^2 \qquad y_N' \to s^1 = s \qquad y_N \to s^0 = 1$$

The resulting expression on the left-hand side of Eq. (9.27) is called the **characteristic polynomial**, and a given circuit has one and only one characteristic polynomial.

As was true of first-order circuits, the characteristic equation determines the value of s for the natural response. But s now appears in a quadratic polynomial that can be factored as

$$s^2 + 2\alpha s + \omega_0^2 = (s - p_1)(s - p_2)$$

Equation (9.27) therefore has *two* solutions, $s = p_1$ and $s = p_2$. The quantities p_1 and p_2 are the **roots** of the characteristic equation, so they are known as the **characteristic values** or **eigenvalues**. Application of the quadratic formula to Eq. (9.27) yields the characteristic values

$$p_1 = -\alpha + \sqrt{\alpha^2 - \omega_0^2} \qquad p_2 = -\alpha - \sqrt{\alpha^2 - \omega_0^2} \tag{9.28}$$

The tidy appearance of these results comes from the factor of 2 multiplying α in Eqs. (9.26) and (9.27).

Since Ae^{st} satisfies Eq. (9.26) with either $s = p_1$ or $s = p_2$, we take the general solution of our second-order differential equation to be the sum

$$y_N(t) = A_1e^{p_1t} + A_2e^{p_2t} \qquad p_2 \neq p_1 \tag{9.29}$$

The two constants A_1 and A_2 reflect the fact that a second-order circuit contains two independent energy-storage elements. Hence, we'll eventually need two constants to incorporate *two initial conditions.* Meanwhile, we should give further attention to the characteristic values p_1 and p_2 because there are three different cases to consider, corresponding to $\alpha^2 > \omega_0^2$, $\alpha^2 < \omega_0^2$ and $\alpha^2 = \omega_0^2$.

A second-order circuit is said to be **overdamped** when its element values are such that

$$\alpha^2 > \omega_0^{\ 2} \tag{9.30}$$

Equation (9.28) then shows that p_1 and p_2 will be *real, unequal*, and *negative*, assuming that $\alpha > 0$. Accordingly, we introduce the time constants

$$\tau_1 = -1/p_1 \qquad \tau_2 = -1/p_2$$

Equation (9.29) then becomes

$$y_N(t) = A_1 e^{-t/\tau_1} + A_2 e^{-t/\tau_2}$$

which consists of *two decaying exponentials*. Figure 9.22 plots a typical overdamped natural response with $\tau_1 > \tau_2$, $A_1 < 0$, and $A_2 > 0$.

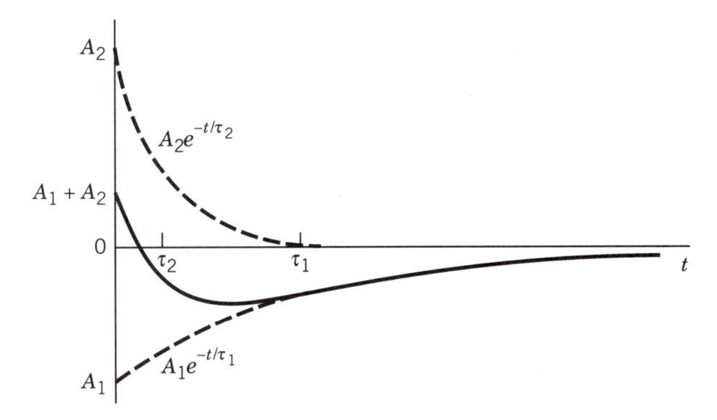

Figure 9.22 Overdamped natural response.

Overdamped behavior *always* occurs when a circuit consists of resistance and either two capacitors or two inductors, but no controlled sources. Other second-order circuits may also be overdamped, provided that the element values are such that $\alpha^2 > \omega_0^{\ 2}$.

Example 9.8 *Natural Response of a Series LRC Circuit*

Suppose the series *LRC* circuit back in Fig. 9.17 has

$$L = 0.1 \text{ H} \qquad R = 14\ \Omega \qquad C = 1/400 \text{ F}$$

We'll find the resulting form of the natural response of the current i_L.

Since the excitation equals zero by definition of natural response, we set $v_s{}' = 0$ in Eq. (9.20) to obtain the homogeneous differential equation

$$i_L'' + (R/L)i_L' + (1/LC)i_L = 0$$

Replacing derivatives of i_L with powers of s gives the characteristic equation

$$s^2 + (R/L)s + (1/LC) = 0$$

Comparison with Eq. (9.27) then shows that

$$\alpha = R/2L = 70 \qquad \omega_0^2 = 1/LC = 4000$$

so $\alpha^2 = 4900 > \omega_0^2$ and the circuit is overdamped.

From Eq. (9.28), the characteristic values are

$$p_1, p_2 = -70 \pm \sqrt{4900 - 4000} = -40, -100$$

Hence,

$$i_L(t) = A_1 e^{-40t} + A_2 e^{-100t}$$

The corresponding time constants are $\tau_1 = -1/p_1 = 25$ ms and $\tau_2 = -1/p_2 = 10$ ms.

Exercise 9.7

Let the circuit in Example 9.6 have $R_x = R$ and $L_1 = L_2 = L$.
(a) Show that this circuit is always overdamped, regardless of the values of R and L.
(b) Calculate τ_1 and τ_2 when $R = 2L$.

Underdamped Response

When a second-order circuit contains both capacitance and inductance, its characteristic equation may yield *complex* roots. Specifically, if

$$\alpha^2 < \omega_0^2 \tag{9.31}$$

then Eq. (9.28) can be rewritten as

$$p_1 = -\alpha + j\omega_d \qquad p_2 = p_1^* = -\alpha - j\omega_d \tag{9.32a}$$

where

$$\omega_d \triangleq \sqrt{\omega_0^2 - \alpha^2} \tag{9.32b}$$

Thus, p_1 and p_2 are *complex conjugates*, with real part $-\alpha$ and imaginary parts $\pm\omega_d$. We say that the circuit is **underdamped** because its natural response exhibits an *oscillatory* behavior.

To show that complex-conjugate roots lead to oscillating time functions, we first insert Eq. (9.32a) into Eq. (9.29) to get

$$y_N(t) = A_1 e^{p_1 t} + A_2 e^{p_2 t} = A_1 e^{(-\alpha + j\omega_d)t} + A_2 e^{(-\alpha - j\omega_d)t}$$
$$= e^{-\alpha t}(A_1 e^{+j\omega_d t} + A_2 e^{-j\omega_d t})$$

Next, we observe that $y_N(t)$ must be a *real* function of time, which will be true only when the constants A_1 and A_2 are also complex conjugates. We therefore assume the polar expressions

$$A_1 = |A_1| e^{j\angle A_1} \qquad A_2 = A_1^* = |A_1| e^{-j\angle A_1}$$

so that

$$y_N(t) = e^{-\alpha t}(|A_1| e^{j\angle A_1} e^{+j\omega_d t} + |A_1| e^{-j\angle A_1} e^{-j\omega_d t})$$
$$= |A_1| e^{-\alpha t}[e^{j(\omega_d t + \angle A_1)} + e^{-j(\omega_d t + \angle A_1)}]$$

Since $e^{j\phi} + e^{-j\phi} = 2 \cos \phi$ for any angle ϕ, we finally obtain

$$y_N(t) = 2|A_1| e^{-\alpha t} \cos (\omega_d t + \angle A_1) \tag{9.33}$$

All imaginary quantities have thereby disappeared, as required for a real time function, but there are still two initial-condition constants — $|A_1|$ and $\angle A_1$.

Figure 9.23 illustrates the underdamped natural behavior described by Eq. (9.33) with $\alpha > 0$. This waveform oscillates at the rate ω_d while its amplitude decays with time constant $\tau = 1/\alpha$. We therefore call ω_d the **damped frequency** and α the **damping coefficient**. A similar decaying oscillation occurs when you strike a bell.

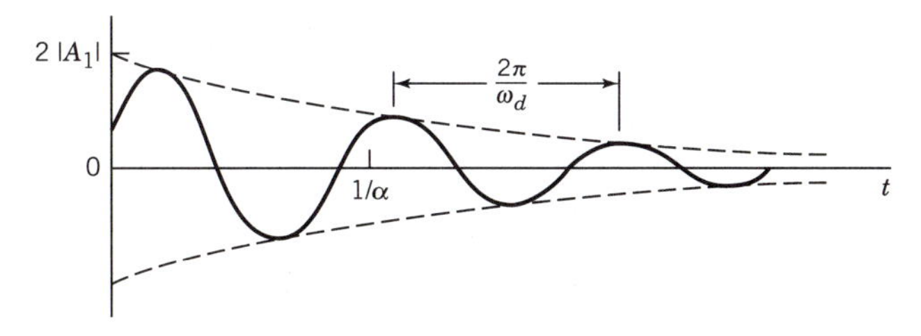

Figure 9.23 Underdamped natural response.

Physically, electrical oscillation results from the exchange of stored energy back and forth between the capacitor and inductor. But any circuit resistance continually dissipates energy and "damps" the oscillation, so the amplitude progressively decreases. If the circuit has zero resistance, then $\alpha = 0$, $\omega_d = \omega_0$, and

$$y_N(t) = 2|A_1| \cos (\omega_0 t + \angle A_1)$$

Thus, the natural response of an ideal lossless *LC* circuit is a *constant-amplitude oscillation* at the resonant frequency ω_0. However, since we cannot build a completely lossless passive circuit, practical oscillators include amplifiers to compensate for dissipation by the resistances.

Example 9.9 ***Natural Response of a Phase-Shift Oscillator***

Suppose the phase-shift oscillator in Example 9.7 has $C = 2\ \mu\text{F}$, $R = 100\ \Omega$, $L = 10$ mH, and variable gain K, so

$$\alpha = \frac{R}{2L}\left(\frac{L}{R^2C} + 1 - K\right) = 5000(2.5 - K) \qquad \omega_0^2 = \frac{2}{LC} = 10^8$$

Let's investigate the natural response of v_{out} with different values of K.

Back in Example 6.10, we used ac phasor analysis to show that this circuit would oscillate at $\omega_{osc} = \sqrt{2/LC} = 10{,}000$ if $K = 1 + (L/CR^2) = 2.5$. Now we see that taking $K = 2.5$ results in $\alpha = 0$, so $\omega_d = \omega_0 = 10{,}000$ and

$$v_{out}(t) = 2|A_1| \cos(10{,}000t + \angle A_1)$$

Thus, the amplifier exactly compensates for resistance dissipation and the natural response is a constant-amplitude sinusoid, as desired. This waveform is started by initial stored energy and lasts indefinitely.

But if the amplifier has insufficient gain, say $K = 2.0$, then

$$\alpha = 2500 \qquad \omega_d = \sqrt{10^8 - 2500^2} = 9680$$

so the characteristic values are

$$p_1, p_2 = -\alpha \pm j\omega_d = -2500 \pm j9680$$

The natural response will thus be the decaying oscillation

$$v_{out}(t) = 2|A_1|e^{-2500t} \cos(9680t + \angle A_1)$$

which rapidly dies away. On the other hand, if the amplifier has too much gain, say $K = 3.0$, then

$$\alpha = -2500 \qquad \omega_d = \sqrt{10^8 - (-2500)^2} = 9680$$

and

$$p_1, p_2 = -\alpha \pm j\omega_d = +2500 \pm j9680$$

Hence, the natural response will be the *growing* oscillation

$$v_{out}(t) = 2|A_1|e^{+2500t} \cos(9680t + \angle A_1)$$

This unstable behavior quickly drives the amplifier into nonlinear saturation.

Exercise 9.8

Let the parallel *CRL* circuit described by Eq. (9.22) have $R = 3\ \Omega$ and $C = 1/24$ F. Find the natural response of $v_C(t)$ when $L = 0.3$ H and when $L = 2$ H. (Save your results for use in Exercises 9.11 and 9.12.)

Critically Damped Response

The dividing line between overdamped and underdamped behavior is known as **critical damping**. Critically damped behavior occurs when the element values exactly satisfy the condition

$$\alpha^2 = \omega_0{}^2 \tag{9.34}$$

and hence

$$p_1 = p_2 = -\alpha \tag{9.35}$$

Consequently, we say that the roots are **repeated**.

Our previous form for $y_N(t)$ does not hold in the case of repeated roots because it would reduce to

$$A_1 e^{p_1 t} + A_2 e^{p_2 t} = (A_1 + A_2)e^{-\alpha t} = A_3 e^{-\alpha t}$$

This expression contains just one initial-condition constant and is therefore incomplete.

To obtain the general solution with critical damping, we assume the modified exponential form

$$y_N(t) = a(t)e^{-\alpha t}$$

so

$$y_N' = (a' - \alpha a)e^{-\alpha t} \qquad y_N'' = (a'' - 2\alpha a' + \alpha^2 a)e^{-\alpha t}$$

Substitution into the homogeneous differential equation with $\omega_0{}^2 = \alpha^2$ then yields

$$(a'' - 2\alpha a' + \alpha^2 a) + 2\alpha(a' - \alpha a) + \alpha^2 a = 0$$

which reduces to $a'' = 0$. Accordingly, $a(t)$ must be

$$a(t) = A_3 + A_4 t$$

where A_3 and A_4 are constants. Hence, the critically damped natural response takes the form

$$y_N(t) = A_3e^{-\alpha t} + A_4te^{-\alpha t} \tag{9.36}$$

You can confirm that Eq. (9.36) satisfies the homogeneous equation if and only if $\omega_0^2 = \alpha^2$.

When α is positive, the first term of Eq. (9.36) is just a decaying exponential. But the second term has the more interesting time variation shown in Fig. 9.24. This waveform starts with a linear rise and goes through the maximum value $A_4/\alpha e$ at $t = 1/\alpha$ before it decays toward zero.

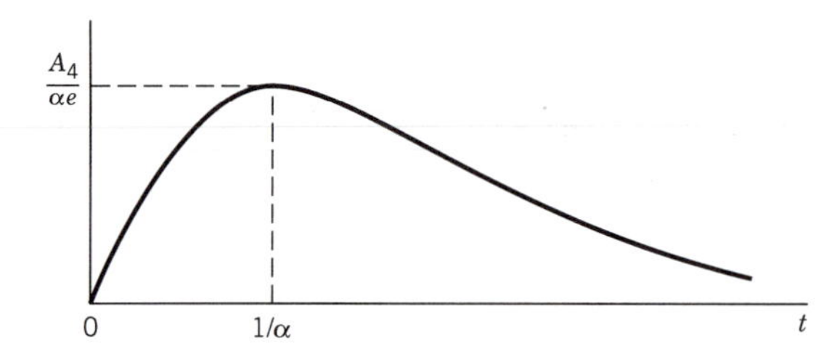

Figure 9.24 Critically damped component $A_4te^{-\alpha t}$.

To summarize our results, Table 9.1 lists the characteristic values and the form of the natural response for each of the three cases. The differences between these cases will be explored more fully in the next section.

Table 9.1 Solutions of $y_N'' + 2\alpha y_N' + \omega_0^2 y_N = 0$

Overdamped case: $\alpha^2 > \omega_0^2$

$$p_1, p_2 = -\alpha \pm \sqrt{\alpha^2 - \omega_0^2}$$
$$y_N(t) = A_1e^{p_1t} + A_2e^{p_2t}$$

Underdamped case: $\alpha^2 < \omega_0^2$

$$p_1, p_2 = -\alpha \pm j\omega_d \qquad \omega_d = \sqrt{\omega_0^2 - \alpha^2}$$
$$y_N(t) = 2|A_1|e^{-\alpha t}\cos(\omega_d t + \angle A_1)$$

Critically damped case: $\alpha^2 = \omega_0^2$

$$p_1 = p_2 = -\alpha$$
$$y_N(t) = A_3e^{-\alpha t} + A_4te^{-\alpha t}$$

Exercise 9.9

If the parallel *CRL* circuit described by Eq. (9.22) has $R = 3\ \Omega$ and $C = 1/24$ F, then what value of L results in critical damping?

9.4 SECOND-ORDER TRANSIENTS

Having studied the natural response of second-order circuits, we're now prepared to investigate second-order transients. First, we'll tackle the problem of initial conditions. Then we'll combine initial conditions with the natural response to obtain transients produced by switched dc sources, including the zero-input response and the step response as special cases.

Initial Conditions

By definition, a second-order circuit has two independent energy-storage elements. Consequently, second-order transients involve *two initial conditions*. We should therefore give special attention to the selection and calculation of initial conditions. We'll take the initial time at $t = 0$, for simplicity, and we'll again let $y(t)$ stand for any voltage or current of interest.

An obvious initial condition on $y(t)$ is its *initial value* $y(0^+)$. If we know the values of the state variables at $t = 0^-$, then $y(0^+)$ can be determined from the familiar continuity relations

$$v_C(0^+) = v_C(0^-) \qquad i_L(0^+) = i_L(0^-) \tag{9.37}$$

For the other initial condition, we draw upon the properties that

$$i_C(t) = Cv_C'(t) \qquad v_L(t) = Li_L'(t)$$

Thus, if we also know the initial values of capacitor currents and inductor voltages, then we can calculate the initial derivatives

$$v_C'(0^+) = i_C(0^+)/C \qquad i_L'(0^+) = v_L(0^+)/L \tag{9.38}$$

Accordingly, we'll take the second initial condition on $y(t)$ to be

$$y'(0^+) \triangleq \left.\frac{dy}{dt}\right|_{t=0^+} \tag{9.39}$$

which is the *initial slope*.

The values of $y(0^+)$ must include the effect of any applied input at $t = 0^+$ as well as continuity of the state variables. Taking account of these factors, a systematic procedure for finding the initial value and slope of any variable $y(t)$ goes as follows:

1. Determine the conditions at $t = 0^-$ needed to evaluate the state variables at $t = 0^+$.
2. Taking $t > 0$, write *equations* for $i_C(t)$, $v_L(t)$, and $y(t)$ in terms of the state variables and the input.
3. Use the results of steps 1 and 2 to evaluate $i_C(0^+)$, $v_L(0^+)$, and $y(0^+)$. Then calculate $v_C'(0^+)$ and $i_L'(0^+)$ via Eq. (9.38).

4. Differentiate the equation for $y(t)$ obtained in step 2, and set $t = 0^+$ to obtain $y'(0^+)$ from the values of $v_C'(0^+)$ and $i_L'(0^+)$.

An example should help clarify this procedure.

Example 9.10 ***Calculating Initial Conditions***

You are given the series *LRC* circuit in Fig. 9.25 with a switched dc source voltage

$$v_s = V_1 \qquad t < 0$$
$$\;\;\; = V_2 \qquad t > 0$$

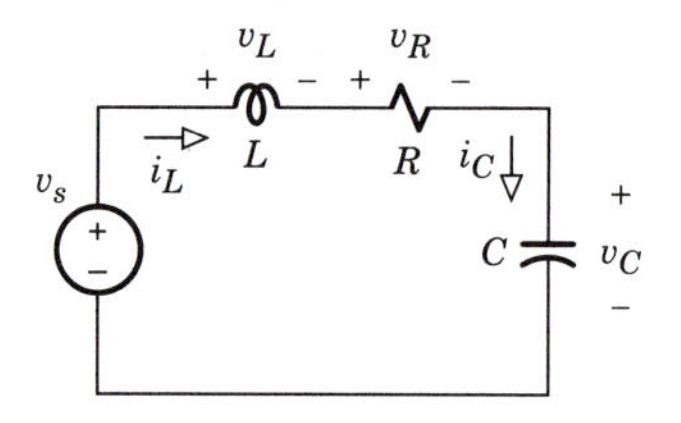

Figure 9.25 Series *LRC* circuit.

You want to find the resulting initial values and slopes of v_L, i_L, v_R, and v_C.

Step 1: You assume the dc steady state with $v_s = V_1$ for all $t < 0$. Since $v_L = 0$ and $i_C = 0$ under dc conditions, continuity requires

$$i_L(0^+) = i_L(0^-) = 0 \qquad v_C(0^+) = v_C(0^-) = V_1$$

Step 2: Noting that $v_s = V_2$ for $t > 0$, you apply Ohm's and Kirchhoff's laws to write the equations

$$v_L(t) = V_2 - Ri_L(t) - v_C(t)$$
$$v_R(t) = Ri_L(t) \qquad i_C(t) = i_L(t)$$

In each case, the right-hand side involves only the state variables and the input voltage.

Step 3: You set $t = 0^+$ in your equations to get the initial values

$$v_L(0^+) = V_2 - Ri_L(0^+) - v_C(0^+) = V_2 - V_1$$
$$v_R(0^+) = Ri_L(0^+) = 0 \qquad i_C(0^+) = i_L(0^+) = 0$$

You then calculate the slopes

$$i_L'(0^+) = \frac{v_L(0^+)}{L} = \frac{V_2 - V_1}{L} \qquad v_C'(0^+) = \frac{i_C(0^+)}{C} = 0$$

Step 4: You differentiate the equations for $v_L(t)$ and $v_R(t)$, and you set $t = 0^+$ to get the remaining slopes

$$v_L'(0^+) = -Ri_L'(0^+) - v_C'(0^+) = -R(V_2 - V_1)/L$$
$$v_R'(0^+) = Ri_L'(0^+) = R(V_2 - V_1)/L$$

Exercise 9.10

Obtain expressions for the initial value and slope of v_C, i_C, i_R, and i_L in Fig. 9.26 when

$$i_s = I_1 \qquad t < 0$$
$$= I_2 \qquad t > 0$$

(Save your results for use in Exercises 9.11 and 9.12.)

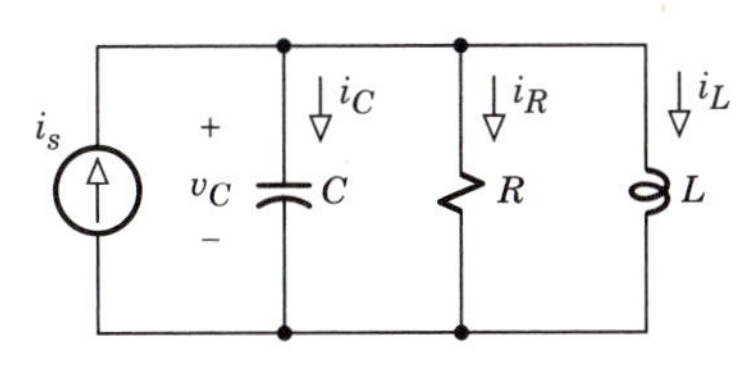

Figure 9.26

Switched DC Transients

Consider a second-order circuit with stored energy at $t = 0$ and a dc excitation for $t > 0$. Since the forcing function is constant for $t > 0$, the behavior of any variable $y(t)$ is governed by a differential equation having the form

$$y'' + 2\alpha y' + \omega_0^2 y = \omega_0^2 Y_{ss} \qquad t > 0 \qquad (9.40)$$

The constant on the right-hand side has been written as $\omega_0^2 Y_{ss}$ to emphasize that Y_{ss} equals the resulting steady-state value of $y(t)$. This interpretation follows from the fact that $y'' = y' = 0$ in the dc steady state, so Eq. (9.40) reduces to $\omega_0^2 y = \omega_0^2 Y_{ss}$ and $y(t) = Y_{ss}$.

The complete solution of Eq. (9.40) consists of the steady-state forced response Y_{ss} plus the natural response. We therefore write our general expression for switched dc transients as

$$y(t) = Y_{ss} + y_N(t) \qquad t > 0 \qquad (9.41)$$

Equation (9.41) includes the *zero-input response* if we let $Y_{ss} = 0$, equivalent to setting the right-hand side of Eq. (9.40) equal to zero. Equation (9.41) also includes the *step response* if we let the initial stored energy be zero.

As we previously learned, the circuit's characteristic values p_1 and p_2 determine the specific components of $y_N(t)$ in Eq. (9.41). However, $y_N(t)$ always contains two constants that must be evaluated to agree with the initial conditions $y(0^+)$ and $y'(0^+)$ dictated by the initial stored energy. We'll pursue these calculations separately for overdamped, underdamped, and critically damped circuits.

Case I: Overdamped Circuits. If $\alpha^2 > \omega_0^2$, then the circuit is overdamped and the characteristic values are

$$p_1 = -\alpha + \sqrt{\alpha^2 - \omega_0^2} \qquad p_2 = -\alpha - \sqrt{\alpha^2 - \omega_0^2}$$

Hence, $y_N(t) = A_1 e^{p_1 t} + A_2 e^{p_2 t}$, and

$$y(t) = Y_{ss} + A_1 e^{p_1 t} + A_2 e^{p_2 t} \qquad t > 0 \tag{9.42}$$

where p_1 and p_2 are real and unequal.

For the purpose of evaluating A_1 and A_2, we first set $t = 0^+$ in Eq. (9.42) to obtain

$$y(0^+) = Y_{ss} + A_1 + A_2$$

Next, we differentiate Eq. (9.42) and again set $t = 0^+$ to get

$$y'(0^+) = p_1 A_1 + p_2 A_2$$

Rearranging these simultaneous equations in matrix form yields

$$\begin{bmatrix} 1 & 1 \\ p_1 & p_2 \end{bmatrix} \begin{bmatrix} A_1 \\ A_2 \end{bmatrix} = \begin{bmatrix} y(0^+) - Y_{ss} \\ y'(0^+) \end{bmatrix} \tag{9.43}$$

Equation (9.43) can then be solved for A_1 and A_2.

Case II: Underdamped Circuits. If $\alpha^2 < \omega_0^2$, then the circuit is underdamped and the roots are complex conjugates, namely,

$$p_1, p_2 = -\alpha \pm j\omega_d \qquad \omega_d = \sqrt{\omega_0^2 - \alpha^2}$$

But we still have *unequal* roots, so Eq. (9.43) still holds for this case by inserting $p_1 = -\alpha + j\omega_d$ and $p_2 = -\alpha + j\omega_d$.

However, the initial-condition constants are complex conjugates, and we only need to evaluate A_1 because we can express the natural response in the form $y_N(t) = 2|A_1|e^{-\alpha t} \cos(\omega_d t + \angle A_1)$. Thus,

$$y(t) = Y_{ss} + 2|A_1|e^{-\alpha t} \cos(\omega_d t + \angle A_1) \qquad t > 0 \tag{9.44}$$

The corresponding requirement that $A_2 = A_1^*$ is guaranteed by Eq. (9.43) when $p_2 = p_1^*$.

Case III: Critically Damped Circuits. If $\alpha^2 = \omega_0 2$, then the circuit is critically damped and

$$p_1 = p_2 = -\alpha$$

Equation (9.43) clearly does not hold in this case because its characteristic determinant is $\Delta = p_2 - p_1$, which equals zero when the roots are repeated.

Instead, we recall that a critically damped circuit has $y_N(t) = A_3 e^{-\alpha t} + A_4 t e^{-\alpha t}$, so

$$y(t) = Y_{ss} + A_3 e^{-\alpha t} + A_4 t e^{-\alpha t} \qquad t > 0 \tag{9.45}$$

Differentiation then yields

$$y'(t) = -\alpha A_3 e^{-\alpha t} + A_4 e^{-\alpha t} - \alpha A_4 t e^{-\alpha t}$$

Hence, the constants A_3 and A_4 must be

$$A_3 = y(0^+) - Y_{ss} \qquad A_4 = y'(0^+) + \alpha A_3 \tag{9.46}$$

which follow directly by setting $t = 0^+$ in $y(t)$ and $y'(t)$.

Example 9.11 ***Underdamped Zero-Input Response***

Let the series *LRC* circuit in Fig. 9.25 have $L = 0.1$ H, $R = 5\ \Omega$, and $C = 1/640$ F. Then $\alpha = R/2L = 25$, $\omega_0{}^2 = 1/LC = 6400$, and the characteristic values are

$$p_1, p_2 = -25 \pm j76$$

We'll find the transient inductor current i_L for t > 0 when

$$\begin{aligned} v_s(t) &= 30 \text{ V} \qquad t < 0 \\ &= 0 \text{ V} \qquad t > 0 \end{aligned}$$

We're thus dealing with an underdamped zero-input response.

For the initial conditions on i_L, we draw upon the results from Example 9.10 with $V_1 = 30$ V and $V_2 = 0$. Hence,

$$i_L(0^+) = 0 \qquad i_L'(0^+) = -30/L = -300 \text{ A/s}$$

The negative initial slope reflects the fact that the capacitor initially discharges back through the inductor. Eventually, of course, i_L goes to the steady-state value $I_{ss} = 0$.

After inserting our numerical values into Eq. (9.43), we have the complex matrix equation

$$\begin{bmatrix} 1 & 1 \\ -25 + j76 & -25 - j76 \end{bmatrix} \begin{bmatrix} A_1 \\ A_2 \end{bmatrix} = \begin{bmatrix} i_L(0^+) - I_{ss} \\ i_L'(0^+) \end{bmatrix} = \begin{bmatrix} 0 \\ -300 \end{bmatrix}$$

The relevant determinants are then found to be $\Delta = -j152$ and $\Delta_1 = 300$, so

$$A_1 = 300/(-j152) = 1.974 \text{ A } \underline{/90°}$$

Equation (9.44) therefore becomes

$$i_L(t) = 3.95e^{-25t} \cos (76t + 90°) \text{ A} \qquad t > 0$$

The waveform shown in Fig. 9.27 illustrates the oscillatory behavior known as **ringing** that always appears in underdamped second-order transients.

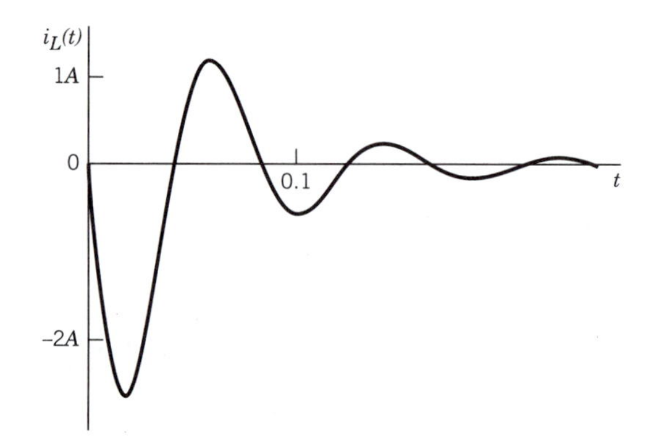

Figure 9.27 Underdamped zero-input response waveform.

Example 9.12 *Step Response with Variable Damping*

For our last example, let's investigate the step response of a series *LRC* circuit. Specifically, we'll find the capacitor voltage $v_C(t)$ produced by

$$v_s(t) = 30u(t) \text{ V} = \begin{cases} 0 & t < 0 \\ 30 \text{ V} & t > 0 \end{cases}$$

We'll let $L = 0.1$ H and $C = 1/640$ F, so

$$\alpha = R/2L = 5R \qquad \omega_0^2 = 1/LC = 6400$$

and we'll explore the damping effect by taking three different values of R.

The initial conditions on v_C can be found using the results of Example 9.10 with $V_1 = 0$ and $V_2 = 30$ V, from which

$$v_C(0^+) = 0 \qquad v_C'(0^+) = 0$$

As $t \rightarrow \infty$, v_C approaches the dc steady-state value $V_{ss} = 30$ V.

Overdamped Response. If we let $R = 34\ \Omega$, then $\alpha = 170$ and

$$p_1, p_2 = -170 \pm \sqrt{170^2 - 6400} = -20, -320$$

Substituting values into Eq. (9.43) we have

$$\begin{bmatrix} 1 & 1 \\ -20 & -320 \end{bmatrix} \begin{bmatrix} A_1 \\ A_2 \end{bmatrix} = \begin{bmatrix} v_C(0^+) - V_{ss} \\ v_C'(0^+) \end{bmatrix} = \begin{bmatrix} -30 \\ 0 \end{bmatrix}$$

Simultaneous solution gives $A_1 = -32$ and $A_2 = 2$, so Eq. (9.42) yields

$$v_C(t) = 30 - 32e^{-20t} + 2e^{-320t}\ \text{V} \qquad t > 0$$

Underdamped Response. If we let $R = 5\ \Omega$, then $\alpha = 25$ and

$$p_1, p_2 = -25 \pm j\sqrt{6400 - 25^2} = -25 \pm j76$$

Thus,

$$\begin{bmatrix} 1 & 1 \\ -25 + j76 & -25 - j76 \end{bmatrix} \begin{bmatrix} A_1 \\ A_2 \end{bmatrix} = \begin{bmatrix} -30 \\ 0 \end{bmatrix}$$

from which

$$A_1 = (750 + j2280)/(-j152) = 15.8\ \underline{/161.8^\circ}$$

Equation (9.44) then yields

$$v_C(t) = 30 + 31.6e^{-25t} \cos(76t + 161.8^\circ)\ \text{V} \qquad t > 0$$

Critically Damped Response. Critical damping occurs when $\alpha^2 = 25R^2 = \omega_0^2 = 6400$, so $R = 6400/25 = 16\ \Omega$ and $\alpha = 80$. Hence, from Eq. (9.46),

$$A_3 = v_C(0^+) - V_{ss} = -30 \qquad A_4 = v_C'(0^+) + \alpha A_3 = -2400$$

so Eq. (9.45) yields

$$v_C(t) = 30 - 30e^{-80t} - 2400te^{-80t}\ \text{V} \qquad t > 0$$

The waveforms from these three cases are plotted together in Fig. 9.28 to facilitate comparisons. All three waveforms start at zero with zero initial slope, and they all approach the steady-state value of 30 V. Not surprisingly, the overdamped response takes the longest time to reach the steady state. The underdamped response has the most rapid initial rise, but the oscillatory ringing produces **overshoot** above the steady-state value. Between these ex-

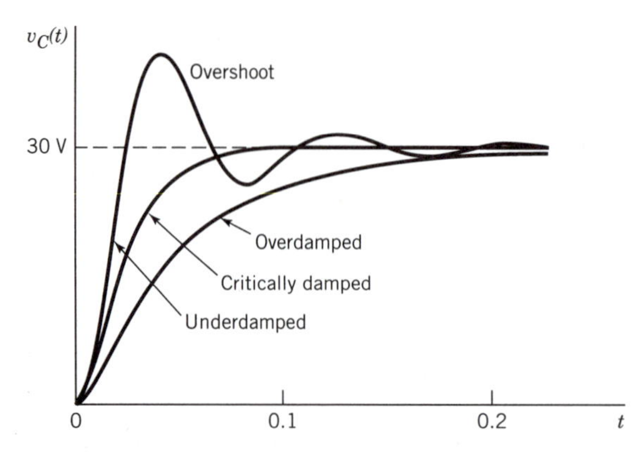

Figure 9.28 Second-order step response waveforms.

tremes, the critically damped response takes the least time to reach the steady state without overshoot.

Incidentally, critical damping is just what you want in an automobile's suspension system, whose dynamic behavior acts essentially like that of a second-order circuit. If the shock absorbers are too soft, meaning insufficient damping, then the riders get bounced up and down each time the tires hit a bump in the road. But if the shock absorbers are too stiff, meaning too much damping, then the chassis may not reach its steady-state position before the next bump.

Exercise 9.11

Use your results fromm Exercises 9.8 and 9.10 to find the transient response of i_L in Fig. 9.26 when $R = 3\ \Omega$, $C = 1/24$ F, and

$$\begin{aligned} i_s &= 2\text{ A} \qquad & t < 0 \\ &= 0 & t > 0 \end{aligned}$$

Do the calculations with $L = 2$ H and $L = 0.3$ H.

Exercise 9.12

Rework Exercise 9.11 with $L = 0.15$ H and

$$\begin{aligned} i_s &= 0 \qquad & t < 0 \\ &= 2\text{ A} & t > 0 \end{aligned}$$

PROBLEMS

Section 9.1 First-Order Transients

(See also PSpice problems B.33-B.37.)

9.1* The switch in Fig. P9.1 has been in the upper position a long time before $t = 0$. Find $v(t)$ and $i(t)$ for $t > 0$ taking element X to be a 5-μF capacitor.

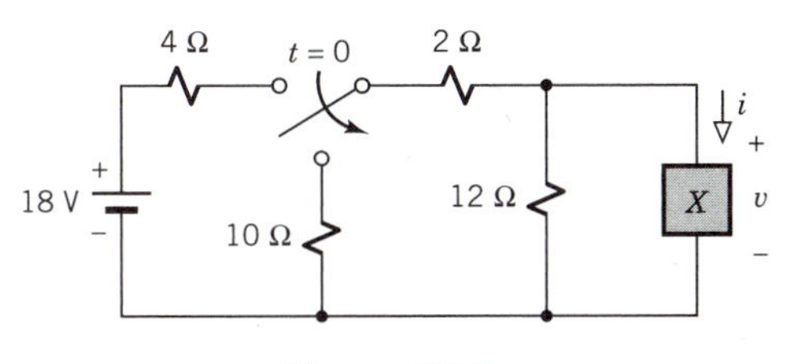

Figure P9.1

9.2 Do Problem 9.1 taking element X to be a 0.3-H inductor.

9.3 The switch in Fig. P9.3 has been in the upper position a long time before $t = 0$. Find $v(t)$, $i(t)$, and $v_1(t)$ for $t > 0$ taking element X to be a 30-μF capacitor.

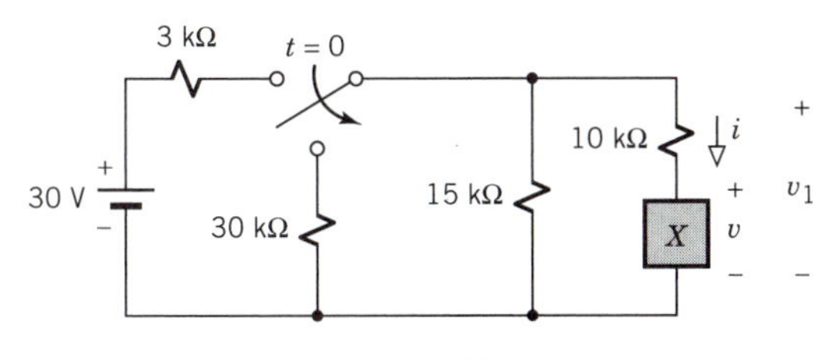

Figure P9.3

9.4 Do Problem 9.3 taking element X to be a 0.2-H inductor.

9.5 Given a 12-V battery with $R_s = 0.6\ \Omega$, design a circuit using a switch, inductor, and resistor to produce a 1000-V voltage spike with $\tau = 20\ \mu$s across a 100-Ω resistance.

9.6 Given a 12-V battery with $R_s = 0.6\ \Omega$, design a circuit using a switch, capacitor, and resistor to produce a 30-A current spike with $\tau = 10\ \mu$s through a 0.1-Ω resistance.

9.7* Let Fig. P9.7 have $R_1 = 25\ \Omega$, $L = 0.1$ H, $R_2 = 1$ kΩ, and $C = 4\ \mu$F. Find $v_C(t)$ and $v_L(t)$ for $t > 0$ when there is no initial stored energy. Hint: The *RL* and *RC* branches act independently.

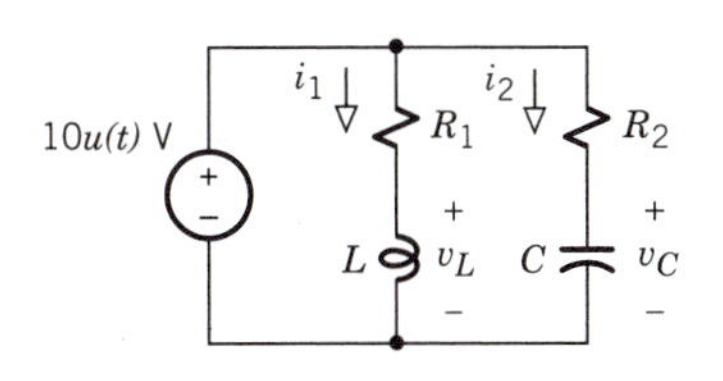

Figure P9.7

9.8 Let Fig. P9.7 have $R_1 = R_2 = 2$ kΩ, $L = 80$ mH, and $C = 0.5\ \mu$F. Find $i_1(t)$ and $i_2(t)$ for $t > 0$ when there is no initial stored energy. Hint: The *RL* and *RC* branches act independently.

9.9 Let Fig. P9.7 have $R_1 = 500\ \Omega$, $L = 50$ mH, $R_2 = 250\ \Omega$, and $C = 0.4\ \mu$F. Find $i_1(t)$ and $i_2(t)$ for $t > 0$ when there is no initial stored energy. Hint: The *RL* and *RC* branches act independently.

9.10 The switch in Fig. P9.10 has been at the upper position a long time before $t = 0$, when it goes to the lower position. The switch returns to the upper position at $t_0 = 1$ s. The resistance values in are in ohms, and element X is a 2-H inductor. (a) Find $i(t)$ for $0 < t < t_0$, and determine when $i(t) = 0$. (b) Find $i(t)$ for $t > t_0$, and sketch the waveform for $0^- \leq t \leq 3$ s.

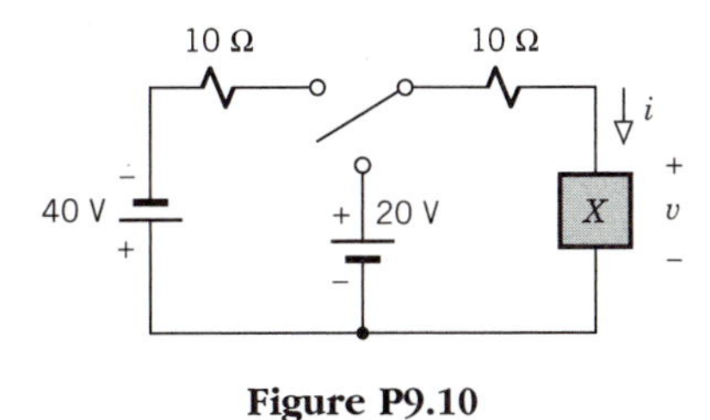

Figure P9.10

9.11 The switch in Fig. P9.10 has been at the upper position a long time before $t = 0$, when it goes to the lower position. The switch returns to the upper position at $t_0 = 1$ s. The resistance values in are in kilohms, and element X is a 20-μF capacitor. (a) Find $v(t)$ for $0 < t < t_0$, and determine when $v(t) = 0$. (b) Find $v(t)$ for $t > t_0$, and sketch the resulting waveform for $0^- \leq t \leq 3$ s.

9.12* The switch in Fig. P9.12 has been at the upper position a long time before $t = 0$, when it goes to the lower position. The switch returns to the upper position at $t_0 = 1$ s. The resistance values in are in ohms, and element X is a 4-H inductor. (a) Find $i(t)$ for $0 < t < t_0$, and determine when $i(t) = 0$. (b) Find $i(t)$ for $t > t_0$, and sketch the waveform for $0^- \leq t \leq 3$ s.

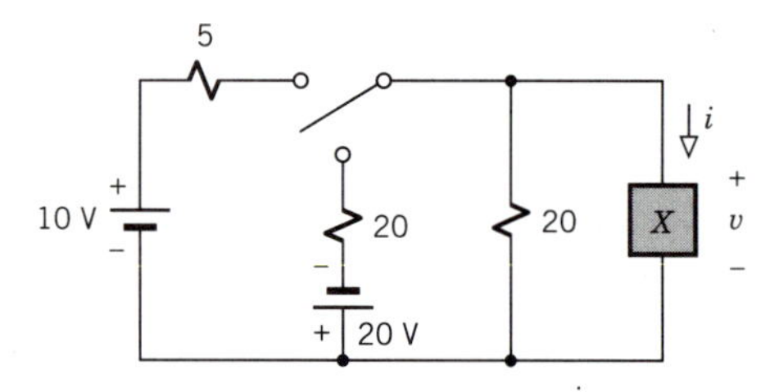

Figure P9.12

9.13 The switch in Fig. P9.12 has been at the upper position a long time before $t = 0$, when it goes to the lower position. The switch returns to the upper position at $t_0 = 1$ s. The resistance values in are in kilohms, and element X is a 50-μF capacitor. (a) Find $v(t)$ for $0 < t < t_0$, and determine when $v(t) = 0$. Find $v(t)$ for $t > t_0$, and sketch the waveform for $0^- \leq t \leq 3$ s.

9.14 The voltage applied to a series RL network is a 10-V rectangular pulse like Fig. 9.11 (p. 414). By considering $i_L(t)$, find and sketch $v_L(t)$ for $t \geq 0^-$ when there is no initial stored energy and $L/R = 2D$.

9.15 Do Problem 9.14 with $L/R = 0.5D$.

9.16 Suppose the double-pulse voltage waveform in Fig. P9.16 is applied to a series RC network with $\tau = 0.5D$. Find $v_C(D)$, $v_C(2D)$, and $v_C(3D)$, and sketch $v_C(t)$.

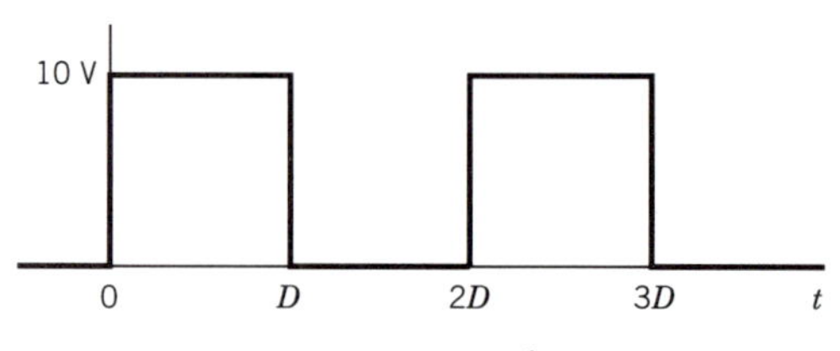

Figure P9.16

9.17 Do Problem 9.16 with $\tau = D$.

9.18 Suppose the double-pulse voltage waveform in Fig. P9.16 is applied to a series RL network with $\tau = D$. By considering $i_L(t)$, find $v_L(t)$ at $t = 0^+, D^-, D^+, 2D^-, 2D^+, 3D^-$, and $3D^+$.

9.19 Do Problem 9.18 with $\tau = 0.5D$.

Section 9.2 Switched AC Transients

9.20* Find $i(t)$ and $v(t)$ for $t > 0$ in Fig. P9.20 when $R = 7\ \Omega$ and

$$v_s(t) = 50 \cos 24t \text{ V} \qquad t < 0$$
$$= 50 \cos (24t + 90°) \text{ V} \qquad t > 0$$

which is a **phase-modulated (PM) signal**.

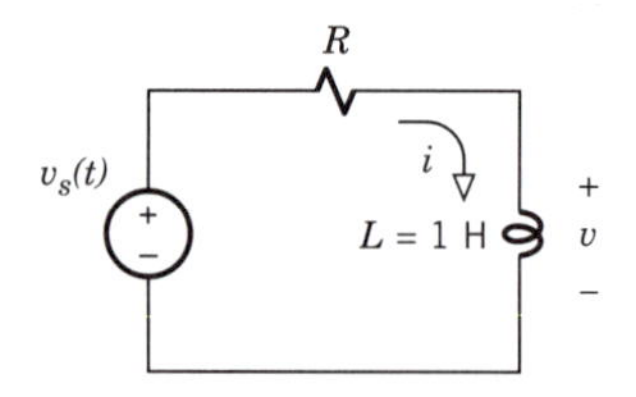

Figure P9.20

9.21 Find $i(t)$ and $v(t)$ for $t > 0$ in Fig. P9.20 when $R = 12\ \Omega$ and

$$v_s(t) = 30 \cos 9t \text{ V} \qquad t < 0$$
$$= 30 \cos 16t \text{ V} \qquad t > 0$$

which is a **frequency-modulated (FM) signal**.

9.22 Find $v(t)$ for $t > 0$ in Fig. P9.22 when $C = 1/300$ F and

$$v_s(t) = 4 \text{ V} \qquad t < 0$$
$$= 11 \cos (5t - 90°) \text{ V} \qquad t > 0$$

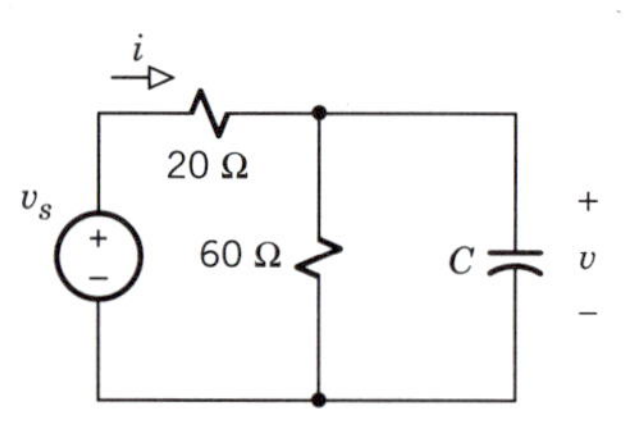

Figure P9.22

9.23 Find $i(t)$ for $t > 0$ in Fig. P9.22 when $C = 1/60$ F and

$$v_s(t) = -16 \text{ V} \qquad t < 0$$
$$= 20 \cos 2t \text{ V} \qquad t > 0$$

9.24 Find $i(t)$ for $t > 0$ in Fig. P9.24 when

$$i_s(t) = 4 \text{ A} \qquad t < 0$$
$$= 4 \cos 10t \text{ A} \qquad t > 0$$

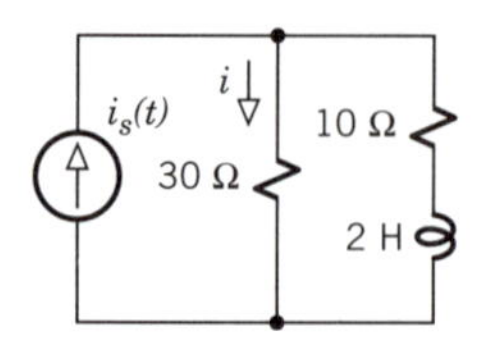

Figure P9.24

9.25 Find $i(t)$ for $t > 0$ in Fig. P9.24 when

$$i_s(t) = 3 \cos 10t \text{ A} \qquad t < 0$$
$$= 8 \cos 10t \text{ A} \qquad t > 0$$

Section 9.3 Second-Order Natural Response

9.26 Derive Eqs. (9.21) and (9.22) directly from Fig. P9.26.

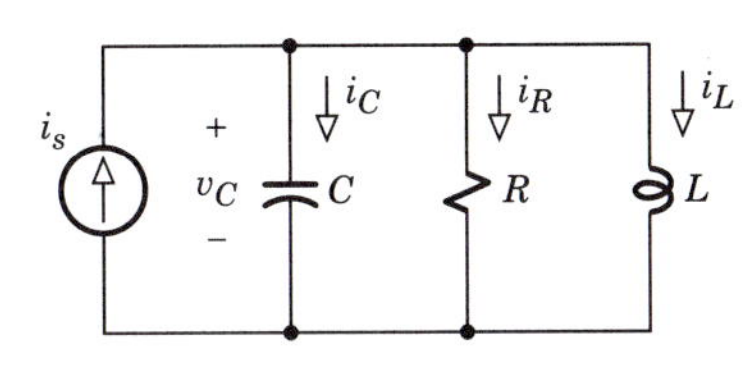

Figure P9.26

9.27* Obtain the differential equation relating i_L to v_s in Fig. P9.27.

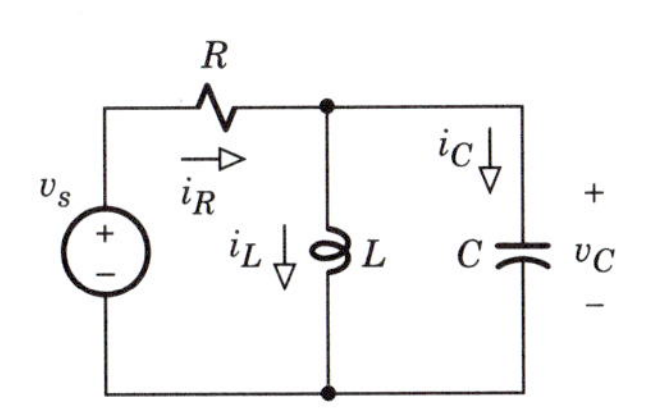

Figure P9.27

9.28 Obtain the differential equation relating v_C to v_s in Fig. P9.27.

9.29 Let $C = 1/12$ F, $R = 4\ \Omega$, and $L = 2$ H in Fig. P9.26. Use Eqs. (9.21) and (9.22) to obtain the differential equation relating i_C to i_s.

9.30 Let $C = 1/10$ F, $R = 2\ \Omega$, $L = 2$ H, and $K = 3$ in Fig. 9.20 (p. 427). Obtain the differential equation governing the behavior of v_1.

9.31 Let $L_1 = L_2 = 1$ H, $R = 4\ \Omega$, and $R_x = 2\ \Omega$ in Fig. P9.31. Obtain the differential equation relating i to i_s.

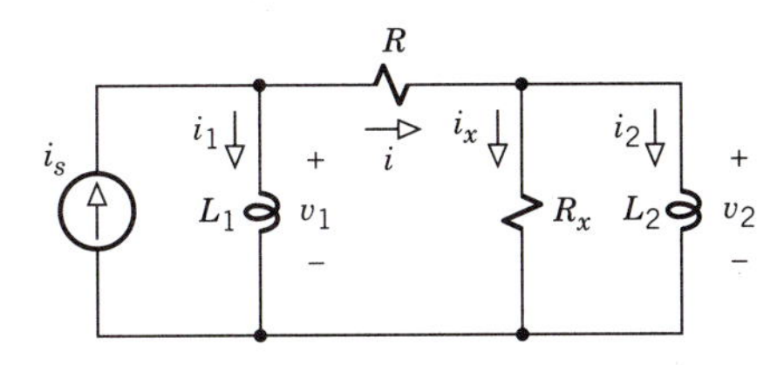

Figure P9.31

9.32* Obtain the solution of

$$2y_N'' + 28y_N' + 80y_N = 0$$

9.33 Obtain the solution of

$$4y_N'' + 4y_N' + y_N = 0$$

9.34 Obtain the solution of

$$3y_N'' + 30y_N' + 150y_N = 0$$

9.35 Obtain the solution of

$$5y_N'' + 160y_N' + 2000y_N = 0$$

9.36 Obtain the solution of

$$y_N'' + 2y_N' - 24y_N = 0$$

9.37 Obtain the solution of

$$y_N'' - 10y_N' + 16y_N = 0$$

9.38 Obtain the solution of

$$y_N'' - 6y_N' + 45y_N = 0$$

9.39 Obtain the solution of

$$y_N'' - 8y_N' - 9y_N = 0$$

9.40 Let $p_1 = -\alpha + j\omega_d$ and $p_2 = p_1^*$ in Eq. (9.29). By writing $A_1 = |A_1|e^{j\phi_1}$ and $A_2 = |A_1|e^{j\phi_2}$, confirm that the imaginary part of $y_N(t)$ equals zero for all t if and only if $A_2 = A_1^*$.

9.41 Show that critical damping in a series *LRC* circuit corresponds to $Q_{ser} = 1/2$, where Q_{ser} is the quality factor for series resonance.

9.42 Show that critical damping in a parallel *CRL* circuit corresponds to $Q_{par} = 1/2$, where Q_{par} is the quality factor for parallel resonance.

9.43 Let $y_N(t) = A \cos(\omega t + \phi)$. Show by substitution that this is the solution of Eq. (9.26) if and only if $\alpha = 0$ and $\omega = \omega_0$.

9.44 Verify the location and value of the maximum of $A_4 te^{-\alpha t}$ as shown in Fig. 9.24 (p. 434).

9.45 Confirm by substitution that the critically damped response in Eq. (9.36) is the solution of Eq. (9.26) if and only if $\alpha^2 = \omega_0^2$.

Section 9.4 Second-Order Transients

(See also PSpice problems B.38–B.42.)

9.46* Find the zero-input response of i_L in Fig. P9.46 when $i_L(0^-) = 3$ A and $v_C(0^-) = 0$. Let $L = 5$ H, $R = 8\ \Omega$, and $C = 1/80$ F, so $\alpha = 5$ and $\omega_0^2 = 16$.

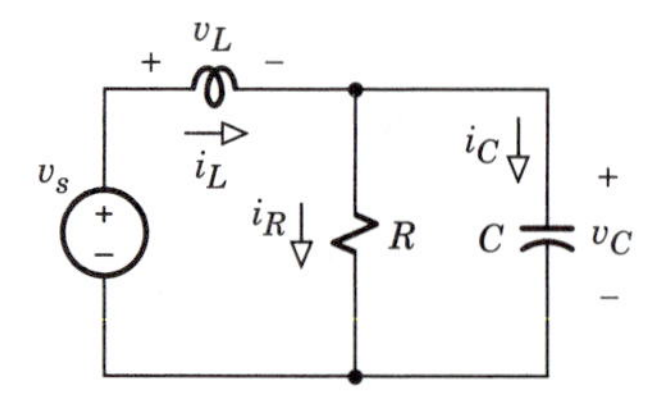

Figure P9.46

9.47 Find the zero-input response of v_C in Fig. P9.26 when $v_C(0^-) = 0$ and $i_L(0^-) = 3$ A. Let $C = 1/200$ F, $R = 25\ \Omega$, and $L = 8$ H, so $\alpha = 4$ and $\omega_0^2 = 25$.

9.48 Find the zero-input response of i_L in Fig. P9.48 when $v_s(t) = 12$ V for $t < 0$. Let $L = 1/2$ H, $R = 20\ \Omega$, and $C = 1/200$ F, so $\alpha = 20$ and $\omega_0^2 = 400$.

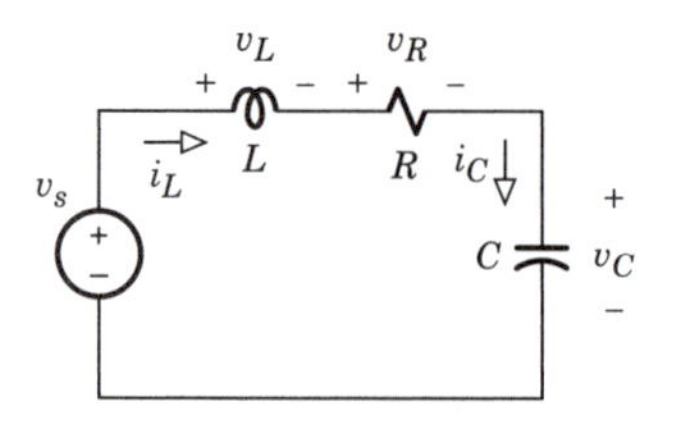

Figure P9.48

9.49 Find the zero-input response of v_C in Fig. P9.48 when $v_s(t) = 8$ V for $t < 0$. Let $L = 1/4$ H, $R = 6\ \Omega$, and $C = 1/100$ F, so $\alpha = 12$ and $\omega_0^2 = 400$.

9.50 Find the zero-input response of i_1 in Fig. P9.31 when $i_s(t) = 8$ A for $t < 0$. Let $L_1 = 6$ H, $R = 12\ \Omega$, $R_x = 3\ \Omega$, and $L_2 = 2$ H, so $\alpha = 2$ and $\omega_0^2 = 3$.

9.51 Consider the phase-shift oscillator in Fig. 9.20 (p. 427). Let $C = 2\ \mu$F, $R = 100\ \Omega$, $L = 10$ mH, and $K = 2.5$, so $\alpha = 0$ and $\omega_0^2 = 10^8$. Find $v_{out}(t)$ for $t > 0$, given that $v_C(0^-) = 4$ V and $i_L(0^-) = 0$.

9.52* Find the step response of i_L in Fig. P9.46 when $v_s(t) = 20u(t)$ V. Let $L = 2$ H, $R = 5\ \Omega$, and $C = 1/50$ F, so $\alpha = 5$ and $\omega_0^2 = 25$.

9.53 Find the step response of i_L in Fig. P9.26 when $i_s(t) = 4u(t)$ A. Let $C = 1/60$ F, $R = 5\ \Omega$, and $L = 3$ H, so $\alpha = 6$ and $\omega_0^2 = 20$.

9.54 Find the step response of v_C in Fig. P9.26 when $i_s(t) = 3u(t)$ A. Let $C = 1/40$ F, $R = 10\ \Omega$, and $L = 2$ H, so $\alpha = 2$ and $\omega_0^2 = 20$.

9.55 Find the step response of i_L in Fig. P9.48 when $v_s(t) = 30u(t)$ V. Let $L = 1/2$ H, $R = 2\ \Omega$, and $C = 1/20$ F, so $\alpha = 2$ and $\omega_0^2 = 40$.

9.56 Find the step response of i_1 in Fig. P9.31 when $i_s(t) = 4u(t)$ A. Let $L_1 = 4$ H, $R = 20\ \Omega$, $R_x = 12\ \Omega$, and $L_2 = 3$ H, so $\alpha = 6$ and $\omega_0^2 = 20$.

9.57* Find v_C for $t > 0$ in Fig. P9.26 when

$$\begin{aligned} i_s(t) &= -3\text{ A} \qquad t < 0 \\ &= 3\text{ A} \qquad t > 0 \end{aligned}$$

Let $C = 1/100$ F, $R = 10\ \Omega$, and $L = 2$ H, so $\alpha = 5$ and $\omega_0^2 = 50$.

9.58 Find v_C for $t > 0$ in Fig. P9.48 when

$$\begin{aligned} v_s(t) &= 12\text{ V} \qquad t < 0 \\ &= -12\text{ V} \qquad t > 0 \end{aligned}$$

Let $L = 4$ H, $R = 56\ \Omega$, and C $1/2500$ F, so $\alpha = 7$ and $\omega_0^2 = 625$.

9.59 Find i_L for $t > 0$ in Fig. P9.48 when

$$\begin{aligned} v_s(t) &= 40\text{ V} \qquad t < 0 \\ &= -20\text{ V} \qquad t > 0 \end{aligned}$$

Let $L = 2$ H, $R = 24\ \Omega$, and $C = 1/22$ F, so $\alpha = 6$ and $\omega_0^2 = 11$.

9.60 Find i_L for $t > 0$ in Fig. P9.46 when

$$\begin{aligned} v_s(t) &= -10\text{ V} \qquad t < 0 \\ &= 30\text{ V} \qquad t > 0 \end{aligned}$$

Let $L = 2$ H, $R = 5\ \Omega$, and $C = 1/50$ F, so $\alpha = 5$ and $\omega_0^2 = 25$.

CHAPTER 10

Network Functions and s-Domain Analysis

Two main threads run throughout circuit analysis, namely, steady-state response and transients. This chapter begins to weave those threads together by developing the related concepts of *complex frequency, generalized impedance,* and *network functions.* These concepts ultimately lead us to *s-domain analysis,* which links natural response and forced response.

We'll show that a network function contains exactly the same information as a circuit's differential equation. We'll also show how network functions can be found with the help of generalized impedance, thereby eliminating the chore of formulating differential equations.

Having obtained the network function for a circuit, we can then apply *s*-domain analysis to predict the behavior of both the forced response and the natural response. For these reasons, the s domain will play a major role in our further investigation of dynamic circuits.

Objectives

After studying this chapter and working the exercises, you should be able to do each of the following:

1. Identify the types of waveforms that can be expressed in terms of complex frequency (Section 10.1).
2. Use generalized impedance to calculate the forced response produced by an input with a complex-frequency waveform (Section 10.1).
3. Apply impedance analysis to obtain the network function relating the excitation and response in a given circuit (Section 10.2).
4. Write the differential circuit equation that corresponds to a given network function, and vice versa (Section 10.2).
5. Find the network function of a circuit with mutual inductance (Section 10.3).†
6. Determine the poles and zeros of a network function, state their significance, and draw the pole-zero pattern (Section 10.4).
7. Use s-plane vectors to evaluate a network function at a particular complex frequency (Section 10.4).
8. Given the poles of a circuit, find the modes in the natural response and determine if the circuit is stable or unstable (Section 10.4).
9. Use magnitude and frequency scaling to simplify network analysis (Section 10.5).†

10.1 COMPLEX FREQUENCY AND GENERALIZED IMPEDANCE

When we investigated forced response in previous chapters, we limited our consideration to circuits with either a constant or sinusoidal excitation. Now we'll expand our scope to include the broader class of input waveforms characterized by *complex frequency*. After studying such waveforms, we'll show that the forced response they produce is easily found by generalizing impedance for complex frequency.

Complex Frequency

Complex frequency pertains to oscillating voltages or currents whose amplitudes may have an exponential variation rather than being constant. We write such waveforms in general as

$$x(t) = X_m e^{\sigma t} \cos(\omega t + \phi_x) \tag{10.1}$$

which consists of an ordinary sinusoid multiplied by $e^{\sigma t}$. The symbol σ (sigma) stands for any real number — positive, negative, or zero — so the combined amplitude term $X_m e^{\sigma t}$ may increase or decrease with time or it may remain constant. Our objective is to express Eq. (10.1) in a form better suited to tasks in circuit analysis.

We begin by noting that $X_m e^{\sigma t}$ is a real quantity and that $\cos(\omega t + \phi_x) = \text{Re}[e^{j(\omega t+\phi_x)}]$. Thus, we can rewrite Eq. (10.1) as

$$\begin{aligned} x(t) &= X_m e^{\sigma t}\text{Re}[e^{j(\omega t+\phi_x)}] = \text{Re}[X_m e^{\sigma t} e^{j(\omega t+\phi_x)}] \\ &= \text{Re}[(X_m e^{j\phi_x})e^{(\sigma+j\omega)t}] \end{aligned}$$

Now we have an expression that looks like the phasor representation of a sinusoid, except that $j\omega$ has been replaced by the complex quantity $\sigma + j\omega$.

To bring this interpretation out more clearly, we define the **complex frequency**

$$s \triangleq \sigma + j\omega \tag{10.2}$$

We also reintroduce our phasor notation

$$\underline{X} \triangleq X_m \underline{/\phi_x} = X_m e^{j\phi_x}$$

Equation (10.1) thereby becomes

$$x(t) = \text{Re}[\underline{X}e^{st}] \tag{10.3}$$

The joint use of phasor notation and complex frequency thus leads to the compact expression in Eq. (10.3) for the waveform $X_m e^{\sigma t}\cos(\omega t + \phi_x)$.

An additional benefit of complex frequency comes from the fact that Eq. (10.3) actually incorporates several other possible waveforms, depending upon the values of σ and ω. For if $\sigma = 0$, then $s = j\omega$ and

$$x(t) = \text{Re}[\underline{X}e^{j\omega t}] = X_m \cos(\omega t + \phi_x) \tag{10.4a}$$

which is a constant-amplitude sinusoid or *ac waveform.* But if $\omega = 0$, then $s = \sigma$ and

$$x(t) = \text{Re}[\underline{X}e^{\sigma t}] = X_m \cos\phi_x\, e^{\sigma t} \tag{10.4b}$$

which is a nonoscillating *exponential waveform.* Going even further, if $\sigma = 0$ and $\omega = 0$, then

$$x(t) = \text{Re}[\underline{X}e^{0t}] = X_m \cos\phi_x \tag{10.4c}$$

which is just a constant or *dc waveform.*

Equations (10.4b) and (10.4c) deserve additional comment because, when $\omega = 0$, the only role of the constant $\cos\phi_x$ is to establish the *sign* of $x(t)$. Since X_m is always a positive quantity, by convention, we either take $\phi_x = 0$ to get $X_m \cos\phi_x = X_m > 0$ or we take $\phi_x = \pm 180°$ to get $X_m \cos\phi_x = -X_m < 0$.

Example 10.1 ***A Complex-Frequency Waveform***

Suppose you want the complex-frequency representation for the current waveform

$$i(t) = 200e^{-5t} \cos(30t + 60°) \text{ mA}$$

Examining this expression term by term, you see that $I_m = 200$ mA, $\sigma = -5$, $\omega = 30$, and $\phi_i = 60°$. Hence, you can write $i(t) = \text{Re}[\underline{I}e^{st}]$ by taking

$$\underline{I} = 200 \text{ mA} \underline{/60°} \qquad s = -5 + j30$$

Exercise 10.1

(a) Find $\underline{V}$ and s such that

$$v(t) = \text{Re}[\underline{V}e^{st}] = -10e^{3t} \text{ V}$$

(b) Write out the waveform $i(t) = \text{Re}[\underline{I}e^{st}]$ when

$$\underline{I} = 0.5 \text{ A} \underline{/-20°} \qquad s = 0 + j500$$

Generalized Impedance and Admittance

Now consider the case of a two-terminal element driven by an input voltage or current with a complex-frequency waveform. The question at hand is: What will be the resulting forced response at the element's terminals?

To address this question, recall that when the input is an ac current such as

$$i(t) = I_m \cos(\omega t + \phi_i) = \text{Re}[\underline{I}e^{j\omega t}]$$

the forced-response voltage will be

$$v(t) = V_m \cos(\omega t + \phi_v) = \text{Re}[\underline{V}e^{j\omega t}]$$

where the associated phasors are

$$\underline{I} = I_m \underline{/\phi_i} = I_m e^{j\phi_i} \qquad \underline{V} = V_m \underline{/\phi_v} = V_m e^{j\phi_v} \tag{10.5}$$

We showed in Section 6.2 that these phasors are related by the ac impedance

$$Z(j\omega) = \underline{V}/\underline{I}$$

Thus, given $i(t)$, we can find $v(t)$ from the phasor equation $\underline{V} = Z(j\omega)\underline{I}$.

Putting the complex frequency s in place of $j\omega$ in the foregoing expressions, it immediately follows that if

$$i(t) = \text{Re}[\underline{I}e^{st}] = I_m e^{\sigma t} \cos(\omega t + \phi_i) \tag{10.6a}$$

then

$$v(t) = \text{Re}[\underline{V}e^{st}] = V_m e^{\sigma t} \cos(\omega t + \phi_v) \tag{10.6b}$$

The phasors $\underline{I}$ and $\underline{V}$ are still as given in Eq. (10.5), but now they are related by

$$Z(s) \triangleq \underline{V}/\underline{I} \tag{10.7}$$

so

$$\underline{V} = Z(s)\underline{I} \tag{10.8}$$

We call $Z(s)$ the **generalized impedance**, as distinguished from the ac impedance $Z(j\omega)$.

By similar reasoning, the input voltage $v(t) = \text{Re}[\underline{V}e^{st}]$ applied to a two-terminal network produces the forced-response current $i(t) = \text{Re}[\underline{I}e^{st}]$ where $\underline{I} = \underline{V}/Z(s)$. We therefore define the **generalized admittance**

$$Y(s) \triangleq 1/Z(s) = \underline{I}/\underline{V} \tag{10.9}$$

Hence,

$$\underline{I} = Y(s)\underline{V} \tag{10.10}$$

which is the dual of Eq. (10.8). The names *generalized impedance* and *generalized admittance* reflect the fact that Eqs. (10.8) and (10.10) hold for the general class of excitation waveforms represented with $s = \sigma + j\omega$.

As implied by the symbols $Z(s)$ and $Y(s)$, the generalized impedance or admittance by replacing $j\omega$ with s. To confirm this assertion, consider an inductor described by $v = L\,di/dt$. If $v = \text{Re}[\underline{V}e^{st}]$ and $i = \text{Re}[\underline{I}e^{st}]$, then

$$\text{Re}[\underline{V}e^{st}] = L\frac{d}{dt}\text{Re}[\underline{I}e^{st}] = \text{Re}\left[L\underline{I}\frac{de^{st}}{dt}\right] = \text{Re}[sL\underline{I}e^{st}]$$

so $\underline{V} = sL\underline{I}$ and $Z_L(s) = \underline{V}/\underline{I} = sL$, as compared to the ac impedance $Z_L(j\omega) = j\omega L$. The derivations for resistors and capacitors follow similar lines.

Table 10.1 lists the expressions for $Z(s)$ and $Y(s)$ for resistors, inductors, and capacitors. The table emphasizes the viewpoint that ac impedance and admittance are special cases of generalized impedance and admittance with $s = j\omega$. Accordingly, we hereafter drop the adjective "generalized" and refer to $Z(s)$ and $Y(s)$ simply as the impedance and admittance.

Going a step further, let the load network in Fig. 10.1*a* consist of two or more elements. Since Kirchhoff's laws hold for phasors, and since the elements are described by their individual impedances, the terminal voltage and current phasors will be related by the network's *equivalent impedance* or *equivalent admittance*. This property is emphasized by the ***s*-domain diagram** in Fig.

TABLE 10.1 Generalized Impedance and Admittance

Element	$Z(s)$	$Y(s)$
Resistor	R	$\frac{1}{R} = G$
Inductor	sL	$\frac{1}{sL}$
Capacitor	$\frac{1}{sC}$	sC

10.1*b*. The *s*-domain diagram reduces to the familiar ac frequency-domain diagram when $s = j\omega$. And as in the ac case, we can find $Z(s)$ or $Y(s)$ using our previous methods for resistive networks with impedance and admittance in place of resistance and conductance.

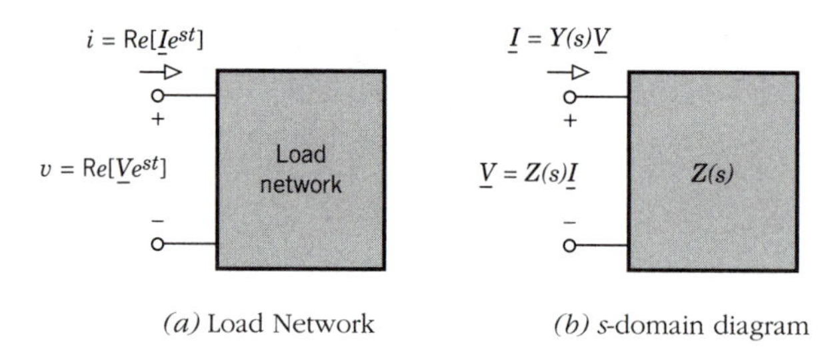

(*a*) Load Network (*b*) *s*-domain diagram

Figure 10.1

Having determined $Z(s)$ or $Y(s)$ for a given network, we can use it to calculate the forced response produced by a complex-frequency excitation with a particular value of s. But there is one exception that should be mentioned here, namely,

> If s happens to equal one of the network's characteristic values, then $Z(s)$ or $Y(s)$ becomes infinitely large and cannot be used by itself to find the forced response.

Further discussion of this special case is deferred to Chapter 13.

Example 10.2 *Calculations with Complex Frequency*

Suppose you want to calculate the forced-response current $i(t)$ in Fig. 10.2*a* when

$$L = 1 \text{ H} \qquad R = 5\ \Omega \qquad C = 1/10 \text{ F}$$

and

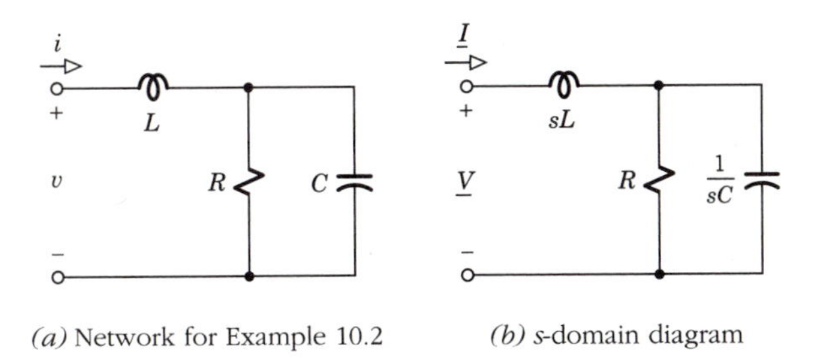

(a) Network for Example 10.2 (b) s-domain diagram

Figure 10.2

$$v(t) = 20e^{-2t} \cos (4t + 90°) \text{ V}$$

To start the analysis, you identify the voltage phasor and complex frequency

$$\underline{V} = 20 \text{ V } \underline{/90°} \qquad s = -2 + j4$$

These are obtained by inspection of the expression for $v(t)$.

Next, you draw the s-domain diagram in Fig. 10.2b, where phasors replace time functions and impedances replace element labels. The terminal impedance is readily determined by series-parallel reduction to be

$$Z(s) = sL + R \left\| \left(\frac{1}{sC} \right) \right. = sL + \frac{R}{sCR + 1}$$

from which

$$Y(s) = \frac{1}{Z(s)} = \frac{RCs + 1}{LRCs^2 + Ls + R}$$

Putting the element values into your expression for $Y(s)$ and setting $s = -2 + j4$ yields

$$Y(-2 + j4) = -0.32 - j0.24 = 0.4 \text{ S } \underline{/-143.1°}$$

Thus, the current phasor is

$$\underline{I} = Y(-2 + j4)\underline{V} = 8 \text{ A } \underline{/-53.1°}$$

so you finally obtain

$$i(t) = 8e^{-2t} \cos (4t - 53.1°) \text{ A}$$

Exercise 10.2

A network consists of $L = 0.5$ H in parallel with $R = 6\ \Omega$. Determine $Z(s)$ and find $v(t)$ when $i(t) = -2e^{-8t}$ A.

Impedance Analysis

Besides numerical calculations of forced response, impedance or admittance analysis often provides useful information about certain properties of networks.

The basic strategy is to find the equivalent impedance or admittance for an arbitrary value of s. Then you use the expression for $Z(s)$ or $Y(s)$ to deduce how the network behaves for specific values of s. We'll illustrate this strategy here with two examples having practical importance.

Example 10.3 *Miller-Effect Capacitance*

Some electronic circuits involve an inverting voltage amplifier with a feedback capacitor connected as shown in Fig. 10.3*a*. This configuration produces effects at the input and output that we'll investigate by finding the admittances $Y_1(s) = \underline{I}_1/\underline{V}_{in}$ and $Y_2(s) = \underline{I}_2/\underline{V}_{out}$.

We see from the s-domain diagram given in Fig. 10.3*b* that $\underline{I}_1 = Y_C(s)(\underline{V}_{in} - \underline{V}_{out})$ where $Y_C(s) = sC$ and $\underline{V}_{out} = -A\underline{V}_{in}$. Thus,

$$\underline{I}_1 = sC[\underline{V}_{in} - (-A\underline{V}_{in})] = sC(1 + A)\underline{V}_{in}$$

so we get

$$Y_1(s) = \frac{\underline{I}_1}{\underline{V}_{in}} = sC_1 \qquad C_1 = (1 + A)C$$

Similar analysis using $\underline{I}_2 = -\underline{I}_1$ and $\underline{V}_{in} = -\underline{V}_{out}/A$ yields

$$Y_2(s) = \frac{\underline{I}_2}{\underline{V}_{out}} = sC_2 \qquad C_2 = \left(1 + \frac{1}{A}\right) C$$

Figure 10.3*c* interprets our results in terms of the equivalent capacitors C_1 and C_2 across the input and output of the amplifier. The multiplication of the feedback capacitance C by the factors $(1 + A)$ and $(1 + 1/A)$ is called the **Miller effect**. This effect may adversely affect the amplifier's performance by decreasing its useful frequency range.

Example 10.4 *Generalized Impedance Converter*

Figure 10.4 diagrams an op-amp network known as a **generalized impedance converter (GIC)**. We'll justify this name by finding $Z_{in}(s) = \underline{V}_{in}/\underline{I}_{in}$ in

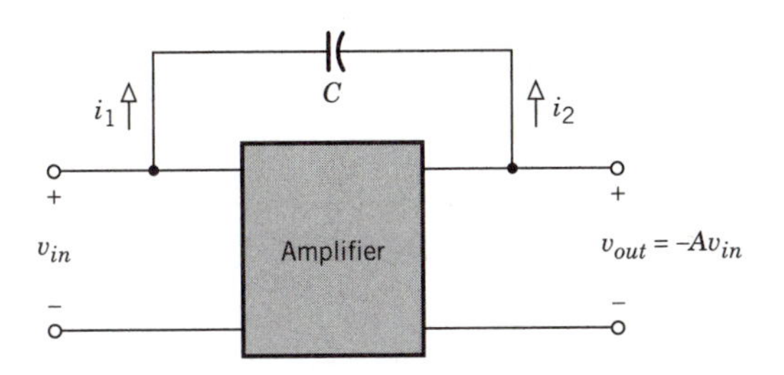

(a) Inverting voltage amplifier with feedback capacitor

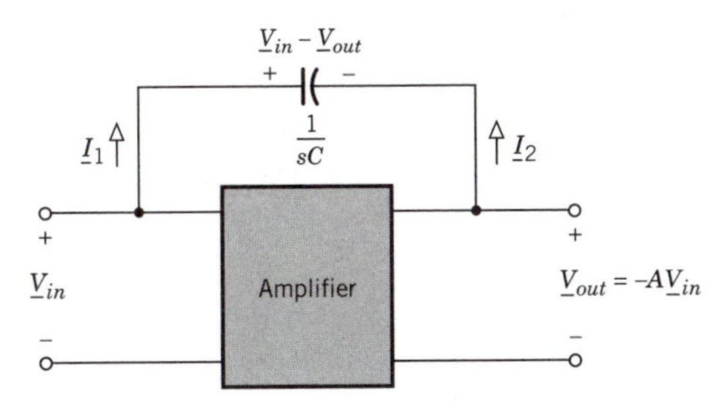

(b) *s*-domain diagram

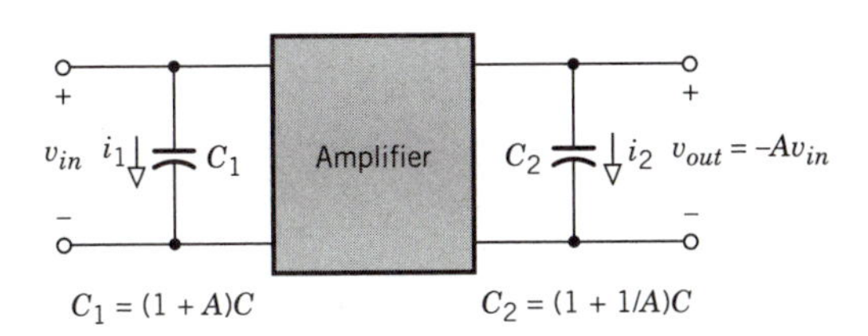

(c) Equivalent Miller-effect capacitances

Figure 10.3

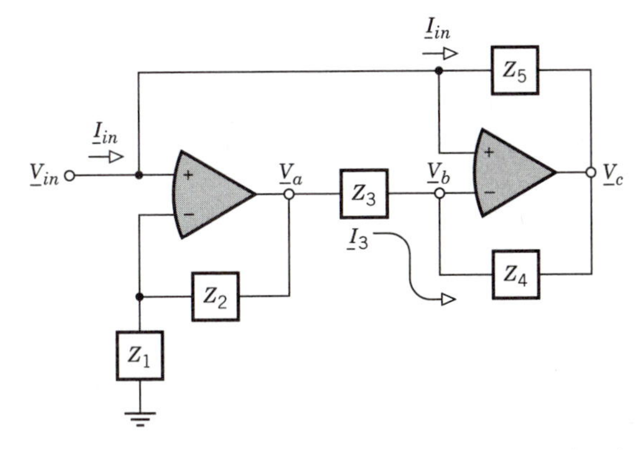

Figure 10.4 Generalized impedance converter (GIC).

terms of the individual impedances labeled $Z_1, Z_2, \ldots, Z_5$. In general, each impedance might be a function of s, so $Z_1 = Z_1(s)$, $Z_2 = Z_2(s)$, etc. Also note that all voltages here are measured with respect to the ground point.

To begin the analysis, recall that the op-amps draw no input current. Consequently, $\underline{I}_{in}$ goes directly through Z_5 and

$$\underline{I}_{in} = (\underline{V}_{in} - \underline{V}_c)/Z_5$$

We therefore need to relate $\underline{V}_c$ to $\underline{V}_{in}$, which we'll do by determining $\underline{V}_a$, $\underline{V}_b$, and $\underline{I}_3$.

The first op-amp just acts as a noninverting amplifier with Z_1 and Z_2 in place of the resistors R_1 and R_F, so

$$\underline{V}_a = \frac{Z_2 + Z_1}{Z_1}\underline{V}_{in} = \left(\frac{Z_2}{Z_1} + 1\right)\underline{V}_{in}$$

The virtual short at the input of the second op-amp results in

$$\underline{V}_b = \underline{V}_{in}$$

Consequently,

$$\underline{I}_3 = \frac{\underline{V}_a - \underline{V}_b}{Z_3} = \frac{Z_2}{Z_1 Z_3}\underline{V}_{in} \qquad \underline{V}_c = \underline{V}_b - Z_4\underline{I}_3 = \left(1 - \frac{Z_2 Z_4}{Z_1 Z_3}\right)\underline{V}_{in}$$

Hence,

$$\underline{I}_{in} = (\underline{V}_{in} - \underline{V}_c)/Z_5 = (Z_2 Z_4/Z_1 Z_3 Z_5)\underline{V}_{in}$$

and

$$Z_{in}(s) = \frac{\underline{V}_{in}}{\underline{I}_{in}} = \frac{Z_1 Z_3 Z_5}{Z_2 Z_4} \qquad (10.11)$$

which is our final result.

To bring out a significant implication of Eq. (10.11), suppose that either Z_2 or Z_4 represents a capacitor's impedance $1/sC$ while all the other impedances are equal resistances of value R. Equation (10.11) then becomes

$$Z_{in}(s) = s(R^2C) = sL_{eq} \qquad L_{eq} = R^2C \qquad (10.12)$$

This means that the GIC "converts" a capacitor and four resistors into an *inductance* — a valuable property because the GIC can be built as an integrated circuit that takes less space and costs less than an actual inductor.

Another interesting conversion occurs when Z_3 and Z_5 represent equal capacitors and the remaining elements are equal resistors, so

$$Z_{in}(s) = 1/s^2C^2R$$

For ac steady-state operation, we have $s = j\omega$ and

$$Z_{in}(j\omega) = -1/\omega^2C^2R = R_{eq}(\omega)$$

The GIC acts like a **frequency-dependent negative resistor**. This effect is exploited in the design of certain electronic filters.

Exercise 10.3

Find and interpret $Z_{in}(s)$ in Fig. 10.4 when Z_2 represents an inductor and all the other impedances are equal resistances.

Exercise 10.4

The network in Fig. 10.5 contains two current-controlled voltage sources. Show that its input impedance has the form

$$Z_{in}(s) = R_{eq} + sL_{eq}$$

Express R_{eq} and L_{eq} in terms of r_m, R, and C.

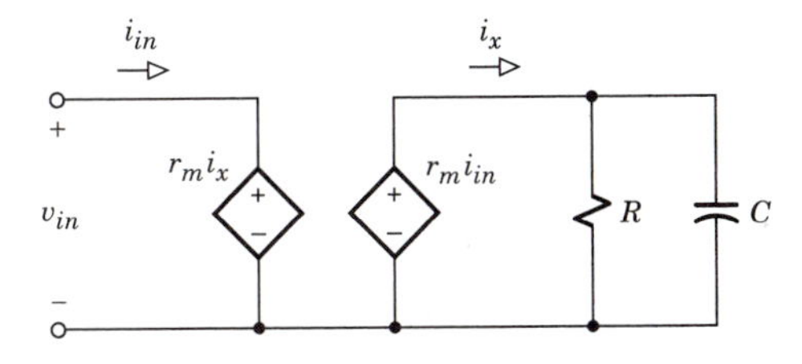

Figure 10.5

10.2 NETWORK FUNCTIONS

This section again considers two-terminal networks driven by an input voltage or current $x(t)$ having a complex-frequency waveform. But now we'll be concerned with the resulting forced response of an arbitrary branch variable $y(t)$. Thus, we seek a general relationship for *any* response forced by a complex-frequency excitation.

The basic relationship between $y(t)$ and $x(t)$ is, of course, the differential equation. Our purpose here is to develop an equivalent and more convenient relationship known as the *network function*.

Network Functions and Circuit Equations

Let the input to a linear two-terminal network be a voltage or current represented by

$$x(t) = X_m e^{\sigma t} \cos (\omega t + \phi_x) = \text{Re}[\underline{X}e^{st}] \tag{10.13}$$

where $s = \sigma + j\omega$ and

$$\underline{X} = X_m \underline{/\phi_x} = X_m e^{j\phi_x}$$

Since the network is linear, the forced response of any branch variable will take the form

$$y(t) = Y_m e^{\sigma t} \cos (\omega t + \phi_y) = \text{Re}[\underline{Y}e^{st}] \tag{10.14}$$

with

$$\underline{Y} = Y_m \underline{/\phi_y} = Y_m e^{j\phi_y}$$

We want to find $y(t)$ given $x(t)$ and the network diagram.

One advantage of the complex-frequency expressions in Eqs. (10.13) and (10.14) is that our task boils down to relating the response phasor $\underline{Y}$ to the excitation phasor $\underline{X}$. Another advantage is the corresponding simplicity of the time derivatives. For instance, the first and second derivatives of $y(t)$ become

$$y' = \frac{d}{dt}\text{Re}[\underline{Y}e^{st}] = \text{Re}\left[\underline{Y}\frac{de^{st}}{dt}\right] = \text{Re}[\underline{Y}se^{st}]$$

$$y'' = \frac{dy'}{dt} = \frac{d}{dt}\text{Re}[\underline{Y}se^{st}] = \text{Re}[\underline{Y}s^2e^{st}]$$

Thus, differentiation results in powers of s multiplying the phasor.

Now suppose the network in question contains two independent energy-storage elements. From our study of second-order circuits in Section 9.3, we know that the forced response of $y(t)$ is governed by a differential equation such as

$$a_2 y'' + a_1 y' + a_0 y = b_2 x'' + b_1 x' + b_0 x \tag{10.15}$$

where the right-hand side allows for the possible inclusion of the derivatives x'' and x' as well as x. Inserting Eqs. (10.13) and (10.14) into Eq. (10.15) yields

$$a_2\text{Re}[\underline{Y}s^2e^{st}] + a_1\text{Re}[\underline{Y}se^{st}] + a_0\text{Re}[\underline{Y}e^{st}]$$
$$= b_2\text{Re}[\underline{X}s^2e^{st}] + b_1\text{Re}[\underline{X}se^{st}] + b_0\text{Re}[\underline{X}e^{st}]$$

But the a's and b's are *real* quantities incorporating element values, so we may regroup terms as

$$\text{Re}[(a_2s^2 + a_1s + a_0)\underline{Y}e^{st}] = \text{Re}[(b_2s^2 + b_1s + b_0)\underline{X}e^{st}]$$

This equality must hold at any instant of time, including $t = 0$. Accordingly, we can find $\underline{Y}$ from the simplified *algebraic* equation

$$(a_2s^2 + a_1s + a_0)\underline{Y} = (b_2s^2 + b_1s + b_0)\underline{X}$$

Solving for $\underline{Y}$ gives

$$\underline{Y} = \frac{b_2s^2 + b_1s + b_0}{a_2s^2 + a_1s + a_0}\underline{X} \qquad (10.16)$$

which is the desired result.

Observe that Eq. (10.16) has the form $\underline{Y} = H(s)\underline{X}$ where $H(s)$ is the phasor ratio defined by

$$H(s) \triangleq \underline{Y}/\underline{X} \qquad (10.17)$$

We therefore call $H(s)$ the **network function** relating the response phasor $\underline{Y}$ to the input phasor $\underline{X}$. Our definition of $H(s)$ extends the concept of equivalent impedance and admittance, and it includes $Z(s)$ and $Y(s)$ as special cases.

Specifically, if $x(t)$ stands for a source current and $y(t)$ is the resulting voltage across the network, then $\underline{X} = \underline{I}$ and $\underline{Y} = \underline{V}$ so $H(s) = \underline{V}/\underline{I} = Z(s)$. Likewise, if $x(t)$ stands for a source voltage and $y(t)$ is the resulting current into the network, then $\underline{X} = \underline{V}$ and $\underline{Y} = \underline{I}$ so $H(s) = \underline{I}/\underline{V} = Y(s)$. Accordingly, we say that $H(s)$ is a **driving-point function** when it relates a network's terminal variables. But our definition of $H(s)$ also permits $y(t)$ to be any voltage or current *within* the network, in which case $H(s)$ is usually called a **transfer function**. A given network may thus be described by several different network functions, one for each response of interest.

The notation $H(s)$ emphasizes the fact that the value of a network function depends upon the *input's* complex frequency s. Nonetheless, all the constants in $H(s)$ represent the *network's* elements as contained in the differential equation. To illustrate these points, consider Eq. (10.16) where

$$H(s) = \underline{Y}/\underline{X} = \frac{b_2s^2 + b_1s + b_0}{a_2s^2 + a_1s + a_0}$$

which was derived from

$$a_2y'' + a_1y' + a_0y = b_2x'' + b_1x' + b_0x$$

By comparison, we clearly see that

> The numerator of $H(s)$ is a polynomial obtained from the right-hand side of the differential equation with derivatives of x replaced by powers of s, while the denominator is a polynomial obtained from the left-hand side with derivatives of y replaced by powers of s.

Consequently, both the numerator and denominator are polynomials in s, and the coefficients come directly from the differential equation.

The foregoing observations remain valid for ***n*th-order networks** containing an arbitrary number of independent energy-storage elements. The forced-response of any variable $y(t)$ is then governed by an nth-order differential equation having the general form

$$a_n \frac{d^n y}{dt^n} + a_{n-1} \frac{d^{n-1} y}{dt^{n-1}} + \cdots + a_1 \frac{dy}{dt} + a_0 y = b_m \frac{d^m x}{dt^m} + b_{m-1} \frac{d^{m-1} x}{dt^{m-1}} + \cdots + b_1 \frac{dx}{dt} + b_0 x \quad (10.18)$$

Inserting Eqs. (10.13) and (10.14) and proceeding as before yields

$$(a_n s^n + a_{n-1} s^{n-1} + \cdots + a_1 s + a_0)\underline{Y} = (b_m s^m + b_{m-1} s^{m-1} + \cdots + b_1 s + b_0)\underline{X}$$

so the network function is

$$H(s) = \underline{Y}/\underline{X} = \frac{b_m s^m + b_{m-1} s^{m-1} + \cdots + b_1 s + b_0}{a_n s^n + a_{n-1} s^{n-1} + \cdots + a_1 s + a_0} \quad (10.19)$$

Hence, we again have a ratio of polynomials that can be written by *inspection* of the differential equation. Polynomial ratios like Eq. (10.19) are referred to in general as **rational functions**.

Example 10.5 ***Series LRC Network Functions***

The series *LRC* circuit in Fig. 10.6 was studied in Section 9.3, where we found that the inductor current i_L and voltage v_L are related to v_s by

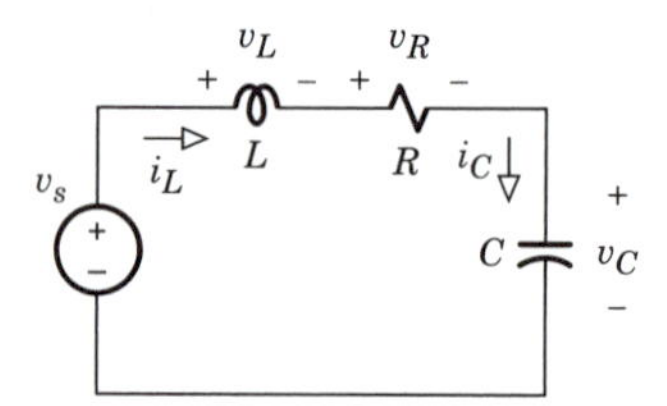

Figure 10.6 Series *LRC* circuit.

$$i_L'' + (R/L)\,i_L' + (1/LC)\,i_L = (1/L)v_s'$$
$$v_L'' + (R/L)v_L' + (1/LC)v_L = v_s''$$

Here, we'll obtain the corresponding driving point function $H_1(s) = \underline{I}_L/\underline{V}_s$ and the transfer function $H_2(s) = \underline{V}_L/\underline{V}_s$.

Replacing time derivatives by powers of s multiplying phasors gives the algebraic equations

$$[s^2 + (R/L)s + (1/LC)]\underline{I}_L = (1/L)s\underline{V}_s$$
$$[s^2 + (R/L)s + (1/LC)]\underline{V}_L = s^2\underline{V}_s$$

Hence,

$$H_1(s) = \frac{\underline{I}_L}{\underline{V}_s} = \frac{(1/L)s}{s^2 + (R/L)s + (1/LC)}$$

$$H_2(s) = \frac{\underline{V}_L}{\underline{V}_s} = \frac{s^2}{s^2 + (R/L)s + (1/LC)}$$

Observe that the same denominator polynomial appears in $H_1(s)$ and $H_2(s)$ because the left-hand side of each differential equation has the same coefficients.

Exercise 10.5

A current source drives a parallel *CRL* network with $C = 1/12$ F, $R = 6\ \Omega$, and $L = 3$ H. Obtain the network function relating v_C to i_s by drawing upon Eq. (9.22) in Section 9.3.

Network Functions and Impedance Analysis

We've seen how a network function can be obtained from the circuit's differential equation. But there's another, easier way to find $H(s)$ that bypasses differential equations entirely.

For this method, you draw the s-domain diagram and apply impedance analysis to get an expression for the response phasor $\underline{Y}$ in terms of the excitation phasor $\underline{X}$. Division then yields the network function $H(s) = \underline{Y}/\underline{X}$ with any value of s.

To carry out the s-domain calculations for $H(s)$, you may invoke any appropriate impedance or admittance techniques from Sections 6.2 and 6.3. Particularly helpful are

- Series-parallel reduction
- Voltage and current dividers
- Proportionality

- Thévenin's theorem
- Source conversions
- Node and mesh equations

The most suitable method for a given circuit depends on its configuration and the variable of interest.

The resulting expression for $H(s)$ can then be checked using the following simple tests:

- The highest power of s should equal the number of independent energy-storage elements.
- The values of $H(0)$ and $H(\infty)$ should agree with direct calculations from the s-domain diagram. In this regard, note that $Z_L = sL$ and $Z_C = 1/sC$ so

$$Z_L = 0 \text{ and } Z_C = \infty \text{ when } s = 0$$
$$Z_L = \infty \text{ and } Z_C = 0 \text{ when } s = \infty$$

- All terms in a polynomial should have the same units. In this regard, note the dimensionless quantities

$$RCs = Z_R/Z_C \qquad (L/R)s = Z_L/Z_R \qquad LCs^2 = Z_L/Z_C$$

These checks for accuracy help detect errors that might have crept in during the algebraic manipulations.

Having found $H(s)$ as a ratio of polynomials, you may work backwards to get the corresponding differential equation if desired. Indeed, impedance analysis usually proves to be the simplest method for obtaining differential equations of complicated circuits. However, since $H(s)$ contains all the relevant information, you seldom actually need to write out the differential equation.

Example 10.6 ***Finding Network Functions***

Given the circuit in Fig. 10.7*a*, we'll obtain the network functions and differential equations relating i_L and v_C to the source voltage v_s. Our methods will be based on impedance analysis of the s-domain diagram in Fig. 10.7*b*.

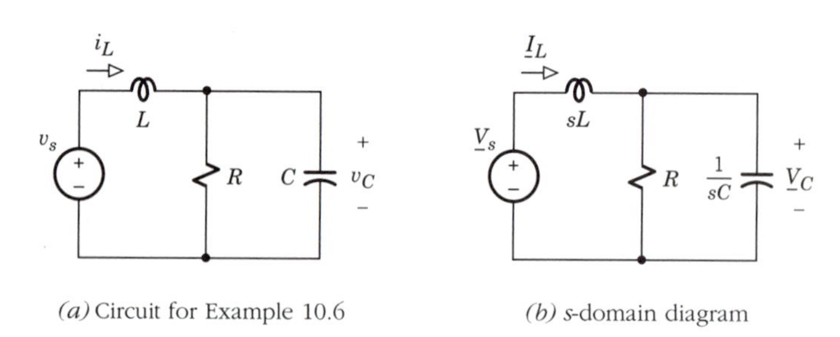

(*a*) Circuit for Example 10.6 (*b*) *s*-domain diagram

Figure 10.7

First we observe that $\underline{I}_L = Y(s)\underline{V}_s$ where $Y(s)$ is the network's input admittance previously determined in Example 10.2. We thus get the driving-point function

$$H_1(s) = \frac{\underline{I}_L}{\underline{V}_s} = Y(s) = \frac{RCs + 1}{LRCs^2 + Ls + R}$$

The s^2 term in the denominator is expected for a circuit with one inductor and one capacitor. Setting $s = 0$ yields $H_1(0) = 1/R$, in agreement with the observation that $\underline{I}_L = \underline{V}_s/R$ under dc steady-state conditions. At the other extreme, we have $H_1(\infty) = 0$ because the inductor becomes an open circuit and $\underline{I}_L = 0$ when $s = \infty$. We also note that both terms in the numerator are dimensionless, while all three terms in the denominator have the units of impedance.

By inspection of $H_1(s)$, the corresponding differential equation is

$$LRCi_L'' + Li_L'' + Ri_L = RCv_s' + v_s$$

Dividing both sides by LRC yields

$$i_L'' + \frac{1}{RC}i_L' + \frac{1}{LC}i_L = \frac{1}{L}v_s' + \frac{1}{LRC}v_s$$

which is the result stated back in Exercise 9.6.

Focusing now on $\underline{V}_C$, a voltage divider or a node equation could be used for the expression relating it to $\underline{V}_s$. But we'll apply the proportionality method by writing

$$\underline{I}_L = sC\underline{V}_C + \underline{V}_C/R = (sCR + 1)\underline{V}_C/R$$

Then

$$\underline{V}_s = sL\underline{I}_L + \underline{V}_C = (s^2LCR + sL + R)\underline{V}_C/R$$

from which we get the transfer function

$$H_2(s) = \frac{\underline{V}_C}{\underline{V}_s} = \frac{R}{LCRs^2 + Ls + R}$$

The corresponding differential equation is $LCRv_C'' + Lv_C' + Rv_C = Rv_s$ or

$$v_C'' + \frac{1}{RC}v_C' + \frac{1}{LC}v_C = \frac{1}{LC}v_s$$

Example 10.7 ***Twin-Tee Network with an Op-Amp***

To further illustrate the value of impedance analysis, consider the op-amp circuit in Fig. 10.8*a*. The op-amp operates in the noninverting mode, so it acts

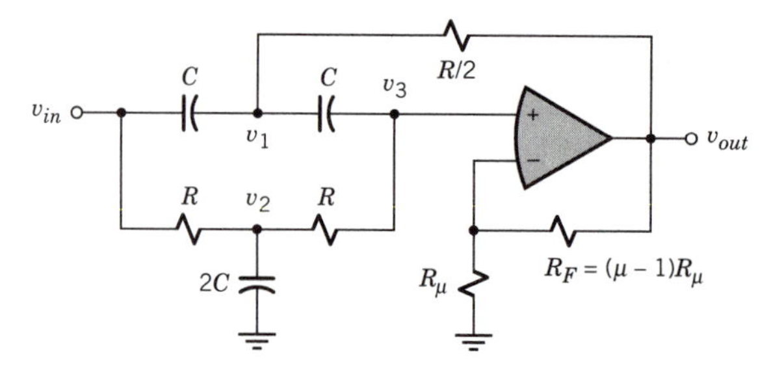

(a) Twin-tee network with an op-amp

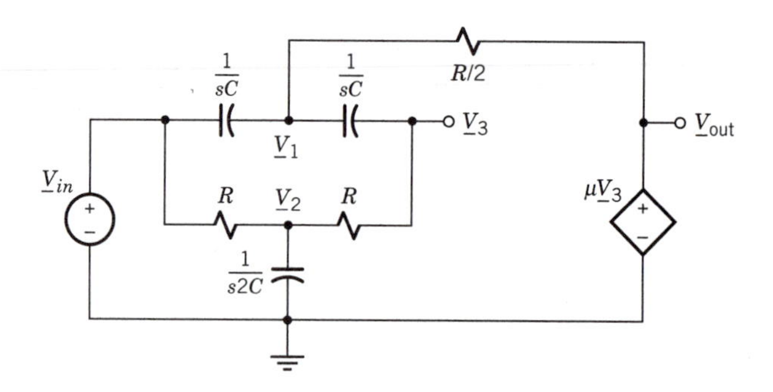

(b) s-domain diagram

Figure 10.8

like a controlled source with $v_{out} = (R_F + R_\mu)v_3/R_\mu = \mu v_3$. But the RC network has a **twin-tee** configuration that makes it difficult to find the differential equation for v_1, v_2, or v_3 by direct means. In contrast, analyzing the s-domain diagram in Fig. 10.8b becomes a fairly routine task with the help of our systematic method for writing matrix equations.

By extension of the ac relations in Section 6.3, we know that the final matrix node equation will have the form

$$[Y(s) - \tilde{Y}(s)][\underline{V}] = [\underline{\tilde{I}}_s]$$

The admittance matrix $[Y(s)]$ can be determined by inspection, whereas $[\tilde{Y}(s)]$ and $[\underline{\tilde{I}}_s]$ are obtained by inserting the constraint equation into the equivalent source-current vector $[\underline{I}_s]$.

Proceeding with these steps, we first let $G = 1/R$ and sum the admittances at each node to get

$$[Y(s)] = \begin{bmatrix} sC + sC + 2G & 0 & -sC \\ 0 & s2C + G + G & -G \\ -sC & -G & sC + G \end{bmatrix}$$

Next, we use the constraint equation $\underline{V}_{out} = \mu\underline{V}_3$ to write the equivalent source-current vector

$$[\underline{I}_s] = \begin{bmatrix} sC\underline{V}_{in} + 2\mu G\underline{V}_3 \\ G\underline{V}_{in} \\ 0 \end{bmatrix} = \begin{bmatrix} sC\underline{V}_{in} \\ G\underline{V}_{in} \\ 0 \end{bmatrix} + \begin{bmatrix} 0 & 0 & 2\mu G \\ 0 & 0 & 0 \\ 0 & 0 & 0 \end{bmatrix} \begin{bmatrix} \underline{V}_1 \\ \underline{V}_2 \\ \underline{V}_3 \end{bmatrix}$$

Then we divide all terms of $[Y(s)]$ and $[\underline{I}_s]$ by C and combine them to obtain

$$\begin{bmatrix} 2(s+a) & 0 & -(s+2\mu a) \\ 0 & 2(s+a) & -a \\ -s & -a & s+a \end{bmatrix} \begin{bmatrix} \underline{V}_1 \\ \underline{V}_2 \\ \underline{V}_3 \end{bmatrix} = \begin{bmatrix} s\underline{V}_{in} \\ a\underline{V}_{in} \\ 0 \end{bmatrix}$$

where

$$a = G/C = 1/RC$$

Application of Cramer's rule now yields the determinants

$$\Delta = 2(s+a)[s^2 + (4 - 2\mu)as + a^2]$$
$$\Delta_1 = 2(s^3 + 2as^2 + a^2 s + \mu a^3]\underline{V}_{in}$$
$$\Delta_2 = 2a[s^2 + (2 - \mu)as + a^2]\underline{V}_{in}$$
$$\Delta_3 = 2(s+a)(s^2 + a^2)\underline{V}_{in}$$

Thus, the node-voltage network functions are

$$H_1(s) = \frac{\underline{V}_1}{\underline{V}_{in}} = \frac{\Delta_1/\Delta}{\underline{V}_{in}} = \frac{s^3 + 2as^2 + a^2 s + \mu a^3}{(s+a)\,[s^2 + (4 - 2\mu)as + a^2]}$$

$$H_2(s) = \frac{\underline{V}_2}{\underline{V}_{in}} = \frac{\Delta_2/\Delta}{\underline{V}_{in}} = \frac{a[s^2 + (2 - \mu)as + a^2]}{(s+a)\,[s^2 + (4 - 2\mu)as + a^2]}$$

$$H_3(s) = \frac{\underline{V}_3}{\underline{V}_{in}} = \frac{\Delta_3/\Delta}{\underline{V}_{in}} = \frac{s^2 + a^2}{s^2 + (4 - 2\mu)as + a^2}$$

Note that the factor $(s + a)$ does not appear in the denominator of $H_3(s)$ because Δ_3 happens to include the same factor and they cancel out. Also note that $\underline{V}_{out} = \mu\underline{V}_3$, so the input-output transfer function is $\underline{V}_{out}/\underline{V}_{in} = \mu\underline{V}_3/\underline{V}_{in} = \mu H_3(s)$.

Exercise 10.6

Use a voltage divider to obtain the network function and differential equation relating v_R to v_s in Fig. 10.6. Then show that your differential equation agrees with the equation relating i_L to v_s in Example 10.5.

Exercise 10.7

Write a single s-domain mesh equation for the circuit in Fig. 10.9 to obtain the driving point function

$$H(s) = \underline{I}/\underline{V} = \frac{s + 4}{s^2 + (6 - 2\beta)s + 24 - 8\beta}$$

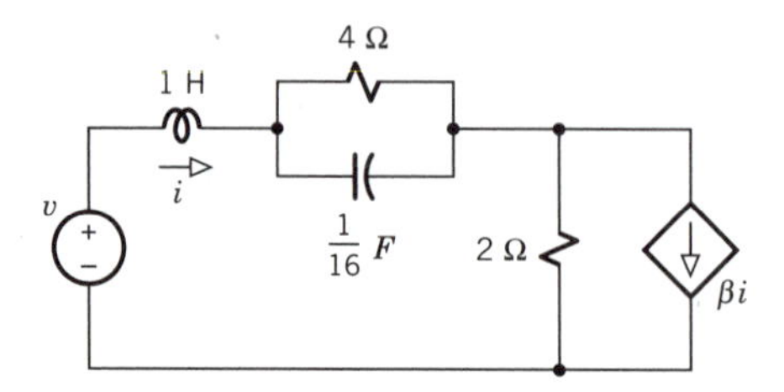

Figure 10.9

10.3 NETWORK FUNCTIONS WITH MUTUAL INDUCTANCE†

Back in Section 8.3 we developed equivalent networks with three inductors to represent the self- and mutual inductances of magnetically coupled coils. Here, we'll invoke an equivalent network to investigate the properties of a network function with mutual inductance. The analysis is straightforward, but the results shed further light on magnetic coupling.

Consider the source-load circuit in Fig. 10.10*a*, where the mutual inductance has been labeled $\pm M$ to account for either of the possible winding senses. This circuit configuration allows us to replace the coils by the tee network from Fig. 8.23*b*, and Fig. 10.10*b* gives the resulting *s*-domain diagram.

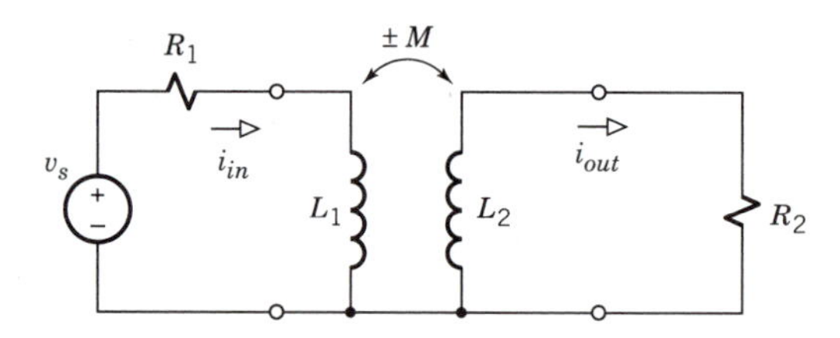

(a) Circuit with mutual inductance

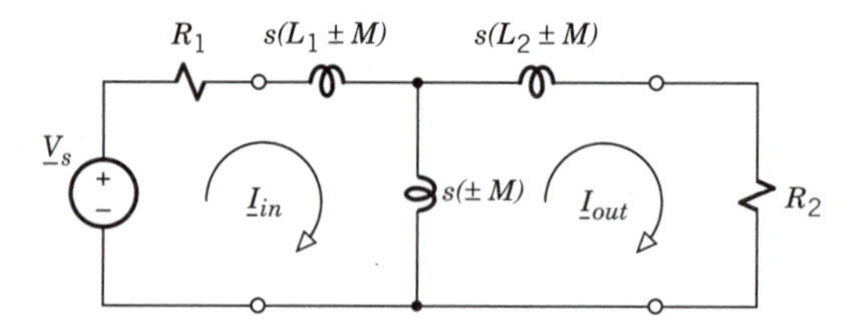

(b) *s*-domain diagram with tee network

Figure 10.10

We'll focus on the two mesh currents, because any other variable of interest can be found from $\underline{I}_{in}$ or $\underline{I}_{out}$.

By direct inspection of Fig. 10.10*b* we obtain the matrix mesh equation

$$\begin{bmatrix} sL_1 + R_1 & \pm sM \\ \pm sM & sL_2 + R_2 \end{bmatrix} \begin{bmatrix} \underline{I}_{in} \\ \underline{I}_{out} \end{bmatrix} = \begin{bmatrix} \underline{V}_s \\ 0 \end{bmatrix}$$

Then, after solving for $\underline{I}_{in}$ and $\underline{I}_{out}$, we obtain the network functions

$$H_1(s) = \frac{\underline{I}_{in}}{\underline{V}_s} = \frac{L_2 s + R_2}{(L_1 L_2 - M^2)s^2 + (L_1 R_2 + L_2 R_1)s + R_1 R_2} \tag{10.20a}$$

$$H_2(s) = \frac{\underline{I}_{out}}{\underline{V}_s} = \frac{\pm Ms}{(L_1 L_2 - M^2)s^2 + (L_1 R_2 + L_2 R_1)s + R_1 R_2} \tag{10.20b}$$

Although the *s*-domain diagram includes *three* inductors, the highest-order term in our network functions is s^2. The reason for this apparent discrepancy is that the three equivalent inductors are not independent — for if you know the currents i_{in} and i_{out} through $L_1 \mp M$ and $L_2 \mp M$, respectively, then you also know the current $i_{in} - i_{out}$ through $\pm M$. Consequently, we're actually dealing with a *second*-order circuit. Moreover, in the case of unity coupling we have $M = \sqrt{L_1 L_2}$ so $L_1 L_2 - M^2 = 0$ and the s^2 term vanishes in Eqs. (10.20a) and (10.20b). This result underscores the fact that unity coupled coils act as a *single* energy-storage element.

Exercise 10.8

Use Eqs. (10.20a) and (10.20b) to find $H_3(s) = \underline{I}_{out}/\underline{I}_{in}$.

10.4 *s*-DOMAIN ANALYSIS

This section expands our discussion of network functions. These functions deserve further attention in view of the following properties:

- With the help of impedance analysis, $H(s)$ is usually easier to find than the corresponding differential equation.
- Network functions simplify the calculation of the *forced* response when the excitation has a generalized complex-frequency waveform, which includes ac and dc steady states as special cases.
- A network function contains all the coefficients from the differential equation, so the form of the *natural* response also can be determined from $H(s)$.

Other useful properties will be developed in later chapters.

The vehicle for our work here is *s-domain analysis*. We'll start with the *poles* and *zeros* of network functions, and we'll use them to deduce the behavior of both forced and natural response.

Poles and Zeros

Consider an nth-order circuit with one input $x(t) = \text{Re}[\underline{X}e^{st}]$, so the forced response of any variable of interest will be $y(t) = \text{Re}[\underline{Y}e^{st}]$. As shown in the Section 10.2, the phasors $\underline{X}$ and $\underline{Y}$ are related by a network function having the general form

$$H(s) = \underline{Y}/\underline{X} = \frac{b_m s^m + b_{m-1} s^{m-1} + \cdots + b_1 s + b_0}{a_n s^n + a_{n-1} s^{n-1} + \cdots + a_1 s + a_0}$$

where the a's and b's are real quantities incorporating element values.

For our present purposes, we draw upon the fact that any rth-order polynomial with real coefficients has r roots that must be real or complex conjugates. We therefore let $K = b_m/a_n$ and write the factored expression

$$H(s) = K\frac{(s - z_1)(s - z_2)\cdots(s - z_m)}{(s - p_1)(s - p_2)\cdots(s - p_n)} \tag{10.21}$$

The n roots of the denominator polynomial are denoted here by $p_1, p_2, \ldots, p_n$, and they are known as the **poles** of $H(s)$. The m roots of the numerator polynomial are denoted here by $z_1, z_2, \ldots, z_m$, and they are known as the **zeros** of $H(s)$. The constant K will be called the **gain factor** since, with $s = 0$,

$$H(0) = K\frac{(-z_1)(-z_2)\cdots(-z_m)}{(-p_1)(-p_2)\cdots(-p_n)}$$

which corresponds to the dc gain.

Although Eq. (10.21) represents an arbitrary network function, the general properties of $H(s)$ always require that:

- The gain factor K is real, but it may be positive or negative;
- All poles and zeros are either real quantities or they occur in complex-conjugate pairs;
- The number of poles equals the number of independent energy-storage elements in the circuit.

Additionally, the network functions for most circuits have $m \leq n$, so the number of zeros does not exceed the number of poles.

The names "poles" and "zeros" come from the behavior of $H(s)$ when s takes on the value of a pole or zero. At one extreme, if z_i stands for any zero of $H(s)$ and if $s = z_i$, then we see from Eq. (10.21) that

$$H(z_i) = 0 \tag{10.22a}$$

The corresponding forced response equals zero since $\underline{Y} = H(z_i)\underline{X} = 0$. At the other extreme, if p_i stands for any pole of $H(s)$ and if $s = p_i$, then

$$H(p_i) = \infty \tag{10.22b}$$

so the network function "blows up" or "climbs a pole." In this special case, you cannot use $H(s)$ directly to find the forced response.

Further examination of Eq. (10.21) reveals that a network function is completely defined by its poles and zeros, together with the gain factor K. Since poles and zeros may be complex numbers, a convenient way of displaying their values is as points in the **s plane** — a complex plane containing all possible values of the complex frequency $s = \sigma + j\omega$. We therefore label the horizontal and vertical axes with the variables $\text{Re}[s] = \sigma$ and $\text{Im}[s] = \omega$, respectively.

Pole locations in the s plane are indicated by small crosses (**x**), while zero locations are indicated by small circles (**o**). The resulting picture is known as the **pole-zero pattern**. By way of illustration, Fig. 10.11a shows the pole-

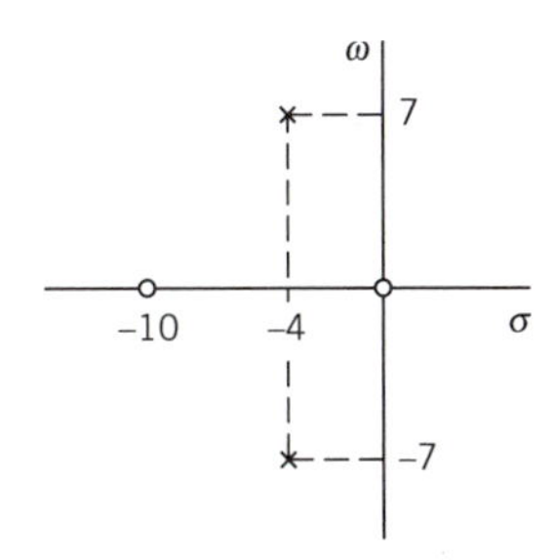

(a) Pole-zero pattern with complex poles

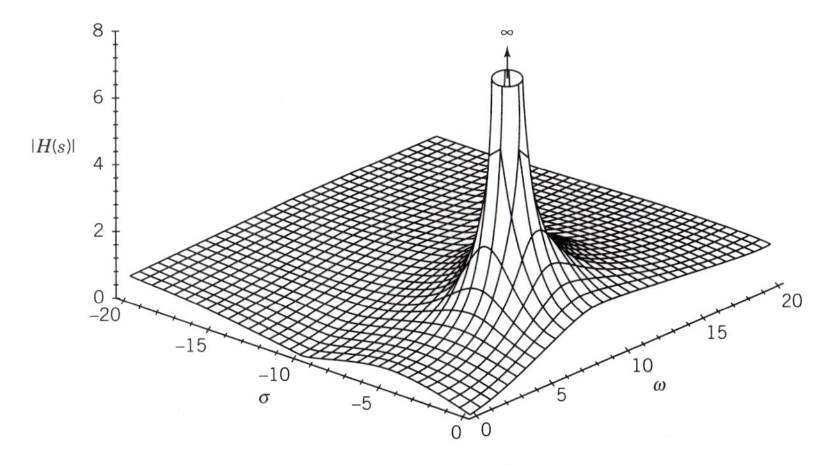

(b) Surface contour |H(s)|

Figure 10.11

zero pattern for a network function with two real zeros and a complex-conjugate pair of poles. Had there been any repeated roots, they would be marked by multiple crosses or circles at the appropriate points.

Having located the poles and zeros, we can visualize the magnitude of $H(s)$ as a three-dimensional surface above the s plane. This surface looks like a thin elastic sheet tacked down to the s plane at each zero location and propped up by an infinitely tall post at each pole location. Thus, the pole-zero pattern in Fig. 10.11*a* leads to the surface contour $|H(s)|$ depicted by Fig. 10.11*b*.

When constructing a pole-zero pattern, the main task is factoring the numerator and denominator of $H(s)$ to obtain the poles and zeros. Drawing on our study of second-order natural response, the roots of quadratic polynomials are most conveniently found by putting them in the standard form $s^2 + 2\alpha s + \omega_0^2$. Higher-order polynomials are more difficult to factor and generally require the use of numerical methods. Consequently, any higher-order polynomials in this text will be given in partially factored form.

Example 10.8 *Pole-Zero Pattern of a Fifth-Order Network*

Consider the partially factored network function

$$H(s) = -5\frac{s^4 + 16s^3 + 164s^2}{(s + 32)(s^2 + 36)(s^2 + 40s + 400)}$$

This function has the form of Eq. (10.21) with $K = -5$ and $m = 4$. We also see that the network in question is fifth order since the highest power of s in the denominator would be $n = 5$ if the three factors were multiplied together.

The network's four zeros are evaluated by writing the numerator polynomial as

$$\begin{aligned} s^4 + 16s^3 + 164s^2 &= s \times s \times (s^2 + 2 \times 8s + 164) \\ &= (s - z_1)(s - z_2)(s - z_3)(s - z_4) \end{aligned}$$

from which

$$z_1 = z_2 = 0 \qquad z_3, z_4 = -8 \pm j\sqrt{164 - 8^2} = -8 \pm j10$$

Thus, there is a repeated or "double" zero at the origin of the s plane, and a complex-conjugate pair of zeros at $s = -8 \pm j10$.

The network's five poles are evaluated by writing the factors of the denominator as

$$\begin{aligned} s + 32 &= s - p_1 \\ s^2 + 36 &= (s - p_2)(s - p_3) \\ s^2 + 2 \times 20s + 400 &= (s - p_4)(s - p_5) \end{aligned}$$

from which

$$p_1 = -32 \qquad p_2, p_3 = \pm j\sqrt{36} = \pm j6 \qquad p_4 = p_5 = -20$$

Thus, p_1 is real, p_2 and p_3 are on the imaginary axis, and p_4 and p_5 are repeated.

Figure 10.12 shows the resulting pole-zero pattern, which has symmetry with respect to the real axis. All pole-zero patterns exhibit that symmetry because the polynomial coefficients are real numbers so any complex poles or zeros always occur in conjugate pairs.

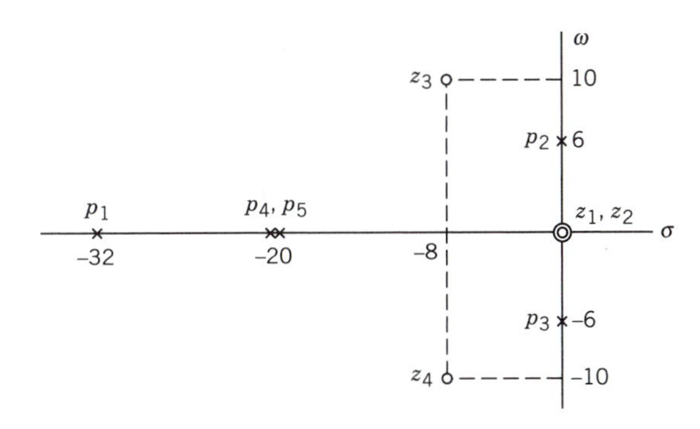

Figure 10.12 Pole-zero pattern of a fifth-order network.

Exercise 10.9

Draw the pole-zero pattern of

$$H(s) = 64(s^3 + 12s^2 + 32s)/(s^2 + 8s + 25)^2$$

(Save your results for use in Exercise 10.10.)

Forced Response and s-Plane Vectors

One important application of pole-zero patterns is the calculation of the forced response $y(t)$ produced by the input $x(t)$ with a specified complex frequency, say $s = s_0$. We know that if $x(t) = \mathrm{Re}[\underline{X}e^{s_0t}]$, and if s_0 does not equal any of the poles of $H(s)$, then $y(t) = \mathrm{Re}[\underline{Y}e^{s_0t}]$ where

$$\underline{Y} = H(s_0)\underline{X} \tag{10.23}$$

Hence, we need to evaluate $H(s)$ at $s = s_0$.

When s_0 is a complex number, $H(s_0)$ will also be complex. We therefore want the polar form of $H(s_0)$ for use in Eq. (10.23). As a preliminary step, we first replace s by s_0 everywhere in Eq. (10.21) to get

$$H(s_0) = K\frac{(s_0 - z_1)(s_0 - z_2)\cdots}{(s_0 - p_1)(s_0 - p_2)\cdots} \tag{10.24}$$

In general, each term of the form $(s_0 - a)$ is a complex quantity having magnitude $|s_0 - a|$ and angle $\angle(s_0 - a)$. Hence, the desired magnitude and angle of $H(s_0)$ are given by

$$|H(s_0)| = |K| \frac{|s_0 - z_1||s_0 - z_2| \cdots}{|s_0 - p_1||s_0 - p_2| \cdots} \tag{10.25a}$$

$$\begin{aligned} \angle H(s_0) = \angle K &+ [\angle(s_0 - z_1) + \angle(s_0 - z_2) + \cdots] \\ &- [\angle(s_0 - p_1) + \angle(s_0 - p_2) + \cdots] \end{aligned} \tag{10.25b}$$

Although K is a real quantity, it may be positive or negative, so either $\angle K = 0$ or $\angle K = \pm 180°$.

Terms such as $|s_0 - p_1|$ and $\angle(s_0 - p_1)$ could be evaluated analytically. But a more satisfying approach involves a graphical picture with **s-plane vectors**. To develop this picture, we observe that s_0 and p_1 are specific points in the s plane that can be defined by vectors from the origin. Furthermore,

$$p_1 + (s_0 - p_1) = s_0$$

so we interpret $s_0 - p_1$ as the vector drawn from pole p_1 to point s_0. Figure 10.13 illustrates these s-plane vectors along with $|s_0 - p_1|$ and $\angle(s_0 - p_1)$.

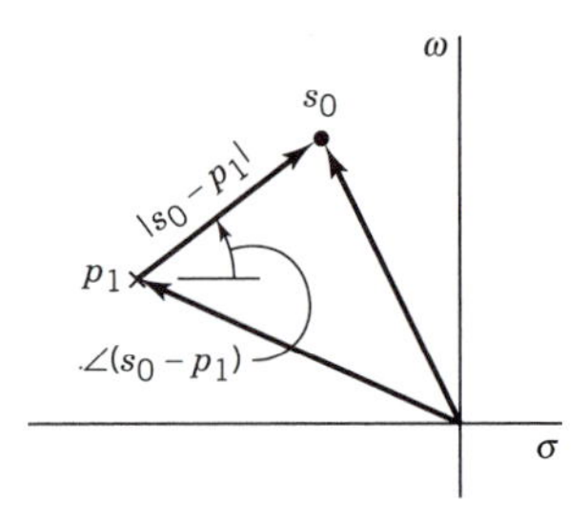

Figure 10.13 s-plane vectors for $s_0 - p_1$.

In like manner, we can obtain the magnitudes and angles for all the complex quantities in $H(s_0)$ by drawing s-plane vectors from all poles and zeros to the point s_0. Such constructions help us avoid numerical errors. Additionally, they bring out the relative influence of each pole and zero on the forced response.

Example 10.9 *Calculations with s-Plane Vectors*

Suppose we are given that

$$H(s) = \frac{-6s}{s^2 + 12s + 45}$$

which has

$$K = -6 = 6 \angle \pm 180^\circ \qquad z_1 = 0 \qquad p_1, p_2 = -6 \pm j3$$

We want to find the forced response $y(t)$ when

$$x(t) = 10e^{-4t} \cos 3t$$

so $\underline{X} = 10$ and $s_0 = -4 + j3$.

After constructing the s-plane vectors in Fig. 10.14, we clearly see that

$$s_0 - z_1 = -4 + j3 = 5 \angle 143.1^\circ$$

Then, from the horizontal and vertical components of the other two vectors, we get

$$s_0 - p_1 = 2 + j0 = 2 \angle 0^\circ$$
$$s_0 - p_2 = 2 + j6 = \sqrt{40} \angle 71.6^\circ$$

Therefore,

$$H(s_0) = 6 \angle \pm 180^\circ \frac{5 \angle 143.1^\circ}{2 \angle 0^\circ \times \sqrt{40} \angle 71.6^\circ} = 2.37 \angle -108.5^\circ$$

where we have taken $\angle K = -180^\circ$. Hence, the forced-response phasor is

$$\underline{Y} = H(s_0)\underline{X} = 23.7 \angle -108.5^\circ$$

and we finally obtain

$$y(t) = 23.7e^{-4t} \cos (3t - 108.5^\circ)$$

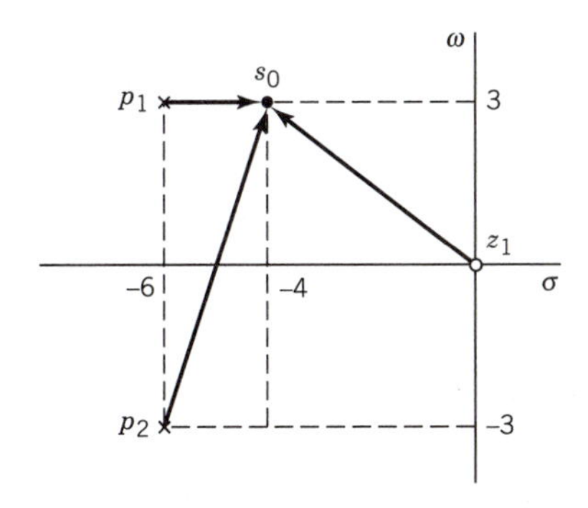

Figure 10.14 s-plane vectors for Example 10.9.

Exercise 10.10

Use s-plane vectors to evaluate the network function in Exercise 10.9 at $s_0 = -4 + j1$.

Natural Response and Stability

We originally defined network functions in conjunction with *forced* response. Even so, we can also use network functions to predict the form of the *natural* response because $H(s)$ contains all the coefficients from the differential equation.

To demonstrate this point, consider the second-order network function

$$H(s) = K\frac{(s - z_1)(s - z_2)}{(s - p_1)(s - p_2)} = K\frac{s^2 - (z_1 + z_2)s + z_1 z_2}{s^2 - (p_1 + p_2)s + p_1 p_2} \tag{10.26}$$

By inspection, the corresponding differential equation is

$$y'' - (p_1 + p_2)y' + p_1 p_2 y = K[x'' - (z_1 + z_2)x' + z_1 z_2 x]$$

If you wanted to find the natural response starting from the differential equation, then you would first let $x = x' = x'' = 0$ and write

$$y_N'' - (p_1 + p_2)y_N' + p_1 p_2 y_N = 0$$

Replacing derivatives of y_N with powers of s yields the characteristic equation

$$s^2 - (p_1 + p_2)s + p_1 p_2 = 0$$

so the characteristic polynomial is

$$P(s) = s^2 - (p_1 + p_2)s + p_1 p_2 = (s - p_1)(s - p_2)$$

The natural response thus depends entirely on the characteristic values p_1 and p_2, which are the roots of $P(s)$.

Now, going back to the network function and comparing it with the characteristic polynomial, we see that

> The denominator of $H(s)$ is identical to $P(s)$, so the poles of $H(s)$ are the characteristic values of the circuit's natural response.

This important conclusion remains valid for a circuit of any order.

Suppose then that all poles of an nth-order circuit are **distinct**, meaning no repeated poles. The general form of the natural response will be

$$y_N(t) = A_1 e^{p_1 t} + A_2 e^{p_2 t} + \cdots + A_n e^{p_n t} \tag{10.27}$$

where $A_1, A_2, \ldots, A_n$ are constants. The values of those constants depend on the initial conditions.

Equation (10.27) still holds when some of the poles have complex values. However, if

$$p_i = \sigma_i + j\omega_i \qquad \omega_i \neq 0$$

then we know that there will be a conjugate pole, say $p_{i+1} = p_i^*$, and we must have $A_{i+1} = A_i^*$ to get a real function. Accordingly, we can group two terms together as

$$\begin{aligned} A_i e^{p_i t} + A_{i+1} e^{p_{i+1} t} &= A_i e^{p_i t} + (A_i e^{p_i t})^* = 2\,\mathrm{Re}[A_i e^{p_i t}] \\ &= 2\,|A_i| e^{\sigma_i t} \cos(\omega_i t + \angle A_i) \end{aligned} \tag{10.28}$$

which is our familiar complex-frequency waveform.

Equations (10.27) and (10.28) indicate that each distinct real pole or complex-conjugate pair contributes a particular waveform component to $y_N(t)$. We call these waveforms the **modes** of the natural response.

Figure 10.15 illustrates the modes associated with distinct poles located on the real axis or in the upper half of the s plane. (We omit the lower half because it only contains the complex conjugates of any poles in the upper half.) This pictorial display emphasizes the properties that:

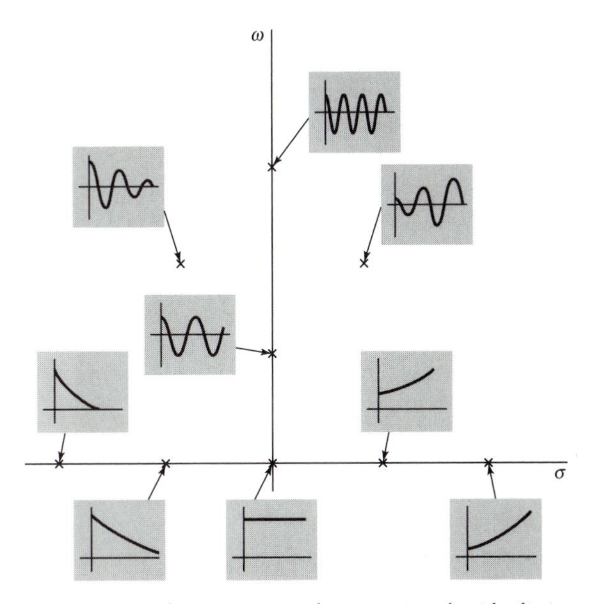

Figure 10.15 Natural-response modes associated with distinct poles.

- Any pole in the left half plane has $\sigma < 0$ so the mode decays with time, while any pole in the right half plane has $\sigma > 0$ so the mode grows with time. The rate of decay or growth increases with the horizontal distance $|\sigma|$ from the imaginary axis.
- Any pole on the imaginary axis has $\sigma = 0$ so the mode is a dc waveform if $\omega = 0$ or an ac waveform if $\omega \neq 0$.
- Any pole located off the real axis has $\omega \neq 0$ and is part of a complex-conjugate pair, so the associated mode oscillates. The rate of oscillation increases with the vertical distance $|\omega|$ from the real axis.

The pole-zero pattern thus tells us a great deal about the natural-behavior modes when the poles are distinct.

But suppose we have a **repeated** real pole, say $p_2 = p_1$. From our study of critically damped second-order circuits, we know that the natural response includes the exponential mode $e^{p_1 t}$ and the modified mode $te^{p_1 t}$. If there are r equal poles, say $p_r = \cdots = p_2 = p_1$, then the general form of the natural response becomes

$$y_N(t) = (A_1 + A_2 t + \cdots + A_r t^{r-1})e^{p_1 t} + \cdots \tag{10.29}$$

Similarly, the two modes associated with a repeated pair of complex-conjugate poles are

$$\begin{aligned}(A_1 + A_2 t)e^{p_1 t} &+ (A_1{}^* + A_2{}^* t)e^{p_1{}^* t} \\ &= 2|A_1|e^{\sigma_1 t}\cos(\omega_1 t + \angle A_1) + 2|A_2|te^{\sigma_1 t}\cos(\omega_1 t + \angle A_2)\end{aligned} \tag{10.30}$$

In short, when poles are repeated, the mode waveforms in Fig. 10.15 are multiplied by t or by powers of t. See, for instance, Fig. 9.24 for the waveform of the mode $te^{-\alpha t}$.

Now observe that the natural response includes a mode that *grows* with time whenever there are poles in the right half of the s plane or repeated poles on the imaginary axis. A network with a growing natural response is said to be **unstable**, and it may cause serious problems when the network in question is part of an electronic circuit or a control system. Furthermore, we cannot meaningfully speak of steady-state conditions in an unstable network because the growing natural response would eventually obscure the forced response.

However, if all poles are in the left half of the s plane, so that

$$\sigma_i = \text{Re}[p_i] < 0 \qquad i = 1, 2, \ldots, n \tag{10.31a}$$

then

$$y_N(t) \rightarrow 0 \text{ as } t \rightarrow \infty \tag{10.31b}$$

Accordingly, we say that

> A network is **stable** when all of its poles fall within the left half of the s plane.

Any network that contains resistance and does not contain controlled sources will be stable, because the resistance dissipates power and controlled sources would be needed to supply the energy for a growing natural response. A network containing controlled sources may or may not be stable, depending on the particular arrangement.

A special situation — neither stable nor unstable — occurs when all poles are in the left half of the s plane, except for a distinct pole at the origin or a conjugate pair of poles on the imaginary axis. The natural response then contains a nondecaying constant or sinusoid. In the latter case, the circuit could serve as an **oscillator**.

Finally, consider a network function in which the value of one of the zeros happens to equal the value of one of the poles. For instance, if $z_2 = p_2$ in Eq. (10.26), then we have **pole-zero cancellation** in that

$$H(s) = K\frac{(s - z_1)(s - p_2)}{(s - p_1)(s - p_2)} = K\frac{s - z_1}{s - p_1}$$

While this reduced expression is valid for the forced response, the natural response still involves both p_1 and p_2 because the circuit has two independent energy-storage elements. In general, the natural response and stability depend on *all* poles, irrespective of pole-zero cancellation. Consequently, you should always check to be sure that the number of poles agrees with the number of independent energy-storage elements.

Example 10.10 ***Natural Responses of a Third-Order Circuit***

In Example 10.7 we analyzed a third-order twin-tee circuit with an op-amp. Two of the network functions obtained there are

$$H_2(s) = \frac{\underline{V}_2}{\underline{V}_{in}} = \frac{a[s^2 + (2 - \mu)as + a^2]}{(s + a)[s^2 + 2(2 - \mu)as + a^2]}$$

$$H_3(s) = \frac{\underline{V}_3}{\underline{V}_{in}} = \frac{s^2 + a^2}{s^2 + 2(2 - \mu)as + a^2}$$

where $a = 1/RC$ and μ is the voltage gain of the op-amp section. The denominator of $H_3(s)$ has been reduced to a second-order polynomial by pole-zero cancellation. Hence, we'll work here with the complete denominator in $H_2(s)$.

For discussion purposes, let $a = 5$ and let μ be variable. The characteristic polynomial becomes

$$P(s) = (s + 5)[s^2 + 2(10 - 5\mu)s + 25]$$

from which we get the poles

$$p_1 = -5 \qquad p_2, p_3 = -(10 - 5\mu) \pm \sqrt{(10 - 5\mu)^2 - 25}$$

The natural-response mode associated with the fixed pole p_1 is just the decaying exponential $A_1 e^{-5t}$. The locations of the other two poles vary with the gain μ.

Table 10.2 lists the values of p_2 and p_3 along with the associated modes for selected values of μ. When the gain falls in the range $1 < \mu < 2$, the mode is a decaying oscillation. Since the circuit contains no inductors, we conclude that this underdamped behavior is produced by the controlled-source effect of the op-amp. Furthermore, when $\mu = 2$ the mode is a nondecaying sinusoid, so the circuit could be used as an oscillator needing no input to drive the output. With $\mu > 2$, the circuit becomes unstable and the mode grows exponentially, with or without oscillations.

TABLE 10.2

μ	p_2, p_3	Mode	Classification
1.2	$-4 \pm j3$	$2\lvert A_2\rvert e^{-4t} \cos(3t + \angle A_2)$	Stable
2.0	$0 \pm j5$	$2\lvert A_2\rvert \cos(5t + \angle A_2)$	Oscillator
2.6	$3 \pm j4$	$2\lvert A_2\rvert e^{3t} \cos(4t + \angle A_2)$	Unstable
3.0	5, 5	$A_2 e^{5t} + A_3 te^{5t}$	Unstable
4.6	1, 25	$A_2 e^{t} + A_3 e^{25t}$	Unstable

Exercise 10.11

Let the circuit in Example 10.10 have $a = 2$. Find $y_N(t)$ when $\mu = 0$ and $\mu = 1$.

Exercise 10.12

Find $y_N(t)$ for the circuit in Exercise 10.7 (p. 465) when $\beta = 3$ and $\beta = 4$.

10.5 NETWORK SCALING†

Many of our examples have involved simple but unrealistic values for elements and frequency — values such as $R = 1\ \Omega$, $C = 1/8$ F, or $\omega = 5$ rad/s. In practice, of course, you encounter actual values such as $R = 4.3 \times 10^3\ \Omega$, $C = 0.22 \times 10^{-6}$ F, or $\omega = 2\pi \times 60$ Hz. Manipulating those awkward numbers is tiresome and easily leads to computational errors. Thankfully, numerical calculations can be simplified by **network scaling**.

The essence of scaling for analysis is to replace the actual network diagram with a scaled version having more convenient element values — somewhat like building a scale model of an aircraft for wind-tunnel tests. Conversely, circuit designers often develop design solutions in the form of prototype networks that can be rescaled to meet the needs of specific applications. In either case, network scaling generally involves both magnitude and frequency.

For **magnitude scaling**, you choose a suitable scale factor k_m and work with a scaled impedance $\hat{Z}$ related to the actual impedance Z by

$$\hat{Z} = k_m Z \qquad k_m > 0 \tag{10.32}$$

All impedances must be increased or decreased by the same factor. Thus, any voltage-to-current phasor ratio $\underline{V}_y/\underline{I}_x$ is multiplied by k_m and the scaled ratio becomes

$$\hat{\underline{V}}_y/\hat{\underline{I}}_x = k_m(\underline{V}_y/\underline{I}_x) \tag{10.33}$$

However, all voltage-to-voltage and current-to-current ratios are dimensionless quantities, so they remain unaffected by magnitude scaling.

For **frequency scaling**, you choose another scale factor k_f and work with a scaled complex frequency $\hat{s}$ related to the unscaled frequency s by

$$\hat{s} = k_f s \qquad k_f > 0 \tag{10.34}$$

Frequency scaling thereby expands or shrinks the entire pole-zero pattern. This property becomes evident by considering the unscaled network function

$$H(s) = K\frac{(s - z_1)\cdots(s - z_m)}{(s - p_1)\cdots(s - p_n)}$$

Replacing s with $\hat{s}/k_f$ produces the frequency-scaled function

$$\begin{aligned}\hat{H}(\hat{s}) &= K\frac{(\hat{s}/k_f - z_1)\cdots(\hat{s}/k_f - z_m)}{(\hat{s}/k_f - p_1)\cdots(\hat{s}/k_f - p_n)}\\ &= Kk_f^{\,n-m}\frac{(\hat{s} - k_f z_1)\cdots(\hat{s} - k_f z_m)}{(\hat{s} - k_f p_1)\cdots(\hat{s} - k_f p_n)}\end{aligned} \tag{10.35}$$

Therefore, each unscaled pole p_i and zero z_i move to

$$\hat{p}_i = k_f p_i \qquad \hat{z}_i = k_f z_i \tag{10.36}$$

Equivalently, frequency scaling corresponds to scaling *time* by $1/k_f$ since $st = (\hat{s}/k_f)t = \hat{s}\hat{t}$ with $\hat{t} = t/k_f$.

With appropriate scale factors, the joint use of magnitude and frequency scaling yields more convenient values for circuit elements. The scaled impedance of any resistor is $k_m Z_R = k_m R$, so the scaled resistance simply becomes

$$\hat{R} = k_m R \tag{10.37a}$$

Likewise, if the circuit contains a current-controlled voltage source or a voltage-controlled current source, then the transresistance r_m or transconductance g_m is scaled as

$$\hat{r}_m = k_m r_m \qquad \hat{g}_m = g_m / k_m \tag{10.37b}$$

Note that resistance or conductance scaling does not involve the frequency-scaling factor k_f. However, the scaled impedance of an inductor is

$$\hat{Z} = k_m Z_L = k_m sL = k_m(\hat{s}/k_f)L = \hat{s}(k_m L/k_f)$$

so the scaled inductance becomes

$$\hat{L} = k_m L / k_f \tag{10.37c}$$

Similarly, a scaled capacitance becomes

$$\hat{C} = C / k_m k_f \tag{10.37d}$$

which follows from $Z_C = 1/sC$.

Table 10.3 summarizes the relationships for magnitude and frequency scaling. The table omits the voltage gain μ of a voltage-controlled voltage source and the current gain β of a current-controlled current source because they are dimensionless ratios and independent of frequency.

TABLE 10.3 Network Scaling Relationships

Unscaled Value	Z	s	R	r_m	g_m	L	C
Scaled Value	$k_m Z$	$k_f s$	$k_m R$	$k_m r_m$	$\dfrac{g_m}{k_m}$	$\dfrac{k_m}{k_f} L$	$\dfrac{C}{k_m k_f}$

Having derived and analyzed a scaled network, you can obtain the poles and zeros of the unscaled network via Eq. (10.36) rewritten as

$$p_i = \hat{p}_i / k_f \qquad z_i = \hat{z}_i / k_f \tag{10.38}$$

Then, for the gain factor K in $H(s)$, you need to take account of both magnitude and frequency scaling. Specifically, if $\hat{H}(\hat{s})$ has gain factor $\hat{K}$, then

$$K = k_m{}^q \hat{K} / k_f^{n-m} \tag{10.39a}$$

where

$$q = \begin{cases} 1 & \text{if } H(s) = \underline{I}_y / \underline{V}_x \\ -1 & \text{if } H(s) = \underline{V}_y / \underline{I}_x \\ 0 & \text{otherwise} \end{cases} \tag{10.39b}$$

Equation (10.39a) combines the results from Eqs. (10.33) and (10.35).

Example 10.11 *Scaling Calculations*

To illustrate the advantage and technique of network scaling, suppose you need the network function $H(s) = \underline{I}_L/\underline{V}_s$ for the circuit in Fig. 10.16*a*. The analysis with symbols is routine, leading to

$$H(s) = \frac{\underline{I}_L}{\underline{V}_s} = \frac{1}{Z(s)} = \frac{1}{sL + \dfrac{1}{sC + 1/R}}$$

However, upon inserting the element values you get

$$H(s) = \frac{1}{0.04s + \dfrac{1}{25 \times 10^{-9}s + 5 \times 10^{-4}}}$$

Rather than manipulating this messy expression, you decide to use network scaling. Your general strategy is to choose k_m and k_f to get more convenient numerical values.

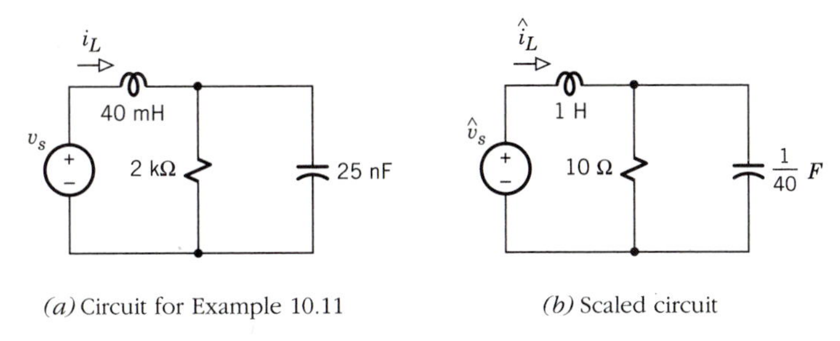

(*a*) Circuit for Example 10.11 (*b*) Scaled circuit

Figure 10.16

Since resistance is affected only by the magnitude scale factor, you might first take

$$k_m = 0.005 = 1/200$$

so that

$$\hat{R} = 0.005 \times 2\ \text{k}\Omega = 10\ \Omega$$

while

$$\hat{L} = \frac{k_m}{k_f} \times 40\ \text{mH} = \frac{2 \times 10^{-4}}{k_f}\ \text{H} \qquad \hat{C} = \frac{25\ \text{nF}}{k_m k_f} = \frac{5 \times 10^{-6}}{k_f}\ \text{F}$$

The expression for the scaled inductance then suggests taking

$$k_f = 2 \times 10^{-4} = 1/5000$$

in which case

$$\hat{L} = 1 \text{ H} \qquad \hat{C} = 0.025 = 1/40 \text{ F}$$

Figure 10.16*b* diagrams the resulting scaled circuit.

The network function for the scaled network is now easily found to be

$$\hat{H}(\hat{s}) = \hat{\underline{I}}_L/\hat{\underline{V}}_s = (\hat{s} + 4)/(\hat{s}^2 + 4\hat{s} + 40)$$

from which

$$\hat{K} = 1 \qquad \hat{z}_1 = -4 \qquad \hat{p}_1, \hat{p}_2 = -2 \pm j6$$

Therefore, the network function for the original circuit is

$$H(s) = \frac{\underline{I}_L}{\underline{V}_s} = K\frac{s + z_1}{(s + p_1)(s + p_2)}$$

where

$$z_1 = -4/k_f = -20{,}000$$
$$p_1, p_2 = (-2 \pm j6)/k_f = -10{,}000 \pm j30{,}000$$

The gain factor is given by Eq. (10.39a) with $q = 1$, $n = 2$, and $m = 1$, so

$$K = k_m/k_f = 25$$

Exercise 10.13

Figure 10.17 is the diagram of a scaled circuit. The unscaled circuit has $R_1 = 2.5\ \text{k}\Omega$ and $L = 50$ mH.

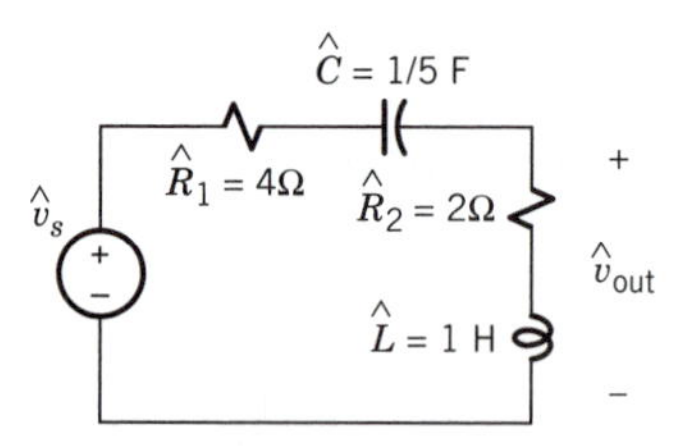

Figure 10.17

(a) What are the other unscaled element values?

(b) Analyze the scaled circuit to find the zeros, poles, and gain factor of the unscaled network function $H(s) = \underline{V}_{out}/\underline{V}_s$.

PROBLEMS

Section 10.1 Complex Frequency and Generalized Impedance

10.1* Let the network in Fig. P10.1 have $R_1 = R_2 = 1\ \Omega$, $L = 1$ H, and $C = 1/20$ F. Show that the input impedance is

$$Z(s) = (2s^2 + 24s + 40)/(s^2 + 4s + 20)$$

Then find the forced response $v(t)$ when $i(t) = 10e^{-5t}$ A.

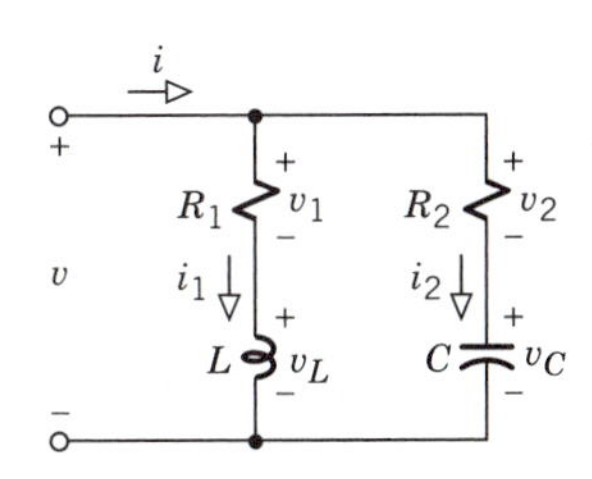

Figure P10.1

10.2 Let the network in Fig. P10.2 have $C = 1/4$ F, $R_1 = 4\ \Omega$, $R_2 = 5\ \Omega$, and $L = 1$ H. Show that the input admittance is

$$Y(s) = (s^2 + 9s)/(4s^2 + 24s + 36)$$

Then find the forced response $i(t)$ when $v(t) = -3e^{-4t}$ V.

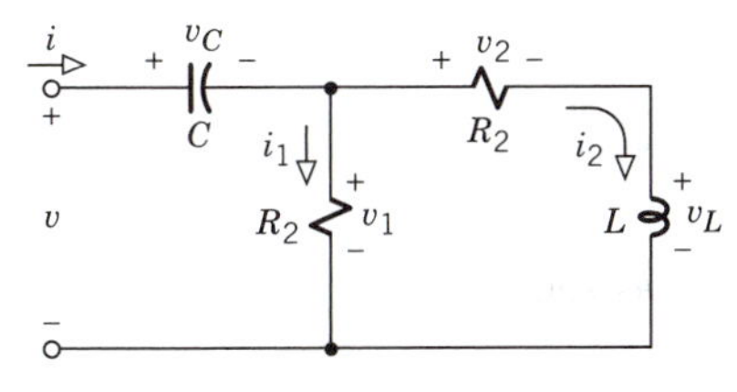

Figure P10.2

10.3 Let the network in Fig. P10.3 have $R_1 = 8\ \Omega$, $L_1 = 2$ H, $R_2 = 4\ \Omega$, and $L_2 = 1$ H. Show that the input impedance is

$$Z(s) = (4s^2 + 16s)/(s^2 + 10s + 16)$$

Then find the forced response $v(t)$ when $i_s(t) = 10 \cos(4t - 90°)$ A.

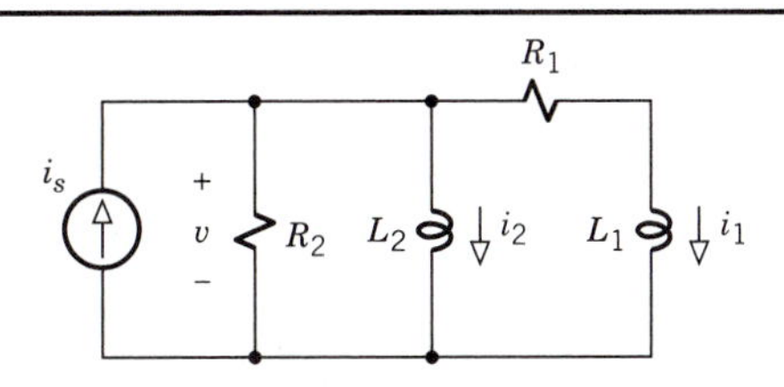

Figure P10.3

10.4 Let the network in Fig. P10.4 have $R_1 = 2\ \Omega$, $C = 1/2$ F, $R_2 = 3\ \Omega$, and $L = 1$ H. Show that the input admittance is

$$Y(s) = (s^2 + 3s + 2)/(2s^2 + 8s + 10)$$

Then find the forced response $i(t)$ when $v(t) = -5 \cos t$ V.

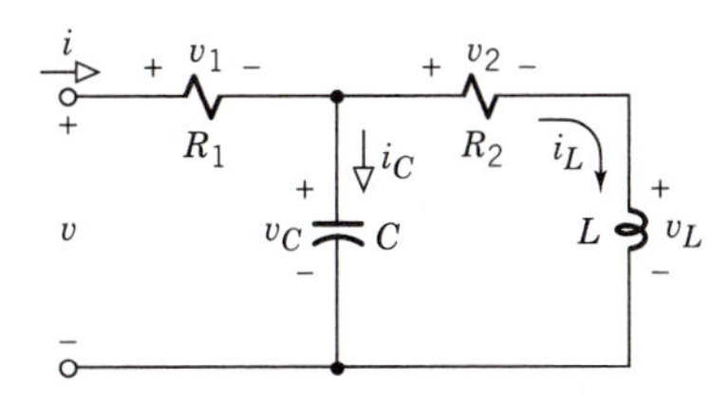

Figure P10.4

10.5* Consider the series LRC circuit in Fig. 9.17 (p. 423). Evaluate $Z(s)$ at $s = -2 + j6$ with $L = 5$ H, $R = 5\ \Omega$, and $C = 1/40$ F. Then find $i_L(t)$ forced by $v_s(t) = 100e^{-2t} \cos(6t + 90°)$ V.

10.6 Consider the parallel CRL circuit in Fig. 9.18 (p. 424). Evaluate $Y(s)$ at $s = -2 + j4$ with $C = 1/40$ F, $R = 20\ \Omega$, and $L = 1$ H. Then find $v_C(t)$ forced by $i_s(t) = 3e^{-2t} \cos(4t - 180°)$.

10.7 Use impedance analysis to show that connecting capacitors C_1 and C_2 in series gives

$$C_{eq} = C_1 C_2/(C_1 + C_2)$$

10.8 Use impedance analysis to show that connecting inductors L_1 and L_2 in parallel gives

$$L_{eq} = L_1 L_2/(L_1 + L_2)$$

10.9* The network in Fig. P10.2 is driven by $v(t) = \text{Re}[\underline{V}e^{st}]$ with $\underline{V} \neq 0$ and the forced response is found to be $i(t) = 0$. Show that there are two possible values of s, and obtain expressions for those values in terms of R_1, R_2, L, and C.

10.10 The network in Fig. P10.1 is driven by $i(t) = \text{Re}[\underline{I}e^{st}]$ with $\underline{I} \neq 0$ and the forced response is found to be $v(t) = 0$. Show that there are two possible values of s, and obtain expressions for those values in terms of R_1, R_2, L, and C.

10.11 The network in Fig. P10.4 is driven by $v(t) = \text{Re}[\underline{V}e^{st}]$ with $\underline{V} \neq 0$ and the forced response is found to be $i(t) = 0$. Show that there are two possible values of s, and obtain expressions for those values in terms of R_1, R_2, L, and C.

Section 10.2 Network Functions

10.12* Obtain $H(s) = \underline{V}_C/\underline{V}_s$ for the circuit in Fig. P10.12 by taking $\underline{V}_C = 1$ and invoking proportionality. Express your result as a ratio of polynomials.

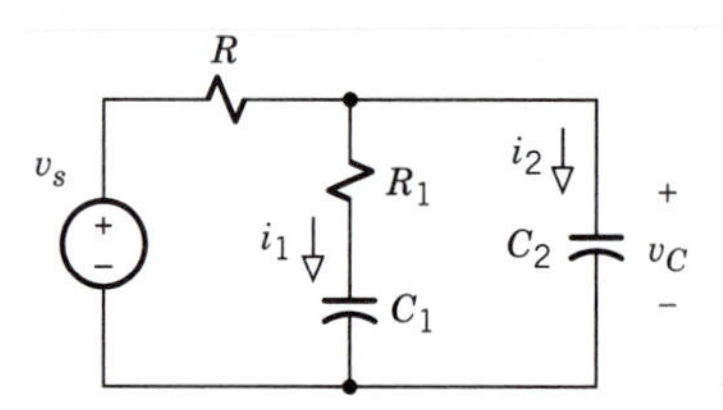

Figure P10.12

10.13 Obtain $H(s) = \underline{I}_2/\underline{I}_s$ for the circuit in Fig. P10.13 by taking $\underline{I}_2 = 1$ and invoking proportionality. Express your result as a ratio of polynomials.

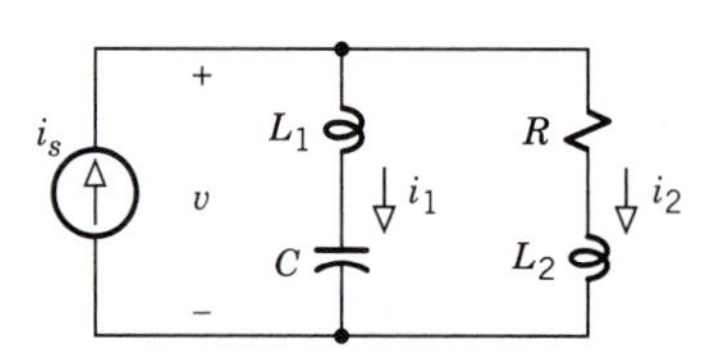

Figure P10.13

10.14 Obtain $H(s) = \underline{I}_R/\underline{V}_s$ for the circuit in Fig. P10.14 by taking $\underline{I}_R = 1$ and invoking proportionality. Express your result as a ratio of polynomials.

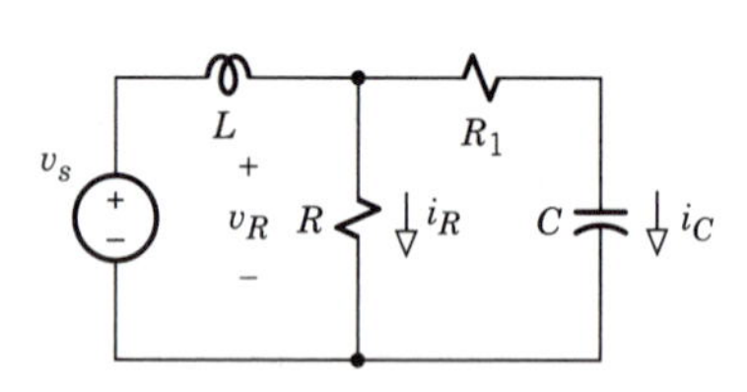

Figure P10.14

10.15 Obtain $H(s) = \underline{I}_1/\underline{I}_s$ for the circuit in Fig. P10.3 by taking $\underline{I}_1 = 1$ and invoking proportionality. Express your result as a ratio of polynomials. Then simplify $H(s)$ for the case when $R_2 \to \infty$.

10.16 Obtain $H(s) = \underline{I}_2/\underline{I}_s$ for the circuit in Fig. P10.3 by writing a single s-domain node equation for v. Express your result as a ratio of polynomials. Then simplify $H(s)$ for the case when $R_1 \to 0$.

10.17 Write a single s-domain node equation for v_C in Fig. P10.17. Then use a voltage divider to obtain

$$H(s) = \underline{V}/\underline{V}_s = 5s/(s+2)(s^2+2s+10)$$

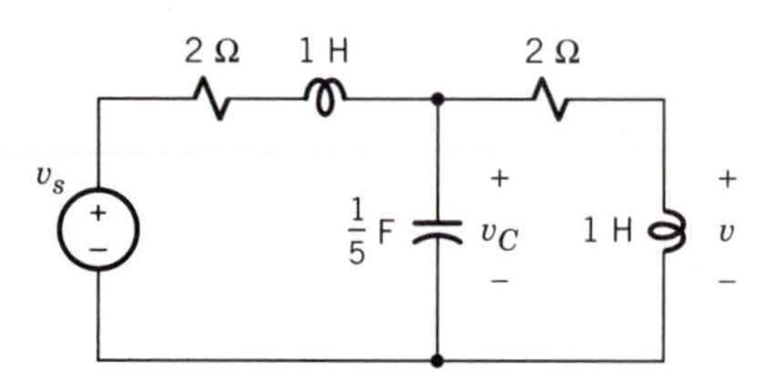

Figure P10.17

10.18 Write a single s-domain mesh equation for i in Fig. P10.18 to obtain

$$H(s) = \frac{\underline{I}}{\underline{I}_s} = \frac{s(s+2-2\mu)}{s^2+(6-2\mu)s+5}$$

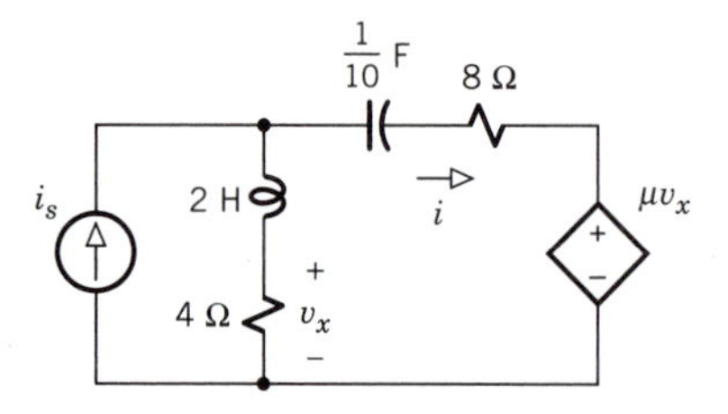

Figure P10.18

10.19 Write a single s-domain node equation for v in Fig. P10.19 to obtain

$$H(s) = \frac{\underline{V}}{\underline{V}_s} = \frac{-4g_m(s+1-1/g_m)}{s^2+(2-4g_m)s+5-4g_m}$$

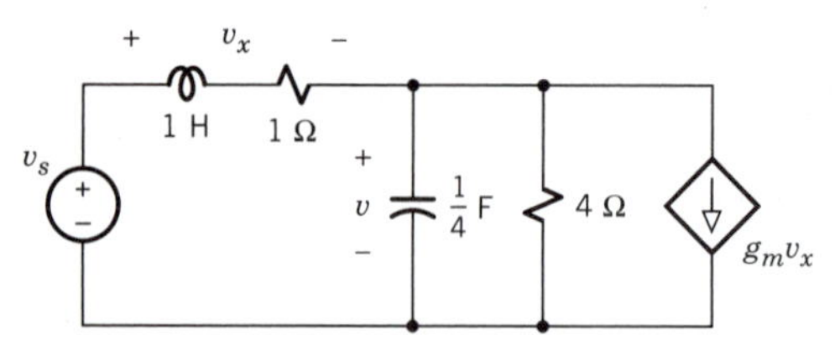

Figure P10.19

10.20 Use *s*-domain matrix mesh analysis for i_1 and i_2 in Fig. P10.20 to obtain

$$H_1(s) = \underline{I}_1/\underline{I}_s = (s^2 + s + 20)/P(s)$$
$$H_2(s) = \underline{I}_2/\underline{I}_s = 20/P(s)$$

where

$$P(s) = s^3 + 5s^2 + 24s + 20 = (s + 1)(s^2 + 4s + 20)$$

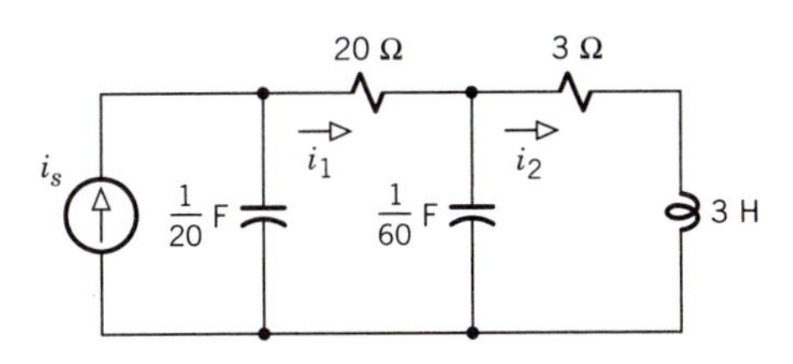

Figure P10.20

10.21 Use *s*-domain matrix node analysis for v_1 and v_2 in Fig. P10.21 to obtain

$$H_1(s) = \underline{V}_1/\underline{V}_s = 24(s^2 + 8s + 24)/P(s)$$
$$H_2(s) = \underline{V}_2/\underline{V}_s = 576/P(s)$$

where

$$P(s) = s^3 + 10s^2 + 64s + 240 = (s + 6)(s^2 + 4s + 40)$$

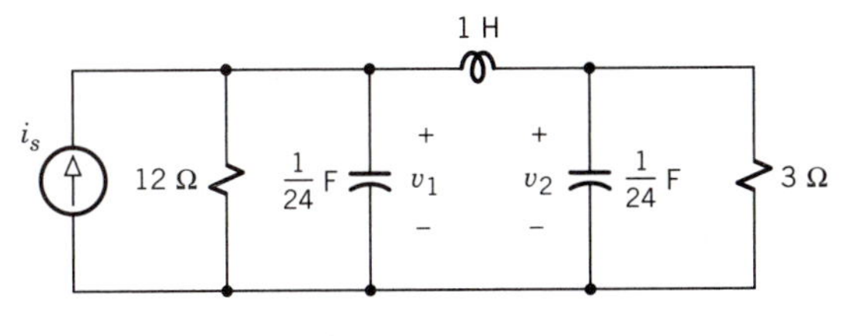

Figure P10.21

10.22 Use *s*-domain matrix mesh analysis for i_1 and i_2 in Fig. P10.22 to obtain

$$H_1(s) = \underline{I}_1/\underline{V}_s = s[(5 - r_m)s + 16]/P(s)$$
$$H_2(s) = \underline{I}_2/\underline{V}_s = (1 - 0.25r_m)s^2/P(s)$$

where

$$P(s) = s^2 + (36 - 4r_m)s + 64$$

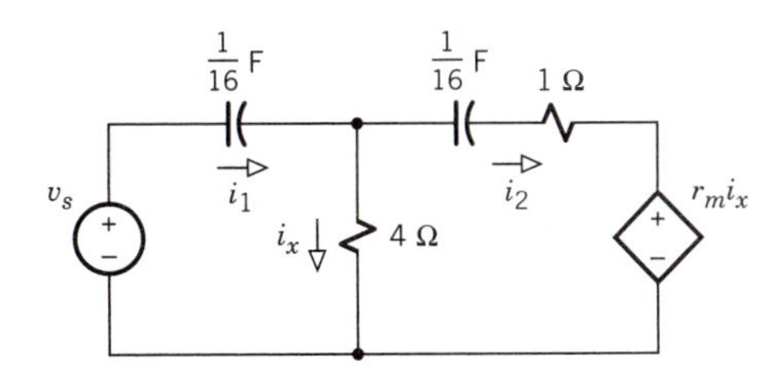

Figure P10.22

10.23 Use *s*-domain matrix node analysis for v_1 and v_2 in Fig. P10.23 to obtain

$$H_1(s) = \underline{V}_1/\underline{I}_s = 20(s + 50)/P(s)$$
$$H_2(s) = \underline{V}_2/\underline{I}_s = 10(\beta s^2 + 100)/P(s)$$

where

$$P(s) = s^2 + (52 - 50\beta)s + 200$$

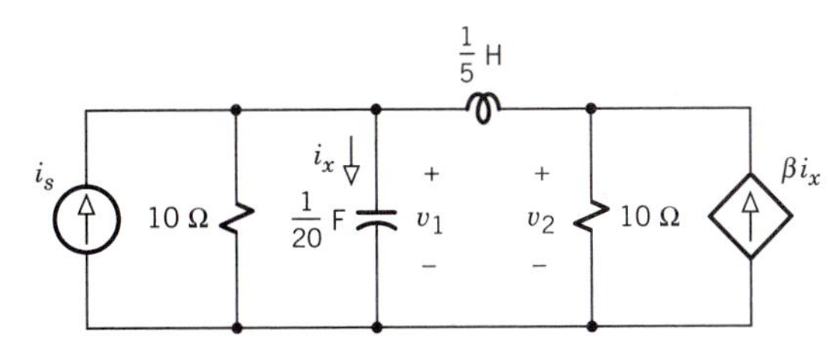

Figure P10.23

Section 10.3 Network Functions with Mutual Inductance

10.24* Let the coils in Fig. P10.24 have $M = 1$ H and the same winding sense. Use series-parallel reduction to find $H(s) = \underline{I}_{in}/\underline{V}_{in} = 1/Z_{in}(s)$.

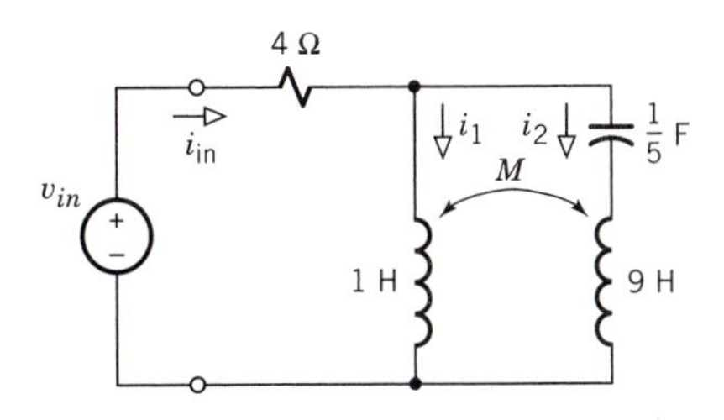

Figure P10.24

10.25 Find $H(s) = \underline{I}_{out}/\underline{I}_{in}$ in Fig. P10.25 when $L_1 = 1$ H, $L_2 = 10$ H, $M = 3$ H, and the load is a 5-Ω resistor.

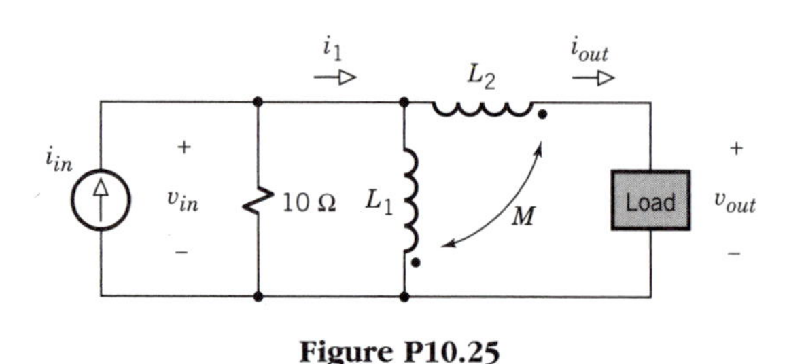

Figure P10.25

10.26 Find $H(s) = \underline{V}_{out}/\underline{I}_{in}$ in Fig. P10.25 when $L_1 = 20$ H, $L_2 = 1$ H, $M = 4$ H, and the load is a 6-Ω resistor in series with a 3-H inductor.

10.27 Let the coils in Fig. P10.24 have $M = 2$ H and the opposite winding sense. Redraw the circuit so that i_1 is a mesh current, and find $H(s) = \underline{I}_1/\underline{V}_{in}$.

10.28 Let the coils in Fig. P10.24 have $M = k\sqrt{L_1 L_2}$ and the same winding sense. Find $H(s) = \underline{I}_2/\underline{V}_{in}$ and simplify your results for the special cases of $k = 0$ and $k = 1$.

10.29 Find $H(s) = \underline{V}_{in}/\underline{I}_{in}$ in Fig. P10.25 when $L_1 = 8$ H, $L_2 = 2$ H, $M = 4$ H, and the load is a 5-Ω resistor in series with a 1/10-F capacitor.

Section 10.4 s-Domain Analysis

10.30* Determine K and draw the pole-zero pattern when $H(s)$ equals $Z(s)$ in Problem 10.1.

10.31 Determine K and draw the pole-zero pattern when $H(s)$ equals $Y(s)$ in Problem 10.2.

10.32 Determine K and draw the pole-zero pattern when $H(s)$ equals $Z(s)$ in Problem 10.3.

10.33 Determine K and draw the pole-zero pattern when $H(s)$ equals $Y(s)$ in Problem 10.4.

10.34* Use s-plane vectors to calculate $H(s_0)$ in polar form when $s_0 = 0 + j8$ and

$$H(s) = 51s^2/(s + 15)(s^2 + 24s + 208)$$

10.35 Use s-plane vectors to calculate $H(s_0)$ in polar form when $s_0 = 0 + j4$ and

$$H(s) = -90(s^2 - 16)/s(s^2 + 8s + 80)$$

10.36 Use s-plane vectors to calculate $H(s_0)$ in polar form when $s_0 = -16 + j12$ and

$$H(s) = 7s(s + 16)/(s + 32)(s^2 + 18s + 225)$$

10.37 Use s-plane vectors to calculate $H(s_0)$ in polar form when $s_0 = -7 + j12$ and

$$H(s) = -26(s^2 + 144)/(s + 12)(s^2 + 14s + 113)$$

10.38 Given the results in Problem 10.20, find the forced response of $i_2(t)$ when $i_s = 8e^{3t}\cos 4t$ A.

10.39 Given the results in Problem 10.21, find the forced response of $v_2(t)$ when $v_s = 5e^{-4t}\cos 3t$ A.

10.40 Given the results in Problem 10.17, find the forced response of $v(t)$ when $v_s = 40e^{3t}\cos 3t$ A.

10.41* Evaluate the poles and zeros of $H(s)$ in Problem 10.18 with $\mu = 0, 2, 3, 4$, and 6. Also write the natural response for each case, and label the ones that are stable, unstable, or oscillatory.

10.42 Evaluate the poles and zeros of $H_1(s)$ in Problem 10.22 with $r_m = 4, 5, 9, 12.2$, and 13. Also write the natural response for each case, and label the ones that are stable, unstable, or oscillatory.

10.43 Evaluate the poles and zeros of $H_2(s)$ in Problem 10.23 with $\beta = 0.38, 0.474, 0.96, 1.44$, and 1.64. Also write the natural response for each case, and label the ones that are stable, unstable, or oscillatory.

10.44 Evaluate the poles and zeros of $H(s)$ in Problem 10.19 with $g_m = 0, 0.5, 1, 1.25$, and 1.5. Also write the natural response for each case, and label the ones that are stable, unstable, or oscillatory.

Section 10.5 Network Scaling

10.45* Let Fig. P10.18 be a prototype circuit with $\mu = 3$ and output i. Taking advantage of the result given in Problem 10.18, obtain a scaled circuit whose poles are at $\pm j2\pi \times 1$ kHz and whose dc input resistance is 1 kΩ. What's the resulting location of the zeros?

10.46 Let Fig. P10.23 be a prototype circuit with $\beta = 1$ and output v_2. Taking advantage of the result given in Problem 10.23, obtain a scaled circuit with a 1-mH inductor whose zeros are at $\pm j2\pi \times 30$ kHz. What's the resulting location of the poles?

10.47 Let Fig. P10.19 be a prototype circuit with $g_m = 0.5$ and output v. Taking advantage of the result given in Problem 10.19, obtain a scaled circuit with a 0.1-μF capacitor and poles at $\pm j2\pi \times 100$ kHz. What's the resulting location of the zeros?

10.48 Let the network in Fig. P10.1 have $R_1 = 100$ Ω, $L = 0.4$ H, $R_2 = 500$ Ω, and $C = 5$ μF. Use a scaled network with $\hat{R}_1 = 1$ Ω and $\hat{L} = 1$ H to find the zeros, poles, and gain factor of $H(s) = 1/Y(s)$.

10.49 Let the network in Fig. P10.2 have $C = 2.5$ nF, $R_1 = 4$ kΩ, $R_2 = 6$ kΩ, and $L = 20$ mH. Use a scaled network with $\hat{R}_1 = 2$ Ω and $\hat{C} = 1/2$ F to find the zeros, poles, and gain factor of $H(s) = 1/Z(s)$.

10.50 Let the network in Fig. P10.1 have $R_1 = 75$ Ω, $L = 10$ mH, $R_2 = 25$ Ω, and $C = 4$ μF. Use a scaled network with $\hat{L} = 1$ H and $\hat{C} = 1$ F to find the zeros, poles, and gain factor of $H(s) = 1/Y(s)$.

10.51 Let the network in Fig. P10.4 have $R_1 = 500$ Ω, $C = 8$ nF, $R_2 = 750$ Ω, and $L = 1$ mH. Use a scaled network with $\hat{C} = 1/2$ F and $\hat{L} = 1$ H to find the zeros, poles, and gain factor of $H(s) = 1/Z(s)$.

CHAPTER 11

Frequency Response and Filters

Audio amplifiers, radio receivers, and television sets are familiar examples of signal-processing systems — the *signals* being electrical waveforms. Such signals may often be viewed as consisting of sinusoidal components at various frequencies. Underscoring this point, Table 11.1 lists the range of significant frequencies (in hertz) for representative signals in communication systems.

A crucial question in the analysis of communication systems and other signal-processing circuits is:

TABLE 11.1 Frequency Ranges of Typical Communication Signals

Signal type	Frequency range
Telephone-quality voice	200 Hz – 3.2 kHz
AM broadcast-quality audio	100 Hz – 5 kHz
High-fidelity audio	20 Hz – 20 kHz
Television video	60 Hz – 4.2 MHz
Broadcast AM on 540-kHz carrier	535 kHz/545 kHz
Broadcast FM on 88.5-MHz carrier	88.3 MHz – 88.7 MHz

How does the circuit respond to the different frequency components of the signal?

A related design question is

How can unwanted frequency components such as "hum" or interference be removed from a signal?

This chapter addresses both questions by considering frequency to be a *variable* so that we can investigate circuit response as a function of frequency.

After an introduction to the concept of frequency response, we'll examine simple frequency-selective networks that act as filters. Then we'll develop the Bode plot as a graphical technique for displaying frequency-response curves and for doing frequency-response design. The closing section discusses more sophisticated types of filters.

Objectives

After studying this chapter and working the exercises, you should be able to do each of the following:

1. Define and interpret the amplitude ratio and phase shift of a network, given its transfer function (Section 11.1).
2. Use *s*-plane vectors to sketch the frequency-response curves of a network (Section 11.1).
3. Distinguish between ideal lowpass, highpass, bandpass, and notch filters (Section 11.2).
4. Identify the characteristics of simple circuits that approximate ideal filters, and calculate the filter parameters (Section 11.2).
5. Design a simple op-amp filter (Section 11.3).†
6. Convert a gain value to or from decibels (Section 11.4).
7. Draw the magnitude and phase asymptotes and add the correction terms to obtain the complete Bode plot for a network function consisting of first-order factors (Section 11.4).
8. Sketch the magnitude Bode plot for a network function with a quadratic factor (Section 11.4).
9. Design an op-amp circuit to achieve a specified dB gain curve consisting of first-order factors (Section 11.5).†

10. Write the factored transfer function of a Butterworth lowpass or highpass filter of specified order (Section 11.6).†
11. Design an op-amp circuit to implement a Butterworth filter (Section 11.6).†

11.1 FREQUENCY RESPONSE

This section introduces frequency response defined in terms of a network's amplitude ratio and phase shift. Then we'll develop a method for sketching frequency-response curves with the help of *s*-plane vectors.

Amplitude Ratio and Phase Shift

Recall from Section 10.2 that an appropriate network function $H(s)$ can be used to find the forced response of any voltage or current in a circuit when the excitation has a complex-frequency waveform. Also recall that a complex-frequency waveform reduces to a sinusoidal ac waveform when $s = j\omega$. This is the case we'll pursue here in greater detail. Throughout, we'll refer to $H(s)$ as the *transfer* function because the response of interest is usually an "output" variable rather than a driving-point variable.

Consider a stable linear network driven by a sinusoidal input written in general as

$$x(t) = X_m \cos(\omega t + \phi_x) = \text{Re}[\underline{X}e^{j\omega t}] \tag{11.1a}$$

Under steady-state conditions, any specified output has the form

$$y(t) = Y_m \cos(\omega t + \phi_y) = \text{Re}[\underline{Y}e^{j\omega t}] \tag{11.1b}$$

The output phasor $\underline{Y}$ is related to the input phasor $\underline{X}$ by the relevant transfer function $H(s)$ with $s = j\omega$, so

$$\underline{Y} = H(j\omega)\underline{X} \tag{11.2}$$

Thus, the output amplitude is $Y_m = |H(j\omega)|X_m$ and the output phase is $\phi_y = \angle H(j\omega) + \phi_x$.

For the study of frequency response, we'll denote the magnitude and angle of $H(j\omega)$ by

$$a(\omega) \triangleq |H(j\omega)| \qquad \theta(\omega) \triangleq \angle H(j\omega) \tag{11.3}$$

We call $a(\omega)$ the network's **amplitude ratio**, and we call $\theta(\omega)$ the **phase shift** in view of the properties

$$Y_m/X_m = a(\omega) \qquad \phi_y - \phi_x = \theta(\omega)$$

Accordingly, we can write the output as

$$y(t) = a(\omega)X_m \cos [\omega t + \theta(\omega) + \phi_x] \tag{11.4}$$

The functional notations $a(\omega)$ and $\theta(\omega)$ emphasize that both the amplitude ratio and phase shift generally depend upon the excitation frequency ω.

Now suppose that the input actually consists of a sum of sinusoids at different frequencies, say

$$x(t) = X_1 \cos (\omega_1 t + \phi_1) + X_2 \cos (\omega_2 t + \phi_2) + \ldots \tag{11.5a}$$

Since the network is linear, we can immediately invoke superposition to obtain the resulting steady-state output as

$$y(t) = a(\omega_1)X_1 \cos [\omega_1 t + \theta(\omega_1) + \phi_1] + a(\omega_2)X_2 \cos [\omega_2 t + \theta(\omega_2) + \phi_2] + \ldots \tag{11.5b}$$

This expression brings out the fact that you may need to evaluate a network's amplitude ratio and phase shift at several different frequencies.

Finally, it should be noted that $a(\omega)$ and $\theta(\omega)$ are meaningful only when the network is *stable*. Otherwise, the output never gets to steady-state conditions because it would include a nondecaying natural response.

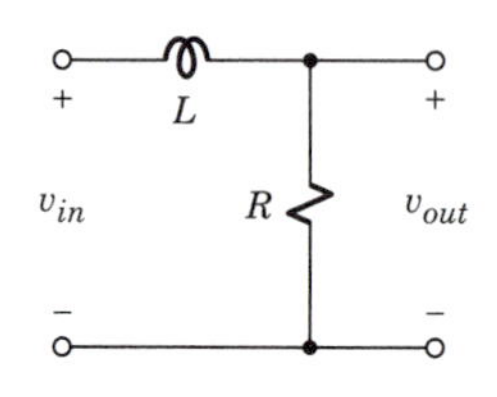

Figure 11.1
A frequency-selective network.

Example 11.1 ***A Frequency-Selective Network***

The network in Fig. 11.1 is driven by

$$v_{in}(t) = 10 \cos 20t + 10 \cos 300t$$

We seek the resulting steady-state output voltage, given that $R = 8\ \Omega$ and $L = 0.2$ H.

First, we let s be arbitrary and calculate the transfer function

$$H(s) = \frac{\underline{V}_{out}}{\underline{V}_{in}} = \frac{R}{sL + R} = \frac{(R/L)}{s + (R/L)}$$

Next, we set $s = j\omega$ and insert $R/L = 40$ to get

$$H(j\omega) = \frac{40}{40 + j\omega}$$

Conversion to polar form then yields the amplitude ratio and phase shift

$$a(\omega) = |H(j\omega)| = \frac{40}{\sqrt{1600 + \omega^2}} \qquad \theta(\omega) = \angle H(j\omega) = -\tan^{-1} \frac{\omega}{40}$$

Evaluating these functions at the input frequencies $\omega = 20$ and $\omega = 300$, we obtain

$$a(20) = 0.894 \qquad \theta(20) = -26.6°$$
$$a(300) = 0.132 \qquad \theta(300) = -82.4°$$

Thus,

$$v_{out}(t) = 8.94 \cos(20t - 26.6°) + 1.32 \cos(300t - 82.4°)$$

A comparison of $v_{out}(t)$ with $v_{in}(t)$ shows that the frequency-selective effect of the network has significantly reduced the amplitude of the output component at the higher frequency.

Exercise 11.1

Rework Example 11.1 with the positions of the resistor and inductor interchanged.

Frequency-Response Curves

The characteristics of frequency-selective networks are often presented as plots of amplitude ratio and phase shift versus frequency. These **frequency-response curves** play a valuable role in the design of audio amplifiers, the performance evaluation of radio circuits, and numerous other applications involving frequency-selective effects. Here, we'll discuss two different methods for obtaining frequency-response curves.

The *analytical method* starts with the expressions for $a(\omega) = |H(j\omega)|$ and $\theta(\omega) = \angle H(j\omega)$, which you often need in subsequent work. However, these expressions become awkward to handle when the transfer function contains complex poles or zeros. Furthermore, $a(\omega)$ and $\theta(\omega)$ must be evaluated at several frequencies so that you can draw curves through the calculated points. A computer could do the tedious calculations, of course, but sheer "number crunching" does not help you understand the factors that account for the shapes of the curves.

The *graphical method* involves the pole-zero pattern of $H(s)$ with s-plane vectors drawn from the poles and zeros to a test point $s = j\omega$ on the imaginary axis. Then, from Eq. (10.25), you can calculate or estimate $a(\omega)$ and $\theta(\omega)$ at any value of ω via

$$a(\omega) = |K| \frac{|j\omega - z_1||j\omega - z_2| \ldots}{|j\omega - p_1||j\omega - p_2| \ldots} \tag{11.6a}$$

$$\theta(\omega) = \angle K + [\angle(j\omega - z_1) + \angle(j\omega - z_2) + \ldots] - [\angle(j\omega - p_1) + \angle(j\omega - p_2) + \ldots] \tag{11.6b}$$

By moving the test point to selected values along the imaginary axis, you can clearly see how the poles and zeros influence the frequency response. Additionally, the s-plane vector constructions resolve phase ambiguities that sometime arise in the analytic method.

Further examination of Eqs. (11.6a) and (11.6b) reveals that the response at very high frequencies depends only on K and the number of poles and zeros. For if there are n poles and $m \leqslant n$ zeros, then letting $\omega \to \infty$ yields

$$a(\infty) = \begin{cases} |K| & m = n \\ 0 & m < n \end{cases} \tag{11.7a}$$

$$\theta(\infty) = \angle K + (m - n) \times 90° \tag{11.7b}$$

In contrast, however, the response at very low frequencies depends on the *locations* of all poles and zeros, as seen by letting $\omega \to 0$ in Eqs. (11.6a) and (11.6b). Thus, finding the low-frequency response generally requires vector constructions with the test point at $\omega = 0^+$. We use $\omega = 0^+$ rather than at $\omega = 0$ to preserve the angle of any vector starting at the origin of the s plane.

Example 11.2 *An All-Pass Network*

Figure 11.2*a* is called an **all-pass network**. We'll justify that name by examining the frequency response.

We first need the transfer function $H(s) = \underline{V}_{out}/\underline{V}_{in}$, obtained by writing the voltage-divider expression

$$\underline{V}_{out} = \underline{V}_C - \underline{V}_x = \frac{1/sC}{R + 1/sC}\underline{V}_{in} - \frac{1}{2}\underline{V}_{in} = \frac{-R + 1/sC}{2(R + 1/sC)}\underline{V}_{in}$$

Rearranging and inserting $a = 1/RC$ yields

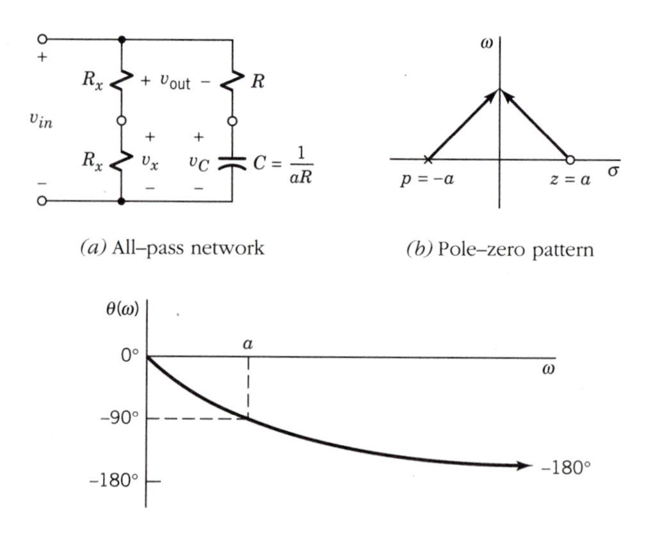

(a) All–pass network

(b) Pole–zero pattern

(c) Phase shift versus ω

Figure 11.2

$$H(s) = \frac{\underline{V}_{out}}{\underline{V}_{in}} = -\frac{1}{2}\frac{s - a}{s + a}$$

so $K = -\frac{1}{2}$, $z = a$, and $p = -a$. The pole-zero pattern in Fig. 11.2*b* brings out the fact that this network has a zero in the right half of the s plane. Even so, the network is stable since its single pole is in the left half. We also see that s-plane vectors drawn from p and z to an arbitrary test point $s = j\omega$ always have the same length. Consequently, we expect that the amplitude ratio $a(\omega)$ will be a constant.

Since we're dealing with a simple function here, we'll pursue the analytical method by setting $s = j\omega$ to get

$$H(j\omega) = -\frac{1}{2}\frac{j\omega - a}{j\omega + a} = \frac{a - j\omega}{2(a + j\omega)}$$

The polar versions of the numerator and denominator are

$$a - j\omega = \sqrt{a^2 + \omega^2}\ \underline{/\tan^{-1}(-\omega/a)}$$
$$2(a + j\omega) = 2\sqrt{a^2 + \omega^2}\ \underline{/\tan^{-1}(\omega/a)}$$

Hence,

$$a(\omega) = |H(j\omega)| = \sqrt{a^2 + \omega^2}/2\sqrt{a^2 + \omega^2} = \tfrac{1}{2}$$
$$\theta(\omega) = \angle H(j\omega) = \tan^{-1}(-\omega/a) - \tan^{-1}(\omega/a) = -2\tan^{-1}(\omega/a)$$

The amplitude ratio is a constant, as expected, while the phase shift varies with ω as plotted in Fig. 11.2*c*.

We call this an *all-pass* network because the amplitudes of all input frequencies remain in the same proportion at the output. Such networks are used to introduce phase shift without altering the relative amplitudes of the frequency components.

Example 11.3 *Frequency Response Calculations*

For a more complicated example, consider the transfer function

$$H(s) = \frac{20(s + 25)}{s^2 + 20s + 500}$$

from which

$$H(j\omega) = \frac{20(j\omega + 25)}{-\omega^2 + 20j\omega + 500} = \frac{20(25 + j\omega)}{(500 - \omega^2) + j20\omega}$$

After converting the numerator and denominator to polar form, we get

$$a(\omega) = \frac{20\sqrt{625 + \omega^2}}{\sqrt{(500 - \omega^2)^2 + 400\omega^2}}$$

$$\theta(\omega) = \tan^{-1}\frac{\omega}{25} - \tan^{-1}\frac{20\omega}{500 - \omega^2} \qquad \omega^2 < 500$$

$$= \tan^{-1}\frac{\omega}{25} \pm 180° + \tan^{-1}\frac{20\omega}{\omega^2 - 500} \qquad \omega^2 > 500$$

Little can be gleaned from such formidable expressions without numerical calculations. Furthermore, the phase shift appears to be ambiguous when $\omega^2 > 500$. We therefore turn to the graphical method.

Since the roots of $P(s) = s^2 + 2 \times 10s + 500$ are $-10 \pm j\sqrt{500 - 10^2}$, $H(s)$ has the factors

$$K = 20 \qquad z_1 = -25 \qquad p_1, p_2 = -10 \pm j20$$

Thus, with $K = 20\ \underline{/0°}$, Eqs. (11.6a) and (11.6b) become

$$a(\omega) = 20|j\omega - z_1|/|j\omega - p_1||j\omega - p_2|$$

$$\theta(\omega) = \angle(j\omega - z_1) - \angle(j\omega - p_1) - \angle(j\omega - p_2)$$

Figure 11.3 shows the s-plane vectors with the test point at $\omega = 0^+$, 10, 20, and 30. These points were chosen for study because p_1 is closest to the imaginary axis at $\omega = 20$.

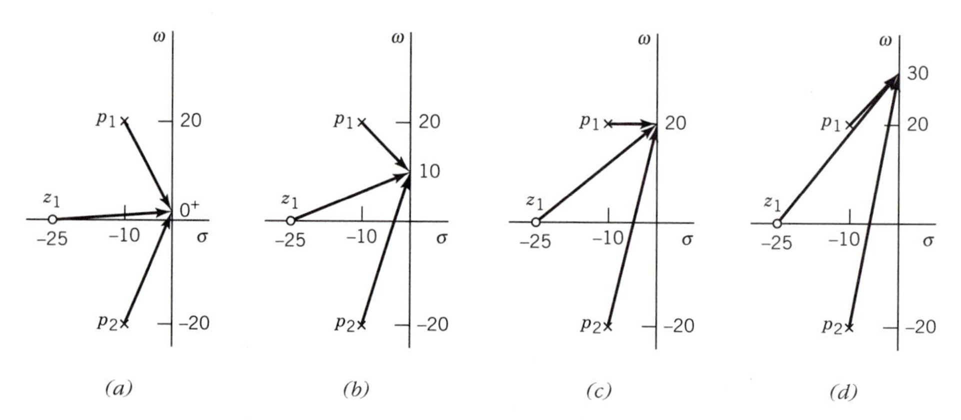

Figure 11.3 s-plane vectors for different test points.

Table 11.2 lists the data obtained from the vector constructions, together with $a(\infty) = 0$ and $\theta(\infty) = -90°$ from Eq. (11.7). The resulting frequency response curves are sketched in Fig. 11.4. An inspection of the tabulated data reveals that p_1 accounts for the "bump" in the amplitude ratio and the rapid phase variation near $\omega = 20$.

TABLE 11.2

ω	0^+	10	20	30	∞
$j\omega - z_1$	$25.0\angle 0°$	$26.9\angle 22°$	$32.0\angle 39°$	$39.1\angle 50°$	
$j\omega - p_1$	$22.4\angle -63°$	$14.1\angle -45°$	$10.0\angle 0°$	$14.1\angle 45°$	
$j\omega - p_2$	$22.4\angle 63°$	$31.6\angle 72°$	$41.2\angle 76°$	$51.0\angle 79°$	
$a(\omega)$	1.00	1.21	1.55	1.09	0
$\theta(\omega)$	0°	−5°	−37°	−74°	−90°

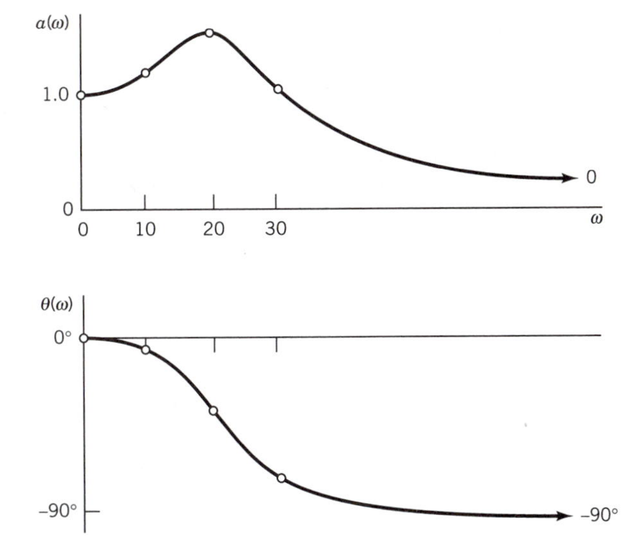

Figure 11.4 Frequency response curves.

Exercise 11.2

Given $H(s) = 20s/(s + 10)^2$, find $a(\omega)$ and $\theta(\omega)$ by first expressing $(10 + j\omega)$ in polar form.

Exercise 11.3

Find $a(\infty)$ and $\theta(\infty)$ when $H(s) = 20s/(s + 10)^2$. Then construct s-plane vectors to evaluate $a(\omega)$ and $\theta(\omega)$ for $\omega = 0^+$, 10, and 20.

11.2 FILTERS

Filters are frequency-selective networks that pass certain frequency components from the input to the output, while rejecting or suppressing other frequency components. The most common filter types are categorized as lowpass, highpass, bandpass, and notch, according to the frequency ranges being passed or rejected.

This section defines the ideal characteristics of basic filters and investigates simple circuits that approximate ideal filtering. For convenience, we'll assume throughout a positive gain constant K, since negative gain just introduces a constant phase shift of $\pm 180°$.

Lowpass and Highpass Filters

Figure 11.5 plots the amplitude-ratio curves of an **ideal lowpass filter** and an **ideal highpass filter**, both with **cutoff frequency** ω_{co}. The lowpass filter has constant gain K over the **passband** from $\omega = 0$ to $\omega = \omega_{co}$, but it totally blocks all frequency components in the **stopband** above ω_{co}. Conversely, the highpass filter blocks all frequency components below ω_{co} and has constant gain for $\omega \geqslant \omega_{co}$. Although these idealized characteristics cannot be achieved exactly, they can be approximated in various ways. Here, we'll study practical lowpass and highpass filter networks containing just one reactive element.

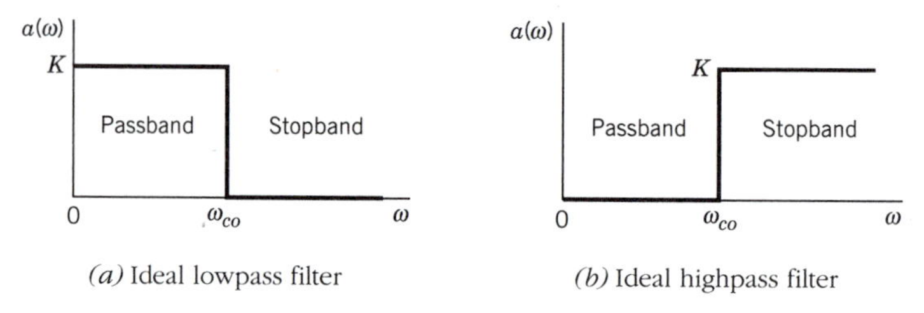

(a) Ideal lowpass filter *(b)* Ideal highpass filter

Figure 11.5 Amplitude-ratio curves.

The simplest frequency-selective network is a **first-order lowpass filter** described by

$$H_{lp}(s) = \frac{K\omega_{co}}{s + \omega_{co}} \tag{11.8}$$

Thus, with $s = j\omega$,

$$H_{lp}(j\omega) = \frac{K\omega_{co}}{j\omega + \omega_{co}} = \frac{K}{1 + j(\omega/\omega_{co})} \tag{11.9}$$

so the amplitude ratio and phase shift are

$$a_{lp}(\omega) = \frac{K}{\sqrt{1 + (\omega/\omega_{co})^2}} \qquad \theta_{lp}(\omega) = -\tan^{-1}\frac{\omega}{\omega_{co}} \tag{11.10}$$

when K is positive.

The lowpass transfer function contains no zeros and just one pole, $p = -\omega_{co}$. The s-plane diagram in Fig. 11.6a readily leads to the frequency

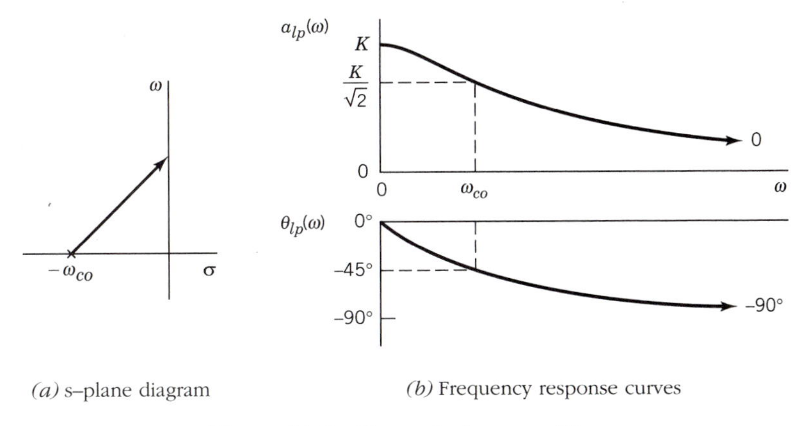

(a) s–plane diagram *(b)* Frequency response curves

Figure 11.6 First-order lowpass filter.

response curves in Fig. 11.6*b*. The amplitude-ratio curve shows that input components at $\omega \ll \omega_{co}$ are passed to the output with **low-frequency gain** K, while high-frequency components at $\omega \gg \omega_{co}$ appear at the output with greatly reduced amplitudes.

However, $a_{lp}(\omega)$ exhibits a gradual transition from passband to stopband rather than an abrupt change at ω_{co}. Lacking a unique landmark, we now define the cutoff frequency somewhat arbitrarily by the property

$$a_{lp}(\omega_{co}) = K/\sqrt{2} = 0.707K \tag{11.11}$$

Thus, the output voltage or current at ω_{co} is reduced by $1/\sqrt{2}$ compared to the maximum output at $\omega \ll \omega_{co}$. This definition of cutoff frequency corresponds to the **half-power point** because the average power in the output signal would be reduced by ½ when the voltage or current is reduced by $1/\sqrt{2}$.

A **first-order highpass filter** differs from a lowpass filter in that it has a zero at the origin of the s plane as well as a pole at $-\omega_{co}$. The corresponding transfer function is

$$H_{hp}(s) = \frac{Ks}{s + \omega_{co}} \tag{11.12}$$

Thus,

$$H_{hp}(j\omega) = \frac{Kj\omega}{j\omega + \omega_{co}} = \frac{K}{1 - j(\omega_{co}/\omega)} \tag{11.13}$$

and

$$a_{hp}(\omega) = \frac{K}{\sqrt{1 + (\omega_{co}/\omega)^2}} \qquad \theta_{hp}(\omega) = \tan^{-1}\frac{\omega_{co}}{\omega} \tag{11.14}$$

These expressions should be compared with the previous lowpass versions in Eqs. (11.8)–(11.10).

Figure 11.7 shows the *s*-plane diagram and the resulting frequency-response curves for our simple highpass filter. The shape of $a_{hp}(\omega)$ confirms that this filter passes frequencies at $\omega \gg \omega_{co}$ with **high-frequency gain** K, while it blocks low frequencies at $\omega \ll \omega_{co}$. The cutoff frequency is again defined by $a_{hp}(\omega_{co}) = K/\sqrt{2}$.

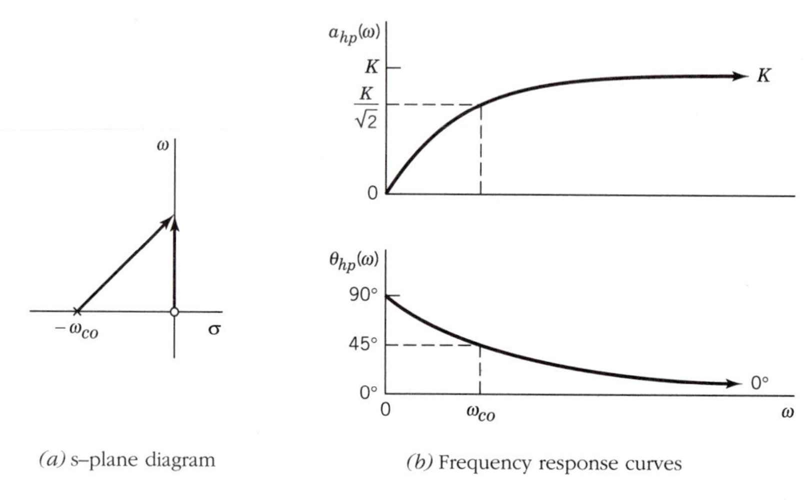

(a) s–plane diagram *(b)* Frequency response curves

Figure 11.7 First-order highpass filter.

We have been working here with *radian* frequency ω, even though engineers usually speak of *cyclical* frequency measured in hertz. But we wrote Eqs. (11.10) and (11.14) in terms of the *ratio* ω/ω_{co}, so they remain valid for cyclical frequency because $\omega = 2\pi f$ and factors of 2π cancel out in the ratios. Thus,

$$\omega/\omega_{co} = f/f_{co}$$

where the cyclical cutoff frequency is $f_{co} = \omega_{co}/2\pi$.

Turning to hardware implementation, Fig. 11.8 diagrams *RC* and *RL* lowpass and highpass filters with voltage inputs and outputs. You can easily confirm that Eq. (11.8) or (11.12) holds for $\underline{V}_{out}/\underline{V}_{in}$ with

$$K = 1 \qquad \omega_{co} = \frac{1}{\tau} \qquad \tau = \begin{cases} RC \\ L/R \end{cases} \tag{11.15}$$

Hence, the cutoff frequency of a first-order filter equals the reciprocal of the time constant τ. Physically, the filtering action results from the capacitor acting as a low-frequency block and a high-frequency short or the inductor acting as a low-frequency short and a high-frequency block. However, any circuitry connected at the output may introduce loading effects and thereby alter the filter's characteristics.

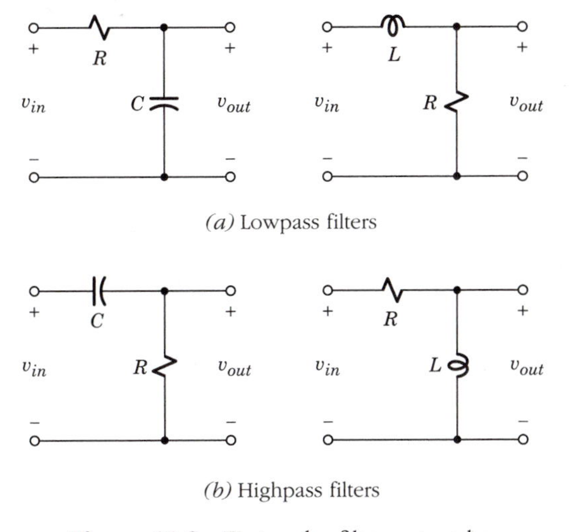

Figure 11.8 First-order filter networks.

Most practical lowpass and highpass filters are built with capacitors to avoid the larger size and cost of inductors. Sophisticated higher-order filter networks include several capacitors arranged to achieve a sharper transition at the cutoff frequency. We'll discuss more selective filters in Section 11.6.

Example 11.4 *Parallel Filter Network*

Problem: Find $H(j\omega)$ for the parallel RC network in Fig. 11.9 when the output is the capacitor current i_C.

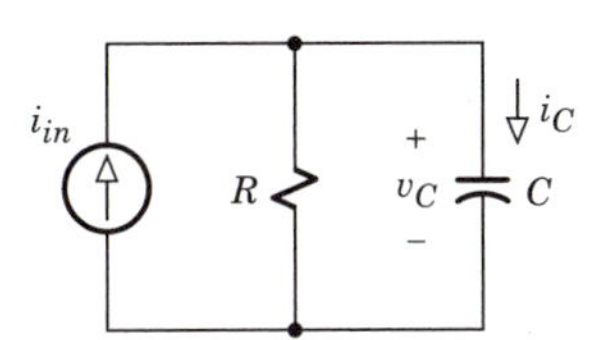

Figure 11.9 Circuit for Example 11.4.

Solution: In the s domain, $\underline{I}_C$ is related to $\underline{I}_{in}$ by the current-divider ratio $R/(R + 1/sC)$, so

$$H(s) = \frac{\underline{I}_C}{\underline{I}_{in}} = \frac{R}{R + 1/sC} = \frac{s}{s + 1/RC}$$

and

$$H(j\omega) = \frac{j\omega}{j\omega + 1/RC} = \frac{1}{1 - j(1/RC\omega)}$$

Hence, from Eq. (11.13), this network acts as a first-order highpass filter with $K = 1$ and $\omega_{co} = 1/RC$.

Example 11.5 *Design of a Lowpass Filter*

Your home intercom system has developed an annoying "whistle" at $f =$ 16 kHz with an amplitude of about 10% of a typical voice signal. Knowing that voice frequency components above 3 kHz are unimportant for intelligibility, you conclude that a lowpass filter with $f_{co} \approx 4$ kHz could quench your whistle while keeping most of the voice signal.

But you also know that the intercom amplifier has a source resistance of 50 Ω, and that the loudspeaker is equivalent to a 200-Ω load resistance at voice frequencies. You therefore decide to obtain lowpass filtering by connecting a capacitor in parallel with the loudspeaker, so the capacitor acts as a short circuit that bypasses high frequencies around the loudspeaker. Figure 11.10 depicts the resulting equivalent circuit in the *s*-domain.

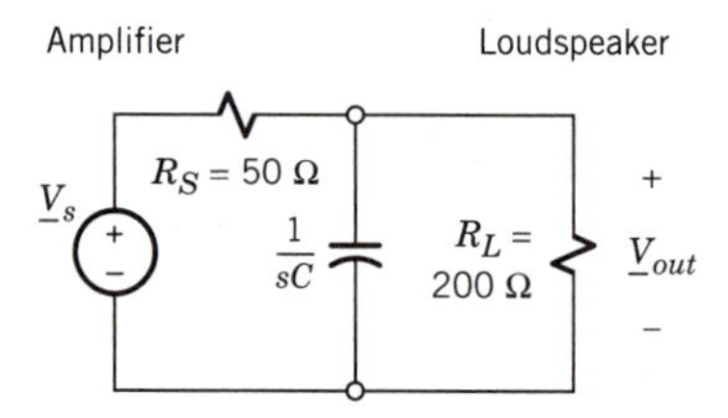

Figure 11.10 Equivalent circuit for Example 11.5.

Since the loudspeaker loads the output of the R_sC network, your first task is to find the transfer function including R_L. Taking $\underline{V}_{out}$ as a node voltage and working with conductances yields

$$(G_s + G_L + sC)\underline{V}_{out} = G_s\underline{V}_s$$

so

$$H(s) = \frac{\underline{V}_{out}}{\underline{V}_s} = \frac{G_s}{sC + G_s + G_L} = \frac{G_s/C}{s + (G_s + G_L)/C} = \frac{K\omega_{co}}{s + \omega_{co}}$$

where

$$\omega_{co} = \frac{G_s + G_L}{C} = \frac{1}{40C} \qquad K = \frac{G_s}{\omega_{co}C} = \frac{G_s}{G_s + G_L} = 0.8$$

Comparison with Eq. (11.8) confirms that $H(s)$ is, indeed, a lowpass filter.

Now you can calculate the desired capacitance via

$$C = 1/(2\pi f_{co} \times 40) \approx 1\ \mu\text{F}$$

By the way, note for this case that $\omega_{co} = 1/\tau$ where

$$\tau = R_{eq}C \qquad R_{eq} = 1/(G_s + G_L) = R_s \| R_L$$

Also note that loading effect reduces the low-frequency gain to $K = 0.8 < 1$.

Finally, you test your design by modeling the voice signal as a 3-kHz sinusoid with $V_m = 5$ V, so the total input signal is

$$v_s(t) = 5 \cos \omega_1 t + 0.5 \cos \omega_2 t$$

where $\omega_1 = 2\pi \times 3$ kHz and $\omega_2 = 2\pi \times 16$ kHz. Using Eq. (11.9) with $K = 0.8$, $\omega_1/\omega_{co} = 3$ kHz/4 kHz $= 0.75$, and $\omega_2/\omega_{co} = 16$ kHz/4 kHz $= 4.0$ gives

$$H(j\omega_1) = 0.64 \angle{-37^\circ} \qquad H(j\omega_2) = 0.19 \angle{-76^\circ}$$

Thus,

$$v_{out}(t) = 3.2 \cos(\omega_1 t - 37^\circ) + 0.095 \cos(\omega_2 t - 76^\circ)$$

which shows that the whistle's amplitude has been cut down to about 3% of the voice signal at the output.

Exercise 11.4

Rework Example 11.4 taking v_C as the output.

Exercise 11.5

Suppose the capacitor is put in series with R_s in Fig. 11.10 rather than in parallel with R_L. Use a voltage divider to show that $H(s)$ will be a highpass function like Eq. (11.12). Then find C to get $f_{co} = 200$ Hz.

Bandpass and Notch Filters

Figure 11.11 plots the amplitude-ratio curves of an **ideal bandpass filter** and an **ideal notch filter**. The bandpass filter passes all frequency components between a **lower cutoff frequency** ω_l and an **upper cutoff frequency** ω_u, so its **bandwidth** is $B = \omega_u - \omega_l$. In contrast, the notch filter blocks all frequency components in the bandwidth between ω_l and ω_u, so it is also known as a **band-reject filter**.

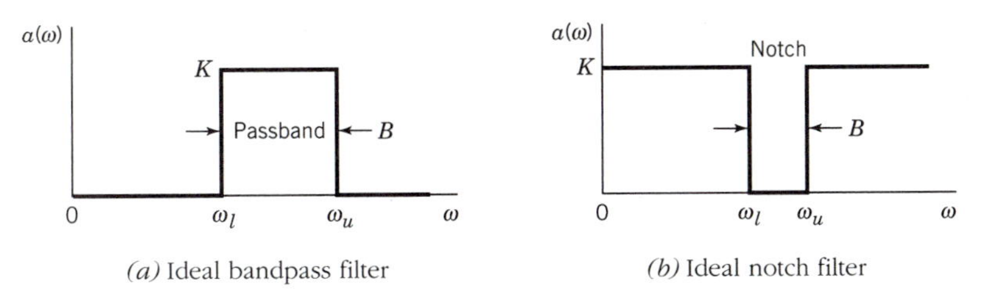

Figure 11.11 Amplitude-ratio curves.

Circuits that act as bandpass or notch filters must include at least two reactive elements in order to have two cutoff frequencies. These circuits play vital roles in radio, TV, radar, and instrumentation systems. Here, we'll study second-order implementations of bandpass and notch filters.

Any **second-order bandpass filter** may be described by the transfer function

$$H_{bp}(s) = \frac{K(\omega_0/Q)s}{s^2 + (\omega_0/Q)s + \omega_0^2} \tag{11.16}$$

The parameter Q stands for the **quality factor** (not to be confused with reactive power), which is related to the damping coefficient α by

$$Q = \omega_0/2\alpha$$

The network is underdamped when $\alpha < \omega_0$ or $Q > ½$, and its poles can be written as

$$p_1, p_2 = -\frac{\omega_0}{2Q} \pm j\omega_0\sqrt{1 - \frac{1}{4Q^2}}$$

Figure 11.12*a* shows the resulting pole-zero pattern, including the zero at the origin. The polar coordinates of the poles are

$$|p_1| = |p_2| = \sqrt{(\omega_0/2Q)^2 + \omega_0^2(1 - 1/4Q^2)} = \omega_0 \tag{11.17a}$$

$$\angle p_1 = 180° - \psi \qquad \angle p_2 = 180° + \psi \tag{11.17b}$$

where

$$\psi = \cos^{-1}(1/2Q) \tag{11.17c}$$

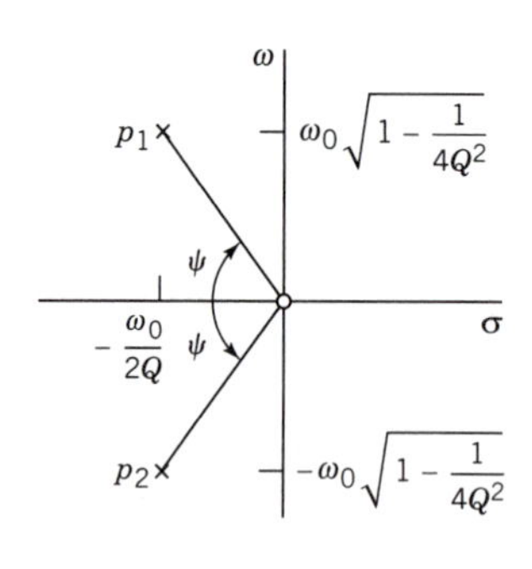

(*a*) Pole–zero pattern for second–order bandpass filter

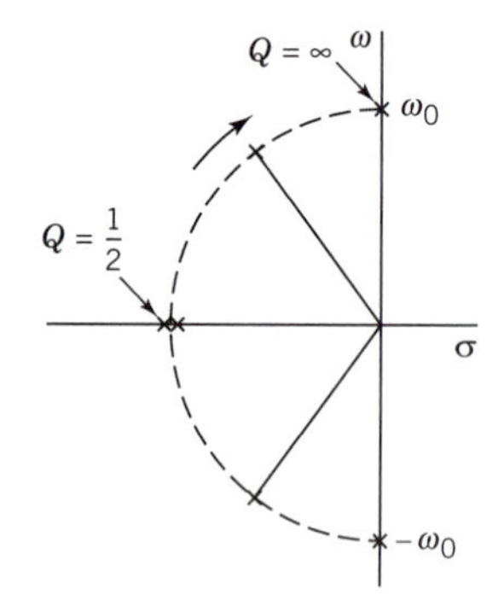

(*b*) Pole movement as Q changes

Figure 11.12

These expressions indicate that if we hold ω_0 constant while varying Q, then the poles move around a semicircle of fixed radius ω_0. Figure 11.12*b* illustrates this effect as the quality factor increases from $Q = ½$ (critical damping) to $Q = \infty$ (zero damping). We'll soon see how Q influences bandpass filtering.

Setting $s = j\omega$ in Eq. (11.16) and performing some manipulations yields

$$H_{bp}(j\omega) = \frac{Kj(\omega_0/Q)\omega}{\omega_0{}^2 - \omega^2 + j(\omega_0/Q)\omega} = \frac{K}{1 + jQ\left(\dfrac{\omega}{\omega_0} - \dfrac{\omega_0}{\omega}\right)} \tag{11.18}$$

from which

$$a_{bp}(\omega) = \frac{K}{\sqrt{1 + Q^2\left(\dfrac{\omega}{\omega_0} - \dfrac{\omega_0}{\omega}\right)^2}}$$
$$\theta_{bp}(\omega) = -\tan^{-1} Q\left(\frac{\omega}{\omega_0} - \frac{\omega_0}{\omega}\right) \tag{11.19}$$

The frequency-response curves are plotted in Fig. 11.13 for $Q = 1$ and $Q = 3$.

The amplitude ratio has the peak value $a_{bp}(\omega_0) = K$, and $a_{bp}(\omega)$ goes to zero as $\omega \to 0$ or $\omega \to \infty$. Thus, the filter blocks very low and very high frequencies while passing frequencies around ω_0 with **midband gain** K. The phase shift in Fig. 11.13 starts at $\theta_{bp}(0) = +90°$ (like a highpass filter), crosses

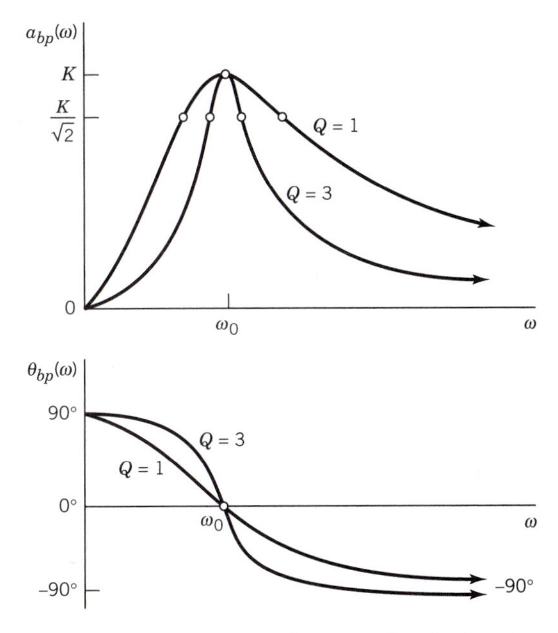

Figure 11.13 Frequency response curves for second-order bandpass filters with $Q = 1$ and $Q = 3$.

$0°$ at ω_0, and goes to $\theta(\infty) = -90°$ (like a lowpass filter). Increasing Q puts pole p_1 closer to ω_0 on the ω axis of the s plane, so the width of the passband decreases and the slope of the phase shift at ω_0 increases.

To calculate the bandwidth B, we again define the cutoff frequencies ω_l and ω_u by the property

$$a_{bp}(\omega_l) = a_{bp}(\omega_u) = K/\sqrt{2}$$

or, equivalently,

$$a_{bp}^2(\omega) = K^2/2 \qquad \omega = \omega_l,\ \omega_u$$

Taking $Q > \frac{1}{2}$ and solving the resulting quadratic from Eq. (11.19) gives

$$\omega_u,\ \omega_l = \omega_0 \sqrt{1 + \frac{1}{4Q^2}} \pm \frac{\omega_0}{2Q} \tag{11.20}$$

Hence,

$$B = \omega_u - \omega_l = \omega_0/Q \tag{11.21}$$

which confirms that B decreases as Q increases. However, the response peak at ω_0 does not fall exactly in the middle of the passband. Instead, upon multiplying ω_l by ω_u we find that

$$\omega_l \omega_u = \omega_0^2 \tag{11.22}$$

so ω_0 is the *geometric mean* of the two cutoff frequencies.

Many applications call for a **narrowband bandpass filter,** meaning that $B \ll \omega_0$. From Eq. (11.21), a narrowband filter must be a "high-Q" network with

$$Q = \omega_0/B \geqslant 10$$

Equation (11.20) then simplifies to

$$\omega_u,\ \omega_l \approx \omega_0 \pm (\omega_0/2Q) = \omega_0 \pm \tfrac{1}{2} B \tag{11.23}$$

Consequently, the narrowband frequency-response curves appear to be symmetrical around ω_0, which may therefore be called the **center frequency**. Figure 11.14 illustrates the amplitude ratio of a high-Q bandpass filter. This type of response would be appropriate for the tuning filter in an AM radio, for instance, since a typical AM radio signal requires a bandwidth of about 15 kHz centered near 1 MHz.

High-Q networks may also function as notch filters. In general, the transfer function of a **second-order notch filter** is

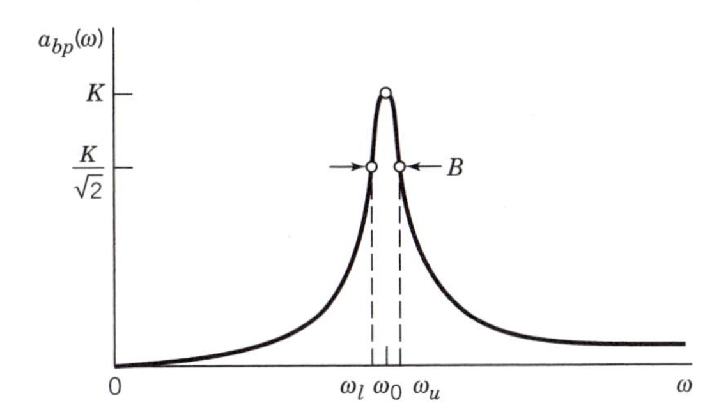

Figure 11.14 Amplitude ratio of a high-Q bandpass filter.

$$H_{no}(s) = \frac{K(s^2 + 2\beta s + \omega_0{}^2)}{s^2 + (\omega_0/Q)s + \omega_0{}^2} \qquad \beta \ll \frac{\omega_0}{2Q} \tag{11.24}$$

The notch effect comes from the quadratic numerator, which results in the pair of complex-conjugate zeros

$$z_1, z_2 = -\beta \pm j\sqrt{\omega_0{}^2 - \beta^2} \approx -\beta \pm j\omega_0$$

The upper half of the pole-zero pattern is given in Fig. 11.15.

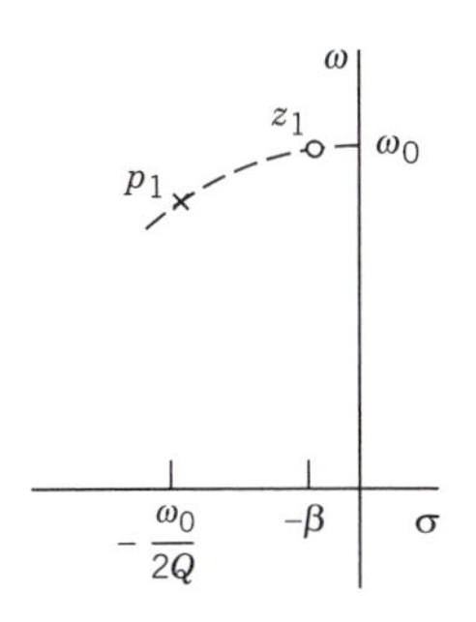

Figure 11.15 Notch filter pole-zero pattern.

Complete rejection at ω_0 requires $\beta = 0$, so z_1 falls on the imaginary axis and forces $a_{no}(\omega_0) = 0$. Figure 11.16 shows the resulting amplitude-ratio curve. The notch width is

$$B = \omega_0/Q \tag{11.25}$$

This type of response is used to remove unwanted narrowband interference from a desired signal. If the filter has $\beta \neq 0$, then $a_{no}(\omega_0) = KQ\beta/\omega_0 \ll K$ and interference at ω_0 is reduced but not totally rejected.

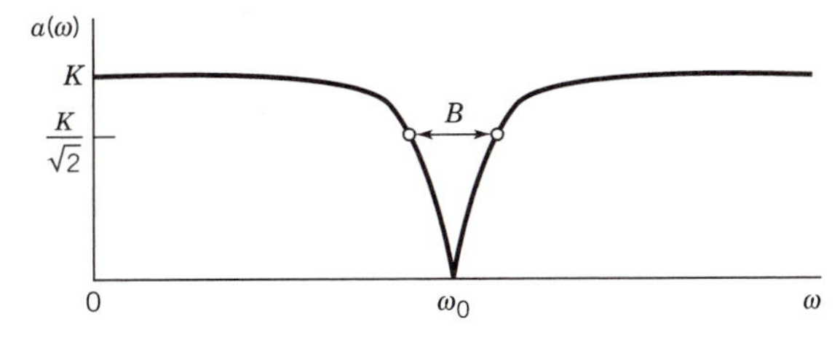

Figure 11.16 Amplitude ratio of a second-order notch filter with $\beta = 0$.

For convenient reference, Table 11.3 lists the transfer functions of second-order bandpass and notch filters, along with first-order lowpass and highpass filters.

TABLE 11.3 Simple Filters

Type	Transfer Function	Properties
Lowpass	$H(s) = \dfrac{K\omega_{co}}{s + \omega_{co}}$	$a(0) = K$ $a(\omega_{co}) = K\sqrt{2}$
Highpass	$H(s) = \dfrac{Ks}{s + \omega_{co}}$	$a(\infty) = K$ $a(\omega_{co}) = K/\sqrt{2}$
Bandpass	$H(s) = \dfrac{K(\omega_0/Q)s}{s^2 + (\omega_0/Q)s + \omega_0^2}$	$a(\omega_0) = K$ $B = \omega_0/Q$
Notch	$H(s) = \dfrac{K(s^2 + 2\beta s + \omega_0^2)}{s^2 + (\omega_0/Q)s + \omega_0^2}$	$a(\omega_0) = KQ\beta/\omega_0$ $B = \omega_0/Q$

Both bandpass and notch filters can be implemented using *resonant* circuits such as the series and parallel circuits in Fig. 11.17. From the low-frequency and high-frequency behavior of capacitors and inductors, it follows that v_{bp} and i_{bp} are bandpass outputs while v_{no} and i_{no} are notch outputs with $\beta = 0$. The respective transfer functions are given by Eqs. (11.16) and (11.24) with

$$K = 1 \qquad \omega_0 = \frac{1}{\sqrt{LC}} \qquad Q = \begin{cases} Q_{ser} \\ Q_{par} \end{cases} \tag{11.26}$$

where Q_{ser} and Q_{par} are the series and parallel quality factors from Section 6.4, namely,

$$Q_{ser} = \frac{\omega_0 L}{R} = \frac{1}{\omega_0 CR} \qquad Q_{par} = \omega_0 CR = \frac{R}{\omega_0 L} \tag{11.27}$$

However, these relations omit the winding resistance R_w associated with real inductors.

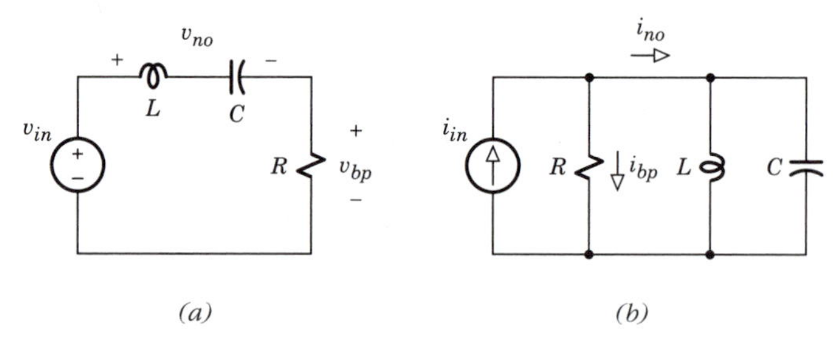

Figure 11.17 Resonant circuits for bandpass and notch filters.

Small winding resistance does not appreciably alter series bandpass filters, but R_w degrades series notch filters by introducing $\beta < 0$. Both of these effects are easily analyzed. For parallel bandpass or notch filtering with winding resistance, consider the network in Fig. 11.18*a*. Recall from Section 6.4 that if $R_w \ll \omega_0 L$, then R_w acts approximately like a large parallel resistance having the value

$$R_{par} = L/CR_w \tag{11.28a}$$

Figure 11.18*b* shows the approximate equivalent network for frequencies near ω_0, so the quality factor becomes

$$Q_{par} = \omega_0 C(R \| R_{par}) = (R \| R_{par})/\omega_0 L \tag{11.28b}$$

Since R_w reduces Q_{par}, the width of the passband or notch will be increased.

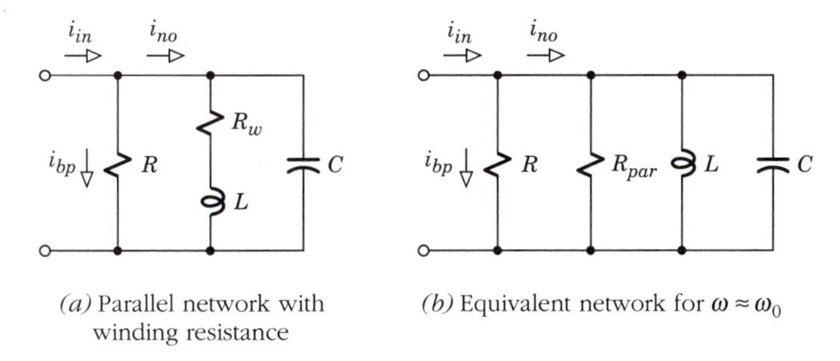

(*a*) Parallel network with winding resistance

(*b*) Equivalent network for $\omega \approx \omega_0$

Figure 11.18

Example 11.6 *Design of a Bandpass Filter*

A certain instrumentation system requires a parallel bandpass filter passing frequencies in the range of 20 kHz ± 250 Hz. You have available for this circuit a 1-mH inductor with $R_w = 1.2\ \Omega$, so you need to find the values of C and R.

You begin by noting that $B = 2\pi \times 2 \times 250$ Hz. Then, assuming that $\omega_0 = 2\pi \times 20$ kHz,

$$Q = \omega_0/B = 20 \text{ kHz}/500 \text{ Hz} = 40$$

This high Q justifies your assumption that ω_0 is indeed the center frequency of the passband. As a further check, you could use Eq. (11.22) with $f_l = \omega_l/2\pi = 19.75$ kHz and $f_u = \omega_u/2\pi = 20.25$ kHz. The resulting geometric mean is

$$f_0 = \omega_0/2\pi = \sqrt{f_l f_u} = 19.998 \text{ kHz}$$

which is virtually identical to the center frequency.

Proceeding with the calculations, you take $\omega_0 = 2\pi \times 20$ kHz and $Q_{par} = Q = 40$ to get

$$C = 1/\omega_0^2 L = 63.3 \text{ nF} \qquad R \| R_{par} = Q\omega_0 L = 5.03 \text{ k}\Omega$$

But $R_{par} = L/CR_w = 13.2 \text{ k}\Omega$, so

$$R = \frac{R_{par}(R \| R_{par})}{R_{par} - (R \| R_{par})} = 8.13 \text{ k}\Omega$$

Had R_{par} been less than the required value of $R \| R_{par}$, then R would be *negative* and the desired values of ω_0 and Q could not be achieved with the available inductor.

Exercise 11.6

The intermediate-frequency (IF) amplifier in an FM tuner has $\omega_0 = 2\pi \times 10.7$ MHz and $B = 2\pi \times 250$ kHz. Find Q_{ser}, R, and L for the corresponding series tuned circuit when $C = 100$ pF and $R_w = 0$.

Exercise 11.7

Let R_w be added in series between L and C in Fig. 11.17*a*. Taking v_{no} as the output, show that $H(s)$ has the form of Eq. (11.24) with β proportional to R_w.

11.3 OP-AMP FILTER CIRCUITS†

Loading effects and other shortcomings of passive filter circuits may be overcome by including an op-amp to make an **active filter**. As a further advantage, op-amps can eliminate the need for inductors in bandpass and notch filters. Here we'll investigate five types of op-amp filter circuits.

Noninverting Lowpass and Highpass Filters. First-order filter circuits with gain $K > 1$ are easily built by connecting the output of an RC network to the input of a noninverting amplifier, as shown in Fig. 11.19 The noninverting

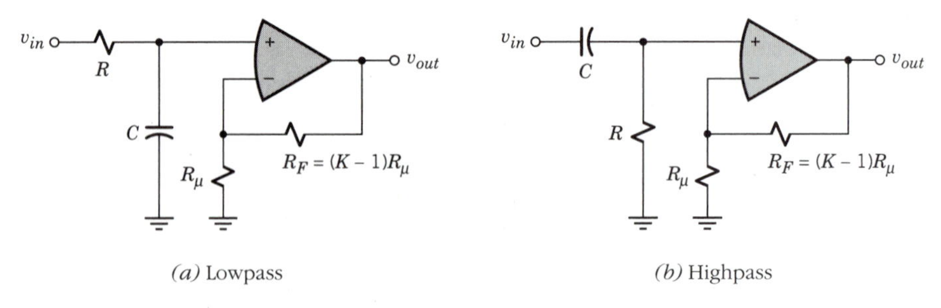

Figure 11.19 Noninverting active filters.

amplifier isolates the *RC* network from the load, so there are no output loading effects.

Inverting Lowpass and Highpass Filters. Inverting filters usually incorporate capacitance directly in the inverting amplifier configuration. Recall that a resistive inverting amplifier has $v_{out}/v_{in} = -R_F/R_1$, where R_1 and R_F are the input and feedback resistances, respectively. Replacing these resistances by the impedances $Z_1(s)$ and $Z_F(s)$, we get the transfer function

$$H(s) = \underline{V}_{out}/\underline{V}_{in} = -Z_F(s)/Z_1(s) \tag{11.29}$$

The lowpass circuit in Fig. 11.20*a* has $Z_1(s) = R_1$ and $Z_F(s) = R_F \| (1/sC_F) = R_F/(1 + sC_F R_F)$. Substitution in Eq. (11.29) therefore yields

$$H(s) = -\frac{R_F}{R_1}\frac{\omega_{co}}{s+\omega_{co}} \qquad \omega_{co} = \frac{1}{R_F C_F} \tag{11.30a}$$

so the resulting low-frequency gain is $K = H(j0) = -R_F/R_1$. The highpass circuit in Fig. 11.20*b* has $Z_1(s) = R_1 + 1/sC_1$ and $Z_F(s) = R_F$, so

$$H(s) = -\frac{R_F}{R_1}\frac{s}{s+\omega_{co}} \qquad \omega_{co} = \frac{1}{R_1 C_1} \tag{11.30b}$$

and the high-frequency gain is $K = H(j\infty) = -R_F/R_1$. Since the inverting amplifier draws input current in both of these circuits, the value of R_1 should include any input source resistance. Alternatively, a voltage follower could be inserted as a buffer between the source and the filter.

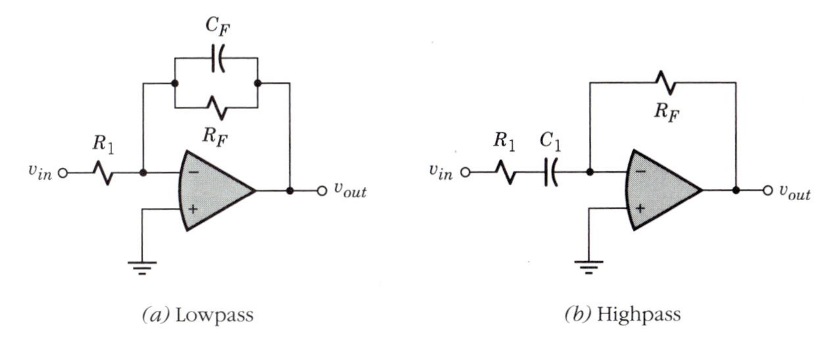

Figure 11.20 Inverting active filters.

Wideband Bandpass Filters. Lowpass and highpass filtering are combined in the circuit of Fig. 11.21*a* to obtain a nonresonant inverting bandpass filter with $Q < ½$. The transfer function is

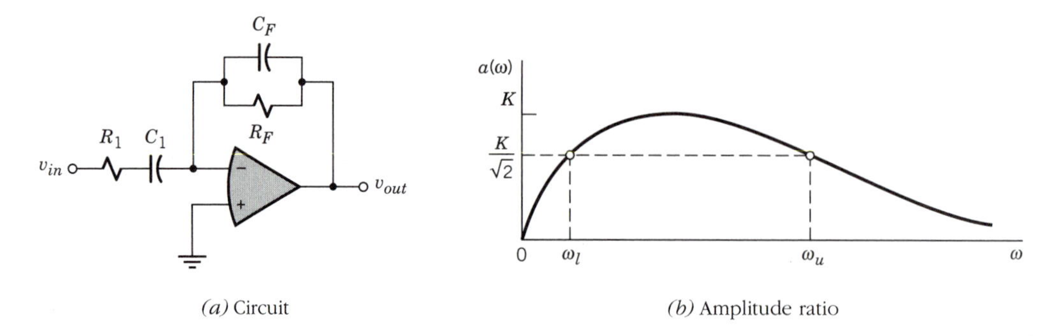

(a) Circuit (b) Amplitude ratio

Figure 11.21 Wideband bandpass filter.

$$H(s) = -K \frac{s}{s + \omega_l} \frac{\omega_u}{s + \omega_u} \qquad (11.31)$$

where

$$K = R_F/R_1 \qquad \omega_l = 1/R_1/C_1 \qquad \omega_u = 1/R_F C_F$$

Hence, there are two real poles, $p_1 = -\omega_l$ and $p_2 = -\omega_u$. This circuit is used only when we want $\omega_1 \ll \omega_u$, in which case

$$a(\omega_l) \approx a(\omega_u) \approx K/\sqrt{2} \qquad \omega_l \ll \omega_u$$

Figure 11.21*b* plots the amplitude ratio curve of a typical wideband filter. Audio amplifiers usually have this type of response, with f_l = 20–100 Hz and f_u = 10–20 kHz.

Narrowband Bandpass Filters. There are several op-amp implementations for resonant bandpass filters, one of them being the **Sallen-Key circuit** diagrammed in Fig. 11.22. The op-amp is in the noninverting configuration, and R_1C_1 form a lowpass section while C_2R_3 form a highpass section. Additional feedback via R_2 produces the necessary high-Q effect.

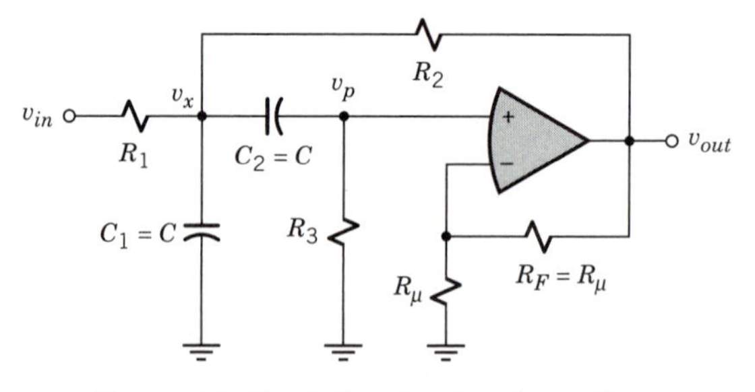

Figure 11.22 Sallen-Key bandpass filter.

Writing the node equations for $\underline{V}_x$ and $\underline{V}_p$ and using the fact that $\underline{V}_{out} = (1 + R_F/R_\mu)\underline{V}_p$ yields our standard bandpass function

$$H(s) = \frac{\underline{V}_{out}}{\underline{V}_{in}} = \frac{K(\omega_0/Q)s}{s^2 + (\omega_0/Q)s + \omega_0^2}$$

with

$$K = \frac{2Q}{R_1 C\omega_0} \qquad \omega_0^2 = \frac{R_1 + R_2}{R_1 R_2 R_3 C^2} \qquad Q = \frac{R_1 R_2 R_3 C\omega_0}{2R_1R_2 + R_2R_3 - R_1R_3} \tag{11.32}$$

Since the analysis results are quite complicated, a more useful procedure is to select C and calculate the resistance values sequentially via

$$R_1 = \frac{2Q}{K\omega_0 C} \quad R_2 = \frac{KR_1}{\sqrt{(K-1)^2 + 8Q^2} - 1} \quad R_3 = \frac{K^2R_1(R_1 + R_2)}{4Q^2R_2} \tag{11.33}$$

These *design equations* follow from algebraic manipulations of Eq. (11.32).

The Sallen-Key circuit has the important advantage of resonant effect without an inductor. But it also has the disadvantage of potential *instability* because R_2 provides *positive* feedback. To bring out that danger, we rewrite the expression for Q in Eq. (11.32) by introducing the conductances $G_1 = 1/R_1$, etc., so

$$Q = C\omega_0/(2G_3 + G_1 - G_2)$$

Thus, if G_2 happens to be too large, then Q may be a *negative* quantity. Returning to Fig. 11.12*a* (p. 502), we see that negative Q moves the poles into the right half of the s plane, meaning that the natural response becomes a growing oscillation. Special care must therefore be taken when designing R_1, R_2, and R_3 to avoid creating an oscillator instead of a filter.

Notch Filters. The simplest op-amp implementation of a notch filter is the twin-tee circuit back in Fig. 10.8*a* (p. 464). From our previous analysis

$$H(s) = \frac{\underline{V}_{out}}{\underline{V}_{in}} = \mu \frac{s^2 + a^2}{s^2 + (4 - 2\mu)as + a^2} \qquad a = \frac{1}{RC} \tag{11.34a}$$

which has the form of $H_{no}(s)$ in Eq. (11.24) with

$$K = \mu \qquad \beta = 0 \qquad \omega_0 = \frac{1}{RC} \qquad Q = \frac{1}{4 - 2\mu} \tag{11.34b}$$

Since the values of K and Q are interrelated here, they cannot be specified independently. However, another amplifier unit could be inserted at the input to achieve the desired gain and eliminate input loading.

As a closing consideration, recall that the input and feedback resistances connected to an op-amp should fall in the range of 1 kΩ to 100 kΩ. Thus, when a resistor and a capacitor determine the cutoff frequency or center frequency of an active filter, the value of the capacitor must be selected with care. A handy rule-of-thumb for this purpose is to choose C such that

$$10^{-6}/f_c < C < 10^{-4}/f_c \tag{11.35}$$

where f_c stands for the cutoff or center frequency in hertz. Equation (11.35) usually results in suitable values for all resistors.

Example 11.7 ***Design of an Active Filter***

Suppose the intercom system in Example 11.5 also suffers from a 60-Hz "hum" and an inadequate gain, along with the 16-kHz whistle. You can cure all of these problems in one fell swoop with an active wideband filter like Fig. 11.21 — provided that the op-amp delivers sufficient output power to drive the loudspeaker. The element values are determined as follows.

The lower cutoff frequency should be above 60 Hz to reduce the hum, but f_l must be below the significant voice frequencies. You therefore select the compromise value $f_l \approx 200$ Hz. You take the upper cutoff frequency to be $f_u \approx 4$ kHz, as before. Since $f_l = 1/2\pi R_1 C_1$ and $f_u = 1/2\pi R_F C_F$, Eq. (11.35) requires

$$5 \text{ nF} < C_1 < 500 \text{ nF} \qquad 0.25 \text{ nF} < C_F < 25 \text{ nF}$$

You also note that $R_F = 1/\omega_u C_F$ and $R_1 = 1/\omega_l C_1$, so

$$K = R_F/R_1 = \omega_l C_1/\omega_u C_F = C_1/20C_F$$

Thus, you can increase the midband voltage gain by taking a large value for C_1 and a small value for C_F.

In light of these considerations, you choose the standard capacitor values

$$C_1 = 100 \text{ nF} \qquad C_F = 1 \text{ nF}$$

The resistors must then be

$$R_1 = 1/2\pi f_l C_1 \approx 8 \text{ k}\Omega \qquad R_F = 1/2\pi f_u C_F \approx 40 \text{ k}\Omega$$

which gives $K = 40/8 = 5$. The value for R_1 also minimizes source loading since $R_s = 50\ \Omega \ll R_1$.

Exercise 11.8

Use Eq. (11.29) to derive Eq. (11.31) from Fig. 11.21*a*.

11.4 BODE PLOTS

In the previous sections we showed plots of amplitude ratio and phase shift on a linear frequency axis. Another useful plotting scheme was devised by the American engineer Hendrick Bode (pronounced Bo-dee). A Bode plot involves a logarithmic measure of the amplitude ratio known as the *decibel,* and the decibel values are plotted along with the phase shift on a logarithmic frequency axis. The logarithmic frequency axis allows for a detailed display of a much greater frequency range.

Bode plots play an important role in the analysis and design of filters, amplifiers, and control systems. Accordingly, various computer programs have been devised to produce plots starting from a transfer function.

But Bode plots also can be sketched by hand, taking advantage of simple asymptotic behavior rather than making extensive numerical calculations. We'll develop that construction technique here, beginning with factored transfer functions. Then we'll examine basic first-order and quadratic functions that serve as building blocks for Bode plots. After using these building blocks to sketch Bode plots by hand, you will be better prepared to interpret computer-generated curves and to design transfer functions for specified frequency-response characteristics.

Factored Functions and Decibels

Given the transfer function of a network, our starting point for constructing its Bode plot is to decompose $H(s)$ into a factored expression of the form

$$H(s) = KH_1(s)H_2(s) \ldots \tag{11.36}$$

The constant K is chosen to be a real number, possibly negative, while $H_1(s)$, $H_2(s)$, . . . , are chosen to be simple functions with known Bode plots. Factoring thereby expresses a complicated function in the form of a more easily handled product.

After factoring $H(s)$ and setting $s = j\omega$, we obtain the total amplitude ratio and phase shift as

$$a(\omega) = |H(j\omega)| = |K|a_1(\omega)a_2(\omega) \ldots \tag{11.37}$$

$$\theta(\omega) = \angle H(j\omega) = \angle K + \theta_1(\omega) + \theta_2(\omega) + \cdots \tag{11.38}$$

in which $a_1(\omega) = |H_1(j\omega)|$, $\theta_1(\omega) = \angle H_1(j\omega)$, and so forth. The angle of K is either 0° or ±180°, depending on the sign of K. The sum of phase shifts in Eq. (11.38) nicely suits our purposes, but the product of amplitude ratios in Eq. (11.37) does not lend itself readily to graphical construction because products are harder to graph than sums.

To recast Eq. (11.37) as a summation, we'll work with gain measured in **decibels (dB)**. Specifically, for a given amplitude ratio $a(\omega)$, the **dB gain** is defined by

$$g(\omega) \triangleq 20 \log a(\omega) \tag{11.39}$$

where log denotes the common or base-10 logarithm. Since the logarithm of a product equals the sum of the logarithms, the total dB gain from Eq. (11.37) becomes

$$\begin{aligned} g(\omega) &= 20 \log |K| + 20 \log a_1(\omega) + 20 \log a_2(\omega) + \cdots \\ &= K_{dB} + g_1(\omega) + g_2(\omega) + \cdots \end{aligned} \tag{11.40}$$

where $K_{dB} = 20 \log |K|$, $g_1(\omega) = 20 \log a_1(\omega)$, and so forth.

Equations (11.38) and (11.40) bring out the underlying strategy here. For if you know the Bode plots of the individual factors $H_1(s)$, $H_2(s)$, . . . , then you can simply *add them together*. Shifting the resulting curves vertically by the constants K_{dB} and $\angle K$ then yields the complete Bode plot of $H(s)$. This approach will be pursued subsequently, after we give further attention to the decibel unit.

We capitalize the B in dB because the decibel is one-tenth of a larger unit named the *bel* to honor Alexander Graham Bell (1847–1922). Before inventing the telephone, Bell conducted experiments on speech and hearing in which he used the logarithm of power ratios. When P_{in} and P_{out} denote average powers, the *power ratio in bels* is $\log (P_{out}/P_{in})$. But average power is proportional to the square of voltage or current, so the output–input power ratio of a network is proportional to $a^2(\omega)$. Correspondingly, the amplitude ratio in bels is $\log a^2(\omega) = 2 \log a(\omega)$, and multiplying by 10 gives the *decibel* gain, as defined in Eq. (11.39).

Being a logarithmic measure, decibels transform powers of 10 into products of 10. In particular, if at some frequency ω we have

$$a(\omega) = 10^m \tag{11.41a}$$

then the dB gain is

$$g(\omega) = 20 \log 10^m = 20 \times m \text{ dB} \tag{11.41b}$$

By taking $m = 0$, we see that if $a(\omega) = 10^0 = 1$ then $g(\omega) = 20 \times 0 = 0$ dB, so unity amplitude ratio corresponds to zero dB gain. Furthermore, if $m < 0$, then $a(\omega) < 1$ and $g(\omega) < 0$ dB, so negative dB gain simply means that $a(\omega)$ is less than unity.

Table 11.4 lists exact or approximate dB values for selected values of amplitude ratio. Other dB values are readily found using Eq. (11.39) and a calculator. Conversion from dB gain back to amplitude ratio is accomplished via

$$a(\omega) = 10^{g(\omega)/20} \tag{11.42}$$

which follows by inversion of Eq. (11.39).

TABLE 11.4 Selected Decibel Values

Amplitude ratio	10	2	$\sqrt{2}$	1	$1/\sqrt{2}$	0.5	0.1
Gain in dB	20	≈ 6	≈ 3	0	≈ -3	≈ -6	-20

Exercise 11.9

Use Table 11.4 and the properties of decibels to evaluate $g(\omega)$ when: **(a)** $a(\omega) = 5 = 10 \times 0.5$; **(b)** $a(\omega) = 0.008 = (2 \times 0.1)^3$; **(c)** $a(\omega) = \sqrt{500} = \sqrt{0.5 \times 1000}$.

First-Order Factors

Many transfer functions of interest consist of products of first-order factors. Accordingly, we'll begin with Bode plots of three basic first-order functions. Then we'll construct Bode plots for transfer functions made up of these building blocks.

Ramp Function. The basic ramp function contains only a zero at the origin of the s-plane. We define this function as

$$H_r(s; W) \triangleq s/W \tag{11.43}$$

where W is any positive real constant. The notation $H_r(s; W)$ stands for a function of s that also involves the *parameter* W, which may take on various values. The need for this special notation will become apparent as we go along.

Setting $s = j\omega$ in Eq. (11.43) and converting to polar form yields

$$H_r(j\omega; W) = j\omega/W = (\omega/W)\ \underline{/90^\circ}$$

from which

$$g_r(\omega; W) = 20 \log (\omega/W) \qquad \theta_r(\omega; W) = 90^\circ \tag{11.44}$$

Hence, the phase shift stays constant at 90°, whereas the gain varies logarithmically with ω. This logarithmic gain variation would be awkward to plot directly versus ω. But we'll plot $g_r(\omega; W)$ versus $\log \omega$ or, equivalently, on the *semilogarithmic graph* shown in Fig. 11.23*a*.

The horizontal axis of our plot has been labeled with values of ω, so equispaced values increase by powers of 10, e.g., $0.1W$, W, $10W$, ... The gain curve thus follows a straight line through -20 dB at $\omega = 0.1W$, through 0 dB at $\omega = W$, and so forth. Since $g_r(\omega; W)$ increases by 20 dB when frequency increases by a factor of 10, we say that this line has a slope of $+20$ dB per **decade**. The slope is also equivalent to about $+6$ dB per **octave**, an octave being a frequency increase by a factor of 2. (The name "octave" comes

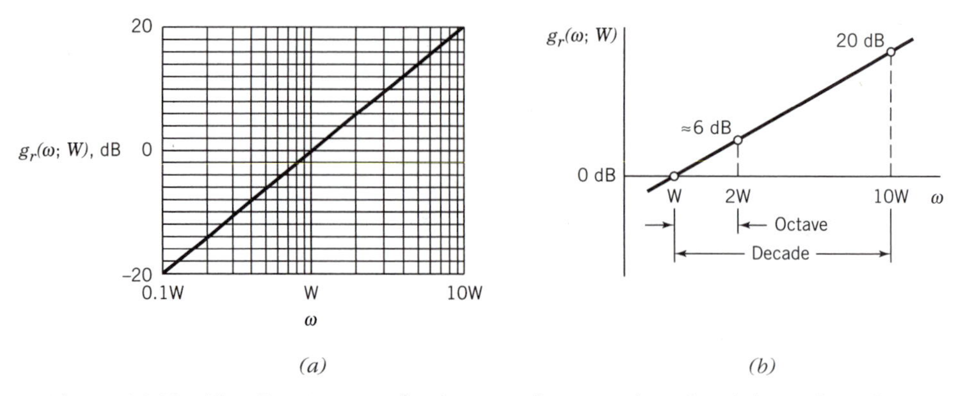

Figure 11.23 The dB gain curve for the ramp function plotted with logarithmic frequency axis.

from the eight notes in a musical scale whose starting and ending frequencies differ by a factor of 2.) Figure 11.23*b* illustrates the decade and octave slopes.

Highpass Function The basic first-order highpass function has $K = 1$, a zero at the origin, and a pole at $s = -W$, where W represents an arbitrary cutoff frequency. Thus,

$$H_{hp}(s; W) \triangleq \frac{s}{s + W} \qquad H_{hp}(j\omega; W) = \frac{j(\omega/W)}{1 + j(\omega/W)} \tag{11.45}$$

The corresponding gain and phase curves are plotted on the semilog graph in Fig. 11.24, but the grid lines have been omitted so you can clearly see the straight-line **asymptotes**.

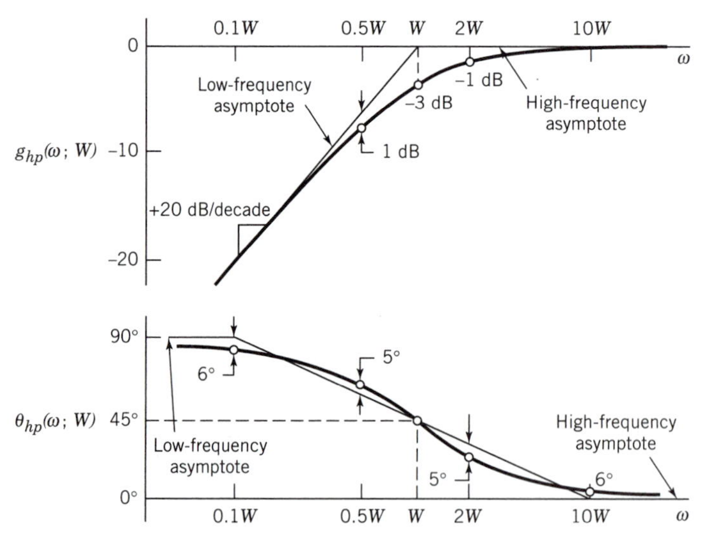

Figure 11.24 Bode plot of the highpass function.

The low-frequency asymptotes embody the approximation that if $\omega/W \ll 1$, then $1 + j(\omega/W) \approx 1$ and $H_{hp}(j\omega; W) \approx j(\omega; W)$. Hence,

$$g_{hp} \approx 20 \log (\omega/W) \qquad \theta_{hp} \approx 90° \qquad \omega < 0.1W \tag{11.46a}$$

Similarly, the high-frequency asymptotes embody the approximation $H_{hp}(j\omega; W) \approx 1$ for $\omega/W \gg 1$. Hence,

$$g_{hp} \approx 0 \text{ dB} \qquad \theta_{hp} \approx 0° \qquad \omega > 10W \tag{11.46b}$$

Between these two extremes at $\omega = W$ we have $H_{hp}(jW; W) = j/(1 + j) = 0.707 \angle 45°$, so

$$g_{hp} \approx -3 \text{ dB} \qquad \theta_{hp} = 45° \qquad \omega = W \tag{11.46c}$$

We call W the **corner frequency** or **break frequency** because the gain approximation initially rises with a 20-dB/decade slope and then "breaks" at $\omega = W$ and becomes a horizontal line at 0 dB. Note, however, that the phase approximation breaks at $0.1W$ and $10W$ and has a diagonal line connecting the horizontal low-frequency and high-frequency asymptotes.

For a given value of W, you can construct the Bode plot shown in Fig. 11.24 by first drawing the straight-line asymptotes on semilog graph paper. Then you plot a few specific points relative to the asymptotes using the "correction" terms from Table 11.5, which were obtained by calculating exact values from Eq. (11.45) at selected frequencies. Finally, you draw smooth curves that go through the plotted points and approach the low-frequency and high-frequency asymptotes.

TABLE 11.5 Asymptote Correction Terms for H_{hp} and H_{lp}

ω/W	0.1	0.5	1	2	10
Δg (db)	0	−1	−3	−1	0
$\Delta\theta$ (°)	−6	+5	0	−5	+6

Lowpass Function The basic first-order lowpass function has $K = 1$, a pole at $s = -W$, and no zero, as defined by

$$H_{lp}(s; W) \triangleq \frac{W}{s + W} \qquad H_{lp}(j\omega; W) = \frac{1}{1 + j(\omega/W)} \tag{11.47}$$

Because $H_{lp}(j\omega; W) \approx 1$ for $\omega/W \ll 1$ while $H_{lp}(j\omega; W) \approx (\omega; W)^{-1} \angle -90°$ for $\omega/W \gg 1$, the low-frequency and high-frequency asymptotic approximations are

$$g_{lp} \approx 0 \text{ dB} \qquad \theta_{lp} \approx 0° \qquad \omega < 0.1W \tag{11.48a}$$

$$g_{lp} \approx -20 \log (\omega/W) \qquad \theta_{lp} \approx -90^\circ \quad \omega > 10W \tag{11.48b}$$

The break-frequency values are

$$g_{lp} \approx -3 \text{ dB} \qquad \theta_{lp} = -45^\circ \qquad \omega = W \tag{11.48c}$$

As shown in Fig. 11.25, the gain curve starts as a horizontal line at 0 dB but falls off with a slope of −20 dB/decade above the break frequency W. Also observe that the phase curve now goes from 0° to −90°. These curves are constructed by the same method as before, using the asymptotes and the correction terms in Table 11.5.

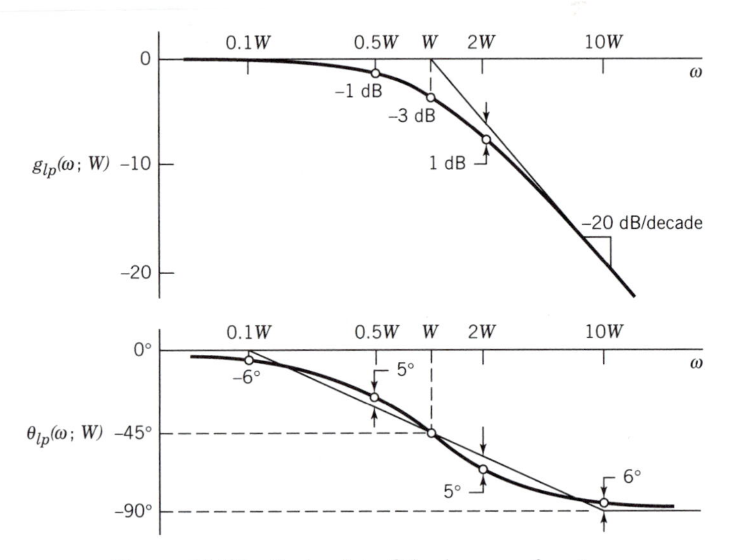

Figure 11.25 Bode plot of the lowpass function.

When a lowpass or highpass function has $K \neq 1$, the gain curve is shifted vertically by the constant K_{dB}. Hence, $g_{max} = K_{\text{dB}}$ and at the break frequency we have

$$g \approx g_{max} - 3 \text{ dB} \qquad \omega = W$$

For this reason, the cutoff frequency of a first-order filter is sometimes called the "3-dB frequency."

Now let $H_x(s)$ stand for any of our three basic functions and consider the case of

$$H(s) = H_x^{\ m}(s) \tag{11.49}$$

where m is a positive or negative integer. If m is negative, then any poles of $H_x(s)$ become zeros of $H(s)$ and vice versa. Expressing $H_x(j\omega)$ in polar form, we obtain

$$H(j\omega) = H_x^m(j\omega) = (a_x \angle\theta_x)^m = a_x^m \angle m\theta_x$$

The gain and phase of $H(s)$ are thus related to the gain and phase of $H_x(s)$ by

$$g(\omega) = m \times g_x(\omega) \qquad \theta(\omega) = m \times \theta_x(\omega) \tag{11.50}$$

Hence, we can obtain the Bode plot of $H(s)$ from the Bode plot of $H_x(s)$ by *multiplying both curves by m.*

Example 11.8 ***An Illustrative Bode Plot***

Suppose you need the Bode plot of the transfer function

$$H(s) = -\frac{(s + 200)^2}{10s^2}$$

Since $(s + 200)/s$ is the reciprocal of the highpass function with $W = 200$, you rewrite $H(s)$ as

$$H(s) = -\frac{1}{10}\left(\frac{s}{s + 200}\right)^{-2} = -0.1H_{hp}^{-2}(s;200)$$

Thus, from Eq. (11.50),

$$g(\omega) = K_{\text{dB}} - 2g_{hp}(\omega;200) \qquad \theta(\omega) = \angle K - 2\theta_{hp}(\omega;200)$$

where $K_{\text{dB}} = 20 \log 0.1 = -20$ dB and $\angle K = \pm 180°$.

Referring to the Bode plot of H_{hp} and Table 11.5, you obtain the straight-line approximations and correction terms for $-2g_{hp}(\omega;200)$ and $-2\theta_{hp}(\omega;200)$, as shown in Fig. 11.26*a*. All dB and angle values have been multiplied by -2, so the low-frequency gain slope becomes -40 dB/decade and the low-frequency phase asymptote becomes $-180°$. All correction terms are likewise doubled and inverted. Adding the constants $K_{\text{dB}} = -20$ dB and $\angle K = +180°$ gives the complete Bode plot for $H(s)$ in Fig. 11.26*b*.

Products of First-Order Factors. With the help of our basic functions and Eq. (11.50), you can construct the Bode plot for any transfer function that consists entirely of first-order factors and powers of first-order factors. Exploiting the additive property of gain and phase, the procedure goes as follows:

1. Factor $H(s)$ in terms of a constant and basic functions, putting the functions in order of increasing break frequencies.

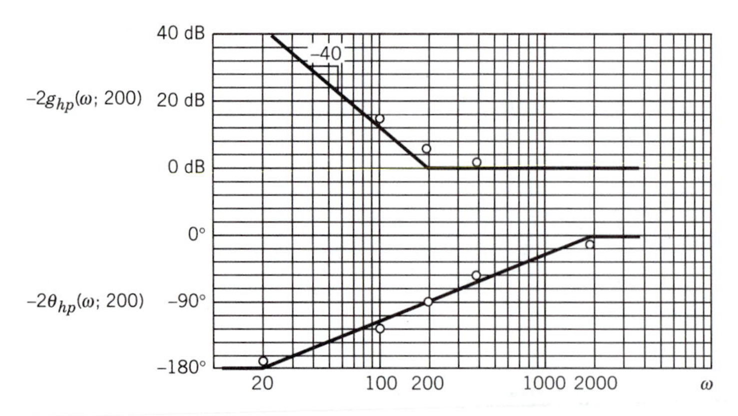

(a) Asymototes and correction terms

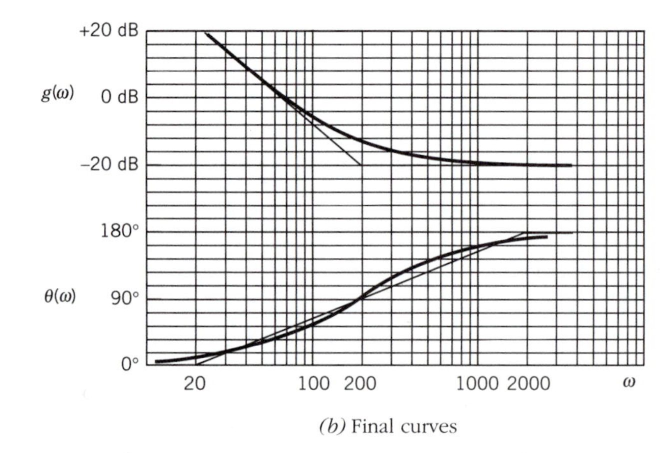

(b) Final curves

Figure 11.26 Bode plot for Example 11.8.

2. Construct the asymptotic gain and phase plots on semilog graph paper by adding the individual straight-line approximations. Convenient vertical scales for this purpose are usually 2 dB and 9° per small division.
3. Plot a few specific points relative to the straight lines by summing correction terms from Table 11.5.
4. Draw smooth curves that go through the plotted points and approach the asymptotes.
5. Shift the asymptotic plots vertically by the constants K_{dB} and $\angle K$ (or just relabel the vertical axes).

The resulting curves will have sufficient accuracy for almost all practical purposes. Sometimes just the straight-line approximations provide enough information, so you can skip the third and fourth steps.

Example 11.9 *Frequency Response of a Bandpass Amplifier*

A certain bandpass amplifier is described by

$$H(s) = \frac{20{,}000s}{(s + 100)(s + 400)}$$

We'll investigate the frequency response by constructing the Bode plot.

Step 1: $H(s)$ consists of a highpass function and a lowpass function, so we factor it as

$$H(s) = \frac{20{,}000}{400} \frac{s}{s + 100} \frac{400}{s + 400} = 50H_1(s)H_2(s)$$

where

$$H_1(s) = \text{H}_{hp}(s;100) \qquad H_2(s) = H_{lp}(s;400)$$

Other fractorizations are possible, but this one contains just two basic functions and leads to the easiest construction.

Step 2: Figure 11.27 shows the sum of the gain and phase asymptotes of $H_1(s)$ and $H_2(s)$. The gain asymptotes are easily added by noting that $g_1 \approx 0$ dB for $\omega \gg 100$ while $g_2 \approx 0$ dB for $\omega \ll 400$. The phase asymptotes are added by summing values at each break frequency and connecting those points with straight lines.

Step 3: Individual correction terms from Table 11.5 and the resulting sums are listed in Table 11.6, where entries marked with a star are estimated values. Selected points have then been plotted on Fig. 11.27.

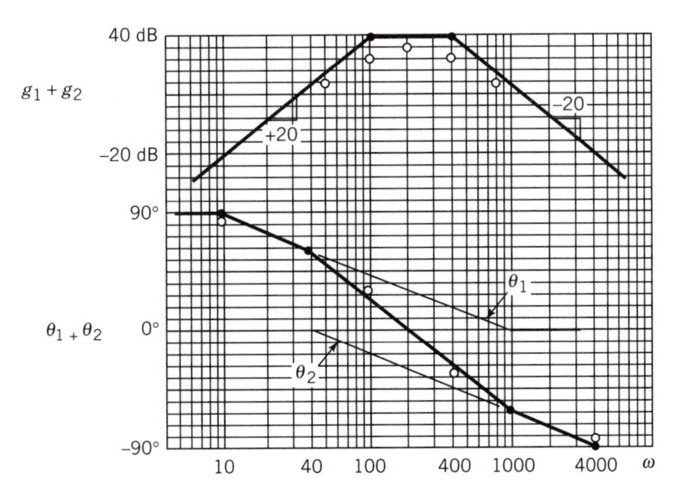

Figure 11.27 Asymptotes and correction terms for Example 11.9.

Steps 4 and 5: Drawing smooth curves through the plotted points and adding $K_{\text{dB}} = 34$ dB yields the complete plot in Fig. 11.28.

TABLE 11.6

ω	10	50	100	200	400	800	4000
Δg_1	0	−1	−3	−1	0	0	0
Δg_2	0	0	0	−1	−3	−1	0
Sum (dB)	0	−1	−3	−2	−3	−1	0
$\Delta\theta_1$	−6	+5	0	−5	−4	+3*	0*
$\Delta\theta_2$	0*	−3*	+4	+5	0	−5	+6
Sum (°)	−6	+2	+4	0	−4	−2	+6

The gain curve reveals that the amplifier has $g_{max} \approx 32$ dB occurring at $\omega \approx 200$ — the same frequency at which $\theta(\omega)$ crosses 0°. The corresponding maximum amplitude ratio is

$$a_{max} \approx 10^{32/20} \approx 40$$

Since a_{max} differs from the gain constant $|K| = 50$, the lower and upper cutoff frequencies do not equal the break frequencies. Instead, they correspond to the points where $g(\omega) = g_{max} - 3$ dB ≈ 29 dB. Hence, $\omega_l \approx 75$ and $\omega_u \approx 550$, so the amplifier's bandwidth is $B = \omega_u - \omega_l \approx 475$ rad/s. These useful results would have been difficult to obtain by any other method short of computer-generated plots.

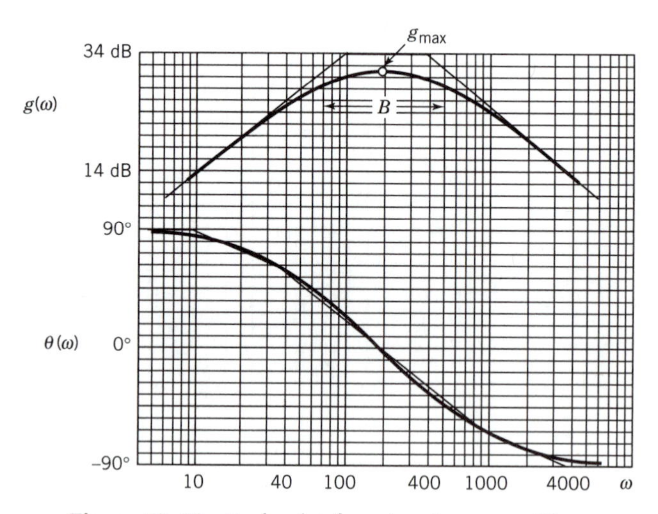

Figure 11.28 Bode plot for a bandpass amplifier.

Exercise 11.10

Construct the dB gain curve for $H(s) = 1000s/(s + 100)^2$ and find g_{max} and a_{max}. Hint: Use a highpass and a lowpass function.

Exercise 11.11

Factor $H(s) = (s + 50)/(s + 200)$ using H_{lp}^{-1} and H_{lp}. Then construct the gain and phase asymptotes for $H(s)$.

Quadratic Factors

Our basic first-order functions take care of all factors of $H(s)$ with poles or zeros at the origin or on the negative real axis of the s plane. We now turn to the case of quadratic functions that produce *complex-conjugate* poles. The reciprocal operation can be invoked to obtain complex-conjugate zeros when needed.

Recall from our discussion of bandpass filters that a second-order circuit with quality factor $Q > \frac{1}{2}$ has complex-conjugate poles, as desired here. However, the bandpass function $H_{bp}(s)$ also includes a zero at the origin. To avoid the unwanted zero, we'll define our basic quadratic function to be

$$H_q(s;\omega_0,Q) \triangleq \frac{\omega_0^2}{s^2 + (\omega_0/Q)s + \omega_0^2} \tag{11.51}$$

which has *two* parameters, ω_0 and Q.

Setting $s = j\omega$ in Eq. (11.51) yields

$$H_q(j\omega;\omega_0,Q) = \frac{1}{1 - (\omega/\omega_0)^2 + j(\omega/Q\omega_0)} \tag{11.52}$$

so

$$\begin{aligned} H_q(j\omega;\omega_0,Q) &\approx 1 & \omega \ll \omega_0 \\ &\approx -(\omega_0/\omega)^2 & \omega \gg \omega_0 \end{aligned}$$

Although the low-frequency and high-frequency behaviors do not involve Q, the gain and phase in the vicinity of ω_0 depend critically on the quality factor. Figure 11.29 gives the resulting Bode plots obtained by numerical calculations with selected values of Q. We see that the gain exhibits a distinct resonance peak at $\omega = \omega_0$ when $Q > 1$, and the phase slope at ω_0 increases as Q increases.

The phase curves in Fig. 11.29 do not lead to simple constructions by hand. Consequently, computer-generated plots are called for when you need accurate phase-shift values of transfer functions with complex-conjugate poles or zeros. However, gain curves can be sketched using the asymptotic approximations

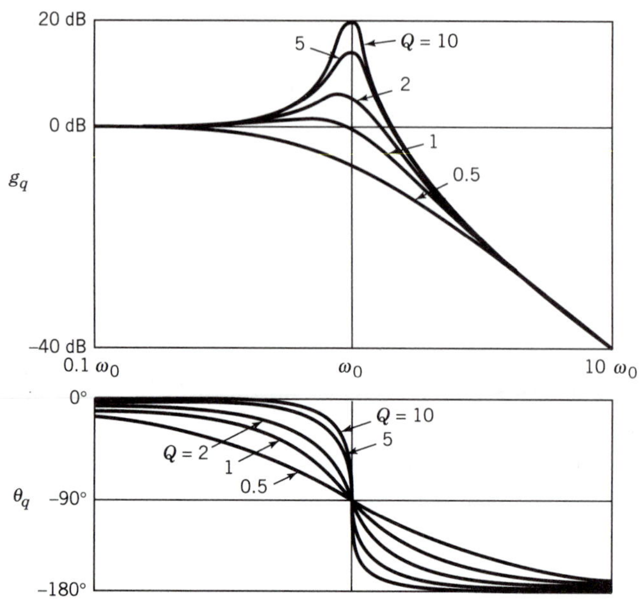

Figure 11.29 Bode plots of the quadratic function for selected values of Q.

$$
\begin{aligned}
g_q &\approx 0 \text{ dB} && \omega < 0.1\omega_0 \\
&\approx -40 \log(\omega/\omega_0) && \omega > 10\omega_0
\end{aligned}
\tag{11.53}
$$

Additionally, based on calculations from Eq. (11.52), three gain points can be plotted using

$$g_q = 20 \log Q = Q_{dB} \qquad \omega = \omega_0 \tag{11.54a}$$

$$\Delta g_q = 10 \log \frac{16}{9 + 4/Q^2} \qquad \frac{\omega}{\omega_0} = 0.5,\ 2 \tag{11.54b}$$

Figure 11.30 shows the asymptotes and typical correction terms.

Finally, it should be mentioned that quadratic factors are also expressed in terms of the **damping ratio ζ** (zeta), rather than the quality factor Q or damping coefficient α. Damping ratio is related to Q and α by

$$\zeta = 1/2Q = \alpha/\omega_0 \tag{11.55}$$

Thus, $\zeta = 1$ corresponds to critical damping, while an underdamped function has $\zeta < 1$.

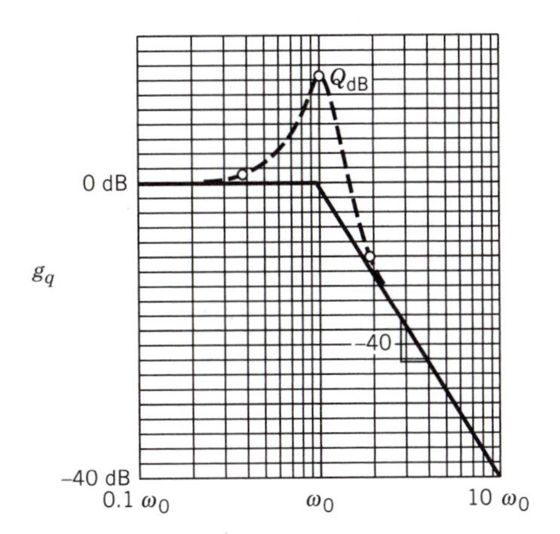

Figure 11.30 Gain asymptotes and correction terms for the quadratic function.

Example 11.10 *Bode Plot of a Narrowband Filter*

Consider a filter described by

$$H(s) = \frac{20s}{s^2 + 20s + 10^4}$$

The quadratic denominator has $\omega_0^{\,2} = 10^4$ and $\omega_0/Q = 20$, so

$$\omega_0 = 100 \qquad Q = \omega_0/20 = 5$$

The corresponding damping ratio is $\zeta = \frac{1}{2}Q = 0.1$

To account for s in the numerator, we'll introduce the ramp function with $W = \omega_0 = 100$ and factor $H(s)$ as

$$\begin{aligned} H(s) &= \frac{20 \times 100}{10^4} \frac{s}{100} \frac{10^4}{s^2 + 20s + 10^4} \\ &= 0.2H_r(s;100)H_q(s;100,5) \end{aligned}$$

Thus,

$$g(\omega) = -14 \text{ dB} + g_r(\omega;100) + g_q(\omega;100,5)$$

The gain curve is sketched in Fig. 11.31, where the inverted vee shape of the asymptotes comes from adding the ramp's slope of 20 dB/decade to the slopes of g_q. The peak gain is $g_{max} = 0$ dB because the ramp has no correction terms while Eq. (11.54a) gives

$$g_q = 20 \log 5 = 14 \text{ dB} \qquad \omega = 100$$

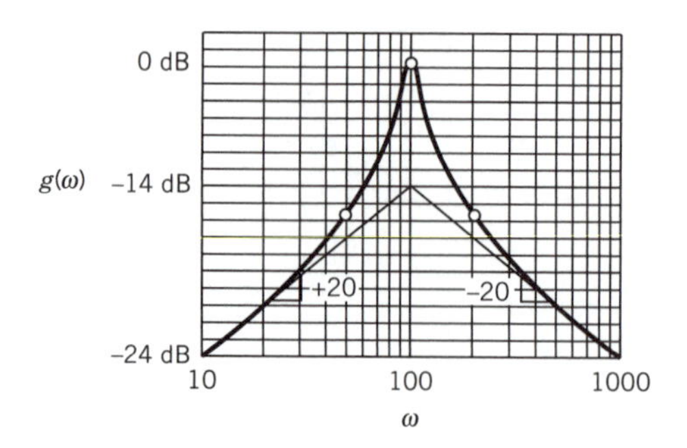

Figure 11.31 Gain plot of a narrowband filter.

which exactly compensates for the downward shift resulting from $K_{dB} = -14$ dB. The other two correction terms from Eq. (11.54b) are

$$\Delta g_q = 10 \log 1.75 = +2.4 \text{ dB} \qquad \omega = 50,\ 200$$

Exercise 11.12

Take $(s/\omega_0)^2$ as a factor to sketch the gain curve for the function $H(s) = 0.2s^2/(s^2 + 5s + 2500)$. What's the value of g_{max}?

11.5 FREQUENCY-RESPONSE DESIGN†

Engineers who design control or communication systems sometimes need frequency-selective networks that differ from standard filter characteristics. When the required response is given as a dB gain curve, without resonant peaks or dips, a convenient design methodology involves working backwards from the Bode plot to the transfer function.

The first task is to construct a straight-line approximation for $g(\omega)$. From our study of Bode plots, we know that the slopes of all lines must be integer multiples of ± 20 dB/decade. Otherwise, the response in question cannot be achieved with a lumped-element network. For the purpose of calculating slopes, we note that the spacing on a logarithmic axis between an arbitrary frequency W_1 and $W_2 > W_1$ is $\log W_2 - \log W_1 = \log W_2/W_1$, so

$$\text{Number of decades} = \log W_2/W_1 \tag{11.56}$$

This expression includes fractional parts of a decade.

Next, we decompose the straight-line approximation into a sum consisting of a constant term and asymptotic plots of basic first-order functions. The gain curve can thus be represented by writing

$$g(\omega) = K_{\text{dB}} + g_1(\omega) + g_2(\omega) + \cdots \tag{11.57}$$

so the required transfer function is found from

$$H(s) = \pm|K|H_1(s)H_2(s)\cdots \tag{11.58}$$

The ambiguous polarity arises here because the sign of K only affects the phase shift, not the dB gain.

Lastly, we need to determine the physical realization of $H(s)$. We generally prefer RC networks, and op-amps may be employed to eliminate interaction and loading effects. Figure 11.32 gives the diagrams of four inverting op-amp circuits that implement various first-order factors useful for frequency-response design. When $H(s)$ has several factors, two or more of the op-amp circuits can be connected in cascade to get the product of the individual functions.

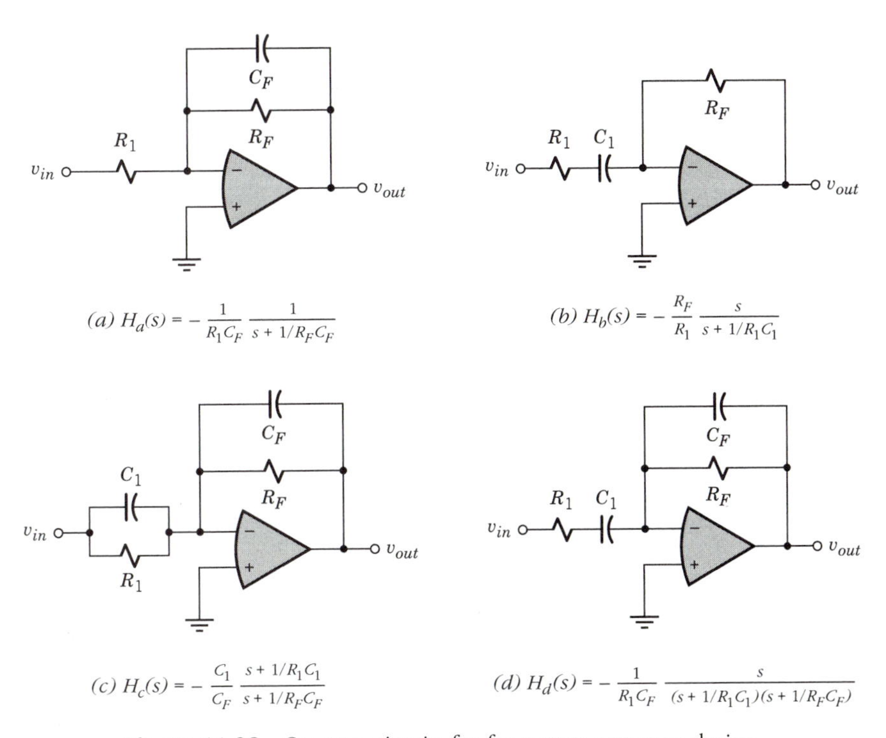

Figure 11.32 Op-amp circuits for frequency-response design.

Example 11.11 *Design of an FM Pre-emphasis Network*

To improve the quality of FM radio reception, the high-frequency components of the audio signal are emphasized before modulation at the transmitter. This strategy allows for a noise-reducing de-emphasis network at the receiver. The gain curve for a suitable FM pre-emphasis network is plotted in Fig. 11.33*a*, along with the straight-line approximation.

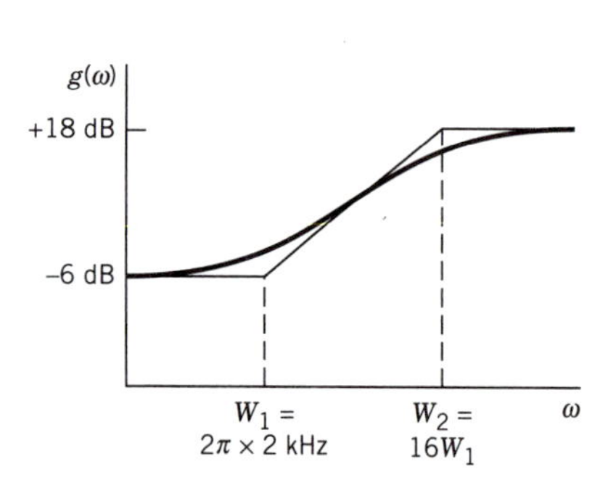

(a) Gain curve for FM premphasis

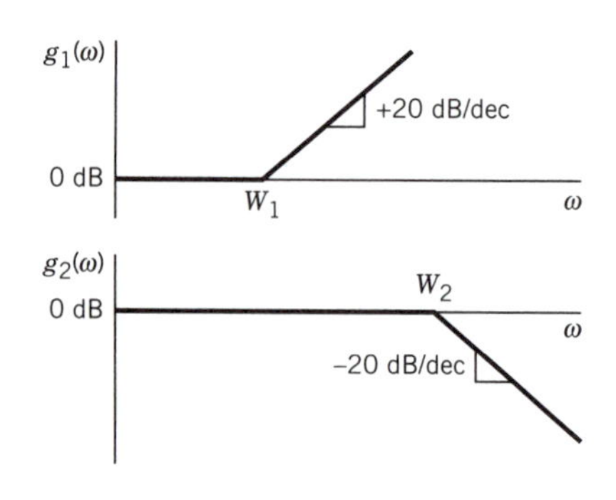

(b) Asymptotic plots

Figure 11.33

We first determine the asymptotic slope in the middle region. The gain increases by $18 - (-6) = +24$ dB when the frequency increases from W_1 to $W_2 = 16W_1$. Since $\log W_2/W_1 = \log 16 = 1.2$ decades, the slope is

$$+24 \text{ dB}/1.2 \text{ decades} = +20 \text{ dB/decade}$$

This calculation confirms the realizability of $g(\omega)$.

Taking $K_{\text{dB}} = -6$ dB, the straight-line approximation can be decomposed into the two asymptotic plots shown in Fig. 11.33*b*. We thereby obtain $g_1(\omega) = g_{lp}^{-1}(\omega; W_1)$ and $g_2(\omega) = g_{lp}(\omega; W_2)$. Thus, inserting $|K| = 10^{-6/20} \approx 0.5$, $W_1 = 2\pi \times 2000 = 12{,}600$ and $W_2 = 16W_1 = 201{,}000$, we have

$$H(s) = \pm 0.5 H_{lp}^{-1}(s; 12{,}600) H_{lp}(s; 201{,}000)$$

$$= \pm 0.5 \frac{s + 12{,}600}{12{,}600} \frac{201{,}000}{s + 201{,}000} = \pm 8 \frac{s + 12{,}600}{s + 201{,}000}$$

If polarity inversion is acceptable, then $H(s)$ can be implemented by the circuit in Fig. 11.32*c* with

$$C_1/C_F = 8 \qquad 1/R_1C_1 = 12{,}600 \qquad 1/R_FC_F = 201{,}000$$

Since the resistances should be in the range of 1 kΩ to 100 kΩ, we'll take $R_F = 10$ kΩ and

$$C_F = 1/201{,}000R_F \approx 5 \times 10^{-10} = 0.5 \text{ nF}$$

Then

$$C_1 = 8C_F = 4 \text{ nF} \qquad R_1 = 1/12{,}600C_1 \approx 20 \text{ k}\Omega$$

which completes the design calculations.

Exercise 11.13

Design an op-amp circuit to get the asymptotic gain curve in Fig. 11.34. Let all capacitors be 1 μF.

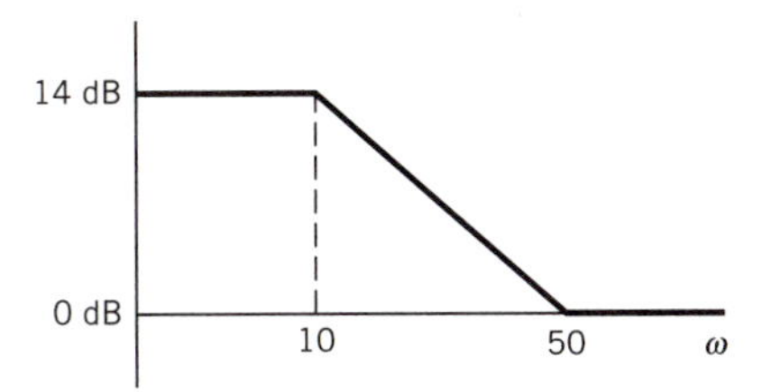

Figure 11.34

11.6 BUTTERWORTH FILTERS†

The simple filters studied in Section 11.2 lack pronounced selectivity because they have rather gradual transitions from passband to stopband. But higher-order filter circuits can provide sharper transitions and act more like ideal filters. For example, Fig. 11.35 shows the amplitude-ratio curves of three different lowpass filters with the same cutoff frequency. Curve A corresponds to our familiar first-order filter, whereas curve B exhibits the improved response of a *fourth-order Butterworth filter*. Curve C is another type of fourth-order response called a *Chebyshev filter*. Compared to the Butterworth response, the Chebyshev response falls off even more rapidly in the stopband. However, the Chebyshev response also has ripples in the passband that may or may not be acceptable.

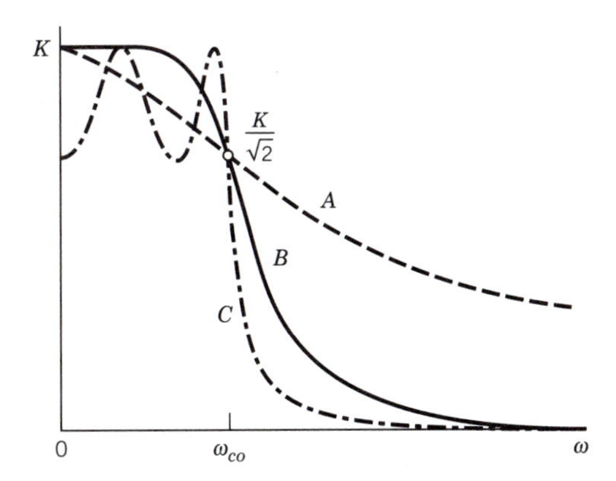

Figure 11.35 Amplitude ratios of lowpass filters: (A) first-order, (B) fourth-order Butterworth, and (C) fourth-order Chebyshev.

As illustrated by Fig. 11.35, the design of higher-order filters involves two general types of tradeoffs. One tradeoff is *performance versus complexity,* in that higher-order circuits may have better characteristics but are more complex and, consequently, more expensive than simple filters. Another tradeoff is *rejection versus ripples,* in that stopband rejection can be increased by allowing ripples in the passband. These and other considerations have lead to the development of several families of higher-order filters.

This section focuses on the family of Butterworth filters, including nth-order lowpass and highpass transfer functions along with op-amp circuit implementations. These filter circuits are useful in many practical applications because they provide improved selectivity without passband ripples. Additionally, the study of Butterworth filters does not require the more advanced mathematical techniques needed for other types such as Chebyshev filters. Even so, we'll limit our consideration to lowpass and highpass filters because higher-order bandpass and notch circuits become appreciably more complicated.

Butterworth Lowpass Filters

Butterworth filters have the distinctive characteristic that the amplitude ratio is as constant or "flat" as possible over the passband. This characteristic comes from the **Butterworth polynomials** denoted by $B_n(s)$, where n indicates the polynomial order. But regardless of n, setting $s = j\omega$ in any Butterworth polynomial yields

$$|B_n(j\omega)| = \sqrt{\omega^{2n} + \omega_{co}{}^{2n}} \tag{11.59}$$

We say that $|B_n(j\omega)|$ is **maximally flat** at $\omega = 0$ because

$$\frac{d^m}{d\omega^m}|B_n(j\omega)|_{\omega=0} = 0 \qquad m = 1, 2, \ldots, 2n-1$$

Additionally, when n is *large,* we see from Eq. (11.59) that

$$\begin{aligned} |B_n(j\omega)| &\approx \omega_{co}{}^n & \omega < \omega_{co} \\ &\approx \omega^n & \omega > \omega_{co} \end{aligned}$$

These properties lend themselves to better approximations of ideal filter characteristics.

An ***n*th-order Butterworth lowpass filter** is a network with n poles arranged to get the transfer function

$$H_{lp}(s) = K\omega_{co}{}^n / B_n(s) \tag{11.60}$$

The corresponding amplitude ratio is

$$a_{lp}(\omega) = \frac{K\omega_{co}{}^n}{|B_n(j\omega)]} = \frac{K}{\sqrt{1 + (\omega/\omega_{co})^{2n}}} \tag{11.61}$$

which assumes $K > 0$ as before. The parameter ω_{co} is the cutoff frequency in the usual sense that

$$a_{lp}(\omega_{co}) = K/\sqrt{2}$$

In fact, if we take $n = 1$, then $a_{lp}(\omega)$ is identical to that of a simple *RC* lowpass filter. But with $n \gg 1$, we obtain the improved response

$$\begin{aligned} a_{lp}(\omega) &\approx K & \omega < \omega_{co} \\ &\approx (\omega_{co}/\omega)^n K \ll K & \omega > \omega_{co} \end{aligned}$$

Figure 11.36 illustrates how a Butterworth lowpass filter approaches ideal filtering as n increases. The almost constant response for $\omega < \omega_{co}$ comes from the maximally flat property of $|B_n(j\omega)|$.

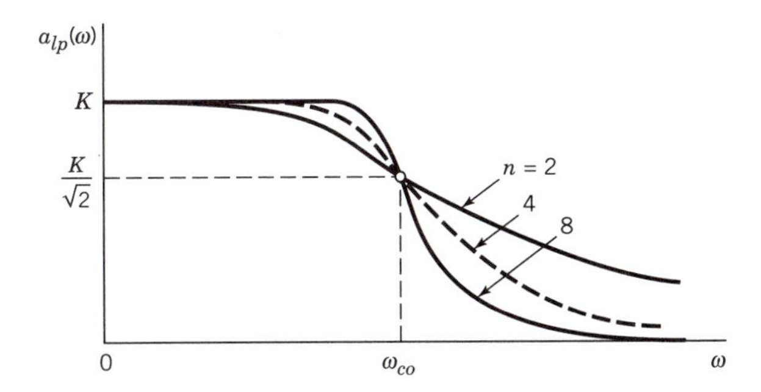

Figure 11.36 Amplitude ratios of Butterworth lowpass filters.

For $\omega > \omega_{co}$, the high-frequency rejection of a Butterworth lowpass filter is better displayed by the dB gain

$$\begin{aligned} g_{lp}(\omega) &= 20 \log a_{lp}(\omega) \\ &\approx K_{\text{dB}} - 20n \log(\omega/\omega_{co}) \qquad \omega > \omega_{co} \end{aligned}$$

The Bode plots in Fig. 11.37 show the resulting high-frequency slope or **rolloff** of $20n$ dB per decade. Increasing n thus increases the rejection above ω_{co}, as well as making the response flatter in the passband. However, the complexity of the circuit also increases with n, so an important design task is finding the *smallest* n that meets the requirements for a particular application.

After determining the value of n, we need the locations of the poles associated with the polynomial $B_n(s)$ to satisfy Eq. (11.59). S. Butterworth tackled this problem in 1930, and he showed in general that

> The poles are uniformly spaced by angles of $180°/n$ around a semicircle of radius ω_{co} in the left half of the s plane.

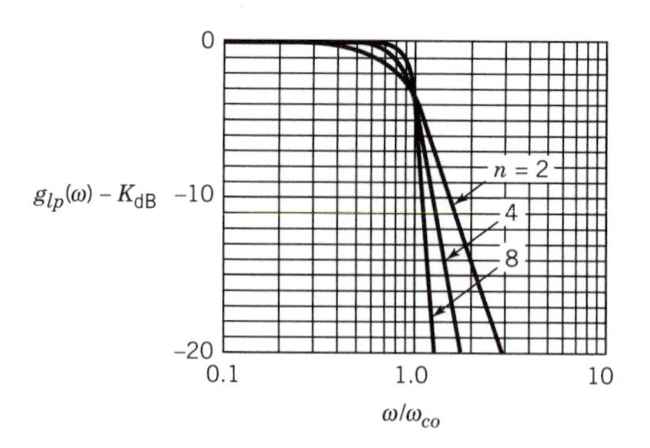

Figure 11.37 Gain curves of Butterworth lowpass filters.

When n is an *even* integer, all poles must be complex-conjugate pairs and we get the pattern in Fig. 11.38*a*. The angular position of the poles in the upper half are

$$\psi_i = \frac{i-1}{n}\,90^\circ \qquad i = 2, 4, \ldots, n$$

When n is an *odd* integer, there must also be a real pole without a complex-conjugate mate. We then get the pattern in Fig. 11.38*b*, where

$$\psi_i = \frac{i-1}{n}\,90^\circ \qquad i = 1, 3, \ldots, n$$

which includes the real pole $p_1 = -\omega_{co}$.

Comparing Fig. 11.38 with Fig. 11.12*a* (p. 502), we see that each complex-conjugate pair of poles corresponds to the roots of a quadratic in the form

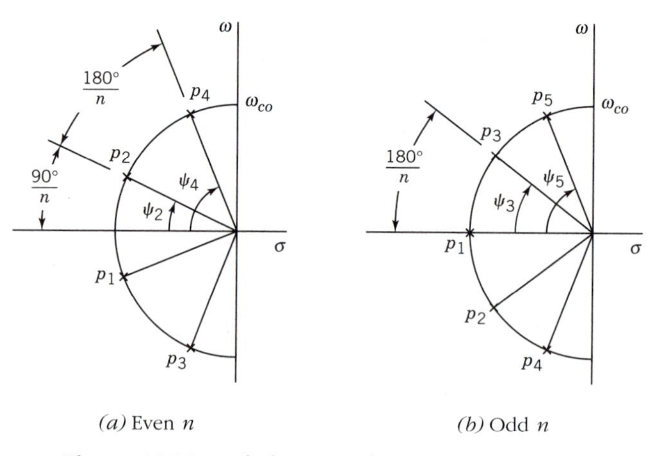

Figure 11.38 Pole locations for Butterworth filters.

$$P_i(s) = s^2 + (\omega_{co}/Q_i)s + \omega_{co}^2 \tag{11.62}$$

with

$$Q_i = \frac{1}{2\cos\psi_i}$$

Therefore, we can write the even-order Butterworth polynomials as the product of $n/2$ quadratics

$$B_n(s) = P_2(s)P_4(s)\ldots P_n(s) \qquad n \text{ even} \tag{11.63a}$$

The odd-order polynomials likewise consist of quadratics and the additional factor $s - p_1 = s + \omega_{co}$, so

$$B_n(s) = (s + \omega_{co})P_3(s)P_5(s)\ldots P_n(s) \qquad n \text{ odd} \tag{11.63b}$$

All quality factors for the $P_i(s)$ are calculated via

$$Q_i = \frac{1}{2\cos[(i-1)90°/n]} \tag{11.64}$$

Table 11.7 lists the resulting values for $n = 2$ through 9.

TABLE 11.7 Quality Factors of Butterworth Polynomials

n	Q_2	Q_4	Q_6	Q_8	n	Q_3	Q_5	Q_7	Q_9
2	0.707				3	1.000			
4	0.541	1.307			5	0.618	1.618		
6	0.518	0.707	1.932		7	0.555	0.802	2.247	
8	0.510	0.601	0.900	2.563	9	0.532	0.653	1.000	2.879

Before moving on, we should emphasize that *pole locations* play a crucial role in filter design. If the poles are equispaced around a semicircle, as in Fig. 11.38, then we have the maximally flat Butterworth response. But if the poles are arranged on an ellipse, then we have a Chebyshev response with ripples in the passband. Figure 11.39 illustrates how the poles of a fourth-order Butterworth filter are moved inward to get a Chebyshev filter. The passband ripples come from the fact that these poles are now closer to the ω axis, so each pole produces a "bump" in the amplitude ratio.

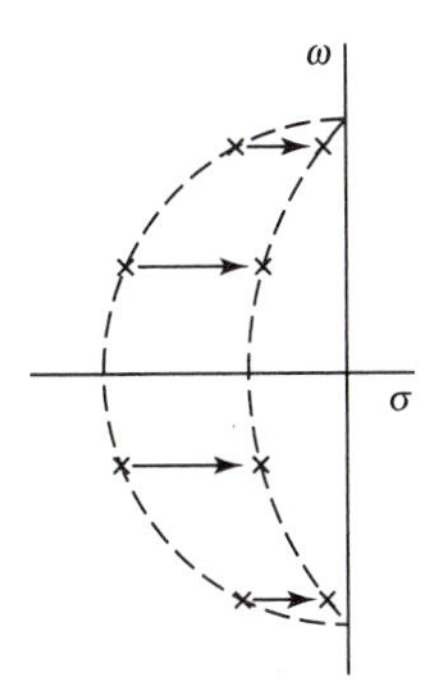

Figure 11.39 Moving the poles of a Butterworth filter to get a Chebyshev response.

Example 11.12 ***FM Stereo Separation Filter***

FM stereo receivers need filters that pass frequency components below 15 kHz and suppress components above 23 kHz. For this purpose, we'll design a Butterworth filter having $f_{co} = 15$ kHz and $a_{lp}(\omega) \leq K/8$ for $f \geq 23$ kHz.

Our first task is to determine the smallest value of n that satisfies the suppression requirement, rewritten as

$$a_{lp}^2(\omega) = \frac{K^2}{1 + (\omega/\omega_{co})^{2n}} \leq \frac{K^2}{8^2} \qquad \omega \geq 2\pi \times 23 \text{ kHz}$$

Since $a_{lp}(\omega)$ decreases as ω increases, we set $\omega = 2\pi \times 23$ kHz along with $\omega_{co} = 2\pi \times 15$ kHz to obtain the condition

$$(23/15)^{2n} \geq 8^2 - 1 = 63$$

from which

$$n \geq \frac{\log 63}{2 \log (23/15)} = 4.85$$

Hence, we round up to the integer value $n = 5$.

Next, we set $n = 5$ in Eq. (11.60) to write the transfer function

$$H_{lp}(s) = K\omega_{co}^{\;5}/B_5(s)$$

The fifth-order Butterworth polynomial is given by Eq. (11.63b) as

$$\begin{aligned} B_5(s) &= (s + \omega_{co})P_3(s)P_5(s) \\ &= (s + \omega_{co})\left(s^2 + \frac{\omega_{co}}{Q_3}s + \omega_{co}^{\;2}\right)\left(s^2 + \frac{\omega_{co}}{Q_5}s + \omega_{co}^{\;2}\right) \end{aligned}$$

The corresponding quality factors are obtained from Table 11.7 in the row for $n = 5$, where

$$Q_3 = 0.618 \qquad Q_5 = 1.618$$

Exercise 11.14

Write the third-order polynomial $B_3(s)$ and show that it satisfies the property in Eq. (11.59).

Butterworth Highpass Filters

Engineers seldom design highpass filters from scratch. Instead, they derive highpass characteristics from existing lowpass designs via the **lowpass-to-highpass transformation**

$$s \rightarrow \omega_{co}^{\;2}/s \tag{11.65}$$

which means that s is replaced wherever it appears by ω_{co}^2/s. For example, applying this transformation to the first-order lowpass function $H_{lp}(s) = K\omega_{co}/(s + \omega_{co})$ yields

$$H_{hp}(s) = H_{lp}(\omega_{co}^2/s) = \frac{K\omega_{co}}{(\omega_{co}^2/s) + \omega_{co}} = \frac{Ks}{s + \omega_{co}}$$

which you should recognize as being our standard first-order highpass function.

Similarly, to obtain the transfer function of an **nth-order Butterworth highpass filter,** we note from Eq. (11.62) that

$$P_i(\omega_{co}^2/s) = (\omega_{co}^2/s)^2 + (\omega_{co}/Q_i)(\omega_{co}^2/s) + \omega_{co}^2 = (\omega_{co}/s)^2 P_i(s)$$

Consequently, $B_n(\omega_{co}^2/s) = (\omega_{co}/s)^n B_n(s)$, and Eq. (11.60) transforms to

$$H_{hp}(s) = Ks^n/B_n(s) \tag{11.66}$$

The resulting amplitude ratio is

$$a_{hp}(\omega) = \frac{K\omega^n}{|B_n(j\omega)|} = \frac{K}{\sqrt{1 + (\omega_{co}/\omega)^{2n}}} \tag{11.67}$$

Plots of $a_{hp}(\omega)$ in Fig. 11.40 again show a maximally flat response over the passband and improved selectivity as n increases. Since Eq.(11.67) is the same as $a_{lp}(\omega)$ with ω_{co}/ω in place of ω/ω_{co}, the dB gain curves look like Fig. 11.37 with the horizontal axis relabeled ω_{co}/ω.

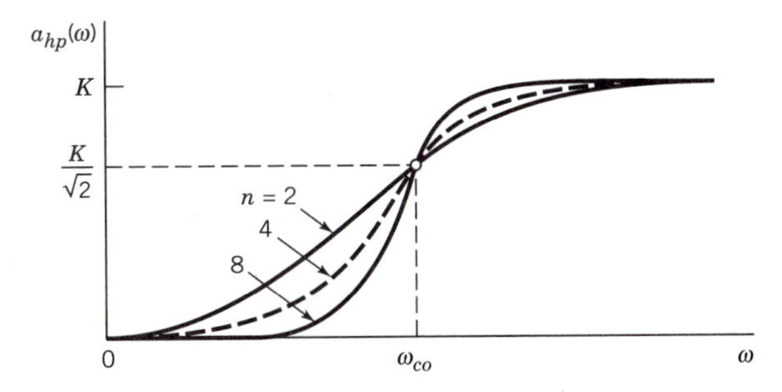

Figure 11.40 Amplitude ratios of Butterworth highpass filters.

Equation (11.66) indicates that we can approximate an ideal highpass filter using n zeros at $s = 0$ along with the n poles from $B_n(s)$. Since the pole locations remain the same as the lowpass case, $B_n(s)$ is still given by Eq. (11.63a) or (11.63b) and Table 11.7 holds for both lowpass and highpass Butterworth filters.

Exercise 11.15

A Butterworth highpass filter with $K = 25$ and $f_{co} = 700$ Hz is to have $a_{hp}(\omega) \leq K/30$ for $f \leq 200$ Hz. Find the required value of n and write the expression for $H_{hp}(s)$. (Save your results for use in Exercise 11.16.)

Op-Amp Implementations

Designing a passive network to achieve the Butterworth lowpass or highpass response is a difficult task because of interaction between the elements. Furthermore, the passive network must contain at least one inductor. But if we write $H(s)$ as a product of simple factors, then the individual factors can be implemented with op-amp stages. The op-amps eliminate inductors and loading interaction between stages, and they allow for gain $K > 1$.

Pursing this approach, we first express the transfer function for an nth-order lowpass or highpass filter as

$$\begin{aligned} H(s) &= H_2(s)H_4(s) \dots H_n(s) \qquad n \text{ even} \\ &= H_1(s)H_3(s) \dots H_n(s) \qquad n \text{ odd} \end{aligned} \tag{11.68}$$

For the lowpass case, Eqs. (11.60) and (11.63) suggest the factors

$$H_{lp1}(s) = \frac{K_1\omega_{co}}{s + \omega_{co}} \qquad H_{lpi}(s) = \frac{K_i{\omega_{co}}^2}{P_i(s)} \qquad i \geq 2 \tag{11.69}$$

For the highpass case, Eqs. (11.66) and (11.63) suggest the factors

$$H_{hp1}(s) = \frac{K_1 s}{s + \omega_{co}} \qquad H_{hpi}(s) = \frac{K_i s^2}{P_i(s)} \qquad i \geq 2 \tag{11.70}$$

In either case, the overall gain is

$$\begin{aligned} K &= K_2K_4 \dots K_n \qquad n \text{ even} \\ &= K_1K_3 \dots K_n \qquad n \text{ odd} \end{aligned} \tag{11.71}$$

so we are free to distribute the gain over the individual stages.

The first-order factors $H_{lp1}(s)$ and $H_{hp1}(s)$ are easily implemented as noninverting circuits like Fig. 11.19 (p. 508) or inverting circuits like Fig. 11.20 (p. 509). Op-amp circuits for second-order functions are diagrammed in Fig. 11.41. These noninverting circuits have the advantage of leading to convenient design equations. If you want an inverting filter with even n, then you can include an additional inverting amplifier.

Node analysis of the noninverting lowpass circuit in Fig. 11.41a confirms that

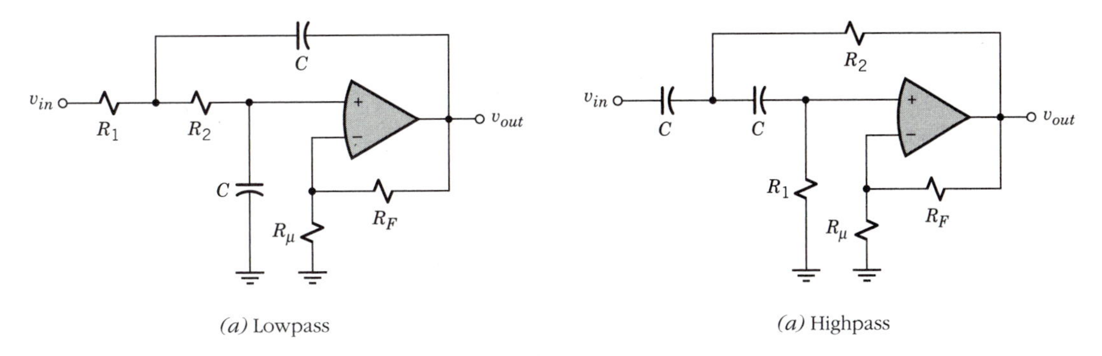

Figure 11.41 Op-amp circuits for second-order functions.

$$H_{lpi}(s) = \frac{\underline{V}_{out}}{\underline{V}_{in}} = \frac{K_i \omega_{co}^2}{s^2 + (\omega_{co}/Q_i)s + \omega_{co}^2} \tag{11.72}$$

with

$$K_i = 1 + \frac{R_F}{R_\mu} \qquad \omega_{co}^2 = \frac{1}{R_1 R_2 C^2} \qquad Q_i = \frac{\sqrt{R_1 R_2}}{R_2 - (K_i - 2)R_1}$$

Solving these expressions for R_1 and R_2 yields the *design equations*

$$R_1 = \frac{2Q_i r}{1 + \sqrt{4Q_i^2(K_i - 2) + 1}} \qquad R_2 = \frac{r^2}{R_1} \tag{11.73}$$

where

$$r = 1/\omega_{co}C$$

which has the units of resistance. The equation for R_1 requires that we have $4Q_i^2(K_i - 2) + 1 \geqslant 0$ or

$$K_i \geqslant 2 - 1/4Q_i^2 \tag{11.74}$$

which sets a lower limit on the stage gain.

Node analysis of the noninverting highpass circuit in Fig. 11.41*b* confirms that

$$H_{hpi}(s) = \frac{\underline{V}_{out}}{\underline{V}_{in}} = \frac{K_i s^2}{s^2 + (\omega_{co}/Q_i)s + \omega_{co}^2} \tag{11.75}$$

with

$$K_i = 1 + \frac{R_F}{R_\mu} \qquad {\omega_{co}}^2 = \frac{1}{R_1 R_2 C^2} \qquad Q_i = \frac{\sqrt{R_1 R_2}}{2R_2 - (K_i - 1)R_1}$$

The corresponding design equations are

$$R_1 = \frac{4Q_i r}{1 + \sqrt{8Q_i^2(K_i - 1) + 1}} \qquad R_2 = \frac{r^2}{R_1} \tag{11.76}$$

where, again, $r = 1/\omega_{co}C$. The expression for R_1 requires $K_i \geq 1 - 1/8Q_i^2$, which is automatically satisfied since $K_i > 1$.

Equations (11.68)–(11.76) together with Table 11.7 provide the information needed to design a lowpass or highpass Butterworth filter with $n \leq 9$. But keep in mind that all capacitor values should be chosen to fall in the range

$$10^{-6}/f_{co} < C < 10^{-4}/f_{co} \tag{11.77}$$

This is the same rule-of-thumb mentioned in Section 11.3.

Example 11.13 ***Op-Amp Circuit for a Lowpass Filter***

Suppose the lowpass filter in Example 11.12 is to have $K = 80$. The op-amp implementation from Eqs. (11.68) and (11.69) calls for one first-order stage and two second-order stages to get

$$H_{lp}(s) = H_{lp1}(s)H_{lp3}(s)H_{lp5}(s)$$

where

$$f_{co} = 15.9 \text{ kHz} \qquad Q_3 = 0.618 \qquad Q_5 = 1.618$$

All stages will be noninverting circuits with gains such that $K_1K_3K_5 = K = 80$.

We'll distribute the total gain more-or-less equally by taking $K_1 = 5$ and $K_3 = K_5 = 4$. These values also ensure that Eq. (11.74) is satisfied for the second-order stages. If we let $R_\mu = 1$ kΩ in all stages, for simplicity, then the feedback resistors are $R_{F1} = (5 - 1)R_\mu = 4$ kΩ and $R_{F3} = R_{F5} = (4 - 1)R_\mu = 3$ kΩ.

We'll also let all capacitors have the same value, which, from Eq. (11.77), should fall between 6.3×10^{-11} F and 6.3×10^{-9} F. An appropriate choice would therefore be the standard value

$$C = 10^{-9} = 1 \text{ nF}$$

Then, since the first-order stage has $\omega_{co} = 1/RC$, the required resistor is

$$R = 1/2\pi f_{co}C = 10.0 \text{ k}\Omega$$

For the second-order stages, we insert $r = 1/\omega_{co}C = 10.0\ \text{k}\Omega$ and $K_i = 4$ in Eq. (11.73) to obtain

$$R_1 = \frac{2Q_i \times 10.0\ \text{k}\Omega}{1 + \sqrt{8Q_i^2 + 1}} \qquad R_2 = \frac{(10.0\ \text{k}\Omega)^2}{R_1}$$

Thus, the resistors for the stage with $Q_i = Q_3 = 0.618$ are

$$R_1 = 4.10\ \text{k}\Omega \qquad R_2 = 24.4\ \text{k}\Omega$$

while the resistors for the stage with $Q_i = Q_5 = 1.618$ are

$$R_1 = 5.69\ \text{k}\Omega \qquad R_2 = 17.6\ \text{k}\Omega$$

The complete circuit is diagrammed in Fig. 11.42, where all capacitances are in nanofarads and all resistances in kilohms.

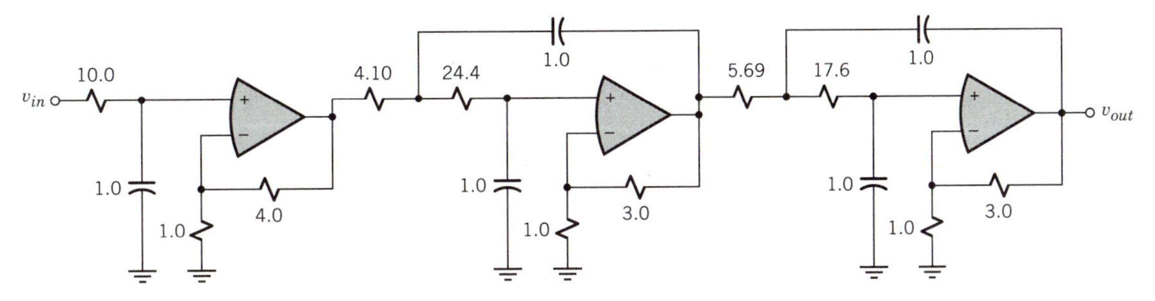

Figure 11.42 Op-amp circuit for Example 11.13 (capacitances in nanofarads and resistances in kilohms).

Exercise 11.16

Obtain the op-amp implementation of the filter in Exercise 11.15. Let all stages have the same gain, and take $R_\mu = 1\ \text{k}\Omega$ and $C = 10$ nF in all stages.

PROBLEMS

Section 11.1 Frequency Response

11.1 Obtain the expressions for $a(\omega)$ and $\theta(\omega)$ given that $H(s) = (s + 20)/5(s + 4)$. Then find the steady-state output $y(t)$ when the input is $x(t) = \cos 2t + \cos 10t + \cos 50t$.

11.2 Do Problem 11.1 with

$$H(s) = 25(s + 4)/(s + 10)^2$$

11.3 Do Problem 11.1 with

$$H(s) = 25s/(s + 5)(s + 20)$$

11.4 Do Problem 11.1 with

$$H(s) = s(s + 20)/(s + 5)(s + 40)$$

11.5 Use s-plane vectors with test points at $\omega = 0^+$, 4, 10, and 20 to sketch the frequency-response curves for $H(s) = (s + 20)/5(s + 4)$.

11.6 Do Problem 11.5 for $H(s) = 25(s+4)/(s+10)^2$.

11.7 Use s-plane vectors with test points at $\omega = 0^+$, 5, 10, and 20 to sketch the frequency-response curves for $H(s) = 20s/(s+10)^2$.

11.8 Do Problem 11.7 for

$$H(s) = 25s/(s+5)(s+20)$$

11.9 Select appropriate test points and use s-plane vectors to sketch the frequency-response curves for

$$H(s) = s^2/(s^2 + 4s + 104)$$

11.10 Do Problem 11.9 for

$$H(s) = (s^2 + 2s + 101)/(s+10)^2$$

11.11 Do Problem 11.9 for

$$H(s) = 10(s^2 + 4s + 104)/(s+10)^3$$

11.12 Do Problem 11.9 for

$$H(s) = s(s+10)/(s^2 + 10s + 50)$$

Section 11.2 Filters

(See also PSpice problems B.43–B.48.)

11.13* Design an LR lowpass filter like the one in Fig. 11.8a (p. 499) such that $f_{co} = 200$ kHz and $|Z_{in}(j\omega_{co})| = 1$ kΩ.

11.14 Design an RL highpass filter like the one in Fig. 11.8b (p. 499) such that $f_{co} = 100$ Hz and $|Z_{in}(j2\omega_{co})| = 100$ Ω.

11.15 Design an RC lowpass filter like the one in Fig. 11.8a (p. 499) such that $f_{co} = 10$ kHz and $|Z_{in}(j\omega_{co})| = 600$ Ω.

11.16 Design a CR highpass filter like the one in Fig. 11.8b (p. 499) such that $f_{co} = 30$ Hz and $|Z_{in}(j\omega_{co}/3)| = 5$ kΩ.

11.17 Obtain the expressions for ω_{co} and K when R_s is added in series with L in the LR lowpass filter in Fig. 11.8a (p. 499).

11.18 Obtain the expressions for ω_{co} and K when R_L is added in parallel with L in the RL highpass filter in Fig. 11.8b (p. 499).

11.19* The circuit in Fig. P11.19 has been proposed as a means of separating the two components of a telemetry signal $v(t) = v_1(t) + v_2(t)$. The component $v_1(t)$ consists of frequencies below $f_1 = 2$ kHz, while $v_2(t)$ consists of frequencies above $f_2 = 20$ kHz. Determine appropriate values for R_a and R_b. Then find $v_a(t)$ and $v_b(t)$ given that $v(t) = 10\cos 0.6\omega_1 t + 10\cos 1.2\omega_2 t$.

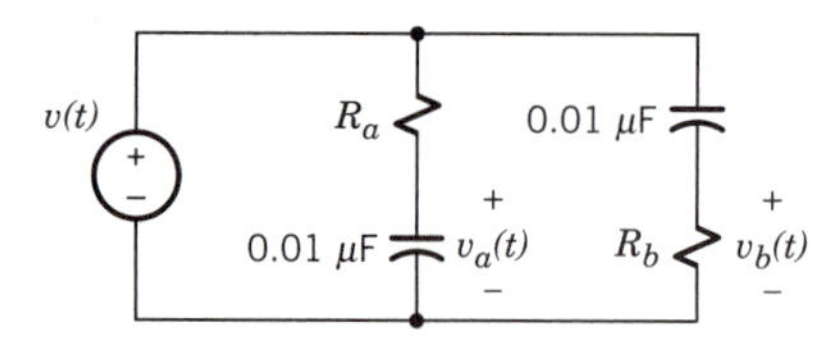

Figure P11.19

11.20 Do Problem 11.19 with $f_1 = 10$ kHz and $f_2 = 50$ kHz.

11.21 Do Problem 11.19 with $f_1 = 500$ Hz and $f_2 = 2$ kHz.

11.22 The ladder network in Fig. P11.22 becomes a second-order lowpass filter when $Z_1 = Z_3 = R$ and $Z_2 = Z_4 = 1/sC$. Take $\underline{V}_{out} = 1$ to find $\underline{V}_{in}$ and show that $H(s) = 1/(R^2C^2s^2 + 3RCs + 1)$. Then let $x = R^2C^2\omega_{co}^2$ and solve the quadratic resulting from $a^2(\omega_{co}) = ½$ to obtain ω_{co} in terms of RC.

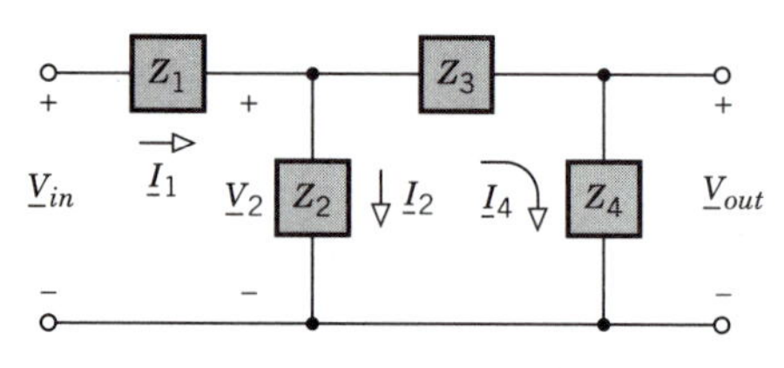

Figure P11.22

11.23 The ladder network in Fig. P11.22 becomes a second order highpass filter when $Z_1 = Z_3 = 1/sC$ and $Z_2 = Z_4 = R$. Take $\underline{V}_{out} = 1$ to find $\underline{V}_{in}$ and show that $H(s) = R^2C^2s^2/(R^2C^2s^2 + 3RCs + 1)$. Then let $x = R^2C^2\omega_{co}^2$ and solve the quadratic resulting from $a^2(\omega_{co}) = ½$ to obtain ω_{co} in terms of RC.

11.24 The ladder network in Fig. P11.22 acts like a highpass filter driving a lowpass filter when $Z_1 = 1/sC_1$, $Z_2 = Z_3 = R$, $Z_4 = 1/sC_4$, and $C_1 \gg C_4$. Take $\underline{V}_{out} = 1$ to find $\underline{V}_{in}$ and $H(s) = \underline{V}_{out}/\underline{V}_{in}$. Then verify that $H(s) \approx [s/(s + \omega_{hp})]\,[\omega_{lp}/(s + \omega_{lp})]$ where $\omega_{hp} = 1/RC_1$ and $\omega_{lp} = 1/RC_4$ so $\omega_{lp} \gg \omega_{hp}$.

11.25* Determine ω_0 and Q for a bandpass filter with $f_l = 1$ kHz and $f_u = 4$ kHz. Then find R and C for implementation as a series resonant circuit with $L = 1$ mH.

11.26 Determine ω_0 and Q for a bandpass filter with $f_l = 40$ kHz and $f_u = 90$ kHz. Then find R and C for implementation as a parallel resonant circuit with $L = 0.1$ mH.

11.27 A low-frequency vibration analyzer requires a parallel bandpass filter like Fig. P11.27 with $R = 10$ kΩ, $\omega_0 = 50$ rad/s, and $B = 2$ rad/s. (a) Find the required values of C and L when R_w is negligible. (b)

Modify the circuit to eliminate the large inductor using the generalized impedance converter from Example 10.4 (p. 454) with a 1-μF capacitor and four equal resistors.

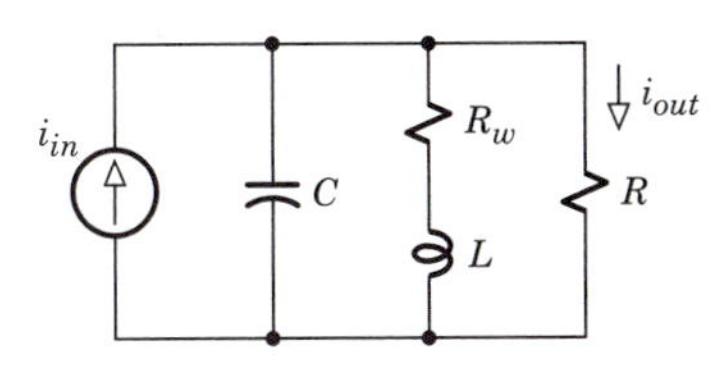

Figure P11.27

11.28 Let the coil in Fig. P11.27 have $L = 100\ \mu$H and $R_w = 0.1\ \Omega$. Find the values of C and R to get $\omega_0 = 200$ krad/s and $B = 2$ krad/s.

11.29 Let the coil in Fig. P11.27 have $L = 100\ \mu$H and $R_w = 0.1\ \Omega$. Find the values of C and R to get $\omega_0 = 50$ krad/s and $B = 1$ krad/s.

11.30 Suppose the coil in Fig. P11.27 has $L = 10\ \mu$H and $R_w = 0.5\ \Omega$, and you want to get $\omega_0 = 500$ krad/s and $B = 25$ krad/s. (a) Find the required values of C and R. (b) Devise a circuit implementation using the negative-resistance converter from Problem 3.39 (p. 129).

11.31 Derive Eq. (11.20) by solving the quadratic that results from Eq. (11.19) when $a_{bp}^2(\omega) = K^2/2$. Hint: Use the facts that $\sqrt{1} = \pm 1$ and $\omega \geqslant 0$.

11.32 Derive from Eq. (11.18) the approximation

$$H_{bp}(j\omega) \approx \frac{K}{1 + j2(\omega - \omega_0)/B} \qquad |\omega - \omega_0| \ll \omega_0$$

Hint: Let $\omega/\omega_0 = 1 + \delta$ with $|\delta| \ll 1$ so $\omega_0/\omega = (1 + \delta)^{-1} \approx 1 - \delta$.

11.33* Find $H(s) = \underline{V}_L/\underline{V}_{in}$ for Fig. P11.33. Express your result in terms of ω_0 and Q_{ser}. Then sketch $a(\omega)$ to show that the circuit acts essentially like a bandpass filter when $Q_{ser} \gg 1$.

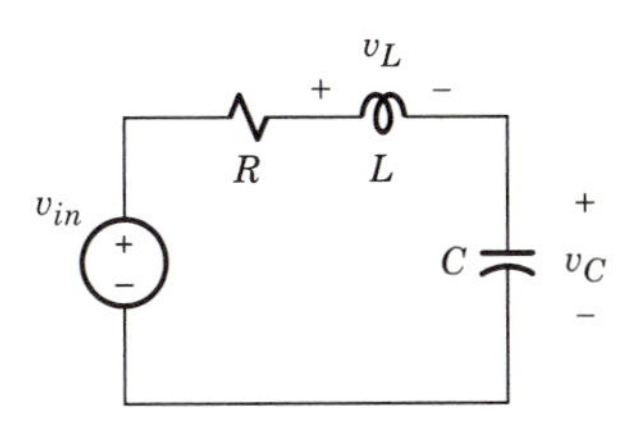

Figure P11.33

11.34 Do Problem 11.33 with $H(s) = \underline{V}_C/\underline{V}_{in}$.

11.35 Consider a notch filter described by Eq. (11.24) with $K = 1$, $\omega_0 = 1$, $\beta = 0$, and $Q \gg 1$. Use s-plane vectors with test points at $\omega = 1 - \frac{1}{2}Q$, 1^-, 1^+, and $1 + \frac{1}{2}Q$ to sketch $a_{no}(\omega)$ and $\theta_{no}\omega)$ for this case. What happens to the phase shift at ω_0?

11.36 Consider a notch filter described by Eq. (11.24) with $K = 1$, $\omega_0 = 1000$, $\beta = 1$, and $Q = 50$. Use s-plane vectors with appropriate test points to sketch $a_{no}(\omega)$ and $\theta_{no}(\omega)$ for this case.

11.37 Do Problem 11.36 with $Q = 25$.

Section 11.3 Op-Amp Filter Circuits

11.38* Design an active wideband filter like Fig. 11.21a (p. 510) to get $f_l = 20$ Hz, $f_u = 1$ kHz, and $K = 4$ with $R_1 \geqslant 5$ kΩ.

11.39 Design an active wideband filter like Fig. 11.21a (p. 510) to get $f_l = 5$ kHz, $f_u = 50$ kHz, and $K = 2$ with $R_1 \geqslant$ kΩ.

11.40 Design an active narrowband filter like Fig. 11.22 (p. 510) to get $f_l = 9.5$ kHz, $f_u = 10.5$ kHz, and $K = 4$ with $R_1 \geqslant 50$ kΩ.

11.41 Design an active narrowband filter like Fig. 11.22 (p. 510) to get $f_l = 490$ Hz, $f_u = 510$ Hz, and $K = 20$ with $R_1 \geqslant 25$ kΩ.

11.42 Design an active narrowband filter like Fig. 11.22 (p. 510) to get $f_l = 2$ kHz, $f_u = 4$ kHz, and $K = 1$ with $R_1 \geqslant 10$ kΩ.

11.43 Let $G_1 = 1/R_1$, etc., in Fig. 11.22 (p. 510) and write the matrix node equation for $\underline{V}_x$ and $\underline{V}_p$. Then solve for $\underline{V}_p$ to derive the results in Eq. (11.32).

11.44 Figure P11.44 diagrams an inverting active bandpass filter. Find $\underline{V}_{out}/\underline{V}_1$ and $\underline{V}_1/\underline{V}_{in}$ to show that

$$H(s) = \underline{V}_{out}/\underline{V}_{in} = -K(\omega_0/Q)s/[s^2 + (\omega_0/Q)s + \omega_0^2]$$

with

$$Q = \frac{\omega_0 R_F C}{2} \qquad K = \frac{R_F}{2R_1} \qquad \omega_0^2 = \frac{R_1 + R_2}{R_1 R_2 F_F C^2}$$

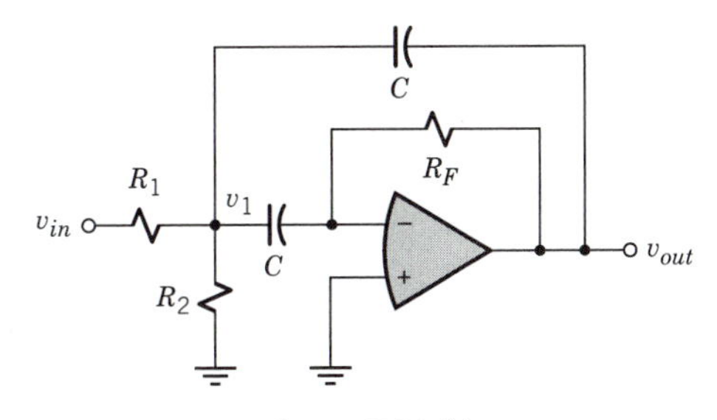

Figure P11.44

11.45 Use the results given in Problem 11.44 to derive the design equations

$$R_F = \frac{2Q}{\omega_0 C} \quad R_1 = \frac{R_F}{2K} \quad R_2 = \frac{R_F}{4Q^2 - 2K}$$

For a given value of Q, what's the upper limit on K?

11.46 Design an active notch filter in the form of Fig. 10.8*a* (p. 464) to get $f_0 = 1$ kHz, $B = 2\pi \times 100$ Hz, and $R \geq 10$ kΩ. What's the resulting value of K?

11.47 Design and active notch filter in the form of Fig. 10.8*a* (p. 464) to get $f_0 = 60$ Hz, $B = 2\pi \times 1$ Hz, and $R \geq 2$ kΩ. What's the resulting value of K?

Section 11.4 Bode Plots

(See also PSpice problems B.49–B.52.)

11.48* Construct the complete Bode plot for

$$H(s) = -20{,}000/(s + 20)(s + 200)$$

11.49 Construct the complete Bode plot for

$$H(s) = 8s^2/(s + 50)\,(s + 500)$$

11.50 Construct the complete Bode plot for

$$H(s) = 0.05(s + 400)/(s + 40)$$

11.51 Construct the complete Bode plot for

$$H(s) = -0.004s(s + 100)/(s + 10)$$

11.52* Draw the asymptotic Bode plot of the gain and phase for

$$H(s) = 8000s/(s + 10)(s + 40)(s + 80)$$

Then add the dB corrections to find the maximum value of $a(\omega)$ and the frequencies ω_l and ω_l at which the gain drops 3 dB below the maximum.

11.53 Do Problem 11.52 for

$$H(s) = -2000s^2/(s + 50)\,(s + 200)\,(s + 400)$$

11.54 Do Problem 11.52 for

$$H(s) = s(s + 800)/(s + 50)\,(s + 200)$$

11.55 Do Problem 11.52 for

$$H(s) = -1000(s + 10)/(s + 100)\,(s + 200)$$

11.56 Construct the asymptotic Bode plot of the gain and phase for $H(s) = 400(s + 10)/(s + 40)^2$.

11.57 Construct the asymptotic Bode plot of the gain and phase for $H(s) = 0.25(s + 80)^2/(s + 10)(s + 40)$.

11.58 Construct the asymptotic Bode plot of the gain and phase for

$$H(s) = 100s(s + 50)/(s + 100)^2(s + 400)$$

11.59 Construct the asymptotic Bode plot of the gain and phase for

$$H(s) = -0.04(s + 50)^3(s + 500)^2/s^3(s + 100)^2$$

11.60* Sketch the Bode plot of $g(\omega)$ when

$$H(s) = 100s/(s + 10)(s^2 + 2s + 100)$$

11.61 Sketch the Bode plot of $g(\omega)$ when

$$H(s) = 20s^2/(s + 20)(s^2 + 10s + 400)$$

11.62 Sketch the Bode plot of $g(\omega)$ when

$$H(s) = 30(s^2 + 10s + 900)/(s + 30)^3$$

11.63 Sketch the Bode plot of $g(\omega)$ when

$$H(s) = s(s^2 + 4s + 1600)/(s + 40)^3$$

11.64 A **stagger-tuned bandpass filter** involves two bandpass functions with the same Q but different resonant frequencies, say ω_1 and ω_2. The transfer function is

$$H(s) = \frac{Ks^2}{(s^2 + \omega_1 s/Q + \omega_1^2)(s^2 + \omega_2 s/Q + \omega_2^2)}$$

Construct the asymptotic Bode plot for $g(\omega)$ taking $\omega_1 = 10$, $\omega_2 = 40$, and $K = \omega_2^2$. Then add correction terms and draw the smooth curves for $Q = 0.5$, 1, and 3. Which value of Q results in the most nearly ideal response, and what are the corresponding values of g_{max}, ω_l, and ω_u?

Section 11.5 Frequency-Response Design

11.65* Design an op-amp circuit to get the asymptotic gain curve in Fig. P11.65 with $\omega_1 = 125$ and $\omega_2 = 4000$. Let all capacitors be 1 μF.

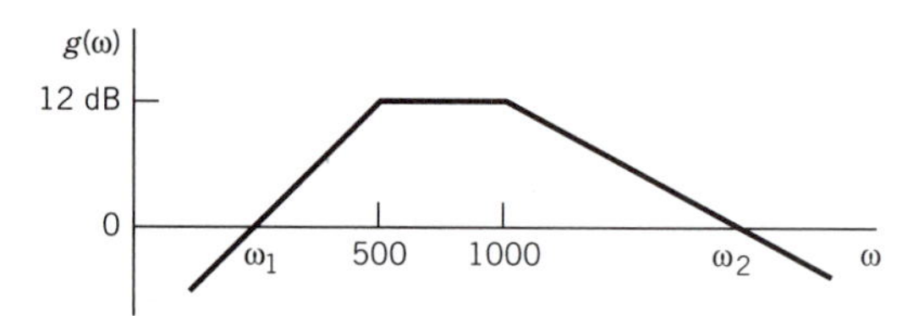

Figure P11.65

11.66 Design an op-amp circuit to get the asymptotic gain curve in Fig. P11.66. Let all capacitors be 0.25 μF.

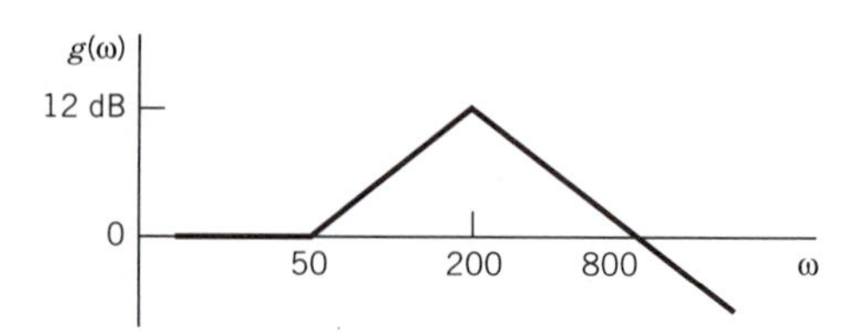

Figure P11.66

11.67 Design an op-amp circuit to get the asymptotic gain curve in Fig. P11.67. Let all capacitors be 0.1 μF.

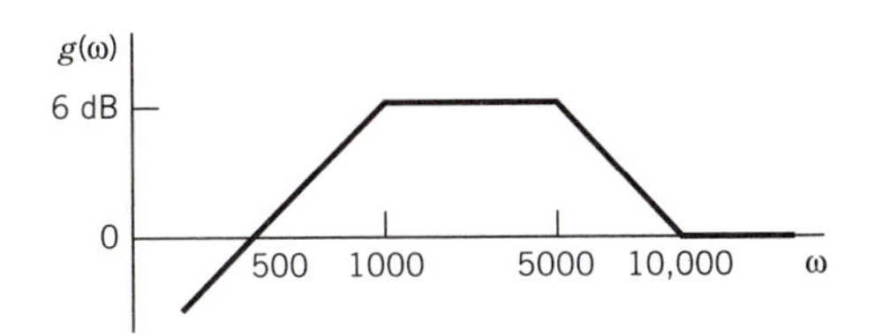

Figure P11.67

11.68 Design an op-amp circuit to get the asymptotic gain curve in Fig. P11.65 with $\omega_1 = 250$ and $\omega_2 = 4000$. Let all capacitors be 1 μF.

11.69 Design an op-amp circuit to get the asymptotic gain curve in Fig. P11.69. Let all capacitors be 0.5 μF.

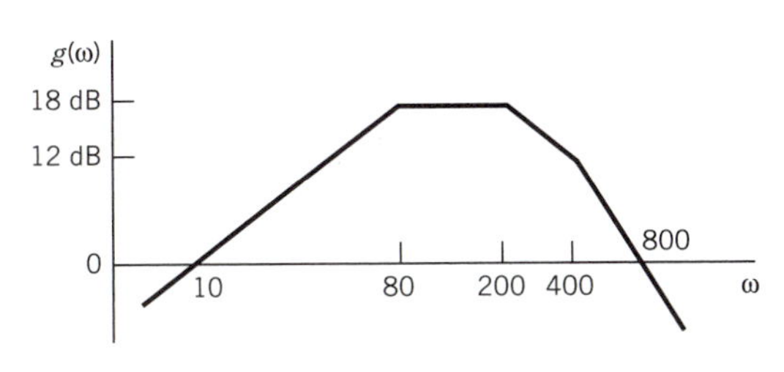

Figure P11.69

11.70 Design an op-amp circuit to get the asymptotic gain curve in Fig. P11.65 with $\omega_1 = 250$ and $\omega_2 = 2000$. Let all capacitors be 1 μF. Hint: Use three op-amps.

Section 11.6 Butterworth Filters

11.71* Find n such that a Butterworth lowpass filter with $f_{co} = 200$ Hz has $a_{lp}(\omega) \leq K/3$ for $f \geq 400$ Hz.

11.72 Find n such that a Butterworth lowpass filter with $f_{co} = 3.2$ kHz has $a_{lp}(\omega) \leq K/10$ for $f \geq 4$ kHz.

11.73 Show that the circuit in Fig. P11.73 becomes a second-order Butterworth lowpass filter with $K = 1$ and $\omega_{co}^2 = 1/LC$ when $R = 1.414\omega_{co}L$.

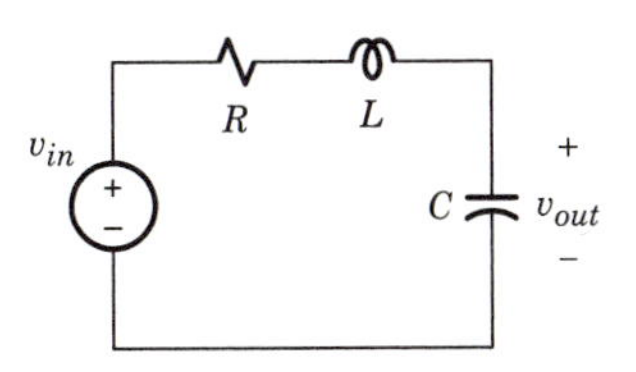

Figure P11.73

11.74 Show that the circuit in Fig. P11.74 becomes a third-order Butterworth lowpass filter with $K = 1$ and $\omega_{co}^2 = 2/3LC$ when $R = 2\omega_{co}L$.

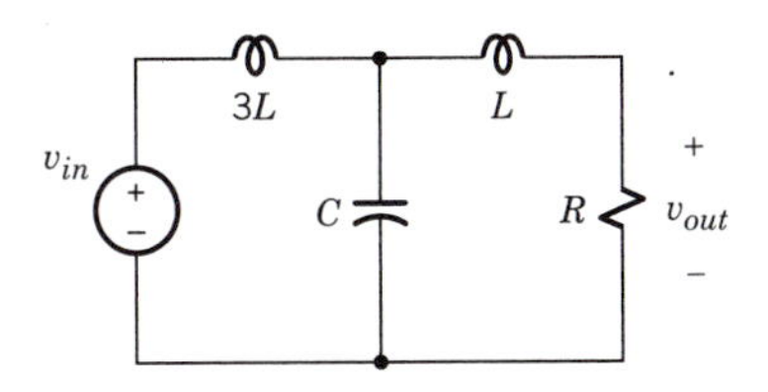

Figure P11.74

11.75 Use Eq. (11.64) to calculate the quality factors for the Butterworth polynomial with $n = 8$.

11.76 Use Eq. (11.64) to calculate the quality factors for the Butterworth polynomial with $n = 9$.

11.77 Find n and the condition on f_{co} such that a Butterworth highpass filter has $a_{hp}(\omega) \geq 0.8K$ for $f \geq 1$ kHz and $a_{hp}(\omega) \leq 0.01K$ for $f \leq 100$ Hz.

11.78 Find n and the condition on f_{co} such that a Butterworth highpass filter has $a_{hp}(\omega) \geq 0.9K$ for $f \geq$ kHz and $a_{hp}(\omega) \leq 0.03K$ for $f \leq 20$ kHz.

11.79 Derive Eq. (11.72) by node analysis of Fig. 11.41*a*.

11.80 Derive Eq. (11.75) by node analysis of Fig. 11.41*b*.

11.81* Design an op-amp circuit with 1-μF capacitors to implement a Butterworth lowpass filter with $n = 2$, $f_{co} = 50$ Hz, and $K = 10$. Take $R_\mu = 1$ kΩ.

11.82 Design an op-amp circuit with 1-nF capacitors to implement a Butterworth lowpass filter with $n = 3$, $f_{co} = 10$ kHz, and $K = 25$. Take $R_\mu = 1$ kΩ, and let all stages have the same gain.

11.83 Design an op-amp circuit with 10-nF capacitors to implement a Butterworth highpass filter with $n = 4$, $f_{co} = 2$ kHz, and $K = 100$. Take $R_\mu = 1$ kΩ, and let all stages have the same gain.

11.84 Design an op-amp circuit with 100-nF capacitors to implement a Butterworth highpass filter with $n = 5$, $f_{co} = 500$ Hz, and $K = 8$. Take $R_\mu = 1$ kΩ, and let all stages have the same gain.

CHAPTER 12
Fourier Series Analysis

Many electrical waveforms are periodic but not sinusoidal. For analysis purposes, such waveforms can be represented in series form based on the original work of Fourier (pronounced Foor-yay). This chapter applies Fourier-series analysis to find the steady-state response of stable circuits driven by periodic excitations. Important practical applications of the Fourier-series method include signal generators, electronic power supplies, and communication circuits.

The essential new technique introduced here is the series expansion of periodic waveforms. Even though the waveforms have nonsinusoidal shapes, the Fourier series consists of sinusoidal components at various frequencies. Drawing upon this property, we'll develop the frequency-domain representation or *spectrum* for periodic waveforms. The spectral concept helps bring out the relationships between time-domain and frequency-domain properties of waveforms. Finally, we combine spectral analysis with frequency response to investigate periodic steady-state response, including waveform distortion and equalization.

Objectives

After studying this chapter and working the exercises, you should be able to do each of the following:

1. Write the expressions for the trigonometric and exponential Fourier series and for the series coefficients (Section 12.1).
2. Calculate the series coefficients of a simple periodic function (Section 12.1).
3. Identify the effects of waveform symmetry on the Fourier series (Section 12.1).
4. Construct the two-sided line spectrum of a waveform represented by rotating phasors (Section 12.2).
5. State the basic relationships between the time-domain and frequency-domain properties of periodic functions (Section 12.2).
6. Find the spectrum of a waveform constructed from simple periodic functions (Section 12.2).
7. Calculate or approximate the periodic steady-state response of a network given its frequency response and the spectrum of the input (Section 12.3).
8. State the frequency-domain conditions for distortionless waveform reproduction, and distinguish between amplitude distortion and delay distortion (Section 12.3).
9. Design an equalizer to correct waveform distortion over a specified frequency range (Section 12.3).

12.1 PERIODIC WAVEFORMS AND FOURIER SERIES

Any "well-behaved" periodic function can be expressed as an infinite series of the type originated by the French scientist Jean Baptiste Joseph Fourier (1768–1830). Fourier first applied series expansions to the study of heat conduction, but his results remain valid for any other types of periodic phenomena. Consequently, Fourier's method appears in many branches of science and engineering, including the analysis of periodic electrical waveforms.

Two versions of the Fourier series are presented here, after an introductory discussion of periodic waveforms. We'll also state the requirements for a "well-behaved" function, and we'll examine the effects of waveform symmetry on the Fourier series.

Of necessity, some of the work involves rather abstract mathematical manipulations. But don't be discouraged, because we'll eventually put our results to practical use in subsequent sections. In fact, Fourier analysis is one of those sophisticated mathematical techniques that engineers have exploited to achieve dramatic advances in modern technology.

Periodic Waveforms

The defining property of a periodic waveform is *repetition over time.* Stated mathematically, any time function $f(t)$ is **periodic** if there exists some repetition interval T such that

$$f(t - T) = f(t) \tag{12.1}$$

Equation (12.1) says that shifting the function T seconds to the right re-creates the original function. For example, the waveform in Fig. 12.1 exhibits the periodic behavior $f(t_i - T) = f(t_i)$ at every instant $t = t_i$. This waveform further illustrates three other periodic properties that follow from Eq. (12.1).

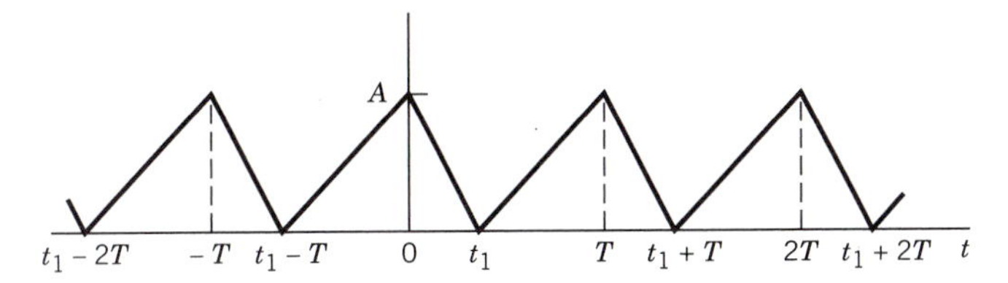

Figure 12.1 Periodic waveform.

First, the waveshape is preserved when we shift $f(t)$ any integer number of periods to the left or right, so

$$f(t + mT) = f(t) \qquad m = 1, 2, 3, \ldots$$

Therefore, if $f(t)$ has period T, then it also repeats itself every mT seconds. To avoid possible confusion, we usually consider T to be the *smallest* constant that satisfies Eq. (12.1). Accordingly, we call T the **fundamental period**.

Second, since Eq. (12.1) must hold for any value of t, a periodic waveform theoretically continues for *all* time — from $t = -\infty$ to $t = \infty$. A real waveform, of course, actually lasts for only a finite amount of time. However, the assumption of true periodicity may be justified when the duration of a repeating waveform is large compared to its period. In fact, we previously took this assumption as the basis of ac steady-state analysis.

Third, the net area under a periodic curve $f(t)$ over any period is independent of where we consider the period to start. Expressing area as an integration and letting t_1 and t_2 be arbitrary starting points, this property means that

$$\int_{t_1}^{t_1+T} f(t)\, dt = \int_{t_2}^{t_2+T} f(t)\, dt \tag{12.2}$$

Such integrals frequently occur in conjunction with Fourier series, so we'll adopt the shorthand notation

$$\int_T f(t)\,dt \triangleq \int_{t_i}^{t_i+T} f(t)\,dt \tag{12.3}$$

The practical implication of Eq. (12.3) is that you are free to choose any starting point t_i that simplifies the integration.

Trigonometric Fourier Series

Now let $f(t)$ stand for any periodic waveform of interest in circuit analysis. Although the waveform is not necessarily sinusoidal, we may still speak of its frequency of repetition, which simply equals $1/T$ periods per second. Correspondingly, we define the **fundamental frequency**

$$\Omega \triangleq 2\pi/T \tag{12.4}$$

so Ω is just the frequency of repetition measured in radians per second.

Fourier made the remarkable discovery that almost any periodic waveform $f(t)$ with fundamental frequency Ω can be expanded as an infinite series in the form

$$\begin{aligned} f(t) = c_0 &+ (a_1 \cos \Omega t + b_1 \sin \Omega t) \\ &+ (a_2 \cos 2\Omega t + b_2 \sin 2\Omega t) \\ &+ (a_3 \cos 3\Omega t + b_3 \sin 3\Omega t) + \cdots \end{aligned}$$

Alternatively, in summation notation,

$$f(t) = c_0 + \sum_{n=1}^{\infty} (a_n \cos n\Omega t + b_n \sin n\Omega t) \tag{12.5}$$

We call Eq. (12.5) the **trigonometric Fourier series**. The constants c_0, a_n, and b_n are the series **coefficients**, and their values depend upon $f(t)$. Some of the coefficients may equal zero for a particular waveform, but there will never be terms other than those indicated here.

Physically, Eq. (12.5) implies that $f(t)$ consists of a constant component c_0 plus the oscillatory components $a_n \cos n\Omega t$ and $b_n \sin n\Omega t$ with $n = 1$, 2, 3, ... All frequencies are integer multiples or **harmonics** of the fundamental frequency Ω — the second harmonic being 2Ω, the third harmonic 3Ω, and so on. The harmonic relationship ensures that each term in Eq. (12.5) repeats every T seconds. For this reason, Fourier expansion is also known as *harmonic analysis.*

Proving the validity of the Fourier series goes beyond our scope, but we'll at least make it plausible by examining some simple sums of harmonically related components. For one example, Fig. 12.2*a* plots the components and resultant sum of the two-term series

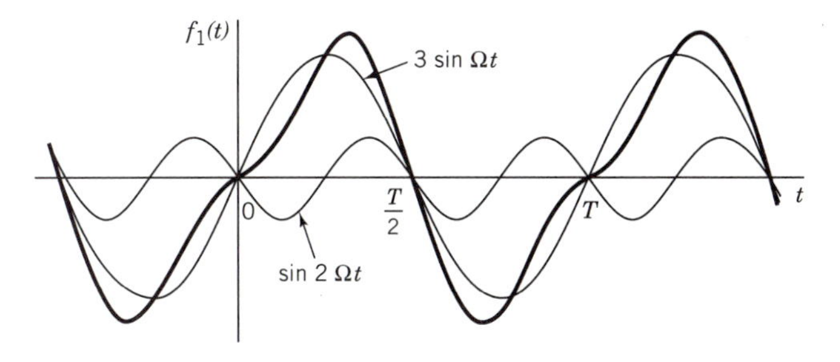

(a) $f_1(t) = 3 \sin \Omega t - \sin 2\,\Omega t$

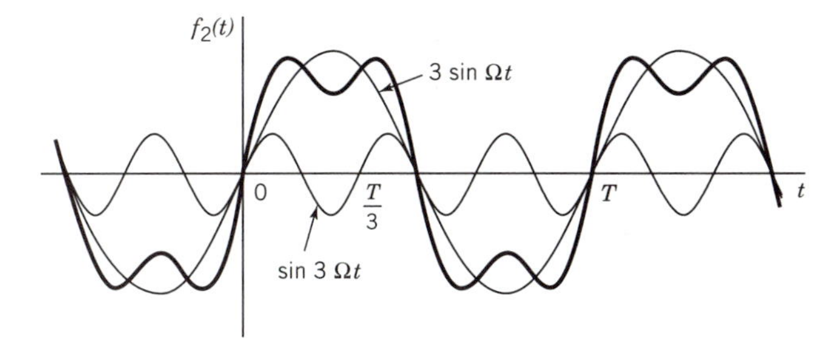

(b) $f_2(t) = 3 \sin \Omega t + \sin 3\,\Omega t$

Figure 12.2

$$f_1(t) = 3 \sin \Omega t - \sin 2\Omega t$$

The second-harmonic component repeats every $T/2$ seconds, and its peaks fall on either side of the peaks of the fundamental-frequency component, so $f_1(t)$ has something like a "sawtooth" shape. For another example, Fig. 12.2*b* plots the components and sum for

$$f_2(t) = 3 \sin \Omega t + \sin 3\Omega t$$

This waveform looks more like a "square wave" because the second harmonic is absent and every third peak of the third harmonic partially cancels a peak of the fundamental-frequency component.

By including an infinite number of harmonics, the Fourier series can represent any "well-behaved" periodic function. In this regard, a well-behaved function is defined by the following **Dirichlet's conditions**:

> The function $f(t)$ must be single valued; it must have a finite number of maxima, minima, and discontinuities per period; and the integral $\int_T |f(t)|\, dt$ must be finite.

All electrical waveforms satisfy these conditions.

When Dirichlet's conditions hold, the infinite series summation converges to the value of $f(t)$ wherever the waveform is continuous. If $f(t)$ is discontinuous at some instant t_0, then the series converges to the midpoint value $\frac{1}{2}[f(t_0^-) + f(t_0^+)]$. These convergence properties will be illustrated after we obtain general formulas for the series coefficients.

Fourier's crucial discovery was that all the terms in Eq. (12.5) are **orthogonal**, meaning that the integral over one period of the product of any two different terms vanishes. This orthogonal property follows from the integral relationships

$$\int_T \cos n\Omega t \, dt = \int_T \sin n\Omega t \, dt = 0 \tag{12.6a}$$

$$\int_T \cos n\Omega t \sin m\Omega t \, dt = 0 \tag{12.6b}$$

$$\int_T \cos n\Omega t \cos m\Omega t \, dt = 0 \qquad n \neq m \tag{12.6c}$$

$$\int_T \sin n\Omega t \sin m\Omega t \, dt = 0 \qquad n \neq m \tag{12.6d}$$

However, when $n = m$, we have

$$\int_T \cos^2 m\Omega t \, dt = \int_T \sin^2 m\Omega t \, dt = \frac{T}{2} \tag{12.7}$$

We'll apply these relationships to the task of finding the Fourier series coefficients.

The constant term c_0 is obtained by integrating both sides of Eq. (12.5) over one period and invoking Eq. (12.6a), so

$$\begin{aligned} \int_T f(t) \, dt &= c_0 \int_T dt + \sum_{n=1}^{\infty} \left(a_n \int_T \cos n\Omega t \, dt + b_n \int_T \sin n\Omega t \, dt \right) \\ &= c_0 \int_T dt = c_0 \int_{t_i}^{t_i+T} dt = c_0 T \end{aligned}$$

Hence,

$$c_0 = \frac{1}{T} \int_T f(t) \, dt \tag{12.8}$$

Equation (12.8) shows that c_0 equals the *average value* of $f(t)$. For simple waveforms, this value can be determined directly by inspection.

To find the cosine coefficient of the mth harmonic, we multiply both sides of Eq. (12.5) by $\cos m\Omega t$ and again integrate over one period so that

$$\int_T f(t)\cos m\Omega t\,dt = c_0 \int_T \cos m\Omega t\,dt + \sum_{n=1}^{\infty} a_n \int_T \cos n\Omega t \cos m\Omega t\,dt + \sum_{n=1}^{\infty} b_n \int_T \sin n\Omega t \cos m\Omega t\,dt$$

But, from Eqs. (12.6a)–(12.6d), all terms on the right vanish except the cosine integral with $n = m$. We are then left with

$$\int_T f(t)\cos m\Omega t\,dt = a_m \int_T \cos^2 m\Omega t\,dt = a_m \frac{T}{2}$$

Letting $m = n$ yields the general result

$$a_n = \frac{2}{T}\int_T f(t)\cos n\Omega t\,dt \tag{12.9}$$

Similarly, the sine coefficient is

$$b_n = \frac{2}{T}\int_T f(t)\sin n\Omega t\,dt \tag{12.10}$$

obtained by multiplying Eq. (12.5) by $\sin m\Omega t$ and integrating.

Example 12.1 *Half-Rectified Sine Wave*

As part of the process of ac to dc conversion, some electronic power supplies generate a **half-rectified sine wave** like Fig. 12.3. This waveform is defined over one period by

$$\begin{aligned} f(t) &= A\sin\Omega t && 0 < t < T/2 \\ &= 0 && T/2 < t < T \end{aligned}$$

Rectification removes the negative excursions from the sine wave $A \sin \Omega t$, so $f(t)$ has a nonzero dc component along with sinusoidal components at the fundamental and harmonic frequencies.

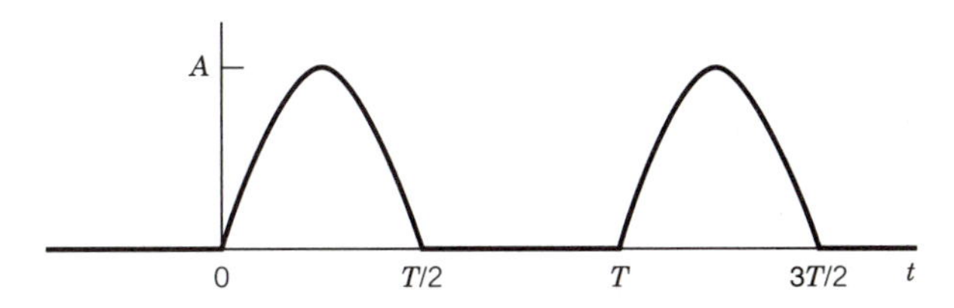

Figure 12.3 Half-rectified sine wave.

To evaluate c_0, we'll take the integration in Eq. (12.8) from $t = 0$ to $t = T$ and insert $\Omega T = 2\pi$. Hence,

$$c_0 = \frac{1}{T}\int_0^{T/2} A \sin \Omega t \, dt + \frac{1}{T}\int_{T/2}^{T} 0 \, dt$$
$$= \frac{-A}{\Omega T}\left(\cos \frac{\Omega T}{2} - 1\right) = \frac{-A}{2\pi}(\cos \pi - 1) = \frac{A}{\pi}$$

The average value of $f(t)$ is thus about one-third of the peak value A.

To find the cosine coefficients, we'll take the same integration range and make the change of variable $\alpha = \Omega t = 2\pi t/T$ in Eq. (12.9). Then, from a standard table of integrals,

$$a_n = \frac{2}{T}\int_0^{T/2} A \sin \Omega t \cos n\Omega t \, dt = \frac{A}{\pi}\int_0^{\pi} \sin \alpha \cos n\alpha \, d\alpha$$
$$= -\frac{A}{2\pi}\left[\frac{1 - \cos (n-1)\pi}{n-1} - \frac{1 - \cos (n+1)\pi}{n+1}\right] \qquad n \neq 1$$

This formidable expression simplifies considerably since $\cos (n \pm 1)\pi = -1$ when n is even whereas $\cos (n \pm 1)\pi = 1$ when n is odd. Thus,

$$a_n = -\frac{2A}{\pi(n^2 - 1)} \qquad n = 2, 4, 6, \ldots$$
$$= 0 \qquad n = 3, 5, 7, \ldots$$

But a separate integration is needed for $n = 1$, which yields

$$a_1 = \frac{A}{\pi}\int_0^{\pi} \sin \alpha \cos \alpha \, d\alpha = 0$$

The sine coefficients are obtained in like manner from Eq. (12.10), the result being

$$b_n = \frac{A}{\pi}\int_0^{\pi} \sin \alpha \sin n\alpha \, d\alpha = A/2 \qquad n = 1$$
$$= 0 \qquad n = 2, 3, 4, \ldots$$

so only b_1 is nonzero.

Substituting our values for c_0, a_n, and b_n into Eq. (12.5) gives the trigonometric series expansion

$$f(t) = \frac{A}{\pi} + \frac{A}{2}\sin \Omega t - \frac{2A}{3\pi}\cos 2\Omega t - \frac{2A}{15\pi}\cos 4\Omega t - \cdots$$

We'll interpret this expression — and gain confidence in the Fourier series — by a partial reconstruction of $f(t)$ from its expansion.

Figure 12.4*a* plots the constant plus fundamental-frequency term along with the second-harmonic component. The peaks of the second-harmonic component have the same polarity as the fundamental-frequency component at $t = T/4, 5T/4, \ldots$, while they have opposite polarity at $t = 3T/4, 7T/4, \ldots$. Consequently, as shown in Fig. 12.4*b*, the sum of the first three terms of the expansion begins to resemble the half-rectified sine wave. The resemblence is greatly improved by summing all terms through the sixth harmonic, which is plotted in Fig. 12.4*c*. Eventually, as higher harmonics are added, the series reconstruction becomes indistinguishable from the original waveform.

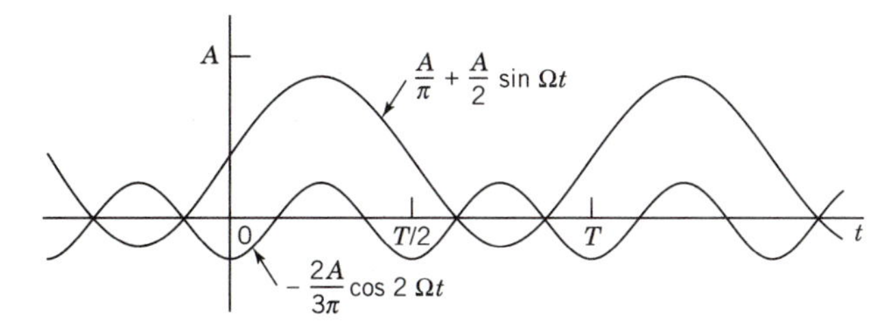

(a) First– and second–harmonic components

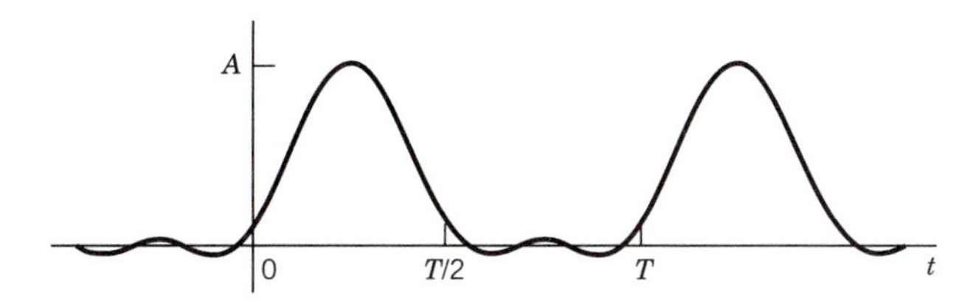

(b) Sum of first three terms

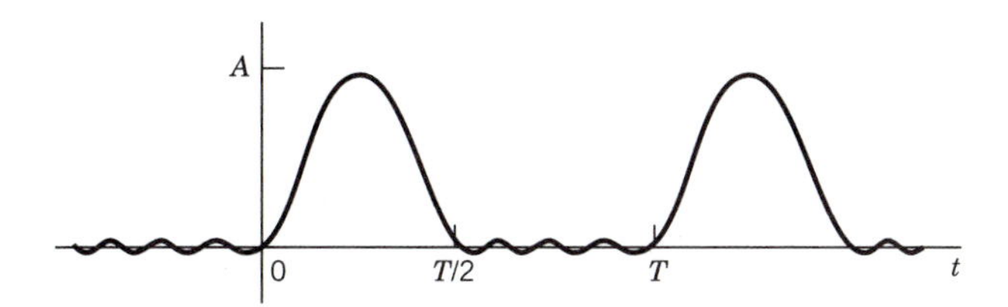

(c) Sum of terms through the sixth harmonic

Figure 12.4 Waveform reconstruction.

Exercise 12.1

Consider the waveform $f(t)$ in Fig. 12.5. Take the range of integration over $-T/2 \le t \le T/2$ to obtain

$$c_0 = A/2 \qquad a_n = 0$$
$$b_n = -2A/n\pi \qquad n = 1, 3, 5, \ldots$$
$$= 0 \qquad n = 2, 4, 6, \ldots$$

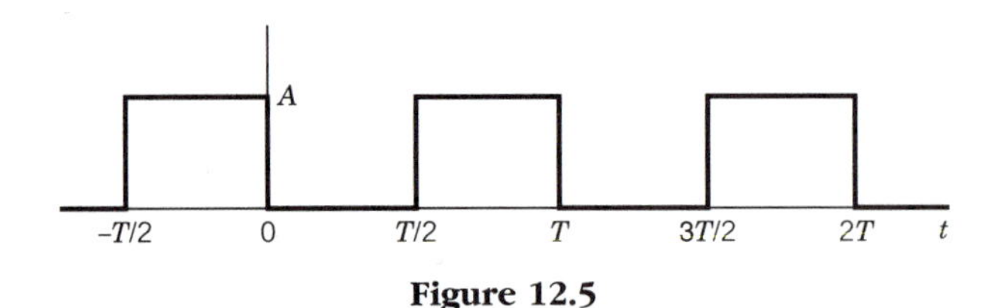

Figure 12.5

Then let $A = 6\pi$ and sketch the approximation for $f(t)$ obtained from the first three nonzero terms of the series.

Exponential Fourier Series

Another version of the Fourier series consists of terms proportional to $e^{\pm jn\Omega t}$ instead of sines and cosines. Since $e^{-jn\Omega t}$ must be added to $e^{jn\Omega t}$ to get a real quantity, this **exponential Fourier series** is written as

$$f(t) = c_0 + \sum_{n=1}^{\infty} (c_n e^{jn\Omega t} + c_{-n} e^{-jn\Omega t})$$

Then, letting $c_0 = c_0 e^{j0t}$ and summing over both positive and negative values of n, we get the compact expression

$$f(t) = \sum_{n=-\infty}^{\infty} c_n e^{jn\Omega t} \qquad (12.11)$$

Although Eq. (12.11) looks more abstract than the trigonometric expansion, it has the advantage that c_n represents *all* the series coefficients. Consequently, we prefer the exponential series for analytical investigations.

Like the trigonometric series, the components of the exponential series are *orthogonal* in the sense that

$$\int_T e^{jn\Omega t} e^{-jm\Omega t}\, dt = 0 \qquad n \neq m \qquad (12.12a)$$

$$= T \qquad n = m \qquad (12.12b)$$

This property quickly yields c_n in terms of $f(t)$ when we multiply both sides of Eq. (12.11) by $e^{-jm\Omega t}$ and integrate over one period. Then

$$\int_T f(t)\, e^{-jm\Omega t}\, dt = \sum_{n=-\infty}^{\infty} c_n \int_T e^{jn\Omega t} e^{-jm\Omega t}\, dt$$

But all the terms with $n \neq m$ vanish per Eq. (12.12a), and the remaining term with $n = m$ reduces to $c_m T$ per Eq. (12.12b), so

$$\int_T f(t)e^{-jm\Omega t}\,dt = c_m T$$

Hence, upon replacing m with n,

$$c_n = \frac{1}{T}\int_T f(t)e^{-jn\Omega t}\,dt \tag{12.13}$$

which is the general result.

Equation (12.13) holds for all values of n, and it reduces to our previous expression for c_0 by letting $n = 0$. However, separate integration with $n = 0$ is usually easier than taking the limit of c_n as $n \to 0$.

For $n \neq 0$, we can insert $e^{-jn\Omega t} = \cos n\Omega t - j \sin n\Omega t$ and Eq. (12.13) becomes

$$c_n = \frac{1}{T}\int_T f(t)\cos n\Omega t\,dt - j\frac{1}{T}\int_T f(t)\sin n\Omega t\,dt \tag{12.14}$$

This expression brings out the fact that the c_n are *complex* quantities. Moreover, changing the sign of n shows that

$$c_{-n} = c_n^* \tag{12.15}$$

Thus, c_n and c_{-n} are *complex conjugates.*

Equation (12.14) also links the exponential series to the trigonometric version by comparison with our previous results

$$a_n = \frac{2}{T}\int_T f(t)\cos n\Omega t\,dt \qquad b_n = \frac{2}{T}\int_T f(t)\sin n\Omega t\,dt$$

We immediately see that

$$a_n = 2\,\mathrm{Re}[c_n] \qquad b_n = -2\,\mathrm{Im}[c_n] \qquad n \geq 1 \tag{12.16}$$

Conversely,

$$c_n = \tfrac{1}{2}(a_n - jb_n) \qquad n \geq 1 \tag{12.17}$$

so $c_{-n} = c_n^* = \tfrac{1}{2}(a_n + jb_n)$. Figure 12.6 interprets these relations in the complex plane.

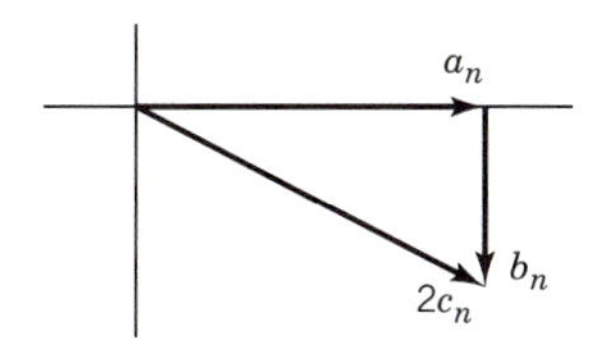

Figure 12.6 Construction showing $a_n - jb_n = 2c_n$.

Example 12.2 *Sinusoidal Waveform*

To bring out the physical meaning of the exponential Fourier series, consider the sinusoidal waveform

$$f(t) = A_m \cos(m\Omega t + \phi_m)$$

The integration for c_n is simplified by the identity

$$\cos(m\Omega t + \phi_m) = ½[e^{j(m\Omega t+\phi_m)} + e^{-j(m\Omega t+\phi_m)}]$$

Then

$$\begin{aligned} c_n &= \frac{1}{T}\int_T ½A_m[e^{j(m\Omega t+\phi_m)} + e^{-j(m\Omega t+\phi_m)}]e^{-jn\Omega t}\,dt \\ &= \frac{A_m}{2T}e^{j\phi_m}\int_T e^{jm\Omega t}\,e^{-jn\Omega t}\,dt + \frac{A_m}{2T}e^{-j\phi_m}\int_T e^{-jm\Omega t}\,e^{-jn\Omega t}\,dt \end{aligned}$$

and the orthogonality property in Eq. (12.12) yields

$$\begin{aligned} c_n &= 0 && n \neq \pm m \\ &= ½A_m e^{j\phi_m} && n = m \\ &= ½A_m e^{-j\phi_m} && n = -m \end{aligned}$$

Thus, the exponential Fourier series is

$$f(t) = ½A_m e^{j\phi_m}\,e^{jm\Omega t} + ½A_m e^{-j\phi_m}\,e^{-jm\Omega t}$$

This expansion consists of just one pair of terms because the waveform contains only one frequency, $m\Omega$, and no constant component.

For ac circuit analysis, we would have represented $f(t)$ in phasor notation as $\text{Re}[A_m\underline{/\phi_m}\,e^{jm\Omega t}]$ so $f(t)$ equals the horizontal projection of a rotating phasor $A_m\,e^{j(m\Omega t+\phi_m)}$. Here, we represent $f(t)$ by summing *two counter-rotating phasors* since

$$f(t) = ½A_m e^{j(m\Omega t+\phi_m)} + ½A_m e^{-j(m\Omega t+\phi_m)}$$

Figure 12.7 shows how these phasors sum to form the sinusoidal function, as compared with taking the horizontal projection of a single rotating phasor.

Example 12.3 *Exponential Series Coefficients*

The nonzero coefficients for the trigonometric expansion of a half-rectified sine wave were found in Example 12.1 to be

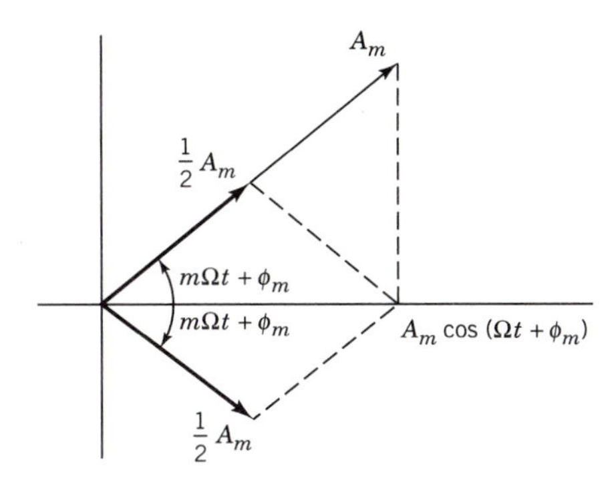

Figure 12.7 Sum of two counter-rotating phasors and projection of one rotating phasor.

$$c_0 = A/\pi \qquad b_1 = A/2$$
$$a_n = -2A/\pi(n^2 - 1) \qquad n = 2, 4, 6, \ldots$$

Now, using $c_n = \frac{1}{2}(a_n - jb_n)$ for $n \geq 1$, we get the exponential series coefficients

$$\begin{aligned} c_n &= A/\pi & n &= 0 \\ &= -jA/4 & n &= 1 \\ &= -A/\pi(n^2 - 1) & n &= 2, 4, 6, \ldots \\ &= 0 & n &= 3, 5, 7, \ldots \end{aligned}$$

The values for negative n then follow from $c_{-n} = c_n^*$.

Exercise 12.2

Verify Eqs. (12.12a) and (12.12b) by carrying out the integrations over $0 \leq t \leq T$ with $n \neq m$ and with $n = m$.

Exercise 12.3

Using Eq. (12.13), show that the waveform in Fig. 12.5 has

$$\begin{aligned} c_n &= jA/n\pi & n &= \pm 1, \pm 3, \pm 5, \ldots \\ &= 0 & n &= \pm 2, \pm 4, \pm 6, \ldots \end{aligned}$$

Then confirm that this result agrees with a_n and b_n in Exercise 12.1.

Waveform Symmetry

Many important periodic waveforms possess symmetry with respect to the time origin. We'll define two types of symmetry here, and we'll examine the effects of symmetry on the Fourier series coefficients.

Even Symmetry A function $f(t)$ is said to have *even* symmetry if it satisfies the condition

$$f(-t) = f(t) \tag{12.18}$$

Thus, an even waveform looks the same when we go either backward or forward from $t = 0$. Examples of even functions include $\cos n\Omega t$ and the **rectangular pulse train** and **triangular wave** shown in Fig. 12.8. This figure illustrates the property that the net area of one period of an even periodic function can be written as

$$\int_T f(t)\, dt = 2\int_0^{T/2} f(t)\, dt \tag{12.19}$$

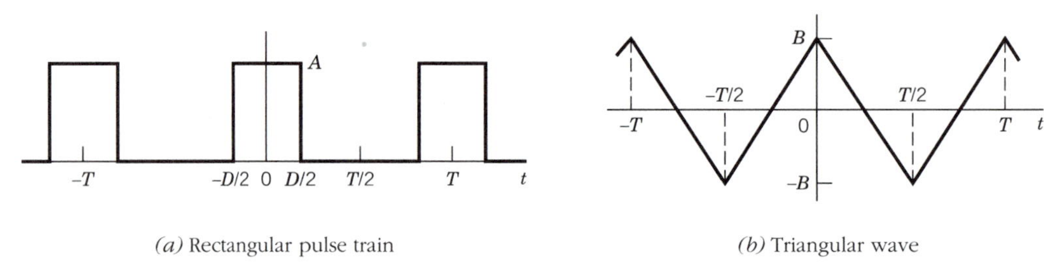

(*a*) Rectangular pulse train (*b*) Triangular wave

Figure 12.8 Even-symmetry waveforms.

Odd Symmetry A function with *odd* symmetry is the "opposite" of an even function in the sense that

$$f(-t) = -f(t) \tag{12.20}$$

Thus, the waveform looks inverted when we go in the negative-time direction. Examples of odd functions include $\sin n\Omega t$ and the **square wave** and **saw-tooth wave** in Fig. 12.9. Clearly, the net area of one period of an odd periodic function equals zero, so

$$\int_T f(t)\, dt = 0 \tag{12.21}$$

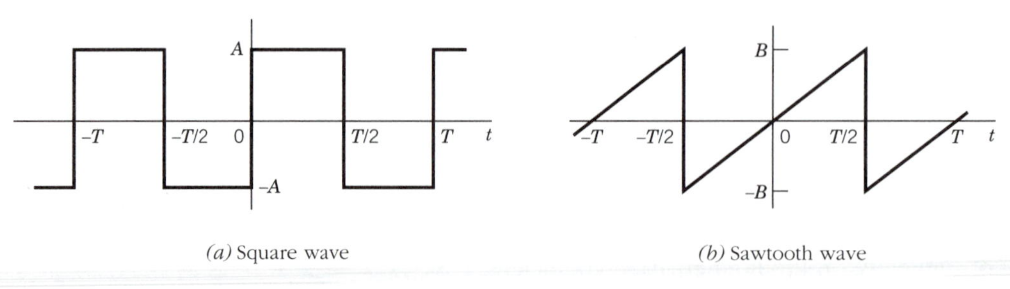

(*a*) Square wave (*b*) Sawtooth wave

Figure 12.9 Odd-symmetry waveforms.

Now consider the product of two symmetric functions, say $g(t) = f_1(t)f_2(t)$. When $f_1(t)$ and $f_2(t)$ are both even, $g(t)$ will be even because

$$g(-t) = f_1(-t)f_2(-t) = f_1(t)f_2(t) = g(t)$$

When $f_1(t)$ and $f_2(t)$ are both odd, $g(t)$ will still be even because

$$g(-t) = f_1(-t)f_2(-t) = [-f_1(t)][-f_2(t)] = g(t)$$

In short, multiplying two functions having the *same symmetry* — even or odd — always results in a function with *even symmetry*. On the other hand, you can easily confirm that when $f_1(t)$ and $f_2(t)$ have *opposite symmetry* — one even, the other odd — the product $f_1(t)f_2(t)$ always has *odd symmetry*.

The foregoing integral and product properties of symmetric functions result in simplifications of the Fourier series coefficients. Specifically, if $f(t)$ is an **even periodic waveform**, then $f(t) \times \cos n\Omega t$ is even while $f(t) \times \sin n\Omega t$ is odd. Equations (12.8) and (12.14) thereby reduce to

$$\begin{aligned} c_0 &= \frac{2}{T}\int_0^{T/2} f(t)\, dt \\ c_n &= \frac{2}{T}\int_0^{T/2} f(t) \cos n\Omega t\, dt \end{aligned} \qquad (12.22)$$

so all the coefficients of the exponential series will be *real* quantities. The trigonometric series then has

$$a_n = 2c_n = \frac{4}{T}\int_0^{T/2} f(t) \cos n\Omega t\, dt \qquad b_n = 0 \qquad (12.23)$$

The expansion thus consists entirely of even-symmetry components, as should be expected.

On the other hand, if $f(t)$ is an **odd periodic waveform**, then $f(t) \times \cos n\Omega t$ is odd while $f(t) \times \sin n\Omega t$ is even. Hence,

$$\begin{aligned} c_0 &= 0 \\ c_n &= -\frac{j2}{T}\int_0^{T/2} f(t) \sin n\Omega t\, dt \end{aligned} \qquad (12.24)$$

so all the coefficients of the exponential series will be *imaginary* quantities. The trigonometric series then has

$$a_n = 0 \qquad b_n = j2c_n = \frac{4}{T}\int_0^{T/2} f(t) \sin n\Omega t\, dt \qquad (12.25)$$

The expansion thus consists entirely of odd-symmetry components.

Example 12.4 *Rectangular Pulse Train*

The rectangular pulse train in Fig. 12.8a consists of periodic pulses with height A and duration $D < T$. We want to perform a harmonic analysis because pulse trains like this appear in digital electronic circuits and pulsed communication systems.

First, noting that each pulse has area $A \times D$, we obtain the dc component

$$c_0 = AD/T$$

Next, for $n \neq 0$, we take advantage of the even symmetry and use Eq. (12.22) to get

$$c_n = \frac{2}{T}\int_0^{D/2} A \cos n\Omega t \, dt = \frac{2A}{n\Omega T} \sin \frac{n\Omega D}{2} = \frac{A}{n\pi} \sin \frac{n\pi D}{T}$$

This expression reduces to c_0 when $n = 0$. However, since n appears in the denominator, we must draw upon the fact that $\sin n\phi \rightarrow n\phi$ when $n\phi \rightarrow 0$. Thus,

$$c_0 = \lim_{n \to 0} \frac{A}{n\pi} \frac{n\pi D}{T} = \frac{AD}{T}$$

which agrees with the average value of the waveform.

Further examination of our result reveals that some of the harmonics may be absent entirely because $c_n = 0$ if nD/T equals an integer. This property is seen more clearly by writing

$$\sin(n\pi D/T) = \sin[(nD/T) \times 180^\circ]$$

where we have converted π radians to 180°.

Suppose, for instance, that $D = T/3$ and $A = 3$. Then $c_0 = 1$ and

$$c_n = (3/n\pi)\sin(n \times 60^\circ) = 0 \qquad n = \pm 3, \pm 6, \pm 9, \ldots$$

Since $a_n = 2c_n$ and $b_n = 0$ per Eq. (12.23), the trigonometric series becomes

$$f(t) = 1 + 1.654 \cos \Omega t + 0.827 \cos 2\Omega t$$
$$- 0.413 \cos 4\Omega t - 0.331 \cos 5\Omega t + \cdots$$

Figure 12.10 plots the partial reconstruction using all terms through the eighth harmonic.

Notice that the series converges to the midpoint value $A/2 = 1.5$ at the pulse edges where the waveform is discontinuous. Also notice the overshoot or "ears" on each side of the discontinuities, an effect known as **Gibb's phenomenon**. This phenomenon occurs in the reconstruction of any periodic waveform having discontinuities.

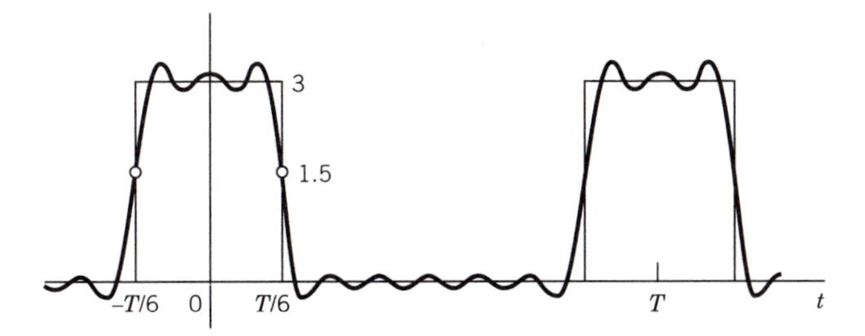

Figure 12.10 Partial reconstruction of a rectangular pulse train showing Gibb's phenomenon.

Exercise 12.4

Let $f_1(t)$ be an odd function and $f_2(t)$ an even function. Show that $g(t) = f_1(t)f_2(t)$ has odd symmetry.

Exercise 12.5

The sawtooth wave in Fig. 12.9*b* is defined by

$$f(t) = 2Bt/T \qquad -T/2 < t < T/2$$

Let $\alpha = n\Omega t = 2\pi nt/T$ in Eq. (12.24) to show that

$$\begin{aligned} c_n &= -jB/n\pi & n &= \pm 1, \pm 3, \pm 5, \ldots \\ &= jB/n\pi & n &= \pm 2, \pm 4, \pm 6, \ldots \end{aligned}$$

Then write out the first three nonzero terms of the trigonometric expansion.

12.2 SPECTRAL ANALYSIS OF PERIODIC WAVEFORMS

This section introduces *spectral analysis* as a means of depicting time-varying voltages or currents in the *frequency domain.* We'll show how to represent periodic functions as plots of amplitude and phase versus frequency. Then we'll examine some key relationships between the time-domain and frequency-domain properties of periodic waveforms.

Line Spectra

Spectral analysis exploits the fact that any sinusoid is completely characterized by three quantities: *frequency, amplitude,* and *phase.* Thus, when a waveform consists entirely of sinusoids, we can convey all the information about it by plotting amplitude and phase versus frequency. This frequency-domain representation of a waveform is called its **spectrum**.

As an introductory example, Fig. 12.11*a* shows the frequency-domain picture of

$$\begin{aligned} f(t) &= 7 + 10 \cos (3t + 30°) + 4 \sin 8t \\ &= 7 \cos (0t + 0°) + 10 \cos (3t + 30°) + 4 \cos (8t - 90°) \end{aligned}$$

The waveform consists of a dc or zero-frequency component plus sinusoids at $\omega = 3$ and $\omega = 8$. The amplitude plot therefore has lines of height 7, 10, and 4 at $\omega = 0$, 3, and 8, respectively. The phase plot likewise has lines of height 0°, 30°, and −90°. Taken together, these two plots constitute the **one-sided line spectrum** of $f(t)$. One-sided amplitude plots can be produced in the laboratory using an instrument called a *spectrum analyzer.*

Another way of displaying the frequency-domain picture relates to *rotating phasors* rather than sinusoids. Drawing upon Example 12.2, any sinusoid may be expanded as

$$\begin{aligned} A \cos (\omega t + \phi) &= \tfrac{1}{2} A e^{j(\omega t+\phi)} + \tfrac{1}{2} A e^{-j(\omega t+\phi)} \\ &= \tfrac{1}{2} A \underline{/\phi}\, e^{j\omega t} + \tfrac{1}{2} A \underline{/-\phi}\, e^{j(-\omega)t} \end{aligned}$$

Our illustrative waveform then becomes

$$\begin{aligned} f(t) = 7\underline{/0°}\, e^{j0t} &+ [5\underline{/30°}\, e^{j3t} + 5\underline{/-30°}\, e^{j(-3)t}] \\ &+ [2\underline{/-90°}\, e^{j8t} + 2\underline{/90°}\, e^{j(-8)t}] \end{aligned}$$

The corresponding frequency-domain picture in Fig. 12.11*b* shows the amplitude and phase of the rotating phasors that make up $f(t)$. We now have a **two-sided line spectrum** because a sinusoid at any frequency $\omega_1 > 0$ is equivalent to the sum of *two* phasors with rotational frequencies ω_1 and $-\omega_1$.

At first glance the two-sided spectrum may appear to be unduly complicated because it includes lines at *negative* frequencies. Observe, however, that

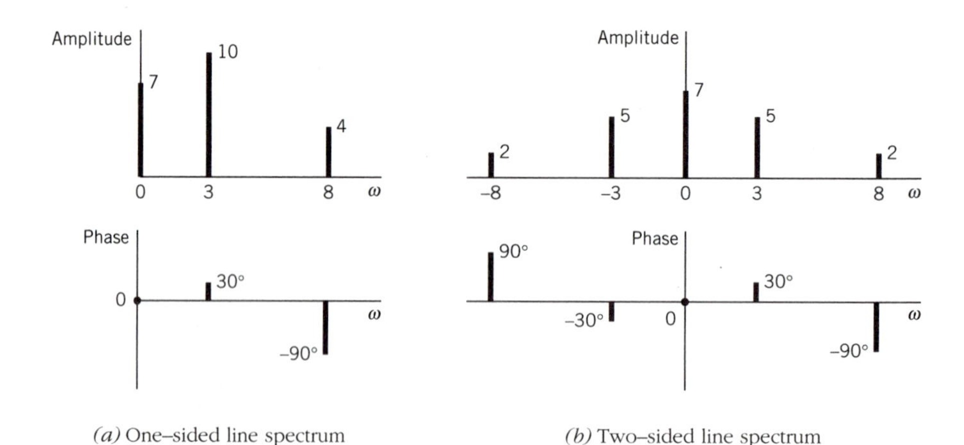

(a) One–sided line spectrum

(b) Two–sided line spectrum

Figure 12.11

the amplitude spectrum has *even symmetry* while the phase spectrum has *odd symmetry*, so the negative-frequency portions are directly related to the positive-frequency portions. We need both negative and positive frequencies to represent a sinusoidal component in terms of rotating phasors.

The advantage of the two-sided spectrum becomes evident when we consider spectral analysis of periodic waveforms. Any periodic electrical waveform with fundamental frequency Ω can be expanded as the exponential Fourier series

$$f(t) = \sum_{n=-\infty}^{\infty} c_n e^{jn\Omega t}$$

This series consists entirely of rotating phasors with length $|c_n|$, phase angle $\angle c_n$, and frequency $\omega = n\Omega$, where $n = 0, \pm 1, \pm 2, \ldots$. The two-sided spectrum therefore contains amplitude lines of height $|c_n|$ and phase lines of height $\angle c_n$, all lines being located at harmonics of Ω.

To emphasize the relationship between spectral analysis and the exponential Fourier series, we'll use the functional notation

$$c(n\Omega) \triangleq c_n \tag{12.25}$$

Consequently, $|c(n\Omega)|$ represents the **amplitude spectrum** and $\angle c(n\Omega)$ represents the **phase spectrum**. Since $c_{-n} = c_n^*$, the amplitude spectrum exhibits the even-symmetry property

$$|c(-n\Omega)| = |c(n\Omega)| \tag{12.26a}$$

while the phase spectrum exhibits the odd-symmetry property

$$\angle c(-n\Omega) = -\angle c(n\Omega) \qquad n \neq 0 \tag{12.26b}$$

A minor exception to the phase symmetry occurs when c_0 is negative, so $\angle c(0) = \pm 180°$.

Example 12.5 ***Spectrum of a Rectangular Pulse Train***

Consider the rectangular pulse train in Example 12.4. From our previous result, the spectrum is described by

$$c(n\Omega) = c_n = \frac{A}{n\pi} \sin \frac{n\pi D}{T}$$

As an aid to plotting the spectrum, we note that $n\pi/T = n\Omega/2$ so

$$c(n\Omega) = \frac{AD}{T} \frac{\sin (n\Omega D/2)}{n\Omega D/2} = \frac{AD}{T} \text{Sa } (n\Omega D/2)$$

where we've introduced the "sine-over-argument" function

$$\text{Sa}\,(\omega D/2) \triangleq \frac{\sin\,(\omega D/2)}{\omega D/2} \tag{12.27}$$

Figure 12.12 shows that Sa $(\omega D/2)$ is an even function of the continuous variable ω, and it has zero crossings uniformly spaced by $2\pi/D$ on both sides of the peak at $\omega = 0$ where Sa $(0) = 1$.

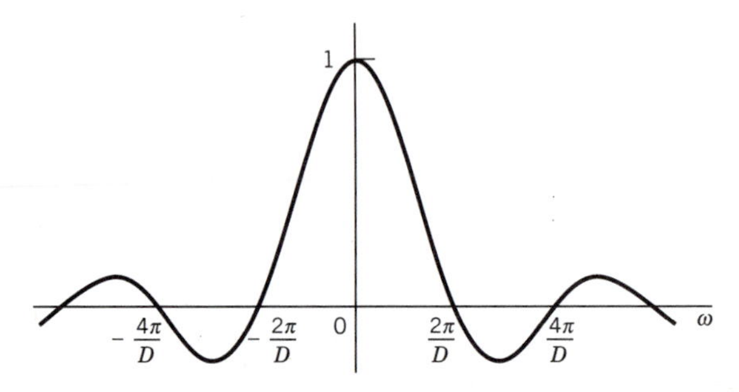

Figure 12.12 Sa $(\omega D/2) = [\sin\,(\omega D/2)]/(\omega D/2)$.

Taking the magnitude and angle of $c(n\Omega)$ for the pulse train yields the amplitude and phase spectrum

$$\begin{aligned} |c(n\Omega)| &= \frac{AD}{T}\,|\text{Sa}\,(\omega D/2)| \\ \angle c(n\Omega) &= \angle\text{Sa}\,(\omega D/2) \end{aligned} \qquad \omega = 0,\ \pm\Omega,\ \pm 2\Omega,\ \ldots$$

Figure 12.13 gives the resulting plots for the case of $D = T/3$, so $3\Omega = 2\pi/D$. The amplitude spectrum has been constructed with the help of Fig. 12.12 by treating the continuous function $(AD/T)|\text{Sa}\,(\omega D/2)|$ as the *envelope*

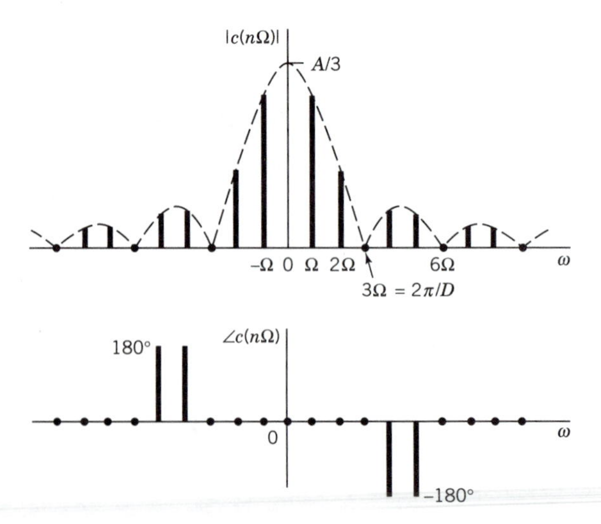

Figure 12.13 Spectrum of a rectangular pulse train with $D = T/3$.

of the lines. All lines appear at harmonics of the fundamental frequency Ω, but every third line is "missing" because multiples of 3Ω happen to fall exactly at the zero crossings of Sa $(\omega D/2)$. The phase spectrum has been constructed by noting that Sa $(\omega D/2)$ is always real but sometimes negative, and thus $\angle$Sa $(\omega D/2) = \pm 180°$ when Sa $(\omega D/2) < 0$. We use both $-180°$ and $+180°$ here to preserve the odd symmetry of the phase spectrum.

Observe from the amplitude spectrum that most of the spectral "content" of the pulse train lies within the range $|\omega| \leq 2\pi/D$. If the pulse duration D increases or decreases while T remains fixed, then the amplitude envelope shrinks or expands relative to the line spacing. Hence, as the pulses get shorter, the spectrum spreads out and the higher frequency components become increasingly important. This **reciprocal spreading effect** is illustrated in Fig. 12.14 with $D = T/2$ and $D = T/5$, omitting the negative-frequency portion of the spectra.

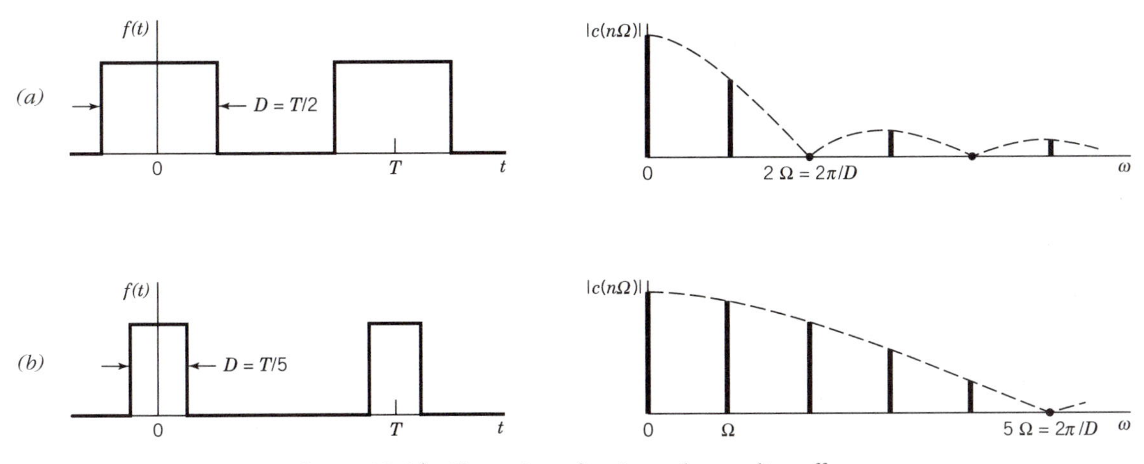

Figure 12.14 Illustration of reciprocal spreading effect.

Exercise 12.6

Construct the amplitude and phase spectra for the sawtooth wave in Exercise 12.5 (p. 561).

Time and Frequency Relations

We now develop some important relationships between the time-domain and frequency-domain properties of periodic waveforms. These relations have practical value because:

- They help us infer spectral characteristics from time-domain information, or vice versa;
- They provide shortcuts for calculating series coefficients and line spectra.

In stating the time and frequency relations, we assume throughout that $f(t)$, $g(t)$, and $z(t)$ are periodic waveforms with the *same* period, and their spectra are represented by $c_f(n\Omega)$, $c_g(n\Omega)$, and $c_z(n\Omega)$, respectively.

Linear Combination Let the waveform $z(t)$ be a linear combination of $f(t)$ and $g(t)$, say

$$z(t) = \alpha f(t) + \beta g(t) \tag{12.28a}$$

where α and β are real constants. Upon substituting the series expansions for $f(t)$ and $g(t)$ we obtain

$$\begin{aligned} z(t) &= \alpha \sum_{n=-\infty}^{\infty} c_f(n\Omega)e^{jn\Omega t} + \beta \sum_{n=-\infty}^{\infty} c_g(n\Omega)e^{jn\Omega t} \\ &= \sum_{n=-\infty}^{\infty} [\alpha c_f(n\Omega) + \beta c_g(n\Omega)]e^{jn\Omega t} \end{aligned}$$

But the series expansion of $z(t)$ has the form

$$z(t) = \sum_{n=-\infty}^{\infty} c_z(n\Omega)e^{jn\Omega t}$$

Therefore,

$$c_z(n\Omega) = \alpha c_f(n\Omega) + \beta c_g(n\Omega) \tag{12.28b}$$

which means that a linear combination in the time domain becomes a linear combination in the frequency domain.

Time Shift Suppose that $z(t)$ has the same shape as $f(t)$ shifted t_0 units in time so that

$$z(t) = f(t - t_0) \tag{12.29a}$$

The time shift could be to the right if t_0 is positive, or to the left if t_0 is negative. Replacing t with $t - t_0$ in the expansion for $f(t)$ gives

$$z(t) = \sum_{n=-\infty}^{\infty} c_f(n\Omega)e^{jn\Omega(t-t_0)} = \sum_{n=-\infty}^{\infty} [c_f(n\Omega)e^{-jn\Omega t_0}]e^{jn\Omega t}$$

from which

$$c_z(n\Omega) = c_f(n\Omega)e^{-jn\Omega t_0} = c_f(n\Omega)e^{-j2\pi n t_0/T} \tag{12.29b}$$

The corresponding amplitude and phase spectra are

$$|c_z(n\Omega)| = |c_f(n\Omega)| \qquad \angle c_z(n\Omega) = \angle c_f(n\Omega) - n\Omega t_0$$

Hence, time shifting introduces a *linear phase shift,* but it does not affect the amplitude spectrum.

Origin Shift Now let the origin of $f(t)$ be shifted A_0 units vertically and t_0 units horizontally to produce

$$z(t) = A_0 + f(t - t_0) \tag{12.30a}$$

The level shift A_0 just adds to the dc component of $f(t)$ and the time shift has no effect when $n = 0$, so

$$\begin{aligned} c_z(n\Omega) &= A_0 + c_f(0) & n &= 0 \\ &= c_f(n\Omega)e^{-j2\pi n t_0/T} & n &\neq 0 \end{aligned} \tag{12.30b}$$

This result simplifies spectral analysis when $z(t)$ possesses inherent symmetry that has been "hidden" by a shift of the origin.

Half-Wave Symmetry A periodic waveform $z(t)$ is said to have *half-wave* symmetry when

$$z(t \pm T/2) = -z(t) \tag{12.31}$$

As illustrated by Fig. 12.15*a*, Eq. (12.31) means that half of each period of $z(t)$ looks like the other half turned upside down. Accordingly, we can write

$$z(t) = z^+(t) - z^+(t - T/2) \tag{12.32a}$$

where $z^+(t)$ stands for the positive portions of $z(t)$ shown in Fig. 12.15*b*. Since Eq. (12.32a) involves both linearity and time shifting, it follows that

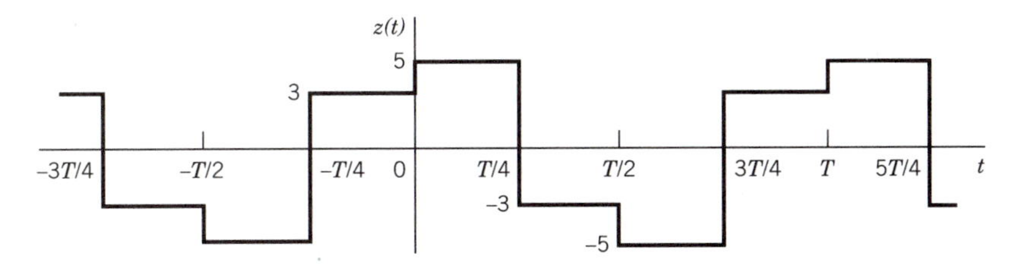

(a) Waveform with half-wave symmetry

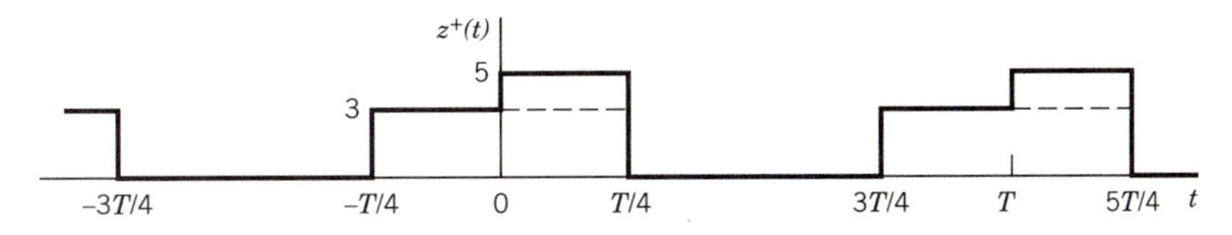

(b) Positive portion of waveform

Figure 12.15

$$c_z(n\Omega) = c_{z+}(n\Omega) - c_{z+}(n\Omega)e^{-jn\Omega T/2} = c_{z+}(n\Omega)(1 - e^{-jn\pi})$$

which simplifies to

$$\begin{aligned} c_z(n\Omega) &= 2c_{z+}(n\Omega) \qquad && n = \pm 1, \pm 3, \ldots \\ &= 0 && n = 0, \pm 2, \pm 4, \ldots \end{aligned} \tag{12.32b}$$

The spectrum of a waveform with half-wave symmetry therefore consists entirely of *odd harmonics*. Even harmonics are absent because they do not satisfy Eq. (12.31).

Our time and frequency relations become particularly helpful in conjunction with various building-block functions that can be used to create new periodic waveforms. For this purpose, Table 12.1 lists the values of $c(n\Omega)$ for selected periodic functions. All of these functions have appeared or will appear in examples or exercises.

TABLE 12.1 Fourier Coefficients of Selected Waveforms

Square wave

$$\begin{aligned} c(n\Omega) &= 0 \qquad && n = 0, 2, 4, \ldots \\ &= -\frac{j2}{n\pi} && n = 1, 3, 5, \ldots \end{aligned}$$

Triangular wave

$$\begin{aligned} c(n\Omega) &= 0 \qquad && n = 0, 2, 4, \ldots \\ &= \frac{4}{n^2\pi^2} && n = 1, 3, 5, \ldots \end{aligned}$$

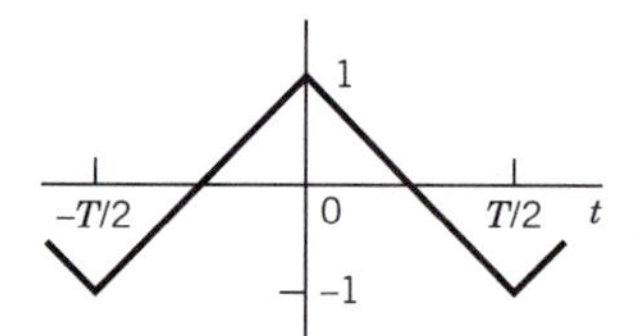

Sawtooth wave

$$\begin{aligned} c(n\Omega) &= 0 \qquad && n = 0 \\ &= -j/n\pi && n = 1, 3, 5, \ldots \\ &= j/n\pi && n = 2, 4, 6, \ldots \end{aligned}$$

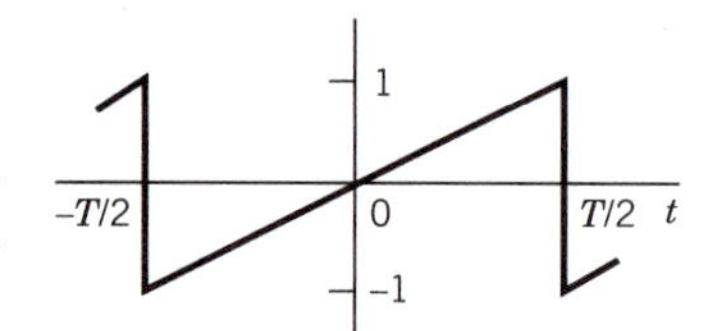

Rectangular pulse train

$$\begin{aligned} c(n\Omega) &= D/T \qquad && n = 0 \\ &= \frac{\sin(n\pi D/T)}{n\pi} && n > 0 \end{aligned}$$

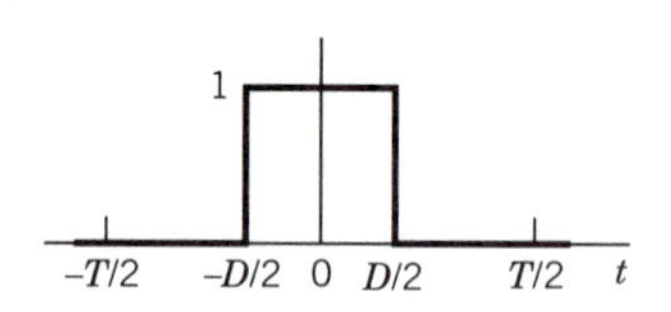

Half-rectified sine wave

$$\begin{aligned} c(n\Omega) &= 1/\pi \qquad && n = 0 \\ &= -j/4 && n = 1 \\ &= \frac{-1}{\pi(n^2 - 1)} && n = 2, 4, 6, \ldots \\ &= 0 && n = 3, 5, 7, \ldots \end{aligned}$$

Example 12.6 *Waveform with Half-Wave Symmetry*

Suppose you want to find $c_z(n\Omega)$ for the waveform in Fig. 12.15*a*. You note that its positive portion in Fig. 12.15*b* can be expressed as the linear combination

$$z^+(t) = 3f(t) + 2g(t)$$

where $f(t)$ and $g(t)$ are the unit-height rectangular pulse trains in Fig. 12.16.

The pulse train $f(t)$ has even symmetry and duration $D = 2 \times T/4 = T/2$. Thus, from Table 12.1, $c_f(n\Omega) = \sin(n\pi/2)/n\pi$. The pulse train labeled $g(t)$ has $D = T/4$ and time shift $t_0 = T/8$, so $n\Omega t_0 = n\pi/4$ and

$$\begin{aligned} c_g(n\Omega) &= \frac{\sin(n\pi/4)}{n\pi} e^{-jn\pi/4} \\ &= \frac{\sin(n\pi/4)}{n\pi} [\cos(n\pi/4 - j\sin(n\pi/4)] \\ &= \frac{\sin(n\pi/2)}{2n\pi} - j\frac{1 - \cos(n\pi/2)}{2n\pi} \end{aligned}$$

Therefore,

$$c_{z+}(n\Omega) = 3c_f(n\Omega) + 2c_g(n\Omega) = \frac{4\sin(n\pi/2)}{n\pi} - j\frac{1 - \cos(n\pi/2)}{n\pi}$$

Finally, from Eq. (12.32b), the nonzero coefficients of $z(t)$ are

$$c_z(n\Omega) = 2c_{z+}(n\Omega) = \frac{8\sin(n\pi/2)}{n\pi} - j\frac{2}{n\pi}$$

which holds for $n = \pm 1, \pm 3, \ldots$

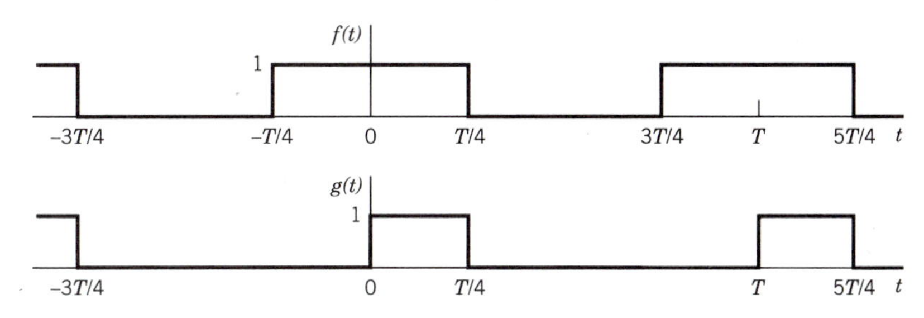

Figure 12.16 Pulse trains for Example 12.6.

Exercise 12.7

Let $z(t)$ be the square wave in Fig. 12.9a (p. 558). Express $z(t)$ as a linear combination of shifted rectangular pulse trains to obtain

$$\begin{aligned} c_z(n\Omega) &= 0 & n &= 0,\ \pm 2,\ \pm 4,\ \ldots \\ &= -j2A/n\pi & n &= \pm 1,\ \pm 3,\ \pm 5,\ \ldots \end{aligned}$$

Differentiation and Integration

Sometimes a periodic waveform is produced by the time-domain operation of differentiation or integration. This might occur, for instance, when an inductor or capacitor carries a periodic current.

To determine the spectral effect of differentiation, let $c_z(n\Omega)$ be the spectrum of a periodic waveform $z(t)$. The derivative of $z(t)$ will then be another periodic waveform, say

$$z'(t) = dz(t)/dt$$

Differentiating the expansion of $z(t)$ gives

$$\frac{dz(t)}{dt} = \sum_{n=-\infty}^{\infty} c_z(n\Omega)\frac{de^{jn\Omega t}}{dt} = \sum_{n=-\infty}^{\infty} [c_z(n\Omega)\,jn\Omega]e^{jn\Omega t}$$

Therefore, it follows that

$$c_{z'}(n\Omega) = jn\Omega\, c_z(n\Omega) = \frac{j2\pi n}{T}\, c_z(n\Omega) \tag{12.33}$$

The dc component is $c_{z'}(0) = 0$, as should be expected since differentiation removes any constant component of $z(t)$.

The reverse of the preceding analysis holds in the case of integration. For if $dz(t)/dt = z'(t)$, then $z(t)$ can be obtained by integrating $z'(t)$ and, if needed, adding a constant term z_{av}. Thus, from Eq. (12.33), the spectrum of $z(t)$ is related to $c_{z'}(n\Omega)$ by

$$c_z(n\Omega) = \frac{c_{z'}(n\Omega)}{jn\Omega} = \frac{T}{j2\pi n}\, c_{z'}(n\Omega) \qquad n \neq 0 \tag{12.34}$$

The dc component must be evaluated separately via $c_z(0) = z_{av}$.

Equation (12.34) shows that integration *decreases* amplitudes at higher frequencies, since $|c_z(n\Omega)| = |c_{z'}(n\Omega)|/|n\Omega|$. Conversely, Eq. (12.33) shows that differentiation *increases* amplitudes at higher frequencies. These effects correspond to the time-domain properties that integration tends to make a waveform "smoother" while diffrentiation tends to make a waveform "rougher."

Example 12.7 *Triangular Pulse Train*

Calculating the coefficients $c_z(n\Omega)$ for the triangular pulse train in Fig. 12.17*a* would be a rather tedious chore. However, differentiating $z(t)$ produces the simpler waveform $z'(t)$ in Fig. 12.17*b*, which consists of two rectangular pulse trains. Accordingly, we can take advantage of several of our time and frequency relations to find the series coefficients for both waveforms.

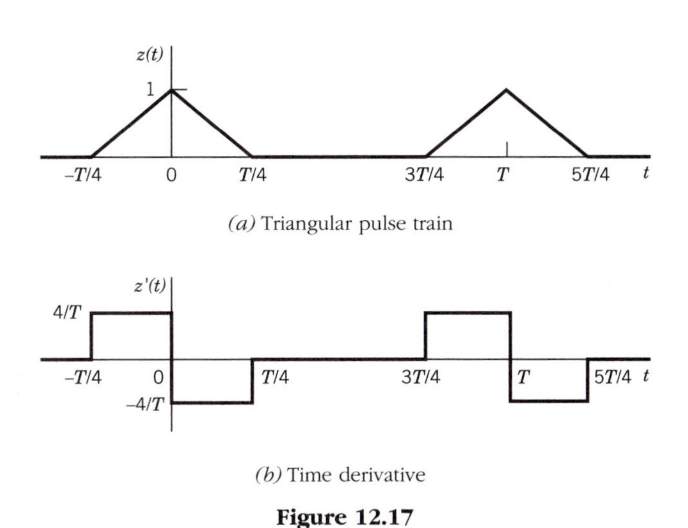

(a) Triangular pulse train

(b) Time derivative

Figure 12.17

First, letting $f(t)$ be a rectangular pulse train with unit height, $D = T/4$, and even symmetry, we write

$$z'(t) = (4/T)f(t + T/8) - (4/T)f(t - T/8)$$

Then, from the linearity and time-shift relations

$$c_{z'}(n\Omega) = (4/T)c_f(n\Omega)e^{jn\pi/4} - (4/T)c_f(n\Omega)e^{jn\pi/4}$$

$$= (4/T)c_f(n\Omega) \times j2 \sin \frac{n\pi}{4} = \frac{j8}{n\pi T} \sin^2 \frac{n\pi}{4}$$

where we have inserted $c_f(n\Omega) = (1/n\pi) \sin (n\pi/4)$. Application of the integration relationship in Eq. (12.34) now yields

$$c_z(n\Omega) = \frac{T}{j2\pi n} c_{z'}(n\Omega) = \frac{4}{n^2\pi^2} \sin^2 \frac{n\pi}{4} \qquad n \neq 0$$

For $n = 0$, inspection of the waveform $z(t)$ reveals that

$$c_z(0) = z_{av} = 1/4$$

Note that $c_z(n\Omega)$ is purely real while $c_{z'}(n\Omega)$ is purely imaginary, in agreement with the even symmetry of $z(t)$ and the odd symmetry of $z'(t)$.

The amplitude spectra $|c_z(n\Omega)|$ and $|c_{z'}(n\Omega)|$ are sketched in Fig. 12.18, omitting the negative-frequency portion. This figure illustrates the increased amplitude at higher frequencies produced by differentiating $z(t)$ to get $z'(t)$.

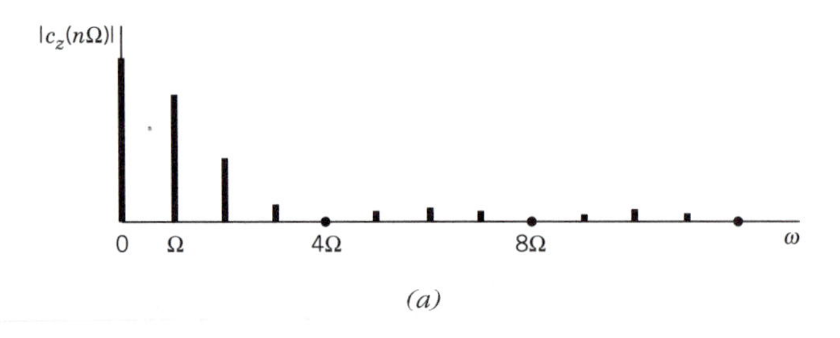

(a)

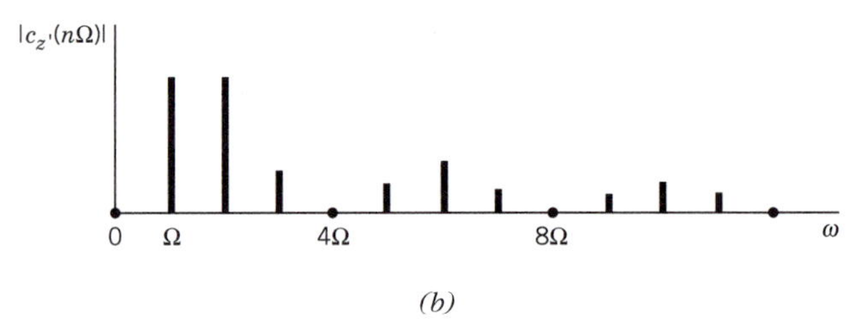

(b)

Figure 12.18 Amplitude spectrum of a triangular pulse train and its derivative.

Exercise 12.8

Sketch $f'(t)$ when $f(t)$ is the triangular wave in Fig. 12.8*b* (p. 558). Then find $c_f(n\Omega)$ using the results of Exercise 12.7.

12.3 SPECTRAL CIRCUIT ANALYSIS

Spectral circuit analysis extends ac steady-state analysis to the more general case of an arbitrary periodic excitation. By taking advantage of the fact that a periodic excitation waveform consists of components at several frequencies, we can determine the resulting periodic steady-state response. Additionally, we can use spectral analysis to investigate the causes of waveform distortion and techniques for reducing unwanted distortion.

Periodic Steady-State Response

We examine here the following type of problem: The input to a stable network is a periodic voltage or current $x(t)$, the transients have died away, and we seek the resulting steady-state response of some output voltage or current $y(t)$. Our approach to this task draws upon spectral analysis together with the network's frequency response.

As preparation, we first develop yet another version of the Fourier series by writing the exponential expansion in the form

$$f(t) = c_0 + \sum_{n=1}^{\infty} (c_n e^{jn\Omega t} + c_{-n} e^{-jn\Omega t})$$

Since c_n is complex, we can introduce the polar notation

$$A_n = 2|c_n| \qquad \phi_n = \angle c_n \tag{12.35}$$

so

$$c_n = \tfrac{1}{2} A_n \underline{/\phi_n} \qquad c_{-n} = c_n^* = \tfrac{1}{2} A_n \underline{/-\phi_n}$$

Adding the pair of exponential terms then yields

$$\begin{aligned} c_n e^{jn\Omega t} + c_{-n} e^{-jn\Omega t} &= (\tfrac{1}{2} A_n \underline{/\phi_n}) e^{jn\Omega t} + (\tfrac{1}{2} A_n \underline{/-\phi_n}) e^{-jn\Omega t} \\ &= \tfrac{1}{2} A_n [e^{j(n\Omega t + \phi_n)} + e^{-j(n\Omega t + \phi_n)}] \\ &= A_n \cos (n\Omega t + \phi_n) \end{aligned}$$

We thereby arrive at the **sinusoidal Fourier series**

$$f(t) = c_0 + \sum_{n=1}^{\infty} A_n \cos (n\Omega t) + \phi_n) \tag{12.36}$$

Relative to circuit analysis, Eq. (12.36) has the advantage that it expresses periodic functions directly in terms of dc and ac components.

Now consider a network driven by a periodic excitation $x(t)$ with spectrum $c_x(n\Omega)$. From Eqs. (12.35) and (12.36), we can write

$$x(t) = c_x(0) + \sum_{n=1}^{\infty} A_{xn} \cos (n\Omega t + \phi_{xn})$$

where, in phasor notation,

$$A_{xn} \underline{/\phi_{xn}} = 2c_x(n\Omega)$$

Having decomposed $x(t)$ into a sum of dc and ac components, we find the resulting output $y(t)$ using the network's frequency-response function $H(j\omega)$.

Each ac input component produces an output component at the same frequency, modified by the amplitude ratio $|H(j\omega)|$ and phase $\angle H(j\omega)$. Thus, the output amplitude and phase of the nth harmonic are

$$\begin{aligned} A_{yn} &= |H(jn\Omega)| A_{xn} = 2|H(jn\Omega)||c_x(n\Omega)| \\ \phi_{yn} &= \phi_{xn} + \angle H(jn\Omega) = \angle c_x(n\Omega) + \angle H(jn\Omega) \end{aligned}$$

But $|c_y(n\Omega)| = \tfrac{1}{2} A_{yn}$ and $\angle c_y(n\Omega) = \phi_{yn}$, so the **output spectrum** is just the product

$$c_y(n\Omega) = H(jn\Omega)c_x(n\Omega) \tag{12.37}$$

Finally, invoking superposition, we get the **periodic steady-state response**

$$y(t) = c_y(0) + \sum_{n=1}^{\infty} A_{yn} \cos (n\Omega t + \phi_{yn}) \tag{12.38}$$

where, from Eq. (12.37),

$$c_y(0) = H(0)c_x(0) \tag{12.39a}$$

$$A_{yn} \underline{/\phi_{yn}} = 2c_y(n\Omega) = 2H(jn\Omega)c_x(n\Omega) \tag{12.39b}$$

Note the factor of 2 that appears in Eq. (12.39b) for the ac components but not in Eq. (12.39a) for the dc component.

At first glance, Eq. (12.38) implies that you must sum an *infinite* number of terms to obtain the output waveform $y(t)$. In many practical applications, however, the network acts as a *filter* that eliminates or greatly reduces all but a few harmonics. A rough but handy rule-of-thumb for determining the *significant* output terms is to

> Ignore any component whose amplitude is less than 10% of the largest output amplitude.

The remaining terms usually constitute a reasonable approximation for $y(t)$.

To summarize, our spectral method for periodic steady-state analysis consists of the following steps:

1. Find the input spectrum $c_x(n\Omega)$.
2. Find the network's frequency-response function $H(jn\Omega)$ relating the input to the output in question.
3. Determine the significant output components by calculating $c_y(0)$ and the largest values of A_{yn}.
4. Insert the significant values into Eq. (12.38) to obtain the approximate expression for $y(t)$.

Example 12.8 *Harmonic Generator*

With the help of electronic switching circuits, square waves are easier to generate than sinusoids. For this reason, some waveform generators produce high-frequency sinusoids by applying a square wave to a bandpass filter. If the filter is tuned to one of the harmonics and has a narrow bandwidth, then the output should be a sinusoid at the selected harmonic frequency.

We'll analyze this scheme when the filter has gain $K = 1$, resonant frequency $\omega_0 = 3\Omega$, and bandwidth $B = \omega_0/10$, so the quality factor is $Q = \omega_0/B = 10$. For simplicity, we'll assume a square wave with height $A = 1$.

Step 1: From the results of Exercise 12.7, the spectrum of a square wave $x(t)$ with $A = 1$ is

$$\begin{aligned} c_x(n\Omega) &= 0 & n &= 0, \pm 2, \pm 4, \ldots \\ &= -j2/n\pi & n &= \pm 1, \pm 3, \pm 5, \ldots \end{aligned}$$

The input amplitude spectrum $|c_x(n\Omega)|$ is plotted in Fig. 12.19*a*

Step 2: From Eq. (11.18), the bandpass filter has

$$H(j\omega) = \left[1 + jQ\left(\frac{\omega}{\omega_0} - \frac{\omega_0}{\omega}\right)\right]^{-1}$$

Thus, with $\omega = n\Omega$, $\omega_0 = 3\Omega$, and $Q = 10$,

$$H(jn\Omega) = \left[1 + j10\left(\frac{n}{3} - \frac{3}{n}\right)\right]^{-1}$$

Figure 12.19*b* shows the corresponding amplitude-ratio curve $|H(j\omega)|$.

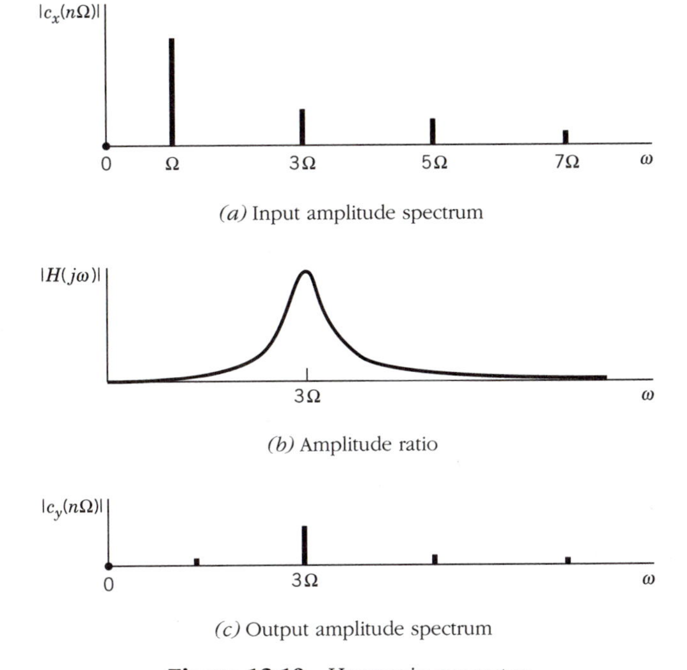

Figure 12.19 Harmonic generator.

Step 3: Multiplying $|H(jn\Omega)|$ and $|c_x(n\Omega)|$ yields the output amplitude spectrum $|c_y(n\Omega)|$ in Fig. 12.19*c*. There is no dc output component, and the third harmonic clearly dominates, as desired. However, there also may be significant components at adjacent harmonic frequencies. For quantitative purposes, Table 12.2 lists the values of $c_x(n\Omega)$, $H(jn\Omega)$, and the phasor $A_{yn} \angle \phi_{yn}$ for $n = 1$, 3, and 5. Since $A_{y5} < 0.1 \times A_{y3}$, we ignore all harmonics except the first and third.

TABLE 12.2

n	$c_x(n\Omega)$	$H(jn\Omega)$	$A_{yn} \angle \phi_{yn}$
1	$0.637 \angle -90°$	$0.038 \angle 88°$	$0.048 \angle -2°$
3	$0.212 \angle -90°$	$1.000 \angle 0°$	$0.424 \angle -90°$
5	$0.127 \angle -90°$	$0.093 \angle -85°$	$0.024 \angle -175°$

Step 4: The significant values from Table 12.2 give the two-term approximation

$$y(t) \approx 0.048 \cos(\Omega t - 2°) + 0.424 \cos(3\Omega t - 90°)$$

This waveform is sketched in Fig. 12.20 along with the input square wave $x(t)$. To generate a "purer" sinusoidal output, the filter should have $Q > 10$ so the bandwidth would be narrower.

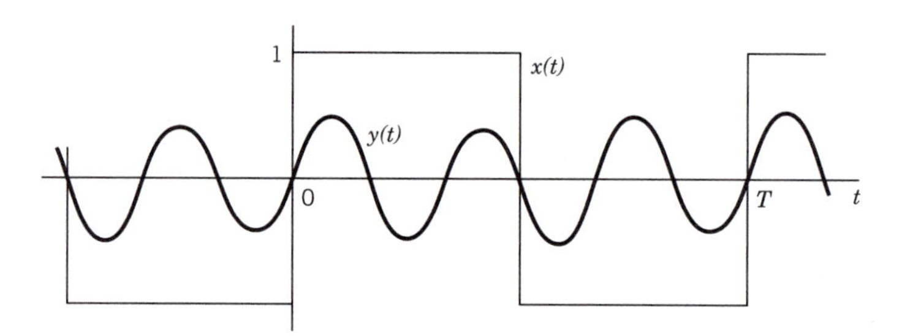

Figure 12.20 Harmonic generator waveforms.

Example 12.9 *Improved Harmonic Generator*

Suppose a harmonic generator is to be designed such that all unwanted output components must have amplitudes less than 1% of the third harmonic. We'll determine the corresponding requirement for the Q of the filter.

As implied by Table 12.2, the largest unwanted component will be the first harmonic. With arbitrary Q, the amplitude ratio of the filter is

$$|H(jn\Omega)| = \left[1 + Q^2\left(\frac{n}{3} - \frac{3}{n}\right)^2\right]^{-1/2}$$

Thus,

$$A_{y1} = 2|H(jn\Omega)||c_x(\Omega)| = 1.274/\sqrt{1 + Q^2(8/3)^2}$$

whereas $A_{y3} = 0.424$. The specifications call for $A_{y1} < 0.01 \times A_{y3}$, so

$$1.274/\sqrt{1 + Q^2(8/3)^2} < 0.00424$$

from which

$$Q > (3/8)/\sqrt{(1.274/0.00424)^2 - 1} \approx 113$$

Exercise 12.9

Obtain an approximation for $y(t)$ when a half-rectified sine wave is applied to a first-order lowpass filter with $K = 1$ and $\omega_{co} = \Omega/4$. Take $A = \pi$ and use the results of Example 12.3 (p. 556).

Waveform Distortion

When an input waveform $x(t)$ drives a linear network, the steady-state output $y(t)$ may have a substantially different shape. In some applications the difference is intentional and desired, as in the case of a harmonic generator producing a sinusoidal output from a square-wave input. In other applications, however, the output waveform is supposed to look like the input, as in the case of an audio amplifier. Any significant difference between the output and input waveshapes is called **linear distortion**. Electronic components may also introduce *nonlinear* distortion, but we limit our attention here to *linear* networks.

To investigate the causes of linear waveform distortion, we'll begin at the opposite extreme. **Distortionless reproduction** occurs only when the network's frequency response has the form

$$H(j\omega) = \alpha e^{-j\omega t_0} \tag{12.40}$$

where α and t_0 are constants. The steady-state response to an input $x(t)$ is then

$$y(t) = \alpha x(t - t_0) \tag{12.41}$$

Thus, a distortionless output $y(t)$ differs from $x(t)$ only by a constant *scale factor* α and a *time shift* t_0. This relationship also allows for polarity inversion, since α could be a negative quantity.

The derivation of Eq. (12.41) is straightforward assuming a periodic input $x(t)$ with spectrum $c_x(n\Omega)$. The steady-state output $y(t)$ will be periodic, and its spectrum is given by

$$c_y(n\Omega) = H(jn\Omega)c_x(n\Omega)$$

But upon setting $\omega = n\Omega$, Eq. (12.40) becomes

$$H(jn\Omega) = \alpha e^{-jn\Omega t_0}$$

so

$$c_y(n\Omega) = \alpha c_x(n\Omega)e^{-jn\Omega t_0}$$

Equation (12.41) now follows from our linearity and time-shift relations in Section 12.2.

Further examination of Eq. (12.40) leads to two separate conditions for distortionless reproduction. Expressing $H(j\omega)$ in polar form and recognizing that α may be positive or negative, we obtain

$$a(\omega) = |H(j\omega)| = |\alpha| \tag{12.42a}$$

$$\theta(\omega) = \angle H(j\omega) = \angle\alpha - \omega t_0 \tag{12.42b}$$

Equation (12.42a) requires the network's amplitude ratio to be constant, while Eq. (12.42b) requires the phase shift to have constant slope. Fortunately, however, these conditions must be satisfied only over the range of frequencies in which the input has significant spectral content — after all, the behavior of $H(j\omega)$ doesn't matter at any frequencies where the input spectrum is negligible.

When $a(\omega)$ is not constant over the input spectrum, per Eq. (12.42a), the amplitudes of some of the output components will be larger or smaller than they should be. This effect is therefore known as **amplitude distortion**. For instance, a series capacitor might act as a dc block and reduce amplitudes at the low end of the spectrum, while a parallel capacitor might act as a high-frequency short and reduce amplitudes at the high end of the spectrum.

As an illustration of amplitude distortion, let the input be the square-wave approximation in Fig. 12.21*a* where

$$x(t) = 15 \sin \Omega t + 5 \sin 3\Omega t + 3 \sin 5\Omega T$$

If this "test" input is applied to a network that acts like an ideal *lowpass* filter and completely removes the high-frequency components at $\omega > 3\Omega$, then

$$y(t) = 15 \sin \Omega t + 5 \sin 3\Omega t$$

The resulting waveform in Fig. 12.21*b* shows that the loss of the high-frequency term makes the corners more rounded than square. On the other hand, if the network acts like an ideal *highpass* filter and completely removes the low-frequency components at $\omega < 3\Omega$, then

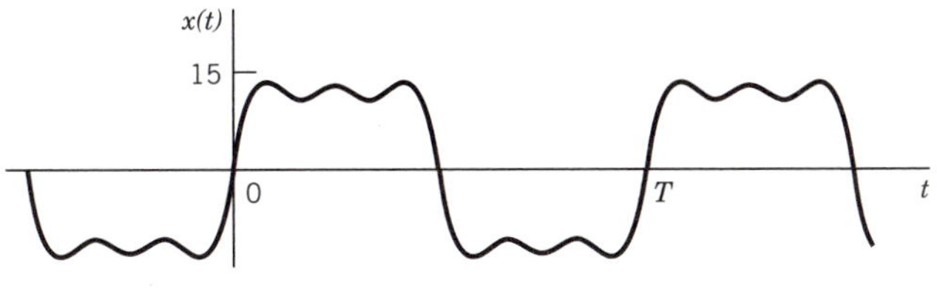

(a) Test waveform

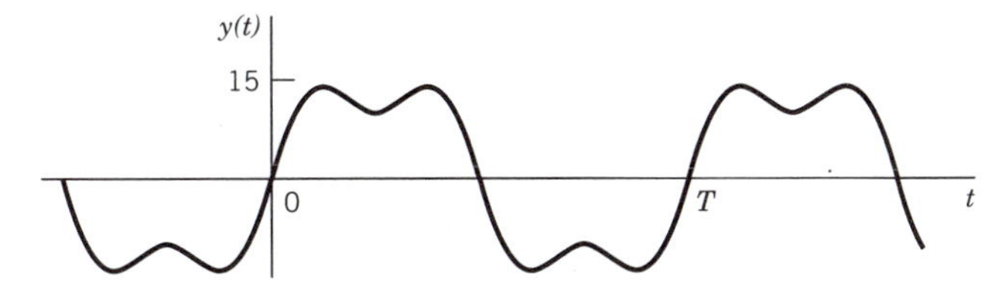

(b) High frequencies removed

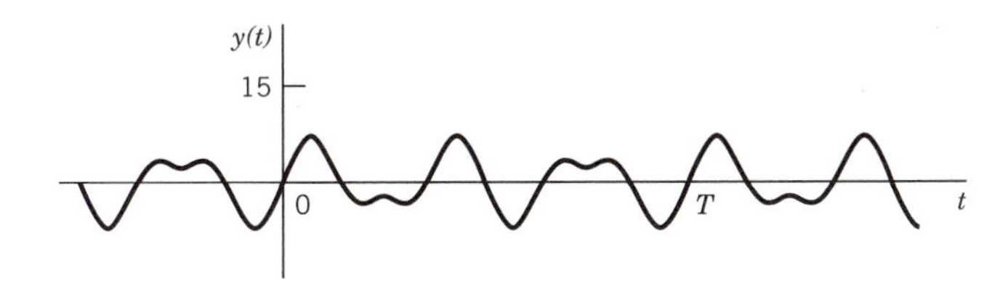

(c) Low frequencies removed

Figure 12.21

$$y(t) = 5 \sin 3\Omega t + 3 \sin 5\Omega t$$

The resulting waveform in Fig. 12.21*c* bears no resemblance to a square wave because it lacks the dominant fundamental-frequency component.

When $\theta(\omega)$ does not have constant slope, per Eq. (12.42b), the various input components appear at the output with different amounts of time delay. This effect is therefore known as **delay distortion**. Note carefully the distinction between *constant time delay* and *constant phase shift.* Any constant phase shift other than $\theta(\omega) = 0°$ or $\theta(\omega) = \pm 180°$ produces delay distortion because each frequency component is shifted in time by a different amount.

As an illustration of delay distortion, let our test input be applied to a network having $\theta(\omega) = -90°$ so

$$\begin{aligned} y(t) &= 15 \sin (\Omega t - 90°) + 5 \sin (3\Omega t - 90°) + 3 \sin (5\Omega t - 90°) \\ &= -15 \cos \Omega t - 5 \cos 3\Omega t - 3 \cos 5\Omega t \end{aligned}$$

The resulting waveform in Fig. 12.22 now looks like a triangular wave instead of a square wave. The peak excursions are larger than before because the peaks of the individual components all add together at $t = 0, \pm T/2, \pm T, \ldots$.

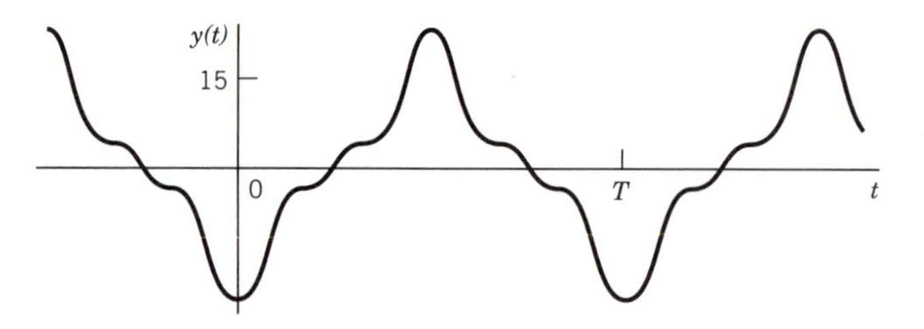

Figure 12.22 Delay distortion of test waveform.

Most amplifiers and other circuits designed for waveform reproduction introduce both amplitude and delay distortion to some degree. The amount of distortion that can be tolerated depends upon the particular application and usually must be determined by experimental studies. By way of example, the amplitude ratio of a high-fidelity audio amplifier should vary no more than about $\pm 10\%$ (or ± 1 dB) over a frequency range of 30 Hz to 15 kHz. Phase shift is less critical in audio amplification because the human ear is not very sensitive to delay distortion — so an acoustical wave like Fig. 12.22 would sound about the same as Fig. 12.21*a*. But delay distortion could be a serious problem in a digital communication system that conveys information via pulsed waveforms.

Example 12.10 *Distortion by a Lowpass Filter*

Recall from Section 11.2 that a simple first-order lowpass filter with low-frequency gain $K > 0$ has

$$a(\omega) = \frac{K}{\sqrt{1 + (\omega/\omega_{co})^2}} \qquad \theta(\omega) = -\tan^{-1}\frac{\omega}{\omega_{co}}$$

Comparisons with Eqs. (12.42a) and (12.42b) indicate that this network would generally be expected to produce both amplitude and delay distortion.

Suppose, however, that the input spectrum is concentrated below some frequency $\omega_{max} \ll \omega_{co}$. With $\omega \ll \omega_{co}$, we can use the approximations $1 + (\omega/\omega_{co})^2 \approx 1$ and $\tan^{-1}(\omega/\omega_{co}) \approx \omega/\omega_{co}$ to write

$$a(\omega) \approx K \qquad \theta(\omega) \approx -\omega/\omega_{co} \qquad \omega \le \omega_{max} \ll \omega_{co}$$

Thus, when the desired input has negligible high-frequency content, the lowpass filter provides essentially distortionless reproduction with $\alpha = K$ and $t_0 = 1/\omega_{co}$.

Exercise 12.10

A certain network completely removes all frequencies above $\omega = 4$ and introduces a phase shift in radians given by $\theta(\omega) = (\pi/12)(7\omega - \omega^2)$. Find and sketch $y(t)$ taking the test input

$$x(t) = 15 \sin t + 5 \sin 3t + 3 \sin 5t$$

Equalization

Linear waveform distortion can often be reduced by **equalization**. This process involves another network called an **equalizer** designed to counteract the effects of the distorting network.

Figure 12.23 depicts the equalization strategy assuming periodic voltage waveforms for discussion purposes. The input $x(t)$ drives a distorting network with frequency response $H_d(j\omega)$, so the spectrum of the distorted waveform $z(t)$ is

$$c_z(n\Omega) = H_d(jn\Omega)c_x(n\Omega)$$

But the final output $y(t)$ is produced by applying $z(t)$ to an equalizer with frequency response $H_{eq}(j\omega)$. Thus,

$$\begin{aligned} c_y(n\Omega) &= H_{eq}(jn\Omega)c_z(n\Omega) = H_{eq}(jn\Omega)H_d(jn\Omega)c_x(n\Omega) \\ &= H(jn\Omega)c_x(n\Omega) \end{aligned}$$

where

$$H(jn\Omega) = H_{eq}(jn\Omega)H_d(jn\Omega)$$

Substituting ω in place $n\Omega$ yields

$$H(j\omega) = H_{eq}(j\omega)H_d(j\omega)$$

which represents the *overall frequency response* relating the input and output spectra. The expression for $H(j\omega)$ assumes that $H_d(j\omega)$ includes any loading introduced by the equalizer.

Since we want $y(t) = \alpha x(t - t_0)$ for distortionless reproduction of $x(t)$, the overall frequency response must have the form $H(j\omega) = H_{eq}(j\omega)H_d(j\omega) = \alpha e^{-j\omega t_0}$. The equalizer should therefore be designed such that

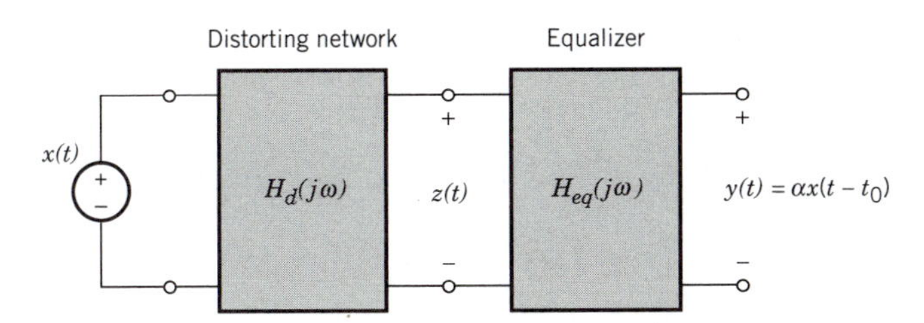

Figure 12.23 Equalization for linear distortion.

$$H_{eq}(j\omega) = \frac{\alpha e^{-j\omega t_0}}{H_d(j\omega)} \tag{12.43}$$

Ideally, Eq. (12.43) should hold for all ω. In practice, we try to design an equalizer that essentially satisfies Eq. (12.43) over the frequency range in which the input has significant spectral content.

For instance, the popular **graphic equalizers** in high-fidelity systems consist of bandpass filters with adjustable gains that allow you to increase or decrease the amplitude ratio over a number of individual frequency bands covering the audio range. As noted before, audio systems do not require phase equalization.

When both amplitude and phase are to be equalized, the *s*-domain interpretation of Eq. (12.43) is helpful. In particular, let $H_d(s)$ have the usual form

$$H_d(s) = K\frac{(s - z_1)(s - z_2)\cdots(s - z_m)}{(s - p_1)(s - p_2)\cdots(s - p_n)}$$

Then, taking $t_0 = 0$ for simplicity, the equalizer's transfer function must be

$$H_{eq}(s) = \frac{\alpha}{H_d(s)} = \frac{\alpha}{K}\frac{(s - p_1)(s - p_2)\cdots(s - p_n)}{(s - z_1)(s - z_2)\cdots(s - z_m)} \tag{12.44}$$

Hence, the *poles* of the equalizer should fall at the *zeros* of $H_d(s)$, and the *zeros* of the equalizer should fall at the *poles* of $H_d(s)$. We would then get complete *pole-zero cancellation,* leaving $H(s) = H_d(s)H_{eq}(s) = \alpha$.

But perfect pole-zero cancellation may be difficult to achieve in practice. Consequently, equalizer design often involves approximating Eq. (12.44) with additional poles or modified pole locations. Op-amp circuits like those back in Fig. 11.32 might be used for this purpose.

Example 12.11 *Equalization of a Distorted Pulse Train*

Suppose the input in Fig. 12.23 is a rectangular pulse train with $D = T/3$, and $H_d(j\omega)$ is a first-order lowpass function with $K = 1$ and $\omega_{co} = \Omega = 2\pi/T$. Figure 12.24*a* compares the resulting distorted waveform $z(t)$ with the original pulse train $x(t)$, taking $t = 0$ at the start of an input pulse. An appropriate equalizer can reduce the distortion and produce an output waveform $y(t)$ like Fig. 12.24*b* — clearly a marked improvement.

The reason why $y(t)$ does not have exactly the same shape as $x(t)$ stems from the fact that $H_d(s) = \Omega/(s + \Omega)$. Consequently, ideal equalization per Eq. (12.44) would require

$$H_{eq}(s) = \alpha/H_d(s) = (\alpha/\Omega)(s + \Omega)$$

which has one zero and no poles. To get a more practical equalizer function, we'll add a pole at $s = \omega_{eq}$ and write

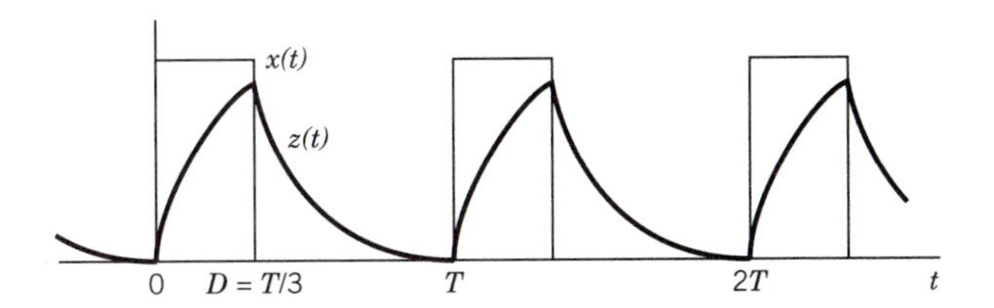

(a) Pulse train and distorted waveform

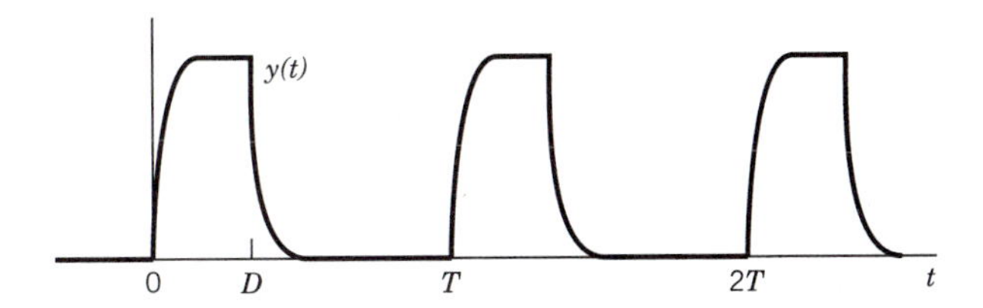

(b) Waveform with equalization

Figure 12.24

$$H_{eq}(s) = K_{eq}\frac{s+\Omega}{s+\omega_{eq}}$$

The overall response then becomes

$$H(s) = H_d(s)H_{eq}(s) = \frac{K_{eq}\Omega}{s+\omega_{eq}}$$

This expression represents another lowpass function with cutoff frequency ω_{eq} and gain $H(0) = K_{eq}\Omega/\omega_{eq}$. Thus, after determining ω_{eq}, we can get $H(0) = 1$ by taking $K_{eq} = \omega_{eq}/\Omega$.

As a guide to selecting a value for ω_{eq}, Fig. 12.25*a* repeats the amplitude spectrum of the pulse train previously found in Example 12.5. The pulse train has significant spectral content up to at least the fifth harmonic, so we can "pass" most of the spectrum by taking $\omega_{eq} = 5\Omega$. The corresponding overall amplitude ratio $a(\omega)$ is plotted in Fig. 12.25*b* along with the unequalized $a_d(\omega)$. With $\omega_{eq} = 5\Omega$ and $K_{eq} = \omega_{eq}/\Omega = 5$, the equalizer itself has

$$H_{eq}(s) = 5\,\frac{s+\Omega}{s+5\Omega}$$

The plot of $a_{eq}(\omega)$ in Fig. 12.25*c* shows that the equalizer enhances high-frequency components that were reduced by $a_d(\omega)$. This equalization yields $y(t)$ back in Fig. 12.24*b*.

Finally, since both $z(t)$ and $y(t)$ consist here of *repeated transients,* we can check our conclusions directly via *time-domain analysis.* Recall in Section 9.1 we found that reasonable pulse reproduction requires a time constant

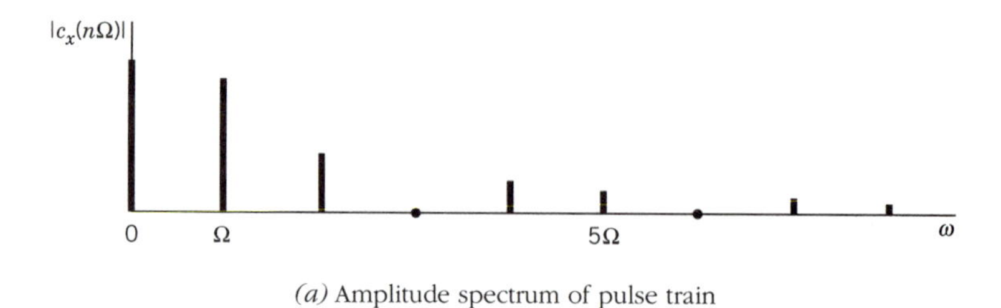

(a) Amplitude spectrum of pulse train

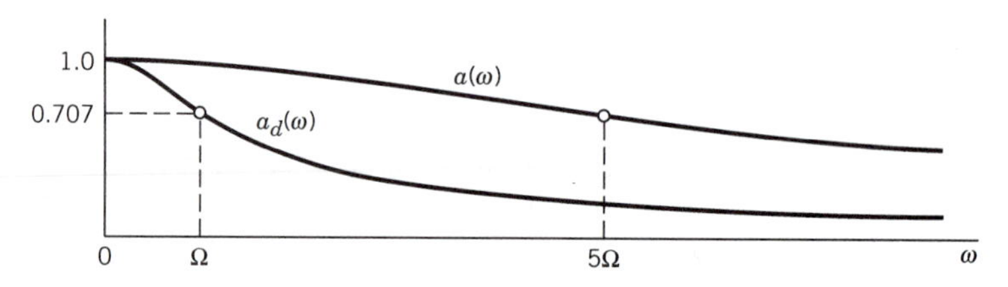

(b) Equalized and unequalized amplitude ratios

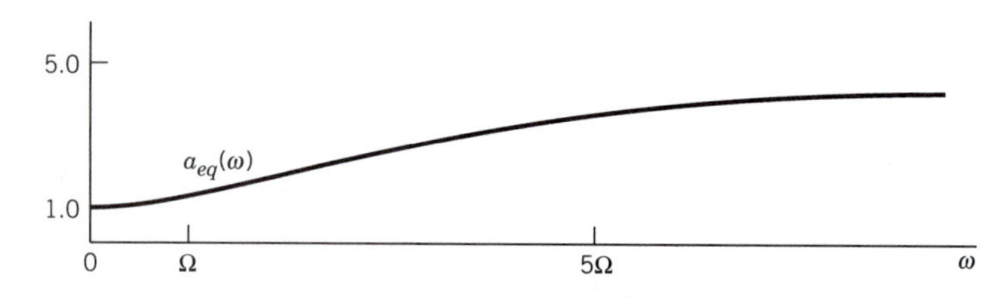

(c) Amplitude ratio of equalizer

Figure 12.25

$\tau \ll D$. Also recall that the time constant of a first-order lowpass function equals the reciprocal of the cutoff frequency. The unequalized function with $\omega_{co} = \Omega$ has $\tau = 1/\omega_{co} = T/2\pi = 3D/2\pi \approx 0.5D$, so the transient response in $z(t)$ is not fast enough. But the equalized function has $\tau = 1/\omega_{eq} = T/10\pi \approx 0.1D$, so $y(t)$ more closely follows $x(t)$.

Exercise 12.11

Let $H_d(s) = s/(s + 10)$ in Fig. 12.23.

(a) Sketch $a_{eq}(\omega)$ for ideal equalization over $5 \le \omega \le 20$ with $\alpha = 1$.

(b) Explain why equalization would be difficult over $0 \le \omega \le 20$.

PROBLEMS

Section 12.1 Periodic Waveforms and Fourier Series

12.1* Find c_n for the periodic waveform defined by $f(t) = e^{-t}$ for $0 < t < T$. Use your result to determine c_0, a_n, and b_n.

12.2 Let $f(t)$ be the periodic waveform in Fig. P12.2. Calculate c_0 and show that

$$c_n = (T/4\pi^2 n^2 D)\,[e^{-jn\Omega D}(jn\Omega D + 1) - 1]$$

for $n \neq 0$. Then find a_n and b_n when $D = T$.

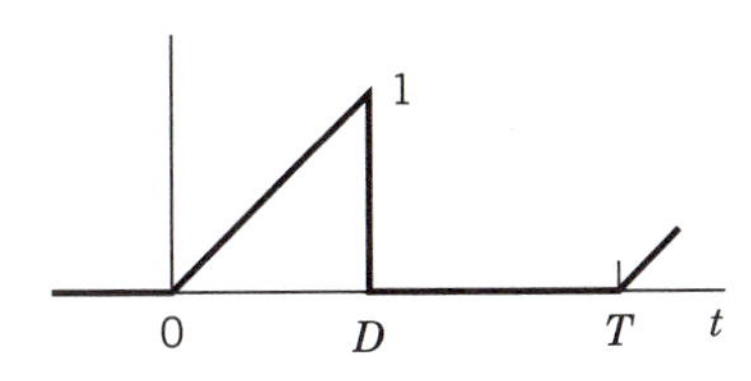

Figure P12.2

12.3 Let $f(t)$ be the periodic waveform in Fig. P12.3 with $B = 3$, $A = 1$, and $D = T/4$. Calculate c_0 and c_n. Then determine the values of n for which $a_n = 0$ and $b_n = 0$.

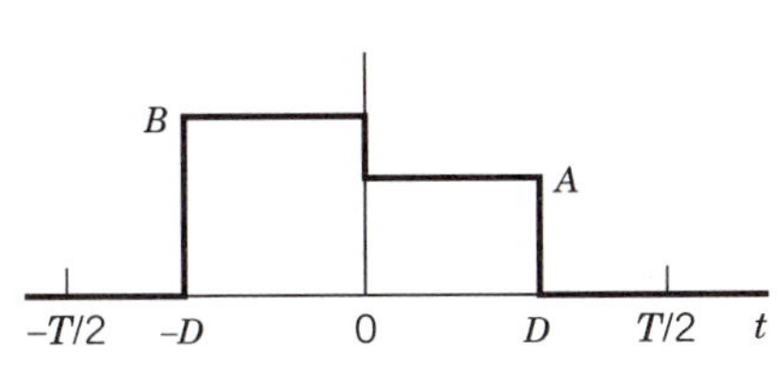

Figure P12.3

12.4 Calculate c_0 and c_n when $f(t)$ is the waveform in Fig. 12.1 (p. 547) with $A = 1$, $t_1 = 1$, and $T = 4$. Hint: Use the property that

$$e^{jn\Omega(T-t_1)} = e^{j2\pi n}\,e^{-jn\Omega t_1} = e^{-jn\Omega t_1}$$

12.5 Do Problem 12.4 with $t_1 = 4$ and $T = 6$.

12.6* Figure P12.6 shows a **full-rectified sine wave** defined by $f(t) = A\,|\sin \Omega t|$. Use symmetry and the results of Example 12.1 to obtain c_0, a_n, and b_n. Then write out the first three nonzero terms of the trigonometric expansion. Why are there only *even* harmonics in this series?

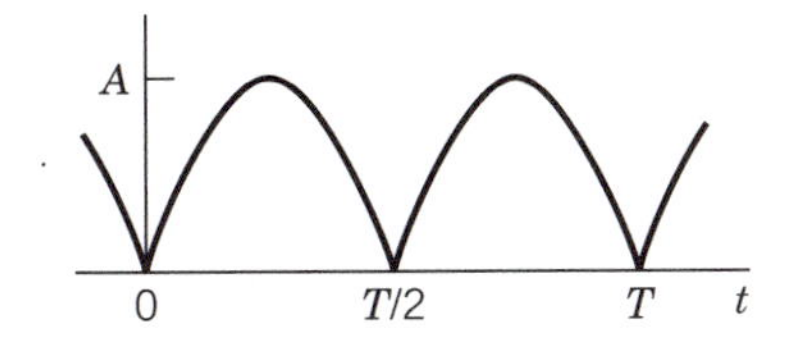

Figure P12.6

12.7 Take advantage of symmetry to find c_0, a_n, and b_n for the periodic waveform in Fig. P12.7. Then let $D = T/3$ and write out the first three nonzero terms of the trigonometric expansion with numerical coefficients.

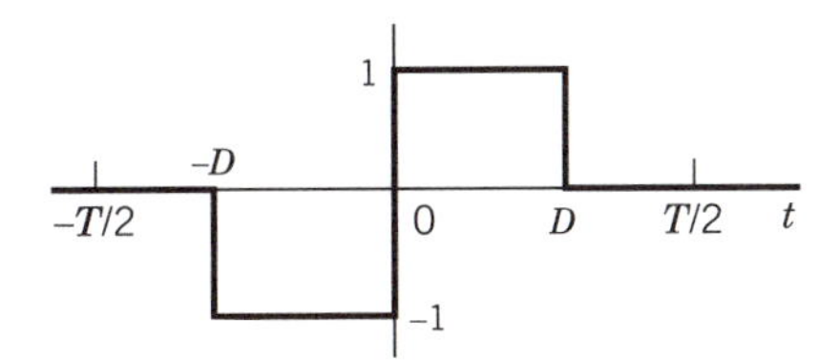

Figure P12.7

12.8 Take advantage of symmetry to find c_0, a_n, and b_n for the periodic waveform defined over one period by

$$f(t) = t/D \qquad |t| < D$$
$$= 0 \qquad D < |t| < T/2$$

Then let $D = T/4$ and write out the first three nonzero terms of the trigonometric expansion with numerical coefficients.

12.9 Take advantage of symmetry to find c_0, a_n, and b_n for the periodic waveform defined over one period by

$$f(t) = |2t/T| \qquad |t| \leq T/2$$

Then write out the first three nonzero terms of the trignometric expansion with numerical coefficients.

12.10 Do Problem 12.9 with

$$f(t) = |4t/T| \qquad |t| < T/4$$
$$= 0 \qquad T/4 < |t| < T/2$$

Section 12.2 Spectral Analysis of Periodic Waveforms

12.11 Sketch and label the two-sided line spectrum of $f(t) = -3 + 4 \sin(5t + 30°)$.

12.12 Sketch and label the two-sided line spectrum of $f(t) = 12 \cos^2(5t + 40°)$.

12.13 Sketch and label the two-sided line spectrum of $f(t) = 8 \cos 2t \cos (5t + 20°)$.

12.14 Sketch and label the two-sided line spectrum of $f(t) = 20 \sin 3t \cos (3t - 60°)$.

12.15 Sketch and label the two-sided line spectrum of $f(t) = 12 \sin (10t + 60°) \cos (5t + 60°)$.

12.16* Construct the waveform in Fig. P12.16 using a triangular wave and find $c_z(n\Omega)$.

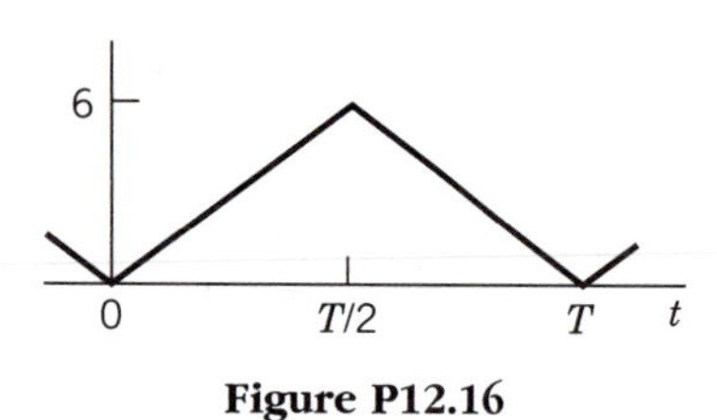

Figure P12.16

12.17 Construct the waveform in Fig. P12.17 using a sawtooth wave and find $c_z(n\Omega)$.

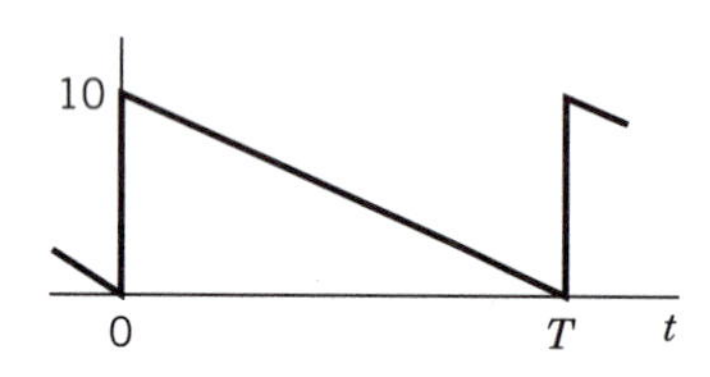

Figure P12.17

12.18 Let the waveform in Fig. P12.3 have $A = 1$, $B = 2$, and $D = T/4$. Obtain $c_z(n\Omega)$ by decomposing this waveform into rectangular pulse trains.

12.19 Let the waveform in Fig. P12.3 have $A = 4$, $B = 6$, and $D = T/3$. Obtain $c_z(n\Omega)$ by decomposing this waveform into rectangular pulse trains.

12.20* Obtain $c_z(n\Omega)$ for the full-rectified sine wave in Fig. P12.6 by decomposing this waveform into half-rectified sine waves.

12.21 Derive $c_z(n\Omega)$ for the waveform in Fig. P12.21 by combining two waveforms from Table 12.1. Simplify your results by considering specific values of n.

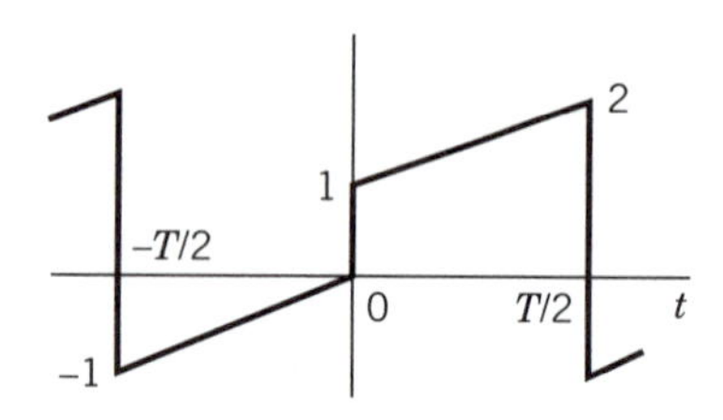

Figure P12.21

12.22 Derive $c_z(n\Omega)$ for the waveform in Fig. P12.22 by combining two waveforms from Table 12.1. Simplify your results by considering specific values of n.

Figure P12.22

12.23 Find $c_z(n\Omega)$ for a waveform with half-wave symmetry when $z(t) = 1$ for $0 < t < T/3$ and $z(t) = 0$ for $T/3 < t < T/2$. What harmonics are absent from the spectrum?

12.24 Draw the waveform $z(t)$ with half-wave symmetry when $z^+(t)$ is as shown in Fig. P12.2 with $D = T/2$. Then use the result in Problem 12.2 to obtain $c_z(n\Omega)$.

12.25 **Quarter-wave symmetry** describes a function with half-wave symmetry and either even or odd symmetry. Let $z(t)$ in Fig. P12.25 be part of a function with even quarter-wave symmetry. Draw the complete waveform for $-T/2 < t < T/2$, and find $c_z(n\Omega)$.

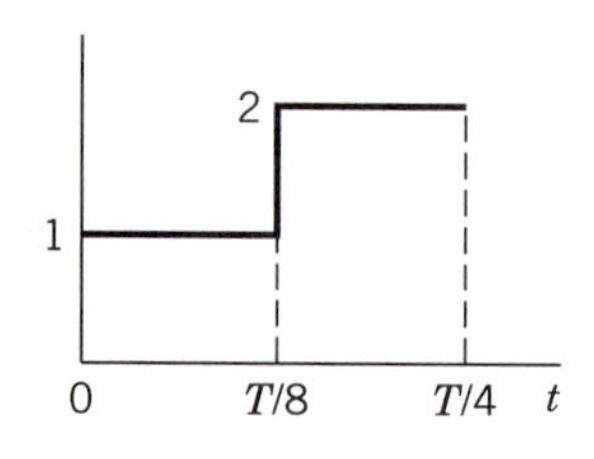

Figure P12.25

12.26 Do Problem 12.25 for the case when $z(t)$ has odd quarter-wave symmetry.

12.27* Let $z(t)$ be the triangular wave in Table 12.1. Sketch $z'(t)$ and find $c_{z'}(n\Omega)$.

12.28 Let $z(t)$ be the half-rectified sine wave in Table 12.1. Sketch $z'(t)$ and find $c_{z'}(n\Omega)$.

12.29 Obtain $c_z(n\Omega)$ by considering $z'(t)$ when $z(t) = 1 - 4t^2/T^2$ for $-T/2 \leq t \leq T/2$.

12.30 Figure P12.30 is a **trapezoidal pulse**. Sketch $z'(t)$ to find $c_z(n\Omega)$ when $t_1 = T/6$ and $t_2 = T/2$.

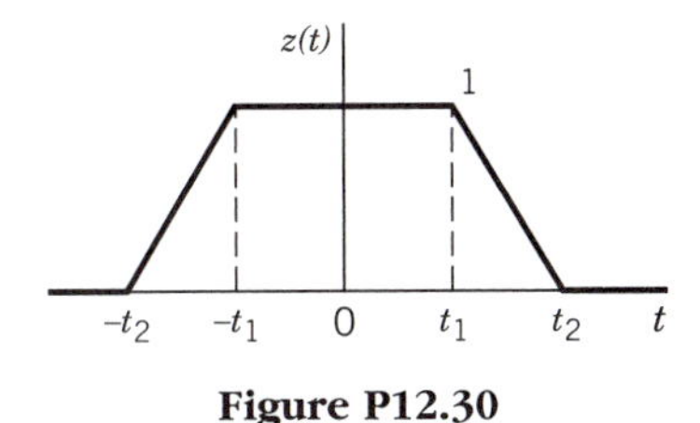

Figure P12.30

12.31 Do Problem 12.30 with $t_1 = T/8$ and $t_2 = T/4$.

Section 12.3 Spectral Circuit Analysis

12.32* The input to a network is a rectangular pulse train having $A = \pi$, $T = 2$ ms, and $D = 1$ ms. The network is an RC lowpass filter with $\omega_{co} = 500\,\pi$. Obtain an approximation for the steady-state output.

12.33 The input to a network is a triangular wave having $T = 0.5$ ms and peak values $\pm\,\pi^2/4$. The network is an RC highpass filter with $\omega_{co} = 10^4\,\pi$. Obtain an approximation for the steady-state output.

12.34 A sawtooth wave with peak values $\pm\,\pi$ is applied to a second-order lowpass filter having

$$H(j\omega) = [1 - (\omega/\omega_{co})^2 + j\sqrt{2}(\omega/\omega_{co})]^{-1}$$

Obtain an approximation for the steady-state output when $\omega_{co} = \Omega$.

12.35 A bandbass filter with $K = 1$, $Q = 2$, and $\omega_0 = 105$ is driven by a rectangular pulse train having $A = \pi$, $T = 60$ ms, and $D = 20$ ms. Obtain an approximation for the steady-state output.

12.36 A harmonic generator consists of a square wave driving a bandpass filter tuned to the fifth harmonic. Obtain an approximation for the steady-state output when the square-wave amplitude is $A = \pi$ and the filter has $Q = 20$.

12.37* Figure P12.37 represents a dc motor connected to a rectifier. The voltage $v(t)$ is a half-rectified sine wave with peak value V_m and $T = 1/60$ s. Given that $R = 10\ \Omega$, determine V_m such that the dc power delivered to the motor is $P_{DC} = 200$ W. Then obtain the condition on L such that

$$i(t) \approx I_{DC} + I_1 \cos(\Omega t + \phi_1)$$

where $I_1 \leq I_{DC}/10$.

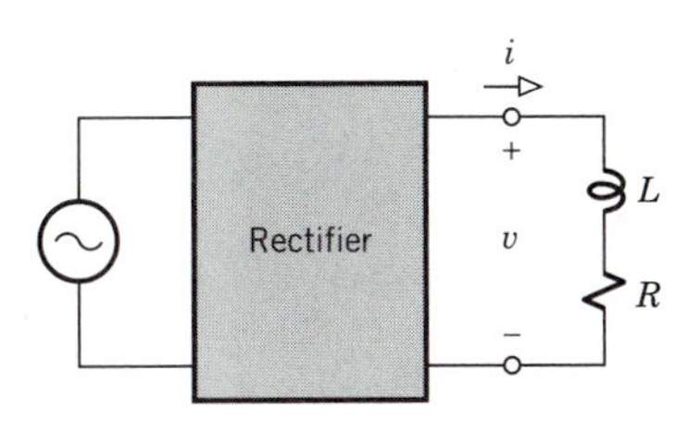

Figure P12.37

12.38 Do Problem 12.37 with $R = 4\ \Omega$, $P_{DC} = 1$ kW, and $I_1 \leq I_{DC}/20$.

12.39 A sinusoidal voltage is to be produced by applying a triangular voltage wave to a series RC network with $R = 1$ kΩ. The triangular wave has $T = 1$ ms and peak value V_m. (a) Find C and V_m such that the first-harmonic amplitude of the voltage across C equals 2 V and the third-harmonic amplitude equals 0.1 V. (b) Show that the third harmonic amplitude cannot be reduced to 1% of the first-harmonic amplitude.

12.40 A sinusoidal voltage is to be produced by applying a square voltage wave to a series RC network with $R = 5$ kΩ. The square wave has $T = 20$ ms and peak value V_m. (a) Find C and V_m such that the first-harmonic amplitude of the voltage across C equals 4 V and the third-harmonic amplitude equals 0.5 V. (b) Show that the third-harmonic amplitude cannot be reduced to 10% of the first-harmonic amplitude.

12.41 If a square wave or rectangular pulse train drives a first-order circuit, then the output consists of exponential transients with time constant τ. (See, for example, Fig. 12.24*a*.) Hence, the **exact steady-state response** can be found by treating the initial value of the transient as an unknown and invoking the periodic property. Use this approach to obtain the steady-state capacitor voltage $v_C(t)$ in a series RC circuit with $R = 1$ kΩ and $C = 1\ \mu$F when the applied voltage is a square wave with $T = 2$ ms and peak values ± 5 V.

12.42 A rectangular pulse train voltage with $A = 12$ V, $D = 1$ ms, and $T = 3$ ms is applied to a series RL circuit having $R = 4\ \Omega$ and $L = 8$ mH. Use the approach outlined in Problem 12.41 to obtain the steady-state current $i(t)$. For convenience, let the pulses start at $t = 0, \pm T, \ldots$

12.43 Use spectral analysis with periodic functions to obtain an expression for $y(t)$ in terms of $x(t)$ when the transfer function of the network is

$$H(s) = 0.2s - 0.5e^{-s/100}$$

12.44 Do Problem 12.43 with

$$H(s) = s(s - 0.1)e^{-s/5}$$

12.45 A transmission line connecting a source to a load introduces time delay and an *echo,* so that

$$y(t) = x(t - t_d) + 0.2x(t - 3t_d)$$

Use spectral analysis with periodic functions to show that the transmission line has

$$H(j\omega) = (1 + 0.2e^{-j2\omega t_d})e^{-j\omega t_d}$$

Then find and sketch $a(\omega)$ to show that there are *ripples* in the amplitude ratio.

12.46 A radio transmission path between a transmitter and receiver introduces time delay and *leading and trailing echos,* so that

$$y(t) = 0.1x(t - t_d + t_0) + x(t - t_d) - 0.1x(t - t_d - t_0)$$

Use spectral analysis with periodic functions to show that the transmission path has

$$H(j\omega) = (1 + 0.1e^{j\omega t_0} - 0.1e^{-j\omega t_0})e^{-j\omega t_d}$$

Then find and sketch $\theta(\omega)$ to show that there are *ripples* in the phase shift.

12.47* Ideal equalization with $\alpha = 1$ and $t_0 = 0.1$ s is desired over $0 \le \omega \le 20$ when

$$H_d(s) = 500/(s + 10)^2$$

Find $H_{eq}(j\omega)$ and sketch its amplitude ratio and phase shift.

12.48 Ideal equalization with $\alpha = 4$ and $t_0 = 0$ is desired over $10 \le \omega \le 30$ when

$$H_d(s) = 200s/(s^2 + 10s + 400)$$

Find $H_{eq}(j\omega)$ and sketch its amplitude ratio and phase shift.

12.49 An equalizer is needed to compensate for $H_d(s) = -(s + 1000)/2(s + 4000)$. Taking $\alpha = 1$ and $t_0 = 0$, design the equalizer using one of the op-amp circuits in Fig. 11.32 (p. 527) with $R_F = 10$ kΩ.

CHAPTER 13

LaPlace Transform Analysis

This chapter considers the broad class of dynamic analysis problems described as follows:

> Given a circuit with some initial state at $t = 0^-$ and an excitation $x(t)$ starting at $t = 0$, find the resulting response of any variable $y(t)$ for $t \geq 0$.

The solution thus includes the transient behavior of $y(t)$ as well as the steady-state response, if one exists. The ability to solve such problems is vital because it helps design engineers predict how circuits will perform in robotic devices, digital computers, and a host of other important applications.

Theoretically, we could tackle circuit dynamics using "classical" time-domain analysis of differential equations. However, as seen in Chapter 9, the time-domain method involves formulating the differential equation, finding the forced response produced by the excitation, and determining the initial-condition constants — any or all of which may be difficult tasks. Here, we'll bypass those troublesome aspects with the help of the Laplace transform.

Figure 13.1 schematically portrays the strategy of Laplace transform analysis, as compared to time-domain analysis. The gist of Laplace analysis follows the same lines as impedance analysis, in that we go to the s domain and work with algebraic equations rather than differential equations. But in place of phasors, we now use Laplace transforms to represent the excitation, to incorporate initial conditions, and to obtain algebraic circuit relations. Finally, after solving the s-domain equations, we return to the time domain via the inverse Laplace transform.

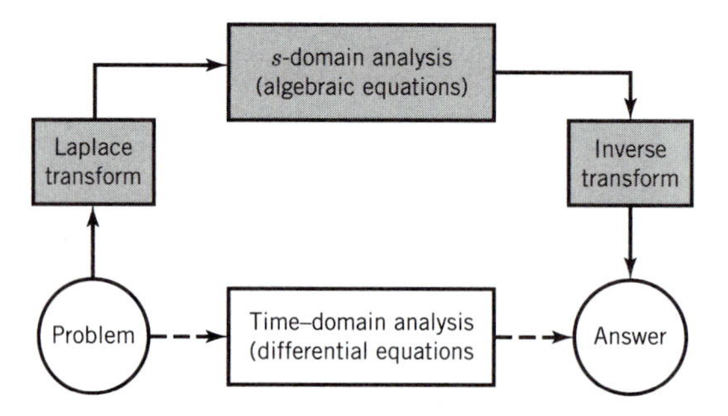

Figure 13.1 Laplace analysis compared with time-domain analysis

Although this strategy appears to be more roundabout than time-domain analysis, it avoids many problems associated with the formulation and solution of differential circuit questions. Additionally, Laplace transforms allow us to handle a much larger group of excitation functions, covering almost all waveforms of practical interest. Extensive tables of Laplace transforms have been compiled for this very purpose.

To develop the transform method, this chapter begins with the basic properties of Laplace transforms and the techniques for transform inversion. We then apply transforms to dynamic circuit analysis, including initial stored energy, the zero-input response, and the zero-state response. The closing section introduces another method for the zero-state response using impulses and convolution, which play a key role in the study of linear systems as well as linear circuits.

Objectives

After studying this chapter and working the exercises, you should be able to do each of the following:

1. Write the defining integral for the Laplace transform, and calculate transforms of simple functions (Section 13.1).

2. Use transform properties to obtain the transform of a function written in terms of simple functions (Section 13.1).
3. Evaluate the coefficients of the partial-fraction expansion to find the inverse Laplace transform of a function containing all distinct poles (Section 13.2).
4. Find the inverse Laplace transform of a function that includes one pair of repeated or complex-conjugate poles (Section 13.2).
5. State and apply the initial-value and final-value theorems (Section 13.2).
6. Use Laplace transforms to calculate the step response or zero-state ac response of a second-order circuit (Section 13.3).
7. Represent initial stored energy in the s-domain to obtain the zero-input response of a circuit with one or two energy-storage elements (Section 13.3).
8. Find the complete response of a first- or second-order circuit whose input has a known Laplace transform (Section 13.3).
9. Apply transform analysis to obtain the transient response of a circuit with mutual inductance (Section 13.4).†
10. State the basic properties of the unit impulse (Section 13.5).†
11. Calculate direct and inverse Laplace transforms involving impulses (Section 13.5).†
12. Find the impulse response of a first- or second-order circuit, and express the zero-state response as a convolution (Section 13.5).†

13.1 LAPLACE TRANSFORMS

The Laplace transform bears the name of the French mathematician Pierre Simon Laplace (1749–1827), who did the theoretical background work. But credit should also go to the English scientist/engineer Oliver Heaviside (1850–1925), whose controversial "operational calculus" pioneered the application of transform methods to engineering problems.

This section states the definition of the Laplace transform and develops basic transform properties. Then we'll show how the Laplace transform can be applied to solve differential equations governing circuit dynamics.

Definition

The complex-frequency variable $s = \sigma + j\omega$ was introduced in Chapter 10 to allow the use of compact phasor notation for a large class of time functions. Specifically, any waveform such as $x(t) = X_m e^{\sigma t} \cos(\omega t + \phi_x)$ could be written as $x(t) = \text{Re}[\underline{X}e^{st}]$ where $\underline{X} = X_m \angle \phi_x$. In essence, we thereby converted the time function $x(t)$ into the s-domain phasor $\underline{X}$.

The Laplace transform also converts time functions into s-domain entities, but with three crucial differences:

- The Laplace transform holds only for $t \geq 0$, and all previous effects are taken into account via the initial conditions at $t = 0^-$.
- The time function being transformed may have almost any reasonable behavior for $t \geq 0$.
- The resulting transform is a function of the variable s.

We'll subsequently see that these properties are ideally suited to dynamic circuit analysis.

For generality of notation, let $f(t)$ represent any circuit time function of interest, be it an excitation or a response. The **Laplace transform** of $f(t)$ is an s-domain function denoted by $F(s)$ or $\mathcal{L}[f(t)]$ and defined by

$$F(s) = \mathcal{L}[f(t)] \triangleq \int_{0^-}^{\infty} f(t)e^{-st}\,dt \tag{13.1}$$

Equation (13.1) is often called the **unilateral** or **one-sided Laplace transform** because it ignores the behavior of $f(t)$ for all $t < 0$. The symbol $\mathcal{L}[f(t)]$ stands for the integration on the right-hand side that produces $F(s)$.

We take the lower limit of the transform integral at $t = 0^-$ to facilitate the treatment of initial conditions. This choice may appear to raise problems when $f(t)$ undergoes a jump discontinuity at $t = 0$ so $f(0^+) \neq f(0^-)$, as illustrated by Fig. 13.2*a*. However, with or without discontinuities, most waveforms have the property that

$$|f(t)| < \infty \qquad 0^- \leq t \leq 0^+ \tag{13.2a}$$

The transform integrand $f(t)e^{-st}$ is then finite over the infinitesimal range $0^- \leq t \leq 0^+$, and Eq. (13.1) becomes

$$\begin{aligned}\mathcal{L}[f(t)] &= \int_{0^-}^{0^+} f(t)e^{-st}\,dt + \int_{0^+}^{\infty} f(t)e^{-st}\,dt \\ &= 0 + \int_{0^+}^{\infty} f(t)e^{-st}\,dt \end{aligned} \tag{13.2b}$$

Hence, the lower limit often may be replaced by 0^+.

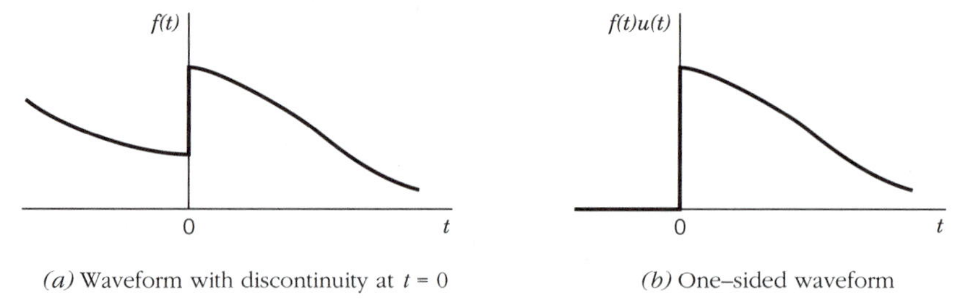

(*a*) Waveform with discontinuity at $t = 0$ (*b*) One–sided waveform

Figure 13.2

To emphasize the "one-sided" nature of the Laplace transform, consider the product $f(t)u(t)$ in Fig. 13.2*b*. We previously defined the **unit step** $u(t)$ as

$$\begin{aligned} u(t) &= 0 \qquad t < 0 \\ &= 1 \qquad t > 0 \end{aligned}$$

so

$$\begin{aligned} f(t)u(t) &= 0 \qquad t < 0 \\ &= f(t) \qquad t > 0 \end{aligned}$$

Correspondingly, the transform of $f(t)u(t)$ is

$$\mathcal{L}[f(t)u(t)] = \int_{0^-}^{\infty} f(t)u(t)e^{-st}\,dt = \int_{0^+}^{\infty} f(t)e^{-st}\,dt$$

We thus see that if $|f(t)| < \infty$ for $0^- \leq t \leq 0^+$, then

$$\mathcal{L}[f(t)u(t)] = \mathcal{L}[f(t)] \tag{13.3}$$

The major exception to Eq. (13.3) occurs when $f(t)$ includes the *impulse* function considered later on in Section 13.5.

Although s appears as a variable in $F(s) = \mathcal{L}[f(t)]$, we hold s constant when performing the integration with respect to t. This is an important feature because the transform of $f(t)$ exists only when the area under $f(t)e^{-st}$ remains finite as $t \to \infty$. A sufficient **existence condition** for the Laplace transform of $f(t)$ is obtained by noting that $f(t)e^{-st} \leq |f(t)e^{-st}| = |f(t)|\,|e^{-st}|$ and that $|e^{-st}| = |e^{-\sigma t}e^{-j\omega t}| = e^{-\sigma t}$. Hence, the area under $f(t)e^{-st}$ remains finite if

$$\lim_{t\to\infty} |f(t)|e^{-\sigma t} = 0 \qquad \sigma > \sigma_c \tag{13.4}$$

where σ_c is some finite real number. We then say that $f(t)$ is of **exponential order**, and σ_c is called the **abscissa of convergence** for the Laplace transform.

Virtually all waveforms of practical interest satisfy Eq. (13.4). Indeed, any function will be of exponential order when $|f(t)| < \infty$ for all $t \geq 0$. Even so, we sometimes encounter cases in which $|f(t)| \to \infty$ as $t \to \infty$. We must then make sure that Eq. (13.4) holds with finite σ_c.

The abscissa of convergence also comes into play in the **inverse Laplace transform**, symbolized by $\mathcal{L}^{-1}[F(s)]$. This operation recovers $f(t)$ from $F(s)$ in the sense that

$$\mathcal{L}^{-1}[F(s)] = f(t) \qquad t \geq 0$$

Formally, the inverse transform is defined as the s-plane integration

$$\mathcal{L}^{-1}[F(s)] \triangleq \frac{1}{2\pi j}\int_{c-j\omega}^{c+j\omega} F(s)e^{st}\,ds \qquad c > \sigma_c \tag{13.5}$$

Since $F(s)$ exists only for $\text{Re}[s] > \sigma_c$, the integration must be carried out along a path within the shaded **region of convergence** illustrated in Fig. 13.3. Although Eq. (13.5) is essential for the theory of Laplace transforms, we seldom use it in circuit analysis. Instead, we'll subsequently develop a more convenient inversion method that draws upon known transforms.

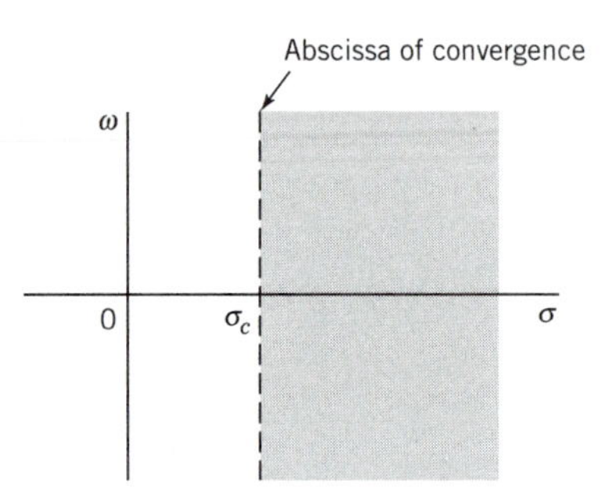

Figure 13.3 Region of convergence.

Example 13.1 *Transforms of* e^{-at} and $u(t)$

Consider the exponential function $f(t) = e^{-at}$, where a has an arbitrary value. This function decays with time when $a > 0$, but it grows without bound when $a < 0$. Nonetheless, $f(t)$ satisfies Eq. (13.4) with *any* value of a because

$$\lim_{t\to\infty} |f(t)|e^{-\sigma t} = \lim_{t\to\infty} e^{-(a+\sigma)t} = 0 \quad \text{if} \quad a + \sigma > 0$$

Hence, the Laplace transform of e^{-at} exists for $\sigma > -a$ and the abscissa of convergence is $\sigma_c = -a$.

To find the transform, we insert $f(t)$ into Eq. (13.1) and perform the integration with fixed s, giving

$$\begin{aligned}\mathcal{L}[e^{-at}] &= \int_{0^-}^{\infty} e^{-at}e^{-st}\,dt = \int_{0^-}^{\infty} e^{-(a+s)t}\,dt \\ &= \left.\frac{e^{-(a+s)t}}{-(a+s)}\right|_{t=0^-}^{t=\infty} = \frac{1}{s+a}\left[1 - \lim_{t\to\infty} e^{-(a+s)t}\right]\end{aligned}$$

But we have established that $e^{-(a+s)t}$ vanishes at $t \to \infty$ when $\text{Re}[s] > \sigma_c = -a$. Therefore,

$$\mathcal{L}[e^{-at}] = \frac{1}{s+a} \tag{13.6}$$

where a may be positive or negative.

Setting $a = 0$ in the foregoing result immediately yields the Laplace transform of a *unit constant*, namely,

$$\mathcal{L}[1] = 1/s \tag{13.7a}$$

Equation (13.7a) may also be written as

$$\mathcal{L}[u(t)] = 1/s \tag{13.7b}$$

which follows from Eqs. (13.3) and (13.6) with $f(t) = = e^{-0t} = 1$.

Example 13.2 *Transform of sin βt*

Now let $f(t) = \sin \beta t$, where β symbolizes a specific frequency as distinguished from the s-domain variable $\omega = \text{Im}[s]$. Since $|\sin \beta t| \leq 1$ for all t, this function is clearly of exponential order and $\sigma_c = 0$.

The Laplace transform is easily found via the exponential expression $\sin \beta t = (e^{j\beta t} - e^{-j\beta t})/j2$. Equation (13.1) then gives

$$\begin{aligned}\mathcal{L}[\sin \beta t] &= \int_{0^-}^{\infty} \frac{1}{j2}(e^{j\beta t} - e^{-j\beta t})e^{-st}\, dt \\ &= \frac{1}{j2}\left[\int_{0^-}^{\infty} e^{-(s-j\beta)t}\, dt - \int_{0^-}^{\infty} e^{-(s+j\beta)t}\, dt\right] \\ &= \frac{1}{j2}\left(\frac{1}{s - j\beta} - \frac{1}{s + j\beta}\right) = \frac{\beta}{s^2 + \beta^2}\end{aligned} \tag{13.8}$$

Exercise 13.1

Consider the **ramp function** $f(t) = t$, which satisfies Eq. (13.4) with $\sigma_c = 0$. Use Eq. (13.1) to show that

$$\mathcal{L}[t] = 1/s^2$$

Transform Properties

Many time functions that occur in circuit analysis may be viewed as the result of performing various operations on other, simpler time functions. This viewpoint often has practical value as a means of finding Laplace transforms without resorting to brute-force integration. We therefore examine here several important time-domain operations and the corresponding properties of the Laplace transform.

Linear Combination. Let $f(t)$ and $g(t)$ be arbitrary time functions with transforms $\mathcal{L}[f(t)] = F(s)$ and $\mathcal{L}[g(t)] = G(s)$. We seek the transform of the linear combination $Af(t) + Bg(t)$, where A and B are constants. From Eq. (13.1),

$$\mathcal{L}[Af(t) + Bg(t)] = \int_{0^-}^{\infty} [Af(t) + Bg(t)]e^{-st}\, dt$$
$$= A\int_{0^-}^{\infty} f(t)e^{-st}\, dt + B\int_{0^-}^{\infty} g(t)e^{-st}\, dt$$

so

$$\mathcal{L}[Af(t) + Bg(t)] = AF(s) + BG(s) \tag{13.9}$$

Thus, a linear combination in the time domain transforms to a linear combination in the s domain. As a special case of Eq. (13.9) we see that

$$\mathcal{L}[Af(t)] = AF(s) = A\mathcal{L}[f(t)]$$

which follows by taking $B = 0$.

Multiplication by e^{-at}. When $e^{-at}f(t)$ replaces $f(t)$ in Eq. (13.1) we have

$$\int_{0^-}^{\infty} e^{-at}f(t)e^{-st}\, dt = \int_{0^-}^{\infty} f(t)e^{-(s+a)t}\, dt = F(s + a)$$

where $F(s) = \mathcal{L}[f(t)]$. Thus,

$$\mathcal{L}[e^{-at}f(t)] = F(s + a) \tag{13.10}$$

so multiplication by e^{-at} corresponds to replacing every s in $F(s)$ by $s + a$.

Multiplication by t. In contrast to Eq. (13.10), multiplying $f(t)$ by t corresponds to differentiating $-F(s)$, in that

$$\mathcal{L}[tf(t)] = -\frac{d}{ds}F(s) \tag{13.11}$$

Equation (13.11) is verified by noting that

$$\frac{d}{ds}F(s) = \frac{d}{ds}\int_{0^-}^{\infty} f(t)e^{-st}\, dt = \int_{0^-}^{\infty} f(t)\frac{de^{-st}}{ds}\, dt$$
$$= \int_{0^-}^{\infty} f(t)(-t)e^{-st}\, dt = -\int_{0^-}^{\infty} tf(t)\, e^{-st}\, dt$$

so $dF(s)/ds = -\mathcal{L}[tf(t)]$.

Time Delay. Since the Laplace transform ignores the behavior of $f(t)$ for $t < 0$, the time-delayed function $f(t - t_0)$ would be undefined for $t < t_0$. However, we can consider the delayed version of $f(t)u(t)$ illustrated in Fig. 13.4 where

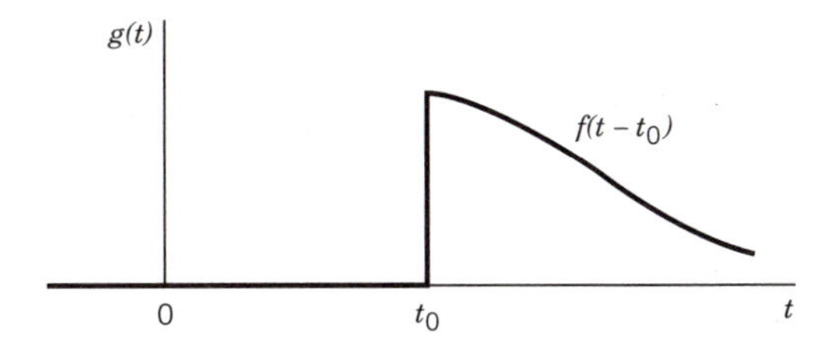

Figure 13.4 Time-delayed, one-sided waveform.

$$g(t) = f(t - t_0)u(t - t_0) = \begin{cases} 0 & t < t_0 \\ f(t - t_0) & t > t_0 \end{cases}$$

The resulting transform is found by making the temporary change of variable $\lambda = t - t_0$ to get

$$\mathcal{L}[g(t)] = \int_{t_0^+}^{\infty} f(t - t_0)e^{-st}\, dt = \int_{0^+}^{\infty} f(\lambda)e^{-s(\lambda + t_0)}\, d\lambda$$

$$= e^{-st_0} \int_{0^+}^{\infty} f(\lambda)e^{-s\lambda}\, d\lambda = e^{-st_0} \int_{0^+}^{\infty} f(t)e^{-st}\, dt$$

Thus,

$$\mathcal{L}[f(t - t_0)u(t - t_0)] = e^{-st_0}\, F(s) \tag{13.12}$$

where $F(s) = \mathcal{L}[f(t)u(t)]$.

Example 13.1-3 *Transform of a Rectangular Pulse*

Consider the rectangular pulse plotted in Fig. 13.5*a*, where

$$\begin{aligned} f(t) &= 1 \qquad 0 < t < D \\ &= 0 \qquad \text{otherwise} \end{aligned}$$

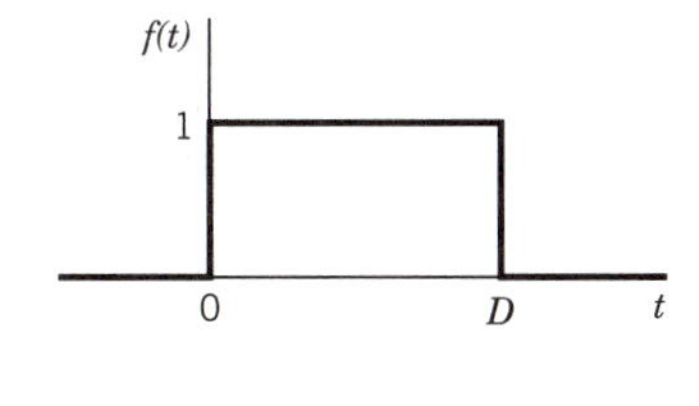

(*a*) Rectangular pulse

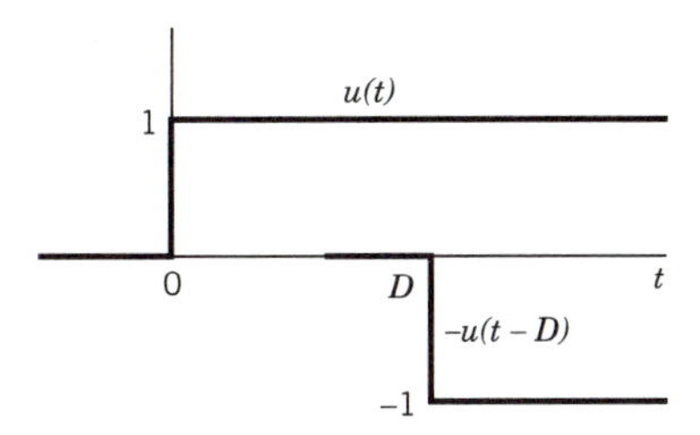

(*b*) Decomposition as two step functions

Figure 13.5

For transform purposes, we can decompose this waveform into an upward step at $t = 0$ plus a downward step at $t = D$, as shown in Fig. 13.5*b*. Thus,

$$f(t) = u(t) - u(t - D) = u(t) + (-1)u(t - D)$$

which is a linear combination with time delay. Since $\mathcal{L}[u(t)] = 1/s$, the transform of $f(t)$ is

$$F(s) = \frac{1}{s} + (-1)e^{-sD}\frac{1}{s} = \frac{1 - e^{-sD}}{s}$$

Exercise 13.2

Show that $\mathcal{L}[te^{-at}] = 1/(s + a)^2$ by taking $f(t) = t$ in Eq. (13.10) and using the result of Exercise 13.1.

Exercise 13.3

Rework Exercise 13.2 by letting $f(t) = e^{-at}$ in Eq. (13.11).

Differentiation and Integration. The Laplace transform of the time derivative $f'(t) = df/dt$ is obtained from Eq. (13.1) using integration by parts, as follows:

$$\begin{aligned}\mathcal{L}[f'(t)] &= \int_{0^-}^{\infty} \frac{df(t)}{dt} e^{-st}\, dt \\ &= f(t)e^{-st}\Big|_{0^-}^{\infty} - \int_{0^-}^{\infty} f(t)\frac{de^{-st}}{dt}\, dt \\ &= -f(0^-) + s\int_{0^-}^{\infty} f(t)e^{-st}\, dt\end{aligned}$$

After rearranging terms we have

$$\mathcal{L}[f'(t)] = sF(s) - f(0^-) \tag{13.13a}$$

Hence, differentiating $f(t)$ corresponds to multiplying $F(s)$ by s and subtracting the initial value $f(0^-)$. Repeating this process for the second derivative yields

$$\mathcal{L}[f''(t)] = s^2F(s) - sf(0^-) - f'(0^-) \tag{13.13b}$$

and so forth for higher derivatives.

Finally, suppose we generate another new time function $g(t)$ from $f(t)$ via the integration operation

$$g(t) = \int_{0^-}^{t} f(\lambda)\, d\lambda$$

The transform of $g(t)$ again involves integration by parts, noting that

$$dg = f(t)\, dt \qquad g(0^-) = 0 \qquad e^{-st}\, dt = d\left(\frac{e^{-st}}{-s}\right)$$

Thus,

$$\begin{aligned}
\mathcal{L}[g(t)] &= \int_{0^-}^{\infty} g(t)\, e^{-st}\, dt = \int_{0^-}^{t} g(t)\, d\left(\frac{e^{-st}}{-s}\right) \\
&= g(t)\left(\frac{e^{-st}}{-s}\right)\Bigg|_{t=0^-}^{t=\infty} - \int_{0^-}^{\infty} \frac{e^{-st}}{-s}\, dg \\
&= [0 - 0] + \frac{1}{s}\int_{0^-}^{\infty} e^{-st} f(t)\, dt
\end{aligned}$$

We thereby obtain the simple result that

$$\mathcal{L}\left[\int_{0^-}^{t} f(\lambda)\, d\lambda\right] = \frac{1}{s}\, F(s) \tag{13.14}$$

Notice that integration corresponds to *dividing* $F(s)$ by s, whereas differentiation corresponds to *multiplying* $F(s)$ by s if $f(0^-) = 0$.

For convenient reference, Table 13.1 summarizes the properties we have discussed here. Application of appropriate transform properties to known Laplace transforms gives the additional transforms listed in Table 13.2. We call this listing a table of **Laplace transform pairs** to reflect the fact that if $\mathcal{L}[f(t)] = F(s)$, then $\mathcal{L}^{-1}[F(s)] = f(t)$ for $t \geq 0$. You should familiarize yourself with these basic transform pairs because they will be used extensively in the rest of this chapter.

TABLE 13.1 Laplace Transform Properties

Operation	Time Function	Laplace Transform
Linear combination	$Af(t) + Bg(t)$	$AF(s) + BG(s)$
Multiplication by e^{-at}	$e^{-at} f(t)$	$F(s + a)$
Multiplication by t	$tf(t)$	$-dF(s)/ds$
Time delay	$f(t - t_0)u(t-t_0)$	$e^{-st_0}F(s)$
Differentiation	$f'(t)$	$sF(s) - f(0^-)$
	$f''(t)$	$s^2F(s) - sf(0^-) - f'(0^-)$
Integration	$\int_{0^-}^{t} f(\lambda)\, d\lambda$	$\frac{1}{s}\, F(s)$

TABLE 13.2 Laplace Transform Pairs

$f(t)$	$F(s)$
A	$\dfrac{A}{s}$
$u(t) - u(t - D)$	$\dfrac{1 - e^{-sD}}{s}$
t	$\dfrac{1}{s^2}$
t^r	$\dfrac{r!}{s^{r+1}}$
e^{-at}	$\dfrac{1}{s + a}$
te^{-at}	$\dfrac{1}{(s + a)^2}$
$t^r e^{-at}$	$\dfrac{r!}{(s + a)^{r+1}}$
$\sin \beta t$	$\dfrac{\beta}{s^2 + \beta^2}$
$\cos (\beta t + \phi)$	$\dfrac{s \cos \phi - \beta \sin \phi}{s^2 + \beta^2}$
$e^{-at} \cos (\beta t + \phi)$	$\dfrac{(s + a) \cos \phi - \beta \sin \phi}{(s + a)^2 + \beta^2}$

Example 13.4 ***Transforms of Cosine Functions***

Given $\mathcal{L}[\sin \beta t] = \beta/(s^2 + \beta^2)$ from Example 13.2, we'll apply transform properties to derive the Laplace transforms of the functions $\cos \beta t$ and $\cos (\beta t + \phi)$.

First, let $f(t) = \sin \beta t$, so $f'(t) = \beta \cos \beta t$ and the differentiation property yields

$$\mathcal{L}[f'(t)] = s \frac{\beta}{s^2 + \beta^2} - \sin (0^-) = \frac{s\beta}{s^2 + \beta^2}$$

Since $\mathcal{L}[f'(t)] = \mathcal{L}[\beta \cos \beta t] = \beta \mathcal{L}[\cos \beta t]$, we now have

$$\mathcal{L}[\cos \beta t] = \frac{s}{s^2 + \beta^2}$$

Next, we invoke the expansion

$$\cos (\beta t + \phi) = \cos \phi \cos \beta t - \sin \phi \sin \beta t$$

from which

$$\mathcal{L}[\cos(\beta t + \phi)] = \cos\phi \frac{s}{s^2 + \beta^2} - \sin\phi \frac{\beta}{s^2 + \beta^2}$$
$$= \frac{s\cos\phi - \beta\sin\phi}{s^2 + \beta^2}$$

Exercise 13.4

Let $F(s) = 1/(s + a)$ in Eq. (13.14) to show that

$$\mathcal{L}^{-1}\left[\frac{1}{s(s+a)}\right] = -\frac{1}{a}(e^{-at} - 1)$$

Solution of Differential Equations

Our motivation for studying the Laplace transform stems from the value of transformation as an efficient way of analyzing circuit dynamics. The transform method will be developed in detail in Section 13.3. Here, we present two circuit examples that illustrate the transform solution of linear differential equations with constant coefficients, including initial conditions.

First-Order Example. As a simple introductory case, consider the switched first-order circuit in Fig. 13.6. The switch has been closed for a long time before it opens at $t = 0$, and we want to find the capacitor voltage $v(t)$ for $t > 0$.

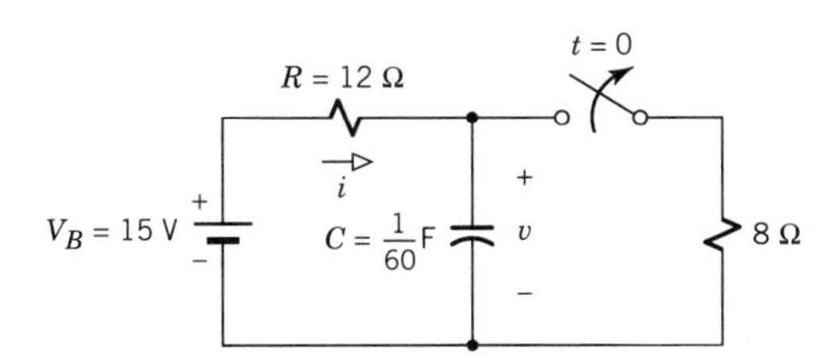

Figure 13.6 Switched first-order circuit.

DC steady-state analysis at $t = 0^-$ gives the initial condition

$$v(0^-) = (8/20)V_B = 6 \text{ V}$$

The behavior for $t > 0$ is then governed by the KVL equation $Ri + v = V_B$ and the element equation $i = Cv'$, which can be combined with the element values as

$$0.2v'(t) + v(t) = 15$$

We must solve this differential equation and incorporate the initial condition to find $v(t)$.

Our first task is to obtain the s-domain function $V(s) = \mathcal{L}[v(t)]$ by transforming the entire differential equation. Since the left-hand side is a linear combination of $v'(t)$ and $v(t)$, we can write

$$0.2\mathcal{L}[v'(t)] + \mathcal{L}[v(t)] = \mathcal{L}[15]$$

But $\mathcal{L}[v'(t)] = s\mathcal{L}[v(t)] - v(0^-) = sV(s) - 6$, and the constant on the right transforms as $\mathcal{L}[15] = 15/s$, so

$$0.2[sV(s) - 6] + V(s) = 15/s$$

Algebraic manipulation then yields

$$V(s) = \frac{1.2 + 15/s}{0.2s + 1} = \frac{6s + 75}{s(s + 5)} = \frac{6}{s + 5} + \frac{75}{s(s + 5)}$$

which is the desired s-domain function.

Our next task is to find the inverse transform $\mathcal{L}^{-1}[V(s)] = v(t)$, which we'll do with the help of known transform properties. To that end, let

$$F_1(s) = \frac{1}{s + 5} \qquad F_2(s) = \frac{1}{s(s + 5)}$$

so

$$V(s) = 6F_1(s) + 75F_2(s)$$

Then

$$v(t) = \mathcal{L}^{-1}[V(s)] = 6\mathcal{L}^{-1}[F_1(s)] + 75\mathcal{L}^{-1}[F_2(s)]$$

where, from Table 13.2 and Exercise 13.4,

$$\mathcal{L}^{-1}[F_1(s)] = e^{-5t} \qquad \mathcal{L}^{-1}[F_2(s)] = -0.2(e^{-5t} - 1)$$

Thus,

$$v(t) = 6e^{-5t} + 75(-0.2)(e^{-5t} - 1) = 15 - 9e^{-5t} \qquad t \geq 0$$

As expected, $v(t)$ has initial value $v(0^+) = 6\text{ V} = v(0^-)$ and steady-state value $v(\infty) = 15\text{ V} = V_B$. Furthermore, the transient part of $v(t)$ is a decaying exponential with time constant $\tau = 1/5 = RC$.

Second-Order Example. The second-order circuit in Fig. 13.7 poses a more challenging problem. We'll seek the inductor current $i(t)$ for $t > 0$, given the switched dc input current

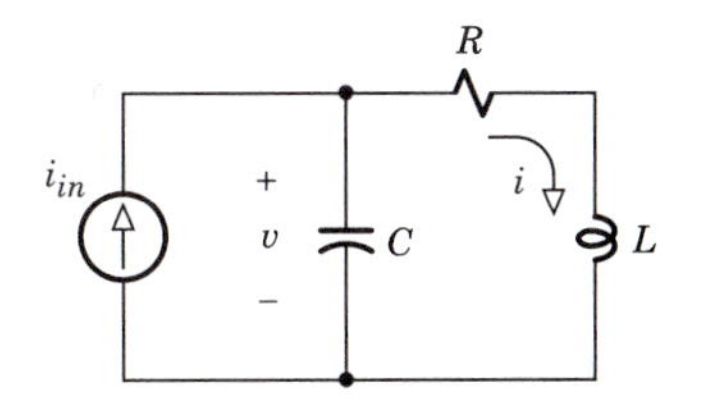

Figure 13.7 Second-order circuit.

$$\begin{aligned} i_{in}(t) &= I_1 \qquad t < 0 \\ &= I_2 \qquad t > 0 \end{aligned}$$

Routine analysis yields the initial conditions

$$i(0^-) = I_1 \qquad v(0^-) = RI_1$$

and the behavior for $t > 0$ is governed by the coupled first-order equations

$$Li'(t) + Ri(t) = v(t) \qquad Cv'(t) + i(t) = I_2$$

Instead of combining these two differential equations, we'll transform them individually to get

$$\begin{aligned} L[sI(s) - I_1] + RI(s) &= V(s) \\ C[sV(s) - RI_1] + I(s) &= I_2/s \end{aligned}$$

where $I(s) = \mathcal{L}[i(t)]$ and $V(s) = \mathcal{L}[v(t)]$. Now we can eliminate $V(s)$ algebraically and solve for $I(s)$ the result being

$$I(s) = \frac{LCsI_1 + RCI_1 + I_2/s}{LCs^2 + RCs + 1} = \frac{I_1 s^2 + (R/L)I_1 s + I_2/LC}{s[s^2 + (R/L)s + 1/LC]}$$

The inverse transform of $I(s)$ involves more than direct table look-ups, so we'll stop here with two observations. First, the denominator of $I(s)$ contains the *characteristic polynomial* $P(s) = s^2 + (R/L)s + 1/LC$. Second, if $I_2 = I_1$, then $I(s) = I_1/s$ and $i(t) = \mathcal{L}^{-1}[I_1/s] = I_1$ for $t \geq 0$ — in agreement with the fact that no switching actually occurs when $I_2 = I_1$.

By way of a summary, our two examples have illustrated the salient features of solving circuit differential equations with the Laplace transform, to wit:

- Transformation automatically incorporates the initial *conditions* via $\mathcal{L}[f'(t)] = sF(s) - f(0^-)$.
- Transformation converts linear *differential* equations into s-domain *algebraic* equations suitable for further manipulation.

- Transformation produces an s-domain function whose denominator includes the *characteristic polynomial.*
- Inverse transformation is needed to go from the s-domain result to the corresponding time function.

We therefore devote the entire next section to techniques for transform inversion.

Exercise 13.5

Find $Y(s)$ given that

$$y'' + 20y' + 100y = 0 \qquad y(0^-) = 0 \quad y'(0^-) = 3$$

Then find $y(t)$ from Table 13.2.

13.2 TRANSFORM INVERSION

Transform inversion is the process of finding the time function $f(t)$ corresponding to a given Laplace transform $F(s)$. As previously stated, the formal expression for $\mathcal{L}^{-1}[F(s)]$ requires integration in the s-plane. Fortunately, however, a simpler method known as *partial-fraction expansion* proves to be sufficient for almost all applications in circuit analysis.

The basic strategy of partial-fraction expansion is to express $F(s)$ as a sum of simple functions whose inverse transforms are known. We'll pursue that strategy here, concentrating primarily on problems of the type associated with first- and second-order circuits.

When higher order circuits are involved, the algebra of partial-fraction expansion often becomes a tiresome chore. Those problems are better handled by a computer program such as Maple, Mathematica, or Matlab that carry out partial-fraction expansions or even complete inverse transforms.

The treatment in this section will prepare you to use computer programs intelligently as well as to do simpler inversions by hand. We'll also discuss the *initial-value* and *final-value theorems* that provide valuable information about $f(t)$ without performing transform inversion.

Partial-Fraction Expansions

When we apply Laplace transforms to investigate circuit dynamics, the s-domain result can usually be written as a ratio of polynomials like

$$F(s) = \frac{N(s)}{D(s)} = \frac{b_m s^m + b_{m-1}s^{m-1} + \cdots + b_1 s + b_0}{s^n + a_{n-1}s^{n-1} + \cdots + a_1 s + a_0} \tag{13.15}$$

Such an expression is known as a **rational function** of s. A rational function is **strictly proper** when

$$m \leq n - 1 \tag{13.16}$$

so the order of the numerator polynomial $N(s)$ is less than the order of the denominator polynomial $D(s)$. This condition will be assumed throughout.

Given a strictly proper rational function $F(s)$, our goal is to expand it as a sum of terms whose individual inverse transforms are known. Accordingly, from a review of Table 13.2, we seek terms proportional to $1/s$, $1/(s + a)$, $1/(s^2 + \beta^2)$, and so forth. The name **partial fraction** aptly describes these terms.

The first expansion step requires factoring the denominator polynomial to obtain

$$D(s) = (s - s_1)(s - s_2) \cdots (s - s_n) \tag{13.17}$$

where $s_1, s_2, \ldots s_n$ are the roots of $D(s)$. We'll refer to these roots as the **poles** of $F(s)$ because $F(s)$ has the same mathematical structure as a network function $H(s)$ with poles $p_1 = s_1$, $p_2 = s_2$, etc. The inversion of $F(s)$ becomes somewhat more complicated when it includes complex or repeated poles, so we focus initially on the case of **distinct real poles**.

When all of the poles of $F(s)$ are distinct, it can be expanded as a sum of partial fractions in the form

$$F(s) = \frac{N(s)}{D(s)} = \frac{A_1}{s - s_1} + \frac{A_2}{s - s_2} + \cdots + \frac{A_n}{s - s_n} \tag{13.18}$$

The constants A_1 through A_n are called the **residues** of $F(s)$ at $s = s_1$ through s_n, respectively. The validity of our expansion rests upon the fact that if we converted the right-hand side of Eq. (13.18) into a ratio of polynomials, then the resulting denominator would equal $D(s)$ and the order of the resulting numerator would be $m \leq n - 1$.

But we kept the partial fractions separate in Eq. (13.18) to take advantage of the known transform $\mathcal{L}[1/(s + a)] = e^{-at}$. Hence, by the property of linear combination,

$$f(t) = A_1 e^{s_1 t} + A_2 e^{s_2 t} + \cdots + A_n e^{s_n t} \qquad t \geq 0 \tag{13.19}$$

The inversion of $F(s)$ therefore boils down to finding the residues $A_1, A_2, \ldots A_n$. Of the several techniques for this task, the most convenient approach involves isolating residues one by one.

Consider, for example, a second-order function and its expansion

$$F(s) = \frac{N(s)}{(s - s_1)(s - s_2)} = \frac{A_1}{s - s_1} + \frac{A_2}{s - s_2}$$

We isolate A_1 by multiplying by $s - s_1$ to get

$$(s - s_1)F(s) = \frac{N(s)}{s - s_2} = A_1 + \frac{s - s_1}{s - s_2}A_2$$

Setting $s = s_1$ then eliminates the term involving A_2 and leaves

$$A_1 = \frac{N(s_1)}{s_1 - s_2} = (s - s_1)F(s)\big|_{s=s_1}$$

Similarly, to isolate A_2, we multiply $F(s)$ by $(s - s_2)$ and set $s = s_2$.

Returning to the nth-order function in Eq. (13.18), we can determine each of the residues of distinct poles via

$$A_i = (s - s_i)F(s)\big|_{s=s_i} \qquad i = 1, 2, \ldots \tag{13.20}$$

Equation (13.20) bears the name **Heaviside's theorem**, but it's more frequently called the **cover-up rule** because multiplying $F(s)$ by $(s - s_i)$ has the same effect as "covering up" $(s - s_i)$ in the factored denominator.

Example 13.5 ***Inversion of a Third-Order Function***

We previously found that the inductor current back in Fig. 13.7 was described by the s-domain function

$$I(s) = \frac{I_1 s^2 + (R/L)I_1 s + I_2/LC}{s[s^2 + (R/L)s + 1/LC]}$$

Now we'll use partial-fraction expansion to obtain $i(t)$ when

$$R = 12\ \Omega \qquad L = 1\text{ H} \qquad C = 1/20\text{ F} \qquad I_1 = -2\text{ A} \qquad I_2 = 2\text{ A}$$

so

$$I(s) = \frac{-2s^2 - 24s + 40}{s(s^2 + 12s + 20)} = \frac{-2s^2 - 24s + 40}{s(s + 2)(s + 10)}$$

We begin by observing that $I(s)$ is a strictly proper rational function with $m = 2$ and $n = 3$. Furthermore, the poles are real and distinct, namely, $s_1 = 0$, $s_2 = -2$, and $s_3 = -10$. The partial-fraction expansion therefore takes the form

$$I(s) = \frac{A_1}{s - s_1} + \frac{A_2}{s - s_2} + \frac{A_3}{s - s_3} = \frac{A_1}{s} + \frac{A_2}{s + 2} + \frac{A_3}{s + 10}$$

Applying the cover-up rule from Eq. (13.20) yields the residues

$$A_1 = sI(s)|_{s=0} = \left.\frac{-2s^2 - 24s + 40}{(s+2)(s+10)}\right|_{s=0} = 2$$

$$A_2 = (s+2)I(s)|_{s=-2} = \left.\frac{-2s^2 - 24s + 40}{s(s+10)}\right|_{s=-2} = -5$$

$$A_3 = (s+10)I(s)|_{s=-10} = \left.\frac{-2s^2 - 24s + 40}{s(s+2)}\right|_{s=-10} = 1$$

To check these values, we combine the partial fractions as the ratio of polynomials

$$\begin{aligned}\frac{A_1}{s} + \frac{A_1}{s+2} + \frac{A_1}{s+10} &= \frac{2}{s} + \frac{-5}{s+2} + \frac{1}{s+10} \\ &= \frac{2(s+2)(s+10) - 5s(s+10) + s(s+10)}{s(s+2)(s+10)} \\ &= \frac{-2s^2 - 24s + 40}{s(s+2)(s+10)} = I(s)\end{aligned}$$

which confirms the validity of our expansion.

Finally, after inserting values for the poles and residues, the inverse transform of $I(s)$ is

$$i(t) = A_1 e^{s_1 t} + A_2 e^{s_2 t} + A_3 e^{s_3 t} = 2 - 5e^{-2t} + e^{-10t}$$

which holds for $t \geq 0$.

Exercise 13.6

Find $f(t)$ when $F(s) = (s - 23)/(s^2 + 3s - 10)$.

Complex Poles

Complex poles always occur in conjugate pairs in the Laplace transform of a real function of time. Partial-fraction expansion remains valid when $F(s)$ has complex poles, but some additional manipulations are needed to put $f(t)$ in a suitable form because it will include one or more sinusoidal components.

To elaborate upon inversion with complex poles, we'll take the denominator of $F(s)$ to be

$$D(s) = (s^2 + 2\alpha s + \omega_0^2)(s - s_3) \cdots (s - s_n)$$

If $\alpha^2 < \omega_0^2$, then the two roots of the quadratic factor are

$$s_1, s_2 = -\alpha \pm j\beta \qquad \beta = \sqrt{\omega_0^2 - \alpha^2} \tag{13.21}$$

Since s_1 and s_2 are complex conjugates, the residues obtained from the cover-up rule will also be complex conjugates.

We therefore write the partial-fraction expansion as

$$F(s) = G(s) + \frac{A_3}{s - s_3} + \cdots + \frac{A_n}{s - s_n} \tag{13.22a}$$

with

$$G(s) = \frac{\frac{1}{2}K}{s + \alpha - j\beta} + \frac{\frac{1}{2}K^*}{s + \alpha + j\beta} \tag{13.22b}$$

Here we have denoted the residue at s_1 by $\frac{1}{2}K$ to simplify further manipulations, so Eq. (13.20) yields

$$K = 2(s + \alpha - j\beta)F(s)\big|_{s=-\alpha+j\beta} \tag{13.23}$$

We don't need a separate equation for the residue at s_2, which equals $\frac{1}{2}K^*$.

Now, assuming that all other poles are real and distinct, the inverse transform of Eq. (13.22a) is

$$f(t) = g(t) + A_3 e^{s_3 t} + \cdots + A_n e^{s_n t}$$

where, from Eq. (13.22b),

$$g(t) = \tfrac{1}{2}Ke^{-(\alpha - j\beta)t} + \tfrac{1}{2}K^* e^{-(\alpha + j\beta)t}$$

We clean up this expression by recognizing that the complex constant K can be put in the polar or exponential notation

$$K = K_m \angle\phi = K_m e^{j\phi} \tag{13.24}$$

Then $K^* = K_m e^{-j\phi}$ and

$$\begin{aligned} g(t) &= \tfrac{1}{2}K_m e^{-\alpha t}[e^{j(\beta t + \omega)} + e^{-j(\beta t + \phi)}] \\ &= K_m e^{-\alpha t} \cos(\beta t + \phi) \end{aligned} \tag{13.25}$$

Thus, a pair of distinct complex poles in $F(s)$ produces a sinusoidal waveform with exponential amplitude variation.

But, returning to Eq. (13.23), we see that the evaluation of K_m and ϕ may entail some tedious complex algebra. An alternative approach follows from the known transform of $g(t)$ in Table 13.2, namely,

$$\mathcal{L}[g(t)] = \frac{(s + \alpha)K_m \cos\phi - \beta K_m \sin\phi}{(s + \alpha)^2 + \beta^2}$$

Accordingly, let's write $G(s)$ as a strictly proper function in the *unfactored* form

$$G(s) = \frac{Bs + C}{s^2 + 2\alpha s + \omega_0^2} \tag{13.26a}$$

$$= \frac{(s + \alpha)B - (\alpha B - C)}{(s + \alpha)^2 + \omega_0^2 - \alpha^2} \tag{13.26b}$$

Since $\omega_0^2 - \alpha^2 = \beta^2$, Eq. (13.26b) equals our expression for $\mathcal{L}[g(t)]$ when $B = K_m \cos \phi$ and $\alpha B - C = \beta K_m \sin \phi$, which can be combined in the complex equation

$$B + j(\alpha B - C)/\beta = K_m \cos \phi + jK_m \sin \phi = K_m e^{j\phi} = K$$

Hence,

$$K = B + j\frac{\alpha B - C}{\beta} \tag{13.27}$$

Compared to Eq. (13.23), Eq. (13.27) is a much easier way of calculating the complex quantity K, provided that we know the values of B and C.

When all or most of the other residues of $F(s)$ can be evaluated by the standard cover-up rule, B and C are most easily found using a variation of the **method of undetermined coefficients**, as follows:

1. Write out the partial-fraction expansion in the desired form.
2. Evaluate all residues that can be found readily from the cover-up rule, and insert those results in the expansion.
3. Evaluate one of the remaining unknowns by manipulating the equation obtained in step 2, using the fact that that this relationship must hold for all values of s. Thus, you may let s be any convenient value (usually 0 or 1) or you may multiply through by s and let $s \to \infty$ to isolate an unknown.
4. Repeat step 3 for each of the remaining unknowns.

The following example illustrates this method.

Example 13.6 ***Inversion with Complex Poles***

Let's find $f(t)$ when

$$F(s) = \frac{15s^2 - 16s - 7}{(s + 2)(s^2 + 6s + 25)}$$

This function has a distinct real pole at $s_1 = -2$, but the quadratic parameters are $\alpha = 3$ and $\omega_0^2 = 25 > \alpha^2$ so

$$\beta = \sqrt{25 - 3^2} = 4 \qquad s_2, s_3 = -3 \pm j4$$

We'll write the partial-fraction expansion as

$$F(s) = \frac{A_1}{s + 2} + G(s)$$

in which

$$G(s) = \frac{\frac{1}{2}K}{s + 3 - j4} + \frac{\frac{1}{2}K^*}{s + 3 + j4} = \frac{Bs + C}{s^2 + 6s + 25}$$

The corresponding time function is $f(t) = 5e^{-2t} + g(t)$, where $g(t)$ has the form of Eq. (13.24).

If we use Eq. (13.23) for K, then we would have to evaluate the complex expression

$$K = 2 \left. \frac{15s^2 - 16s - 7}{(s + 2)(s + 3 + j4)} \right|_{s=-3+j4}$$

Instead, we'll apply the method of undetermined coefficients to find B and C.

Step 1: We write out the expansion as

$$F(s) = \frac{15s^2 - 16s - 7}{(s + 2)(s^2 + 6s + 25)} = \frac{A_1}{s + 2} + \frac{Bs + C}{s^2 + 6s + 25}$$

Step 2: The standard cover-up rule for A_1 yields

$$A_1 = \left. \frac{15s^2 - 16s - 7}{s^2 + 6s + 25} \right|_{s=-2} = 5$$

Thus, we now have

$$\frac{15s^2 - 16s - 7}{(s + 2)(s^2 + 6s + 25)} = \frac{5}{s + 2} + \frac{Bs + C}{s^2 + 6s + 25}$$

Step 3: To isolate and evaluate B, we multiply both sides by s so

$$sF(s) = \frac{15s^3 - \cdots}{s^3 + \cdots} = \frac{5s}{s + 2} + \frac{Bs^2 + Cs}{s^2 + 6s + 25}$$

and letting $s \to \infty$ gives

$$\lim_{s\to\infty} sF(s) = 15 = 5 + B \quad \Rightarrow \quad B = 15 - 5 = 10$$

Step 4: With $B = 10$, the expansion becomes

$$\frac{15s^2 - 16s - 7}{(s+2)(s^2+6s+25)} = \frac{5}{s+2} + \frac{10s + C}{s^2 + 6s + 25}$$

Examination then suggests setting $s = 0$ to get

$$F(0) = \frac{-7}{2 \times 25} = \frac{5}{2} + \frac{C}{25} \quad \Rightarrow \quad C = \frac{-7 - 25 \times 5}{2} = -66$$

Now we can invoke Eq. (13.27) to find the complex quantity

$$K = 10 + j\frac{3 \times 10 - (-66)}{4} = 10 + j24 = 26 \angle 67.4°$$

Thus, Eq. (13.24) becomes

$$g(t) = 26e^{-3t} \cos (4t + 67.4°)$$

and

$$f(t) = 5e^{-2t} + 26e^{-3t} \cos (4t + 67.4°) \qquad t \geq 0$$

Exercise 13.7

Use Eq. (13.27) and the method of undetermined coefficients to find $f(t)$ when

$$F(s) = \frac{s^2 + 14s + 25}{s^3 + 8s^2 + 25s}$$

Repeated Poles Repeated poles at $s = s_i$ arise when the denominator of $F(s)$ has the factor $(s - s_i)^r$. The partial-fraction expansion must then include terms proportional to $1/(s - s_i)^r$, $1/(s - s_i)^{r-1}$, . . . , along with $1/(s - s_i)$. These modified partial fractions are needed so that the sum still consists of n *different* terms.

Suppose, in particular, that $F(s)$ has a **double pole** corresponding to the factor $(s - s_i)^2$. If all other poles are distinct, then we write

$$F(s) = G(s) + \frac{A_3}{s - s_3} + \cdots + \frac{A_n}{s - s_n} \tag{13.28a}$$

with

$$G(s) = \frac{A_{i1}}{s - s_i} + \frac{A_{i2}}{(s - s_i)^2} \tag{13.28b}$$

The inversion of $G(s)$ gives

$$g(t) = A_{i1}e^{s_i t} + A_{i2}te^{s_i t} \tag{13.29}$$

where we have drawn upon the known inverse transform $\mathcal{L}^{-1}[1/(s + a)^2] = te^{-at}$.

Our usual cover-up rule remains valid for the residues at the nonrepeated poles, but not for A_{i1} and A_{i2}. The difficulty here is that the second term in the expansion of $G(s)$ blows up if we multiply it by $s - s_i$ and set $s = s_i$. Instead, we multiply Eq. (13.28a) by $(s - s_i)^2$ and rearrange as

$$(s - s_i)A_{i1} + A_{i2} = (s - s_i)^2F(s) - \frac{(s - s_i)^2}{s - s_3}A_3 - \cdots$$

Setting $s = s_i$ now yields

$$A_{i2} = (s - s_i)^2F(s)|_{s=s_i} \tag{13.30a}$$

which differs from the standard cover-up formula in that $F(s)$ is multiplied by $(s - s_i)^2$. Furthermore, since $(s - s_i)A_{i1}$ appears in the preceding expression, differentiation with respect to s eliminates all other residues and leaves

$$A_{i1} = \frac{d}{ds}[(s - s_i)^2F(s)]|_{s=s_i} \tag{13.30b}$$

Notice that the differentiation is performed *before* setting $s = s_i$. Functions with two or more pairs of double poles are handled in the same fashion.

But suppose that the denominator of $F(s)$ includes $(s - s_i)^3$, so we have a **triple pole** at s_i. The partial-fraction expansion for the triple pole is

$$G(s) = \frac{A_{i1}}{s - s_i} + \frac{A_{i2}}{(s - s_i)^2} + \frac{A_{i3}}{(s - s_i)^3}$$

from which

$$g(t) = A_{i1}e^{s_i t} + A_{i2}te^{s_i t} + \frac{1}{2}A_{i3}t^2e^{s_i t} \tag{13.31}$$

By analysis similar to the double-pole case, the residue formulas are

$$A_{i3} = [(s - s_i)^2F(s)]|_{s=s_i} \tag{13.32a}$$

$$A_{i2} = \frac{d}{ds}[(s - s_i)^3F(s)]|_{s=s_i} \tag{13.32b}$$

$$A_{i1} = \frac{1}{2!}\frac{d^2}{ds^2}[(s - s_i)^3F(s)]|_{s=s_i} \tag{13.32c}$$

Alternatively, to avoid the need for differentiation, A_{i2} and A_{i1} may be found by the method of undetermined coefficients.

The foregoing expressions may be extrapolated for four or more identical poles. Additionally, Eqs. (13.29)–(13.32) hold when s_i is a *complex repeated root,* in which case the residues will be complex-conjugate pairs. However, the algebra required for such cases is better done by computer than by hand.

Example 13.7 ***Inversion with a Triple Pole***

Consider the inversion of the function

$$F(s) = \frac{-s^2 - 2s + 14}{(s+4)^3(s+5)} = \frac{A_{11}}{s+4} + \frac{A_{12}}{(s+4)^2} + \frac{A_{13}}{(s+4)^3} + \frac{A_4}{s+5}$$

We'll evaluate A_4 and A_{13} via cover-up formulas. Then we'll use the method of undetermined coefficients to find A_{11} and A_{12}.

From Eqs. (13.20) and (13.32a),

$$A_4 = (s+5)F(s)|_{s=-5} = 1$$
$$A_{13} = (s+4)^3 F(s)|_{s=-4} = 6$$

so

$$\frac{-s^2 - 2s + 14}{(s+4)^3(s+5)} = \frac{A_{11}}{s+4} + \frac{A_{12}}{(s+4)^2} + \frac{6}{(s+4)^3} + \frac{1}{s+5}$$

We then observe that

$$\lim_{s\to\infty} sF(s) = 0 = A_{11} + 0 + 0 + 1 \quad \Rightarrow \quad A_{11} = -1$$

$$F(0) = \frac{14}{320} = \frac{A_{11}}{4} + \frac{A_{12}}{16} + \frac{6}{64} + \frac{1}{5} \quad \Rightarrow \quad A_{12} = -4A_{11} - 4 = 0$$

Hence,

$$F(s) = \frac{-1}{s+4} + \frac{0}{(s+4)^2} + \frac{6}{(s+4)^3} + \frac{1}{s+5}$$

and

$$f(t) = -1e^{-4t} + 0te^{-4t} + \frac{1}{2}6t^2e^{-4t} + 1e^{-5t}$$
$$= -e^{-4t} + 3t^2e^{-4t} + e^{-5t} \qquad t \geq 0$$

Exercise 13.8

Use Eqs. (13.30a) and (13.30b) to find $f(t)$ when

$$F(s) = (2s - 4)/(s + 3)^2$$

(Save your results for use in Exercise 13.10.)

Exercise 13.9

Recalculate both residues needed in Exercise 13.8 via the method of undetermined coefficients.

Time Delay

The Laplace transform automatically accounts for any switching that takes place at $t = 0$. But if switching also occurs at a later time, or if the circuit is driven by a pulsed waveform, then $F(s)$ explicitly includes a time-delay term. We examine here the corresponding inversion.

Recall that the Laplace transform of the delayed time function $g(t - t_0)u(t - t_0)$ is $\mathcal{L}[g(t)]e^{-st_0}$. Consequently, a typical s-domain result with a single time delay has the form

$$F(s) = \frac{N_1(s) + N_2(s)e^{-st_0}}{D(s)} = F_1(s) + F_2(s)e^{-st_0} \qquad (13.33a)$$

where $F_1(s) = N_1(s)/D(s)$ and $F_2(s) = N_2/D(s)$. The inverse transform is then

$$f(t) = f_1(t) + f_2(t - t_0)u(t - t_0) \qquad t \geq 0 \qquad (13.33b)$$

where $f_1(t) = \mathcal{L}^{-1}[F_1(s)]$ and $f_2(t) = \mathcal{L}^{-1}[F_2(s)]$. If $F_1(s)$ and $F_2(s)$ are strictly proper, then partial-fraction expansion may be applied to each function separately to obtain $f_1(t)$ and $f_2(t)$.

Example 13.8 ***Inversion with Time Delay***

The excitation of a certain unstable circuit is the stepped waveform $x(t) = 20u(t) - 40u(t - 3)$, and time-domain analysis for the response $y(t)$ yields the differential equation

$$y'(t) - 5y(t) = -x(t) = -20u(t) + 40u(t - 3)$$

We want to find $Y(s)$ and $y(t)$ when the circuit has zero initial state, so $y(0^-) = 0$.

After taking the Laplace transform of both sides, the differential equation becomes the algebraic relation

$$sY(s) - 5Y(s) = (-20 + 40e^{-3s})/s$$

Solving for $Y(s)$ gives

$$Y(s) = \frac{-20 + 40e^{-3s}}{s(s - 5)} = F_1(s) - 2F_1(s)e^{-3s}$$

where

$$F_1(s) = \frac{-20}{s(s - 5)} = \frac{A_1}{s} + \frac{A_2}{s - 5}$$

The residues of $F_1(s)$ are

$$A_1 = sF_1(s)|_{s=0} = 4 \qquad A_2 = (s - 5)F_1(s)|_{s=5} = -4$$

so

$$f_1(t) = \mathcal{L}^{-1}[F_1(s)] = 4 - 4e^{5t}$$

Then, from our expansion of $Y(s)$,

$$\begin{aligned} y(t) &= f_1(t) - 2f_1(t - 3)u(t - 3) \\ &= 4 - 4e^{5t} - [8 - 8e^{5(t-3)}]u(t - 3) \qquad t \geq 0 \end{aligned}$$

The growing exponentials e^{5t} and $e^{5(t-3)}$ are a result of the unstable nature of the circuit.

Exercise 13.10

Using the result of Exercise 13.8, find $f(t)$ when

$$F(s) = \frac{5(s + 3) - (2s - 4)e^{-8s}}{(s + 3)^2}$$

Initial and Final Values

Sometimes we need to know how $f(t)$ behaves at $t = 0^+$ or as $t \rightarrow \infty$, but we don't care about the entire time function. The desired information can be obtained directly from $F(s)$, without inversion, by applying the simple relations presented here.

Consider a typical transient problem with initial condition $f(0^-)$. The value of $f(0^+)$ may or may not equal $f(0^-)$, depending upon the excitation and continuity conditions. In any case, for a given $F(s)$, the **initial-value theorem** states that

$$f(0^+) = \lim_{s \to \infty} sF(s) \qquad (13.34)$$

provided that the limit exists. This limit always exists for any strictly proper rational function $F(s) = N(s)/D(s)$, since the order of $sN(s)$ will not exceed the order of $D(s)$.

The proof of Eq. (13.34) must account for a possible jump discontinuity at $t = 0$, as illustrated in Fig. 13.8*a*. We temporarily remove the discontinuity by defining the *continuous* waveform in Fig. 13.8*b* where

$$f_c(t) = f(t) - [f(0^+) - f(0^-)]u(t)$$

The transforms of $f_c(t)$ and $f(t)$ are thus related by

$$F_c(s) = F(s) - [f(0^+) - f(0^-)]/s$$

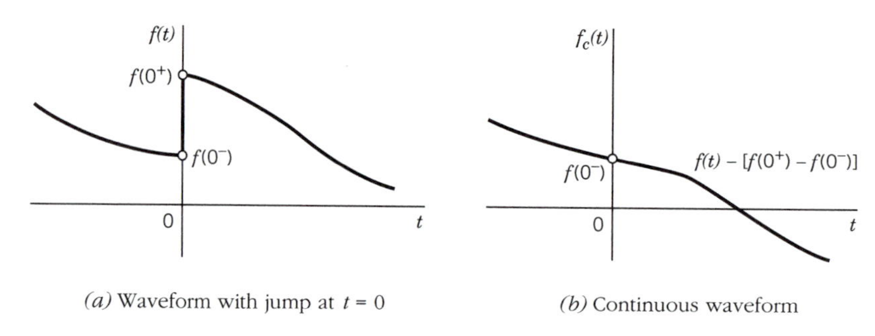

(*a*) Waveform with jump at $t = 0$ (*b*) Continuous waveform

Figure 13.8

Now recall that the transform of the derivative $f_c{}'(t) = df_c/dt$ is given by

$$\mathcal{L}[f_c'(t)] = sF_c(s) - f_c(0^-) = \int_{0^-}^{\infty} f_c'(t)e^{-st}\,dt$$

This transform exists when $f_c{}'(t)$ is of exponential order, and we have defined $f_c(t)$ such that $|f_c{}'(t)| < \infty$ over $0^- \leq t \leq 0^+$. Consequently,

$$\lim_{s \to \infty} \int_{0^-}^{\infty} f_c'(t)e^{-st}\,dt = \int_{0^+}^{\infty} \left[\lim_{s \to \infty} f_c'(t)e^{-st} \right] dt = 0$$

from which

$$\lim_{s \to \infty} sF_c(s) - f_c(0^-) = 0$$

But $sF_c(s) = sF(s) - [f(0^+) - f(0^-)]$ and $f_c(0^-) = f(0^-)$, so

$$\lim_{s \to \infty} sF(s) - f(0^+) = 0$$

and rearrangement yields Eq. (13.34).

The initial-value theorem may also be invoked to find the **initial slope** $f'(0^+)$, provided that you know the initial condition $f(0^-)$. Since

$$\mathcal{L}[f'(t)] = sF(s) - f(0^-) = \frac{sN(s)}{D(s)} - f(0^-)$$

application of Eq. (13.34) gives

$$f'(0^+) = \lim_{s\to\infty} s\,\frac{sN(s) - f(0^-)D(s)}{D(s)} \tag{13.35}$$

Note carefully that Eq. (13.35) involves $f(0^-)$, whereas Eq. (13.34) yields $f(0^+)$.

Turning to the behavior as $t \to \infty$, the **final-value theorem** states that if $f(t)$ approaches a unique finite value, then

$$f(\infty) = \lim_{s\to 0} sF(s) \tag{13.36}$$

This theorem *cannot* be used when $F(s)$ includes:

- Poles in the right half of the s-plane;
- Multiple poles at $s = 0$;
- Poles on the imaginary axis.

A unique final value does not exist in these cases because $f(t)$ either becomes unbounded as $t \to \infty$ or it contains a nondecaying oscillation.

The proof of Eq. (13.36) comes from the transform of $f'(t)$ written as

$$\mathcal{L}[f'(t)] = sF(s) - f(0^-) = \int_{0^-}^{\infty} f'(t)e^{-st}\,dt$$

Thus,

$$\lim_{s\to 0} sF(s) - f(0^-) = \int_{0^-}^{\infty} f'(t)\,dt = f(\infty) - f(0^-)$$

and Eq. (13.36) follows directly.

Example 13.9 *Calculating Initial and Final Values*

Suppose you have a dynamic circuit problem with the initial condition

$$f(0^-) = 5$$

Further suppose that your analysis yields the s-domain result

$$F(s) = \frac{N(s)}{D(s)} = \frac{5s^3 - 1600}{s(s^3 + 18s^2 + 90s + 800)}$$

The denominator $D(s)$ is fourth order, so you would have to evaluate four residues to carry out the inversion for $f(t)$.

However, noting that $F(s)$ is strictly proper, you immediately apply the initial-value theorem to obtain

$$f(0^+) = \lim_{s\to\infty} sF(s) = \lim_{s\to\infty} \frac{5s^3 - \cdots}{s^3 + \cdots} = 5$$

Since $f(0^+) = f(0^-)$, you conclude that $f(t)$ has continuity at $t = 0$. You next determine the initial slope of $f(t)$ from

$$\frac{sN(s) - f(0^-)D(s)}{D(s)} = \frac{-90s^3 - 450s^2 - 5600s}{s(s^3 + 18s^2 + 90s + 800)}$$

which is also strictly proper. Hence,

$$f'(0^+) = \lim_{s\to\infty} s\,\frac{-90s^3 - \cdots}{s^4 + \cdots} = -90$$

so $f(t)$ initially decreases.

Application of the final-value theorem entails a little more work because you must be check the pole locations. You see by inspection of $F(s)$ that $s_1 = 0$, which is at the origin but not repeated. Then, using your calculator, numerical analysis of the cubic in $D(s)$ gives

$$s_2, s_3 = -4 \pm j8 \qquad s_4 = -10$$

which are all within the left half of the s plane. Therefore, the limit exists and

$$f(\infty) = \lim_{s\to 0} sF(s) = \lim_{s\to 0} \frac{\cdots - 1600}{\cdots + 800} = -2$$

The nonzero value of $f(\infty)$ agrees with $s_1 = 0$, because the inverse transform would include a constant and the other poles correspond to decaying functions.

Exercise 13.11

Let $f(0^-) = 0$ and let $F(s)$ be given by Eq. (13.15). Determine if the limit exists and, if so, evaluate $f(0^+)$ and $f'(0^+)$ when $m = n - 1$ and $m = n - 2$.

13.3 TRANSFORM CIRCUIT ANALYSIS

Having established the required background, we now apply Laplace transforms to circuit analysis. The general class of problems considered here is as follows:

Given a circuit with some initial state at $t = 0^-$ and an excitation $x(t)$ starting at $t = 0$, find the resulting behavior of any voltage or current $y(t)$ for $t \geq 0$.

We'll first investigate the simpler case of the zero-state response, including a re-examination of the natural and forced components. Then we'll incorporate nonzero initial conditions to obtain the zero-input response and complete response.

Zero-State Response

By definition, the zero-state response occurs when a circuit contains no stored energy at $t = 0^-$. We may therefore assume that the excitation equals zero for $t < 0$. We'll also assume a *single* input, since superposition can be invoked to handle multiple inputs.

Consider a circuit with n energy-storage elements and input $x(t)$. The behavior of any voltage or current $y(t)$ is governed by a differential equation such as

$$a_n \frac{d^n y}{dt^n} + a_{n-1} \frac{d^{n-1} y}{dt^{n-1}} + \cdots + a_1 \frac{dy}{dt} + a_0 y$$
$$= b_m \frac{d^m x}{dt^m} + b_{m-1} \frac{d^{m-1} x}{dt^{m-1}} + \cdots + b_1 \frac{dx}{dt} + b_0 x \quad (13.37)$$

We'll go from the time domain to the s domain by taking the Laplace transform of Eq. (13.37). To that end, let

$$Y(s) = \mathcal{L}[y(t)] \qquad X(s) = \mathcal{L}[x(t)] \quad (13.38)$$

so

$$\mathcal{L}\left[\frac{d^k y}{dt^k}\right] = s^k Y(s) \qquad \mathcal{L}\left[\frac{d^k x}{dt^k}\right] = s^k X(s) \quad (13.39)$$

No initial-condition constants appear in these derivative relations because we have assumed zero-state conditions at $t = 0^-$ and zero input for $t < 0$.

After using Eq. (13.39) for the transform on both sides of Eq. (13.37), we collect terms and get the algebraic equation

$$(a_n s^n + a_{n-1} s^{n-1} + \cdots + a_1 s + a_0) Y(s)$$
$$= (b_m s^m + b_{m-1} s^{m-1} + \cdots + b_1 s + b_0) X(s)$$

from which

$$Y(s) = \frac{b_m s^m + b_{m-1} s^{m-1} + \cdots + b_1 s + b_0}{a_n s^n + a_{n-1} s^{n-1} + \cdots + a_1 s + a_0} X(s)$$

But the ratio of polynomials here is identical to the *network function* $H(s)$ obtained from Eq. (13.37). Hence, we can write the compact relationship

$$Y(s) = H(s)X(s) \tag{13.40}$$

which is the desired result.

Equation (13.40) shows that the Laplace transform of the zero-state response equals the product of the network function and the Laplace transform of the input. Furthermore, the network function $H(s)$ is exactly the same as first introduced in Chapter 10. Accordingly, we can bypass the differential equation entirely and find $y(t)$ by the following procedure:

1. Draw the s-domain circuit diagram labeled with the symbols $X(s)$ and $Y(s)$, and use standard analysis methods to obtain $H(s) = Y(s)/X(s)$.
2. Take the Laplace transform of $x(t)$ to get the expression for $X(s)$.
3. Multiply $H(s)$ by $X(s)$ to get the expression for $Y(s)$.
4. Take the inverse transform of $Y(s)$ to get $y(t)$.

The final inverse transformation generally involves partial-fraction expansion and deserves elaboration.

We know from our previous work that a network function can always be expressed in the form

$$H(s) = N_H(s)/P(s)$$

where $P(s)$ is the network's *characteristic polynomial.* We've also seen that the transforms of most input time functions have the form

$$X(s) = N_X(s)/D_X(s)$$

where $D_X(s)$ is another polynomial. Consequently,

$$Y(s) = H(s)X(s) = \frac{N_H(s)N_X(s)}{P(s)D_X(s)} \tag{13.41}$$

Equation (13.41) demonstrates that the poles of $Y(s)$ consist of the network's poles $p_1, p_2, \ldots$ together with additional poles equal to the roots of $D_X(s)$. We'll draw upon this property to discuss two special but important cases.

Step Response. The step response is produced when $x(t) = u(t) = 1$ for $t > 0$, so $X(s) = 1/s$. Equation (13.41) then becomes

$$Y(s) = H(s) \times \frac{1}{s} = \frac{N_H(s)}{sP(s)} \tag{13.42}$$

Hence, partial-fraction expansion for the step response involves a pole at the origin and the network's poles.

Example 13.10 *Step Response*

The circuit in Fig. 13.9*a* is driven by the voltage $v_s(t) = u(t)$ and we want to find the step response of the current $i(t)$.

Step 1: From the *s*-domain diagram in Fig. 13.9*b* we see that $I(s) = V_s(s)/Z(s)$ where

$$Z(s) = s + 2\|(16/s) = s + \frac{32}{2s + 16}$$

Thus,

$$H(s) = \frac{I(s)}{V_s(s)} = \frac{1}{Z(s)} = \frac{2s + 16}{2s^2 + 16s + 32} = \frac{s + 8}{(s + 4)^2}$$

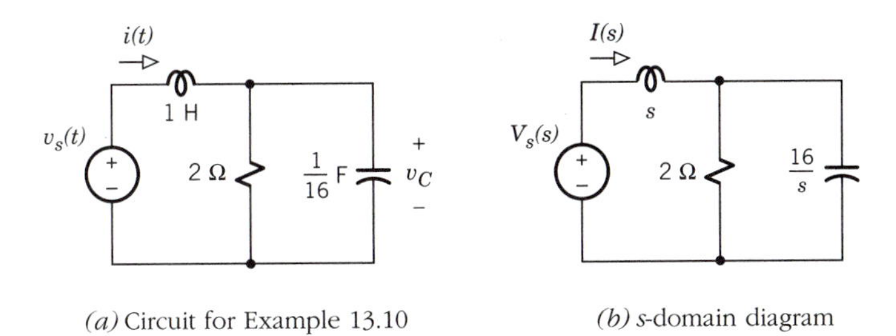

(*a*) Circuit for Example 13.10

(*b*) *s*-domain diagram

Figure 13.9

Step 2: With $v_s(t) = u(t)$, $V_s(s) = \mathcal{L}[u(t)] = 1/s$.

Step 3: The transform of the step response is

$$I(s) = H(s) \times \frac{1}{s} = \frac{s + 8}{s(s + 4)^2}$$

Before taking the inverse transform, we can check our *s*-domain result by applying the initial-value and final-value theorems. The initial value is

$$i(0^+) = \lim_{s \to \infty} sI(s) = 0$$

which agrees with the continuity of inductor current since $i(0^+) = i(0^-) = 0$. The final value is

$$i(\infty) = \lim_{s \to 0} sI(s) = 0.5$$

which agrees with the dc steady-state condition $i = 1 \text{ V}/2\ \Omega = 0.5$ A.

Step 4: To obtain the entire step-response waveform, we write the partial-fraction expansion

$$I(s) = \frac{s + 8}{s(s + 4)^2} = \frac{A_1}{s} + \frac{A_{21}}{s + 4} + \frac{A_{22}}{(s + 4)^2}$$

The residues are then found to be

$$A_1 = 0.5 \qquad A_{21} = -0.5 \qquad A_{22} = -1$$

Hence,

$$i(t) = 0.5 - 0.5e^{-4t} - te^{-4t} \qquad t \geq 0$$

Inspecting this expression, we see that $i(t)$ consists of the steady-state forced response $i_F(t) = 0.5$ plus the decaying natural response $i_N(t) = -0.5e^{-4t} - te^{-4t}$.

Exercise 13.12

Find the step response of $y(t)$ when $H(s) = 2s/(s + 3)$.

Zero-State AC Response. Let the input applied at $t = 0$ be a sinusoid, say

$$x(t) = X_m \cos(\beta t + \phi_x)$$

The Laplace transform of any constant-amplitude ac excitation at frequency β has $D_X(s) = s^2 + \beta^2$, so Eq. (13.41) becomes

$$Y(s) = \frac{N_H(s)N_X(s)}{P(s)(s^2 + \beta^2)} \tag{13.43}$$

Although we could take the inverse of this entire expression, we'll gain insight by writing

$$Y(s) = Y_N(s) + Y_F(s) \tag{13.44a}$$

where

$$Y_N(s) = \frac{N_N(s)}{P(s)} \qquad Y_F(s) = \frac{N_F(s)}{s^2 + \beta^2} \tag{13.44b}$$

Since $Y_N(s)$ contains all of the network's poles but none of the poles from $X(s)$, we conclude that $Y_N(s)$ corresponds to the *natural-response component* $y_N(t)$. Conversely, since $Y_F(s)$ contains all of the poles from $X(s)$ but none of the network's poles, we conclude that $Y_F(s)$ corresponds to the *force-response component* $y_F(t)$. The total response is then given by the familiar expression

$$y(t) = y_N(t) + y_F(t) \tag{13.45}$$

provided that we can determine $y_N(t)$ and $y_F(t)$.

The natural-response component can be found from Eq. (13.43) by evaluating the residues of $Y(s)$ at the poles of $P(s)$, so we don't actually need the numerator polynomial $N_N(s)$ in $Y_N(s)$. We could also find the forced-response component by evaluating residues, but standard phasor analysis usually proves to be easier. In particular, if we let $X = \underline{X_m \angle \phi_x}$, then

$$\underline{Y} = H(j\beta)X = Y_m \angle \phi_y \tag{13.46a}$$

and

$$y_F(t) = Y_m \cos(\beta t + \phi_y) \tag{13.46b}$$

This phasor method takes the place of inverting $Y_F(s)$, so we don't need the other numerator $N_F(s)$ or the residues of $Y_F(s)$.

Example 13.11 ***Zero-State AC Response***

Suppose the circuit in Fig. 13.9*a* has zero initial stored energy and is driven by the ac source voltage

$$v_s(t) = 50 \cos 8t \qquad t \geq 0$$

We previously found that the network function is $H(s) = I(s)/V_s(s) = (s + 8)/(s + 4)^2$. Now we'll determine the zero-state ac response of $i(t)$ by finding and adding the natural and forced components.

For the natural-response component, we take the Laplace transform of $v_s(t)$ to get

$$V_s(s) = 50\mathcal{L}[\cos 8t] = 50s/(s^2 + 64)$$

Multiplying $H(s)$ by $V_s(s)$ and expanding yields

$$\begin{aligned} I(s) = H(s)V_s(s) &= \frac{50s^2 + 400s}{(s + 4)^2(s^2 + 64)} \\ &= \frac{A_{11}}{s + 4} + \frac{A_{12}}{(s + 4)^2} + \frac{N_F(s)}{s^2 + 64} \end{aligned}$$

where

$$A_{11} = -1 \qquad A_{12} = -10$$

Hence,

$$I_N(s) = \frac{-1}{s+4} + \frac{-10}{(s+4)^2}$$

and

$$i_N(t) = -e^{-4t} - 10te^{-4t}$$

For the forced-response component, we write the input phasor $\underline{V}_s = 50 \angle 0°$ and we set $s = j\beta = j8$ in $H(s)$ to calculate

$$H(j8) = \frac{j8 + 8}{(j8)^2 + 8 \times j8 + 16} = 0.141 \angle -81.9°$$

Then

$$\underline{I} = H(j8)\underline{V}_s = 7.07 \angle -81.9°$$

and

$$i_F(t) = 7.07 \cos (8t - 81.9°)$$

Lastly, we obtain $i(t)$ by adding $i_N(t)$ and $i_F(t)$, the final result being

$$i(t) = -(1 + 10t)e^{-4t} + 7.07 \cos (8t - 81.9°) \qquad t \geq 0$$

As a check of this result, we calculate the initial value

$$i(0^+) = -1 + 7.07 \cos (-81.9°) = 0$$

which again agrees with the continuity condition $i(0^+) = i(0^-) = 0$. The plot of $i(t)$ in Fig. 13.10 shows the rather brief transient behavior from the initial value to the ac steady-state condition $i(t) = i_F(t)$.

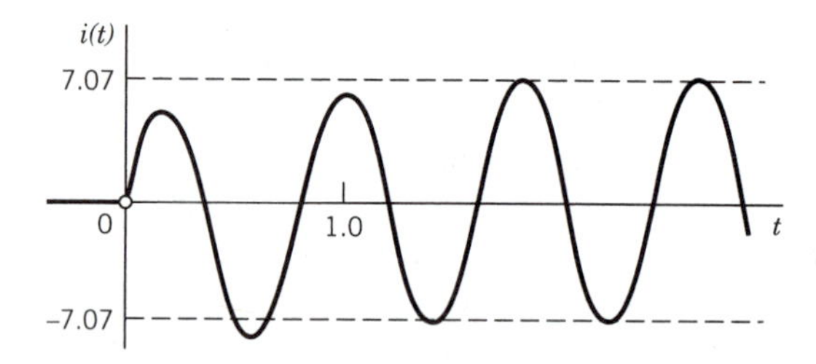

Figure 13.10 Zero-state ac response waveform.

Exercise 13.13

Given $H(s) = 2s/(s + 3)$, find the zero-state response when $x(t) = 25 \sin 4t$ for $t \geq 0$.

Natural Response and Forced Response

The zero-state response generally consists of a natural-response component plus a forced-response component. In Chapter 10, we showed that the modes in the natural response could be predicted from the poles of $H(s)$. We also showed how the forced response produced by a complex-frequency excitation of the form $x(t) = \text{Re}[\underline{X}e^{s_0 t}]$ could be found from the phasor relation $\underline{Y} = H(s_0)\underline{X}$, provided that s_0 does not equal any of the poles of $H(s)$. Here we'll apply transform analysis to further investigate the natural and forced components of the zero-state response with a complex-frequency excitation.

For purposes of discussion, let the network function in question be

$$H(s) = \frac{1}{(s+1)(s+2)}$$

This circuit's poles are $p_1 = -1$ and $p_2 = -2$, so its natural response has the form $y_N(t) = A_1 e^{-t} + A_2 e^{-2t}$. Also let the input be the exponential function

$$x(t) = 10e^{-s_0 t}$$

where s_0 is real. We'll consider what happens when s_0 does not and does equal one of the circuit's poles.

Case I: $s_0 = -3$. Since s_0 does not equal either pole of $H(s)$, we could find the forced response by the method from Chapter 10. Introducing the phasor $\underline{X} = 10\,\underline{/0^\circ}$ to write $x(t) = 10e^{-3t} = \text{Re}[\underline{X}e^{-3t}]$, we calculate the response phasor via $\underline{Y} = H(-3)\underline{X} = 5\,\underline{/0^\circ}$ and

$$y_F(t) = \text{Re}[\underline{Y}e^{-3t}] = 5e^{-3t}$$

But to find the entire zero-state response by transform analysis, we note that $X(s) = \mathcal{L}[10e^{-3t}] = 10/(s+3)$ so

$$Y(s) = H(s)X(s) = \frac{10}{(s+1)(s+2)(s+3)} = \frac{5}{s+1} + \frac{-10}{s+2} + \frac{5}{s+3}$$

and

$$y(t) = 5e^{-t} - 10e^{-2t} + 5e^{-3t}$$

We see that the first two terms of $y(t)$ constitute the natural response, while the third term is identical to the previously determined forced response $y_F(t)$. Thus, $y(t) = y_N(t) + y_F(t)$ as expected.

Case II: $s_0 = -2$. Since s_0 equals a pole of $H(s)$, we call this case **excitation at a natural frequency**. Now we cannot use the phasor method to find the forced response because $H(-2) = \infty$ and the response phasor $\underline{Y}$ would be undefined. Turning directly to the transform method, we note that $X(s) = \mathcal{L}[10e^{-2t}] = 10/(s+2)$ so $Y(s)$ will have a *repeated* pole. Specifically,

$$Y(s) = H(s)X(s) = \frac{10}{(s+1)(s+2)^2} = \frac{10}{s+1} + \frac{-10}{s+2} + \frac{-10}{(s+2)^2}$$

and

$$y(t) = 10e^{-t} - 10e^{-2t} - 10te^{-2t}$$

The first term here is clearly a natural-response mode while the last term must be a forced-response component since $H(s)$ by itself does not have a repeated pole. But the middle term has an ambiguous interpretation because it could be either a natural or forced response. In any case, we see that excitation at a natural frequency causes the forced response to be multiplied by t (or powers of t).

Exercise 13.14

Obtain an expression for the zero-state response $y(t)$ given that $H(s) = K/(s + a)$ and $x(t) = X_m e^{-at}$. Comment on your result in terms of natural and forced response.

Zero-Input Response

By definition, the zero-input response occurs when the excitation equals zero for $t \geq 0$ but the circuit contains stored energy at $t = 0^-$. As previously illustrated in Section 13.1, we could analyze this situation by starting with the differential equation and use the Laplace transform to incorporate the initial conditions. But a more efficient *s-domain method* simplifies analysis with initial stored energy by focusing on capacitor voltages and inductor currents. These *state variables* fully determine the stored energy at any time, and their initial values can be represented directly in the *s*-domain diagram.

Consider first the capacitor in Fig. 13.11*a*. The current–voltage relationship is

$$i_C(t) = Cv_C'(t)$$

which transforms to

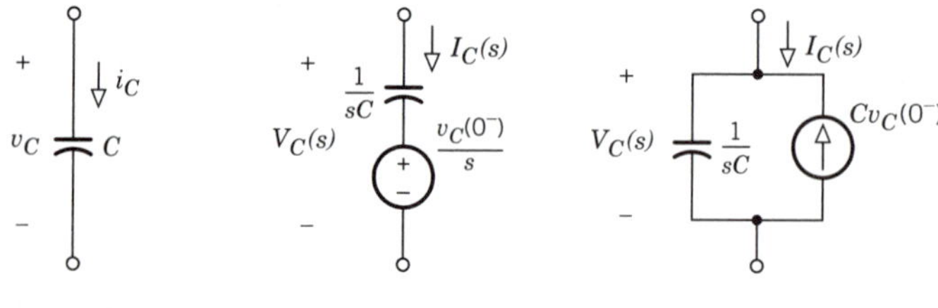

(*a*) Capacitor in the time domain

(*b*) *s*-domain Thévenin model of initial voltage

(*c*) *s*-domain Norton model of initial voltage

Figure 13.11

$$I_C(s) = C[sV_C(s) - v_C(0^-)]$$

Rearrangement then yields

$$V_C(s) = \frac{v_C(0^-)}{s} + \frac{1}{sC} + \frac{1}{sC} I_C(s) \tag{13.47a}$$

The s-domain representation of this expression is diagrammed in Fig. 13.11b. Here we have the usual impedance $Z_C = 1/sC$, but a fictitious voltage source of value $v_C(0^-)/s$ has been added in series to account for the initial stored energy. Alternatively, we may rewrite Eq. (13.47a) as

$$I_C(s) = sCV_C(s) - Cv_C(0^-) \tag{13.47b}$$

which leads to the representation in Fig. 13.11c. Now Z_C is in parallel with a fictitious current source of value $Cv_C(0^-)$. We'll call this parallel structure the **Norton model**, as distinguished from the **Thévenin model** in Fig. 13.11b. Of course, either model may be obtained from the other by source conversion.

Next, consider the inductor in Fig. 13.12a. The voltage–current relationship is

$$v_L(t) = Li_L'(t)$$

which transforms to

$$V_L(s) = L[sI_L(s) - i_L(0^-)]$$

Rearrangement then yields

$$I_L(s) = \frac{i_L(0^-)}{s} + \frac{1}{sL} V_L(s) \tag{13.48a}$$

$$V_L(s) = sLI_L(s) - Li_L(0^-) \tag{13.48b}$$

Figures 13.12b and 13.12c give the corresponding Norton and Thévenin models. As you might have expected, these inductance networks are the duals of the capacitance networks in Fig. 13.11.

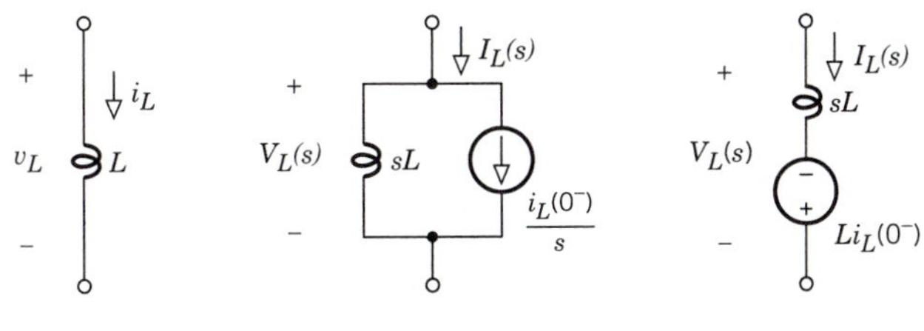

(a) Inductor in the time domain

(b) s-domain Norton model of initial current

(c) s-domain Thévenin model of initial current

Figure 13.12

Observe carefully that our initial-value sources are integral parts of the s-domain representation of capacitance and inductance. Consequently, when you use a Thévenin model, the time-domain voltage across the element in question corresponds to the transformed voltage across the series combination of impedance *and* initial-value voltage source. Similarly, when you use a Norton model, the time-domain current through the element in question corresponds to the transformed current through the parallel combination of impedance *and* initial-value current source.

Our s-domain models for capacitance and inductance directly incorporate initial stored energy, so the zero-input response of any variable can be found using s-domain analysis, as follows:

1. Calculate the value of all capacitor voltages and inductor currents at $t = 0^-$.
2. Draw the s-domain circuit diagram for $t \geq 0$ with the initial values represented by appropriate fictitious sources, taking care to match the source polarities with the polarities of the initial voltages and currents.
3. Find $Y(s)$ by analyzing the s-domain diagram.
4. Take the inverse transform of $Y(s)$ to get $y(t)$.

Since each energy-storage element may be represented by either a Thévenin or a Norton model, you should select the type that expedites systematic circuit analysis — usually Norton models for node equations and Thévenin models for mesh equations.

Example 13.12 ***Calculating a Zero-Input Response***

The circuit in Fig. 13.13a has $i_s(t) = 6$ A for all $t < 0$, and the input source is turned off at $t = 0$. We want to find the resulting zero-input response of i_L.

Step 1: DC steady-state analysis for $t < 0$ yields the initial values

$$i_L(0^-) = 6 \text{ A} \qquad v_C(0^-) = 20\ \Omega \times i_L(0^-) = 120 \text{ V}$$

Step 2: For $t \geq 0$, the input current source is an open circuit, leaving a

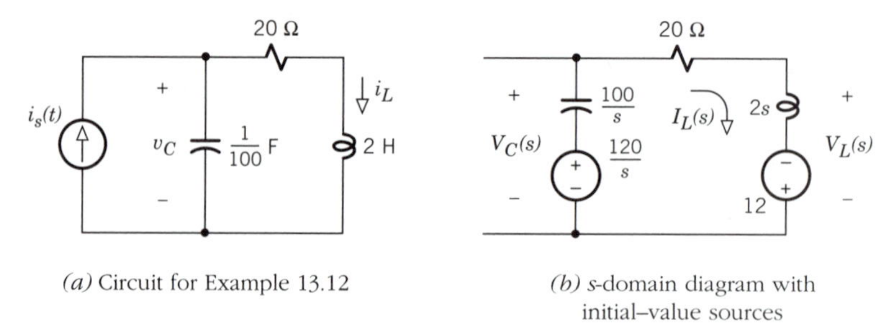

(a) Circuit for Example 13.12

(b) s-domain diagram with initial–value sources

Figure 13.13

CRL series loop. We therefore choose to represent both initial values by voltage sources. Since $Li_L(0^-) = 12$, our *s*-domain diagram becomes as shown in Fig. 13.13*b*. The diagram also includes labels for $V_C(s)$ and $V_L(s)$, which could be found after calculating $I_L(s)$.

Step 3: By inspection of Fig. 13.13*b*, the mesh equation for $I_L(s)$ is

$$(2s + 20 + 100/s)I_L(s) = 12 + 120/s$$

so

$$I_L(s) = \frac{12 + 120/s}{2s + 20 + 100/s} = \frac{6s + 60}{s^2 + 10s + 50}$$

Step 4: $I_L(s)$ has the form $(Bs + C)/(s^2 + 2\alpha s + \omega_0{}^2)$ with

$$B = 6 \qquad C = 60 \qquad \alpha = 5 \qquad \omega_0{}^2 = 50$$

We thus find that

$$\beta = 5 \qquad K = 6 - j6 = 6\sqrt{2}\,\underline{/-45^\circ}$$

and

$$\begin{aligned} i_L(t) &= K_m e^{-\alpha t} \cos(\beta t + \phi) \\ &= 6\sqrt{2}e^{-5t} \cos(5t - 45^\circ) \text{ A} \qquad t \geq 0 \end{aligned}$$

Exercise 13.15

Find $V_C(s)$ for the zero-input response of $v_C(t)$ in Fig. 13.9*a* when $v_s(t) =$ 8 V for all $t < 0$. Use current sources to represent the initial conditions.

Complete Response

When a circuit has nonzero stored energy at $t = 0^-$ *and* a nonzero excitation for $t \geq 0$, the resulting complete response consists of the zero-input response plus the zero-state response. Thus, you could find $Y(s)$ by first assuming zero excitation to calculate the zero-input response, and then assuming zero initial stored energy to calculate the zero-state response.

But a more efficient strategy leads directly to the sum of both components by representing the initial conditions and the excitation in the *s*-domain. This strategy combines our zero-input and zero-state methods as follows;

1. Calculate the values of all capacitor voltages and inductor currents at $t = 0^-$.
2. Take the Laplace transform of $x(t)$ to get the expression for $X(s)$.

3. Draw the s-domain circuit diagram for $t \geq 0$ with the input represented by $X(s)$ and initial values represented by appropriate fictitious sources.
4. Find $Y(s)$ by analyzing the s-domain diagram.
5. Take the inverse transform of $Y(s)$ to get $y(t)$.

As in the case of the zero-input response, you may choose either a voltage or current source to represent each initial condition. Additionally, a source conversion of the input may expedite the s-domain analysis.

Example 13.13 ***Calculating a Complete Response***

We want to find $v_C(t)$ for $t \geq 0$ in Fig. 13.14*a*, given that

$$v_s(t) = 20 \text{ V} \qquad t < 0$$
$$= -20 \text{ V} \qquad t \geq 0$$

Step 1: The dc steady state for $t < 0$ produces the initial values

$$i_L(0^-) = 4 \text{ A} \qquad v_C(0^-) = 20 \text{ V}$$

Step 2: The Laplace transform of the input for $t \geq 0$ is

$$V_s(s) = \mathcal{L}[-20] = -20/s$$

Step 3: Since v_C is a node voltage, we'll represent its initial value in the s domain by the parallel source current $Cv_C(0^-) = 0.5$. However, noting that the inductor will be in series with $V_s(s)$, we'll represent its initial current as another series source voltage $Li_L(0^-) = 2$. Figure 13.14*b* shows the resulting s-domain diagram with three sources.

Step 4: Inspection of Fig. 13.14*b* yields the systematic node equation

$$\left(\frac{s}{40} + \frac{1}{5} + \frac{1}{0.5s}\right) V_C(s) = \frac{2 - 20/s}{0.5s} + 0.5$$

from which

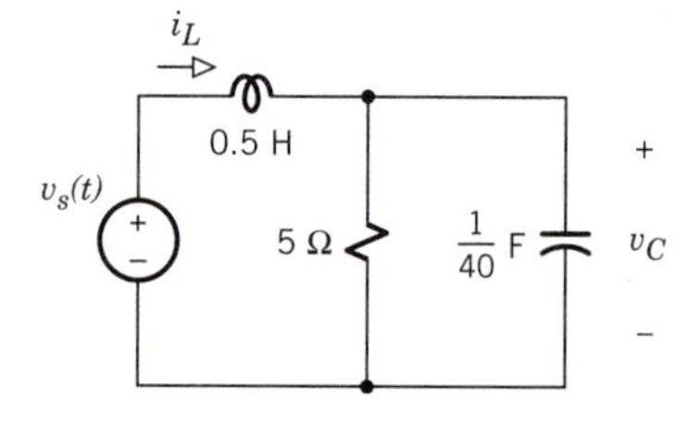

(*a*) Circuit for Example 13.13

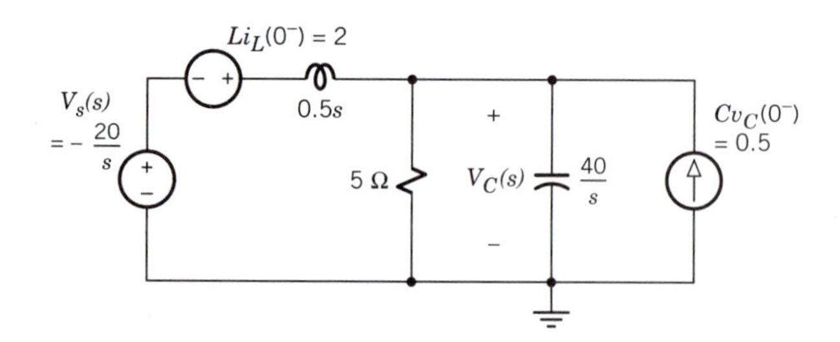

(*b*) s-domain diagram with input and initial–value sources

Figure 13.14

$$V_C(s) = \frac{20(s^2 + 8s - 80)}{s(s^2 + 8s + 80)} = \frac{A_1}{s} + \frac{Bs + C}{s^2 + 8s + 80}$$

Step 5: The inversion of $V_C(s)$ is a routine task, the final result being

$$v_C(t) = -20 + 20\sqrt{5}e^{-4t} \cos (8t - 26.6°) \text{ V} \qquad t \geq 0$$

The plot of $v_C(t)$ in Fig. 13.15 illustrates the overshoot of an underdamped circuit before it reaches the new steady-state condition at $v_C(t) = -20$ V.

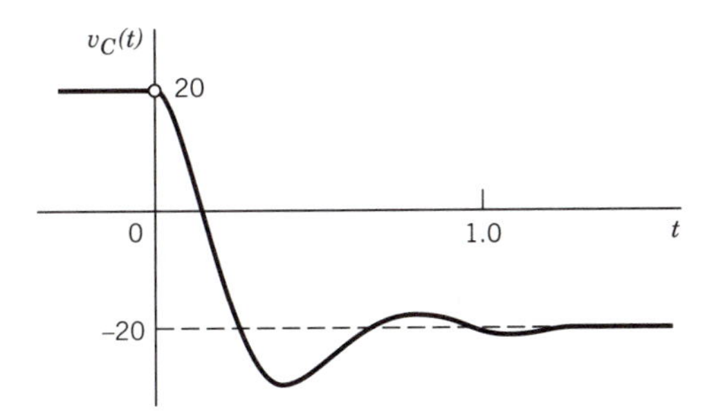

Figure 13.15 Complete-response waveform.

Exercise 13.16

By modifying the analysis in Example 13.12 (p. 628), find $I_L(s)$ when

$$i_s(t) = 3 \text{ A} \qquad t < 0$$
$$= -1 \text{ A} \qquad t \geq 0$$

Hint: Use source conversions to obtain a series circuit with two voltage sources.

13.4 TRANSFORM ANALYSIS WITH MUTUAL INDUCTANCE†

Recall from Section 8.3 that the combined effects of self- and mutual inductance in magnetically coupled coils usually can be represented by an equivalent three-inductor network. In particular, when the coils have the same winding sense as shown in Fig. 13.16*a*, the tee equivalent network takes the form of Fig. 13.16*b*. (The case of opposite winding sense is handled by substituting $-M$ for M.) This tee network expedites transform analysis of transients in circuits with mutual inductance.

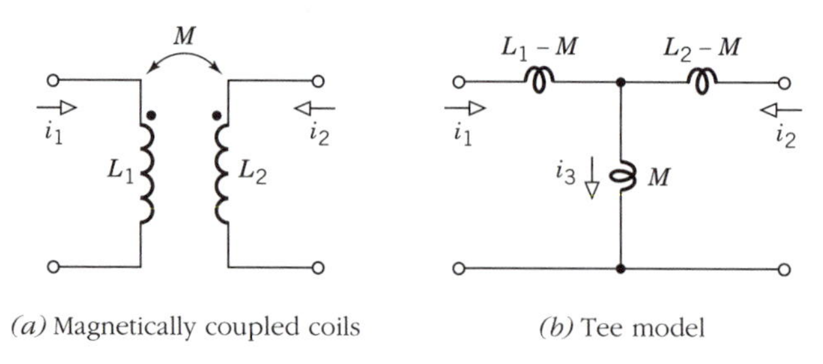

(a) Magnetically coupled coils (b) Tee model

Figure 13.16

If both the primary current i_1 and the secondary current i_2 equal zero at $t = 0^-$, then $i_3(0^-)$ also equals zero and the analysis is straightforward. You just replace the three inductors by their s-domain impedances and proceed as before to get the response in question.

But if there is initial stored energy, then you must include fictitious sources to account for $i_3(0^-)$ as well as $i_1(0^-)$ and $i_2(0^-)$. Figure 13.17 diagrams the resulting s-domain model with voltage sources representing the initial conditions. Note carefully that the inductors have values $L_1 - M$, $L_2 - M$, and M, so the initial-condition source voltages are $(L_1 - M)i_1(0^-)$, $(L_2 - M)i_2(0^-)$, and $Mi_3(0^-)$. We choose voltage sources here since the tee configuration suggests the use of mesh analysis. Source conversions may be performed, if necessary, for node analysis.

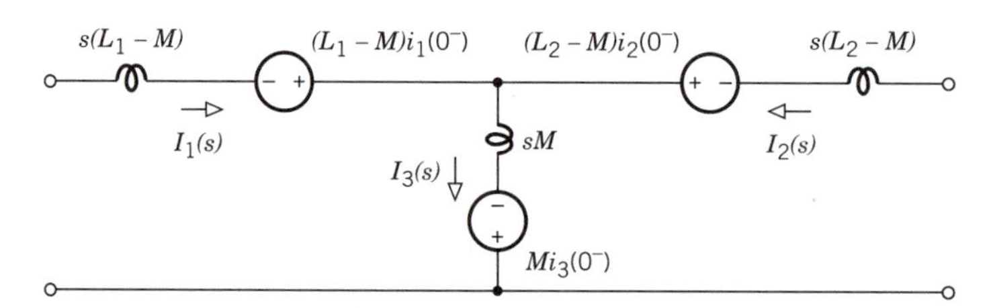

Figure 13.17 s-domain tee model with initial-value sources.

When working with the tee network, you should keep in mind that the three inductors are *interdependent* because i_3 equals $i_1 + i_2$. Consequently, in absence of other energy-storage elements, the characteristic polynomial will be *second* order rather than third order. Furthermore, in the case of *unity coupling* ($k = 1$), the primary and secondary coils act like just *one* element with respect to energy storage. Thus, the currents may undergo discontinuous jumps while the terminal voltages remain finite. The following example illustrates these effects.

Example 13.14 *Complete Response of a Transformer Circuit*

Figure 13.18*a* depicts a transformer connecting a source to a load. We'll find the resulting currents i_1 and i_2 for $t \geq 0$ when

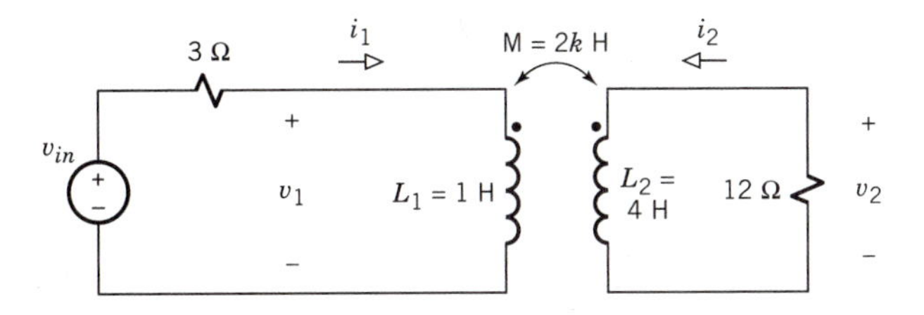

(a) Transformer circuit

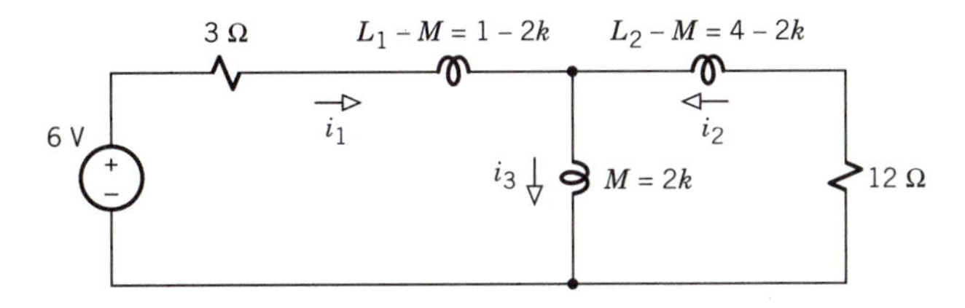

(b) Diagram with tee model for $t < 0$

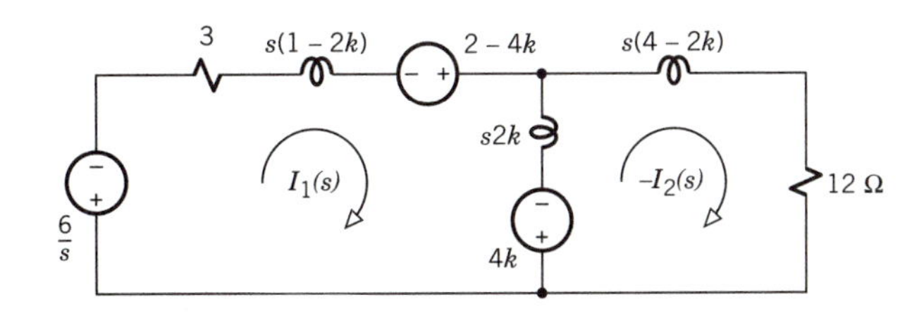

(c) s–domain diagram for $t \geq 0$ with input and initial-value sources

Figure 13.18

$$\begin{aligned} v_{in} &= +6 \text{ V} \qquad t < 0 \\ &= -6 \text{ V} \qquad t > 0 \end{aligned}$$

To explore the coupling effect, we'll take

$$M = k\sqrt{L_1 L_2} = 2k \text{ H}$$

where $0 \leq k \leq 1$.

For $t < 0$, the time-domain conditions with the tee network are as shown in Fig. 13.18*b*. Since the inductors behave as short circuits in the dc steady state, the initial-condition currents are

$$i_1(0^-) = i_3(0^-) = v_{in}/3 = 2 \text{ A} \qquad i_2(0^-) = 0$$

Thus,

$$(L_1 - M)i_1(0^-) = 2 - 4k \qquad Mi_3(0^-) = 4k \qquad (L_2 - M)i_2(0^-) = 0$$

Figure 13.18c gives the corresponding s-domain diagram for $t \geq 0$, including $\mathcal{L}[v_{in}(t)] = \mathcal{L}[-6] = -6/s$ and mesh currents $I_1(s)$ and $-I_2(s)$.

By inspection of this diagram we get the matrix mesh equation

$$\begin{bmatrix} s+3 & -2ks \\ -2ks & 4s+12 \end{bmatrix} \begin{bmatrix} I_1(s) \\ -I_2(s) \end{bmatrix} = \begin{bmatrix} 2-6/s \\ -4k \end{bmatrix}$$

Clearing the fractions and using Cramer's rule yields

$$I_1(s) = \frac{(8-8k^2)s^2 - 72}{sP(s)} \qquad I_2(s) = \frac{24k}{P(s)}$$

where

$$P(s) = (4-4k^2)s^2 + 24s + 36$$

Further analysis now requires a specific value of the coupling coefficient k.

If $k = \frac{1}{2}$, then $P(s) = 3s^2 + 24s + 36 = 3(s^2 + 8s + 12)$ and

$$I_1(s) = \frac{2s^2 - 24}{s(s^2 + 8s + 12)} \qquad I_2(s) = \frac{4}{s^2 + 8s + 12}$$

Application of the initial-value theorem gives

$$i_1(0^+) = 2 = i_1(0^-) \qquad i_2(0^+) = 0 = i_2(0^-)$$

so the currents are continuous at $t = 0$. The waveforms for $t \geq 0$ in this case are

$$i_1(t) = -2 + 2e^{-2t} + 2e^{-6t} \qquad i_2(t) = e^{-2t} - e^{-6t}$$

which follow from partial-fraction expansion.

If the coils have unity coupling, then $k = 1$ and $P(s) = 24s + 36 = 24(s + 1.5)$ — a *first*-order polynomial. Thus,

$$I_1(s) = \frac{-3}{s(s+1.5)} \qquad I_2(s) = \frac{1}{s+1.5}$$

and the initial-value theorem now gives

$$i_1(0^+) = 0 \neq i_1(0^-) \qquad i_2(0^+) = 1 \neq i_2(0^-)$$

This result means that both currents jump discontinuously at $t = 0$. The waveforms for $t \geq 0$ are

$$i_1(t) = -2 + 2e^{-1.5t} \qquad i_2(t) = e^{-1.5t}$$

Returning to Fig. 13.18a, we see that $v_1(t) = v_{in} - 3i_1(t)$ and $v_2(t) = -12i_2(t)$ for all t, so the voltages remain finite despite the current jumps.

Exercise 13.17

Let the circuit in Example 13.14 have $k = 1$, and let capacitance $C = 1/10$ F be inserted between the 3-Ω resistor and the transformer. Find $I_1(s)$ and compare $i_1(0^+)$ with $i_1(0^-)$.

13.5 IMPULSES AND CONVOLUTION†

Taking a broader viewpoint, linear circuits constitute a subset of the general category of **linear systems**. Such systems may contain a variety of physical devices — electrical, mechanical, hydraulic, etc. — provided that they all obey the superposition principle.

Transform methods play an important role in the study of linear systems, along with more sophisticated methods involving *impulses* and *convolution*. This section introduces impulses and convolution specifically in the context of circuit analysis.

Impulses

To develop the impulse concept, let's begin with what appears to be a simple problem: A constant voltage V_0 is applied directly to an uncharged capacitor at $t = 0$, and we want to find the resulting current. Nothing that $v_C = 0$ for $t < 0$ and $v_C = V_0$ for $t > 0$, the capacitor voltage is described by

$$v_C(t) = V_0 u(t)$$

where $u(t)$ is the unit step. Differentiation then yields the current

$$i_C(t) = C\, dv_C/dt = CV_0\, du/dt$$

which involves the derivative of the unit step. But du/dt is undefined in the ordinary sense because $u(t)$ has a *discontinuity* at $t = 0$. We must therefore employ a limiting process to investigate the properties of du/dt.

Let $u_\epsilon(t)$ be the *continuous* function plotted in Fig. 13.19*a*. Clearly, $u_\epsilon(t)$ approaches the unit step as $\epsilon \rightarrow 0$, as shown in Fig. 13.19*b*, so

$$\lim_{\epsilon \rightarrow 0} u_\epsilon(t) = u(t)$$

Hence, it follows that

$$\lim_{\epsilon \rightarrow 0} du_\epsilon/dt = du/dt$$

With $\epsilon \neq 0$, the derivative of $u_\epsilon(t)$ is the rectangular pulse plotted in Fig. 13.19*c*. The pulse has width 2ϵ, height $1/2\epsilon$, and area $2\epsilon \times 1/2\epsilon = 1$. As $\epsilon \rightarrow 0$, the pulse becomes vanishingly narrow and infinitely high, while its

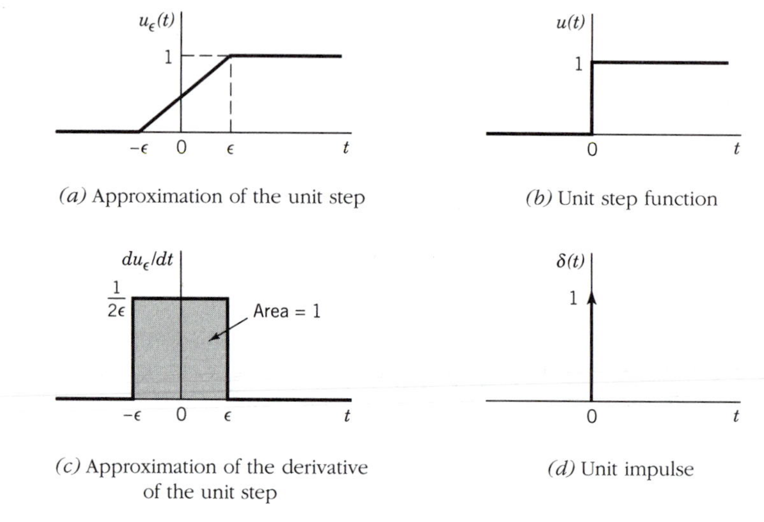

(*a*) Approximation of the unit step

(*b*) Unit step function

(*c*) Approximation of the derivative of the unit step

(*d*) Unit impulse

Figure 13.19

area remains constant at unity. We call this result the **unit impulse**, symbolized by $\delta(t)$ and represented pictorially in Fig. 13.19*d*. The number "1" next to the arrowhead denotes the unit area.

Based on our limiting process, we now write

$$du/dt = \delta(t)$$

where

$$\delta(t) = 0 \qquad t \neq 0 \tag{13.49a}$$

$$\int_{-\infty}^{\infty} \delta(t)\, dt = \int_{0^-}^{0^+} \delta(t)\, dt = 1 \tag{13.49b}$$

Taken together, Eqs. (13.49a) and (13.49b) mean that $\delta(t)$ has unit area concentrated at the instant $t = 0$. More generally, an impulse with area A concentrated at $t = t_0$ would be written as $A\delta(t - t_0)$ so that

$$A\delta(t - t_0) = 0 \qquad t \neq t_0$$

$$\int_{-\infty}^{\infty} A\delta(t - t_0)\, dt = \int_{t_0^-}^{t_0^+} A\delta(t - t_0)\, dt = A$$

Figure 13.20 shows the pictorial representation.

Although the impulse concept demands quite a stretch of the imagination, it does agree with physical reasoning in extreme situations. For instance, if a capacitor voltage jumps from zero to V_0 at $t = 0$, then the total charge transferred must be $q = CV_0$ so

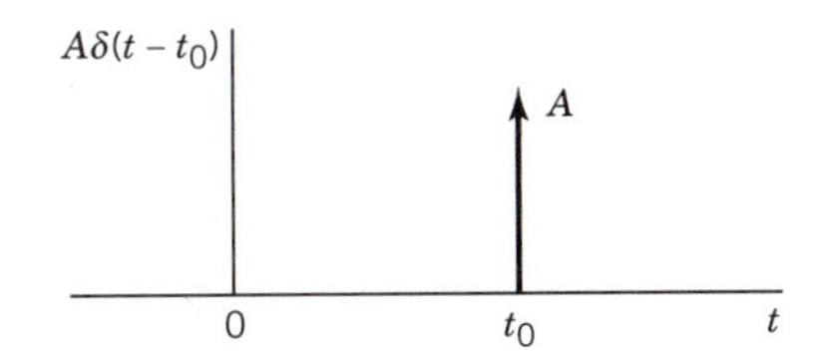

Figure 13.20 Delayed impulse with area A.

$$\int_{0^-}^{0^+} i_C(t)\, dt = CV_0$$

This expression implies that

$$i_C(t) = CV_0\delta(t)$$

which corresponds to $i_C(t) = C\, dv_C/dt = CV_0\, du/dt$ when $v_C(t) = V_0 u(t)$. Of course, an actual circuit always has some nonzero resistance preventing the current impulse needed to get an instantaneous voltage jump. Nonetheless, impulsive current serves as a useful *idealized model* when capacitor voltage changes in a very short time interval. Similar observations hold for impulsive voltages when inductor currents undergo abrupt changes.

Returning to Eqs. (13.49a) and (13.49b), observe that these properties do not adequately define $\delta(t)$ as a mathematical function. In fact, the impulse belongs to the special family of **generalized functions** or **distributions**, which are defined indirectly in terms of processes rather than direct equations. Specifically, taking $f(t)$ to be any ordinary function with a unique value at $t = t_0$, $\delta(t)$ is defined by the process

$$\int_{-\infty}^{\infty} f(t)\delta(t - t_0)\, dt = f(t)|_{t=t_0} = f(t_0) \tag{13.50}$$

Equation (13.50) is known as the **sampling process** because it extracts from $f(t)$ the sample value $f(t_0)$. As an immediate consequence of Eq. (13.50), we can write

$$f(t)\delta(t - t_0) = f(t_0)\delta(t - t_0) \tag{13.51}$$

which justifies our previous assertion that $\delta(t - t_0) = 0$ for $t \neq t_0$.

All other impulse properties also stem from the sampling process. In particular, consider the finite-range integral

$$\int_{t_1}^{t_2} f(t)\delta(t - t_0)\, dt \qquad t_1 < t_2$$

We recast this expression with infinite limits by letting

$$f_{12}(t) = f(t) \qquad t_1 < t < t_2$$
$$= 0 \qquad \text{otherwise}$$

so, from Eq. (13.50),

$$\int_{t_1}^{t_2} f(t)\delta(t - t_0)\, dt = \int_{-\infty}^{\infty} f_{12}(t)\delta(t - t_0)\, dt = f_{12}(t_0)$$

Thus,

$$\int_{t_1}^{t_2} f(t)\delta(t - t_0)\, dt = f(t_0) \qquad t_1 < t_0 < t_2$$
$$= 0 \qquad \text{otherwise} \tag{13.52}$$

Equation (13.52) says that sampling still occurs when the impulse "location" t_0 falls within the range of integration. Setting $t_1 = 0^-$, $t_2 = 0^+$, $f(t) = 1$, and $t_0 = 0$ takes us back to the unit-area property in Eq. (13.49b).

Finally, let λ be a dummy variable of integration and consider the integral

$$\int_{-\infty}^{t} \delta(\lambda - t_0)\, d\lambda$$

The impulse is located at $\lambda = t_0$, so it falls within the range of integration only when $t > t_0$. Hence,

$$\int_{-\infty}^{t} \delta(\lambda - t_0)\, d\lambda = 0 \qquad t < t_0$$
$$= 1 \qquad t > t_0$$

or, more compactly,

$$\int_{-\infty}^{t} \delta(\lambda - t_0)\, d\lambda = u(t - t_0) \tag{13.53}$$

Differentiating both sides of Eq. (13.53) then yields

$$\frac{d}{dt} u(t - t_0) = \delta(t - t_0) \tag{13.54}$$

which supports our initial statement that $du/dt = \delta(t)$.

Example 13.15 ***Discontinuous Inductor Current***

Suppose the current through an inductor has the waveform in Fig. 13.21*a*, where

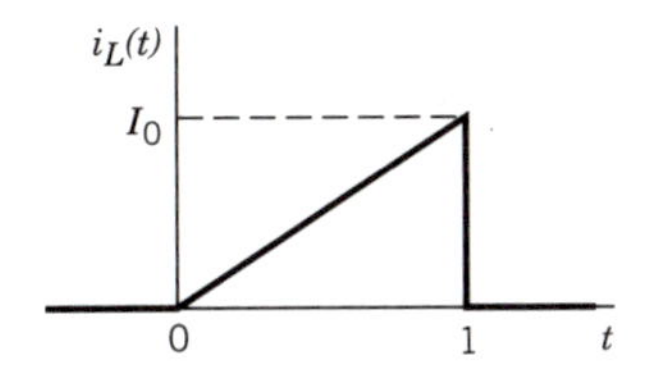

(a) Inductor current waveform

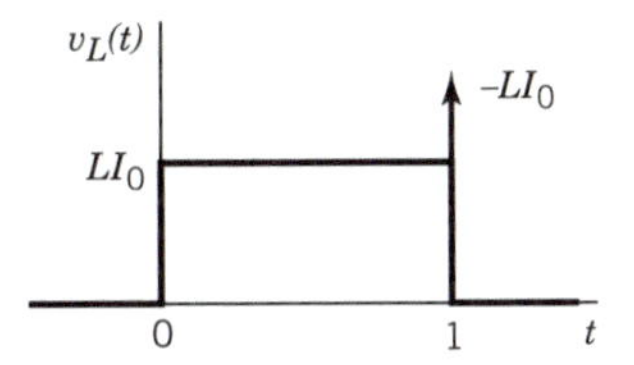

(b) Voltage waveform with an impulse

Figure 13.21

$$\begin{aligned} i_L(t) &= I_0 t \qquad 0 < t < 1 \\ &= 0 \qquad \text{otherwise} \end{aligned}$$

We want to find the resulting voltage $v_L(t)$ across the inductor.

We begin by writing the current for all t as

$$i_L(t) = I_0 t[u(t) - u(t-1)]$$

Then, using Eq. (13.54) and the chain rule for differentiation,

$$\begin{aligned} v_L(t) &= L\frac{d}{dt}\{I_0 t[u(t) - u(t-1)]\} \\ &= LI_0[u(t) - u(t-1)] + LI_0 t[\delta(t) - \delta(t-1)] \end{aligned}$$

But, from Eq. (13.51),

$$\begin{aligned} LI_0 t\delta(t) &= LI_0 t|_{t=0} \times \delta(t) = 0 \\ LI_0 t\delta(t-1) &= LI_0 t|_{t=1} \times \delta(t-1) = LI_0\delta(t-1) \end{aligned}$$

Thus,

$$v_L(t) = LI_0[u(t) - u(t-1)] - LI_0\delta(t-1)$$

which is plotted in Fig. 13.21*b*. The single impulse at $t = 1$ reflects the fact that $i_L(t)$ is continuous everywhere else.

Exercise 13.18

Obtain the results of the following expressions:

(a) $\dfrac{d}{dt}[t^3 u(t-2)]$ **(b)** $\displaystyle\int_1^{\infty} e^t[\delta(t) - \delta(t-2)]\,dt$

(c) $\displaystyle\int_{-\infty}^{t} [3\delta(\lambda) - 5\delta(\lambda - 2)]\,d\lambda$

Transforms with Impulses

The Laplace transform of the unit impulse is obtained from the defining integral

$$\mathcal{L}[\delta(t)] = \int_{0^-}^{\infty} \delta(t)e^{-st}\,dt$$

The lower limit at $t = 0^-$ ensures that the range of integration includes the impulse located at $t = 0$, so the sampling process applies here and

$$\mathcal{L}[\delta(t)] = e^{-st}|_{t=0} = 1 \tag{13.55}$$

Hence, the transform of $\delta(t)$ equals the constant 1. More generally,

$$\mathcal{L}[A\delta(t - t_0)] = Ae^{-st_0} \tag{13.56}$$

which follows from basic transform properties.

For an independent confirmation of Eq. (13.55), we use $\delta(t) = du/dt$ together with the differentiation property of the Laplace transform, which yields

$$\mathcal{L}[\delta(t)] = \mathcal{L}[du/dt] = s\mathcal{L}[u(t)] - u(0^-)$$

But $\mathcal{L}[u(t)] = 1/s$ and $u(0^-) = 0$, so

$$\mathcal{L}[\delta(t)] = s(1/s) - 0 = 1$$

Equation (13.55) is thus consistent with our previous results.

As the converse of Eq. (13.55), the inverse Laplace transform of an s-domain constant must be a time-domain impulse. Specifically, for any constant A.

$$\mathcal{L}^{-1}[A] = A\delta(t) \tag{13.57}$$

This result allows us to take the inverse transform of a rational function when the order of the numerator polynomial equals the order of the denominator polynomial.

Consider, for example, the inverse transform of $F(s) = 3s/(s + 2)$. This function can be reduced to a constant plus a strictly proper function by dividing the denominator into the numerator as follows:

$$\begin{array}{r|l} & \;\;\;\;\;\;\;3 \\ \hline s + 2 & 3s \\ & \underline{3s + 6} \\ & \;\;\;\; - 6 \end{array}$$

Upon putting the remainder over the denominator, we have

$$F(s) = \frac{3s}{s+2} = 3 + \frac{-6}{s+2}$$

Thus,

$$f(t) = \mathcal{L}^{-1}[F(s)] = 3\delta(t) - 6e^{-2t}$$

To generalize on our simple example, let $F(s)$ be any ratio of polynomials with $m = n$, say

$$F(s) = \frac{N(s)}{D(s)} = \frac{b_n s^n + \cdots + b_1 s + b_0}{s^n + \cdots + a_1 s + a_0}$$

Long division then yields

$$F(s) = b_n + G(s) \qquad G(s) = \frac{R(s)}{D(s)} \tag{13.58a}$$

where the remainder polynomial $R(s)$ has order $m_R \leq n - 1$. Since $G(s)$ is strictly proper, standard inversion methods hold for $g(t) = \mathcal{L}^{-1}[G(s)]$, and

$$f(t) = b_n\delta(t) + g(t) \tag{13.58b}$$

When $f(t)$ contains an impulse at $t = 0$, as in Eq. (13.58b), the value of $f(0)$ is undefined. However, $\delta(0^+) = 0$ so we can meaningfully speak of the initial value just after the impulse, namely

$$f(0^+) = g(0^+)$$

Therefore, if desired, $f(0^+)$ may be obtained by applying the initial-value theorem to $G(s)$.

Example 13.16 ***Impulsive Zero-State Response***

Suppose you want to find the zero-state response of $i(t)$ in Fig. 13.22 when $v(t) = 80u(t)$ V. You suspect that $i(t)$ will include an impulse because the voltage step appears directly across two uncharged capacitors.

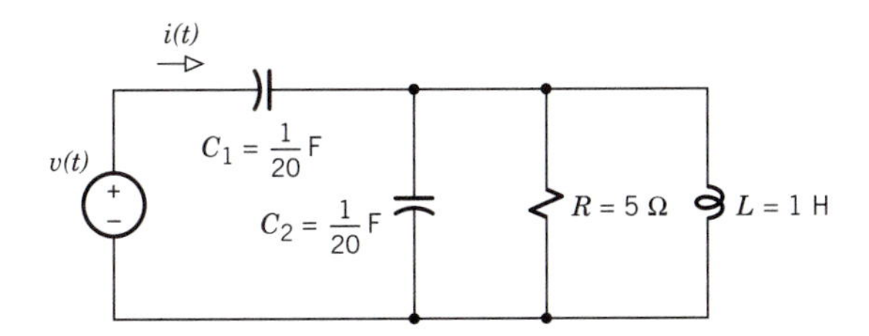

Figure 13.22 Circuit for Example 13.16.

Proceeding with the analysis, you take advantage of the zero-state condition and calculate the impedance

$$Z(s) = \frac{V(s)}{I(s)} = \frac{20}{s} + \frac{1}{\frac{s}{20} + \frac{1}{5} + \frac{1}{s}} = \frac{40s^2 + 80s + 400}{s(s^2 + 4s + 20)}$$

Then, with $V(s) = 80/s$,

$$I(s) = \frac{V(s)}{Z(s)} = \frac{2s^2 + 8s + 40}{s^2 + 2s + 10}$$

or, after long division,

$$I(s) = 2 + G(s) \qquad G(s) = \frac{4s + 20}{s^2 + 2s + 10}$$

Inverse transformation now gives

$$i(t) = 2\delta(t) + 4.80e^{-t} \cos(3t - 33.7°)$$

The initial value just after the impulse is

$$i(0^+) = 4.80 \cos(-33.7°) = 4.00$$

Alternatively, you might note that

$$g(0^+) = \lim_{s \to \infty} sG(s) = 4$$

so $i(0^+) = g(0^+) = 4$.

Exercise 13.19

Use the differentiation property to show that

$$\mathcal{L}[A\delta'(t)] = As$$

Exercise 13.20

Rework Example 13.16 with $C_1 = C_2 = 1/8$ F and $R = 1\ \Omega$, so $Z(s) = (16s^2 + 64s + 64)/s(s^2 + 8s + 8)$.

Convolution and Impulse Response

We conclude this section with an introduction to the study of circuit dynamics using *convolution.* Convolution is an integration process that can be done by hand in simple cases. Moreover, it particularly suits the task of obtaining complicated response waveforms with the help of numerical integration routines on a computer.

Convolution analysis focuses on the *zero-state response* of a circuit driven by a single input. We'll show that the circuit can be represented in the time domain by its *impulse response* $h(t)$ and that the zero-state response $y(t)$ equals the convolution of $h(t)$ and the input $x(t)$. Thus, the analysis can be performed entirely in the *time domain.* But before we get to the circuit applications, we must first discuss the general definition and properties of convolution.

Let $f(t)$ and $g(t)$ be any two time functions. The **convolution** of $f(t)$ and $g(t)$ is an integral operation defined

$$f(t) * g(t) \triangleq \int_{-\infty}^{\infty} f(\lambda) g(t-\lambda)\, d\lambda \tag{13.59}$$

The notation $f(t) * g(t)$ stands for the integration on the right in Eq. (13.59), and the asterisk should not be confused with complex conjugation. The integration is always performed with respect to a dummy variable, such as λ. Thus, although convolving $f(t)$ with $g(t)$ produces another function of time, t acts as a constant insofar as the integration is concerned.

Since λ is a dummy variable in Eq. (13.59), we may make the change of variable $\mu = t - \lambda$. Then $\lambda = t - \mu$ and $d\lambda = -d\mu$, so

$$\begin{aligned}\int_{-\infty}^{\infty} f(\lambda) g(t-\lambda)\, d\lambda &= -\int_{+\infty}^{-\infty} f(t-\mu) g(\mu)\, d\mu \\ &= \int_{-\infty}^{\infty} g(u) f(t-\mu)\, d\mu = g(t) * f(t)\end{aligned}$$

We have thereby established the **commutative property** that

$$f(t) * g(t) = g(t) * f(t) \tag{13.60}$$

Equation (13.60) means that we are free to interchange the two functions being convolved, which may simplify the integration.

Another convolution relationship is the **distributive property** obtained by setting $g(t) = g_1(t) + g_2(t)$ in Eq. (13.59). Regrouping terms then yields

$$f(t) * [g_1(t) + g_2(t)] = [f(t) * g_1(t)] + [f(t) * g_2(t)] \tag{13.61}$$

Convolution with a sum of functions may therefore be treated as a sum of convolutions.

A third important relationship is the **replication property** when $g(t) = \delta(t)$. Using the commutative property of convolution and the sampling property of the impulse we get

$$\begin{aligned} f(t) * \delta(t) &= \delta(t) * f(t) \\ &= \int_{-\infty}^{\infty} \delta(\lambda) f(t-\lambda)\, d\lambda = f(t-\lambda)\big|_{\lambda=0} \end{aligned}$$

Thus,

$$f(t) * \delta(t) = f(t) \tag{13.62}$$

which says that convolving a unit impulse with an ordinary function $f(t)$ simply reproduces $f(t)$.

When convolution involves step functions, a *graphical interpretation* proves to be very helpful. By way of example, take the functions in Fig. 13.23*a* where

$$f(t) = 5e^{-t}u(t) \qquad g(t) = (3 + t)u(t)$$

so

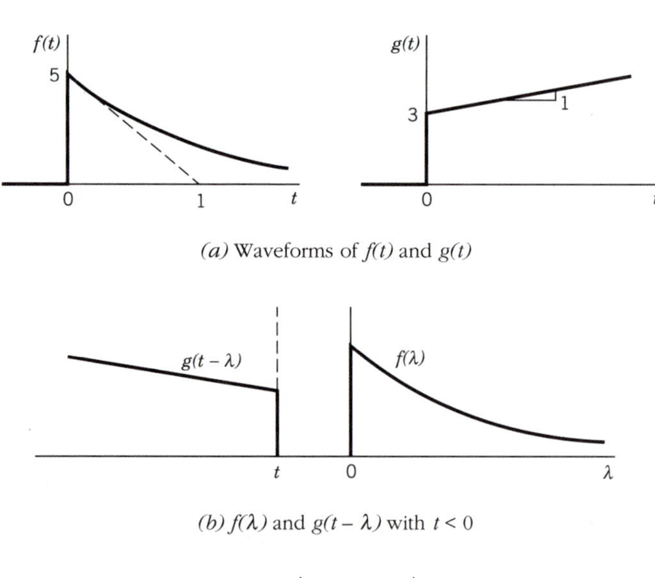

(a) Waveforms of $f(t)$ and $g(t)$

(b) $f(\lambda)$ and $g(t-\lambda)$ with $t < 0$

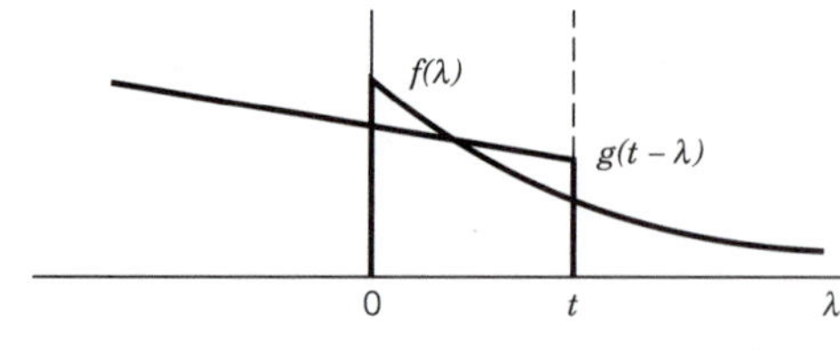

(c) $f(\lambda)$ and $g(t-\lambda)$ with $t > 0$

Figure 13.23 Graphical interpretation of convolution.

$$f(t) * g(t) = \int_{-\infty}^{\infty} 5e^{-\lambda}\, u(\lambda)(3 + t - \lambda)u(t - \lambda)\, d\lambda$$

Since the two step functions equal either zero or one, we can account for their effect by sketching $f(\lambda)$ and $g(t - \lambda)$ versus λ. Clearly, $f(\lambda)$ has the same shape as $f(t)$, but the sketch of $g(t - \lambda)$ requires two steps:

1. Reverse $g(t)$ in time and replace t with λ to get $g(-\lambda)$;
2. Shift $g(-\lambda)$ by t units to the right to get $g[-(\lambda - t)] = g(t - \lambda)$ for any particular value of t.

Figure 13.23*b* shows $f(\lambda)$ together with $g(t - \lambda)$ when $t < 0$. The value of t always equals the distance from the origin of $f(\lambda)$ to the shifted origin of $g(t - \lambda)$ indicated by the dashed line. Incidentally the name convolution comes from the German word for *folding*, which describes the reversed shape of $g(t - \lambda)$ in this figure.

To evaluate $f(t) * g(t)$ over $-\infty < t < \infty$, $g(t - \lambda)$ must slide from left to right with respect to $f(\lambda)$. Accordingly, the convolution integrand may change with t. Indeed, we see in Fig. 13.23*b* that the functions don't "overlap" when $t < 0$, so $f(\lambda)g(t - \lambda) = 0$ and

$$f(t) * g(t) = 0 \qquad t < 0$$

However, for $t > 0$, Fig. 13.23*c* shows that the functions overlap over the range $0 < \lambda < t$. The limits of integration thus become 0 to t, and

$$\begin{aligned} f(t) * g(t) &= \int_0^t 5e^{-\lambda}\,(3 + t - \lambda)\, d\lambda \\ &= 5(3 + t)\int_0^t e^{-\lambda}\, d\lambda - 5\int_0^t \lambda e^{-\lambda}\, d\lambda \\ &= 10 - 10e^{-t} + 5t \qquad t > 0 \end{aligned}$$

The complete result plotted in Fig. 13.24 illustrates that convolution has a *smoothing* effect because the discontinuities in $f(t)$ and $g(t)$ do not appear in $f(t) * g(t)$.

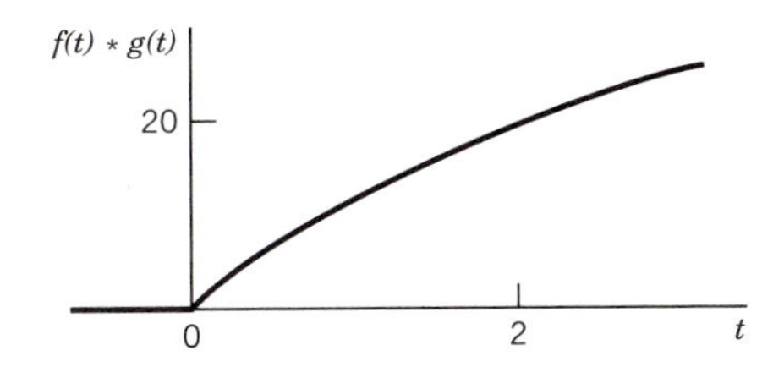

Figure 13.24 Waveform produced by convolution.

Our graphical example also illustrates that certain simplifications occur when the convolved functions start at finite instants of time. In particular, suppose both $f(t)$ and $g(t)$ are **causal functions**, defined by

$$f(t) = g(t) = 0 \qquad t < 0 \tag{13.63}$$

Since $f(\lambda) = 0$ for $\lambda < 0$ while $g(t - \lambda) = 0$ for $\lambda > t$, the convolution integral simplifies to

$$f(t) * g(t) = \int_{0^-}^{\infty} f(\lambda)g(t - \lambda)\, d\lambda = \int_{0^-}^{t} f(\lambda)g(t - \lambda)\, d\lambda \tag{13.64a}$$

Additionally, from the graphical interpretation,

$$f(t) * g(t) = 0 \qquad t < 0 \tag{13.64b}$$

The convolution of two causal functions therefore produces another causal function. Equations (13.64a) and (13.64b) have particular importance in conjunction with Laplace transforms because the transform ignores all time behavior before $t = 0$, equivalent to assuming causal time functions.

Now consider the Laplace transform of the convolution of causal functions. From Eq. (13.64a) and the definition of the transform,

$$\begin{aligned}\mathcal{L}[f(t) * g(t)] &= \int_{0^-}^{\infty} \left[\int_{0^-}^{\infty} f(\lambda)g(t - \lambda)\, d\lambda\right] e^{-st}\, dt \\ &= \int_{0^-}^{\infty} f(\lambda) \left[\int_{0^-}^{\infty} g(t - \lambda)e^{-st}\, dt\right] d\lambda\end{aligned}$$

where we have interchanged the order of integration. Next, let $\mu = t - \lambda$ so $t = \mu + \lambda$ and

$$\begin{aligned}\mathcal{L}[f(t) * g(t)] &= \int_{0^-}^{\infty} f(\lambda) \left[\int_{-\lambda}^{\infty} g(\mu)e^{-s(\mu+\lambda)}\, d\mu\right] d\lambda \\ &= \int_{0^-}^{\infty} f(\lambda)e^{-s\lambda} \left[\int_{-\lambda}^{\infty} g(\mu)e^{-s\mu}\, d\mu\right] d\lambda\end{aligned}$$

But the causal nature of $g(t)$ means that $g(\mu) = 0$ for $\mu < 0$. The double integration thus separates into a product of two transform integrals since

$$\begin{aligned}\mathcal{L}[f(t) * g(t)] &= \int_{0^-}^{\infty} f(\lambda)e^{-s\lambda} \left[\int_{0^-}^{\infty} g(\mu)e^{-s\mu}\, d\mu\right] d\lambda \\ &= \int_{0^-}^{\infty} f(\lambda)e^{-s\lambda}\, d\lambda \times \int_{0^-}^{\infty} g(\mu)e^{-s\mu}\, d\mu \\ &= \mathcal{L}[f(t)] \times \mathcal{L}[g(t)]\end{aligned}$$

Finally, writing $\mathcal{L}[f(t)] = F(s)$ and $\mathcal{L}[g(t)] = G(s)$, we obtain the simple result

$$\mathcal{L}[f(t) * g(t)] = F(s)G(s) \tag{13.65a}$$

Hence, convolution in the time domain because *multiplication* in the s domain. Conversely, the inverse transform of Eq. (13.65a) yields

$$\mathcal{L}^{-1}[F(s)G(s)] = f(t) * g(t) \tag{13.65b}$$

These results complete our preparation for the application of convolution to circuit analysis.

Now let $y(t)$ be any zero-state response of a linear circuit with input $x(t)$ applied at $t = 0$. If we know the appropriate network function $H(s)$, then the Laplace transforms $Y(s) = \mathcal{L}[y(t)]$ and $X(s) = \mathcal{L}[x(t)]$ are related by

$$Y(s) = H(s)X(s)$$

Figure 13.25a portrays this relationship in the form of an **s-domain block diagram**.

We return to the time domain by taking the inverse transform of $Y(s)$. Drawing upon Eq. (13.65b) we get

$$y(t) = \mathcal{L}^{-1}[H(s)X(s)] = \mathcal{L}^{-1}[H(s)] * \mathcal{L}^{-1}[X(s)]$$

or, more compactly,

$$y(t) = h(t) * x(t) \tag{13.66}$$

where

$$h(t) = \mathcal{L}^{-1}[H(s)] \tag{13.67}$$

Figure 13.25b shows the corresponding **time-domain block diagram**. The circuit's properties are represented here by $h(t)$ rather than $H(s)$ since Eq. (13.67) implies that $H(s)$ can be obtained from $h(t)$ via $H(s) = \mathcal{L}[h(t)]$.

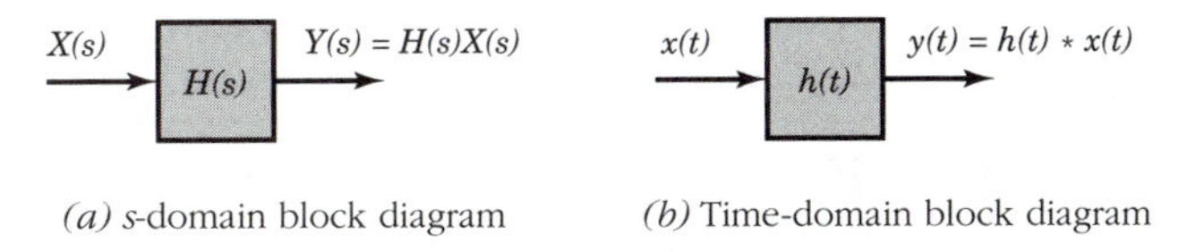

(*a*) s-domain block diagram (*b*) Time-domain block diagram

Figure 13.25

Clearly, all the information available from $H(s)$ is also contained in $h(t)$. But what's the physical interpretation of this important time function? To answer that question, recall that $f(t) * \delta(t) = f(t)$ for any $f(t)$. Thus, with $x(t) = \delta(t)$, Eq. (13.66) becomes

$$y(t) = h(t) * \delta(t) = h(t)$$

so $h(t)$ equals the zero-state response produced by a unit impulse at the input.

We therefore call $h(t)$ the **impulse response**. The impulse response of any real circuit always has the causal property

$$h(t) = 0 \qquad t < 0$$

After all, the response to $\delta(t)$ cannot appear before the impulse occurs at $t = 0$.

Another interpretation of the impulse response involves the *step response*, which we'll symbolize by $y_u(t)$. Since $y_u(t)$ is produced when $x(t) = u(t)$, and since $u(t) = 0$ for $t < 0$, the step response of a real circuit must also be causal. Furthermore, the transform of $y_u(t)$ is

$$Y_u(s) = H(s) \times \mathcal{L}[u(t)] = H(s)/s$$

from which

$$H(s) = sY_u(s)$$

But, nothing that $y_u(0^-) = 0$, we can write

$$H(s) = sY_u(s) = sY_u(s) - y_u(0^-) = \mathcal{L}[y_u'(t)]$$

The inverse transform then yields

$$h(t) = y_u'(t) \tag{13.68}$$

Hence, the impulse response of a circuit equals the *derivative* of the step response.

In the general study of linear systems, particular systems are often represented by block diagrams consisting of several functional blocks. The system's impulse response can then be found by assuming a unit impulse or step at the input and determining the resulting output. In circuit analysis, however, we usually work with s-domain circuit diagrams and we find $h(t)$ by taking the inverse transform of $H(s)$.

Example 13.17 ***Impulse Response and Pulse Response***

Figure 13.26*a* shows a critically damped *LRC* circuit with zero-state conditions $i(0^-) = v_C(0^-) = 0$, so the s-domain diagram takes the form of Fig. 13.26*b*. We'll use convolution to obtain $i(t)$ when

$$v(t) = \frac{1}{D}[u(t) - u(t - D)]$$

which is a rectangular pulse with height $1/D$, duration D, and unit area.

Convolution analysis requires that we first determine the impulse response. Since the output in question is $i(t)$, the relevant network function is

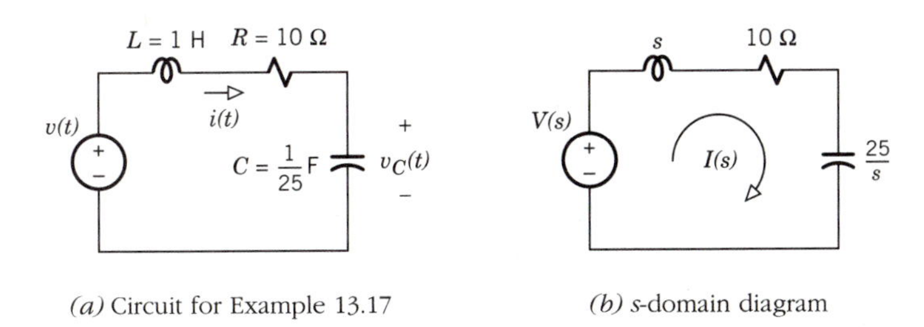

(*a*) Circuit for Example 13.17

(*b*) *s*-domain diagram

Figure 13.26

$$H(s) = \frac{I(s)}{V(s)} = \frac{1}{Z(s)} = \frac{1}{s + 10 + 25/s}$$

Alternatively, we could have started with $I(s) = V(s)/Z(s)$ and assumed that $v(t) = \delta(t)$, so $i(t) = h(t)$, $V(s) = 1$, and $H(s) = I(s) = 1/Z(s)$. After rearrangement and expansion, the network function becomes

$$H(s) = \frac{s}{s^2 + 10s + 25} = \frac{s}{(s + 5)^2} = \frac{1}{s + 5} + \frac{-5}{(s + 5)^2}$$

Thus, for t > 0,

$$h(t) = \mathcal{L}^{-1}[H(s)] = (1 - 5t)e^{-5t}$$

and we know that $h(t) = 0$ for $t < 0$. This impulse response is sketched in Fig. 13.27*a*.

Proceeding with the convolution, the graphical interpretation of $i(t) = h(t) * v(t)$ in Fig. 13.27*b* leads to three distinct cases, as follows: For $t < 0$,

$$i(t) = 0$$

For $0 < t < D$,

$$i(t) = \int_0^t (1 - 5\lambda)e^{-5\lambda} \frac{1}{D}\, d\lambda = \frac{t}{D} e^{-5t}$$

For $t > D$,

$$i(t) = \int_{t-D}^t (1 - 5\lambda)e^{-5\lambda} \frac{1}{D}\, d\lambda = \frac{t}{D} e^{-5t} - \frac{t - D}{D} e^{-5(t-D)}$$

The complete pulse response is plotted in Figs. 13.27*c* and 13.27*d* taking $D = 0.5$ and 0.1, respectively.

These drawings show that the pulse response $i(t)$ begins to look like the impulse response $h(t)$ when the duration of the input pulse gets small. In fact, since the input pulse has unit area, $v(t)$ approaches $\delta(t)$ as $D \to 0$ and, correspondingly, $i(t)$ approaches $h(t)$. But even with a nonzero value of D, we

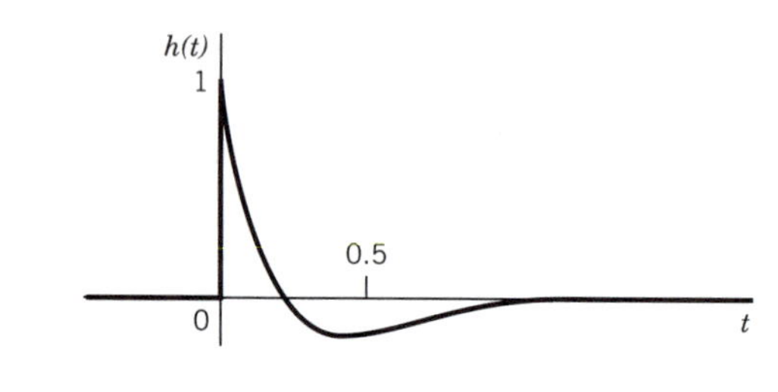

(a) Impulse response

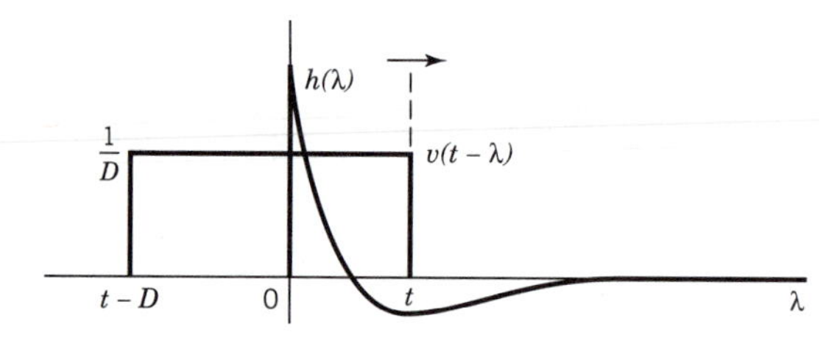

(b) Convolution with rectangular pulse

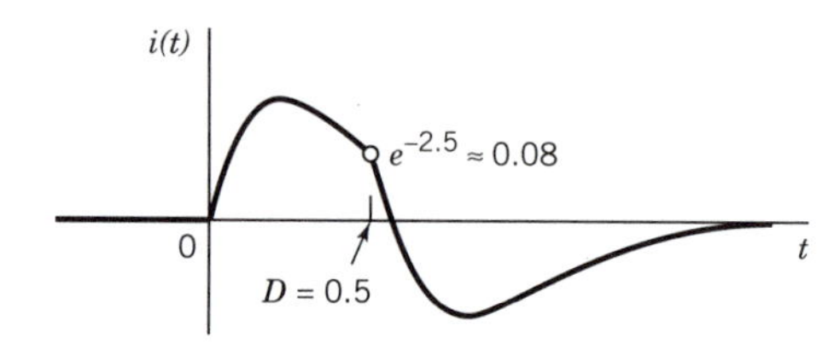

(c) Pulse response with $D = 0.5$

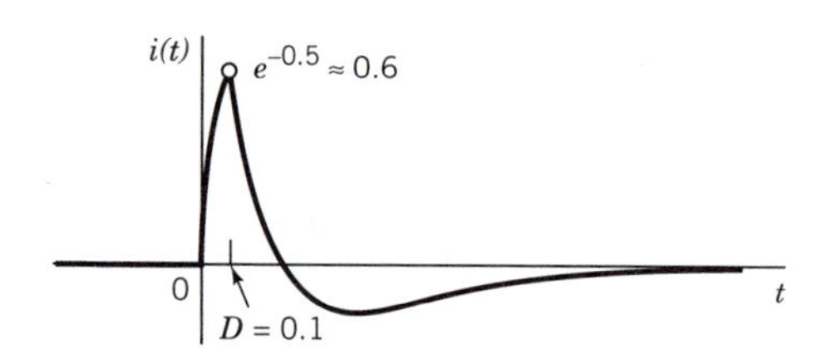

(d) Pulse response with $D = 0.1$

Figure 13.27

get $i(t) \approx h(t)$ if D is small enough. Under that condition, the input pulse has essentially the same effect as an impulse.

Exercise 13.21

Use the graphical interpretation of convolution to obtain $f(t) * g(t)$, as shown in Fig. 13.28.

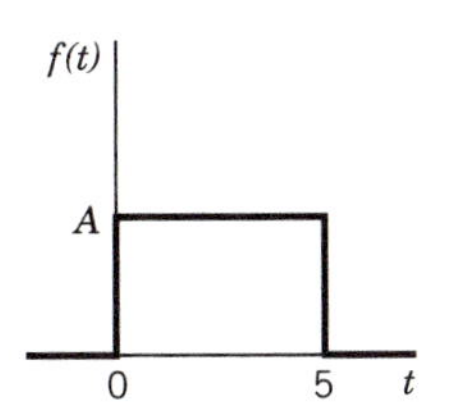

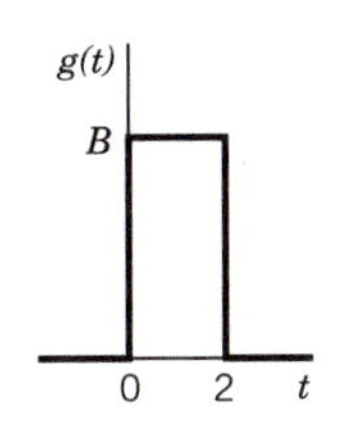

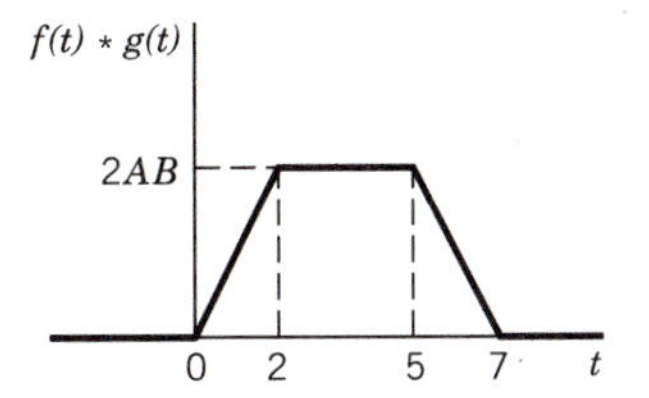

Figure 13.28

Exercise 13.22

Find $h(t)$ when $v_C(t)$ is the response of interest in Fig. 13.26*a*. Then find $v_C(t) = h(t) * v(t)$ when $v(t) = u(t - 2)$.

PROBLEMS

Section 13.1 Laplace Transforms

13.1* Use Eq. (13.1) to find $F(s)$ when

$$f(t) = 1 - t \qquad 0 < t < 1$$
$$= 0 \qquad \text{otherwise}$$

13.2 Use Eq. (13.1) to find $F(s)$ when

$$f(t) = t \qquad 0 < t < 1$$
$$= 1 \qquad 1 < t < 2$$
$$= 0 \qquad \text{otherwise}$$

13.3 Use Eq. (13.1) to find $F(s)$ when

$$f(t) = \sin \pi t \qquad 0 < t < 1$$
$$= 0 \qquad \text{otherwise}$$

13.4 Use Eq. (13.1) to find $F(s)$ when

$$f(t) = t \cos \pi t \qquad 0 < t < 1$$
$$= 0 \qquad \text{otherwise}$$

13.5 Use Eq. (13.1) to find $F(s)$ when

$$f(t) = t \qquad 0 < t < 1$$
$$= 2 - t \qquad 1 < t < 2$$
$$= 0 \qquad \text{otherwise}$$

13.6 Derive from Eq. (13.1) the **time-scaling property**

$$\mathcal{L}[f(t/\tau)] = \tau F(s\tau)$$

13.7 Let $g(t) = f'(t)$ to derive Eq. (13.13b) from Eq. (13.13a).

13.8 Use Eq. (13.13b) to obtain the expression for $\mathcal{L}[d^3f/dt^3]$.

13.9 Replace $F(s)$ with the transform integral to derive the property

$$\int_s^{\infty} F(s)\, ds = \mathcal{L}[t^{-1}f(t)]$$

13.10 Consider the **semiperiodic function**

$$g(t) = f(t) + f(t - T)u(t - T)$$
$$+ f(t - 2T)u(t - 2T) + \cdots$$

where $f(t) = 0$ outside $0 < t < T$. Show that

$$\mathcal{L}[g(t)] = F(s)/(1 - e^{-sT})$$

Hint: Use the fact that $1 + a + a^2 + \ldots = (1 - a)^{-1}$.

13.11 Use Tables 13.1 and 13.2 to find the transform of the **hyperbolic functions**

$$\sinh at = \tfrac{1}{2}(e^{at} - e^{-at}) \qquad \cosh at = \tfrac{1}{2}(e^{at} + e^{-at})$$

13.12* Given that $\mathcal{L}[t] = 1/s^2$, use an appropriate transform property to find $\mathcal{L}[t^2]$ and $\mathcal{L}[t^3]$. Extrapolate your results for $\mathcal{L}[t^r]$.

13.13 Find $\mathcal{L}[t \cos \beta t]$ using the known transform of $\cos \beta t$.

13.14 Find $\mathcal{L}[\sin^2 at]$ using Table 13.2 and a trigonometric identity.

13.15 Express $f(t)$ in Fig. P13.15 as a linear combination of simple functions to obtain $F(s)$.

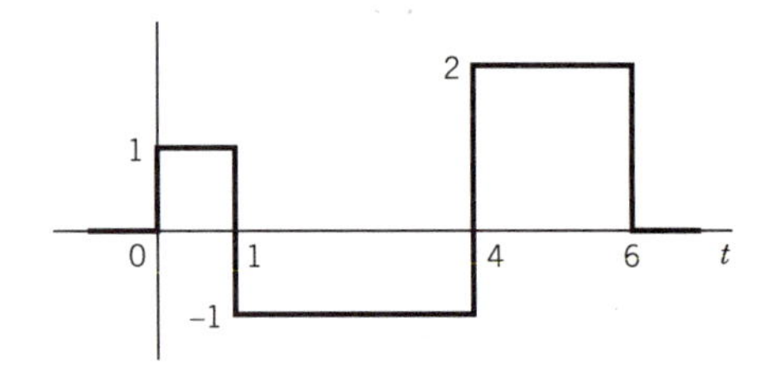

Figure P13.15

13.16 Express $f(t)$ in Fig. P13.16 as a linear combination of simple functions to obtain $F(s)$.

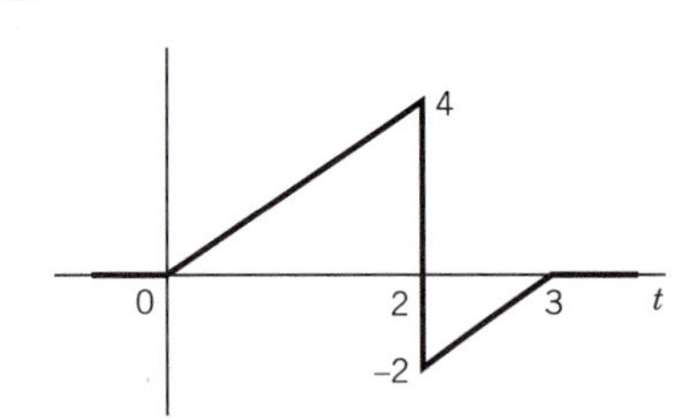

Figure P13.16

13.17 Express $f(t)$ in Fig. P13.17 as a linear combination of simple functions to obtain $F(s)$.

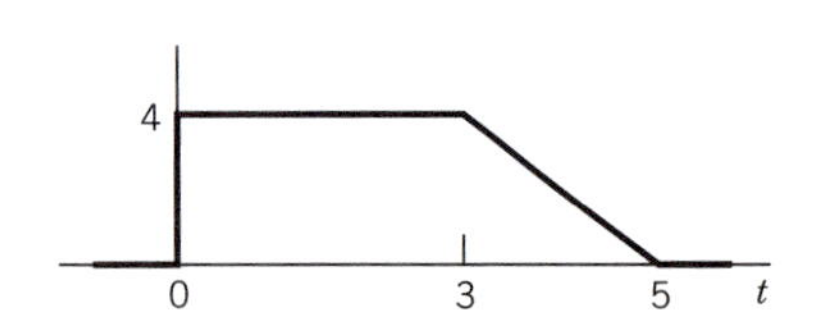

Figure P13.17

13.18 Express $f(t)$ in Fig. P13.18 as a linear combination of simple functions to obtain $F(s)$.

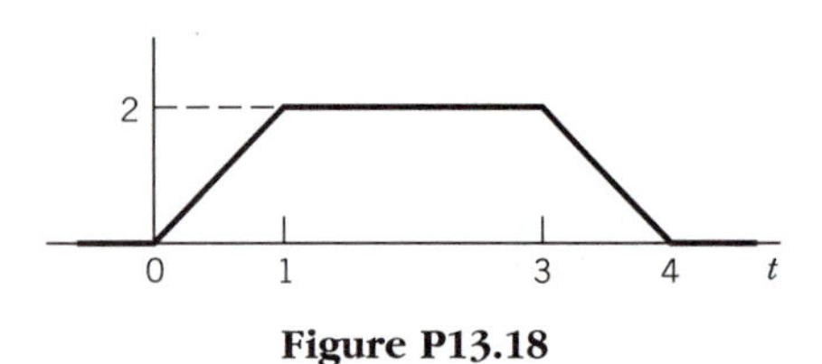

Figure P13.18

13.19* Use transforms and Table 13.2 to find $Y(s)$ and $y(t)$ for $t \geq 0$, given that

$$y' + 4y = 10te^{-4t}u(t) \qquad y(0^-) = -3$$

13.20 Do Problem 13.19 with

$$y'' + 2y' + 10y = 0 \quad y(0^-) = 6 \quad y'(0^-) = -6$$

13.21 Do Problem 13.19 with

$$2y' + 6y = x \qquad x' + x = -2y$$
$$y(0^-) = 0 \quad x(0^-) = -10$$

13.22 Do Problem 13.19 with

$$y' + 3y = x \qquad x' = 3y' - 16y$$
$$y(0^-) = 10 \qquad x(0^-) = 30$$

Section 13.2 Transform Inversion

13.23* Obtain the partial-fraction expansion and inverse transform of

$$F(s) = \frac{5s^2 + 10s + 24}{s^3 + 6s^2 + 8s}$$

13.24 Obtain the partial-fraction expansion and inverse transform of

$$F(s) = \frac{27s + 138}{(s - 1)(s^2 + 14s + 40)}$$

13.25 Obtain the partial-fraction expansion and inverse transform of

$$F(s) = \frac{-s^3 - 4s^2 - 12s - 96}{(s^2 + 4s)(s^2 + 8s + 12)}$$

13.26 Use Eq. (13.27) and the method of undetermined coefficients to find $f(t)$ when

$$F(s) = \frac{-2s^2 + 58s - 616}{(s + 3)(s^2 + 4s + 104)}$$

13.27 Do Problem 13.26 with

$$F(s) = \frac{5s^2 - 100s - 770}{(s + 4)(s^2 + 12s + 61)}$$

13.28 Do Problem 13.26 with

$$F(s) = \frac{8s^3 + 82s^2 + 240s + 200}{s(s + 4)(s^2 + 6s + 10)}$$

13.29 Do Problem 13.26 with

$$F(s) = \frac{-14s^2 + 92s - 50}{(s + 1)(s + 2)(s^2 + 25)}$$

13.30 Use the method of undetermined coefficients to find $f(t)$ when

$$F(s) = (2s^2 + 30s + 180)/s(s + 6)^2$$

13.31 Use the method of undetermined coefficients to find $f(t)$ when

$$F(s) = (4s^2 + 12s - 10)/(s + 3)^2(s + 4)$$

13.32* Determine $f(t)$ by applying Eq. (13.30) to

$$F(s) = (12s^2 - 250)/s^2(s + 5)^2$$

13.33 Determine $f(t)$ by applying Eq. (13.30) to

$$F(s) = (-2s + 2)/(s + 1)^2(s + 2)^2$$

13.34 Determine $f(t)$ by applying Eq. (13.30) to

$$F(s) = (16s + 64)/(s^2 + 16)^2$$

13.35 Determine $f(t)$ by applying Eq. (13.30) to

$$F(s) = (-2s^2 - 4s + 6)/(s^2 + 2s + 5)^2$$

13.36 Using Eq. (13.32), find $f(t)$ when

$$F(s) = (3s - 12)/(s + 2)^3(s + 5)$$

13.37 Using Eq. (13.32), find $f(t)$ when

$$F(s) = (3s + 2)/s(s + 1)^3(s + 2)$$

13.38 Derive the extension of Eq. (13.32) for the case when $F(s)$ includes a quadruple pole.

13.39* Obtain the partial-fraction expansion and inverse transform of

$$F(s) = \frac{-3s + 15 + (2s^2 + 6s)e^{-4s}}{s^3 + 8s^2 + 15s}$$

13.40 Obtain the partial-fraction expansion and inverse transform of

$$F(s) = \frac{4s^2 + 40s + (10s + 20)e^{-3s}}{s^3 + 12s^2 + 20s}$$

13.41 Obtain the partial-fraction expansion and inverse transform of

$$F(s) = \frac{10s + 40 - (20s + 40)e^{-s}}{s^3 + 6s^2 + 8s}$$

13.42 Derive from the initial-value theorem an expression for $f''(0^+)$ in terms of $F(s) = N(s)/D(s)$ and the initial conditions.

13.43* Evaluate $f(0^+)$ and $f'(0^+)$ given

$$F(s) = \frac{-5s^2 + 12}{s(s + 1)^2(s + 3)} \qquad f(0^-) = 0$$

Also evaluate $f(\infty)$ or state why the limit does not exist.

13.44 Do Problem 13.43 with

$$F(s) = \frac{3s^2 + 20s - 4}{(s + 2)(s + 5)^2} \qquad f(0^-) = 3$$

13.45 Do Problem 13.43 with

$$F(s) = \frac{-3s^2 + 2}{(s^2 + 4)(s^2 + 6s + 5)} \qquad f(0^-) = 0$$

13.46 Do Problem 13.43 with

$$F(s) = \frac{2s^3 + 9s^2 + 16}{s(s - 1)(s + 4)^2} \qquad f(0^-) = 2$$

Section 13.3 Transform Circuit Analysis

13.47* Find the zero-state response $y(t)$ when

$$x(t) = 2 \qquad t \geq 0$$
$$H(s) = (7s + 5)/(s^2 + 7s + 10)$$

13.48 Find the zero-state response $y(t)$ when

$$x(t) = 10 \qquad t \geq 0$$
$$H(s) = (s^2 + 30s)/(s + 10)(s^2 - 100)$$

13.49 Find the zero-state response $y(t)$ when

$$x(t) = 20 \qquad t \geq 0$$
$$H(s) = (4s + 50)/s^2 + 4s + 200)$$

13.50 Find the zero-state response $y(t)$ when

$$x(t) = e^{-t} \qquad t \geq 0$$
$$H(s) = (s^2 + 2s - 2)/s(s + 1)^2$$

13.51 Find the zero-state response $y(t)$ when

$$x(t) = 1 - e^{-3t} \qquad t \geq 0$$
$$H(s) = (2s + 30)/1(s^2 + 8s + 15)$$

13.52 Find the zero-state response $y(t)$ when

$$x(t) = u(t) - u(t - 4)$$
$$H(s) = -3s/(s^2 + 2s + 10)$$

13.53 Find the zero-state response $y(t)$ when

$$x(t) = 3u(t) + u(t - 1)$$
$$H(s) = (s^2 + 18s)/(s^2 - 6s - 40)$$

13.54* Calculate the zero-state ac response $y(t)$ given

$$x(t) = 20 \cos (5t + 90°) \qquad t \geq 0$$
$$H(s) = (s + 10)/s(s + 5)$$

13.55 Calculate the zero-state ac response $y(t)$ given

$$x(t) = 10 \cos (4t - 36.9°) \qquad t \geq 0$$
$$H(s) = 13s/(s + 3)(s + 6)$$

13.56 Calculate the zero-state ac response $y(t)$ given

$$x(t) = 4 \cos 12t \qquad t \geq 0$$
$$H(s) = 10(s^2 - 36)/s(s + 9)(s + 16)$$

13.57 Calculate the zero-state ac response $y(t)$ given

$$x(t) = 5 \cos (3t + 90°) \qquad t \geq 0$$
$$H(s) = (s^2 + 9)/s(s^2 + 2s + 5)$$

13.58 Write a matrix mesh equation to obtain the zero-input response of $i_1(t)$ in Fig. P13.58 when $i_s(t) = 6$ A for $t < 0$.

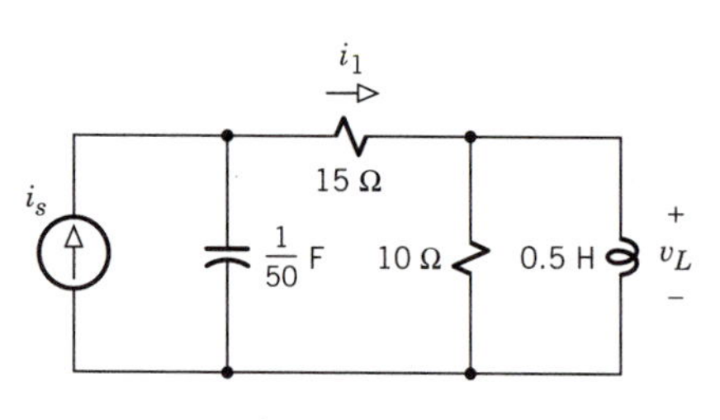

Figure P13.58

13.59 Write a single node equation to obtain the zero-input response of $v_1(t)$ in Fig. P13.59 when $v_s(t) = -8$ V for $t < 0$.

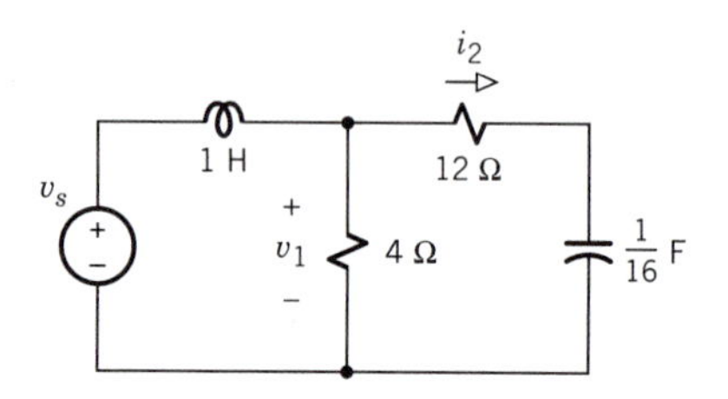

Figure P13.59

13.60 Write a matrix mesh equation to obtain the zero-input response of $i_2(t)$ in Fig. P13.59 when $v_s(t) = 16$ V for $t < 0$.

13.61 Write a single node equation to obtain the zero-input response of $v_C(t)$ in Fig. P13.61 when $v_s(t) = 10$ V for $t < 0$.

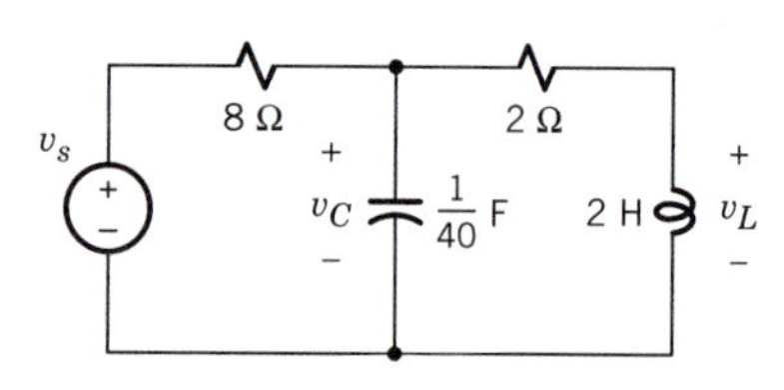

Figure P13.61

13.62 Write a single node equation to obtain the zero-input response of $v_C(t)$ in Fig. P13.62 when $\mu = 4$ and $i_s(t) = 1$ A for $t < 0$.

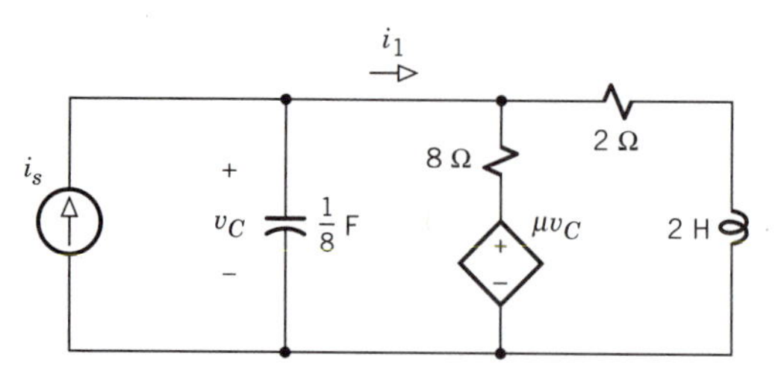

Figure P13.62

13.63 Write a matrix node equation to obtain the zero-input response of $v_1(t)$ in Fig. P13.63 when $\beta = 9$ and $v_s(t) = 25$ V for $t < 0$.

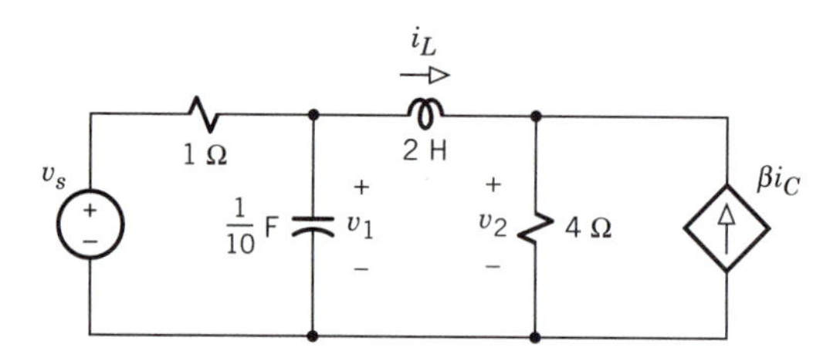

Figure P13.63

13.64* Let the circuit in Fig. P13.59 have

$$v_s(t) = 40 \text{ V} \quad t < 0$$
$$= -40 \text{ V} \quad t > 0$$

Find $I_2(s)$ and evaluate $i_2(0^+)$ and $i_2{}'(0^+)$.

13.65 Let the circuit in Fig. P13.58 have

$$i_s(t) = 1 \text{ A} \quad t < 0$$
$$= 2 \text{ A} \quad t > 0$$

Find $V_L(s)$ and evaluate $v_L(0^+)$ and $v_L{}'(0^+)$.

13.66 Let the circuit in Fig. P13.61 have

$$v_s(t) = -10 \text{ V} \quad t < 0$$
$$= 20e^{-t} \text{ V} \quad t > 0$$

Find $V_L(s)$ and evaluate $v_L(0^+)$ and $v_L{}'(0)$.

13.67 Let the circuit in Fig. P13.61 have

$$v_s(t) = 10 \text{ V} \quad t < 0$$
$$= \sin t \text{ V} \quad t > 0$$

Find $V_C(s)$ and evaluate $v_C(0^+)$ and $v_C{}'(0^+)$.

13.68 Let the circuit in Fig. P13.63 have $\beta = 6$ and

$$v_s(t) = -5 \text{ V} \quad t < 0$$
$$= 5t \text{ V} \quad t > 0$$

Find $I_L(s)$ and evaluate $i_L(0^+)$ and $i_L{}'(0^+)$.

13.69 Let the circuit in Fig. P13.62 have $\mu = 2$ and

$$i_s(t) = -3 \text{ A} \qquad t < 0$$
$$= 1 \text{ A} \qquad t > 0$$

Find $I_1(s)$ and evaluate $i_1(0^+)$ and $i_1{}'(0^+)$.

Section 13.4 Transform Analysis with Mutual Inductance

13.70* Given the transformer circuit in Fig. P13.70, calculate the zero-input response $i_2(t)$ given that $R = 12\ \Omega$, $R_1 = 0$, and $v(t) = 36$ V for $t < 0$.

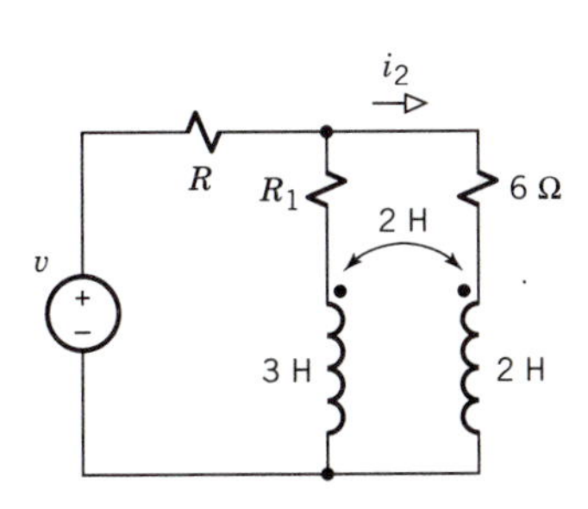

Figure P13.70

13.71 Given the transformer circuit in Fig. P13.70, calculate the zero-input response $i_2(t)$ when $R = 0$, $R_1 = 4\ \Omega$, and $v(t) = 132$ V for $t < 0$.

13.72 Let $M = 4$ H in Fig. P13.72. Find $I_1(s)$ for $t > 0$ and evaluate $i_1(0^+)$ and $i_1{}'(0^+)$ when

$$i_s(t) = 2 \text{ A} \qquad t < 0$$
$$= -1 \text{ A} \qquad t > 0$$

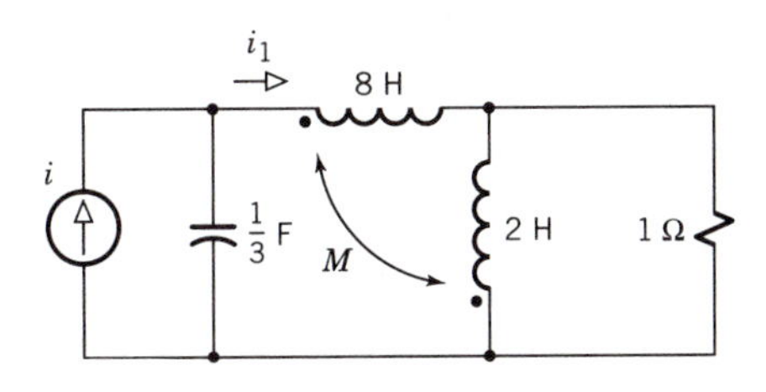

Figure P13.72

13.73 Do Problem 13.72 with $M = 3$ H.

Section 13.5 Impulses and Convolution

13.74* Suppose $f(t)$ is such that its derivative $f'(t)$ has a "simple" shape so you can easily find $\mathcal{L}[f'(t)]$. The Laplace transform of $f(t)$ may then be determined via the differentiation property written as

$$F(s) = \{\mathcal{L}[f'(t)] + f(0^-)\}/s$$

Use this method to obtain $F(s)$ when $f(t)$ is the waveform in Fig. 13.21*a* (p. 639).

13.75 Do Problem 13.74 for the waveform in Fig. P13.17.

13.76 Do Problem 13.74 for the waveform in Fig. P13.15.

13.77 Do Problem 13.74 for the waveform in Fig. P13.16.

13.78* Find the zero-state response of $i_2(t)$ and $v_1(t)$ in Fig. P13.78 when $R = 10\ \Omega$ and $i_s(t) = 5u(t)$ A.

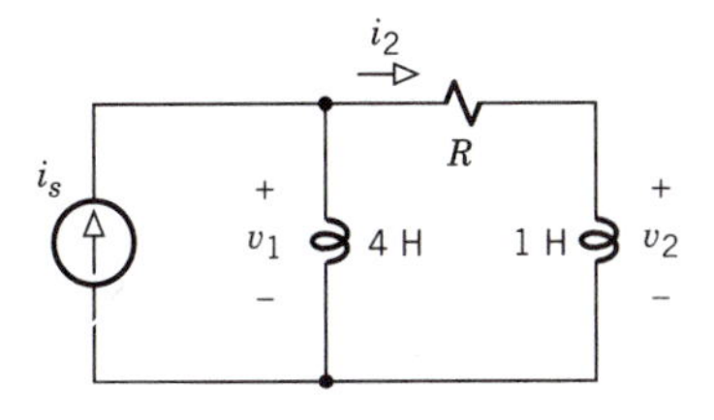

Figure P13.78

13.79 Find the zero-state response of $i_1(t)$ and $v_2(t)$ in Fig. P13.79 when $R = 1\ \Omega$ and $v_s(t) = 50u(t)$ V.

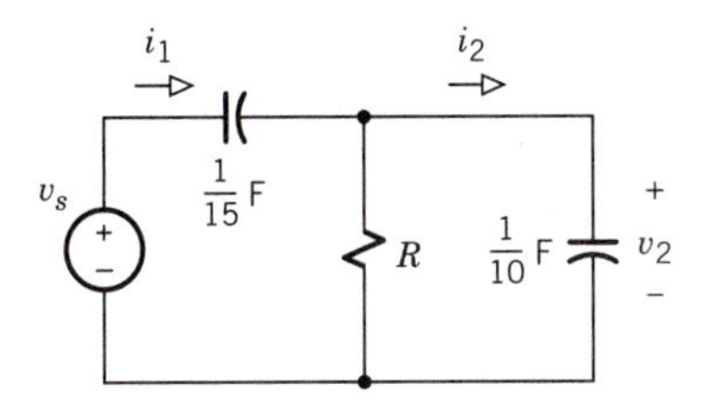

Figure P13.79

13.80 Find the zero-input response of $i_2(t)$ and $v_2(t)$ in Fig. P13.78 when $R = 5\ \Omega$ and $i_s(t) = 2.5$ A for $t < 0$.

13.81 Find the zero-input response of $v_2(t)$ and $i_2(t)$ in Fig. P13.79 when $R = 2\ \Omega$ and $v_s(t) = 75$ V for $t < 0$.

13.82 Use a single node equation to obtain the zero-state response of $i_2(t)$, $i_3(t)$, and $i_1(t)$ in Fig. P13.82 when $R = 5\ \Omega$ and $v_s(t) = 48u(t)$ V.

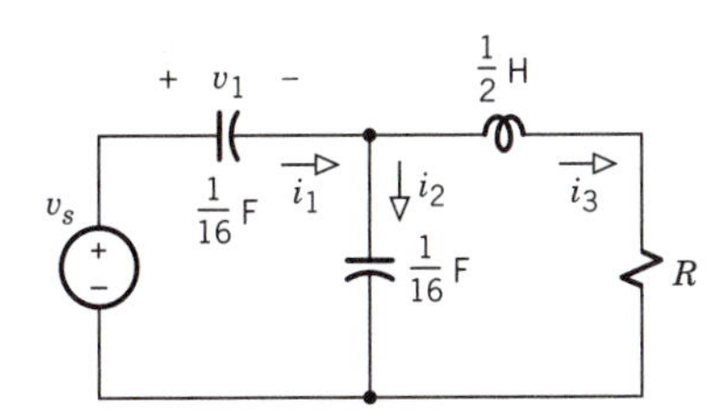

Figure P13.82

13.83 Use a single mesh equation to obtain the zero-state response of $v_2(t)$, $v_3(t)$, and $v_1(t)$ in Fig. P13.83 when $R = 40\ \Omega$ and $i_s(t) = 2u(t)$ A.

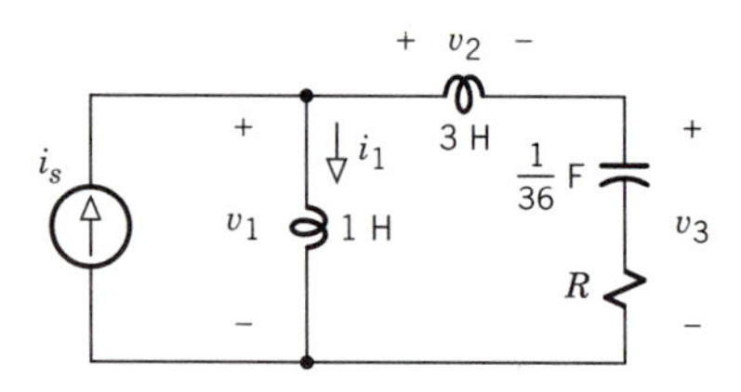

Figure P13.83

13.84 Use a single node equation to obtain the zero-input response of $i_2(t)$ and $i_3(t)$ in Fig. P13.82 when $R = 4\ \Omega$ and $v_s(t) = -32$ V for $t < 0$. Then find $v_1(t)$ to confirm that $v_1(\infty) = 0$.

13.85 Use a single mesh equation to obtain the zero-input response of $v_2(t)$ and $v_3(t)$ in Fig. P13.83 when $R = 24\ \Omega$ and $i_s(t) = 4$ A for $t < 0$. Then find $i_1(t)$ to confirm that $i_1(\infty) = 0$.

13.86* Apply Eq. (13.65a) to find $f(t)$ for $t > 0$, given that $f(t)$ is causal and

$$f(t) * [\cos 2t\ u(t)] = 3u(t)$$

13.87 Do Problem 13.86 with

$$f(t) + 2f(t) * [e^{-3t}u(t)] = 10\delta(t)$$

13.88 Do Problem 13.86 with

$$f(t) + f(t) * [9tu(t)] = 2\delta(t-4)$$

13.89 Do Problem 13.86 with

$$f'(t) + 10f(t) + f(t) * [50u(t)] = 4\delta(t)$$

13.90* Obtain $y(t)$ by convolution when $h(t) = te^{-t}u(t)$ and $x(t) = u(t) - u(t-D)$.

13.91 Obtain $y(t)$ by convolution when $h(t) = e^{-t}u(t)$ and $x(t) = t[u(t) - u(t-D)]$.

13.92 Given the step response $y_u(t) = e^{-t}u(t)$, calculate the impulse response $h(t)$ and the zero-state response $y(t)$ produced by $x(t) = \sin \beta t\ u(t)$.

13.93 Given the step response $y_u(t) = \cos \beta t$, calculate the impulse response $h(t)$ and the zero-state response $y(t)$ produced by $x(t) = tu(t)$.

13.94 Let i_2 be the response of interest in Fig. P13.78 with $R = 20\ \Omega$. Find $H(s)$ and $h(t)$. Then use convolution to obtain the zero-state response when

$$i_s(t) = 5[u(t) - u(t-2)] \text{ A}$$

13.95 Let v_2 be the response of interest in Fig. P13.79 with $R = 3\ \Omega$. Find $H(s)$ and $h(t)$. Then use convolution to obtain the zero-state response when

$$v_s(t) = 10t[u(t) - u(t-1)] \text{ V}$$

CHAPTER 14

Two-Port Networks

At the very beginning of this book, we noted that every electrical circuit consists of two essential parts, a source and a load. Subsequently, we extended that notion by speaking of source networks and load networks — each network having one external pair of terminals forming an input or output port.

But many applications involve an additional network inserted between the source and load for purposes such as amplification, impedance matching, or filtering. The additional network must have two pairs of terminals and is therefore named a *two-port network,* or *two-port* for short.

This chapter introduces the basic properties of two-port networks and develops their description in terms of *two-port parameters.* These parameters lead to compact models of two-ports, akin to our models for source networks in terms of Thévenin parameters. We then use the two-port parameters to analyze circuits comprising one or more two-ports. The viewpoint and techniques presented here provide the generality needed for the analysis or design of sophisticated filters and electronic amplifiers.

Objectives

After studying this chapter and working the exercises, you should be able to do each of the following:

1. Distinguish between a four-terminal network, a two-port network, and a three-terminal network (Section 14.1).
2. State the properties of a reciprocal network (Section 14.1).
3. Write matrix two-port equations in terms of the z, y, h, or $ABCD$ parameters (Sections 14.1 and 14.2).
4. Calculate z, y, h, or $ABCD$ parameters using open-circuit and/or short-circuit conditions (Sections 14.1 and 14.2).
5. Convert one set of parameters into another set (Section 14.2).
6. Analyze a circuit consisting of a two-port connected between a source and a load (Section 14.3).
7. Find the equivalent parameters when two or more two-ports are connected in cascade, series, or parallel (Section 14.3).

14.1 TWO-PORTS AND IMPEDANCE PARAMETERS

This section begins with the basic concepts of two-port networks. We then develop two-port equations and the corresponding model of a two-port in terms of *impedance parameters*. We'll also look at the special case of *reciprocal networks*.

Two-Port Concepts

A two-port network has four external terminals, as illustrated by Fig. 14.1. Ordinarily, the treatment of any network with four terminals would involve eight variables — four currents and four voltages. However, a two-port must be structured such that one pair of terminals serves as an **input port** and the other pair serves as an **output port**. In order to speak of unique input and output currents, we impose the condition that

$$i_3 = i_1 \qquad i_4 = i_2$$

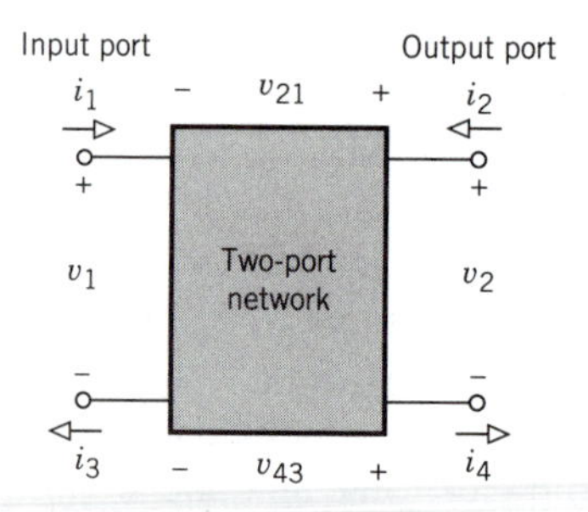

Figure 14.1 Two-port network.

We also note that the voltages v_{21} and v_{43} usually hold no interest because they do not appear across the input or output ports. Consequently, our study of two-ports will focus on just two pairs of variables: the input voltage and current, v_1 and i_1, and the output voltage and current, v_2 and i_2.

Figure 14.2 depicts a two-port interfacing a source and a load, an arrangement that guarantees unique input and output currents. For if we apply KCL to a supernode enclosing the two-port and load, then the current i_1 entering the upper terminal of the two-port must equal the current leaving the lower terminal. Likewise, if we apply KCL to a supernode enclosing the source and two-port, then the current i_2 entering the upper terminal of the two-port must equal the current leaving the lower terminal. By standard convention, both current reference arrows point inward at the upper terminals of the two-port.

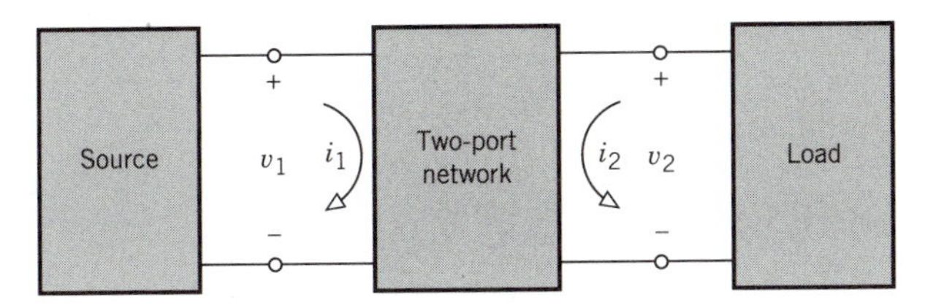

Figure 14.2 Two-port interfacing a source and load.

The usual role of a two-port is to "process" the input voltage or current coming from the source. Accordingly, we'll assume throughout that

> A two-port network contains no independent sources, although it may contain controlled sources.

The load network also contains no independent sources, by definition, so the source sees a modified load looking into the input terminals of a two-port with attached load. Conversely, the load sees a modified source looking back into the output terminals of the two-port.

Many two-ports are actually **three-terminal networks**, represented by Fig. 14.3*a*. The three-terminal configuration is important because most transistors and amplifier circuits have three rather than four distinct external terminals. A three-terminal network becomes a two-port when one terminal connects to both the input and output side, per Fig. 14.3*b*. The lower input

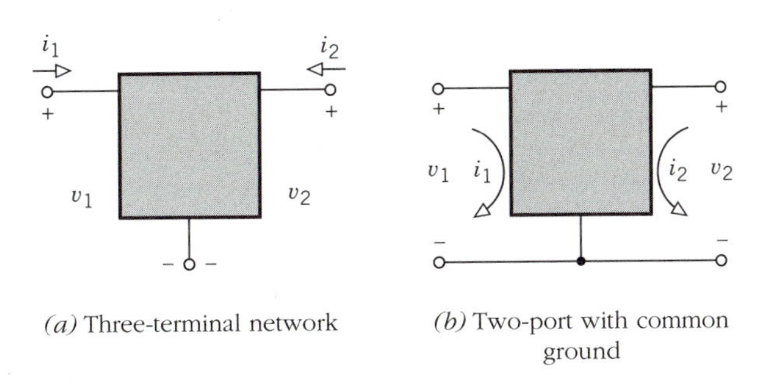

(a) Three-terminal network

(b) Two-port with common ground

Figure 14.3

and output terminals are thus at the same potential, and the two-port is said to have a **common ground**.

Now recall that we previously developed models for source and load networks by studying the voltage–current equations at the terminals. Models of two-port networks are derived in a similar fashion. The major difference here is that a two-port has *two* pairs of terminals.

For generality, we'll work with *s*-domain networks using the phasor notation in Fig. 14.4. Our objective is to obtain two equations relating the four phasors $\underline{V}_1$, $\underline{I}_1$, $\underline{V}_2$, and $\underline{I}_2$. These equations will ultimately lead to network functions for cicrcuits comprising two-port networks. The network functions can then be used to carry out specific analysis or design tasks — impedance matching, stability tests, frequency response, etc.

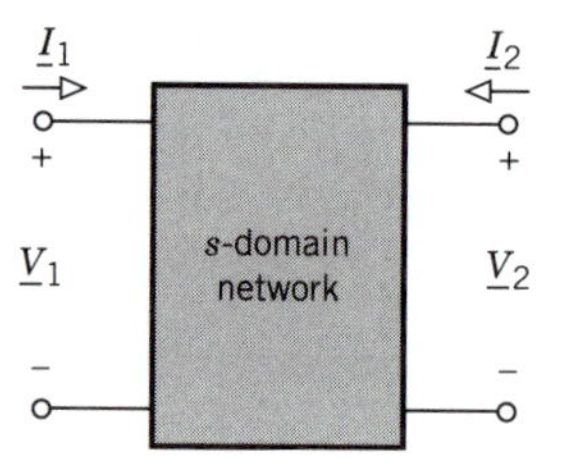

Figure 14.4 Two-port in the *s* domain.

Impedance Parameters

One approach to modeling two-ports treats $\underline{I}_1$ and $\underline{I}_2$ as source currents and assumes a common ground, so $\underline{V}_1$ and $\underline{V}_2$ are unknown node voltages. Figure 14.5 depicts this approach, noting that there may be a total of N unknown node voltages of which $N - 2$ are internal to the network. Upon writing and solving an $N \times N$ matrix node equation, we'll get equations for $\underline{V}_1$ and $\underline{V}_2$ in terms of $\underline{I}_1$ and $\underline{I}_2$.

Since $\underline{I}_1$ and $\underline{I}_2$ are the only independent sources, the matrix node equation takes the form

$$\begin{bmatrix} Y_{11} & -Y_{12} & \cdots & -Y_{1N} \\ -Y_{21} & Y_{22} & \cdots & -Y_{2N} \\ -Y_{31} & -Y_{32} & \cdots & -Y_{3N} \\ \vdots & \vdots & & \vdots \\ -Y_{N1} & -Y_{N2} & \cdots & Y_{NN} \end{bmatrix} \begin{bmatrix} \underline{V}_1 \\ \underline{V}_2 \\ \underline{V}_3 \\ \vdots \\ \underline{V}_N \end{bmatrix} = \begin{bmatrix} \underline{I}_1 \\ \underline{I}_2 \\ 0 \\ \vdots \\ 0 \end{bmatrix} \tag{14.1}$$

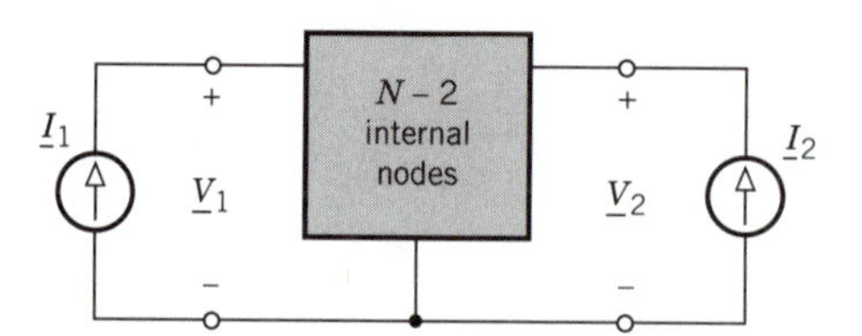

Figure 14.5 Common ground two-port with current sources.

in which Y_{11}, Y_{12}, . . . are the usual node admittances. Solving for $\underline{V}_1$ and $\underline{V}_2$ via Cramer's rule yields

$$\underline{V}_1 = \frac{\Delta_{11}}{\Delta}\underline{I}_1 + \frac{\Delta_{21}}{\Delta}\underline{I}_2 \qquad \underline{V}_2 = \frac{\Delta_{12}}{\Delta}\underline{I}_1 + \frac{\Delta_{22}}{\Delta}\underline{I}_2 \tag{14.2}$$

where Δ is the determinant of the admittance matrix $[Y]$ and Δ_{ij} stands for the cofactor of the ijth element of $[Y]$.

To express Eq. (14.2) more compactly, we observe that the coefficients Δ_{ij}/Δ must have the units of impedance. We therefore introduce four **impedance parameters**

$$z_{11} = \frac{\Delta_{11}}{\Delta} \qquad z_{12} = \frac{\Delta_{21}}{\Delta} \qquad z_{21} = \frac{\Delta_{12}}{\Delta} \qquad z_{22} = \frac{\Delta_{22}}{\Delta} \tag{14.3}$$

These quantities are also called the ***z* parameters**, and lowercase letters are used to avoid confusion with mesh impedances such as $Z_{11}(s)$. Nonetheless, each z parameter may depend upon s when the two-port contains energy-storage elements, so $z_{11} = z_{11}(s)$, etc.

After inserting the z parameters, Eq. (14.2) becomes our desired set of relations for the terminal variables in the form

$$\begin{aligned} \underline{V}_1 &= z_{11}\underline{I}_1 + z_{12}\underline{I}_2 \\ \underline{V}_2 &= z_{21}\underline{I}_1 + z_{22}\underline{I}_2 \end{aligned} \tag{14.4}$$

Or, using matrix notation, we have

$$\begin{bmatrix} \underline{V}_1 \\ \underline{V}_2 \end{bmatrix} = [z] \begin{bmatrix} \underline{I}_1 \\ \underline{I}_2 \end{bmatrix}$$

where

$$[z] \triangleq \begin{bmatrix} z_{11} & z_{12} \\ z_{21} & z_{22} \end{bmatrix} \tag{14.5}$$

which is the z-parameter matrix.

Although we derived Eq. (14.4) by assuming source currents for $\underline{I}_1$ and $\underline{I}_2$, the voltage–current relationships obviously remain valid when the ports are connected to arbitrary external circuitry. Closer examination of Eq. (14.4) then suggests a way of measuring z parameters with *open-circuit* conditions at the input or output ports. Setting $\underline{I}_2 = 0$ or $\underline{I}_1 = 0$, we see that

$$\begin{aligned} z_{11} &= \underline{V}_1/\underline{I}_1\big|_{\underline{I}_2=0} \qquad & z_{12} &= \underline{V}_1/\underline{I}_2\big|_{\underline{I}_1=0} \\ z_{21} &= \underline{V}_2/\underline{I}_1\big|_{\underline{I}_2=0} \qquad & z_{22} &= \underline{V}_2/\underline{I}_2\big|_{\underline{I}_1=0} \end{aligned} \tag{14.6}$$

These expressions serve as alternative definitions of the z parameters, and they lead to physical interpretations in that

$$z_{11} = \text{input impedance with open output}$$
$$z_{21} = \text{forward transfer impedance with open output}$$
$$z_{12} = \text{reverse transfer impedance with open input}$$
$$z_{22} = \text{output impedance with open input}$$

The open-circuit conditions explain why the z parameters are also known as the **open-circuit impedance parameters**.

Drawing upon Eqs. (14.4) and (14.6), any two-port whose z parameters are known can be modeled by the equivalent s-domain diagram in Fig. 14.6. Each side of the model looks like a Thévenin network, except that the voltage source is controlled by the current on the other side to account for the transfer effect. The dashed line connecting the input and output sides indicates that the bottom nodes will be at the same potential when the two-port has a common ground.

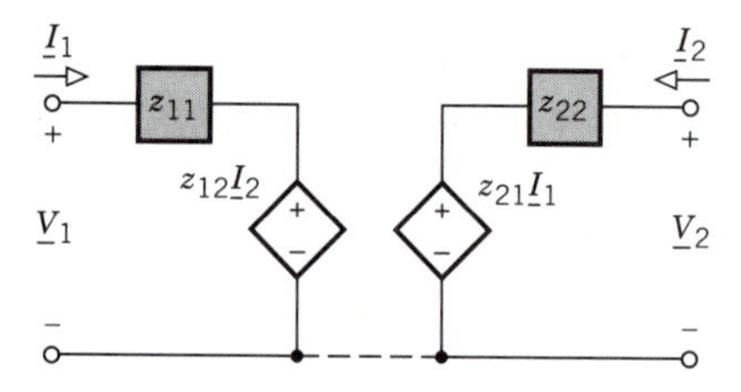

Figure 14.6 z-parameter model.

Clearly, the essential initial task in two-port modeling is determining the z parameters from the actual diagram of a particular network. Based on Eq. (14.4), an **indirect method** for finding z parameters goes as follows:

> Treat $\underline{I}_1$ and $\underline{I}_2$ as source currents and use any standard analysis techniques to obtain $\underline{V}_1$ and $\underline{V}_2$ in terms of $\underline{I}_1$ and $\underline{I}_2$.

But a more **direct method** based on Eq. (14.6) involves two steps:

1. Open the output port so $\underline{I}_2 = 0$, find $\underline{V}_1$ and $\underline{V}_2$ in terms of $\underline{I}_1$, then calculate $z_{11} = \underline{V}_1/\underline{I}_1$ and $z_{21} = \underline{V}_2/\underline{I}_1$.
2. Open the input port so $\underline{I}_1 = 0$, find $\underline{V}_1$ and $\underline{V}_2$ in terms of $\underline{I}_2$, then calculate $z_{12} = \underline{V}_1/\underline{I}_2$ and $z_{22} = \underline{V}_2/\underline{I}_2$.

The parameter subscripts serve as a guide in this method, because the first subscript of each parameter identifies the voltage involved and the second subscript identifies the nonzero current.

Finally, we point out that not all two-ports possess meaningful z parameters. Going back to the original definitions in Eq. (14.3), we see that the z parameters would become undefined if the network configuration happens to yield $\Delta = 0$. A simple and general **existence test** comes from Eq. (14.4), which implies that

> The z parameters exist only when independent current sources may be connected to the input and output ports without violating KCL.

If it turns out that the z parameters do not exist, then the two-port in question cannot be modeled by z parameters.

Example 14.1 ***z Parameters by the Indirect Method***

The two-port in Fig. 14.7*a* has a common ground, and v_1 and v_2 can be taken as the only node voltages. We can thus get the z parameters via node analysis of the *s*-domain diagram in Fig. 14.7*b*.

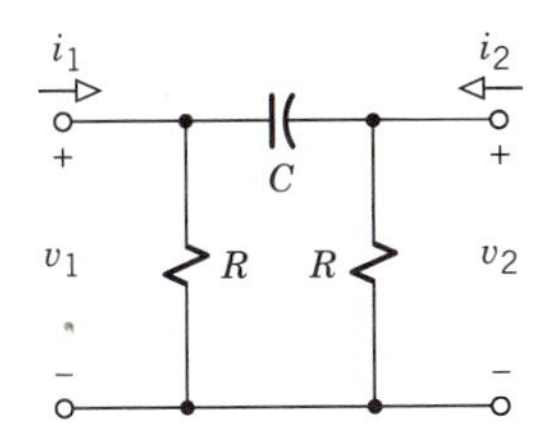

(*a*) Network for Example 14.1

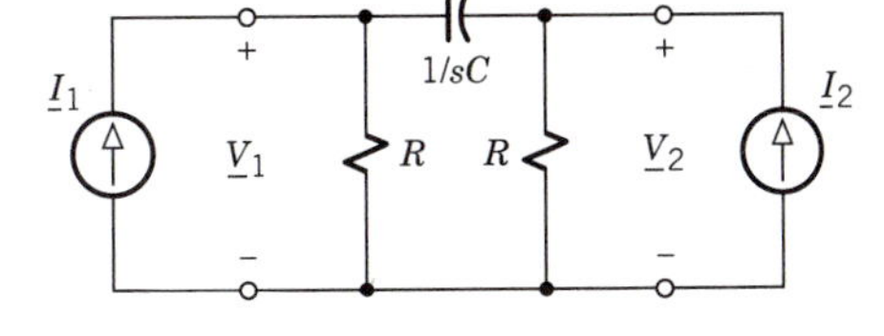

(*b*) *s*-domain diagram with current sources

Figure 14.7

By inspection, the matrix node equation is

$$\begin{bmatrix} sC + 1/R & -sC \\ -sC & sC + 1/R \end{bmatrix} \begin{bmatrix} V_1 \\ V_2 \end{bmatrix} = \begin{bmatrix} I_1 \\ I_2 \end{bmatrix}$$

Applications of Cramer's rule yields

$$\Delta = |[Y]| = (2sCR + 1)/R^2$$

$$V_1 = \frac{1}{\Delta}\begin{vmatrix} I_1 & -sC \\ I_2 & sC + 1/R \end{vmatrix} = \frac{sC + 1/R}{\Delta} I_1 + \frac{sC}{\Delta} I_2$$

$$V_2 = \frac{1}{\Delta}\begin{vmatrix} sC + 1/R & I_1 \\ -sC & I_2 \end{vmatrix} = \frac{sC}{\Delta} I_1 + \frac{sC + 1/R}{\Delta} I_2$$

Hence, by comparison with Eq. (14.4),

$$z_{11} = z_{22} = \frac{sC + 1/R}{\Delta} = \frac{sCR^2 + R}{2sCR + 1} \qquad z_{12} = z_{21} = \frac{sC}{\Delta} = \frac{sCR^2}{2sCR + 1}$$

But consider what happens when $R \rightarrow \infty$, so both vertical branches become open circuits. Although we still have a plausible network, all of the z parameters "blow up" and therefore do not exist. The physical reason underlying this effect is that the currents cannot come from independent sources because $i_2 = -i_1$ when $R \rightarrow \infty$.

Example 14.2 *z Parameters by the Direct Method*

Suppose you want to model the resistive two-port in Fig. 14.8*a*, which includes a VCVS. The interior node and floating voltage source would make node analysis awkward, so you select the direct method. The absence of energy-storage elements allows you to work here with instantaneous variables rather than phasors.

Step 1: You set $i_2 = 0$ and draw the open-output diagram, Fig. 14.8*b*. Here you see that $v_1 = (10 + 50)\,i_1$, $v_x = 50\,i_1$, and $v_2 = v_x - 3\,v_x = -2 \times 50\,i_1$. Thus,

$$z_{11} = v_1/i_1 = 60\ \Omega \qquad z_{21} = v_2/i_1 = -100\ \Omega$$

Step 2: Now you set $i_1 = 0$ and draw the open-input diagram, Fig. 14.8*c*, from which you see that $v_1 = v_x = 50\,i_2$ and $v_2 = v_x - 3\,v_x = -2 \times 50\,i_2$. Thus,

$$z_{21} = v_2/i_1 = 50\ \Omega \qquad z_{22} = v_2/i_2 = -100\ \Omega$$

The results from both steps are summarized in the z-parameter matrix

$$[z] = \begin{bmatrix} 60\ \Omega & 50\ \Omega \\ -100\ \Omega & -100\ \Omega \end{bmatrix}$$

An illustration of the meaning of $[z]$ is provided by assuming that the output port is left open and $v_1 = 12$ V. Then, from Eq. (14.4), $i_1 = v_1/z_{11} = 12/60 = 0.2$ A and $v_2 = z_{21}\,i_1 = -100 \times 0.2 = -20$ V.

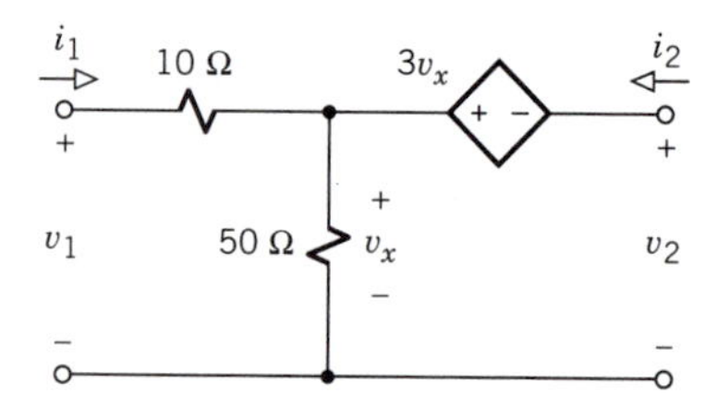

(a) Network for Example 14.2

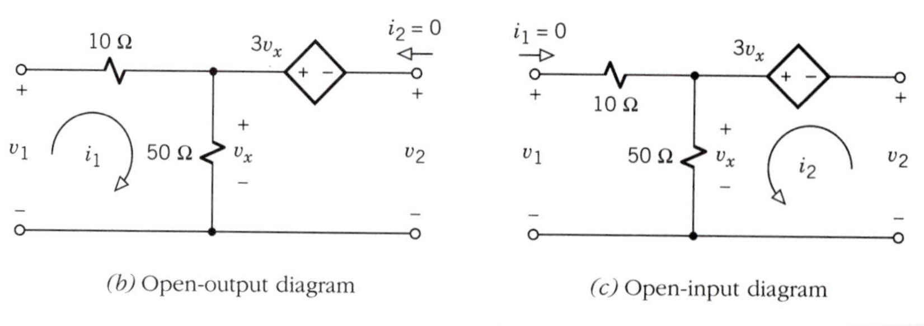

(b) Open-output diagram

(c) Open-input diagram

Figure 14.8

Exercise 14.1

The ladder network in Fig. 14.9*a* consists of two arbitrary branch impedances, Z_a and Z_b. Find $[z]$ by writing two *loop* equations.

Exercise 14.2

Consider the two-port in Fig. 14.9*b*.
(**a**) Use the direct method to obtain $[z]$ when R has an arbitrary value.
(**b**) Explain why $[z]$ does not exist when $R \to \infty$.

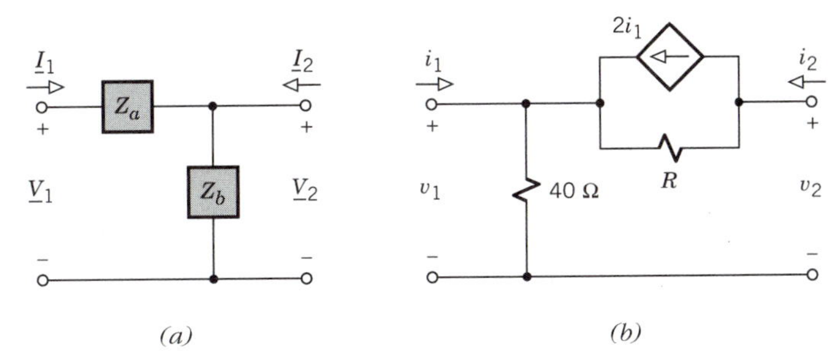

Figure 14.9

Reciprocal Networks

As illustrated by Examples 14.1 and 14.2, some of the four z parameters of a two-port may turn out to have the same value. A case of particular interest occurs when

$$z_{21} = z_{12} \tag{14.7}$$

We then say that the two-port is a **reciprocal network**.

One implication of Eq. (14.7) is the **reciprocity theorem**, which states that

> In a reciprocal network, interchanging a single current source and an open circuit does not alter the open-circuit voltage, and interchanging a single voltage source and a short circuit does not alter the short-circuit current.

Figure 14.10 illustrates these interchanges.

We prove the first part of the reciprocity theorem by applying the z-parameter equations to Fig. 14.10*a*, so either $\underline{V}_{1oc} = z_{12}\underline{I}_2$ when $\underline{I}_1 = 0$ or $\underline{V}_{2oc} = z_{21}\underline{I}_1$ when $\underline{I}_2 = 0$. Thus, if $z_{21} = z_{12}$, then $\underline{V}_{2oc} = \underline{V}_{1oc}$ when $\underline{I}_1 = \underline{I}_2$. The two cases in Fig. 14.10*b* are slightly more complicated because we

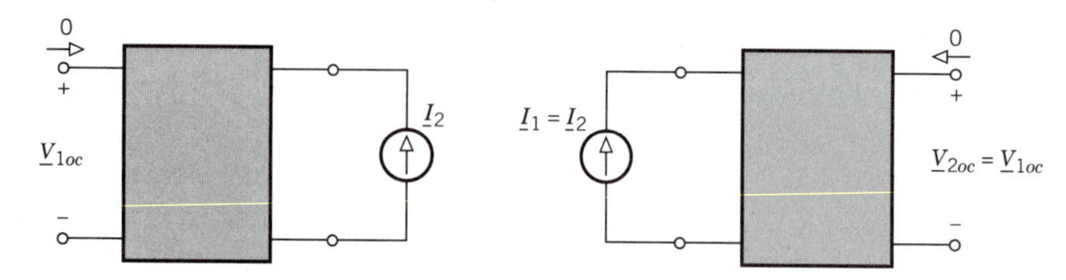

(a) Interchanging open circuit and current source

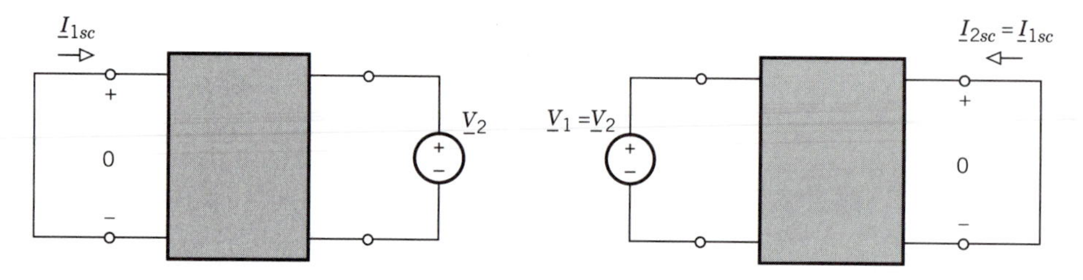

(b) Interchanging short circuit and voltage source

Figure 14.10 Reciprocity conditions.

must eliminate $\underline{I}_2$ or $\underline{I}_1$. With the input shorted, $z_{11}\underline{I}_{1sc} + z_{12}\underline{I}_2 = \underline{V}_1 = 0$ and $\underline{V}_2 = z_{21}\underline{I}_{1sc} + z_{22}\underline{I}_2$, from which

$$\underline{I}_{1sc} = \underline{V}_2/(z_{21} - z_{11}z_{22}/z_{12})$$

Similarly, with the output shorted we find that

$$\underline{I}_{2sc} = \underline{V}_1/(z_{12} - z_{11}z_{22}/z_{21})$$

Thus, if $z_{21} = z_{12}$, then $\underline{I}_{2sc} = \underline{I}_{1sc}$ when $\underline{V}_1 = \underline{V}_2$.

Another implication of Eq. (14.7) is that the model of a reciprocal two-port involves just three parameter values and does not require controlled sources. For when $z_{21} = z_{12}$, the z-parameter equations become

$$\underline{V}_1 = z_{11}\underline{I}_1 + z_{12}\underline{I}_2 = (z_{11} - z_{12})\underline{I}_1 + z_{12}(\underline{I}_1 + \underline{I}_2)$$
$$\underline{V}_2 = z_{12}\underline{I}_1 + z_{22}\underline{I}_2 = z_{12}(\underline{I}_1 + \underline{I}_2) + (z_{22} - z_{12})\underline{I}_2$$

which correspond to the *tee network* in Fig. 14.11. Consequently, any reciprocal two-port whose z parameters exist is equivalent at its terminals to a tee network consisting of three branch impedances.

Frequently, you can identify a reciprocal two-port simply by inspecting its actual diagram, because

> Any linear network that contains no controlled sources will be reciprocal.

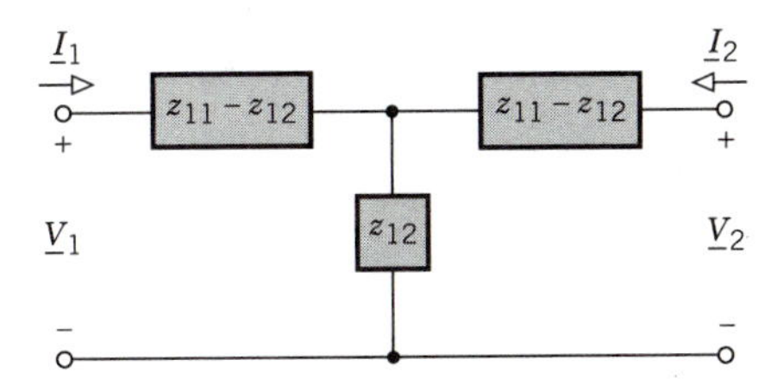

Figure 14.11 Tee model of a reciprocal two-port.

This assertion follows from our original equations

$$z_{12} = \Delta_{21}/\Delta \qquad z_{21} = \Delta_{12}/\Delta$$

Referring to Eq. (14.1), one of the cofactors in question is

$$\Delta_{12} = -\begin{vmatrix} -Y_{21} & -Y_{23} & \cdots & -Y_{2N} \\ -Y_{31} & Y_{33} & \cdots & -Y_{3N} \\ \vdots & \vdots & & \vdots \\ -Y_{N1} & -Y_{N3} & \cdots & Y_{NN} \end{vmatrix}$$

The other cofactor retains its value when we interchange rows and columns of the determinant, so

$$\Delta_{21} = -\begin{vmatrix} -Y_{12} & -Y_{13} & \cdots & -Y_{1N} \\ -Y_{32} & Y_{33} & \cdots & -Y_{3N} \\ \vdots & \vdots & & \vdots \\ -Y_{N2} & -Y_{N3} & \cdots & Y_{NN} \end{vmatrix} = -\begin{vmatrix} -Y_{12} & -Y_{32} & \cdots & -Y_{N2} \\ -Y_{13} & Y_{33} & \cdots & -Y_{N3} \\ \vdots & \vdots & & \vdots \\ -Y_{1N} & -Y_{3N} & \cdots & Y_{NN} \end{vmatrix}$$

Now recall that the admittance matrix has symmetry about the main diagonal when there are no controlled sources. Since $Y_{12} = Y_{21}$, $Y_{32} = Y_{23}, \ldots,$ we see that $\Delta_{21} = \Delta_{12}$ and therefore $z_{12} = z_{21}$.

Networks containing controlled sources may or may not be reciprocal. Reciprocity must then be tested by calculation. For instance, the two-port back in Example 14.2 is not reciprocal because we found that $z_{12} = 50\ \Omega$ while $z_{21} = -100\ \Omega$.

Example 14.3 *z Parameters of a Tee Network*

Suppose a two-port actually consists of three tee-connected impedances, as shown in Fig. 14.12. The absence of controlled sources guarantees that such a network is reciprocal. Thus, by comparison with Fig. 14.11,

$$z_{11} - z_{12} = Z_a \qquad z_{12} = Z_c \qquad z_{22} - z_{12} = Z_b$$

so

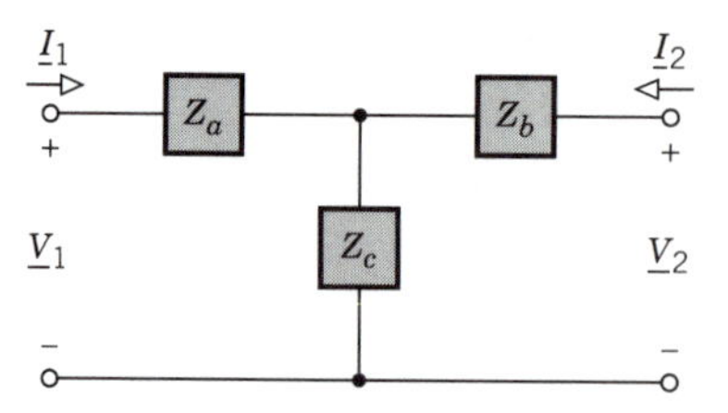

Figure 14.12 Tee network.

$$[z] = \begin{bmatrix} Z_a + Z_c & Z_c \\ Z_c & Z_b + Z_c \end{bmatrix}$$

Exercise 14.3

Show that the network in Fig. 14.13 is reciprocal by finding the z parameters via the direct method. Then draw and label the equivalent tee network.

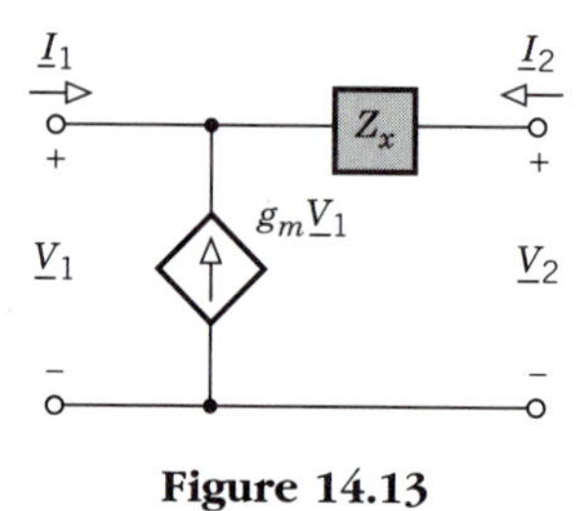

Figure 14.13

14.2 ADMITTANCE, HYBRID, AND TRANSMISSION PARAMETERS

Besides the impedance-parameter equations, there are several other useful ways of modeling a two-port network. This section defines and discusses three additional models using admittance parameters, hybrid parameters, and transmission parameters.

One reason for studying other types of two-port parameters is that they may be well defined even though the impedance parameters do not exist. More importantly, each of the models considered here has practical significance for specific circuit applications. To facilitate circuit analysis with two-ports, we'll also show how to convert one set of parameters into another set.

Admittance Parameters

The *dual* of the impedance parameters are the **admittance parameters** or $\boldsymbol{y}$ **parameters**. These parameters relate the terminal currents to the voltages in the form

$$\underline{I}_1 = y_{11}\underline{V}_1 + y_{12}\underline{V}_2$$
$$\underline{I}_2 = y_{21}\underline{V}_1 + y_{22}\underline{V}_2 \tag{14.8}$$

or

$$\begin{bmatrix} \underline{I}_1 \\ \underline{I}_2 \end{bmatrix} = [y] \begin{bmatrix} \underline{V}_1 \\ \underline{V}_2 \end{bmatrix}$$

where

$$[y] \triangleq \begin{bmatrix} y_{11} & y_{12} \\ y_{21} & y_{22} \end{bmatrix} \tag{14.9}$$

All of the admittances in this y-parameter matrix may depend on s when the two-port contains energy-storage elements.

Equation (14.8) defines the y parameters indirectly. For more direct definitions, we set $\underline{V}_2 = 0$ or $\underline{V}_1 = 0$ to get

$$y_{11} = \underline{I}_1/\underline{V}_1|_{\underline{V}_2=0} \qquad y_{12} = \underline{I}_1/\underline{V}_2|_{\underline{V}_1=0}$$
$$y_{21} = \underline{I}_2/\underline{V}_1|_{\underline{V}_2=0} \qquad y_{22} = \underline{I}_2/\underline{V}_2|_{\underline{V}_1=0} \tag{14.10}$$

These expressions lead to the physical interpretations

y_{11} = input admittance with shorted output

y_{21} = forward transfer admittance with shorted output

y_{12} = reverse transfer admittance with shorted input

y_{22} = output admittance with shorted input

Consequently, the y parameters are also known as the **short-circuit admittance parameters**.

Since short-circuit conditions are involved here, the individual y parameters are not the reciprocals of the respective open-circuit z parameters. In fact, the y parameters may not even exist when a network has well-defined z parameters, or vice versa. Closer examination of Eq. (14.8) reveals that

> The y parameters exist only when independent voltage sources may be connected to the input and output ports without violating KVL.

This existence test is, of course, the dual of the test for z parameters. Two-ports that contain ideal op-amps may not have y parameters because the output of an ideal op-amp inherently comes from a controlled voltage source, so an independent voltage source connected to the output port would violate KVL.

Any two-port whose y parameters exist can be modeled by the equivalent s-domain diagram in Fig. 14.14*a*. As might have been expected from duality, each side of the model looks like a Norton network with a controlled current source to account for the transfer effect. The impedances in parallel with the

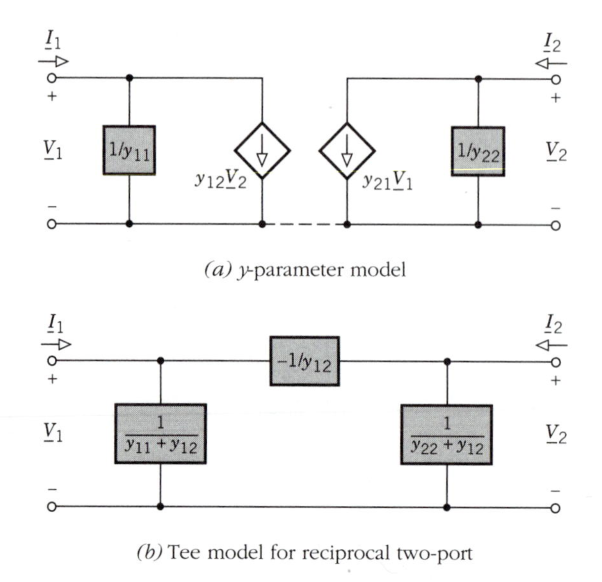

(a) y-parameter model

(b) Tee model for reciprocal two-port

Figure 14.14

current sources are labeled $1/y_{11}$ and $1/y_{22}$ because y_{11} and y_{22} have the units of admittance.

If the two-port in question happens to be a *reciprocal network,* then $\underline{I}_{2sc}/\underline{V}_1 = \underline{I}_{1sc}/\underline{V}_2$. But Fig. 14.14*a* shows that $\underline{I}_{2sc} = y_{21}\underline{V}_1$ and $\underline{I}_{1sc} = y_{12}\underline{V}_2$. Hence, reciprocity requires

$$y_{21} = y_{12} \tag{14.11}$$

The y-parameter model thereby reduces to the equivalent *pi network* in Fig. 14.14*b*.

Returning to Eq. (14.8), we see that an indirect method for determining the y parameters of a given two-port goes as follows:

> Treat $\underline{V}_1$ and $\underline{V}_2$ as source voltages and use any standard analysis techniques to obtain $\underline{I}_1$ and $\underline{I}_2$ in terms of $\underline{V}_1$ and $\underline{V}_2$.

Alternatively, a direct method based on Eq. (14.10) involves two steps:

1. Short the output port so $\underline{V}_2 = 0$, find $\underline{I}_1$ and $\underline{I}_2$ in terms of $\underline{V}_1$, then calculate $y_{11} = \underline{I}_1/\underline{V}_1$ and $y_{21} = \underline{I}_2/\underline{V}_1$.
2. Short the input port so $\underline{V}_1 = 0$, find $\underline{I}_1$ and $\underline{I}_2$ in terms of $\underline{V}_2$, then calculate $y_{12} = \underline{I}_1/\underline{V}_2$ and $y_{22} = \underline{I}_2/\underline{V}_2$.

The first subscript of each parameter now identifies the current involved and the second subscript identifies the nonzero voltage.

Example 14.4 *y Parameters by the Indirect Method*

The z parameters of the network in Fig. 14.15*a* were calculated in Example 14.1, where we found that $z_{11} = z_{22}$ and $z_{12} = z_{21}$. We also found that the z parameters did not exist if $R \to \infty$. Here, we'll use the indirect method to investigate the y parameters.

As the s-domain diagram in Fig. 14.15*b* shows, we can connect independent voltage sources at both ports without violating KVL. The resulting currents are

$$\underline{I}_1 = \underline{V}_1/R + (\underline{V}_1 - \underline{V}_2)/(1/sC) = (G + sC)\underline{V}_1 - sC\underline{V}_2$$
$$\underline{I}_2 = \underline{V}_2/R + (\underline{V}_2 - \underline{V}_1)/(1/sC) = -sC\underline{V}_1 + (G + sC)\underline{V}_2$$

where $G = 1/R$. Hence, by comparison with Eq. (14.8),

$$y_{11} = y_{22} = G + sC \qquad y_{12} = y_{21} = -sC$$

Since the network is reciprocal, it could be represented by either of the models in Fig. 14.14. Furthermore, the y parameters exist when $G \to 0$, corresponding to $R \to \infty$, because we can still have $\underline{V}_2 \neq \underline{V}_1$ even though $\underline{I}_2 = -\underline{I}_1$.

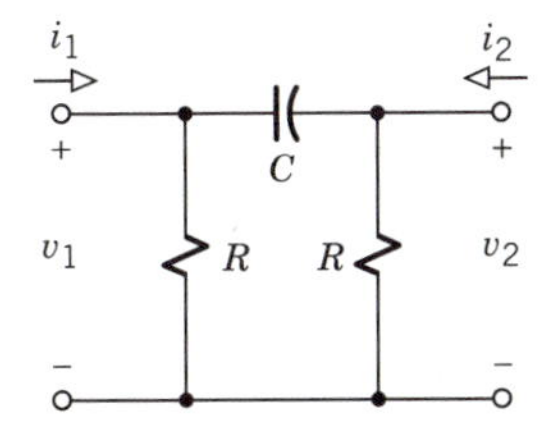

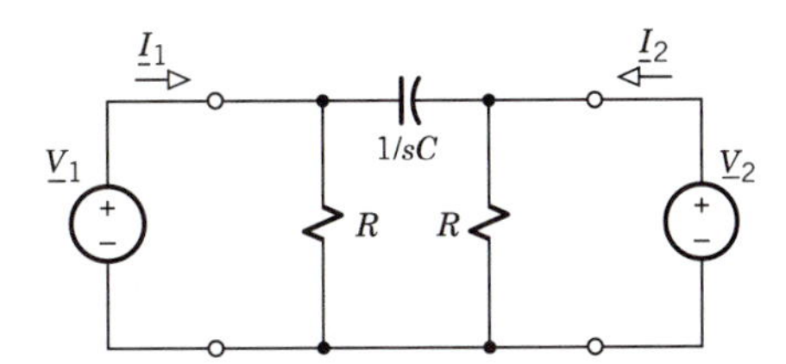

(*a*) Network for Example 14.4 (*b*) s–domain diagram with voltage sources

Figure 14.15

Example 14.5 *y Parameters by the Direct Method*

Suppose you need the y parameters for the two-port in Fig. 14.16*a*. This network configuration also allows independent voltage sources connected at either port, so you know that the y parameters will exist.

Applying the direct method, you seek the currents in terms of the voltages from the shorted-port s-domain diagrams in Fig. 14.16*b* and *c*. With $\underline{V}_2 = 0$, $\underline{I}_1 = (s/40)\underline{V}_1$ and $\underline{I}_2 = 3\underline{I}_1 - \underline{I}_1 = 2(s/40)\underline{V}_1$ so

$$y_{11} = \underline{I}_1/\underline{V}_1 = s/40 \qquad y_{21} = \underline{I}_2/\underline{V}_1 = s/20$$

With $\underline{V}_1 = 0$, $\underline{I}_1 = -(s/40)\underline{V}_2$ and $\underline{I}_2 = 3\underline{I}_1 - \underline{I}_1 + \underline{V}_2/10$ so

$$y_{12} = \underline{I}_1/\underline{V}_2 = -s/40 \qquad y_{22} = \underline{I}_2/\underline{V}_2 = (2 - s)/20$$

Thus,

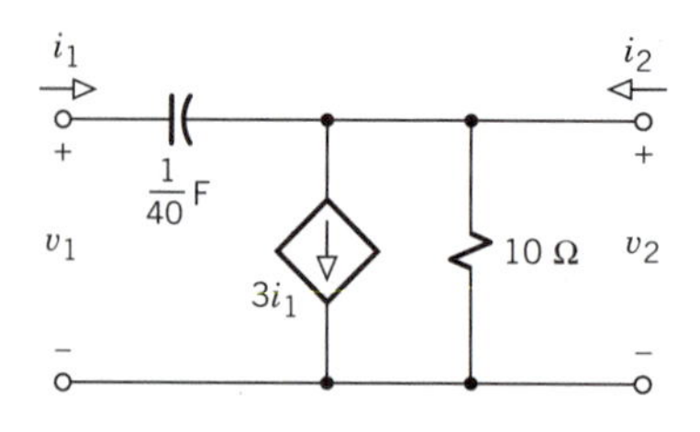

(a) Network for Example 14.5

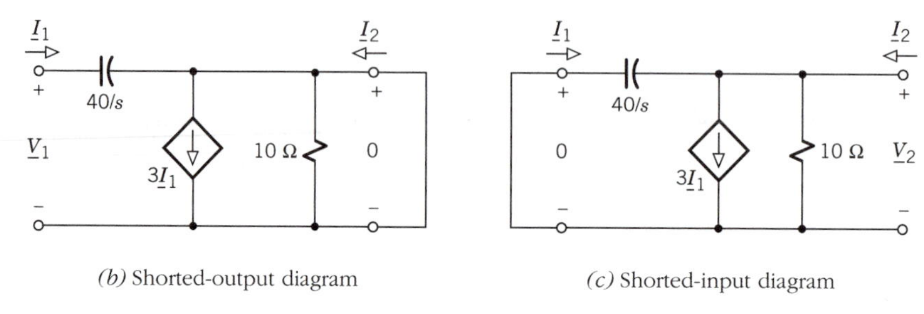

(b) Shorted-output diagram

(c) Shorted-input diagram

Figure 14.16

$$[y] = \begin{bmatrix} s/40 & -s/40 \\ s/20 & (2-s)/20 \end{bmatrix}$$

You might also put these parameters values on the equivalent diagram in Fig. 14.14*a*. However, since $y_{21} \neq y_{12}$, the two-port is not reciprocal and does not have an equivalent pi network.

Exercise 14.4

Consider the ladder network in Fig. 14.9*a* (p. 665). Use the direct method to obtain $[y]$ in terms of $Y_a = 1/Z_a$ and $Y_b = 1/Z_b$.

Exercise 14.5

By finding $\underline{I}_1$ and $\underline{I}_2$ in terms of $\underline{V}_1$ and $\underline{V}_2$, confirm that the pi network in Fig. 14.14*b* represents the *y*-parameter equations of a reciprocal two-port.

Hybrid Parameters

The **hybrid parameters** or ***h*** parameters involve a mixture of impedance and admittance concepts. These parameters relate $\underline{V}_1$ and $\underline{I}_2$ to $\underline{V}_2$ and $\underline{I}_1$ via

$$\begin{aligned} \underline{V}_1 &= h_{11}\underline{I}_1 + h_{12}\underline{V}_2 \\ \underline{I}_2 &= h_{21}\underline{I}_1 + h_{22}\underline{V}_2 \end{aligned} \qquad (14.12)$$

Thus,

$$\begin{bmatrix} \underline{V}_1 \\ \underline{I}_2 \end{bmatrix} = [h] \begin{bmatrix} \underline{I}_1 \\ \underline{V}_2 \end{bmatrix}$$

where

$$[h] \triangleq \begin{bmatrix} h_{11} & h_{12} \\ h_{21} & h_{22} \end{bmatrix} \tag{14.13}$$

which is the h-parameter matrix.

The particular merit of the h parameters is in conjunction with *transistors,* whose properties are conveniently determined experimentally from measurements with the output shorted or the input open. Setting $\underline{V}_2 = 0$ or $\underline{I}_1 = 0$ in Eq. (14.12) yields

$$\begin{aligned} h_{11} &= \underline{V}_1/\underline{I}_1|_{\underline{V}_2=0} & h_{12} &= \underline{V}_1/\underline{V}_2|_{\underline{I}_1=0} \\ h_{21} &= \underline{I}_2/\underline{I}_1|_{\underline{V}_2=0} & h_{22} &= \underline{I}_2/\underline{V}_2|_{\underline{I}_1=0} \end{aligned} \tag{14.14}$$

Correspondingly, the direct method for measuring or calculating h parameters goes as follows:

1. Short the output port so $\underline{V}_2 = 0$, find $\underline{V}_1$ and $\underline{I}_2$ in terms of $\underline{I}_1$, then calculate $h_{11} = \underline{V}_1/\underline{I}_1$ and $h_{21} = \underline{I}_2/\underline{I}_1$.
2. Open the input port so $\underline{I}_1 = 0$, find $\underline{V}_1$ and $\underline{I}_2$ in terms of $\underline{V}_2$, then calculate $h_{12} = \underline{V}_1/\underline{V}_2$ and $h_{22} = \underline{I}_2/\underline{V}_2$.

Again, the parameter subscripts guide you through the procedure.

The relations in Eq. (14.14) indicate that h_{11} has the units of impedance, h_{22} has the units of admittance, and h_{12} and h_{21} are dimensionless ratios. We therefore say that

$$\begin{aligned} h_{11} &= \text{input impedance with shorted output} \\ h_{21} &= \text{forward current ratio with shorted output} \\ h_{12} &= \text{reverse voltage ratio with open input} \\ h_{22} &= \text{output admittance with open input} \end{aligned}$$

The mixed units here account for the name "hybrid" parameters. To emphasize the physical interpretations, the h parameters of transistors are usually denoted by h_i, h_f, h_r, and h_o rather than h_{11}, h_{21}, h_{12}, and h_{22}.

The mixed nature of the h parameters also manifests itself in the existence test, which involves both KVL and KCL. Specifically,

> The h parameters exist only when an independent current source may be connected to the input port and an independent voltage source may be connected to the output port without violating Kirchhoff's laws.

This test follows from Eq. (14.12).

Any two-port whose h parameters exist can be modeled by the equivalent s-domain diagram in Fig. 14.17. The input side looks like a Thévenin or z-parameter network, while the output side looks like a Norton or y-parameter network. However, going back to the interpretations of the z and y parameters, we find that $h_{11} = 1/y_{11} \neq z_{11}$ and $h_{22} = 1/z_{22} \neq y_{22}$, while h_{12} and h_{21} have no direct counterparts. The hybrid structure of Fig. 14.17 thus precludes simplification for the case of a reciprocal two-port.

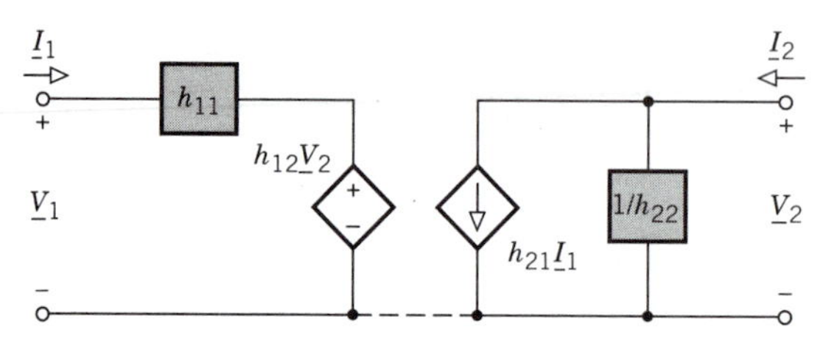

Figure 14.17 h-parameter model.

Example 14.6 *Calculating h Parameters*

The two-port from Example 14.5 has been redrawn in Fig. 14.18 in order to calculate the h parameters. The parameter subscripts serve as a guide when we first short the output port and then open the input port.

With $\underline{V}_2 = 0$, $\underline{V}_1 = (40/s)\underline{I}_1$ and $\underline{I}_2 = 3\underline{I}_1 - \underline{I}_1$ so

$$h_{11} = \underline{V}_1/\underline{I}_1 = 40/s \qquad h_{21} = \underline{I}_2/\underline{I}_1 = 2$$

With $\underline{I}_1 = 0$, $\underline{V}_1 = \underline{V}_2$ and $\underline{V}_2 = 10\underline{I}_2$ so

$$h_{12} = \underline{V}_1/\underline{V}_2 = 1 \qquad h_{22} = \underline{I}_2/\underline{V}_2 = 0.1$$

Thus, including the impedance and admittance units,

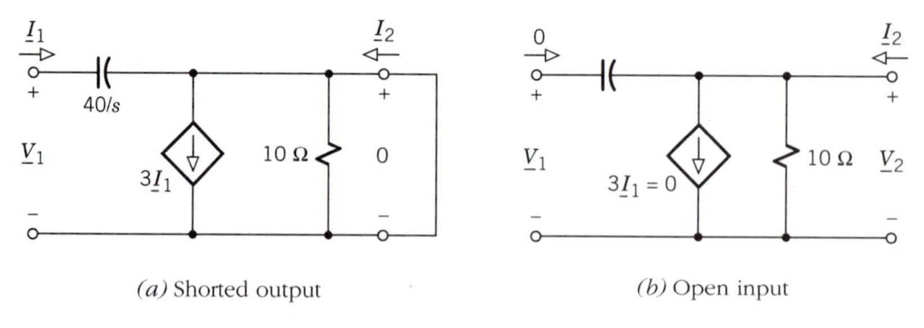

Figure 14.18 Diagrams for Example 14.6.

$$[h] = \begin{bmatrix} 40/s\ \Omega & 1 \\ 2 & 0.1\ \text{S} \end{bmatrix}$$

Exercise 14.6

Consider the two-port in Fig. 14.8*a* (p. 664). Use the fact that $v_2 = -2v_x$ to obtain

$$[h] = \begin{bmatrix} 10\ \Omega & -0.5 \\ -1 & -0.01\ \text{S} \end{bmatrix}$$

Transmission Parameters

Transmission parameters or ***ABCD* parameters** differ in several ways from the other parameters we have discussed. The *ABCD* parameters relate the input variables to the output variables via

$$\begin{aligned} \underline{V}_1 &= A\underline{V}_2 - B\underline{I}_2 \\ \underline{I}_1 &= C\underline{V}_2 - D\underline{I}_2 \end{aligned} \tag{14.15}$$

Note carefully the minus signs in these equations. Also note that the parameters A, B, C, and D are written without subscripts.

Double subscripts are not used in Eq. (14.15) because both variables on the right-hand side bear the same subscript. Nonetheless, we can still define the transmission matrix as

$$[T] \triangleq \begin{bmatrix} A & B \\ C & D \end{bmatrix} \tag{14.16}$$

so that

$$\begin{bmatrix} \underline{V}_1 \\ \underline{I}_1 \end{bmatrix} = [T] \begin{bmatrix} \underline{V}_2 \\ -\underline{I}_2 \end{bmatrix}$$

The minus sign is attached to $\underline{I}_2$ rather than to B and D because the *ABCD* parameters originated in conjunction with transmission lines and it was desirable to take the same reference direction for the input and output currents. The significance of *ABCD* parameters in circuit analysis will emerge in the next section when we analyze interconnected two-ports.

Meanwhile, for the physical interpretation of the *ABCD* parameters, we observe from Eq. (14.15) that

$$A = \underline{V}_1/\underline{V}_2|_{I_2=0} \qquad B = -\underline{V}_1/\underline{I}_2|_{V_2=0}$$
$$C = \underline{I}_1/\underline{V}_2|_{I_2=0} \qquad D = -\underline{I}_1/\underline{I}_2|_{V_2=0} \tag{14.17}$$

Hence,

A = reverse voltage ratio with open output
C = reverse transfer admittance with open output
$-B$ = reverse transfer impedance with shorted output
$-D$ = reverse current ratio with shorted output

Further consideration of Eq. (14.17) leads to three other points. First, with respect to existence,

> The *ABCD* parameters exist only if $\underline{V}_{2oc} \neq 0$ and $\underline{I}_{2sc} \neq 0$.

Second, when the transmission parameters exist, the direct method for determining their values goes as follows:

1. Open the output port so $\underline{I}_2 = 0$, find $\underline{V}_1$ and $\underline{I}_1$ in terms of $\underline{V}_2$, then calculate $A = \underline{V}_1/\underline{V}_2$ and $C = \underline{I}_1/\underline{V}_2$.
2. Short the output port so $\underline{V}_2 = 0$, find $\underline{V}_1$ and $\underline{I}_1$ in terms of $\underline{I}_2$, then calculate $B = -\underline{V}_1/\underline{I}_2$ and $D = -\underline{I}_1/\underline{I}_2$.

Third, since all of the *ABCD* parameters represent *transfer* ratios, an equivalent *s*-domain diagram would consist entirely of controlled sources. Such a diagram provides little help in circuit analysis, so we work instead with the transmission-parameter equations.

Example 14.7 ***Calculating ABCD Parameters***

Let's find the *ABCD* parameters for the two-port considered in Examples 14.5 and 14.6.

With $\underline{I}_2 = 0$, Fig. 14.19*a* shows that $\underline{I}_1 - 3\underline{I}_1 = \underline{V}_2/10$ so $\underline{I}_1 = -0.05\underline{V}_2$ and $\underline{V}_1 = \underline{V}_2 + (40/s)\underline{I}_1 = (1 - 2/s)\underline{V}_2$. Thus,

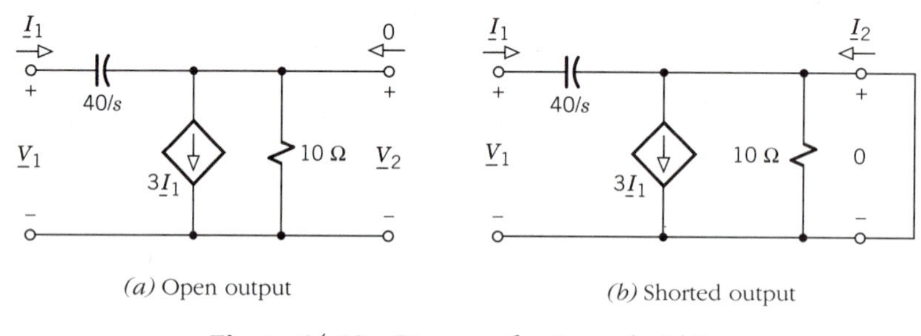

Figure 14.19 Diagrams for Example 14.7.

$$A = \underline{V}_1/\underline{V}_2 = 1 - 2/s \qquad C = \underline{I}_1/\underline{V}_2 = -0.05$$

With $\underline{V}_2 = 0$, Fig. 14.19*b* shows that $\underline{I}_1 - 3\underline{I}_1 = -\underline{I}_2$ so $\underline{I}_1 = 0.5\underline{I}_2$ and $\underline{V}_1 = (40/s)\underline{I}_1 = (20/s)\underline{I}_2$. Thus,

$$B = -\underline{V}_1/\underline{I}_2 = -20/s \qquad D = -\underline{I}_1/\underline{I}_2 = -0.5$$

Summarizing in matrix form, we have

$$[T] = \begin{bmatrix} 1 - 2/s & -20/s\ \Omega \\ -0.05\ \text{S} & -0.5 \end{bmatrix}$$

Exercise 14.7

Again consider the two-port in Fig. 14.8*a* (p. 664). Show that

$$[T] = \begin{bmatrix} -0.6 & 10\ \Omega \\ -0.01\ \text{S} & 1 \end{bmatrix}$$

Parameter Conversion

Having found one set of parameters for a particular two-port, you can obtain any other set by parameter conversion, provided that the other parameters exist. We'll develop parameter-conversion formulas here with reference to Table 14.1, which summarizes the two-port equations for the z, y, h, and $ABCD$ parameters.

TABLE 14.1 Two-port Equations

Impedance Parameters	Admittance Parameters
$\begin{bmatrix} \underline{V}_1 \\ \underline{V}_2 \end{bmatrix} = \begin{bmatrix} z_{11} & z_{12} \\ z_{21} & z_{22} \end{bmatrix} \begin{bmatrix} \underline{I}_1 \\ \underline{I}_2 \end{bmatrix}$	$\begin{bmatrix} \underline{I}_1 \\ \underline{I}_2 \end{bmatrix} = \begin{bmatrix} y_{11} & y_{12} \\ y_{21} & y_{22} \end{bmatrix} \begin{bmatrix} \underline{V}_1 \\ \underline{V}_2 \end{bmatrix}$
Hybrid Parameters	**Transmission Parameters**
$\begin{bmatrix} \underline{V}_1 \\ \underline{I}_2 \end{bmatrix} = \begin{bmatrix} h_{11} & h_{12} \\ h_{21} & h_{22} \end{bmatrix} \begin{bmatrix} \underline{I}_1 \\ \underline{V}_2 \end{bmatrix}$	$\begin{bmatrix} \underline{V}_1 \\ \underline{I}_1 \end{bmatrix} = \begin{bmatrix} A & B \\ C & D \end{bmatrix} \begin{bmatrix} \underline{V}_2 \\ -\underline{I}_2 \end{bmatrix}$

We begin by considering the impedance-parameter equations

$$\begin{bmatrix} \underline{V}_1 \\ \underline{V}_2 \end{bmatrix} = [z] \begin{bmatrix} \underline{I}_1 \\ \underline{I}_2 \end{bmatrix}$$

Solving for $\underline{I}_1$ and $\underline{I}_2$ in terms of $\underline{V}_1$ and $\underline{V}_2$ yields

$$\begin{bmatrix} \underline{I}_1 \\ \underline{I}_2 \end{bmatrix} = [z]^{-1} \begin{bmatrix} \underline{V}_1 \\ \underline{V}_2 \end{bmatrix} = [y] \begin{bmatrix} \underline{V}_1 \\ \underline{V}_2 \end{bmatrix}$$

Hence,

$$[y] = [z]^{-1} = \begin{bmatrix} z_{22}/\Delta_z & -z_{12}/\Delta_z \\ -z_{21}/\Delta_z & z_{11}/\Delta_z \end{bmatrix}$$

where we have introduced the determinant

$$\Delta_z = |[z]| = z_{11}z_{22} - z_{12}z_{21}$$

Conversely, it follows that

$$[z] = [y]^{-1} = \begin{bmatrix} y_{22}/\Delta_y & -y_{12}/\Delta_y \\ -y_{21}/\Delta_y & y_{11}/\Delta_y \end{bmatrix}$$

where

$$\Delta_y = |[y]| = y_{11}y_{22} - y_{12}y_{21}$$

In short, conversion between the z and y parameters simply requires inverting the parameter matrix.

Conversions involving hybrid or transmission parameters are less direct because of the mixed variables. The general strategy employs algebraic manipulation to put one set of equations in the form of the other set. Although these manipulations do not involve matrices, they always produce one term that equals the determinant of the original parameter matrix.

For instance, to find the $\boldsymbol{h}$ parameters from the z parameters, we start with the z-parameter equations

$$\underline{V}_1 = z_{11}\underline{I}_1 + z_{12}\underline{I}_2 \qquad \underline{V}_2 = z_{21}\underline{I}_1 + z_{22}\underline{I}_2$$

Since the $\boldsymbol{h}$ parameters relate $\underline{V}_1$ and $\underline{I}_2$ to $\underline{I}_1$ and $\underline{V}_2$, we can rewrite the second z-parameter equation in the desired form as

$$\underline{I}_2 = (-z_{21}/z_{22})\underline{I}_1 + (1/z_{22})\underline{V}_2$$

Inserting this relation into the first z-parameter equation yields

$$\begin{aligned} \underline{V}_1 &= z_{11}\underline{I}_1 + z_{12}[(-z_{21}/z_{22})\underline{I}_1 + (1/z_{22})\underline{V}_2] \\ &= [(z_{11}z_{22} - z_{12}z_{21})/z_{22}]\underline{I}_1 + (z_{12}/z_{22})\underline{V}_2 \end{aligned}$$

Thus, after putting these equations in matrix form and recognizing that $z_{11}z_{22} - z_{12}z_{21} = \Delta_z$, we have

$$\begin{bmatrix} V_1 \\ I_2 \end{bmatrix} = \begin{bmatrix} \Delta_z/z_{22} & z_{12}/z_{22} \\ -z_{21}/z_{22} & 1/z_{22} \end{bmatrix} \begin{bmatrix} I_1 \\ V_2 \end{bmatrix}$$

from which

$$[h] = \begin{bmatrix} \Delta_z/z_{22} & z_{12}/z_{22} \\ -z_{21}/z_{22} & 1/z_{22} \end{bmatrix}$$

Other conversions follow similar lines of attack.

Table 14.2 lists all the conversion relationships for the z, y, h, and $ABCD$ parameters. The symbols Δ_z, Δ_y, Δ_h, and Δ_T stand for the determinants of the respective parameter matrices. This table also provides information about impossible conversions — for if any of the denominator terms happens to equal zero, then the converted parameters will not exist for that particular two-port. By way of illustration, suppose a certain transistor has $h_{22} = 0$ and the other h parameters are nonzero, so $\Delta_h \neq 0$. Scanning down the third column of Table 14.2, we see that $[y]$ and $[T]$ can be calculated, but the z parameters are undefined when $h_{22} = 0$.

TABLE 14.2 Parameter Conversions

$[z]$	$\begin{matrix} z_{11} & z_{12} \\ z_{21} & z_{22} \end{matrix}$	$\begin{matrix} \frac{y_{22}}{\Delta_y} & \frac{-y_{12}}{\Delta_y} \\ \frac{-y_{21}}{\Delta_y} & \frac{y_{11}}{\Delta_y} \end{matrix}$	$\begin{matrix} \frac{\Delta_h}{h_{22}} & \frac{h_{12}}{h_{22}} \\ \frac{-h_{21}}{h_{22}} & \frac{1}{h_{22}} \end{matrix}$	$\begin{matrix} \frac{A}{C} & \frac{\Delta_T}{C} \\ \frac{1}{C} & \frac{D}{C} \end{matrix}$
$[y]$	$\begin{matrix} \frac{z_{22}}{\Delta_z} & \frac{-z_{12}}{\Delta_z} \\ \frac{-z_{21}}{\Delta_z} & \frac{z_{11}}{\Delta_z} \end{matrix}$	$\begin{matrix} y_{11} & y_{12} \\ y_{21} & y_{22} \end{matrix}$	$\begin{matrix} \frac{1}{h_{11}} & \frac{-h_{12}}{h_{11}} \\ \frac{h_{21}}{h_{11}} & \frac{\Delta_h}{h_{11}} \end{matrix}$	$\begin{matrix} \frac{D}{B} & \frac{-\Delta_T}{B} \\ \frac{-1}{B} & \frac{A}{B} \end{matrix}$
$[h]$	$\begin{matrix} \frac{\Delta_z}{z_{22}} & \frac{z_{12}}{z_{22}} \\ \frac{-z_{21}}{z_{22}} & \frac{1}{z_{22}} \end{matrix}$	$\begin{matrix} \frac{1}{y_{11}} & \frac{-y_{12}}{y_{11}} \\ \frac{y_{21}}{y_{11}} & \frac{\Delta_y}{y_{11}} \end{matrix}$	$\begin{matrix} h_{11} & h_{12} \\ h_{21} & h_{22} \end{matrix}$	$\begin{matrix} \frac{B}{D} & \frac{\Delta_T}{D} \\ \frac{-1}{D} & \frac{C}{D} \end{matrix}$
$[T]$	$\begin{matrix} \frac{z_{11}}{z_{21}} & \frac{\Delta_z}{z_{21}} \\ \frac{1}{z_{21}} & \frac{z_{22}}{z_{21}} \end{matrix}$	$\begin{matrix} \frac{-y_{22}}{y_{21}} & \frac{-1}{y_{21}} \\ \frac{-\Delta_y}{y_{21}} & \frac{-y_{11}}{y_{21}} \end{matrix}$	$\begin{matrix} \frac{-\Delta_h}{h_{21}} & \frac{-h_{11}}{h_{21}} \\ \frac{-h_{22}}{h_{21}} & \frac{-1}{h_{21}} \end{matrix}$	$\begin{matrix} A & B \\ C & D \end{matrix}$

Another observation from Table 14.2 concerns reciprocity. We already know that a reciprocal network has $z_{22} = z_{12}$ and/or $y_{21} = y_{12}$. The first and second rows of the table now indicate that if $z_{21} = z_{12}$ or $y_{21} = y_{12}$, then

$$h_{21} = -h_{12} \qquad \Delta_T = AD - BC = 1 \tag{14.18}$$

Hence, you can also test for reciprocity using h or $ABCD$ parameters.

Example 14.8 *Parameters of a Tee Network*

The tee network in Fig. 14.20 was previously found to have

$$[z] = \begin{bmatrix} Z_a + Z_c & Z_c \\ Z_c & Z_b + Z_c \end{bmatrix}$$

The first column of Table 14.2 now yields

$$[y] = \begin{bmatrix} (Z_b + Z_c)/\Delta_z & -Z_c/\Delta_z \\ -Z_c/\Delta_z & (Z_a + Z_c)/\Delta_z \end{bmatrix}$$

$$[h] = \begin{bmatrix} \Delta_z/(Z_b + Z_c) & Z_c/(Z_b + Z_c) \\ -Z_c/(Z_b + Z_c) & 1/(Z_b + Z_c) \end{bmatrix}$$

$$[T] = \begin{bmatrix} 1 + Z_a/Z_c & \Delta_z/Z_c \\ 1/Z_c & 1 + Z_b/Z_c \end{bmatrix}$$

where $\Delta_z = Z_aZ_b + Z_aZ_c + Z_bZ_c$.

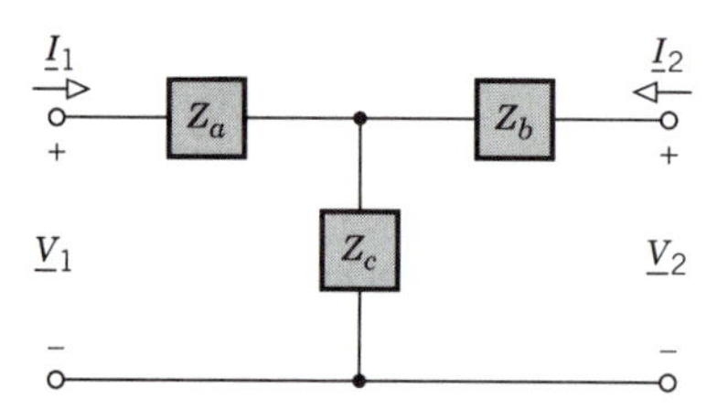

Figure 14.20 Tee network.

Exercise 14.8

A reciprocal two-port is known to have $A = 5$, $C = -0.1$ S, and $D = 0.2$. Evaluate B and find the other parameter sets that exist. Include all units.

14.3 CIRCUIT ANALYSIS WITH TWO-PORTS

Armed with the background from the previous sections, we're prepared at last to analyze complete circuits containing two-ports. The results developed here have important applications in the analysis and design of amplifiers and filters.

We'll begin with the basic case of a single two-port having terminations at the input and output. Then we'll consider two or more two-ports interconnected in various ways.

Terminated Two-Ports

Usually, a two-port interfaces a source and a load. The load acts as an equivalent impedance Z_L, and the source network may be replaced by a Thévenin or Norton model with parameters $\underline{V}_s$, $\underline{I}_s$, and Z_s. Figure 14.21 shows the general *s*-domain diagram of a terminated two-port, taking the Thévenin model for the source.

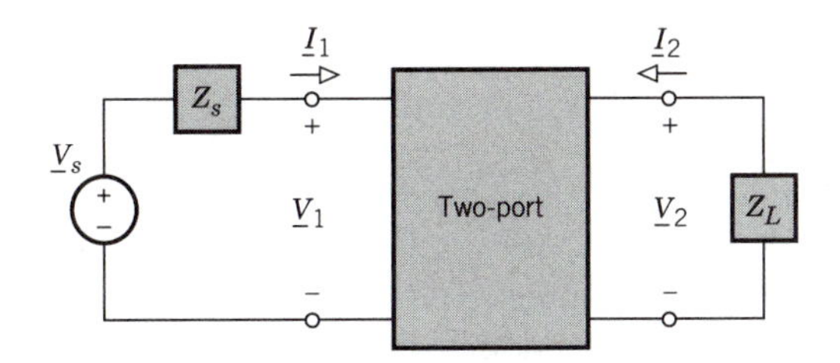

Figure 14.21 Two-port terminated with source and load.

Our objective here is to analyze this complete circuit using two-port parameters so the results will hold for *any* two-port. The particular network functions we seek are the

$$
\begin{aligned}
&\text{Voltage transfer function} && H_v(s) \triangleq \underline{V}_2/\underline{V}_1 \\
&\text{Current transfer function} && H_i(s) \triangleq \underline{I}_2/\underline{I}_1 \\
&\text{Equivalent input impedance} && Z_i(s) \triangleq \underline{V}_1/\underline{I}_1 \\
&\text{Equivalent output impedance} && Z_o(s) \triangleq \underline{V}_2/\underline{I}_2\big|_{\underline{V}_s=0}
\end{aligned}
\tag{14.19}
$$

Other quantities involving terminal variables can be obtained from these four basic ratios.

We begin the analysis by assuming that the z parameters exist, so

$$\underline{V}_1 = z_{11}\underline{I}_1 + z_{12}\underline{I}_2 \qquad \underline{V}_2 = z_{21}\underline{I}_1 + z_{22}\underline{I}_2$$

But the load impedance across the output port imposes a constraint on $\underline{V}_2$ and $\underline{I}_2$, namely

$$\underline{V}_2 = -Z_L\underline{I}_2$$

Replacing $\underline{V}_2$ with $-Z_L\underline{I}_2$ in the second z-parameter equation yields the current transfer function

$$H_i(s) = \frac{\underline{I}_2}{\underline{I}_1} = \frac{-z_{21}}{z_{22} + Z_L}$$

Next, we obtain the voltage transfer function and the input impedance by substituting $\underline{I}_2 = H_i(s)\underline{I}_1$ in both z-parameter equations, the results being

$$H_v(s) = \frac{\underline{V}_2}{\underline{V}_1} = \frac{z_{21}Z_L}{z_{11}z_{22} - z_{12}z_{21} + z_{11}Z_L} = \frac{z_{21}Z_L}{\Delta_z + z_{11}Z_L}$$

$$Z_i(s) = \frac{\underline{V}_1}{\underline{I}_1} = \frac{z_{11}z_{22} - z_{12}z_{21} + z_{11}Z_L}{z_{22} + Z_L} = \frac{\Delta_z + z_{11}Z_L}{z_{22} + Z_L}$$

Finally, for the output impedance, we use the fact that $\underline{V}_1 = -Z_s\underline{I}_1$ when $\underline{V}_s = 0$. Eliminating both $\underline{V}_1$ and $\underline{I}_1$ from the z-parameter equations then yields

$$Z_o(s) = \left.\frac{\underline{V}_2}{\underline{I}_2}\right|_{\underline{V}_s=0} = \frac{\Delta_z + z_{22}Z_s}{z_{11} + Z_s}$$

which completes the analysis.

By similar manipulations, the desired expressions can be obtained with other two-port parameters. The results are widely known, of course, and they are listed in Table 14.3 for convenient reference. Some of these relations are written in terms of the external admittances $Y_s = 1/Z_s$ and $Y_L = 1/Z_L$.

TABLE 14.3 Relations for Terminated Two-Ports

$H_v(s)$	$\dfrac{z_{21}Z_L}{\Delta_z + z_{11}Z_L}$	$\dfrac{-y_{21}}{y_{22} + Y_L}$	$\dfrac{-h_{21}Z_L}{h_{11} + \Delta_h Z_L}$	$\dfrac{Z_L}{AZ_L + B}$
$H_i(s)$	$\dfrac{-z_{21}}{z_{22} + Z_L}$	$\dfrac{y_{21}Y_L}{\Delta_y + y_{11}Y_L}$	$\dfrac{h_{21}Y_L}{h_{22} + Y_L}$	$\dfrac{-1}{CZ_L + D}$
$Z_i(s)$	$\dfrac{\Delta_z + z_{11}Z_L}{z_{22} + Z_L}$	$\dfrac{y_{22} + Y_L}{\Delta_y + y_{11}Y_L}$	$\dfrac{\Delta_h + h_{11}Y_L}{h_{22} + Y_L}$	$\dfrac{AZ_L + B}{CZ_L + D}$
$Z_o(s)$	$\dfrac{\Delta_z + z_{22}Z_s}{z_{11} + Z_s}$	$\dfrac{y_{11} + Y_s}{\Delta_y + y_{22}Y_s}$	$\dfrac{h_{11} + Z_s}{\Delta_h + h_{22}Z_s}$	$\dfrac{B + DZ_s}{A + CZ_s}$

Example 14.9 ***Calculating a Transfer Function***

Suppose we want to study the behavior of i_2 when $Z_s = 0$ back in Fig. 14.21 and we know the *ABCD* parameters of the two-port. Since $\underline{V}_1 = \underline{V}_s$ when $Z_s = 0$, the desired information can be obtained from the transfer function $H(s) = \underline{I}_2/\underline{V}_1$.

Although $\underline{I}_2/\underline{V}_1$ is not one of our basic ratios, we observe that $\underline{V}_1 = Z_i(s)\underline{I}_1$. Thus,

$$H(s) = \frac{\underline{I}_2}{\underline{V}_1} = \frac{\underline{I}_2}{Z_i(s)\underline{I}_1} = \frac{H_i(s)}{Z_i(s)} = \frac{-1}{AZ_L + B}$$

where we have used Table 14.3 to express $H_i(s)/Z_i(s)$ in terms of the *ABCD* parameters.

For further illustration, let the load be a 2.5-H inductor, so $Z_L = 2.5s$, and let the two-port be the one considered in Example 14.7, so $A = 1 - 2/s$ and $B = -20/s$. Substitution into $H(s)$ yields

$$H(s) = \frac{-1}{2.5s - 5 - 20/s} = \frac{-0.4s}{s^2 - 2s - 8}$$

The characteristic polynomial is $P(s) = s^2 - 2s - 8$, whose roots are $p_1 = -2$ and $p_2 = 4$. Consequently, the natural behavior of i_2 will include a decaying mode proportional to e^{-2t} and a growing mode proportional to e^{4t}. The growing mode means, of course, that this circuit is unstable.

Example 14.10 *A Mid-Frequency Transistor Amplifier*

Figure 14.22 represents a simple transistor amplifier operating at midrange frequencies, so all quantities are independent of s. The dc biasing circuitry has been omitted here to focus on the "small-signal" performance.

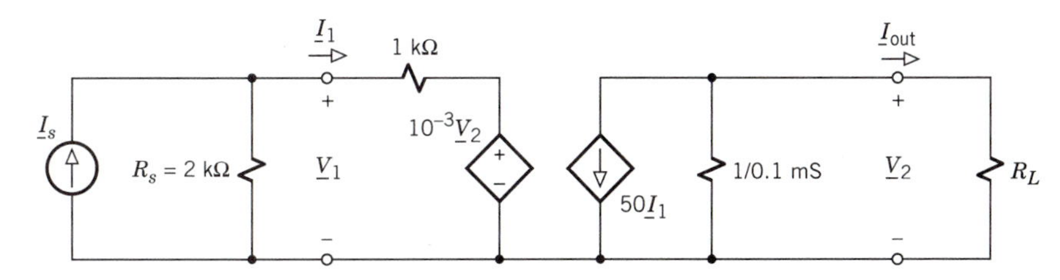

Figure 14.22 Midfrequency model of a transistor amplifier.

The transistor is modeled by the indicated small-signal $\boldsymbol{h}$ parameters, namely,

$$[h] = \begin{bmatrix} 1\ \text{k}\Omega & 10^{-3} \\ 50 & 0.1\ \text{mS} \end{bmatrix}$$

so

$$\Delta_h = h_{11}h_{22} - h_{12}h_{21} = 0.05$$

We want to find the load resistance R_L needed to get overall current gain $A_i = \underline{I}_{out}/\underline{I}_s = -25$, where the negative value reflects the fact that $\underline{I}_{out} = -\underline{I}_2$.

Loading effect at the input causes $\underline{I}_1$ to differ from $\underline{I}_s$ by the current-divider ratio $R_s/[R_s + Z_i(s)]$. Consequently, the magnitude of $\underline{I}_{out}/\underline{I}_s$ will be less than the magnitude of $H_i(s) = \underline{I}_2/\underline{I}_1$. Furthermore, as seen in Table 14.3, both $Z_i(s)$ and $H_i(s)$ vary with $Z_L = R_L$. Using the given $\boldsymbol{h}$ parameters, and working with impedances in kilohms and admittances in millisiemens, we obtain

$$Z_i(s) = \frac{0.05R_L + 1}{0.1R_L + 1} \qquad H_i(s) = \frac{50}{0.1R_L + 1}$$

Thus, over the range $0 \le R_L \le \infty$, $Z_i(s)$ goes from 1 kΩ to 0.5 kΩ and $H_i(s)$ goes from 50 to 0.

We account for both varying effects by writing the chain expansion

$$A_i = \frac{-\underline{I}_2}{\underline{I}_s} = -\frac{\underline{I}_1}{\underline{I}_s} \times \frac{\underline{I}_2}{\underline{I}_1} = -\frac{R_s}{R_s + Z_i(s)} H_i(s)$$

Inserting $R_s = 2\ \text{k}\Omega$ and our expressions for $Z_i(s)$ and $H_i(s)$ yields

$$A_i = \frac{-100}{3 + 0.25R_L}$$

where, again, R_L is in kilohms. The specification $A_i = -25$ thus requires $R_L = 4\ \text{k}\Omega$.

Exercise 14.9

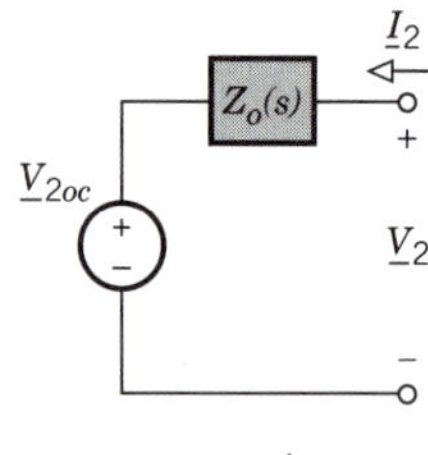

Figure 14.23

Figure 14.23 is the Thévenin model looking back into the output terminals of a two-port with a source connected at the input. Use z-parameter relations with $Z_L = \infty$ to show that the open-circuit output voltage may be expressed in terms of source voltage by

$$\underline{V}_{2oc} = \frac{z_{21}}{Z_s + z_{11}} \underline{V}_s$$

Exercise 14.10

Find $H_v(s)$ when $Z_L = 1/sC$ and

$$[y] = \begin{bmatrix} 0.1 + 1/s & -1/s \\ -1/s & 0.2 + 1/s \end{bmatrix}$$

What value of C results in critical damping?

Interconnected Two-Ports

Some circuits contain more than one two-port, or they may be viewed that way for purposes of analysis or design. We therefore conclude our investigation by studying interconnected two-ports. Throughout, we'll assume that

> When two-ports are connected together, the interconnections do not alter the properties of the individual two-ports.

Otherwise, the two-port analysis method would be invalid.

Cascade connection Probably the most obvious way of combining two-ports is the *cascade* diagrammed in Fig. 14.24. Here, the output terminals of the first two-port are connected directly to the input terminals of the second. The currents and voltages at the connection points are distinguished by the subscripts a and b, and we see that $\underline{V}_{1b} = \underline{V}_{2a}$ whereas $\underline{I}_{1b} = -\underline{I}_{2a}$. This type of interconnection always preserves the individual properties of the two-ports.

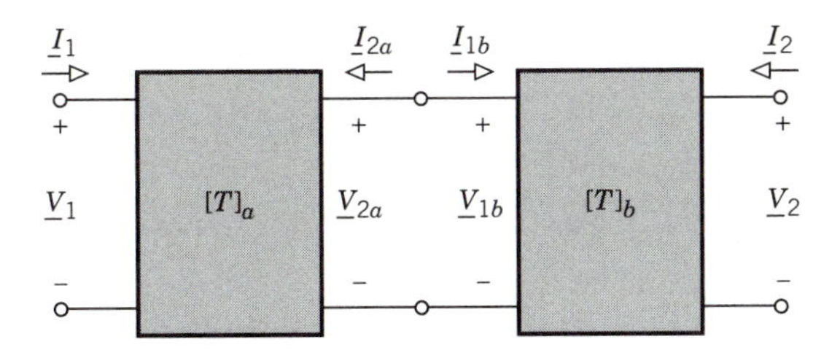

Figure 14.24 Cascade connection of two-ports.

The two-ports themselves have been labeled with their transmission matrices $[T]_a$ and $[T]_b$ because the cascade acts like a single equivalent two-port with

$$[T]_{cas} = [T]_a[T]_b \tag{14.20}$$

This valuable result comes from the two-port equations

$$\begin{bmatrix} \underline{V}_1 \\ \underline{I}_1 \end{bmatrix} = [T]_a \begin{bmatrix} \underline{V}_{2a} \\ -\underline{I}_{2a} \end{bmatrix} \qquad \begin{bmatrix} \underline{V}_{2a} \\ -\underline{I}_{2a} \end{bmatrix} = \begin{bmatrix} \underline{V}_{1b} \\ \underline{I}_{1b} \end{bmatrix} = [T]_b \begin{bmatrix} \underline{V}_2 \\ -\underline{I}_2 \end{bmatrix}$$

Hence,

$$\begin{bmatrix} \underline{V}_1 \\ \underline{I}_1 \end{bmatrix} = [T]_a[T]_b \begin{bmatrix} \underline{V}_2 \\ -\underline{I}_2 \end{bmatrix}$$

which confirms that $[T]_{cas} = [T]_a[T]_b$.

Equation (14.20) readily generalizes for cascades of three or more two-ports. The overall equivalent transmission matrix simply equals the product of the individual transmission matrices.

Example 14.11 *A Cascade Amplifier*

In Example 14.10 we studied a transistor amplifier that had $A_i = \underline{I}_{out}/\underline{I}_s = -25$ when $R_s = 2\ \text{k}\Omega$ and $R_L = 4\ \text{k}\Omega$. Now let's replace the single transistor with two identical transistors in cascade, as depicted by Fig. 14.25. Analysis of this arrangement will show that it produces much more current gain.

For cascade analysis, we must first obtain the *ABCD* parameters of the transistors. Converting the previously given h parameters yields

$$[T]_a = [T]_b = \begin{bmatrix} -10^{-3} & -20\ \Omega \\ -2\ \mu\text{S} & -0.02 \end{bmatrix}$$

Then, by matrix multiplication,

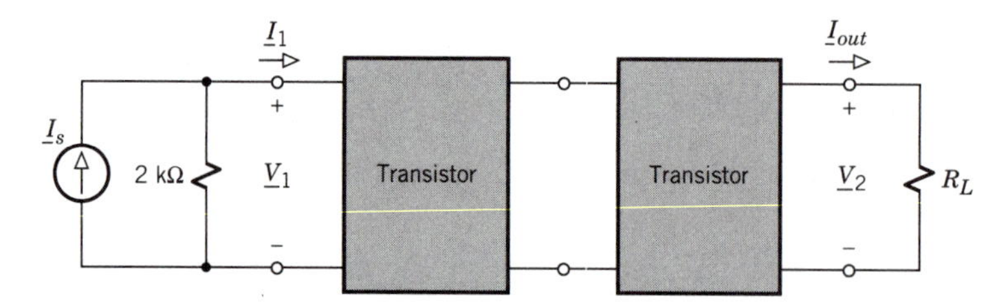

Figure 14.25 Cascade amplifier.

$$[T]_{cas} = [T]_a[T]_b = \begin{bmatrix} 41 \times 10^{-6} & 0.42\ \Omega \\ 0.042\ \mu\text{S} & 440 \times 10^{-6} \end{bmatrix}$$

Next, we use the *ABCD* parameters of the cascade along with $R_L = 4$ kΩ to calculate

$$Z_i(s) = \frac{AR_L + B}{CR_L + D} = 961\ \Omega \qquad H_i(s) = \frac{-1}{CR_L + D} = -1{,}645$$

Finally, from the expression derived in Example 14.10,

$$A_i = \frac{\underline{I}_{out}}{\underline{I}_s} = -\frac{R_s}{R_s + Z_i(s)} H_i(s) \approx 1{,}100$$

The value of A_i is now positive, as well as larger in magnitude, because each stage acts as an inverting amplifier.

Exercise 14.11

Suppose both two-ports in Fig. 14.24 have

$$[T] = \begin{bmatrix} 2 & 5\ \text{k}\Omega \\ -1\ \text{mS} & 3 \end{bmatrix}$$

Find v_1 and i_2 when $i_1 = 8$ mA and the load is a short circuit.

Series and Parallel Connections Figure 14.26 shows two other configurations of practical interest. If you cover up both output ports or both input ports in each diagram, then you'll see why Fig. 14.26*a* is called a *series* connection and Fig. 14.26*b* is called a *parallel* connection.

The equivalent *z*-parameter matrix of the series connection is

$$[z]_{ser} = [z]_a + [z]_b \tag{14.21}$$

while the equivalent *y*-parameter matrix of the parallel connection is

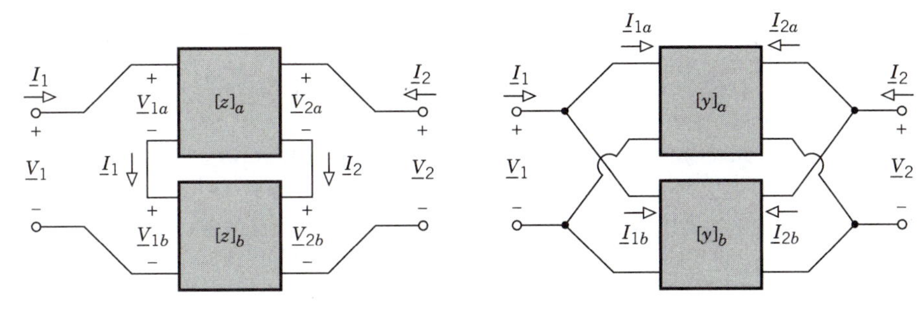

(a) Series connection of two-ports *(b)* Parallel connection of two-ports

Figure 14.26

$$[y]_{par} = [y]_a + [y]_b \tag{14.22}$$

Similarly, for three or more two-ports connected in series or parallel, you add the individual z or y parameters term by term.

The derivation of Eq. (14.21) starts with the observation that the series connection in Fig. 14.26*a* has

$$\begin{bmatrix} \underline{V}_{1a} \\ \underline{V}_{2a} \end{bmatrix} = [z]_a \begin{bmatrix} \underline{I}_1 \\ \underline{I}_2 \end{bmatrix} \qquad \begin{bmatrix} \underline{V}_{1b} \\ \underline{V}_{2b} \end{bmatrix} = [z]_b \begin{bmatrix} \underline{I}_1 \\ \underline{I}_2 \end{bmatrix}$$

But $\underline{V}_1 = \underline{V}_{1a} + \underline{V}_{1b}$ and $\underline{V}_2 = \underline{V}_{2a} + \underline{V}_{2b}$, so

$$\begin{bmatrix} \underline{V}_1 \\ \underline{V}_2 \end{bmatrix} = \begin{bmatrix} \underline{V}_{1a} \\ \underline{V}_{2a} \end{bmatrix} + \begin{bmatrix} \underline{V}_{1b} \\ \underline{V}_{2b} \end{bmatrix} = \{[z]_a + [z]_b\} \begin{bmatrix} \underline{I}_1 \\ \underline{I}_2 \end{bmatrix}$$

from which $[z]_{ser} = [z]_a + [z]_b$. Equation (14.22) then follows by duality, since the parallel connection in Fig. 14.26*b* is the dual of the series connection.

Unlike a cascade, a series or parallel connection may alter the properties of the individual two-ports, especially when one of them has a common ground. Figure 14.27 illustrates two such cases, using a three-terminal network for one two-port. The resistive networks are altered because the common

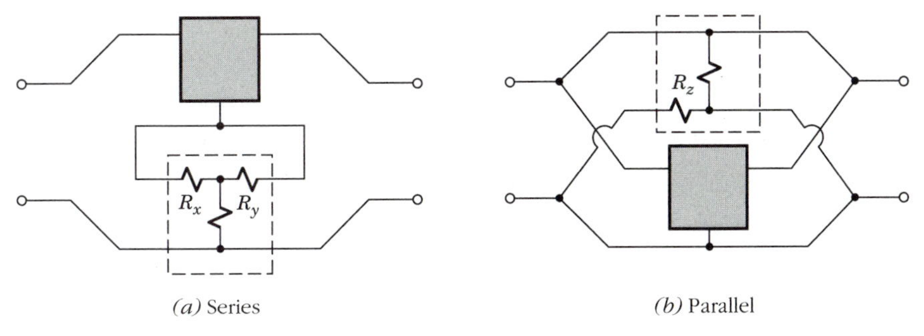

(a) Series *(b)* Parallel

Figure 14.27 Two-port connections that alter properties.

ground in Fig. 14.27*a* puts R_x and R_y in parallel, and the common ground in Fig. 14.27*b* puts a short around R_z. Although formal tests for valid interconnections exist, you can usually spot an invalid one by examining the diagram.

Parallel combinations of two or more common-ground two-ports are always valid if all common-ground terminals get tied together. A simple but significant case of this type is shown in Fig. 14.28*a*, where impedance Z_F bridges across the top of a three-terminal network. When redrawn as in Fig. 14.28*b*, we more clearly see that the bridging impedance may be viewed as a common-ground two-port in parallel with the three-terminal network.

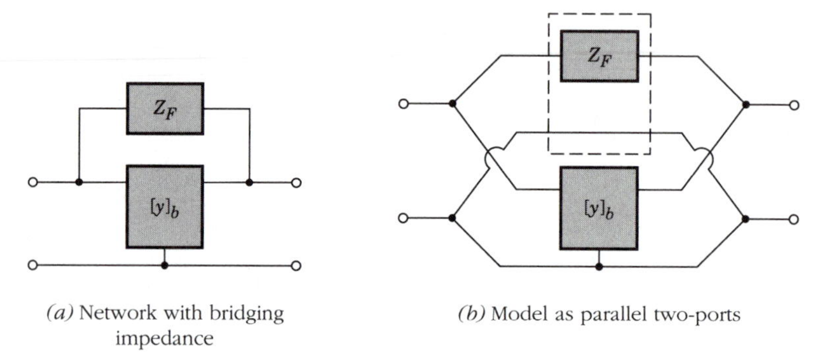

(*a*) Network with bridging impedance

(*b*) Model as parallel two-ports

Figure 14.28

Letting $Y_F = 1/Z_F$, the y parameters of the upper two-port are readily found to be

$$[y]_a = \begin{bmatrix} Y_F & -Y_F \\ -Y_F & Y_F \end{bmatrix}$$

Hence, the parallel connection has

$$[y] = \begin{bmatrix} Y_F + y_{11\,b} & -Y_F + y_{12\,b} \\ -Y_F + y_{21\,b} & Y_F + y_{22\,b} \end{bmatrix} \tag{14.23}$$

This result plays an important role when $[y]_b$ represents an electronic amplifier and Z_F provides *feedback* from output to input.

Example 14.12 *A High-Frequency Amplifier*

When high frequencies are to be amplified, the circuit model of an amplifier must include small stray capacitances that would be ignored at lower frequencies. For instance, Fig. 14.29 gives the simplified model of a high-frequency transistor voltage amplifier with one capacitance. We'll analyze this circuit to find the frequency response from the voltage transfer function $H_v(s) = \underline{V}_2/\underline{V}_1$ when load resistance R_L is connected at the output.

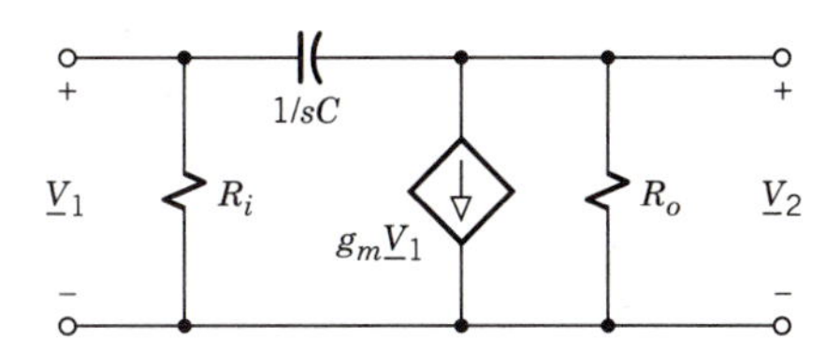

Figure 14.29 High-frequency model of a transistor amplifier.

Closer examination of the diagram reveals that the capacitance acts as a bridging impedance, like Fig. 14.28*a* with $Z_F = 1/sC$. Moreover, by inspection of the remaining resistive network,

$$[y]_b = \begin{bmatrix} 1/R_i & 0 \\ g_m & 1/R_o \end{bmatrix}$$

Thus, using Eq. (14.23),

$$[y] = \begin{bmatrix} sC + 1/R_i & -sC \\ -sC + g_m & sC + 1/R_o \end{bmatrix}$$

From Table 14.3, the corresponding voltage transfer function with $Y_L = 1/R_L$ is

$$H_v(s) = \frac{-y_{21}}{y_{22} + Y_L} = \frac{sC - g_m}{sC + 1/R_o + 1/R_L} = \frac{s - g_m/C}{s + G/C}$$

where $G = 1/R_o + 1/R_L = 1/(R_o \| R_L)$.

Our result shows that $H_v(s)$ has a pole at $p = -G/C$ and a zero at $z = +g_m/C$. This right-hand plane zero does not imply instability. Rather, it accounts for the low-frequency voltage inversion since $H_v(0) = -g_m/G$. Figure 14.30 plots the amplitude response $|H_v(j\omega)|$ assuming $g_m/G \gg 1$. The lower cutoff frequency is $\omega_{co} \approx G/C$, and the gain drops to unity for $\omega \gg g_m/C$.

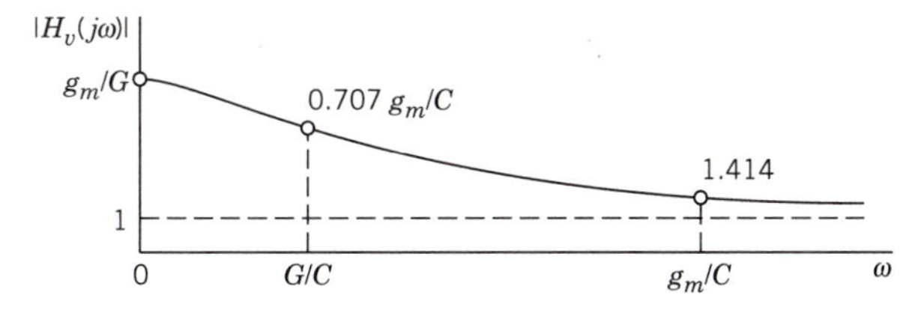

Figure 14.30 Amplitude response of a high-frequency amplifier.

Exercise 14.12

The arrangement in Fig. 14.31 is called a **series-parallel connection**. Show that the equivalent h-parameter matrix is $[h] = [h]_a + [h]_b$.

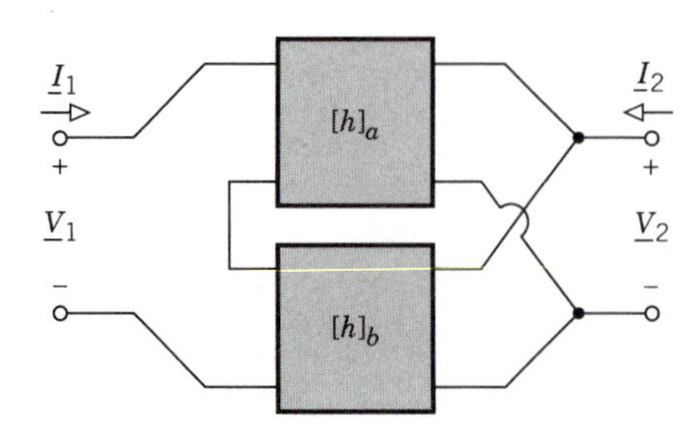

Figure 14.31 Series-parallel connection.

PROBLEMS

Section 14.1 Two-Ports and Impedance Parameters

14.1* Use the direct method to obtain $[z]$ for Fig. P14.1 when $i_x = 3i_2$.

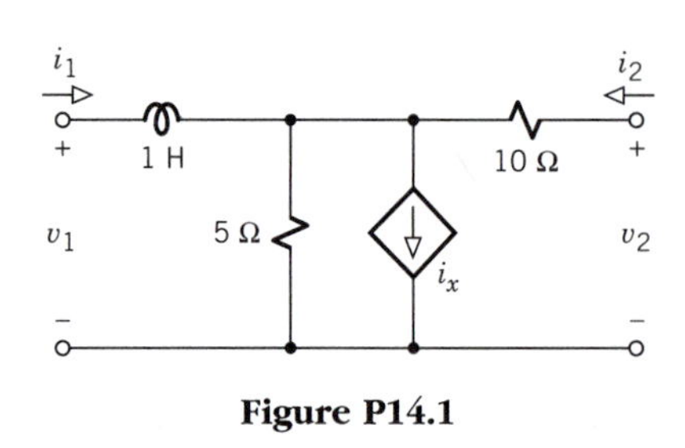

Figure P14.1

14.2 Obtain $[z]$ for Fig. P14.2 by the direct method.

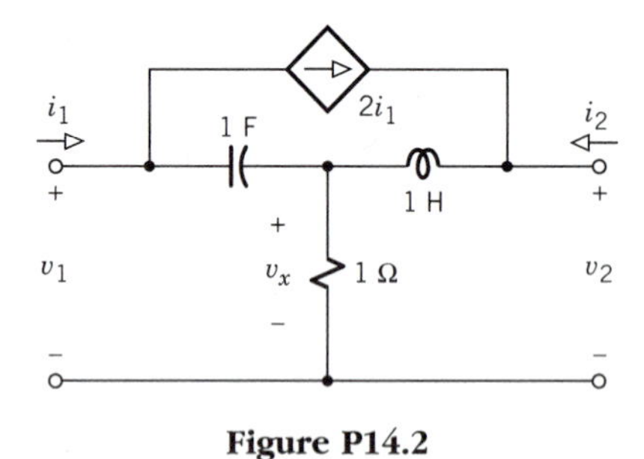

Figure P14.2

14.3 Obtain $[z]$ for Fig. P14.3 by the direct method.

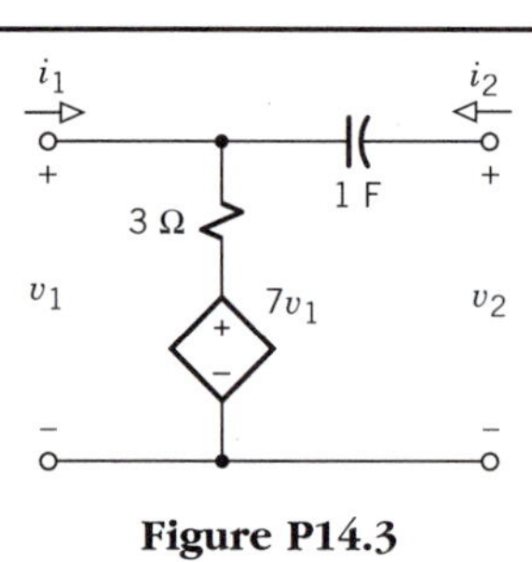

Figure P14.3

14.4 Take advantage of the symmetry of Fig. P14.4 to find the z parameters by direct method.

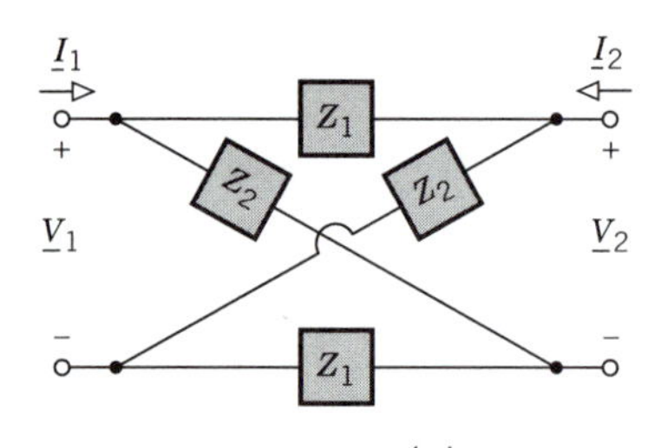

Figure P14.4

14.5 Using the direct method, show that the pi network in Fig. P14.5 has

$$[z] = \begin{bmatrix} Z_a(Z_c + Z_b)/\Sigma Z & Z_a Z_b/\Sigma Z \\ Z_a Z_b/\Sigma Z & Z_b(Z_c + Z_a)/\Sigma Z \end{bmatrix}$$

where $\Sigma Z = Z_a + Z_b + Z_c$.

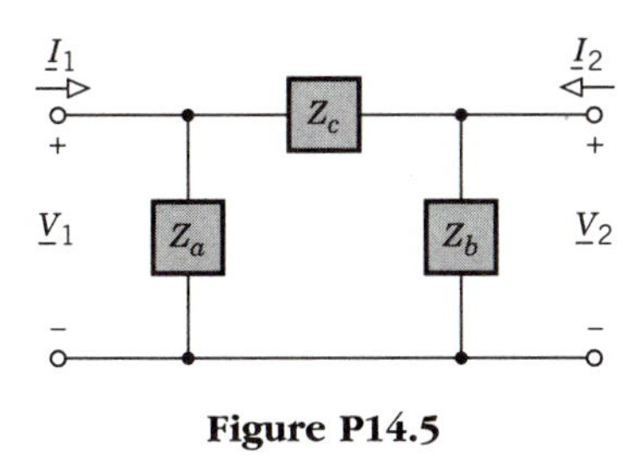

Figure P14.5

14.6* Find the z parameters for Fig. P14.6 by the following method: attach current sources for $\underline{I}_1$ and $\underline{I}_2$, write a matrix node equation, and solve for $\underline{V}_1$ and $\underline{V}_2$.

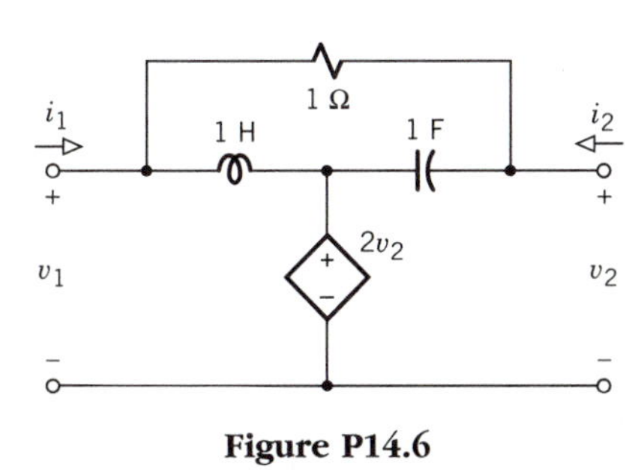

Figure P14.6

14.7 Consider the ladder network in Fig. P14.7. Letting all impedances equal Z, write mesh equations with unknown currents $\underline{I}_1$, $\underline{I}_x$, and $-\underline{I}_2$. Then eliminate $\underline{I}_x$ to get the z-parameter equations.

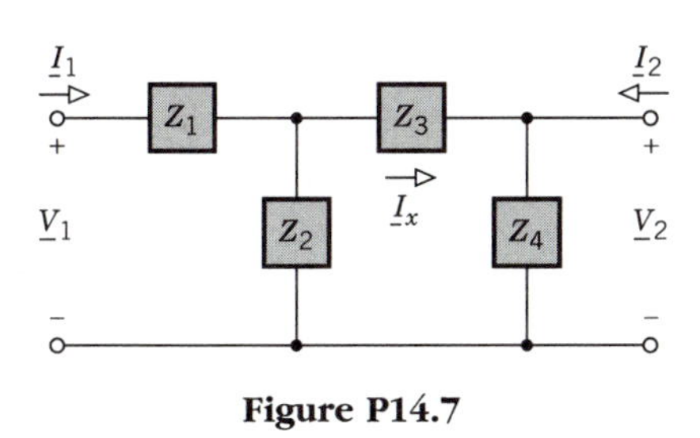

Figure P14.7

14.8 Do Problem 14.7 for the bridged-tee network in Fig. P14.8.

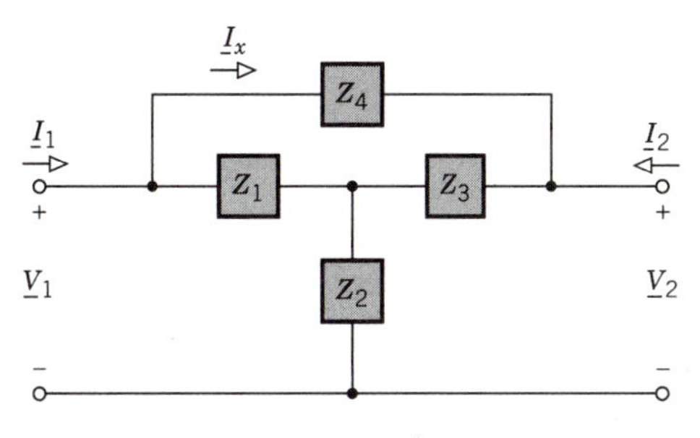

Figure P14.8

14.9 The equivalent tee network in Fig. 14.11 (p. 667) can be modified to represent a nonreciprocal two-port by adding an element in series at the output. Determine the required element and draw the modified network.

14.10 The tee network in Fig. 14.12 (p. 668) can be modified to represent a nonreciprocal two-port by adding an element in series at the input. Determine the required element and the expressions for Z_a, Z_b, and Z_c in terms of the z parameters. Then draw the modified network.

14.11 Draw and label the simplest possible equivalent network for a resistive two-port given the following dc measurement results:

Output Open	*Input Open*
$v_1 = 20$ V	$v_1 = 10$ V
$v_2 = 20$ V	$v_2 = 20$ V
$i_1 = 4$ mA	$i_2 = 2$ mA

14.12 Do Problem 14.1 given:

Output Open	*Input Open*
$v_1 = 30$ V	$v_1 = 12$ V
$v_2 = 18$ V	$v_2 = 30$ V
$i_1 = 3$ A	$i_2 = 2$ A

14.13 Draw and label the simplest possible equivalent impedance network given the following ac measurement results:

Output Open	*Input Open*
$\underline{V}_1 = 34\text{ V }\angle -28.1°$	$\underline{V}_1 = 30\text{ V }\angle -60°$
$\underline{V}_2 = 30\text{ V }\angle 0°$	$\underline{V}_2 = 60\text{ V }\angle 0°$
$\underline{I}_1 = 2\text{ A }\angle 0°$	$\underline{I}_2 = 2\text{ A }\angle -60°$

14.14 Do Problem 14.13 given:

Output Open	*Input Open*
$\underline{V}_1 = 20\text{ V }\angle 0°$	$\underline{V}_1 = 26\text{ V }\angle 67.4°$
$\underline{V}_2 = 13\text{ V }\angle 30.5°$	$\underline{V}_2 = 34\text{ V }\angle 61.9°$
$\underline{I}_1 = 1\text{ A }\angle -36.9°$	$\underline{I}_2 = 2\text{ A }\angle 0°$

Section 14.2 Admittance, Hybrid, and Transmission Parameters

14.15* Obtain $[y]$ for Fig. P14.5 by comparison with Fig. 14.14*b* (p. 670).

14.16 Use the direct method to find $[y]$ for Fig. P14.6.

14.17 Determine the y parameters for Fig. P14.2 by the following method: attach sources for $\underline{V}_1$ and $\underline{V}_2$, write a matrix mesh equation with unknowns $\underline{I}_1$ and $-\underline{I}_2$, and solve for $\underline{I}_1$ and $\underline{I}_2$.

14.18 Use the direct method to find $[h]$ for Fig. P14.1 when $i_x = 16 i_1$.

14.19 Use the direct method to find $[h]$ for Fig. P14.4. Hint: Consider the voltages across the top and bottom branches when $\underline{V}_2 = 0$.

14.20* Determine the h parameters for Fig. P14.2 by the following method: attach sources for $\underline{I}_1$ and $\underline{V}_2$, write and solve a matrix node equation for $\underline{V}_1$ and $\underline{V}_x$, then use your results to find $\underline{I}_2$.

14.21 The **g parameters** are defined by

$$\underline{I}_1 = g_{11}\underline{V}_1 + g_{12}\underline{I}_2 \qquad \underline{V}_2 = g_{21}\underline{V}_1 + g_{22}\underline{I}_2$$

Draw the corresponding two-port model and develop a direct method for calculating g_{11}, g_{12}, g_{21}, and g_{22}. What's the relationship between the g parameters and the h parameters?

14.22 Use the direct method to find $[T]$ for Fig. P14.3.

14.23* Use the direct method to find $[T]$ for Fig. P14.1 when $i_x = v_2/(2\ \Omega)$.

14.24 Use the direct method to find $[T]$ for Fig. P14.4. Hint: Consider the voltages across the top and bottom branches with $\underline{V}_2 = 0$.

14.25 Consider the ideal transformer in Fig. P14.25. Obtain $[T]$ by inspection. Then calculate all the other parameter matrices that exist.

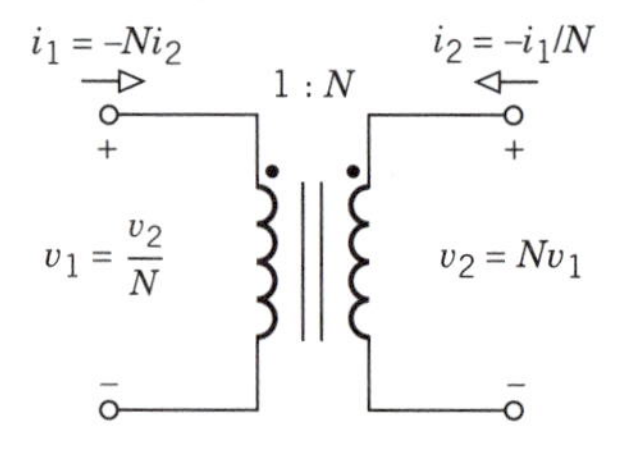

Figure P14.25

14.26 Consider the ideal inverting op-amp circuit in Fig. P14.26. Let $v_{in} = v_1$ and $v_{out} = v_2$ to obtain $[T]$ by inspection. Then calculate all the other parameter matrices that exist.

14.27 Consider the voltage amplifier model in Fig. P14.27. Let $v_{in} = v_1$ and $v_{out} = v_2$ to obtain $[z]$ by inspection. Then calculate all the other parameter matrices that exist.

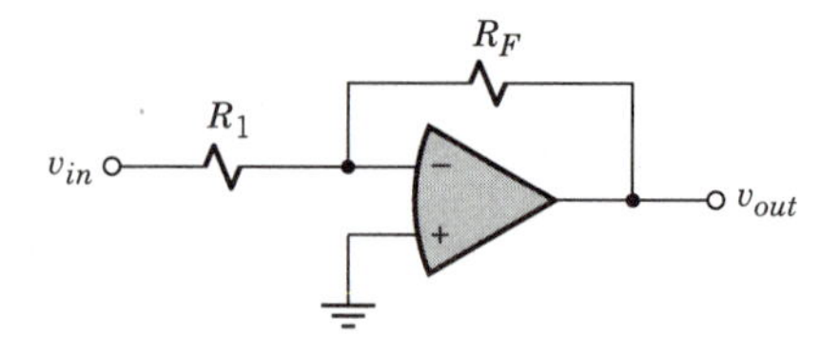

Figure P14.26

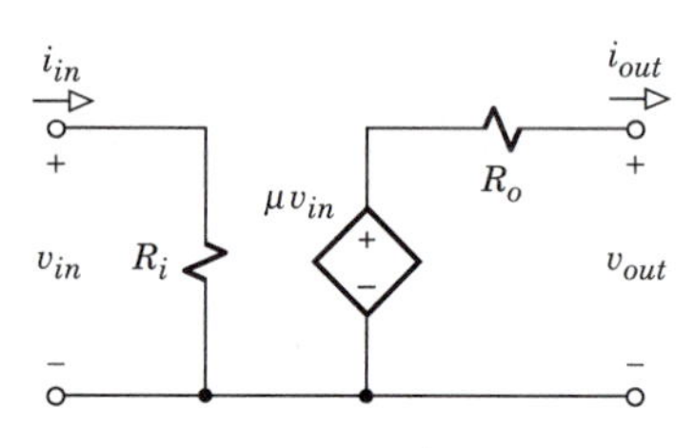

Figure P14.27

14.28 Let $R = 60\ \Omega$ in the network in Fig. 14.9b (p. 665). Find $[h]$ by the direct method. Then calculate all the other parameter matrices that exist.

14.29 Use the direct method to find $[y]$ for Fig. P14.3. Then calculate all the other parameter matrices that exist.

14.30 Derive the formulas for conversions between $[y]$ and $[h]$.

14.31 Derive the formulas for conversions between $[h]$ and $[T]$.

14.32 Derive the formulas for conversions between $[y]$ and $[T]$.

14.33 Derive the formulas for conversions between $[z]$ and $[T]$.

Section 14.3 Circuit Analysis with Two-Ports

14.34* The resistive two-port in Fig. 14.8a was found to have

$$[z] = \begin{bmatrix} 60\ \Omega & 50\ \Omega \\ -100\ \Omega & -100\ \Omega \end{bmatrix}$$

If a load resistance R_L is attached at the output such that $Z_i(s) = 0$, then what are the corresponding values of R_L and $H_i(s)$?

14.35 Derive an expression for $A_v = \underline{V}_2/\underline{V}_s$ in terms of R_L when a resistive load is attached to the two-port in Problem 14.34 and the input source has $R_s = 40\ \Omega$. Then find the values of R_L such that $A_v = 4$ and $A_v = -6$.

14.36 The two-port in Fig. 14.16*a* was found to have

$$[y] = \begin{bmatrix} s/40 & -s/40 \\ s/20 & (2 - s)/20 \end{bmatrix}$$

Find Y_L to maximize $|H_v(s)|$ when $s = j40$. Then calculate $H_v(j40)$ in polar form.

14.37 The two-port in Fig. 14.16*a* was found to have

$$[h] = \begin{bmatrix} 40/s & 1 \\ 2 & 0.1 \end{bmatrix} \qquad [T] = \begin{bmatrix} 1 - 2/s & -20/s \\ -0.05 & -0.5 \end{bmatrix}$$

Using the *ABCD* parameters, find i_1 and i_2 in the steady state when $v_1 = 30 \cos 2t$ V and $Z_L = 10\ \Omega$.

14.38 The tee network in Fig. 14.12 was found to have

$$[z] = \begin{bmatrix} Z_a + Z_c & Z_c \\ Z_c & Z_b + Z_c \end{bmatrix}$$

Determine the poles and zeros of $H_v(s)$ when $Z_a = 1$, $Z_b = 2s$, $Z_c = 2/s$, and the output is terminated with $Z_L = 4$.

14.39 Do Problem 14.38 with $Z_a = 8/s$, $Z_b = s$, $Z_c = 8$, and $Z_L = 7$.

14.40* When operated as a current amplifier, a certain transistor can be represented by

$$[h] = \begin{bmatrix} 10^5/(s + 5000) & 0 \\ 10^6/(s + 5000) & 10^{-3} \end{bmatrix}$$

Show that if the load is a resistor, then $H_i(s)$ is a lowpass filter with $\omega_{co} = 5000$. Then find Z_L such that $H_i(s) = 100\,\omega_{co}/(s + \omega_{co})$ where $\omega_{co} = 10{,}000$.

14.41 Given the transistor amplifier in Problem 14.40, find Z_L such that $\underline{V}_2/\underline{I}_1 = -10^9/(s + \omega_{co})$ where $\omega_{co} = 25{,}000$.

14.42 An ac source with $v_s = V_m \cos 4t$ and $Z_s = 20\ \Omega$ is connected to a load impedance Z_L via a two-port having

$$[z] = \begin{bmatrix} -20 + 40/s & 10 \\ -20 & 10 \end{bmatrix}$$

Find Z_L to achieve maximum power transfer to the load. Then find V_m such that the load receives $P_L = 1$ W.

14.43 Do Problem 14.42 with a two-port having

$$[h] = \begin{bmatrix} 5s & 0 \\ -10 & 0.04 - 0.12/s \end{bmatrix}$$

14.44 Derive the formulas for $H_v(s)$, $H_i(s)$, and $Z_i(s)$ in terms of the *ABCD* parameters.

14.45 Derive the formulas for $H_v(s)$, $H_i(s)$, and $Z_i(s)$ in terms of the y parameters.

14.46 Derive the formulas for $H_v(s)$, $H_i(s)$, and $Z_i(s)$ in terms of the $\mathbf{h}$ parameters.

14.47* Consider a two-stage cascade built with the two-port in Problem 14.34. Obtain an expression for $H_i(s)$ in terms of $Z_L = R_L$. Then find the values of R_L such that $H_i(s) = -10$ and $H_i(s) = +10$.

14.48 Suppose a third identical transistor is inserted in the cascade amplifier in Example 14.11 (p. 000). Use the given transmission matrices to calculate $Z_i(s)$, $H_i(s)$, and A_i.

14.49 A two-stage cascade is built with the two-port in Problem 14.37. Evaluate $H_v(s)$ and $Z_i(s)$ in polar form when $s = j2$ and $Z_L = 10\ \Omega$.

14.50 Consider a two-stage amplifier using the transistor in Problem 14.40 and having $Z_L = 1\ \text{k}\Omega$. Letting $s + 5000 = a$ for convenience, show that $H_v(s)$ has the form of a first-order lowpass filter. Then evaluate $H_v(0)$ and ω_{co}.

14.51 The ladder network in Fig. P14.7 may be viewed as a cascade of two tee networks, each having $Z_b = 0$. Take this approach to find the poles and zeros of $H_v(s)$ when $Z_1 = Z_4 = 20/s$ and $Z_2 = Z_3 = Z_L = 10$. You may use the matrices in Example 14.8 (p. 000).

14.52 An amplifier with $Z_L = 1\ \text{k}\Omega$ is constructed by the series two-port connection of a tee network and a transistor. The transistor has $z_{11} = 2\ \text{k}\Omega$, $z_{12} = 8\ \text{k}\Omega$, and $z_{21} = z_{22} = 0$. The tee network has $Z_a = 0$ and $Z_b = Z_c = R$. Working with all values in kilohms, obtain expressions for $H_v(s)$ and $Z_i(s)$ in terms of R. Then evaluate $H_v(s)$ and $Z_i(s)$ for $R = 0$, 1 kΩ, and 4 kΩ.

14.53 Consider the amplifier in Problem 14.52 when the tee network has $Z_b = 1000$ and $Z_c = 5000/s$. (a) Let $s' = 1000s$ so you can work with values in kilohms to obtain an expression for $H_v(s)$. (b) What are the poles and zeros of this function?

14.54 Two networks are described by the g parameters defined in Problem 14.21. Find and draw the interconnection that results in $[g] = [g]_a + [g]_b$.

14.55 The bridged-tee network in Fig. P14.8 may be viewed as a particular case of Fig. 14.28 (p. 000). Use this approach to determine the poles and zeros of $H_i(s)$ when $Z_L = 0$, $Z_1 = Z_3 = R$, $Z_2 = 1/sC$, $Z_4 = 1/sC_F$, $RC = 1$, and $RC_F = ½$. (Note that you only need to find two of the combined y parameters for this problem.)

14.56 Suppose a feedback resistance $Z_F = 1/G$ bridges across the transistor in Problem 14.40. (a) Letting $s + 5000 = a$ for convenience, show that $H_i(s)$ has the form of a first-order lowpass filter when $Z_L = 0$. (Note that you only need to find two of the combined y parameters for this purpose.) (b) Obtain expressions for $H_i(0)$, ω_{co}, and the "gain-bandwidth product" $H_i(0)\omega_{co}$ in terms of G. Make a table listing the values of these quantities for $Z_F = 10\ \Omega$, $100\ \Omega$, and $1\ k\Omega$.

14.57 A twin-tee network used for notch filtering consists of two parallel-connected tee networks. One tee network has $Z_a = Z_b = R$ and $Z_c = ½sC$, while the other has $Z_a = Z_b = 1/sC$ and $Z_c = R/2$. (a) Obtain the resulting expression for $H_v(s)$ when $Z_L = \infty$. (Note that you only need to find two of the combined y parameters for this purpose.) (b) Draw the pole-zero pattern.

14.58 A certain amplifier consists of a capacitative pi network connected in parallel with a transistor. The pi network is like Fig. P14.5 with $Z_a = Z_b = Z_c = 1/sC$. The transistor has $y_{11} = G_i$, $y_{12} = 0$, $y_{21} = g_m$, and $y_{22} = G_o$. (a) Obtain an expression for $H_i(s)$ with an arbitrary load resistance $Z_L = 1/G$. (b) What's the condition on g_m for stability?

CHAPTER 15
State Variable Analysis

We previously observed that capacitor voltages and inductor currents determine the *state* of the energy stored in a circuit. More generally, the state of any nth-order linear circuit or system can be defined by the appropriate choice of n time-varying physical quantities known as *state variables.*

This chapter introduces state variables as an alternative way of analyzing dynamic circuits and obtaining transfer functions. We'll show how to formulate circuit state equations, and how to solve those equations with the help of the Laplace transform.

Along the way we'll see that the state-variable approach offers distinctive advantages over other methods, especially when the circuit has several inputs and/or outputs. Because some of those advantages stem from the use of matrix relations, a review of Appendix A may be helpful as preparation.

Objectives

After studying this chapter and working the exercises, you should be able to do each of the following:

1. Define the state of a circuit and identify the properties of the state variables (Section 15.1).
2. Write the standard form of the matrix state and output equations for an arbitrary circuit (Section 15.1).
3. Distinguish between proper and improper circuits (Section 15.2).
4. Obtain the state and output equations for a proper circuit, without or with controlled sources (Section 15.2).
5. Identify the modifications required for state-variable analysis when the circuit is improper (Section 15.3).†
6. Obtain the characteristics polynomial of a circuit from its state equation (Section 15.4).
7. Use the resolvent matrix to obtain the zero-input response or complete response of a second-order circuit (Section 15.4).
8. Calculate the transfer-function matrix for a second-order circuit with multiple inputs/outputs (Section 15.4).

15.1 INTRODUCTION TO STATE VARIABLES

Our treatment of state-variable analysis begins here with the properties of state variables. The we'll develop the description of a circuit in terms of two matrix equations, called the *state equation* and the *output equation.* The matrix formulation leads to several advantages for circuit analysis.

State Variables

Of the several voltages and currents in an arbitrary circuit, the state variables have special significance. Taken collectively, the state variables constitute the **state** of the circuit, as defined by the following two properties:

> The state and inputs at any instant t_0 uniquely determine the values of all circuit variables at t_0.
> The state at any later time $t_1 > t_0$ can be determined uniquely from the state at t_0 and the inputs for $t > t_0$.

In short, the state incorporates all information about the past needed to calculate present values and to predict future behavior.

Any minimal set of mutually independent variables satisfying the foregoing properties may be selected as the state variables for a particular circuit. However, in order to represent the circuit's instantaneous stored energy, the state variables are always equal to or related to capacitor voltages and inductor currents. Furthermore, the number of state variables never exceeds the number of energy-storage elements. In fact, the number of state variables equals the number of *independent* energy-storage elements.

Because the state variables provide information about future behavior, and because future behavior depends on the present rate of change, there is an important connection between the state variables and their *derivatives*. Specifically,

> The first derivative of any state variable depends only upon the values of the state variables and the inputs.

As a consequence of this property, we can write the **state equation** as a set of coupled first-order differential equations relating the state variables and the inputs. The following example illustrates these concepts of state variables and state equations.

Example 15.1 *Testing State Variables*

The circuit in Fig. 15.1 contains seven different voltages and currents in addition to the input voltage. But there are just two energy-storage elements, suggesting that the state is defined by the capacitor voltage v_1 and the inductor current i_2. We want to confirm this assumption.

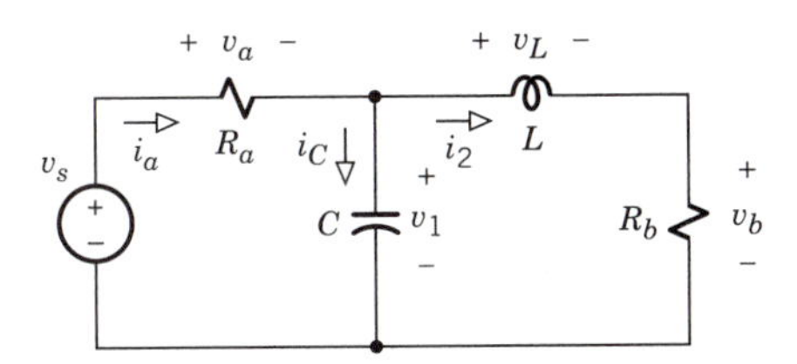

Figure 15.1 Circuit for Example 15.1.

If v_1 and i_2 are appropriate state variables, then we should be able to evaluate all other variables in terms of v_1 and i_2, together with the input v_s. To test our hypothesis, we observe from the diagram that

$$v_a = v_s - v_1 \qquad i_a = (v_s - v_1)/R_a \qquad v_b = R_b i_2$$

so

$$i_C = i_a - i_2 = \frac{v_s - v_1}{R_a} - i_2 \qquad v_L = v_1 - v_b = v_1 - R_b i_2$$

Hence, v_1 and i_2 satisfy the present-value property of state variables.

To predict future behavior, recall that $i_C = C\,dv_1/dt = Cv_1'$ and $v_L = L\,di_2/dt = Li_2'$. Then, using the previous expressions for i_C and v_L, we have

$$Cv_1{}' = i_C = \frac{v_s}{R_a} - \frac{v_1}{R_a} - i_2 \qquad Li_2{}' = v_L = v_1 - R_b i_2$$

As required, these first-order differential equations involve only the state variables and the input.

But the differential equations are *coupled* in the sense that $v_1{}'$ depends on i_2 as well as v_1, while $i_2{}'$ depends on v_1 as well as i_2. We emphasize the coupling by rewriting the equations in the matrix form

$$\begin{bmatrix} v_1{}' \\ i_2{}' \end{bmatrix} = \begin{bmatrix} -1/R_aC & -1/C \\ 1/L & -R_b/L \end{bmatrix} \begin{bmatrix} v_1 \\ i_2 \end{bmatrix} + \begin{bmatrix} 1/R_aC \\ 0 \end{bmatrix} v_s$$

Which is the *matrix state equation.*

Exercise 15.1

Confirm that all other variables in Fig. 15.2 can be found from i_1, v_2, i_s, and v_s. Then write equations for $Li_1{}'$ and $Cv_2{}'$ in terms of i_1, v_2, i_s, and v_s. (Save your results for use in Exercise 15.2.)

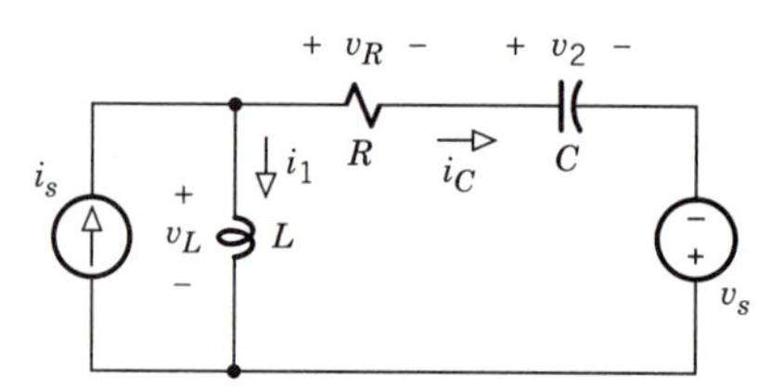

Figure 15.2

State and Output Equations

State-variable analysis starts with the relationship between the state variables and the inputs. A particular circuit may have several inputs coming from both voltage and current sources. Thus, as in the past, we'll represent the inputs by x_1, x_2, . . . The state variables may likewise be a mix of voltages and currents. For generality, we'll denote them by q_1, q_2, . . . , so their derivatives are $q_1{}'$, $q_2{}'$, . . . (These q's should not be confused with the symbol for charge.)

A circuit with n independent energy-storage elements is described by n state variables. If the circuit has k inputs, then the matrix state equation takes the form

$$\begin{bmatrix} q_1{}' \\ q_2{}' \\ \vdots \\ q_n{}' \end{bmatrix} = \begin{bmatrix} a_{11} & a_{12} & \cdots & a_{1n} \\ a_{21} & a_{22} & \cdots & a_{2n} \\ \vdots & \vdots & & \vdots \\ a_{n1} & a_{n2} & \cdots & a_{nn} \end{bmatrix} \begin{bmatrix} q_1 \\ q_2 \\ \vdots \\ q_n \end{bmatrix} + \begin{bmatrix} b_{11} & b_{12} & \cdots & b_{1k} \\ b_{21} & b_{22} & \cdots & b_{2k} \\ \vdots & \vdots & & \vdots \\ b_{n1} & b_{n2} & \cdots & b_{nk} \end{bmatrix} \begin{bmatrix} x_1 \\ x_2 \\ \vdots \\ x_k \end{bmatrix} \qquad (15.1)$$

The a's and b's here are constants incorporating the value of the circuit components.

A more compact version of Eq. (15.1) involves the n-element **state vector** $[q]$ and its derivative $[q']$, defined as

$$[q] \triangleq \begin{bmatrix} q_1 \\ q_2 \\ \vdots \\ q_n \end{bmatrix} \qquad [q'] \triangleq \begin{bmatrix} q_1' \\ q_2' \\ \vdots \\ q_n' \end{bmatrix} \tag{15.2}$$

Similarly, we define the k-element **input vector**

$$[x] \triangleq \begin{bmatrix} x_1 \\ x_2 \\ \vdots \\ x_k \end{bmatrix} \tag{15.3}$$

In general, all the elements of $[q]$, $[q']$, and $[x]$ are functions of time.

Using the state and input vectors, the matrix state equation can be written symbolically as

$$[q'] = [A][q] + [B][x] \tag{15.4}$$

where

$$[A] \triangleq \begin{bmatrix} a_{11} & a_{12} & \cdots & a_{1n} \\ a_{21} & a_{22} & \cdots & a_{2n} \\ \vdots & \vdots & & \vdots \\ a_{n1} & a_{n2} & \cdots & a_{nn} \end{bmatrix} \qquad [B] \triangleq \begin{bmatrix} b_{11} & b_{12} & \cdots & b_{1k} \\ b_{21} & b_{22} & \cdots & b_{2k} \\ \vdots & \vdots & & \vdots \\ b_{n1} & b_{n2} & \cdots & b_{nk} \end{bmatrix}$$

The circuit components are thus contained in the $n \times n$ coefficient matrix $[A]$ and the $n \times k$ coefficient matrix $[B]$. Indeed, we see from Eq. (15.4) that the first main task in state-variable analysis is to find the $[A]$ and $[B]$ matrices.

The solution of the state equation for a particular circuit gives the behavior of the state variables $q_1 = q_1(t)$, $q_2 = q_2(t)$, etc. But we may also be concerned with voltages and currents that are not state variables. All such responses of interest will be called "outputs" and denoted by $y_1, y_2, \ldots$ When there is a total of r outputs under consideration, we define the r-element **output vector**

$$[y] \triangleq \begin{bmatrix} y_1 \\ y_2 \\ \vdots \\ y_r \end{bmatrix} \tag{15.5}$$

If some of the responses of interest happen to be state variables, then they would be included in both the state and output vectors.

To obtain the desired expression for the outputs, we draw upon the property that the state and inputs uniquely determine the instantaneous value of *any* circuit variable. Accordingly, we write the **output equation** in matrix form as

$$[y] = [C][q] + [D][x] \tag{15.6}$$

Here $[C]$ is an $r \times n$ coefficient matrix and $[D]$ is an $r \times k$ coefficient matrix. These matrices are found by analysis of the circuit in question.

In a few circuits, some of the outputs also depend on derivatives of some of the inputs. The output equation must then be written as

$$[y] = [C][q] + [D][x] + [E][x'] \tag{15.7}$$

where $[E]$ is another $r \times k$ coefficient matrix and $[x']$ is the derivative of the input vector. Equation (15.7) reduces to Eq. (15.6) when $[E] = [0]$, meaning that all coefficients in the $[E]$ matrix equal zero.

Example 15.2 *Matrix State and Output Equations*

The circuit in Fig. 15.3 has two inputs (i_s and v_s), and obvious choices for the state variable are i_1 and v_2. The specified responses of interest are v_a, i_1, and i_b. Hence, we take the state, input, and output vectors to be

$$[q] = \begin{bmatrix} i_1 \\ v_2 \end{bmatrix} \qquad [x] = \begin{bmatrix} i_s \\ v_s \end{bmatrix} \qquad [y] = \begin{bmatrix} v_a \\ i_1 \\ i_b \end{bmatrix}$$

Our task is to find the coefficient matrices for the state and output equations.

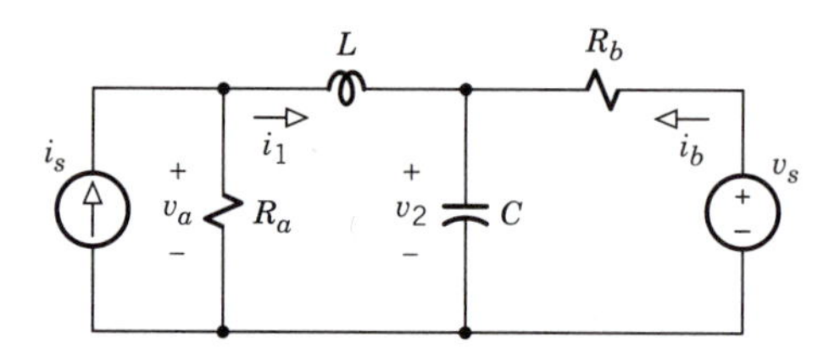

Figure 15.3 Circuit for Example 15.2.

Application of Kirchhoff's laws yields

$$v_a = R_a(i_s - i_1) \qquad i_b = (v_s - v_2)/R_b$$

$$Li_1' = v_a - v_2 \qquad Cv_2' = i_1 + i_b$$

from which

$$i_1' = (-R_a/L)i_1 + (-1/L)v_2 + (R_a/L)i_s + 0 \times v_s$$
$$v_2' = (1/C)i_1 + (-1/R_bC)v_2 + 0 \times i_s + (1/R_b)v_s$$

We thereby obtain the matrix state equation

$$\begin{bmatrix} i_1' \\ v_2' \end{bmatrix} = \begin{bmatrix} -R_a/L & -1/L \\ 1/C & -1/R_bC \end{bmatrix} \begin{bmatrix} i_1 \\ v_2 \end{bmatrix} + \begin{bmatrix} R_a/L & 0 \\ 0 & 1/R_bC \end{bmatrix} \begin{bmatrix} i_s \\ v_s \end{bmatrix}$$

Comparison with Eq. (15.4) then shows that

$$[A] = \begin{bmatrix} -R_a/L & -1/L \\ 1/C & -1/R_bC \end{bmatrix} \qquad [B] = \begin{bmatrix} R_a/L & 0 \\ 0 & 1/R_bC \end{bmatrix}$$

These matrices are both 2×2 because there are $n = 2$ states and $k = 2$ inputs.

For the output equation we note that

$$v_a = (-R_a)i_1 + 0 \times v_2 + R_a i_s + 0 \times v_s$$
$$i_1 = 1 \times i_1 + 0 \times v_2 + 0 \times i_s + 0 \times v_s$$
$$i_b = 0 \times i_1 + (-1/R_b)v_2 + 0 \times i_s + (1/R_b)v_s$$

Thus, in matrix form,

$$\begin{bmatrix} v_a \\ i_1 \\ i_b \end{bmatrix} = \begin{bmatrix} -R_a & 0 \\ 1 & 0 \\ 0 & -1/R_b \end{bmatrix} \begin{bmatrix} i_1 \\ v_2 \end{bmatrix} + \begin{bmatrix} R_a & 0 \\ 0 & 0 \\ 0 & 1/R_b \end{bmatrix} \begin{bmatrix} i_s \\ v_s \end{bmatrix}$$

This output equation has the form of Eq. (15.6) with

$$[C] = \begin{bmatrix} -R_a & 0 \\ 1 & 0 \\ 0 & -1/R_b \end{bmatrix} \qquad [D] = \begin{bmatrix} R_a & 0 \\ 0 & 0 \\ 0 & 1/R_b \end{bmatrix}$$

These matrices are both 3×2 because there are $r = 3$ outputs, $n = 2$ states, and $k = 2$ inputs.

Exercise 15.2

Take the state, input, and output vectors for the circuit in Fig. 15.2 to be

$$[q] = \begin{bmatrix} i_1 \\ v_2 \end{bmatrix} \qquad [x] = \begin{bmatrix} i_s \\ v_s \end{bmatrix} \qquad [y] = \begin{bmatrix} v_L \\ v_R \end{bmatrix}$$

Use the results of Exercise 15.1 to find the coefficient matrices $[A]$, $[B]$, $[C]$, and $[D]$.

Advantages of State-Variable Analysis

Thanks to matrix notation, the use of state variables provides a convenient way of tackling complicated circuit problems. The specific advantages of state-variable analysis are as follows:

- State variables lead to a better understanding of physical behavior because they are directly associated with energy storage.
- State equations may be easier to obtain than node or mesh equations.
- The state and output equations provide the most compact descriptions of a circuit with multiple inputs and outputs.
- State equations can be solved analytically in the s domain via the Laplace transform.
- Transformed state equations can be combined with output equations to obtain transfer functions relating all outputs to all inputs.

These advantages will be developed in the remainder of this chapter.

Additionally, state equations can be solved numerically in the time domain with the help of computer programs. Furthermore, state equations can be written even when a circuit contains nonlinear or time-varying components. These features make state-variable analysis a powerful tool for more advanced circuit and system problems beyond the scope of this book.

15.2 CIRCUIT STATE EQUATIONS

The first step in state-variable circuit analysis is writing the state and output equations. This section presents a simple and systematic way of obtaining these equations for the case of *proper* circuits. As essential background, we begin with the distinction between proper and improper circuits.

Proper and Improper Circuits

A circuit is said to be **proper** when all capacitors and inductors act independently with respect to energy storage. A necessary condition for such independence is the absence of configurations that produce direct relationships between capacitor voltages or inductor currents.

Figure 15.4 shows two examples of capacitors connected to form the configuration we'll call a ***C*-*v* loop**. In general,

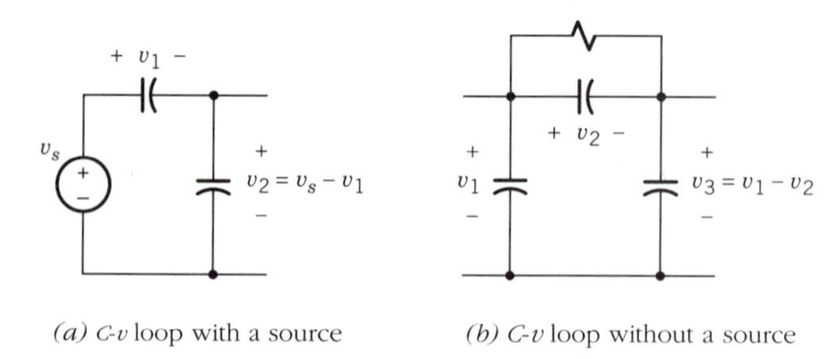

(a) C-v loop with a source *(b) C-v* loop without a source

Figure 15.4

> A C-v loop is a closed path consisting entirely of capacitors or capacitors and voltage sources.

From KVL, it follows that one of the capacitor voltages equals an algebraic sum of the other loop voltages. Hence, the energy stored by one capacitor cannot be independent of the others.

The dual of a C-v loop is an ***L*-*i* cutset**, illustrated by two examples in Fig. 15.5. In general,

> An L-i cutset is a node or supernode whose currents consist entirely of inductor currents or inductor currents and source currents.

From KCL, it follows that one of the inductor currents equals an algebraic sum of the other cutset currents. Hence, the energy stored by one inductor cannot be independent of the others.

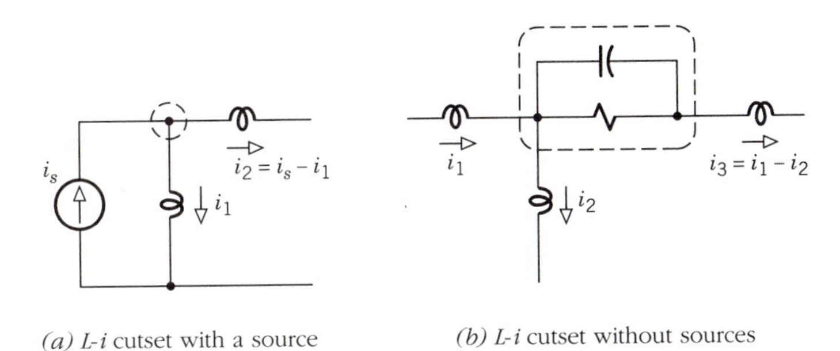

(a) L-i cutset with a source (b) L-i cutset without sources

Figure 15.5

Clearly, a circuit that contains a C-v loop or L-i cutset is **improper**. Even in absence of C-v loops or L-i cutsets, a circuit with a controlled source will be improper if the controlled source establishes a direct relationship between energy-storage elements. Consequently, we can say that

> A circuit is proper when it contains no C-v loops, no L-i cutsets, and no controlled sources.

If the circuit contains controlled sources, then it may be proper but this must be confirmed by further analysis.

Equations for Proper Circuits

When a circuit is proper, the number of state variables equals the number of energy-storage elements. Because the capacitor voltages and inductor currents completely determine the stored energy at any instant of time, we conclude that

> An appropriate set of state variables for a proper circuit consists of all capacitor voltages and inductor currents.

Furthermore, the output equation of a proper circuit always has the form

$$[y] = [C][q] + [D][x] \tag{15.8}$$

which does not include a term proportional to $[x']$.

As implied by Eq. (15.8), any variable in a proper circuit depends only upon the values of the capacitor voltages, inductor currents, and input sources at any instant of time. To emphasize this property, we'll treat the capacitors and inductors as *fictitious sources.*

Figure 15.6*a* shows capacitor C viewed as a voltage source of value v_C. The capacitor's terminal i-v relationship is retained here by the current arrow drawn in accordance with the passive convention and labeled Cv_C'. Similarly, Fig. 15.6*b* shows an inductor L viewed as a current source i_L with the passive-convention voltage labeled Li_L'.

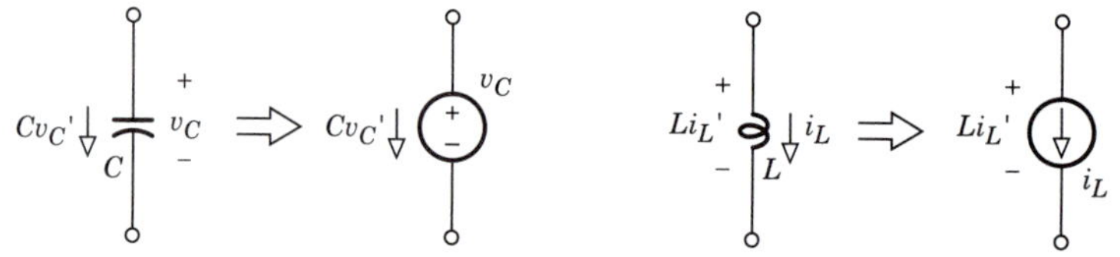

(*a*) Capacitor represented by a voltage source

(*b*) Inductor represented by a current source

Figure 15.6

With appropriate labeling, the fictitious sources still represent the actual behavior of the energy-storage elements. Thus, for purposes of analysis, we are free to put such sources in place of all capacitors and inductors in the circuit diagram. Having done so, the diagram reduces to a *resistive circuit.* And if the resulting resistive circuit consists mostly of independent sources (real or fictitious), then it can easily be analyzed by *superposition.*

Based on these observations, a systematic procedure for obtaining the state and output equations of a proper circuit goes as follows:

1. Take the state variables to be all capacitor voltages and inductor currents, and mark their reference polarities on the circuit diagram.
2. Replace each capacitor with a fictitious voltage source and each inductor with a fictitious current source, labeled as in Fig. 15.6.
3. Apply superposition by suppressing all but one of the independent sources to find the contribution from that source to each Cv_C' current, each Li_L' voltage, and each output of interest. Repeat for all other fictitious sources and independent input sources.

4. Sum the contributions to get complete expressions for each Cv_C' current and Li_L' voltage. Dividing these expressions by L or C yields the state equations.
5. Obtain the output equations by summing the contributions to each output variable.

A tabular arrangement of the superposition results expedites the last two steps.

Example 15.3 *State Analysis without Controlled Sources*

Figure 15.7a repeats the circuit diagram analyzed nonsystematically back in Example 15.2. Now we recognize the circuit as being proper since it contains no *C-v* loops, *L-i* cutsets, nor controlled sources. Accordingly, we'll apply our systematic procedure to obtain the matrix state equation. We'll also obtain the matrix output equation for voltage v_a, current i_1, and current i_b.

Step 1: We take the inductor current i_1 and the capacitor voltage v_2 as the state variables.

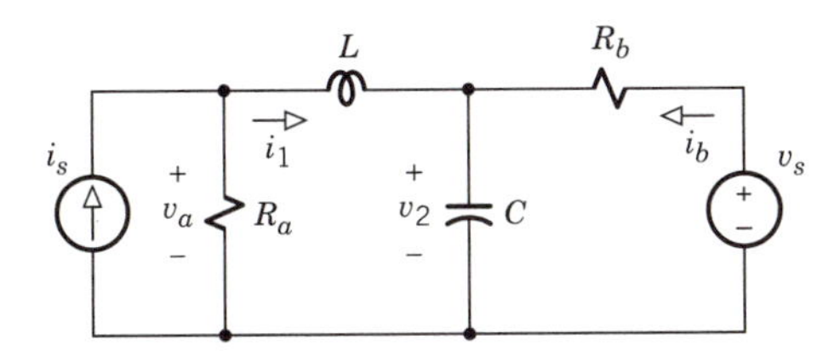

(a) Circuit for Example 15.3

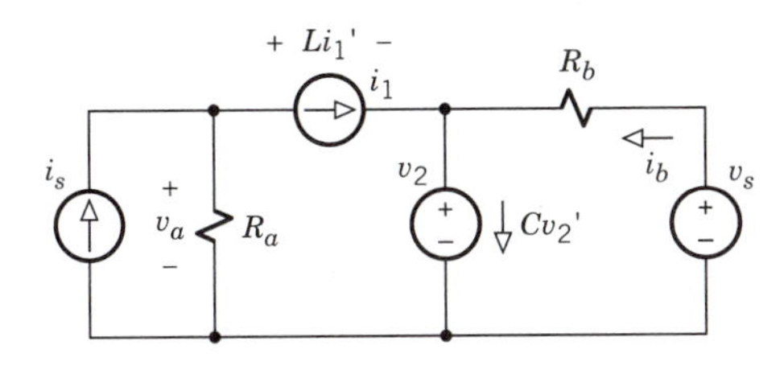

(b) Equivalent resistive circuit

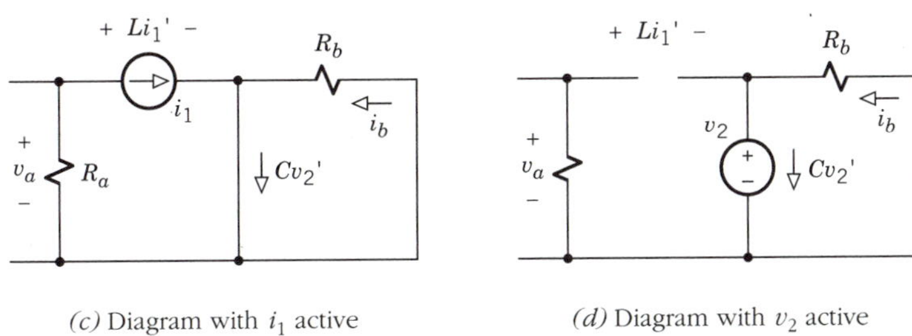

(c) Diagram with i_1 active

(d) Diagram with v_2 active

Figure 15.7

Step 2: Replacing the inductor and capacitor with fictitious sources gives the resistive circuit in Fig. 15.7*b*, which includes the labels for the inductor voltage Li_1' and capacitor current Cv_2'.

Step 3: To find the contributions from i_1, we suppress all other sources as shown in Fig. 15.7*c* where

$$Li_1' = v_a = -R_a i_1 \qquad Cv_2' = i_1 \qquad i_b = 0$$

Next, with v_2 as the only active source, the circuit reduces to Fig. 15.7*d* and we see that

$$Li_1' = -v_2 \qquad Cv_2' = i_b = -v_2/R_b \qquad v_a = 0$$

Similarly, the reduced circuit with only i_s active yields

$$Li_1' = v_a = R_a i_s \qquad Cv_2' = i_b = 0$$

while the reduced circuit with only v_s active yields

$$Li_1' = 0 \qquad Cv_2' = i_b = v_s/R_b \qquad v_a = 0$$

These results are summarized in Table 15.1.

TABLE 15.1

Active Source:	i_1	v_2	i_s	v_s
Li_1'	$-R_a$	-1	R_a	0
Cv_2'	1	$-1/R_b$	0	$1/R_b$
v_a	$-R_a$	0	R_a	0
i_1	1	0	0	0
i_b	0	$-1/R_b$	0	$1/R_b$

Table 15.1 has the active sources labeled across the top, with the state variables in subscript order followed by the inputs. The first two rows of the table are for the contributions to the state-variable derivatives, while the remaining three rows are for the outputs, including the trivial case of i_1. The dashed lines divide the table into sections corresponding to the matrix equations.

Each column of the table shows the *coefficient* of the contribution from one active source to each quantity under consideration. The column entries can often be determined by inspection of the corresponding reduced circuit, together with a few simple calculations.

Step 4: Summing across the rows of Table 15.1 yields the desired superposition totals. Hence, after dividing the first row by L and the second row by C, we get the matrix state equation

$$\begin{bmatrix} i_1' \\ v_2' \end{bmatrix} = \begin{bmatrix} -R_a/L & -1/L \\ 1/C & -1/R_bC \end{bmatrix} \begin{bmatrix} i_1 \\ v_2 \end{bmatrix} + \begin{bmatrix} R_a/L & 0 \\ 0 & 1/R_bC \end{bmatrix} \begin{bmatrix} i_s \\ v_s \end{bmatrix}$$

Step 5: The output equation is then obtained directly from the bottom three rows as

$$\begin{bmatrix} v_a \\ i_1 \\ i_b \end{bmatrix} = \begin{bmatrix} -R_a & 0 \\ 1 & 0 \\ 0 & -1/R_b \end{bmatrix} \begin{bmatrix} i_1 \\ v_2 \end{bmatrix} + \begin{bmatrix} R_a & 0 \\ 0 & 0 \\ 0 & 1/R_b \end{bmatrix} \begin{bmatrix} i_s \\ v_s \end{bmatrix}$$

Exercise 15.3

Use the systematic procedure to obtain the matrix state and output equations for Fig. 15.2 with state variable i_1 and v_2 and output v_R.

Circuits with Controlled Sources. When a circuit contains no *C-v* loops or *L-i* cutsets but does include one or more controlled sources, you cannot tell immediately if it is proper or improper. Such cases are best approached by *assuming* that the circuit is proper and applying the previous systematic procedure. If the analysis yields a valid matrix state equation, then the assumption of independent energy storage is confirmed.

Example 15.4 *State Analysis with a Controlled Source*

Figure 15.8*a* contains a controlled source with an arbitrary parameter μ. We must therefore test the assumption that this is a proper circuit. If so, than i_1 and i_2 are appropriate state variables. We'll take the voltage v_R as the only output.

Figure 15.8*b* shows the reduced resistive circuit with i_1 being the active independent source. (As always, the controlled source is never suppressed.) We immediately see that $v_R = 5i_1$, and KVL yields

$$v_x - \mu v_x = -(3 + 5)i_1$$

Thus, assuming that $1 - \mu \neq 0$, we get

$$0.2i_1' = v_x = 8i_1/(\mu - 1)$$

$$0.5i_2' = 5i_1 - \mu v_x = -(3\mu + 5)i_1/(\mu - 1)$$

Similar calculations with i_2 active and v_s active yield the remaining results listed in Table 15.2

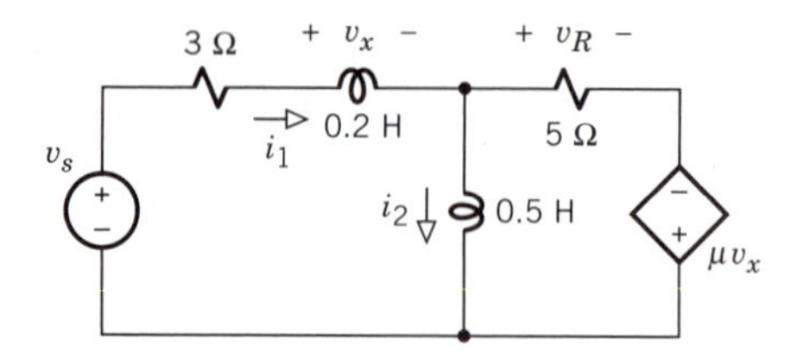

(a) Circuit for Example 15.4

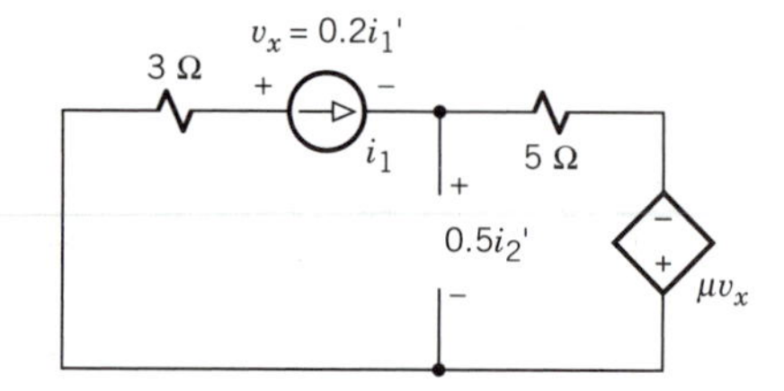

(b) Equivalent resistive circuit with i_1 active

Figure 15.8

TABLE 15.2

Active Source:	i_1	i_2	v_s
$0.2i_1'$	$\dfrac{8}{\mu - 1}$	$\dfrac{-5}{\mu - 1}$	$\dfrac{-1}{\mu - 1}$
$0.5i_2'$	$-\dfrac{3\mu + 5}{\mu - 1}$	$\dfrac{5}{\mu - 1}$	$\dfrac{\mu}{\mu - 1}$
v_R	5	−5	0

Upon dividing the first two rows by 0.2 and 0.5, respectively, and factoring the common term $1/(\mu - 1)$, we get the matrix equation

$$\begin{bmatrix} i_1' \\ i_2' \end{bmatrix} = \frac{1}{\mu - 1}\begin{bmatrix} 40 & -25 \\ -6\mu - 10 & 10 \end{bmatrix}\begin{bmatrix} i_1 \\ i_2 \end{bmatrix} + \frac{1}{\mu - 1}\begin{bmatrix} -5 \\ 2\mu \end{bmatrix} v_s$$

This is a standard state equation with two state variables and one input. The last row of the table yields the corresponding matrix output equation

$$v_R = [5 \quad -5]\begin{bmatrix} i_1 \\ i_2 \end{bmatrix}$$

which happens to have $[D] = 0$ so v_s does not appear.

Since we were able to write a standard state question, we conclude that the circuit is, in fact, proper — provided that $\mu \neq 1$. But if $\mu = 1$, then we no longer have a valid state equation because the term $1/(\mu - 1)$ "blows up." Hence, the circuit is *improper* only in the special case when $\mu = 1$.

To bring out that effect more clearly, we set $\mu = 1$ and apply KVL around the outer loop in Fig. 15.8*a*, getting

$$v_s - 3i_1 - v_x - 5(i_1 - i_2) + v_x = 0$$

from which

$$8i_1 - 5i_2 = v_s$$

This expression shows that the controlled source establishes a direct relationship between the currents i_1 and i_2, so the two inductors are not independent with respect to energy storage.

Exercise 15.4

Let the voltage source in Fig. 15.2 be controlled by v_L such that $v_s = 3v_L$. Assume the circuit is still proper, and find that state equation with state variables i_1 and v_2.

15.3 IMPROPER CIRCUITS†

An improper circuit has the distinguishing feature that the number of state variables is *less* then the number of energy storage elements. Consequently, such circuits are also said to be **degenerate**.

The degeneracy clearly emerges when we put our fictitious sources in place of capacitors in a *C-v* loop or inductors in an *L-i* cutset. For example, the capacitor loop in Fig. 15.9*a* has $v_1 - v_2 - v_3 = 0$. If we represent v_1 and v_2 by fictitious independent sources, then the remaining loop voltage must be represented by a *controlled* source with $v_3 = v_1 - v_2$. Similarly, the *L-i* cutset in Fig., 15.9*b* has $i_2 = i_1 + i_s$, so one of the three cutset currents must be represented by a controlled source.

Since controlled sources are never suppressed when we apply superposition, our previous superposition method for proper circuits does not lead to helpful simplifications of improper circuits. Instead, we must resort to direct use of KCL and KVL. Furthermore, the fictitious controlled sources cause the appearance of unwanted derivative terms when we attempt to write the state equations. There are two distinct cases to consider.

Case 1: *C-v* Loops and *L-i* Cutsets Without Sources. If the *C-v* loops and the *L-i* cutsets do not involve input sources, then the expressions for Cv_C' currents and/or Li_L' voltages cannot be written without including *derivatives of state variables.* These expressions must therefore be manipulated to obtain the state equations in standard form.

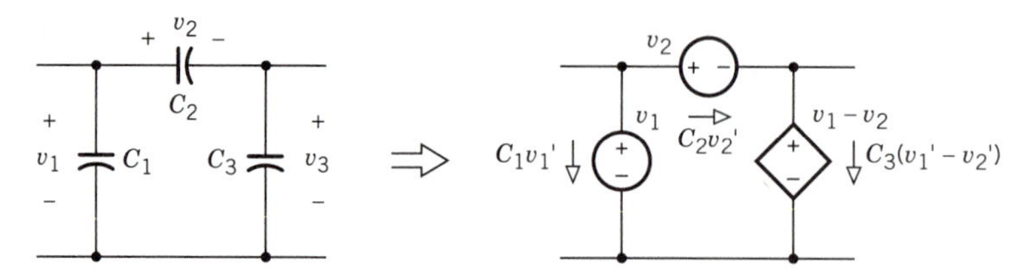

(a) *C-v* loop represented by fictitious sources

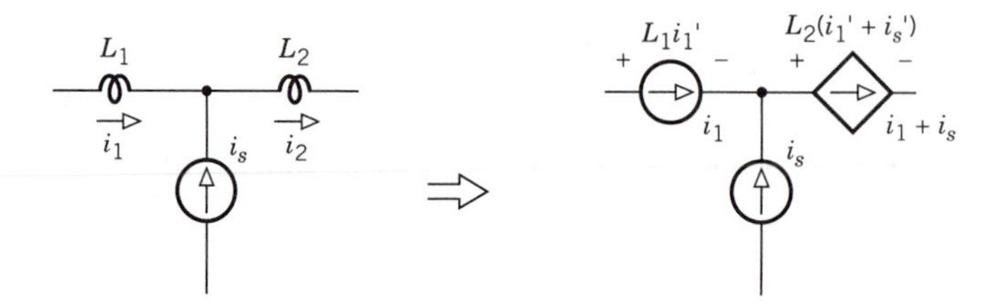

(b) *L-i* cutset represented by fictitious sources

Figure 15.9

Example 15.5 *L-i Cutset Without Input Sources*

The transformer in Fig. 15.10*a* has mutual inductance that can be represented by the tee model in Fig. 15.10*b*. We thus have an improper circuit since the tee model forms an *L-i* cutset without applied current sources. Figure 15.10*c* shows the corresponding resistive circuit for analysis, taking i_1 and i_2 as the

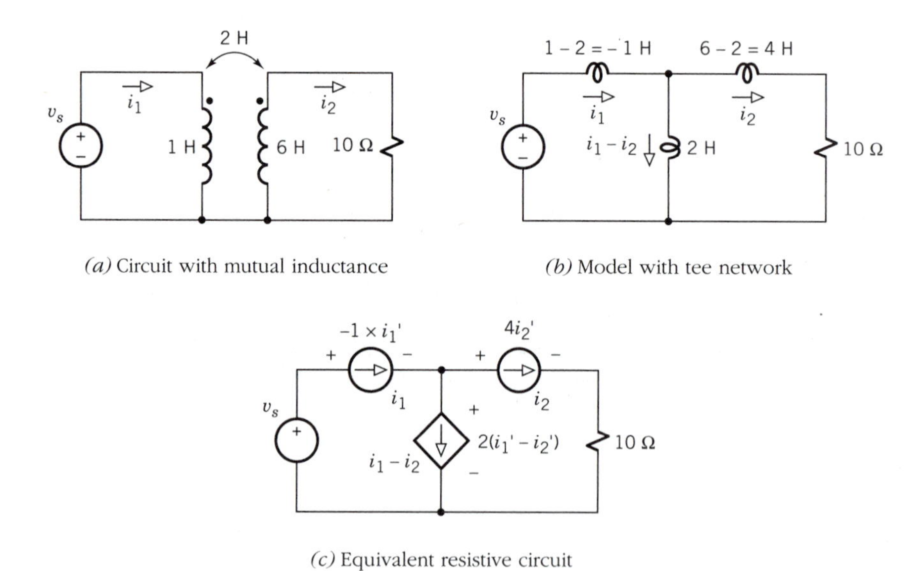

(a) Circuit with mutual inductance

(b) Model with tee network

(c) Equivalent resistive circuit

Figure 15.10

state variables. The third inductor current is not an independent variable, so we use a controlled source to represent $i_3 = i_1 - i_2$.

Upon applying KVL to Fig. 15.10*c*, we find that all loops contain two Li_L' terms. In particular, from the outer loop and the left-hand loop we get

$$-i_1' + 4i_2' = v_s - 10i_2 \qquad -i_1' + 2(i_1' - i_2') = v_s$$

Because we need separate equations for the state-variable derivatives i_1' and i_2', we treat the above expressions as a pair of linear equations rewritten as

$$\begin{bmatrix} -1 & 4 \\ 1 & -2 \end{bmatrix}\begin{bmatrix} i_1' \\ i_2' \end{bmatrix} = \begin{bmatrix} v_s - 10i_2 \\ v_s \end{bmatrix}$$

Simultaneous solution then yields

$$i_1' = -10i_2 + 3v_s \qquad i_2' = -5i_2 + v_s$$

Hence, we arrive at the state equations

$$\begin{bmatrix} i_1' \\ i_2' \end{bmatrix} = \begin{bmatrix} 0 & -10 \\ 0 & -5 \end{bmatrix}\begin{bmatrix} i_1 \\ i_2 \end{bmatrix} + \begin{bmatrix} 3 \\ 1 \end{bmatrix} v_s$$

Any other variable considered to be an output is easily expressed in terms of i_1, i_2, and v_s.

Exercise 15.5

Let the improper circuit in Fig. 15.11 have $C_1 = C_2 = C_3 = 1/12$ F, $L = 0$, and $R = 2\ \Omega$. Find the state equation taking v_2 and v_3 as the state variables.

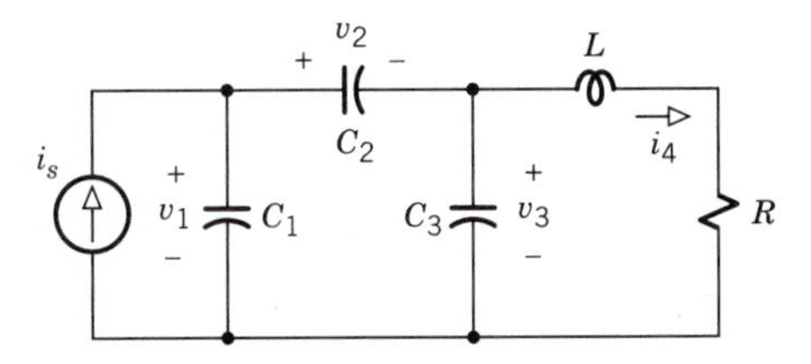

Figure 15.11

Case II: *C-v* Loops and *L-i* Cutsets with Sources. If a *C-v* loop and/or *L-i* cutset does involve input sources, then the expressions for Cv_C' currents and/or Li_L' voltages cannot be written without including *derivatives of input sources.* These derivatives must therefore be eliminated by defining one or more new state variables to get a standard state equation. As a consequence, the resulting output equation may take the form

$$[y] = [C][q] + [D][x] + [E][x'] \tag{15.9}$$

which includes the derivatives of the input vector.

Example 15.6 *C-v Loop with an Input Source*

The improper circuit in Fig. 15.12a has a C-v loop with an input source. Figure 15.12b is the corresponding resistive circuit taking $v_3 = v_s - v_1$. The presence of two independent fictitious sources implies that we need two state variables, but not necessarily v_1 and i_2.

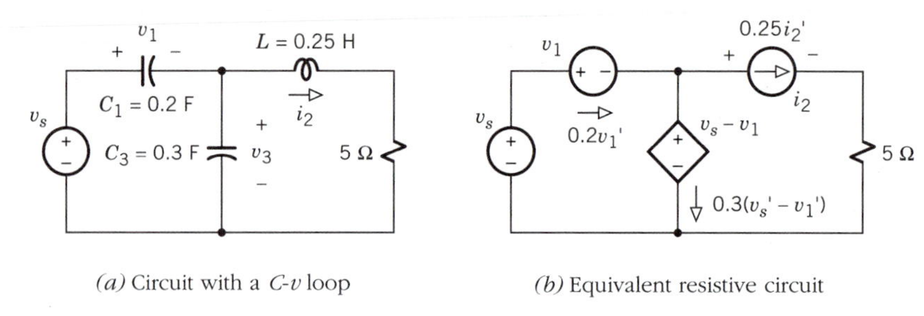

(a) Circuit with a C-v loop (b) Equivalent resistive circuit

Figure 15.12

Because the controlled source automatically satisfies KVL around the left-hand loop, the only independent combining equations come from KCL at the upper or lower node and KVL around the outer loop. We therefore start by writing

$$0.2v_1' = 0.3(v_s' - v_1') + i_2$$
$$0.25i_2' = v_s - v_1 - 5i_2$$

The equation for i_2' has the required form, but the equation for v_1' also includes the source derivatives v_s'.

To eliminate that unwanted derivative, we rewrite the first equation as

$$v_1' - 0.6v_s' = 2i_2$$

This expression suggests defining the new state variable

$$q_1 = v_1 - 0.6v_s$$

so the first state equation becomes

$$q_1' = 2i_2$$

Now the equation for i_2' must be modified to get rid of v_1. From the definition of q_1 we have

$$v_1 = q_1 + 0.6v_s$$

Substitution then yields

$$i_2' = 4(v_s - v_1 - 5i_2) = -4q_1 - 20i_2 + 1.4v_s$$

Thus, in terms of q_1 and i_2, the matrix state equation is

$$\begin{bmatrix} q_1' \\ i_2' \end{bmatrix} = \begin{bmatrix} 0 & 2 \\ -4 & -20 \end{bmatrix}\begin{bmatrix} q_1 \\ i_2 \end{bmatrix} + \begin{bmatrix} 0 \\ 1.6 \end{bmatrix} v_s$$

Although we have eliminated v_s' to obtain a standard state equation, the output equation may still involve v_s'. In particular, suppose the responses of interest are the inductor voltage $v_a = 0.25i_2'$ and the capacitor current $i_b = 0.3(v_s' - v_1')$. From the state equation and the expression for v_1 in terms of q_1 we get

$$\begin{aligned} v_a &= 0.25i_2' = 0.25(-4q_1 - 20i_2 + 1.6v_s) \\ i_b &= 0.3(v_s' - v_1') = 0.3v_s' - 0.3(q_1' + 0.6v_s') \\ &= -0.3q_1' + 0.12v_s' = -0.3(2i_2) + 0.12v_s' \end{aligned}$$

Hence,

$$\begin{bmatrix} v_a \\ i_b \end{bmatrix} = \begin{bmatrix} -1 & -5 \\ 0 & -0.6 \end{bmatrix}\begin{bmatrix} q_1 \\ i_2 \end{bmatrix} + \begin{bmatrix} 0.4 \\ 0 \end{bmatrix} v_s + \begin{bmatrix} 0 \\ 0.12 \end{bmatrix} v_s'$$

Controlled Sources. Finally, recall that circuits containing "nonfictitious" controlled sources may be improper even in absence of *C-v* loops and *L-i* cutsets. The analysis of such circuits generally follows the line of one of the two previous cases.

Exercise 15.6

Suppose the controlled source in Fig. 15.8*a* (p. 708) has $\mu = 1$, so the circuit becomes improper. Apply KVL around the left-hand loop and the outer loop to obtain the state equation with one variable $q = i_1 - 0.1v_s$.

15.4 TRANSFORM SOLUTION OF STATE EQUATIONS

Having shown how to obtain the state-variable description of a circuit, we finally turn to solution methods. We'll start with the zero-state response, which leads us to the *resolvent matrix.* Then we'll consider the complete response and the *transfer-function matrix.*

Although state variables were developed primarily for computer-aided numerical solution of higher order circuits, we'll focus on second-order circuits whose state and output equations can be solved by hand via Laplace transforms. Nonetheless, the methods presented here also hold for more complicated cases — perhaps with the aid of a computer algebra program such as Maple or Mathematica to carry out the matrix manipulations.

Zero-Input Response

The matrix state and output equations for any linear circuit have the general form

$$[q'] = [A][q] + [B][x]$$
$$[y] = [C][q] + [D][x] + [E][x']$$

Initially we'll concentrate on the zero-input response, so $[x] = [x'] = 0$ for $t \geq 0$ and

$$[q'] = [A][q] \qquad (15.10)$$
$$[y] = [C][q] \qquad (15.11)$$

We want to solve Eq. (15.10) for the state variables. Then we can use Eq. (15.11) to compute the outputs.

Since Eq. (15.10) involves derivatives, the Laplace transform solution requires *initial conditions.* Specifically, for each state variable $q_i(t)$ with initial condition $q_i(0^-)$, the Laplace transform of the derivative is

$$\mathcal{L}[q_i'(t)] = sQ_i(s) - q_i(0^-)$$

where $Q_i(s) = \mathcal{L}[q_i(t)]$. We therefore define the n-element vectors

$$[Q(s)] \triangleq \begin{bmatrix} Q_1(s) \\ Q_2(s) \\ \vdots \\ Q_n(s) \end{bmatrix} \qquad [q(0^-)] \triangleq \begin{bmatrix} q_1(0^-) \\ q_2(0^-) \\ \vdots \\ q_n(0^-) \end{bmatrix} \qquad (15.12)$$

Note that $[Q(s)]$ stands for the element-by-element transform of the state vector $[q]$.

The transform of the matrix equation $[q'] = [A]\,[q]$ thus becomes $s[Q(s)] - [q(0^-)] = [A]\,[Q(s)]$ or

$$s[Q(s)] - [A][Q(s)] = [q(0^-)]$$

Since $[Q(s)]$ comprises the unknowns to be found, we introduce the $n \times n$ *identity matrix* $[I]$ to write $s[Q(s)] = s[I]\,[Q(s)] = [sI]\,[Q(s)]$ and

$$[sI][Q(s)] - [A][Q(s)] = [q(0^-)]$$

Factoring $[Q(s)]$ now yields

$$[sI - A][Q(s)] = [q(0^-)] \tag{15.13}$$

The new matrix $[sI - A]$ is given in general by

$$[sI - A] = \begin{bmatrix} s - a_{11} & -a_{12} & \cdots & -a_{1n} \\ -a_{21} & s - a_{22} & \cdots & -a_{2n} \\ \vdots & \vdots & & \vdots \\ -a_{n1} & -a_{n2} & \cdots & s - a_{nn} \end{bmatrix} \tag{15.14}$$

where the a's are the respective coefficients in the $[A]$ matrix. Note that s comes from $[sI]$ and thus appears only on the main diagonal.

Solving Eq. (15.13) for $[Q(s)]$ requires premultiplying both sides by the inverse of $[sI - A]$. Thus, for convenience, we'll define the **resolvent matrix**

$$[\Phi(s)] \triangleq [sI - A]^{-1} \tag{15.15}$$

The Laplace transform of the state vector for the zero-input response is then

$$[Q(s)] = [sI - A]^{-1}[q(0^-)] = [\Phi(s)][q(0^-)] \tag{15.16}$$

We return to the time domain via

$$[q(t)] = \mathcal{L}^{-1}\{[Q(s)]\} = \mathcal{L}^{-1}\{[\Phi(s)][q(0^-)]\} \tag{15.17}$$

which is the symbolic version of our desired result.

Equation (15.17) reveals that we need to calculate the resolvent matrix $[\Phi(s)]$ and form the partial-fraction expansion of each term in the matrix product $[\Phi(s)]\,[q(0^-)]$. To expedite both tasks, recall that the inverse of the square matrix $[sI - A]$ can be written as

$$[sI - A]^{-1} = \text{adj}[sI - A]/|[sI - A]|$$

where $\text{adj}[sI - A]$ is the *adjoint matrix* and $|[sI - A]|$ is the *determinant.*

But Eq. (15.14) implies that the determinant of $[sI - A]$ will be a *polynomial* with variable s. We therefore let

$$P(s) \triangleq |[sI - A]| \tag{15.18}$$

so

$$[\Phi(s)] = [sI - A]^{-1} = \frac{1}{P(s)}\,\text{adj}[sI - A] \tag{15.18}$$

Because $P(s)$ will appear in the denominator of every term in $[Q(s)]$, the roots of $P(s)$ determine all the modes of the zero-input response. Hence, as our notation suggests, $P(s)$ is the circuit's *characteristic polynomial.* Furthermore, the state-variable analysis confirms that

> A given circuit has one and only one characteristic polynomial $P(s)$.

This property was previously observed in Chapters 9 and 10.

To elaborate further, consider an arbitrary second-order circuit. The 2×2 $[A]$ matrix has the form

$$[A] = \begin{bmatrix} a_{11} & a_{12} \\ a_{21} & a_{22} \end{bmatrix}$$

so

$$[sI - A] = \begin{bmatrix} s - a_{11} & -a_{12} \\ -a_{21} & s - a_{22} \end{bmatrix}$$

The corresponding resolvent matrix is

$$[\Phi(s)] = \frac{1}{P(s)} \begin{bmatrix} s - a_{22} & +a_{12} \\ +a_{21} & s - a_{11} \end{bmatrix} \tag{15.20a}$$

with

$$\begin{aligned} P(s) = \|[sI - A]\| &= \begin{vmatrix} s - a_{11} & -a_{12} \\ -a_{21} & s - a_{22} \end{vmatrix} \\ &= s^2 - (a_{11} + a_{22})s + a_{11}a_{22} - a_{12}a_{21} \end{aligned} \tag{15.20b}$$

Equation (15.20b) thus expresses the characteristic polynomial for *any* second-order circuit in terms of the coefficients of its $[A]$ matrix.

Example 15.7 ***Calculating the Zero-Input Response***

Suppose you're given a circuit with $n = 2$ states and $r = 2$ outputs related by the matrices

$$[A] = \begin{bmatrix} -8 & -5 \\ 3 & 0 \end{bmatrix} \qquad [C] = \begin{bmatrix} 6 & 10 \\ 0 & -2 \end{bmatrix}$$

You want to find the zero-input response produced by the initial conditions $q_1(0^-) = 2$ and $q_2(0^-) = -4$, so

$$[q(0^-)] = \begin{bmatrix} 2 \\ -4 \end{bmatrix}$$

To begin the calculations, you first form the matrix $[sI - A]$ and its adjoint

$$[sI - A] = \begin{bmatrix} s+8 & 5 \\ -3 & s \end{bmatrix} \qquad \text{adj}[sI - A] = \begin{bmatrix} s & -5 \\ 3 & s+8 \end{bmatrix}$$

Next, you calculate the characteristic polynomial

$$P(s) = \|[sI - A]\| = s^2 + 8s + 15 = (s+3)(s+5)$$

Then, since the resolvent matrix is $[\Phi(s)] = \text{adj}[sI - A]/P(s)$, you obtain the transform of the state vector

$$\begin{aligned} [Q(s)] &= [\Phi(s)][q(0^-)] \\ &= \frac{1}{P(s)}\begin{bmatrix} s & -5 \\ 3 & s+8 \end{bmatrix}\begin{bmatrix} 2 \\ -4 \end{bmatrix} = \frac{1}{P(s)}\begin{bmatrix} 2s+20 \\ -4s-26 \end{bmatrix} \end{aligned}$$

Writing out the two elements of $[Q(s)]$ and taking the partial-fraction expansion yields

$$\begin{aligned} Q_1(s) &= \frac{2s+20}{(s+3)(s+5)} = \frac{7}{s+3} + \frac{-5}{s+5} \\ Q_2(s) &= \frac{-4s-26}{(s+3)(s+5)} = \frac{-7}{s+3} + \frac{3}{s+5} \end{aligned}$$

By inverse transformation,

$$q_1(t) = 7e^{-3t} - 5e^{-5t} \qquad q_2(t) = -7e^{-3t} + 3e^{-5t}$$

which describes the zero-input response of the state variables.

Finally, you calculate the resulting outputs via

$$[y] = [C][q] = \begin{bmatrix} 6 & 10 \\ 0 & -2 \end{bmatrix}\begin{bmatrix} q_1 \\ q_2 \end{bmatrix} = \begin{bmatrix} 6q_1 + 10q_2 \\ -2q_2 \end{bmatrix}$$

Thus,

$$\begin{aligned} y_1(t) &= 6q_1(t) + 10q_2(t) = -28e^{-3t} \\ y_2(t) &= -2q_2(t) = 14e^{-3t} - 6e^{-5t} \end{aligned}$$

Note that the natural-response mode e^{-5t} cancels out and does not appear in $y_1(t)$.

Exercise 15.7

A second-order circuit with two inputs and one output is described by the matrices

$$[A] = \begin{bmatrix} 0 & 1 \\ 0 & -4 \end{bmatrix} \qquad [B] = \begin{bmatrix} 1 & -1 \\ 0 & 4 \end{bmatrix}$$
$$[C] = [2 \quad -1] \qquad [D] = [0 \quad 4] \qquad [E] = [0]$$

Find the zero-state response of $q_1(t)$, $q_2(t)$, and $y(t)$ for $t \geq 0$ when $q_1(0^-) = 0$ and $q_2(0^-) = 12$. (Save your work for use in Exercises 15.8 and 15.9.)

Complete Response

The complete response with nonzero input and nonzero initial state is governed by the matrix state equation

$$[q'] = [A][q] + [B][x]$$

which includes the input vector $[x]$. Since the initial state $[q(0^-)]$ takes account of all past inputs, we assume that $[x] = [x'] = 0$ for $t < 0$.

For $t \geq 0$, we'll need the transformed input vector

$$[X(s)] \triangleq \mathscr{L}\{[x(t)]\} = \begin{bmatrix} X_1(s) \\ X_2(s) \\ \vdots \\ X_k(s) \end{bmatrix} \tag{15.21}$$

Taking the Laplace transform of both sides of the state equation then gives

$$s[Q(s)] - [q(0^-)] = [A][Q(s)] + [B][X(s)]$$

or, upon rearrangement,

$$[sI - A][Q(s)] = [q(0^-)] + [B][X(s)]$$

Premultiplying both sides by $[\Phi(s)] = [sI - A]^{-1}$ yields

$$[Q(s)] = [\Phi(s)][q(0^-)] + [\Phi(s)][B][X(s)] \tag{15.22}$$

We recognize the first part of Eq. (15.22) as the transform of the zero-input response, so the second part represents the zero-state response.

But we're concerned here with the *complete* response, which equals the sum of the zero-input and zero-state responses. To expedite that calculation, let's factor the common term $[\Phi(s)] = \text{adj}[sI - A]/P(s)$ in Eq. (15.22) and write

$$[Q(s)] = \frac{1}{P(s)} \operatorname{adj}[sI - A]\{[q(0^-)] + [B][X(s)]\} \tag{15.23}$$

The inverse transform of Eq. (15.23) gives the complete response of the state vector $[q]$. The resulting outputs can then be calculated algebraically from the output equation

$$[y] = [C][q] + [D][x] + [E][x'] \tag{15.24}$$

Most circuits of interest are proper, so the term $[E]$ $[x']$ disappears from Eq. (15.24). The compact matrix notation allows us to carry along the $[E]$ matrix for the sake of generality.

Example 15.8 ***Calculating the Complete Response***

Suppose the circuit considered in Example 15.7 has $k = 2$ inputs and

$$[B] = \begin{bmatrix} -8 & 0 \\ 3 & 1 \end{bmatrix} \qquad [D] = \begin{bmatrix} 0 & 0 \\ 2 & 0 \end{bmatrix} \qquad [E] = [0]$$

We previously found that $P(s) = (s + 3)(s + 5)$ and

$$\operatorname{adj}[sI - A] = \begin{bmatrix} s & -5 \\ 3 & s + 8 \end{bmatrix}$$

Now we'll find the complete response when

$$q_1(0^-) = 2 \qquad q_2(0^-) = -4 \qquad x_1(t) = 10t \qquad x_2(t) = 0$$

so

$$[q(0^-)] = \begin{bmatrix} 2 \\ -4 \end{bmatrix} \qquad [x] = \begin{bmatrix} 10t \\ 0 \end{bmatrix} \qquad [X(s)] = \begin{bmatrix} 10/s^2 \\ 0 \end{bmatrix}$$

Using Eq. (15.23), the transformed state vector is

$$\begin{aligned}
[Q(s)] &= \frac{1}{P(s)} \begin{bmatrix} s & -5 \\ 3 & s + 8 \end{bmatrix} \left\{ \begin{bmatrix} 2 \\ -4 \end{bmatrix} + \begin{bmatrix} -8 & 0 \\ 3 & 1 \end{bmatrix} \begin{bmatrix} 10/s^2 \\ 0 \end{bmatrix} \right\} \\
&= \frac{1}{P(s)} \begin{bmatrix} s & -5 \\ 3 & s + 8 \end{bmatrix} \begin{bmatrix} 2 - 80/s^2 \\ -4 + 30/s^2 \end{bmatrix} \\
&= \frac{1}{P(s)} \begin{bmatrix} (2s^3 + 20s^2 - 80s - 150)/s^2 \\ (-4s^2 - 26s + 30)/s \end{bmatrix}
\end{aligned}$$

Thus,

$$Q_1(s) = \frac{2s^3 + 20s^2 - 80s - 150}{s^2(s+3)(s+5)} \qquad Q_2(s) = \frac{-4s^2 - 26s + 30}{s(s+3)(s+5)}$$

and inverse transformation yields

$$q_1(t) = -10t + 12e^{-3t} - 10e^{-5t} \qquad q_2(t) = 2 - 12e^{-3t} + 6e^{-5t}$$

Since $[E] = [0]$, the resulting outputs are given by

$$[y] = [C][q] + [D][x] = \begin{bmatrix} 6 & 10 \\ 0 & -2 \end{bmatrix}\begin{bmatrix} q_1 \\ q_2 \end{bmatrix} + \begin{bmatrix} 0 & 0 \\ 2 & 0 \end{bmatrix}\begin{bmatrix} 10t \\ 0 \end{bmatrix}$$

from which

$$y_1(t) = 6q_1(t) + 10q_2(t) = 20 - 10t - 48e^{-3t}$$
$$y_2(t) = -2q_2(t) + 20t = -4 + 20t + 24e^{-3t} - 12e^{-5t}$$

Exercise 15.8

Find $q_1(t)$, $q_2(t)$, and $y(t)$ for $t \geq 0$ when the circuit in Exercise 15.7 has $q_1(0^-) = 0$, $q_2(0^-) = 12$, $x_1(t) = 0$, and $x_2(t) = 8u(t)$.

Transfer-Function Matrix

In principle, the zero-state response is just a special case of the complete response with $[q(0^-)] = [0]$. In practice, however, we gain a computational advantage by combining the state and output equations into one expression for the transform of the zero-state output vector. An important result therefrom is the *transfer-function matrix.*

Consider the Laplace transform of the matrix output equation

$$[y] = [C][q] + [D][x] + [E][x']$$

Since $[x(0^-)] = [0]$, we have $\mathcal{L}\{[x'(t)]\} = s[X(s)]$ and

$$[Y(s)] = [C][Q(s)] + [D][X(s)] + [E]s[X(s)]$$

Under zero-state conditions, the transformed state vector in Eq. (15.22) reduces to $[Q(s)] = [\Phi(s)][B]\,[X(s)]$. Thus,

$$\begin{aligned}[Y(s)] &= [C][\Phi(s)][B][X(s)] + [D][X(s)] + [E]s[X(s)] \\ &= \{[C][\Phi(s)][B] + [D + sE]\}[X(s)]\end{aligned}$$

which has the form

$$[Y(s)] = [H(s)][X(s)] \tag{15.25}$$

where

$$[H(s)] \triangleq [C][\Phi(s)][B] + [D + sE] \tag{15.26}$$

We call $[H(s)]$ the **transfer-function matrix** because it incorporates the zero-state s-domain relationships between *all* the inputs and *all* the outputs under consideration.

To bring out the meaning of $[H(s)]$, we rewrite Eq. (15.25) as the set of r output equations

$$Y_1(s) = H_{11}(s)X_1(s) + H_{12}(s)X_2(s) + \cdots + H_{1k}(s)X_k(s)$$
$$Y_2(s) = H_{21}(s)X_1(s) + H_{22}(s)X_2(s) + \cdots + H_{2k}(s)X_k(s)$$
$$\vdots$$
$$Y_r(s) = H_{r1}(s)X_1(s) + H_{r2}(s)X_2(s) + \cdots + H_{rk}(s)X_k(s)$$

This display clearly shows that each element of $[H(s)]$ is a transfer function relating one of the outputs to one of the inputs. Consequently, when you're interested in the zero-state behavior of a state variable, you should include that variable among the outputs so you can obtain the corresponding transfer function.

Another important property of the transfer-function matrix becomes apparent after inserting $[\Phi(s)] = \text{adj}[sI - A]/P(s)$ and rewriting Eq. (15.26) as

$$P(s)[H(s)] = [C]\text{adj}[sI - A][B] + P(s)[D + sE] \tag{15.27}$$

We infer from this expression that every element of $[H(s)]$ is a ratio of polynomials and the denominator of each one is the characteristic polynomial $P(s)$. Thus,

> In absence of pole-zero cancellations, the poles of all transfer functions of a given circuit equal the roots of $P(s)$.

This is another general property observed in Chapter 10 that is now confirmed by state-variable analysis. Additionally, Eq. (15.27) is a convenient form for computing $[H(s)]$ by hand.

Example 15.9 *Calculating the Zero-State Response*

Our previous examples focused on the zero-state response and complete response of a second-order circuit with two inputs and two outputs. Here, we'll find the transfer-function matrix and the zero-state response at both outputs when $x_1(t) = u(t)$ and $x_2(t) = 0$.

Substituting the various matrices into Eq. (15.27) gives the transfer-function matrix via

$$\begin{aligned} P(s)[H(s)] &= [C]\text{adj}[sI - A][B] + P(s)[D] \\ &= \begin{bmatrix} 6 & 10 \\ 0 & -2 \end{bmatrix}\begin{bmatrix} s & -5 \\ 3 & s+8 \end{bmatrix}\begin{bmatrix} -8 & 0 \\ 3 & 1 \end{bmatrix} + \begin{bmatrix} 0 & 0 \\ 2P(s) & 0 \end{bmatrix} \\ &= \begin{bmatrix} -18s - 90 & 10s + 50 \\ -6s & -2s + 16 \end{bmatrix} + \begin{bmatrix} 0 & 0 \\ 2s^2 + 16s + 30 & 0 \end{bmatrix} \\ &= \begin{bmatrix} -18s - 90 & 10s + 50 \\ 2s^2 + 10s + 30 & -2s + 16 \end{bmatrix} \end{aligned}$$

Since $P(s) = s^2 + 8s + 15 = (s + 3)(s + 5)$, the zero-state input-output relations in the s domain are

$$H_{11}(s) = \frac{-18s - 90}{s^2 + 8s + 15} = \frac{-18(s+5)}{(s+3)(s+5)} = \frac{-18}{s+3}$$

$$H_{12}(s) = \frac{10s + 50}{s^2 + 8s + 15} = \frac{10(s+5)}{(s+3)(s+5)} = \frac{10}{s+3}$$

$$H_{21}(s) = \frac{2s^2 + 10s + 30}{s^2 + 8s + 15} = \frac{2s^2 + 10s + 30}{(s+3)(s+5)}$$

$$H_{22}(s) = \frac{-2s + 16}{s^2 + 8s + 15} = \frac{-2s + 16}{(s+3)(s+5)}$$

Note that $H_{11}(s)$ and $H_{12}(s)$ simplify by virtue of pole-zero cancellations.

To obtain the zero-state outputs produced by $x_1(t) = u(t)$ and $x_2(t) = 0$, we set $X_1(s)$ and $X_2(s) = 0$ in the matrix equation

$$[Y(s)] = [H(s)][X(s)] = \begin{bmatrix} H_{11}(s) & H_{12}(s) \\ H_{21}(s) & H_{22}(s) \end{bmatrix}\begin{bmatrix} 1/s \\ 0 \end{bmatrix}$$

from which

$$Y_1(s) = \frac{-18}{s(s+3)} \qquad Y_2(s) = \frac{2s^2 + 10s + 30}{s(s+3)(s+5)}$$

Taking the inverse transform then yields

$$y_1(t) = -6 + 6e^{-3t} \qquad y_2(t) = 2 - 3e^{-3t} + 3e^{-5t}$$

The mode e^{-5t} does not appear in $y_1(t)$ because of the pole-zero cancellation in $H_1(s)$.

Example 15.10 ***Universal State-Variable Filter***

Most of this chapter has been rather theoretical, so our closing example applies state-variable analysis to the important practical circuit in Fig. 15.13. This important practical circuit in Fig. 15.13. This second-order circuit consists of four op-amp stages whose output voltages v_1, v_2, v_3, and v_4 are related to the input voltage by different filter functions. Two of the outputs equal the capacitor voltages, labeled q_1 and q_2, and the circuit is called a *universal state-*

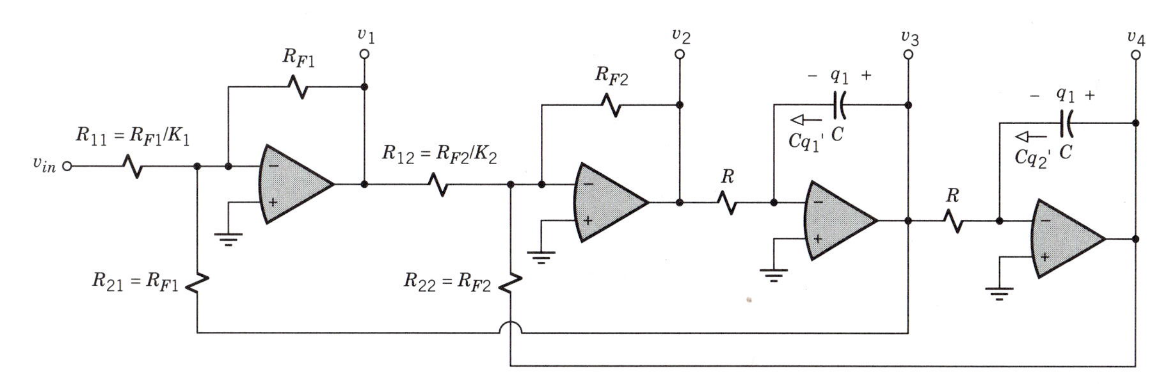

Figure 15.13 Universal state-variable filter.

variable filter. We'll justify the name by finding the four components of the transfer-function matrix $[H(s)]$. To that end, we first need the coefficient matrices for the state and output equations.

The four op-amps are assumed to be ideal, so each of the inverting input terminals is at virtual ground and draws no current. Furthermore, the first two stages act as inverting summing amplifiers. Consequently,

$$v_1 = -(R_{F1}/R_{11})v_{in} - (R_{F1}/R_{21})v_3 = -K_1 v_{in} - v_3$$
$$v_2 = -(R_{F2}/R_{12})v_1 - (R_{F2}/R_{22})v_4 = -K_2 v_1 - v_4$$
$$v_3 = q_1 \qquad Cq_1' = -v_2/R$$
$$v_4 = q_2 \qquad Cq_2' = -v_3/R$$

Then, after applying superposition which q_1 and q_2 treated as fictitious voltage sources, we get the components listed in Table 15.3. The resulting coefficient matrices are

$$[A] = \begin{bmatrix} -K_2/RC & 1/RC \\ -1/RC & 0 \end{bmatrix} \qquad [B] = \begin{bmatrix} -K_2K_1/RC \\ 0 \end{bmatrix}$$
$$[C] = \begin{bmatrix} -1 & 0 \\ K_2 & -1 \\ 1 & 0 \\ 0 & 1 \end{bmatrix} \qquad [D] = \begin{bmatrix} -K_1 \\ K_2K_1 \\ 0 \\ 0 \end{bmatrix}$$

TABLE 15.3

Active Source:	q_1	q_2	v_{in}
Cq_1'	$-K_2/R$	$1/R$	$-K_2K_1/R$
Cq_2'	$-1/R$	0	0
v_1	-1	0	$-K_1$
v_2	K_2	-1	K_2K_1
v_3	1	0	0
v_4	0	1	0

We next observe that

$$[sI - A] = \begin{bmatrix} s + K_2/RC & -1/RC \\ 1/RC & s \end{bmatrix}$$

The corresponding characteristic polynomial is

$$P(s) = \|[sI - A]\| = s^2 + (K_2/RC)s + 1/R^2C^2 = s^2 + (\omega_0/Q)s + \omega_0^{\,2}$$

where

$$\omega_0/Q = K_2/RC \qquad \omega_0 = 1/RC$$

so

$$Q = RC\omega_0/K_2 = 1/K_2$$

Finally, the matrix calculations in Eq. (15.27) yield

$$P(s)[H(s)] = \begin{bmatrix} -K_1(s^2 + \omega_0^{\,2}) \\ K_1 s^2/Q \\ -K_1\omega_0 s/Q \\ K_1\omega_0^{\,2}/Q \end{bmatrix}$$

Thus, the first component of $[H(s)]$ is

$$H_1(s) = \frac{\underline{V}_1}{\underline{V}_{in}} = \frac{-K_1(s^2 + \omega_0^{\,2})}{s^2 + (\omega_0/Q)s + \omega_0^{\,2}}$$

which is an *inverting notch filter* with gain $K = -K_1$. The second component is

$$H_2(s) = \frac{\underline{V}_2}{\underline{V}_{in}} = \frac{(K_1/Q)s^2}{s^2 + (\omega_0/Q)s + \omega_0^{\,2}}$$

which is a *second-order highpass filter* with $K = K_1/Q$, and a Butterworth response is obtained by taking $Q = 0.707$. The third component is

$$H_3(s) = \frac{\underline{V}_3}{\underline{V}_{in}} = \frac{-K_1(\omega_0/Q)s}{s^2 + (\omega_0/Q)s + \omega_0^{\,2}}$$

which is an *inverting bandpass filter* with $K = -K_1$. The fourth component is

$$H_4(s) = \frac{\underline{V}_4}{\underline{V}_{in}} = \frac{(K_1/Q)\omega_0^{\,2}}{s^2 + (\omega_0/Q)s + \omega_0^{\,2}}$$

which is a *second-order lowpass filter* with $K = K_1/Q$, and a Butterworth response is obtained by taking $Q = 0.707$.

This circuit is called a *universal* filter because you can select an appropriate output to get any one of the four basic filter functions, lowpass, highpass, bandpass, or notch. The filter parameters K, Q, and ω_0 are adjusted via the values of K_1, K_2, and RC. Complete units like Fig. 15.13 are commercially available as integrated circuits.

Exercise 15.9

Find $[H(s)]$ and the individual transfer functions $H_{11}(s)$ and $H_{12}(s)$ for the circuit in Exercise 15.7.

PROBLEMS

Section 15.2 Circuit State Equations

15.1* Obtain the matrix state and output equations for the circuit in Fig. P15.1. Take v_1 and i_2 as the state variables and i_a, v_b, and v_c as the outputs.

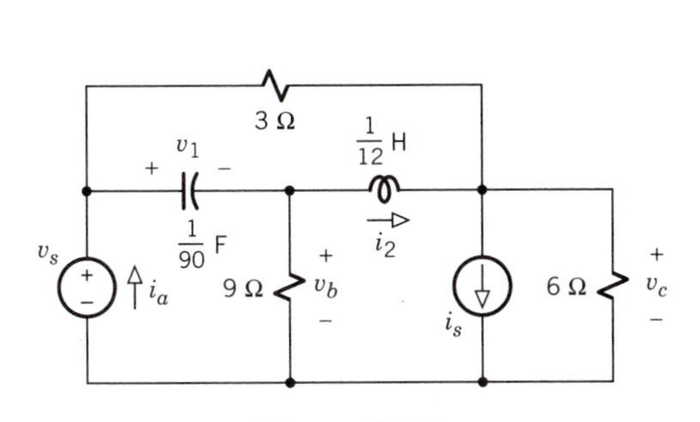

Figure P15.1

15.2 Obtain the matrix state and output equations for the circuit in Fig. P15.2. Take i_1 and v_2 as the state variables and v_a and i_b as the outputs.

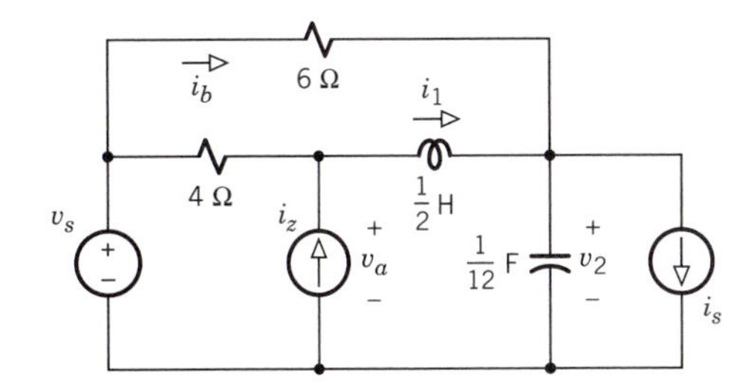

Figure P15.2

15.3 Obtain the matrix state and output equations for the circuit in Fig. P15.3. Take v_1 and i_2 as the state variables and i_a, v_b, and i_c as the outputs.

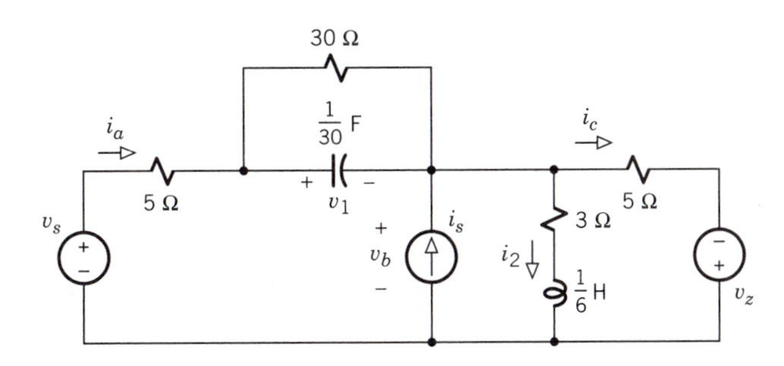

Figure P15.3

15.4 Obtain the matrix state and output equations for the circuit in Fig. P15.4. Take v_1, v_2, and i_3 as the state variables and i_a as the output.

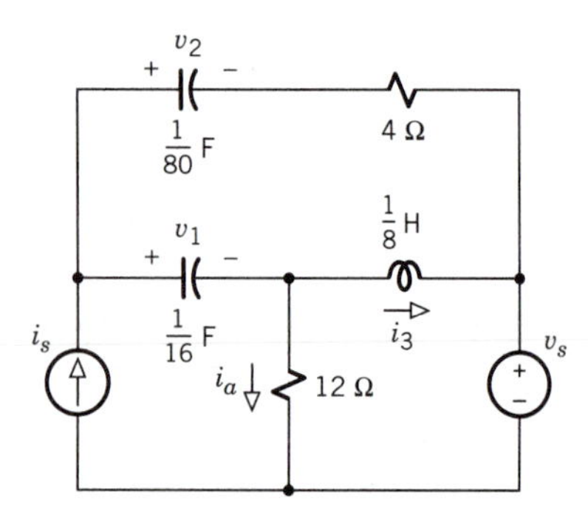

Figure P15.4

15.5* Let v_s in Fig. P15.1 be a controlled source such that $v_s = 2v_b$. Obtain the matrix state and output equations taking v_1 and i_2 as the state variables and i_a, v_b, and v_c as the outputs.

15.6 Let i_z in Fig. P15.2 be a controlled source such that $i_z = 3i_b$. Obtain the matrix state and output equations taking i_1 and v_2 as the state variables and v_a and i_b as the outputs.

15.7 Let v_z in Fig. P15.3 be a controlled source such that $v_z = 4v_b$. Obtain the matrix state and output equations taking v_1 and i_2 as the state variables and i_a and v_b as the outputs.

15.8 Let i_s in Fig. P15.4 be a controlled source such that $i_s = 6i_a$. Obtain the matrix state and output equations taking v_1, v_2, and i_3 as the state variables and i_a as the output.

Section 15.3 Improper Circuits

15.9* Let the circuit in Fig. 15.11 (p. 711) have $C_1 = C_3 = 1/10$ F, $C_2 = 1/5$ F, $L = ½$ H, and $R = 4\ \Omega$. Obtain the state equation with variables v_1, v_2, and i_4.

15.10 Let the circuit in Fig. 15.11 (p. 711) have $C_1 = C_2 = 1/10$ F, $C_3 = 1/5$ F, $L = 1/4$ H, and $R = 3\ \Omega$. Obtain the state equation with variables v_1, v_3, and i_4.

15.11 Let the circuit in Fig. P15.11 have $C_1 = C_2 = 1/10$ F, $R = 5\ \Omega$, and $L = ½$ H. Treat C_2 as a controlled source to obtain the state equation with variables $q_1 = 2v_1 - v_s$ and i_3. Then write the matrix output equation for i_a and v_2.

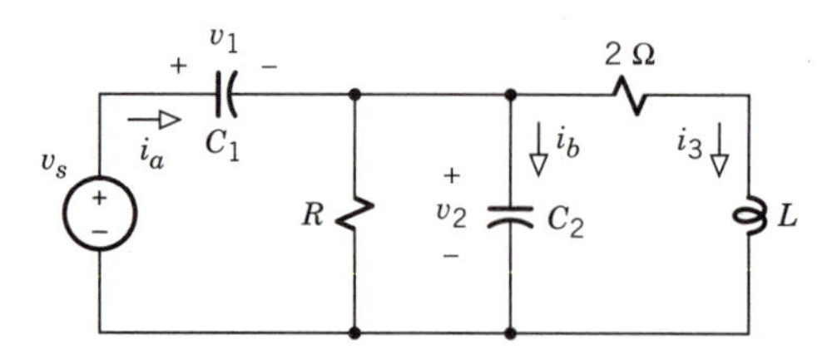

Figure P15.11

15.12 Let the improper circuit in Fig. P15.11 have $C_1 = 1/12$ F, $C_2 = 1/6$ F, $R = 4\ \Omega$, and $L = 1/3$ H. Treat C_1 as a controlled source to obtain the state equation with variables $q_1 = v_s - 3v_2$ and i_3. Then write the matrix output equations for v_1 and i_b.

15.13* Let the improper circuit in Fig. P15.13 have $L_1 = 3$ H, $L_2 = 1$ H, and $R = 12\ \Omega$. Treat L_1 as a controlled source to obtain the state equation with variables $q_1 = 4i_2 - 3i_s$ and v_3. Then write the matrix output equation for q_1 and v_b.

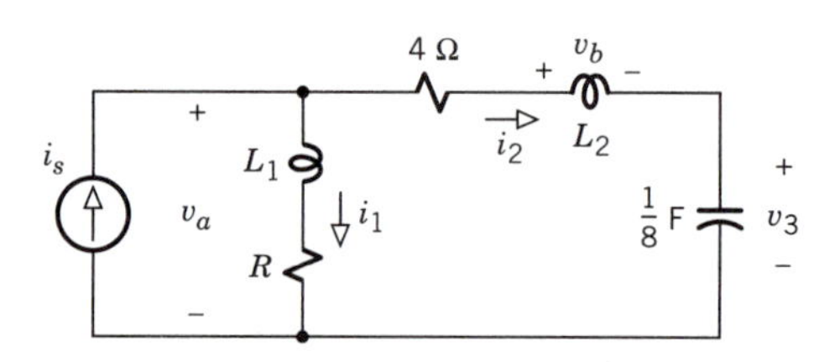

Figure P15.13

15.14 Let the improper circuit in Fig. P15.13 have $L_1 = L_2 = 2$ H and $R = 6\ \Omega$. Treat L_2 as a controlled source to obtain the state equation with variables $q_1 = i_s - 2i_1$ and v_3. Then write the matrix output equation for v_a and i_2.

15.15 Let the improper circuit in Fig. P15.15 have $L = 0$. Obtain the state equation with one variable $q = 5v_1 - 2v_s$. Then write the output equation for i_a.

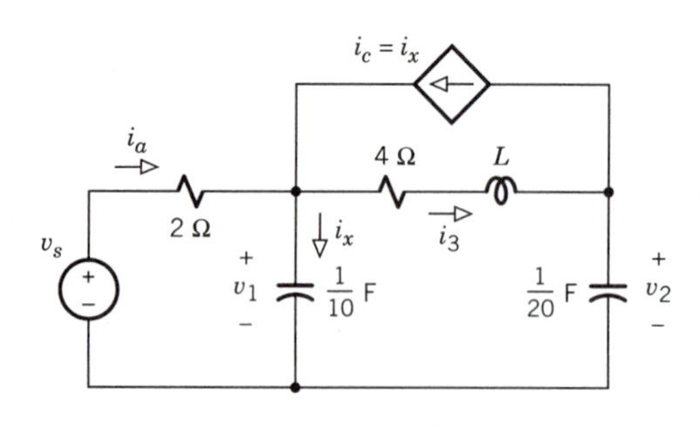

Figure P15.15

15.16 Let the improper circuit in Fig. P15.15 have $L = 1$ H. Obtain the state equation with variables $q_1 = v_2 + 2v_s$ and i_3.

Section 15.4 Transform Solution of State Equations

15.17* A second-order circuit with three inputs and two outputs is described by the following coefficient matrices:

$$[A] = \begin{bmatrix} 2 & 1 \\ 6 & 3 \end{bmatrix} \quad [B] = \begin{bmatrix} 1 & 1 & 0 \\ 0 & 3 & 0 \end{bmatrix}$$
$$[C] = \begin{bmatrix} 2 & 1 \\ -3 & 1 \end{bmatrix} \quad [D] = \begin{bmatrix} -2 & 0 & 0 \\ 1 & 0 & -1 \end{bmatrix}$$
$$[E] = [0]$$

(a) Show that the characteristic polynomial is $P(s) = s(s - 5)$. (b) Find the outputs for $t \geq 0$ given that $[x] = 0$, $q_1(0^-) = 5$, and $q_2(0^-) = 0$.

15.18 A second-order circuit with two inputs and two outputs is described by the following coefficient matrices:

$$[A] = \begin{bmatrix} -4 & -2 \\ 2 & 0 \end{bmatrix} \quad [B] = \begin{bmatrix} -1 & 1 \\ 2 & -2 \end{bmatrix}$$
$$[C] = \begin{bmatrix} -1 & 1 \\ 0 & 3 \end{bmatrix} \quad [D] = [0] \quad [E] = \begin{bmatrix} 0 & 0 \\ 0 & 1 \end{bmatrix}$$

(a) Show that the characteristic polynomial is $P(s) = (s + 2)^2$. (b) Find the outputs for $t \geq 0$ given that $[x] = 0$, $q_1(0^-) = 5$ and $q_2(0^-) = -3$.

15.19 A second-order circuit with two inputs and three outputs is described by the following coefficient matrices:

$$[A] = \begin{bmatrix} 0 & 1 \\ -2 & -3 \end{bmatrix} \quad [B] = \begin{bmatrix} 1 & 0 \\ 0 & -1 \end{bmatrix}$$
$$[C] = \begin{bmatrix} 0 & 2 \\ -1 & 0 \\ 1 & 1 \end{bmatrix} \quad [D] = \begin{bmatrix} 2 & 0 \\ 0 & 0 \\ 0 & 1 \end{bmatrix} \quad [E] = [0]$$

(a) Show that the characteristic polynomial is $P(s) = (s + 1)(s + 2)$. (b) Find the outputs for t ≥ 0 given that $[x] = 0$, $q_1(0^-) = 10$ and $q_2(0^-) = 0$.

15.20 A second-order circuit with two inputs and one output is described by the following coefficient matrices:

$$[A] = \begin{bmatrix} -4 & 5 \\ -1 & 0 \end{bmatrix} \quad [B] = \begin{bmatrix} 0 & 2 \\ -1 & 0 \end{bmatrix}$$
$$[C] = [0 \quad 4] \quad [D] = [1 \quad 0] \quad [E] = [0]$$

(a) Show that the characteristic polynomial is $P(s) = s^2 + 4s + 5$. (b) Find the output for $t \geq 0$ given that $[x] = 0$ and $q_1(0^-) = q_2(0^-) = 2$.

15.21 A second-order circuit with two inputs and two outputs is described by the following coefficient matrices:

$$[A] = \begin{bmatrix} -1 & 4 \\ -1 & -1 \end{bmatrix} \quad [B] = \begin{bmatrix} -2 & 0 \\ 3 & 0 \end{bmatrix}$$
$$[C] = \begin{bmatrix} 1 & 0 \\ 0 & -2 \end{bmatrix} \quad [D] = \begin{bmatrix} 0 & -3 \\ 2 & 1 \end{bmatrix} \quad [E] = [0]$$

(a) Show that the characteristic polynomial is $P(s) = s^2 + 2s + 5$. (b) Find the outputs for $t \geq 0$ given that $[x] = 0$ and $q_1(0^-) = q_2(0^-) = 4$.

15.22* Find the complete response of the outputs of the circuit in Problem 15.17 when $q_1(0^-) = 0$, $q_2(0^-) = 9$, $x_2(t) = x_3(t) = 0$, and $x_1(t) = 3e^{-t}$ for $t \geq 0$.

15.23 Find the complete response of the outputs of the circuit in Problem 15.21 when $q_1(0^-) = 6$, $q_2(0^-) = 0$, $x_1(t) = 0$, and $x_2(t) = 5$ for $t \geq 0$.

15.24 Find the complete response of the outputs of the circuit in Problem 15.18 when $q_1(0^-) = -5$, $q_2(0^-) = 2$, and $x_1(t) = x_2(t) = 10t$ for $t \geq 0$.

15.25 Find the complete response of the outputs of the circuit in Problem 15.19 when $q_1(0^-) = 1$, $q_2(0^-) = 0$, $x_1(t) = 0$, and $x_2(t) = 3e^{-t}$ for $t \geq 0$.

15.26 Find the complete response of the output of the circuit in Problem 15.20 when $q_1(0^-) = 0$, $q_2(0^-) = 2$, $x_1(t) = 0$, and $x_2(t) = 5$ for $t \geq 0$.

15.27* Obtain the elements of the transfer-function matrix for the circuit in Problem 15.20.

15.28 Obtain the elements of the transfer-function matrix for the circuit in Problem 15.21.

15.29 Obtain the elements of the transfer-function matrix for the circuit in Problem 15.18.

15.30 Obtain the elements of the transfer-function matrix for the circuit in Problem 15.17.

15.31 Obtain the elements of the transfer-function matrix for the circuit in Problem 15.19.

APPENDIX A
Matrix Algebra

Circuit analysis frequently involves sets of linear equations relating a group of dependent variables to a group of independent variables. Matrix notation provides a compact way of representing such equations. Additionally, when linear equations have been written in matrix form, they can be solved for the unknowns through the use of Cramer's rule and determinants.

This appendix presents those aspects of matrix algebra needed for efficient circuit analysis. The first section covers basic matrix concepts and techniques employed starting in Chapter 4 and used regularly thereafter. The second section deals with the somewhat more advanced matrix operations that come into play in Chapters 14 and 15. Both sections emphasize applications to sets of two or three equations, because problems with four or more simultaneous equations are better handled by calculator or computer programs.

Objectives

After studying this appendix and working the exercises, you should be able to do each of the following:

1. Distinguish between a matrix and a vector, and carry out matrix-vector multiplication (Section A.1).
2. Write a set of linear equations in matrix form (Section A.1).
3. Calculate the cofactors and determinants of a 2×2 or 3×3 matrix (Section A.1)
4. Determine when a set of two or three simultaneous equations will yield a unique solution, and use Cramer's rule to obtain the numeric or symbolic solution (Section A.1).
5. State the conditions for and basic properties of matrix addition and multiplication, and carry out the addition, subtraction or multiplication of two matrices (Section A.2).
6. Identify the properties of the null matrix and the identity matrix (Section A.2).
7. Find the transpose, adjoint, and inverse of a 2×2 or 3×3 matrix (Section A.2).
8. Use the inverse matrix to solve a set of two or three simultaneous equations (Section A.2).

A.1 MATRIX EQUATIONS AND DETERMINANTS

This section summarizes the basics of matrix algebra essential for representing and solving sets of linear equations. The topics covered are matrix notation and equations, determinants, and Cramer's rule.

Matrix Notation and Equations

A **matrix** is a rectangular array of quantities. We'll use the symbol $[A]$ to stand for an arbitrary matrix with m rows and n columns. The quantities that make up $[A]$ are called its **elements**, and a_{ij} denotes the element in the ith row and jth column, Thus, in general,

$$[A] \triangleq \begin{bmatrix} a_{11} & a_{12} & \cdots & a_{1n} \\ a_{21} & a_{22} & \cdots & a_{2n} \\ \vdots & \vdots & & \vdots \\ a_{m1} & a_{m2} & \cdots & a_{mn} \end{bmatrix} \tag{A.1}$$

This array is said to be of **order** $m \times n$ (pronounced "m by n"). The row designation always comes before the column designation, just like the subscripts of the elements.

An array having only one column is known as a **vector**. We distinguish this special case using lowercase letters, so an arbitrary nth-order vector is

$$[x] = \begin{bmatrix} x_1 \\ x_2 \\ \vdots \\ x_n \end{bmatrix}$$

Note that each vector element needs only a single subscript.

Matrix/vector notation allows compact representation of sets of linear equations. The key ingredient for that purpose is multiplication of a matrix $[A]$ and a vector $[x]$. In this regard

> The product $[A][x]$ is defined only when the number of rows in $[x]$ equals the number of columns in $[A]$.

If $[A]$ is of order $m \times n$ and $[x]$ is nth order, then $[A][x]$ is an mth-order vector given by

$$\begin{bmatrix} a_{11} & a_{12} & \cdots & a_{1n} \\ a_{21} & a_{22} & \cdots & a_{2n} \\ \vdots & \vdots & & \vdots \\ a_{m1} & a_{n2} & \cdots & a_{mn} \end{bmatrix} \begin{bmatrix} x_1 \\ x_2 \\ \vdots \\ x_n \end{bmatrix} \triangleq \begin{bmatrix} a_{11}x_1 + a_{12}x_2 + \cdots + a_{1n}x_n \\ a_{21}x_1 + a_{22}x_2 + \cdots + a_{2n}x_n \\ \vdots \qquad \vdots \qquad \vdots \\ a_{m1}x_1 + a_{m2}x_2 + \cdots + a_{mn}x_n \end{bmatrix} \tag{A.2}$$

For a helpful picture of this product operation, turn $[x]$ sideways and put its elements directly above the elements in the ith row of $[A]$. Term-by-term multiplication and addition then yields the ith element of the product, as follows:

$$\begin{array}{cccc} x_1 & x_2 & \cdots & x_n \\ a_{i1} & a_{i2} & \cdots & a_{in} \\ \hline \multicolumn{4}{c}{a_{i1}x_1 + a_{i2}x_2 + \cdots + a_{in}x_n} \end{array}$$

This display also emphasizes that the number of *rows* in $[x]$ must match up with the number of *columns* in $[A]$.

Equation (A.2) shows that multiplying $[A]$ times $[x]$ produces a new column vector, say $[y]$. Accordingly, we can write the **matrix equation**

$$[y] = [A][x] \tag{A.3a}$$

The fundamental property of this or any other matrix equation is that each element on one side must equal the corresponding element on the other side. Thus, from Eq. (A.2), the elements of $[y]$ are

$$y_i = a_{i1}x_1 + a_{i2}x_2 + \cdots + a_{in}x_n \tag{A.3b}$$

which holds for $i = 1, 2, \ldots, m$. In short, each element of $[y]$ equals a *sum of row-by-column products.*

More often, however, we must work with indirect relationships between the dependent variables in $[y]$ and the independent variables in $[x]$. In particular, consider the set of n simultaneous linear equations

$$\begin{aligned} a_{11}y_1 + a_{12}y_2 + \cdots + a_{1n}y_n &= x_1 \\ a_{21}y_1 + a_{22}y_2 + \cdots + a_{2n}y_n &= x_2 \\ \vdots \qquad \vdots \qquad \vdots \quad &\quad \vdots \\ a_{n1}y_1 + a_{n2}y_2 + \cdots + a_{nn}y_n &= x_n \end{aligned} \tag{A.4}$$

This set may also be written in matrix form as

$$[A][y] = [x] \tag{A.5}$$

Here, both $[x]$ and $[y]$ are nth order vectors. Furthermore, $[A]$ has n rows and n columns, so we call it an nth-order **square matrix**. We'll subsequently discuss determinants of square matrices as a means of solving for $[y]$ from $[A][y] = [x]$.

Example A.1 *Matrix-Vector Multiplication*

Suppose two dependent variables are directly related to three independent variables by the set of equations

$$y_1 = 12x_1 - 7x_2$$
$$y_2 = -3x_1 + 10x_2 + 4x_3$$

This set may be written in matrix form as $[y] = [A][x]$ where

$$[y] = \begin{bmatrix} y_1 \\ y_2 \end{bmatrix} \qquad [A] = \begin{bmatrix} 12 & -7 & 0 \\ -3 & 10 & 4 \end{bmatrix} \qquad [x] = \begin{bmatrix} x_1 \\ x_2 \\ x_3 \end{bmatrix}$$

If $x_1 = 5$, $x_2 = -2$, and $x_3 = 8$, then

$$[y] = \begin{bmatrix} 12 & -7 & 0 \\ -3 & 10 & 4 \end{bmatrix}\begin{bmatrix} 5 \\ -2 \\ 8 \end{bmatrix}$$
$$= \begin{bmatrix} (12)(5) + (-7)(-2) + (0)(8) \\ (-3)(5) + (10)(-2) + (4)(8) \end{bmatrix} = \begin{bmatrix} 74 \\ -3 \end{bmatrix}$$

Thus, $y_1 = 74$ and $y_2 = -3$.

Exercise A.1

Find y_1, y_2, and y_3, given that $[y] = [A][x]$ with

$$[A] = \begin{bmatrix} 0 & -3 \\ 4 & 10 \\ 2 & 2 \end{bmatrix} \qquad [x] = \begin{bmatrix} 5 \\ -2 \end{bmatrix}$$

Determinants

Associated with any square matrix $[A]$ is a single quantity called its **determinant**, denoted symbolically by

$$\Delta = |[A]| \tag{A.6}$$

Despite the notation $\|[A]\|$, the determinant of $[A]$ may be either positive or negative.

The determinant of an arbitrary 2×2 matrix is computed via

$$\Delta = \begin{vmatrix} a_{11} & a_{12} \\ a_{21} & a_{22} \end{vmatrix} = a_{11}a_{22} - a_{12}a_{22} \tag{A.7}$$

Thus, you take the product along the **main diagonal** (upper left to lower right) and subtract the product along the other diagonal. To exploit the simplicity of Eq. (A.7), determinants of larger matrices can be evaluated by expansions involving reduced arrays.

For each element a_{ij} in the square matrix $[A]$, there is a **cofactor** defined by

$$\Delta_{ij} \triangleq (-1)^{i+j}|[A]_{ij}| \tag{A.8}$$

where $[A]_{ij}$ stands for the reduced array obtained from $[A]$ by deleting row i and column j. The term $(-1)^{i+j}$ in Eq. (A.8) just produces alternating signs, since $(-1)^{i+j} = \pm 1$ according to whether $i + j$ is even or odd.

To calculate the determinant of an $n \times n$ matrix, **Laplace's expansion** along any row i yields

$$\Delta = a_{i1}\Delta_{i1} + a_{i2}\Delta_{i2} + \cdots + a_{in}\Delta_{in} \tag{A.9a}$$

Alternatively, expansion down any column j yields

$$\Delta = a_{1j}\Delta_{1j} + a_{2j}\Delta_{2j} + \cdots + a_{nj}\Delta_{nj} \tag{A.9b}$$

It follows from these two equivalent expansions that interchanging the rows and columns of $[A]$ does not alter the value of Δ.

Applying Laplace's expansion to a 3×3 matrix reduces all cofactors to 2×2 determinants, which are easily evaluated via Eq. (A.7). Determinants of larger matrices may be handled by further expansions of the cofactors until all arrays have been reduced to 2×2. When a particular row or column of $[A]$ contains some zero elements, expanding along that row or column saves labor because you only need the cofactors of the nonzero elements.

Example A.2 ***Calculating a Determinant***

Let's calculate the determinant of the third-order matrix

$$[A] = \begin{bmatrix} 1 & 2 & 3 \\ 4 & 5 & 6 \\ 0 & 8 & 9 \end{bmatrix}$$

Since a zero element appears in the third row of the first column, we'll do the expansion along the third row.

Referring to Eq. (A.9a) with $i = 3$, $n = 3$, and $a_{31} = 0$, we see that we need the two cofactors

$$\Delta_{32} = (-1)^{3+2} \begin{vmatrix} 1 & \text{X} & 3 \\ 4 & \text{X} & 6 \\ \text{X} & \text{X} & \text{X} \end{vmatrix} = -\begin{vmatrix} 1 & 3 \\ 4 & 6 \end{vmatrix}$$

$$\Delta_{33} = (-1)^{3+3} \begin{vmatrix} 1 & 2 & \text{X} \\ 4 & 5 & \text{X} \\ \text{X} & \text{X} & \text{X} \end{vmatrix} = \begin{vmatrix} 1 & 2 \\ 4 & 5 \end{vmatrix}$$

Thus

$$\begin{aligned} \Delta &= a_{31}\Delta_{31} + a_{32}\Delta_{32} + a_{33}\Delta_{33} \\ &= 0 - 8(6 - 12) + 9(5 - 8) = 21 \end{aligned}$$

In practice, this expansion is usually written out directly by inspection as

$$\Delta = 0 - 8\begin{vmatrix} 1 & 3 \\ 4 & 6 \end{vmatrix} + 9\begin{vmatrix} 1 & 2 \\ 4 & 5 \end{vmatrix} = 0 - 8(-6) + 9(-3) = 21$$

Exercise A.2

Given $[A]$ in Example A.2, evaluate $|[A]|$ by: **(a)** expansion down the first column; **(b)** expansion along the second row.

Cramer's Rule

Now we're prepared to solve sets of simultaneous linear equations having the form

$$\begin{aligned} a_{11}y_1 + a_{12}y_2 + \cdots + a_{1n}y_n &= x_1 \\ a_{21}y_1 + a_{22}y_2 + \cdots + a_{2n}y_n &= x_2 \\ \vdots \qquad \quad \vdots \qquad \qquad \quad \vdots \quad &\quad \vdots \\ a_{n1}y_1 + a_{n2}y_2 + \cdots + a_{nn}y_n &= x_n \end{aligned}$$

We begin by putting the coefficients in a square matrix $[A]$ and calculating its determinant Δ and cofactors Δ_{ij}. The solution for the ith unknown is then given by **Cramer's rule** as

$$y_i = \frac{1}{\Delta}(\Delta_{1i}x_i + \Delta_{2i}x_2 + \cdots + \Delta_{ni}x_n) \tag{A.10}$$

where $i = 1, 2, \ldots, n$. Note carefully the cofactor subscripts, which change in *row* sequence rather than column sequence. Also note that Eq. (A.10) requires $\Delta = |[A]| \neq 0$. For this reason Δ is called the **characteristic determinant** in the context of simultaneous equations.

If the elements of $[A]$ happen to be such that $\Delta = 0$, then Eq. (A.10) becomes invalid. Accordingly, we say that $[A]$ is **singular** when $|[A]| = 0$. A singular characteristic determinant means that the simultaneous equations are interdependent and therefore do not provide sufficient information for unique solution.

Another version of Cramer's rule involves combining the matrix $[A]$ and the vector $[x]$. Specifically, let Δ_j stand for the determinant of $[A]$ with the jth column replaced by the independent variables from $[x]$, so

$$\Delta_j \triangleq \begin{vmatrix} a_{11} & \cdots & x_1 & \cdots & a_{1n} \\ a_{21} & \cdots & x_2 & \cdots & a_{2n} \\ \vdots & & \vdots & & \vdots \\ a_{n1} & \cdots & x_n & \cdots & a_{nn} \end{vmatrix} \quad (\text{jth column at } x) \tag{A.11}$$

Expansion down the jth column then yields

$$\Delta_j = x_1\Delta_{1j} + x_2\Delta_{2j} + \cdots + x_n\Delta_{nj}$$

Hence, upon comparison with Eq. (A.10), we get Cramer's rule in the more compact expression

$$y_j = \Delta_j/\Delta \qquad j = 1, 2, \ldots, n \tag{A.12}$$

This expression is particular handy when we have numerical values for the elements of $[x]$.

Two final points deserve mention here. First, Cramer's rule is used primarily for theoretical work or for numerical solution of two or three simultaneous equations. But the number crunching becomes quite tedious for $n \geq 4$. Consequently, those problems are better handled by calculator routines or computer programs that exploit more efficient methods such as the popular technique of gaussian elimination.

Second, when carrying out numerical evaluation of second- or third-order determinants, you may want to do some preliminary manipulations to simplify the numbers. As a case in point, suppose that $[A]$ consists mostly of large numbers or of fractions. Multiplying all elements of $[A]$ by some constant λ yields

$$\lambda[A] = [\lambda A] = \begin{bmatrix} \lambda a_{11} & \cdots & \lambda a_{1n} \\ \vdots & & \vdots \\ \lambda a_{m1} & \cdots & \lambda a_{mn} \end{bmatrix} \tag{A.13}$$

from which $|[\lambda A]| = \lambda^n|[A]|$. Thus, you can evaluate the determinant $|[\lambda A]|$ and obtain $|[A]|$ via

$$|[A]| = |[\lambda A]|/\lambda^n \tag{A.14}$$

In the same vein, suppose that awkward numbers appear in different rows of the matrix equation $[A][y] = [x]$. You might then multiply each row by a different constant to get the tidier equation

$$\begin{bmatrix} \lambda_1 a_{11} & \cdots & \lambda_1 a_{1n} \\ \vdots & & \vdots \\ \lambda_n a_{n1} & \cdots & \lambda_n a_{nn} \end{bmatrix} \begin{bmatrix} y_1 \\ \vdots \\ y_n \end{bmatrix} = \begin{bmatrix} \lambda_1 x_1 \\ \vdots \\ \lambda_n x_n \end{bmatrix}$$

Of course, the λ's must now be incorporated into the calculations for Cramer's rule.

Example A.3 *Solving Simultaneous Equations*

Consider the set of equations represented by

$$\begin{bmatrix} -5 & 0 & 2 \\ -500 & -125 & 0 \\ 0 & 1/3 & 2/5 \end{bmatrix} \begin{bmatrix} y_1 \\ y_2 \\ y_3 \end{bmatrix} = \begin{bmatrix} v_1 \\ v_2 \\ v_3 \end{bmatrix}$$

We'll solve for the y's in terms of the v's via Eq. (A.10). Then we'll use Eq. (A.12) to evaluate the y's, given values of the v's.

Because the coefficient matrix includes some awkward numbers, a wise preliminary step is to divide the second row by 125 and multiply the third row by 15. We then get

$$\begin{bmatrix} -5 & 0 & 2 \\ -4 & -1 & 0 \\ 0 & 5 & 6 \end{bmatrix} \begin{bmatrix} y_1 \\ y_2 \\ y_3 \end{bmatrix} = \begin{bmatrix} v_1 \\ v_2/125 \\ 15 v_3 \end{bmatrix} = \begin{bmatrix} x_1 \\ x_2 \\ x_3 \end{bmatrix}$$

which has the form $[A][y] = [x]$ with

$$x_1 = v_1 \qquad x_2 = v_2/125 \qquad x_3 = 15 v_3$$

The characteristic determinant is easily found to be

$$\Delta = |[A]| = -10$$

so we know that the equations have a unique solution.

To apply our first version of Cramer's rule, we first calculate all nine cofactors of $[A]$, obtaining

$$\begin{array}{lll} \Delta_{11} = -6 & \Delta_{12} = 24 & \Delta_{13} = -20 \\ \Delta_{21} = 10 & \Delta_{22} = -30 & \Delta_{23} = 25 \\ \Delta_{31} = 2 & \Delta_{32} = -8 & \Delta_{33} = 5 \end{array}$$

Thus, from Eq. (A.10),

$$y_1 = \frac{1}{-10}(-6x_1 + 10x_2 + 2x_3) = 0.6v_1 - 0.008v_2 - 3v_3$$

$$y_2 = \frac{1}{-10}(24x_1 - 30x_2 - 8x_3) = -2.4v_1 = 0.024v_2 + 12v_3$$

$$y_3 = \frac{1}{-10}(-20x_1 + 25x_2 + 5x_3) = 2v_1 - 0.02v_2 - 7.5v_3$$

Had we initially been given the values $v_1 = 12$, $v_2 = 1000$, and $v_3 = 0.6$, then $x_1 = 12$, $x_2 = 8$, and $x_3 = 9$ so

$$\begin{bmatrix} -5 & 0 & 2 \\ -4 & -1 & 0 \\ 0 & 5 & 6 \end{bmatrix} \begin{bmatrix} y_1 \\ y_2 \\ y_3 \end{bmatrix} = \begin{bmatrix} 12 \\ 8 \\ 9 \end{bmatrix}$$

The shorter version of Cramer's rule is probably more direct for this all-numerical case, and Eq. (A.12) yields

$$y_1 = \frac{1}{\Delta} \begin{vmatrix} 12 & 0 & 2 \\ 8 & -1 & 0 \\ 9 & 5 & 6 \end{vmatrix} = \frac{26}{-10} = -2.6 \quad (\text{1st column})$$

$$y_2 = \frac{1}{\Delta} \begin{vmatrix} -5 & 12 & 2 \\ -4 & 8 & 0 \\ 0 & 9 & 6 \end{vmatrix} = \frac{-24}{-10} = 2.4 \quad (\text{2nd column})$$

$$y_3 = \frac{1}{\Delta} \begin{vmatrix} -5 & 0 & 12 \\ -4 & -1 & 8 \\ 0 & 5 & 9 \end{vmatrix} = \frac{5}{-10} = -0.5 \quad (\text{3rd column})$$

Exercise A.3

Given that

$$\begin{bmatrix} 3/8 & 1/8 \\ -1/5 & 1/10 \end{bmatrix} [y] = \begin{bmatrix} v_1 \\ v_2 \end{bmatrix}$$

(a) Solve for y_1 and y_2 when $v_1 = 30$ and $v_2 = -16$.
(b) Find y_1 and y_2 in terms of v_1 and v_2.

A.2 MATRIX OPERATIONS

This section introduces the additional matrix operations needed for the study of two-port networks in Chapter 14 and state-variable circuit analysis in Chapter 15.

Addition and Multiplication

Matrix addition is defined only when the matrices involved have the same order. Each element of the sum then equals the sum of the respective elements. For example, if $[A]$ and $[B]$ are both of order $m \times n$, and if

$$[C] = [A + \lambda B] = [A] + \lambda[B] \tag{A.15a}$$

then

$$c_{ij} = a_{ij} + \lambda b_{ij} \tag{A.15b}$$

Obviously, the sum matrix $[C]$ also has order $m \times n$.

The multiplying constant λ has been included in Eq. (A.15a) to allow generalization for **matrix subtraction**. Thus, upon letting $\lambda = -1$, we get $[C] = [A] - [B]$ with $c_{ij} = a_{ij} - b_{ij}$. Furthermore, all of the elements of $[C]$ equal zero when $[C] = [A] - [A]$, so

$$[A] - [A] = [0] \tag{A.16}$$

We call $[0]$ the **null matrix** of order $m \times n$.

Based upon Eq. (A.15), you can easily see that matrix addition exibits the properties

$$\begin{aligned}
&[A] + [0] = [A] \\
&[A] + [B] = [B] + [A] \\
&\lambda([A] + [B]) = \lambda[A] + \lambda[B] \\
&[A] + ([B] + [C]) = ([A] + [B]) + [C]
\end{aligned} \tag{A.17}$$

These properties are the same as ordinary addition, bearing in mind that the matrices must be of the same order.

Matrix multiplication is quite another matter, as implied by our previous definition of a matrix-vector product. To start from that simpler case, consider the following two products with the indicated orders:

$$\underset{m\times 1}{[y]} = \underset{m\times k}{[A]}\ \underset{k\times 1}{[v]} \qquad \underset{k\times 1}{[v]} = \underset{k\times n}{[B]}\ \underset{n\times 1}{[x]}$$

Replacing $[v]$ in the first equation with $[B][x]$ from the second yields

$$[y] = [A][B][x] = [C][x]$$

where

$$\underset{m\times n}{[C]} = \underset{m\times k}{[A]}\ \underset{k\times n}{[B]} \tag{A.18a}$$

We thereby see that

> The product $[A][B]$ is defined only when the number of rows in $[B]$ equals the number of columns in $[A]$.

The elements of the resulting matrix $[C]$ are given by

$$c_{ij} = a_{i1}b_{1j} + a_{i2}b_{2j} + \cdots + a_{ik}b_{kj} \tag{A.18b}$$

This expression is obtained by carrying out the multiplication $[v] = [B][x]$ and substituting into $[y] = [A][v]$.

Equation (A.18b) describes a *sum of row-by-column products,* like matrix-vector multiplication. Now, however, the operation must be performed with each column of $[B]$ multiplying each row of $[A]$. Thus, the pattern for a typical element c_{ij} of the matrix product $[A][B]$ is illustrated by

$$\begin{bmatrix} & & \\ a_{i1} & \cdots & a_{ik} \\ & & \end{bmatrix} \begin{bmatrix} & b_{1j} & \\ & \vdots & \\ & b_{kj} & \end{bmatrix} = \begin{bmatrix} & & \\ & c_{ij} & \\ & & \end{bmatrix} \tag{A.19}$$

As a consequence of Eq. (A.18), matrix multiplication with appropriate orders exhibits the familiar properties

$$\begin{aligned} [A]([B] + [C]) &= [A][B] + [A][C] \\ [A]([B][C]) &= ([A][B])[C] \end{aligned} \tag{A.20}$$

But is is *not* true in general that $[A][B]$ equals $[B][A]$, so we must distinguish between **premultiplication** and **postmultiplication**. Moreover, the result $[A][B] = [0]$ does not necessarily imply that $[A] = [0]$ or $[B] = [0]$, and $[A][B] = [B][C]$ does not necessarily imply that $[A] = [C]$.

A matrix having special interest in multiplication is the **identity matrix** $[I]$, a square matrix of arbitrary order with 1's on the main diagonal and 0's elsewhere. Thus, for $n = 2$ and $n = 3$,

$$[I] = \begin{bmatrix} 1 & 0 \\ 0 & 1 \end{bmatrix} \qquad [I] = \begin{bmatrix} 1 & 0 & 0 \\ 0 & 1 & 0 \\ 0 & 0 & 1 \end{bmatrix} \tag{A.21}$$

When the order of $[I]$ is chosen for multiplication with $[A]$, Eq. (A.18b) yields

$$[A][I] = [I][A] = [A] \tag{A.22}$$

This relationship expedites factoring in matrix equations.

Consider, for instance, the nth-order square matrix $[A]$ and vector $[x]$ combined to form

$$[y] = \lambda[x] + [A][x]$$

Taking $[I]$ to be nth order, we write $\lambda[x] = \lambda[I][x] = [\lambda I][x]$ and obtain

$$[y] = [\lambda I][x] + [A][x] = [\lambda I + A][x]$$

where $[\lambda I + A] = [\lambda I] + [A]$.

Example A.4 ***Performing Matrix Operations***

Suppose we want to simplify $[A][B] - [A][C]$ given

$$[A] = \begin{bmatrix} 0 & 1 \\ -2 & 3 \\ 4 & -5 \end{bmatrix} \qquad [B] = \begin{bmatrix} 7 & -1 & -4 \\ 5 & 0 & -3 \end{bmatrix} \qquad [C] = \begin{bmatrix} 8 & -3 & -5 \\ 2 & 6 & 0 \end{bmatrix}$$

Since $[A][B] - [A][C] = [A]([B] - [C])$, we first perform the routine subtraction

$$[B] - [C] = \begin{bmatrix} 7-8 & -1+3 & -4+5 \\ 5-2 & 0-6 & -3-0 \end{bmatrix} = \begin{bmatrix} -1 & 2 & 1 \\ 3 & -6 & -3 \end{bmatrix}$$

Multiplication per Eq. (A.19) then yields the 3×3 array

$$\begin{bmatrix} 0 & 1 \\ -2 & 3 \\ 4 & -5 \end{bmatrix}\begin{bmatrix} -1 & 2 & 1 \\ 3 & -6 & -3 \end{bmatrix} = \begin{bmatrix} 0+3 & 0-6 & 0-3 \\ 2+9 & -4-18 & -2-9 \\ -4-15 & 8+30 & 4+15 \end{bmatrix}$$

Thus,

$$[A]([B] - [C]) = \begin{bmatrix} 3 & -6 & -3 \\ 11 & -22 & -11 \\ -19 & 38 & 19 \end{bmatrix}$$

Exercise A.4

Taking $[A]$ and $[B] - [C]$ as in Example A.4, carry out the multiplication to show that $([B] - [C])[A] = [0]$, despite the fact that $[B] - [C] \neq [0]$, $[A] \neq [0]$, and $[A]([B] - [C]) \neq [0]$.

Inverse and Adjoint Matrices

We have not yet mentioned matrix division simply because that operation has no meaning. However, we may speak of the **inverse** of $[A]$, denoted $[A]^{-1}$ and defined by the property

$$[A]^{-1}[A] = [A][A]^{-1} = [I] \tag{A.23}$$

Careful examination of Eq. (A.23) reveals that $[A]$ must be a *square* matrix. Additionally, as shown later, $[A]$ must be *nonsingular* in the sense that $\Delta = |[A]| \neq 0$.

The inverse matrix plays an important role in the study of matrix-vector equations such as

$$[A][y] = [x] \tag{A.24a}$$

Premultiplying both sides by $[A]^{-1}$ yields $[A]^{-1}[A][y] = [A]^{-1}[x]$. But $[A]^{-1}[A][y] = [I][y] = [y]$, so

$$[y] = [A]^{-1}[x] \tag{A.24b}$$

which provides an analytical expression for $[y]$ in terms of $[x]$.

Letting α_{ij} represent the elements of $[A]^{-1}$, we next seek the relationship between α_{ij} and the elements of $[A]$. To that end, we observe from Eq. (A.24b) that the ith element of $[y]$ will be

$$y_i = \alpha_{i1}x_1 + \alpha_{i2}x_2 + \cdots + \alpha_{in}x_n$$

Hence, upon comparison with Cramer's rule as stated in Eq. (A.10), we find that

$$\alpha_{ij} = \Delta_{ji}/\Delta \tag{A.25}$$

where Δ_{ji} is the cofactor of element a_{ji} in $[A]$. Equation (A.25) indicates that $\alpha_{ij} \neq 1/a_{ij}$, and it clearly shows that $[A]^{-1}$ exists only when $\Delta \neq 0$.

Further consideration of Eq. (A.25) leads to an expression for $[A]^{-1}$ involving the cofactors of $[A]$ with the rows and columns interchanged or *transposed* — transposition being needed because α_{ij} is proportional to Δ_{ji}. We therefore introduce the matrix **transpose** $[A]^T$ whose elements are

$$a^T_{ij} = a_{ji}$$

We also introduce the **adjoint** of $[A]$, symbolized by adj$[A]$ and defined as

$$\text{adj}[A] \triangleq [\Delta_{ij}]^T = \begin{bmatrix} \Delta_{11} & \Delta_{21} & \cdots & \Delta_{n1} \\ \Delta_{12} & \Delta_{22} & \cdots & \Delta_{n2} \\ \vdots & \vdots & & \vdots \\ \Delta_{1n} & \Delta_{2n} & \cdots & \Delta_{nn} \end{bmatrix} \tag{A.26}$$

This transposed array of cofactors can be obtained by calculating the cofactors of $[A]^T$.

Finally, multiplying adj$[A]$ by $1/\Delta$ gives our desired matrix equation

$$[A]^{-1} = \frac{1}{\Delta}\,\text{adj}[A] \tag{A.27}$$

Like Cramer's rule, Eq. (A.27) is used primarily for theoretical work or for numerical solution with $n = 2$ or 3.

For an arbitrary 2×2 matrix, we obtain $[A]^{-1}$ by starting with the transpose

$$[A]^T = \begin{bmatrix} a_{11} & a_{21} \\ a_{12} & a_{22} \end{bmatrix}$$

Inspection of $[A]^T$ then yields

$$\text{adj}[A] = \begin{bmatrix} a_{22} & -a_{12} \\ -a_{21} & a_{11} \end{bmatrix}$$

Hence, from Eq. (A.27),

$$[A]^{-1} = \begin{bmatrix} a_{22}/\Delta & -a_{12}/\Delta \\ -a_{21}/\Delta & a_{11}/\Delta \end{bmatrix}$$

where $\Delta = a_{11}a_{22} - a_{12}a_{21}$.

Example A.5 ***Solving a Matrix Equation***

Suppose you want to solve the third-order set of equations represented by $[A][y] = [x]$ with

$$[A] = \begin{bmatrix} -5 & 0 & 2 \\ -4 & -1 & 0 \\ 0 & 5 & 6 \end{bmatrix}$$

First, you pivot the elements of $[A]$ around the main diagonal to get

$$[A]^T = \begin{bmatrix} -5 & -4 & 0 \\ 0 & -1 & 5 \\ 2 & 0 & 6 \end{bmatrix}$$

Second, you find the cofactors of $[A]^T$ by covering up rows and columns, from which

$$\text{adj}[A] = \begin{bmatrix} (-6 - 0) & -(0 - 10) & (0 + 2) \\ -(-24 - 0) & (-30 - 0) & -(0 + 8) \\ (-20 - 0) & -(-25 - 0) & (5 - 0) \end{bmatrix}$$

Third, you calculate $\Delta = |[A]| = -10$, so

$$[A]^{-1} = -0.1 \text{ adj}[A] = \begin{bmatrix} 0.6 & -1.0 & -0.2 \\ -2.4 & 3.0 & 0.8 \\ 2.0 & -2.5 & -0.5 \end{bmatrix}$$

Fourth, since $[y] = [A]^{-1}[x]$, your solution is

$$\begin{bmatrix} y_1 \\ y_2 \\ y_3 \end{bmatrix} = \begin{bmatrix} 0.6x_1 - 1.0x_2 - 0.2x_3 \\ -2.4x_1 + 3.0x_2 + 0.8x_3 \\ 2.0x_1 - 2.5x_2 - 0.5x_3 \end{bmatrix}$$

Exercise A.5

Find adj$[A]$ and $[A]^{-1}$ for $[A]$ in Example A.2 (p. 733).

APPENDIX B
Circuit Analysis with PSpice

Many computer programs have been devised to aid circuit analysis and design work. The most powerful and versatile of these originated at the University of California, Berkeley, and was named SPICE. The name stands for *Simulation Program with Integrated Circuit Emphasis* because the program contains many aspects specifically for integrated electronic circuits. Nonetheless, SPICE has all the capabilities needed to analyze any linear circuit of the type considered in this text.

PSpice® is a commercial software product based on the SPICE algorithms. But PSpice includes additional features that make the program more flexible and user-friendly. Notable among those features is the graphics postprocessor Probe™, which acts like a "software oscilloscope" capable of displaying waveforms and much more. For these reasons, PSpice has become one of the most popular circuit simulation programs in industry and education.

This appendix introduces the basic techniques of linear circuit analysis using PSpice. Following an overview of PSpice, the sequence of topics parallels that of the text itself — going from resistive circuits to ac steady state, transients, and frequency response. Corresponding PSpice simulation problems are at the end of the appendix.

By doing PSpice simulations, you will gain a fuller understanding of circuit behavior under various conditions. Indeed, the interpretation of simulation results is just as important as the results per se. Additionally, the background gained by simulating linear circuits makes you better prepared to apply PSpice to the nonlinear devices and circuits encountered in the study of electronics.

Three important points regarding our treatment of PSpice should be stated at the outset:

- Many specialized or advanced features have been purposely omitted in the interest of simplicity;
- Some of the operational details depend upon the particular computer platform as well as the version of PSpice;
- All of the features covered here are included in the evaluation edition or student edition of PSpice version 4.03 or later.

Earlier versions of PSpice may be used, but they lack some of the convenient enhancements.

PSpice versions 5.1 and later are usually part of a larger software package called *The Design Center,* which also simulates digital electronic circuits. The current evaluation edition of PSpice is available for several platforms. Educators may obtain the software with installation instructions by contacting MicroSim Corporation, 20 Fairbanks, Irvine, CA 92718, (800) 245-3022 or (714) 770-3022, Fax (714) 455-0554.

For convenient reference, a concise summary of PSpice and Probe information is included at the back of this text.

Objectives

After studying this appendix and working the exercises, you should be able to do each of the following:

1. Write and run a PSpice circuit file including `.param`, `.print`, and `.dc` statements (Section B.1).
2. Invoke Probe and obtain plots of variables (Section B.1).
3. Use PSpice and the `.tf` command to analyze a resistive circuit, possibly including controlled sources (Section B.2).
4. Vary a parameter value via the `.step` command (Section B.2).
5. Simulate a single-phase or three-phase ac circuit, and sweep the frequency with a `.ac` command (Section B.3).

6. Write a PSpice description of an ideal transformer or mutual inductance (Section B.3).
7. Use PSpice and Probe to obtain transient waveforms using initial conditions, time-varying sources, and the `.tran` command (Section B.4).
8. Write PSpice statements to actuate a switch for transient analysis (Section B.4).
9. Use PSpice and Probe to obtain frequency-response curves or Bode plots (Section B.5).

B.1 INTRODUCTION TO PSPICE

As an introductory overview of PSpice, this section describes and discusses the simulation of a simple resistive circuit. We'll also present the essential details needed to run PSpice and Probe.

PSpice Basics

PSpice is a computer program itself, so you don't have to do "programming" to use it. Rather, you prepare a **circuit file** that tells PSpice what you want done. The circuit file consists of several statements that

- describe the circuit to be simulated,
- specify the analysis to be performed,
- control the output format and variables.

PSpice does the simulation called for in the circuit file and stores the results in an **output file**. The circuit file may also invoke Probe, which allows you to display results graphically in various ways.

The type of simulation performed by PSpice depends on the source specifications and control statements. We'll subsequently discuss five important simulation types:

- **DC analysis** with fixed source values
- **DC sweep analysis** with varying source or element values
- **AC analysis** with fixed source frequency
- **AC sweep analysis** with varying source frequency
- **Transient analysis** with particular source waveforms

Probe operates in conjunction with sweep or transient analyses that generate an independent variable needed for plotting.

PSpice works fundamentally with *node voltages*, and the circuit description requires **node numbers**. Every circuit must contain a **reference node** numbered 0 (zero), and all other nodes are identified by unique positive integers. (PSpice also allows nodes identified by *text names*, which we'll reserve for special purposes.) If two or more connection points are joined by a zero-resistance path, then that combination is treated as a single node. For example, Fig. B.1 shows the diagram of a circuit labeled with node numbers 0, 1, and 2. These numbers happen to be sequential, but this is not required.

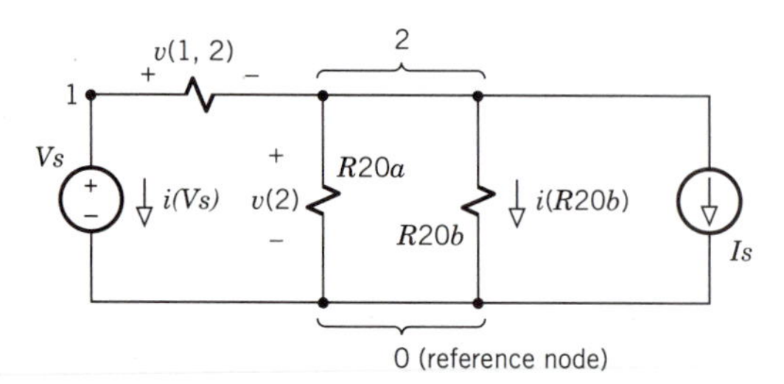

Figure B.1 Circuit diagram with PSpice notation.

Each element in a PSpice circuit must connect to nodes that specify the element's location. Consequently, even a series connection point like node 1 in Fig. B.1 requires identification. Each element must also have a **device name**, such as *Vs, Is,* or *R20 b*, with an appropriate key initial letter. You may use both uppercase and lowercase letters, because PSpice is insensitive to case in circuit files. However, PSpice does not allow subscripts or superscripts.

The basic PSpice **variables** are node voltages and branch currents. A node voltage is specified by a node number, so $v(2)$ in Fig. B.1 denotes the potential difference between node 2 and the reference node. You can also use two node numbers to specify a difference voltage, so $v(1,2)$ stands for $v(1) - v(2)$. A branch current is specified by the element through which it passes, such as $i(Vs)$ or $i(R20\,b)$.

The reference polarity for node voltages is always positive relative to node 0. Other references are dictated by the property that

> PSpice employs the passive convention for all elements, with the first-named node taken to be at the higher potential.

Accordingly, the reference direction of any current goes from the first-named node (higher potential) to the second (lower potential). Thus, for instance, the reference arrow for $i(Vs)$ in Fig. B.1 points from node 1 to node 0 — so you might expect $i(Vs)$ to have a negative value.

Circuit Files

Circuit files contain five types of statements, some of them being optional. The statement types are

- The title, which must be the *first* line. This title line will appear on any output, so you may want to include your name for identification purposes.

- Device statements, which specify circuit elements and have a *key letter* in the first column.
- Control statements, which have a period (.) in the first column. Some control statements pertain to devices, while others specify the type of simulation or output format.
- Comment lines, which have an asterisk (*) in the first column. Comments may also be appended to other lines after a semicolon (;).
- The `.end` statement, which must be the *last* line.

Any statement that is too long for one line can be carried over on a continuation line having a plus (+) in the first column.

Except for the title and end line, statements may appear in almost any order. For clarity, however, you should group the device statements together in a logical order. Comments can be interspersed as needed to provide explanations and documentation.

Each **device statement** includes a user-supplied name, the numbers of the nodes to which the element is connected, and the element's value. Independent dc voltage and current sources are described by statements having the form

```
Vname n+ n- dc value
Iname n+ n- dc value
```

Here, the required key letters (V or I) are capitalized, while user-supplied items such as *name* and *value* are italicized. The higher- and lower-potential node numbers are denoted, respectively, by n^+ (first named) and n^- (second named). Hence, the reference current direction goes from node n^+ to node n^-. The value may be positive, negative, or zero.

The device statement syntax for a resistor is

```
Rname n+ n- value
```

The order of node numbers affects only the reference direction of its branch current $i(Rname)$. The resistance value cannot be zero or infinite, but negative values are allowed.

The names of any elements in these statements are arbitrary, provided that they begin with the proper key letter and that all elements in a given circuit have different names. Meaningful names like *Vs* and *Rload* should be used when appropriate. Otherwise, the name may just echo the node numbers, as in $R12$ for the resistance between nodes 1 and 2.

PSpice automatically assigns the corresponding SI unit for each element based on the key letter of its name, so units need not be written after element values. The spacing between entries on a given statement line is arbitrary, and entries may be separated by one or more blank spaces, commas, or parentheses. Thus, you may insert optional spaces, commas, and parentheses to make the statements more readable.

Numerical values are expressed in either fixed-point or floating-point form. The PSpice floating-point notation for $x \times 10^m$ is xEm, where x is any number and m is a positive or negative integer. Alternatively, a number may be followed immediately by one of the symbolic scale factors in Table B.1. You may enter these scale factors as uppercase or lowercase letters. (However, Probe is case sensitive and it reserves M for 10^6 and m for 10^{-3}.)

TABLE B.1 PSpice Scale Factors

Symbol	Value	Symbol	Value
k	10^3	m	10^{-3}
Meg	10^6	u	10^{-6}
G	10^9	n	10^{-9}
T	10^{12}	p	10^{-12}

To illustrate several of the foregoing points, consider the circuit diagram in Fig. B.2. The nodes have been numbered, and each element has been labeled with both a name and a value. The device statements might be written as

```
Vs    1 0 dc 15
R12   1 2    500
R20a  2 0    1.2k ; 1200-ohm resistor
R20b  2 0    2E3  ; 2000-ohm resistor
Is    2 0 dc 10m  ; 10-mA current from node 2 to node 0
```

The last three of these statements include appended comments, and you should take a moment to relate each device statement to the circuit diagram. Also note that the order of node numbers for $R20a$ and $R20b$ is such that the directions of branch currents $i(R20a)$ and $i(R20b)$ will correspond to Ohm's law with voltage $v(2)$.

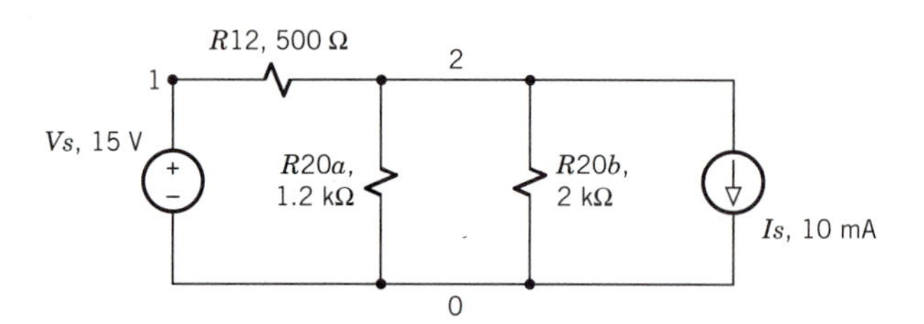

Figure B.2 Diagram for PSpice simulation.

Numerical values in device statements may also be specified via **expressions**. The value is then replaced by {expression}, where the curly braces enclose a mathematical relation with operation symbols and parentheses as needed. The basic mathematical operations are symbolized by

+ (addition) − (subtraction or negation)
* (multiplication) / (division)

Thus, for instance, a value of two and one-third could be written in a device statement as {2 + 1/3} or as {7/3}.

Turning to **control statements**, the only essential one is the required last line

```
.end
```

Additional `.end` statements may appear within a circuit file for two or more simulations in one PSpice run. Each simulation must then start with a title line.

When a circuit file contains dc sources and no explicit output statements, PSpice produces its **default output** consisting of

- A table of node–voltage values
- The value of the current through each independent voltage source
- The total power dissipation supplied by independent voltage sources, but not by current sources

This output may be sufficient information in some analysis problems. Other statements controlling output variables will be discussed shortly.

The syntax of the **options control statement** that affects output format is

```
.options ⟨list⟩ ⟨node⟩ ⟨nopage⟩
```

Here, each key word in angle brackets may be included or omitted. When included, they have the following effects on the output file:

`list` produces a summary of elements

`node` produces a summary of connections

`nopage` suppresses paging at the start of each section.

You should always include `.options nopage` to conserve paper in printed output. The `list` and `node` options may be of help for debugging errors in a large circuit file.

Example B.1 ***Simulation of a Simple Resistive Circuit***

Again consider the labeled circuit diagram in Fig. B.2. If the only variables of interest are node voltages and voltage–source currents, then the complete circuit file could be:

```
Example B.1, My Name
*Resistive circuit with two sources
```

```
Vs   1 0 dc 15
R12  1 2    500
R20a 2 0    1.2k
R20b 2 0    2E3
Is   2 0 dc 10m
.options nopage
.end
```

After PSpice completes the simulation, the results appear in the output file under the title line. The first output section echoes the circuit file, without the title line. The remaining output section is as follows.

```
**************************************************************
 **** SMALL SIGNAL BIAS SOLUTION TEMPERATURE = 27.00 DEG C

 NODE   VOLTAGE  NODE   VOLTAGE NODE VOLTAGE
(    1) 15.0000 (    2)   6.0000

   VOLTAGE SOURCE CURRENTS
   NAME     CURRENT
   Vs        -1.800E-02

   TOTAL POWER DISSIPATION   2.70E-01 WATTS
**************************************************************
```

The heading `SMALL SIGNAL BIAS SOLUTION TEMPERATURE = 27.00 DEG C` has meaning primarily for electronic circuits and can be ignored in our work. The significant values shown here are

$$\begin{aligned} v(1) &= 15\text{ V} && \text{(established by } Vs) \\ v(2) &= 6\text{ V} && \text{(simulation result)} \\ i(Vs) &= -18\text{ mA} && \text{(simulation result, passive convention)} \end{aligned}$$

The `TOTAL POWER DISSIPATION` corresponds to the power supplied only by the voltage source, so the value is $Vs \times [-i(Vs)] = 270$ mW. The additional power supplied by the current source may, of course, be calculated by hand via $[-v(2)] \times Is = -60$ mW. Thus, the power actually dissipated by the resistors in this circuit is $270 + (-60) = 210$ mW.

Exercise B.1

Write a complete circuit file to find v_1 and v_2 in Fig. B.3 when $i_s = 3$ mA and $v_s = 27$ V. (Save your file for use in Exercises B.2 and B.3.)

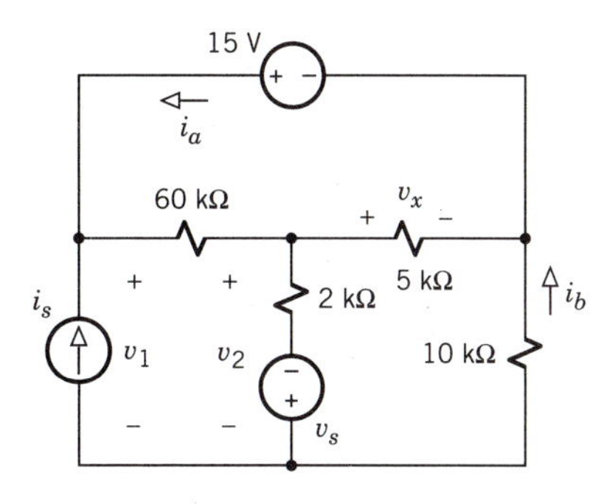

Figure B.3

PSpice Output

When the PSpice default output does not provide enough information, the circuit file must be augmented to obtain values of other variables of interest. There are several different ways to handle this situation, depending upon the additional information desired.

If you only want the value of a branch current, then you may take advantage of the fact that the default output includes the currents through all independent voltage sources. Accordingly, you can put a *current-sensing zero-voltage source* in the appropriate branch. Such a source acts like a short and does not alter the circuit's behavior, but it does require an additional node number. For instance, Fig. B.4 shows how node 2 from Fig. B.2 could be split into two nodes to insert a current-sensing source for $i(V23)$. The device statement might read

```
V23 2 3 dc 0 ; current-sensing voltage source
```

Other device statements must then be modified to account for the new node number.

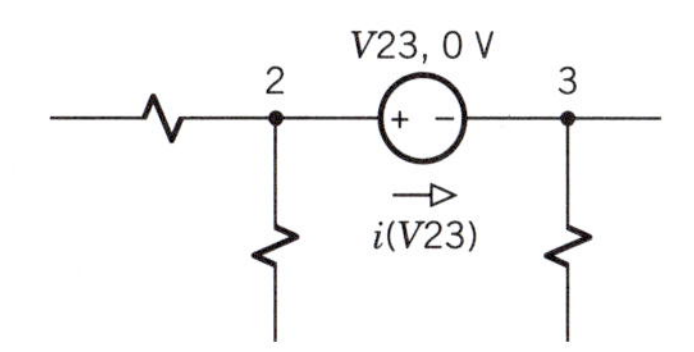

Figure B.4 Current-sensing voltage source.

To get several branch currents and/or voltages, you can use a **print control statement** having the general form

```
.print dc var1 var2 . . .
```

where *var1*, *var2*, . . . constitute *all* the variables of interest. These variables are specified by the usual PSpice notation for node voltages, difference voltages, and branch currents. The default output is not supplied when the circuit file includes a `.print dc` statement.

The `.print dc` statement must always be accompanied by a **source control statement** governing the value of an independent voltage or current source. One form of the statement for dc sources is

```
.dc [V|I]name list val1 . . .
```

where [*V*|*I*]*name* stands for either *Vname* or *Iname*. The list entry *val1* . . . gives the source value or a sequence of values in ascending or descending order. This specification overrides the value given in the device statement for [*V*|*I*]*name*, and the simulation is redone with each value when more than one is listed.

Example B.2 *Simulation with Three Source Values*

The following file for the circuit back in Fig. B.2 includes a source control statement that sets Vs = 15 V, 0 V, and −7.5 V. The print control statement calls for four outputs.

```
Example B.2, My Name
* Illustrating .dc list and .print dc
Vs   1 0 dc 15
R12  1 2    500
R20a 2 0    1.2k
R20b 2 0    2E3
Is   2 0 dc 10m
* Simulate with Vs = 15 V, 0 V, and -7.5 V
.dc Vs list 15 0 -7.5
* Output variables to be printed
.print dc v(2) v(1,2) i(R12) i(R20a)
.options nopage
.end
```

The resulting output shown as follows is a table giving values of the specified variables for each value of Vs. Note that $v(2) = Vs$ in the last line of the table, so we would expect to get $v(1,2) = 0$ and $i(R12) = v(1,2)/500 = 0$. However, round-off errors in the numerical simulation yield very small but nonzero values for these two quantities.

```
*************************************************************
Example B.2, My Name

 ****            DC TRANSFER CURVES            TEMPERATURE = 27.00 DEG C

     Vs           V(2)          V(1,2)          I(R12)         I(R20a)
  1.500E+01    6.000E+00     9.000E+00      1.800E-02      5.000E-03
```

```
 0.000E+00  -3.000E+00   3.000E+00   6.000E-03  -2.500E-03
-7.500E+00  -7.500E+00  -2.250E-09  -4.500E-12  -6.250E-03
**************************************************************
```

Exercise B.2

Modify your circuit file from Exercise B.1 to obtain values for i_a, i_b, and v_x in Fig. B.3 with $v_s = 27$ V while $i_s = 0$, 3 mA, and 6 mA. Hint: Enter the value of the 15-V source as a *negative* quantity to get the desired reference direction for i_a.

Probe Output with dc Sweep Another way of changing a source voltage is the **dc sweep statement**

```
.dc lin [V|I]name start, stop, step
```

where optional commas have been included for clarity. This statement sweeps the value of *Vname* or *Iname* linearly from *start* to *stop* in increments of *step.*

When a source is swept over several values, you probably want the results depicted graphically. For simple graphical output, PSpice provides a `.plot` statement that produces rather crude graphs by line plotting with characters. But much more satisfactory are plots created with the graphics postprocessor Probe.

The **Probe control statement** that invokes Probe from the circuit file is

```
.probe ⟨var1 var2 . . .⟩
```

The list of output variables usually should be omitted, in which case *all* voltages and currents will be available. The accompanying sweep statement should generate a sufficient number of points for plotting purposes. Keep in mind, however, that processing time increases with the number of sample points.

Sometimes PSpice invokes Probe automatically, even when the circuit file does not include `.probe`. You then press ⟨Esc⟩ if you do not want graphical output.

Example B.3 *Simulation with Probe Output*

The following file for the circuit back in Fig. B.2 includes a dc sweep statement that sweeps the source current from 0 to 40 mA with steps of 1 mA. Graphics output is obtained via Probe, so `.options nopage` is not needed.

```
Example B.3, My Name
* Illustrating dc sweep and probe
```

```
Vs   1 0 dc 15
R12  1 2    500
R20a 2 0    1.2k
R20b 2 0    2E3
Is   2 0 dc 10m
* Sweep current Is linearly from 0 to 40 mA in 1-mA steps
.dc lin Is 0, 40m, 1m
* Invoke Probe for plotting with all variables available
.probe
.end
```

Figure B.5 reproduces Probe output in the form of a *split-screen display* with current *Is* as the independent variable along the horizontal axis (*X*-axis). The upper plot of $v(2)$ versus *Is* corresponds to the v-i curve of the equivalent source network connected to the current source. Thus, we see from this plot that the open-circuit voltage of the network is $v_{oc} = 9$ V and the short-circuit current is $i_{sc} = 30$ mA, so its Thévenin resistance is $R_t = 9/30 = 0.3$ kΩ. The lower plot of $-v(2)*Is$ represents the power supplied by *Is*, so the lower vertical axis is labeled with the unit mW (milliwatts). This plot shows that the current source actually absorbs power when $0 < Is < 30$ mA.

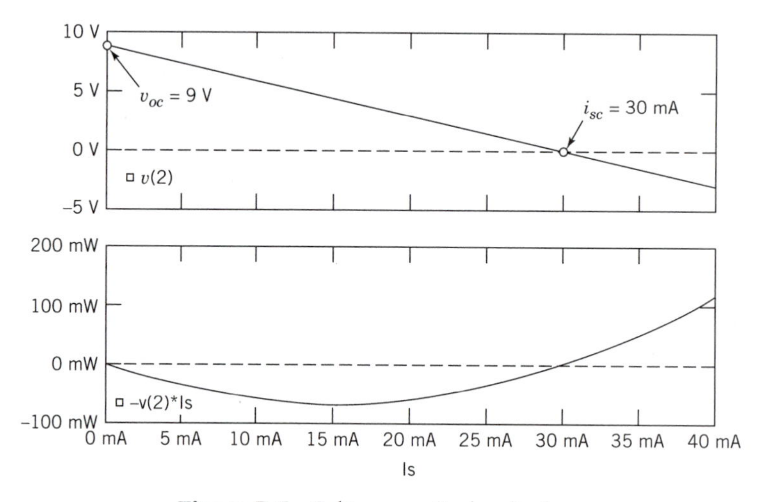

Figure B.5 Split-screen Probe display.

Since the split-screen format reduces vertical size, the same information can be put in the larger *dual-trace display* of Fig. B.6. Here, the power curve has been multiplied by 50 to make its vertical deflection comparable to $v(2)$.

Exercise B.3

Modify your circuit file from Exercise B.1 to generate Probe plots with about 100 points while v_s varies from -20 V to $+20$ V, keeping i_s fixed at 3 mA.

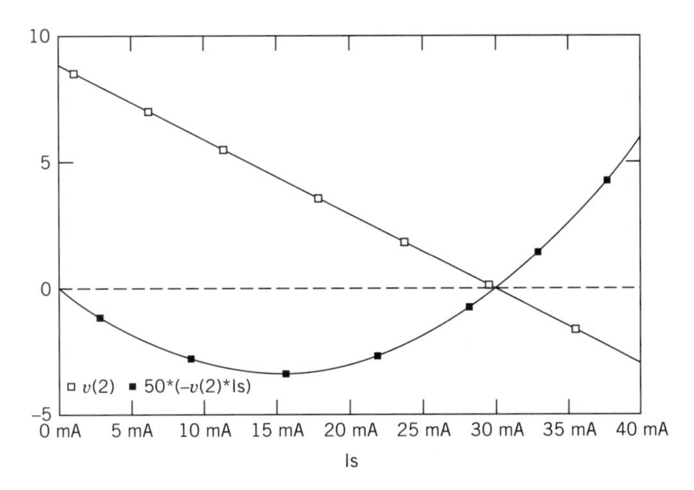

Figure B.6 Dual-trace Probe display.

Running PSpice

As preparation for a PSpice simulation, you should always make a circuit diagram labeled with node numbers, device names, and element values. You should also include reference polarity marks for any branch variables of interest and label them with PSpice notation.

PSpice versions 5.0 and later include a graphical program called *Schematics* that allows you to draw and label your circuit diagram on the screen. The program then automatically generates a circuit file from your schematic diagram. This option is particularly convenient for complicated electronic circuits, but it should be used only *after* you have mastered the basics of PSpice by writing your own circuit files.

Circuit files that you write can be entered using either a word processor or the text editor under the control shell of PSpice versions without *Schematics.* The following instructions for both methods assume a computer with the DOS operating system. Similar procedures hold for other operating systems.

Using the Control Shell The control shell is accessed by typing *ps.* Each heading at the top of the menu screen has a drop-down submenu. Menu selections are made using the initial letters or arrow keys and the ⟨*Enter*⟩ key. Support for mouse selection is also provided, but the mouse pointer does not appear on the screen. The ⟨*Esc*⟩ key cancels an unwanted selection and/or takes you back to the main menu.

To enter your circuit file, select **Files** + **Current File** and type your *filename,* which can be any alphanumeric sequence with no more than eight characters. Then select **Files** + **Edit** to begin typing your file text. Press *F1* for editor instructions as needed. Press ⟨*Esc*⟩ after you have completed your file.

If your circuit file has syntax errors, then some of them may be detected at this point. Select **Edit** and press *F6* to get the error messages, which are given by line number excluding the title line. Press ⟨*Esc*⟩ to edit the errors and press ⟨*Esc*⟩ again.

Now select **Analysis** + **Run PSpice** to start the simulation. Additional errors may be detected by PSpice, and they will be indicated in the output file. You can view this file on the monitor by selecting **Files** + **Browse Output**. Return to **Edit** to correct the errors, and then restart the simulation. After browsing the results in the final output, you exit the control shell by selecting **Quit**.

Using a Word Processor Word processors afford somewhat more convenience than the text editor, but the circuit file must be saved as plain ASCII text. Be careful to avoid a blank line after the final end statement, because PSpice will consider it to be the title line of another circuit description.

After completing your circuit file, you must save it under a name in the form *filename.cir,* where *filename* is an alphanumeric sequence with no more than eight characters. PSpice is then invoked by typing the command *pspice filename,* and output is written to a file named *filename.out.*

You can view the output file on your monitor using the word processor or a file browser. If your circuit file contains errors, then they will be indicated in the output file. Any errors should be corrected by editing the circuit file — not the output file — and running PSpice again.

Printing Output After a successful PSpice run, you should obtain a printed copy of the output file for documentation. (Recall that this file also includes the statements from your circuit file.) To get the printed output, key in the DOS command *type filename.out.*

Running Probe

Probe is a menu-driven interactive graphics program, so you can easily try various plots and formats. You can also define traces by expressions involving constants and variables, and the cursor option allows you to read numerical values accurately from the traces. The best way to become familiar with Probe's many features is by hands-on experimentation!

If your circuit file includes multiple simulations, then Probe first presents a start-up menu for selecting the case or cases to be displayed. Usually you just press ⟨*Enter*⟩ so that all cases will be available.

The main Probe menu then appears with a blank plotting area. Select a menu command by arrow key, initial letter, or mouse pointer. You can use the menu and submenu commands to generate one or more plots with single or multiple traces. The most useful menu items for our work are explained here.

Add_trace Select this to produce one or more traces by entering variables or expressions with variables. For instance, you can generate a multitrace plot showing the node voltage $v(2)$, the scaled current $i(R)/5$, and the product of the difference voltage $v(1,2)$ times $i(R)$ by typing

$$v(2),\ i(R)/5,\ v(1,2)^*\, i(R)\ \langle \textit{Enter} \rangle$$

Pressing ⟨*F4*⟩ produces a listing of all the available variables. You may also display a constant, such as 3.142, or a function of the horizontal variable, such as 4* *Is*.

If your circuit file generates two or more simulations, then entering a variable name such as $v(2)$ produces traces for *all* cases. A *single* trace is obtained by appending @ and the case number to the name of the variable. Thus, the entry $v(2)$@2 generates a single trace for the second case alone.

Remove_trace Select this to remove one or more traces from the display.

Plot_control Select this to add or remove plotting areas or plotting symbols. Adding split-screen plotting areas reduces the vertical size of existing plots because new plots appear across the top, as previously illustrated in Fig. B.5. The current active plot is indicated on the screen by SEL⟩⟩.

X_axis and Y_axis Select either of these to invoke submenus controlling the horizontal or vertical axis. You may switch between linear or logarithmic display, or set a new range for the axis, or return to automatic ranging.

The **X_axis** submenu also includes the **X_variable** option that allows you to enter a new horizontal plotting variable. The new variable is specified in the same manner as for **Add_trace**, and it may involve an expression. For instance, when a dc sweep of a source current makes *Is* the "normal" horizontal variable, you could change it to the product $v(2)$***Is*. If the circuit file generates multiple cases, then the horizontal variable must be from just *one* case, such as *Is*@2.

Cursor Select this to obtain movable cursors that provide accurate readouts of points on traces. Cursor C1 is moved along a trace using the arrow keys, and it is jumped to the next trace by pressing ⟨*Ctrl*⟩ and an arrow key. C2 is moved and jumped in the same fashion by holding down ⟨*Shift*⟩. A display box gives the readouts of the cursors' horizontal and vertical coordinates.

Hard_copy Select this followed by **1_page_long** to obtain a printed copy of the display for documentation. You can then modify the display and print other hard copies.

Exit Select this to end the Probe session.

B.2 RESISTIVE CIRCUITS

This section covers additional PSpice statements and techniques for analyzing resistive circuits, including those with controlled sources. Many of these statements and techniques are also applicable in other simulation problems.

Although we continue to use dc sources, the terminology may be somewhat misleading. For if a circuit contains only sources and resistors, then the "dc" source values could actually represent *instantaneous* values of time-varying sources.

Controlled Sources and Op-Amps

PSpice provides device statements for all four types of controlled sources: VCVS, VCCS, CCCS, and CCVS. Additionally, the VCVS can be used to model a real op-amp or to approximate an ideal op-amp.

The device statements for voltage-controlled sources are similar to those for independent sources, with the additional specification of the control voltage. The control voltage must be a difference voltage such as $v(nc^+, nc^-)$. However, you can let $nc^- = 0$ for the node voltage $v(nc^+)$, or you can let $nc^+ = 0$ for $-v(nc^-)$. Using optional parentheses for clarity, the statement syntax is

```
[E|G]name n+ n- (nc+, nc-) value
```

where

Ename denotes a VCVS with $\mu = value$

Gname denotes a VCCS with $g_m = value$

Accordingly, the source voltage across *Ename* is $value \times v(nc^+, nc^-)$ while the source current through *Gname* is $value \times v(nc^+, nc^-)$. As before, the reference current direction goes from n^+ to n^-.

The specification for a current-controlled source requires that the control current pass through an independent voltage source, say *Vcname,* to sense the control current i(*Vcname*). Consequently, *Vcname* may need to be an added zero-voltage source. Again using optional parentheses, the statement syntax is

```
[F|H]name n+ n- (Vcname) value
```

where

Fname denotes a CCCS with $\beta = value$

Hname denotes a CCVS with $r_m = value$

Accordingly, the source current through *Fname* is $value \times i$(*Vcname*), while the source voltage across *Hname* is $value \times i$(*Vcname*).

By way of example, Fig. B.7 diagrams a rather unlikely circuit fragment containing all four types of controlled sources. The corresponding device statements are

```
E12 1 2 (2, 3) 50    ; VCVS = 50 * v(2,3)
H20 2 0 (V43)  4k    ; CCVS = 4 kilohms * i(V43)
R23 2 3        1k
G24 2 4 (4, 0) 0.2m ; VCCS = 0.2 mS * v(4)
V43 4 3 dc     0     ; senses i(V43)
F40 4 0 (V43)  20    ; CCCS = 20 * i(V43)
```

Operational amplifiers can be simulated via the model in Fig. B.8, which includes the internal op-amp resistances r_i and r_o along with a VCVS having

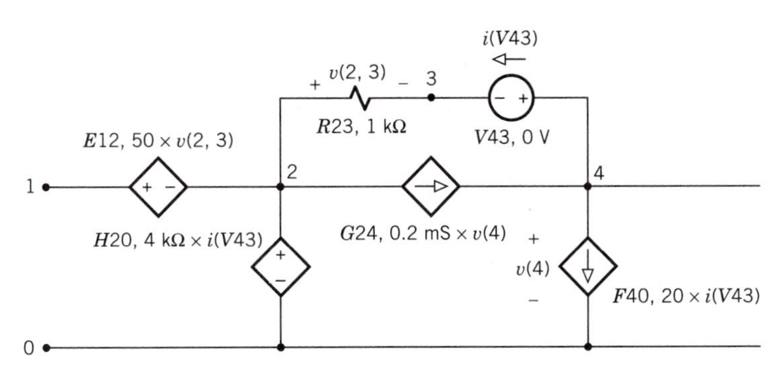

Figure B.7 Diagram with four controlled sources.

$\mu = A$. Typical parameter values for a real op-amp are $r_i \approx 1\ \text{M}\Omega$, $r_o \approx 100\ \Omega$, and $A = 10^5$. For an ideal op-amp, the input resistance r_i can be replaced by an open circuit and the output resistance r_o can be replaced by a short circuit. However, the infinite gain must be approximated as a large but finite value such as $A = 10^9$.

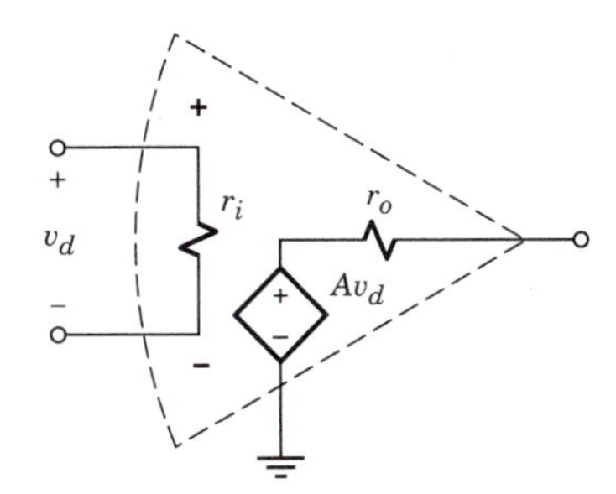

Figure B.8 Op-amp model.

Transfer Function Analysis

PSpice has a transfer-function control statement tailor made for the analysis of resistive amplifier circuits and other circuits with controlled sources. The general form is

```
.tf outvar [V|I]name
```

The output variable *outvar* must be a node voltage, a difference voltage, or the current through an independent voltage source. For meaningful results, [*V*|*I*]*name* should be the only input source, and its value should be constant.

The `.tf` command causes PSpice to calculate the following three quantities:

- The *gain,* defined as the value of *outvar* divided by the value of the specified source [*V*|*I*]*name.*

- The *input resistance* seen by [V|I]*name* with any other independent sources suppressed. (This value can also be calculated from the values of the input voltage and current.)
- The *output resistance,* defined as the Thévenin resistance between the nodes associated with *outvar.*

When *outvar* is the voltage at the terminals of a source network, the gain and output resistance can be used to obtain the network's Thévenin or Norton model.

Example B.4 *Thévenin Model of an Inverting Amplifier*

Figure B.9*a* is a simple inverting amplifier built with a nonideal op-amp. We'll treat the output terminal and ground as the terminals of a source network to find the Thévenin model with any input voltage.

Figure B.9*b* gives the simulation diagram assuming the op-amp has r_i = 1 MΩ, r_o = 100 Ω, and $A = 10^5$. The ground point has been taken as the reference node, and v_{in} is arbitrarily set at 1 V. The corresponding circuit file is as follows:

```
Example B.4, Inverting Amplifier
Vin 1 0 dc      1
R1   1 2        8k
RF   2 4        40k
* Op-amp model with Vd = v(0, 2)
ri   2 0        1Meg
ro   3 4        100
Eoa 3 0 (0, 2) 1E5
* Vout = v(4)
.tf v(4) Vin
.end
```

Simulation results from the `.tf` command are

```
V(4)/Vin = -5.000E+00
```

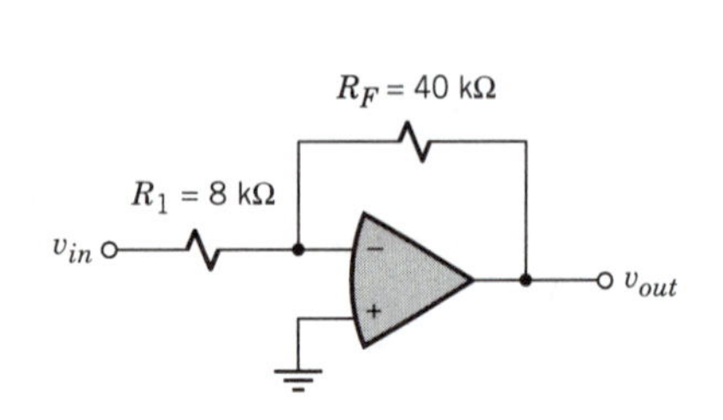

(*a*) Inverting amplifier

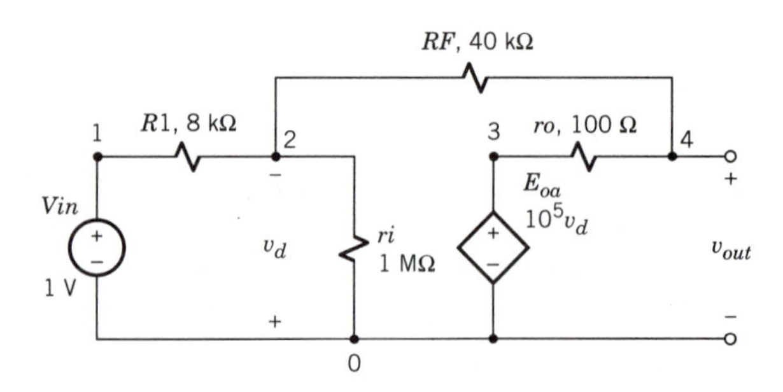

(*b*) Simulation diagram

Figure B.9

```
INPUT RESISTANCE AT Vin = 8.000E+03
OUTPUT RESISTANCE AT V(4) = 6.040E-03
```

The output resistance equals the Thévenin resistance, so $R_t = 6.040$ mΩ. And since $v_{out}/v_{in} = -5.000$, we conclude from proportionality that $v_{out-oc} = -5v_{in}$ for any input voltage.

Exercise B.4

Write the device and control statements needed to find the Thévenin parameters when the source network in Fig. B.10 has $r_m = 10$ kΩ.

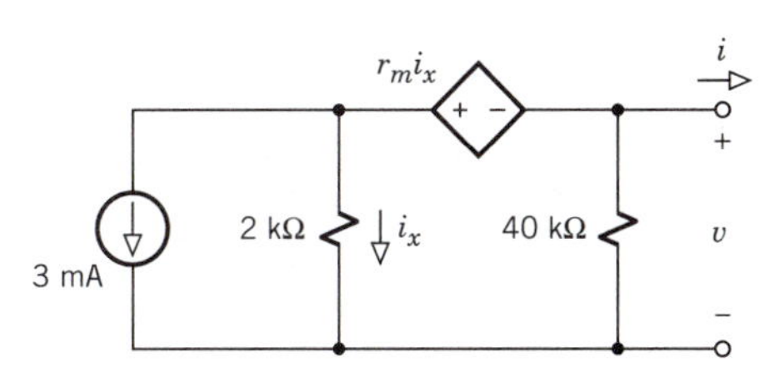

Figure B.10

Varying Parameters

PSpice becomes particularly helpful when you want to see what happens when some circuit value changes. We've already discussed how to vary source values. Now we'll consider varying two kinds of parameter values, namely, resistance and controlled-source gain.

Stepping Resistance A convenient way of varying resistance involves the parameter and step statements. A **parameter** is a global constant whose value is defined by a control statement having the format

```
.param paramname = nomval
```

This statement defines the nominal value for a parameter named *paramname,* which may or may not be the same as a device name. The parameter value is invoked in a device statement by replacing *value* with either {*paramname*} or an expression involving *paramname.*

A **step** statement can then be used to override *nomval* and vary the parameter. Two forms of this control statement are

```
.step param     paramname list val1 . . .
.step lin param paramname start, stop, step
```

The `list` and `lin` operations are just like those of the `.dc` statements. Thus, for instance, a resistor named *Rvar* can be stepped through the values 2, 4, and 8 Ω by the statements

```
Rvar    n+ n-  {Rvar}
.param        Rvar = 2 ; nominal value of Rvar is 2 ohms
.step   param Rvar list  2 4 8 ; Rvar = 2, 4, 8
```

The simulation in which this appears would be repeated three times, once for each value of *Rvar.*

The values of two or more parameters may also be varied in relationship by using different expressions with the same *paramname.* Suppose, for example, that a 10-Ω potentiometer is represented by resistors $R12$ and $R23$. You could step $R12$ through the values 1, 3, 5, 7, and 9 Ω while keeping the potentiometer resistance fixed at $R12 + R23 = 10\ \Omega$ via the statements

```
R12     1 2 {R12}
R23     2 3 {10 − R12}; R23 = 10 − R12
.param       R12 = 2
.step lin param R12 1, 9, 2 ; R12 = 1, 3, . . . , 9
```

Because zero resistance is not allowed, you should avoid setting $R12 = 0$ or setting $R12 = 10\ \Omega$ so $R23 = 0$.

Only one `.step` statement is allowed in a circuit file. Furthermore, `.param` and `.step` cannot be used to vary the value of a controlled source.

Example B.5 *Analysis of a Transistor Amplifier*

Figure B.11 gives the PSpice model of a transistor amplifier with a CCCS controlled by *Ib.* We want to find the voltage gain, input resistance, and output resistance when $Re = 0.2$, 0.5, 1, and 2 kΩ. For that purpose, we arbitrarily take $Vin = 1$ V and use the following circuit file.

```
Example B.5
* Transfer-function analysis with stepped parameter Re
```

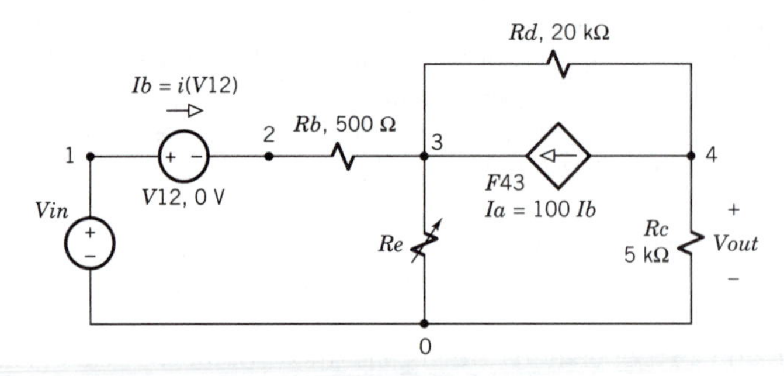

Figure B.11 PSpice model of a transistor amplifier.

```
Vin 1 0 dc     1
V12 1 2 dc     0 ; senses i(V12) = Ib
Rb  2 3        500
Re  3 0        {Re}
F43 4 3 (V12) 100 ; CCCS controlled by Ib
Rd  3 4        20k
Rc  4 0        5k
.param            Re = 1k
.step param Re list .2k .5k 1k 2k ; four values of Re
* Transfer-function analysis for each Re, Vout = v(4)
.tf v(4) Vin
.end
```

The simulation results are tabulated as follows. Scanning across each line reveals that the gain and input resistance depend significantly upon the value of *Re*, while the output resistance stays nearly constant.

PARAM RE	200	500	1.00E+03	2.00E+03
V(4)/Vin	−2.39E+01	−9.75E+00	−4.90E+00	−2.45E+00
INPUT RESISTANCE AT Vin	1.65E+04	4.02E+04	7.38E+04	1.50E+05
OUTPUT RESISTANCE AT V(4)	4.95E+03	4.97E+03	4.98E+03	4.98E+03

Exercise B.5

Let $R12a$ and $R12b$ be in parallel. Write the device and control statements to keep $R12a \| R12b$ constant at 4 kΩ while $R12b$ varies over 5, 6, . . . , 20 kΩ. Hint: Use Eq. (2.10) for the "reverse parallel" formula.

Sweeping Resistance The `.step` statement does not generate a "swept" parameter suitable for Probe plots. Instead, to get a graphical display with varying resistance, you must sweep the resistance using a `.dc` statement with a parameter. The syntax for the device and control statements are as follows:

```
Rname n+ n- {paramname}
.param paramname = nomval
.dc lin param paramname start, stop, step
.probe
```

Probe then takes *paramname* as the X-axis plotting variable.

If a large resistance range is under consideration, then a **logarithmic sweep** may be advantageous. The sweep statement form is

```
.dc dec param paramname start, stop, npoints
```

This statement generates *npoints* per *decade*, which is an increase by a factor of 10. The X-axis of Probe can then be set for a logarithmic display.

When sweeping resistance, either linearly or logarithmically, there should be no other sweep statements in the circuit file. If a `.tf` statement is included, it will be executed before the sweep using only the nominal value of the resistance.

Varying Controlled Sources PSpice does not have direct provision for varying the gain parameter of a controlled source. However, you can accomplish the same end with the help of an auxiliary controlled current source and a stepped or swept resistor arranged such that the new control variable becomes the voltage across the variable resistance.

To elucidate this strategy, Fig. B.12*a* shows part of a circuit containing a VCVS whose gain μ is to be varied. The corresponding simulation diagram in Fig. B.12*b* includes an auxiliary VCCS generating $i_x = v_c/1\ \Omega$. This source and the variable resistor are connected between the reference node and an auxiliary node to produce $v_x = R_{var}i_x$. The control relation for the original controlled source has been changed so that

$$v_a = v_x = R_{var}i_x = R_{var}v_c/1\ \Omega = \mu v_c$$

where

$$\mu = R_{var}/1\ \Omega$$

Hence, varying R_{var} is equivalent to varying μ. And since no current goes from node x to the original circuit, the two auxiliary elements have no other effect on the simulation.

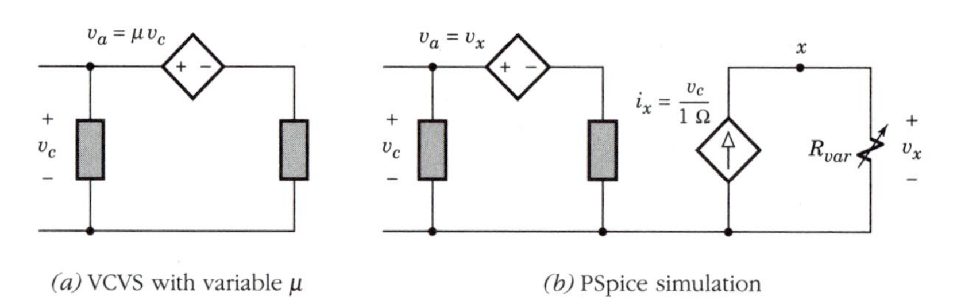

(*a*) VCVS with variable μ

(*b*) PSpice simulation

Figure B.12

Example B.6 ***Sweeping a CCCS***

Figure B.13 repeats the simulation diagram for the transistor amplifier from Example B.5. But this time we'll analyze the performance while sweeping the gain of the CCCS from 10 to 200, keeping *Re* fixed at 1 kΩ.

To sweep the gain β, two modifications have been made to the diagram:

1. An auxiliary CCCS named *Faux* and a variable resistor *Rvar* have been added to produce the new control variable $Vx = Rvar \times Ix$ with $Ix = Ib$;

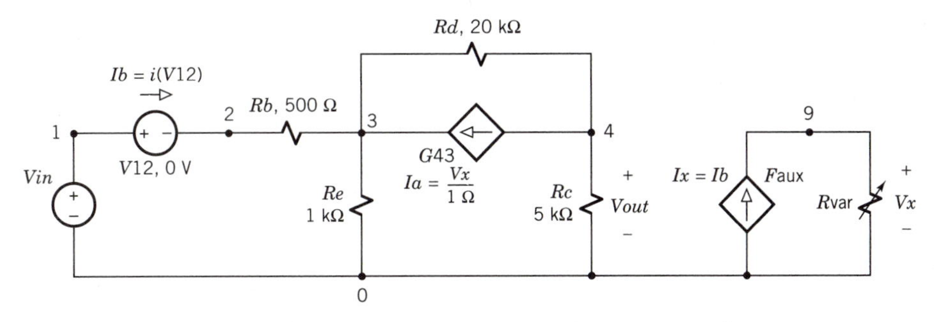

Figure B.13 A simulation diagram for Example B.6.

2. The original CCCS has been changed to a VCCS, named $G43$, such that

$$Ia = Vx/1\ \Omega = Rvar(Ix/1\ \Omega) = (Rvar/1\ \Omega)Ib = \beta \times Ib$$

where

$$\beta = Rvar/1\ \Omega$$

Thus, in the following circuit file, we use *Beta* as the parameter name for *Rvar* so Probe's *X*-axis will be labeled *Beta*.

```
Example B.5, Sweeping a CCCS
Vin    1 0  dc      1
V12    1 2  dc      0    ; senses i(V12) = Ib
Rb     2 3          500
Re     3 0          1k
G43    4 3  (9, 0)  1    ; Ia = v(9)
Rd     3 4          20k
Rc     4 0          5k
Faux   0 9  (V12)   1    ; Ix = Ib
Rvar   9 0     {Beta}    ; v(9) = Beta*Ib
.param Beta = 10
* Sweep Beta from 10 to 200 in steps of 5
.dc lin param Beta 10, 200, 5
.probe
.end
```

Figure B.14 reproduces the results of our simulation. The upper trace shows the voltage gain $Vout/Vin = v(4)/v(1)$, which approaches -5 as *Beta* increases. The lower trace shows the input resistance $Vin/Iin = v(1)/i(V12)$, which increases linearly with *Beta*.

Exercise B.6

Consider the op-amp model in Fig. B.9*b*. Write the simulation statements needed to sweep the gain A logarithmically from 10^2 to 10^6, with a total of 80 points.

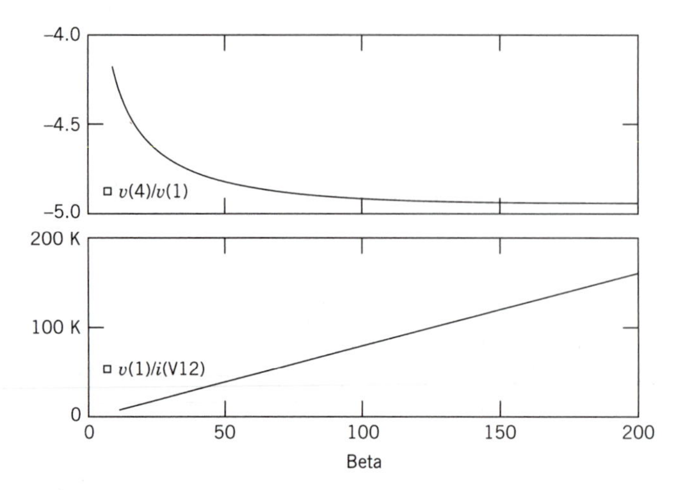

Figure B.14 A Probe display of voltage gain and input resistance versus Beta.

Subcircuits

When a circuit contains two or more building-block networks having the same structure, the circuit file may be simplified by defining a *subcircuit* for the repeated network. This technique is handy, for example, when analyzing circuits with several op-amps.

The definition of a subcircuit named *subcktname* requires three or more statements, as follows:

```
.subckt subcktname nx1 . . . ⟨params: param1 = value . . .⟩
(device statements for the subcircuit)
.ends ⟨subcktname⟩
```

Nodes *nx1*, . . . identify the external terminals of the subcircuit, excluding the reference node which is global. Using *text names* for these nodes helps prevent confusion. The device statements describe the elements within the subcircuit. The `.ends` statement concludes the subcircuit description, and it may include *subcktname* for identification.

The `params` option makes it possible to alter some parameter values in the calling statement. However, the gains of any controlled sources must have numerical values assigned by device statements in the subcircuit. Separate parameter statements are not allowed in subcircuits, nor are any other control statements.

A subcircuit is inserted into the main circuit by the calling statement

```
Xname na1 . . . subcktname ⟨params: param1 = newvalue . . . ⟩
```

Nodes *na1*, . . . , are the circuit nodes to which the respective subcircuit terminals are attached. A separate calling statement is required for each insertion of the subcircuit.

Example B.7 *Circuit with Two Op-Amps*

Figure B.15*a* diagrams a circuit with two op-amps, which are represented by the subcircuit model in Fig. B.15*b*. The resistance values for op-amp #1 are ri = 500 kΩ and ro = 100 Ω while op-amp #2 has ri = 2 MΩ and ro = 500 Ω. If the subcircuit model is defined using the same parameters as op-amp #1, then the relevant PSpice statements could read as follows:

```
* Calling statement for op-amp #1
Xoa1 1 2 3 opamp ; same parameters as subcircuit
* Calling statement for op-amp #2
Xoa2 0 4 5 opamp params: ri = 2 Meg ro = 500
* Subcircuit for op-amp model
.subckt opamp pos neg out
+ params: ri = 500k ro = 100
ri pos neg {ri}
Eoa int 0 (pos,neg) 1E5
ro int out {ro}
.ends opamp
```

Note that `params` is not needed in the first calling statement because op-amp #1 has the same values as the subcircuit. (All `params` could be omitted if both op-amps had the same values as the subcircuit.) Also note in the second calling statement that the first node number is 0 because op-amp #2 has a grounded noninverting terminal. The node named *int* is internal to the subcircuit and does not appear in either calling statement.

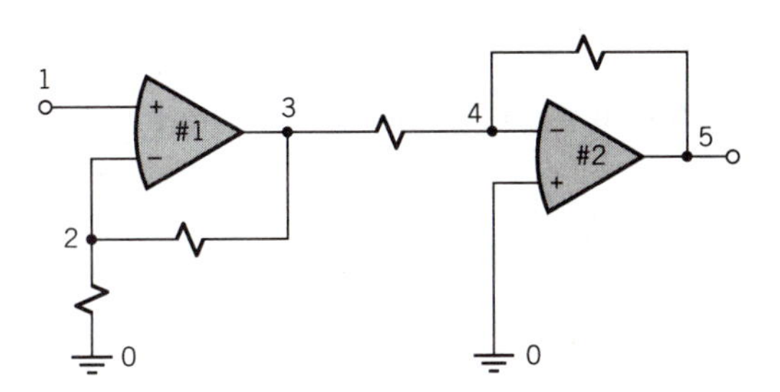

(a) Circuit with two op-amps

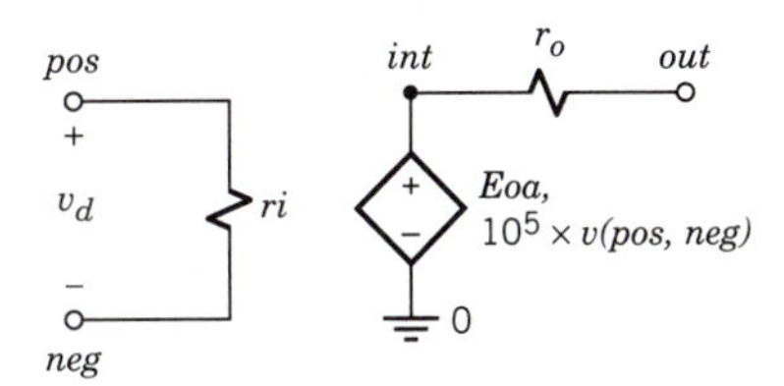

(b) Subcircuit model for op-amps

Figure B.15

Exercise B.7

Consider the op-amp circuit in Fig. 4.45 (p. 180). Write a complete circuit file using `.subckt` and `.step` to find v_{out} for $K_2 = 0.5$ and 1.5 when $v_{in} = 1$ V, $R_1 = R_2 = R_F = 1$ kΩ, and both op-amps have $r_i = 1$ MΩ, $r_o = 0$, and $A = 10^6$.

B.3 AC CIRCUITS

This section presents the methods for PSpice analysis of circuits operating under ac steady-state conditions. Also included are the statements and techniques for simulating capacitors, inductors, frequency sweep, three-phase circuits, ideal transformers, and mutual inductance.

AC Circuit Files

PSpice simulation of ac circuits differs from resistive circuits in three major respects:

1. AC circuits generally include capacitors and/or inductors.
2. AC sources are specified in terms of their phasor parameters as well as frequency.
3. The output variables are phasors.

Although all ac voltages and currents vary sinusoidally, PSpice simulates the corresponding phasors rather than the time-varying waveforms.

Capacitors and Inductors Resistors in ac circuits are described by the same device statements used for dc circuits. The device statements for capacitors and inductors follow the same pattern, namely,

```
[C|L]name n+ n- value
```

You may also replace *value* with an expression or a parameter name and use a `.step` statement to vary capacitance or inductance. However, any `.dc` sweep is executed before the ac simulation and has no effect on the ac results.

When any circuit contains capacitors or inductors, PSpice imposes the following requirements:

- A dc path must exist from the reference node to every other node when all sources are suppressed.
- A closed path consisting entirely of voltage sources and/or inductors is not allowed.
- A node or supernode whose impinging branches consist entirely of current sources and/or capacitors is not allowed.

To satisfy these requirements, you may need to put a very large additional resistor in parallel with a capacitor or a very small additional resistor in series with an inductor.

Sometimes you know the ac impedances rather than element values. You must then convert any reactance X into the corresponding series inductance or capacitance via $L = X/\omega$ for $X > 0$ or $C = 1/\omega|X|$ for $X < 0$.

AC Sources and Print Statements The device statements for ac sources have the general form

```
[V|I]name n+ n- ac amplitude ⟨phase⟩
```

The *phase* value is expressed in *degrees,* and the default is 0° when the value is omitted. The *amplitude* may be either the peak or rms value, as long as all sources are specified the same way.

All ac sources in a given circuit file are at the same frequency, which must be expressed in *hertz* rather than rad/s. The frequency or frequency sweep is specified by the separate statement

```
.ac lin npoints, fstart, fstop
```

where *npoints* is the number of sample frequencies in the range *fstart* to *fstop.* (Note that the syntax differs from a `.dc` sweep statement in that *npoints* appears before the start and stop values.) For single-frequency analysis, use *npoints* = 1 and *fstart* = *fstop.* To obtain a particular frequency *step* in a linear sweep, take *npoints* = 1 + (*fstart* − *fstop*)/*step.*

The output variables in an ac simulation are phasor voltages or currents, and we'll emphasize that fact by using capital letters. The print control statement is

```
.print ac var1 var2 . . .
```

where *var1 var2* . . . are the polar or rectangular components of phasor quantities. The desired forms are specified as follows:

Amplitude − `[V|I]M(. . .)`	Phase (°) − `[V|I]P(. . .)`
Real part − `[V|I]R(. . .)`	Imaginary part − `[V|I]I(. . .)`

Thus, for instance, *VM*(3) means the amplitude of the voltage at node 3, while *IP*($C12$) means the phase of the current through capacitor $C12$. The same notation is used to specify traces in Probe when you sweep the frequency. If rms values have been used for source amplitudes, then the amplitude, real part, and imaginary part of any output are also rms values.

Example B.8 ***Single-Frequency Analysis***

For the circuit in Fig. B.16, you want to find i_{in} and v_3 given that $v_{in} = 90 \cos 5000t$ V. You also want to find the input impedance at the source frequency.

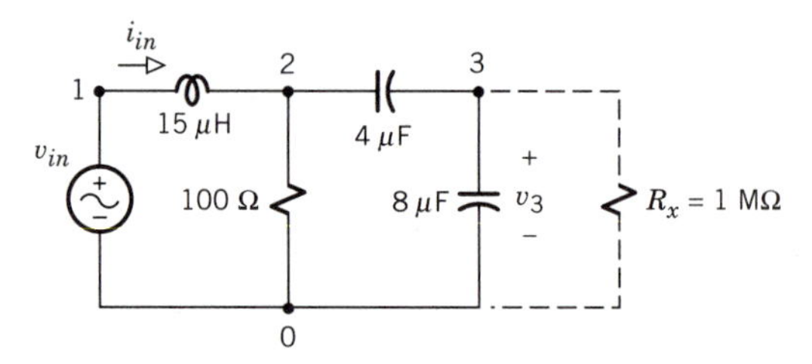

Figure B.16 A circuit for Example B.8.

But the conditions at node 3 violate two PSpice requirements, because that node does not have a dc path to the reference node and the impinging branches consist entirely of capacitors. You therefore add a resistor R_x as shown, and you take $R_x = 1\ \text{M}\Omega$ so its impedance is very large compared to the impedance of the parallel 8-μF capacitor, which has $|Z| = 1/\omega C = 25\ \Omega$. Finally, noting that the source frequency in hertz is $f = 5000/2\pi = 795.8$, your circuit file is

```
Example B.8
Vin  1 0  ac  90 ; phase equals 0 by default
L12  1 2      15m
R20  2 0      100
C23  2 3      4u
C30  3 0      8u
Rx   3 0      1Meg
.ac lin 1, 795.8, 795.8 ; single frequency, omega = 5000
.print ac IM(L12) IP(L12) VM(3) VP(3) ; Iin = I(L12)
.end
```

PSpice produces the following results in the output file:

```
****************************************************************
FREQ          IM(L12)        IP(L12)        VM(3)          VP(3)
7.958E+02     2.000E+00     -3.687E+01     4.000E+01     -9.000E+01
****************************************************************
```

Thus, $\underline{I}_{in} = 2\ \text{A}\ \angle{-36.9°}$ and $\underline{V}_3 = 40\ \text{V}\ \angle{-90°}$, so the waveforms are $i_{in} = 2\cos(5000t - 36.9°)$ and $v_3 = 40\cos(5000t - 90°)$. Finally, you calculate the input impedance by hand via $Z_{in} = \underline{V}_{in}/\underline{I}_{in} = 45\ \Omega\ \angle 36.9° = 36 + j27\ \Omega$.

Example B.9 ***Frequency-Sweep Analysis***

Resonance effects in Fig. B.17 are to be investigated using frequency sweep and Probe. This circuit differs from standard parallel resonance in that R is in series with L. However, the resonant frequency is still expected to be around $1/2\pi\sqrt{LC} \approx 5$ kHz. The source current phasor is specified as $\underline{I}_s = 1\ \text{A}\ \angle 0°$ to facilitate displays of impedance quantities. The circuit file is as follows:

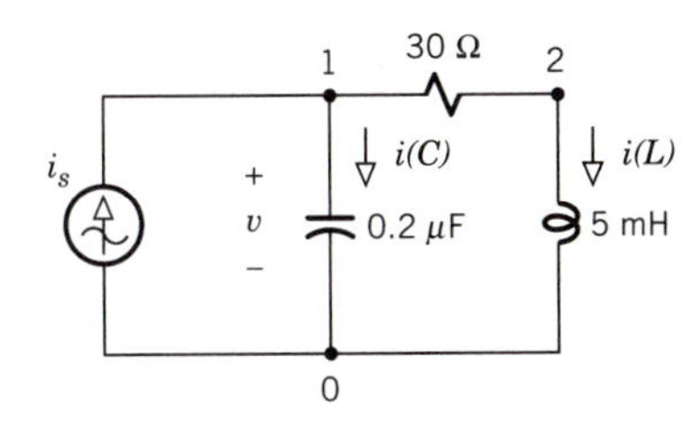

Figure B.17 Circuit for Example B.9.

```
Example B.8
Is   0 1   ac   1     ; so Z = V(1)/Is = V(1)
C    1 0        .2u
R    1 2        30
L    2 0        5m
.ac lin 200, 3k, 7k ; sweep 200 points, 3 kHz to 7 kHz
.probe
.end
```

Several resulting Probe plots are reproduced in Fig. B.18. Note that a trace of the constant 0 has been added in the lower plot of Fig. B.18*a* to clearly show where $\angle Z$ changes sign. Numerical values obtained using Probe's cursor are given in Table B.2. These results indicate that the resonant condition is ambiguous because the frequency at which $\angle Z = 0$ does not equal the frequency at which $|Z|$ is maximum. The quality factor is also ambiguous because $|\underline{I}_C|_{max} \neq |\underline{I}_L|_{max}$.

TABLE B.2

Quantity	$\|Z\|_{max}$	$\angle Z$	$\text{Re}[Z]_{max}$	$\|\underline{I}_C\|_{max}$	$\|\underline{I}_L\|_{max}$
Value	848	0°	841	5.38	5.29
Frequency	5020	4941	4980	5061	4980

Exercise B.8

Consider the circuit in Fig. B.19 with

$$i = \sin \omega t \text{ A} \qquad v = 30 \cos(\omega t - 45°) \text{ V}$$

Write a circuit file to find the polar components of $\underline{I}_1$ when $\omega = 10$, 30, and 50 rad/s.

Three-Phase Circuits

PSpice easily handles any balanced or unbalanced three-phase circuit driven by a wye-connected source. The source is described using three ac source

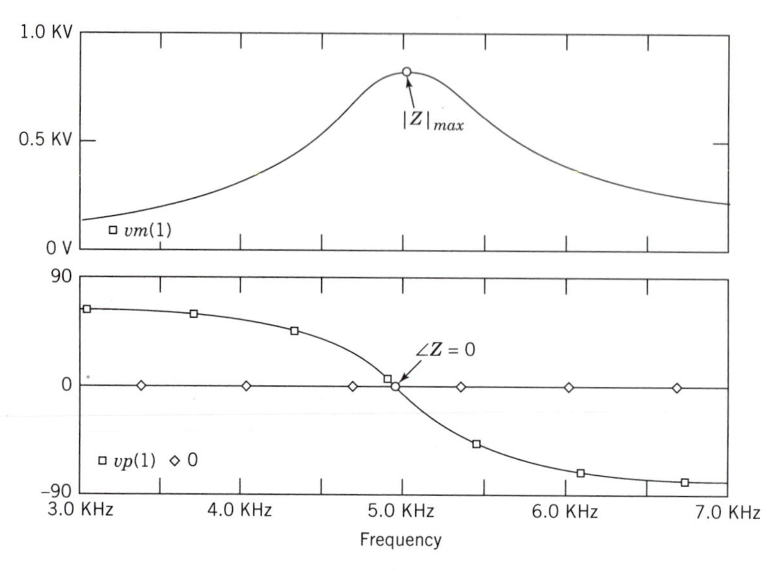

(a) Magnitude and angle of input impedance

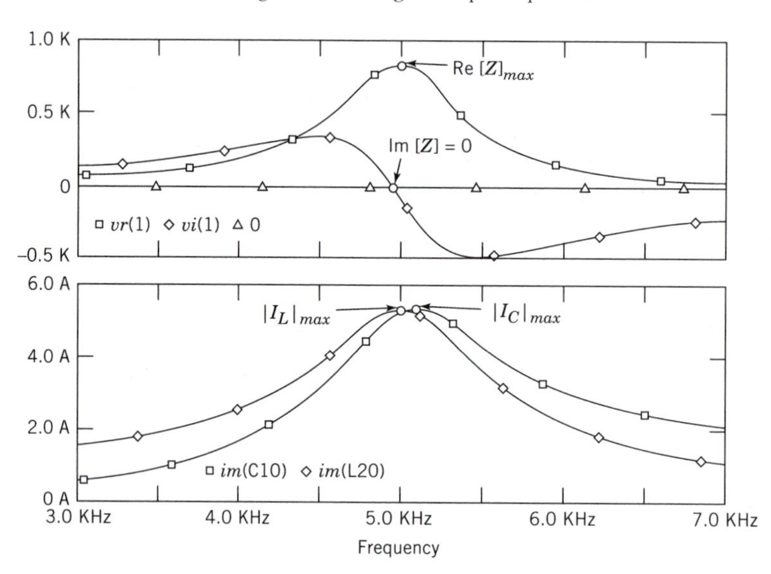

(b) Input impedance and current amplitudes

Figure B.18 Swept-frequency Probe displays.

statements with appropriate phase angles (e.g., 0°, −120°, and −240°). The neutral point at the source is a convenient reference node, and the load may have a separate neutral point if wye connected.

Ideal delta-connected sources cannot be simulated because they have a closed path consisting entirely of voltage sources. Such sources can be represented instead by an equivalent wye connection, or small resistances may be inserted in series with the delta sources to satisfy the PSpice requirement.

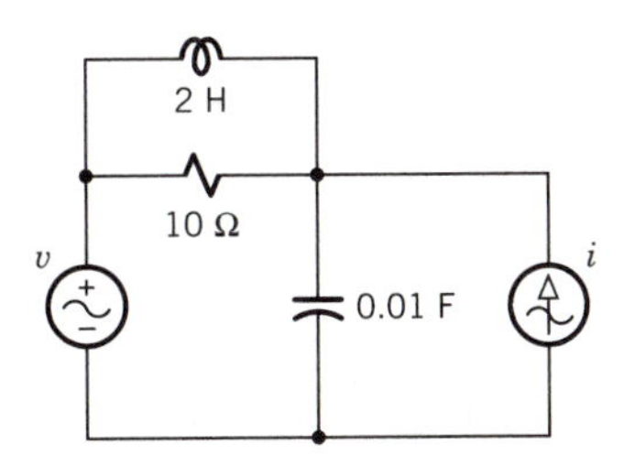

Figure B.19

Exercise B.9

Write the device statements to describe a three-phase circuit having the following specifications: wye-connected generator with *a-b-c* phase sequence, $\underline{V}_a = 300$ V(rms) $\angle 0°$, and $f = 60$ Hz; delta-connected balanced load with branch impedances of $10 + j20\ \Omega$.

Transformers and Mutual Inductance

Ideal transformers are modeled in PSpice using controlled sources, as discussed in Section 8.1. Both a VCVS and a CCCS are required, and an independent voltage source must be included to sense the control current. For instance, the ideal transformer in Fig. B.20*a* could be simulated by Fig. B.20*b*. Note that the polarities of the controlled sources must agree with the dot markings.

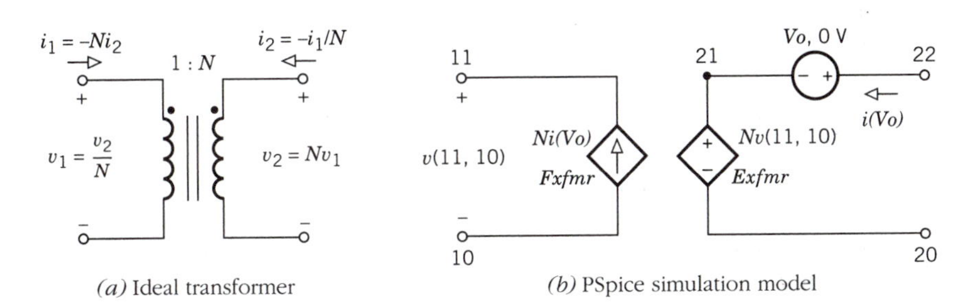

Figure B.20

Mutual inductance can be represented by any of the equivalent inductor networks derived in Section 8.3, provided that the equivalent network suits the circuit in question. Alternatively, PSpice models magnetically coupled inductors directly in terms of the **coupling coefficient**

$$k = M/(L_1 L_2)^{1/2} < 1$$

Here, L_1 and L_2 are the associated self-inductances defined by separate device statements. The statement syntax for inductor coupling is

```
Kname Lname1 Lname2 value
```

The n^+ nodes of L_1 and L_2 must correspond to the leads marked with dots.

Since ideal transformers and coupled inductors often lack direct electrical connections, care must be taken to ensure that all nodes have a dc path to the reference node. If necessary, a very large additional resistance may be included for this purpose. Circuits with coupled inductors may also need an additional small resistance to eliminate a closed path of voltage sources and inductors.

Example B.10 *Inductor Coupling*

Consider the simple circuit with coupled inductors in Fig. B.21. The large resistor R_x has been added to provide the required dc path to nodes 2 and 3.

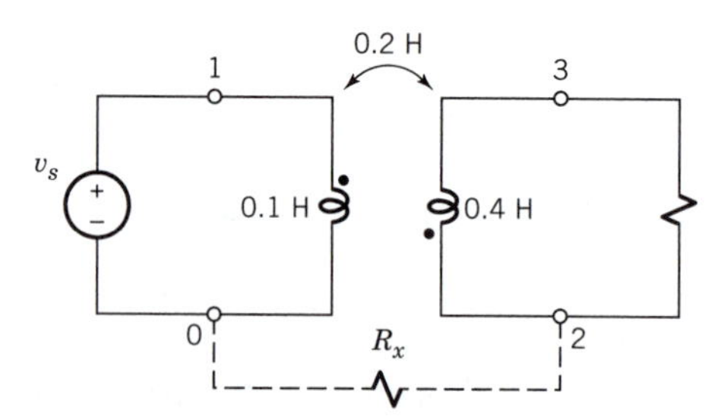

Figure B.21 Circuit for Example B.10.

From the given element values, the coupling coefficient is $k = 0.2/(0.1 \times 0.4)^{1/2} = 1$. However, PSpice requires $k < 1$, so we might take $k = 0.999$. The corresponding device statements are

```
L10  1 0   0.1          ; dot at node 1
L23  2 3   0.4          ; dot at node 2
K12  L10   L23   0.999  ; approximates k = 1
```

Exercise B.10

Write device statements for the ideal transformer in Fig. B.20*a* with $N = 25$ and the following circuitry: A source network connected to the primary has $R_s = 2\ \Omega$ and $\underline{V}_s = 10\text{ V}\ \underline{/0^\circ}$ at $\omega = 10^5$; a 1-kΩ load resistor is across the secondary. Use the current through v_s as the control current for i_2.

B.4 TRANSIENTS

This section presents the methods for PSpice analysis of transient behavior, including initial conditions, time-varying sources, and switches. The use of

Probe is assumed throughout, because graphical pictures of transient waveforms are the most valuable and informative simulation results.

Transient Circuit Files

PSpice simulation of transients involves three additional considerations:

1. Initial conditions of capacitors and inductors
2. Control statements for transient analysis
3. Description of any time-varying input sources or switches.

We'll begin here with the first two of these.

The initial-condition voltage across a capacitor or current through an inductor may be incorporated in the device statement as follows:

```
[C|L]name n+ n- value IC = value
```

The node numbers n^+ and n^- establish the polarity of the initial capacitor voltage or inductor current.

PSpice transient simulation always starts at $t = 0$ and ends at the user-specified time *tfinal.* The control statement syntax is

```
.tran tstep tfinal ⟨td ⟨maxstep⟩⟩ ⟨uic⟩
```

where *tstep* is the time step for any printed output. For Probe output without printing, you should take *tstep* = *tfinal.*

Specifying *td* suppress printed or Probe output prior to $t =$ *td*, although the simulation still starts at $t = 0$. The default is *td* = 0.

Specifying *maxstep* sets an upper limit on the internal time increment used for PSpice calculations. The default value is the smaller of *tstep* or *tfinal*/50. In most cases, you should let PSpice determine the time increment since a small value of *maxstep* significantly increases computation time. However, a smooth plot of an oscillatory waveform at frequency f requires

$$maxstep \leq 0.05/f$$

If *maxstep* is specified, then a value must be included for *td* even if *td* = 0.

The option `uic` (use initial conditions) invokes the IC values given in capacitor and inductor statements. When `uic` is included, any unspecified initial value defaults to zero. When `uic` is not included, any stated IC values are ignored and PSpice calculates all initial conditions via dc steady-state analysis at $t = 0$. Consequently, to be on the safe side, you should *always* specify IC values and include `uic` in the `.tran` statement.

Example B.11 ***Second-order Step Response***

Suppose you want to simulate the step response of v_C in Fig. B.22 when $R_s = 10$, 100, and 1000 Ω. You can do this by taking v_s to be a 1-V dc source

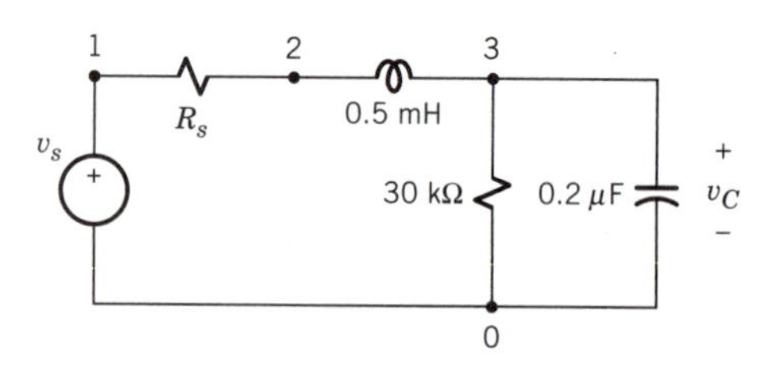

Figure B.22 Circuit for Example B.11.

and specifying zero initial conditions for the inductor and capacitor. Thus, the simulated circuit has $i_L(0) = v_C(0) = 0$ and $v_s = 1$ V for $t > 0$.

Since the transient duration is not easily estimated, you try a few runs with different values of *tfinal* until you get a suitable Probe plot with *tfinal* = 500 μs. Your circuit file then reads

```
Example B.11, Step response
Vs  1 0  dc    1
Rs  1 2       {Rs}
L   2 3       0.5m  IC = 0
R   3 0        30k
C   3 0       0.2u  IC = 0
.param Rs = 100
.step param Rs list 10 100 1k
.tran 500u 500u uic
.probe
.end
```

The resulting Probe plot of $v(3) = v_C$ in Fig. B.23a reveals that this second-order circuit is overdamped when $R_s = 1000\ \Omega$, underdamped when $R_s = 10\ \Omega$, and at or near critical damping when $R_s = 100\ \Omega$. The underdamped response exhibits the distinctive overshoot with oscillations at $f \approx 16$ kHz, but the waveform appears to be "kinky" because PSpice's internal time increment is too large.

For a better plot of the underdamped case, you want to have *maxstep* ≤ 0.05/(16 kHz) ≈ 3.13 μs. You therefore rerun the simulation with R_s fixed at 100 Ω and with the `.tran` statement modified to read

```
.tran 200u 200u 0 3u uic
```

so *td* = 0 and *maxstep* = 3 μs. The plot of $v(3) - 1$ in Fig. B.23b corresponds to the natural response alone, because the forced response equals 1 V. You can now estimate the damping coefficient α and oscillation frequency ω_d by using the cursor to measure the value and location of two peaks, as marked on the figure.

You know that the general form of an underdamped second-order natural response is

$$y_N(t) = Ae^{-\alpha t}\cos(\omega_d t + \phi)$$

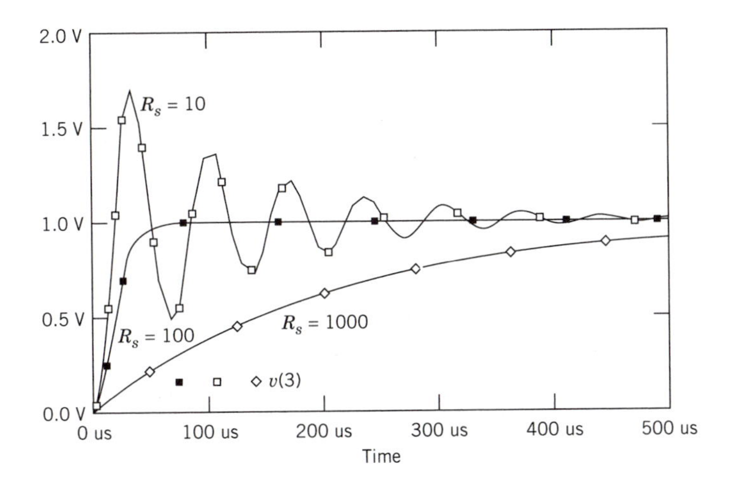

(*a*) With three values of R_S

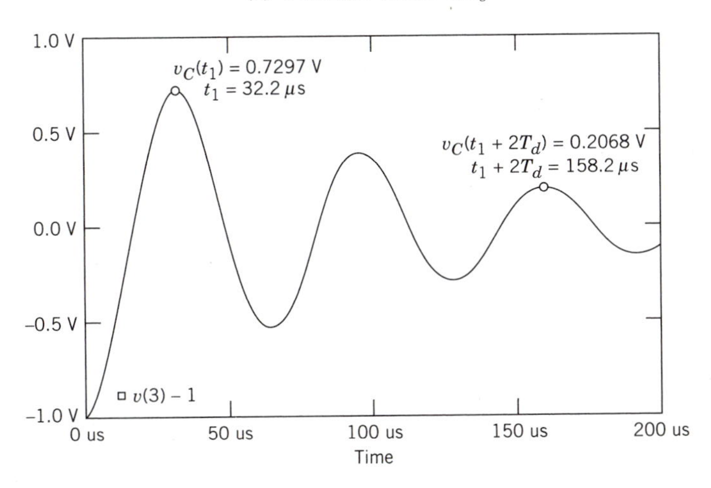

(*b*) Expanded view of underdamped waveform

Figure B.23 Probe display of step response.

which has peaks when $\cos(\omega_d t + \phi) = 1$. The two labeled peaks in Fig. B.23*b* are spaced by $2T_d \approx 126$ μs, so $T_d \approx 63$ μs and $\omega_d = 2\pi/T_d \approx 100{,}000$. To calculate the value of α, you note that the peaks at t_1 and $t_1 + 2T_d$ correspond to

$$v_C(t_1) = Ae^{-\alpha t_1} \qquad v_C(t_1 + 2T_d) = Ae^{-\alpha(t_1+2T_d)}$$

Hence, $v_C(t_1)/v_C(t_1 + 2T_d) = e^{\alpha 2T_d}$ and

$$\alpha = \frac{1}{2T_d} \ln \frac{v_C(t_1)}{v_C(t_1 + 2T_d)} \approx 10{,}000$$

Accordingly, when $R_s = 100\ \Omega$, the roots of the characteristic polynomial of this circuit are $p_1, p_2 = -\alpha + j\omega_d \approx -10{,}000 \pm j100{,}000$.

Exercise B.11

The circuit in Fig. B.24 is in the steady state with $v_s = 10$ V for $t < 0$. Given that $v_s = 0$ for $t > 0$, write a circuit file for transient analysis using Probe with $R_s = 5$ kΩ and 20 kΩ. Take *tfinal* $= 3\tau_{max}$ and specify the IC value by an {*expression*}.

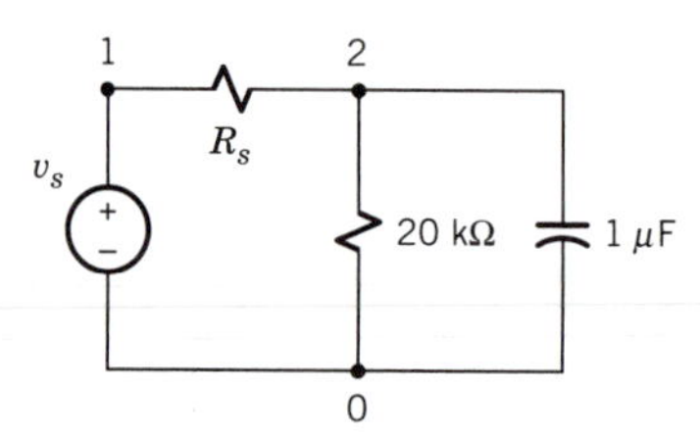

Figure B.24

Time-Varying Sources

PSpice provides several time-varying voltage or current sources for use in transient analysis. The device statements and waveform properties for the most useful ones are detailed here. Except where otherwise indicated, any source parameter may be changed via `.param` and `.step` statements.

Piecewise Linear Source This source produces a waveform consisting of straight-line segments, as illustrated by Fig. B.25. The statement syntax is

```
[V|I]name n+ n- pwl t0,X0 t1,X1 t2,X2 . . .
```

where $0 \leq t0 < t1 < t2$ . . . The values of *t0,X0* . . . , cannot involve parameters.

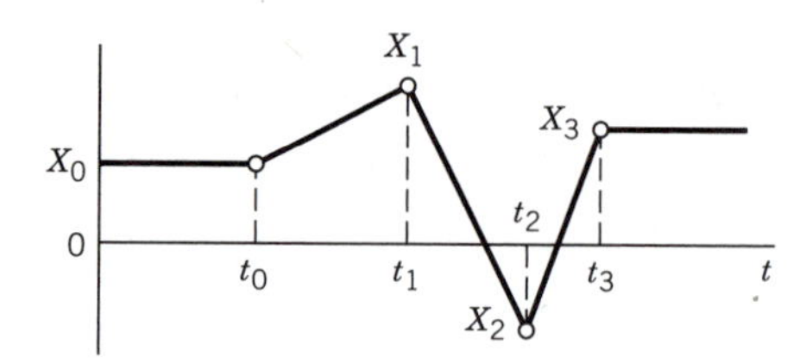

Figure B.25 Waveform of piecewise linear source.

Although a perfect vertical jump is not possible, you can get a brief risetime or falltime by taking two time points close together. Thus, for instance, a voltage step of height 5 V at $t = 0$ might be approximated with 1-μs risetime by writing

```
Vin 1 0 pwl 0,0 1u,5
```

But keep in mind that the shorter the transition time, the longer it takes PSpice to do the simulation.

Pulse Source This source produces a periodic train of pulses, as illustrated by Fig. B.26. The statement syntax is

```
[V|I]name n+ n- pulse Xo Xp ⟨td ⟨tr ⟨tf ⟨pw ⟨per⟩⟩⟩⟩⟩
```

The nested angle brackets for the optional entries mean that *td* must be specified when *tr* is specified, *td* and *tr* must be specified when *tf* is specified, and so forth. Default values are

$$td = 0 \quad tr = tstep \quad tf = tstep \quad pw = tfinal \quad per = tfinal$$

Thus, for instance, a single 5-V pulse starting at $t = 0$ and ending at $t = 8$ ms with 1-ms transition times would be written as

```
Vin 1 0 pulse 0 5 0 1m 1m 6m
```

The same waveform could also be produced by a `pwl` source, but the `pulse` source is more convenient for multiple pulses.

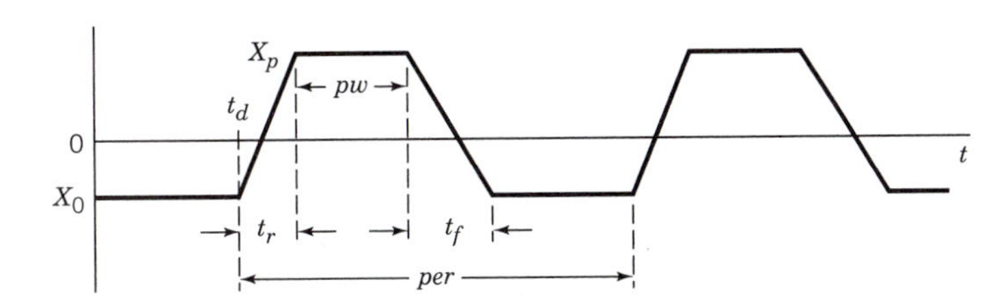

Figure B.26 Waveform of pulse source.

Damped Sinusoidal Source This source produces a delayed sinusoid with a dc offset and an exponential decay, as illustrated in Fig. B.27. The waveform is described mathematically by

$$\begin{aligned} x(t) &= X_0 + X_m \sin \phi & 0 < t < t_d \\ &= X_0 + X_m\, d(t) \sin [2\pi f(t - t_d) + \phi] & t > t_d \end{aligned}$$

with

$$\phi = 2\pi \frac{phase}{360} \qquad d(t) = e^{-alpha(t-t_d)}$$

The source statement is

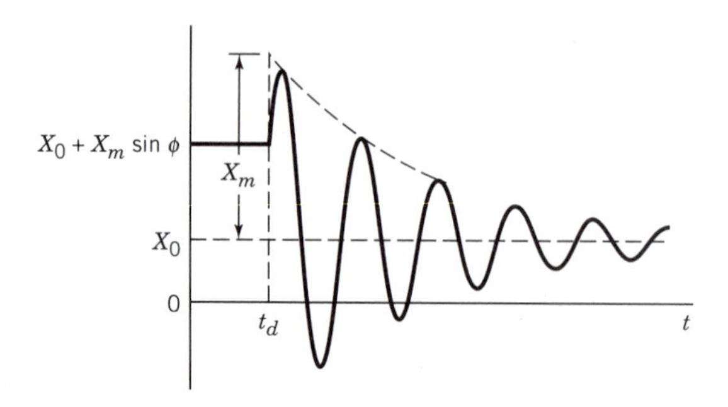

Figure B.27 Waveform of damped sinusoidal source.

```
[V|I]name n+ n- sin Xo Xm f ⟨td ⟨alpha ⟨phase⟩⟩⟩
```

where f is in hertz and *phase* is in degrees. Default values are $td = 0$, $alpha = 0$, and $phase = 0$.

An important special case is the ordinary constant-amplitude sinusoidal waveform

$$x(t) = X_m \sin(2\pi ft + \phi) \qquad t > 0$$

for which the device statement reads

```
[V|I]name n+ n- sin 0 Xm f 0 0 phase
```

Such sources are needed for the study of ac transients prior to steady-state conditions. (The standard PSpice ac source is only for the steady state *after* transients have died away.)

Example B.12 ***AC Transient Behavior***

Suppose the second-order circuit back in Fig. B.22 has no stored energy at $t = 0$, and the input is $v_s(t) = \sin(2\pi ft + \phi)$ for $t > 0$. Taking $R_s = 100\ \Omega$, we'll investigate the critically damped transition of v_C to ac steady-state conditions at $f = 50$ kHz with $\phi = 0°$ and $90°$.

From Fig. B.23*a*, we expect the transient to last about 50 μs, so we'll take *tfinal* = 100 μs. For a smooth plot of the 50-kHz oscillations, we'll take *maxstep* = $0.05/f = 1\ \mu$s. The corresponding circuit file is

```
Example B.12
Vs 1 0 sin  0 1 50k 0 0 {phase} ; variable phase
Rs 1 2     100
L  2 3     0.5m IC = 0
R  3 0      30k
C  3 0     0.2u IC = 0
.param phase = 0
```

```
.step param phase list 0 90 ; phase = 0, 90 degrees
.tran 100u 100u 0 1u uic    ; maxstep = 1 microsecond
.probe
.end
```

The resulting Probe plots in Fig. B.28*a* show that the value of ϕ significantly affects the transient behavior. However, in the steady-state conditions after about 60 μs, the waveforms differ only by the phase shift.

Another interesting Probe display is reproduced in Fig. B.28*b*. Here, we have plotted v_C versus v_s with $\phi = 0°$. The corresponding PSpice variables are entered as $v(3)@1$ and $v(1)@1$ since the case of $\phi = 0°$ was simulated first. The trace starts at $t = 0$ and follows a spiraling path during the transient period. The final steady-state oval pattern is called a **Lissajous figure**.

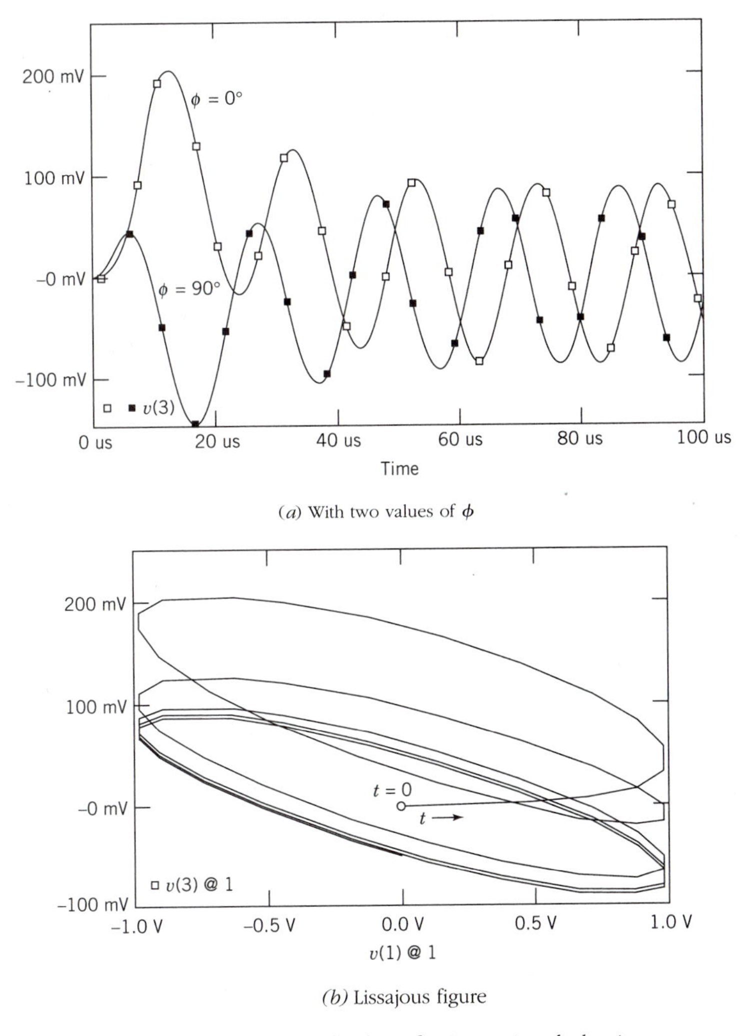

(*a*) With two values of ϕ

(*b*) Lissajous figure

Figure B.28 Probe display of AC transient behavior.

Exercise B.12

Write device statements for the source current from node 0 to node 1 to get the following waveforms:

(a) $i_s(t) = 5000t \qquad 0 < t < 2\ \mu s$
$= 0.02 - 5000t \qquad 2\ \mu s < t < 4\ \mu s$
$= 0 \qquad t > 4\ \mu s$

(b) $i_s(t) = 0.002 + 0.003e^{-10t}\cos 100t$

Switches

Switching at $t = 0$ is easily handled in PSpice using initial conditions. For switching after $t = 0$, PSpice provides **voltage-controlled switches**.

The switch description requires three statements. The first two are the device and model statements, for which the syntax is

```
Sname n+ n- (nc+, nc-) modname
.model modname vswitch ⟨Ron = value⟩ ⟨Roff = value⟩
+ ⟨Von = value⟩ ⟨Voff = value⟩
```

The switch is closed and has resistance *Ron* between nodes n^+ and n^- when $v(nc^+, nc^-) \geq Von$. The swith is open and has resistance *Roff* when $v(nc^+, nc^-) \leq Voff$. Default values are

$$Ron = 1\ \Omega \quad Roff = 1\ \text{M}\Omega \quad Von = 1\ \text{V} \quad Voff = 0$$

When other resistance values are specified, the ratio should not exceed $Roff/Ron = 10^{12}$.

We also need a third statement for the time-varying control voltage $v(nc^+, nc^-)$, which is usually a piecewise linear or pulse source. The risetime or falltime must be short to approximate instantaneous switching, but transition times should not be much less than *tfinal*/1000. If the control voltage has no other function in the circuit, then the source should connect from the reference node to a separate node, together with an arbitrary resistance.

Example B.13 *Switched RC Circuit*

To illustrate the use of PSpice switches, consider the circuit in Fig. B.29. We are given that this circuit is in the steady state for $t < 0$, with switch *S*25 closed and switch *S*45 open. At $t = 0$, *S*25 opens and *S*45 closes; then at $t = 50$ ms, *S*25 closes and *S*45 opens. We want to simulate this circuit for $t \geq 0$.

The switching at $t = 0$ is taken care of by noting that the initial capacitor voltage will be 10 V at node 5. For the switching at $t = 50$ ms, we need a control voltage *V*25 that jumps up to close *S*25 and another control voltage *V*45 that jumps down to open *S*45. Both switches can have the same `.model`

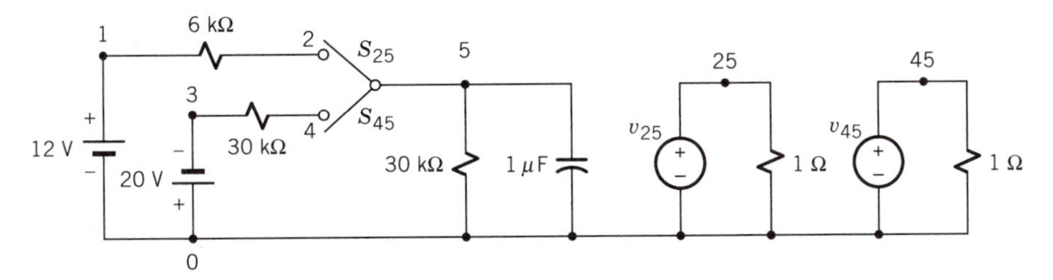

Figure B.29 Simulation diagram of switched *RC* circuit.

statement, and the default values for *Von, Voff,* and *Ron* are appropriate here. However, we'll take *Roff* = 10^9 Ω to better approximate an open circuit relative to the other resistances. The corresponding circuit file with *tfinal* = 100 ms is

```
Example B.13
V10  1 0  DC  12
R12  1 2       6k
V30  0 3  DC  20
R34  3 4      30k
R50  5 0      30k
C    5 0       1u IC = 10
S25  2 5  (25,0) switch
S45  4 5  (45,0) switch
.model switch vswitch Roff = 1G
V25 25 0 pwl 0,0 49.9m,0 50m,1 ; risetime = 0.1 ms
V45 45 0 pwl 0,1 49.9m,1 50m,0 ; falltime = 0.1 ms
.tran 100m 100m uic
.probe
.end
```

Figure B.30 reproduces the Probe plots of the capacitor voltage and current. The current jump at t = 50 ms appears to be instantaneous because the switch transition time is only 0.1 ms.

Exercise B.13

Write the statements needed to actuate two switches with default parameter values and *tfinal* = 10 s. Switch *S*12 is to be closed for $2 < t < 3$ and $6 < t < 7$; otherwise, it is open. Switch *S*34 is to be open when *S*12 is closed, and vice versa. Take nodes 0 and 10 for the control voltage, and use a pulse source with *Xo* = −1 V and *Xp* = +1 V.

B.5 FREQUENCY RESPONSE

The Probe output of PSpice provides a very convenient way of displaying frequency response. Furthermore, the interactive capabilities of Probe allow

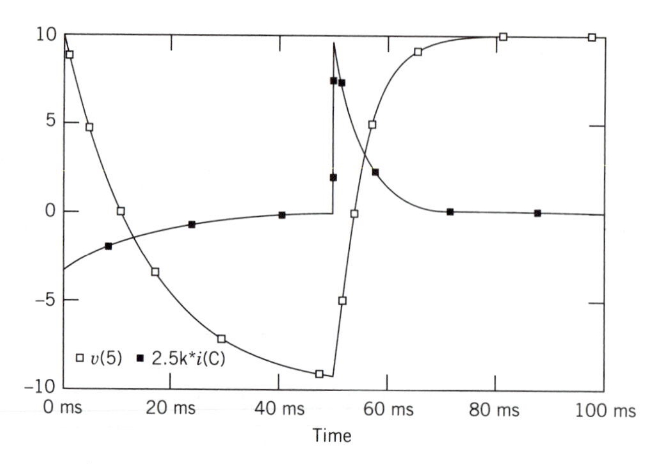

Figure B.30 Probe display of switched transients.

you to generate informative plots that would be tedious to obtain by other means.

The circuit under investigation should have just one independent source, so you can define appropriate transfer functions. The `.ac` statement should sweep frequency through a sufficient number of points to obtain meaningful plots. As a rule of thumb, 50 to 200 points usually gives satisfactory results. After viewing the Probe output, you may want to edit your sweep statement and rerun the circuit file.

When an `.ac` statement is used, Probe's default *X*-axis variable is frequency in hertz. However, you can change the variable to $\omega = 2\pi f$ by selecting the *X*-variable option and entering the expression 6.283**frequency*.

If the frequency range is large, then a **logarithmic sweep** may be desirable. The statement form is

```
.ac [dec|oct] npoints, fstart, fstop
```

This statement generates *npoints* per decade (`dec`) or per octave (`oct`). The *X*-axis of Probe may then be set for either a linear or logarithmic display. The logarithmic display may also be used with the `.ac lin` sweep statement.

Bode plots are obtained using the logarithmic *X*-axis and specifying magnitudes of output variables in **decibels**. The dB specification is [*V*|*I*]*DB*(. . .), which may also be used in print statements.

Example B.14 *Frequency-Response Analysis*

Suppose you want to investigate the frequency response of the circuit in Fig. B.31, taking $H_2(j\omega) = \underline{V}_2/\underline{V}_{in}$ and $H_4(j\omega) = \underline{V}_4/\underline{V}_{in}$. You let $\underline{V}_{in} = 1\text{ V} \angle 0°$, so $H_2(j\omega) = V(2)$ and $H_4(j\omega) = V(4)$. To determine the sweep range, you note that the first *CR* section by itself should act like a high-pass filter with $f_{co} = 1/2\pi RC \approx 1.6$ kHz, while the *LC* section has at $f_0 = 1/(2\pi\sqrt{LC}) \approx 16$ kHz. You therefore sweep well beyond those points using the following circuit file.

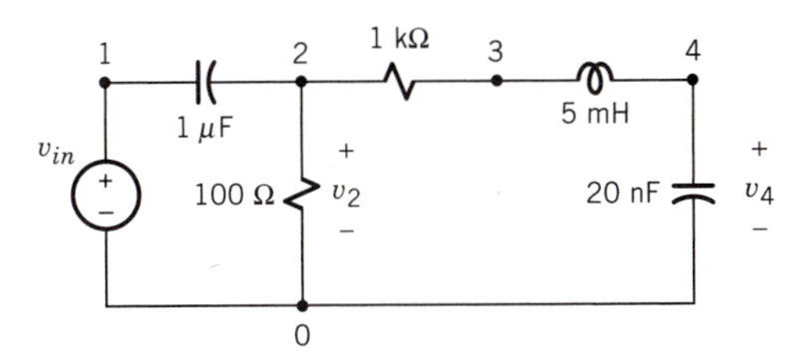

Figure B.31 Circuit for Example B.14.

```
Example B.14
Vin  1 0  ac   1    ; H2 = V(2), H4 = V(4)
C12  1 2        1u
R20  2 0     100
R23  2 3        1k
L34  3 4        5m
C40  4 0      20n
* Sweep from 1E2 to 1E6 hertz, 4 decades, 50 points/decade
.ac dec 50, 100, 1Meg
.probe
.end
```

Fig. B.32*a* shows the resulting curves of amplitude ratio and phase shift for $H_2(j\omega) = V(2)$ and $H_4(j\omega) = V(4)$ with a linear frequency axis. Also included are the curves for the *RLC* section by itself, obtained via $|\underline{V}_4/\underline{V}_2| = VM(4)/VM(2)$ and $\angle(\underline{V}_4/\underline{V}_2) = VP(4) - VP(2)$. Probe's *Set_Range* option has been used so the *X*-axis goes from 0 to 30 kHz.

The Bode plot of $g_4(\omega)$ versus log ω is shown in Fig. B.32*b*, where $g_4(\omega) = VDB(4)$ and the *X*-axis variable has been changed to 6.283**frequency*. By constructing straight-line asymptotes on this plot, you conclude that the transfer function $H_4(s)$ has a zero at $s = 0$, a pole at $s \approx -8$ krad/s, and repeated poles at $s \approx -100$ krad/s.

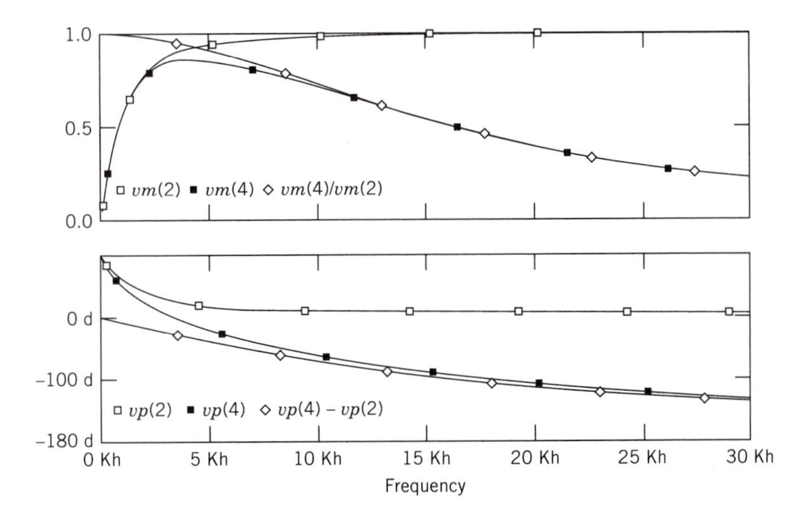

(*a*) Amplitude ratio and phase shift

Figure B.32 Probe displays of frequency response.

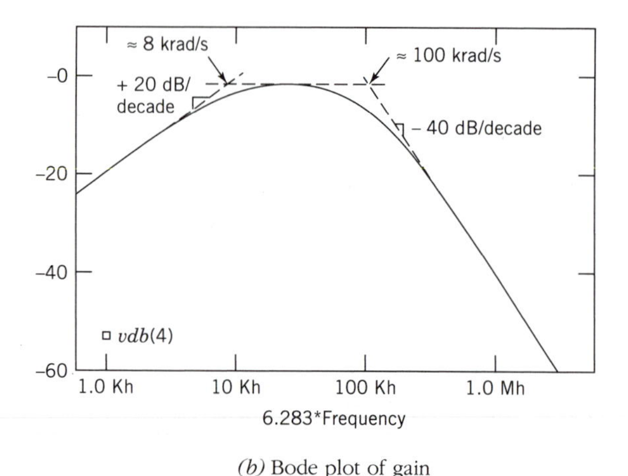

(b) Bode plot of gain

Figure B.32 *(continued)*

PROBLEMS

B.2 Resistive Circuits

B.1* Simulate the circuit in Fig. P1.34 (p. 37) with $R_3 = 8\ \Omega$. Use `.print` and `.dc` statements to find v_a, v_b, i_2, and the power supplied by each source when $v_s = 12$ V, 6 V, and -6 V.

B.2 Simulate the circuit in Fig. P1.40 (p. 38) with $v_s = -40$ V. Use `.print` and `.dc` statements to find v_2, v_4, i_1, i_3, and the power supplied by each source when $i_s = -1$ mA, 1 mA, and 2 mA.

B.3 Simulate the circuit in Fig. P2.35 (p. 81). Obtain Probe plots of v_4 and $R_{eq} = 48\ \text{V}/i_1$ versus R_1 with R_1 swept from 5 Ω to 50 Ω. Use the cursor to find the values of v_4 and R_{eq} when $R_1 = 10\ \Omega$, 20 Ω, and 40 Ω.

B.4 Simulate the circuit in Fig. P2.55 (p. 82). Take $v_s = 30$ V and $\mu = 6$, and replace the 4-Ω resistor with a variable resistor R_{var}. Obtain Probe plots of v_2 and $R_{eq} = v/i$ versus R_{var} with R_{var} swept from 4 Ω to 12 Ω. Use the cursor to find the values of v_2 and R_{eq} when $R_{var} = 4\ \Omega$, 8 Ω, and 12 Ω.

B.5* This simulation explores the properties of the amplifier circuit in Fig. PB.5 with $g_m = 60$ mS. Use a `.tf` statement to find the input resistance and the output Thévenin parameters with $R_f = 30$ kΩ and 60 kΩ.

B.6 Do Problem B.5 for the amplifier circuit in Fig. PB.6 with $r_m = 120$ kΩ.

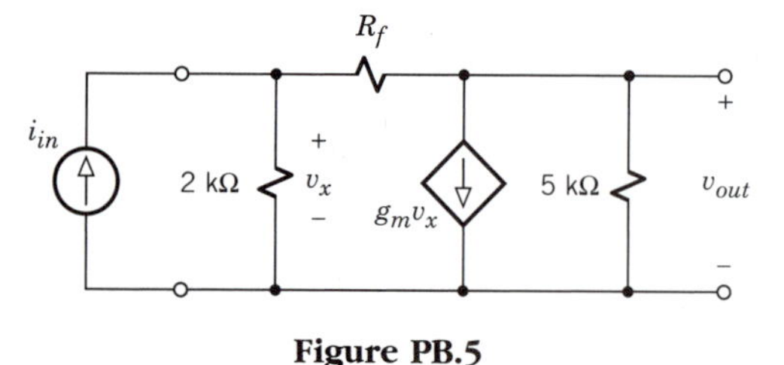

Figure PB.5

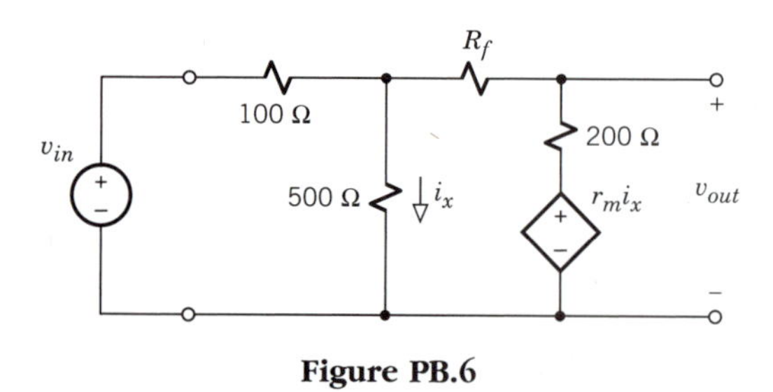

Figure PB.6

B.7 This simulation illustrates the effect of finite op-amp gain on the noninverting amplifier circuit in Fig. 3.16*a* (p. 101). Take $v_{in} > 0$ and $R_1 = 1$ kΩ, and let $R_F = 24$ kΩ, 49 kΩ, and 99 kΩ. Model the op-amp as in Fig. B.8 with $r_i = \infty$, $r_o = 0$, and variable gain A, and sweep A logarithmically from 10^2 to 10^7. Obtain a multitrace Probe plot showing v_{out} versus A for each value of R_F, and use the cursor to determine the minimum value of A in each case such that $|v_{out}| \geq 0.99|v_{out}|_{max}$. Also obtain a multitrace plot showing $v_d = v_p - v_n$, and use the cursor to determine the minimum value of A in each case such that $|v_d| \leq |v_{in}|/1000$.

B.8 Do Problem B.7 for the inverting amplifier circuit in Fig. 3.21a (p. 107). Take $v_{in} > 0$ and $R_1 =$ 1 kΩ, and let $R_F =$ 25 kΩ, 50 kΩ, and 100 kΩ.

B.9 This simulation illustrates the effect of internal op-amp resistances on the noninverting amplifier circuit in Fig. 3.16a (p. 101). Take $v_{in} =$ 1 V, $R_1 =$ 1 kΩ, and $R_F =$ 9 kΩ. (a) Model the op-amp model as in Fig. B.8 with $r_i =$ 100 kΩ, $r_o =$ 500 Ω, and variable gain A. Use a `.tf` statement to find v_{out}/v_{in} and the input and output resistance for $A = 10^2$, 10^4, and 10^6. (b) Repeat part (a) with a better op-amp having $r_i =$ 1 MΩ and $r_o =$ 100 Ω.

B.10 Do Problem B.9 for the inverting amplifier circuit in Fig. 3.21a (p. 107). Take $v_{in} =$ 1 V, $R_1 =$ 1 kΩ, and $R_F =$ 10 kΩ.

B.11 Simulate the amplifier circuit in Fig. PB.5 with $g_m =$ 60 mS. Add an independent voltage source at the output so you can sweep v_{out} from −10 V to +10 V. Generate a dual-trace Probe plot of the output current versus v_{out} for $R_f =$ 30 kΩ and 60 kΩ. Use cursor measurements to find the output Thévenin parameters for both cases.

B.12 Do Problem B.11 for the amplifier circuit in Fig. PB.6 with $r_m =$ 120 kΩ.

B.13* Simulate the circuit with two op-amps in Fig. P4.81 (p. 192). Model both op-amps by subcircuit with $r_i =$ 1 MΩ, $r_o =$ 100 Ω, and $A = 10^5$. Take $R =$ 1 kΩ and use a `.tf` statement to find the gain and input and output resistances with $K =$ 1, 2, and 4.

B.14 Simulate the circuit with two op-amps in Fig. P4.84 (p. 192). Model both op-amps by a subcircuit with $r_i =$ 1 MΩ, $r_o =$ 100 Ω, and $A = 10^5$. Take $R =$ 1 kΩ and $R_a = (K-1)R/2$, and use a `.tf` statement to find the gain and input and output resistances with $K =$ 2, 3, and 5.

B.3 AC Circuits

B.15* Simulate the circuit in Fig. P6.26 (p. 000) under steady-state conditions with $v =$ 40 cos 50,000t V, $L =$ 0.2 H, $R_1 =$ 5 Ω, $R_2 =$ 10 Ω, and $C =$ 2 μF. Include a print statement to find $v_2(t)$, $i_C(t)$, and the input impedance $Z = \underline{V}/\underline{I}$. Hint: You will need to add a small resistance.

B.16 Simulate the circuit in Fig. P6.24 (p. 297) under steady-state conditions with $i =$ 4 cos 20,000t A, $C =$ 10 μF, $R_1 =$ 10 Ω, $R_2 =$ 5 Ω, and $L =$ 0.25 mH. Include a print statement to find $v_C(t)$, $i_L(t)$, and the input impedance $Z = \underline{V}/\underline{I}$. Hint: You will need to add a large resistance.

B.17 The source network in Fig. PB.17 is operating in the steady state with $v =$ 30 cos 1000t V and $r_m =$ 40 Ω. Find the ac Thévenin parameters by simulation with a load resistance that steps from a very large to a very small value. Is the network capacitive or inductive?

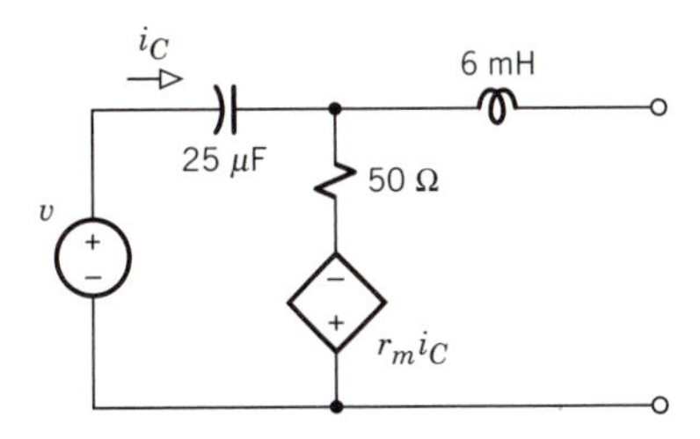

Figure PB.17

B.18 The source network in Fig. PB.18 is operating in the steady state with $i =$ 6 cos $10^5 t$ mA and $r_m =$ 1 kΩ. Find the ac Thévenin parameters by simulation with a load resistance that steps from a very large to a very small value. Is the network capacitive or inductive?

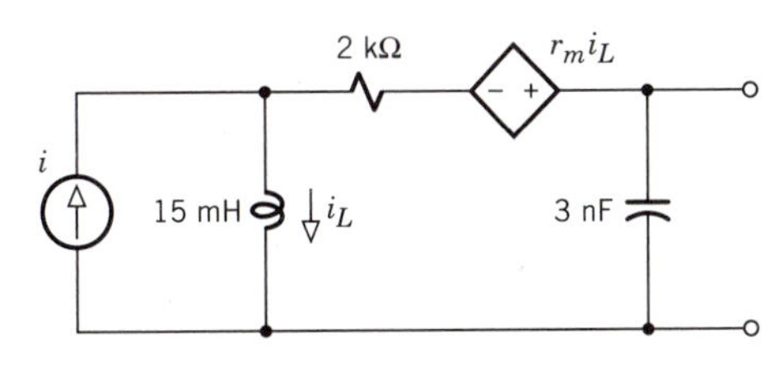

Figure PB.18

B.19* Consider the modified resonant circuit in Fig. PB.19 with $v_{in} =$ cos ωt and $R =$ 100 Ω. Calculate $f_{LC} = 1/2\pi\sqrt{LC}$ and do a linear frequency-sweep simulation from about $0.5f_{LC}$ to $1.5f_{LC}$. Invoke Probe to obtain the following: (a) A split-screen display of $|Z_{in}|$ and $\angle Z_{in}$, where $Z_{in} = \underline{V}_{in}/\underline{I}_{in}$. Use the cursor to find the frequency at which $\angle Z_{in} = 0$ and the maximum or minimum value of $|Z_{in}|$ and the corresponding frequency. (b) Dual-trace displays of $|\underline{I}_C|$ and $|\underline{I}_L|$ and of $|\underline{V}_C|$ and $|\underline{V}_L|$. Use the cursor to determine the maximum value of each of these four quantities and the frequencies at which they occur. In view of your results from (a) and (b), what can you say about the type of resonance (series or parallel) and the resonant frequency?

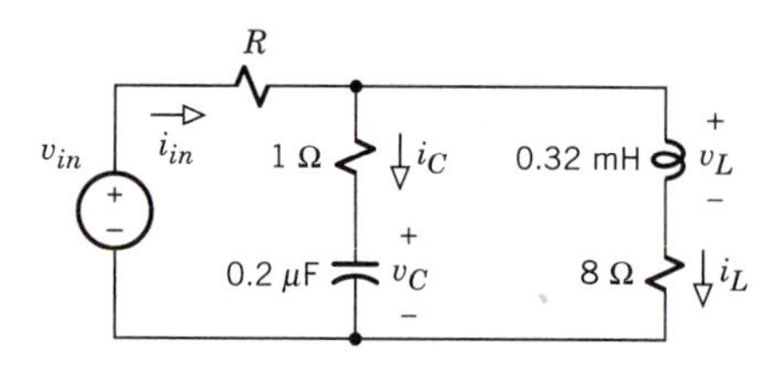

Figure PB.19

B.20 Do Problem B.19 for the circuit in Fig. PB.20 with $v_{in} = \cos \omega t$ and $R = 6\ \Omega$.

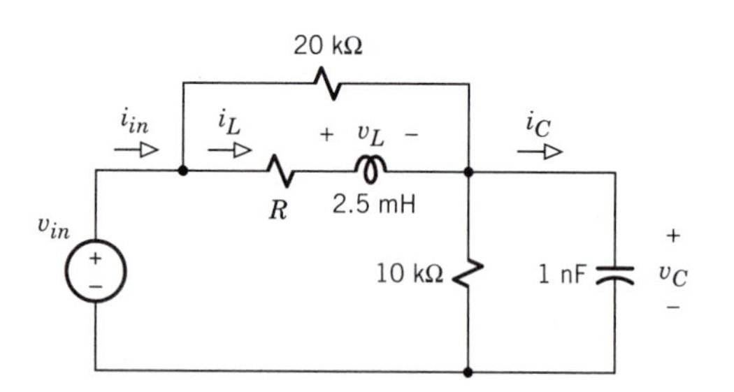

Figure PB.20

B.21 Let the circuit in Fig. PB.21 have $R_x = 20\ \Omega$, $L_x = 25$ mH, $R_y = 4\ \Omega$, and $L_y = 10$ mH, so it exhibits both series and parallel resonance effects. Do a linear frequency-sweep simulation from 10 Hz to 10 kHz and invoke Probe to obtain the following: (a) A split-screen display of $|Z_{in}|$ and $\angle Z_{in}$, where $Z_{in} = \underline{V}_{in}/\underline{I}_{in}$; (b) A dual-trace display of $|\underline{I}_x|$ and $|\underline{I}_y|$. Use the cursor to find the frequencies at which $|Z_{in}|$, $|\underline{I}_x|$, and $|\underline{I}_y|$ have local maximum or minimum values and the frequencies at which $\angle Z_{in} = 0$. In view of your results, what can you say about the resonance effects?

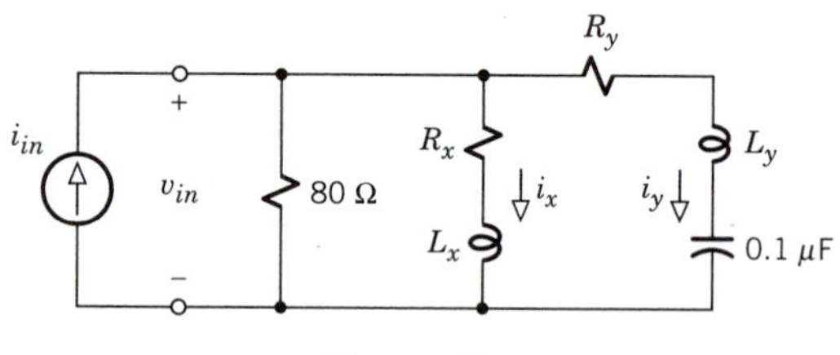

Figure PB.21

B.22 Do Problem B.21 for the circuit in Fig. PB.22.

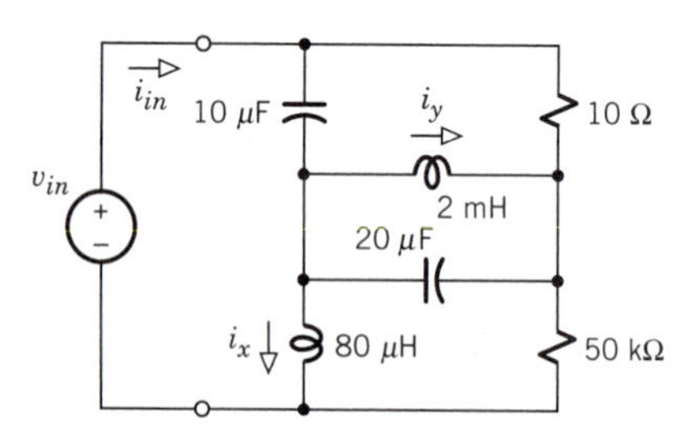

Figure PB.22

B.23 The ac power system in Fig. PB.23 has load impedances $Z_1 = 2 - j12\ \Omega$ and $Z_2 = 1.5 + j4\ \Omega$. Use a simulation to find $\underline{I}$. Then calculate P, Q, P_L/P, and the load's power factor.

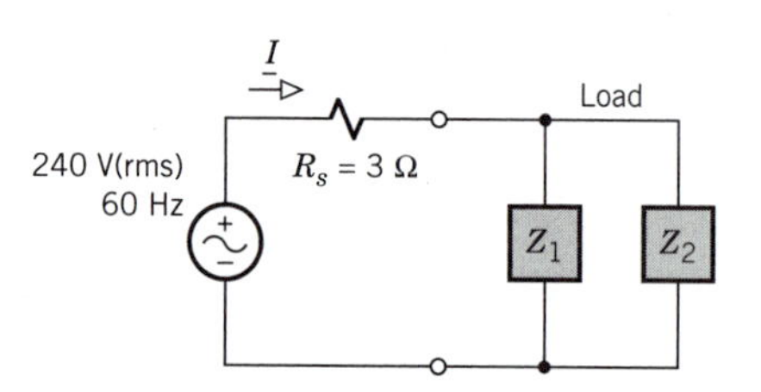

Figure PB.23

B.24 Do Problem B.23 with $Z_1 = 2.5 - j8\ \Omega$ and $Z_2 = 4 + j10\ \Omega$.

B.25* Simulate the balanced three-phase circuit in Fig. 7.22*a* (p. 336) with *a-b-c* phase sequence, $\underline{V}_a = 5$ kV(rms) $\angle 0°$, $f = 60$ Hz, $Z_l = 0.3\ \Omega$, and a delta load having 4 Ω and 10 mH in series in each branch. Include a print statement to obtain $\underline{V}_{ab}$, $\underline{I}_a$, $\underline{I}_b$, $\underline{I}_c$, and $\underline{V}_{AB}$.

B.26 Simulate the three-phase circuit in Fig. 7.25 (p. 340) with *a-b-c* phase sequence, $\underline{V}_a = 580$ V(rms) $\angle 0°$, $f = 60$ Hz, and a balanced load having 7 Ω and 64 mH in series in each branch. Include a print statement with appropriate outputs so you can calculate the wattmeter readings P_1 and P_2.

B.27 Simulate both of the unbalanced three-phase circuits in Fig. 7.27 (p. 344) with: *a-b-c* phase sequence, $\underline{V}_a = 120$ V(rms) $\angle 0°$, $f = 60$ Hz, $Z_a = 3\ \Omega$, $Z_b = 4\ \Omega$, and $Z_c = 3 + j4\ \Omega$. Include a print statement to obtain all phasor line currents in each case. Hint: You will need to add a small resistor for the neutral wire.

B.28 Simulate the unbalanced three-phase circuit in Fig. 7.29*a* (p. 346) with *a-b-c* phase sequence,

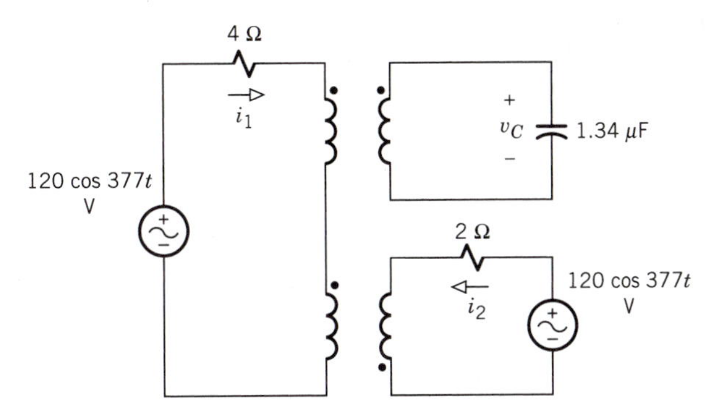

Figure PB.30

V_p = 580 V(rms), f = 60 Hz, $Z_{aA} = Z_{bB} = Z_{cC}$ = 2 Ω, $Z_{AB} = Z_{BC}$ = 25 Ω, and Z_{CA} = 7 + j24 Ω. Include a print statement to obtain the rms values of all line currents and the phase voltages across the delta load.

B.29 Simulate the ideal-transformer circuit in Fig. P8.10 (p. 398) with N_A = 24, Z = 1152 + j576 Ω, N_B = 0.05, and Z_{out} = 28.8 − j36 Ω. Use any convenient frequency, and include a print statement to find $\underline{I}_{in}$, $\underline{V}_{in}$, $\underline{V}_a$, $\underline{I}$, $\underline{V}_b$, and $\underline{V}_{out}$.

B.30 Use a simulation to find $\underline{I}_1$, $\underline{I}_2$, and $\underline{V}_C$ in Fig. PB.30 when both transformers are ideal with their primaries on the left and turns ratios 1:6 (upper) and 1:0.5 (lower). Hint: You will need to add two or three large resistors.

B.31 Simulate the coupled-inductor circuit in Fig. P8.43 (p. 400) with L_1 = 0.1 H, L_2 = 0.2 H, and a 10-Ω load resistor. (a) Include `.step` and `.print` statements to calculate $|\underline{V}_{out}|/|\underline{V}_{in}|$ and $|\underline{I}_{out}|/|\underline{I}_{in}|$ for k = 0.5, 0.9, and 0.999. (b) Repeat (a) with the dot moved to the other end of one coil.

B.32 Let both transformers in Fig. PB.30 be coupled inductors with $L_1 = L_2$ = 0.1 H and k = 0.95. Do a simulation with two additional large resistors to find $\underline{I}_1$, $\underline{I}_2$, and $\underline{V}_C$.

B.4 Transients

B.33* The RL circuit in Fig. PB.33 has L = 50 mH, $i_L(0^-)$ = 0, and

$$v_s(t) = 1000t \qquad 0 \le t < 1 \text{ ms}$$
$$= 1 \qquad t > 1 \text{ ms}$$

Do a transient simulation over $0 \le t \le 8$ ms with R = 50 Ω, 100 Ω, and 200 Ω. Obtain dual-trace Probe plots of $v_R(t)$ and $v_L(t)$ for each case and use the cursor to find the maximum value of $v_L(t)$. Also estimate the time t_{ss} when the circuit has essentially reached the steady state. Put your results in a table.

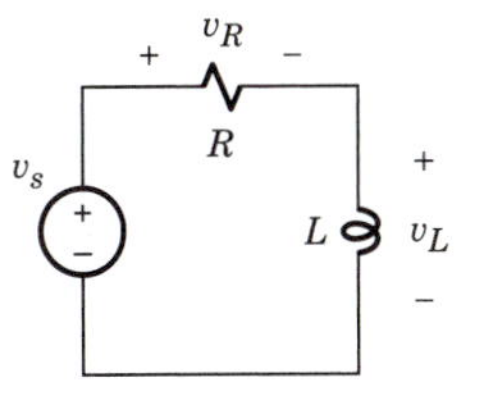

Figure PB.33

B.34 This simulation illustrates transient behavior leading into **periodic steady-state conditions**. Let $v_s(t)$ in Fig. PB.33 be a periodic train of rectangular pulses starting at t = 0 with 1-V amplitude, 2-ms duration, and 3-ms spacing. Simulate this circuit over $0 \le t \le 30$ ms, taking L = 800 mH, $i_L(0^-)$ = 0, and R = 200 Ω, 400 Ω, and 800 Ω. Obtain separate Probe plots of $v_R(t)$ for each case, estimate the time t_{ss} at which the response essentially reaches the periodic steady state, and use the cursor to find the maximum and minimum values of v_R under this condition. Put your results in a table.

B.35 Simulate the circuit in Fig. PB.35 over the range $0 \le t \le 6$ s with the following switching events: S_1 is at position a and S_2 at position d for all $t < 0$; S_1 goes to b at t = 0; S_1 goes back to a and S_2 goes to e at t = 3 s. Obtain a split-screen Probe plot showing $v(t)$ and $i(t)$, and use the cursor to determine their extreme values.

B.36 Simulate the circuit in Fig. PB.35 over the range $0 \le t \le 8$ s with the following switching events: S_1 is at position a and S_2 at position e for all $t < 0$; S_2 goes to f at t = 0; S_1 goes to b at t = 3 s; S_2 goes back to e at t = 5 s. Obtain a split-screen Probe plot showing $v(t)$ and $i(t)$, and use the cursor to determine their extreme values.

B.37 Simulate the circuit in Fig. PB.35 over the range $0 \le t \le 10$ s with the following switching events: S_1

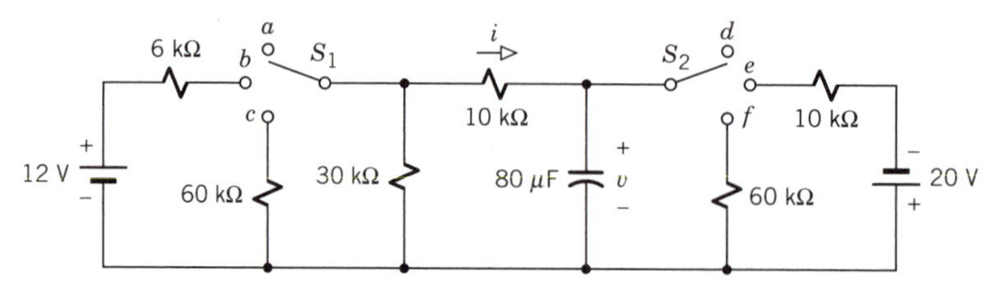

Figure PB.35

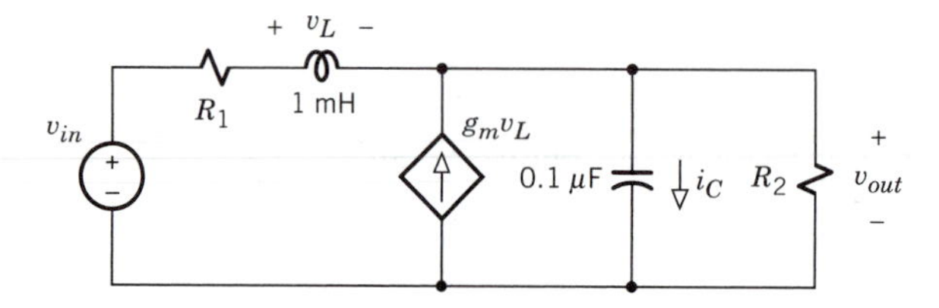

Figure PB.40

is at position b and S_2 at position d for all $t < 0$; S_1 goes to c at $t = 0$; S_2 goes to e at $t = 3$ s; S_1 goes back to b at $t = 5$ s; S_2 goes back to d at $t = 6$ s. Obtain a split-screen Probe plot showing $v(t)$ and $i(t)$, and use the cursor to determine their extreme values.

B.38 Let the circuit in Fig. PB.19 have $v_{in} = 0$, $i_L(0) = 0$, and $v_C(0) = 1$ V. Simulate the resulting natural response over $0 \leq t \leq 100$ μs with $R = 10$, 20, 50, and 100 Ω. Generate multitrace Probe plots of $v_C(t)$ and $i_L(t)$ to determine which value(s) of R corresponds to overdamping, underdamping, and at or near critical damping. Also use the cursor to estimate ω_d for the case with the least damping.

B.39 Simulate the step response of the circuit in Fig. PB.20 over $0 \leq t \leq 20$ μs when $R = 1$ kΩ, 2 kΩ, 5 kΩ, and 10 kΩ. Generate multitrace Probe plots of $v_C(t)$ and $i_C(t)$ to determine which value(s) of R corresponds to overdamping, underdamping, and at or near critical damping. Also use the cursor to estimate $i_C(0^+)$ and the value of ω_d for the case with the least damping.

B.40 The circuit in Fig. PB.40 has $g_m = 5$ mS, $R_1 = b$, and $R_2 = 100b$. Simulate the step response over $0 \leq t \leq 100$ μs with $b = 20$, 100, and 500. Generate a multitrace Probe plot of $i_C(t)$ to determine which value(s) of b corresponds to overdamping, underdamping, and at or near critical damping. Also use cursor measurements to estimate the locations of the network's poles for the case with the least damping.

B.41 The controlled source in Fig. PB.41 represents a noninverting op-amp circuit with input v_x. Simulate the step response over $0 \leq t \leq 2$ ms with $R_1 = 10^3 b$ and $R_2 = 10^5/b$, taking $b = 1$, 5, and 25. Generate a multitrace Probe plot of $v_{out}(t)$ to determine which value(s) of b corresponds to overdamping, underdamping, and at or near critical damping. Also use cursor measurements to estimate the locations of the network's poles for the case with the least damping.

B.42 This simulation explores transient effects in a resonant circuit. Consider a series RLC circuit with $\omega_0 = 2\pi \times 1$ kHz, $C = 1$ μF, $L = 1/\omega_0^2 C$, $R = 1/\omega_0 C Q_{ser}$, $v_C(0^-) = i_L(0^-) = 0$, and input voltage

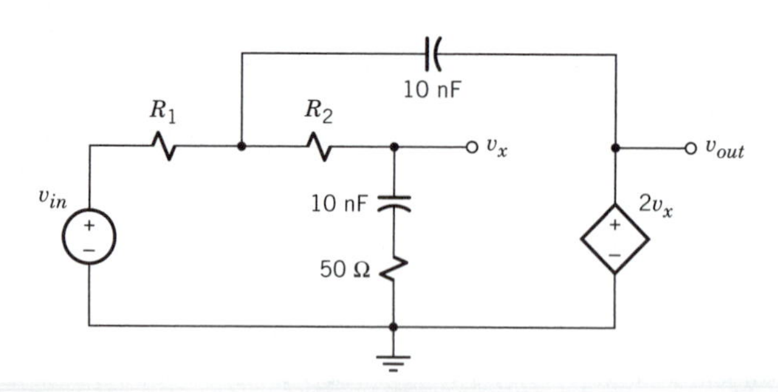

Figure PB.41

$v_s(t) = \sin 2\pi ft$ for $t \geq 0$. Simulate the transient response over $0 \leq t \leq 20$ ms for (a) $f = 1$ kHz with $Q_{ser} = 1, 5,$ and 10; (b) $Q_{ser} = 5$ with $f = 800$ Hz and 1 kHz. Obtain multitrace Probe plots of $v_C(t)$ and $v_C(t) + v_L(t)$ for parts (a) and (b), and estimate from each trace of $v_C(t)$ the time t_{ss} when the circuit essentially reaches steady-state conditions. Also, obtain the Lissajous figures for part (a) by plotting $v_C(t)$ versus $v_s(t)$. (If you have separate colors for each trace, then a multitrace Lissajous figure produces an interesting display.)

B.5 Frequency Response

B.43* Simulate the bandpass frequency response $H(j\omega) = \underline{I}_1/\underline{I}_{in}$ of the circuit in Fig. PB.43 with $C = 0.2\ \mu$F and $L = 1$ mH. Obtain plots of $a(\omega)$ and $\theta(\omega)$ over the linear frequency range $0 \leq f \leq 20$ kHz for $R_w = 0$, 2 Ω, and 10 Ω. (Hint: Take a very small but nonzero value to approximate $R_w = 0$.) Use cursor measurements to find a_{max}, the frequency f_0 at which $a(\omega) = a_{max}$, the frequencies f_l and f_u at which $a(\omega) = a_{max}/\sqrt{2}$, and the frequency f_p at which $\theta(\omega) = 0°$. Put your results in a table along with calculated values of the bandwidth $B = f_u - f_l$, quality factor $Q = f_0/B$, and geometric mean frequency $\sqrt{f_l f_u}$.

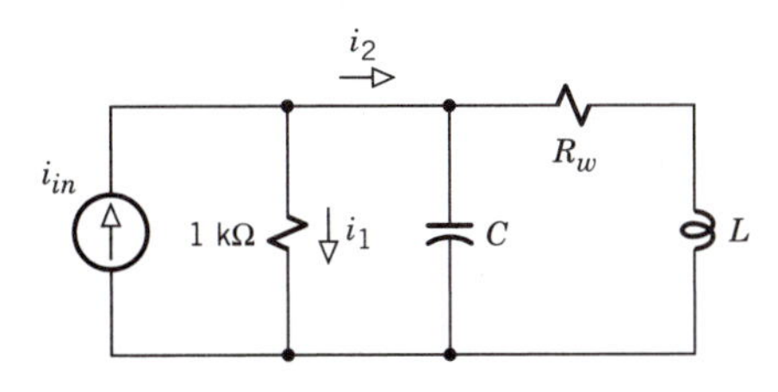

Figure PB.43

B.44 Simulate the band-reject frequency response $H(j\omega) = \underline{I}_2/\underline{I}_{in}$ of the circuit in Fig. PB.43 with $C = 0.4\ \mu$F and $L = 2.5$ mH. Obtain plots of $a(\omega)$ and $\theta(\omega)$ over the linear frequency range $0 < f \leq 10$ kHz for $R_w = 0$, 2 Ω, and 5 Ω. (Hint: Take a very small but nonzero value to approximate $R_w = 0$.) Use cursor measurements to find a_{min}, the frequency f_0 at which $a(\omega) = a_{min}$, the frequencies f_l and f_u at which $a(\omega) = a_{max}/\sqrt{2}$, and the frequency f_p at which $\theta(\omega) = 0°$. Put your results in a table along with calculated values of the bandwidth $B = f_u - f_l$, quality factor $Q = f_0/B$, and geometric mean frequency $\sqrt{f_l f_u}$.

B.45 Figure PB.45 is a simplified op-amp model including high-frequency effects. The value of R_m depends upon the quality (and cost) of the op-amp. (a) Use this model with $R_m = 800$ kΩ to simulate a standard noninverting amplifier having $R_1 = 1$ kΩ and $R_F = 4$ kΩ, 9 kΩ, and 19 kΩ. Let $H(j\omega) = \underline{V}_{out}/\underline{V}_{in}$ and obtain a multitrace plot of $a(\omega)$ over the linear frequency range $0 < f \leq f_{max}$, where $f_{max} = 2 \times 10^{11}/R_m$. Take cursor measurements to find a_{max} and the frequency f_{co} at which $a(\omega) = a_{max}/\sqrt{2}$ for each value of R_F. Then calculate the **gain-bandwidth product** $a_{max}f_{co}$. (b) Repeat (a) with $R_m = 40$ kΩ, which corresponds to a more expensive op-amp. What conclusions can you draw from your results?

B.46 Consider an active inverting lowpass filter like Fig. 11.20*a* (p. 509). The filter has been designed assuming an ideal op-amp, taking $R_1 = 1$ kΩ, $R_F = 20$ kΩ, and $C_F = 1/2\pi f_{co}R_F$, where f_{co} is the desired cutoff frequency. (a) Simulate this circuit for $f_{co} = 500$ Hz, 5 kHz, and 50 kHz, using the high-frequency op-amp model in Fig. PB.45 with $R_m = 800$ kΩ. Obtain a multitrace plot of $a(w)$ over the linear frequency range $0 < f \leq 50$ kHz and use cursor measurements to determine the actual cutoff frequency for each value of C_F. (b) Repeat (a) with $R_m = 40$ kΩ, which corresponds to a more expensive op-amp. What conclusions can you draw from your results?

B.47 Figure PB.47 is a **stagger-tuned amplifier**, which can better approximate an ideal bandpass filter by using two resonant sections with resonant frequencies slightly above and below the desired center frequency. Simulate this circuit with $\beta = 1$, $R_2 = 1$ kHz, and $L_1 = 200\ \mu\text{H} - \Delta L$ and $L_2 = 200\ \mu\text{H} + \Delta L$, where $\Delta L = 0$, 5 μH, 10 μH, and 15 μH. Obtain a multitrace plot of $|\underline{I}_2/\underline{I}_{in}|$ over the linear frequency range 40 kHz $\leq f \leq$ 60 kHz. Use cursor measurements to determine the lower and upper cutoff frequencies f_l and f_u for

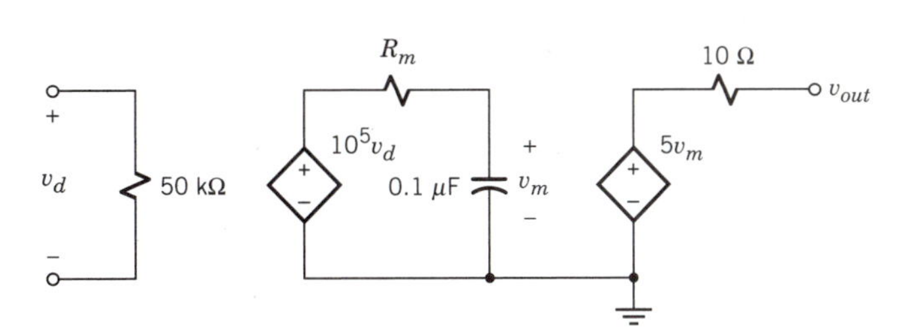

Figure PB.45

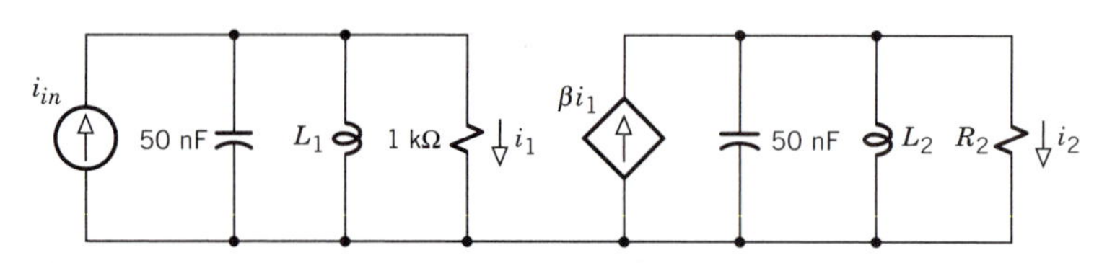

Figure PB.47

each trace. Then calculate the bandwidth $B = f_u - f_l$ and the center frequency $f_c = (f_l + f_u)/2$. Also, to illustrate the stagger-tuning effect, take the case of $\Delta L = 10\ \mu$H and obtain a multitrace plot showing $|\underline{I}_1/\underline{I}_{in}|$, $|\underline{I}_2/\underline{I}_1|$, and $|\underline{I}_2/\underline{I}_{in}|$.

B.48 Consider the stagger-tuned amplifier described in Problem B.47 with $\beta = 1$. (a) Simulate this circuit taking $L_1 = 215\ \mu$H, $L_2 = 185\ \mu$H, and $R_2 = 800, 850, 900, 950$, and $1000\ \Omega$. Obtain a multitrace plot showing $|\underline{I}_2/\underline{I}_{in}|$ over the linear frequency range 40 kHz $\le f \le$ 60 kHz. Determine which value of R_2 gives the best bandpass response, and use cursor measurements to find the lower and upper cutoff frequencies f_l and f_u, the bandwidth $B = f_u - f_l$, and the center frequency $f_c = (f_l + f_u)/2$. (b) Repeat (a) with $L_1 = 220\ \mu$H and $L_2 = 180\ \mu$H. Also, to illustrate the stagger-tuning effect, take the best value of R_2 and obtain a multitrace plot showing $|\underline{I}_1/\underline{I}_{in}|$, $|\underline{I}_2/\underline{I}_1|$, and $|\underline{I}_2/\underline{I}_{in}|$.

B.49* Do a decade frequency-sweep simulation over $1 \le \omega \le 10^9$ for the circuit in Fig. PB.41 with $R_1 = 1$ kΩ and $R_2 = 10$ kΩ. Invoke Probe to obtain the Bode plot of $H(s) = \underline{V}_{out}/\underline{V}_{in}$, and construct asymptotes for $g(\omega)$ to estimate the values of the poles and zeros.

B.50 Let the circuit in Fig. PB.40 have $g_m = 5$ mS, $R_1 = b$, and $R_2 = 100b$. Do a decade frequency-sweep simulation over $10 \le \omega \le 10^{10}$ with $b = 20$ and 2000, and invoke Probe to obtain the Bode plots of $H(s) = \underline{V}_{out}/\underline{V}_{in}$ for both cases. Then construct asymptotes for $g(\omega)$ with $b = 2000$ to estimate the values of the poles and zeros in this case. What can you conclude from your results about the pole locations when $b = 20$?

B.51 Let the circuit in Fig. PB.21 have $R_x = 10\ \Omega$, $L_x = 1$ mH, $R_y = 200\ \Omega$, and $L_y = 20\ \mu$H. Do a decade frequency-sweep simulation over $10^2 \le \omega \le 10^9$ and invoke Probe to obtain the Bode plot of $H(s) = \underline{I}_y/\underline{I}_{in}$. Construct asymptotes for $g(\omega)$ to estimate the values of the poles and zeros.

B.52 Do a decade frequency-sweep simulation over $10 \le \omega \le 10^7$ for the circuit in Fig. PB.21 with $R_x = 0.5\ \Omega$, $L_x = 0.5$ mH, $R_y = 1$ kΩ, and $L_y = 10$ mH. Invoke Probe to obtain a dual-trace plot showing $g_1(\omega)$ and $g_2(\omega)$, where $H_1(s) = \underline{I}_y/\underline{I}_{in}$ and $H_2(s) = \underline{V}_{in}/\underline{I}_{in}$. Construct asymptotes for both traces to estimate the values of the poles and zeros. What do you conclude about $H_2(s)$ from your results?

Tables of Mathematical Relations

The following tables summarize most of the mathematical relationships used in the text and the problems.

Basic Derivatives and Integrals

$$\frac{d}{dt} t^n = nt^{n-1} \qquad \int t^n \, dt = \frac{t^{n+1}}{n+1} \qquad n \neq -1$$

$$\frac{d}{dt} e^{st} = se^{st} \qquad \int e^{st} \, dt = \frac{1}{s} e^{st}$$

$$\frac{d}{dt} \cos \omega t = -\omega \sin \omega t \qquad \int \cos \omega t \, dt = \frac{1}{\omega} \sin \omega t$$

$$\frac{d}{dt} \sin \omega t = \omega \cos \omega t \qquad \int \sin \omega t \, dt = -\frac{1}{\omega} \cos \omega t$$

$$\frac{d}{dx} \frac{u}{v} = \frac{v \, du/dx - u \, dv/dx}{v^2} \qquad \int u \frac{dv}{dx} dx = uv - \int v \frac{du}{dx} dx$$

Complex Numbers

$$A = a_r + ja_i = |A| \angle \phi_a = |A| e^{j\phi_a}$$

$$\mathrm{Re}[A] = a_r = |A| \cos \phi_a$$

$$\mathrm{Im}[A] = a_i = |A| \sin \phi_a$$

$$|A| = \sqrt{a_r^2 + a_i^2}$$

$$\phi_a = \tan^{-1}(a_i/a_r) \qquad a_r > 0$$

$$= \pm 180° - \tan^{-1}[a_i/(-a_r)] \qquad a_r < 0$$

$$A^* = a_r - ja_i = |A| \underline{/-\phi_a} = |A|e^{-j\phi_a}$$

$$AA^* = |A|^2 = a_r^2 + a_i^2 \qquad A + A^* = 2a_r$$

$$e^{\pm j\alpha} = \cos\alpha \pm j\sin\alpha = 1 \underline{/\pm\alpha}$$

$$e^{j\alpha} + e^{-j\alpha} = 2\cos\alpha \qquad e^{j\alpha} - e^{-j\alpha} = j2\sin\alpha$$

$$j = 0 + j1 = 1 \underline{/90^\circ} = e^{j90^\circ}$$

$$j^2 = -1 \qquad 1/j = -j$$

Exponential and Logarithmic Functions

$$e^\alpha e^\beta = e^{(\alpha+\beta)} \qquad e^\alpha / e^\beta = e^{(\alpha-\beta)}$$

$$\log xy = \log x + \log y \qquad \log(x/y) = \log x - \log y$$

$$\log x^n = n\log x$$

Trigonometric Functions

$$\sin(\alpha \pm 90^\circ) = \pm\cos\alpha \qquad \cos(\alpha \pm 90^\circ) = \mp\sin\alpha$$

$$\sin(\alpha \pm 180^\circ) = -\sin\alpha \qquad \cos(\alpha \pm 180^\circ) = \cos\alpha$$

$$\cos^2\alpha = \tfrac{1}{2}(1 + \cos 2\alpha) \qquad \sin^2\alpha = \tfrac{1}{2}(1 - \cos 2\alpha)$$

$$\sin(\alpha \pm \beta) = \sin\alpha\cos\beta \pm \cos\alpha\sin\beta$$

$$\cos(\alpha \pm \beta) = \cos\alpha\cos\beta \mp \sin\alpha\sin\beta$$

$$\sin\alpha\sin\beta = \tfrac{1}{2}\cos(\alpha - \beta) - \tfrac{1}{2}\cos(\alpha + \beta)$$

$$\sin\alpha\cos\beta = \tfrac{1}{2}\sin(\alpha - \beta) + \tfrac{1}{2}\sin(\alpha + \beta)$$

$$\cos\alpha\cos\beta = \tfrac{1}{2}\cos(\alpha - \beta) + \tfrac{1}{2}\cos(\alpha + \beta)$$

$\cos(\omega t + \phi) = \cos\omega(t + t_0)$ where

$$t_0 = -\frac{\phi(\text{rad})}{\omega} = -\frac{\phi(\text{deg})}{360^\circ}\frac{2\pi}{\omega}$$

Series Expansions and Approximations

$$(1 + x)^n = 1 + nx + \frac{n(n-1)}{2!}x^2 + \cdots$$

$$\approx 1 + nx \qquad x^2 \ll 1$$

$$(1 + x)^{-n} = 1 - nx + \frac{n(n+1)}{2!}x^2 - \cdots$$

$$\approx 1 - nx \qquad x^2 \ll 1$$

$$e^\alpha = 1 + \alpha + \frac{1}{2!}\alpha^2 + \cdots$$

$$\approx 1 + \alpha \qquad \alpha^2 \ll 1$$

$$\sin\alpha = \alpha - \frac{1}{3!}\alpha^3 + \frac{1}{5!}\alpha^5 - \cdots$$
$$\approx \alpha \qquad \alpha^2 \ll 1$$

$$\cos\alpha = 1 - \frac{1}{2!}\alpha^2 + \frac{1}{4!}\alpha^4 - \cdots$$
$$\approx 1 \qquad \alpha^2 \ll 1$$

$$\tan^{-1} x = x - \frac{1}{3}x^3 + \frac{1}{5}x^5 - \cdots \qquad x^2 < 1$$
$$\approx x \qquad x^2 \ll 1$$

Trigonometric and Exponential Integrals

$$\int \cos^2\alpha\, d\alpha = \frac{\alpha}{2} + \frac{\sin 2\alpha}{4} \qquad \int \sin^2\alpha\, d\alpha = \frac{\alpha}{2} - \frac{\sin 2\alpha}{4}$$

$$\int \alpha\cos\alpha\, d\alpha = \cos\alpha + \alpha\sin\alpha$$

$$\int \alpha\sin\alpha\, d\alpha = \sin\alpha - \alpha\cos\alpha$$

$$\int \alpha^2\cos\alpha\, d\alpha = 2\alpha\cos\alpha + (\alpha^2 - 2)\sin\alpha$$

$$\int \alpha^2\sin\alpha\, d\alpha = 2\alpha\sin\alpha - (\alpha^2 - 2)\cos\alpha$$

$$\int \cos at \cos bt\, dt = \frac{\sin(a-b)t}{2(a-b)} + \frac{\sin(a+b)t}{2(a+b)} \qquad a^2 \neq b^2$$

$$\int \sin at \sin bt\, dt = \frac{\sin(a-b)t}{2(a-b)} - \frac{\sin(a+b)t}{2(a+b)} \qquad a^2 \neq b^2$$

$$\int \sin at \cos bt\, dt = -\frac{\cos(a-b)t}{2(a-b)} - \frac{\cos(a+b)t}{2(a+b)} \qquad a^2 \neq b^2$$

$$\int t\, e^{at}\, dt = \frac{e^{at}}{a^2}(at - 1)$$

$$\int t^2 e^{at}\, dt = \frac{e^{at}}{a^3}(a^2t^2 - 2at + 2)$$

$$\int \cos bt\, e^{at}\, dt = \frac{e^{at}}{a^2 + b^2}(a\cos bt + b\sin bt)$$

$$\int \sin bt\, e^{at}\, dt = \frac{e^{at}}{a^2 + b^2}(a\sin bt - b\cos bt)$$

Exercise Solutions

Chapter 1

1.1 $dq = -7 \cdot 10^{15}|q_e| + 2 \cdot 10^{15}|q_e| = -5 \cdot 10^{15}|q_e|$
$i = dq/dt = -5 \cdot 10^{15}|q_e|/5 \cdot 10^{-3} = -0.16$ A

1.2 $q_1 = -8 \text{ A} \times 0.2 \text{ s} = -1.6$ C
$q_2 = 2 \text{ A} \times (0.5 - 0.2) \text{ s} = 0.6$ C
$q_T = q_1 + q_2 = -1.0$ C
$i_{av} = -1.0 \text{ C}/0.5 \text{ s} = -2$ A

1.3 (a) $v = 4.5 \text{ W}/0.5 \text{ A} = 9$ V
(b) $v = -60 \text{ W}/0.5 \text{ A} = -120$ V

1.4 $i = 60 \text{ W}/12 \text{ V} = 5$ A
$iT = q_{stored} = 116 \text{ Ah} \Rightarrow T = 116/5 = 23.2$ hours

1.5 (a) $p = 5 \cdot 10^3 \times 0.4 \cdot 10^{-6} = 2 \cdot 10^{-3} = 2$ mW
(b) $p = 5 \cdot 10^3 \times 10 \cdot 10^{-3} = 50$ W
(c) $p = 5 \cdot 10^3 \times 300 = 1.5 \cdot 10^6 = 1.5$ MW

1.6

v	0	± 10	± 20	± 30
i	0	± 0.01	± 0.08	± 0.27

1.7 (a) $i = 10^{-5}(-20)^3 = -0.08$ A, $p = vi = 1.6$ W
(b) $v = (10^5 \times 0.01)^{1/3} = 10$ V
$p = vi = 0.1 \text{ W} = 100$ mW

1.8 $R = 20 \text{ V}/50 \text{ mA} = 0.4 \text{ k}\Omega$
$G = 50 \text{ mA}/20 \text{ V} = 2.5$ mS

1.9 $i = 12 \text{ V}/5 \text{ k}\Omega = 2.4$ mA
$p = (12 \text{ V})^2/5 \text{ k}\Omega = 28.8$ mW

1.10 $R = 10^{-6}\ \Omega\text{-m} \times 1 \text{ m}/\pi(2 \cdot 10^{-5} \text{ m})^2 = 796\ \Omega$
$\Delta R = 2 \times 796 \times 0.001 = 1.59\ \Omega$

1.11 $i_2 = i_x - i_1 = 6$ A, $i_3 = i_x - i_4 = -2$ A
$i_6 = i_3 + i_5 - i_2 = 0$

1.12 The voltage values must be equal, and the higher-potential terminals must be connected to the same node.

1.13 $v_y = v_5 - v_4 = -9$ V, $v_2 = v_3 + v_y = 11$ V

1.14 (a) $v_R = 9 - 12 = -3$ V
$i_R = -3/60 = -0.05$ A
(b) $v_R = 9 - 9 = 0$, $i_R = 0$
(c) $v_R = 9 - (-9) = 18$ V, $i_R = 18/60 = 0.3$ A

1.15 $i_R = i_s + i_x = 6$ mA, $v_s = v = Ri_R = 150$ V

1.16 If the current source is in series with X, then $i_x = 3 \text{ A} \neq 2.5$ A. If the voltage source is in parallel with X, then $v_x = 12 \text{ V} \neq 10$ V.

1.17 $v_1 = 6i_1 = 30$ V, $v_2 = 50 - v_1 = 20$ V
$i_2 = v_2/20 = 1$ A, $v_3 = v_2 = 20$ V
$i_3 = i_1 - i_2 = 4$ A, $R_3 = v_3/i_3 = 5\ \Omega$

1.18 Typical design

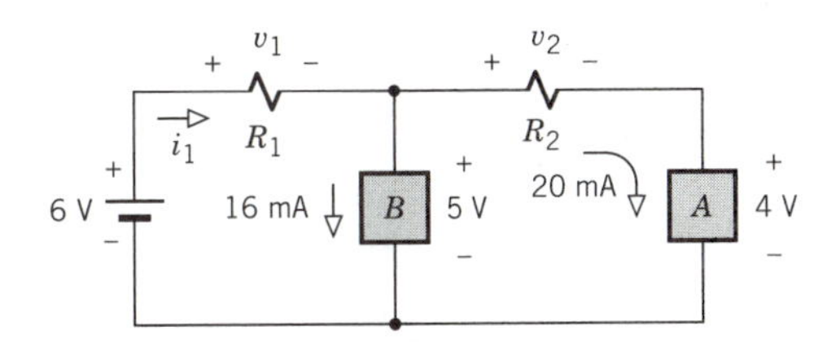

$i_1 = 16 + 20 = 36$ mA, $v_1 = 6 - 5 = 1$ V

$R_1 = v_1/i_1 = 0.0278\ \text{k}\Omega = 27.8\ \Omega$
$v_2 = 5 - 4 = 1\ \text{V}$
$R_2 = v_2/20\ \text{mA} = 0.05\ \text{k}\Omega = 50\ \Omega$

Chapter 2

2.1 $v_1/v = R_1/R_{ser} = 0.2 \Rightarrow R_1 = 0.2 \times 40 = 8\ \Omega$
$v_2/v = R_2/R_{ser} = 0.5 \Rightarrow R_2 = 0.5 \times 40 = 20\ \Omega$
$R_{ser} = R_1 + R_2 + R_3 \Rightarrow R_3 = 40 - (8 + 20) = 12\ \Omega$

2.2 $R_{par} = (1/4 + 1/36 + 1/18)^{-1}$
$= 36/(9 + 1 + 2) = 3\ \Omega$
$4\|(36\|18) = 4\|12 = 3\ \Omega$
$(4\|36)\|18 = 3.6\|18 = 3\ \Omega$

2.3 $p_{max} = (120\ \text{V})^2/(60\|R_2) = 720\ \text{W}$ so $60\|R_2 = 20\ \Omega$
$R_2 = (60 \times 20)/(60 - 20) = 30\ \Omega$
$p_1 = 120^2/60 = 240\ \text{W}$
$p_2 = 120^2/30 = 480\ \text{W}$
$p_{min} = 120^2/(60 + 30) = 160\ \text{W}$

2.4 Let $R_2 = (4 + 8)\|(3 + 9) = 6\ \Omega$
$R_{eq} = 4\|(6 + R_2) = 3\ \Omega$
$v = R_{eq} \times 4\ \text{A} = 12\ \text{V}$, $p = 12\ \text{V} \times 4\ \text{A} = 48\ \text{W}$
$v_2 = [R_2/(6 + R_2)] \times 12\ \text{V} = 6\ \text{V}$

2.5 $R = (1\ \text{S})^{-1} = 1\ \Omega$
$R_1 = (20\ \text{S})^{-1} = 0.05\ \Omega$
$R_2 = (5\ \text{S})^{-1} = 0.2\ \Omega$

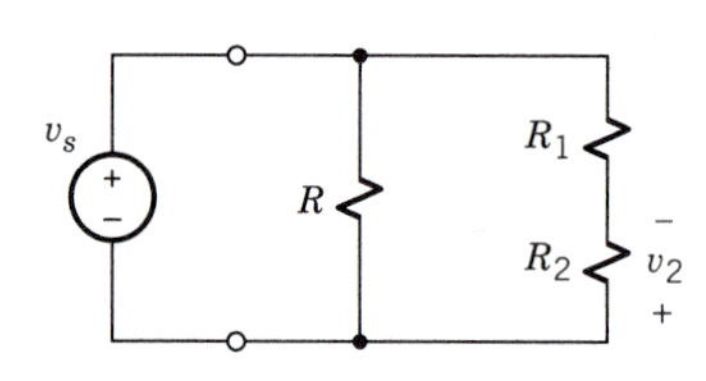

$G_{eq} = 1/1 + 1/(0.05 + 0.2) = 5\ \text{S}$, $i = 5v_s$
$v_2 = [0.2/(0.05 + 0.2)](-v_s) = -0.8v_s$
For the original circuit
$R_{eq} = 1 + 20\|5 = 5\ \Omega$, $v = 5i_s$
$i_2 = [20/(20 + 5)](-i_s) = -0.8i_s$

2.6 $v = 5i_1$
$i_b = i - i_1$, $v = 2(i_b + i_c) = 20i_b = 20i - 20i_1$
$5i_1 = 20i - 20i_1 \Rightarrow i_1 = (20/25)i = 0.8i$

2.7 $i = -i_c = -g_m v$ so $R_{eq} = v/i = -1/g_m$

2.8 $i_1 = 0.8i$, $v = 5\ \text{k}\Omega \times i_1 = 4\ \text{k}\Omega \times i$
$R_{eq} = v/i = 4\ \text{k}\Omega$

2.9 $f(v) = G \times v$
$f(K_a v_a + K_b v_b) = G \times (K_a v_a + K_b v_b)$
$= K_a G v_a + K_b G v_b$
$= K_a f(v_a) + K_b f(v_b)$

2.10 $i_b = i_c/9 = 1\ \text{mA}$, $v = 2(i_b + i_c) = 20\ \text{V}$
$i_1 = v/5 = 4\ \text{mA}$, $i = i_b + i_1 = 5\ \text{mA}$
$K = \hat{i}/i = 3/5 = 0.6$
$\hat{v} = Kv = 0.6 \times 20 = 12\ \text{V}$

2.11 $R\|8 = 6\ \Omega$
4-A source active: $v_{1-1} = (12\|6) \times 4 = 16\ \text{V}$
24-V source active:

$$v_{1-2} = \frac{12}{12 + 6}(-24) = -16\ \text{V}$$

$v_1 = v_{1-1} + v_{1-2} = 0\ \text{V}$

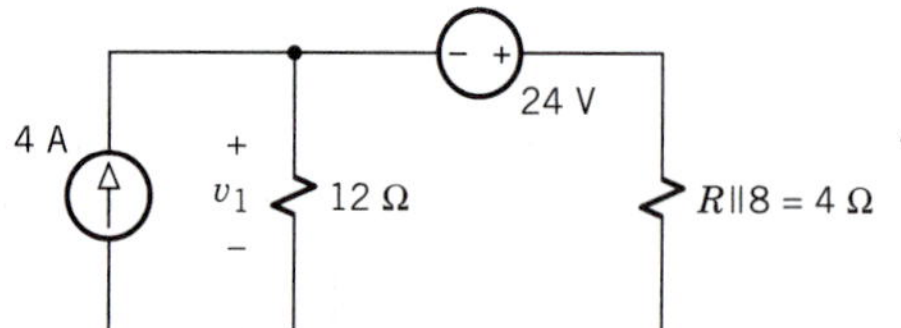

2.12 Superposition would be of no help because there is just one independent source, and the controlled source must not be suppressed when v_s is active.

2.13 $v_{oc} - R_t(5\ \text{mA}) = 30\ \text{V}$ and $v_{oc} - R_t(-15\ \text{mA}) = -30\ \text{V}$ so $v_{oc} = 15\ \text{V}$, $R_t = -3\ \text{k}\Omega$
$i_{sc} = v_{oc}/R_t = -5\ \text{mA}$

2.14 $i = 0 \Rightarrow v_{oc} = 10\ \Omega \times 6\ \text{A} = 60\ \text{V}$
$v = 0 \Rightarrow i_{sc} = [10/(2 + 10)] \times 6\ \text{A} = 5\ \text{A}$
$R_t = 60\ \text{V}/5\ \text{A} = 12\ \Omega$
$i = 60/(12 + R_L) = 4 \Rightarrow R_L = 3\ \Omega$

2.15 With $i = 0$: $-i_x - 3i_x = 0 \Rightarrow i_x = 0$ so $v_{oc} = 10\ \text{V}$
With source suppressed: $i_x = v_t/2$
$i_t = i_x + 3i_x = 4(v_t/2) = 2v_t$
so $R_t = v_t/i_t = ½ = 0.5\ \text{k}\Omega$
Then $i_{sc} = v_{oc}/R_t = 10/0.5 = 20\ \text{mA}$
With $v = 0$: $i_x = -10/2 = -5\ \text{mA}$
$i_{sc} = -i_x - 3i_x = -4(-5\ \text{mA}) = 20\ \text{mA}$

2.16

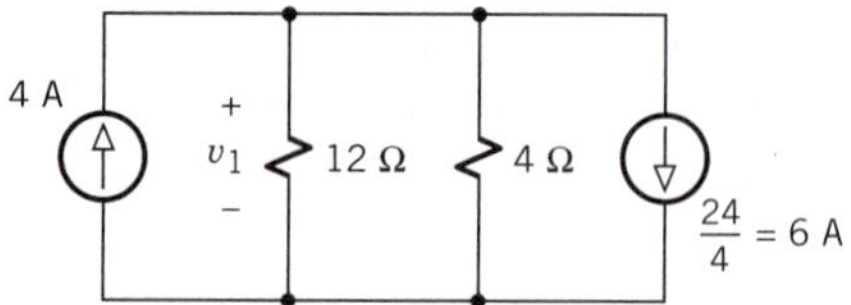

$12\|4 = 3\ \Omega$, $v_1 = 3(4 - 6) = -6\ \text{V}$

2.17

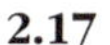

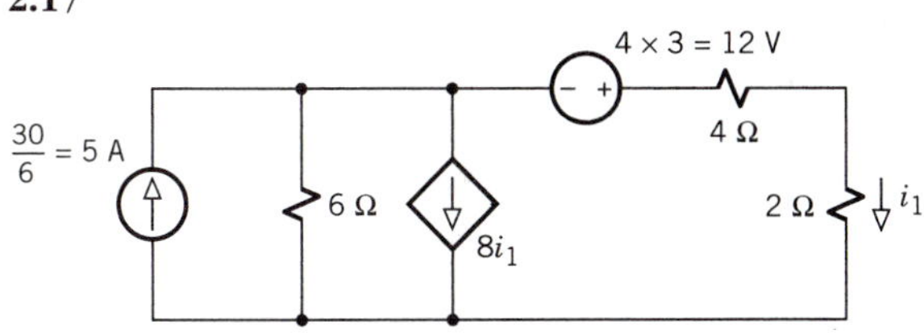

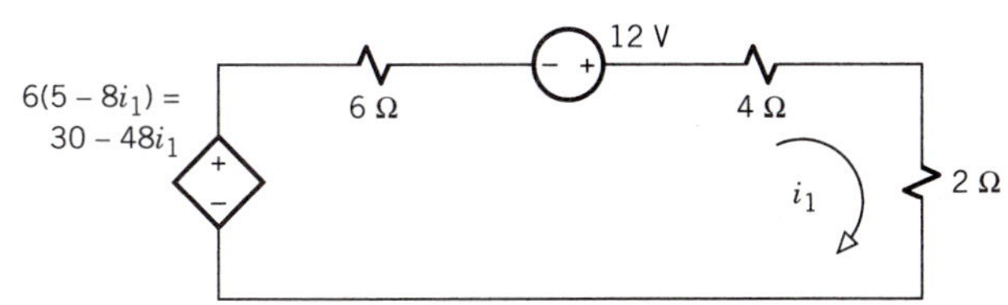

$(30 - 48i_1) + 12 - (6 + 4 + 2)i_1 = 0$
so $i_1 = 0.7$ A

Chapter 3

3.1 $R_L = 0 \Rightarrow v = 0$ so $i_s = 100$ mA
With $R_L = 1$ kΩ,

$$i = \frac{R_s}{R_s + 1\text{ k}\Omega} \times 100\text{ mA}$$

$$= 80\text{ mA} \Rightarrow R_s = 4\text{ k}\Omega$$

3.2 With 0.4 Ω and 0.2 Ω in series with the voltage sources
$i = (12 - 6)\text{V}/(0.4 + 0.2)\Omega = 10$ A
$v = 12 - 0.4i = 0.2i + 6 = 8$ V

3.3 $v_s = v_{oc} = 40$ V, $R_s = R_t = 4\ \Omega$ so $R_L = 4\ \Omega$
$p_{max} = 40^2/(4 \times 4) = 100$ W
$i = 40/(4 + 4) = 5$ A
$v = 40 - 4 \times 5 = 20$ V
$p_s = v^2/20 + (50 - v)^2/5 = 200$ W
Eff $= 100/(100 + 200) \approx 33\%$

3.4 $$A_i = \frac{i_{out}}{i_s} = \frac{i_{in}}{i_s} \times \frac{i_c}{i_{in}} \times \frac{i_{out}}{i_c}$$

$$= \frac{R_s}{R_s + R_i} \times \beta \times \frac{R_o}{R_o + R_L}$$

3.5

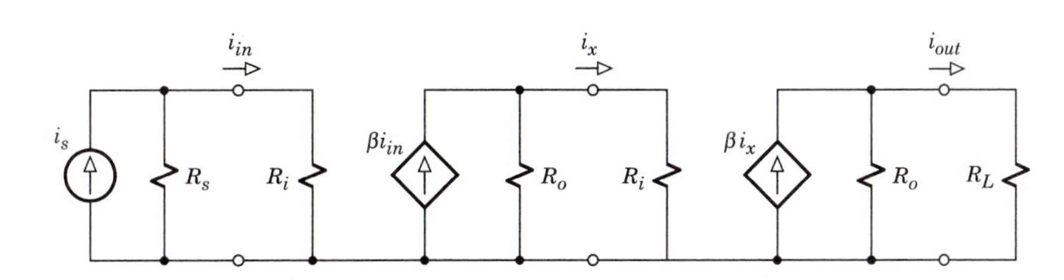

$$A_i = \frac{i_{in}}{i_s} \times \frac{\beta i_{in}}{i_{in}} \times \frac{i_x}{\beta i_{in}} \times \frac{\beta i_x}{i_x} \times \frac{i_{out}}{\beta i_x}$$

$$= \frac{R_s}{R_s + R_i} \times \beta \times \frac{R_o}{R_o + R_i} \times \beta \times \frac{R_o}{R_o + R_L}$$

$A_i \approx \beta^2$ when $R_i \ll R_s$, $R_i \ll R_o$, and $R_o \gg R_L$

3.6 $v_{max} = 25/10^5 = 0.25$ mV

v_n	0	0.2 mV	0.4 mV	0.6 mV	0.8 mV	1.0 mV
v_d	0.5 mV	0.3 mV	0.1 mV	−0.1 mV	−0.3 mV	−0.5 mV
v_{out}	25 V	25 V	10 V	−10 V	−25 V	−25 V

3.7 (a) $B = 4/(76 + 4) = 1/20$
$|v_{out}|_{max} \approx |v_{in}|_{max}/B = 16\text{ V} \Rightarrow V_{PS} > 16$ V
(b) $A = 10^4 \Rightarrow AB = 500$

$$\frac{v_d}{v_{in}} = \frac{1}{501} = 1.99601 \cdot 10^{-3}$$

$$\frac{v_{out}}{v_{in}} = \frac{10^4}{501} = 19.9601$$

$A = 10^6 \Rightarrow AB = 50{,}000$

$$\frac{v_d}{v_{in}} = \frac{1}{50{,}001} = 1.99996 \cdot 10^{-5}$$

$$\frac{v_{out}}{v_{in}} = \frac{10^6}{50{,}001} = 19.9996$$

3.8 $v_n = \dfrac{R_1}{R_1 + R_F} v_{out}$ and $v_{in} = v_p = v_n$ so
$v_{out}/v_{in} = v_{out}/v_n = (R_1 + R_F)/R_1$

3.9 $R_i = R_1$, $R_F/R_1 = 20$, and $R_F \leq 100$ kΩ, so
$R_{i-max} = R_{F-max}/20 = 5$ kΩ, $R_1 = 5$ kΩ
$R_F = 100$ kΩ

3.10 $|v_{out}|_{max} = K(3 \times 0.2 + 0.6) \leq 12$ V so
$K_{max} = 12/1.2 = 10$ and
$v_{out} = -(30v_1 + 10v_2)$
$= -[(R_F/R_1)v_1 + (R_F/R_2)v_2]$
Thus $R_F = 30R_1 = 30$ kΩ, $R_2 = R_F/10 = 3$ kΩ

3.11 v_1 active $\Rightarrow v_p = 0 \Rightarrow$ inverting amplifier with
$v_{out-1} = -(R_F/R_1)v_1$
v_2 active $\Rightarrow R_1$ grounded $\Rightarrow$ noninverting amplifier with
$v_p = R_3v_2/(R_3 + R_2) = v_2/(1 + R_2/R_3)$
$= v_2/(1 + R_1/R_F)$

$$v_{out-2} = (R_1 + R_F)v_p/R_1 = (R_1 + R_F)v_2/R_1(1 + R_1/R_F) = (R_F/R_1)v_2$$

$$v_{out} = v_{out-2} + v_{out-1} = (R_F/R_1)(v_2 - v_1)$$

3.12 $B = 1/20,\ 1 + AB = 501$
$R_i \approx 501 \times 3{\cdot}10^6 = 1{,}503\ \text{M}\Omega$
$R_o = 100/501 = 0.1996\ \Omega$

3.13 $I = [R_1/(R_1 + R_m)]I_u$ so
$I_{ufs} = [(R_1 + R_m)/R_1]I_{fs} = (1 + R_m/R_1)I_{fs}$

3.14 $V_{fs} = 15\ \Omega \times 2\ \text{mA} = 30\ \text{mV}$

$1 + R_v/15 = 5/0.03 \Rightarrow R_v = 2485\ \Omega$

$1 + 15/R_a = 30/2 \ \Rightarrow R_a = 1.071\ \Omega$

3.15 $I_u = 1.60\ \text{mA} \pm 0.05 \times 1.60\ \text{mA} = 1.52\text{–}1.68\ \text{mA}$
$R_{am} = 200\ \Omega\text{–mA}/2\ \text{mA} = 100\ \Omega$ and $R \geq 1\ \text{k}\Omega$ so $1 + R_{am}/R \leq 1.1$
$I_{act} \leq 1.1 \times 1.68 = 1.85\ \text{mA}$
$I_{act} \geq 1.0 \times 1.52 = 1.52\ \text{mA}$

3.16 With R_a removed, $R_a \| R_m \Rightarrow R_m$, so half-scale deflection corresponds to $R_u = R_r + R_m \geq 3\ \text{k}\Omega$. Thus, values of R_u much smaller than 3 kΩ could not be read accurately.

3.17 (a) $R = R_u/(R_2/R_1) \approx 50/(R_2/R_1)$
Since we must have $1 \leq R \leq 1000$, the bridge can be balanced with $R_2/R_1 = 0.1$, 1, or 10.
(b) Maximum accuracy, with three significant figures, occurs when $100 \leq R \leq 999$, so $R_2/R_1 = 0.1$.

Chapter 4

4.1 $G_{11} = \dfrac{1}{15} + \dfrac{1}{30} + \dfrac{1}{10} = \dfrac{1}{5}\ \text{S}$

$$i_{s1} = 6 - \frac{20}{10} = 4\ \text{A}$$

$$\frac{1}{5}v_1 = 4 \Rightarrow v_1 = 20\ \text{V}$$

$$i_a = \frac{(-20) - v_1}{10} = -4\ \text{A}$$

4.2
$$[G] = \begin{bmatrix} \frac{1}{5} + \frac{1}{1} + \frac{1}{12+8} & -\frac{1}{12+8} \\ -\frac{1}{12+8} & \frac{1}{20} + \frac{1}{12+8} + \frac{1}{10} \end{bmatrix}$$

$$[i_s] = \begin{bmatrix} \frac{20}{1} + 6 \\ \frac{20}{20} - 6 \end{bmatrix}$$

$$\begin{bmatrix} 1.25 & -0.05 \\ -0.05 & 0.2 \end{bmatrix}\begin{bmatrix} v_1 \\ v_2 \end{bmatrix} = \begin{bmatrix} 26 \\ -5 \end{bmatrix}$$

$$v_1 = 20\ \text{V},\ v_2 = -20\ \text{V}$$

4.3 (a) R_x is connected between node 1 and the reference node.
(b) R_x is connected between node 1 and a voltage source.

4.4 The current leaving node 3 and entering node 1 still equals 3 mA, regardless of the value of R in series.

4.5
$$[G] = \begin{bmatrix} 1 + 0.4 + 0.1 & -1 & -0.4 \\ -1 & 1 + 0.5 & 0 \\ -0.4 & 0 & 0.4 + 0.2 \end{bmatrix}$$

$$[i_s] = \begin{bmatrix} -0.4v_x \\ i_z - 0.5v_w \\ 0.4v_x - i_z + i_y \end{bmatrix}$$

4.6
$$\frac{v_1 - (-12)}{4} + \frac{v_1}{6} + \frac{v_1 - 36}{12} + \frac{(v_1 - 36) - 24}{3} = 0$$

$$v_1 = 24\ \text{V},\ v_2 = v_1 - 36 = -12\ \text{V}$$

4.7 $5i_1 + (4 + 6)(i_1 + 3) - 60 = 0 \Rightarrow 15i_1 = 30$
$i_1 = 2\ \text{A},\ v_a = 4(i_1 + 3) = 20\ \text{V}$

4.8
$$\begin{bmatrix} 2 + 4 + 8 & -8 \\ -8 & 8 + 7 + 6 \end{bmatrix}\begin{bmatrix} i_1 \\ i_2 \end{bmatrix} = \begin{bmatrix} 20 - 4 \times 3 \\ -7 \times 3 \end{bmatrix}$$

$$i_1 = 0,\ i_2 = -1\ \text{A}$$

4.9 (a) R_x is on the perimeter of mesh 2.
(b) R_x is the branch between mesh 2 and a current source.

4.10 The branch voltage still equals 20 V, regardless of the value of R in parallel.

4.11 $(5 + 4 + 6)i_1 = 4 \times 3 - 6 \times 2 \Rightarrow i_1 = 0$

4.12
$$\begin{bmatrix} 9 + 4 + 5 & 0 & 0 \\ 0 & 3 + 6 & -6 \\ 0 & -6 & 8 + 6 + 2 \end{bmatrix}\begin{bmatrix} i_1 \\ i_2 \\ i_3 \end{bmatrix} = \begin{bmatrix} -v_z - 5i_x \\ v_z - 6i_y \\ 6i_y - 8i_x \end{bmatrix}$$

4.13 Supermesh has i_1 in upper half and $i_1 + i_y$ in lower half, so
$4i_1 + 2(i_1 - i_2) + 5(i_1 + i_y - i_2) - v_x = 0$
Mesh 2 has
$8i_2 + 5[i_2 - (i_1 + i_y)] + 2(i_2 - i_1) = 0$

$$\begin{bmatrix} 4 + 2 + 5 & -2 - 5 \\ -2 - 5 & 8 + 5 + 2 \end{bmatrix}\begin{bmatrix} i_1 \\ i_2 \end{bmatrix} = \begin{bmatrix} v_x - 5i_y \\ 5i_y \end{bmatrix}$$

4.14
$$[R] = \begin{bmatrix} 12 & -2 \\ -2 & 9 \end{bmatrix} \qquad i_a = 3 + i_1$$

$$[v_s] = \begin{bmatrix} 20i_a - 6 \times 3 \\ -20i_a \end{bmatrix} = \begin{bmatrix} 42 + 20i_1 \\ -60 - 20i_1 \end{bmatrix}$$

$$= \begin{bmatrix} 42 \\ -60 \end{bmatrix} + \begin{bmatrix} 20 & 0 \\ -20 & 0 \end{bmatrix}\begin{bmatrix} i_1 \\ i_2 \end{bmatrix}$$

$$\begin{bmatrix} 12 - 20 & -2 - 0 \\ -2 - (-20) & 9 - 0 \end{bmatrix}\begin{bmatrix} i_1 \\ i_2 \end{bmatrix} = \begin{bmatrix} 42 \\ -60 \end{bmatrix}$$

4.15 $[R]$ is the same as in Exercise 4.14.
$v_b = 2(i_1 - i_2)$

$$[v_s] = \begin{bmatrix} -6 \times 3 + 6 \times 2(i_1 - i_2) \\ -6 \times 2(i_1 - i_2) \end{bmatrix} = \begin{bmatrix} -18 \\ 0 \end{bmatrix} + \begin{bmatrix} 12 & -12 \\ -12 & 12 \end{bmatrix}\begin{bmatrix} i_1 \\ i_2 \end{bmatrix}$$

$$\begin{bmatrix} 12 - 12 & -2 - (-12) \\ -2 - (-12) & 9 - 12 \end{bmatrix}\begin{bmatrix} i_1 \\ i_2 \end{bmatrix} = \begin{bmatrix} -18 \\ 0 \end{bmatrix}$$

$i_2 = -1.8$ A

4.16 $[G] = \dfrac{1}{16}\begin{bmatrix} 14 & -10 \\ -10 & 11 \end{bmatrix} \qquad v_a = v_1 - 12$

$$[i_s] = \begin{bmatrix} 12/4 + 3v_a/2 \\ -3v_a/2 \end{bmatrix} = \begin{bmatrix} -15 + 1.5v_1 \\ 18 - 1.5v_1 \end{bmatrix} = \begin{bmatrix} -15 \\ 18 \end{bmatrix} + \frac{1}{16}\begin{bmatrix} 24 & 0 \\ -24 & 0 \end{bmatrix}\begin{bmatrix} v_1 \\ v_2 \end{bmatrix}$$

$$\frac{1}{16}\begin{bmatrix} 14 - 24 & -10 - 0 \\ -10 - (-24) & 11 - 0 \end{bmatrix}\begin{bmatrix} v_1 \\ v_2 \end{bmatrix} = \begin{bmatrix} -15 \\ 18 \end{bmatrix}$$

4.17 Use current source i and mesh current i_b.
Then $i_a = i - i_b$

$(20 + 5)i_b = 20i - 50(i - i_b) \Rightarrow i_b = 1.2i$

$v = 10i + 20(i - i_b) = 6i \Rightarrow R_{eq} = v/i = 6\ \Omega$

Or use voltage source v and node voltage v_a.
Then $i_a = v_a/20$

$$\left(\frac{1}{10} + \frac{1}{20} + \frac{1}{5}\right)v_a = \frac{v}{10} + \frac{50}{5}\frac{v_a}{20} \Rightarrow v_a = -\frac{2}{3}v$$

$i = (v - v_a)/10 = 5v/30 \Rightarrow R_{eq} = v/i = 6\ \Omega$

4.18 $i_x = (v - 10)/2$
$(1/2)v = 10/2 - 3i_x - i$
$\quad = 20 - 1.5v - i$
$v = 10 - 0.5i$
$v_{oc} = 10$ V, $R_t = 0.5$ kΩ

4.19 $$\left(\frac{1}{R_1 + R_2} + \frac{1}{R_3} + \frac{1}{R_4}\right)v_1 = \frac{v_{out}}{R_4}$$

$$v_{out} = \left(\frac{R_4}{R_1 + R_2} + \frac{R_4}{R_3} + \frac{R_4}{R_4}\right)v_1 = (K_2 + 2)v_1$$

Virtual short requires $v_n = v_{in}$ so

$$v_{in} = v_n = [R_1/(R_1 + R_2)]v_1 = v_1/K_1$$

Thus, $v_1 = K_1 v_{in}$ and $v_{out} = K_1(K_2 + 2)v_{in}$

4.20 $R_{ab} = 12$, $R_{bc} = 8$, $R_{ca} = 4$

$$R_a = \frac{12 \times 4}{24} = 2,\ R_b = \frac{8 \times 12}{24} = 4,$$

$$R_c = \frac{4 \times 8}{24} = \frac{4}{3}$$

$R_{eq} = R_b + (R_a + 3)\|(R_c + 2)$
$\quad = 4 + 5\|(10/3) = 6\ \Omega$

Chapter 5

5.1 $i(t) = 200 \cdot 10^{-12}\, dv/dt$
$\quad = 0.4$ mA $\quad 0 < t < 50\ \mu s$
$\quad = -2$ mA $\quad 50 < t < 60\ \mu s$

$w(t) = \tfrac{1}{2}\, 200 \cdot 10^{-12} v^2 \qquad p(t) = v(t)i(t)$

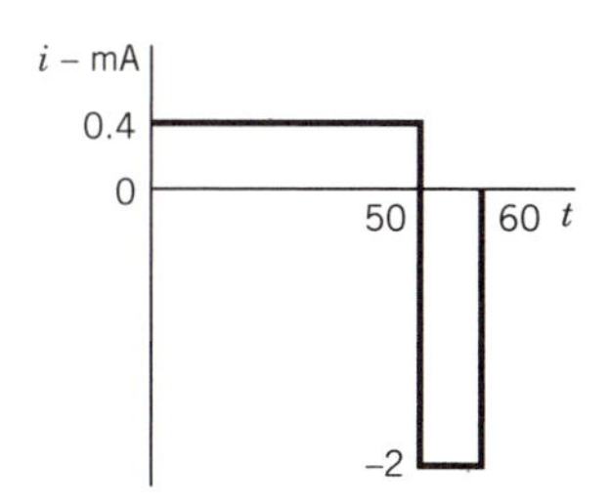

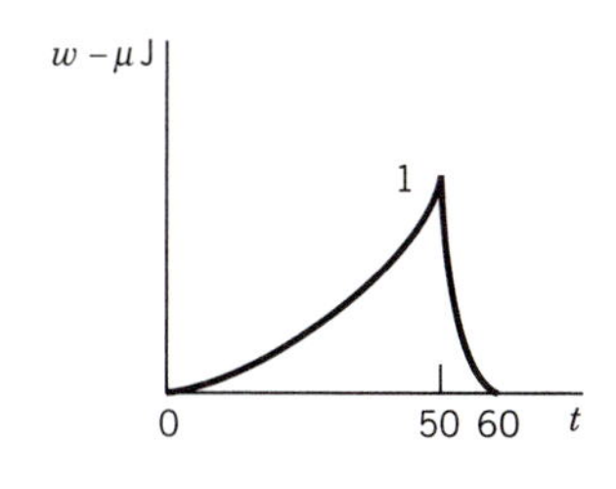

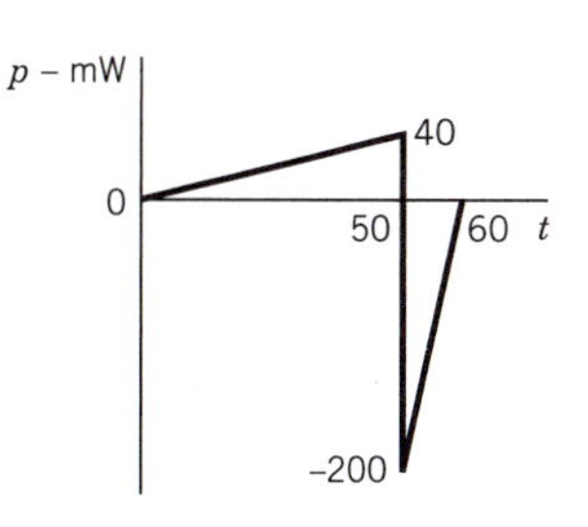

5.2 Use Eq. (5.9) with $t_1 = 0$:

$$v(t) = 0 + \frac{5 \cdot 10^{-3}}{10 \cdot 10^{-6}}t = 200 \Rightarrow t = 0.4\ \text{s}$$

5.3 $$v(t) = 0 + \frac{1}{C}\int_0^t Ie^{-a\lambda}\, d\lambda = \frac{I}{aC}(1 - e^{-at}) \quad t > 0$$

$v(0^+) = (I/aC)(1 - 1) = 0$

5.4 $i = C_{par}\, dv/dt \Rightarrow dv/dt = i/C_{par}$
$i_n = C_n\, dv/dt = (C_n/C_{par})i$

5.5 $C_2 = 8 + 12 = 20\ \mu$F
$C_{eq} = (5 \times 20)/(5 + 20) = 4\ \mu$F
$v(0) = v_1(0) + v_2(0) = 10$ V
From Eq. (5.16) with $C_{ser} = C_{eq}$:

$v_1(t_1) = 0 + (4/5)(0 - 10) = -8$ V

$v_2(t_1) = 10 + (4/20)(0 - 10) = +8$ V

5.6 $v_L = 0$ and $i_C = 0$ so $v_C = 0$ and
$i_L = 30/20 = 1.5$ A
$w = w_L = \tfrac{1}{2}\, 16 \cdot 10^{-3} \times 1.5^2 = 18$ mJ

5.7 $dw = p\,dt = (L\,di/dt)i\,dt = Li\,di$
$w = L \int i\,di = \frac{1}{2}Li^2 + w_0$
But $w = 0$ when $i = 0$ so $w_0 = 0$ and $w = \frac{1}{2}Li^2$

5.8 $v = V \qquad t_1 < t < t_2$
$= 0 \qquad t > t_2$
$i = i(t_1) + (V/L)t \qquad t_1 < t < t_2$
$= i(t_2) \qquad t > t_2$

5.9 $1/L_{eq} = 1/0.4 + 1/0.4 + 1/(0.1 + 0.5)$ so
$L_{eq} = 0.15$ H
$v = 0.15\,di/dt = 30 \cos 50t$ V
From Eq. (5.30):

$$i_a(t) = 2 + (0.15/0.6)[4 \sin 50t - 0]$$
$$= 2 + \sin 50t$$

5.10 $i = C\,dv_C/dt$ and $L\,di/dt + Ri + v_C = v$ so

$$LC\frac{d^2 v_C}{dt^2} + RC\frac{dv_C}{dt} + v_C = v$$
$$LC\frac{d^3 v_C}{dt^3} + RC\frac{d^2 v_C}{dt^2} + \frac{dv_C}{dt} = \frac{dv}{dt}$$

But $dv_C/dt = i/C$, so $C\,d^2 v_C/dt^2 = di/dt$ and

$$L\frac{d^2 i}{dt^2} + R\frac{di}{dt} + \frac{1}{C}i = \frac{dv}{dt}$$

5.11 $L\,di/dt + Ri = 0 \Rightarrow Ls + R = 0$ so
$s = -R/L = -10$
$A = i_N(0^+) = i(0^-) = 2$ A
$i_N(t) = 2e^{-10t}$ A $= 0.736$ A $\qquad t = 0.1$
$= 13.5$ mA $\qquad t = 0.5$
$= 90.8$ μA $\qquad t = 1$

5.12 $10^{-4}\,dv/dt + 2{\cdot}10^{-4}v = i$
$10^{-4}\,dv_F/dt + 2{\cdot}10^{-4}v_F = 0.002t$
From Table 5.3: $v_F = K_1 t + K_0 \Rightarrow dv_F/dt =$
$K_1 10^{-4} K_1 + 2{\cdot}10^{-4} K_1 t + 2{\cdot}10^{-4} K_0 = 0.002t$
$K_1 = 0.002/2{\cdot}10^{-4} = 10$
$K_0 = -10^{-4}K_1/2{\cdot}10^{-4} = -5$
$v_F(t) = 10t - 5$ V

5.13 $10^{-4}\,dv_N/dt + 2{\cdot}10^{-4}v_N = 0$ and $v_N = Ae^{st}$ so
$10^{-4}sAe^{st} + 2{\cdot}10^{-4}Ae^{st} = 0$
$s + 2 = 0 \Rightarrow v_n = Ae^{-2t}$
$10^{-4}\,dv_F/dt + 2{\cdot}10^{-4}v_F = 0.005e^{-2t}$ so
$v_F = K_2 te^{-2t}$
$10^{-4}(1 - 2t)K_2 e^{-2t} + 2{\cdot}10^{-4}K_2 te^{-2t} =$
$0.005e^{-2t}$
$10^{-4}K_2 e^{-2t} = 0.005e^{-2t}$ so
$K_2 = 0.005/10^{-4} = 50$
$v_F(t) = 50te^{-2t}$ V

5.14 $v_F(t) = 10t - 5$ from Exercise 5.12
$v_N(t) = Ae^{-2t}$ from Exercise 5.13
$v(t) = v_F(t) + v_N(t) = 10t - 5 + Ae^{-2t}$
$v(0) = -5 + A = -30 \Rightarrow A = -25$
$v(t) = 10t - 5 - 25e^{-2t}$
$v_N(t) \approx 0$ when $e^{-2t} < e^{-5}$, so the duration of the transient interval is about 2.5 s.

Chapter 6

6.1 (a) $\sin(\omega t - 60°) = \cos(\omega t - 60° - 90°)$ so
$\underline{V} = 100 \angle -150°$
(b) $-\cos(\omega t + 80°) = \cos(\omega t - 80° \pm 180°)$ so
$\underline{I} = 0.2 \angle 260°$ or $\underline{I} = 0.2 \angle -100°$

6.2 (a) $T = 1/(250{\cdot}10^3) = 4$ μs
$t_0 = -(90/360)T = -1$ μs
(b) $T = 2\pi/100\pi = 20$ ms
$t_0 = -(-144/360)T = +8$ ms

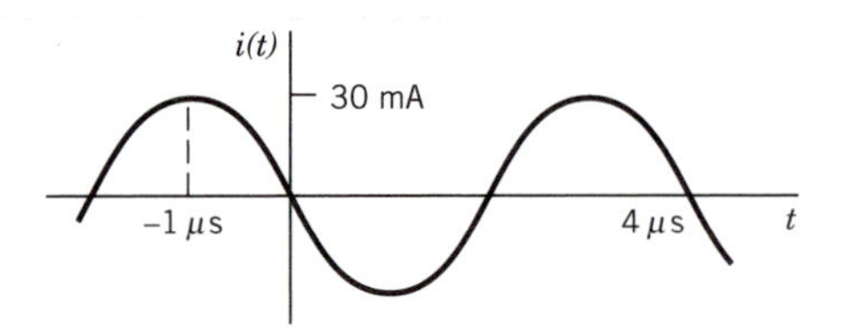

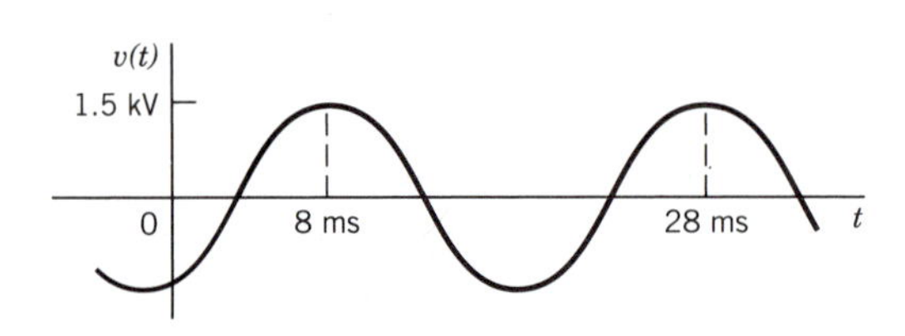

6.3 $A - A^* = (-4 + j3) - (-4 - j3) = 0 + j6$
$= \sqrt{0^2 + 6^2} \angle \tan^{-1}(6/0) = 6 \angle 90°$

$jA = j(-4) + j^2 3 = -3 - j4$
$= \sqrt{(-3)^2 + (-4)^2} \angle \pm 180° - \tan^{-1}(-4/3)$
$= 5 \angle -126.9°$

$$A + \frac{25}{A} = A + \frac{25A^*}{AA^*} = -4 + j3 + \frac{25(-4 - j3)}{(-4)^2 + 3^3}$$
$= -8 + j0$
$= \sqrt{(-8)^2 + 0^2} \angle \pm 180° - \tan^{-1}(0/8)$
$= 8 \angle \pm 180°$

6.4 (a) $A = \sqrt{3^2 + 4^2} \angle \tan^{-1}(4/3) = 5 \angle 53.1°$
$A^2 = 5 \times 5 \angle 53.1° + 53.1° = 25 \angle 106.2°$
$1/A = (1/5) \angle 0° - 53.1° = 0.2 \angle -53.1°$
(b) $A^*/B^* = |A| \angle -\phi_a / |B| \angle -\phi_b$
$= (|A|/|B|) \angle -\phi_a + \phi_b$
$= [(|A|/|B|) \angle \phi_a - \phi_b]^* = (A/B)^*$

6.5 $i_R(t) = (60/5) \sin 8000t$
$= 12 \cos(8000t - 90°)$ A
so $\underline{I}_R = 12$ A $\angle -90° = 0 - j12$
$i_C(t) = 25{\cdot}10^{-6}(8000 \times 60 \cos 8000t)$
$= 12 \cos 8000t$ A

so $\underline{I}_C = 12\ \text{A}\ \angle 0° = 12 + j0$
$\underline{I} = 12 - j12 = 17.0\ \text{A}\ \angle -45°$
$i(t) = 17.0 \cos(8000t - 45°)\ \text{A}$

6.6 $dv/dt = \text{Re}[\underline{V}\, d(e^{j\omega t})/dt] = \text{Re}[\underline{V} j\omega e^{j\omega t}]$
$\text{Re}[\underline{I}e^{j\omega t}] = C\,\text{Re}[\underline{V} j\omega e^{j\omega t}] = \text{Re}[j\omega C\underline{V}e^{j\omega t}]$
so $\underline{I} = j\omega C\underline{V} \Rightarrow \underline{V} = \underline{I}/j\omega C$

6.7 $Z_L = j800 \times 25{\cdot}10^{-3} = j20\ \Omega = 20\ \Omega\ \angle 90°$
$Y_L = 1/Z_L = -j0.05\ \text{S} = 0.05\ \text{S}\ \angle -90°$
$\underline{V} = Z_L\underline{I} = (20\ \Omega\ \angle 90°)(4\ \text{A}\ \angle -50°)$
$= 80\ \text{V}\ \angle 40°$
$v(t) = 80 \cos(800t + 40°)\ \text{V}$

6.8 $Z_L = j20,\ Y_L = -j0.05$

$$Z = \frac{Z_R Z_L}{Z_R + Z_L} = \frac{10 \times j20}{10 + j20}$$

$$= \frac{4000 + j2000}{10^2 + 20^2} = 8 + j4\ \Omega$$

$R(\omega) = 8\ \Omega,\ X(\omega) = 4\ \Omega$
$Y = Y_R + Y_L = 0.1 - j0.05$ so
$G(\omega) = 0.1\ \text{S},\ B(\omega) = -0.05\ \text{S}$

6.9 $$Z(j\omega) = \frac{R(j\omega L)}{R + j\omega L}\,\frac{R - j\omega L}{R - j\omega L} = \frac{(\omega L)^2 R + j\omega L R^2}{R^2 + (\omega L)^2}$$

$$R(\omega) = \frac{(\omega L)^2 R}{R^2 + (\omega L)^2} \qquad X(\omega) = \frac{\omega L R^2}{R^2 + (\omega L)^2}$$

6.10 $Z_L = j14,\ Z_C = -j40,\ (-j40)\|80 = 16 - j32$
$Z = (8 + j14) + (16 - j32) = 24 - j18$
$= 30\ \Omega\ \angle -36.9°$
$\underline{I} = \underline{V}/Z = 2\ \text{A}\ \angle 36.9°$
$i(t) = 2 \cos(1000t + 36.9°)\ \text{A}$
$\underline{I}_C = [80/(80 - j40)]\underline{I} = 0.8 - j1.6$
$= 1.79\ \text{A}\ \angle 63.4°$
$i_C(t) = 1.79 \cos(1000t + 63.4°)\ \text{A}$

6.11 $\underline{I}_2 = 1,\ \underline{I}_C = 80/(-j40) = j2$
$\underline{I} = \underline{I}_2 + \underline{I}_C = 1 + j2$
$\underline{V} = (8 + j14)\underline{I} + \underline{V}_2 = 60 + j30$
$\underline{V}_2/\underline{V} = 1.19\ \angle -26.6°$

6.12 $\omega_{osc} = 2\pi f_{osc} = 62.8{\cdot}10^3$
$C = 2/L\omega_{osc}^2 = 25.3\ \text{nF}$
$K = 1 + (L/CR^2) = 1.790$

6.13 $Z_L = j5,\ Z_C = -j10$
$Z_t = 12 + j5 - j10 = 13\ \Omega\ \angle -22.6°$
$\underline{V}_{oc} = (12 + j5) \times 2 = 26\ \text{V}\ \angle 22.6°$
$\underline{I}_{sc} = [(12 + j5)/(12 + j5 - j10)] \times 2$
$= 2\ \text{A}\ \angle 45.2°$
$Z_t\underline{I}_{sc} = 26\ \text{V}\ \angle 22.6° = \underline{V}_{oc}$

6.14 $Y = j/5 + 1/5 + (1 - j2)/5 = (2 - j1)/5$
$\underline{V}_x = \underline{V}_1 - 30$

$$\underline{I}_s = \frac{30}{-j5} + \frac{1 - j2}{5}\,2\underline{V}_x$$

$$= \frac{-60 + j150}{5} + \frac{2 - j4}{5}\,\underline{V}_1$$

$[2 - j1 - (2 - j4)]\underline{V}_1 = -60 + j150$
$\underline{V}_1 = 50 + j20 = 53.9\ \text{V}\ \angle 21.8°$

6.15 $$[Z] = \begin{bmatrix} 5 - j5 & -5 \\ -5 & 6 + j2 \end{bmatrix} \qquad \underline{V}_x = -(-j5)\underline{I}_1$$

$$[\underline{V}_s] = \begin{bmatrix} 30 \\ -2\underline{V}_x \end{bmatrix} = \begin{bmatrix} 30 \\ 0 \end{bmatrix} + \begin{bmatrix} 0 & 0 \\ -j10 & 0 \end{bmatrix}\begin{bmatrix} \underline{I}_1 \\ \underline{I}_2 \end{bmatrix}$$

$$\begin{bmatrix} 5 - j5 & -5 \\ -5 + j10 & 6 + j2 \end{bmatrix}\begin{bmatrix} \underline{I}_1 \\ \underline{I}_2 \end{bmatrix} = \begin{bmatrix} 30 \\ 0 \end{bmatrix}$$

6.16 $\theta = 0 \Rightarrow \angle\underline{I} = \angle\underline{V} = 0$
$\underline{I}_R + \underline{I}_L = \underline{I}$
$\underline{I}_R = \underline{V}_2/4,\ \underline{I}_L = \underline{V}_2/j4$
$\angle\underline{I}_L = \angle\underline{I}_R - 90°,\ |\underline{I}_R| = |\underline{I}_L|$
$\underline{V}_1 + \underline{V}_2 = \underline{V}$
$\underline{V}_1 = -j2\underline{I},\ \underline{V}_2 = 4\underline{I}_R$
$\angle\underline{V}_1 = \angle\underline{I} - 90°,\ \angle\underline{V}_2 = \angle\underline{I}_R$

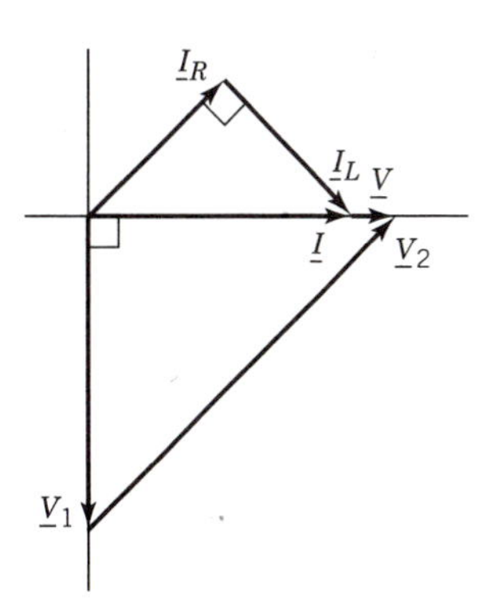

6.17 $\omega_0 = 1/\sqrt{LC} = 2{\cdot}10^5,\ Q_{ser} = \omega_0 L/R = 4$
$\omega = 1.6{\cdot}10^5$
$\underline{V}_L = j\omega L\underline{I} = 32\ \text{V}\ \angle 90°$
$\underline{V}_C = \underline{I}/j\omega C = 50\ \text{V}\ \angle -90°$
$\underline{V}_x = \underline{V}_L + \underline{V}_C = 18\ \text{V}\ \angle -90°$
$\underline{V}_R = R\underline{I} = 10\ \text{V}\ \angle 0°$
$\underline{V} = \underline{V}_R + \underline{V}_x = 20.6\ \text{V}\ \angle -60.9°$

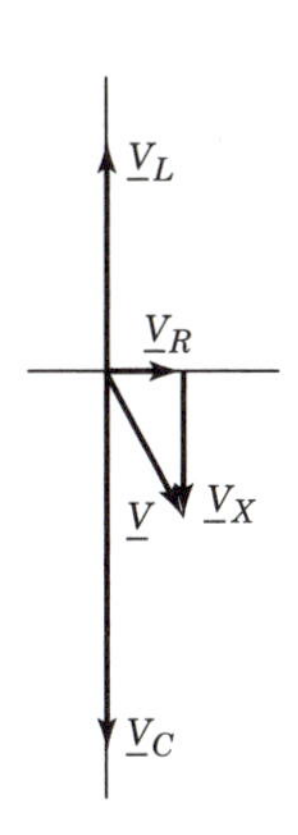

6.18 $Z(j\omega_0) = 1/Y(j\omega_0) = R,\ \underline{V} = R\underline{I}$
$\underline{I}_C = j\omega_0 C\underline{V} = j\omega_0 CR\underline{I} = jQ_{par}\underline{I}$
$\underline{I}_L = \underline{V}/j\omega_0 L = -j(R/\omega_0 L)\underline{I} = -jQ_{par}\underline{I}$

6.19 With current source active, $\omega = 5$ as in Exercise 6.12, so

$$\underline{I}_1 = \underline{I}_{sc} = 2\text{ A }\angle 45.2^\circ$$

With voltage source active, $\omega = 10$, $Z_L = j10$ and $Z_C = -j5$ so

$$\underline{I}_2 = -\frac{52\ \angle 0^\circ}{12 + j10 - j5} = \frac{52\ \angle \pm 180^\circ}{13\ \angle 22.6^\circ} = 4\text{ A }\angle 157.4^\circ$$

$$i = i_1 + i_2 = 2\cos(5t + 45.2^\circ) + 4\cos(10t + 157.4^\circ)$$

6.20 At any one frequency, the net reactance of Z_u depends on both L_u and C_u so their effects cannot be evaluated separately.

6.21 $R_u + 1/j\omega C_u = R_2(R_3 + 1/j\omega C_3)/R_1 = R_2R_3/R_1 + R_2/j\omega C_3 R_1$
so $R_u = (R_2/R_1)R_3$
and $C_u = (R_1/R_2)C_3$

Chapter 7

7.1 $Z = 10\|j5 = 2 + j4\ \Omega = 4.47\ \Omega\ \angle 63.4^\circ$
$P = \frac{1}{2}I_m^2 \times 2\ \Omega = 25\text{ W}$
$p_{max} = \frac{1}{2}|Z|I_m^2(\cos\theta + 1) = 80.9\text{ W}$
$p_{min} = \frac{1}{2}|Z|I_m^2(\cos\theta - 1) = -30.9\text{ W}$

7.2 $\cos^2(\omega t + \phi_v) = \frac{1}{2} + \frac{1}{2}\cos(2\omega t + 2\phi_v)$

$$\int_{t_1}^{t_1+T} \cos(2\omega t + 2\phi_v)\,dt = 0$$

$$V_{rms}^2 = \frac{V_m^2}{2T}\int_{t_1}^{t_1+T} dt = \frac{V_m^2}{2}$$

7.3
$$\frac{dP}{dR} = \frac{(R_s + R)^2 + (X_s + X)^2 - R2(R_s + R)}{[(R_s + R)^2 + (X_s + X)^2]^2} \times V_{rms}^2 = \frac{R_s^2 + (X_s + X)^2 - R^2}{[(R_s + R)^2 + (X_s + X)^2]^2} V_{rms}^2$$

$dP/dR = 0 \Rightarrow R^2 = R_s^2 + (X_s + X)^2$

7.4 $Z = 2 + j12.5 + Z_{RC} = 6 + j4.5 = 7.5\ \Omega\ \angle 36.9^\circ$
$I_{rms} = |\underline{I}| = 300/7.5 = 40\text{ A}$
$P = 6 \times 40^2 = 9.6\text{ kW}$
$Q = 4.5 \times 40^2 = 7.2\text{ kVAr}$
$P_L/P = 4/(2 + 4) \approx 67\%$

7.5 $|S| = 200\text{ V} \times 85\text{ A} = 17\text{ kVA}$
pf $= 8\text{ kW}/17\text{ kVA} = 0.471$ lagging

$Q = +\sqrt{17^2 - 8^2} = +15\text{ kVAr}$

$X = Q/I_{rms}^2 = +2.35\ \Omega$

$R = P/I_{rms}^2 = 1.11\ \Omega,\ Z = 1.11 + j2.35\ \Omega$

7.6 For heater: $I_{rms} = 200/8 = 25\text{ A}$
$P = RI_{rms}^2 = 5\text{ kW}$
For motor and heater:

$|S| = \sqrt{(8 + 5)^2 + 15^2} = 19.8\text{ kVA}$
$I_{rms} = |S|/V_{rms} = 99.3\text{ A}$

For plant: $S = 13 + j0,\ I_{rms} = |S|/V_{rms} = 65\text{ A}$

Load	P(kW)	Q(kVAr)	$\|S\|$(kVA)	I_{rms}(A)
Motor	8	+15	17	85
Heater	5	0	5	25
	13	+15	19.8	99.3
Capacitor	0	−15	15	75
Plant	13	0	13	65

$C = -Q_C/\omega|V_C|^2 = 995\ \mu\text{F}$

7.7 $i_i(t) \approx i(t)$
$\underline{I}_v = \underline{V}/j\omega L_M = (|\underline{V}|/\omega L_M)\ \angle \phi_v - 90^\circ$
Replacing ϕ_v by $\phi_v - 90^\circ$ and R_M by ωL_M in Eq. (7.32) gives
$y_{ss} = (K_M/\omega L_M)\ |\underline{V}|\ |\underline{I}| \cos(\phi_v - 90^\circ - \phi_i)$
But $\phi_v - 90^\circ + \phi_i = \theta - 90^\circ$ and
$\cos(\theta - 90^\circ) = \sin\theta$, so
$y_{ss} = (K_M/\omega L_M)\ |\underline{V}|\ |\underline{I}| \sin\theta = (K_M/\omega L_M)Q$

7.8 $\cos(\alpha \pm 120^\circ) = \cos\alpha\cos 120^\circ \mp \sin\alpha\sin 120^\circ$
$\cos(\alpha - 120^\circ) + \cos(\alpha + 120^\circ) = 2\cos 120^\circ\cos\alpha = -\cos\alpha$
$\cos\alpha + \cos(\alpha - 120^\circ) + \cos(\alpha + 120^\circ) = 0$

7.9 $V_\phi = 52/\sqrt{3} = 30.0\text{ kV}$ and $\angle\underline{V}_a = \angle\underline{V}_{ab} - 30^\circ$
$\underline{V}_a = 30\text{ kV }\angle -30^\circ,\ \underline{V}_b = 30\text{ kV }\angle -150^\circ$
$\underline{V}_c = 30\text{ kV }\angle 90^\circ$

7.10 $\underline{V}_a = 30\text{ kV }\angle -30^\circ$ from Exercise 7.9
$Z_Y = \underline{V}_a/\underline{I}_a = 30\ \Omega\ \angle 60^\circ = 15 + j26\ \Omega$
$P = 3 \times 15\ |\underline{I}_a|^2 = 45\text{ MW}$
$Q = 3 \times 26\ |\underline{I}_a|^2 = 78\text{ MVAr}$

7.11 $Z_Y = Z_\Delta/3 = 1.5 + j2 = 2.5\ \Omega\ \angle 53.1^\circ$
Before pf correction: $I_l = 500\text{ V}/2.5\ \Omega = 200\text{ A}$,
$P_\phi = 1.5 \times 200^2 = 60\text{ kW}$
$Q_\phi = 2 \times 200^2 = 80\text{ kVAr}$
After pf correction:
$3P_\phi = |S| = \sqrt{3}(\sqrt{3} \times 500)I_l$,
$I_l = 60\text{ kW}/500\text{ V} = 120\text{ A}$
$Q_C = -\omega C_\Delta|\sqrt{3} \times 500|^2 = -80\text{ kVAr}$ so
$C_\Delta = 340\ \mu\text{F}$

7.12 $\underline{I}_a^* = 1\text{ kA }\angle +90^\circ,\ \underline{V}_{bc} = 52\text{ kV }\angle -120^\circ$,
$\underline{I}_b^* = (1\ \angle -90^\circ - 120^\circ)^* = 1\text{ kA }\angle +210^\circ$
$\underline{V}_{ac} = -\underline{V}_{ca} = 52\ \angle 120^\circ \pm 180^\circ = 52\text{ kV }\angle -60^\circ$
$P_1 = \text{Re}[52\text{ MVA }\angle -60^\circ + 90^\circ] = 52\cos 30^\circ = 45\text{ MW}$

$P_2 = \text{Re}[52 \text{ MVA } \angle -120° + 210°]$
$= 52 \cos 90° = 0$
$P_1 + P_2 = 45 \text{ MW} = P$
$\sqrt{3}|P_2 - P_1| = \sqrt{3} \times 45 = 78 \text{ MVAr} = |Q|$

7.13 $600 \text{ V}/12\ \Omega = 50 \text{ A}$
$\underline{I}_a = 50 \text{ A} \angle 0°,\ \underline{I}_b = 50 \text{ A} \angle -120°$
$\underline{I}_C = 50 \text{ A} \angle 120°$
$\underline{I}_{cn} = (600 \angle +120°)/(5 \angle -90°) = 120 \angle 210°$
$\underline{I}_c = \underline{I}_C + \underline{I}_{cn} = 130 \text{ A} \angle -173°$
$\underline{I}_n = -\underline{I}_{cn} = 120 \text{ A} \angle 30°$
$P_\phi = 12|\underline{I}_a|^2 = 30 \text{ kW},\ Q_\phi = 0$
$P = 3P_\phi = 90 \text{ kW},\ Q = (-5)|\underline{I}_{cn}|^2 = -72 \text{ kVAr}$
$\text{pf} = 90/\sqrt{90^2 + 72^2} = 0.78 \text{ leading}$

7.14 $V_\phi = 208/\sqrt{3} = 120 \text{ V}$
$\underline{I}_a = (120 \angle -30°)/(10 \angle 90°)$
$= 12 \angle -120° = -6 - j10.39$
$\underline{I}_b = (120 \angle -150°)/(10 \angle 53.1°)$
$= 12 \angle -203.1° = -11.04 + j4.71$
$\underline{I}_c = (120 \angle 90°)/(10 \angle 0°) = 12 \angle 90° = 0 + j12$
$\underline{I}_n = -(\underline{I}_a + \underline{I}_b + \underline{I}_c)$
$= 17.04 - j6.32 = 18.17 \angle -20.3°$

Chapter 8

8.1 $N = 20/50 = 0.4,\ j\omega L = j4$
$\underline{I}_2 = (25 - 10)/(3 + j4) = 3 \text{ A} \angle -53.1°$
$\underline{I}_1 = -0.4\underline{I}_2 = 1.2 \text{ A} \angle 126.9°$

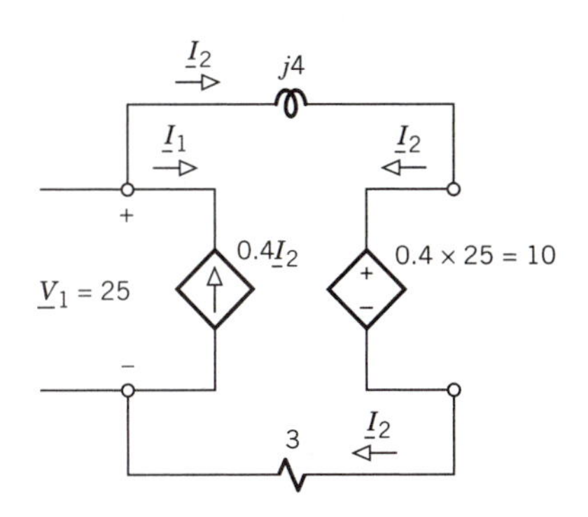

8.2 $Z/N^2 = 3 - j2.25\ \Omega$
$\underline{I}_{in} = 25/(3 + j4 + 3 - j2.25) = 4 \text{ A} \angle -16.3°$
$\underline{V}_{in} = (3 - j2.25)\underline{I}_{in} = 15 \text{ V} \angle -53.1°$
$\underline{I}_{out} = \underline{I}_{in}/2 = 2 \text{ A} \angle -16.3°$
$\underline{V}_{out} = 2\underline{V}_{in} = 30 \text{ V} \angle -53.1°$

8.3 $N\underline{V}_s = 50 \text{ V} \angle 0°,\ N^2 Z_s = 12 + j16\ \Omega$
$\underline{I}_{out} = 50/(12 + j16 + 12 - j9) = 2 \text{ A} \angle -16.3°$
$\underline{V}_{out} = (12 - j9)\underline{I}_{out} = 30 \text{ V} \angle -53.1°$
$\underline{I}_{in} = 2\underline{I}_{out} = 4 \text{ A} \angle -16.3°$
$\underline{V}_{in} = \underline{V}_{out}/2 = 15 \text{ V} \angle -53.1°$

8.4 $Z = 500 + j0,\ |Z_s| = 179\ \Omega$
$|Z|/N^2 = |Z_s| \Rightarrow N^2 = 500/179,\ N = 1.67$
$|\underline{I}_{in}| = 18/|80 - j160 + 179| = 59.1 \text{ mA}$
$P = \tfrac{1}{2}|\underline{I}_{in}|^2 \times 179 = 313 \text{ mW}$

8.5 $L_1 - 2M + L_2 \geq L_1 - 2M_{max} + L_2$
$M_{max} = \sqrt{L_1 L_2}$
$L_1 - 2M_{max} + L_2 = L_1 - 2\sqrt{L_1 L_2} + L_2$
$= (\sqrt{L_1} - \sqrt{L_2})^2 \geq 0$
so $L_1 - 2M + L_2 \geq 0$

8.6 $v_2 = M\dfrac{di_1}{dt} + L_2\dfrac{di_2}{dt} = 0 \Rightarrow \dfrac{di_2}{dt} = -\dfrac{M}{L_2}\dfrac{di_1}{dt}$

$v_1 = L_1\dfrac{di_1}{dt} + M\left(-\dfrac{M}{L_2}\dfrac{di_1}{dt}\right) = \left(L_1 - \dfrac{M^2}{L_2}\right)\dfrac{di_1}{dt}$

$M^2 = k^2 L_1 L_2$ so $L_{eq} = L_1 - M^2/L_2 = (1 - k^2)L_1$

8.7 $dw/di_2 = Mi_1 + L_2 i_2 = 0 \Rightarrow i_2 = (-M/L_2)i_1$
Then $w_{min} = \tfrac{1}{2}L_1 i_1^2 + Mi_1(-M/L_2)i_1$
$+ \tfrac{1}{2}L_2(-M/L_2)^2 i_1^2$
$= (L_1 L_2 - M^2)i_1^2/2L_2$

8.8 $\omega M = 0.75\sqrt{8 \times 2} = 3\ \Omega$ and $\underline{V}_1 = \underline{V}_2 = \underline{V}$, so
$j8\underline{I}_1 + j3\underline{I}_2 = j3\underline{I}_1 + j2\underline{I}_2 \Rightarrow \underline{I}_2 = -5\underline{I}_1$
$\underline{V} = \underline{V}_1 = j8\underline{I}_1 + j3\underline{I}_2 = j[8 + 3(-5)]\underline{I}_1 = -j7\underline{I}_1$
$\underline{I} = \underline{I}_1 + \underline{I}_2 = (1 - 5)\underline{I}_1 = -4\underline{I}_1$
$Z_{eq} = \underline{V}/\underline{I} = j7/4 = j1.75\ \Omega$

8.9 $Z_{in} = j\omega L_1 + \dfrac{(\omega M)^2}{R + j(\omega L_2 + X)}\dfrac{R - j(\omega L_2 + X)}{R - j(\omega L_2 + X)}$

$= \dfrac{(\omega M)^2 R}{R^2 + (\omega L_2 + X)^2} + j\omega L_1$

$- j\dfrac{(\omega M)^2(\omega L_2 + X)}{R^2 + (\omega L_2 + X)^2}$

$\text{Re}[Z_{in}] = 0 \Rightarrow R = 0$
$\text{Im}[Z_{in}] < 0 \Rightarrow$
$\omega L_1 < (\omega M)^2(\omega L_2 + X)/[R^2 + (\omega L_2 + X)^2]$
so $(\omega M)^2 > \omega L_1(\omega L_2 + X)$

8.10 $L_{eq} = (L_1 - M)\|(L_2 - M) + M$

$= \dfrac{(L_1 - M)(L_2 - M)}{L_1 + L_2 - 2M} + M$

$= \dfrac{L_1 L_2 - M^2}{L_1 + L_2 - 2M}$

8.11 $L_1 L_2 - M^2 = 0.01,\ \omega L_{22} = 100/40$
$\omega L_{12} = 100/(-30)$
$(-j0.4 + j0.3 + 1/50)\underline{V}_{out} = -j0.4\underline{V}_{in}$
$\underline{V}_{out}/\underline{V}_{in} = -j0.4/(0.02 - j0.1)$
$= 3.92 \angle -11.3°$

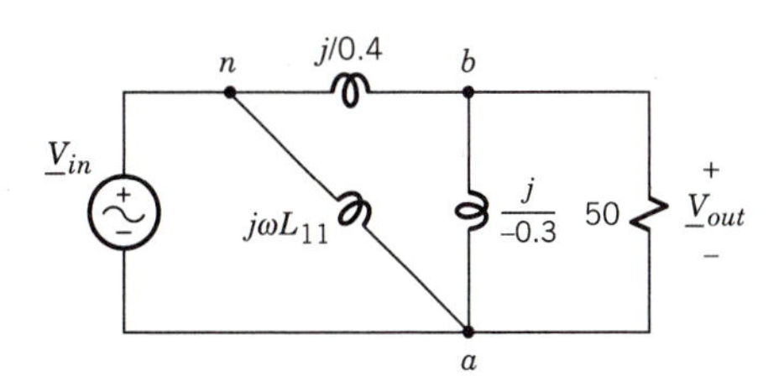

8.12 $v_1 = \left(L_1 \mp \frac{M}{N_0}\right)\frac{di_1}{dt} \pm \frac{M}{N_0}\frac{d}{dt}(i_1 + N_0 i_2)$

$= L_1 \frac{di_1}{dt} \pm M\frac{di_2}{dt}$

$\frac{v_2}{N_0} = \left(\frac{L_2}{N_0^2} \mp \frac{M}{N_0}\right)\frac{d}{dt}(N_0 i_2)$

$\pm \frac{M}{N_0}\frac{d}{dt}(i_1 + N_0 i_2)$

$= \pm (M/N_0)\, di_1/dt + (L_2/N_0)\, di_2/dt$

so $v_2 = \pm M\, di_1/dt + L_2\, di_2/dt$

8.13 $Z/N^2 = 2.32 + j5\ \Omega$

$5\underline{I}_{out} = 240/(2.4 + j5.5) = 40.0\ \angle{-66.4°}$

$\underline{V}_{out} = (58 + j125) \times \underline{I}_{out} = 1100\text{ V}\ \angle{-1.3°}$

$P_{out} = 58 \times |\underline{I}_{out}|^2 = 3712\text{ W}$

$\underline{I}_c = 240/40 = 6\text{ A}$

$P_{dis} = 40 \times 6^2 + 0.08 \times 40^2 = 1568\text{ W}$

$\text{Eff} = 3712/5280 \approx 70\%$

8.14 $P_{sc} = |\underline{V}_{in}|^2/R_c + R_w|N\underline{I}_{out}|^2$ so

$R_w = (P_{sc} - |\underline{V}_{in}|^2/R_c)/N^2|\underline{I}_{out}|^2$

$Q_{sc} = |\underline{V}_{in}|^2/X_m + X_l|N\underline{I}_{out}|^2$ so

$X_l = (Q_{sc} - |\underline{V}_{in}|^2/X_m)/N^2|\underline{I}_{out}|^2$

Chapter 9

9.1 $i(0^-) = 0,\ v_C(0^+) = v_C(0^-) = 25\text{ V}$

$v(0^+) = [40/(40 + 10)]v_C(0^+) = 20\text{ V}$

$\tau = (40 + 10)\ \Omega \times 2\ \mu\text{F} = 100\ \mu\text{s}$

$v(t) = 20e^{-t/\tau}\text{ V} \qquad t > 0$

$i(t) = -v(t)/(40\ \Omega) = -0.5e^{-t/\tau}\text{ A} \qquad t > 0$

9.2 $I_{ss} = 12\text{ V}/3\text{ k}\Omega = 4\text{ mA},\ R_{eq} = 2\text{ k}\Omega,$

$\tau = 20\text{ mH}/2\text{ k}\Omega = 10^{-5}\text{ s}$

$i(t) = 4(1 - e^{-100{,}000t})\text{ mA} \qquad t > 0$

$v(t) = 8\text{ V} - 2\text{ k}\Omega \times i(t) = 8e^{-100{,}000t}\text{ V}$

for $t > 0$

9.3 $R_{eq} = 2\text{ k}\Omega,\ D = 0.4\text{ s}$

(a) $\tau = R_{eq}C \le D/5$ so

$C \le 0.4/(5 \times 2000) = 40\ \mu\text{F}$

(b) $\tau = R_{eq}C = 0.4\text{ s} = D,\ V_{ss} = 8\text{ V}$

$v(t)$ has the same shape as Fig. 9.12b with $D = 0.4$ s and $v(D) = 8(1 - e^{-0.4/0.4}) = 5.06$ V

9.4 $i_L(0^+) = i_L(0^-) = [30/(10 + 30)](-4) = -3\text{ A}$

$\tau = 2\ H/(10 + 30)\ \Omega = 1/20\text{ s}$

$I_0 = i(0^+) = i_s(0^+) - i_L(0^+) = 5\text{ A}$

$I_{ss} = [10/(10 + 30)] \times 2 = 0.5\text{ A}$

$i(t) = 0.5 + 4.5e^{-20t}\text{ A} \qquad t > 0$

9.5 $i_L(0^+) = -3$ A and $\tau = 1/20$ s from Exercise 9.4

$I_0 = i(0^+) = i_s(0^+) - i_L(0^+) = 7\text{ A}$

$\underline{I} = [(10 + j20)/(30 + 10 + j20)]\underline{I}_s = 2\ \angle{36.9°}$

$i_F(t) = 2\cos(10t + 36.9°),\ i_F(0^+) = 1.6\text{ A}$

$i(t) = 2\cos(10t + 36.9°) + 5.4e^{-20t}\text{ A} \qquad t \ge 0$

9.6 $i_R = v_C/R,\ Cv_C' = i_C = i_L - i_R = i_L - v_C/R$

$Li_L' = v_L = v_s - v_C \Rightarrow v_C = v_s - Li_L'$

Thus, $C(v_s' - Li_L'') = i_L - (v_s - Li_L')/R$

so $i_L'' + (1/RC)i_L' + (1/LC)i_L =$

$(1/L)v_s' + (1/LRC)v_s$

9.7 (a) $\alpha = (R/2L) + (R + R)/2L = 3R/2L$

$\omega_0^2 = R^2/L^2$

so $\alpha^2 = 9R^2/4L^2 = 2.25(R^2/L^2) > \omega_0^2$

(b) $\alpha = 3,\ \omega_0^2 = 4 \Rightarrow p_1, p_2 = -3 \pm \sqrt{3^2 - 4}$

$= -0.764, -5.24$

so $\tau_1 = -1/p_1 = 1.31,\ \tau_2 = -1/p_2 = 0.191$

9.8 $\alpha = 1/2RC = 4,\ \omega_0^2 = 1/LC = 24/L$

$L = 0.3 \Rightarrow \omega_0^2 = 80 > \alpha^2$

$\omega_d = \sqrt{80 - 4^2} = 8$ so

$p_1, p_2 = -4 \pm j8$ and

$v_C(t) = 2|A_1|e^{-4t}\cos(8t + \angle A_1)$

$L = 2 \Rightarrow \omega_0^2 = 12 < \alpha^2$ so

$p_1, p_2 = -4 \pm \sqrt{4^2 - 12} = -2, -6$

$v_C(t) = A_1e^{-2t} + A_2e^{-6t}$

9.9 $\alpha = 1/2RC = 4,\ \omega_0^2 = 1/LC = 24/L$

$\omega_0^2 = \alpha^2 \Rightarrow 24/L = 16,\ L = 1.5\text{ H}$

9.10 $v_C(0^+) = v_C(0^-) = 0,\ i_L(0^+) = i_L(0^-) = I_1$

$i_R(t) = v_C(t)/R,\ i_R(0^+) = 0$

$i_C(t) = I_2 - i_L(t) - v_C(t)/R,\ i_C(0^+) = I_2 - I_1$

$v_L(t) = v_C(t),\ v_L(0^+) = 0$

$v_C'(0^+) = (I_2 - I_1)/C,\ i_L'(0^+) = 0$

$i_R'(0^+) = v_C(0^+)/R = (I_2 - I_1)/RC$

$i_C'(0^+) = -i_L'(0^+) - v_C'(0^+)/R$

$= -(I_2 - I_1)/RC$

9.11 $i_L(0^+) = 2,\ i_L'(0^+) = 0,\ I_{ss} = 0$

$L = 2 \Rightarrow p_1 = -2,\ p_2 = -6$

$$\begin{bmatrix} 1 & 1 \\ -2 & -6 \end{bmatrix}\begin{bmatrix} A_1 \\ A_2 \end{bmatrix} = \begin{bmatrix} 2 \\ 0 \end{bmatrix} \Rightarrow \begin{matrix} A_1 = 3 \\ A_2 = -1 \end{matrix}$$

$i_L(t) = 3e^{-2t} - e^{-6t}\text{ A}$

$L = 0.3 \Rightarrow p_1, p_2 = -4 \pm j8$

$$\begin{bmatrix} 1 & 1 \\ -4 + j8 & -4 - j8 \end{bmatrix}\begin{bmatrix} A_1 \\ A_2 \end{bmatrix} = \begin{bmatrix} 2 \\ 0 \end{bmatrix}$$

$A_1 = (-8 - j16)/(-j16) = 1.12\ \angle{-26.2°}$

$i_L(t) = 2.24e^{-4t}\cos(8t - 26.2°)\text{ A}$

9.12 $i_L(0^+) = 0,\ i_L'(0^+) = 0,\ I_{ss} = 2,\ L = 1.5$ so

$p_1 = p_2 = -4$

$A_3 = 0 - 2 = -2,\ A_4 = 0 + 4(-2) = -8$

$i_L(t) = 2 - 2e^{-4t} - 8te^{-4t}\ \text{A} \qquad t > 0$

Chapter 10

10.1 (a) $\underline{V} = -10 = 10\ \text{V}\ \angle \pm 180^\circ,\ s = 3 + j0$

(b) $i(t) = 0.5e^{0t}\cos(500t - 20^\circ) = 0.5\cos(500t - 20^\circ)\ \text{A}$

10.2 $Z(s) = 0.5s \| 6 = 3s/(0.5s + 6)$

$\underline{I} = 2\ \text{A}\ \angle \pm 180^\circ,\ s = -8$

$Z(-8) = -12 = 12\ \Omega\ \angle \pm 180^\circ$

$\underline{V} = 12\ \angle \pm 180^\circ \times 2\ \angle \pm 180^\circ = 24\ \text{V}\ \angle 0^\circ$

$v(t) = 24e^{-8t}\ \text{V}$

10.3 $Z_{in}(s) = R^3/(sLR) = 1/[s(L/R^2)] = 1/sC_{eq}$ with $C_{eq} = L/R^2$

10.4 $\underline{I}_x = \underline{I}_R + \underline{I}_C = r_m\underline{I}_{in}/R + r_m\underline{I}_{in}/(1/sC)$

$\underline{V}_{in} = r_m\underline{I}_x = r_m^2\underline{I}_{in}(1/R + sC)$

$Z_{in}(s) = \underline{V}_{in}/\underline{I}_{in} = r_m^2/R + sr_m^2C$

$R_{eq} = r_m^2/R,\ L_{eq} = r_m^2C$

10.5 $v_C'' + \dfrac{12}{6}v_C' + \dfrac{12}{3}v_C = 12i_s'$

$$H(s) = \frac{\underline{V}_C}{\underline{I}_s} = \frac{12s}{s^2 + 2s + 4}$$

10.6 $\underline{V}_R = \dfrac{R}{sL + R + 1/sC}\underline{V}_s$

$$H(s) = \frac{\underline{V}_R}{\underline{V}_s} = \frac{RCs}{LCs^2 + RCs + 1}$$

so $LCv_R'' + RCv_R' + v_R = RCv_s'$

$v_R = Ri_L \Rightarrow LCRi_L'' + R^2Ci_L' + Ri_L = RCv_s'$

so $i_L'' + (R/L)i_L' + (1/LC)i_L = (1/L)v_s'$

10.7 $4\|(16/s) = 16/(s + 4)$

$[s + 16/(s + 4) + 2]\underline{I} = \underline{V} + 2 \times \beta\underline{I}$

$[s + 16/(s + 4) + 2 - 2\beta]\underline{I} = \underline{V}$

$$H(s) = \frac{\underline{I}}{\underline{V}} = \frac{s + 4}{s^2 + (6 - 2\beta)s + 24 - 8\beta}$$

10.8 $H_3(s) = \dfrac{\underline{I}_{out}}{\underline{I}_{in}} = \dfrac{\underline{I}_{out}}{\underline{V}_s} \times \dfrac{\underline{V}_s}{\underline{I}_{in}} = \dfrac{H_2(s)}{H_1(s)} = \dfrac{\pm Ms}{L_2s + R_2}$

10.9 $s^3 + 12s^2 + 32s = s(s^2 + 2 \times 6s + 32)$ so

$z_1 = 0,\ z_2, z_3 = -6 \pm \sqrt{6^2 - 32} = -4, -8$

$(s^2 + 8s + 25)^2 = (s^2 + 2 \times 4s + 25)^2$ so

$p_1, p_2 = -4 \pm j\sqrt{25 - 4^2} = -4 \pm j3,$

$p_3 = p_1,\ p_4 = p_2$

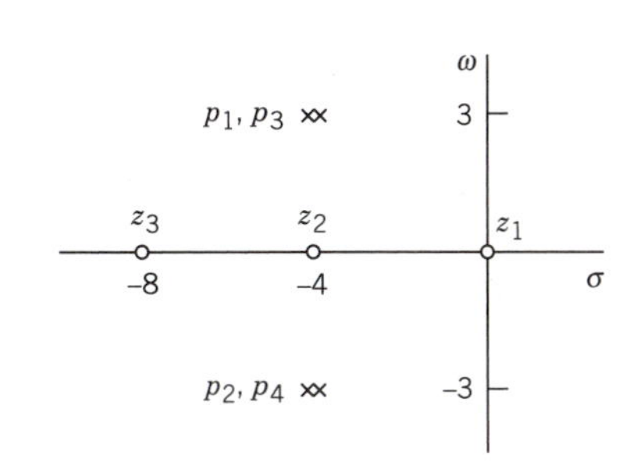

10.10

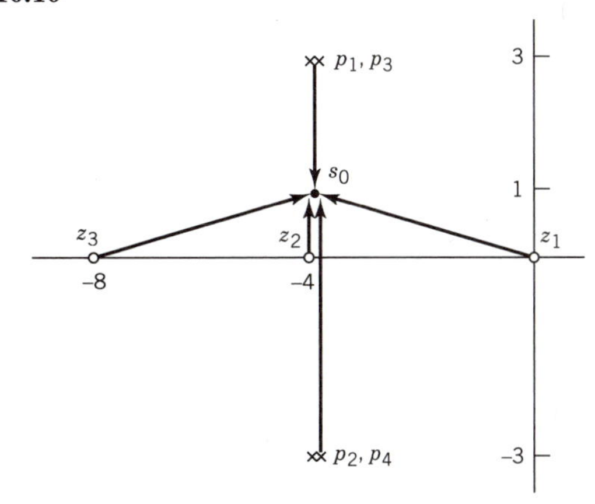

$s_0 - z_1 = -4 + j1 =$

$s_0 - z_2 = 0 + j1 = 1\ \angle 90^\circ$

$s_0 - z_3 = 4 + j1 = \sqrt{17}\ \angle 14.0^\circ$

$s_0 - p_1 = s_0 - p_3 = 0 - j2 = 2\ \angle -90^\circ$

$s_0 - p_2 = s_0 - p_4 = 0 + j4 = 4\ \angle 90^\circ$

$$H(s_0) = \frac{64\sqrt{17}\angle 166.0^\circ\ 1\angle 90^\circ\ \sqrt{17}\angle 14.0^\circ}{2\angle -90^\circ\ 4\angle 90^\circ\ 2\angle -90^\circ\ 4\angle 90^\circ} = 17\ \angle 270^\circ$$

10.11 $p_1 = -a = -2$

$p_2, p_3 = -(4 - 2\mu) \pm \sqrt{(4 - 2\mu)^2 - 4}$

$\mu = 0 \Rightarrow p_2, p_3 = -4 \pm \sqrt{12} = -0.536, -7.46$

$y_N(t) = K_1e^{-2t} + K_2e^{-0.536t} + K_3e^{-7.46t}$

$\mu = 1 \Rightarrow p_2, p_3 = = -2 \pm 0$ so

$p_1 = p_2 = p_3 = -2$

$y_N(t) = K_1e^{-2t} + K_2te^{-2t} + K_3t^2e^{-2t}$

10.12 $\alpha = (6 - 2\beta)/2,\ \omega_0^2 = 24 - 8\beta$

$p_1, p_2 = -(3 - \beta) \pm \sqrt{(3 - \beta)^2 - (24 - 8\beta)}$

$\beta = 3 \Rightarrow p_1, p_2 = -0 \pm 0$ so $p_1 = p_2 = 0$

$y_N(t) = K_1 + K_2t$

$\beta = 4 \Rightarrow p_1, p_2 = +1 \pm 3 = 4, -2$

$$y_N(t) = K_1 e^{4t} + K_2 e^{-2t}$$

10.13 (a) $\hat{R}_1 = k_m \times 2500 = 4 \Rightarrow k_m = 1.6 \cdot 10^{-3}$
$= (k_m/k_f) \times 0.05 = 1 \Rightarrow k_f = 8 \cdot 10^{-5}$
so $R_2 = 2/k_m = 1.25$ kΩ,
$C = k_m k_f \times 0.2 = 25.6$ nF
(b) $\hat{H}(\hat{s}) = (2 + \hat{s})/(4 + \hat{s}/5 - 2 + \hat{s})$
$= \hat{s}(\hat{s} + 2)/(\hat{s}^2 + 6\hat{s} + 5)$
$\hat{K} = 1,\ \hat{z}_1 = 0,\ \hat{z}_2 = -2,\ \hat{p}_1 = -1,\ \hat{p}_2 = -5$
$z_1 = 0,\ z_2 = -2/k_f = -25{,}000$
$p_1 = -1/k_f = -12{,}500$
$p_2 = -5/k_f = -62{,}500$
$q = 0$ and $m = n = 0 \Rightarrow K = \hat{K} = 1$

Chapter 11

11.1 $H(s) = \dfrac{sL}{sL + R} = \dfrac{s}{s + R/L}$

$H(j\omega) = \dfrac{j\omega}{40 + j\omega}$
$a(\omega) = \omega/\sqrt{1600 + \omega^2}$
$\theta(\omega) = 90° - \tan^{-1}(\omega/40)$
$a(20) = 0.447 \qquad \theta(20) = 63.4°$
$a(300) = 0.991 \qquad \theta(300) = 7.6°$
$v_{out}(t) = 4.47 \cos(20t + 63.4°)$
$+ 9.91 \cos(300t + 7.6°)$

11.2 $H(j\omega) = j20\omega/(10 + j\omega)^2$
$(10 + j\omega) = (100 + \omega^2)^{1/2} \angle \tan^{-1}(\omega/10)$ so
$(10 + j\omega)^2 = (100 + \omega^2) \angle 2\tan^{-1}(\omega/10)$
$a(\omega) = 20\omega/(100 + \omega^2)$
$\theta(\omega) = 90° - 2\tan^{-1}(\omega/10)$

11.3 $K = 20,\ z_1 = 0,\ p_1 = p_2 = -10$
$m = 1,\ n = 2 \Rightarrow a(\infty) = 0,\ \theta(\infty) = -90°$
$a(\omega) = 20|j\omega - z_1|/|j\omega - p_1|^2$
$\theta(\omega) = \angle(j\omega - z_1) - 2 \times \angle(j\omega - p_1)$

	$\omega = 0^+$	$\omega = 10$	$\omega = 20$
$j\omega - z_1$	$0\angle 90°$	$10\angle 90°$	$20\angle 90°$
$j\omega - p_1$	$10\angle 0°$	$14.1\angle 45°$	$22.4\angle 63°$
$a(\omega)$	0	1.0	0.80
$\theta(\omega)$	90°	0°	−36°

11.4 $\underline{V}_C = [R \| (1/sC)]\underline{I}_{in}$
$H(s) = \underline{V}_C/\underline{I}_{in} = (R/sC)/(R + 1/sC)$
$= (1/C)/(s + 1/RC)$

$$H(j\omega) = \frac{1/C}{j\omega + 1/RC} = \frac{R}{1 + j(RC\omega)}$$

From Eq. (11.9), the network acts as a first-order lowpass filter with $K = R$ and $\omega_{co} = 1/RC$.

11.5 $H(s) = \dfrac{\underline{V}_{out}}{\underline{V}_{in}} = \dfrac{R_L}{R_s + R_L + 1/sC}$
$= \dfrac{[R_L/(R_L + R_s)]s}{s + 1/(R_L + R_s)C}$

$K = R_L/(R_L + R_s),\ \omega_{co} = 1/(R_L + R_s)C$
$C = 1/2\pi f_{co}(R_L + R_s) = 3.18\ \mu$F

11.6 $Q_{ser} = \omega_0/B = 42.8,\ R = 1/Q\omega_0 C = 3.48\ \Omega$,
$L = 1/\omega_0^2 C = 2.21\ \mu$H

11.7 $\dfrac{\underline{V}_{no}}{\underline{V}_{in}} = \dfrac{sL + R_w + 1/sC}{sL + R_w + 1/sC + R}$
$= \dfrac{s^2 + (R_w/L)s + 1/LC}{s^2 + [(R_w + R)/L]s + 1/LC}$
so $K = 1,\ \omega_0^2 = 1/LC,\ \beta = R_w/2L$,
$Q = \omega_0 L/(R_w + R)$

11.8 $Z_1(s) = R_1 + 1/sC_1$
$Z_F(s) = R\|(1/sC_F) = R_F/(1 + sC_F R_F)$

$H(s) = \dfrac{-R_F}{(R_1 + 1/sC_1)(1 + sC_F R_F)}$
$= \dfrac{-R_F s/R_1 R_F C_F}{(s + 1/R_1 C_1)(s + 1/R_F C_F)}$
$= -\dfrac{R_F}{R_1}\dfrac{s}{s + 1/R_1 C_1}\dfrac{1/R_F C_F}{s + 1/R_F C_F}$

11.9 (a) $g(\omega) = 20 \log 10 + 20 \log 0.5$
$\approx 20 - 6 = 14$ dB
(b) $g(\omega) = 20 \log 2^3 + 20 \log 0.1^3$
$\approx 3(6) + 3(-20) = -42$ dB
(c) $g(\omega) = 20 \log 0.5^{1/2} + 20 \log 10^{3/2}$
$\approx (1/2)(-6) + (3/2)(20) = 27$ dB

11.10 $H(s) = 10\dfrac{s}{s + 100}\dfrac{100}{s + 100}$
$= 10 H_{hp}(s;100) H_{lp}(s;100)$
$g(\omega) = 20$ dB $+ g_{hp}(\omega;100) + g_{lp}(\omega;100)$

ω	50	100	200
Δg_{hp}	−1	−3	−1
Δg_{lp}	−1	−3	−1
Sum	−2	−6	−2

$g_{max} = 14$ dB
$a_{max} = 10^{14}/20 \approx 5$

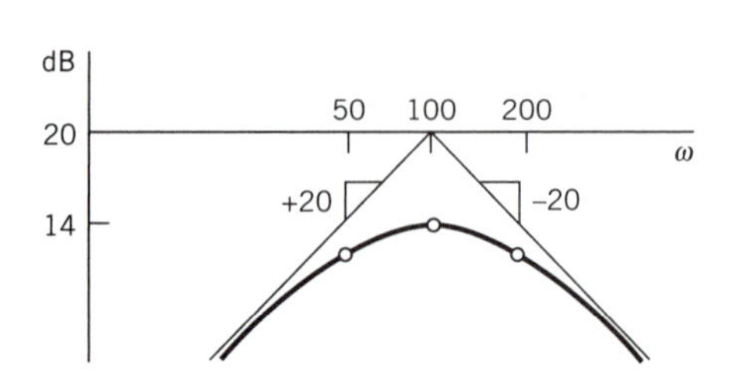

11.11 $H(s) = \dfrac{50}{200}\dfrac{s+50}{50}\dfrac{200}{s+200}$
$= 0.25 H_{lp}^{-1}(s;50)\, H_{lp}(s;200)$

$g(\omega) = -12 \text{ dB} - g_{lp}(\omega;50) + g_{lp}(\omega;200)$

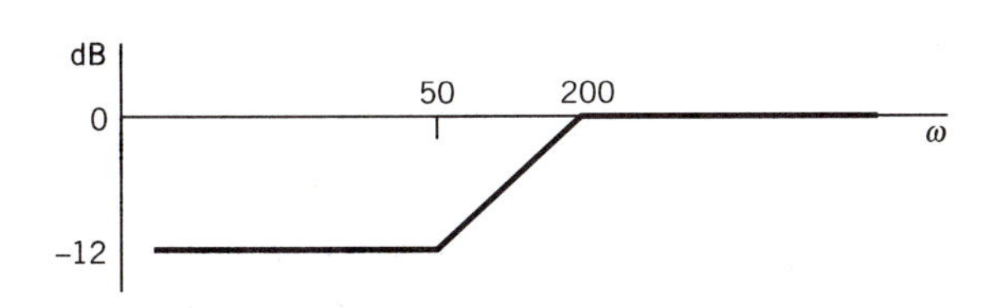

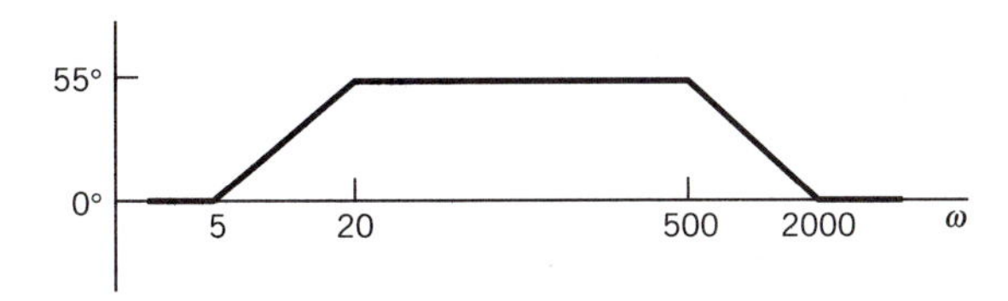

11.12 $\omega_0 = 50,\ Q = \omega_0/5 = 10$

$$H(s) = \frac{0.2 \times 50^2}{2500}\frac{s^2}{50^2}\frac{2500}{s^2 + 5s + 2500}$$
$$= 0.2 H_r^2(s;50)\, H_q(s;50,10)$$

$g(\omega) = -14 \text{ dB} + 2g_r(\omega;50) + g_q(\omega;50,10)$
$Q_{dB} = +20 \text{ dB},\ g_{max} = 20 - 14 = 6 \text{ dB}$

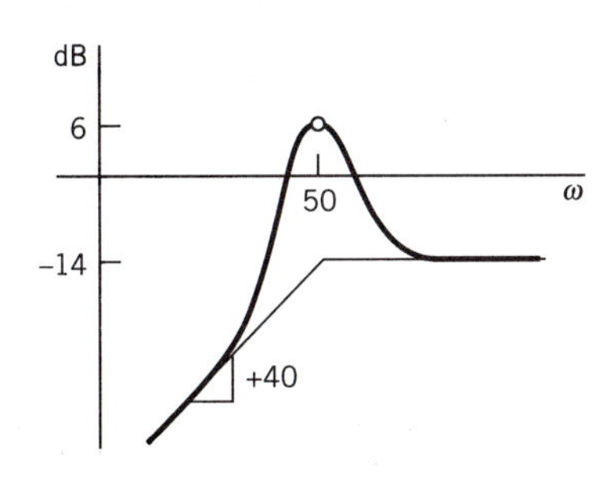

11.13 $\log 50/10 = 0.70$
$(-14 \text{ dB})/(0.7 \text{ decades}) = -20 \text{ dB/decade}$
$g(\omega) = 14 \text{ dB} + g_{lp}(\omega;10) + g_{lp}^{-1}(\omega;50)$

$$H(s) = \pm 5 \frac{10}{s+10}\frac{s+50}{50} = \pm 1 \frac{s+50}{s+10}$$

Use Fig. 11.32*c* with $C_1 = C_F = 1\ \mu\text{F}$ so $R_1 = 1/50C_1 = 20\ \text{k}\Omega$ and $R_F = 1/10C_F = 100\ \text{k}\Omega$.

11.14 $Q_3 = 1.000$
$B_3(s) = (s + \omega_{co})P_3(s)$
$= (s + \omega_{co})(s^2 + \omega_{co}s + \omega_{co}^2)$
$B_3(j\omega) = (j\omega + \omega_{co})(-\omega^2 + j\omega_{co}\omega + \omega_{co}^2)$
$|B_3(j\omega)|^2 = (\omega^2 + \omega_{co}^2)[(\omega_{co}^2 - \omega^2)^2 + (\omega_{co}\omega)^2]$
$= \omega^6 + \omega_{co}^6$

$|B_3(j\omega)| = \sqrt{\omega^{2n} + \omega_{co}^{2n}},\ n = 3$

11.15 $\omega_{co} = 2\pi \times 700$
$a_{hp}^2(\omega) = 25^2/[1 + (\omega_{co}/\omega)^{2n}]$
$\leq (25/30)^2$ for $\omega \leq 2\pi \times 200$ so
$(700/200)^{2n} \geq 30^2 - 1 = 899$
$n \geq (\log 899)/(2 \log 3.5) = 2.71 \Rightarrow n = 3$

$Q_3 = 1.000$

$$H_{hp}(s) = \frac{25s^3}{(s + \omega_{co})(s^2 + \omega_{co}s + \omega_{co}^2)}$$

11.16 $n = 3,\ \omega_{co} = 2\pi \times 700$
$K = K_1K_3 = 25 \Rightarrow K_1 = K_3 = 5$

$$H_{hp}(s) = \frac{5s}{s + \omega_{co}}\frac{5s}{P_3(s)} \qquad Q_3 = 1.000$$

First-order stage is like Fig. 11.19*b* with

$R = 1/\omega_{co}C = 22.7\ \text{k}\Omega,\ R_F = (K_1 - 1) \times 1\ \text{k}\Omega$
$= 4\ \text{k}\Omega$

Second-order stage is like Fig. 11.41*b* with

$R_1 = (4 \times 22.7\ \text{k}\Omega)/[1 + \sqrt{8(4-1) + 1}]$
$= 15.2\ \text{k}\Omega$
$R_2 = (22.7\ \text{k}\Omega)^2/R_1 = 34.1\ \text{k}\Omega,\ R_F = 4\ \text{k}\Omega$

Chapter 12

12.1 $c_0 = \dfrac{1}{T}\displaystyle\int_{-T/2}^{0} A\, dt = \dfrac{A}{2}$

$$a_n = \frac{1}{T}\int_{-T/2}^{0} A \cos n\Omega t\, dt = \frac{2A}{n\Omega T}\sin n\pi = 0$$

$$b_n = \frac{1}{T}\int_{-T/2}^{0} A \sin n\Omega t\, dt$$
$$= \frac{2A}{n\Omega T}(\cos n\pi - 1)$$
$= -2A/n\pi \qquad n = 1, 3, 5, \ldots$
$= 0 \qquad n = 2, 4, 6, \ldots$
$f(t) \approx 3\pi - 12 \sin \Omega t - 4 \sin 3\Omega t$

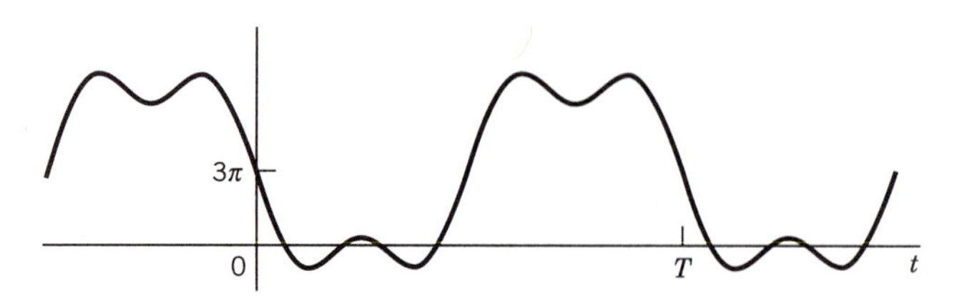

12.2 $\int_T e^{jn\Omega t}e^{-jm\Omega t}\,dt = \int_0^T e^{j(n-m)\Omega t}\,dt$

$= [e^{j(n-m)2\pi} - 1]/j(n-m)\Omega$

$= 0 \qquad n \neq m$

$\int_T e^{jm\Omega t}e^{-jm\Omega t}\,dt = \int_0^T dt = T$

12.3 $c_n = \frac{1}{T}\int_{-T/2}^{0} A\, e^{-jn\Omega t}dt = \frac{A}{-j2\pi n}(1 - e^{-jn\pi})$

$= jA/n\pi \qquad n = \pm 1, \pm 3, \ldots$

$= 0 \qquad n = \pm 2, \pm 4, \ldots$

$a_n = 2\,\mathrm{Re}[c_n] = 0$

$b_n = -2\,\mathrm{Im}[c_n]$

$= -2A/n\pi \qquad n = 1, 3, \ldots$

$= 0 \qquad n = 2, 4, \ldots$

12.4 $g(-t) = f_1(-t)f_2(-t) = [-f_1(t)]f_2(t)$

$= -f_1(t)f_2(t) = -g(t)$

12.5 $c_n = \frac{-j2}{T}\int_0^{T/2} \frac{2Bt}{T}\sin n\Omega\, dt$

$= \frac{-jB}{n^2\pi^2}\int_0^{n\pi} \alpha \sin \alpha\, d\alpha$

$= (jB/n\pi)\cos n\pi$

$= -jB/n\pi \qquad n = \pm 1, \pm 3, \ldots$

$= jB/n\pi \qquad n = \pm 2, \pm 4, \ldots$

$b_n = -2\,\mathrm{Im}[c_n] = 2B/n\pi \qquad n = 1, 3, 5, \ldots$

$= -2B/n\pi \qquad n = 2, 4, 6, \ldots$

$f(t) = (2B/\pi)\sin \Omega t - (B/\pi)\sin 2\Omega t + (2B/3\pi)\sin 3\Omega t + \cdots$

12.6 $c_n = 0 \qquad n = 0$

$= -jB/n\pi \qquad n = \pm 1, \pm 3, \ldots$

$= jB/n\pi \qquad n = \pm 2, \pm 4, \ldots$

12.7 $z(t) = f(t) - g(t)$ where

$f(t)$ has $D = T/2$ and $t_0 = +T/4$,

$g(t)$ has $D = T/2$ and $t_0 = -T/4$

$c_z(n\Omega) = (A/n\pi)\sin(n\pi/2)(e^{-jn\pi/2} - e^{jn\pi/2})$

$= (A/n\pi)\sin(n\pi/2)(-j2)\sin(n\pi/2)$

$= 0 \qquad n = 0, \pm 2, \pm 4, \ldots$

$= -j2A/n\pi \qquad n = \pm 1, \pm 3, \ldots$

12.8 $f'(t)$ is a square wave with $A = -2B/(T/2) = -4B/T$, so

$c_{f'}(n\Omega) = 0 \qquad n = 0, \pm 2, \pm 4, \ldots$

$= j8B/n\pi T \qquad n = \pm 1, \pm 3, \ldots$

Since $f_{av} = 0$,

$c_f(n\Omega) = (T/j2\pi n)c_{f'}(n\Omega)$

$= 0 \qquad n = 0, \pm 2, \pm 4, \ldots$

$= 4B/n^2\pi^2 \qquad n = \pm 1, \pm 3, \ldots$

12.9 $c_x(n\Omega) = 1 \qquad n = 0$

$= -j\pi/4 \qquad n = 1$

$= -1/(n^2 - 1) \qquad n = 2, 4, \ldots$

$= 0 \qquad n = 3, 5, \ldots$

$H(j\omega) = 1/[1 + j(\omega/\omega_{co})]$

$H(jn\Omega) = 1/(1 + j4n)$

$c_y(0) = 1 \times 1 = 1$

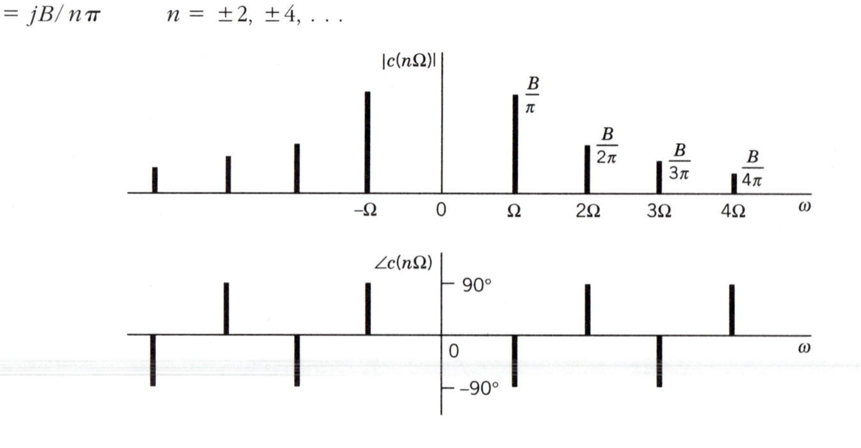

n	$c_x(n\Omega)$	$H(jn\Omega)$	$A_{yn}\,\underline{/\phi_{yn}}$
1	$0.785\underline{/-90°}$	$0.243\underline{/-76°}$	$0.382\underline{/-166°}$
2	$0.333\underline{/\pm 180°}$	$0.124\underline{/-83°}$	$0.083\underline{/97°}$

$y(t) \approx 1 + 0.382 \cos(\Omega t - 166°)$

12.10 $\theta(1) = (\pi/12)(7 - 1) = \pi/2 = 90°$

$\theta(3) = (\pi/12)(21 - 9) = \pi = 180°$

$y(t) = 15 \sin(t + 90°) + 5 \sin(3t + 180°)$

$= 15 \cos t - 5 \sin 3t$

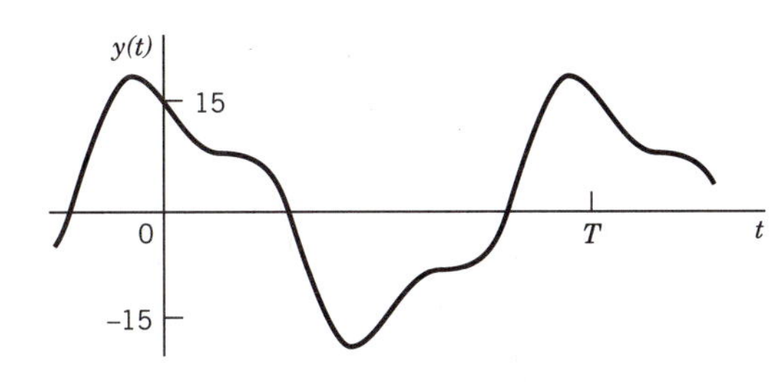

12.11 $H_{eq}(s) = \dfrac{s + 10}{s}$

$H_{eq}(j\omega) = \dfrac{j\omega + 10}{j\omega},$

$a_{eq}(\omega) = \dfrac{\sqrt{\omega^2 + 100}}{\omega}$

(a)

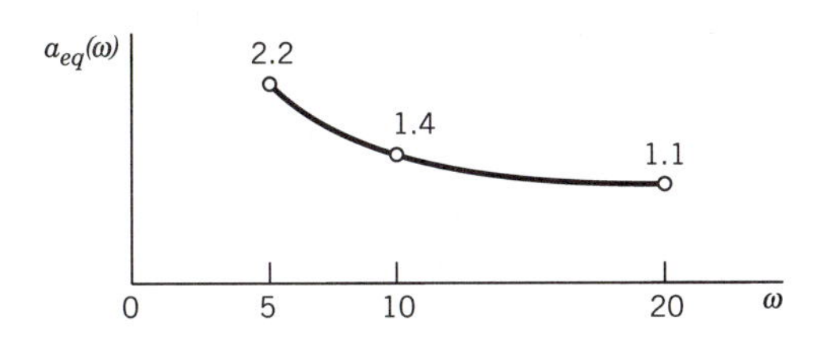

(b) As $\omega \to 0$, $a_{eq}(\omega) \to \infty$, so the equalizer would have to have infinite gain at $\omega = 0$.

Chapter 13

13.1 $\mathcal{L}[t] = \displaystyle\int_0^\infty te^{-st}\,dt = \frac{e^{-st}}{s^2}(-st - 1)\Bigg|_{0^-}^{\infty}$

$= \dfrac{1}{s^2} - \displaystyle\lim_{t\to\infty} \frac{te^{-st}}{s} = \frac{1}{s^2} \quad$ for $\sigma > 0$

13.2 $F(s) = \mathcal{L}[t] = 1/s^2$ so

$\mathcal{L}[te^{-at}] = F(s + a) = 1/(s + a)^2$

13.3 $\mathcal{L}[e^{-at}] = 1/(s + a)$

$\mathcal{L}[te^{-at}] = -d(s + a)^{-1}/ds$

$= -(-1)(s + a)^{-2} = 1/(s + a)^2$

13.4 $F(s) = 1/(s + a) \Rightarrow f(t) = e^{-at}$

$\mathcal{L}^{-1}[F(s)/s] = \displaystyle\int_{0^-}^{t} e^{-a\lambda}\,d\lambda = -(e^{-at} - 1)/a$

13.5 $[s^2 Y(s) - sy(0^-) - y'(0^-)] +$

$20[sY(s) - y(0^-)] + 100\,Y(s) = 0$

$(s^2 + 20s + 100)\,Y(s) =$

$sy(0^-) + y'(0^-) + 20y(0^-) = 3$

$Y(s) = 3/(s^2 + 20s + 100) = 3/(s + 10)^2$

$y(t) = 3\mathcal{L}^{-1}[1/(s + 10)^2] = 3te^{-10t} \qquad t \geq 0$

13.6 $s^2 + 3s - 10 = (s - 2)(s + 5)$ so

$F(s) = \dfrac{A_1}{s - 2} + \dfrac{A_2}{s + 5}$

$A_1 = \dfrac{s - 23}{s + 5}\Bigg|_{s=2} = -3, \; A_2 = \dfrac{s - 23}{s - 2}\Bigg|_{s=-5} = 4$

$f(t) = -3e^{2t} + 4e^{-5t} \qquad t \geq 0$

13.7 $D(s) = s(s^2 + 8s + 25)$

$F(s) = \dfrac{s^2 + 14s + 25}{s(s^2 + 8s + 25)} = \dfrac{A_1}{s} + \dfrac{Bs + C}{s^2 + 8s + 25}$

$A_1 = sF(s)|_{s=0} = 1$

$\displaystyle\lim_{s\to\infty} sF(s) = 1 = A_1 + B \Rightarrow B = 0$

$F(1) = 40/34 = A_1 + (B + C)/34 \Rightarrow C = 6$

$\alpha = 4,\; \beta = \sqrt{25 - 4^2} = 3$

$K = 0 + j\dfrac{-6}{3} = -j2 = 2\,\underline{/-90°}$

$f(t) = 1 + 2e^{-4t}\cos(3t - 90°)$

13.8 $F(s) = \dfrac{2s - 4}{(s + 3)^2} = \dfrac{A_{11}}{s + 3} + \dfrac{A_{12}}{(s + 3)^2}$

$A_{12} = (s + 3)^2 F(s)|_{s=-3} = -10$

$A_{11} = \dfrac{d}{ds}[(s + 3)^2 F(s)]|_{s=-3} = 2|_{s=-3} = 2$

$f(t) = 2e^{-3t} - 10te^{-3t} \qquad t \geq 0$

13.9 $\dfrac{2s - 4}{(s + 3)^2} = \dfrac{A_{11}}{s + 3} + \dfrac{A_{12}}{(s + 3)^2}$

$\displaystyle\lim_{s\to\infty} sF(s) = 2 = A_{11} + 0 \Rightarrow A_{11} = 2$

$F(0) = -4/9 = A_{11}/3 + A_{12}/9 \Rightarrow A_{12} = -10$

13.10 $F(s) = F_1(s) - F_2(s)e^{-8s}$

$F_1(s) = 5(s + 3)/(s + 3)^2 = 5/(s + 3)$

$f_1(t) = 5e^{-3t}$

$$F_2(s) = (2s - 4)/(s + 3)^2$$
$$f_2(t) = 2e^{-3t} - 10te^{-3t}$$
$$f(t) = f_1(t) - f_2(t - 8)u(t - 8)$$
$$= 5e^{-3t} - [2 - 10(t - 8)]e^{-3(t-8)}u(t - 8)$$

13.11 $m = n - 1$

$$f(0^+) = \lim_{s\to\infty} \frac{b_{n-1}s^n + \cdots}{s^n + \cdots} = b_{n-1}$$

$$f'(0^+) = \lim_{s\to\infty} \frac{b_{n-1}s^{n+1} + \cdots}{s^n + \cdots}$$

does not exist

$m = n - 2$

$$f(0^+) = \lim_{s\to\infty} \frac{b_{n-2}s^{n-1} + \cdots}{s^n + \cdots} = 0$$

$$f'(0^+) = \lim_{s\to\infty} \frac{b_{n-2}s^n + \cdots}{s^n + \cdots} = b_{n-2}$$

13.12 $Y(s) = H(s)\dfrac{1}{s} = \dfrac{2}{s + 3}$

$y(t) = 2e^{-3t} \quad t \ge 0$

13.13 $X(s) = 25 \times 4/(s^2 + 4^2)$

$$Y(s) = \frac{200s}{(s + 3)(s^2 + 16)} = \frac{A_1}{s + 3} + \frac{N_F(s)}{s^2 + 16}$$

$A_1 = (s + 3)\,Y(s)|_{s=-3} = -24$ so

$y_N(t) = -24e^{-3t}$

$x = 25 \sin 4t = 25 \cos(4t - 90°)$ so

$\underline{X} = 25\,\underline{/-90°}$

$H(j4) = j8/(j4 + 3) = 1.6\,\underline{/36.9°}$

$\underline{Y} = H(j4)\underline{X} = 40\,\underline{/-53.1°}$ so

$y_F(t) = 40 \cos(4t - 53.1°)$

$y(t) = -24e^{-3t} + 40 \cos(4t - 53.1°)$

13.14 $X(s) = X_m/(s + a)$

$Y(s) = H(s)X(s) = KX_m/(s + a)^2$

$y(t) = KX_m te^{-at}$

There is no natural-response component in $y(t)$, and the forced-response has been multiplied by t.

13.15 $i_L(0^-) = 4,\ v_C(0^-) = 8,\ Cv_C(0^-) = 0.5$

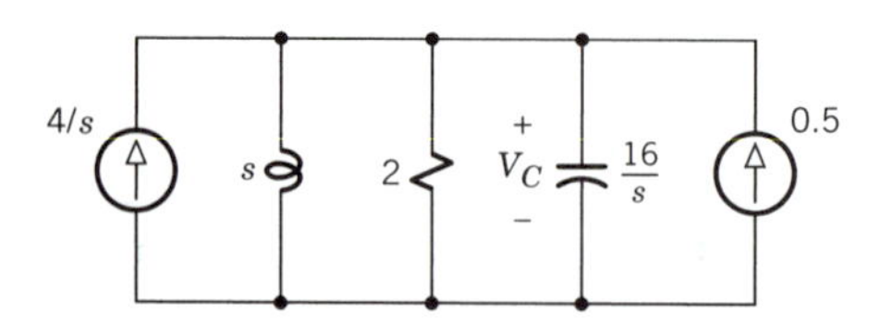

$$(s/16 + 1/2 + 1/s)V_C(s) = 0.5 + 4/s$$

$$V_C(s) = \frac{8s + 64}{s^2 + 8s + 16}$$

13.16 $I_s(s) = -1/s,\ Cv_C(0^-) = 0.6$

$(100/s)(0.6 - 1/s) = 60/s - 100/s^2$

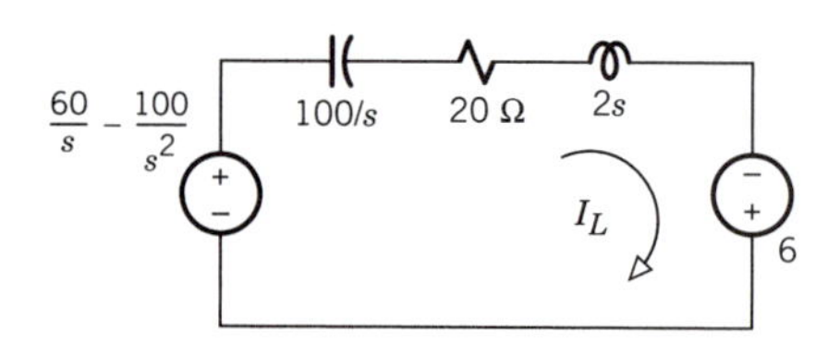

$$(2s + 20 + 100/s)I_L(s) = 6 + 60/s - 100/s^2$$

$$I_L(s) = \frac{3s^2 + 30s - 50}{s(s^2 + 10s + 50)}$$

13.17 At $t = 0^-$, $i_1 = i_2 = i_3 = 0$, $v_C = 6$

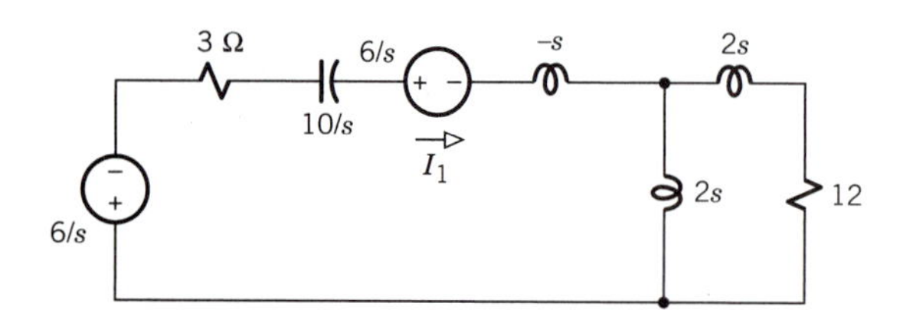

$I_1(s) = -(6/s + 6/s)/Z(s)$ where

$$Z(s) = 3 + 10/s - s + 2s\|(2s + 12)$$
$$= \frac{6s^2 + 19s + 30}{s(s + 3)}$$

$$I_1(s) = \frac{-12(s + 3)}{6s^2 + 19s + 30},\quad i_1(0^+) = -2 \ne i_1(0^-)$$

13.18 (a) $3t^2u(t - 2) + t^3\delta(t - 2) =$
$3t^2u(t - 2) + 8\delta(t - 2)$

(b) $0 - e^t|_{t=2} = -e^2$ (c) $3u(t) - 5u(t - 2)$

13.19 $\mathcal{L}[\delta(t)] = 1,\ \delta(0^-) = 0$ so
$\mathcal{L}[A\delta'(t)] = sA - A \times 0 = As$

13.20 $I(s) = \dfrac{80/s}{Z(s)} = \dfrac{5s^2 + 40s + 40}{s^2 + 4s + 4} = 5 + G(s)$

$$G(s) = \frac{20s + 20}{s^2 + 4s + 4} = \frac{20}{s+2} + \frac{-20}{(s+2)^2}$$
$i(t) = 5\delta(t) + 20e^{-2t} - 20te^{-2t}$

13.21 $t < 0 \quad f(t) * g(t) = 0$

$0 < t < 2 \quad f(t) * g(t) = \int_0^t AB\,d\lambda = ABt$

$2 < t < 5 \quad f(t) * g(t) = \int_{t-2}^t AB\,d\lambda = 2AB$

$5 < t < 7 \quad f(t) * g(t) = \int_{t-2}^5 AB\,d\lambda$
$= AB(7 - t)$

$t > 7 \quad f(t) * g(t) = 0$

13.22 $H(s) = \dfrac{V_C(s)}{V(s)} = \dfrac{25/s}{s + 10 + 25/s} = \dfrac{25}{(s+5)^2}$

$h(t) = 25te^{-5t} \quad t > 0$
$t < 2 \quad v_C(t) = 0$
$t > 2 \quad v_C(t) = \int_0^{t-2} 25\lambda e^{-5\lambda}\,d\lambda$
$= 1 - (5t - 9)e^{-5(t-2)}$

Chapter 14

14.1 $\underline{V}_1 = Z_a\underline{I}_1 + Z_b(\underline{I}_1 + \underline{I}_2) = (Z_a + Z_b)\underline{I}_1 + Z_b\underline{I}_2$
$\underline{V}_2 = Z_b(\underline{I}_1 + \underline{I}_2) = Z_b\underline{I}_1 + Z_b\underline{I}_2$
Thus,
$$[z] = \begin{bmatrix} Z_a + Z_b & Z_b \\ Z_b & Z_b \end{bmatrix}$$

14.2 (a) $i_2 = 0 \Rightarrow v_1 = 40i_1,$
$v_2 = v_1 + R(-2i_1) = (40 - 2R)i_1$
$i_1 = 0 \Rightarrow v_1 = 40i_2$
$v_2 = v_1 + Ri_2 = (40 + R)i_2$
Thus,
$$[z] = \begin{bmatrix} 40 & 40 \\ 40 - 2R & 40 + R \end{bmatrix}$$
(b) With $R = \infty$, $i_2 = 2i_1$ so i_1 and i_2 cannot come from independent current sources.

14.3 With $i_2 = 0$: $g_m\underline{V}_1 = -\underline{I}_1$ so
$\underline{V}_2 = \underline{V}_1 = -(1/g_m)\underline{I}_1$
$z_{11} = \underline{V}_1/\underline{I}_1 = -1/g_m,\ z_{21} = \underline{V}_2/\underline{I}_1 = -1/g_m$
With $i_1 = 0$: $g_m\underline{V}_1 = -\underline{I}_2$ so
$\underline{V}_1 = -(1/g_m)\underline{I}_2$
$\underline{V}_2 = \underline{V}_1 + Z_x\underline{I}_2 = (Z_x - 1/g_m)\underline{I}_2$
$z_{12} = \underline{V}_1/\underline{I}_2 = -1/g_m$
$z_{22} = \underline{V}_2/\underline{I}_2 = Z_x - 1/g_m$

Thus, $z_{21} = z_{12}$, $z_{11} - z_{12} = 0$, and $z_{22} - z_{12} = Z_x$.

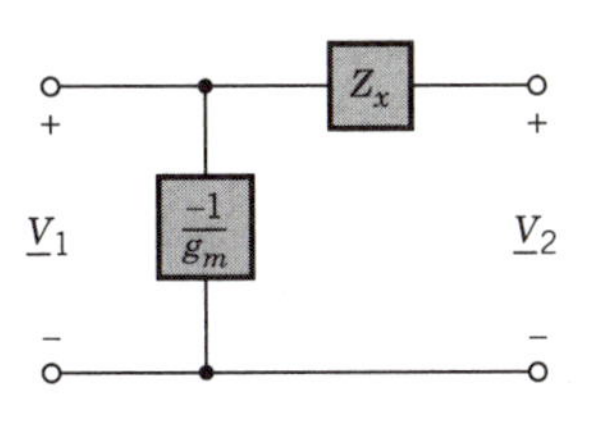

14.4 With $\underline{V}_2 = 0$: $\underline{I}_1 = \underline{V}_1/Z_a,\ \underline{I}_2 = -\underline{V}_1/Z_a$
$y_{11} = Y_a,\ y_{21} = -Y_a$
With $\underline{V}_1 = 0$: $\underline{I}_1 = -\underline{V}_2/Z_a$
$\underline{I}_2 = \underline{V}_2/Z_a + \underline{V}_2/Z_b$
$Y_{12} = -Y_a,\ y_{22} = Y_a + Y_b$
Thus,
$$[y] = \begin{bmatrix} Y_a & -Y_a \\ -Y_a & Y_a + Y_b \end{bmatrix}$$

14.5 $\underline{I}_1 = (y_{11} + y_{12})\underline{V}_1 + (-y_{12})(\underline{V}_1 - \underline{V}_2)$
$= y_{11}\underline{V}_1 + y_{12}\underline{V}_2$
$\underline{I}_2 = (y_{22} + y_{12})\underline{V}_2 + (-y_{12})(\underline{V}_2 - \underline{V}_1)$
$= y_{12}\underline{V}_1 + y_{22}\underline{V}_2$
$= y_{21}\underline{V}_1 + y_{22}\underline{V}_2$ when $y_{21} = y_{12}$

14.6 $v_2 = v_x - 3v_x = -2v_x$
With $v_2 = 0$: $v_x = 0,\ i_2 = -i_1$
$v_1 = 10i_1 = -10i_2$
$h_{11} = v_1/i_1 = 10,\ h_{21} = i_2/i_1 = -1$
With $i_1 = 0$: $v_1 = v_x = 50i_2$
$v_2 = -2v_1 = -100i_2$
$h_{12} = v_1/v_2 = -0.5,\ h_{22} = i_2/v_2 = -0.01$

14.7 $v_2 = v_x - 3v_x = -2v_x$
With $v_2 = 0$: $v_x = 0,\ i_2 = -i_1$
$v_1 = 10i_1 = -10i_2$
$B = -v_1/i_2 = 10$
$D = -i_1/i_2 = 1$
With $i_2 = 0$: $v_x = 50i_1,\ v_1 = 60i_1$
$v_2 = -2v_x = -100i_1 = -(100/60)v_1$
$A = v_1/v_2 = -60/100,$
$C = i_1/v_2 = -1/100$

14.8 $\Delta_T = 5 \times 0.2 - B(-0.1) = 1 \Rightarrow B = 0\ \Omega$
$[y]$ does not exist
$$[z] = \begin{bmatrix} -50\ \Omega & -10\ \Omega \\ -10\ \Omega & -2\ \Omega \end{bmatrix}$$
$$[h] = \begin{bmatrix} 0\ \Omega & 5 \\ -5 & -0.5\ \text{S} \end{bmatrix}$$

14.9 With $Z_L = \infty$, $Z_i(s) = z_{11}$ and
$H_v(s) = \underline{V}_{2oc}/\underline{V}_1 = z_{21}/z_{11}$
$\underline{V}_1 = \{Z_i(s)/[Z_s + Z_i(s)]\}\underline{V}_s$

$$\underline{V}_{2oc} = \frac{\underline{V}_{2oc}}{\underline{V}_1}\frac{\underline{V}_1}{\underline{V}_s}\underline{V}_s = \frac{H_v(s)Z_i(s)}{Z_s + Z_i(s)}\underline{V}_s$$

$$= \frac{z_{21}}{Z_s + z_{11}}\underline{V}_s$$

14.10 $H_v(s) = \dfrac{1/s}{0.2 + 1/s + sC}$

$$= \frac{1/C}{s^2 + 0.2s/C + 1/C}$$

$\alpha = 0.1/C$ and $\omega_0^2 = 1/C$ so
$\alpha^2 = \omega_0^2 \Rightarrow C = 0.01$

14.11 $[T]_{cas} = \begin{bmatrix} 2 & 5 \\ -1 & 3 \end{bmatrix}\begin{bmatrix} 2 & 5 \\ -1 & 3 \end{bmatrix}$

$$= \begin{bmatrix} -1 & 25\text{ k}\Omega \\ -5\text{ mS} & 4 \end{bmatrix}$$

With $Z_L = 0$, $Z_i(s) = B/D$ and $H_i(s) = -1/D$.
$v_1 = Z_i(s)i_1 = (25/4) \times 8 = 50$ V
$i_2 = H_i(s)i_1 = (-1/4) \times 8 = -2$ mA

14.12 $\underline{I}_{1a} = \underline{I}_{1b} = \underline{I}_1$, $\underline{V}_{1a} + \underline{V}_{1b} = \underline{V}_1$
$\underline{V}_{2a} = \underline{V}_{2b} = \underline{V}_2$, $\underline{I}_{2a} + \underline{I}_{2b} = \underline{I}_2$

$$\begin{bmatrix} \underline{V}_{1a} \\ \underline{I}_{2a} \end{bmatrix} = [h]_a \begin{bmatrix} \underline{I}_1 \\ \underline{V}_2 \end{bmatrix} \qquad \begin{bmatrix} \underline{V}_{1b} \\ \underline{I}_{2b} \end{bmatrix} = [h]_b \begin{bmatrix} \underline{I}_1 \\ \underline{V}_2 \end{bmatrix}$$

$$\begin{bmatrix} \underline{V}_1 \\ \underline{I}_2 \end{bmatrix} = \begin{bmatrix} \underline{V}_{1a} \\ \underline{I}_{2a} \end{bmatrix} + \begin{bmatrix} \underline{V}_{1b} \\ \underline{I}_{2b} \end{bmatrix}$$

$$= \{[h]_a + [h]_b\} \begin{bmatrix} \underline{I}_1 \\ \underline{V}_2 \end{bmatrix}$$

Chapter 15

15.1 $i_C = i_s - i_1$, $v_R = Ri_C = Ri_s - Ri_1$
$v_L = v_R + v_2 - v_s = Ri_s - Ri_1 + v_2 - v_s$
$Li_1' = v_L = Ri_s - Ri_1 + v_2 - v_s$
$Cv_2' = i_C = i_s - i_1$

15.2 $i_1' = (-R/L)i_1 + (1/L)v_2 + (R/L)i_s + (-1/L)v_s$
$v_2' = (-1/C)i_1 + 0 \times v_2 + (1/C)i_s + 0 \times v_s$

$$[A] = \begin{bmatrix} -R/L & 1/L \\ -1/C & 0 \end{bmatrix} \quad [B] = \begin{bmatrix} R/L & -1/L \\ 1/C & 0 \end{bmatrix}$$

$v_L = (-R)i_1 + 1 \times v_2 + R \times i_s + (-1)v_s$
$v_R = (-R)i_1 + 0 \times v_2 + R \times i_s + 0 \times v_s$

$$[C] = \begin{bmatrix} -R & 1 \\ -R & 0 \end{bmatrix} \quad [D] = \begin{bmatrix} R & -1 \\ R & 0 \end{bmatrix}$$

15.3

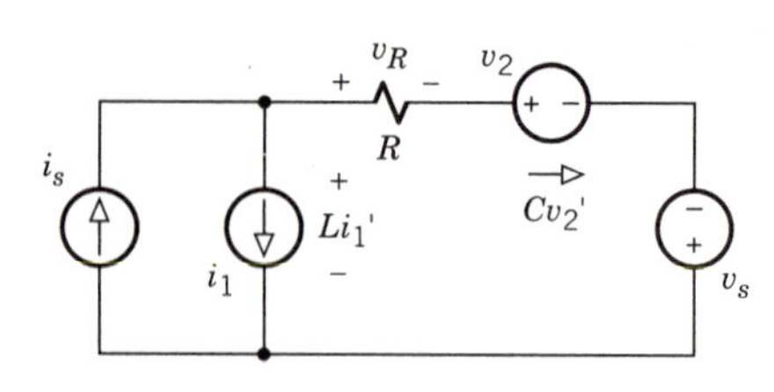

Active Source:	i_1	v_2	i_s	v_s
Li_1'	$-R$	1	R	-1
Cv_2'	-1	0	1	0
v_R	$-R$	0	R	0

$$\begin{bmatrix} i_1' \\ v_2' \end{bmatrix} = \begin{bmatrix} -R/L & 1/L \\ -1/C & 0 \end{bmatrix}\begin{bmatrix} i_1 \\ v_2 \end{bmatrix} + \begin{bmatrix} R/L & -1/L \\ 1/C & 0 \end{bmatrix}\begin{bmatrix} i_s \\ v_s \end{bmatrix}$$

$$v_R = [-R \quad 0]\begin{bmatrix} i_1 \\ v_2 \end{bmatrix} + [R \quad 0]\begin{bmatrix} i_s \\ v_s \end{bmatrix}$$

15.4 i_1 active:
$Cv_2' = -i_1$
$v_L + 3v_L = R(-i_1)$

v_2 active:
$Cv_2' = 0$
$v_L + 3v_L = v_2$

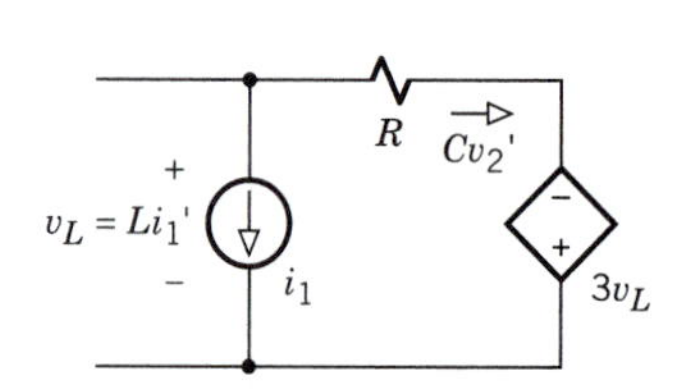

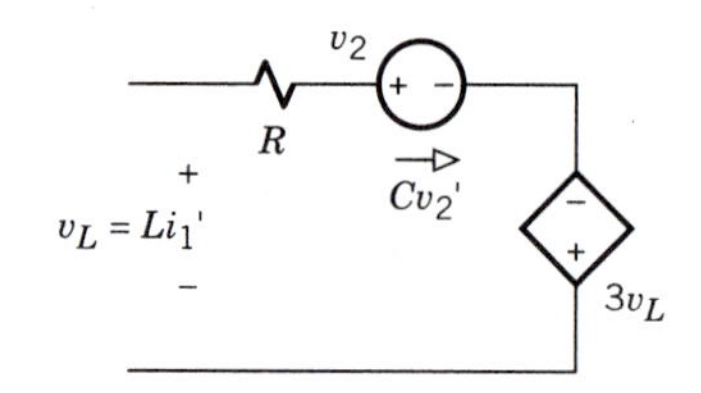

i_s active: Like i_1 active with $i_1 \rightarrow -i_s$

Active Source:	i_1	v_2	i_s
Li_1'	$-R/4$	$1/4$	$R/4$
Cv_2'	-1	0	1

$$\begin{bmatrix} i_1' \\ v_2' \end{bmatrix} = \begin{bmatrix} -R/4L & 1/4L \\ -1/C & 0 \end{bmatrix}\begin{bmatrix} i_1 \\ v_2 \end{bmatrix} + \begin{bmatrix} R/4L \\ 1/C \end{bmatrix} i_s$$

15.5 $v_1 = v_2 + v_3$
$C_1(v_2' + v_3') + C_2v_2' = i_s$
$C_2v_2' - C_3v_3' = v_3/R$

$$\begin{bmatrix} 2/12 & 1/12 \\ 1/12 & -1/12 \end{bmatrix}\begin{bmatrix} v_2' \\ v_3' \end{bmatrix} = \begin{bmatrix} i_s \\ v_3/2 \end{bmatrix}$$

$$\begin{bmatrix} v_2' \\ v_3' \end{bmatrix} = \begin{bmatrix} 0 & 2 \\ 0 & -4 \end{bmatrix}\begin{bmatrix} v_2 \\ v_3 \end{bmatrix} + \begin{bmatrix} 4 \\ 4 \end{bmatrix} i_s$$

15.6 $0.2i_1' + 0.5i_2' = v_s - 3i_1$
$v_s - 3i_1 - v_x - 5(i_1 - i_2) + v_x = 0$
$5i_2 = -v_s + 8i_1 \Rightarrow 0.5i_2' = -0.1v_s' + 0.8i_1'$

Thus, $0.2i_1' - 0.1v_s' + 0.8i_1' = v_s - 3i_1$

$i_1' - 0.1v_s' = v_s - 3i_1$

Let $q = i_1 - 0.1v_s \Rightarrow i_1 = q + 0.1v_s$

$q' = v_s - 3i_1 = v_s - 3q - 0.3v_s$

$q' = -3q + 0.7v_s$

15.7 $[sI - A] = \begin{bmatrix} s & -1 \\ 0 & s+4 \end{bmatrix} \qquad P(s) = s^2 + 4s$

$$[\Phi(s)] = \frac{1}{P(s)}\begin{bmatrix} s+4 & 1 \\ 0 & s \end{bmatrix}$$

$$[Q(s)] = \frac{1}{P(s)}\begin{bmatrix} s+4 & 1 \\ 0 & s \end{bmatrix}\begin{bmatrix} 0 \\ 12 \end{bmatrix} = \frac{1}{P(s)}\begin{bmatrix} 12 \\ 12s \end{bmatrix}$$

$$Q_1(s) = \frac{12}{s(s+4)} = \frac{3}{s} - \frac{3}{s+4}$$

$$Q_2(s) = \frac{12s}{s(s+4)} = \frac{12}{s+4}$$

$$q_1(t) = 3 - 3e^{-4t} \qquad q_2(t) = 12e^{-4t}$$

$$y(t) = [2 \quad -1]\begin{bmatrix} q_1 \\ q_2 \end{bmatrix} = 6 - 18e^{-4t}$$

15.8
$$[Q(s)] = \frac{1}{P(s)}\begin{bmatrix} s+4 & 1 \\ 0 & s \end{bmatrix}\left\{\begin{bmatrix} 0 \\ 12 \end{bmatrix} + \begin{bmatrix} 1 & -1 \\ 0 & 4 \end{bmatrix}\begin{bmatrix} 0 \\ 8/s \end{bmatrix}\right\}$$

$$= \frac{1}{P(s)}\begin{bmatrix} s+4 & 1 \\ 0 & s \end{bmatrix}\begin{bmatrix} -8/s \\ 12 + 32/s \end{bmatrix}$$

$$= \frac{1}{P(s)}\begin{bmatrix} 4 \\ 12s + 32 \end{bmatrix}$$

$$Q_1(s) = \frac{4}{s(s+4)} = \frac{1}{s} - \frac{1}{s+4}$$

$$Q_2(s) = \frac{12s+32}{s(s+4)} = \frac{8}{s} + \frac{4}{s+4}$$

$$q_1(t) = 1 - e^{-4t} \qquad q_2(t) = 8 + 4e^{-4t}$$

$$y(t) = [2 \quad -1]\begin{bmatrix} q_1 \\ q_2 \end{bmatrix} + [0 \quad 4]\begin{bmatrix} 0 \\ 8 \end{bmatrix}$$

$$= 2q_1 - q_2 + 32 = 26 - 6e^{-4t}$$

15.9
$$P(s)[H(s)] = [2 \quad -1]\begin{bmatrix} s+4 & 1 \\ 0 & s \end{bmatrix}\begin{bmatrix} 1 & -1 \\ 0 & 4 \end{bmatrix} + P(s)[0 \quad 4]$$

$$= [2 \quad -1]\begin{bmatrix} s+4 & -s \\ 0 & 4s \end{bmatrix} + [0 \quad (s^2+4s)4]$$

$$= [2s + 8 \quad -6s] + [0 \quad 4s^2 + 16s]$$

$$[H(s)] = \frac{1}{s(s+4)}[2(s+4) \quad s(4s+10)]$$

$$H_{11}(s) = 2/s \qquad H_{12}(s) = (4s+10)/(s+4)$$

APPENDIX A

A.1 $y_1 = (0))(5) + (-3)(-2) = -6$

$y_2 = (4)(5) + (10)(-2) = 0$

$y_3 = (2)(5) + (2)(-2) = 6$

A.2 (a) $\Delta = 1\begin{vmatrix} 5 & 6 \\ 8 & 9 \end{vmatrix} - 4\begin{vmatrix} 2 & 3 \\ 8 & 9 \end{vmatrix} + 0$

$= 1(-3) - 4(-6) = 21$

(b) $\Delta = -4\begin{vmatrix} 2 & 3 \\ 8 & 9 \end{vmatrix} + 5\begin{vmatrix} 1 & 3 \\ 0 & 9 \end{vmatrix} - 6\begin{vmatrix} 1 & 2 \\ 0 & 8 \end{vmatrix}$

$= -4(-6) + 5(9) - 6(8) = 21$

A.3 $\begin{bmatrix} 3 & 1 \\ -2 & 1 \end{bmatrix}[y] = \begin{bmatrix} 8v_1 \\ 10v_2 \end{bmatrix} = \begin{bmatrix} x_1 \\ x_2 \end{bmatrix}$

$\Delta = 5$

(a) $y_1 = \dfrac{1}{5}\begin{vmatrix} 240 & 1 \\ -160 & 1 \end{vmatrix} = 80$

$y_2 = \dfrac{1}{5}\begin{vmatrix} 3 & 240 \\ -2 & -160 \end{vmatrix} = 0$

(b) $\Delta_{11} = 1,\ \Delta_{12} = 2,\ \Delta_{21} = -1,\ \Delta_{22} = 3$

$y_1 = (x_1 - x_2)/5 = 1.6v_1 - 2v_2$

$y_2 = (2x_1 + 3x_2)/5 = 3.2v_1 + 6v_2$

A.4
$$\begin{bmatrix} -1 & 2 & 1 \\ 3 & -6 & -3 \end{bmatrix}\begin{bmatrix} 0 & 1 \\ -2 & 3 \\ 4 & -5 \end{bmatrix}$$

$$= \begin{bmatrix} 0 - 4 + 4 & -1 + 6 - 5 \\ 0 + 12 - 12 & 3 - 18 + 15 \end{bmatrix}$$

$$= \begin{bmatrix} 0 & 0 \\ 0 & 0 \end{bmatrix}$$

A.5 $[A]^T = \begin{bmatrix} 1 & 4 & 0 \\ 2 & 5 & 8 \\ 3 & 6 & 9 \end{bmatrix} \qquad \text{adj}[A] = \begin{bmatrix} -3 & 6 & -3 \\ 36 & 9 & 6 \\ 32 & -8 & -3 \end{bmatrix}$

$|[A]| = 21$

$$[A]^{-1} = \begin{bmatrix} -1/7 & 2/7 & -1/7 \\ 12/7 & 3/7 & 2/7 \\ 32/21 & -8/21 & -1/7 \end{bmatrix}$$

APPENDIX B

B.1
```
Exercise B.1, My Name
Is    0 1   dc   3m
* current from node 0 to node 1
R12   1 2        60k
R23   2 3        2k
Vs    0 3   dc   27
* node 0 is at higher potential
```

```
R24  2 4     5k
R40  4 0     10k
V14  1 4  dc  15
.options nopage
.end
```

B.2 Modified and additional statements are

```
R04  0 4    10k ; node order so ib = i(R04)
V41  4 1 dc -15 ; node order so ia = i(V41)
.dc  Is  list 0 3m 6m
.print dc  i(V41) i(R04) v(2,4) ; v(2,4) = vx
```

B.3 Modified and additional statements are

```
* Sweep Vs linearly from
* −20 V to +20 V in 0.25-V steps
.dc lin Vs -20  20 0.25
.probe
.end
```

B.4

```
Is  1 0 dc 3m
R12 1 2       2k
V20 2 0 dc    0      ; senses i(V20) = ix
H13 1 3 (V20)    10k ; CCVS = 10 kohm * ix
R30 3 0      40 k
* Rt = output resistance,
* Voc = [v(3)/Is] * 3 mA
.tf v(3) Is
```

B.5

```
R12a   1 2 {10k*R12b/(R12b - 10k)}
R12b   1 2 {R12b}
.param      R12b = 5k
.step lin param R12b 5k 20k 1k
```

B.6

```
Eoa  3 0 (5,0) 1 ; Va = v(5)
Gaux 0 5 (0,2) 1 ; Ix = Vd = v(0,2)
Rvar 5 0 {A} ; v(5) = A * Ix
.param A = 1E2
.dc dec param A 1E2, 1E6, 20
* 4 * 20 = 80 points
```

B.7

```
Op-amp circuit from Fig. 4.45
Vin  1 0   dc 1
Xoa1 2 1 3 opamp
R2   2 3   1k
R3   2 5   {K2*1k}
R1   3 4   1k
RF   4 5   1k
Xoa2 0 4 5 opamp
.subckt opamp pos neg out
ri pos neg 1meg
Eoa out 0 (pos,neg) 1E6
.ends opamp
.param K2 = 0.5
.step param K2 list 0.5 1.5
.end
```

B.8

```
Exercise B.8
Vs  0 1 ac 30 -45
V12 1 2 ac 0          ; senses i1 = i(V12)
R23 2 3    10
L23 2 3    2
C30 3 0    .01
Is  0 3 ac 1 -90      ; sin x = cos (x - 90)
.ac lin 3 1.59 7.96 ; 10, 30, 50 rad/s
.print ac im(V12) ip(V12)
.end
```

B.9

```
* Use rms values for all sources and variables
Va 1 0 ac 300       ; node 1 = a
Vb 2 0 ac 300 -120 ; node 2 = b
Vc 3 0 ac 300 -240 ; node 3 = c
* omega = 377, L = 20/omega = 53.1 mH
R14 1 4      10
L42 4 2      53.1m
R25 2 5      10
L53 5 3      53.1m
R36 3 6      10
L61 6 1      53.1m
.ac lin 1 60 60
```

B.10

```
Vs     1 0 ac     10
Rs     1 2        2
Exfmr  2 0 (4,3)  .04
* v1 = v2/N = 0.04v(4,3)
Fxfmr  4 3 (Vs)   .04
* i2 = -i1/N = 0.04i(Vs)
Rload  4 3        1k
Rx     3 0        1meg
* DC path to nodes 3 and 4
.ac lin 1 15.9k 15.9k
```

B.11 $v_C(0) = Rv_s(0^-)/(R + R_s)$

$\tau_{max} = (R \| R_s)_{max} C = 10$ ms

```
Exercise B.11
Rs 1 0 {Rs}
R  1 0  20k
C  1 0   1u IC = {200k/(20k+Rs)}
.param Rs = 20k
.step param Rs list 5k 20k
.tran 30m 30m uic
.probe
.end
```

B.12 (a)

t	0	2 μs	4 μs
i_s	0	10 mA	0

```
Is 0 1 pwl 0,0 2u,10m 4u,0
```

(b) $\cos 100t = \sin(2\pi ft + \phi)$ so

$f = 50/\pi = 15.92,\ \phi = 90°$

```
Is 0 1 sin 2m 3m 15.92 0 10 90
```

B.13

```
S12 1 2 10  0 switch
S34 3 4 0  10 switch
.model switch vswitch
V10 10 0 pulse -1 1 2 0.01 0.01 0.99 4
```

Answers to Selected Problems

1.1 $q = 60$ C, $i_{av} = 3$ A

1.6 Rating = 0.2 Ah

1.17 kA, Ω; μA, μW

1.20 $v = 20$ V, $i = 5$ A when $R = 4$ Ω

1.29 $v_3 = 6$ V, $i_2 = 3$ A, $i_s = 6$ A

1.37 $v_{BE} = 0.8$ V, $i_B = 0.4$ mA, $v_{CE} = 15.2$ V

1.40 $p_i = 2880$ mW, $p_v = 5400$ mW

2.3 $R_2 = 24$ Ω, $p_{min} = 450$ W

2.9 $R_x = 60$ Ω

2.15 $R_{eq} = 9$ kΩ, $i_2 = 2$ mA, $p = 576$ mW

2.23 $i_x = 1$ A

2.29 $R_{eq} = -5$ Ω

2.37 $K = 2.5$, $i_1 = 1.6$ mA

2.46 $v_{a-1} = v_{a-2} = 18$ V

2.57 $v_{oc} = 27$ V, $i_{sc} = 9$ A

2.64 $R_t = -60$ Ω, $i_{sc} = i_s$

2.71 $i_1 = 1$ mA

3.1 $p_{max} = 1200$ W, $R_L \geq 33$ Ω

3.9 $p_L = 75$ W, Eff = 35.7%

3.16 $R_i' = 4$ kΩ, $v_{out}/v_s = 8$

3.24 $B = 0.0399$, $R_1 = 4.16$ kΩ

3.46 $v_{out}/v_{in} = -AR_F/[(A + 1)R_1 + R_F]$

3.50 $v_d \approx 10$ mV, $R_o \approx 0.1$ Ω

3.54 $R_i = 600$ Ω

3.62 $i_R = 0.5$ mA, $R = 12$ kΩ

4.1 $v_1 = 16$ V, $v_2 = -24$ V, $v_3 = 16$ V

4.6 $v_2 = v_4 = 36$ V, $i_1 = 2.4$ mA

4.11 $v_1 = 30$ V, $v_2 = v_3 = 20$ V

4.15 $v_1 = 12$ V, $v_2 = 24$ V

4.23 $i_1 = 5$ A, $i_2 = 6$ A, $i_3 = -1$ A

4.29 $i_1 = 1$ mA, $i_3 = -1$ mA

4.38 $i_1 = 2$ A, $v_2 = 16$ V

4.44 $i_x = 5$ mA, $i_z = 3$ mA

4.52 $v_x = 12$ V, $v_z = 0$

4.62 $R_{eq} = 5$ Ω

4.72 $v_{oc} = 6$ V, $R_t = 32$ kΩ

4.76 $v_{out} = K(R_2 v_1 + R_1 v_2)/(R_1 + R_2)$

4.85 $R_{eq} = 5$ Ω

5.1 $i = 0.4$ mA

5.5 $i = 5$ mA

5.12 $C_{eq} = 9\ \mu\text{F}$

5.16 $v_1(t_1) = -10$ V, $v_2(t_1) = v_3(t_1) = 5$ V

5.24 $w = 10.83$ J, $v = 0$

5.30 $w_L = 10$ mJ, $w_C = 5$ mJ

5.39 $i_1(t) = -4 + 5t^2$

5.42 $(R + R_1)L/R_1\ di_L/dt + Ri_L = v$

5.48 $C\,d^2v_L/dt^2 + (CR/L)\,dv_L/dt + v_L/L = di/dt$

5.54 $i_F = 0.01t - 0.02$

5.62 $i(t) = 2t - 0.1 + 0.5e^{-20t}$

6.1 (a) $3\ \angle 180^\circ - \alpha$, (b) $2y\ \angle 30^\circ - \phi$, (c) $(1/y)\ \angle 90^\circ + \phi$

6.12 $v = 106\cos(\omega t + 45^\circ)$ V

6.18 $\omega = 20$, $L = 2.5$ H

6.22 $Z = 10\ \Omega\ \angle 0^\circ$, $i_1 = 5.66\cos(\omega t - 45^\circ)$ A

6.28 $X(\omega) = (10\omega - 30/\omega^2)/[1 + (\omega - 2/\omega)^2]$

6.32 $C = 57.7\ \mu\text{F}$

6.38 $\underline{I}_L/\underline{I} = 0.5\ \angle{-90^\circ}$, $\underline{V}/\underline{I} = 26\ \Omega\ \angle{-22.6^\circ}$

6.50 $\underline{V}_1 = j6$

6.56 $\underline{I} = 3 - j1$

6.62 $\underline{V}_1 = 5 + j5 = \underline{V}_L + \underline{V}_2$, $\underline{V}_L = j2R_2I_2$, $\underline{V}_2 = R_2I_2$

6.68 $C = 0.2\ \mu\text{F}$, $|\underline{V}_L| = 20$ V, $|\underline{V}_R + \underline{V}_L| = 21.5$ V

6.83 $i_1(t) = 4 + 1.41\cos(20{,}000t - 45^\circ)$ mA

6.90 $R_u = (C_1/C_3)R_2$, $L_u = C_1R_2R_3$

7.1 $V_{rms} = 47.5$ V, $P = 90$ W

7.10 $V_{rms} = 36.1$ V

7.16 $P_{max} = 25$ mW

7.26 (a) $Q = 7.68$ kVAr, (b) $|\underline{I}| = 24$ A, (c) $|\underline{I}| = 66.7$ A

7.29 (a) $|\underline{I}| = 347$ A, (b) $Q_C = -151$ kVAr

7.41 (a) $\underline{I}_a = 100\text{ A}\ \angle{-90^\circ}$, (b) $C = 1920\ \mu\text{F}$

7.48 $|\underline{I}_a| = 20$ A, $P_2 = 3.6$ kW

7.53 $Z_y = 5.2 + j3.9\ \Omega$

7.59 $\underline{I}_a = 24\text{ A}\ \angle{-53.1^\circ}$, $\underline{I}_n = 115\text{ A}\ \angle 30^\circ$

7.67 $\underline{I}_a = 60\text{ A}\ \angle{-30^\circ}$, $P = 27$ kW

8.3 $v_{in}/i_{in} = 144\ \Omega$

8.10 $\underline{I} = 2.4\text{ A}\ \angle{-16.3^\circ}$, $\underline{V}_{out} = 480\text{ V}\ \angle 161.4^\circ$

8.18 $\underline{I} = 0.6\text{ A}\ \angle{-36.9^\circ}$, $\underline{V}_b = 0$

8.21 $N = 3$, $R = 468\ \Omega$

8.25 $N_1 = 40$, $N_2 = 99$, $M = 19$ mH

8.33 $\underline{V}_1 = 0$, $\underline{I}_2 = 24\text{ A}\ \angle 0^\circ$

8.40 $\underline{I}_1 = 3\text{ A}\ \angle{-90^\circ}$, $\underline{I}_2 = 2.12\text{ A}\ \angle{-45^\circ}$

8.46 $\underline{I}_{out}/\underline{I}_{in} = 2\ \angle{-143.1^\circ}$, $Z_{in} = 12 + j9\ \Omega$

8.57 $\underline{V}_{out} = 300\text{ V}\ \angle{-53.1^\circ}$, $P_{in} = 5760$ W

9.1 $\tau = 30\ \mu\text{s}$, $v(t) = 12e^{-t/\tau}$ V

9.7 $v_C(t) = 10(1 - e^{-t/RC})$ V, $RC = R_2C = 4$ ms

9.12 (a) $i(t) = -1 + 3e^{-2.5t}$, (b) $i(t) = 2 - 2.75e^{-(t-1)}$

9.20 $i(t) = 2\cos(24t + 16.3^\circ) - 1.36e^{-7t}$ A

9.27 $i_L'' + i_L'/RC + i_L/LC = v_s/RLC$

9.32 $y_N(t) = A_1e^{-7t} + A_2e^{-10t}$

9.46 $i_L(t) = 4e^{-2t} - e^{-8t}$

9.52 $i_L(t) = 4 - 4e^{-5t} - 10te^{-5t}$

9.57 $v_C(t) = 120e^{-5t}\cos(5t - 90^\circ)$

10.1 $v(t) = -12e^{-5t}$

10.5 $Z = 25\ \angle 106.3^\circ$, $i_L(t) = 4e^{-2t}\cos(6t - 16.3^\circ)$

10.9 $s = 0$ or $s = -(R_1 + R_2)/L$

10.12 $\dfrac{C_1R_1s + 1}{C_1C_2RR_1s^2 + (C_1R + C_2R + C_1R_1)s + 1}$

10.24 $H(s) = (6s^2 + 5)/(5s^3 + 24s^2 + 5s + 20)$

10.30 $K = 2$, $z_1, z_2 = -2, -10$, $p_1, p_2 = -2 \pm j4$

10.34 $H(s_0) = 0.8\ \angle 98.8^\circ$

10.41 $\mu = 4$, $y_N(t) = 2|A_1|e^t\cos(2t + \angle A_1)$, unstable

10.45 $k_f = 2{,}810$, $k_m = 250$

11.13 $R = 707\ \Omega$, $L = 563\ \mu\text{H}$

11.19 $R_a = 1/2\pi f_1C = 7.96\ \text{k}\Omega$, $R_b = 1/2\pi f_2C = 796\ \Omega$

11.25 $Q = 2/3,\ R = 18.8\ \Omega,\ C = 6.33\ \mu F$

11.33 $a(\omega) = \omega^2/[(\omega_o^2 - \omega^2)^2 + (\omega_o\omega/Q_{ser})^2]^{1/2}$

11.38 Take $C_1 = 1\ \mu F,\ C_F = 5$ nF, $R_1 = 7.96\ k\Omega$, $R_F = 31.8\ k\Omega$

11.48 $H(s) = -5\ H_{lp}(s;20)\ H_{lp}(s;200)$

11.52 $a_{max} \approx 2,\ \omega_l \approx 7,\ \omega_u \approx 45$

11.60 $H(s) = H_{hp}(s;10)\ H_q(s;10,5)$

11.65 $H(s) = (R_{Fb}/R_{1b}R_{1a}C) \times s/(s + 1/R_{1b}C)(s + 1/R_{Fa}C)$

11.71 $n \geq 1.50$

11.81 Fig. 11.41*a*, $R_F = 9\ k\Omega,\ R_1 = 0.878\ k\Omega$, $R_2 = 11.5\ k\Omega$

12.1 $c_n = (1 - e^{-T})(1 - jn\Omega)/(1 + n^2\Omega^2)T$

12.6 $f(t) = (2A/\pi) - (4A/3\pi)\cos 2\Omega t - (4A/15\pi)\cos 4\Omega t - \ldots$

12.16 $c_z(n\Omega) = -12/n^2\pi^2$ for $n = 1, 3, 5, \ldots$

12.20 $c_z(n\Omega) = -2/\pi(n^2 - 1)$ for $n = 2, 4, 6, \ldots$

12.27 $c_z(n\Omega) = j8/n\pi T$ for $n = 1, 3, 5, \ldots$

12.32 $y(t) \approx 1.57 + 0.894\cos(\Omega t - 63°)$

12.37 $V_m = 140$ V, $L \geq 0.416$ H

12.47 $H_{eq}(j\omega) = (10 + j\omega)^2 e^{-j0.1\omega}/500$

13.1 $F(s) = (s - 1 + e^{-s})/s^2$

13.12 $\mathcal{L}[t^r] = r!/s^{r+1}$

13.19 $y(t) = 5t^2e^{-4t} - 3e^{-4t}$

13.23 $f(t) = 3 - 6e^{-2t} + 8e^{-4t}$

13.32 $f(t) = -10t + 4 + 2te^{-5t} - 4e^{-5t}$

13.39 $f(t) = -1 + 3e^{-3t} + 2e^{-5(t-4)}u(t - 4)$

13.43 $f(0^+) = 0,\ f(\infty) = 2,\ f'(0^+) = -5$

13.47 $y(t) = 1 + 3e^{-2t} - 4e^{-5t}$

13.54 $y(t) = -8 + 2e^{-5t} + 6.33\cos(5t - 18.4°)$

13.64 $i_2(0^+) = 0,\ i_2'(0^+) = -20$

13.70 $i_2(t) = -2e^{-3t} + 2e^{-12t}$

13.74 $F(s) = I_o(1 - e^{-s} - se^{-s})/s^2$

13.78 $v_1(t) = 4\,\delta(t) + 32e^{-2t}$

13.86 $f(t) = 3\delta(t) + 12tu(t)$

13.90 $y(t) = [(1 - e^{-D})(t + 1) - D]e^{-(t-D)},\ t \geq D$

14.1 $z_{11} = s + 5,\ z_{12} = -10,\ z_{21} = 5,\ z_{22} = 0$

14.6 $z_{11} = (s^2 - s)/(s^2 + s + 1)$

14.15 $y_{11} = Y_a + Y_c,\ y_{21} = y_{12} = -Y_c$, $y_{22} = Y_b + Y_c$

14.20 $h_{11} = -(s^2 + s + 1)/(s^2 + s)$

14.23 $A = 0.7s + 1,\ B = 3s + 10,\ C = 0.7,\ D = 3$

14.34 $R_L = 16.7\ \Omega,\ H_i = -1.2$

14.40 $Z_L = 5{\cdot}10^6/(s + 5000)$

14.47 $H_i = -10/(-0.04R_L + 9)$

15.1

$$[A] = \begin{bmatrix} -10 & 90 \\ -12 & -24 \end{bmatrix} \qquad [B] = \begin{bmatrix} 10 & 0 \\ 4 & 24 \end{bmatrix}$$

$$[C] = \begin{bmatrix} -1/9 & 1/3 \\ -1 & 0 \\ 0 & 2 \end{bmatrix} \qquad [D] = \begin{bmatrix} 2/9 & 2/3 \\ 1 & 0 \\ 2/3 & -2 \end{bmatrix}$$

15.5

$$[A] = \begin{bmatrix} 10 & 90 \\ -4 & -24 \end{bmatrix} \qquad [B] = \begin{bmatrix} 0 \\ 24 \end{bmatrix}$$

$$[C] = \begin{bmatrix} 1/3 & 1/3 \\ 1 & 0 \\ 4/3 & 2 \end{bmatrix} \qquad [D] = \begin{bmatrix} 2/3 \\ 0 \\ -2 \end{bmatrix}$$

15.9

$$\begin{bmatrix} v_1' \\ v_2' \\ i_4' \end{bmatrix} = \begin{bmatrix} 0 & 0 & -4 \\ 0 & 0 & 2 \\ 2 & -2 & -8 \end{bmatrix} \begin{bmatrix} v_1 \\ v_2 \\ i_4 \end{bmatrix} + \begin{bmatrix} 6 \\ 2 \\ 0 \end{bmatrix} i_s$$

15.13

$$\begin{bmatrix} q_1' \\ v_3' \end{bmatrix} = \begin{bmatrix} -4 & -1 \\ 2 & 0 \end{bmatrix} \begin{bmatrix} q_1 \\ v_3 \end{bmatrix} + \begin{bmatrix} 0 \\ 6 \end{bmatrix} i_s$$

15.17 (a) $P(s) = s^2 - 5s$, (b) $q_1(t) = 3 + 2e^{5t}$

15.22 $y_1(t) = -7e^{-t} + 10e^{5t}$

15.27 $H_{11}(s) = (s^2 - 11)/(s^2 + 4s + 5)$

B.1 $v_s = 12$ V: $v_a = 6$ V, $p_a = 6$ W, $i_2 = 1.5$ A

B.5 $R_f = 30\ k\Omega$: $R_i = 109.9\ \Omega,\ v_{oc} = -2.82$ V, $R_t = 251.2\ \Omega$

B.13 $K = 1$: $v_{out}/v_{in} = 5.0,\ R_i = 50\ G\Omega$, $R_o = 1.05\ \mu\Omega$

B.15 $Z = 10\ \Omega\ \angle 36.9°$

B.19 $|Z_{in}|_{max} = 281\ \Omega$ at 19.8 kHz, $\angle Z_{in} = 0$ at 19.5 kHz

B.25 $\underline{I}_a = 2.43$ kA $\angle -37.6°$

B.29 $\underline{V}_{out} = 154$ V $\angle 159°$

B.33 $R = 50\ \Omega$: $v_{Lmax} = 0.633$ V, $t_{ss} \approx 6$ ms

B.43 $R_w = 0$: $a_{max} = 0.999$, $B = 0.8$ kHz, $Q = 14.1$

B.49 $p_1 \approx -1{,}300$, $p_2 \approx -63{,}000$, $z \approx -2{,}500{,}000$

Index

Reference Guide to PSpice

This summary provides a concise reference guide to PSpice and Probe for analyzing linear circuits. Further details are given in Appendix B of the accompanying text.

GENERAL INFORMATION

Circuit Requirements

- There must be a reference node numbered 0 (zero).
- Two or more elements must be connected at each node.
- A dc path is required from the reference node to all other nodes.
- PSpice does not allow the following configurations:
  - —A closed path that consists entirely of voltage sources and/or inductors
  - —A node or supernode whose impinging branches consist entirely of current sources and/or capacitors

Circuit Files

- The first line is the title, and it will appear on any output.
- The last line must be `.end`. Additional `.end` statements may also appear within the file to separate multiple simulations, each of which must also have a title line.
- Comments may be included as an entire line with an asterisk (*) in the first column, or after a semicolon (;) in other lines.
- Continuation of a line is indicated by a plus (+) in the first column.

- Uppercase and/or lowercase letters may be used, but PSpice is insensitive to case in circuit files. However, Probe is case sensitive in that labels appear as entered in the circuit file.
- Entries on a given line may be separated by blank spaces, commas, or parentheses.

Polarity Convention And Notation

- PSpice employs the passive convention for all elements, with the first-named node taken to be at the higher reference potential. The reference direction for a current goes from the first-named node to the second.
- Node voltages are identified by node number in the form *v(nodenum)*. The voltage difference between two nodes may be written such as *v(node1,node2)*, which stands for *v(node1)* − *v(node2)*.
- Branch currents are identified by an element through which they pass, written in the form *i(elname)*.

Numerical Values

Numbers may be entered in fixed-point or floating-point form. The floating-point notation for $x \times 10^n$ is xEn, where x is a positive or negative number and n is a positive or negative integer. Alternatively, a number in a circuit file may be followed immediately by one of the following symbolic scale factors (either uppercase or lowercase):

Symbol	Value	Symbol	Value
k	10^3	m	10^{-3}
Meg	10^6	u	10^{-6}
G	10^9	n	10^{-9}
T	10^{12}	p	10^{-12}

However, Probe is case sensitive and uses M for 10^6 and m for 10^{-3}.

Expressions

Mathematical expressions for circuit files or Probe are formed using operators, constants, and parentheses as needed. The operator symbols are

+ (addition) − (subtraction or negtation)
* (multiplication) / (division)

A device expression must not be interrupted by the plus sign for a continuation line.

DEVICE DESCRIPTIONS

Required key letters are shown here in uppercase type, and the notation $[X|Y]$ means that either X or Y is a required part of the statement. User-supplied names and numbers are in *italics*. Higher- and lower-potential nodes are denoted by n^+ and n^-, respectively. Optional items are within angle brackets ⟨ ⟩, which may be nested. Lists of arbitrary length are denoted by ellipses (. . .) after one or two entries.

Except where otherwise indicated, any numerical value for a device may be replaced by {*paramname*} or {*expression*}.

Resistors

```
Rname n+ n- value
```

The *value* cannot be zero or infinite, but it may be negative.

Capacitors and Inductors

```
[C|L]name n+ n- value ⟨IC = value⟩
```

The IC value is the initial capacitor voltage or inductor current.

Inductor Coupling

```
Kname Lname1 Lname2 value
```

The value of the coupling coefficient is related to the mutual and self-inductances by $k = M/(L_1 L_2)^{1/2} < 1$. The polarity dots are assumed to be at the n^+ nodes of L_1 and L_2.

Controlled Sources

```
[E|G]name n+ n- (nc+, nc-) value
```

E = VCVS with μ = *value*; G = VCCS with g_m = *value*. The control voltage is $v(nc^+, nc^-)$.

```
[F|H]name n+ n- Vcname value
```

F = CCCS with β = *value*; H = CCVS with r_m = *value*. The control current is *i*(*Vcname*).

The *value* of a controlled source cannot involve a parameter. However, the effect of parameter variation can be simulated using an auxiliary controlled current source driving a stepped or swept resistance.

DC Steady-State Sources

```
[V|I]name n+ n- dc value
```

The *value* may be positive, negative, or zero.

AC Steady-State Sources (used only with .ac)

```
[V|I]name n+ n- ac amplitude ⟨phase⟩
```

The *phase* is in degrees, and the default is

$$phase = 0$$

The frequency is specified in a `.ac` statement.

Damped Sinusoidal Sources (used only with `.tran`)

```
[V|I]name n+ n- sin Xo Xm f ⟨td ⟨alpha ⟨phase⟩⟩⟩
```

Frequency f is in hertz, and *phase* is in degrees. The waveform is

$$x(t) = X_o + X_m \sin\theta \qquad 0 < t < t_d$$
$$= X_o + X_m\, d(t) \sin[2\pi f(t - t_d) + \theta] \qquad t > t_d$$

with

$$\theta = 2\pi\frac{phase}{360} \qquad d(t) = e^{-alpha(t - t_d)}$$

Default values are

$$td = 0 \quad alpha = 0 \quad phase = 0$$

Pulse Sources (used only with `.tran`)

```
[V|I]name n+ n- pulse Xo Xp ⟨td ⟨tr ⟨tf ⟨pw ⟨per⟩⟩⟩⟩⟩
```

See the waveform in Fig. B.26. Default values are

$$td = 0 \quad tr = tstep \quad tf = tstep \quad pw = tfinal \quad per = tfinal$$

Piecewise Linear Sources (used only with `.tran`)

```
[V|I]name n+ n- pwl t0,X0 t1,X1 t2,X2 . . .
```

where $0 \leq t0 < t1 < t2 \ldots$ See the waveform in Fig. B.25. The values of $t0, X0, \ldots,$ cannot involve parameters.

Voltage-Controlled Switches (used only with `.tran`)

```
Sname n+ n- (nc+, nc-) modname
.model modname vswitch ⟨Ron = value⟩ ⟨Roff = value⟩
+ ⟨Von = value⟩ ⟨Voff = value⟩
```

The switch is closed and has resistance *Ron* between nodes n^+ and n^- when $v(nc^+, nc^-) \geq Von$. The switch is open and has resistance *Roff* when $v(nc^+, nc^-) \leq Voff$. Default values are

$$Ron = 1\ \Omega \quad Roff = 1\ \text{M}\Omega \quad Von = 1\ \text{V} \quad Voff = 0$$

The resistance ratio should not exceed $Roff/Ron = 10^{12}$. The transition time of the control voltage should not be much less than $tfinal/1000$.

SIMULATION AND OUTPUT COMMANDS

The notation [*xx*|*yy*|*zz*] means that either *xx* or *yy* or *zz* is a required part of the statement. User-supplied names and numbers are in *italics*. Optional entries are indicated within angle brackets ⟨ ⟩. Lists of arbitrary length are denoted by ellipses (. . .) after one or two entries.

AC Sweep (used for ac sources)

```
.ac [lin|dec|oct] npoints, fstart, fstop
```

Sweeps the frequency of all independent ac sources from *fstart* to *fstop* (hertz).

In a linear sweep, *npoints* is the total number of sample frequencies. For single-frequency analysis, use *npoints* = 1 and *fstart* = *fstop*.

In a logarithmic sweep, *npoints* is the number of sample frequencies per decade or octave.

DC Sweep (used for dc sources and resistances)

```
.dc ⟨lin⟩ [V|I]name start, stop, step
.dc [dec|oct] [V|I]name start, stop, npoints
.dc [V|I]name list val1 val2 . . .
```

Sweeps [*V*|*I*]*name* linearly or logarithmically from the *start* value to the *stop* value or over the list of values. In a logarithmic sweep, *npoints* is the number of values per decade or octave.

A resistance whose value is a parameter may be swept via

```
.dc lin param paramname start, stop, step
.dc dec param paramname start, stop, npoint
```

This sweep affects only the dc simulation.

Options

```
.options ⟨list⟩ ⟨node⟩ ⟨nopage⟩
```

`list` — produces a summary of elements.

`node` — produces a summary of connections.

`nopage` — suppresses paging at the start of each section.

Parameters

```
.param paramname = [value|{expression}]
```

Defines the name and nominal value of a global parameter, which may be used in device statements.

Parameter values may be varied in a `.ac`, `.dc`, or `.tran` simulation via

```
.step param paramname start, stop, step
.step [dec|oct] param paramname start, stop, npoints
.step param paramname list val1 val2 . . .
```

For a single element, use `.step` with the following statements:

```
[R|L|C]name n+ n- {paramname}
.param paramname = nomval
```

The values of two or more elements may be varied in relationship using the following statements:

```
[R|L|C]name1 n1+ n1- {expression1}
[R|L|C]name2 n2+ n2- {expression2}
```

where *expression1* and *expression2* involve the same *paramname* defined by a parameter statement.

Printer Output (used with `.ac`, `.dc`, or `.tran`)

```
.print [dc|ac|tran] var1 . . .
```

Produces a table of specified output variables, which may be node voltages, difference voltages, and/or element currents. AC phasor quantities are specified as follows:

Amplitude — $[V|I]M(\ldots)$
Phase (°) — $[V|I]P(\ldots)$
Real part — $[V|I]R(\ldots)$
Imaginary part — $[V|I]I(\ldots)$
Decibel value of magnitude $[V|I]DB(\ldots)$

Subcircuits

```
.subckt subcktname nx1 . . . ⟨params: param1 = value . . .⟩
(device statements for the subcircuit)
.ends ⟨subcktname⟩
```

Nodes *nx1*, . . . , are dummy numbers for the external terminals of the subcircuit, not including the reference node, which is global.

A subcircuit is inserted into the main circuit by the calling statement

```
Xname na1 . . . subcktname ⟨params: param1 = newvalue . . .⟩
```

Nodes *na1*, . . . , are the circuit nodes to which the respective subcircuit terminals are attached. Parameter values may be omitted if they are the same as in the subcircuit statements.

Transfer Functions (used for resistive circuits)

```
.tf outvar [V|I]name
```

Where *outvar* must be a node voltage, a difference voltage, or an element current. Calculates the gain *outvar*/[*V*|*I*]*name*, the input resistance seen by [*V*|*I*]*name*, and the output resistance between the nodes associated with *outvar*.

Transient Simulation

```
.tran tstep tfinal ⟨td ⟨maxstep⟩⟩ ⟨uic⟩
```

Performs transient simulation from $t = 0$ to $t = tfinal$, and prints data points separated by *tstep*. For Probe output without printing, take *tstep* = *tfinal*.

Specifying *td* suppresses output prior to $t = td$. The default is $td = 0$.

Specifying *maxstep* sets an upper limit on the internal time increment used for calculations. The default value is the smaller of *tstep* or *tfinal*/50. Satisfactory plots of oscillatory waveforms at frequency *f* require $maxstep \leq 0.05/f$.

The option `uic` (use initial conditions) invokes the IC values given in capacitor and inductor statements. When `uic` is included, any unspecified initial value defaults to zero. When `uic` is not included, any stated IC values are ignored and PSpice calculates all initial conditions via dc steady-state analysis at $t = 0$.

PROBE

Probe is invoked from the circuit file by the statement

```
.probe ⟨var1 var2 . . .⟩
```

Output variables for plotting are specified as in a `.print` statement. When no variables are specified, all voltages and currents will be available.

Probe can display plots of any quantities involving node voltages and/or element currents as functions of time, frequency, dc input, or parameter value. It can also plot one variable versus another. The circuit file should therefore contain a `.dc`, `.ac`, or `.tran` statement that generates a sufficient number of points for plotting.

If the circuit file generates multiple simulations, then Probe first presents a start-up menu for selecting the case or cases to be displayed. The main menu then appears under a blank plotting area. The most useful menu items are summarized here.

Add_trace

Adds one or more traces to the display.

Traces are defined by typing variable names (as in `.print` statements) or expressions involving constants, variables and operators. Symbolic scale factors are the same as in PSpice, except that Probe uses M for 10^6 and m for 10^{-3}.

Trace expressions may also contain an existing trace (denoted by typing *#tracenum*), or the *X*-axis variable (denoted by the name that appears under the horizontal axis).

If two or more cases have been selected from the start-up menu, then entering *outvar* produces traces for all available cases. Individual cases are selected by entering *outvar@casenum*.

Remove_trace

Appears only when traces are present, and goes to a submenu to remove one or all of them.

X_axis

Controls the horizontal axis via the following submenu:

Linear/Log — toggles the type of axis
Set_range — allows entry of lower and upper limits
Auto_range — restores full range
X_variable — allows entry of a new variable

The new variable for the **X_variable** option is specified in the same manner as **Add_trace**, and it may involve an expression. This option thus allows plotting one variable versus another. When multiple cases are available, you must specify a single case for the new *X*-axis variable.

Y_axis

Controls the vertical axis via the following submenu:

Linear/Log — toggles the type of axis
Set_range — allows entry of lower and upper limits
Auto_range — restores full range

Plot_control

Goes to a submenu to add or remove plots or plotting symbols. New plots are added across the top, reducing the vertical size of existing plots. The selected active plot is indicated by SEL⟩⟩.

Hard_copy

Goes to a submenu to produce hard-copy output of the display. Select **1_page_long** for normal output. All outputs will bear the title line from the circuit file.

Cursor

Activates cursors C1 and C2, which provide accurate readouts of points on traces. A display box gives the read-outs of the cursors' horizontal and vertical coordinates.

C1 is moved along a trace using the arrow keys, and it is jumped to the next trace by pressing the control key and an arrow key. C2 is moved and jumped in the same fashion by holding down the shift key.